The Oceans and Marine Geochemistry

Edited by

H. Elderfield
University of Cambridge, UK

TREATISE ON GEOCHEMISTRY
Volume 6

Executive Editors

H. D. Holland
Harvard University, Cambridge, MA, USA

and

K. K. Turekian
Yale University, New Haven, CT, USA

ELSEVIER

AMSTERDAM – BOSTON – HEIDELBERG – LONDON – NEW YORK – OXFORD
PARIS – SAN DIEGO – SAN FRANCISCO – SINGAPORE – SYDNEY – TOKYO

Elsevier
The Boulevard, Langford Lane, Kidlington, Oxford OX5 1GB, UK
Radarweg 29, PO Box 211, 1000 AE Amsterdam, The Netherlands

First edition 2006

British Library Cataloguing in Publication Data
A catalogue record for this book is available from the British Library

Library of Congress Cataloging-in-Publication Data
A catalog record for this book is available from the Library of Congress

ISBN-13: 978-0-08-045101-5
ISBN-10: 0-08-045101-2

For information on all Elsevier publications
visit our website at books.elsevier.com

Printed and bound in Italy

06 07 08 09 10 10 9 8 7 6 5 4 3 2 1

The Oceans and Marine Geochemistry

Citations

Please use the following example for citations:

Millero F.J. (2003) Physicochemical controls on seawater, pp. 1–21. In *The Oceans and Marine Geochemistry* (ed. H. Elderfield) Vol. 6 *Treatise on Geochemistry* (eds. H.D. Holland and K.K. Turekian), Elsevier–Pergamon, Oxford.

Cover photo: Scanning electron microscope images of marine calcifying organisms. (Left) The coccolithophore *Emiliania huxleyi* (diameter 5 μm) and (right) the foraminifer *Globigerina bulloides* (diameter 400 μm). Reprinted with permission from H. Elderfield, Science, vol. 296, p. 1618. Copyright 2002 AAAS. (The *Emiliania huxleyi* image is reprinted with permission from P. Ziveri and S. Kars).

DEDICATED
TO

HARMON CRAIG
(1926–2003)

JOHN EDMOND
(1943–2001)

CESARE EMILIANI
(1922–1995)

Photograph provided by Scripps Institution of Oceanography Archives
Photograph provided by Family of John Edmond
Photograph provided by Rosensteil School of Marine and Atmospheric Sciences, University of Miami

Contents

Executive Editors' Foreword

H. D. Holland

Harvard University, Cambridge, MA, USA

and

K. K. Turekian

Yale University, New Haven, CT, USA

Geochemistry has deep roots. Its beginnings can be traced back to antiquity, but many of the discoveries that are basic to the science were made between 1800 and 1910. The periodic table of elements was assembled, radioactivity was discovered, and the thermodynamics of heterogeneous systems was developed. The solar spectrum was used to determine the composition of the Sun. This information, together with chemical analyses of meteorites, provided an entry to a larger view of the universe.

During the first half of the twentieth century, a large number of scientists used a variety of methods to determine the major-element composition of the Earth's crust, and the geochemistries of many of the minor elements were defined by V. M. Goldschmidt and his associates using the then new technique of emission spectrography. V. I. Vernadsky founded biogeochemistry. The crystal structures of most minerals were determined by X-ray diffraction techniques. Isotope geochemistry was born, and age determinations based on radiometric techniques began to define the absolute geologic timescale. The intense scientific efforts during World War II yielded new analytical tools and a group of people who trained a new generation of geochemists at a number of universities. But the field grew slowly. In the 1950s, a few journals were able to report all of the important developments in trace-element geochemistry, isotopic geochronometry, the exploration of paleoclimatology and biogeochemistry with light stable isotopes, and studies of phase equilibria. At the meetings of the American Geophysical Union, geochemical sessions were few, none were concurrent, and they all ranged across the entire field.

Since then the developments in instrumentation and the increases in computing power have been spectacular. The education of geochemists has been broadened beyond the old, rather narrowly defined areas. Atmospheric and marine geochemistry have become integrated into solid Earth geochemistry; cosmochemistry and biogeochemistry have contributed greatly to our understanding of the history of our planet. The study of Earth has evolved into "Earth System Science," whose progress since the 1940s has been truly dramatic.

Major ocean expeditions have shown how and how fast the oceans mix; they have demonstrated the connections between the biologic pump, marine biology, physical oceanography, and marine sedimentation. The discovery of hydrothermal vents has shown how oceanography is related to economic geology. It has revealed formerly unknown oceanic biotas, and has clarified the factors that today control, and in the past have controlled the composition of seawater.

Seafloor spreading, continental drift and plate tectonics have permeated geochemistry. We finally understand the fate of sediments and oceanic crust in subduction zones, their burial and their exhumation. New experimental techniques at temperatures and pressures of the deep Earth interior have clarified the three-dimensional structure of the mantle and the generation of magmas.

Moon rocks, the treasure trove of photographs of the planets and their moons, and the successful search for planets in other solar systems have all revolutionized our understanding of Earth and the universe in which we are embedded.

Geochemistry has also been propelled into the arena of local, regional, and global anthropogenic problems. The discovery of the ozone hole came as a great, unpleasant surprise, an object lesson for optimists and a source of major new insights into the photochemistry and dynamics of the atmosphere. The rise of the CO_2 content of the atmosphere due to the burning of fossil fuels and deforestation has been and will continue to be at the center of the global change controversy, and will yield new insights into the coupling of atmospheric chemistry to the biosphere, the crust, and the oceans.

The rush of scientific progress in geochemistry since World War II has been matched by organizational innovations. The first issue of *Geochimica et Cosmochimica Acta* appeared in June 1950. The Geochemical Society was founded in 1955 and adopted *Geochimica et Cosmochimica Acta* as its official publication in 1957. The International Association of Geochemistry and Cosmochemistry was founded in 1966, and its journal, *Applied Geochemistry*, began publication in 1986. *Chemical Geology* became the journal of the European Association for Geochemistry.

The Goldschmidt Conferences were inaugurated in 1991 and have become large international meetings. Geochemistry has become a major force in the Geological Society of America and in the American Geophysical Union. Needless to say, medals and other awards now recognize outstanding achievements in geochemistry in a number of scientific societies.

During the phenomenal growth of the science since the end of World War II an admirable number of books on various aspects of geochemistry were published. Of these only three attempted to cover the whole field. The excellent *Geochemistry* by K. Rankama and Th.G. Sahama was published in 1950. V. M. Goldschmidt's book with the same title was started by the author in the 1940s. Sadly, his health suffered during the German occupation of his native Norway, and he died in England before the book was completed. Alex Muir and several of Goldschmidt's friends wrote the missing chapters of this classic volume, which was finally published in 1954.

Between 1969 and 1978 K. H. Wedepohl together with a board of editors (C. W. Correns, D. M. Shaw, K. K. Turekian and J. Zeman) and a large number of individual authors assembled the *Handbook of Geochemistry*. This and the other two major works on geochemistry begin with integrating chapters followed by chapters devoted to the geochemistry of one or a small group of elements. All three are now out of date, because major innovations in instrumentation and the expansion of the number of practitioners in the field have produced valuable sets of high-quality data, which have led to many new insights into fundamental geochemical problems.

At the Goldschmidt Conference at Harvard in 1999, Elsevier proposed to the Executive Editors that it was time to prepare a new, reasonably comprehensive, integrated summary of geochemistry. We decided to approach our task somewhat differently from our predecessors. We divided geochemistry into nine parts. As shown below, each part was assigned a volume, and a distinguished editor was chosen for each volume. A tenth volume was reserved for a comprehensive index:

(i) *Meteorites, Comets, and Planets*: Andrew M. Davis

(ii) *Geochemistry of the Mantle and Core*: Richard Carlson

(iii) *The Earth's Crust*: Roberta L. Rudnick

(iv) *Atmospheric Geochemistry*: Ralph F. Keeling

(v) *Freshwater Geochemistry, Weathering, and Soils*: James I. Drever

(vi) *The Oceans and Marine Geochemistry*: Harry Elderfield

(vii) *Sediments, Diagenesis, and Sedimentary Rocks*: Fred T. Mackenzie

(viii) *Biogeochemistry*: William H. Schlesinger

(ix) *Environmental Geochemistry*: Barbara Sherwood Lollar

(x) *Indexes*

The editor of each volume was asked to assemble a group of authors to write a series of chapters that together summarize the part of the field covered by the volume. The volume editors and chapter authors joined the team enthusiastically. Altogether there are 155 chapters and 9 introductory essays in the Treatise. Naming the work proved to be somewhat problematic. It is clearly not meant to be an encyclopedia. The titles *Comprehensive Geochemistry* and *Handbook of Geochemistry* were finally abandoned in favor of *Treatise on Geochemistry*.

The major features of the Treatise were shaped at a meeting in Edinburgh during a conference on Earth System Processes sponsored by the Geological Society of America and the Geological Society of London in June 2001. The fact that the Treatise is being published in 2003 is due to a great deal of hard work on the part of the editors, the authors, Mabel Peterson (the Managing Editor), Angela Greenwell (the former Head of Major Reference Works), Diana Calvert (Developmental Editor, Major Reference Works),

Bob Donaldson (Developmental Manager), Jerome Michalczyk and Rob Webb (Production Editors), and Friso Veenstra (Senior Publishing Editor). We extend our warm thanks to all of them. May their efforts be rewarded by a distinguished journey for the Treatise.

Finally, we would like to express our thanks to J. Laurence Kulp, our advisor as graduate students at Columbia University. He introduced us to the excitement of doing science and convinced us that all of the sciences are really subdivisions of geochemistry.

Contributors to Volume 6

R. F. Anderson
Lamont-Doherty Earth Observatory of Columbia University, Palisades, NY, USA

D. Archer
University of Chicago, IL, USA

M. P. Bacon
Woods Hole Oceanographic Institution, MA, USA

W. S. Broecker
Columbia University, Palisades, NY, USA

K. W. Bruland
University of California at Santa Cruz, CA, USA

H. Cheng
University of Minnesota, Minneapolis, MN, USA

K. B. Cutler
University of Minnesota, Minneapolis, MN, USA

C. L. de la Rocha
University of Cambridge, UK

R. L. Edwards
University of Minnesota, Minneapolis, MN, USA

T. I. Eglinton
Woods Hole Oceanographic Institution, MA, USA

S. Emerson
University of Washington, Seattle, WA, USA

C. D. Gallup
University of Minnesota, Duluth, MN, USA

C. R. German
Southampton Oceanography Centre, Southampton, UK

S. L. Goldstein
Columbia University, Palisades, NY, USA

G. H. Haug
Geoforschungszentrum Potsdam, Germany

J. Hedges
University of Washington, Seattle, WA, USA

S. R. Hemming
Columbia University, Palisades, NY, USA

T. D. Herbert
Brown University, Providence, RI, USA

H. D. Holland
Harvard University, Cambridge, MA, USA

W. J. Jenkins
Woods Hole Oceanographic Institution, MA, USA

D. W. Lea
University of California, Santa Barbara, CA, USA

P. S. Liss
University of East Anglia, Norwich, UK

M. C. Lohan
University of California at Santa Cruz, CA, USA

J. Lynch-Stieglitz
Columbia University, Palisades, NY, USA

F. J. Millero
University of Miami, FL, USA

A. J. Milligan
Princeton University, NJ, USA

F. M. M. Morel
Princeton University, NJ, USA

P. D. Nightingale
Plymouth Marine Laboratory, Devon, UK

G. E. Ravizza
University of Hawaii, Manoa, HI, USA

D. J. Repeta
Woods Hole Oceanographic Institution, MA, USA

M. A. Saito
Princeton University, NJ, USA

D. M. Sigman
Princeton University, NJ, USA

K. K. Turekian
Yale University, New Haven, CT, USA

K. L. Von Damm
University of New Hampshire, Durham, NH, USA

J. C. Zachos
University of California, Santa Cruz, CA, USA

Volume Editor's Introduction

H. Elderfield

University of Cambridge, UK

The chemical composition of the oceans defines the impact of the other geochemical reservoirs on the oceanic reservoir leading to the addition of elements to, or removal from, seawater. Superimposed on this balance is the interplay of ocean chemistry with physics and biology that controls the internal geochemical cycling of elements and isotopes in the oceans. Thus, seawater chemistry is a remarkable integration of processes key to how the Earth has evolved, on a range of timescales equivalent to the oceanic residence times of its constituent chemical elements.

The question "why is the sea salty?" has long fascinated natural philosophers from Aristotle, Pliny the Elder and Mas'udi, through to the seventeenth- to nineteenth-century scientists (Lavoisier, Gay-Lussac, Davy, Marcet) and leading to the expedition of HMS Challenger and the work of W. Dittmar, setting the scene for modern scientific research on ocean chemistry (a brief review of this history is given by Riley, 1965).

A rather better question "How have sea water and air got their present compositions?" was posed by L.G. Sillén (1967) among others. One answer to the former is simply "P–E" but the answer from geochemists is much more interesting. The manner in which the chemistry of the oceans reflects its sources and sinks (the continental crust, the oceanic crust and mantle, the atmosphere, extra-terrestrial inputs) means that the archive of this composition contains in a decipherable code a record of the chemical evolution of the Earth and its atmosphere.

The scientific community has evolved from the difficult task of "simply" thinking about modern ocean composition to the even more difficult task of what was this composition in the past and how has it changed through time. The first edition of *Chemical Oceanography* (Riley and Skirrow, 1965), an enormously successful predecessor to this volume, was concerned almost exclusively with the modern oceans. The perceptive comment of Broecker (2002) that "The paleoclimate record challenges us by demonstrating that much more must be learned about the Earth's climate system before we can have any confidence in model-based predictions of the changes which will result from the buildup of CO_2 and other greenhouse gases in the atmosphere" is a stark reminder that the reason for understanding the modes of operation of past climate (which means past oceans) is more than that of intellectual curiosity. Geochemistry is a major key to this understanding.

These remarks do not mean, at all, that we can tick off modern ocean chemistry/geochemistry/biogeochemistry as a job well done and focus exclusively on records of the past. It is clear from the chapters in this volume that this is far from the truth. An excellent example of this is understanding of the trace-metal chemistry of seawater (Chapters 6.02 and 6.05). The decade from the mid-1970s saw the initial major advance in this field when the first reliable sets of vertical profiles of a host of oceanic trace metals were obtained, the criterion for reliability being what John Edmond called "oceanographic consistency." Advances in metal speciation (Chapter 6.01) and in understanding the biological roles of trace metals (Chapters 6.04 and 6.05) are leading to a second major growth of this research field.

The twenty-one chapters following this introduction contain many examples of how dramatically marine geochemistry has advanced since the last attempt was made to provide a "authoritative and comprehensive coverage of the many facets of this rapidly expanding subject."

What has led to such a dramatic expansion of understanding in this field? One answer, of course, is data. Technological advance is a great driver in subjects like geochemistry. Coupled with the healthy aphorism that data survive its (initial) interpretation, simply measuring the chemistry of the oceans has itself driven forward understanding. There are numerous examples here, but the most striking must be the discovery of and documentation of the chemistry of hot springs on the seafloor (Chapter 6.07).

But equally important have been concepts (maybe "ideas" is a better description). One example, included in the introductory text above, is that of oceanic residence time. The concept of the oceanic residence time of the elements was "initiated" by Barth (1952) in a book on a completely different subject. Residence time (τ) is obtained from the ratio of the total amount of an element in the oceans (A) and its rate of input (flux) to the oceans F_{in}, i.e.,

$$\tau = \frac{A}{F_{in}}$$

This simple equation, when applied to the elements whose concentrations were known in seawater, revealed a staggering range of values from 10^8 yr for sodium, through 10^6 yr for calcium, to values less than 10^3 yr, the mixing time of the oceans, for iron. This exercise showed that even those elements thought to be conservative, major constituents in seawater, exhibited a range in residence times. This was summarized by Broecker (1974), who stated that for every unit of difficulty the ocean has in getting rid of calcium, it has five units of difficulty in ridding itself of potassium, 25 units for magnesium and 125 units for sodium. This led to an upsurge in research into the oceanic sinks for the elements. It was obvious, for example, that the oceanic sink for calcium is the burial of biogenic carbonates in oceanic sediments, but what is the oceanic sink for magnesium? (Chapter 6.07). Drever (1974) posed the "Mg problem," that it was impossible to balance inputs with known outputs, a few years before the first discovery of submarine hot springs at Galapago. It is now recognized that removal of magnesium from seawater by reaction with oceanic crust provides the major oceanic magnesium sink (Edmond *et al.*, 1979). Barth's (1952) work was strongly influenced by a remarkable paper. Conway (1943) developed a number of approaches that are now called "Global Geochemical Cycles" or "Earth System Science" which attempt to provide an integrated view of the chemical evolution of the oceans (Chapter 6.21).

The other striking observation was that trace metals in seawater are not present in trace concentrations because they are present in low concentrations in rivers but because of their high reactivity in seawater. This set the scene for what is now called "biogeochemical cycling" and involves topics such applications of U-series isotopes for studies of particle reactivity, speciation, and bioavailability of trace metals (Chapters 6.01, 6.02, 6.04, and 6.09).

Another consequence is that implicit in the definition of residence time is the assumption of steady state (flux in = flux out) and that τ is a reciprocal first-order rate constant, such that for non-steady state (C = concentration and V = volume)

$$\frac{dC}{dt} = \frac{F_{in}}{V} - \frac{C}{\tau}$$

This equation is the fundamental basis of an important component of the marine geochemist's tool box, the box model, which has played a major role in understanding the role of the oceans in changing atmospheric CO_2 (Chapters 6.10 and 6.18).

It also highlights the importance of time. Defining the rates of oceanic geochemical processes and understanding the underlying causes has been pivotal in the evolution of marine geochemistry. Two other equations are relevant here. One is the decay equation:

$$-\frac{dN}{dt} = \lambda N$$

(N is the number of radioactive atoms present and λ is the decay constant), the application of which to the U- and Th-series isotopes and the cosmogenic isotopes has led to one of the most exciting developments, that of determining ages of water masses (Chapter 6.08) and understanding the interaction of dissolved and particulate forms of elements over a wide range of time constants (Chapters 6.09 and 6.12).

The other is a rate equation of sorts:

$$CaC^{18}O_3 + H_2{}^{16}O \leftrightarrow CaC^{16}O_3 + H_2{}^{18}O$$

Building on the quantification of the equilibrium fractionation of oxygen between calcite and seawater and its temperature sensitivity by H. Urey and S. Epstein, the demonstration of a climate signal for $\delta^{18}O$ in deep-sea cores by Emiliani (1955) and Shackleton (1967) has had a profound influence on marine geochemistry, providing the lynch pin for the discipline of paleoceanography (Chapters 6.13–6.17).

The penultimate concept I shall mention is that of element stoichiometry, illustrated notably by

$$(CH_2O)_{106}(NH_3)_{16}(H_3PO_4)_1 + 138O_2 = 106CO_2 + 16HNO_3 + 1H_3PO_4 + 122H_2O$$

or perhaps $(C_{106}N_{16}P_1)_{1000}Fe_8Mn_4Zn_{0.8}Cu_{0.4}Co_{0.2}Cd_{0.2}$ (Chapter 6.05). The Redfield ratio of the elements assimilated from seawater by phytoplankton (Redfield, 1934) provided the basis for defining biological transformation in water masses during lateral transport, with ramifications for physical and chemical as well as biological oceanography and sediment diagenesis (Chapters 6.02, 6.04, 6.05, and 6.11). Now, it is deviations from Redfield stoichiometry that are proving of greater significance (Chapters 6.04 and 6.05).

Finally, as marine geochemistry has increasingly become concerned with the past, present, and future status (health?) of the ocean–atmosphere system, the concept of "proxy" has become an increasing focus for geochemists. The list is long and becoming longer: $\delta^{18}O$, $\delta^{13}C$, Mg/Ca, Sr/Ca, Cd/Ca, Ba/Ca, Zn/Ca, δ^6Li, $\delta^{10}B$, εNd, $^{87}Sr/^{86}Sr$, *U37K* (Chapters 6.14–6.17 and 6.21).

A special focus in this work is the carbon cycle and its link to changes in greenhouse gases in the atmosphere and implications for climate.

The first section (Chapters 6.01–6.06) discusses the contemporary ocean composition, dealing with physicochemical controls on seawater, controls on trace metals in seawater, gases in seawater, the biological pump, marine bio-inorganic chemistry, and marine organic geochemistry.

The second section (Chapters 6.07–6.11) deals with transport processes in the ocean: hydrothermal processes, tracers of ocean mixing, chemical tracers of particle transport, biological fluxes in the ocean and atmospheric p_{CO_2}, and benthic fluxes and early diagenesis.

The third section (Chapters 6.12–6.17) moves to ocean history and covers the area of paleoclimatology and paleoceanography from marine deposits: geochronometry of marine deposits, geochemical evidence for quaternary sea-level changes, elemental and isotopic proxies of past ocean temperatures, alkenones as paleotemperature indicators, tracers of ocean mixing in the past, and application of long-lived radiogenic isotopic tracers to paleoceanography.

The fourth and final section (Chapters 6.18–6.21) discusses the evolution of seawater composition: the biological pump in the past, the ocean calcium carbonate cycle, Cenozoic ocean chemistry—records from multiple proxies, and the early history of seawater.

As the editor, it has been a pleasure to invite leading scientists to contribute to this volume. I am grateful to them for their outstanding contributions. It is also a pleasure to acknowledge Diana Calvert, Angela Greenwell and Friso Veenstra at Elsevier, Mabel Peterson at Yale, and Sandra Last at Cambridge for their efficient editorial and organizational support and very hard work, "without which this volume would not have been completed" (I have read these words in edited volumes before, but now I fully appreciate their meaning). In particular, thanks go to the Editors-in Chief, Dick Holland and Karl Turekian, for their invitation for me to act as editor, for their helpful advice throughout the project, and for providing two of the chapters that follow.

The three photographs preceding this introduction are of colleagues to whom this volume is dedicated: Harmon Craig (1926–2003), John Edmond (1943–2001), and Cesare Emiliani (1922–1995). Each made outstanding and enduring contributions to this field of endeavor.

REFERENCES

Barth T. W. (1952) *Theoretical Petrology*. Wiley, New York.

Broecker W. S. (1974) *Chemical Oceanography*. Harcourt Brace, Jovanovitch.

Broecker W. S. (2002) *The Glacial World According to Wally*, 3rd edn. Eldigio Press, Palisades, NY.

Conway E. J. (1943) The chemical evolution of the ocean. *Roy. Irish Acad. Proc.* **48B**, 161–212.

Drever J. I. (1974) The magnesium problem. In *The Sea* (ed. E. D. Goldberg). Wiley-Interscience, New York, vol. 5, pp. 337–358.

Edmond J. M., Measures C. I., McDuff R. E., Chan L. H., Collier R., and Grant B. (1979) Ridge crest hydrothermal activity and the balances of the major and minor elements in the ocean: the Galapagos data. *Earth Planet. Sci. Lett.* **46**, 1–18.

Emiliani C. (1955) Pleistocene temperatures. *J. Geol.* **63**, 538–578.

Redfield A. C. (1934) On the proportions of organic derivatives in sea water and their relation to the composition of plankton. In *James Johnstone Memorial Volume* (ed. R. J. Daniel). Univeristy Press of Liverpool, pp. 177–192.

Riley J. (1965) Historical introduction. In *Chemical Oceanography* (eds. J. P. Riley and G. Skirrow). Academic Press, London and New York, vol. 1, chap. 1, 1p.

Riley J. P. and Skirrow G. (eds.) (1965) Academic Press, London and New York, 2 vols.

Shackleton N. (1967) Oxygen isotope analyses and Pleistocene temperatures re-assessed. *Nature* **215**, 15–17.

Sillen L. G. (1967) How have seawater and air got their present compositions? *Chem. Brit.* **3**, 291–297.

6.01
Physicochemical Controls on Seawater

F. J. Millero

University of Miami, FL, USA

NOMENCLATURE

Cl	chlorinity
g_{Est}	grams of salts in an estuary
g_{R}	grams of sea salts in a river
I	molal ionic strength
K	thermodynamic constant
K^*	stoichiometric constant
m	molality
P	physical property of solutions
pK	$= -\log K$
P^0	physical property of water
S	salinity
S_{A}	absolute salinity
S_X	Debye–Huckel limiting law slope
TCO_2	total inorganic carbon
α	molar fraction of a species
β	stability constants for the formation of an ion pair
γ	activity coefficient
σ	ionic strength function
ϕ	apparent molal property of a salt
Φ	apparent molar property for a mixture

The physical–chemical controls on seawater can be attributed to the effect of composition of the major components on the thermodynamic and kinetics of processes in the oceans. In this chapter I will review the experimental and modeling work that has been done on how the composition of the major components of seawater (Na^+, Mg^{2+}, Ca^{2+}, K^+, Sr^{2+}, Cl^-, SO_4^{2-}, HCO_3^-, Br^-, CO_3^{2-}, $B(OH)_3$, $B(OH)_4^-$, F^-, CO_2) controls the rates and equilibria of processes in the oceans. The effect of the major components on (i) the physical–chemical properties of seawater; (ii) the carbonate system in the oceans; (iii) the solubility of iron in seawater; and (iv) the redox reactions of iron and copper in natural waters will be examined.

6.01.1 COMPOSITION OF SEAWATER

Early workers (Culkin, 1965; Millero, 1996) showed that the relative compositions of major components (Na^+, Mg^{2+}, Ca^{2+}, K^+, Sr^{2+}, Cl^-, SO_4^{2-}, HCO_3^-, Br^-, CO_3^{2-}, $B(OH)_3$, $B(OH)_4^-$, F^-) of seawater were constant. This led to the concept of salinity, which is a measure of the total dissolved salts in seawater. The compositions of the major components were initially characterized relative to the chlorinity, which was defined as the mass of the total Cl^- and Br^- in 1 kg of seawater (usually expressed in parts per thousand, ‰). Values of the grams of the major components of seawater as a function of the chlorinity are given in Table 1 (Millero, 1996). The sum of the masses of the individual components of the solution as a function of the chlorinity can be obtained from the individual values. The product of this sum and the chlorinity of a given sample of seawater can be used to define the true or absolute salinity (again normally expressed in parts per thousand, ‰):

$$S_A = 1.815\text{Cl} \tag{1}$$

Since it was difficult to determine the composition of a given sample of seawater, the relationship between salinity and chlorinity was determined on natural samples of seawater (Knudsen, 1901). The salinity of each sample of known chlorinity was determined by heating to dryness and weighing. This led to the so-called Knudsen equation,

$$S = 0.03 + 1.8065\text{Cl} \tag{2}$$

The salinity determined from this equation (e.g., $S = 35.00$ at $\text{Cl} = 19.374$) is lower than the value determined from the composition (Equation (1)), due to the loss of CO_2, HCl, and $B(OH)_3$ on heating. The salinity determined from Equation (2) was used for over 60 years to examine the properties of seawater.

Table 1 The composition of average seawater ($\text{Cl} = 19.374$, $\text{pH} = 8.1$, and $t = 25\,°\text{C}$).[a]

Solute	g_i/Cl (‰)	*mol/(kg solution)*
Na^+	0.55661	0.46907
Mg^{2+}	0.06626	0.05282
Ca^{2+}	0.02127	0.01028
K^+	0.02060	0.01021
Sr^{2+}	0.00041	0.00009
Cl^-	0.99891	0.54588
SO_4^{2-}	0.14000	0.02824
HCO_3^-	0.00552	0.00175
Br^-	0.00347	0.00084
CO_3^{2-}	0.00083	0.00027
$B(OH)_4^-$	0.000415	0.00010
F^-	0.000067	0.00068
$B(OH)_3$	0.001002	0.00031
$\sum =$	1.81540	

[a] Source: Millero (1996).

In more recent years the salinity of seawater is determined from conductivity measurements using the practical salinity scale (PSS) (Lewis, 1978, 1980). The practical salinity of seawater is related to the conductivity of seawater relative to KCl. The salinity of seawater is defined as 35.000 (no units) if the conductance ratio is 1.0000 to a KCl solution containing a mass of 32.4356 g of KCl in a mass of 1 kg of solution at 15 °C (Culkin and Smith, 1980; Poisson, 1980). The practical salinity of seawater as a function of temperature and salinity was characterized by making conductance measurements on seawater of known salinity ($S = 35$) and chlorinity (19.374) by weight evaporation or dilution with distilled water (Dauphinee *et al.*, 1980). Since the scale was set up using real seawater of known chlorinity and salinity, it has an initial connection to the original definition of salinity by Knudsen ($S = 1.80655\text{Cl}$ when $S = 35$ the $\text{Cl} = 19.374$). This relationship between Cl and S, however, will not exist in the future since the conductivity may change when Cl is constant. The practical salinity is valid from $S = 2$ to 42. Extensions of the scale to lower salinities are also available (Hill *et al.*, 1986). Care must be taken in using the PSS outside of the range of the equations and for waters of different composition than seawater (Millero, 2000a).

The composition of seawater with a chlorinity of 19.374 or practical salinity of 35.000 at 25 °C and $\text{pH} = 8.1$ is given in Table 1. The molality (m, mol (kg $H_2O)^{-1}$) and ionic strength (I, mol (kg $H_2O)^{-1}$) for sea salt of salinity $S = 1.80655\text{Cl}$ are given by

$$m = 16.011S/(1,000 - 1.00488S) \tag{3}$$

$$I = 19.9243S/(1,000 - 1.00488S) \tag{4}$$

The molecular weight of sea salt is $M = 62.793\ \text{g mol}^{-1}$ (Millero, 1996). The molalities (m_i) of the individual components of seawater can also be defined by equations similar to Equation (3).

Since the salinity is related to the total salts in a given sample of seawater, it can be used to express the changes in the physical chemical properties of seawater as a function of composition. In essence, one treats seawater as a sea salt similar to an individual salt such as NaCl. Most of the early physical–chemical properties of seawater were made on diluted or evaporated seawater of a known chlorinity (19.374 or $S = 35.000$). By multiplying the values of Cl by

1.80655, one can convert the equations to a function of salinity.

6.01.1.1 Causes of Major Components Not Being Conservative

Although the major components of seawater are relatively constant, a number of factors can cause the ocean and estuarine waters to be non-conservative. These processes occur in estuaries, anoxic basins and sediments, hydrothermal vents, and evaporation basins and include precipitation, dissolution, evaporation, freezing, and oxidation processes. Some examples will be discussed briefly.

The finding of hydrothermal vents has led to the discovery that a number of elements can be added (Ca^{2+}, Cu^{2+}, Zn^{2+}, Mn^{2+}, SiO_2) and taken out (Mg^{2+}, SO_4^{2-}) of ocean waters. The loss of Mg^{2+} at hydrothermal vents in vent fluids of high SiO_2 is shown in Figure 1 (Millero, 1996). The waters coming out of hydrothermal vents at high temperature are devoid of Mg^{2+}. This is related to the formation of magnesium silicates when the seawater reacts with molten basalt. As shown in Figure 2, this deficiency of Mg^{2+} changes the amount of Mg^{2+} in deep Pacific waters (De Villiers and Nelson, 1999).

The nature of the rocks being weathered and the soil types yield groundwaters of different chemical composition and control the concentrations of salts in rivers . The total solids in most rivers are less than 200 ppm or $S = 0.2‰$, and are mostly composed of Mg^{2+}, Ca^{2+}, and HCO_3^-. The major river cation is Ca^{2+} and HCO_3^- is the major anion. The major components of world river water (Ca^{2+} and HCO_3^-) come from the weathering of $CaCO_3$. The mixing of river waters with different salts can result in an estuary that has a composition different from seawater's. This mixing will result in a linear equation with intercepts equal to the values for conservative river ions, as is shown for Mg^{2+}, Ca^{2+}, HCO_3^-, and SO_4^{2-} in the Baltic estuary in Figure 3 (Millero, 1996). The total grams of salts or ions in an estuary (g_{Est}) are given by

$$g_{Est} = g_R + [(35.171 - g_R)/19.374]Cl \qquad (5)$$

where g_R is the grams of river salts. The values of g_R for conservative elements can be obtained from

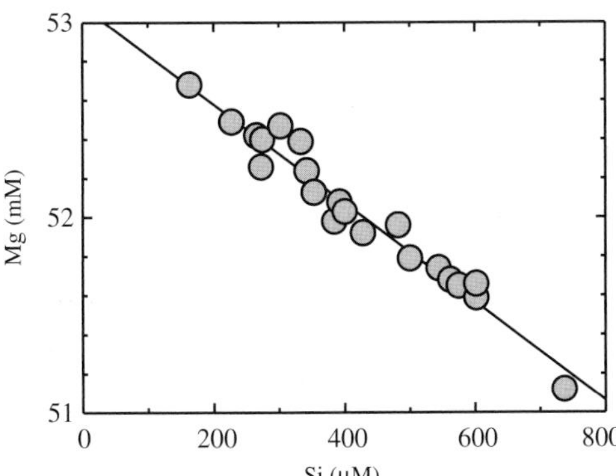

Figure 1 The concentration of Mg^{2+} versus SiO_2 for hydrothermal vent waters (source Millero, 1996).

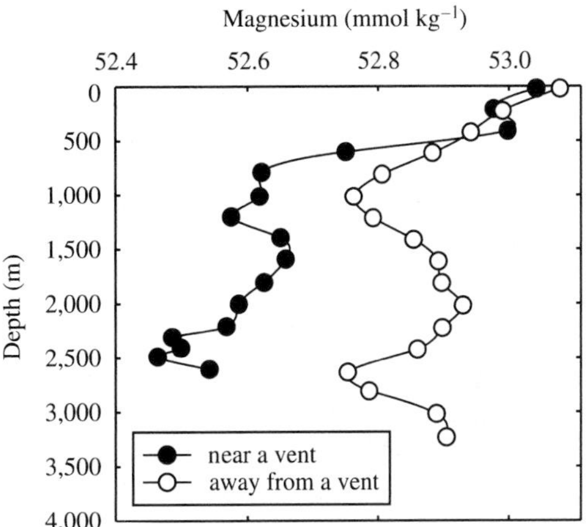

Figure 2 The concentration of Mg^{2+} in deep waters of the Pacific (source De Villiers and Nelson, 1999).

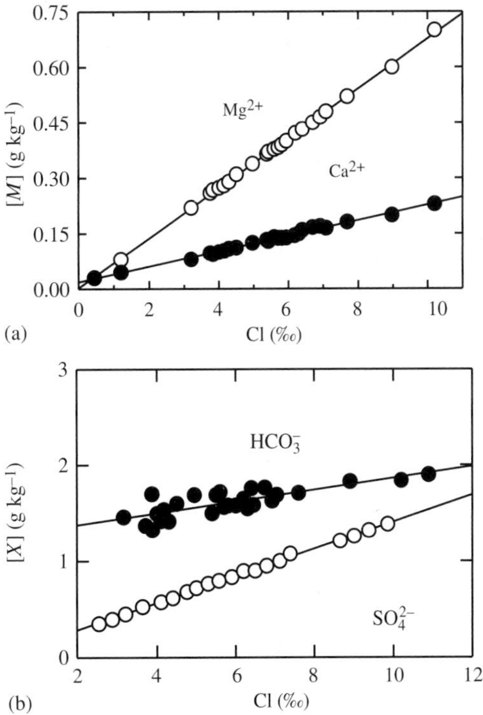

Figure 3 (a) The concentration of Mg^{2+} and Ca^{2+} and (b) HCO_3^- and SO_4^{2-} Baltic estuary as a function of chlorinity (source Millero, 1996).

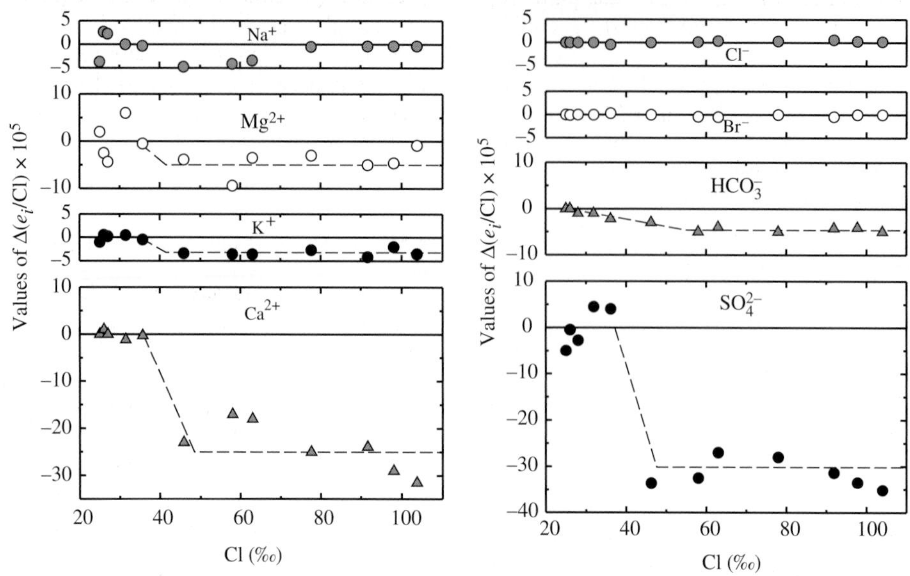

Figure 4 The changes in the cations and anions during the evaporation of Mexican lagoon waters (source Fernandez *et al.*, 1982).

the measured values in the estuary (g_E) as a function of the chlorinity or salinity by rearranging Equation (5).

Since the composition of brines can be different from that of seawater, its mixtures with seawater will have a different composition. The evaporation of seawater in isolated basins can also change the composition due to the precipitation of a number of salts. This is shown in Figure 4 for the evaporation of a Mexican lagoon (Fernandez *et al.*, 1982). Ca^{2+}, K^+, HCO_3^-, and SO_4^{2-} are lost from the solution during the initial evaporation at values of Cl near 40 ($S \approx 74$). This is due to the initial precipitation of $CaCO_3$ and later precipitation of $CaSO_4$. The loss of K^+ may be related to its co-precipitation with $CaCO_3$ or $CaSO_4$. As will be discussed later, the changes in the composition of seawater due to mixing, precipitation, and biochemical reactions in any natural water of known composition can be estimated from ionic interaction models. These models can give reasonable approximations of the rates and equilibrium for most processes occurring in natural waters.

6.01.2 PHYSICAL AND CHEMICAL PROPERTIES OF SEAWATER

6.01.2.1 Pressure–Volume–Temperature Properties

Equations are available for the pressure–volume–temperature (*PVT*) properties of seawater as a function of salinity (0–40), temperature (−2 °C to 40 °C), and pressure (0–1,000 bar) (Millero and Poisson, 1981; Millero, 1982, 1983a, 2001). The 1 atm part of the equation of state is based on the density measurements of Millero *et al.* (1976a) and Poisson *et al.* (1980a). The high-pressure equation of state (Millero *et al.*, 1980) is based on three independent studies: (i) the sound-derived specific volumes of water and seawater (Wilson, 1959, 1960a,b; Fine and Millero, 1973; Wang and Millero, 1973; Fine *et al.*, 1974; Millero and Kubinski, 1975; Chen and Millero, 1978; Chen *et al.*, 1977); (ii) the direct measurements of the specific volume of seawater using a magnetic float method (Chen and Millero, 1976b); and (iii) the direct measurements of the thermal expansion and compressibility of water and seawater by dilatometry (Bradshaw and Schleicher, 1970, 1976, 1986).

Dushaw *et al.* (1993) have shown that the equations for sound speed given by Del Grosso (1974) give better predictions of acoustic arrival patterns than those computed using the equations of Chen and Millero (1978). These findings are in agreement with the earlier calculations of Spiesberger and Metzger (1991a,b). Both studies strongly suggest that similar equations by Chen and Millero (1977) are in error between 0 °C and 15 °C and above a pressure of 200 bar. Millero and Li (1994) have discussed the possible causes of these differences. They showed that the sound speeds of Chen and Millero (1978) were in error because they (Chen and Millero, 1976a) converted their relative measurements to absolute values using the pure-water sound speeds of Wilson (1959). These measurements appear to be in error at low temperatures and high pressures. They gave a corrected equation that gives an adequate representation of the sound speed from 0 °C to 40 °C, $S = 0$–35, and $P = 0$–1,000 bar. They also showed that the use of unreliable sound speeds for water and seawater (Chen and Millero, 1978; Chen *et al.*, 1977) does not affect the equation of state of seawater at low temperatures and high pressures. This is because the equation of state was determined by fitting the differences in the *PVT* properties of seawater and water rather than the absolute values. These results have led to equations that can be used to determine the thermal expansion, adiabatic temperature gradient, and potential of ocean waters (Bryden, 1973).

The *PVT* properties of seawater have been fitted to equations of the form (Wirth, 1940)

$$P = P^0 + AS + BS^{3/2} + CS^2 \qquad (6)$$

where P^0 is the property for pure water, and A, B, and C are functions of temperature and pressure. This equation is related to the functional form of the Debye–Hückel equation (Millero, 2001) for the thermodynamic properties of electrolyte solutions.

6.01.2.2 Thermochemical Properties

The thermochemical properties of seawater from 0 °C to 40 °C and $S = 0$–40 have been determined by a number of workers (Millero, 1983a, 2001). Results over a wider range of temperatures (0–200 °C) and concentrations ($S = 0$–300) are also available (Bromley *et al.*, 1970, 1974). The thermochemical properties include the heat capacity (Cox and Smith, 1959; Bromley *et al.*, 1967, 1970; Bromley, 1968a,b; Millero *et al.*, 1973a), enthalpy (Connors, 1970; Bromley, 1968a; Singh and Bromley, 1973; Millero *et al.*, 1973b), vapor pressure (Robinson, 1954; Rush and Johnson, 1966), and freezing point (Miyake, 1939; Murray and Murray, as quoted by Riley and Chester, 1971; Doherty and Kester, 1974; Fujino *et al.*, 1974). The apparent molal properties (ϕ_X) of seawater have been fitted (Millero and Leung, 1976; Bromley *et al.*, 1967) to equations of the form

$$\phi_X = S_X I^{0.5}[I/(1+I^{0.5}) - \sigma/3] + B_X I + C_X I^{1.5} \qquad (7)$$

where S_X is the Debye–Hückel limiting law slope (Millero, 2001), I is the molal ionic strength, σ is a function of ionic strength, and B_X and C_X are functions of temperature (X = heat capacity, enthalpy, free energy, etc.). More recent studies (Pierrot and Millero, 2000; Millero and Pierrot, submitted) have fit the apparent molal *PVT* and thermal chemical properties of seawater to the Pitzer equations over a wide range of temperature (0–200 °C), salinity (0–200), and pressure (0–1,000 bar). These equations allow one to calculate all the thermodynamic properties by differentiation or integration. The thermochemical equations can be used to determine the free energy, entropy, osmotic pressure, and activity of water in seawater.

6.01.2.3 Other Properties of Seawater

Several workers have determined a number of other physical properties of seawater (Seidler and Peters, 1986; Millero, 2001). These include the boiling point (Stoughton and Lietzke, 1967), the diffusion coefficient (Caldwell, 1974a,b), surface tension (Loglio *et al.*, 1978), thermal conductivity (Caldwell, 1974b, 1978; Jamieson and Tudhoe, 1970), viscosity (Chen *et al.*, 1973; Horne and Johnson, 1966; Stanley and Batten, 1969), dielectric constant (Ho and Hall, 1973), and refractive index (Utterback *et al.*, 1934; Eisenberg, 1958; Cox, 1965; Mehu and Johannin-Gilles, 1969; Rusby, 1967; Leyendekkers and Hunter, 1976; Stanley, 1971). The conductivity of seawater as a function of temperature and salinity (discussed earlier and used to determine the PSS) was determined by Poisson (1980) and Dauphinee and Klein, 1977.

6.01.2.4 Application to other Natural Waters

Although the equations developed for the physical–chemical properties of seawater are strictly valid for ocean waters, studies (Millero *et al.*, 1976a,b; Poisson *et al.*, 1980b, Millero, 2000a) have shown that for dilute natural waters the physical properties are nearly equal to the values for seawater at the same absolute salinity (Poisson *et al.*, 1981; Millero, 2000a). Finally, it should be pointed out that the physical properties of seawater change in deep waters due to the oxidation of organic material (Brewer and Bradshaw, 1975; Millero *et al.*, 1976a,b,c, 1978; Millero, 2000b). This oxidation adds nitrate, phosphate, carbonate, and silicate to the waters (Brewer and Bradshaw, 1975; Millero, 2000b) that change the physical properties such as density, but do not change the practical salinity by the appropriate amount. Direct measurements of the density (ρ) of seawater (Millero *et al.*, 1976a,b,c, 1978) have shown that the values of σ_T ($=10^3(\rho - 1)$) could be 0.025 higher due to this oxidation of organic carbon in deep Pacific waters. This correction can be made by increasing the salinity of the water by the additions of the nutrient salts or by accounting for the changes in the density due to the addition of HCO_3^-, NO_3^-, and SiO_2 to the waters (Brewer and Bradshaw, 1975; Millero *et al.*, 1976c, 1978; Millero, 2000b). The latter technique is the preferred method, since oceanographers do not like to manipulate the practical salinity. As will be discussed later, reasonable estimates of all the physical–chemical properties of natural waters of known composition can be made using ionic interaction models.

6.01.3 THERMODYNAMIC EQUILIBRIA IN SEAWATER

Although the thermodynamic equilibrium of processes in seawater have been studied for many systems, only two systems of wide interest will be examined: (i) the carbonate equilibrium in seawater and (ii) the solubility of Fe(III) in seawater.

6.01.3.1 Carbonic Acid Equilibria in Seawater

Measurements of a number of acids have been made in seawater, and equations are available to determine the effect of salinity, temperature, and pressure on the dissociation constants (Millero, 2001). The most widely studied acids in seawater are those related to the carbonate system. Most of these studies have been made in artificial seawater that contains the major ionic components (Na^+, Mg^{2+}, Ca^{2+}, K^+, Cl^-, SO_4^{2-}, F^-). It was thought that constants determined in this artificial seawater could be used in real seawater. Although this is the case for most acid–base systems, as will be discussed below, this is not the case for the carbonate system (Mojica and Millero, 2002).

The interest in the carbonate system is related to attempts to understand the uptake of fossil fuel produced CO_2 by the oceans. The carbonate system can be studied by measuring pH, total alkalinity (TA), total inorganic carbon (TCO_2), and the fugacity of CO_2 (f_{CO_2}). At least two of these variables are needed (Park, 1969) to characterize the CO_2 system in the oceans. Reliable stoichiometric constants (K^*) for the carbonate system are needed to determine the concentration, mol (kg solution)$^{-1}$, of the components of the CO_2 system ($[HCO_3^-]$, $[CO_2]$, $[CO_3^{2-}]$) and the saturation state of $CaCO_3$ as a function of salinity, temperature, and pressure (Culberson and Pytkowicz, 1968; Ingle, 1975; Millero, 1995, 2001). This includes constants for the solubility of CO_2 in seawater (Weiss, 1974)

$$CO_2(g) = CO_2(l): \quad K_0^* = [CO_2]/f_{CO_2} \qquad (8)$$

constants for the dissociation of carbonic acid (Mehrbach *et al.*, 1973; Hansson, 1973; Dickson and Millero, 1987; Goyet and Poisson, 1989; Roy *et al.*, 1993a; Mojica and Millero, 2002)

$$H_2O(l) + CO_2(l) = H^+ + HCO_3^- \qquad K_1^* = [H^+][HCO_3^-]/[CO_2] \qquad (9)$$

$$HCO_3^- = H^+ + CO_3^{2-} \qquad K_2^* = [H^+][CO_3^{2-}]/[HCO_3^-] \qquad (10)$$

and the solubility of calcium carbonate (Ingle, 1975; Mucci, 1983)

$$Ca^{2+} + CO_3^{2-} = CaCO_3(s) \qquad K_{sp}^* = [Ca^{2+}][CO_3^{2-}] \qquad (11)$$

It is also necessary to know the dissociation constants of boric acid (Dickson, 1990a; Roy *et al.*, 1993b), water (Culberson and Pytkowicz, 1973; Dickson and Riley, 1979a; Millero, 1995), phosphoric acid (Kester and Pytkowicz, 1967; Johansson and Wedborg, 1979; Dickson and Riley, 1979b; Yao and Millero, 1995), hydrogen fluoride (Dickson and Riley, 1979a), hydrogen sulfate (Khoo *et al.*, 1977a; Dickson, 1990b; Campbell *et al.*, 1993), and silicic acid (Busey and Mesmer, 1977; Sjöberg *et al.*, 1981; Millero, 1995) for open-ocean studies, and ammonium (Khoo *et al.*, 1977b; Johansson and Wedborg, 1980; Yao and Millero, 1995) and hydrogen sulfide (Millero *et al.*, 1988; Yao and Millero, 1995) in anoxic waters. The equations needed for these calculations are summarized elsewhere (Millero, 1995; Millero, 2001). A program is also available to calculate the parameters of the CO_2 system in seawater as a function of the input parameters, of pH, TA, TCO_2, f_{CO_2}, salinity, phosphate, silicate, temperature, and pressure (Lewis and Wallace, 1998).

Four groups have made measurements of pK_1^* and pK_2^* for carbonic acid: Hansson (1973), Mehrbach *et al.* (1973), Goyet and Poisson (1989), and Roy *et al.* (1993a). The measurements of Mehrbach *et al.* (1973) were made in real seawater, while the other studies were made in artificial seawater. The standard errors of the fits of the measured values of pK_1^* and pK_2^* to functions of temperature and salinity by the authors are within 0.007 and 0.011, respectively. Comparisons of the values of pK_1^*, pK_2^*, and $pK_2^* - pK_1^*$ measured in artificial seawater with the results in seawater at different temperatures are shown in Figures 5 and 6. Near 25 °C the measurements of pK_1^* in seawater are 0.01 lower than the measurements in artificial seawater, whereas the measurements of pK_2^* in seawater are 0.04 higher than the measurements in artificial seawater. The measured values of $pK_2^* - pK_1^*$ in seawater are 0.05 higher than the measurements in artificial seawater near 25 °C. These differences can lead to large errors in the calculated

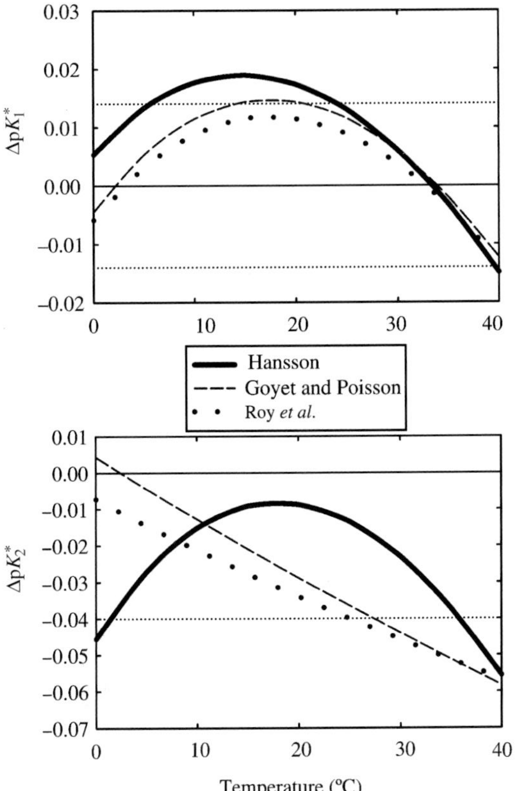

Figure 5 A comparison of the effect of temperature on pK_1^* and pK_2^* at $S = 35$ of various workers to the results of Mehrbach *et al.* (1973) (Δ = others − Mehrbach) (source Mojica and Millero, 2002).

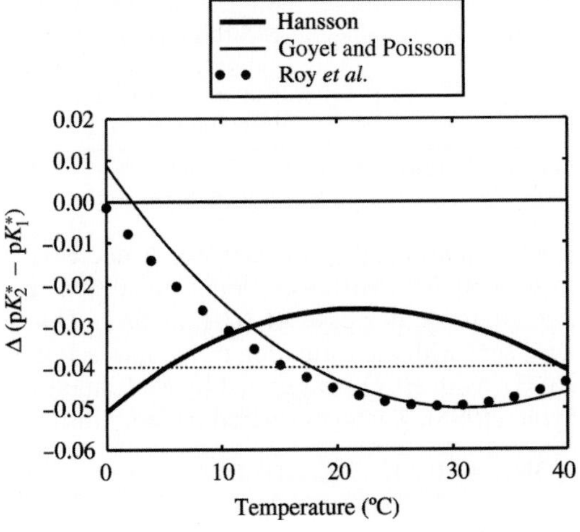

Figure 6 A comparison of the effect of temperature on $pK_2^* - pK_1^*$ at $S = 35$ of various workers to the results of Mehrbach *et al.* (1973) (Δ = others − Mehrbach) (source Mojica and Millero, 2002).

components of the CO_2 system. For example, an input of pH and TA or TCO_2 can lead to errors in the calculated values of f_{CO_2} of 10 μatm due to errors in pK_2^*. Calculations of the CO_2 parameters using an input of TA and TCO_2, f_{CO_2} and TCO_2, or f_{CO_2} and TA require reliable values of $pK_2^* - pK_1^*$ (Lee *et al.*, 2000). For example, an error of 0.04 in $pK_2^* - pK_1^*$ with an input of TA and TCO_2 can lead to errors in the calculated values of f_{CO_2} of 26 μatm.

A number of internal consistency tests using measurements made in the laboratory (Lee *et al.*, 1996; Lueker *et al.*, 2000) and at sea (Lee *et al.*, 1997, 2000; Wanninkhof *et al.*, 1999; Millero *et al.*, 2002) indicate that the values of $pK_2^* - pK_1^*$ of Mehrbach *et al.* (1973) are more reliable than the results of other workers (Hansson, 1973; Goyet and Poisson, 1989; Roy *et al.*, 1993a,b). These studies showed that measurements of TA and TCO_2 of ocean waters gave more reliable predictions of f_{CO_2} when the values of $pK_2^* - pK_1^*$ determined by Mehrbach *et al.* (1973) were used. Since the values of pK_1^* determined by various workers are in reasonable agreement (0.01), these results suggest that the values of pK_2^* determined by Mehrbach *et al.* (1973) are more reliable.

Recent studies (Millero *et al.*, 2002) have examined the carbonate constants that could be determined from the measurements of pH, TA, TCO_2, and f_{CO_2} on surface and deep waters in the Atlantic, Indian, Southern, and Pacific oceans made during the WOCE cruises. The results of this study on samples where f_{CO_2} was determined at 4 °C and 20 °C are shown in Table 2. These results clearly indicate that the carbonate constants of Mehrbach *et al.* (1973) are more reliable than other workers.

The recent pK_1^* and $pK_1^* + pK_2^*$ measurements (Mojica and Millero, 2002) on seawater confirm the reliability of the measurements of Mehrbach *et al.* (1973). Their measured values of (1/2) ($pK_1^* + pK_2^*$) are in good agreement (0.005) with the results of Mehrbach *et al.* (1973). Values of the pK_1^* in seawater were also in agreement (0.005) with the results of Mehrbach *et al.* between 20 °C and 30 °C than the studies made in artificial seawater. The results of Mojica and Millero (2002) and Mehrbach *et al.* (1973) have been combined to yield equations of pK_1^* ($\sigma = 0.0056$) and pK_2^* ($\sigma = 0.010$) that are valid over a wide range of temperature (0–45 °C) and salinity (5–42).

Both studies indicate that the values of pK_1^* (seawater) > pK_1^* (artificial seawater) by ~0.01 and pK_2^* (seawater) < pK_2^* (artificial seawater) by ~0.05 near 25 °C. Mojica and Millero (2002) made measurements of $pK_1^* + pK_2^*$ and pK_1^* in artificial seawater at 25 °C and found that the results were in good agreement with the measurements of Hansson (1973), Goyet and Poisson (1989), and Roy *et al.* (1993a). Mojica and Millero (2002) also made $pK_1^* + pK_2^*$ and pK_1^* measurements in artificial seawater with boric acid and found that the results were in reasonable agreement with the measurements in real seawater (see Table 3). These results indicate that the offsets

Table 2 Calculated values of pK_1, pK_2, and $pK_2 - pK_1$ ($S = 35$) at various temperatures (on the seawater pH scale).

Data	*Waters*	pK_1^*	pK_2^*	$pK_2^* - pK_1^*$
4 °C	Deep[a]	6.062 ± 0.009	9.321 ± 0.020	3.257 ± 0.020
	Surface[b]	6.057 ± 0.005	9.324 ± 0.008	3.258 ± 0.008
	Lee *et al.*[c]	6.058 ± 0.006	9.303 ± 0.011	3.245 ± 0.011
	Lueker *et al.*[c]			3.241 ± 0.013
20 °C	Deep[d]	5.888 ± 0.010	9.035 ± 0.019	3.148 ± 0.019
	Surface[b]	5.879 ± 0.005	9.032 ± 0.008	3.152 ± 0.008
	Lee *et al.*[c]	5.888 ± 0.006	9.011 ± 0.011	3.139 ± 0.014
	Lueker *et al.*			3.141 ± 0.013

Source: Millero *et al.* (2002).
[a] Adjustments of 0.002 in the pK_1 and 0.006 in pK_2 were made to correct the values from an average salinity of 34.5 ± 0.3 to 35.00 using Equations (5) and (6). [b] At $S = 35$. [c] Values at low f_{CO_2}. [d] Adjustments of 0.0005 in the pK_1 and 0.001 in pK_2 to correct for an average salinity of 34.9 ± 0.6 to 35.00.

Table 3 A comparison of the values of pK_1^* and pK_2^* for artificial seawater with and without boric acids ($S = 34.855$ and $t = 25$ °C) with other workers.[a]

Constant	*Without*	*With*	*Reference*
pK_1^*	5.849	5.836	Mojica and Millero (2002)
		5.836[b]	Mojica and Millero (2002)
		5.837[b]	Mehrbach *et al.* (1973)
	5.849		Goyet and Poisson (1989)
	5.847		Roy *et al.* (1993a)
	5.850		Hansson (1973)
pK_2^*	8.90	8.94	Mojica and Millero (2002)
		8.96[b]	Mojica and Millero (2002)
		8.96[b]	Mehrbach *et al.* (1973)
	8.92		Goyet and Poisson (1989)
	8.92		Roy *et al.* (1993a)
	8.94		Hansson (1973)
$(1/2)(pK_1^* + pK_2^*)$	7.36	7.39	Mojica and Millero (2002)
		7.40[b]	Mojica and Millero (2002)
		7.40[b]	Mehrbach *et al.* (1973)
	7.38		Goyet and Poisson (1989)
	7.38		Roy *et al.* (1993a)
	7.40		Hansson (1973)

[a] All the constants have been adjusted to the SW pH scale. [b] Measurements made in real seawater.

between seawater and artificial seawater are due to interactions of boric acid with HCO_3^- and CO_3^{2-} and can be attributed to changes in the activity coefficients of HCO_3^- and CO_3^{2-}. The stoichiometric constants are related to the thermodynamic values in water (K_1^0 and K_2^0), the activity (a_i), and the activity coefficients (γ_i) by

$$K_1^* = [H^+][HCO_3^-]/[CO_2] = K_1^0 \alpha_{H_2O} \gamma_{CO_2} / \gamma_H \gamma_{HCO_3} \quad (12)$$

$$K_2^* = [H^+][CO_3^{2-}]/[HCO_3^-] = K_2^0 \gamma_{HCO_3} / \gamma_H \gamma_{CO_3} \quad (13)$$

The increase in K_1^* and the decrease in K_2^* can be related to a decrease in γ_{HCO_3} and an increase in γ_{CO_3}. This implies that the interactions of HCO_3^- and CO_3^{2-} with $B(OH)_3$ or $B(OH)_4^-$ cause γ_{HCO_3} to decrease and the γ_{CO_3} to increase. A decrease in $\gamma_{HCO_3} = 0.70$ (Millero, 1996) to 0.684 can account for a decrease of 0.01 in pK_1^*. This is consistent with the formation of the interaction of $B(OH)_3$ with HCO_3^- suggested by McElligott and Byrne (1998). They considered the equilibrium

$$B(OH)_3 + HCO_3^- = B(OH)_2CO_3^- + H_2O \quad (14)$$

This equilibrium (Equation (14)) will lower the concentration of HCO_3^- and decrease γ_{HCO_3}, which is consistent with the observed behavior. Unfortunately, the value they found for the stability constant for the formation of $B(OH)_2CO_3^-$ is not large enough to change the values of K_1^* by the amount observed.

Since the changes in K_2^* are much larger than in K_1^*, the interactions of $B(OH)_3$ or $B(OH)_4^-$ with CO_3^{2-} must be greater than with HCO_3^-. An increase of γ_{CO_3} from 0.020 (Millero, 1996) to

0.0215 would result in an increase of 0.03 in pK_2^* (the interactions of HCO_3^- with $B(OH)_3$ would account for 0.01 in the increase in pK_2^*). Since the formation of a complex between CO_3^{2-} and $B(OH)_3$ would lower the γ_{CO_3}, this is unlikely to be the case. The increase in γ_{CO_3} is probably related to anion–anion interactions (CO_3^{2-} with $B(OH)_4^-$), similar to that found for other anions (Pitzer, 1991). New studies are needed to examine these interactions in more detail.

That the effect of boric acid on K_1^* and K_2^* appears to be a function of temperature is also surprising. At 0 °C, the values of pK_1^* and pK_2^* in seawater and artificial seawater are in reasonable agreement. The offsets in pK_1^* and pK_2^* just occur at 10–30 °C. These results indicate that the interactions between HCO_3^- and CO_3^{2-} with $B(OH)_3$ or $B(OH)_4^-$ are strong functions of temperature. This is not commonly found for anion–anion interactions (Millero, 2001).

Laboratory studies (Lee *et al.*, 1996; Lueker *et al.*, 2000) and field measurements (Millero *et al.*, 2002) also indicate that the values of $pK_2^* - pK_1^*$ of seawater appear to decrease as the p_{CO_2} increases above ~650 μatm. This is shown in Figure 7 (Lee *et al.*, 1996). It is not clear that this effect is related to borate–carbonate interactions or organic acids that affect the TA in the oxygen minimum zone.

6.01.3.2 Solubility of Fe(III) in Seawater

The chemistry of iron in natural waters has received a lot of attention (Chisholm and Morel, 1991; Bruland and Wells, 1995). Much of this interest comes from its limitations in high-nutrient low-chlorophyll waters (Martin and Fitzwater, 1988). Although iron is one of the most abundant elements on Earth, its concentration in seawater is very low (Johnson *et al.*, 1997). The need of iron as a micronutrient for the growth of phytoplankton has been demonstrated in recent IRONEX studies in high-nitrate low-chlorophyll open-ocean waters (Martin *et al.*, 1994; Coale *et al.*, 1996) and in some coastal upwelling areas (Hutchins and Bruland, 1998). Redox reactions, inorganic and organic complexation, adsorption, and precipitation control the concentration of iron in natural waters. Dissolved iron can exist in seawater in two oxidation states, Fe(II) and Fe(III), free or complexed with inorganic and organic ligands. The availability of iron in ocean waters is predominately limited by the solubility of Fe(III) (Johnson *et al.*, 1997). Recent authors (Johnson *et al.*, 1997; Millero *et al.*, 1995; Millero, 1998; Kuma *et al.*, 1992, 1996, 1998a,b) have discussed the factors that control the solubility of iron in natural waters. Iron hydroxide complexes dominate the inorganic speciation of Fe(III) in the pH range of most natural waters (Byrne and Kester, 1976; Byrne and Luo, 2000; Millero *et al.*, 1995; Kuma *et al.*, 1996). The formation of organic complexes is important at a pH near 8 (Johnson *et al.*, 1997; Kuma *et al.*, 1996; Millero, 1998; van den Berg, 1995). These studies point out that an accurate knowledge of the Fe(III) hydrolysis chemistry and the formation of organic complexes is critical to the understanding of the behavior of iron in natural waters (Byrne and Kester, 1976; Bruland and Wells, 1995; Millero *et al.*, 1995; Kuma *et al.*, 1996).

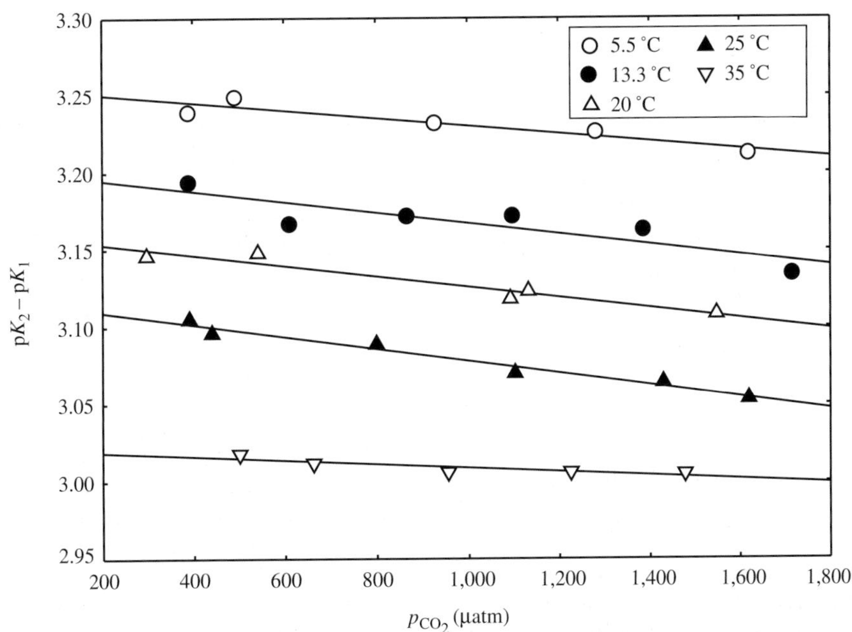

Figure 7 Values of $pK_2^* - pK_1^*$ determined from the measurements of Lee *et al.* (1996) as a function of p_{CO_2}.

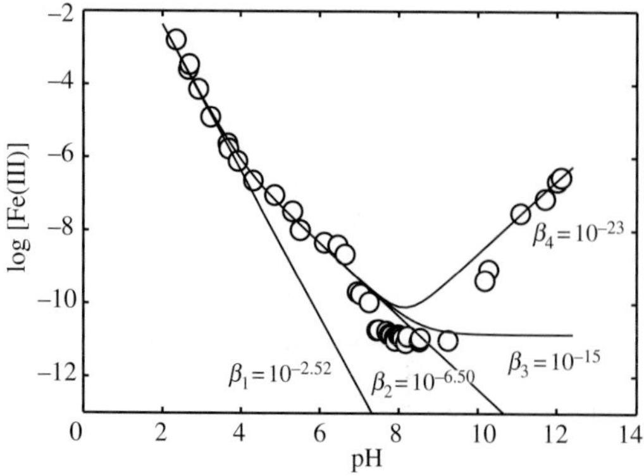

Figure 8 The solubility of Fe(III) in 0.7 *m* NaCl at 25 °C (source Liu and Millero, 1999).

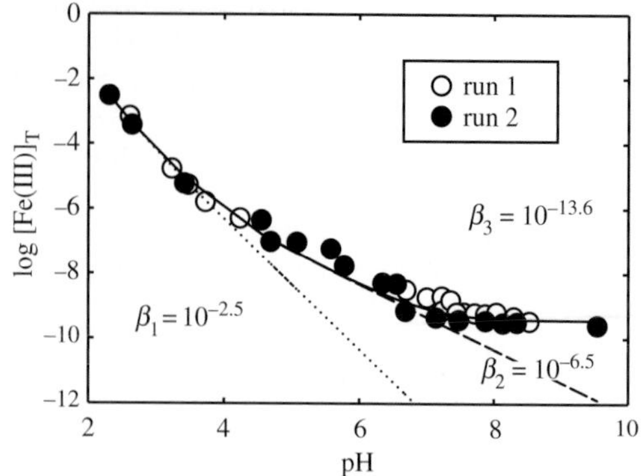

Figure 9 The solubility of Fe(III) in seawater ($S = 36$) at 25 °C (source Liu and Millero, 2002).

Liu and Millero (1999) have determined the solubility of Fe(III) in NaCl solutions as a function of pH, temperature, and ionic strength. These results clearly show that the solubility in NaCl (10 pM) near pH = 8 is much higher than in seawater (200–600 pM) at the same temperature and ionic strength. These results have been used to determine the hydrolysis constants for the formation of Fe(III) complexes with hydroxide (Figure 8) as a function of temperature and ionic strength. The cumulative stoichiometric hydrolysis constants β_j^* for reactions are given by

$$Fe^{3+} + H_2O \stackrel{\beta_1^*}{=} Fe(OH)^{2+} + H^+ \quad (15)$$

$$Fe^{3+} + 2H_2O \stackrel{\beta_2^*}{=} Fe(OH)_2^+ + 2H^+ \quad (16)$$

$$Fe^{3+} + 3H_2O \stackrel{\beta_3^*}{=} Fe(OH)_3^0 + 3H^+ \quad (17)$$

$$Fe^{3+} + 4H_2O \stackrel{\beta_4^*}{=} Fe(OH)_4^- + 4H^+ \quad (18)$$

More recent measurements (Liu and Millero, 2002) of the solubility of Fe(III) in seawater have been made as a function of pH (2–9), temperature (5–50 °C), and salinity (0.2–36). The results as a function of pH at 25 °C and $S = 36$ are shown in Figure 9. These results have been used to derive equations that can be used to determine the solubility in ocean waters. As shown in Figure 10, the solubilities of Fe(III) in seawater (200–300 pM) near a pH of 8 are much higher than the results in 0.7 *m* NaCl (10 pM). Measurements of the solubility in sea salts, and artificial and UV irradiated seawater were the same as in NaCl at the same ionic strength (Figure 11). The increased solubilities in seawater are due to the complexation of Fe(III) with natural organic ligands. The solubility in seawater was higher at low temperatures and low pH similar to NaCl. These effects could partly be responsible for the higher concentrations of soluble Fe(III) in deep waters (Martin and Gordon, 1988; Johnson *et al.*, 1997; Wu *et al.*, 2001).

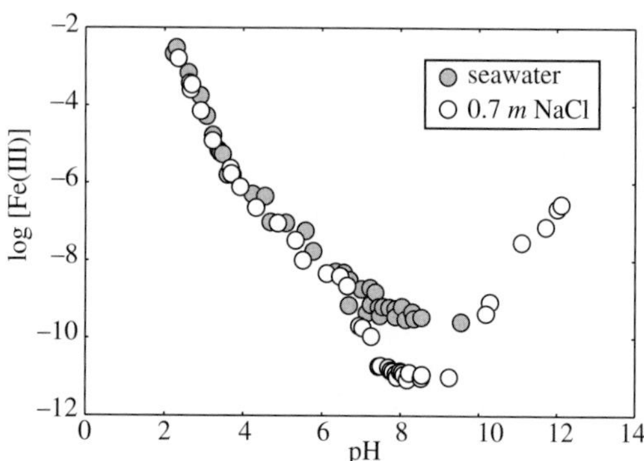

Figure 10 Comparison of the solubility of Fe(III) in seawater ($S = 36$) and 0.7 *m* NaCl at 25 °C (source Liu and Millero, 2002).

The measurements of the solubility of Fe(III) in seawater point out the importance of how low concentrations of natural organic material can affect the speciation and solubility of metals in seawater. This organic material does not have an effect on the physical and chemical properties of seawater and the activity of the major components of seawater, but can change the reactivity of most trace metals in the oceans. These effects are discussed in more detail in other chapters.

6.01.4 KINETIC PROCESS IN SEAWATER

The major components of seawater can also affect the rates of ionic reactions of metals and nonmetals in seawater and other natural waters (Millero, 1985, 1989, 2001). The rates of oxidation of metals can be affected by the anions (Cl^-, OH^-, SO_4^{2-}, HCO_3^-) in aqueous solutions. For example, the formation of the ion pairs

$$Cu^+ + Cl^- = CuCl^0 \quad (19)$$

$$Fe^{2+} + SO_4^{2-} = FeSO_4^0 \quad (20)$$

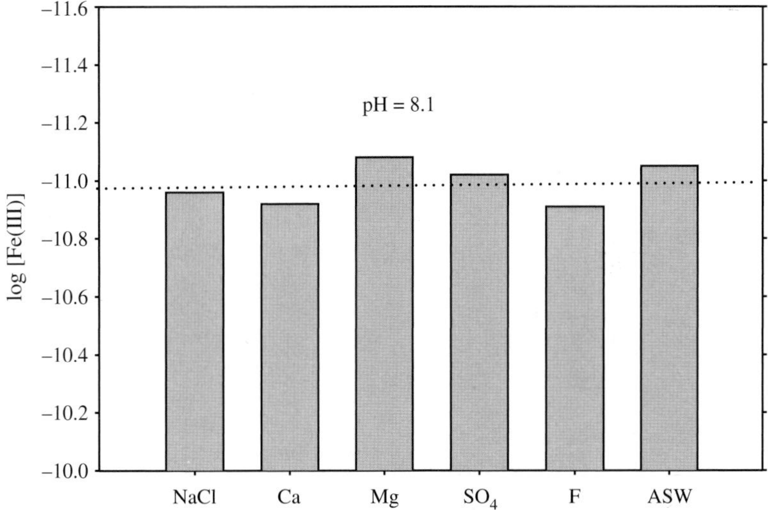

Figure 11 The solubility of Fe(III) in NaCl, NaCl plus the major cations and anions, and artificial seawater (source Liu and Millero, 2002).

causes the rates of oxidation of Cu(I) and Fe(II) with O_2 and H_2O_2 to decrease. The formation of the ion pairs

$$Cu^{2+} + CO_3^{2-} = CuCO_3^0 \qquad (21)$$

$$Fe^{2+} + OH^- = FeOH^+ \qquad (22)$$

causes the rates of oxidation of Cu(I) and Fe(II) to increase. The rates of oxidation of nonmetals can be affected by the cations (H^+, Mg^{2+}, Ca^{2+}, etc.). For example, the formation of the iron complexes

$$Fe^{2+} + HS^- = FeHS^+ \qquad (23)$$

causes the rates of oxidation of H_2S to increase at very low levels of iron. In the next section, the effect of the major anions on the oxidation of Fe(II) and Cu(I) with O_2 will be used as an example of how rates can be affected by ionic interactions.

6.01.4.1 Oxidation of Fe(II) with O_2

Reduced forms of iron, Fe(II), can be produced in natural waters in anoxic basins and hydrothermal waters and by the photoreduction of Fe(III) (Waite and Morel, 1984; Wells and Mayer, 1991; King *et al.*, 1995; Johnson *et al.*, 1994; Miller *et al.*, 1995; Voelker *et al.*, 1997). Once this reduced iron is formed, its lifetime will be related to its rate of oxidation. The kinetics of oxidation of Fe(II) with O_2 have been studied by a number of workers (Stumm and Lee, 1961; Kester *et al.*, 1975; Murray and Gill, 1978; Sung and Morgan, 1980; Davison and Seed, 1983; Roekens and Van Grieken, 1983, 1984; Waite and Morel, 1984; Millero *et al.*, 1987, 1990; Millero and Izaguirre, 1989; King *et al.*, 1995). In this section we will review some of the results that indicate how some of the components of natural waters can affect the rates of this oxidation.

The overall rate equation for the oxidation of Fe(II) with O_2 is given by

$$-d[Fe(II)]/dt = k[Fe(II)][O_2] \qquad (24)$$

Under pseudo-first-order conditions, $\{[O_2] \gg [Fe(II)]\}$, the rate equation can be simplified to

$$-d[Fe(II)]/dt = k_1'[Fe(II)] \qquad (25)$$

where $k_1' = k/[O_2]$. Measurements of oxidation under these conditions have been made in seawater and sea salts by our group (Millero *et al.*, 1987, 1990; Millero and Izaguirre, 1989; King *et al.*, 1995). The effect of ionic strength on the oxidation of Fe(II) in NaCl and seawater (Millero and Izaguirre, 1989) solutions is shown in Figure 12. The values in seawater are lower than the results in NaCl at the same ionic strength. This decrease is related to the formation of Fe(II) ion pairs,

$$Fe^{2+} + Cl^- = FeCl^+ \qquad (26)$$

$$Fe^{2+} + SO_4^{2-} = FeSO_4^0 \qquad (27)$$

that are less reactive to oxidation. To investigate this decrease, measurements have been made as a function of composition at a constant pH and ionic strength ($0.7m$). The results are shown in Figure 13. The addition of HCO_3^- causes the rate to increase, whereas the addition of SO_4^{2-} and $B(OH)_4^-$ causes the rate to decrease. The large increase due to the addition of HCO_3^- can be attributed to the formation of $FeCO_3^0$, which has a

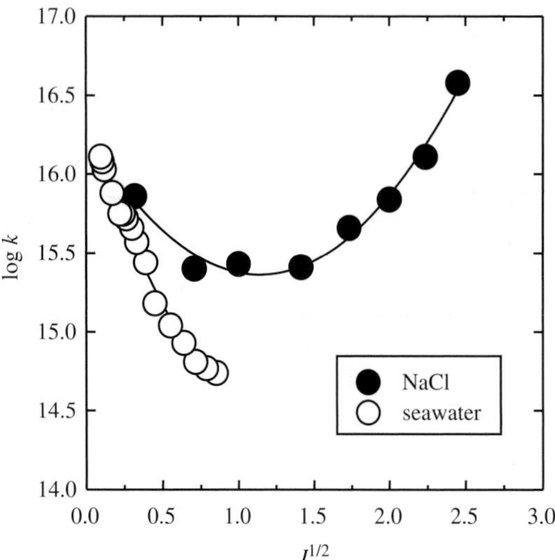

Figure 12 A comparison of the rates of oxidation of Fe(II) with O_2 in NaCl and seawater ($S = 35$) as a function of the square root of ionic strength (pH = 8 and $t = 25$ °C) (sources Millero *et al.*, 1987; Millero and Izaguirre, 1989).

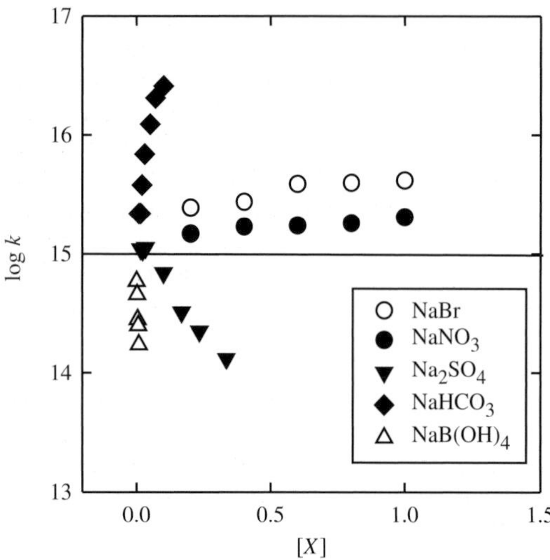

Figure 13 The effect of anions on the oxidation of Fe(II) with O_2 in NaCl + NaX solutions at $I = 0$, pH = 8, and 25 °C (source Millero and Izaguirre, 1989).

faster rate of

$$Fe^{2+} + CO_3^{2-} = FeCO_3^0 \qquad (28)$$

oxidation than $Fe(OH)_2^0$. King (1998) has suggested that the effect of pH on oxidation of Fe(II) is due to the formation of $FeCO_3^0$ rather than $Fe(OH)_2^0$ at pH above 7. The slight increases due to the addition of the anions Br^-, NO_3^-, and ClO_4^- are due to the weaker interaction of these anions with Fe^{2+} compared to Cl. The strong decrease in the rate due to the addition of SO_4^{2-} and $B(OH)_4^-$ can be attributed to the formation of the $FeSO_4^0$ and $FeB(OH)_4^+$ ion pairs. Similar ionic interaction effects have been found for the oxidation of Fe(II) with H_2O_2 in natural waters (Millero and Sotolongo, 1989).

6.01.4.2 Oxidation of Cu(I) with O_2

Our interest in the oxidation of Cu(I) in natural waters arose from the suggestion of Moffett and Zika (1983) that dynamic nonequilibria in surface seawater may lead to the formation of Cu(I). Their later measurements (Moffett and Zika, 1988) demonstrated that measurable amounts of Cu(I) were present in surface seawater. The formation of Cu(I) as discussed above may be caused by the reaction of Cu(II) with reducing agents (O_2^- and H_2O_2) formed by photo-oxidation of dissolved organic matter. One might also expect that Cu(I) may occur in anoxic waters and sediments. Once Cu(I) is formed in natural waters, its lifetime will be related to its rate of oxidation with O_2 to the more stable Cu(II).

The overall rate equation for the oxidation of Cu(I) with O_2 is given by

$$-d[Cu(I)]/dt = k[Cu(I)][O_2] \qquad (29)$$

Under pseudo-first-order conditions, $\{[O_2] \gg [Cu(I)]\}$, the rate equation can be simplified to

$$-d[Cu(I)]/dt = k_1'[Cu(I)] \qquad (30)$$

where $k_1' = k/[O_2]$. Moffett and Zika (1983) were the first to measure the rate of oxidation of Cu(I) in seawater. These measurements showed a large Cl^- dependence and were related to the differences in the rates of oxidation of the various forms of Cu(I) chloride complexes (Millero, 1985). The rates of oxidation of Cu(I) with O_2 have been made in seawater, NaCl, and sea salts (Sharma and Millero, 1988a,b,c,d, 1989) as a function of pH (5–9), temperature (5–45 °C), ionic strength (0.1–6 m), and composition. A comparison of the results for seawater and NaCl solution is shown in Figure 14. The rates decrease with an increase in ionic strength due to the interactions of Cu(I) with Cl^-. The lower rates in seawater at the same ionic strength are due to the lower concentration of Cl^- in seawater. This chloride dependence can be analyzed by assuming that the various Cu(I) chloro-complexes have different rates of oxidation (Millero, 1985). This gives

$$Cu^+ + O_2 \overset{k_0}{\Rightarrow} \text{products} \qquad (31)$$

$$CuCl^0 + O_2 \overset{k_1}{\Rightarrow} \text{products} \qquad (32)$$

$$CuCl_2^- + O_2 \overset{k_2}{\Rightarrow} \text{products} \qquad (33)$$

$$CuCl_3^{2-} + O_2 \overset{k_3}{\Rightarrow} \text{products} \qquad (34)$$

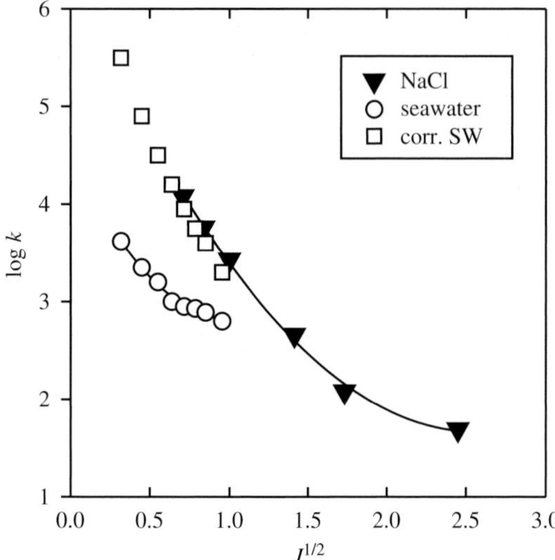

Figure 14 Comparison of the rates of oxidation of Cu(I) with O_2 in NaCl and seawater ($S = 35$) as a function of the square root of ionic strength at 25 °C (source Sharma and Millero, 1989).

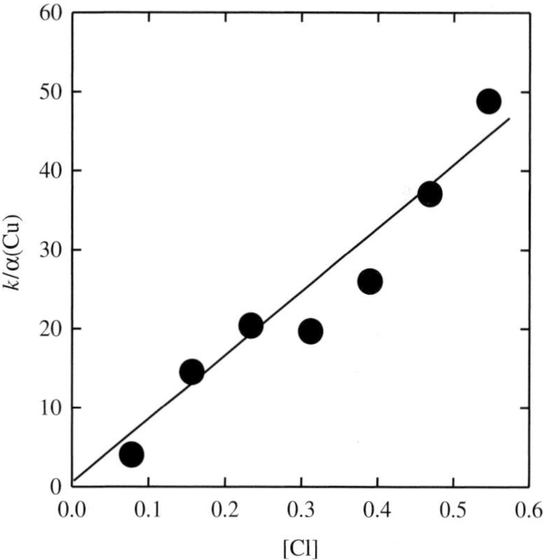

Figure 15 Plot of $k/\alpha(Cu^+)$ for the oxidation of Cu(I) with O_2 in seawater (source Sharma and Millero, 1988a).

The observed rate constant is given by

$$k = k_0\alpha_{Cu} + k_1\alpha_{CuCl} + k_2\alpha_{CuCl_2} + k_3\alpha_{CuCl_3} \quad (35)$$

where k_i is the rate constant and α_i the molar fraction of species i. The substitution of the stepwise stability constants, β_i, for the formation of the various ion pairs gives

$$k/\alpha_{Cu} = k_0 + k_1\beta_1[Cl^-] + k_2\beta_2[Cl^-]^2 + \cdots \quad (36)$$

Values of k/α_{Cu} for measurements in seawater as a function of $[Cl^-]$ are shown in Figure 15. The linear behavior indicates that Cu^+ and $CuCl^0$ are the reactive species. Similar results were found for the oxidation of Cu(I) with H_2O_2 (Sharma and Millero, 1989). Measurements made over a wider range of ionic strengths indicate that the $CuCl_2^-$ species becomes important at high levels of Cl^-.

Measurements in NaBr and NaI solutions indicate that the ion pairs $CuBr^0$ and CuI^0 have rates of oxidation similar to $CuCl^0$ (Sharma and Millero, 1988d). The differences of the overall rates of oxidation of Cu(I) in Cl^-, Br^-, and I^- solutions ($k_{Cl} > k_{Br} > k_I$) are related to the differences in the stability constants ($\beta_{CuCl} > \beta_{CuBr} > \beta_{CuI}$), not the inherent rates.

As mentioned earlier, the results for the oxidation of Cu(I) in seawater at the same Cl^- concentration are lower than the values in NaCl (see Figure 14). To elucidate these differences, measurements have been made in various sea-salt mixtures (Sharma and Millero, 1988b,c). The addition of SO_4^{2-} shows no effect, whereas that of Mg^{2+} and Ca^{2+} causes the rates to decrease, and the addition of HCO_3^- causes the rate to increase. A solution containing seawater concentrations of Na^+, Mg^{2+}, Cl^-, and HCO_3^- gives rates in agreement with the seawater results. Measurements of the effects of added Mg^{2+} and HCO_3^- have been made over a wide range of ionic strengths (Sharma and Millero, 1988b,c). Even though the cause of how Mg^{2+}, Ca^{2+}, and HCO_3^- affect the rates of oxidation of Cu(I) is uncertain, the correction of the seawater results for these effects yields rate constants that agree with the measured values in NaCl solutions (see Figure 14). The effects of Cl^- on the rates of oxidation of Cu(I) with H_2O_2 are similar to those with O_2 (Sharma and Millero, 1989).

In conclusion, studies of the rates of oxidation of Fe(II) and Cu(I) with O_2 and H_2O_2 clearly demonstrate how the formation of ion complexes can affect the rates. The oxidation of other metals is discussed in more detail elsewhere (Millero, 2001).

6.01.5 MODELING THE IONIC INTERACTIONS IN SEAWATER AND OTHER NATURAL WATERS

The ionic interactions in a mixed electrolyte solution like seawater can affect the physical properties (density, heat capacity, etc.). Since the composition of natural waters can be quite different, it is useful to have models that can be used to describe how the ionic components affect the physical properties. This requires knowledge of ionic interactions in the solutions of interest. Over the years, a great deal of progress has been made in interpreting and modeling the physical–chemical properties of mixed electrolyte solutions (Millero, 2001). This has led to the development of models that can be used to estimate the properties of natural waters of known composition. These models consider the changes that occur due to ion–water interactions in dilute solutions and the resultant ion–ion interactions as one moves to more concentrated solutions.

6.01.5.1 Physical–Chemical Properties

The physical–chemical properties of natural waters are determined from the apparent molal properties (Φ) of the components of the mixture. The apparent molal property is related to the change that occurs when a salt is added to water. The apparent molal property is defined by

$$\Phi = \Delta P/n = (P - P^0)/n_T \qquad (37)$$

where n_T is the number of moles or equivalents of added salt, P is the property of the solution, and P^0 is the property of water. The apparent molal property for a mixed electrolyte solution is nearly equal to the weighted sum of the component electrolytes. This additivity, called Young's rule (Young, 1951; Young and Smith, 1954), is given by

$$\Phi = \sum_M \sum_X E_M E_X \, \phi(MX) \qquad (38)$$

where E_M and E_X are the equivalent fractions of cations (M) and anions (X), and $\phi(MX)$ is the apparent property of electrolyte MX at the ionic strength of the mixture. For the major components of seawater (SW in the equation), this sum can be made three ways:

$$\text{(i)} \quad \Phi(\text{SW}) = E_{\text{Na}}E_{\text{Cl}}\,\phi(\text{NaCl}) + E_{\text{Mg}}E_{\text{SO}_4}\,\phi(\text{MgSO}_4)$$

$$\text{(ii)} \quad \Phi(\text{SW}) = E_{\text{Na}}E_{\text{SO}_4}\,\phi(\text{Na}_2\text{SO}_4) + E_{\text{Mg}}E_{\text{Cl}}\,\phi(\text{MgCl}_2)$$

$$\text{(iii)} \quad \Phi(\text{SW}) = E_{\text{Na}}E_{\text{Cl}}\phi(\text{NaCl}) + E_{\text{Na}}E_{\text{SO}_4}\,\phi(\text{MgSO}_4) + E_{\text{Mg}}E_{\text{Cl}}\,\phi(\text{MgCl}_2) + E_{\text{Mg}}E_{\text{SO}_4}\,\phi(\text{MgSO}_4)$$

Experimentally, it is found that the third summation works best because it considers the weighted sum of all the possible cation–anion interactions in the solution. These plus–minus interactions represent the major ionic interactions that occur in the mixture. Once the Φ for the mixture is estimated, a given physical property can be determined from

$$P = P^0 + \Phi n_T \qquad (39)$$

For seawater, Equation (38) can be broken down into terms for the individual major cations in the solution:

$$\Phi(\text{SW}) = E_{\text{Na}}\phi(\text{Na}\textstyle\sum \text{X}_i) + E_{\text{Mg}}\phi(\text{Mg}\sum \text{X}_i) + E_{\text{Ca}}\phi(\text{Ca}\sum \text{X}_i) + E_{\text{K}}\phi(\text{K}\sum \text{X}_i) + E_{\text{Sr}}\phi(\text{Sr}\sum \text{X}_i) \qquad (40)$$

The individual terms are given by

$$\phi(\text{Na}\textstyle\sum \text{X}_i) = E_{\text{Cl}}\phi(\text{NaCl}) + E_{\text{SO}_4}\phi(\text{Na}_2\text{SO}_4) + E_{\text{HCO}_3}\phi(\text{NaHCO}_3) + E_{\text{Br}}\phi(\text{NaBr}) + E_{\text{CO}_3}\phi(\text{Na}_2\text{CO}_3) + E_{\text{B(OH)}_4}\phi(\text{Na(BOH)}_4) + E_{\text{F}}\phi(\text{NaF}) \qquad (41)$$

Similar equations can be written for $E_{\text{Mg}}\phi(\text{Mg}\sum X_i)$, $E_{\text{Ca}}\phi(\text{Ca}\sum X_i)$, etc. The value of $\Phi(\text{SW})$ can be separated into three terms as given by

$$\Phi(\text{SW}) = \Phi^0(\text{SW}) + AI^{1/2} + BI \qquad (42)$$

where $\Phi^0(\text{SW}) = \sum_M \sum_X E_M E_X \phi^0(MX)$, $A = \sum_M \sum_X E_M E_X a(MX)$, and $B = \sum_M \sum_X E_M E_X b(MX)$. The values of $a(MX)$ and $b(MX)$ are the adjustable parameters for the individual components fitted to equations of the form

$$\phi(MX) = \phi^0(MX) + a(MX)I^{1/2} + b(MX)I \qquad (43)$$

The superscript zero is used to denote the values at infinite dilution or extrapolations to pure water and is only due to ion–water interactions. The A and B terms are related to the ion–ion interactions in the solution. This gives

$$P = P^0 + \textstyle\sum \text{ion–water} + \sum \text{ion–ion interactions} \qquad (44)$$

Thus, any physical property of a solution at a given ionic strength is equal to the property of pure water plus a term related to the weighted ion–water and ion–ion interactions.

These equations have been used to estimate the properties of seawater (Millero, 2001), seas (Millero, 1978; Millero and Chetirkin, 1980; Millero *et al.*, 1982), lakes (Effler *et al.*, 1986; Millero, 2000a), rivers (Millero, 1974; Millero *et al.*, 1976b, 1978), estuaries (Millero, 1974; Millero and Kremling, 1976; Millero *et al.*, 1976a,b), and brines (Millero *et al.*, 1982). This includes the estimates of sound speeds (Millero *et al.*, 1977), heat capacities (Millero, 1974; Millero *et al.*, 1973a), enthalpies (Millero, 1974; Duer *et al.*, 1976), freezing points (Millero, 1974), densities (Millero, 1974), expansibilities (Millero, 1974), and compressibilites (Millero, 1974). At higher ionic strengths, the estimates are not as good due to excess-mixing parameters (ΔP_{EX}) due to the interactions of ions of the same sign (Na^+–Mg^{2+}, Cl^-–SO_4^{2-}). These excess-mixing properties can be studied by mixing two electrolyte solutions at a constant ionic strength. For the mixing of the major sea salts, there are six possible mixtures that can be represented by the so-called cross-square diagram:

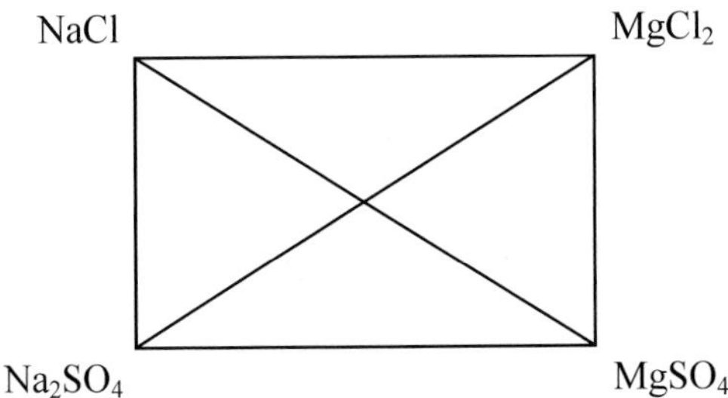

The mixing of the salts around the sides of this diagram has either a common cation or anion during mixing. A number of studies by Young *et al.* (1957) have shown that the values of ΔP_{EX} follow some simple rules. The addition to Young's simple rule gives

$$\phi = \sum_M \sum_X E_M E_X \phi(MX) + \Delta P_{EX}/n_T \quad (45)$$

For most physical properties in dilute solutions, the estimates can be made without the ΔP_{EX} term. When adding the excess-mixing terms to the equations, one is dividing the summation $\sum$ion–ion into three terms:

$$\sum \text{ion–ion} = \text{DH} + \sum \text{binary} + \sum \text{ternary} \quad (46)$$

where DH is a Debye–Hückel contribution, the $\sum$binary term is related to the interactions of ions of opposite (Na–Cl, Mg–Cl) and like signs (Na–Na, Cl–Cl, Na–Mg), and the $\sum$ternary term is related to triplet interactions (Na–Mg–Cl, Cl–SO_4–Na). The Pitzer equations (Pitzer, 1991) attribute the binary (Na–Na, Cl–Cl) and ternary interactions (Na–Na–Cl) for single electrolytes (NaCl) to three terms β^0, β^1, and C^ϕ. The binary (Na–Mg) interactions for mixtures (NaCl + $MgCl_2$) are related to Θ and the ternary interactions (Na–Mg–Cl) are related to Ψ. The Pitzer's equations thus incorporate Young's rule in all of its formulations. This general approach, although somewhat complicated, can account for all the possible interactions in a stepwise manner. Computer codes have been written that can be used to estimate the physical–chemical properties of natural waters over a wide range of temperatures (0–100 °C) and ionic strengths (0–6 m) (Millero, 2001).

6.01.5.2 Estimating Activity Coefficients

Thermodynamic equilibria in natural waters can be affected by ionic interactions in mixed electrolyte solutions. This can be demonstrated by considering the solubility of siderite in natural waters

$$FeCO_3(s) = Fe^{2+} + CO_3^{2-} \quad (47)$$

The thermodynamic solubility constant in pure water ($K^0_{CaCO_3}$) is given by

$$\begin{aligned} K^0_{FeCO_3} &= a_{Fe} a_{CO_3} / a_{FeCO_3} \\ &= [Fe^{2+}]_T [CO_3^{2-}]_T \{\gamma^T_{Fe} \gamma^T_{CO_3}\} \\ &= K^*_{FeCO_3} \{\gamma^T_{Fe} \gamma^T_{CO_3}\} \end{aligned} \quad (48)$$

where a_{FeCO_3} is assumed to be equal to 1.0, a_i is the activity, $[i]$ is the total concentration, and γ_i^T is the total activity of species i. The desired stoichiometric constant, $K^*_{FeCO_3}$, can be determined from the thermodynamic value provided that one can determine the total activity coefficients of Fe^{2+} and CO_3^{2-} in the solution of interest.

The desired total activity coefficients can be determined by assuming that the activities of Fe^{2+} (a_{Fe}) or CO_3^{2-} (a_{CO_3}) are equal to the product of the total concentration and activity coefficient, and the free concentration and activity coefficient. For Fe^{2+} and CO_3^{2-} this gives

$$\alpha_{Fe} = \gamma^T_{Fe}[Fe^{2+}]_T = \gamma^F_{Fe}[Fe^{2+}]_F \quad (49)$$

$$a_{CO_3} = \gamma^T_{CO_3}[CO_3^{2-}]_T = \gamma^F_{CO_3}[CO_3^{2-}]_F \quad (50)$$

The total activity coefficients for Fe^{2+} and CO_3^{2-} are given by

$$\gamma^T_{Fe} = \gamma^F_{Fe}\{[Fe^{2+}]_F/[Fe^{2+}]_T\} \quad (51)$$

$$\gamma^T_{CO_3} = \gamma^F_{CO_3}\{[CO_3^{2-}]_F/[CO_3^{2-}]_T\} \quad (52)$$

The molar fractions of the free cation (α_{Fe}) and anion (α_{CO_3}) are determined from

$$\begin{aligned} \alpha_{Fe} &= [Fe^{2+}]_F/[Fe^{2+}]_T \\ &= 1/(1 + \sum K^*_{FeXi}[X_i]_F) \end{aligned} \quad (53)$$

$$\begin{aligned} \alpha_{CO_3} &= [CO_3^{2-}]_F/[CO_3^{2-}]_T \\ &= 1/(1 + K^*_{MiCO_3}[M_i]_F) \end{aligned} \quad (54)$$

where $X_i = OH^-$, CO_3^{2-}, $B(OH)_4^-$, and $M_i = Mg^{2+}$, Ca^{2+} are the major anions and cations. If the stability constants for the formation of complexes of metals (K^*_{FeL}) with natural organic ligands ([L]) are available, one needs to add a $\{K^*_{FeL}$ [L]$\}$ term inside the bracket in Equation (53). The values of α_{Fe} and α_{CO_3} are determined by a series of iterations using calculated stability constants for the formation of the ion pairs. The fractions of Fe^{2+} and CO_3^{2-} in most natural waters are affected by the formation of carbonate

Table 4 Concentration and stability constants for metal organic complexes in seawater

Metal	*M* (nM)	*L_1* (nM)	*L_2* (nM)	*log K_1*	*log K_2*
Cu(II)[a]	1–10	2–60	10–300	11–12	8.5–10.2
Cd(II)[b]	0.002–0.8	0.1		12	
Zn(II)[c]	0.1–2	1.2		11	
Pb(II)[d]	0.017–0.049	0.2–0.5		9.7	
Ni(II)[e]					
Co(II)[f,g]	0.009–0.083			16.3, 17.3–18.7	
Fe(III)[h]	0.2–8	0.4–13		19–23	

[a] Coale and Bruland (1988), Moffett and Zika (1987), Hering *et al.* (1987), Sunda and Hanson (1987), Sunda and Ferguson (1983), Van den Berg (1984), Anderson *et al.* (1984), Sunda *et al.* (1984), Kramer and Duinker (1984), and Van den Berg (1982). [b] Bruland (1992).
[c] Bruland (1989). [d] Capodaglio *et al.* (1990). [e] Van den Berg and Nimmo (1986). [f] Zhang *et al.* (1990). [g] Saito and Moffett (2001).
[h] Gledhill and van den Berg (1994), Wu and Luther (1995), Rue and Bruland (1995), van den Berg (1995), and Witter *et al.* (2000).

complexes (Silva *et al.*, 2002):

$$\mathrm{Fe^{2+} + CO_3^{2-} = FeCO_3^0} \tag{55}$$

$$\mathrm{Mg^{2+} + CO_3^{2-} = MgCO_3^0} \tag{56}$$

$$\mathrm{Ca^{2+} + CO_3^{2-} = CaCO_3^0} \tag{57}$$

The stability constants for the formation of these complexes can be determined from

$$K^*_{\mathrm{FeCO_3}} = K_{\mathrm{FeCO_3}}\{\gamma^{\mathrm{F}}_{\mathrm{Fe}}\gamma^{\mathrm{F}}_{\mathrm{CO_3}}/\gamma^{\mathrm{F}}_{\mathrm{FeCO_3}}\} \tag{58}$$

where $K_{\mathrm{FeCO_3}}$ is the thermodynamic constant and γ_i^{F} is free-activity coefficient of species *i*.

The activity coefficients of ions in mixed electrolyte solutions need to be estimated to determine equilibria in natural waters of known composition. Activity coefficients of ions in natural waters can be determined by using the ion-pairing model (Garrels and Thompson, 1962; van Breeman, 1973; Dickson and Whitfield, 1981; Turner *et al.*, 1981; Millero and Schreiber, 1982; Truesdell and Jones, 1969) and the specific interaction model (Harvie and Weare, 1980; Harvie *et al.*, 1984; Millero, 1982; Felmy and Weare, 1986; Greenberg and Møller, 1989; Campbell *et al.*, 1993; Clegg and Whitfield, 1991, 1995; Millero and Roy, 1997; Millero and Pierrot, 1998). The recent progress in using these models to estimate the activity of ionic solutes and speciation of metals is described elsewhere (Millero, 2001). The specific interaction model as formulated by Pitzer (1979, 1991) is used to estimate the activity coefficients of free ions in our model (Millero and Pierrot, 1998). This model accounts for all the interactions of ions with the major cations and anions of the solution that do not have strong interactions or form ion pairs. The model is able to account for cation–anion, cation–cation, and anion–anion interactions in a stepwise fashion. It also determines the activity coefficients in electrolyte mixtures using an equation of the form

$$\ln\gamma_i = \mathrm{DH} + \sum_{ij} m_i m_j B^{\gamma}_{ij} + \sum_{ijk} m_i m_j m_k C^{\gamma}_{ijk} \tag{59}$$

The parameters B^{γ}_{ij} and C^{γ}_{ijk} are related to the binary (ions *i* and *j*) and ternary (ions *i*, *j*, and *k*) interactions and can be a function of ionic strength (Pitzer, 1991). The activity coefficients determined from this equation can account for all the interactions of an ion with all the major components of natural water. The model has been shown to produce reliable activity coefficients for the major sea-salt ions (Millero and Pierrot, 1998) and the dissociation of acids in seawater (Millero, 1983b). Computer codes are available to determine the activity coefficient and speciation of divalent and trivalent metals (Millero *et al.*, 1995; Millero and Pierrot, 1998) over a wide range of ionic strengths (0–6 *m*) and temperatures (0–50 °C). The addition of humic ligands (Mantoura *et al.*, 1978) and natural organic ligands (see Table 4) of known strength and concentration that can compete with the inorganic ligands can be analyzed with the program.

In this chapter the effects of the major components of natural waters on the physical properties, ionic equilibria, and rates of reactions have been reviewed briefly. For natural waters of know composition, ionic interaction models can be used to estimate the physical properties and equilibria from 0 °C to 50 °C and $I = 0$–$6\,m$. Measurements of stability constants for the formation of metal complexes are needed to extend the present models to wider range of temperatures (200 °C).

ACKNOWLEDGMENTS

This work was supported by a grant from the oceanographic section of the National Science Foundation.

REFERENCES

Anderson D. M., Lively J. S., and Vaccaro R. F. (1984) Copper complexation during spring phytoplankton blooms in coastal waters. *J. Mar. Res.* **4**, 677–695.

Bradshaw A. L. and Schleicher K. E. (1970) Direct measurement of thermal expansion of seawater under pressure. *Deep-Sea Res.* **17**, 691–706.

Bradshaw A. L. and Schleicher K. E. (1976) Compressibility of distilled water and seawater. *Deep-Sea Res.* **23**, 583–593.

Bradshaw A. L. and Schleicher K. E. (1986) An empirical equation of state for pure water in the oceanographic region of temperature and pressure determined from direct measurements. *J. Chem. Eng. Data* **31**, 189–194.

Brewer P. G. and Bradshaw A. (1975) The effect of non-ideal composition of seawater on salinity and density. *J. Mar. Res.* **33**, 157–175.

Bromley L. A. (1968a) Relative enthalpies of sea salt solutions at 25 °C. *J. Chem. Eng. Data* **13**, 399–402.

Bromley L. A. (1968b) Heat capacity of seawater solutions: partial and apparent values for salts and water. *J. Chem. Eng. Data* **13**, 60–62.

Bromley L. A., Desaussure V. A., Clipp J. C., and Wright J. S. (1967) Heat capacities of seawater solutions at salinities of 1 to 12 and temperatures of 2 to 80 °C. *J. Chem. Eng. Data* **12**, 202–206.

Bromley L. A., Diamond A. E., Salami E., and Wilkins D. G. (1970) Heat capacities and enthalpies of sea salt solutions to 200 °C. *J. Chem. Eng. Data* **15**, 246–253.

Bromley L. A., Singh D., Ray P., Sridhar A., and Read S. M. (1974) Thermodynamics properties of sea salt solutions. *AIChE J.* **20**, 325–335.

Bruland K. W. (1989) Complexation of zinc by natural organic ligands in the central North Pacific. *Limnol. Oceanogr.* **34**, 269–285.

Bruland K. W. (1992) Complexation of cadmium by natural organic ligands in the central North Pacific. *Limnol. Oceanogr.* **37**, 1008–1017.

Bruland K. W. and Wells M. (1995) The Chemistry of iron in seawater and its interaction with phytoplankton. *Spec. Issue Mar. Chem.* **50**, 1–241.

Bryden H. G. (1973) New polynomials for thermal expansion, adiabatic temperature gradient and potential of seawater. *Deep-Sea Res.* **20**, 401–408.

Busey R. and Mesmer R. E. (1977) Ionization equilibria of silicic acid and polysilicate formation in aqueous sodium chloride solution to 300 °C. *Inorg. Chem.* **16**, 2444–2450.

Byrne R. H. and Kester D. R. (1976) Solubility of hydrous ferric oxide and iron speciation in sea water. *Mar. Chem.* **4**, 255–274.

Byrne R. H. and Luo Yu-Ran (2000) Direct observations of nonintegral hydrous ferric oxide solubility products: $K^{*}so = [Fe^{3+}][H^{+}]-2.86$. *Geochim. Cosmochim. Acta* **64**, 1873–1877.

Caldwell D. R. (1974a) The effect of pressure on thermal and Fickian diffusion of sodium chloride. *Deep-Sea Res.* **21**, 369–375.

Caldwell D. R. (1974b) Thermal conductivity of sea water. *Deep-Sea Res.* **21**, 131–137.

Caldwell D. R. (1978) The maximum density points of pure and saline water. *Deep-Sea Res.* **25**, 175–181.

Campbell D. M., Millero F. J., Roy R., Roy L., Lawson M., Vogel K. M., and Moore C. P. (1993) The standard potential for the hydrogen–silver, silver chloride electrode in synthetic seawater. *Mar. Chem.* **44**, 221–234.

Capodaglio G., Coale K. H., and Bruland K. W. (1990) Lead speciation in surface waters of the eastern North Pacific. *Mar. Chem.* **29**, 221–233.

Chen C.-T. and Millero F. J. (1976a) Reevaluation of Wilson's sound-speed measurements for pure water. *J. Acoust. Soc. Am.* **60**, 1270–1273.

Chen C.-T. and Millero F. J. (1976b) The specific volume of seawater at high pressures. *Deep-Sea Res.* **23**, 595–612.

Chen C.-T. and Millero F. J. (1977) Speed of sound in seawater at high pressures. *J. Acoust. Soc. Am.* **62**, 1129–1135.

Chen C.-T. and Millero F. J. (1978) The equation of state of seawater determined from sound speeds. *J. Mar. Res.* **36**, 657–691.

Chen C.-T., Fine R. A., and Millero F. J. (1977) The equation of state of pure water determined from sound speeds. *J. Chem. Phys.* **66**, 2142–2144.

Chen S. F., Chan R. C., Read S. M., and Bromley L. A. (1973) Viscosity of seawater solutions. *Desalination* **13**, 37–51.

Chisholm S. W. and Morel F. M. M. (1991) What controls phytoplankton production in nutrient-rich areas of the open sea? *Spec. Issue Limnol. Oceanogr.* **36**, 1507–1965.

Clegg S. L. and Whitfield M. (1991) Activity coefficients in natural waters. In *Activity Coefficients in Electrolyte Solutions* (ed. K. S. Pitzer). CRC, Boca Raton, FL, pp. 279–434.

Clegg S. L. and Whitfield M. (1995) A chemical model of seawater including dissolved ammonia and the stoichiometric dissociation constant of ammonia in estuarine water and seawater from − 2 to 40 °C. *Geochim. Cosmochim. Acta* **59**, 2403–2421.

Coale K. H. and Bruland K. W. (1988) Copper complexation in the northeast Pacific. *Limnol. Oceanogr.* **33**, 1084–1101.

Coale K. H., Johnson K. S., Fitzwater S. E., Gordon R. M., Tanner S., Chavez F., Ferioli L., Sakamoto C., Rogers P., Millero F. J., Steinberg P., Nightingale P., Cooper D., Cochlan W., Landry M., Constantinou J., Rollwagen G., and Travina A. (1996) A massive phytoplankton blooms induced by an ecosystem iron fertilization experiment in the equatorial Pacific Ocean. *Nature* **383**, 495–501.

Connors D. M. (1970) On the enthalpy of seawater. *Limnol. Oceanogr.* **15**, 587–594.

Cox R. A. (1965) The physical properties of seawater. In *Chemical Oceanography* (eds. J. P. Riley and G. Skirrow). Academic Press, London, vol. 1, pp. 73–120.

Cox R. A. and Smith N. D. (1959) The specific heat of seawater. *Proc. Roy. Soc. A* **252**, 51–62.

Culberson C. and Pytkowicz R. M. (1968) Effect of pressure on carbonic acid, boric acid and the pH in seawater. *Limnol. Oceanogr.* **13**, 403–417.

Culberson C. and Pytkowicz R. M. (1973) Ionization of water in seawater. *Mar. Chem.* **1**, 309–316.

Culkin F. (1965) The major constituents of seawater. In *Chemical Oceanography* (eds. J. P. Riley and G. Skirrow). Academic Press, London, vol. 1, pp. 121–161.

Culkin F. and Smith N. D. (1980) Determination of the concentration of potassium chloride solution having the same electrical conductivity at 15 °C and infinite frequency as standard seawater of salinity 35.000% (Chlorinity 19.37394). *IEEE J. Ocean. Eng.* **OE-5**, 22–23.

Davison W. and Seed G. (1983) The kinetics of oxidation ferrous iron in synthetic and natural waters. *Geochim. Cosmochim. Acta* **47**, 67–79.

Dauphinee T. M. and Klein H. P. (1977) The effect of temperature on the electrical conductivity of seawater. *Deep-Sea Res.* **24**, 891–902.

Dauphinee T. M., Ancsin J., Klein H. P., and Phillips M. J. (1980) The effect of concentration and temperature on the conductivity ratio of potassium chloride solution to standard seawater for salinity 35% (Cl 19.3740) at 15 °C and 24 °C. *IEEE J. Oceanic Eng.* **OE-5**, 17–21.

Del Grosso V. A. (1974) Speed of sound in seawater. *J. Acoust. Soc. Am.* **56**, 1084–1091.

De Villiers S. and Nelson B. K. (1999) Detection of low temperature hydrothermal fluxes by seawater Mg and Ca anomalies. *Science* **285**, 721–723.

Dickson A. G. (1990a) Thermodynamics of the dissociation of boric acid in synthetic seawater from 273.15 to 318.15 K. *Deep-Sea Res.* **37**, 755–766.

Dickson A. G. (1990b) Standard potential of the reaction: $AgCl(s) + 1.2H_2(g) = Ag(s) + HCl\ (aq)$, and the standard acidity constant of the ion HSO_4^- in synthetic sea water from 273.15 to 318.15. *J. Chem. Thermodyn.* **22**, 113–127.

Dickson A. G. and Millero F. J. (1987) A comparison of the equilibrium constants for the dissociation of carbonic acid in seawater media. *Deep-Sea Res.* **34**, 1733–1743.

Dickson A. G. and Riley J. P. (1979a) The estimation of acid dissociation constants in seawater from potentiometric titrations with strong base: I. The ion product of water—K_w. *Mar. Chem.* **7**, 89–99.

Dickson A. G. and Riley J. P. (1979b) The estimation of acid dissociation constants in seawater from potentiometric titrations with strong base: II. The dissociation of phosphoric acid. *Mar. Chem.* **7**, 101–109.

Dickson A. G. and Whitfield M. (1981) An ion-association model for estimating acidity constants (at 25 °C and 1 atm total pressure) in electrolyte mixtures related to seawater (ionic strength <1 mol Kg^{-1} H_2O). *Mar. Chem.* **10**, 315–333.

Duer W. C., Leung W. H., Oglesby G. B., and Millero F. J. (1976) Seawater. A test for multicomponent electrolyte solution theories: II. Enthalpy of mixing and dilution of the major sea salts. *J. Solut. Chem.* **5**, 509–528.

Doherty B. T. and Kester D. R. (1974) Freezing point of seawater. *J. Mar. Res.* **32**, 285–300.

Dushaw B. D., Worcester P. F., Cornuelle B. D., and Howe B. M. (1993) On equation for the speed of sound in seawater. *J. Acoust. Soc. Am.* **93**, 255–275.

Effler S. W., Schimel K., and Millero F. J. (1986) Salinity, ionic strength, and chloride relationships in ion polluted Onondaga Lake, NY. *Water Air Soil Pollut.* **27**, 169–180.

Eisenberg H. (1958) Equation for the refractive index of water. *J. Chem. Phys.* **43**, 3887–3892.

Felmy A. R. and Weare J. H. (1986) The prediction of borate mineral equilibria in natural waters: application to Searles Lake, California. *Geochim. Cosmochim. Acta* **50**, 2771–2783.

Fernandez H., Vazquez F., and Millero F. J. (1982) The density and composition of hypersaline waters of a Mexican Lagoon. *Limnol. Oceanogr.* **27**, 315–321.

Fine R. A. and Millero F. J. (1973) The compressibility of water as a function of temperature and pressure. *J. Chem. Phys.* **59**, 5529–5536.

Fine R. A., Wang D. P., and Millero F. J. (1974) The equation of state of seawater. *J. Mar. Res.* **32**, 433–456.

Fujino K., Lewis E. L., and Perkin R. G. (1974) The freezing point of seawater at pressures up to 100 bars. *J. Geophys. Res.* **79**, 1792–1797.

Garrels R. M. and Thompson M. E. (1962) A chemical model for seawater at 25 °C and one atmosphere total pressure. *Am. J. Sci.* **260**, 57–66.

Gledhill M. and van den Berg C. M. G. (1994) Determination of complexation of iron (III) with natural organic complexing ligands in seawater using cathodic stripping voltammetry. *Mar. Chem.* **47**, 41–54.

Goyet C. and Poisson A. (1989) New determination of carbonic acid dissociation constants in seawater as a function of temperature and salinity. *Deep-Sea Res.* **36**, 1635–1654.

Greenberg J. P. and Møller N. (1989) The prediction of mineral solubilities in natural waters: a chemical equilibrium model for the Na–K–Ca–Cl–SO_4–H_2O system to high concentration from 0 to 250 °C. *Geochim. Cosmochim. Acta* **53**, 2503–2518.

Hansson I. (1973) A new set of acidity constants for carbonic acid and boric acid in seawater. *Deep-Sea Res.* **20**, 461–478.

Harvie C. E. and Weare J. H. (1980) The prediction of mineral solubilities in natural waters: the Na–K–Mg–Ca–SO_4–Cl–H_2O system from zero to high concentration at 25 °C. *Geochim. Cosmochim. Acta* **44**, 981–997.

Harvie C. E., Møller N., and Weare J. H. (1984) The prediction of mineral solubilities in natural waters: the Na–K–Mg–Ca–H–Cl–SO_4–OH–HCO_3–CO_3–CO_3–H_2O system to high ionic strengths at 25 °C. *Geochim. Cosmochim. Acta* **48**, 723–752.

Hering J. G., Sunda W. G., Ferguson R. L., and Morel F. M. M. (1987) A field comparison of two methods for the determination of copper complexation: bacterial bioassay and fixed-potential amperometry. *Mar. Chem.* **20**, 299–312.

Hill K. D., Dauphinee T. M., and Woods D. J. (1986) The extension of the practical salinity scale 1978 to low salinities. *IEEE J. Oceanic Eng.* **OE-11**, 109–112.

Ho W. and Hall W. F. (1973) Measurements of the dielectric properties of seawater and NaCl solutions at 2.65 GHz. *J. Geophys. Res.* **78**, 6301–6315.

Horne R. A. and Johnson D. S. (1966) The viscosity of water under pressure. *J. Phys. Chem.* **70**, 2182–2190.

Hutchins D. A. and Bruland K. (1998) Iron-limited diatom growth and Si:N uptake ratio in a coastal upwelling regime. *Nature* **393**, 561–564.

Ingle S. E. (1975) Solubility of calcite in the ocean. *Mar. Chem.* **3**, 301–319.

Jamieson D. T. and Tudhoe J. S. (1970) Physical properties of sea water solutions: thermal conductivity. *Desalination* **8**, 393–401.

Johansson O. and Wedborg M. (1979) Stability constants of phosphoric acid in seawater of 5–40‰ salinity and temperature of 5–25 °C. *Mar. Chem.* **8**, 57–69.

Johansson O. and Wedborg M. (1980) The ammonia–ammonium equilibrium in seawater at temperatures between 5 and 25 °C. *J. Solut. Chem.* **9**, 37–44.

Johnson K. S., Coale K. H., Elrod V. A., and Tindale N. W. (1994) Iron photochemistry in seawater from the equatorial Pacific. *Mar. Chem.* **46**, 319–334.

Johnson K. S., Gordon R. M., and Coale K. H. (1997) What controls dissolved iron concentrations in the world ocean. *Mar. Chem.* **57**, 137–161.

Kester D. R. and Pytkowicz R. M. (1967) Determination of the apparent dissociation constants of phosphoric acid in seawater. *Limnol. Oceanogr.* **12**, 243–252.

Kester D. R., Byrne R. H., and Liang Y. J. (1975) Redox reactions and solution complexes of iron in marine systems. In *Marine Chemistry in the Coastal Environment*, ACS Symposium 18 (ed. T. M. Church), pp. 56–79.

Khoo K. H., Ramette R. W., Culberson C. H., and Bates R. G. (1977a) Determination of hydrogen ion concentrations in seawater from 5 to 40 °C: standard potentials at salinities from 20 to 45‰. *Anal. Chem.* **49**, 29–34.

Khoo K. H., Culberson C. H., and Bates R. G. (1977b) Thermodynamics of ammonium ion in seawater from 5 to 40 °C. *J. Solut. Chem.* **6**, 281–290.

King D. W. (1998) Role of carbonate speciation on the oxidation rate of Fe(II) in aquatic systems. *Environ. Sci. Technol.* **32**, 2997–3003.

King D. W., Lounsbury H. A., and Millero F. J. (1995) Rates and mechanism of Fe(II) oxidation at nanomolar total iron concentrations. *Environ. Sci. Technol.* **29**, 818–824.

Knudsen M. (1901) *Hydrographical Tables* (according to the measurings of Carl Forch, P. Jacobsen, Martin Knudsen and S. P. L. Sorensen). G. E. C. Gad, Copenhagen, Williams Norgate, London, 63pp..

Kramer C. J. M. and Duinker J. C. (1984) Complexation capacity and conditional stability constants for copper of sea and estuarine waters, sediment extracts and colloids. In *Complexation of Trace Metals in Natural Waters* (eds. C. J. M. Kramer and J. C. Duinker). Nijhoff/Junk, The Hague, The Netherlands, pp. 217–228.

Kuma K., Nakabayashi S., Suzuki Y., and Matsunaga K. (1992) Dissolution rate and solubility of colloidal hydrous ferric oxide in seawater. *Mar. Chem.* **38**, 133–143.

Kuma K., Nishioka J., and Matsunaga K. (1996) Controls on iron(III) hydroxide solubility in seawater: the influence of pH and natural organic chelators. *Limnol. Oceanogr.* **41**, 396–407.

Kuma K., Katsumoto A., Kawakami H., Takatori F., and Matsunaga K. (1998a) Spatial variability of Fe(III) hydroxide solubility in the water column of the northern North Pacific Ocean. *Deep-Sea Res. I* **45**, 91–113.

Kuma K., Katsumoto A., Nishioka J., and Matsunaga K. (1998b) Size-fractionated iron concentrations and Fe(III) hydroxide solubilities in various coastal waters. *Estuar. Coast. Shelf Sci.* **47**, 275–283.

Lee K., Millero F. J., and Campbell D. M. (1996) The reliability of the thermodynamic constants for the dissociation of carbonic acid in seawater. *Mar. Chem.* **55**, 233–246.

Lee K., Millero F. J., and Wanninkhof R. (1997) The carbon dioxide system in the Atlantic Ocean. *J. Geophys. Res.* **102**, 15696–15707.

Lee K., Millero F. J., Byrne R. H., Feely R. A., and Wanninkhof R. (2000) The recommended dissociation constants for carbonic acid in seawater. *Geophys. Res. Lett.* **27**, 229–232.

Lewis E. L. (1980) The practical salinity scale 1978 and its antecedents. *IEEE J. Ocean. Eng.* **OE-5**, 3–8.

Lewis E. L. and Perkin R. G. (1978) Salinity: its definition and calculation. *J. Geophys. Res.* **83**, 466–478.

Lewis E. and Wallace D. W. R. (1998) Basic program for CO_2 system in seawater. ORNL/CDIAC-105, Oak Ridge National Laboratory.

Leyendekkers J. V. and Hunter R. J. (1976) The Tammann-Tait-Gibson model for aqueous electrolyte solutions: application to the refractive index. *J. Phys. Chem.* **81**, 1657–1663.

Liu X. and Millero F. J. (1999) The solubility of iron in sodium chloride solutions. *Geochim. Cosmochim. Acta* **63**, 3487–3497.

Liu X. and Millero F. J. (2002) The solubility of iron in seawater. *Mar. Chem.* **77**, 43–54.

Loglio G., Ficalbi A., and Cini R. (1978) A new evaluation of the surface tension temperature coefficients for water. *J. Colloid Interface Sci.* **64**, 198.

Lueker T. J., Dickson A. G., and Keeling C. D. (2000) Ocean pCO_2 calculated from dissolved inorganic carbon, alkalinity, and equations for K_1 and K_2: validation based on laboratory measurements of CO_2 in gas and seawater at equilibrium. *Mar. Chem.* **70**, 105–119.

Mantoura R. F. C., Dickson A., and Riley J. P. (1978) The complexation of metals with humic materials in natural waters. *Estuar. Coast. Mar. Sci.* **6**, 387–408.

Martin J. H. and Fitzwater S. E. (1988) Iron deficiency limits phytoplankton growth in the north-east Pacific sub-Arctic. *Nature* **331**, 341–343.

Martin J. H. and Gordon R. M. (1988) Northeast Pacific iron distributions in relation to phytoplankton productivity. *Deep-Sea Res.* **35**, 177–196.

Martin J. H., Coale K., Johnson K. S., Fitzwater S. E., Gordon R. M., Tanner S. J., Hunter C. N., Elrod V., Norwicki J., Coley T., Barber R., Lindley S., Watson A., Van Scoy K., Law C., Ling R., Stanton T., Stockel J., Collins C., Anderson A., Onrusek R. M., Latasa M. M., Millero F. J., Lee K., Yao W., Zhang J. Z., Friederich G., Sakamoto C., Chavez C., Buck K., Kolber S., Green R., Falkowski P., Chisholm S. W., Hoge F., Swift B., Turner S., Nightingale P., Liss P., and Tindale N. (1994) Testing the iron hypothesis in ecosystems of the equatorial Pacific Ocean. *Nature* **371**, 123–129.

McElligott S. and Byrne R. H. (1998) Interaction of $B(OH)_3^0$ and HCO_3^- in seawater: formation of $B(OH)_2CO_3^-$. *Aquat. Geochem.* **3**, 345–356.

Mehu A. and Johannin-Gilles A. (1969) Variation de la réfraction spécifique de l'eau de mer étalon de Copehague et de ses dilutions en fonction de longueur d'onde, de la temperature et de al chlorinate. *Deep-Sea Res.* **16**, 605–611.

Mehrbach C., Culberson C. H., Hawley J. E., and Pytkowicz R. N. (1973) Measurement of the apparent dissociation constants of carbonic acid in seawater at atmospheric pressure. *Limnol. Oceanogr.* **18**, 897–907.

Miller W. L., King D. W., Lin J., and Kester D. R. (1995) Photochemical redox cycling of iron in coastal waters. *Mar. Chem.* **50**, 63–77.

Millero F. J. (1974) Seawater as a multicomponent. In *The Sea, Ideas and Observations* (ed. E. D. Goldberg). Interscience Div., Wiley, New York, NY, vol. 5, chap. I, pp. 3–80.

Millero F. J. (1978) The physical chemistry of Baltic Sea waters. *Thalassia Jugoslavica* **14**, 1–46.

Millero F. J. (1982) The thermodynamics of seawater: Part I. The PVT properties. *Ocean Sci. Eng.* **7**, 403–460.

Millero F. J. (1983a) Thermodynamics of seawater: Part II. Thermochemical properties. *Ocean Sci. Eng.* **8**, 1–40.

Millero F. J. (1983b) The estimation of the pK_{HA}^* of acids in seawater using the Pitzer equations. *Geochim. Cosmochim. Acta* **47**, 2121–2129.

Millero F. J. (1985) The effect of ionic interactions on the oxidation of metals in natural waters. *Geochim. Cosmochim. Acta* **49**, 547–553.

Millero F. J. (1989) Effect of ionic interactions on the oxidation of Fe(II) and Cu(I) in natural waters. *Mar. Chem.* **28**, 1–18.

Millero F. J. (1995) The thermodynamics of the carbonic acid system in the oceans. *Geochim. Cosmochim. Acta* **59**, 661–667.

Millero F. J. (1996) *Chemical Oceanography*. CRC Press, Boca Raton, FL, 469pp.

Millero F. J. (1998) Solubility of Fe(III) in seawater. *Earth Planet. Sci. Lett.* **154**, 323–330.

Millero F. J. (2000a) The equation of state of lake waters. *Aquat. Geochem.* **6**, 1–17.

Millero F. J. (2000b) Effect of changes in the composition of seawater on the density–salinity relationship. *Deep-Sea Res. I* **47**, 1583–1590.

Millero F. J. (2001) *Physical Chemistry of Natural Waters*. Wiley-Interscience, NY, 654pp.

Millero F. J. and Chetirkin P. V. (1980) The density of Caspian Sea waters. *Deep-Sea Res.* **27**, 265–271.

Millero F. J. and Izaguirre M. (1989) Effect of ionic strength and ionic interactions on the oxidation of Fe(II). *J. Solut. Chem.* **18**, 585–599.

Millero F. J. and Kremling K. (1976) The densities of Baltic Sea waters. *Deep-Sea Res.* **23**, 1129–1138.

Millero F. J. and Kubinski T. (1975) Speed of sound in seawater as a function of temperature and salinity at one atmosphere. *J. Acoust. Soc. Am.* **57**, 312–319.

Millero F. J. and Leung W. H. (1976) The thermodynamics of seawater at one atmosphere. *Am. J. Sci.* **275**, 1035–1077.

Millero F. J. and Li X. (1994) Comments on "On equations from the speed of sound in seawater". *J. Acoust. Soc. Am.* **95**, 2757–2759.

Millero F. J. and Pierrot D. (1998) A chemical equilibrium model for natural waters. *Aquat. Geochem.* **4**, 153–199.

Millero F. J. and Poisson A. (1981) International one-atmosphere equation of state of seawater. *Deep-Sea Res.* **28**, 625–629.

Millero F. J. and Roy R. (1997) A chemical model for the carbonate system in natural waters. *Croatia Chem. Acta* **70**, 1–38.

Millero F. J. and Schreiber D. R. (1982) Use of the ion pairing model to estimate activity coefficients of the ionic components of natural waters. *Am. J. Sci.* **282**, 1508–1540.

Millero F. J. and Sotolongo S. (1989) The oxidation of Fe(II) with H_2O_2 in seawater. *Geochim. Cosmochim. Acta* **53**, 1867–1873.

Millero F. J., Perron G., and Desnoyers J. E. (1973a) Heat capacity of seawater solutions from 5 to 35 °C and 0.5 to 22‰ chlorinity. *J. Geophys. Res.* **78**, 4499–4507.

Millero F. J., Hansen L. D., and Hoff E. V. (1973b) The enthalpy of seawater from 0 to 30 °C and from 0 to 40‰ salinity. *J. Mar. Res.* **31**, 21–39.

Millero F. J., Gonzalez A., and Ward G. K. (1976a) The density of seawater solutions at one atmosphere as a function of temperature and salinity. *J. Mar. Res.* **34**, 61–93.

Millero F. J., Lawson D., and Gonzalez A. (1976b) The density of artificial river and estuarine waters. *J. Geophys. Res.* **81**, 1177–1179.

Millero F. J., Gonzalez A., Brewer P. G., and Bradshaw A. (1976c) The density of North Atlantic and North Pacific deep waters. *Earth Planet. Sci. Lett.* **32**, 468–472.

Millero F. J., Ward G. K., and Chetirkin P. V. (1977) Relative sound velocities of sea salts at 25 °C. *J. Acoust. Soc. Am.* **61**, 1492–1498.

Millero F. J., Forsht D., Means D., Gieskes J., and Kenyon K. (1978) The density of North Pacific Ocean waters. *J. Geophys. Res.* **83**, 2359–2364.

Millero F. J., Chen C.-T., Bradshaw A., and Schleicher K. (1980) A new high pressure equation of state for seawater. *Deep-Sea Res.* **27**, 255–264.

Millero F. J., Mucci A., Zullig J., and Chetirkin P. (1982) The density of Red Sea brines. *Mar. Chem.* **11**, 463–475.

Millero F. J., Sotolongo S., and Izaguirre M. (1987) The oxidation kinetics of Fe(II) in seawater. *Geochim. Cosmochim. Acta* **51**, 793–801.

Millero F. J., Plese T., and Fernandez M. (1988) The dissociation of hydrogen sulfide in seawater. *Limnol. Oceanogr.* **33**, 269–274.

Millero F. J., Izaguirre M., and Sharma V. K. (1990) The effect of ionic interactions on the rates of oxidation in natural waters. *Mar. Chem.* **22**, 179–191.

Millero F. J., Yao W., and Aicher J. A. (1995) The speciation of Fe(II) and Fe(III) in natural waters. *Mar. Chem.* **50**, 21–39.

Millero F. J., Pierrot D., Lee K., Wanninkhof R., Feely R., Sabine C. L., and Key R. M. (2002) Dissociation constants for carbonic acid determined from field measurements. *Deep-Sea Res. I* **49**, 1705–1723.

Miyake Y. (1939) Chemical studies of the western Pacific Ocean: III. Freezing point, osmotic pressure, boiling point, and vapor pressure of seawater. *Chem. Soc. Japan Bull.* **14**, 58–62.

Moffett J. W. and Zika R. G. (1983) Oxidation kinetics of Cu(I) in seawater: implications for the existence in the marine environment. *Mar. Chem.* **13**, 239–251.

Moffett J. W. and Zika R. G. (1987) Solvent extraction of copper acetylacetonate in studies of copper (II) speciation in seawater. *Mar. Chem.* **21**, 301–313.

Moffett J. W. and Zika R. G. (1988) Measurement of copper(I) in surface waters of the subtropical Atlantic and Gulf of Mexico. *Geochim. Cosmochim. Acta* **52**, 1849.

Mojica P. F. and Millero F. J. (2002) The determination of $pK_1 + pK_2$ in seawater as a function of temperature and salinity. *Geochim. Cosmochim. Acta* **66**, 2529–2540.

Mucci A. (1983) The solubility of calcite and aragonite in seawater at various salinities, temperatures and one atmosphere total pressure. *Am. J. Sci.* **283**, 780–799.

Murray J. W. and Gill G. (1978) The geochemistry of iron in Puget Sound. *Geochim. Cosmochim. Acta* **42**, 9–19.

Park K. (1969) Oceanic CO_2 system: an evaluation of ten methods of investigation. *Limnol. Oceanogr.* **14**, 179–186.

Pierrot D. and Millero F. J. (2000) The apparent molal volume and compressibility of seawater fit to the Pitzer equations. *J. Solut. Chem.* **29**, 719–742.

Pitzer K. S. (1979) Theory: ion interaction approach. In *Activity Coefficients in Electrolyte Solutions* (ed. R. M. Pytkowicz). CRC Press, Boca Raton, FL, vol. I, pp. 157–208.

Pitzer K. S. (1991) Theory: ion interaction approach—theory and data collection. In *Activity Coefficients in Electrolyte Solutions*, 2nd edn. (ed. K. S. Pitzer). CRC Press, Boca Raton, FL, vol. I, pp. 75–153.

Poisson A. (1980) Conductivity/salinity/temperature relationship of diluted and concentrated standard seawater. *IEEE J. Oceanic Eng.* **OE-5**, 41–50.

Poisson A., Brunet C., and Brun-Coltan J. C. (1980a) Density of standard seawater solutions at atmospheric pressure. *Deep-Sea Res.* **27**, 1013–1028.

Poisson A., Lebel J., and Brunet C. J. C. (1980b) Influence of local variations in the ionic ratios on the density of seawater in the St. Lawrence area. *Deep-Sea Res.* **27**, 763–781.

Poisson A., Lebel J., and Brunet C. J. C. (1981) The densities of western Indian Ocean, Red Sea and eastern Mediterranean surface waters. *Deep-Sea Res.* **28**, 1161–1172.

Riley J. P. and Chester R. (1971) *Introduction to Marine Chemistry*. Academic Press, New York, 35p.

Robinson R. A. (1954) The vapor pressure and osmotic equivalence of seawater. *J. Mar. Biol. Assoc. Unit. Kingdom* **33**, 449–455.

Roekens E. J. and Van Grieken R. E. (1983) Kinetics of iron (II) oxidation in seawater of various pH. *Mar. Chem.* **13**, 195–202.

Roekens E. J. and Van Grieken R. E. (1984) Kinetics of iron (II) oxidation in seawater of various pH. *Mar. Chem.* **15**, 281–284.

Roy R. N., Vogel K. M., Moore C. P., Pearson T., Roy L. N., Johnson D. A., Millero F. J., and Campbell D. M. (1993a) The dissociation constants of carbonic acid in seawater at salinities 5 to 45 and temperatures 0 to 45 °C. *Mar. Chem* **44**, 249–267.

Roy R. N., Roy L. N., Lawson M., Vogel K. M., Moore C. P., Davis W., and Millero F. J. (1993b) Thermodynamics of the dissociation of boric acid in seawater at $S = 35$ from 0 to 55 °C. *Mar. Chem.* **44**, 243–248.

Rue E. L. and Bruland K. W. (1995) Complexation of iron (III) by natural organic ligands in the central North Pacific as determined by a new competitive ligand equilibration/adsorptive cathodic stripping voltammetric method. *Mar. Chem.* **50**, 117–138.

Rusby J. S. M. (1967) Measurements of the refractive index of seawater relative to Copenhagen Standard Sea Water. *Deep-Sea Res.* **14**, 427–439.

Rush R. M. and Johnson J. S. (1966) Osmotic coefficients of synthetic seawater solutions at 25 °C. *J. Chem. Eng. Data* **11**, 590–592.

Saito M. A. and Moffet J. W. (2001) Complexation of cobalt by natural organic ligands in the Sargasso Sea as determined by a new high-sensitivity electrochemical colbalt speciation method suitable for open ocean water. *Mar. Chem.* **75**, 49–68.

Seidler G. and Peters H. (1986) Physical properties of sea water. In *Landolt-Börnstein*, Numerical Data and Functional Relationships in Science and Technology, New Series, Oceanography (ed. J. Sündermann), V/3a, pp. 233–264.

Sharma V. K. and Millero F. J. (1988a) Oxidation of copper(I) in seawater. *Environ. Sci. Technol.* **22**, 768–771.

Sharma V. K. and Millero F. J. (1988b) The oxidation of Cu(I) in electrolyte solution. *J. Solut. Chem.* **17**, 581–599.

Sharma V. K. and Millero F. J. (1988c) Effect of ionic interactions on the rates of oxidation of Cu(I) with O_2 in natural waters. *Mar. Chem.* **25**, 141–161.

Sharma V. K. and Millero F. J. (1988d) Determining the stability constant of copper (I) halide complexes from kinetic measurements. *Inorg. Chem.* **27**, 3256–3259.

Sharma V. K. and Millero F. J. (1989) The oxidation of Cu(I) with H_2O_2 in natural waters. *Geochim. Cosmochim. Acta* **53**, 2269–2276.

Silva C. A. R., Liu X., and Millero F. J. (2002) Solubility of siderite ($FeCO_3$) in NaCl solutions. *J. Solut. Chem.* **31**, 97–108.

Singh D. and Bromley L. A. (1973) Relative enthalpies of sea salt solutions at 0 and 70 °C. *J. Chem. Eng. Data* **18**, 174–181.

Sjöberg S., Nordin A., and Ingri N. (1981) Equilibrium and structural studies of silicon(IV) and aluminum(III) in aqueous solution. *Mar. Chem.* **10**, 521–532.

Spiesberger J. L. and Metzger K. (1991a) New estimates of sound speed in water. *J. Acoust. Soc. Am.* **89**, 1697–1700.

Spiesberger J. L. and Metzger K. (1991b) A new algorithm for sound speed in seawater. *J. Acoust. Soc. Am.* **89**, 2677–2688.

Stanley E. M. (1971) Refractive index of pure water for wavelength of 6,328 Å at high pressure and moderate temperature. *J. Chem. Eng. Data* **16**, 454–457.

Stanley E. M. and Batten R. C. (1969) Viscosity of sea water at moderate temperatures and pressures. *J. Geophys. Res.* **74**, 3415–3420.

Stoughton R. W. and Lietzke M. H. (1967) Thermodynamic properties of sea salt solutions. *J. Chem. Eng. Data* **12**, 101–104.

Stumm W. and Lee G. F. (1961) Oxygenation of ferrous iron. *Indust. Eng. Chem.* **53**, 143–146.

Sunda W. G. and Ferguson R. L. (1983) Sensitivity of natural bacterial communities to additions of copper and to cupric ion activity: a bioassay of copper complexation in seawater. In *Trace Metals in Seawater* (eds. C. S. Wong, E. Boyle, K. W. Bruland, J. D. Burton, and E. D. Goldberg). Plenum, New York, pp. 871–891.

Sunda W. G. and Hanson A. K. (1987) Measurement of free cupric ion concentration in seawater by a ligand competition technique involving copper sorption onto C18 SEP-PAK cartridges. *Limnol. Oceanogr.* **32**, 537–551.

Sunda W. G., Klaveness D., and Palumbo A. V. (1984) Bioassays of cupric ion activity and copper complexation. In *Complexation of Trace Metals in Natural Waters* (eds. C. J. M. Kramer and J. C. Duinker). Nijhoff/Junk, The Hague, The Netherlands, pp. 399–409.

Sung W. and Morgan J. J. (1980) Kinetics and product of ferrous oxygenation in aqueous solutions. *Environ. Sci. Technol.* **14**, 561–568.

Truesdell A. H. and Jones B. F. (1969) Ion association of natural brines. *Chem. Geol.* **4**, 51–62.

Turner D. R., Whitfield M., and Dickson A. G. (1981) The equilibrium speciation of dissolved components in freshwater and seawater at 25 °C and 1 atm pressure. *Geochim. Cosmochim. Acta* **45**, 855–881.

Utterback C. L., Thompson T. G., and Thomas B. A. (1934) Refractivity–chlorinity temperature relationship of ocean waters. *J. Cons. Perm. Int. Explor. Mer.* **9**, 35–38.

van Breeman N. (1973) Calculation of activity coefficients in natural waters. *Geochim. Cosmochim. Acta* **37**, 101–107.

van den Berg C. M. G. (1982) Determination of copper complexation with natural organic ligands in seawater by equilibration with MnO_2: II. Experimental procedures and application to surface seawater. *Mar. Chem.* **11**, 323–342.

van den Berg C. M. G. (1984) Determination of the complexing capacity and conditional stability constants of complexes of copper(II) with natural organic ligands in seawater by cathodic stripping voltammetry of copper-catechol complexions. *Mar. Chem.* **15**, 1268–1274.

van den Berg C. M. G. (1995) Evidence for organic complexation of iron in seawater. *Mar. Chem.* **50**, 139–157.

van den Berg C. M. G. and Nimmo M. (1986) Determination of interaction of nickel with dissolved organic material in seawater using cathodic stripping voltammetry. *Sci. Total Environ.* **60**, 185–195.

Voelker B. M., Morel F. M. M., and Sulzberger B. (1997) Iron redox cycling in surface waters: effects of humic substances and light. *Environ. Sci. Technol.* **31**, 1004–1017.

Waite T. D. and Morel F. M. M. (1984) Photoreductive dissolution of colloidal iron oxides in natural waters. *Environ. Sci. Technol.* **18**, 860–868.

Wang D.-P. and Millero F. J. (1973) Precise representation of the $P-V-T$ properties of water and seawater determined from sound speeds. *J. Geophys. Res.* **78**, 7122–7128.

Wanninkhof R., Lewis E., Feely R. A., and Millero F. J. (1999) The optimal carbonate dissociation constants for determining surface water pCO_2 from alkalinity and total inorganic carbon. *Mar. Chem.* **65**, 291–301.

Wells M. L. and Mayer L. M. (1991) The photoconversion of colloidal iron oxyhydroxides in seawater. *Deep-Sea Res.* **38**, 1379–1395.

Weiss R. F. (1974) Carbon dioxide in water and seawater: the solubility of a non-ideal gas. *Mar. Chem.* **2**, 203–215.

Wilson W. D. (1959) The speed of sound in distilled water as a function of temperature and pressure. *J. Acoust. Soc. Am.* **31**, 1067–1072.

Wilson W. D. (1960a) The speed of sound in seawater as a function of temperature, pressure and salinity. *J. Acoust. Soc. Am.* **32**, 641–644.

Wilson W. D. (1960b) Equation for the speed of sound in seawater. *J. Acoust. Soc. Am.* **32**, 1357.

Wirth H. E. (1940) The problem of the density of seawater. *J. Mar. Res.* **3**, 230–247.

Witter A. E., Hutchins D. A., Butler A., and Luther G. W., III (2000) Determination of conditional stability constants and kinetic constants for strong model Fe-binding ligands in seawater. *Mar. Chem.* **69**, 1–17.

Wu J. and Luther G. W. (1995) Complexation of Fe(III) by natural organic ligands in the North west Atlantic Ocean by a competitive ligand equilibration method and a kinetic approach. *Limnol. Oceanogr.* **50**, 177–1119.

Wu J., Boyle E., Sunda W., and Wen L.-S. (2001) Soluble and colloidal iron in the oligotrophic North Atlantic and North Pacific. *Science* **293**, 847–849.

Yao W. and Millero F. J. (1995) The chemistry of anoxic waters in the Framvaren Fjord, Norway. *Aquat. Geochem.* **1**, 53–88.

Young T. F. (1951) Recent developments in the study of interactions between molecules and ions, and of equilibrium in solutions. *Rec. Chem. Progr.* **12**, 81–95.

Young T. F. and Smith M. B. (1954) Thermodynamic properties of mixtures of electrolytes in aqueous solutions. *J. Phys. Chem.* **58**, 716–724.

Young T. F., Wu Y.-C., and Krawetz A. A. (1957) Thermal effects of interaction between ions of like charge. *Discuss. Faraday Soc.* **24** 37–42, 77–80.

Zhang J., van den Berg C. M. G., and Wollast R. (1990) The determination of interactions of cobalt(II) with organic compounds in seawater using cathodic stripping voltammetery. *Mar. Chem.* **28**, 285–300.

6.02
Controls of Trace Metals in Seawater

K. W. Bruland and M. C. Lohan

University of California at Santa Cruz, CA, USA

6.02.1 INTRODUCTION

Since the early 1970s, marine chemists have gained a first-order understanding of the concentrations, distributions, and chemical behaviors of trace metals in seawater. Important factors initiating this quantum leap in knowledge were major advances in modern analytical chemistry and instrumentation, along with the development and adoption of clean techniques. An instrumental development in the mid-1970s that spurred the early research on trace metals was the availability of the sensitive graphite furnace as the sample introduction system to an atomic absorption spectrometer. More recently, the appearance of inductively coupled plasma (ICP) mass spectrometers has provided an even more sensitive and powerful instrumental capability to the arsenal of marine chemists. In addition to these instruments back in shore-based laboratories, there has been the development of sensitive shipboard methods such as stripping voltammetry and flow injection analysis (FIA) systems with either chemiluminescence or catalytically enhanced spectrophotometric detection. Along with the development of these highly sensitive analytical techniques came a recognition and appreciation of the importance of handling contamination issues by using clean techniques during all phases of sampling and analysis. This is necessary due to low concentrations of trace metals in seawater relative to the ubiquitousness of metals on a ship or in a laboratory (e.g., dust, steel hydrowire, rust, paint with copper and zinc antifouling agents, brass fittings, galvanized material, sacrificial zinc anodes, etc.). As a result, seawater concentrations of most trace metals have now been accurately determined in at least some parts of the oceans, and their oceanic distributions have been found to be consistent with oceanographic processes.

The concentrations and distributions of trace metals in seawater are controlled by a combination of processes. These processes include external sources of trace metals delivered by rivers along ocean boundaries, by wind-blown dust from arid and semi-arid regions of the continents, and by hydrothermal circulation at mid-ocean ridges. Processes removing trace metals from seawater include active biological uptake or passive scavenging onto either living or nonliving particulate material. Much of this particulate material (along with its associated trace metals) is internally recycled either in the water column or in surficial sediments. The ultimate sink of trace metals is generally marine sediments. These various sources and sinks are superimposed on the general circulation and mixing of the oceans, resulting in the characteristic distributions of each trace metal. One of the first examples of the emergence of oceanographically consistent vertical profiles was for the trace-metal cadmium (Boyle *et al.*, 1976; Martin *et al.*, 1976; Bruland *et al.*, 1978a). These studies demonstrated that the distribution of dissolved cadmium in the sea follows a pattern similar to that of the nutrients phosphate and nitrate. Sparked by these surprising results, several investigators during the following two decades were able to obtain excellent data sets on a wide variety of trace metals. This chapter will attempt to provide a basic overview of what is known about the controls of the concentrations and distributions of trace metals in the open ocean. Subtleties in their distributions will not be presented. The distributions of trace metals in coastal regions are more dynamic and complicated and will not be discussed in this chapter.

The bulk of the data for vertical profiles of trace metals in seawater are from papers published in the 1980s and 1990s and most of the profiles are from either the North Pacific or North Atlantic. There is a paucity of vertical profiles from the South Atlantic and South Pacific. It has recently been argued that a new "GEOSECS"-type trace-metal program needs to be in place in order to provide appropriate global coverage of trace metals. Much of the impetus for such a program comes from the recognition of iron as an important micronutrient influencing global biogeochemical cycles in the oceans (Moore *et al.*, 2002) and the potential role of other trace metals such as zinc. In particular, there is a pressing need for an expansion of the global database of dissolved iron distributions in the oceans. These measurements are needed to both initiate and verify models and to identify processes not contained in existing models.

There have been a number of reviews of trace elements in seawater that form a foundation for this chapter. Among them are: Bruland (1983) on oceanographically consistent data sets; Burton and Statham (1990) on trace metals in seawater; and Donat and Bruland (1995) on trace elements in oceans. There are two reviews that deal with more of the biological role of trace metals: Bruland *et al.* (1991) on interactive influence of bioactive trace metals on biological production in ocean waters; and Hunter *et al.* (1997) on biological roles of trace metals in natural waters. A highly complementary chapter in this Treatise that deals with the influence of essential trace metals on biological processes has been written by Morel *et al.* (Chapter 6.05). Turning to "on-line" sources of information, Nozaki has done an excellent job perusing the available literature and compiling vertical profiles from the North Pacific for each element in a periodic table that makes an excellent figure (http://www.agu.org/eos_elec/97025e.html). Ken Johnson, a marine chemist at the Monterey Bay Aquarium Research Institute (MBARI), has a web site with a periodic table of the elements containing a brief review of information on each element (http://www.mbari.org/chemsensor/pteo.htm).

6.02.1.1 Concentrations

Concentrations of trace metals in seawater fall within a range bounded by ~10 μmol kg^{-1} at the upper end and extending through fmol kg^{-1} at the lower end (μ(micro) $= 10^{-6}$, n (nano) $=10^{-9}$, p (pico) $= 10^{-12}$, and f (fempto) $= 10^{-15}$). The concentrations of metals in seawater range over 15 orders of magnitude from sodium, the most abundant cation, at a concentration close to 0.5 mol kg^{-1} to iridium at a concentration as low as 0.5 fmol kg^{-1}. A concentration of 10 μmol kg^{-1} (~1 ppm by weight) is chosen as the concentration separating trace metals from the major and minor metals in seawater. The mean concentration of a trace metal is strongly influenced by its deep-water value, particularly that found in the deep waters of the Pacific Ocean with its large volume. Table 1 presents the range of concentrations and their mean for many of the elements in seawater along with an estimate of the major inorganic form or species found in seawater.

Trace metals in seawater can exist in a variety of physical and chemical forms. The simplest physical distinction is particulate versus dissolved forms. This is somewhat of an operational definition with 0.4 μm or 0.2 μm pore size filters generally providing this separation. Particulate forms include those metals adsorbed onto particle surfaces, incorporated within particles of biogenic origin and incorporated in the matrix of aluminosilicate minerals or co-precipitated in other authigenic minerals. Dissolved metals include various soluble complexes of the trace metals and potential colloidal forms. The redox chemistry of the particular metal and its environment dictate the oxidation state and form of the parent species. For trace metals, the parent species can be the simple mono-, di-, or trivalent cation such as Zn^{2+}, or for metals existing in higher

Table 1 A periodic table of the elements in seawater indicating the element number, the dominant inorganic species predicted to be found in the oceans, the range of concentrations observed in the open ocean, and an estimate of the element's mean concentration. The mean concentration is shown on the left-hand column in parenthesis and on the right-hand column as a thick line. Hydrogen, noble gases, lanthanides, and elements after lead are omitted.

Element and species	Concentration (fmol kg^{-1} 10^{-15} · pmol kg^{-1} 10^{-12} · nmol kg^{-1} 10^{-9} · μmol kg^{-1} 10^{-6} · mmol kg^{-1} 10^{-3} · mol kg^{-1} 10^{0})
3 Li Li^+	*(25.9 μmol kg^{-1})*
4 Be $Be(OH)^+$, $Be(OH)_2^0$	4 to 30 pmol kg^{-1} *(23)*
5 B H_3BO_3	*(416 μmol kg^{-1})*
6 C HCO_3^-	1.9 to 2.5 mmol kg^{-1} *(2.25)*
7 N N_2	350 to 610 μmol kg^{-1} *(590)*
7 N NO_3^-	0.01 to 45 μmol kg^{-1} *(30)*
8 O O_2	1 to 350 μmol kg^{-1} *(175)*
9 F F^-	*(68 μmol kg^{-1})*
11 Na Na^+	*(468 mmol kg^{-1})*
12 Mg Mg^{2+}	*(53 mmol kg^{-1})*
13 Al $Al(OH)_4^-$, $Al(OH)_3^0$	0.3 to 40 nmol kg^{-1} *(~2)*
14 Si H_4SiO_4	0.5 to 180 μmol kg^{-1} *(100)*
15 P HPO_4^{2-}	0.001 to 3.5 μmol kg^{-1} *(2.3)*
16 S SO_4^{2-}	*(28 mmol kg^{-1})*
17 Cl Cl^-	*(546 mmol kg^{-1})*
19 K K^+	*(10.2 mmol kg^{-1})*
20 Ca Ca^{2+}	10.1 to 10.3 mmol kg^{-1} *(10.3)*
21 Sc $Sc(OH)_3^0$	8 to 20 pmol kg^{-1} *(16)*
22 Ti $Ti(OH)_4^0$, $TiO(OH)_2^0$	6 to 250 pmol kg^{-1} *(160)*
23 V HVO_4^{2-}	30 to 36 nmol kg^{-1}
24 Cr CrO_4^{2-}	3 to 5 nmol kg^{-1} *(4)*
25 Mn Mn^{2+}	0.08 to 5 nmol kg^{-1} *(0.3)*
26 Fe $Fe(OH)_2^+$, $Fe(OH)_3^0$	0.02 to 2 nmol kg^{-1} *(0.5)*
27 Co Co^{2+}	4 to 300 pmol kg^{-1} *(20)*
28 Ni Ni^{2+}	2 to 12 nmol kg^{-1} *(8)*
29 Cu $CuCO_3^0$	0.5 to 4.5 nmol kg^{-1} *(3)*
30 Zn Zn^{2+}	0.05 to 9 nmol kg^{-1} *(5)*
31 Ga $Ga(OH)_4^-$	12 to 30 pmol kg^{-1}
32 Ge H_4GeO_4	1 to 100 pmol kg^{-1} *(70)*
32 Ge MethylGe	*(400 pmol kg^{-1})*
33 As $HAsO_4^{2-}$	20 to 25 nmol kg^{-1} *(23)*
34 Se SeO_4^{2-}	0.5 to 2.3 nmol kg^{-1} *(1.7)*
35 Br Br^-	*(0.84 mmol kg^{-1})*

(continued)

Table 1 (continued).

Concentration scale: fmol kg^{-1} (10^{-15}) | pmol kg^{-1} (10^{-12}) | nmol kg^{-1} (10^{-9}) | µmol kg^{-1} (10^{-6}) | mmol kg^{-1} (10^{-3}) | mol kg^{-1} (10^{0})

37 Rb	Rb^+	*(1.4 µmol kg−1)*
38 Sr	Sr^{2+}	*(90 µmol kg−1)*
39 Y	$Y(CO_3)^+$, $Y(OH)^{2+}$	60 to 300 pmol kg−1 *(200)*
40 Zr	$Zr(OH)_5^-$, $Zr(OH)_4^0$	12 to 300 pmol kg−1 *(200)*
41 Nb	$Nb(OH)_6^-$, $Nb(OH)_5^0$	≤50 pmol kg−1
42 Mo	MoO_4^{2-}	*(105 nmol kg−1)*
43 Tc		No stable isotope
44 Ru	$Ru(OH)_n^{4-n}$	< 50 fmol kg−1 ?
45 Rh	$Rh(OH)_n^{3-n}$, $RhCl_n^{3-n}$	0.4 to 1 pmol kg−1 *(0.8)*
46 Pd	$PdCl_4^{2-}$	0.2 to 0.7 pmol kg−1 *(0.6)*
47 Ag	$AgCl_2^-$, $AgCl_3^{2-}$	1 to 35 pmol kg−1 *(20)*
48 Cd	$CdCl_2^0$	1 to 1000 pmol kg−1 *(600)*
49 In	$In(OH)_3^0$	40 to 100 fmol kg−1 *(70)*
50 Sn	$SnO(OH)_3^-$, $Sn(OH)_4^0$	1 to 20 pmol kg−1 *(4)* ?
51 Sb	$Sb(OH)_6^-$	*(1.6 nmol kg−1)*
52 Te	$Te(OH)_6^0$	0.5 to 1.2 pmol kg−1 *(0.6)*
53 I	IO_3^-	400 to 460 nmol kg−1 *(450)*
55 Cs	Cs^+	*(2.2 nmol kg−1)*
56 Ba	Ba^{2+}	30 to 150 nmol kg−1 *(110)*
57–71	Lanthanides	
71 Lu	$Lu(CO_3)^+$, $Lu(OH)^{2+}$	0.3 to 1.5 pmol kg−1 *(1)*
72 Hf	$Hf(OH)_4^0$, $Hf(OH)_5^-$	100 to 800 fmol kg−1 *(700)*
73 Ta	$Ta(OH)_5^0$	60 to 220 fmol kg−1 *(200)*
74 W	WO_4^{2-}	*(60 pmol kg−1)*
75 Re	ReO_4^-	*(40 pmol kg−1)*
76 Os	$H_3OsO_6^-$	15 to 60 fmol kg−1 *(50)* ?
77 Ir	$Ir(OH)_3^0$?	0.5 to 1 fmol kg−1
78 Pt	$PtCl_4^{2-}$?	0.2 to 1.5 pmol kg−1 *(0.25)* ?
79 Au	$AuCl_2^-$?, $Au(OH)_3^0$?	10 to 100 fmol kg−1 *(<100)*
80 Hg	$HgCl_4^{2-}$	0.2 to 2 or 2 to 10 pmol kg−1 ? *(1)*?
81 Tl	Tl^+, $TlCl^0$	60 to 80 pmol kg−1 *(70)*
82 Pb	$PbCO_3^0$	5 to 150 pmol kg−1 *(10)*

oxidation states, it can be an oxycation such as UO_2^{2+} or an oxyanion such as MoO_4^{2-}. For a trace metal such as mercury, the parent species can be Hg^{2+}, CH_3Hg^+, or Hg^0. Cationic parent species can form complexes with a variety of both inorganic and organic ligands in seawater. For example, Hg^{2+} will complex with chloride and exist primarily as $Hg(Cl)_4^{2-}$, and UO_2^{2+} will complex with carbonate and exist as the uranyl carbonate complex ($UO_2(CO_3)_3^{4-}$).

The inorganic speciation presented in Table 1 is primarily taken from the compilations of Turner *et al.* (1981) and Byrne *et al.* (1988). Turner *et al.* (1981) used a database of stability constants for more than 500 metal complexes to calculate the inorganic speciation for 58 trace elements in model seawater at pH 8.2, 25 °C, and 1 atm. Byrne *et al.* (1988) extended this work by considering the influence of temperature and pH on speciation. The free hydrated divalent cation dominates the dissolved inorganic speciation of Zn(II) and the first transition series metals Mn(II), Co(II), and Ni(II). Strongly hydrolyzed trace metals include Be(II), Al(III), Fe(III), and Ga(III). Trace metals whose dissolved speciation is dominated by chloride complexation include Cd(II), Hg(II), Ag(I), and Pd(II).

The complexation or chelation of trace metals with organic ligands will be discussed separately at the end of this chapter. As we will see, the chemistry and behavior of many trace metals in the water column is dominated by complexation, biological assimilation at uptake sites on cell surfaces, and adsorption on surface sites of suspended particles. All three of these processes, which are particularly important in surface waters where biological activity is most intense, are controlled by similar coordination mechanisms (Hering and Morel, 1990).

Starting at the beginning of the periodic table, beryllium (element #4) is the first trace metal, and is the only trace metal with an atomic number less than 12. Beryllium has a concentration range in the oceans of 4–30 pmol kg^{-1} (Measures and Edmond, 1982) and exists as Be(II) with the hydrolysis species $Be(OH)^+$ and $Be(OH)_2^0$ as the major inorganic species. Aluminum (element #13) is the next trace metal encountered in the periodic table. Although aluminum is very abundant in the Earth's crust, it is a trace metal in the open ocean with concentrations ranging between 0.3 nmol kg^{-1} and 40 nmol kg^{-1} (Hydes, 1983; Orians and Bruland, 1986). Dissolved aluminum exists as Al(III), with the hydrolysis species $Al(OH)_3^0$ and $Al(OH)_4^-$ dominating its inorganic speciation.

All of the first-row transition metals are trace metals with concentrations ranging by a factor of 10^4—from a low of 4 pmol kg^{-1} up to 36 nmol kg^{-1}. The least abundant first-row transition metal is cobalt, thought to exist as Co(II), with a concentration range of 4–300 pmol kg^{-1}, while the most abundant is vanadium, existing as the vanadate oxyanion (HVO_4^{2-}) at a concentration between 30 nmol kg^{-1} and 36 nmol kg^{-1}. Interestingly, the most abundant transition metal in the oceans is molybdenum (element #42), existing at ~105 nmol kg^{-1} (Sohrin *et al.*, 1989) as Mo(VI) in the form of the molybdate oxyanion (MoO_4^{2-}). The least abundant trace metal in seawater (excluding radioactive elements) is thought to be the platinum group metal iridium (element #77). Iridium is extremely rare in the Earth's crust and its seawater concentration is reported to be between 0.5 fmol kg^{-1} and 1 fmol kg^{-1} (Anbar *et al.*, 1996). It is thought to exist as Ir(III) primarily as the hydrolysis species $Ir(OH)_3^0$. The analytical challenges of determining the concentration of an element at sub-femptomolar concentrations ($<10^{-15}$ M) are impressive indeed. In this case it was determined by isotope dilution using thermal ionization mass spectrometry after concentration with anion exchange resins.

6.02.1.2 Distributions

Trace metals exhibiting a relatively narrow range of concentrations in seawater tend to exist either as oxyanions (MoO_4^{2-}, WO_4^{2-}, ReO_4^{2-}) or as larger monovalent cations (Cs^+, Rb^+). Some of these metals such as molybdenum have a significant biological requirement; however, they tend to exist in seawater at relatively high concentrations relative to their requirement by the biota. Those trace metals exhibiting the greatest range in concentrations tend to be intimately involved in the major biogeochemical cycles and are actively assimilated by phytoplankton in surface waters. These include trace metals such as iron, zinc, and cadmium. Trace metals have been grouped into three principal categories reflecting their distributions and chemical behavior in seawater: conservative, nutrient, and scavenged.

6.02.1.2.1 Conservative-type distributions

Conservative-type trace metals interact only weakly with particles, have oceanic residence times greater than 10^5 yr (much greater than the mixing time of the oceans), and have concentrations that maintain a relatively constant ratio to salinity. Trace metals with conservative-type distributions in seawater such as molybdenum, antimony, tungsten, rhenium, caesium, and rubidium are involved in the major biogeochemical cycles of particle formation and destruction, but this is negligible relative to their concentration in seawater. Molybdenum is probably the best example of a conservative-type trace metal. It exists at an average concentration of 105 nmol kg^{-1} as the oxyanion molybdate, MoO_4^{2-}, and has an oceanic residence time of ~8×10^5 yr (Emerson and Huested, 1991). It exhibits an almost uniform distribution in the oceans with only a slight depletion at the surface. Although molybdenum is required as an essential metal co-factor in a number of enzymes such as nitrogenase, this requirement is small relative to the amount of molybdenum available and does not impact its distribution appreciably. In addition, negatively charged anions such as MoO_4^{2-} have a relatively low particle affinity at the slightly basic pH of seawater. Other trace metals existing as oxyanions that exhibit relatively conservative-type behavior include tungstate (WO_4^{2-}) and perrhenate (ReO_4^-).

There is interest in the use of some of these oxyanions as potential paleogeochemical proxies of the oxygen content or the redox state of deep waters (Emerson and Huested, 1991; Morford and Emerson, 1999). For example, under anoxic conditions molybdenum can be reduced from the +6 oxidation state as an oxyanion to insoluble MoS_2(s) or converted to particle-reactive thiomolybdates (Vorlick and Helz, 2002).

Other examples of conservative-type trace metals include caesium and rubidium. Caesium(I) exists at an average concentration of 2.2 nmol kg^{-1} (Brewer *et al.*, 1972) as the relatively unreactive monovalent cation Cs^+. Its oceanic residence time has been estimated to be $\sim 3 \times 10^5$ yr (Broecker and Peng, 1982). Rubidium exists as the monovalent cation Rb^+ at a concentration of 1.4 μmol kg^{-1} (Spencer *et al.*, 1970) with an oceanic residence time estimated to be 3 Myr.

6.02.1.2.2 Nutrient-type distributions

Trace metals with nutrient-type distributions are significantly involved with the internal cycles of biologically derived particulate material. Their distributions are dominated by the internal cycle of assimilation by plankton in surface waters and the export production or transport of part of this material out of the surface layer followed by oxidation and remineralization of the bulk of this material in deeper waters. Consequently, their concentrations are lowest in surface waters where they are assimilated by phytoplankton and/or adsorbed by biogenic particles, and increase in the subsurface waters as sinking particles undergo decomposition or dissolution. In addition, nutrient-type metals exhibit a relatively low level of scavenging in the deep sea and thus their concentrations increase along the flow path of water in the world's oceans as the water ages. Oceanic residence times of nutrient-type, recycled elements are intermediate (a few thousand to one hundred thousand years).

Zinc is perhaps the most striking example of a trace metal with a nutrient-type distribution in the oceans. Bruland *et al.* (1978b) reported the first accurate zinc distribution in seawater and demonstrated its strong correlation with silicic acid. Subsequently, Bruland (1980), Bruland and Franks (1983), Martin *et al.* (1989, 1993), Morley *et al.* (1993), Bruland *et al.* (1994), and Lohan *et al.* (2002) have provided other consistent profiles. Characteristic vertical profiles of dissolved zinc and silicic acid from high latitudes of the North Atlantic and North Pacific are presented in Figure 1, demonstrating the strong interbasin fractionation that exists for this nutrient-type trace metal. The concentration data available for deep waters of the oceans yield a linear relationship between dissolved zinc (nmol kg^{-1}) and silicic acid (μmol kg^{-1}): $[Zn] = 0.05[H_4SiO_4] + 0.8$. Figure 2(a) presents the distribution of silicic acid at 3,000 m depth in the world's oceans (from the NODC data set) and, using the above relationship between silicic acid and dissolved zinc, Figure 2(b) illustrates the estimated dissolved zinc concentration at 3,000 m depth. Silicic acid increases by a factor of 10 from a concentration of 20 μmol kg^{-1} at a depth of 3,000 m at high latitudes of the North Atlantic to 200 μmol kg^{-1} at high latitudes of the North Pacific. Similarly, dissolved zinc increases by a factor of 5 from 2 nmol kg^{-1} in the young waters of the western North Atlantic to 10 nmol kg^{-1} in

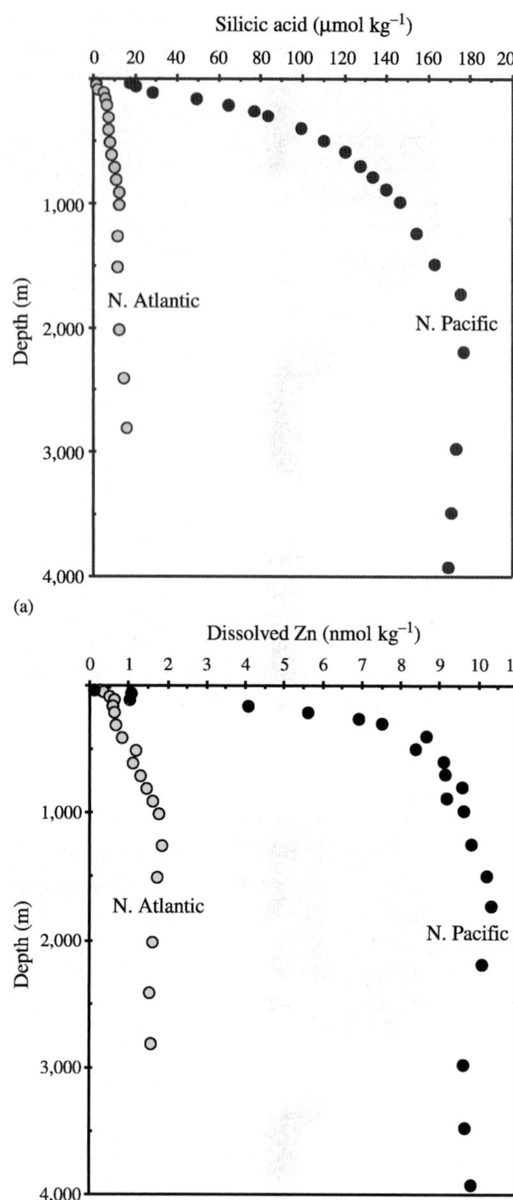

Figure 1 Vertical profiles of (a) silicic acid and (b) dissolved zinc observed at high latitudes of the North Atlantic (○) (59° 30′ N, 20° 45′ W; data from Martin *et al.*, 1993) and the North Pacific (●) (50° N, 145° W; data from Martin *et al.*, 1989).

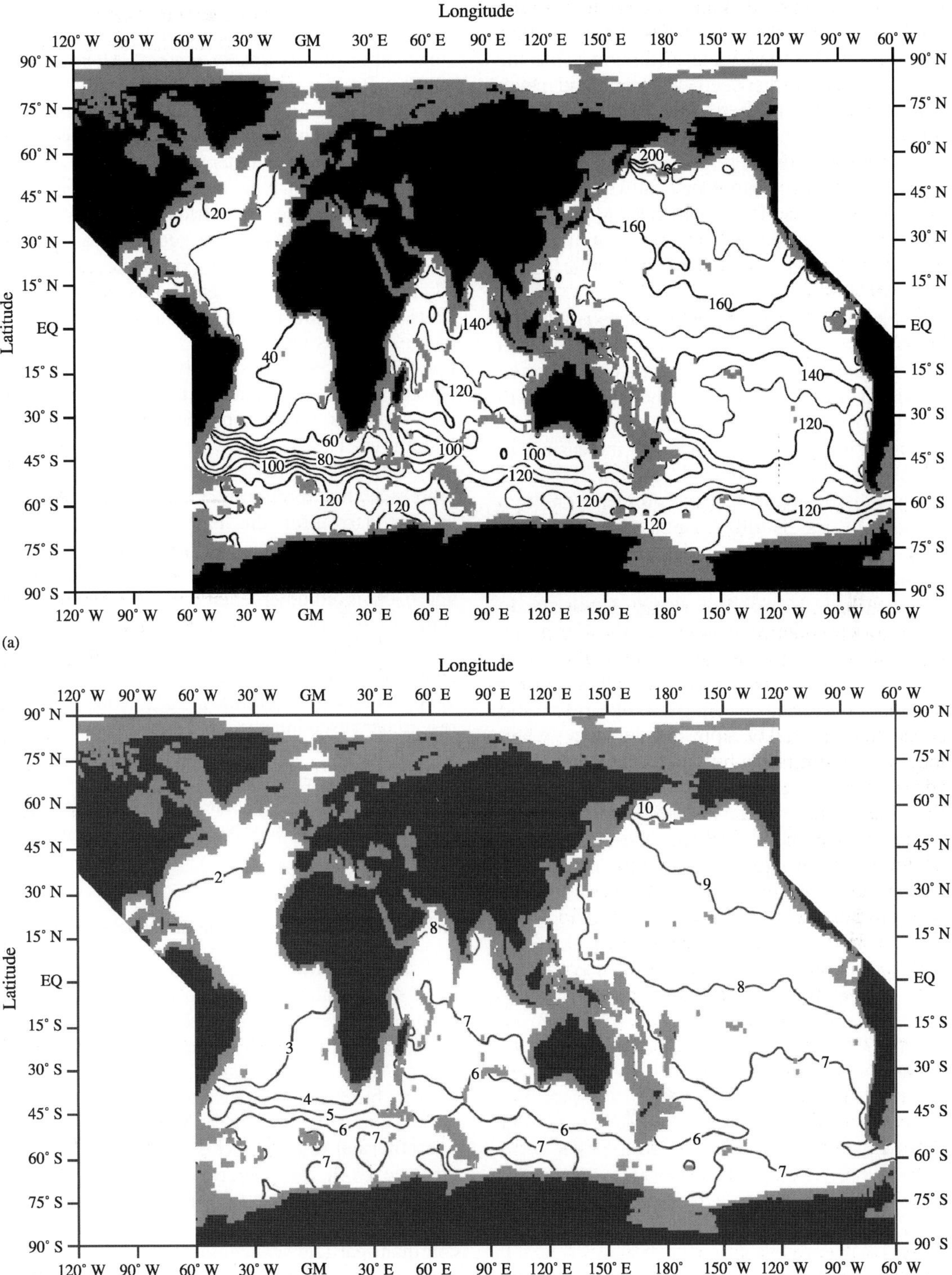

Figure 2 Horizontal gradients of the annual mean concentration of (a) silicic acid (μmol kg^{-1}) at 3,000 m depth in the world's oceans (source NODC), and (b) dissolved zinc (nmol kg^{-1}) at a depth of 3,000 m based upon the deep-water relationship ($[Zn] = 0.05[H_4SiO_4] + 0.8$) between silicic acid and dissolved zinc from stations in the North Atlantic (Bruland and Franks, 1983; Martin *et al.*, 1993), North Pacific (Bruland, 1980; Martin *et al.*, 1989; Bruland *et al.*, 1994), and Southern Ocean (Martin *et al.*, 1990).

the old deep waters found at high latitudes of the North Pacific. Similar figures can be produced for the nutrient-type trace metal cadmium using the strong correlation between cadmium and phosphate or nitrate in deep waters of the oceans. Barium also exhibits a nutrient-type distribution with concentrations ranging from 35 nmol kg^{-1} in surface waters to 56 nmol kg^{-1} at 3,000 m depth in the North Atlantic, and increasing along the flow path of deep water to concentrations of 150 nmol kg^{-1} in the deep North Pacific (Chan *et al.*, 1976, 1977).

Silver provides another interesting example of a nutrient-type trace metal with a strong interbasin fractionation. The first accurate data were reported by Martin *et al.* (1983) for the eastern North Pacific where they observed values of 0.4 pmol kg^{-1} in surface waters increasing to 23 pmol kg^{-1} at a depth of 2,300 m. Figure 3 presents dissolved silver depth profiles from high latitudes of the North Atlantic (Rivera-Duarte *et al.*, 1999) and high latitudes of the Northwest Pacific (Zhang *et al.*, 2001). Vertical profiles of dissolved silver exhibit a strong similarity to silicic acid and the deep waters of the Northwest Pacific are enriched by slightly more than a factor of 10 for silicic acid and slightly less than a factor of 10 for silver (Figure 3). This interbasin fractionation for silver represents an even greater fractionation than that observed for zinc. It is unclear why silver would exhibit such a nutrient-type profile, since there is no known biological requirement for silver. Perhaps the nutrient metal acquisition sites are not selective enough to discriminate against silver and it is mistakenly assimilated into phytoplankton by the zinc (or some other metal) uptake system. Alternatively, silver may be passively adsorbed to selected surface sites of biogenic particles and transported to depth where it is remineralized as the biogenic particulate material undergoes oxidation or dissolution. Nutrient-type trace metals have been used as paleoproxies for nutrient concentrations or ages of deep waters in the geologic past. Good direct fossil records of the phosphate, nitrate, or silicic acid content of deep waters do not exist. Thus, the fossil record of nutrient-type trace metals such as cadmium or zinc whose concentrations mimic and are strongly correlated with the macronutrients phosphate, nitrate, or silicic acid, can serve as indirect proxies of past nutrient conditions. The best known example involves the use of cadmium (Boyle, 1988), where the Cd/Ca ratio in benthic foraminifera in sediment cores can be used to infer the past concentrations of cadmium in deep seawater, and by correlation, phosphate in the overlying deep waters. This approach is complementary to the use of more standard tracers such as $\delta^{13}C$. The nutrient-type distribution of zinc has also been used as a paleoproxy for the age and nutrient content of deep waters (Marchitto *et al.*, 2002). These researchers used increases in benthic foraminiferal Zn/Ca and Cd/Ca ratios as evidence for a greatly increased presence of nutrient-rich Southern Ocean water in the glacial North Atlantic versus relatively nutrient-poor North Atlantic deep water.

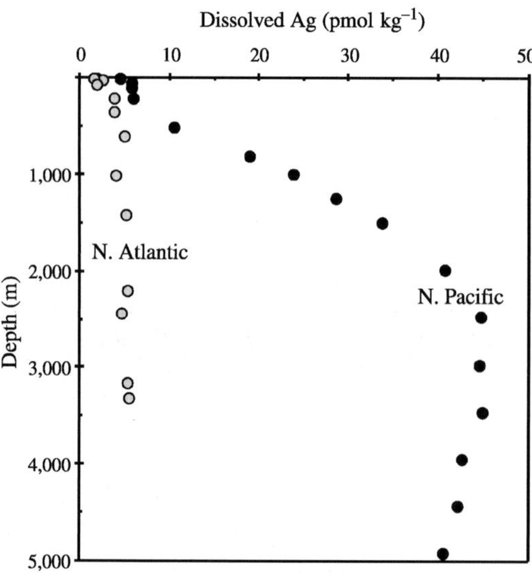

Figure 3 Vertical profiles of dissolved silver in the North Atlantic (○) (composite of two stations 54.5° N, 48.5° W, and 52.7° N, 35° W; data from Rivera-Duarte *et al.*, 1999) and the western North Pacific (●) (40° N, 145° W; data from Zhang *et al.*, 2001).

6.02.1.2.3 Scavenged-type distributions

Trace metals with scavenged-type distributions have strong interactions with particles and short oceanic residence times (~100–1,000 yr), residence times that are less than the ventilation or mixing time of the oceans. Their concentrations tend to be maximal near major sources such as rivers, atmospheric dust, bottom sediments, and hydrothermal vents. Concentrations decrease with distance from the sources and, in general, the concentrations of the scavenged metals tend to decrease along the flow path of deep water due to continual particle scavenging.

Aluminum is the best illustration of a trace metal with a scavenged-type distribution in the oceans. The major external input of aluminum is from the partial dissolution of atmospheric dust delivered to the surface ocean. Vertical profiles in the Mediterranean, the North Atlantic, and the North Pacific are presented in Figure 4. Extremely elevated concentrations of dissolved aluminum are observed in the Mediterranean Sea (Hydes *et al.*, 1988), a region that receives a high atmospheric input of dust. Concentrations in

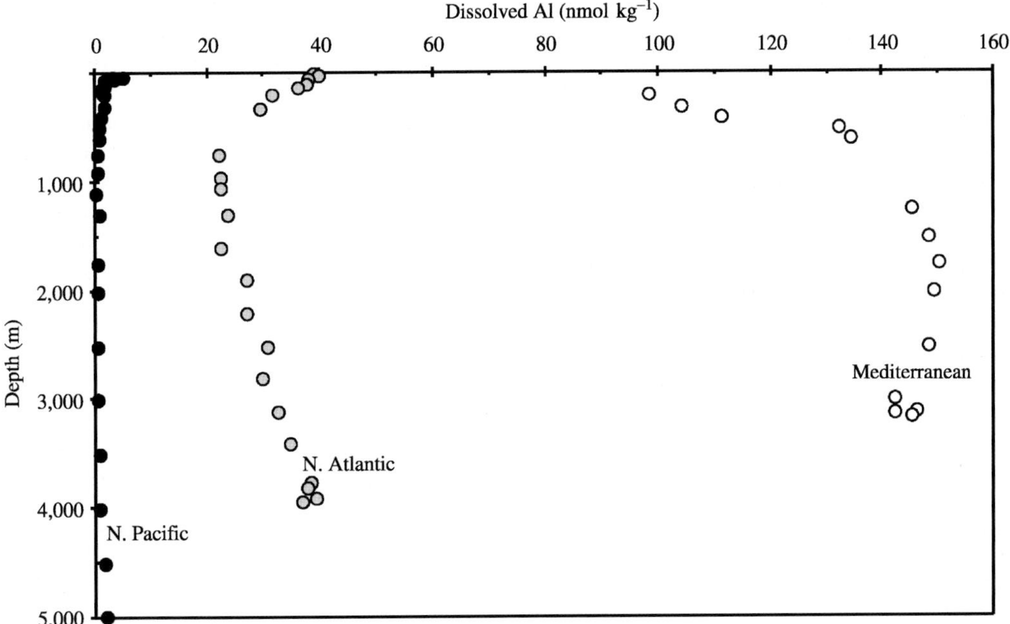

Figure 4 Vertical profiles of dissolved aluminum in the Mediterranean Sea (○) (34° 18′ N, 20° 02′ W; data from Hydes *et al.*, 1988), the North Atlantic (**○**) (40° 51′ N, 64° 10′ W; data from Hydes, 1979), and the North Pacific (●) (28° 15′ N, 155° 07′ W; data from Orians and Bruland, 1986).

the North Atlantic (Hydes, 1983) are maximal in the surface waters and elevated throughout the relatively young deep waters of the North Atlantic. The surface waters of the North Pacific, a region that receive less dust input than the Atlantic, exhibit aluminum concentrations that are correspondingly lower (Orians and Bruland, 1985, 1986). The old deep waters of the North Pacific have dissolved aluminum concentrations that are (8–40)-fold lower than in the North Atlantic and ~100-fold lower than those observed in the Mediterranean. This decrease along the deep-water flow path is consistent with an oceanic residence time for dissolved aluminum of only ~200 yr (Orians and Bruland, 1986). This marked difference between the North Pacific and North Atlantic deep waters is the reverse of that shown by the nutrient-type trace metals and the greatest interbasin fractionation of any trace metal. Unfortunately, it does not appear that this marked interbasin fractionation observed for dissolved aluminum can be utilized as a paleogeochemical tracer.

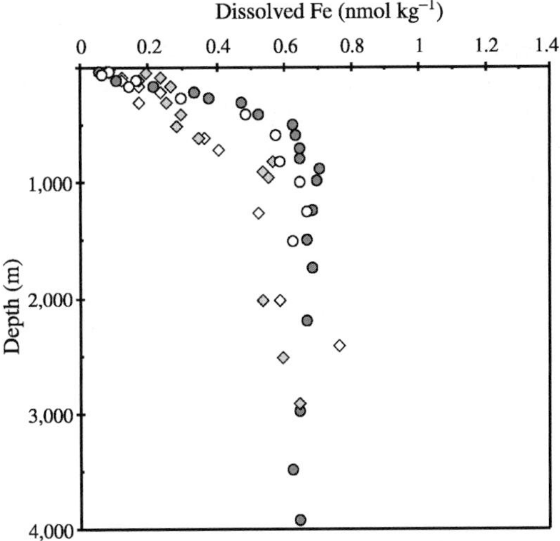

Figure 5 Vertical profiles of dissolved iron from high latitudes of the North Atlantic (◇, ◇) (59° 30′ N, 20° 45′ W and 47° N, 20° W; data from Martin *et al.*, 1993) and the North Pacific (**○**, ○) (50° N, 145° W and 45° N, 142° 52′ W; data from Martin *et al.*, 1989).

6.02.1.2.4 Hybrid distributions

Some trace metals, such as iron and copper, have distributions that are strongly influenced by both recycling and relatively intense scavenging processes. Like nutrient-type elements, dissolved iron is observed to be depleted in remote oceanic surface waters such as high-nutrient, low-chlorophyll (HNLC) regimes, and appears to be regenerated with depth (Figure 5) (Martin and Gordon, 1988; Johnson *et al.*, 1997). In less productive waters of the oligotrophic central gyres, particularly in areas of high dust inputs, dissolved iron can exhibit surface-water maxima more indicative of scavenged elements (Bruland *et al.*, 1994; Measures *et al.*, 1995; Johnson *et al.*, 1997). While nutrient-type metals, with their relatively long oceanic residence times, tend to increase in concentration in the deep waters of the ocean as the latter age, the residence time of iron in deep waters is estimated to be ~200 yr and does not exhibit this trend. Figure 5 presents vertical profiles of dissolved iron in remote high-latitude regions of the North Atlantic and North Pacific oceans. The concentration of iron at depths greater than 1,000 m is not significantly different, which is in marked contrast to profiles of nutrients or nutrient-type trace metals. Instead, the deep-water concentration of dissolved iron appears to be controlled by a balance of remineralization from the rain of particulates from above and particulate scavenging (Johnson *et al.*, 1997).

6.02.1.2.5 Mixed distributions

There are also trace metals that exist in more than one chemical form with substantially differing distributions. A fascinating example involves the trace element germanium. Germanium (element #32) is located just beneath silicon in the periodic table. Inorganic germanic acid behaves similar to silicic acid in seawater (Froelich and Andreae, 1981). It is assimilated at a molar ratio Ge : Si of $\sim 0.7 \times 10^{-6}$ into the siliceous tests of diatoms and other planktonic organisms that make tests of biogenic opal. When the tests dissolve, the germanium is released in the same ratio and as a result there is a tight correlation between the distribution of germanic acid (H_4GeO_4) and silicic acid (H_4SiO_4) in seawater (Froelich and Andreae, 1981). Unlike silicon, however, germanium is also found to exist as the methylated forms $CH_3Ge(OH)_3^0$ and $(CH_3)_2Ge(OH)_2^0$ (Lewis *et al.*, 1989), that appear so stable to degradation that they have been called the "Teflon of the sea." The remarkable stability of these methylated species is reflected in their conservative vertical profiles (concentrations of dimethylgermanium are 100 pmol kg^{-1} and monomethylgermanium are 310–330 pmol kg^{-1}) (Figure 6). This conservative distribution is in marked contrast to the nutrient-type distribution of germanic acid (2–120 pmol kg^{-1}). There are a number of other examples of methylated compounds that comprise a significant fraction of trace metals and metalloids in seawater (e.g., arsenic, selenium, mercury, tin).

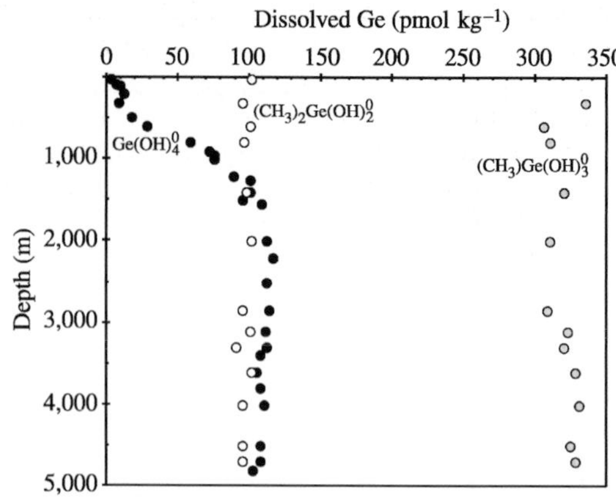

Figure 6 Vertical profiles of dissolved germanium species from the North Pacific: inorganic germanium (●) (25° N, 170° 05′ E; data from Froelich and Andreae, 1981); methyl-germanium (**O**, ○) (data from Lewis *et al.*, 1985).

6.02.2 EXTERNAL INPUTS OF TRACE METALS TO THE OCEANS

6.02.2.1 Rivers

For the major ions in seawater, the input from rivers is generally the dominant source. The historical approach to estimate the river flux of major elements is to measure their concentrations in both dissolved and particulate forms in the river and multiply these concentrations by the river discharge rate, thus arriving at the input of both forms of the elements. For trace metals, however, estimating the river flux is more difficult. There are major problems due to under-sampling of representative river systems. Although large rivers dominate the global river input to the oceans, such rivers are located in remote regions and are insufficiently sampled to allow adequate fluxes of dissolved trace-metal concentrations in relation to season and flow to be determined (Jickells, 1995). Most of the historical river data for trace metals are not accurate, and the development of trace-metal clean techniques also needs to be applied to river sampling (Shiller and Boyle, 1991; Windom *et al.*, 1991; Kim *et al.*, 1999).

Not only are accurate data for trace metals in rivers sparse, there are complications that exist at the river–sea interface. The increase in salinity occurring at the river–sea water interface, with its concomitant increase in the concentrations of the major seawater cations, can lead to flocculation and sedimentation of trace metals such as iron (Boyle *et al.*, 1978; Sholkovitz and Copeland, 1983) or to desorption from suspended riverine particles of trace metals such as barium (Edmond *et al.*, 1978). In organic-rich rivers a major fraction of dissolved trace metals can exist in physiochemical association with colloidal humic acids. Sholkovitz and Copeland (1983) used "product-mode" mixing experiments on filtered Scottish river water, and observed that iron removal was almost complete due to the flocculation of strongly associated iron-humic acid colloids in the presence of the increased concentrations of Ca^{2+} and Mg^{2+} found in estuaries. Copper and nickel were also removed to an appreciable extent. This removal within estuarine mixing zones is not as important in rivers with lower dissolved organic carbon content. Nonconservative behavior within estuaries makes it difficult to obtain realistic estimates of the actual river input of trace metals to the oceans.

6.02.2.2 Atmosphere

Using a sparse network of field measurements of atmospheric aerosols, Duce and Tindale (1991) were able to provide some of the first global estimates of atmospheric dust input to the oceans. More recently, satellites have provided estimates of the global distribution of atmospheric aerosols allowing model estimates of dust deposition (Tegen and Fung, 1995; Mahowald *et al.*, 1999). Figure 7 presents the results of such models (Moore *et al.*, 2002). The atmospheric input of trace metals varies markedly spatially and temporally, and is of a similar magnitude as the riverine input. However, aeolian fluxes impact directly on the oceanic euphotic zone, while fluvial inputs are subjected to considerable modification in estuarine and coastal waters (Jickells, 1995).

Aluminum is a major component of continental materials and is present in seawater at low concentrations in regions devoid of large dust deposition due to the short residence time of aluminum in surface waters (3–5 yr; Orians and Bruland, 1986). Therefore, dissolved aluminum concentrations are an excellent tracer of atmospheric inputs to the ocean. A study by Vink and Measures (2001), using both a model of dust deposition (Measures and Brown, 1996) and dissolved aluminum concentrations in the Atlantic, has demonstrated that surface-water aluminum concentrations can be utilized to study the spatial and interannual variations of aeolian input. Concomitant with the aluminum concentrations, dissolved iron measurements were also carried out. At the interface between the canary current and the south equatorial current, a similar trend of maxima and minima in both the iron and aluminum concentrations was observed, implying a common atmospheric source. Although iron concentrations in the oceans are influenced by atmospheric inputs, low iron values in the surface waters of South Atlantic were observed, where aluminum concentrations suggested high dust deposition indicating that other factors such as, solubility and biological removal are more important in controlling iron distributions.

The maximum concentration of iron in oceanic surface seawater is controlled by the solubility of inorganic forms and the availability of organic complexing ligands to promote higher solubilities (Zhu *et al.*, 1997; Jickells and Spokes, 2001; Vink and Measures, 2001). Perhaps the greatest uncertainty in estimating the impact of this atmospheric input on the oceans is the estimate of the percentage of trace metals associated with dust that is soluble upon entering the ocean. For iron, recent estimates are between 1% and 10%. Jickells and Spokes (2001) in a review suggest

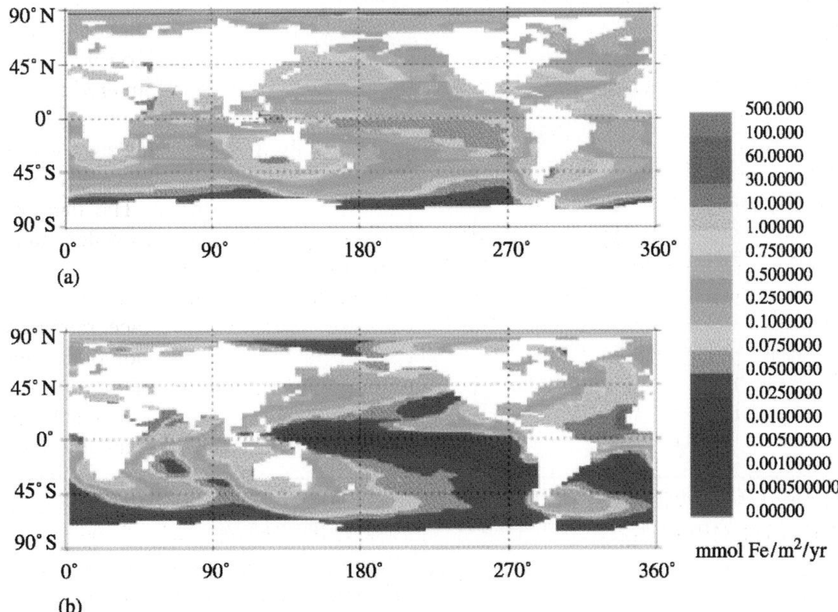

Figure 7 Modeled global estimates of aeolian iron deposition from: (a) Tegen and Fung (1994, 1995) and (b) Mahowald *et al.* (1999) (source Moore *et al.*, 2002).

the mean solubility of iron from atmospheric dust is ~2%. New estimates indicate that ~40% of the world's oceans are limited by the trace metal iron. These regions are primarily located in remote regions where the supply of nutrients is high. In the remote subtropical gyres of the oceans, the source of nutrients and iron from vertical mixing or upwelling is small and, therefore, the atmospheric supply of iron is the dominant input. In the North Pacific, the input of Asian dust is at a maximum in the spring and this input may significantly increase the primary production during this time. The atmospheric input of iron to the central gyres can also be important with respect to nitrogen fixation, a process that requires iron as a metal co-factor (see Chapter 6.05). The variability in atmospheric input of trace metals such as iron can be extreme—the variability ranges from rapid day-to-day changes as a result of dust storms in Asia, to seasonal changes, to decadal changes (Jickells, 1995; Zhu *et al.*, 1997). Less is known about the mean solubility of other biologically important trace metals such as zinc and manganese.

An excellent example of the atmospheric input of a trace metal strongly influencing its surface-water concentration is lead. Tetraethyl lead, an anti-knock gasoline additive, was used extensively in the 1960s and 1970s, with its usage peaking in the late 1970s and then markedly declining in the 1980s and 1990s as a result of actions taken under the Clean Air Act of 1970 (Nriagu, 1989). This extensive use of leaded gasoline resulted in a large anthropogenic lead signal in atmospheric dust, particularly downwind of industrialized nations. Pioneering work of Patterson and co-workers (Schaule and Patterson, 1981, 1983; Flegal and Patterson, 1983) on the distribution of lead in the major central gyres of the Atlantic and Pacific oceans graphically point out how this anthropogenic atmospheric lead input markedly perturbed the distribution of lead in the different ocean basins at the peak of its input in the late 1970s and early 1980s (Figure 8). Due to the prevailing wind pattern, the North Atlantic received the brunt of the US lead input to the oceans and contained markedly elevated concentrations of lead in the upper 1,000 m. In contrast, the remote South Pacific central gyre had surface lead concentrations over an order-of-magnitude lower. Subsequent to this work, Boyle and co-workers (Wu and Boyle, 1997) presented results from a 16-year time series of lead concentrations in the western North Atlantic

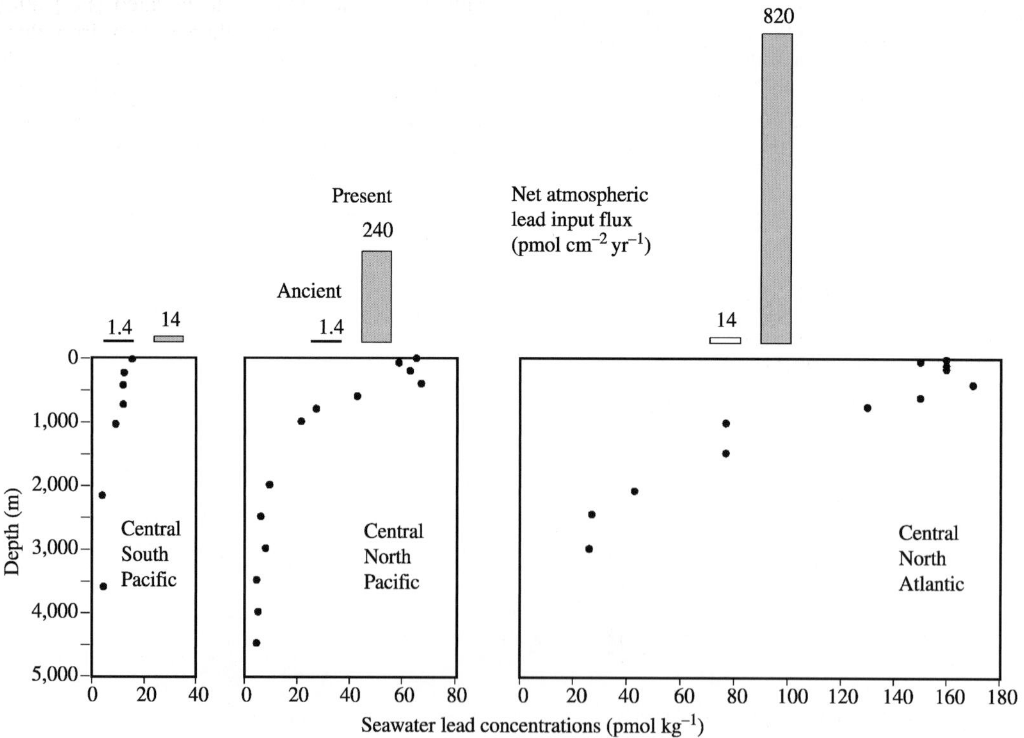

Figure 8 Vertical profiles of dissolved lead in the central North Atlantic (34° 15′ N, 66° 17′ W; data from Schaule and Patterson, 1983), the central North Pacific (32° 41′ N, 145° W; data from Schaule and Patterson, 1981), and the central South Pacific (20° S, 160° W; source Flegal and Patterson, 1983). Estimates of the atmospheric input at the time of sampling and in ancient times prior to the large anthropogenic lead input are also shown (Flegal and Patterson, 1983).

showing that lead concentrations decreased markedly during the decade of the 1980s and that this decrease can be attributed to the phasing out of leaded gasoline in the US.

6.02.2.3 Hydrothermal

It has been estimated, using arguments based upon the ^{3}He anomaly and heat flux, that the entire ocean mixes through hydrothermal vent systems, undergoing high-temperature interaction with fresh oceanic basalt every 8–10 Myr, leading to the production of high temperature (~350 °C), acidic (pH~3.5), reducing, sulfide- and metal-rich hydrothermal fluids (Edmond *et al.*, 1979; von Damm *et al.*, 1985). This ridge crest hydrothermal activity has proven to be the major oceanic sink for the major ions, magnesium, and sulfate, and to be a major source for trace metals such as iron and manganese (see Chapter 6.08). Iron and manganese concentrations in the 350 °C vent waters can be a million-fold higher than in the surrounding seawater. Much of the iron, however, is rapidly precipitated; initially either as iron sulfides and then oxidized to iron oxyhydroxide precipitates, or rapidly oxidized from the soluble Fe(II) form to insoluble Fe(III), and deposited as sediments over the mid-ocean ridges. The dissolved manganese can advect further away from the vent source prior to its microbially mediated oxidation and precipitation (Cowen *et al.*, 1998). The hydrothermal input of iron and manganese, however, is essentially all scavenged and removed in the deep sea prior to having a chance to mix back into the surface waters.

6.02.3 REMOVAL PROCESSES

6.02.3.1 Active Biological Uptake in the Surface Waters

6.02.3.1.1 Lessons from laboratory studies

A great deal of insight has been gained from well-defined laboratory studies of the effects of trace metals on phytoplankton growth rates, which in turn has provided knowledge on the control of trace metals in the upper water column by biological processes (see Chapter 6.05). Culture media have been designed in which the concentration and speciation of trace metals are controlled by the use of strong chelating ligands such as EDTA (Morel *et al.*, 1979; Price *et al.*, 1988/89; Sunda, 1988/89). The rate of uptake of a trace metal (M) is usually proportional to its free-metal concentration [M^{2+}] or its unchelated concentration, [M′] (defined as the sum of the kinetically labile inorganic species of M). The [M′] and [M^{n+}] are related by their inorganic side reaction coefficient ($\alpha = [M']/[M^{n+}]$). The EDTA chelated metal can act as a metal ion buffer that maintains [M^{n+}] and [M′] at constant values in the media. By judiciously varying the EDTA and total metal concentrations, the experimental [M^{2+}] and [M′] can be controlled over a wide range.

Diatoms are particularly important in biogeochemical cycles because of their role as major players in new and export production (Smetacek, 1999). Along with the assimilation and export of carbon, nitrogen, phosphorus, and silicon, diatoms also play an important role in the export of bioactive trace metals from surface waters. Sunda and Huntsman (1995a) have carried out extensive laboratory studies on iron uptake by coastal and oceanic diatoms. Figure 9 presents the cellular Fe/C (μmol mol^{-1}) ratio in diatoms as a function of the estimated [Fe′]. These diatoms exhibit an increasing cellular Fe/C ratio as the [Fe′] in the media increases. At elevated [Fe′] in the media, the diatoms exhibit luxury uptake with the two coastal species reaching cellular Fe/C values that were 20–30 times higher than those estimated to be required for maximum growth (Sunda and Huntsman, 1995a). It has been suggested that this high uptake rate and storage capacity in the diatoms at elevated [Fe′] allows these species to accumulate excess iron during periods of high iron availability, that can then be passed on to their progeny and utilized when the dissolved iron may later be drawn down to concentrations limiting growth rates (Sunda and Huntsman, 1995a). This can be an effective strategy for diatom blooms in coastal upwelling regimes.

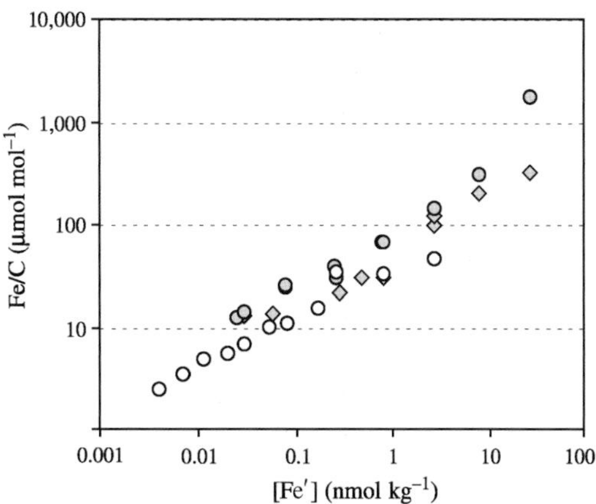

Figure 9 The relationship between the intracellular Fe/C ratio as a function of the inorganic iron concentration, [Fe′], for three diatoms species, *Thalassiosira oceanica* (○) an oceanic species (**○**), and two coastal species *Thalassiosira weissflogii*, and *Thalassiosira pseudonana* (◇) (source Sunda and Huntsman, 1995a).

6.02.3.1.2 Non-Redfieldian assimilation

Figure 9 is important with respect to removal of iron from the surface ocean. Nitrogen and phosphorus are assimilated and removed from the surface ocean at a ratio within about a factor of 2 of the Redfield ratio (Falkowski, 2000; Karl, 2002). Differences are observed at time-series stations in the North Pacific and North Atlantic (HOTS and BATS) that vary depending upon the source of nitrogen; however, they generally vary by less than a factor of 2. In contrast, there is no constant Redfield ratio of Fe/C; the Fe/C ratio in diatoms can vary by a factor of 100 depending upon the availability of iron and whether they are oceanic or coastal species. Interestingly, oceanic species have evolved to "get by" at lower iron concentrations than coastal species (Sunda and Huntsman, 1995a). Diatoms are also the type of phytoplankton most responsive to episodic changes in iron and nutrient inputs as demonstrated in each of the novel mesoscale iron-enrichment experiments (Martin *et al.*, 1994; Coale *et al.*, 1996; Boyd *et al.*, 2000). Their ability to maximize iron uptake and to store luxury uptake, can be a major factor in controlling iron concentrations in surface seawater. Bruland *et al.* (2001) examined this process in coastal diatoms under upwelling conditions and argued that iron is removed preferentially to nitrate in coastal upwelling regimes and tends to drive the system towards iron limitation. The amount of uptake of silicic acid by diatoms relative to the assimilation of nitrate and phosphate has been shown to depend on iron availability (Hutchins and Bruland, 1998; Takeda, 1998). Under iron-replete conditions, the silicon and nitrogen uptake is roughly equal, while under low-iron conditions diatoms exhibit Si/N uptake ratios of ~3. Under low iron concentrations the suppression of iron-containing enzymes such as nitrate or nitrite reductase has been confirmed for the chain-forming diatoms in the laboratory (de Baar *et al.*, 2000), resulting in reduced uptake of nitrogen relative to phosphorus and N/P ratios of 4–6 as compared to ~12–14 at adequate iron supply.

Studies have also been carried out on other essential trace metals such as zinc (Sunda and Huntsman, 1992, 1995b), and similar relationships between Zn/C and $[Zn^{2+}]$ have been observed. In studies with the diatom *Thalassiosira oceanica*, the Zn/C (μmol mol^{-1}) ratio varied from 0.2 at $[Zn^{2+}]$ of 10^{-13} M to ~40 at $[Zn^{2+}]$ of 10^{-9} M (Figure 10). As with iron, there is more than 100-fold variation in the Zn/C ratio in diatoms as a function of $[Zn^{2+}]$. The uptake of cadmium by diatoms is somewhat unique and is dependent not only upon the $[Cd^{2+}]$, but also on $[Zn^{2+}]$ (Sunda and Huntsman, 1998). At elevated $[Zn^{2+}]$, cadmium assimilation and Cd/C ratios in diatoms are low. Under zinc depletion, however, the uptake of cadmium and Cd/C ratios markedly increase. Morel and co-workers (Price and Morel, 1990; Lee and Morel, 1995) have shown that cadmium can functionally replace zinc and that diatoms have a carbonic anhydrase enzyme utilizing cadmium instead of zinc as a metal co-factor. Thus, under low-zinc conditions, the assimilation of cadmium markedly increases and results in the depletion of cadmium in surface water. This is an example of where the cellular uptake of one metal, cadmium, responds to a complex matrix of other metals such as zinc. Cullen *et al.* (2003) provided evidence that the effect of iron limitation on resident diatoms in the Southern Ocean is to decrease growth rates, leading to elevated cellular cadmium content. In this case the assimilation of carbon, nitrogen, and phosphorus was markedly decreased as a result of iron limitation, but cadmium assimilation continued, leading to enhanced cellular Cd/P ratios.

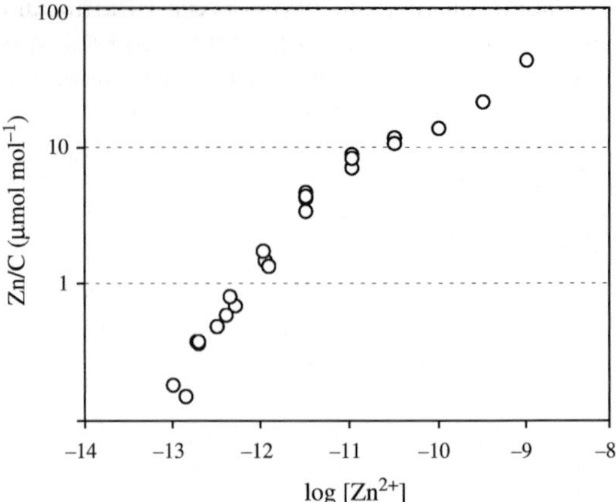

Figure 10 The relationship between the intracellular Zn/C ratio as a function of the free ionic zinc concentration, $[Zn^{2+}]$, for an oceanic diatom species *Thalassiosira oceanica* (source Sunda and Huntsman, 1992).

For copper, the cellular metal/carbon ratios vary in more of a sigmoidal fashion, with what appears to be a region of varying $[Cu^{2+}]$ with the Cu/C ratio somewhat constant and regulated (Sunda and Huntsman, 1995c). The Cu/C ratio in diatoms, however, still varies by roughly a factor of 100 over a wide range of concentrations of $[Cu^{2+}]$ and has been implicated as an important factor in controlling the distribution of copper in the oceans (Sunda and Huntsman, 1995c).

Trace metals for which active biological assimilation may be an important factor in controlling surface-water concentrations and distributions include the first-row transition metals iron, zinc, manganese, copper, nickel, and cobalt, along with cadmium. Bruland *et al.* (1991) compiled data on the composition of plankton in the Pacific taken from two sources (Martin and Knauer, 1973; Martin *et al.*, 1976) and excluded only data with an aluminum content >4 μmol g^{-1} (>100 μg g^{-1}) dry weight in an attempt to minimize the contribution from aluminosilicates minerals. Metal/carbon ratios for iron were close to 50 μmol mol^{-1}, Zn/C ranged from 8 μmol mol^{-1} to 17 μmol mol^{-1}, while Mn/C ratios averaged 3.6 μmol mol^{-1}. These M/C values for iron and zinc in diatom samples from the field off central California lie within the range expected for these region based upon laboratory studies. Biological assimilation of these metals at such M/C ratios with subsequent export of a fraction of this material to the deeper water column is particularly important in influencing the oceanic distributions of iron and zinc. These are two of the trace elements that exhibit marked surface depletion due to their involvement in this biological cycle.

6.02.3.2 Passive Scavenging

6.02.3.2.1 Adsorption/desorption processes

In addition to the role of active assimilation of required trace metals by phytoplankton, there is also passive scavenging of trace metals onto the wide variety of relatively high affinity surface sites on both living and dead particulate material existing in the surface waters. The combined process of surface adsorption, followed by particle settling, is termed scavenging (Goldberg, 1954; Turekian, 1977). Such binding is effectively "passive," in contrast with the active uptake of essential trace metals. Examples of trace metals implicated in such scavenging from surface waters include lead, aluminum, gallium, and the radioactive isotopes of thorium.

6.02.3.2.2 Lessons from radionuclides

Thorium, with its four different radioactive isotopes of greatly differing half-lives, is an excellent tracer which provides insight into the rates and the process of scavenging. Thorium isotopes and their half-lives are: ^{232}Th, $\tau_{1/2} = 1.4 \times 10^{10}$ yr (essentially ^{232}Th can be considered a stable isotope); ^{230}Th, $\tau_{1/2} = 7.54 \times 10^4$ yr with ^{234}U as a parent; ^{228}Th, $\tau_{1/2} = 1.91$ yr with ^{228}Ra as a parent; and ^{234}Th, $\tau_{1/2} = 24.1$ d with ^{238}U as a parent. Two conceptual models that have been used to model particle reactive thorium data are presented in Figure 11. Figure 11(a) incorporates reversible exchange and remineralization of the particles, while Figure 11(b) uses a net scavenging rate constant to examine the net scavenging of thorium. Figure 12 presents the distribution of the net scavenging rate constant of ^{234}Th in surface waters of the Pacific. The scavenging intensity of ^{234}Th varies dramatically between oligotrophic gyres of the North Pacific and South Pacific and productive regions such as the subarctic Pacific, the upwelling regime off central California and the equatorial Pacific. Values of the net scavenging rate constant in the surface mixed layer of the oligotrophic gyres were between 0.009 d^{-1} and 0.003 d^{-1} (yielding a mean life of dissolved ^{234}Th with respect to particle scavenging of 100–300 d). In contrast, in productive regions such as the subarctic Alaskan Gyre and intense coastal upwelling regimes off central California, the net scavenging rate constants were 0.11–0.10 d^{-1} (mean life of 9–10 d) and 0.175–0.125 d^{-1} (mean life of 6–8 d). The net scavenging rate constant has been argued to be proportional to new or export production (Coale and Bruland, 1987) and has been observed to be most intense in regimes of high particle production with substantial export of this material from the surface waters, and least intense in regions of lower primary production where regenerated production dominates.

Aluminum and gallium are two trace metals with hydrolysis chemistry similar to that of thorium and the affinity of a metal cation to form these hydrolysis species has been suggested as an important parameter in models for adsorption and scavenging processes. Orians and Bruland (1986, 1988) have presented data for aluminum and gallium (Figure 13) at some of the same stations where net thorium scavenging rates were determined (Figure 12). The surface-water concentrations of aluminum and gallium should be a function of the magnitude of sources relative to their scavenging removal rate. Both of these metals exhibit higher concentrations in the surface waters of the oligotrophic central gyres, with much lower values observed towards the north and eastern boundaries where rates of export production and ^{234}Th scavenging are higher. Undoubtedly, since Fe(III) is another trace metal with similar hydrolysis chemistry, such scavenging would also take place for it. It is, however, difficult to separate such passive scavenging from active assimilation that also occurs with iron.

Thorium isotopes have also been used to gain insight into scavenging in deep waters. Using a combination of thorium isotopes, investigators have been able to determine the dynamics of thorium scavenging (Bacon and Anderson, 1982; Nozaki *et al.*, 1987; Clegg *et al.*, 1991; Murname, 1994; Roy-Barman *et al.*, 1996). These studies have provided evidence for a dynamic system whereby thorium isotopes appear to be reversibly scavenged from the deep sea onto fine particles, these particles are packaged into larger particles that sink and then either disaggregate or are

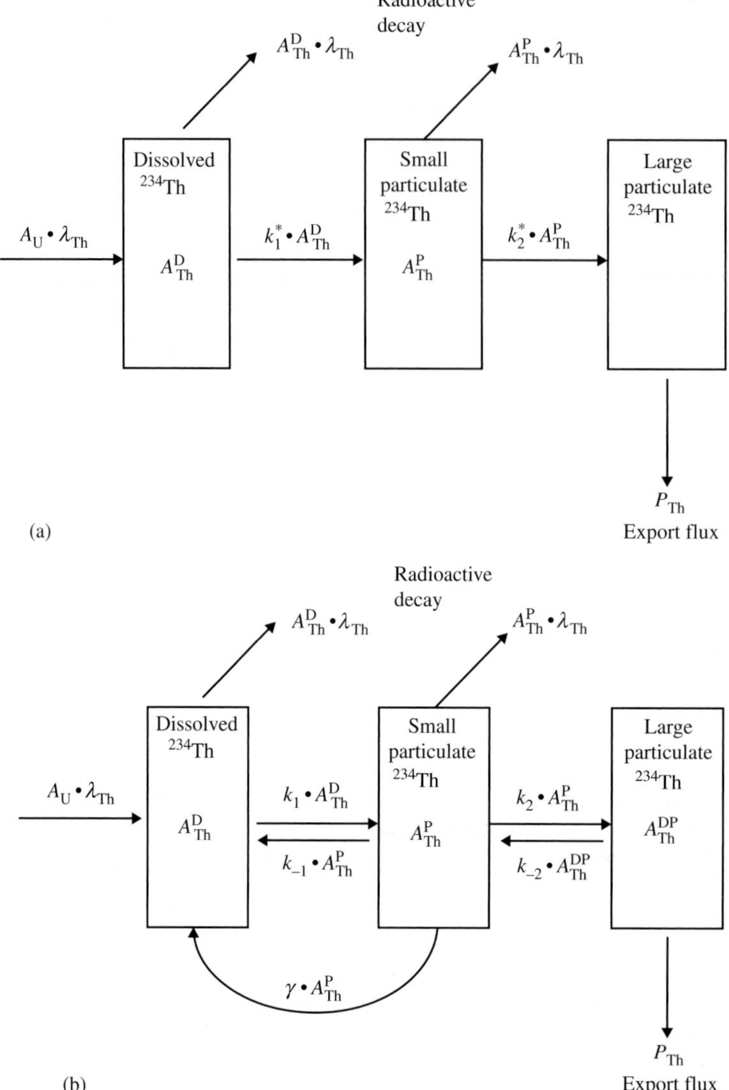

Figure 11 Conceptual models of thorium scavenging (Coale and Bruland, 1985; Bruland and Coale, 1986; Clegg *et al.*, 1991). (a) The surface water ^{234}Th net scavenging model. This model incorporates two different size classes of particles, small suspended particles and large sinking particles with the various sources and sinks for the activity (A) of ^{234}Th depicted. λ_{Th} is the decay constant of ^{234}Th, k^*_1 is the net rate transfer of ^{234}Th from dissolved to suspended particles and k^*_2 is the net rate of transfer of ^{234}Th from small suspended particles to large sinking particles. (b) A reversible scavenging model including desorption, particle disaggregation and remineralization for the deep sea. ^{230}Th and ^{234}Th can be both modeled to yield estimates of rate constants. Rate constants: k_1, adsorption onto small suspended particles and k_{-1}, desorption from small suspended particles; k_2, aggregation or packaging rate of small suspended particles into large sinking particles and k_{-2} is the disaggregation of large sinking particles and γ remineralization of carrier phases including respiration of organic matter or dissolution.

remineralized, and the thorium is either desorbed or released back into solution. In this manner, an individual thorium isotope might spend a total of 20–50 yr in the deep sea before burial in the sediments (see Chapter 6.09) and during this time undergo numerous reversible adsorption/desorption exchanges with particles that are continually aggregating into larger particles, sinking and disaggregating or being remineralized. Estimates of residence times of thorium in various forms within the deep sea suggest that an isotope might spend close to a year in the dissolved form prior to adsorption onto the surface of a small particle where it may reside for another four months prior to aggregating to a larger, more rapidly sinking particle. It might spend a few days sinking a few hundred meters prior to disaggregation and then a few months residing on the small particle prior to desorption or remineralization back into the dissolved phase

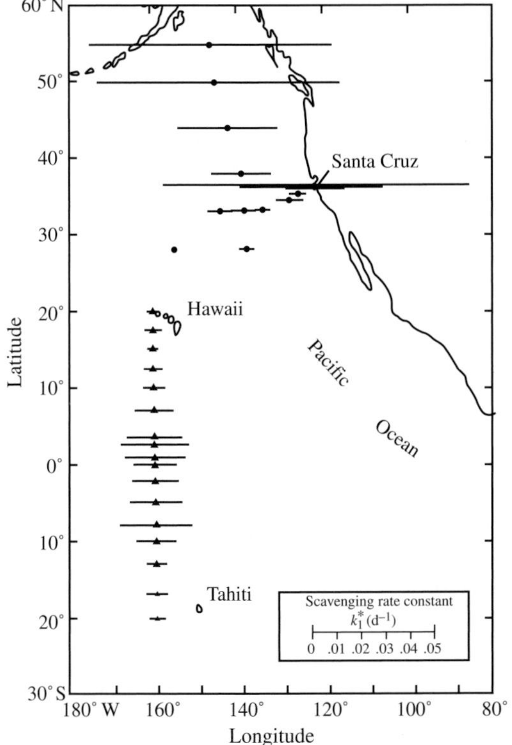

Figure 12 The net scavenging rate constant, k_1^* (d^{-1}), of ^{234}Th from the surface mixed layer of the Pacific Ocean (sources Bruland and Coale, 1986; Bruland and Beals, unpublished).

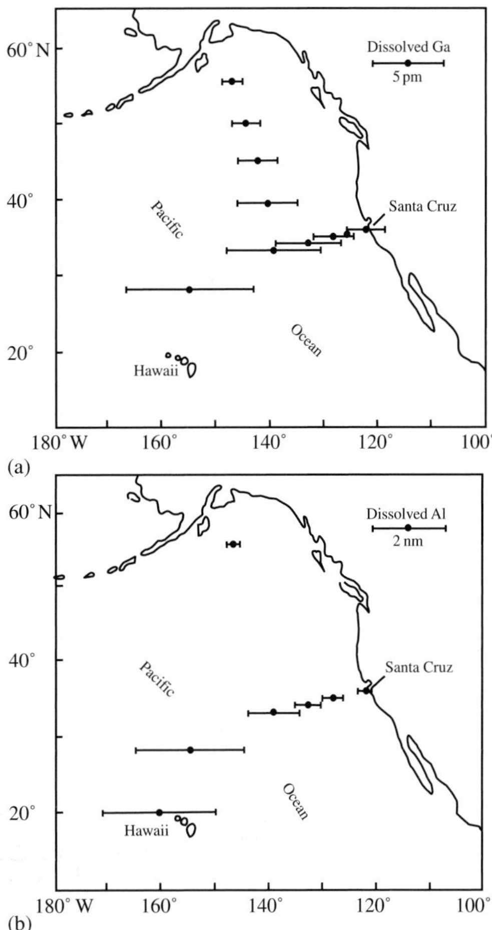

Figure 13 Concentrations of (a) dissolved gallium (data from Orians and Bruland, 1988) and (b) dissolved aluminum (data from Orians and Bruland, 1986) in the surface waters of the eastern North Pacific at stations for which ^{234}Th net scavenging rate data exist (see Figure 12).

(Murname, 1994). Similar dynamic reversible exchange processes in the deep sea may be occurring for other trace metals such as iron, aluminum, gallium, and titanium.

6.02.4 INTERNAL RECYCLING

6.02.4.1 Recycling within the Water Column

Internal recycling is particularly relevant for nutrient-type trace metals that, like the macronutrients nitrate, phosphate, and silicic acid, undergo multiple cycles of assimilation into biogenic particulate material within surface waters and release or remineralization at depth. In this manner an element can undergo many internal cycles within the ocean prior to ultimate burial in sediments. For example, imagine an individual zinc ion delivered to the surface waters by upwelling or vertical mixing. It can be assimilated into a phytoplankton cell, where it might reside a day or two prior to being grazed. In the open ocean the bulk of this phytoplankton zinc and other nutrients will be remineralized in the surface waters and available to undergo perhaps 5 or 10 such assimilation/remineralization cycles within the surface layer (Hutchins *et al.*, 1993; Hutchins and Bruland, 1994, 1995) prior to removal as an export flux in the form of a fecal pellet excreted from a zooplankter or a larger aggregate of particles. This particulate zinc can be remineralized in the deep sea as the fecal pellet serves as a source of food and nutrition for heterotrophic organisms in the deep sea. Remineralization of particulate zinc can occur at depth by the degradation of organic matter or the dissolution of the inorganic carrier phases (metal oxides, opal, calcite). Then perhaps a few hundred to a thousand years later this same zinc ion can be mixed back up into the surface waters. Numerous such cycles can occur during its $(10-50) \times 10^4$ yr odyssey in the oceans. Once finally removed via burial in the sediments, the zinc atom may have to wait ~100 Myr to be tectonically uplifted onto a continent and then another 100 Myr before it is exposed to continental weathering and makes its way via rivers or dust input back to the ocean.

6.02.4.2 Benthic Inputs

Upon settling to the surface sediments, trace metals can be recycled back into the dissolved phase and act as a source to the deep ocean. Pore-water concentrations of trace metals can be significantly higher than that observed in the overlying water column. Elevated pore-water concentrations suggested a potential for benthic fluxes of dissolved metals out of the sediment, which has been verified by direct measurements (Elderfield *et al.*, 1981; Westerlund *et al.*, 1986).

Trace metals in marine sediments are frequently associated with iron and manganese hydroxides and changes in the sediment redox chemistry near the sediment water interface can lead to alternating periods of reductive dissolution and oxidation of these phases. Depth profiles in the shelf waters off the coast of the Falkland Islands and SW Africa indicate a significant supply of dissolved iron (2–38 nmol kg^{-1}) to the overlying water column through reductive benthic processes (Bowie *et al.*, 2002). Iron concentrations just above the sediment interface were 12-fold higher than those observed at the surface. Benthic inputs of iron, cobalt, and manganese have been observed in the highly productive coastal waters of the North Sea (Tappin *et al.*, 1995). Trace-metal recycling through benthic inputs has the potential to supply trace metals back into surface waters. The dominant source of iron to waters off the coast of California is through sediment resuspension followed by upwelling (Johnson *et al.*, 1999; Bruland *et al.*, 2001).

6.02.5 COMPLEXATION WITH ORGANIC LIGANDS

Studies using electrochemical techniques have demonstrated that in surface seawater a major fraction of many trace metals, particularly the bioactive trace metals such as iron, zinc, copper, cobalt, and cadmium, are present as chelates with strong metal-binding organic ligands. These electrochemical methods employ sensitive stripping voltammetric analysis. Anodic stripping voltammetry (ASV), using a thin mercury film (TMF), rotating glassy carbon disk electrode (RGCDE) in the differential pulse (DP) mode, has been used in open ocean studies of copper (Coale and Bruland, 1988, 1990), zinc (Bruland, 1989), cadmium (Bruland, 1992), and lead (Capodaglio *et al.*, 1990). This method directly measures the kinetically labile [M′] (inorganic complexes and free metal), while the metal chelated with strong organic ligands is kinetically inert with respect to being detected by this method. ASV involves a deposition or concentration step that can be ~10 min or 20 min in duration where M′ is continually reduced and concentrated into the mercury amalgam. The DP stripping step involves ramping the potential in a positive direction and the measurement of the anodic stripping current as the metals are oxidized back into solution. Titrations of a sample with the metal of interest and determining [M′] at each titration point allow both the concentrations of the M-binding ligands and binding strengths (conditional stability constants) of the metal–ligand complexes to be determined. The number of trace metals that can be determined by ASV, however, is limited to those that can be reduced (and reoxidized) at appropriate potentials and that are soluble in a mercury amalgam.

A second powerful voltammetric approach that is amenable to a far broader group of trace metals is adsorptive cathodic stripping voltammetry (AdCSV). The application of AdCSV to speciation studies involves the addition of a well-characterized added ligand (AL) that sets up a competitive equilibrium with the natural ligands for the metal of interest (van den Berg, 1988). Most methods involve formation of a neutral biscomplex with the AL, $M(AL)_2^0$ (Bruland *et al.*, 2000). The ALs generally form planer biscomplexes with the metal of interest that have a strong tendency to adsorb on the surface of a hanging mercury drop electrode, whereas the natural metal–ligand complexes do not. After an appropriate adsorption or accumulation time period, the potential is ramped in a negative direction (in either linear, DP or Osteryoung square wave (SW) mode) and the cathodic stripping current is measured as the metal (and sometimes also the AL) is reduced at the electrode surface. The use of these AdCSV methods does not require the metal to be soluble in a mercury amalgam and thus can be used for a wide variety of trace metals. Metal titrations and the determination of the $M(AL)_2^0$ concentration at each titration point allows the determination of natural, strong metal-binding ligand concentrations and their conditional stability constants. The use of AdCSV methods has been applied to studies of the speciation of iron (van den Berg, 1995; Rue and Bruland, 1995; Bruland and Rue, 2001), copper (van den Berg, 1984; Moffett *et al.*, 1990; Donat and van den Berg, 1992), zinc (van den Berg, 1985; Donat and Bruland, 1990; Ellwood and van den Berg, 2000; Lohan *et al.*, 2002), and cobalt (Saito and Moffett, 2001; Ellwood and van den Berg, 2001). It should be noted that "detection windows" of voltammetric techniques examine stability constants of metal-binding ligands specified by the technique, and that natural metal-binding organic ligands are not a single entity but rather classes of ligands with average values assigned.

Results of such voltammetric studies have demonstrated that complexation of trace metals with relatively specific and strong metal-binding organic ligands is important in oceanic surface waters. Greater than 99% of Fe(III) in surface waters is complexed with strong Fe(III)-binding organic ligands existing at sub-nanomolar concentrations in slight excess of the dissolved iron (Rue and Bruland, 1995; van den Berg, 1995; Wu and Luther, 1995; Powell and Donat, 2001; Boye *et al.*, 2003). Greater than 99% of copper exists as organic complexes (Coale and Bruland, 1988; Moffett, 1990). Approximately 98% of dissolved zinc in surface waters is complexed with organic ligands (Bruland, 1989; Donat and Bruland, 1990; Ellwood and van den Berg, 2000; Lohan *et al.*, 2002). Greater than 90% of cobalt exists complexed to strong cobalt-binding organic ligands (Saito and Moffett, 2001; Ellwood and van den Berg, 2001) and ~80% of cadmium in surface waters is complexed with organic ligands (Bruland, 1992).

We know little about the chemical structure or architecture of the organic ligands involved in binding metals in seawater. There have been recent advances, however, into the structure and function of marine siderophores. The conditional stability constants of the marine siderophores so far examined are similar to the stability constants of the natural Fe(III)-binding ligands found in seawater (Barbeau *et al.*, 2001, 2003) and siderophores appear to constitute a significant fraction of the natural Fe(III)-binding organic ligands in seawater (Macrellis *et al.*, 2001). Siderophores are defined as low molecular weight organic chelators with a very high and specific affinity for Fe(III), the biosynthesis of which is regulated by iron levels, and whose function is to mediate iron uptake by microbial cells.

The selectivity of siderophores for Fe(III) is achieved through optimal selection of metal-binding groups, the number of binding units, and their stereochemical arrangement (Boukhalfa and Crumbliss, 2002). Most siderophores are hexadentate and incorporate hydroxamate, catecholate, and/or α-hydroxy carboxylate binding subunits arranged in different architectures. Barbeau *et al.* (2003) have presented a summary of many of the marine siderophores groups produced by both heterotrophic and photosynthetic marine bacteria and characterized the photochemical reactivity of the different Fe(III)-binding functional groups. Hydroxamate groups are photochemically resistant regardless of Fe(III) complexation. Catecholates are susceptible to photo-oxidation in the uncomplexed form, but stabilized against photo-oxidation when ferrated. α-Hydroxy carboxylate groups are stable as the uncomplexed acid, but when coordinated to Fe(III) these moieties undergo light-induced ligand oxidation and reduction of Fe(III) to Fe(II). These photochemical properties appear to determine the reactivity and fate of Fe(III)-binding siderophores in ocean surface waters (Barbeau *et al.*, 2003).

Other possible candidates of natural Fe(III)-binding ligands in seawater are porphyrin-type ligands released as degradation products of cytochrome-containing systems (Rue and Bruland, 1997). As yet, little is known about this possibility. In addition, little is known about the structure, functional groups, or architecture of other metal-binding organic ligands in seawater. What is known is that there appear to be small concentrations of strong and relatively specific metal-binding organic ligands that play an important role in the chemical speciation of quite a few of the bioactive trace metals in the sea.

6.02.5.1 Copper

Copper provides an interesting example of a trace metal that is an essential, required element, but that can be toxic at relatively low concentrations. It can be considered the "Goldilocks" metal. Surface-water concentrations in the open ocean are ~1 nmol kg^{-1}. Without organic complexation, the free copper concentration would be approximately a factor of 20 lower than the total dissolved concentration, with $Cu(CO_3)^0$ predicted to be the dominant species. This would yield [Cu^{2+}] concentrations ~0.5×10^{-10} M or $10^{-10.3}$ M. This concentration would be toxic to many oceanic phytoplankton, particularly the prokaryotic photosynthetic bacteria such as synechococcus (Brand *et al.*, 1986). Figure 14(b) presents the actual vertical distribution of [Cu^{2+}] in the upper 500 m of the Northeast Pacific. Coale and Bruland (1988, 1990) observed a slight excess of a strong copper-binding class of organic ligands, called L_1. This class of strong copper-binding ligands was found to occur in surface waters at concentrations of 1–2 nmol kg^{-1} and its presence led to greater than 99.8% of the copper being chelated to this class of ligands. As a result, [Cu^{2+}] was reduced by close to a factor of 1,000 and exists at concentrations ~10^{-13} M, which is a concentration not toxic to phytoplankton. Moffett and Brand (1996) have shown that cyanobacteria when stressed with slightly elevated [Cu^{2+}] can produce a ligand with a similar conditional stability constant. It appears that, somehow, the phytoplankton of the open ocean, particularly the prokaryotic phytoplankton, are controlling the external concentration of free copper by producing a strong copper-binding ligand that reduces the [Cu^{2+}] to levels that are no longer toxic. As a result, the [Cu^{2+}] in surface waters is buffered by the L_1 class of ligands at a concentration that is "not too little" and "not too much," but "just right"; thus, the "Goldilocks example." This buffering of [Cu^{2+}] also influences its distribution

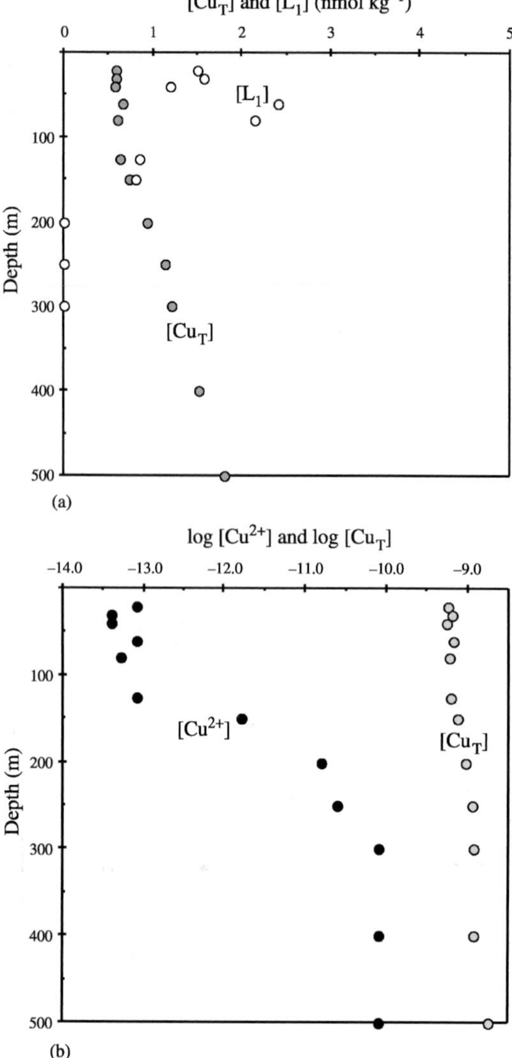

Figure 14 Concentration profiles of (a) total dissolved copper, $[Cu_T]$, and strong copper-binding organic ligands, $[L_1]$, and (b) free $[Cu^{2+}]$ and $[Cu_T]$ in the upper 500 m of the North Pacific (source Coale and Bruland, 1988).

and, because this chelated form is unavailable biologically, the dissolved copper is not depleted to a great degree in oceanic surface waters.

6.02.5.2 Iron

Dissolved iron provides an example of a bioactive, essential trace metal that is depleted in oceanic surface waters to such an extent that it has been estimated to be the limiting nutrient in ~40% of the world's oceans (Moore *et al.*, 2002). In particular, iron has been shown to be the limiting nutrient in the HNLC regions of the Southern Ocean, the equatorial Pacific, and the subarctic Pacific (Martin *et al.*, 1994). Not only does iron exist at extremely low concentrations in surface waters, its chemical speciation is dominated by complexation with Fe(III)-binding organic ligands. The initial evidence that dissolved iron is strongly chelated by natural Fe(III)-binding organic ligands in seawater was provided by van den Berg (1995), Rue and Bruland (1995), and Wu and Luther (1995). Results from surface waters of the equatorial and North Pacific oceans (Rue and Bruland, 1995, 1997) indicate that two classes of Fe(III)-binding ligands exist. There appears to be 0.3–0.4 nmol kg^{-1} of particularly strong ligands (an L_1 class) with a conditional stability constant (with respect to Fe(III)′) of $10^{12.5}$–10^{13} M^{-1}. In addition, there is a more variable class of weaker ligands (an L_2 class) existing at concentrations of 0.2–1.5 nmol kg^{-1} with a conditional stability constant of $10^{11.5}$–$10^{11.8}$ M^{-1}. This is a case where it appears that microorganisms are essentially carrying out "chemical warfare" in their attempts to acquire this metal.

Marine siderophores produced by both heterotrophic and photosynthetic bacteria that have been so far examined have conditional stability constants consistent with the classes of natural organic ligands observed in seawater (Barbeau *et al.*, 2001, 2003). For marine siderophores to be competitive for binding Fe(III) in surface seawater, they would have to be this strong. Barbeau *et al.* (2001) have also shown that photodegradation products of the original siderophores have weaker conditional stability constants similar to the L_2 class of ligands observed by Rue and Bruland (1995, 1997). It is known that many microorganisms are able to not only utilize their own siderophores, but in addition, can assimilate numerous other bacteria's siderophores (Wilhelm and Trick, 1994). Eukaryotic diatoms are thought to not have the receptor sites to assimilate Fe(III)–siderophores directly. There is evidence, however, that diatoms can utilize cell-surface reductase systems to reduce the Fe(III) bound to the siderophore, and Fe(II) can dissociate and become available either as Fe(II) or can be reoxidized to Fe(III) and become available for assimilation (Maldonado and Price, 1999). Barbeau *et al.* (2001) have shown that iron associated with the more weakly held photoproduct of marine siderophores is more readily available than the original Fe(III)–siderophore. Hutchins *et al.* (1999) have also addressed the issue of the availability of Fe(III) bound to various siderophores and other ligands and have observed differences in availability. Fe(III)–porphyrin complexes seem to be more readily available to the eukaryotic diatoms, while Fe(III)–siderophores are more readily available by the prokaryotic phytoplankton community.

In the iron-limited regions of the ocean, iron is cycled through the planktonic community so rapidly that an individual iron atom may be in a different form each day. It may exist dissolved as an Fe(III)–siderophore one day, be part of the intracellular photosynthetic machinery of a photosynthetic bacteria the next day, be regenerated as an iron–porphyrin cell lysis product the next day, and then rapidly reassimilated by a diatom. In this case, the iron that is biologically available is changing on a day-to-day basis. Of interest to this chapter is the eventual export of iron from the surface water as the diatom is grazed by a copepod and a part of the undigested residue of the diatom and its associated iron is removed from the surface layer as a fecal pellet to be transported into the deep sea where it can be remineralized and either scavenged or eventually mixed back into the surface layer once again. Interestingly, it has been argued that Fe(III)-binding organic ligands observed in the deep sea (Rue and Bruland, 1995) play an important role in allowing the dissolved iron in the deep sea to exist at concentrations on the order of a nmol kg^{-1} (Johnson *et al.*, 1997). This is a higher concentration than would be expected from estimates of inorganic solubility (Liu and Millero, 2002).

6.02.5.3 Zinc

Field studies have revealed that there are nanomolar concentrations of strong zinc-binding organic ligands in surface waters (Bruland, 1989; Donat and Bruland, 1990; Ellwood and van den Berg, 2000; Lohan *et al.*, 2002) that play an important role in chelating zinc in surface waters. Figures 15(a) and (b) presents data from the North Pacific (Bruland, 1989) and it is apparent that ~98% of the dissolved zinc in surface waters exists in a chelated form with organic ligands. Although we have insight into the concentrations and conditional stability constants of these ligands, we know little or nothing about their functional character or molecular architecture. There are no immediate solubility constraints on the zinc concentration and it does not undergo redox cycling so that the benefits conferred by complexation upon iron availability do not apply to zinc. Unlike copper, zinc is not toxic to phytoplankton at concentrations observed in the open ocean. There appears to be no immediate advantage to the phytoplankton community in reducing $[Zn^{2+}]$ through organic complexation.

One interesting idea is that this might be an example of "smart banking" by the plankton community. By having the bulk of the zinc chelated and presumably less bioavailable, the removal rate of zinc by the biota will be less (see Figure 10). As discussed previously in Section 6.02.3.2.1, the Zn/C ratio in diatoms is dependent upon the $[Zn^{2+}]$. Therefore, by markedly decreasing $[Zn^{2+}]$ with the production of a zinc-binding

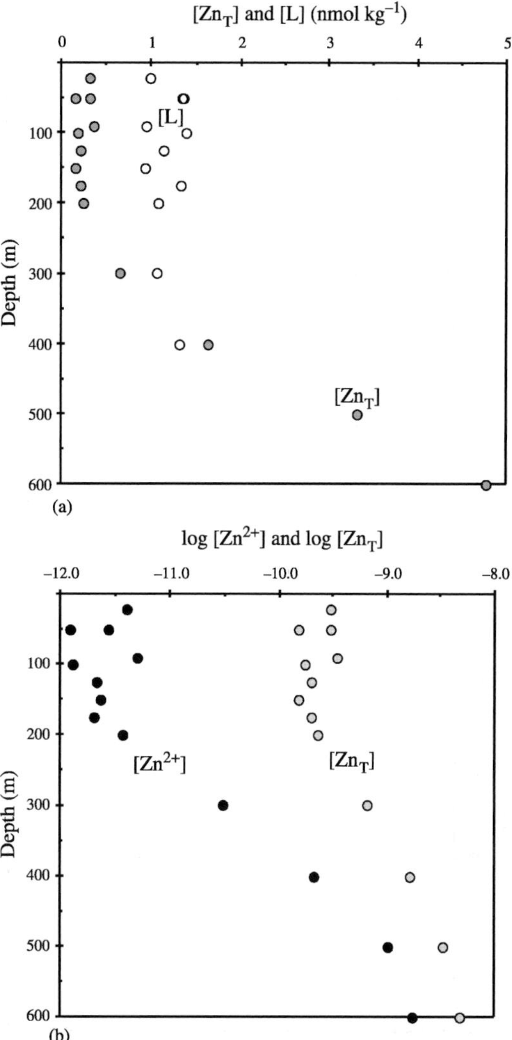

Figure 15 Concentration profiles of (a) total dissolved zinc, $[Zn_T]$, and strong zinc-binding organic ligands, [L] and (b) free $[Zn^{2+}]$ and $[Zn_T]$ in the upper 600 m of the North Pacific (source Bruland, 1989).

organic ligand, the removal rate of zinc will be decreased and its residence time in the surface water increased, e.g., the capital is preserved. Conversely, some of the zinc-binding ligands may be a zinc siderophore-type compound whereby certain microorganisms are attempting to gain an advantage for this potentially biolimiting element.

Voltammetric techniques have provided a powerful tool for measuring the speciation of trace metals at low concentrations. Most of the bioactive trace metals have now been analyzed using these techniques. Knowledge of the inorganic, organic, and free metal forms in the dissolved phase is critical as the different forms are involved in very different biological and geochemical interactions and hence the cycling of trace metals within the ocean. Studies of trace-metal speciation have influenced ideas about the role of trace metals in biological systems. The previous conceptions that trace metals bound to these ligands were not bioavailable has undergone a complete change since the discovery that iron bound to siderophores can be readily accessed by phytoplankton. It may be that the strong cobalt-binding ligands observed by Saito and Moffett (2001) and Ellwood and van den Berg (2001) are a type of "cobalophore" and play a role similar to siderophores with iron. The production of highly specific metal-binding ligands by phytoplankton has led to many interesting questions which are yet to be resolved.

New analytical techniques are emerging such as electrospray ionization mass spectroscopy (ESI-MS) that may prove useful in gaining insight into the structure and functional groups of metal-binding organic ligands in seawater. First, however, the low concentrations of metal-binding ligands must be concentrated from the seawater matrix and isolated or separated from the rest of the dissolved organic matter. This will hopefully allow a better mechanistic understanding of the production and regulation of these ligands in the upper water column and hence increase the understanding of the cycling of dissolved trace metals in the oceans.

In summary, we have presented a general overview of the major controls of trace metals in seawater developed from the extensive research on trace metals over the last few decades. We wanted the reader to gain a first order understanding and not present a comprehensive review on the distribution of each trace metal. Each of the trace metals discussed will undoubtedly prove to have unique characteristics and subtle differences from this simplified version, yet the comparison with these simplified characteristics will serve as a good spring board to a more complete understanding. We have gained an insight into many of the processes affecting trace-metal cycling within the oceans, which has revealed many interesting questions still to be answered. Developments in isolating and characterizing metal-binding organic ligands should elucidate further the controls of trace metals in the upper water column. While in the mesopelagic zone the characterization of trace-metal fluxes and the processes involved in remineralization will greatly enhance our understanding of trace-metal cycling both within this region and exchanges between the benthos and the upper photic zone.

REFERENCES

Anbar A. D., Wasserburg G. J., Papanastassiou D. A., and Anderson P. S. (1996) Iridium in natural waters. *Science* **273**, 1524–1528.

Bacon M. P. and Anderson R. F. (1982) Distribution of thorium isotopes between dissolved and particulate forms in the deep-sea. *J. Geophys. Res.* **87**, 2045–2050.

Barbeau K., Rue E. L., Bruland K. W., and Butler A. (2001) Photochemical cycling of iron in the surface ocean mediated by microbial iron(III)-binding ligands. *Nature* **413**, 409–413.

Barbeau K., Rue E. L., Trick C. G., Bruland K. W., and Butler A. (2003) The photochemical reactivity of siderophores produced by marine heterotrophic bacteria and cyanobacteria, based on characteristic iron(III)-binding groups. *Limnol. Oceanogr.* **48**, 1069–1078.

Boukhalfa H. and Crumbliss A. L. (2002) Chemical aspects of siderophore mediated iron transport. *Biometals* **15**, 325–339.

Bowie A., Maldonado M., Frew R. D., Croot P. L., Achterberg E. P., Mantoura R. F. C., Worsfold P. J., Law C. S., and Boyd P. W. (2001) The fate of added iron during a mesoscale fertilisation experiment in the Southern Ocean. *Deep-Sea Res. II* **48**, 2703–2743.

Boyd P. W., Watson A. J., Law C. S., Abraham E. R., Trull T., Murdoch R., Bakker D. G. E., Bowie A., Buesseler K. O., Chan H., Charette M. A., Croot P. L., Downing K., Frew R., Gall M. P., Hadfield M., Hall J., Harvey M., Jameson J., LaRoche J., Liddicoat M., Ling R., Maldonado M. T., McKay R. M., Nodder S., Pickmere S., Pridmore R., Rintoul S., Safi K., Sutton P., Strzepek R., Tanneberger K., Turner S., Waite A., and Zeldis J. (2000) A mesoscale phytoplankton bloom in the Polar Southern Ocean stimulated by iron fertilization. *Nature* **407**, 695–702.

Boye M. B., Aldrich A. P., van den Berg C. M. G., deJong J. T. M., Veldhuis M. J. W., and de Baar H. J. W. (2003) Horizontal gradient of the chemical speciation of iron in surface waters of NE Atlantic Ocean. *Mar. Chem.* **50**, 129–143.

Boyle E. A. (1988) Cadmium: chemical tracer of deep water oceanography. *Paleoceanography* **3**, 471–489.

Boyle E. A., Sclater F. R., and Edmond J. M. (1976) On the marine geochemistry of cadmium. *Nature* **263**, 42–44.

Boyle E. A., Edmond J. M., and Sholkovitz E. R. (1978) The mechanism of iron removal in estuaries. *Geochim. Cosmochim. Acta* **41**, 1313.

Brand L. E., Sunda W. G., and Guillard R. R. L. (1986) Reduction in marine phytoplankton reproduction rates by copper and cadmium. *J. Exp. Mar. Biol. Ecol.* **96**, 225–250.

Brewer P. G., Spenser D. W., and Robertson D. E. (1972) Trace element profiles from the GEOSECS II test station in the Sargasso Sea. *Earth Planet. Sci. Lett.* **16**, 111–116.

Broecker W. S. and Peng T. H. (1982) Tracers in the Sea. Eldigo Press, Palisades, NY.

Bruland K. W. (1980) Oceanographic distributions of Cd, Zn, Cu, and Ni in the North Pacific. *Earth Planet. Sci. Lett.* **47**, 176–198.

Bruland K. W. (1983) Trace elements in sea water. In *Chemical Oceanography* (eds. J. P. Riley and R. Chester). Academic Press, London, vol. 8, pp. 157–220.

Bruland K. W. (1989) Complexation of zinc by natural organic ligands in the central North Pacific. *Limnol. Oceanogr.* **34**, 269–285.

Bruland K. W. (1992) Complexation of cadmium by natural organic ligands in the central North Pacific. *Limnol. Oceanogr.* **37**, 1008–1017.

Bruland K. W. and Coale K. H. (1986) Surface water $^{234}Th/^{238}U$ disequilibria: spatial and temporal variations of scavenging rates within the Pacific Ocean. In *Dynamic Processes in the Chemistry of the Upper Ocean*, NATO Conference Series IV: Marine Sciences (eds. J. D. Burton, P. G. Brewer, and R. Chesselet). Plenum, New York, pp. 159–172.

Bruland K. W. and Franks R. P. (1983) Mn, Ni, Cu, Zn, and Cd in the western North Atlantic. In *Trace Metals in Seawater* (eds. C. S. Wong, E. A. Boyle, K. W. Bruland, J. D. Burton, and E. D. Goldberg). Plenum, New York, pp. 395–414.

Bruland K. W. and Rue E. L. (2001) Analytical methods for determination of concentrations and speciation of iron. In *The Biogeochemistry of Iron in Seawater* (eds. D. R. Turner and K. A. Hunter). Wiley, Chichester, pp. 255–289.

Bruland K. W., Knauer G., and Martin J. (1978a) Cadmium in Northeast Pacific waters. *Limnol. Oceanogr.* **23**, 618–625.

Bruland K. W., Knauer G., and Martin J. (1978b) Zinc in Northeast Pacific waters. *Nature* **271**, 741–743.

Bruland K. W., Donat J. R., and Hutchins D. T. (1991) Interactive influences of bioactive trace metals on biological production in oceanic waters. *Limnol. Oceanogr.* **36**, 1555–1577.

Bruland K. W., Orians K. J., and Cowen J. P. (1994) Reactive trace metals in the stratified central North Pacific. *Geochim. Cosmochim. Acta* **58**, 3171–3182.

Bruland K. W., Rue E. L., Donat J. R., Skabal S., and Moffett J. W. (2000) An intercomparison of voltammetric approaches to determine the chemical speciation of dissolved copper in a coastal seawater sample. *Anal. Chim. A* **405**, 99–113.

Bruland K. W., Rue E. L., and Smith G. J. (2001) The influence of iron and macronutrients in coastal upwelling regimes off central California: implications for extensive blooms of large diatoms. *Limnol. Oceanogr.* **46**, 1661–1674.

Burton J. D. and Statham P. J. (1990) Trace metals in seawater. In *Heavy Metals in the Marine Environment* (eds. P. S. Rainbow and R. W. Furness). CRC Press, Boca Raton, FL.

Byrne R. H., Kump L. R., and Cantrell K. J. (1988) The influence of temperature and pH on trace metal speciation in seawater. *Mar. Chem.* **25**, 163–181.

Capodaglio G., Coale K. H., and Bruland K. W. (1990) Lead speciation in surface waters of the eastern North Pacific. *Mar. Chem.* **29**, 221–238.

Chan L. H., Edmond J. M., Stallard R. G., Broecker W. S., Chung Y., Weiss R. F., and Ku T. L. (1976) Radium and barium at GEOSECS stations in the Atlantic and Pacific. *Earth Planet. Sci. Lett.* **32**, 258–267.

Chan L. H., Drummond D., Edmond J. M., and Grant B. (1977) On the barium data from the Atlantic GEOSECS expedition. *Deep-Sea Res.* **24**, 613–649.

Clegg S. L., Bacon M. P., and Whitfield M. (1991) Application of a generalized scavenging model to thorium isotope and particle data at equatorial and high latitude sites in the Pacific Ocean. *J. Geophys. Res.* **96**, 20665–20670.

Coale K. H. and Bruland K. W. (1985) $^{234}Th/^{238}U$ disequilibria within the California Current. *Limnol. Oceanogr.* **30**, 22–33.

Coale K. H. and Bruland K. W. (1987) Oceanic stratified euphotic zone as elucidated by $^{234}Th/^{238}U$ disequilibria. *Limnol. Oceanogr.* **32**, 189–200.

Coale K. H. and Bruland K. W. (1988) Copper complexation in the northeast Pacific. *Limnol. Oceanogr.* **33**, 1084–1101.

Coale K. H. and Bruland K. W. (1990) Spatial and temporal variability in copper complexation in the North Pacific. *Deep-Sea Res.* **37**, 317–336.

Coale K. H., Johnson K. S., Fitzwater S. E., Gordon R. M., Tanner S., Chavez F. P., Ferioli L., Sakamoto C., Rogers P., Millero F., Steinberg P., Nightingale P., Cooper D., Cochlan W. P., Landry M. R., Constantinou J., Rollwagen G., Trasvina A., and Kudela R. (1996) A massive phytoplankton bloom induced by a ecosystem-scale iron fertilisation experiment in the equatorial Pacific Ocean. *Nature* **383**, 495–501.

Cowen J. P., Betram M. A., Baker G. T., Feely R. A., Massoth G. J., and Summit M. (1998) Geomicrobial transformations of manganese in Gorda Ridge event plumes. *Deep-Sea Res. II* **45**, 2713–2737.

Cullen J. T., Chase Z., Coale K. H., Fitzwater S. E., and Sherrell R. M. (2003) Effect of iron limitation on the cadmium to phosphorus ratio on natural phytoplankton assemblages from the Southern Ocean. *Limnol. Oceanogr.* **48**, 1079–1087.

de Baar H. J. W., Croot P. L., Stoll M. H. C., Kattner G., Pickmere S., Freyer U., Boyd P., and Smetacek V. (2000) Nutrient anomalies of *Fragilariopsis kerguelensis* blooms revisited. Abstract Southern Ocean JGOFS Symposium, Brest, July 2000.

Donat J. R. and Bruland K. W. (1990) A comparison of two voltammetric techniques for determining zinc speciation in northeast Pacific Ocean waters. *Mar. Chem.* **28**, 301–323.

Donat J. R. and Bruland K. W. (1995) Trace elements in the oceans. In *Trace Elements in Natural Waters* (eds. E. Steinnes and B. Salbu). CRC Press, Boca Raton, FL, pp. 247–280.

Donat J. R. and van den Berg C. M. G. (1992) A new cathodic stripping voltammetric method for determining organic complexation of copper in seawater. *Mar. Chem.* **38**, 69–90.

Duce R. A. and Tindale N. W. (1991) Atmospheric transport of iron and its deposition on the ocean. *Limnol. Oceanogr.* **36**, 1715–1726.

Edmond J. M., Boyle E. A., Drummond D., Grant B., and Mislick T. (1978) Desorption of barium in the plume of the Zaire (Congo) River. *Netherlands J. Sea Res.* **12**, 324–328.

Edmond J. M., Measures C. I., Mangum B., Grant B., Sclater F. R., Collier R., Hudson A., Gordon L. I., and Corliss J. B. (1979) Formation of metal-rich deposits at ridge crests. *Earth Planet. Sci. Lett.* **46**, 19–30.

Elderfield H., Luedke N., McCaffery R. J., and Bender M. (1981) Benthic flux studies in Narragansett Bay. *Am. J. Sci.* **281**, 768–787.

Ellwood M. J. and van den Berg C. M. G. (2000) Zinc speciation in the Northeastern Atlantic Ocean. *Mar. Chem.* **68**, 295–306.

Ellwood M. J. and van den Berg C. M. G. (2001) Determination of organic complexation of cobalt in seawater by cathodic stripping voltammetry. *Mar. Chem.* **75**, 49–68.

Emerson S. R. and Huested S. S. (1991) Ocean anoxia and the concentration of molybdenum and vanadium in seawater. *Mar. Chem.* **34**, 177–196.

Falkowski P. G. (2000) Rationalizing elemental ratios in unicellular algae. *J. Phycol.* **36**, 3–6.

Flegal A. R. and Patterson C. C. (1983) Vertical concentration profiles of lead in the central Pacific at 15° N and 20° S. *Earth Planet. Sci. Lett.* **64**, 19–32.

Froelich P. N. and Andreae M. O. (1981) The marine geochemistry of germanium: Ekasilicon. *Science* **213**, 205–207.

Goldberg E. D. (1954) Marine Geochemistry: 1. Chemical scavengers of the sea. *J. Geol.* **62**, 249–265.

Hering J. G. and Morel F. M. M. (1990) Kinetics of trace-metal complexation-ligand-exchange reactions. *Environ. Sci. Technol.* **24**, 242–252.

Hunter K. A., Kim J. P., and Croot P. L. (1997) Biological role of trace metals in natural waters. *Environ. Monitor. Assess.* **44**, 103–147.

Hutchins D. A. and Bruland K. W. (1994) Grazer-mediated regeneration and assimilation of Fe, Zn, and Mn from planktonic prey. *Mar. Ecol. Prog. Ser.* **11**, 259–269.

Hutchins D. A. and Bruland K. W. (1995) Fe, Zn, Mn and N transfer between size classes in a coastal phytoplankton community: trace metal and major nutrient recycling compared. *J. Mar. Res.* **53**, 1–18.

Hutchins D. A. and Bruland K. W. (1998) Iron-limited diatom growth and Si : N uptake in a coastal upwelling regime. *Nature* **393**, 561–564.

Hutchins D. A., DiTullio G. R., and Bruland K. W. (1993) Iron and regenerated production: evidence for biological iron recycling. *Limnol. Oceanogr.* **38**, 1242–1255.

Hutchins D. A., Witter A. E., Butler A., and Luther G. W. (1999) Competition among marine phytoplankton for different chelated iron species. *Nature* **400**, 858–861.

Hydes D. J. (1979) Aluminum in seawater: control by inorganic processes. *Science* **205**, 1260–1262.

Hydes D. J. (1983) Distribution of aluminum in waters of the Northeast Atlantic 25-degrees-N to 35-degrees-N. *Geochim. Cosmochim. Acta* **47**, 967–973.

Hydes D. J., De Lange G. T., and De Baar H. J. W. (1988) Dissolved aluminum in the Mediterranean. *Geochim. Cosmochim. Acta* **52**, 2107–2114.

Jickells T. M. (1995) Atmospheric inputs of metals and nutrients to oceans: their magnitude and effects. *Mar. Chem.* **48**, 199–214.

Jickells T. M. and Spokes L. J. (2001) Atmospheric iron inputs to the oceans. In *The Biogeochemistry of Iron in Seawater* (eds. D. R. Turner and K. Hunter). Wiley, Chichester, pp. 85–121.

Johnson K. S., Gordon R. M., and Coale K. H. (1997) What controls dissolved iron concentrations in the world ocean? *Mar. Chem.* **57**, 137–161.

Johnson K. S., Chavez F. P., and Freiderich G. E. (1999) Continental-shelf sediments as a primary source of iron for coastal phytoplankton. *Nature* **398**, 697–700.

Karl D. M. (2002) Nutrient dynamics in the deep blue sea. *Trends Microbiol.* **10**, 410–418.

Kim J. P., Hunter K. A., and Reid M. R. (1999) Geochemical processes affecting the major ion composition of river in the South Island, New Zealand. *Mar. Freshwater Res.* **50**, 699–707.

Lee J. G. and Morel F. M. M. (1995) Replacement of zinc by cadmium in marine phytoplankton. *Mar. Ecol. Prog. Ser.* **127**, 305–309.

Lewis B. L., Froelich P. N., and Andreae M. O. (1985) Methylgermanium in natural waters. *Nature* **313**, 303–305.

Lewis B. L., Andreae M. O., and Froelich P. N. (1989) Sources and sinks of methylgermanium in natural waters. *Mar. Chem.* **27**, 179–200.

Liu X. W. and Millero F. J. (2002) The solubility of iron in seawater. *Mar. Chem.* **77**, 43–54.

Lohan M. C., Statham P. J., and Crawford D. W. (2002) Dissolved zinc in the upper water column of the subarctic North East Pacific. *Deep-Sea Res. II* **49**, 5793–5808.

Macrellis H. M., Trick C. G., Rue E. L., Smith G. J., and Bruland K. W. (2001) Collection and detection of natural iron-binding ligands from seawater. *Mar. Chem.* **76**, 175–187.

Maldonado M. T. and Price N. M. (1999) Utilization of iron bound to strong organic ligands by plankton communities in the subarctic Pacific Ocean. *Deep-Sea Res. II* **46**, 2447–2473.

Mahowald N., Kohfeld K., Hannson M., Balkanski Y., Harrison S. P., Prentice I. C., Schulz M., and Rodhe H. (1999) Dust sources during the last glacial maximum and current climate: a comparison of model results with paleodata from ice cores and marine sediments. *J. Geophys. Res.* **104**, 15895–15916.

Marchitto T. M., Oppo D. W., and Curry W. B. (2002) Paired benthic foraminiferal Cd/Ca and Zn/Ca evidence for a greatly increased presence of Southern Ocean Water in the glacial North Atlantic. *Paleoceanography* **17**, 1038.

Martin J. H. and Gordon R. M. (1988) Northeast Pacific iron distributions in relation to phytoplankton productivity. *Deep-Sea Res.* **35**, 177–196.

Martin J. H. and Knauer G. A. (1973) The elemental composition of plankton. *Geochim. Cosmochim. Acta* **37**, 1639–1653.

Martin J., Bruland K. W., and Broenkow W. (1976) Cadmium transport in the California current. In *Marine Pollutant Transfer* (eds. H. L. Windom and R. A. Duce). Lexington Books, Toronto, pp. 84–159.

Martin J. H., Knauer G. A., and Gordon R. M. (1983) Silver distributions and fluxes in the North-east Pacific waters. *Nature* **305**, 306–309.

Martin J. H., Gordon R. M., Fitzwater S., and Broenkow W. W. (1989) VERTEX: phytoplankton/iron studies in the Gulf of Alaska. *Deep-Sea Res.* **36**, 649–680.

Martin J. H., Fitzwater S. E., Gordon R. M., Hunter C. N., and Tanner S. J. (1993) Iron, primary production and carbon–nitrogen flux studies during the JGOFS North Atlantic Bloom Experiment. *Deep-Sea Res.* **40**, 115–134.

Martin J. H., Coale K. H., Johnson K. S., Fitzwater S. E., Gordon R. M., Tanner S. J., Hunter C. N., Elrod V. A., Nowicki J. L., Coley T. L., Barber R. T., Lindley S., Watson A. J., Vanscoy K., Law C. S., Liddicoat M. I., Ling R., Stanton T., Stockel J., Collins C., Anderson A., Bidigare R., Ondrusek M., Latasa M., Millero F. J., Lee K., Yao W., Zhang J. Z., Friederich G., Sakamoto C., Chavez F., Buck K., Kolber Z., Greene R., Falkowski P., Chisholm S. W., Hoge F., Swift R., Yungel J., Turner S., Nightingale P., Hatton A., Liss P., and Tindale N. W. (1994) Testing the iron hypothesis in ecosystems of the equatorial Pacific-Ocean. *Nature* **371**, 123–129.

Measures C. I. and Brown E. T. (1996) Estimating dust input to the Atlantic Ocean using surface water Al concentrations. In *The Impact of African Dust across the Mediterranean* (eds. S. Guerzoni and R. Chester). Kluwer, Dordrecht, 398pp.

Measures C. I. and Edmond J. M. (1982) Beryllium in the water column of the central North Pacific. *Nature* **257**, 51–53.

Measures C. I., Yuan J., and Resing J. A. (1995) Determination of iron in seawater by flow injection-analysis using in-line preconcentration and spectrophotometric detection. *Mar. Chem.* **50**, 3–12.

Moffett J. W. and Brand L. E. (1996) Production of strong, extracellular Cu chelators by marine cyanobacteria in response to Cu stress. *Limnol. Oceanogr.* **41**, 388–395.

Moffett J. W., Brand L. E., and Zika R. G. (1990) Distribution and potential sources and sinks of copper chelators in the Sargasso Sea. *Deep-Sea Res.* **37**, 27–36.

Moore J. K., Doney S. C., Glover D. M., and Fung I. Y. (2002) Iron cycling and nutrient limitation patterns in surface waters of the world ocean. *Deep-Sea Res. II* **49**, 463–507.

Morel F. M. M., Reuter J., Anderson D., and Guillard R. (1979) Aquil: a chemically defined phytoplankton culture medium for trace metal studies. *Limnol. Oceanogr.* **36**, 27–36.

Morford J. L. and Emerson S. S. (1999) The geochemistry of redox sensitive trace metals in sediments. *Geochim. Cosmochim. Acta* **63**, 1735–1750.

Morley N. H., Statham P. J., and Burton J. D. (1993) Dissolved trace metals in the southwestern Indian Ocean. *Deep-Sea Res.* **30**, 1043–1062.

Murname R. J. (1994) Determination of thorium and particulate matter cycling parameters at Station P: a reanalysis and comparison of least squares techniques. *J. Geophys. Res.* **99**, 3393–3405.

Nozaki Y., Yang H. S., and Yamada M. (1987) Scavenging of thorium in the ocean. *J. Geophys. Res.* **92**, 772–778.

Nriagu J. (1989) The rise and fall of leaded gasoline. *Sci. Total Environ.* **92**, 13–28.

Orians K. J. and Bruland K. W. (1985) Dissolved aluminum in the central North Pacific. *Nature* **316**, 427–429.

Orians K. J. and Bruland K. W. (1986) The biogeochemistry of aluminum in the Pacific Ocean. *Earth Planet. Sci. Lett.* **78**, 397–410.

Orians K. J. and Bruland K. W. (1988) The marine geochemistry of dissolved gallium: a comparison with dissolved aluminum. *Geochim. Cosmochim. Acta* **52**, 1–8.

Powell R. T. and Donat J. R. (2001) Organic complexation and speciation of iron in the South and equatorial Atlantic. *Deep-Sea Res. II* **48**, 2877–2893.

Price N. M. and Morel F. M. M. (1990) Cadmium and cobalt substitution for zinc in a marine diatom. *Nature* **344**, 658–660.

Price N. M., Harrison G. I., Hering J. G., Hudson R. J., Nirel P., Palenik B., and Morel F. M. M. (1988/89) Preparation and chemistry of the artifical algal culture medium Aquil. *Biol. Oceanogr.* **5**, 43–46.

Rivera-Duarte I., Flegal A. R., Sanudo-Wilhelmy S. A., and Veron A. J. (1999) Silver in the far North Atlantic Ocean. *Deep-Sea Res. II* **46**, 979–990.

Roy-Barman M., Chen J. H., and Wasserburg G. J. (1996) ^{230}Th–^{232}Th systematics in the central Pacific Ocean: the sources and fates of thorium. *Earth Planet. Sci. Lett.* **139**, 315–363.

Rue E. L. and Bruland K. W. (1995) Complexation of iron(III) by natural organic ligands in the central North Pacific as determined by a new competitive ligand equilibration/adsorptive cathodic stripping voltammetry method. *Mar. Chem.* **50**, 117–138.

Rue E. L. and Bruland K. W. (1997) The role of organic complexation on ambient iron chemistry in the equatorial Pacific Ocean and the response of a mesoscale iron addition experiment. *Limnol. Oceanogr.* **42**, 901–910.

Saito M. A. and Moffett J. W. (2001) Complexation of cobalt by natural organic ligands in the Sargasso Sea as determined by a new high-sensitivity electrochemical cobalt speciation method suitable for open ocean work. *Mar. Chem.* **75**, 69–88.

Schaule B. K. and Patterson C. C. (1981) Lead concentrations in the northeast Pacific: evidence for global anthropogenic perturbations. *Earth Planet. Sci. Lett.* **54**, 97–116.

Schaule B. K. and Patterson C. C. (1983) Perturbations of the natural lead depth profile in the Sargasso Sea by industrial lead. In *Trace Metals in Seawater* (eds. C. S. Wong, E. Boyle, K. W. Bruland, J. D. Burton, and E. D. Goldberg). Plenum, New York, pp. 487–503.

Shiller A. M. and Boyle E. A. (1991) Trace elements in the Mississippi river-delta outflow region-behaviour at high discharge. *Geochim. Cosmochim. Acta* **55**, 3241–3251.

Sholkovitz E. R. and Copeland D. (1983) The coagulation, solubility and adsorption properties of Fe, Mn, Cu, Ni, Cd, Co and humic acids in river water. *Geochim. Cosmochim. Acta* **45**, 181–189.

Smetacek V. (1999) Diatoms and the ocean carbon cycle. *Protist* **150**, 25–32.

Sohrin Y., Isshiki K., Nakayama E., and Matsui M. (1989) Simultaneous determination of tungsten and molybdenum in sea-water by catalytic current polarography after preconcentration on a resin column. *Anal. Chim. A* **218**, 25–35.

Spencer D. W., Robertson D. E., Turekian K. K., and Folsom T. M. (1970) Trace element calibrations and profiles at the GEOSECS test station in the northeast Pacific Ocean. *J. Geophys. Res.* **75**, 7688.

Sunda W. G. (1988/89) Trace metal interactions with marine phytoplankton. *Biol. Oceanogr.* **6**, 411–442.

Sunda W. G. and Huntsman S. A. (1992) Feedback interactions between zinc and phytoplankton in seawater. *Limnol. Oceanogr.* **37**, 25–40.

Sunda W. G. and Huntsman S. A. (1995a) Iron uptake and growth limitation in oceanic and coastal phytoplankton. *Mar. Chem.* **50**, 189–206.

Sunda W. G. and Huntsman S. A. (1995b) Cobalt and zinc interreplacement in marine phytoplankton: biological and geochemical implications. *Limnol. Oceanogr.* **40**, 1404–1407.

Sunda W. G. and Huntsman S. A. (1995c) Regulation of copper concentration in the oceanic nutricline by phytoplankton uptake and regeneration cycles. *Limnol. Oceanogr.* **40**, 132–137.

Sunda W. G. and Huntsman S. A. (1998) Control of Cd concentrations in a coastal diatom by interactions among free ionic Cd, Zn, and Mn in seawater. *Environ. Sci. Technol.* **32**, 2961–2968.

Takeda S. (1998) Influence of iron availability on nutrient consumption ratio of diatoms in oceanic waters. *Nature* **393**, 774–777.

Tappin A. D., Millward G. E., Statham P. J., Burton J. D., and Morris A. W. (1995) Trace metal in the central and southern North Sea. *Estuar. Coast. Shelf Sci.* **41**, 275–323.

Tegen I. and Fung I. (1995) Contribution to the atmospheric mineral aerosol load from land-surface modification. *J. Geophys. Res.* **100**, 18707–18726.

Turekian K. K. (1977) The fate of metals in the oceans. *Geochim. Cosmochim. Acta* **41**, 1139–1144.

Turner D. R., Whitfield M., and Dickson A. G. (1981) The equilibrium speciation of dissolved components in freshwater and seawater at 25 °C and 1 atm pressure. *Geochim. Cosmochim. Acta* **45**, 855–881.

van den Berg C. M. G. (1984) Determination of copper in seawater by cathodic stripping voltammetry of complexes with catechol. *Anal. Chim. A* **164**, 195–207.

van den Berg C. M. G. (1985) Determination of the zinc complexing capacity in seawater by cathodic stripping voltammetry of zinc-APDC complex ions. *Mar. Chem.* **16**, 121–130.

van den Berg C. M. G. (1988) Adsorptive cathodic stripping voltammetry and chronopotentiometry of trace metals in sea water. *Anal. Proc.* **25**, 265–266.

van den Berg C. M. G. (1995) Evidence for organic complexation of iron in seawater. *Mar. Chem.* **50**, 139–157.

Vink S. and Measures C. I. (2001) The role of dust deposition in determining surface water distributions of Al and Fe in the South west Atlantic. *Deep-Sea Res.* **48**, 2787–2809.

von Damm K. L., Edmond J. M., Measures C. L., and Grant B. (1985) Chemistry of submarine hydrothermal solutions at 21° N, East Pacific Rise. *Geochim. Cosmochim. Acta* **49**, 2197–2220.

Vorlicek T. P. and Helz G. R. (2002) Catalysis by mineral surfaces: implications for Mo geochemistry in anoxic environments. *Geochim. Cosmochim. Acta* **66**, 3679–3692.

Westerlund S. F. G., Anderson L. G., Hall P. O. J., Iverfeldt A., Vanderloff M. M. R., and Sundby B. (1986) Benthic flues of cadmium, copper, nickel and lead in coastal environments. *Geochim. Cosmochim. Acta* **50**, 1289–1296.

Wilhelm S. W. and Trick C. G. (1994) Iron-limited growth of cyanobacteria: multiple siderophore production is a common response. *Limnol. Oceanogr.* **39**, 197–1984.

Windom H., Byrd J., Smith R., Hungspreugs M., Dharmvanii S., Thumtrakul W., and Yeats P. (1991) Trace-metal nutrients relationships in estuaries. *Mar. Chem.* **32**, 177–194.

Wu J. and Boyle E. A. (1997) Low blank preconcentration technique for the determination of lead, copper and cadmium in small-volume seawater samples by isotope dilution ICPMS. *Anal. Chem.* **69**, 2464–2470.

Wu J. F. and Luther G. W. (1995) Complexation of Fe(III) by natural organic ligands in the northwest Atlantic Ocean by a competitive ligand equilibration method and a kinetic approach. *Mar. Chem.* **50**, 159–177.

Zhang Y., Amakawa H., and Nozaki Y. (2001) Oceanic profiles of dissolved silver; precise measurements in the basins of western North Pacific, Sea of Okhotsk, and the Japan Sea. *Mar. Chem.* **75**, 151–162.

Zhu X. R., Prospero J. M., and Millero F. J. (1997) Diel variability of soluble Fe(II) and soluble total Fe in North African dust in the trade winds near Barbados. *J. Geophys. Res.* **102**, 21297–21305.

6.03
Gases in Seawater

P. D. Nightingale
Plymouth Marine Laboratory, Devon, UK

and

P. S. Liss
University of East Anglia, Norwich, UK

6.03.1 INTRODUCTION

The annual gross and net primary productivity of the surface oceans is similar in size to that on land (IPCC, 2001). Marine productivity drives the cycling of gases such as oxygen (O_2), dimethyl sulfide (DMS), carbon monoxide (CO), carbon dioxide (CO_2), and methyl iodide (CH_3I) which are of fundamental importance in studies of marine productivity, biogeochemical cycles, atmospheric chemistry, climate, and human health, respectively. For example, ~30% of the world's population (1,570 million) is thought to be at risk of iodine-deficiency disorders that impair mental development (WHO, 1996). The main source of iodine to land is the supply of volatile iodine compounds produced in the ocean and then transferred to the atmosphere via the air–surface interface. The flux of these marine iodine species to the atmosphere is also thought to be important in the oxidation capacity of the troposphere by the production of the iodine oxide radical (Alicke *et al.*, 1999). A further example is that the net flux of CO_2 from the atmosphere to the ocean, $\sim 1.7 \pm 0.5$ Gt C yr^{-1}, represents ~30% of the annual release of anthropogenic CO_2 to the atmosphere (IPCC, 2001). This net flux is superimposed on a huge annual flux (90 Gt C yr^{-1}) of CO_2 that is cycled "naturally" between the ocean and the atmosphere. The long-term sink for anthropogenic CO_2 is recognized as transfer to the ocean from the atmosphere. A final example is the emission of volatile sulfur, in the form of DMS, from the oceans. Not only is an oceanic flux from the oceans needed to balance the loss of sulfur (a bioessential element) from the land via weathering, it has also been proposed as having a major control on climate due to the formation of cloud condensation nuclei (Charlson *et al.*, 1987). Indeed, the existence of DMS and CH_3I has been used as evidence in support of the Gaia hypothesis (Lovelock, 1979).

There are at least four main processes that affect the concentration of gases in the water column: biological production and consumption, photochemistry, air–sea exchange, and vertical mixing. We will not discuss the effect of vertical mixing on gases in seawater and instead refer the reader to Chapter 6.08. Nor will we consider the deeper oceans as this region is discussed in chapters on benthic fluxes and early diagenesis (Chapter 6.11), the biological pump (Chapter 6.04), and the oceanic calcium carbonate cycle (Chapter 6.19) all in this volume. We will discuss the cycling of gases in surface oceans, including the thermocline, and in particular concentrate on the exchange of various volatile compounds across the air–sea interface.

As we will show, while much is known about the cycling of gases such as CO_2 and DMS in the water column, frustratingly little is known about many of the chemical species for which the ocean is believed to be a significant source to the atmosphere. We suspect the passage of time will reveal that the cycling of volatile compounds containing selenium and iodine may well prove as complex as that of DMS. Early studies of DMS assumed that it was produced from a precursor compound, dimethylsulfoniopropionate (DMSP), known to be present in some species of phytoplankton, and that the main sink in the water column was exchange across the air–sea interface. We now know that DMSP and DMS are both rapidly cycled in water column by a complex interaction between phytoplankton, microzooplankton, bacteria, and viruses (see Figure 1). Some detailed process experiments have revealed that only ~10% of the total DMS produced (and less than 1.3% of the DMSP produced) is transferred to the atmosphere, with the bulk of the DMS and DMSP, either being recycled in the water column or photo-oxidized (Archer *et al.*, 2002b).

This chapter is split into two main sections. The first deals with air–sea gas exchange mechanisms and measurements, the second with the processes of production and consumption that control the distributions of gases and includes estimates of

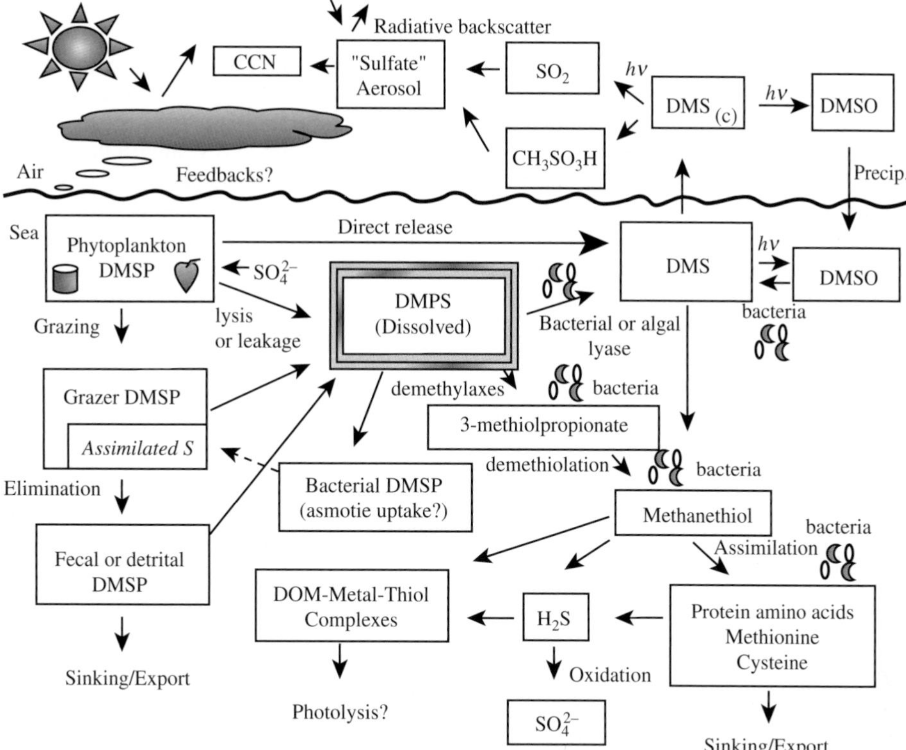

Figure 1 A conceptual model illustrating the biogeochemical cycle of DMS and DMSP (Kiene *et al.*, 2000) (reproduced by permission of Elsevier from *J. Sea Res.*, **2000**, *43*, 209–224).

the size of the oceanic source (or sink) of the compound under discussion.

6.03.2 AIR–SEA GAS EXCHANGE

Despite the central role that air–sea gas exchange plays in studies of marine productivity, biogeochemical cycles, atmospheric chemistry, and climate, it has proved extremely difficult to measure air–sea gas fluxes *in situ*. Only in 2001 were believable direct measurements of oceanic CO_2 fluxes reported in the literature (McGillis *et al.*, 2001a). In this section we examine the various models that have been proposed to understand the basic processes that control gas exchange mechanisms, describe results from laboratory experiments, and discuss the various techniques that have been developed to try to measure gas transfer rates *in situ*. Finally, we describe the development of wind speed (U) based parametrizations and assess their impact on computation of air–sea gas fluxes.

6.03.2.1 Air–Sea Gas Exchange Models

6.03.2.1.1 The two-film model of gas exchange

The simplest model of air–sea gas transfer is the two-film model (Liss and Slater, 1974; Whitman, 1923) which is illustrated in Figure 2. Although the representation of the sea surface as a flat, solid boundary and the presence of two uniform thin layers on either side of it is physically unrealistic, it is a useful starting point to visualize

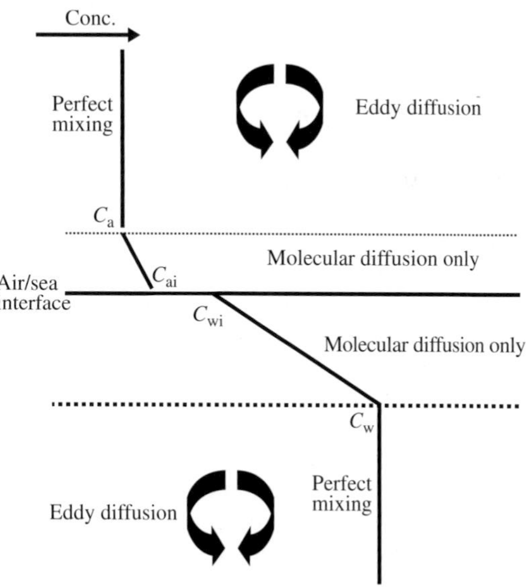

Figure 2 The two-film model of air–sea gas exchange (reproduced by permission of Nature Publishing Group from *Nature*, **1974**, *247*, 181–184).

how gas transfer between the two fluids may occur. The model assumes that the main bodies of air and water are well mixed such that the concentration of any particular gas is uniform in both phases. This does not mean that the gas is inert but that the rate of mixing is greater than the rate of any production or destruction process that may be operating (see Section 6.03.3.1). Transfer through the two thin films is by molecular diffusion only. Note that this also assumes that any production or destruction processes that may occur in the thin films are slow compared to molecular diffusion. The net flux of gas through one film is then given by the product of the concentration difference across the film that drives the flux and a kinetic (or rate) term known as the gas transfer coefficient (k). The gas transfer coefficient is also known as a piston velocity, or more commonly transfer velocity, as it has dimensions of length per time. Given the assumptions above, then

$$F = k_w(C_{wi} - C_w) = k_a(C_a - C_{ai}) \quad (1)$$

where C_a is the concentration in the bulk air, C_w is the concentration in the bulk seawater, and C_{ai} and C_{wi} represent the concentrations at the interface on the air and seawater sides, respectively. If the gas obeys Henry's law, then the relationship between C_{ai} and C_{wi} is given by

$$C_{ai} = H/C_{wi} \quad (2)$$

It can then be shown that

$$F = K_w(C_a/H - C_w) = K_a(C_a - HC_w) \quad (3)$$

where

$$\begin{aligned} 1/K_w &= 1/k_w + 1/Hk_a \\ 1/K_a &= H/k_w + 1/k_a \end{aligned} \quad (4)$$

The flux of gas across the air–water interface is therefore given by the concentration difference between the bulk air and bulk seawater (ΔC) after correcting for solubility (i.e., the degree of disequilibrium between the two phases) and an overall transfer velocity that is itself dependent on the individual transfer velocities in the air and water. In practice, for most sparingly soluble gases (i.e., high H) the rate-limiting step (or main resistance) is transfer through the water-side thin film, as molecular diffusion through water is considerably slower than in air. Examples of these gases are O_2, CO_2, methane (CH_4), methyl bromide (CH_3Br), and sulfur hexafluoride (SF_6). In these cases, the term $1/k_w$ therefore dominates and Equation (3) then simplifies to the more familiar expression for estimating air/sea gas fluxes, i.e.,

$$F = k_w(C_a/H - C_w) \quad (5)$$

For some gases that either react with water, or are highly soluble, the $1/Hk_a$ term dominates. These gases include hydrogen chloride, sulfur dioxide, water, and probably ammonia (NH_3). In these cases the main resistance to transfer is in the air-side thin film and K_w can be approximated by the term Hk_a.

There are relatively few gases where Equation (3) does not simplify, although recent laboratory experiments with DMS suggest that both k_a and k_w may have to be included in air–sea flux estimates at low temperatures or moderate wind speeds (McGillis *et al.*, 2000), as well as for some organic gases, e.g., PCBs and some pesticides (Duce *et al.*, 1991). As there is a good body of information on the term k_a from research into air–sea fluxes of water vapor, and as the transfer of most gases of environmental interest is limited by the water-side resistance, the remainder of this section will concentrate on the determination of k_w.

6.03.2.1.2 Chemical enhancement

As outlined above, the model assumes that no significant loss or production mechanisms are present in the thin film. For some gases this may be incorrect. For example, the transfer rate of CO_2 (k_{CO_2}) may be enhanced by reaction with hydroxide ions (OH^-) within the thin film (Hoover and Berkshire, 1969), i.e.,

$$CO_{2(aq)} + OH^-_{(aq)} = HCO^-_{3(aq)} \quad (6)$$

This effect is pH dependent (Emerson, 1975) and is most important at low wind speeds and high water temperatures (e.g., equatorial regions). Modeling studies have indicated that chemical enhancement may increase k_{CO_2} by up to 8% in these regions (Boutin and Etcheto, 1995).

Additionally, there may be a biological enhancement to k_{CO_2}. It is reasonably well established that some species of phytoplankton possess an enzyme (carbon anhydrase) that catalyzes the above reaction inside algal cells (Raven, 1995). Although early experiments indicated that bovine-derived carbonic anhydrase increased the exchange rates of CO_2 (Berger and Libby, 1969), later laboratory investigations with natural seawater failed to find any enhancement (Goldman and Dennett, 1983). However, recent laboratory studies using a circular gas exchange tank showed that there is indeed a considerable enhancement to k_{CO_2} in the presence of various algal species (Matthews, 2000). This enhancement was species dependent and increased significantly as CO_2 concentrations decreased. The lack of a response observed by Goldman and Dennet could well be due to the short lifetime of the enzyme is seawater. Modeling of these results suggested that biological enhancements could be globally significant (Matthews, 2000). We should caution, however, that no experiment has yet shown that either chemical or biological enhancements of k_{CO_2} operate at sea, although chemical enhancement for sulfur dioxide is well known (see Section 6.03.3.4.2).

6.03.2.1.3 The Schmidt number

It is clearly important to know how to scale k_w for different gases and under different physical conditions. If k_w is known for any particular gas under a particular set of conditions, then k_w can be derived for a different gas using Equation (7) below:

$$k_{w1}/k_{w2} = (Sc_1/Sc_2)^{-n} \quad (7)$$

where Sc (the Schmidt number) is temperature dependent and determined from the ratio of the kinematic viscosity of seawater (ν) and molecular diffusivity (D) of the gas in seawater. Gas transfer velocities are typically normalized to a common Sc (either 600 or 660 the values for CO_2 in freshwater and seawater at 20 °C, respectively) in order to aid comparison under different physical conditions. For the remainder of this section we will use a normalization of k_w to an Sc of 600, i.e., k_{600} where appropriate. Note that the two-film model implies that k_w is proportional to D to the power of unity, i.e.,

$$k_w = D/Z \quad (8)$$

where Z represents the thickness of the film.

6.03.2.1.4 Surface renewal model

Although molecular (diffusive) process will become progressively more important as the interface is approached, the existence of a stagnant film whose thickness, for a given degree of turbulence, is invariant with time and space is clearly physically unrealistic. This led to the development of a variety of surface renewal models in which the stagnant fluid close to the interface is replaced periodically by material from the bulk (Danckwerts, 1951; Higbie, 1935; Ledwell, 1984). The physical process, and its mathematical description, by which liquid is envisaged as being transferred from the bulk to near surface is somewhat different in each of these versions of the surface renewal model. However, for present purposes, the important point is that all the variants predict that k_w will vary with $D^{0.5}$ (i.e., $Sc^{-0.5}$). Similarly, the so-called large- (Fortescue and Pearson, 1967) and small-eddy models (Lamont and Scott, 1970), in which transfer in the near-surface water is described in terms of a series of cells of rotating fluid, also lead to the conclusion that k_w will be proportional to $Sc^{-0.5}$.

Although physically more realistic than the film model, surface renewal and eddy models have been used rather little, particularly in environmental investigations. This is mainly because of the difficulty of specifying the film/fluid replacement rate, except in well-defined laboratory systems. However, the dependence of k_w on $D^{0.5}$ predicted by this family of models is generally taken as effective under wave covered (but not necessarily for bubbled—see Section 6.03.2.1.6) surfaces in the marine environment.

6.03.2.1.5 Micrometeorological models

There is a great deal of theoretical and experimental information from micrometeorological research on the transfer of momentum, heat, and mass at solid and liquid surfaces and across their associated air boundary layers (hence the term "boundary layer models" for relationships arising from this approach). Based on the analogy between transfer of momentum and mass, it has been shown that k_w is proportional to the friction velocity in air (u^*) and that k_w is also proportional to $Sc^{-2/3}$. Apart from an assumption that the surface was smooth and rigid, it was also necessary to assume continuity of stress across the interface in order to convert the velocity profile in air to the equivalent profile in the water (Deacon, 1977). The relationship developed by Deacon is as follows:

$$k_w = 0.082Sc^{-2/3}(r_a/r_w)^{1/2}u^* \quad (9)$$

where r_a and r_w are the densities of air and water, respectively. The friction velocity is related to U at a particular height (z) by the following equation:

$$u_z = (u^*/V)\ \ln(z/z_0) \quad (10)$$

where V = von Karman's constant and z_0 is the roughness length.

The Deacon model is found to work well when the water surface is unruffled by waves, as might be predicted from the rigid wall assumption, but underestimates k_w when the surface ceases to be smooth. To counter this deficiency, the micrometeorological approach was extended (Kerman, 1984) by using an earlier treatment of heat and mass transfer to a rough surface (Yaglom and Kader, 1974). By estimating the amount of surface roughening caused by patches of breaking waves, Kerman was able to predict k_w for a nonsmooth surface.

6.03.2.1.6 Bubble-mediated gas transfer

Bubbles have long been thought to play an important role in air–sea gas exchange. The main mechanism behind bubble formation is the entrainment of air in breaking waves. There are at least three ways in which bubble formation may enhance air–sea gas transfer rates above those predicted by the gas transfer theories already

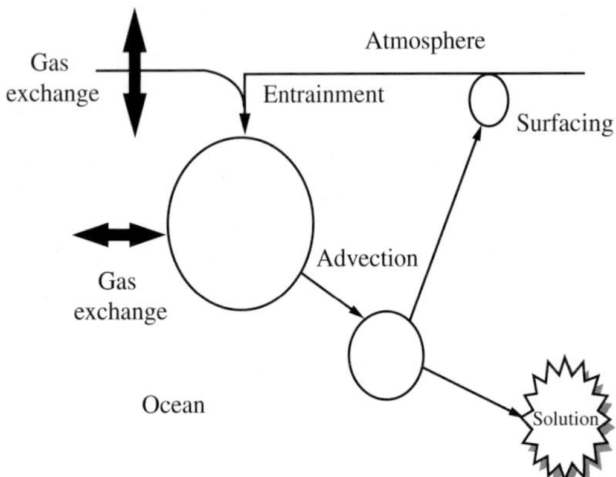

Figure 3 A schematic of bubble evolution after entrainment into the upper ocean (Woolf, 1997) (reproduced by permission of Cambridge University Press from *The Sea Surface and Global Change*, **1997**, pp. 173–205).

discussed and these are illustrated in Figure 3 (Woolf, 1997). The initial injection of the bubble can enhance air–sea gas transfer by disrupting the microlayer. This enhancement can be thought of as temporarily reducing or removing the thin film that acts as a barrier to air–sea gas exchange or as encouraging surface renewal. The air inside the bubble can then exchange directly with the bulk seawater, i.e., air–water gas transfer that bypasses the air–sea interface, and the rate of exchange is then governed by transfer through the microlayer surrounding the bubble. However, the capacity of the bubble to either supply or receive gas may also be limiting, in contrast to the case of transfer across the air–sea interface, where the volume of the atmosphere is effectively unlimited, at least for major atmospheric constituents. There may, therefore, be a solubility effect as the bubble will approach equilibrium with the bulk seawater more quickly for soluble gases (e.g., CO_2) compared to less soluble gases (e.g., SF_6), i.e., bubble-mediated gas exchange is likely to be more effective for less soluble gases (Woolf, 1997). The bubble then either dissolves completely or rises to the surface where it again disrupts the microlayer causing surface renewal.

Although laboratory experiments have shown that bubbles are extremely efficient in enhancing k_w (see Section 6.03.2.2.3), it has so far proved impossible to determine directly the importance of bubble-mediated gas transfer in the field. Models are, therefore, essential in order to try to assess the contribution of bubble-mediated gas transfer to the total. Typically, these models include a bubble-mediated contribution (k_b) to the direct gas transfer rate (k_d) to arrive at the total gas transfer rate (k_w) (Keeling, 1993; Woolf, 1997, 1993). However, the situation is slightly more complicated due to the fact that the concentration of a particular gas inside the bubble is likely to be different from its concentration in the atmosphere. This is in part due to the hydrostatic pressure experienced by a submerged bubble as well as effects due to bubble curvature. As a result, even when a dissolved gas is at equilibrium with respect to the atmosphere, there may be a net transfer of that gas from bubbles to seawater (Woolf, 1997; Woolf and Thorpe, 1991). It is therefore possible that, even under equilibrium conditions, seawater can be slightly supersaturated in a particular gas if significant populations of bubbles are present. Additionally, the composition of the bubble will change as a result of gas exchange between the bubble and the bulk seawater. The net air–sea flux (F) can be represented by the following modification to Equation (5) (Woolf, 1997):

$$F = (k_d + k_b)(C_a(1 + \Lambda)/H - C_w) \quad (11)$$

where Λ represents the equilibrium supersaturation at which there will be no net air–sea transfer of gas. Both k_b and Λ are dependent on solubility and D. The term k_d is predicted to increase approximately with U^2, whereas k_b is predicted to be approximately proportional to white-cap coverage (or U^3) (Woolf, 1997). These models of bubble-mediated gas transfer correctly predict supersaturations of a few percent in dissolved O_2 and nitrogen (N_2) levels at higher wind speeds (Woolf, 1997). Although they predict that the bubble-induced contribution to the total global mean k_{CO_2} may be as high as 30%, many of the parameters in the model are too poorly defined by field measurements for us to be confident of how much bubbles and breaking waves really contribute to gas exchange at sea.

6.03.2.2 Laboratory Studies of Air–Water Gas Exchange

The vast majority of investigations designed to understand the mechanisms behind the process of air–sea gas exchange and to identify variables that influence k_w have been conducted in the laboratory and most commonly in wind/wave tunnels. These vary greatly in scale and sophistication from simple circular tanks with a diameter of less than a meter and just a rotating paddle to enhance stirring (Matthews, 2000)—to even large national facilities up to 50 m in length that may have the ability to mechanically generate waves and bubbles, manipulate air and water temperatures, and possess a return flow for both the air and water phases such that the whole system is closed (de Leeuw *et al.*, 2002).

6.03.2.2.1 The effect of wind

Early experiments in wind/wave tunnels quickly showed that the most obvious parameter that influenced k_w was U. Note that as U is dependent on both height and on atmospheric stability, k_w is more typically plotted against u^*. However, it equally became apparent that values for k_{600} varied for any given U or u^*, depending on the facility being used and on other conditions (Liss, 1983). Considerable field data also exist to show that U has a major (but not necessarily direct) influence on k_w (see Section 6.03.2.3).

6.03.2.2.2 The effect of waves

The presence of wind-induced ripples was quickly noted as being coincident with an enhancement in k_w (Kanwisher, 1963). Careful experiments with various pairs of gases were used to identify n at different U in order to identify which, if any, of the models discussed previously might best represent air–water gas transfer (Jahne *et al.*, 1987). These experiments showed that the thin-film model is too simple and that k_w varied with $Sc^{-2/3}$ at low U as predicted by boundary layer models (see Section 6.03.2.1.5). However, once wind-induced waves were observed on the surface of the water, k_w was found to vary with $Sc^{-1/2}$, in agreement with surface renewal models (see Section 6.03.2.1.4). The exact U at which this regime changes was found to vary with the facility being used but was typically $\sim$2 m s^{-1} ($u^* = 0.3$ cm s^{-1}) (Jahne *et al.*, 1987).

A much tighter correlation was found between k_w and the mean square wave slope of shorter wind waves than there was with U or u^* (Jahne *et al.*, 1987). This led to the conclusion that a wave-related mechanism was controlling k_w at intermediate U and that enhancements were due to increased turbulence generated by waves close to the water surface. More recently, the presence of microscale wave breaking has been speculated to be a fundamental mechanism underlying this increase in turbulence and hence the enhancement in k_w (Zappa *et al.*, 2001). There are surprisingly few, if any, field experiments that show a direct link between wave properties and k_w.

6.03.2.2.3 The effect of bubbles

Experiments at even higher U indicated another enhancement in k_w coincident with the onset of wave breaking and the formation of bubbles (Broecker and Siems, 1984). Large breaking waves are known to be associated with high levels of near-surface turbulence and with the injection of bubbles. The magnitude of the enhancement in k_w was found to depend on the solubility of the gas (Broecker and Siems, 1984; de Leeuw *et al.*, 2002; Merlivat and Memery, 1983). As discussed in Section 6.03.2.1.6, bubble-mediated gas exchange is the most likely explanation for the observations of a dependence on solubility. Again, the wind speeds at which the onset of the enhancement was observed varied, typically between 10 m s^{-1} and 14 m s^{-1}. It is, however, difficult to separate out the effects of increased turbulence due to wave breaking and bubble surfacing from direct bubble-mediated gas exchange, without the use of the models discussed earlier. Indeed, it has been proposed that large breaking waves might suppress the occurrence of microscale breakers at higher U (Csanady, 1990). It has not proved possible to measure the contribution of bubbles to k_w in the field, partly due to a lack of suitable techniques to determine bubble densities and size distributions close to the sea surface (Woolf, 1997).

6.03.2.2.4 Temperature, humidity, and rain effects

There are several ways in which heat and water transfer across the air–sea interface might affect the exchange of trace gases. First, it has been proposed that Equation (1) gives an incomplete description of the gas exchange process, since the irreversible thermodynamic coupling between heat and mass is not included (Phillips, 1994, 1997). Basically, the idea is that heat exchanges across the interface will affect the thermodynamics of gas dissolution, and vice versa. The concept has been criticized by Doney (1995) on the grounds of inconsistencies in Phillips' treatment of the molecular boundary layer and irreversible effects near the air–water interface, and the matter does not appear to have been definitively settled. However, what has been agreed is that, at least under oceanic conditions where the temperature gradients, and thus heat exchanges across the sea surface, are quite small, any effect is very limited and is invariably neglected.

A second and better-recognized effect arises from the fact that evaporation and other heat exchanges at the sea surface will lead to temperature changes, e.g., a cooling of the water very close to the interface if evaporation is dominant. The magnitude of this "cool skin" effect is typically 0.1–0.3 °C. Since the solubility of almost all gases increases with decreasing water temperature, an error is made if the equilibrium solubility is assumed to be at the bulk water temperature. The effect can be of significance particularly for gases such as CO_2 whose global net exchange with the oceans is composed of the sum of uptake minus release, much of which is driven by concentration differences that are close to equilibrium (see Section 6.03.3.3.1). Inclusion of the skin effect into calculations of the net uptake of CO_2 by the oceans via Equation (5) has been estimated to lead to an enhancement by 0.7 Gt C yr^{-1}

(Robertson and Watson, 1992), which is of significance given that the global net flux is ~1.7 Gt C yr^{-1}. Although this does not affect uptake estimates derived from mathematical models, it does help to resolve discrepancies between these estimates and those derived from a combination of CO_2 concentration fields and estimates of k_w (see Section 6.03.3.3.1).

It can also be envisaged that heat exchanges across an air–water interface might lead to alterations in the value of k_w. For example, the "cool skin" effect will lead to instability in the water close to the interface, with ensuing increased turbulence and enhancement of the transfer velocity. This idea has been tested in a wind-tunnel study in which k_w was measured as a function of the evaporation and condensation of water molecules (Liss *et al.*, 1981). The results showed that under evaporative conditions there was no measurable enhancement in k_w due to destabilization of the near-surface water, any effect being masked by mechanically generated mixing. However, under condensing conditions there was up to a 30% decrease in k_w due to increased stratification. The importance of this result in the environment is likely to be limited since evaporative conditions are typically found over the oceans. However, there are situations in which condensation is the dominant process, e.g., in coastal upwelling regions where cold, deep water is brought to the surface and may then be in contact with warm, humid air. Since such areas are often biologically rich, the importance of lowered k_w under condensing conditions could be some significance for the air–sea exchange of biologically active trace gases.

Finally, a significant enhancement in k_w has been noted for SF_6 due to the influence of rainfall (Ho *et al.*, 1997). A first-order relationship between k_w and rain rate was observed in laboratory simulations and in simple field experiments. However, it is not clear how this may apply to the oceans where rainfall may also cause salinity gradients and disrupt surfactant layers.

6.03.2.2.5 The effect of surfactants

The effect of artificial surfactants in reducing k_w has long been recognized. Early experiments showed that reductions in k_w of up to 60% could be observed in wind/wave tanks for a given wind speed (Broecker *et al.*, 1978), and contamination of the air–water interface by surface films was a common problem, particularly in circular tanks where there was no "beach" at the end of the tunnel for the film to collect (Jahne *et al.*, 1987). The presence of concentrated insoluble surfactant films (slicks) acts as a barrier to gas exchange, either by forming a condensed insoluble monolayer on the sea surface (Springer and Pigford, 1970), or by providing an additional liquid phase that provides a resistance to mass transfer (Liss and Martinelli, 1978). However, this effect is believed to be important only at low wind speeds, as slicks are easily dispersed by wind and waves (Liss, 1983). The main effect of surface-active material is believed to be due to the presence of soluble surfactants that alter the hydrodynamic properties of the sea surface (Frew, 1997). Indeed, Frew has argued that soluble surfactants can reduce k_w even in the presence of breaking waves. One of the implications of this is that air–sea gas transfer is not purely a physically driven process but that surface chemistry may also play an important role.

Even more interesting are observations that k_w could vary with biological activity (Goldman *et al.*, 1988) due to the exudation of soluble surface active material by phytoplankton (Frew *et al.*, 1990). More recently, supporting evidence from laboratory experiments using seawater collected on a transect from USA to Bermuda, has shown that the decrease in k_w correlates inversely with bulk-water chlorophyll, dissolved organic carbon (DOC) and colored dissolved organic matter (CDOM) (see Figure 4). The importance of the relationship with CDOM is the possibility that this parameter could be determined remotely by satellite. The implication of this work is that U may not be the best parameter with which to parametrize k_w in the oceans, particularly in biologically productive regions. A reasonable correlation with the total mean square wave slope for both filmed and film-free surfaces, particularly of shorter wind waves, suggested that this parameter, although difficult to measure at sea, might be a more useful predictor of k_w (Jahne *et al.*, 1987).

A note of caution is that although surfactants have been shown to be so important in laboratory-based experiments, the impact of surface films on k_w in the oceans has yet to be evaluated directly. A deliberate tracer experiment (see Section 6.03.2.3.2(iii)) conducted as part of an open ocean iron enrichment experiment in the equatorial Pacific found that there was no measurable reduction in k_w during the development of a large algal bloom (Nightingale *et al.*, 2000a), even though chlorophyll *a* levels increased 30-fold (see Figure 5). However, no measurements of surfactants were made, so it is not known whether there was any change in this class of compounds during the algal bloom. Indeed, it is not inconceivable that much of the world's oceans are already covered by soluble surfactant films.

6.03.2.2.6 Summary of laboratory experiments

To summarize, it seems reasonable to assume that there are several mechanisms, some of them

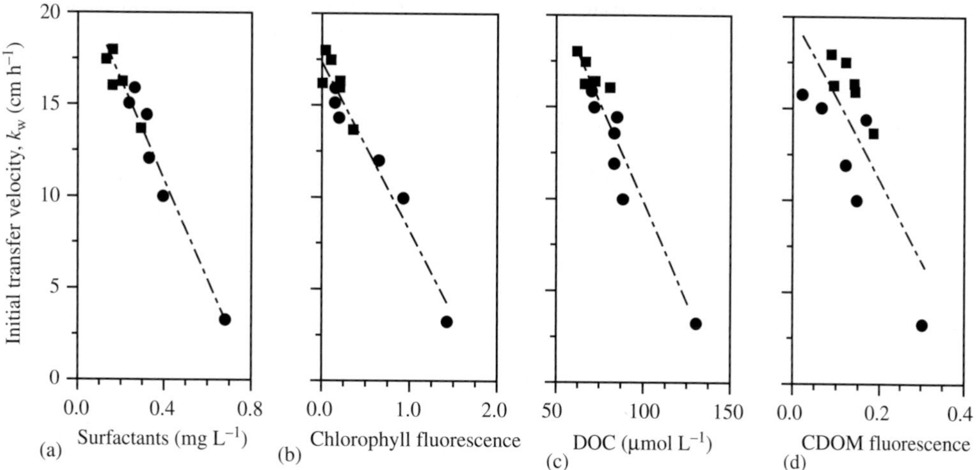

Figure 4 Correlations of k_w with (a) surfactant concentration, (b) *in situ* chlorophyll fluorescence, (c) dissolved organic carbon (DOC), and (d) colored dissolved organic matter (CDOM) for seawater samples collected from Monterey Bay, USA (■) and on a transect from Narragansett USA to Bermuda (●) (Frew, 1997) (reproduced by permission of Cambridge University Press from *The Sea Surface and Global Change*, **1997**, pp. 121–172).

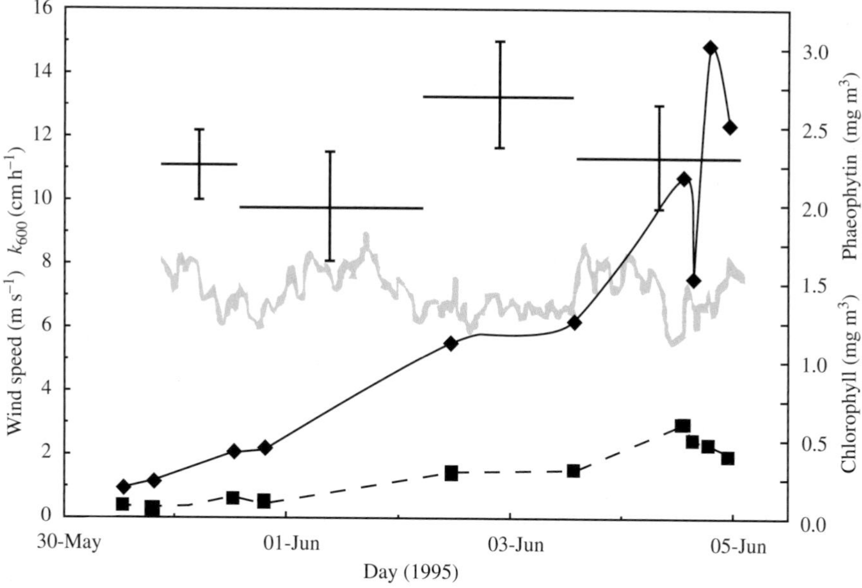

Figure 5 Gas transfer rates during an open-ocean iron enrichment experiment measured by release of deliberate tracers. Estimates of k_w are given by the horizontal black lines. The gray shaded line represents the wind, speed, the dashed line represents levels of the pigment Phaeophytin, the solid line represents concentrations of chlorophyll *a* (Nightingale *et al.*, 2000a) (reproduced by permission of American Geophysical Union from *Geophys. Res. Lett.*, **2000**, *27*, 2117–2120).

competing, that influence k_w in the oceans. Variables that have been found to influence k_w include wind speed, bulk-water temperature, molecular diffusivity, wave field, humidity and temperature gradients, rain fall, surfactant concentration and composition, boundary layer stability, microscale breakers, breaking waves, and bubbles. However, it has proved extremely difficult to extrapolate results to coastal seas and the oceans from laboratory experiments due to the limited fetch, a lack of gusting and artifacts including boundary conditions, the use of "clean" freshwater in most laboratory experiments and the artificial manner in which some of the variables above were manipulated. Although wind and wave tank experiments suggest that there are three clear regimes (smooth, ripples, and breaking waves) in which k_w can be predicted from wind speed, such prediction is much more difficult in a system as complex and variable as the ocean where wind is neither directly nor immediately linked to wave field.

6.03.2.3 Field Studies of Air–Sea Gas Transfer

A major and long-standing problem in studies of air–sea gas exchange is the obvious practical limitation of experimentation at the air–sea interface. Not only is it a formidable task to try to obtain *in situ* measurements close to such a dynamic region as the sea surface and particularly in the microlayer (typically less than 1 mm), it has also proved extremely difficult to actually measure the flux of gas across the air–sea interface directly. It is not straightforward, for obvious reasons, to manipulate experimental conditions in order to probe the mechanisms that might control gas transfer. Various ingenious techniques have therefore been used to derive air–sea gas exchange rates indirectly, typically by estimating air- and/or water-based budgets.

6.03.2.3.1 Large-scale techniques

(i) *Radiocarbon*. Radiocarbon (^{14}C) is produced naturally in the atmosphere and then transferred across the air–sea interface into the oceans as $^{14}CO_2$ (Broecker and Peng, 1974). Assuming that the system is at steady state, the mass of ^{14}C from the atmosphere must be balanced by decay of ^{14}C in the water column. An estimate of the global annual mean value for k_{CO_2} of 21 $\pm$ 5 cm h^{-1} can therefore be produced by a budgeting technique (Broecker *et al*., 1986). Similarly, by measuring the increased ^{14}C present in the oceans as a result of atmospheric nuclear weapon tests in the middle of the twentieth century, an independent estimate of k_{CO_2} of 22 $\pm$ 3 cm h^{-1} can be derived (Broecker *et al*., 1986), in very good agreement with the natural ^{14}C measurement. However, there are some inconsistencies in the global atmospheric ^{14}C budget, yet to be fully explained, that suggest that the oceanic uptake of ^{14}C may have been overestimated (Hesshaimer *et al*., 1994). Although the ^{14}C method allows a reasonable estimate of the globally averaged CO_2 uptake by the oceans, it yields little information on how k_w varies in time and space or indeed how to calculate k_w for other gases.

(ii) *Oxygen/nitrogen ratios*. A more recent technique for deriving regional estimates of k_w is based on high-precision measurements of atmospheric O_2/N_2 ratios from baseline sites situated around the globe (Keeling *et al*., 1998). The technique is dependent on the use of CO_2 data to correct for the effects of land/atmosphere fluxes on O_2/N_2 ratios, and on an atmospheric transport model to simulate oceanic fluxes. Annual values for k_{O_2} of 24 $\pm$ 6 cm h^{-1} and 29 $\pm$ 12 cm h^{-1} were calculated for ocean areas north of 30° N and south of 30° S, respectively, areas with above average winds (Keeling *et al*., 1998). When normalized to k_{600}, these values are ~25% higher than estimates predicted from a ^{14}C-based parametrization of k_{600} with U (Wanninkhof, 1992; see Section 6.03.2.4.1), but are just within measurement uncertainties.

6.03.2.3.2 Local techniques

(i) *Mass balance*. This technique involves time-series measurements of a gas (typically O_2) that is out of equilibrium with the atmosphere, in order to try to obtain the flux of the gas across the air–sea interface by using mass budget. Initial data from O_2 budgeting in the Gulf of Maine (Redfield, 1948) gave monthly mean gas exchange estimates that were subsequently found to be consistent with wind/wave tank experiments (Liss, 1983). Later results from Funka Bay in Japan indicated a clear increase in k_{O_2} in winter compared to the summer and that the increase with wind speed was greater than linear (Tsunogai and Tanaka, 1980).

There are two main difficulties with the technique. The first is to accurately budget the other production and/or removal processes such that the remainder represents the air–sea flux. The second is to overcome the problem of advection and dispersion at the sampling site. Use of Lagrangian techniques may help to overcome this. An estimate of k_{600} of 19 $\pm$ 11 cm h^{-1} has been obtained in a month long budgeting study of CO_2 in an SF_6 labeled tracer patch in the North Atlantic Ocean (Feely *et al*., 2002). This is in agreement with predictions of parametrizations of k_{600} based on the ^{14}C technique and on deliberate tracers (see Section 6.03.2.4.1) and suggests that the mass budget technique ought to be more widely used.

An alternative approach that has been used is to construct an atmospheric mass budget based on measurements of DMS in air and water off Cape Grim, Australia in combination with an atmospheric model to estimate the air–sea flux of DMS (Gabric *et al*., 1995). Their results were consistent with a parametrization of k_w with U based on deliberate tracers (Liss and Merlivat, 1986) but somewhat lower than the ^{14}C-derived values (see Section 6.03.2.4.1).

(ii) *Radon*. Radioactive decay of natural radium-226 to the gas radon-222 occurs within the water column and results in a loss of radon from the surface mixed layer to the atmosphere. A mass budget can be made of the "missing" radon by assuming steady state with deeper waters and hence a value for k_w derived

(Peng *et al.*, 1979). The mean value for k_{600} obtained using this technique is $\sim$14 cm h^{-1} and is somewhat lower than the ^{14}C-derived global mean. Further measurements indicated that mean value for k_{600} was $\sim$18 cm h^{-1} for the tropical Atlantic (Smethie *et al.*, 1985). All the radon data showed a large amount of scatter with U and the technique has some well-documented shortcomings in that it does not allow for the effect of vertical mixing on the radon profile and the long averaging time imposed by the several-day half-life of radon-222 (Liss, 1983). This was overcome by repeated measurements of the vertical profiles of radon at the same station in the subarctic Pacific resulting in a reasonable correlation with U (Emerson *et al.*, 1991). Again estimates tended to be lower than those predicted from ^{14}C-based parametrizations (see Section 6.03.2.4.1).

(iii) *Deliberate tracers experiments.* Inert volatile tracers, particularly SF_6, have been deliberately added to water bodies in order to determine k_w via water-based mass budgeting techniques. The lack of production or removal processes for SF_6 in the water column makes the technique simpler and far more precise than those discussed above. The technique was originally utilized in enclosed lakes and a good correlation was observed between k_w and U (Upstill-Goddard *et al.*, 1990; Wanninkhof *et al.*, 1985, 1987). A major improvement to this technique was to co-release helium-3 (^{3}He) in lake experiments (Watson *et al.*, 1991). This allowed the first determination of n (see Equation (7)) to be made *in situ*, the value of 0.51 being in excellent agreement with the predictions of surface renewal models.

In order to use the deliberate tracer technique at sea, two tracers are required in order to correct for their dilution due to horizontal and vertical mixing. These tracers are usually SF_6 and ^{3}He, and values for k_w can then be calculated from the change in the ratio of the two tracers over time (^{3}He diffuses more rapidly across the air–water interface), and subsequently correlated with environmental variables (Watson *et al.*, 1991). The main drawback to this dual-tracer technique is that the time interval over which measurements are made is typically 8–72 h and that only a small number of data can be obtained per experiment. Results from studies in coastal regions (Asher and Wanninkhof, 1998; Nightingale *et al.*, 2000b; Wanninkhof *et al.*, 1997) and the Atlantic Ocean (McGillis *et al.*, 2001b) and Pacific Ocean (Nightingale *et al.*, 2000a) are shown in Figure 6, and fall between existing U-based parametrizations of k_w. A fit to the dual-tracer data combined with global wind speeds gave a mean value for k_{600} of 18 cm h^{-1}, within the uncertainty associated with the ^{14}C values (Nightingale *et al.*, 2000b). Other nonvolatile tracers (e.g., bacterial

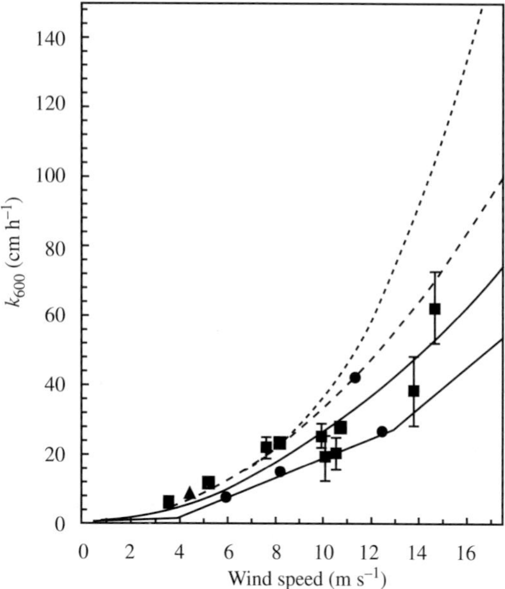

Figure 6 Parametrizations of k_w and deliberate tracer data: parametrizations of k_w with wind speed, LM86 (black line), W92 (black dashed line), WM99 (gray dashed line), and N2000 (gray solid line)—see text for details. Also shown are deliberate tracer data from the southern North Sea (■), the Florida Shelf (▲), and the Georges Bank (●) (Nightingale *et al.*, 2000a) (reproduced by permission of American Geophysical Union from *Geophys. Res. Lett.*, **2000**, *27*, 2117–2120).

spores) have been used in combination with SF_6 and ^{3}He, allowing a first measurement of n at sea (Nightingale *et al.*, 2000b). The value determined is close to 0.5, again in agreement with that predicted by surface renewal models (see Section 6.03.2.1.4).

(iv) *Direct covariance.* The direct covariance or eddy correlation technique is based upon very fast and accurate measurement of atmospheric gas concentrations and correlation with the simultaneously measured vertical wind velocity (Jones and Smith, 1977). Early measurements were controversial because fluxes were orders of magnitude greater than predicted by all of the other techniques described so far (Broecker *et al.*, 1986). However, an experiment based on a stable platform (thereby minimizing problems with flow distortion and motion contamination) in the southern North Sea showed that the gap between more established techniques such as deliberate tracers and direct covariance was narrowing (Jacobs *et al.*, 2002), mostly due to improvements in the sensitivity of CO_2 analyzers (Jacobs *et al.*, 1999). A breakthrough occurred with measurements that were made from a ship in a region of the North Atlantic where there was a large air–sea flux of CO_2 (McGillis *et al.*, 2001a) (see Figure 7). As of early 2000s, only this one study using direct

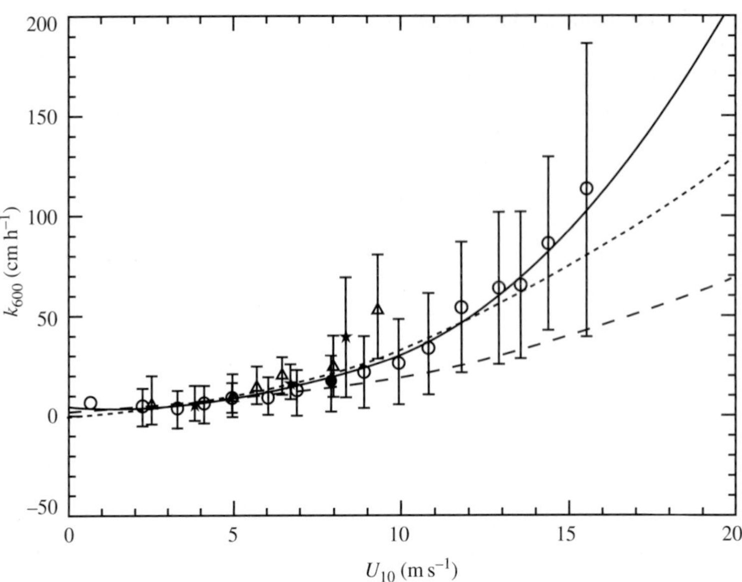

Figure 7 Results from a field comparison of k_w derived from direct covariance (○), deliberate tracers (●), and the atmospheric profile technique for CO_2 (△) and DMS (★) plotted against wind speed. Also shown are the parametrizations of LM86 (dashed line), W92 (dotted line), and a cubic fit to the data (solid line) (reproduced by permission of Elsevier from *Mar. Chem.*, **2001b**, *75*, 267–280).

covariance measurements over the oceans has been published and measurements of k_{CO_2} support a parametrization of k_w with U^3 (Wanninkhof and McGillis, 1999) (see Section 6.03.2.4.1). Only fluxes of CO_2 can be measured using this technique, although it has been successfully used for the measurement of gases where k_a dominates, e.g., ozone (Lenschow *et al.*, 1982) (see Section 6.03.3.4.1).

(v) *Relaxed eddy accumulation.* The relaxed eddy accumulation, or conditional sampling technique, is similar in concept to that of the eddy correlation technique. The principle behind it is to accumulate air samples associated with updrafts and downdrafts at a constant flow rate into two chambers that can be analyzed later for a wide range of compounds (Businger and Oncley, 1990). The flux integration is based upon the duration of the updrafts and downdrafts, rather than the vertical wind velocity, and the concentration difference between the two chambers (Pattey *et al.*, 1993). The technique is extremely attractive as it overcomes the need for fast response sensors and therefore ought to allow the air–sea flux of a wide range of gases to be determined. Although the technique has been regularly used over land (Moncrieff *et al.*, 1998; Xu *et al.*, 2002), it has proved very difficult to implement over the oceans. The only published study is for DMS fluxes measured from a stationary platform in the coastal North Atlantic. Estimates of k_{600} derived showed a high degree of scatter when plotted against U but were in reasonable agreement with the parametrizations shown in Figures 6 and 7 (Zemmelink *et al.*, 2002).

(vi) *Atmospheric profiles.* This technique depends on accurately measuring small gradients in the vertical concentration profile of gases such as DMS in the boundary layer above the air–sea interface. There are considerable uncertainties in the application of this technique, commonly used over land, to the oceans (McGillis *et al.*, 2001b). Measurements of k_w based on profile measurements of CO_2 and DMS have been reported from just one study but were somewhat higher than simultaneous estimates derived from the direct-covariance technique (McGillis *et al.*, 2001b), although within experimental uncertainties (see Figure 7).

6.03.2.3.3 Summary of field data

There are now a considerable number of measurements of k_w at sea. We have a good idea of the global mean value of k_{CO_2}. Significant data sets of k_w have been obtained using the radon and deliberate tracer techniques; the latter in particular indicates that there is a reasonable correlation of k_w with U although the obvious scatter in the data (see Figure 6) suggests that processes other than U also influence k_w. It is noteworthy that there is not a significant difference in measurements of k_w made in coastal seas and open oceans when plotted against U. As yet, there are no field measurements that directly link k_w to any other variables. Advances in the direct-covariance technique probably offer the best likelihood of improving this. Finally, although it is reassuring that deliberate tracer data and the direct-covariance data are in good agreement at U up to $10\ \mathrm{m\ s^{-1}}$, there are still some differences at high U where bubbles and breaking waves are likely to be increasingly important and where few data are available.

6.03.2.4 Parametrizations of Air–Sea Gas Transfer

6.03.2.4.1 Wind speed based parametrizations

A large number of parametrizations of k_w have been proposed. The great majority of these are based upon U, temperature, and D, mostly because they are fairly simple variables to measure and because field experiments have shown that U has a major influence on k_w. Here we will focus on four parametrizations of k_w that have commonly been used in estimates of air–sea gas fluxes (see Figure 6).

One of the first parametrizations of k_w was developed from the original lake experiment of Wanninkhof *et al.* (1985) (see Section 6.03.2.3.2(iii)) for low to intermediate U and combined with data from wind tunnels (Broecker *et al.*, 1978) to obtain estimates at high U (Liss and Merlivat, 1986; henceforth LM86). The relationship was essentially predictive in nature as few accurate oceanic measurements of k_w had been made. At wind speeds of less than $3.6\ \mathrm{m\ s^{-1}}$, LM86 assumed that the value for n was 0.67 (i.e., in agreement with boundary layer models—see Section 6.03.2.1.5) but that at higher wind speeds n was 0.5 (i.e., surface renewal mechanisms dominated—see Section 6.03.2.1.4). The LM86 parametrization predicts a similar global mean ($12\ \mathrm{cm\ h^{-1}}$) to that obtained from the radon data ($14\ \mathrm{cm\ h^{-1}}$—see Section 6.03.2.3.2(ii)) but is lower than the ^{14}C global mean value of $21\ \mathrm{cm\ h^{-1}}$ (see Section 6.03.2.3.1(i)).

An alternative relationship is a curve such that, when used with an approximation of global U distribution, it is in agreement with the global mean estimate derived from the ^{14}C technique (Wanninkhof, 1992; henceforth W92). Although this is a sensible approach, the main difficulty is in deciding the appropriate shape of the curve. Wanninkhof used a dependence of k_w on U^2 based on results from wind tunnel studies and n was assumed to be 0.5.

More recently, it has been proposed that k_w might depend on U^3 (Wanninkhof and McGillis, 1999; henceforth WM99). This was justified by the retardation of k_{600} at low U due to the possible presence of surfactants (see Section 6.03.2.2.5) and enhancements to k_{600} at higher winds due to the presence of bubbles. A cubic relationship is also in good agreement with results from a direct-covariance study (see Section 6.03.2.3.2(iv)). The parametrization was again chosen such that, when averaged over the same approximation of global winds, it was in agreement with the mean estimate of k_{CO_2} derived from the ^{14}C technique. However, although W92 and WM99 are both calibrated against the ^{14}C technique, they can give very different gas flux estimates (see Section 6.03.2.4.2).

A fourth relationship that has been used recently is a quadratic developed with the aim of comparing *in situ* measurements of k_w derived from deliberate tracer experiments (see Section 6.03.2.3.2(iii)) with the mean estimate derived from ^{14}C (Nightingale *et al.*, 2000b; henceforth N2000). This is simply a statistical best fit to the deliberate tracer data and it was derived with regard to the wind speed variability during measurement periods. The relationship is in reasonable agreement with the ^{14}C mean, given experimental uncertainties. Both WM99 and N2000 assume that n is 0.5.

6.03.2.4.2 Use and comparison of different wind-based parametrizations

The nonlinearity of these parametrizations means that care has to be taken to avoid averaging effects and different relationships have been proposed for short- and long-term wind data (Wanninkhof, 1992; Wanninkhof and McGillis, 1999). Despite this, the relationships are often used indiscriminately by biogeochemists to estimate air–sea fluxes. A study using DMS measurements from a Lagrangian study in the North Sea showed that the use of hourly wind speed data can lead to flux estimates ~3 times those obtained from daily averages (Archer *et al.*, 2002a).

Although the different parametrizations can be compared fairly easily using simulated U, it is more instructive to use real data, preferably on a global scale. One way is to use a year of satellite-derived U measurements combined with all four relationships for k_w and combine this with global CO_2 concentration fields (Boutin *et al.*, 2002). The authors found that the annual estimates of the oceanic uptake of anthropogenic CO_2 from the same Δp_{CO_2} dataset varied from $1.2\ \mathrm{Gt\ C\ yr^{-1}}$ to $2.7\ \mathrm{Gt\ C\ yr^{-1}}$ (see Table 1). Of real interest are the results for W92 and WM99. Although both parametrizations are calibrated with the same global mean k_{CO_2}, the global annual uptake of anthropogenic CO_2 increases from $2.2\ \mathrm{Gt\ C\ yr^{-1}}$ to $2.7\ \mathrm{Gt\ C\ yr^{-1}}$. This compares with an increase of $1.4–2.2\ \mathrm{Gt\ C\ yr^{-1}}$ when using a modeled fit for

Table 1 Estimates of the annual oceanic uptake of carbon dioxide (Gt C yr^{-1}) derived from measurements of carbon dioxide and gas transfer velocity parametrizations.

Parametrization of k_w	*Annual oceanic uptake of carbon dioxide* (Gt C yr^{-1})		
	Boutin et al. (2002)	*Wanninkhof and McGillis (1999)*	*Takahashi et al. (2002)*
LM86	1.2		
W92	2.2	1.4	2.0
WM99	2.7	2.2	3.7
N2000	1.7		

global wind speeds (Wanninkhof and McGillis, 1999). More recently, the same Δp_{CO_2} data set used by Boutin *et al.* (2002) has been combined with the same global wind speeds as Wanninkhof and McGillis (1999) and the use of WM99 instead of W92 has been shown to increase the oceanic sink from 2.0 Gt C yr^{-1} to 3.7 Gt C yr^{-1} (Takahashi *et al.*, 2002). We can draw two conclusions. The first is that the form of the relationship of k_w with U is of prime importance. This is due to a correlation between wind speeds and direction of CO_2 flux, i.e., windy areas tend to be sinks of CO_2 and calm areas tend to be sources of CO_2. The second is the large impact of the different U distributions on the net global CO_2 flux.

Finally, we should caution that all the originators of these parametrizations recognize they have substantial limitations. First is the assumption of values for n. Although a value of 0.5 was measured *in situ* for the $^3He/SF_6$ tracer pair (Nightingale *et al.*, 2000b), this is unlikely to be correct at high U or for other gases, due to solubility effects (see Section 6.03.2.1.6). Indeed, a wind speed based parametrization has been proposed that predicts bubble-enhanced gas exchange from white-cap coverage and includes solubility (Asher and Wanninkhof, 1998). Additionally, we should caution that if surfactants influence k_w to a similar extent as in laboratory experiments, then parametrizations of k_w based on U are unlikely to satisfactorily represent a considerable proportion of the global ocean where there is significant biological activity. Indeed, surface films have been proposed as the cause of some of the scatter amongst the data in Figure 6 (Frew, 1997).

6.03.2.5 Remote Sensing and Estimation of Transfer Velocity

As is clear from several of the earlier sections, estimation of k_w in the field is technically difficult and results are spatially and temporally limited. Some efforts have therefore been made to investigate whether remote-sensing techniques can be applied, with the potential advantage of near-global coverage, as well as the ability to obtain repetitive measurements of a frequency dependent on the orbiting time of the satellites.

The most straightforward example of this approach is the use of scatterometer data to obtain estimates of U that are then combined with a parametrization relating k_w to U (see Section 6.03.2.4.2) to obtain large-scale and repetitive fields of k_w. One of the first applications of this approach was to combine LM86 with wind estimates from the Seasat scatterometer (Etcheto and Merlivat, 1988). The approach was extended by using ERS1 scatterometer wind speeds, to try to assess the importance of chemical enhancement of k_{CO_2} (Boutin and Etcheto, 1995) (see Section 6.03.2.1.2). More recently this technique has been used to derive CO_2 flux estimates for four common parametrizations of k_w (see Section 6.03.2.4.2). A typical map of global annual k_{CO_2} values derived from satellite observations is shown in Figure 8.

Another possible approach uses the laboratory finding that k_w is linearly related to the fractional area of the water surface covered by white caps (Asher *et al.*, 1998). Since it is potentially possible to estimate white-cap coverage from the microwave brightness temperature of the sea surface, which can itself be measured by satellite or aircraft-borne radiometers, this may represent a way forward for the estimation of k_w on the wide range of scales available by using these various platforms.

Finally, we discuss an approach which relies on the relationship discussed earlier between the mean square slope of surface waves and k_w (Glover *et al.*, 2002) (see Section 6.03.2.2.2). Although the mean square wave slope is not a simple parameter to measure at sea, it is possible to obtain mean square wave slopes from microwave backscatter measurements. Given a relationship between mean square wave slope and k_w, it ought to be feasible to obtain k_w from satellite radar sensors. Figure 9 shows the results obtained using this technique by Glover *et al.* (2002). The dual frequency altimeter on the TOPEX satellite is used to obtain k_w over a 6 yr time period (1993–1998) at 9 JGOFS

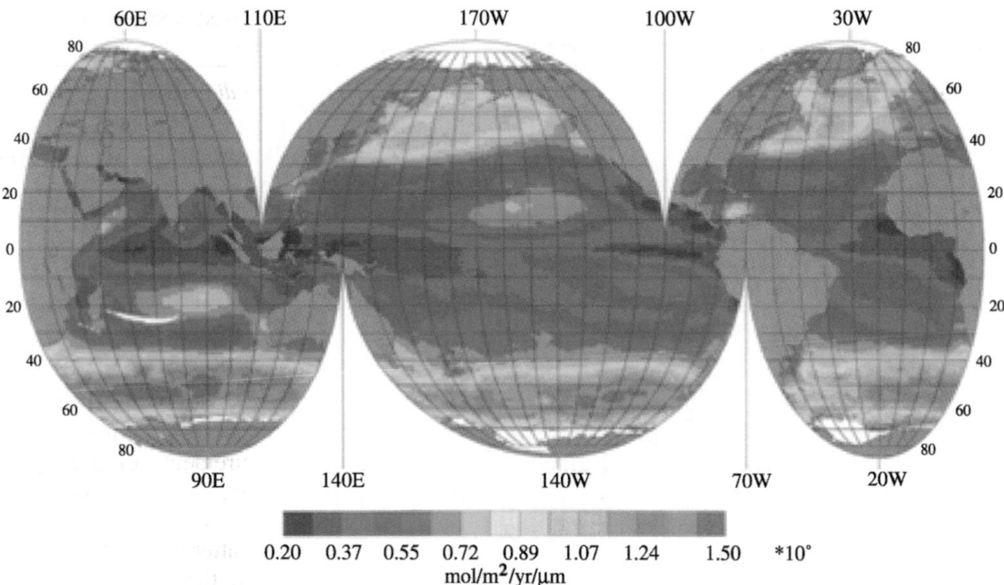

Figure 8 A global map of the annual gas exchange coefficient for carbon dioxide for the year 2001. The map was derived from a combination of wind speeds derived from the QSCAT satellite sensor and the W92 parametrization of k_w. Note that k_w has been corrected for the solubility of CO_2 in this figure (reproduced by permission of Dr. J. Boutin, Laboratorie d'Oceanography Dynamique et de Climatologie (LODYC), Universite Pierre et Marie Curie, Paris, France).

(Joint Global Ocean Flux Study) time-series stations. The results in Figure 9 compare k_w from the TOPEX satellite with the predictive parametrizations of LM86 and W92 (see Section 6.03.2.4.1). The results are in general agreement (which is not necessarily to be expected, since the three estimations are based on very different principles), but the satellite-derived k_w values are more in line with the LM86 relationship than with that of W92 over most of the globe.

6.03.3 THE CYCLING OF GASES IN SEAWATER

As discussed earlier, the flux of gas across the air–sea interface is typically estimated from the product of the concentration difference driving the flux (ΔC) and k_w. So far we have dealt in detail with k_w; the following section covers the ΔC term. Gases of environmental significance are predominantly biogenic in origin but some are also anthropogenic. Concentrations of relevant compounds in seawater and air are generally routine to determine and are not described here. However, as most gases of interest are produced and/or destroyed within the ocean or atmosphere, there is a great deal of spatial and temporal variability in their concentration fields. We briefly discuss the processes that control this variability and include an estimate of the marine contribution to the total atmospheric flux of the gas under discussion. The section is split between those gases for which the ocean acts as a source to the atmosphere, those for which it acts as a sink from the atmosphere, and those for which it can be both.

6.03.3.1 Introduction

Two of the key assumptions of the thin-film model (see Section 6.03.2.1.1) are that the main bodies of air and water are well mixed, i.e., that the concentration of gas at the interface between the thin film and the bulk fluid is the same as in the bulk fluid itself, and that any production or removal processes in the thin film are slow compared to transport across it. It is quite likely that there are near-surface gradients in concentrations of many photochemically active gases. Little research has been published, although the presence of near-surface gradients (10 cm to 2.5 m) in levels of CO during the summer in the Scheldt estuary has been reported (Law *et al.*, 2002). Gradients may well exist for other compounds either produced or removed photochemically, e.g., di-iodomethane, nitric oxide, or carbonyl sulfide (COS). Hence, a key assumption made in most flux calculations that concentrations determined from a typical sampling depth of 4–8 m are the same as immediately below the microlayer may well often be incorrect.

There has been relatively little research into the microlayer, mainly due to the difficulty in defining and sampling it. There seems little doubt that it exists; it can be measured in terms of temperature difference (the skin effect discussed

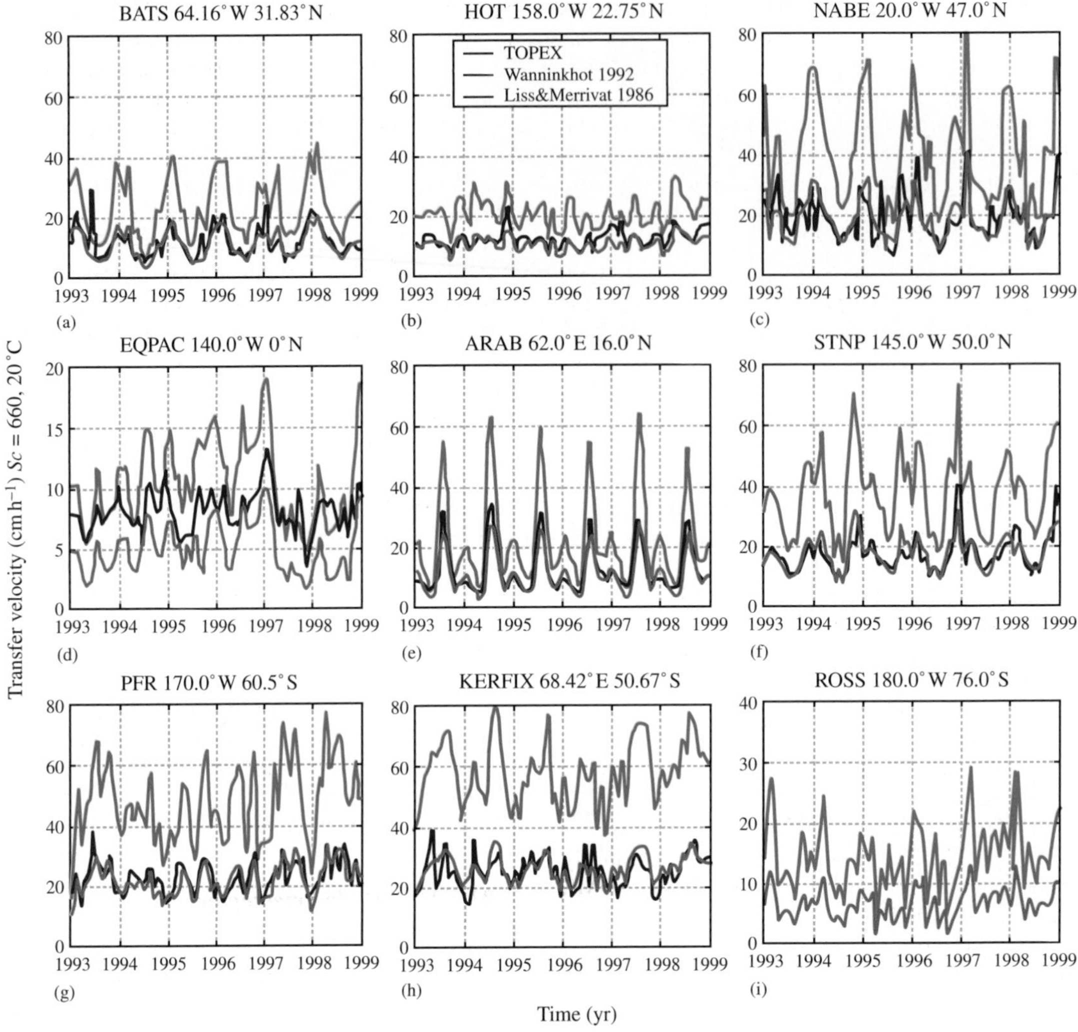

Figure 9 A 6 yr time series of k_w at nine time-series sites around the world. Values derived from the TOPEX algorithm are shown as a blue line, those from W92 as a green line, and those from LM86 as a red line (Glover *et al.*, 2002) (reproduced by permission of American Geophysical Union from *Geophysical Monographs*, **2002**, *127*, 325–332).

in Section 6.03.2.2.4), a distinct biological community (Gladyshev, 1997), and its chemical properties (Hunter, 1997). It is usually operationally defined as the top 1 mm of the ocean surface, although its depth depends in part on the property being studied (i.e., for gases its depth is probably less than 50 μm) and can probably best be thought of as a series of sublayers (Hardy *et al.*, 1997). It is as yet unclear whether there are significant production or removal mechanisms operating in the microlayer for many of the gases discussed here. It is not hard to envisage that the high levels of UV that penetrate into the microlayer may well result in production or removal of photochemically sensitive gases (Plane *et al.*, 1997). There are few data published, although enrichments of DMS in the sea surface microlayer as compared to the bulk water have been reported (Yang *et al.*, 2001).

6.03.3.2 The Oceans as a Source of Gases to the Atmosphere

6.03.3.2.1 Dimethyl sulfide

Sulfur is an element that is essential to life. It had been speculated for many years that there must be a major source of volatile sulfur from the oceans to the land via the atmosphere in order to balance the loss of sulfur from the land via weathering (Eriksson, 1959). Although this source was originally believed to be air–sea transfer of hydrogen sulfide, oceanic measurements subsequently indicated that DMS was the dominant volatile organic sulfur species in the oceans (Lovelock *et al.*, 1972). Indeed, Lovelock had predicted that this compound was the missing link in the biogeochemical cycle of sulfur and he has used the existence of DMS and other methylated compounds produced by marine organisms as evidence in support of the Gaia hypothesis (Lovelock, 1979). DMS is of great interest as one of its oxidation products in the troposphere is sulfur dioxide that can itself be further oxidized to sulfuric acid and then form aerosol sulfate (Plane, 1989). DMS is therefore thought to be an important determinant of atmospheric acidity, particularly in remote areas away from anthropogenic influence (Keene *et al.*, 1998) and may also act as a source of cloud condensation nuclei (Andreae *et al.*, 1995). This link led to the CLAW hypothesis (named after the four authors of Charlson *et al.*, 1987), the idea that phytoplankton may influence global climate. The hypothesis has stimulated much research but remains controversial and has yet to be successfully tested.

DMS is produced from DMSP, a compound that is thought to regulate osmotic pressure in the cells of some species of phytoplankton (Kirst *et al.*, 1991). There may be an antigrazing function for DMSP (Kirst *et al.*, 1991) as the cleavage of DMSP to DMS also produces acrylic acid, a compound believed to have antimicrobial properties (Sieburth, 1960). This cleavage is known to be catalyzed by a DMS lyase enzyme (Steinke *et al.*, 1998). Finally, DMSP may also play a role in polar regions as a cryoprotectant (Kirst *et al.*, 1991). The presence of DMSP in phytoplankton is limited to specific groups of algae, particularly prymnesiophytes (Keller *et al.*, 1989). DMS production is enhanced by microzooplankton grazing (Dacey and Wakeham, 1986), viral infection (Malin *et al.*, 1998), senescence (Nguyen *et al.*, 1988), and bacterial conversion (Kiene and Bates, 1990). Recent detailed process experiments have revealed that DMSP and DMS are both rapidly cycled in the water column by a complex interaction between phytoplankton, microzooplankton, bacteria, and viruses (Archer *et al.*, 2002b). Only ~10% of the total DMS produced (and less than 1.3% of the DMSP produced) was transferred to the atmosphere, with the bulk of the DMS being recycled in the water column or photo-oxidized (Archer *et al.*, 2002a).

DMS is therefore highly variable in both space and time, and it is particularly difficult to characterize global concentration fields. An interpolated global map containing most of the published data on DMS (Kettle *et al.*, 1999) is shown in Figure 10 together with sample points. The highest concentrations are found in coastal upwelling regions (e.g., equatorial Pacific Ocean) and seasonally in high-latitude areas. The Southern Ocean appears to represent a major source of DMS to the atmosphere although, as data from this region are sparse, this conclusion is somewhat tentative. The global marine flux of DMS to the atmosphere has been estimated to be between 15 Tg S yr^{-1} and 33 Tg S yr^{-1} (Kettle and Andreae, 2000), representing ~80% of the global source of DMS (Watts, 2000) and ~50% of the global organic sulfur flux to the atmosphere (see Table 2).

6.03.3.2.2 Methyl mercaptan

Methyl mercaptan (CH_3SH), also known as methanethiol, is also produced from DMSP in a pathway that competes with DMS production and may even be the dominant product

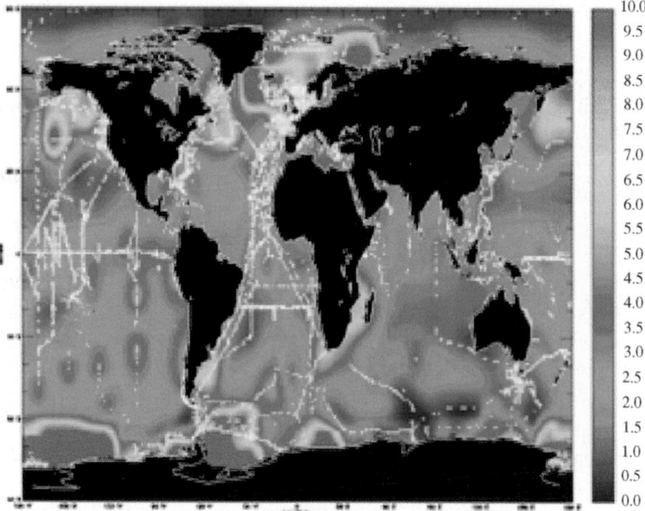

Figure 10 An interpolated map of DMS concentrations. The map is based upon 15,617 DMS measurements from 150 cruises. Concentrations are in nM. Sampling points are represented by white circles indicating that many areas are sparsely sampled (Kettle *et al.*, 1999) (reproduced by permission of American Geophysical Union from *Glob. Biogeochem. Cycles* **1999**, *13*, 399–444).

Table 2 Summary of gases in seawater and the importance of their global atmosphere/ocean flux.

Gas	*Atmospheric role*	*Main production mechanism*	*Net annual flux to the atmosphere*	*% of atmospheric source*
DMS	Cloud formation and acidity	Phytoplankton	15–33 Tg S	80
CH_3SH	Cloud formation and acidity	Phytoplankton	15–33 Tg S	?
COS	Cloud formation and acidity	Photochemistry	−0.1–0.3 Tg	40
CS_2	Cloud formation and acidity	Photochemistry	0.13–0.24 Tg	20–35
H_2S	?	Hydrolysis	1.8 Tg	25
CH_3I	Oxidation capacity	Phytoplankton? Photochemistry?	0.13–0.36 Tg	?
CH_3Cl	Ozone depletion	Phytoplankton? Chemical?	0.2–0.4 Tg	7–14
$CHCl_3$	?	?	0.1–0.35 Tg	25–70
N_2O	Greenhouse gas, ozone depletion	Microbial (de)nitrification	11–17 Tg N	60–90
CH_4	Greenhouse gas, oxidation capacity	Bacteria	15–24 Tg	3–5
CO	Oxidation capacity	Photochemistry	10–650 Tg C	3–20
NMHC	Oxidation capacity	Photochemistry?	2.1 Tg	0.2
Alkyl nitrates	Oxidation capacity	Photochemistry?	~30 Gg	Significant
Oxygenated organics	Oxidation capacity	Photochemistry?	?	?
H_2	Oxidation capacity	Biological photochemistry	1 Tg	5
Mercury	Pollution	Biological	1 Gg	20
Se	Geochemical cycling	Biological	5–8 Gg	45–77
Po	?	Biological?	?	?
CO_2	Greenhouse gas	Respiration	−1.7 +/− 0.5 Pg C	−30
CH_3Br	Ozone depletion	Phytoplankton?	−11 to −20 Gg	−9–17%
$CHBr_3$	Oxidation capacity	Macroalgae	0.22 Tg	70
NH_4	Aerosol formation	Biological	20 Tg	Significant
O_3	Oxidation capacity	NA	−500 Tg	?
SO_2	Aerosol formation	NA	?	?
$HCN + CH_3CN$	Oxidation capacity	NA	−1.3 Tg N	−80
CFCs	Ozone depletion	NA	?	?

Sources are described in the relevant sections in the text. NA = not applicable.

(Kiene *et al.*, 2000). However, CH_3SH is thought to be rapidly removed from the water column due to assimilation into proteins by marine bacteria (Kiene, 1996) and reaction with dissolved organic matter (Kiene *et al.*, 2000). CH_3SH may also have a small photochemical sink in seawater (Flock and Andreae, 1996). Seawater concentrations are assumed to be considerably lower than for DMS, although few measurements have been made (Kiene, 1996). However, recent data indicate that levels of CH_3SH in the Atlantic Ocean ranged from 150 pM to 1,500 pM and were typically ~10% those of DMS (Kettle *et al.*, 2001). Open-ocean values were close to 300 pM, but increased dramatically in upwelling and coastal regions, such that the mean concentration of CH_3SH was as high as 20% that of DMS. In some areas the ratio of CH_3SH to DMS was unity. This would make CH_3SH the second-most dominant volatile sulfur compound in seawater and suggests that CH_3SH ought to be considered in future studies of DMS production or DMSP degradation and in estimates of the flux of biogenic sulfur from the oceans.

6.03.3.2.3 Carbonyl sulfide

COS is produced photochemically from the interaction of UV light with dissolved organic matter (Uher and Andreae, 1997) and its principal loss mechanism in the water column is hydrolysis (Andreae and Ferek, 1992). Typical atmospheric concentrations are ~500 ppt(v), while seawater concentrations vary from 1 pM to 100 pM (Kettle *et al.*, 2001). A diurnal cycle for COS has been reported in surface waters with the ocean acting as a sink of COS late at night and in early morning, but as a source to the atmosphere for the remainder of the day (e.g., Kettle *et al.*, 2001). The presence of significant undersaturations suggests that *in situ* degradation rates must be high. The net flux from the oceans to the air is thought to be ~0.1 Tg COS yr^{-1} (Watts, 2000), although there is considerable uncertainty in this number. A recent modeling exercise suggests that the flux could be between −0.1 Tg COS yr^{-1} and +0.2 Tg COS yr^{-1} (Kettle *et al.*, 2002). As CDOM levels are higher in coastal areas, fluxes from the coastal seas could be up to 0.2 Tg COS yr^{-1} (Watts, 2000), although the data set on which this estimate is based is somewhat sparse. A total flux of 0.3 Tg COS yr^{-1} represents ~40% of the total atmospheric source.

6.03.3.2.4 Carbon disulfide

Carbon disulfide (CS_2) is known to have a photochemical source from CDOM (Xie *et al.*, 1998), although a biological source has also been reported (Xie *et al.*, 1999). CS_2 has no significant sink in the water column other than transfer to the atmosphere, although a small diurnal signal has been observed (Kettle *et al.*, 2001). There are very few published oceanic measurements. Seawater concentrations are typically ~5 pM S, although higher levels (up to 150 pM) have been reported from upwelling areas (Kettle *et al.*, 2001). The global oceanic flux to the atmosphere has been estimated to be 0.13–0.24 Tg CS_2 yr^{-1} (Xie and Moore, 1999), ~20–35% of the total atmospheric source (Watts, 2000).

6.03.3.2.5 Hydrogen sulfide

Hydrogen sulfide (H_2S) was originally thought to be the dominant volatile sulfur compound in the oceans but is now known to make a rather minor contribution to the total marine flux of sulfur to the atmosphere (Watts, 2000). It is produced mainly as a product of the hydrolysis of COS (Elliott, 1989) and from particulate organic material, although there is some evidence for a direct algal source (Andreae *et al.*, 1991). H_2S is rapidly oxidized in surface waters (2–50 h) and exhibits a diurnal cycle with a maximum before dawn (Andreae, 1990). The oceanic source strength has been estimated at 1.8 Tg yr^{-1} ~25% of the total source to the atmosphere (Watts, 2000).

6.03.3.2.6 Methyl iodide

The main source of iodine to land is from the supply of volatile iodine compounds produced in the oceans and transferred to the atmosphere via the air–sea interface (Compos *et al.*, 1996). The supply of these marine iodine compounds to the atmosphere is thought to be important in the oxidation capacity of the troposphere via the production of the iodine oxide (IO) radical (Alicke *et al.*, 1999). The atmospheric chemistry of halogenated compounds is discussed in detail in Chapter 4.02. CH_3I is substantially supersaturated in almost all surface waters and throughout the year (Lovelock, 1975; Moore and Groszko, 1999; Nightingale, 1991; Reifenhauser and Heumann, 1992), although undersaturations have been reported in the Greenland and Norwegian seas during fall (Happell and Wallace, 1996). High levels of CH_3I in coastal seas and in the Southern Ocean have been associated with blooms of the algae *Phaeocystis* (Nightingale, 1991; also Nightingale, 1992, unpublished data). This, together with evidence from laboratory culture experiments that some diatoms release CH_3I (Tokarczyk and Moore, 1994), has led to the supposition that the main source of oceanic CH_3I is production by phytoplankton. However, the presence of supersaturated levels of CH_3I in the tropical Atlantic Ocean, an area of low biological activity, has been used to argue for a photochemical source (Happell and Wallace, 1996). Little is known about possible removal mechanisms in the water column although a temperature-dependent chemical loss has been proposed via hydrolysis, reaction with chloride ions (Cl^-) (Zafiriou, 1975), and photolysis (Zika *et al.*, 1984). The global ocean-to-atmosphere flux has been estimated at between 0.13 Tg yr^{-1} and 0.36 Tg yr^{-1} (Moore and Groszko, 1999). A modeling study has recently predicted that the sea to air flux of CH_3I is 0.21 Tg yr^{-1}, in good agreement with the earlier estimate (Bell *et al.*, 2002). Known terrestrial sources of CH_3I are insignificant.

A series of other halogenated compounds have been identified in seawater, including ethyl iodide, propyl iodide, bromoiodomethane, chloroiodomethane, and di-iodomethane (Carpenter *et al.*, 2000; Klick and Abrahamsson, 1992). Little is known about their production mechanisms. Loss mechanisms are likely to include photolysis and reaction with chloride and hydroxide ions. Information is too limited to be used to derive global fluxes for these compounds, although the data available indicate that a reasonable case can be made that the iodine flux from these compounds is similar to that from CH_3I.

6.03.3.2.7 *Methyl chloride*

Early measurements of CH_3Cl suggested that the ocean was a major source of this compound to the atmosphere (Lovelock, 1975; Singh *et al.*, 1983). Initially, it was thought that the main source of CH_3Cl was via reaction of Cl^- with CH_3I (Zafiriou, 1975). However, this now seems unlikely, given the relatively long reaction time and the two orders of magnitude greater concentrations of CH_3Cl compared to CH_3I (Singh *et al.*, 1983). Although some laboratory cultures of phytoplankton are associated with CH_3Cl production, release rates are too low to explain levels of CH_3Cl in the oceans (Scarratt and Moore, 1998). The main production mechanism of CH_3Cl in the oceans is therefore unknown.

Measurements of CH_3Cl have been reported in the Labrador Sea that were at or below saturation, while warmer waters south of the Gulf Stream and all Pacific samples were consistently supersaturated (Moore *et al.*, 1996a). No correlation was observed with chlorophyll *a*. The authors concluded that earlier estimates of the sea to air flux were overestimates and that there was a global efflux of 0.4–0.6 Tg yr^{-1} to the atmosphere supplied by warm waters, and a high-latitude influx of 0.1–0.3 Tg yr^{-1}. The net global ocean source to the atmosphere was estimated at 0.2–0.4 Tg yr^{-1} compared with a total global source to the atmosphere of 2.8 Tg yr^{-1} (Keene *et al.*, 1999). The dominant source of CH_3Cl to the atmosphere remains unknown (Keene *et al.*, 1999).

6.03.3.2.8 *Chloroform*

Measurements of chloroform ($CHCl_3$) are very rare in seawater. Virtually the only published study of oceanic $CHCl_3$ consists of ~12 measurements in the Atlantic Ocean and Pacific Ocean (Khalil and Rasmussen, 1983. These authors concluded that the annual air–sea flux was ~0.35 Tg $CHCl_3$ yr^{-1}. Nightingale (1991) reported measurements from the southern North Sea on monthly cruises from February through October. The main sources of $CHCl_3$ were coastal and subsequent investigations revealed substantial production by macroalgae (Nightingale *et al.*, 1995). It is assumed that production is related to the presence of haloperoxidase enzymes in algae, but virtually nothing is known about production or removal mechanisms in seawater. Extrapolation of flux estimates from samples in the central area of the North Sea (i.e., away from coastal influences) suggested a lower flux of 0.1 Tg $CHCl_3$ yr^{-1} (Nightingale, 1991) compared to land-based emissions of 0.22 Tg $CHCl_3$ yr^{-1} (Khalil *et al.*, 1999). Emissions from macroalgae appear to be insignificant globally (Nightingale *et al.*, 1995).

6.03.3.2.9 *Nitrous oxide*

Nitrous oxide (N_2O) is a long-lived (120 yr) trace component of the atmosphere (Prinn *et al.*, 1990). It is a climate-active gas as it has a radiative forcing 300 times that of CO_2, although N_2O presently contributes only 5% to the total "greenhouse effect" (Schimel, 1996). N_2O also acts as a source of nitric oxide in the stratosphere and therefore participates in the catalytic removal of ozone (Crutzen, 1970). It is produced as a reaction intermediate in both microbial denitrification and nitrification processes and at greater rates under conditions of low O_2 (Law and Owens, 1990) (see Chapter 6.11 by Emerson and Hedges for more details).

Most of the world's oceans are close to equilibrium with respect to the atmosphere (300 ppbv) and the mean oceanic supersaturation has been calculated at only 4% (Nevison *et al.*, 1995). However, there are significant oceanic areas of N_2O production that include the northwestern

Indian Ocean, the eastern tropical Pacific Ocean, the subtropical North Pacific and some coastal regions (Bange *et al.*, 1996, 2001; Dore *et al.*, 1998; Law and Owens, 1990; Naqvi *et al.*, 2000). It was originally thought the main source of N_2O was in and below the pycnocline where oxygen levels are often reduced. More recently, estimates of diffusive supply of N_2O through the thermocline suggest that there are probably bacterial nitrification processes occurring in the mixed layer that also contribute significantly to the flux of N_2O (Dore and Karl, 1996; Law and Ling, 2001). The total global flux of N_2O to atmosphere has been estimated at ~18 Tg N yr^{-1} (Kroeze *et al.*, 1999) of which the oceans may contribute 2–7 Tg N yr^{-1} (Nevison *et al.*, 1995; Suntharalingam *et al.*, 2000). The total marine source could increase to between 11 Tg N yr^{-1} and 17 Tg N yr^{-1} when coastal areas, particularly estuaries, are included (Bange *et al.*, 1996).

6.03.3.2.10 Methane

CH_4 is another long-lived atmospheric trace gas that is radiatively active, it presently contributes ~15% of the current greenhouse forcing (IPCC, 2001), and is important in the oxidative capacity of the atmosphere (Crutzen, 1991). Typical atmospheric concentrations are ~1,750 ppbv. CH_4 is thought to be formed in seawater by methanogenic bacteria present in anoxic micro-environments, e.g., in zooplankton guts and detritus (Oremland, 1979; Sieburth *et al.*, 1987). CH_4 production is greatest in biologically productive areas. Saturations of up to 200% have been observed in upwelling regions (Owens *et al.*, 1991), but are close to equilibrium in the open oceans. Production appears to dominate at the base of the mixed layer with concentrations as high as 400% saturated (Upstill-Goddard *et al.*, 1999). CH_4 levels up to 20,000% supersaturated have been observed in estuaries and linked to turbidity maxima (Upstill-Goddard *et al.*, 2000). The impact of estuarine sources on global CH_4 fluxes has been estimated at 1–3 Tg C yr^{-1} (Middelburg *et al.*, 2002). The dominant flux of marine methane, 11–18 Tg C yr^{-1} (75%), to the atmosphere is thought to be from shelf regions (Bange *et al.*, 1994), although even this flux is minor compared to anthropogenic fluxes estimated at ~450 Tg C yr^{-1} (IPCC, 2001).

6.03.3.2.11 Carbon monoxide

CO is an important atmospheric gas due to its role in the oxidation capacity of the atmosphere. It is known to be produced photochemically from CDOM (Valentine and Zepp, 1993) and hence its main source is likely to be the upper few meters of the ocean surface. CO also has a marine sink in that it can be oxidized to CO_2 by some microbial groups (Johnson and Bates, 1996), although this process is probably only significant in coastal regions. As with other rapidly cycled gases, CO exhibits a diurnal cycle (Johnson and Bates, 1996). The oceanic flux to the atmosphere has been estimated at anything from 10 Tg yr^{-1} to 240 Tg yr^{-1} (Bates *et al.*, 1995; Cicerone, 1988). Most CO measurements in surface seawater are typically taken from underway measurement systems whose intakes are usually between 4 m and 7 m. Modeling of atmospheric CO concentration profiles close to the sea surface indicates that the oceanic flux could be at the higher end of the estimates above (Springer-Young *et al.*, 1996). Much of the uncertainty is due to poor seasonal and spatial coverage but there may be a systematic underestimation as, since the main production mechanism is photochemical, CO concentrations may be greatest close to the air–sea interface, implying that fluxes have been significantly underestimated (Erickson *et al.*, 2000). Additionally, this compound is likely to have large sources from coastal areas and estuaries where DOC levels are greatest and large supersaturations have been measured (Law *et al.*, 2002). Coastal regions and shelf seas have been tentatively estimated to be a source of 300–400 Tg CO yr^{-1} to the atmosphere (Zou and Jones, 1998), although this is based on a limited number of measurements. Indeed, Law *et al.* (2002) have proposed a total estuarine source of just 0.035–0.095 Tg CO yr^{-1}. Although the marine flux may be small compared to anthropogenic sources, the oceans represent a major source of CO to the marine boundary layer and are therefore likely to have a significant influence on its chemistry (Erickson *et al.*, 2000).

6.03.3.2.12 Non-methane hydrocarbons

Non-methane hydrocarbons (NMHCs such as ethane, ethene, propane, propene, and isoprene) are trace atmospheric constituents that play an important role in both providing a sink for hydroxyl radicals and in controlling ozone concentrations (Donahue and Prinn, 1990). The oceans are known to be a source of NMHCs to the atmosphere, although globally they are significantly smaller than terrestrial sources. However, the main marine-produced NMHCs, ethane and propene, may have an important local impact on atmospheric photochemistry (Plass-Dulmer *et al.*, 1995), particularly in ocean areas far from land. The production mechanisms for NMHCs in seawater are poorly understood, although there is some evidence that the alkenes are produced by photochemical reaction with dissolved organic matter in surface waters (Ratte *et al.*, 1993). In contrast, the highly reactive NMHC, isoprene, has been shown to exhibit a significant correlation with chlorophyll content of seawater, so that its emission from seawater is strongly seasonally dependent (Broadgate *et al.*, 1997). It seems likely that there are distinct production processes for individual NMHCs or groups of them. It has been speculated that marine algae may use NMHCs as signaling compounds (Steinke *et al.*, 2002).

6.03.3.2.13 Alkyl nitrates

Like the NMHCs, alkyl (methyl, ethyl, and propyl) nitrates are important compounds in the oxidant chemistry of the atmosphere. In particular, they are a significant component of its "odd nitrogen" chemistry, which plays an important role in regulating levels of tropospheric ozone in marine areas, particularly those remote from land. There is evidence from measurements in the marine troposphere (Atlas *et al.*, 1993; Blake *et al.*, 1999), a coastal site on Antarctica (Fischer *et al.*, 2002; Jones *et al.*, 1999) as well as from an Antarctic ice core (McIntyre, 2001) that a significant oceanic source of these compounds must exist. This idea is supported by measurements of the oceanic emission flux of alkyl nitrates to the atmosphere (Chuck *et al.*, 2002). In this paper the authors give the results of measurements of methyl and ethyl nitrates in seawater and air made on two research cruises spanning a large latitudinal band of the North and South Atlantic. Surface waters were supersaturated with respect to atmospheric concentrations over large areas of the Atlantic, with supersaturations reaching up to 800% in equatorial waters, thus confirming the oceanic source. The authors present results from a simple box model that suggest the equatorial marine source is likely to be a major component of the regional atmospheric alkyl nitrate budget. There is little evidence available for the mode of formation of alkyl nitrates in the oceans with both biological and photochemical processes being invoked (Chuck *et al.*, 2002; Dahl *et al.*, 2003).

6.03.3.2.14 Oxygenated organic compounds

The oxygenated organics comprise compounds such as low molecular weight carbonyls, alcohols, and peroxides. As with the NMHCs and alkyl nitrates discussed above, these compounds play an important role in controlling the oxidizing capacity and ozone-forming potential of the atmosphere. There are major sources of these compounds from terrestrial sources (Kesselmeier and Staudt, 1999), although substantial marine emissions have to be invoked to explain their observed concentrations in marine air over, for example, the tropical Pacific Ocean (Singh *et al.*, 2001). There is evidence from laboratory experiments for the photochemical formation of some of these compounds (e.g., low molecular weight carbonyls) from organic precursors, with highly elevated concentrations reported for the sea surface microlayer (Zhou and Mopper, 1997), and for the production of acetone by marine microbes (Nemecek-Marshall *et al.*, 1995). However, as with the NMHCs and alkyl nitrates discussed earlier, our knowledge of how these compounds are formed in the oceans is rudimentary.

6.03.3.2.15 Hydrogen

Molecular hydrogen (H_2) is rather stable in the atmosphere with a residence time of 6–8 yr. It is formed in surface ocean waters by microbial processes, so that surface seawater is generally found to be supersaturated by 2–3 times with respect to the concentration of the gas in the overlying atmosphere (Herr and Barger, 1978). These concentrations imply a flux from sea to air globally of $\sim 10^{12}$ g yr^{-1}, which accounts for only a few percent of estimates of the total global H_2 production rate, most of which is thought to be of anthropogenic origin (Schmidt, 1974). As is implied by the early publication dates of the references given above, little or no research on this topic appears to have been carried out in recent years.

6.03.3.2.16 Mercury

Oceanic volatile forms of mercury include mercury metal (Hg^0) and to a lesser extent the fully di-methylated form (Me_2Hg). There is evidence that these species are formed biologically (Fitzgerald, 1989) and, in the case of Hg^0, by photochemical processes in seawater (Costa and Liss, 1999). Their emission from marine waters constitutes a significant component of the total global cycling of mercury (Fitzgerald, 1989; Lamborg *et al.*, 2002). This is important both biogeochemically and from the pollution perspective, since volatilization from seawaters will not only decrease the lifetime of mercury in the water column, but will also bring about its wider geographical dispersion to other environments.

The latter means that mercury from pollution sources transported to the oceans by rivers and through the atmosphere can be recycled back to terrestrial environments.

6.03.3.2.17 Selenium and polonium

There are considerable similarities between the oceanic (bio)geochemical behavior of the elements sulfur (see Section 6.03.3.2.1) and selenium. This is not surprising in view of their proximity as adjacent members of group VIb of the periodic table. For example, both elements require a significant source of a volatile form(s) to be emitted from the oceans to the atmosphere in order for their geochemical budgets to achieve balance (Mosher and Duce, 1987). In the case of sulfur the carrier molecule is DMS, although, until recently, the equivalent volatile form of selenium was unknown. However, measurements of volatile selenium species in samples from the North Atlantic have been reported (Amouroux *et al.*, 2001). The dominant volatile forms were dimethyl selenide (DMSe—the direct analogue of DMS) and the mixed S/Se compound dimethyl selenyl sulfide (DMSeS). Amouroux *et al.* calculate that there is more than enough flux of volatile selenium from the global oceans to the atmosphere to balance the geochemical budget of this element. The mode of formation of these volatile selenium species is assumed to be by analogous routes to those discussed earlier (see Section 6.03.3.2.1) for DMS which is supported by the positive correlation found by Amouroux *et al.* between the seawater concentrations of DMSe and DMS. Once emitted to the atmosphere, DMSe (and DMSeS) are likely to be subject to similar atmospheric transformations as described previously for DMS, although the four orders of magnitude smaller size of the selenium compared with the sulfur sea-to-air flux means that their importance for atmospheric properties will be insignificant. However, the recycling of selenium from the sea to land via the atmosphere may have important implications for human health (Rayman, 2000) in a cycle analogous to that already discussed for iodine (see Section 6.03.3.2.6).

Extension of the group VIb similarity to later members of the group (tellurium and polonium) is problematical. There are measurements of polonium isotopes in aerosols measured at a coastal site with both onshore and offshore winds (Kim *et al.*, 2000), where the higher "excess" ^{210}Po in samples from air masses which have traveled over the oceans is attributed to volatilization of polonium from the sea surface. This could be by a process analogous to that already described for sulfur and selenium, although the evidence is only circumstantial. In the case of tellurium, there are no known marine measurements of volatile forms of the element due to its extremely low concentration in seawater, although a similar cycle could well exist for it.

6.03.3.3 The Oceans as a Source and a Sink of Volatile Compounds

6.03.3.3.1 Carbon dioxide

It is well known that there has been an increase in atmospheric CO_2 from ~280 ppm in 1800 (and several thousand years before) to ~370 ppm at the present time. This increase is due to a combination of fossil fuel burning and deforestation that presently represents a supply of CO_2 to the atmosphere equivalent to ~7 Gt C yr^{-1} (IPCC, 2001). The observed annual increase in atmospheric CO_2 represents ~3.2 Gt C, the balance being removed from the atmosphere and taken up by the oceans and land. The ocean sink of this anthropogenic CO_2 has been estimated to be 1.7 ± 0.5 Gt C yr^{-1} (IPCC, 2001).

The oceans are strongly buffered with respect to CO_2 and much of the anthropogenic CO_2 absorbed by the oceans is converted to other forms of dissolved inorganic carbon (DIC). At a typical seawater pH of 8, the dominant DIC species is HCO_3^- with only 1% in the form of dissolved CO_2. The long-term sink for anthropogenic CO_2 is therefore thought to be the ocean via transfer from the atmosphere (Archer *et al.*, 1997). The rate-limiting step in the long-term oceanic uptake of anthropogenic CO_2 is not air–sea gas exchange, but the mixing of the surface waters with the deep ocean (Sarmiento and Sundquist, 1992).

Large, multinational research programs have tried to determine CO_2 levels across the world's oceans and in different seasons in order to try to quantify the ΔC term of Equation (5) so that an estimate of the global air–sea flux can be derived (see Figure 11 based on 9.4×10^5 measurements complied by Takahashi *et al.* (1997) and extended in Takahashi *et al.* (2000). The transfer velocity term is derived from a combination of U-based parametrizations and global maps of U (see Section 6.03.2.4.1 and Figure 8). Estimates of the oceanic uptake of anthropogenic CO_2 using these techniques vary from 0.6 Gt C yr^{-1} to 3.7 Gt C yr^{-1}, much of this variability depending on the parametrization of k_w employed in the calculation and on the wind-speed distribution (see Section 6.03.2.4.2 and Table 1). This approach has previously tended to give lower uptake estimates than those derived from

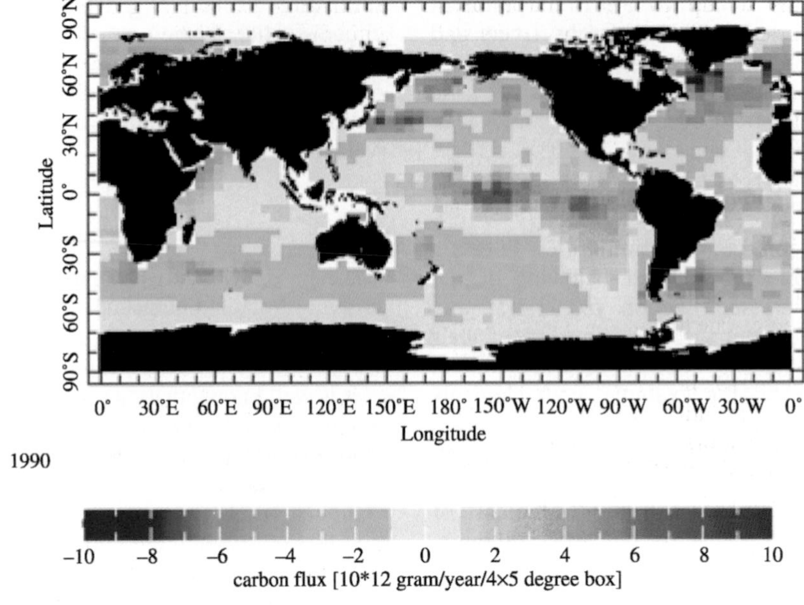

Figure 11 The annual flux of carbon dioxide between the oceans and the atmosphere (Takahashi *et al.*, 1997) (reproduced by permission of National Academy of Sciences, USA from *Proc. Natl. Acad. Sci. USA*, **1997**, *94*, 8292–8299).

global circulation models and atmospheric isotopic measurements (Battle *et al.*, 2000; Keeling *et al.*, 1996; Sarmiento *et al.*, 2000), but following the use of enhanced Δp_{CO_2} data sets (Takahashi *et al.*, 2002) there is now generally good agreement, except when the parametrization of WM99 is used.

There is also a large natural annual flux of CO_2 between the ocean and the atmosphere of almost 90 Gt C yr^{-1} that, pre-1800, was believed to be almost in balance. There was probably a net flux from ocean to atmosphere of ~0.6 Gt C yr^{-1} that balanced the supply of DIC to the oceans via rivers (Sarmiento and Sundquist, 1992). This huge influx and efflux is due to a combination of marine productivity (the biological pump) and ocean circulation (the solubility pump). The biological pump is discussed in detail by de la Rocha in Chapter 6.04. Ocean circulation also results in air–sea exchange of CO_2 as the solubility of CO_2 is temperature dependent. Warming decreases the solubility such that CO_2 becomes supersaturated with respect to the atmosphere and there is a net transfer of CO_2 to the atmosphere, whereas cooling results in a flux from the atmosphere to the ocean. Almost all of the *anthropogenic* CO_2 is thought to be taken up by the solubility pump (IPCC, 2001), as CO_2 availability does not normally limit biological productivity in the world's oceans. Indeed, a recent modeling study has indicated that including present biology in oceanic global circulation models increases the oceanic sink of anthropogenic CO_2 by only 4.9% (Orr *et al.*, 2001).

However, the observation that the net uptake of *anthropogenic* CO_2 is only ~2% of the CO_2 cycled annually across the air–sea interface ought to be of major concern as this suggests that the net flux is extremely sensitive to changes in the cycling of CO_2. Any changes in ocean circulation or the biogeochemistry of the mixed layer could have a major impact on the magnitude (or even sign) of the present sink of CO_2 and hence on the Earth's climate. For example, the flux of anthropogenic CO_2 into the oceans will reduce the pH of surface waters. Changes in pH have been shown to affect biogenic calcification (Riebesell *et al.*, 2000) and could affect the community structure within the water column, thus altering the biological pump.

6.03.3.3.2 Methyl bromide

CH_3Br is a toxic compound commonly used in agriculture as a fumigant whose use is currently limited by international agreement due to its role on stratospheric ozone destruction (Butler, 1995). CH_3Br is removed by reaction with the hydroxyl radical in the troposphere but the reaction is slow enough that a significant fraction is mixed into the stratosphere (Solomon, 1992). Indeed, CH_3Br is thought to be the main source of bromine to the stratosphere where it can participate in the catalytic removal of ozone (Schauffler *et al.*, 1993). Bromine is thought to be 40–100 times more efficient than chlorine in removing ozone (Penkett *et al.*, 1995). Typical marine atmospheric concentrations of CH_3Br are ~10–15 pptv.

Super-saturations of up to 300% have been observed in the northern Atlantic and in coastal regions (Baker *et al.*, 1999; King *et al.*, 2000), although undersaturations have been measured in the tropical Atlantic Ocean and Pacific Ocean (King *et al.*, 2000; Lobert *et al.*, 1995) and the Southern Ocean (Lobert *et al.*, 1997). CH_3Br is known to have significant loss mechanisms in the water column via reaction with Cl^- and biological uptake. The former is strongly temperature dependent (King and Saltzman, 1997), while the latter is thought to be due to consumption by some classes of marine bacteria (Goodwin *et al.*, 1997). The source(s) of CH_3Br in surface waters are not well known but a correlation has been established with pigment indicators of prymnesiophytes, a class of phytoplankton (Baker *et al.*, 1999). Laboratory incubation studies have also shown that some other species of phytoplankton release CH_3Br but at rates that were too low to support the observed oceanic concentrations (Scarratt and Moore, 1996). However, studies into the cycling of CH_3Br in the water column are still in their infancy. Present research suggests that the ocean acts as a net sink for atmospheric CH_3Br of 11–20 Gg yr^{-1} (King *et al.*, 2000).

6.03.3.3.3 Bromoform

Early measurements of bromoform, or tribromomethane ($CHBr_3$), from the Arctic Ocean suggested that the main sources of this compound were coastal (Dyrssen and Fogelqvist, 1981; Fogelqvist, 1985). Although there is evidence of significant production in the open Arctic Ocean away from coastal influence (Krysell, 1991), subsequent work in the southern North Sea (Nightingale, 1991) and the Atlantic Ocean (Moore and Tokarczyk, 1993) has indicated that the main sources of this compound are indeed coastal. Laboratory incubation and rock-pool experiments have shown that macroalgae are prolific in their release of polybrominated compounds and $CHBr_3$ in particular (Geschwend *et al.*, 1985; Manley *et al.*, 1992; Nightingale *et al.*, 1995). The main production mechanism is believed to be due to a haloperoxidase enzyme that is able to convert bromide ions into hypobromous acid (Wever, 1988). The presence of this enzyme may be related to oxidative stress (Mtolera *et al.*, 1996). Laboratory incubations of phytoplankton cultures have indicated that some marine diatoms produce $CHBr_3$ (Moore *et al.*, 1996b). Release was linked to the presence of haloperoxidase enzymes, but the production of $CHBr_3$ by microalgae does not appear to be common. A recent reassessment of the marine $CHBr_3$ budget has concluded that macroalgae are responsible for a total global source strength estimate of 2.2×10^{11} g $CHBr_3$ yr^{-1} (Carpenter and Liss, 2000). This is ~70% of the total source of $CHBr_3$ to the atmosphere. Open-ocean areas may well act as a sink for $CHBr_3$ (Nightingale, 1991).

6.03.3.3.4 Ammonia and methylamines

NH_3 and to a lesser extent mono-, di-, and trimethylamines are the only significant gaseous bases in the atmosphere, and there has been considerable interest in whether the oceans are a source or sink of these gases. Early attempt to assess the air–sea flux from concentration measurements are probably suspect because of the ease with which sample contamination can occur during laboratory processing and analysis. It should be noted here that due to its high solubility (low value of Henry's law constant), the air–water transfer of NH_3 (and the methylamines for the same reason) is under gas phase control (see Section 6.03.2.1.1). The first reliable measurements were probably from the North and South Pacific and indicated that the flux of NH_3 from sea to air is of a size similar to that for emission of DMS (Quinn *et al.*, 1990, 1988). Indeed, the authors showed that this similarity was mirrored in the molar ratio of (non-sea-salt) sulfate to ammonium (1.3 ± 0.7) in atmospheric aerosol particles collected on the cruise, indicating that for clean marine air remote from terrestrial sources, the emission of DMS and NH_3 from the sea appears to control the composition of the aerosol.

Gibb *et al.* (1999) have reported shipboard measurements in the Arabian Sea confirming the emission of NH_3 from these waters. These authors also measured methylamines in air and seawater and found that seawater concentrations of NH_3 were 10–100 times greater than those of the methylamines, of which monomethylamine was the most abundant. They found that whilst inshore waters acted as sources and sinks for methylamines, offshore waters were a consistent sink.

6.03.3.4 The Oceans as a Sink for Atmospheric Gases

6.03.3.4.1 Ozone

The oceans acts as a one-way sink for atmospheric ozone due to the high reactivity of the gas with components in the surface water. The dominant reaction at the surface appears to be with iodide ions and unidentified organic surfactants (Garland *et al.*, 1980). Given that this reactivity makes the sea a perfect sink, calculation of the downward flux involves only knowledge of the atmospheric concentration and the appropriate deposition velocity. Early estimates of the deposition flux give values of $\sim 5 \times 10^{14}$ g yr^{-1}. More recent estimates are in agreement with this. It has proved possible to check these values using the micrometeorological technique of eddy correlation (see Section 6.03.2.3.2(iv)) applied from an aircraft, which appears to work well in this case (Lenschow *et al.*, 1982).

6.03.3.4.2 Sulfur dioxide

Like ozone, sulfur dioxide is subject to deposition into the oceans, with no re-emission. This arises from the high reactivity of the gas in seawater, which ensures its rapid destruction in the water and effective zero surface-water concentration driving the one-way flux (Liss, 1971). The high solubility and aqueous reactivity of SO_2 makes its exchange subject to gas phase control (see Section 6.03.2.1.1).

6.03.3.4.3 Hydrogen cyanide and methyl cyanide

Hydrogen cyanide (HCN) and methyl cyanide (CH_3CN) are trace gases in the atmosphere occurring at the 100–200 pptv level. Their main source is biomass burning, with smaller contributions from automobiles and industry. Concentration measurements over the oceans show lower amounts in the marine boundary layer, and this has been attributed to uptake by the oceans (Singh *et al.*, in press). These authors have used a simple box model to try to quantify the uptake and deduced that ocean surface waters have to be $\sim$20% undersaturated to achieve balance. Since there are no measurements of HCN or CH_3CN in seawater, it is currently not possible to verify these undersaturations, nor is anything known about the destruction processes for the gases in seawater. The box model suggests a deposition to the global ocean of 1.3 Tg N yr^{-1} HCN and CH_3CN, which is $\sim$4% of the total yearly amount of nitrogen entering the oceans via atmospheric deposition (mainly in the form of nitrate and ammonium in rain and particles).

6.03.3.4.4 Synthetic organic compounds

Synthetic organic compounds are a vast group of man-made chemicals; here we consider only a small subset including the polychlorinated biphenyls (PCBs) and various chlorinated organic pesticides (e.g., DDT, chlordane, and dieldrin), for which there are particular environmental concerns. They are emitted to the atmosphere during use or disposal and are found dispersed throughout the environment. Deposition to the oceans occurs by both wet and dry processes, in varying proportions for the different compounds, although gaseous deposition is always a significant route (25–85% of total) (Duce *et al.*, 1991). Re-emission is also possible where the concentration gradient changes sign, either because of reduced air concentrations or elevated water concentrations, or both. From the point of view of gas exchange, many of these compounds are interesting since both gas and liquid phase resistances are significant for their air–sea exchange, which is not the situation for most gases where one or other resistance is dominant (see Section 6.03.2.1.1).

6.03.3.4.5 Chlorofluorocarbons

Measurements of methyl chloroform or 1,1,1-trichloroethane (CH_3CCl_3) showed that this compound was significantly unsaturated in the equatorial Pacific Ocean (Butler *et al.*, 1991). Loss rates were supported roughly by known hydrolysis rates and the authors calculated that $\sim$6% of atmospheric CH_3CCl_3 is removed by consumption in the oceans. With this exception, most of the chlorofluorocarbons were originally thought to be stable in the water column.

The first evidence that this assumption was incorrect was provided by observations of carbon tetrachloride (CCl_4) removal in the Baltic Sea under anoxic conditions (Krysell *et al.*, 1994). A later investigation in the Black Sea found that reductions in CCl_4, $CHCl_3$, CH_3CCl_3, dibromomethane and dibromochloromethane, and bromodichloromethane were related to oxygen/hydrogen sulfide concentrations (Tanhua *et al.*, 1996). Most of the CCl_4 was transformed to $CHCl_3$ as an intermediate product. Subsequent work in a fjord in Norway showed that CFC-11 was also removed in anoxic waters (Shapiro *et al.*, 1997). Loss rates of both CCl_4 and F11 in anoxic waters are probably due to biological rather than chemical removal (Lee *et al.*, 1999). It also seems likely that some of the chlorofluorocarbons are removed in fully oxygenated surface waters. Observations show that there is a deficit of CCl_4 in the Antarctic surface and bottom waters (Meredith *et al.*, 1996). Finally, fluorinated compounds such as CFC-113 are degraded in warm surface waters of the temperate North Atlantic, the tropical western Pacific, the Eastern Mediterranean, and even the Weddell Sea (Roether *et al.*, 2001). CFC-113 depletions were $\sim$3% yr^{-1}, with possibly accelerated rates in the mixed layer or near the surface.

Selected dechlorination of chlorinated compounds by soil bacteria has long been recognized

(Vogel *et al.*, 1987). It seems likely to us that there is a biological transformation of these compounds by marine bacteria, particular as marine bacteria can transform CH_3Br (see Section 6.03.3.3.2). Not only are these compounds likely to be removed from oceanic and coastal waters under anoxic and suboxic conditions, but given that compounds such as CH_4 and N_2O are thought to be produced in suboxic micro-environments within the water column (see Section 6.03.3.2.9), it seems reasonable to assume that the same sites might be areas of chlorofluorocarbon removal.

6.03.4 POSSIBLE EFFECTS OF CLIMATE CHANGE ON GASES IN THE OCEANS

In this brief and somewhat speculative section, we review some potential effects of climate and other global changes on the exchange of trace gases between the atmosphere and oceans. First the effects of climate change on k_w are discussed, followed by some examples of how trace gas concentration fields and ocean/atmosphere fluxes may be affected.

6.03.4.1 Effects on Air–Sea Gas Transfer

The factors controlling k_w have been discussed in detail in Section 6.03.2. Wind is clearly the main driver of k_w, either directly or indirectly via its role in the formation of waves and bubbles. Thus, in order to predict how change in climate may alter k_w, knowledge of how the wind field may vary is a prerequisite. Models of a warmer world show greater warming at high (particularly northern) latitudes than at lower latitudes (IPCC, 2001), and since it is the equator–pole temperature gradients which ultimately drive the wind systems, warming should result in reduced windiness and thence lower k_w. This may be counteracted, to some extent, by the possibility of greater storminess in the tropics (Knutson and Tuleya, 1999). However, better prediction of k_w in the future (or the past) is arguably more dependent on improved knowledge of winds and storminess than it is on the current uncertainties in the relationship between k_w and wind (see Section 6.03.2.4). A second factor that could affect k_w is that of temperature change on the physical terms D, ν, and H (see Section 6.03.2). The temperature dependence of these terms is reasonably well understood and can be straightforwardly incorporated into predictive models as required. The effects are likely to be quite small in comparison with other changes and uncertainties. Prediction of the role of surface films in affecting k_w is difficult in part because of our present lack of understanding of their importance in the field (see Section 6.03.2.2.5), but also because any temporal change is likely to be driven by the availability of suitable surface-activity organic material, which in turn will be dependent on levels of marine biological activity.

6.03.4.2 Effects on Dissolved Gas Concentrations

Attempts to quantify changes in gas fluxes under altered climates are often undertaken using coupled atmosphere–ocean models. These often start with a purely physical approach. Our ability to incorporate biological processes and, even more so, changes to them is currently rudimentary. Early attempts have shown the effects of climate change on CO_2 uptake to be potentially significant and often apparently contradictory (Sarmiento *et al.*, 1998; Sarmiento and LeQuere, 1996). A recent coupled model for DMS emissions incorporating a biological scheme, and with "normal" and doubled atmospheric CO_2, shows that the changes due to climate are maybe quite small for this gas on average, but with significant regional excursions from the mean, both positive and negative (Bopp *et al.*, 2003).

However, even the most sophisticated model available does not incorporate many important factors (themselves subject to climatic and other global changes) that could cause major alterations in trace gas formation (and destruction) in the oceans. For example, altered solar radiation entering the oceans will potentially affect the concentrations of those trace gases (such as COS—see Section 6.03.3.2.3) for which photochemistry is established as a significant mode of formation. Elevated CO_2 uptake by the oceans will itself lead to significant decrease in the pH of surface waters, which is likely to adversely affect phytoplankton, particularly those with calcium carbonate structures (Riebesell *et al.*, 2000). The effect of such changes on trace-gas production is presently unknown, but is likely to be significant. Inputs of iron and nitrogen from the atmosphere, or deeper waters, affect biological activity in the surface oceans and may affect trace-gas fluxes, e.g., DMS (Turner *et al.*, 1996). Again, the impact of such change on trace-gas formation and destruction is largely unknown but is the subject of increasing attention, e.g., in the SOLAS (Surface Ocean–Lower Atmosphere Study) project (www.solas-int.org).

REFERENCES

Alicke B., Hebestreit K., Stutz J., and Platt U. (1999) Iodine oxide in the marine boundary layer. *Nature* **397**, 572–573.

Amouroux D., Liss P. S., Tessier E., Hamren-Larsson M., and Donard O. F. X. (2001) Role of oceans as biogenic sources of selenium. *Earth Planet. Sci. Lett.* **189**, 277–283.

Andreae M. O. (1990) Ocean–atmosphere interactions in the global biogeochemical sulfur cycle. *Mar. Chem.* **30**, 1–29.

Andreae M. O. and Ferek R. J. (1992) Photochemical production of carbonyl sulfide in seawater and its emission to the atmosphere. *Global Biogeochem. Cycles* **6**, 173–175.

Andreae M. O., Elbert W., and Demora S. J. (1995) Biogenetic sulfur emissions and aerosols over the tropical South-Atlantic: 3. Atmospheric dimethylsulfide, aerosols and cloud condensation nuclei. *J. Geophys. Res.* **100**, 11335–11356.

Andreae T. W., Cutter G. A., Hussain N., Radford-Knoery J., and Andreae M. O. (1991) Hydrogen-sulfide and radon in and over the western North-Atlantic Ocean. *J. Geophys. Res.* **96**, 18753–18760.

Archer D., Kheshgi H., and MaierReimer E. (1997) Multiple timescales for neutralization of fossil fuel CO_2. *Geophys. Res. Lett.* **24**, 405–408.

Archer S. D., Gilbert F. J., Nightingale P. D., Zubkov M. V., Taylor A. H., Smith G. C., and Burkill P. H. (2002a) Transformation of dimethylsulphoniopropionate to dimethyl sulphide during summer in the North Sea with an examination of key processes via a modelling approach. *Deep-Sea Res. II* **49**, 3067–3101.

Archer S. D., Smith G. C., Nightingale P. D., Widdicombe C. E., Tarran G. A., Rees A. P., and Burkill P. H. (2002b) Dynamics of particulate dimethylsulphoniopropionate during a Lagrangian experiment in the northern North Sea. *Deep-Sea Res. II* **49**, 2979–2999.

Asher W. E. and Wanninkhof R. (1998) The effect of bubble-mediated gas transfer on purposeful dual-gaseous tracer experiments. *J. Geophys. Res.* **103**, 10555–10560.

Asher W., Wang Q., Monahan E. C., and Smith P. M. (1998) Estimation of air–sea gas transfer velocities from apparent microwave brightness temperature. *Mar. Tech. Soc. J.* **32**, 32–40.

Atlas E., Pollock W., Greenberg J., Heidt L., and Thompson A. M. (1993) Alkyl nitrates, nonmethane hydrocarbons, and halocarbon gases over the equatorial Pacific-Ocean during Saga-3. *J. Geophys. Res.* **98**, 16933–16947.

Baker J. M., Reeves C. E., Nightingale P. D., Penkett S. A., Gibb S. W., and Hatton A. D. (1999) Biological production of methyl bromide in the coastal waters of the North Sea and open ocean of the northeast Atlantic. *Mar. Chem.* **64**, 267–285.

Bange H. W., Bartell U. H., Rapsomanikis S., and Andreae M. O. (1994) Methane in the Baltic and North seas and a reassessment of the marine emissions of methane. *Global Biogeochem. Cycles* **8**, 465–480.

Bange H. W., Rapsomanikis S., and Andreae M. O. (1996) Nitrous-oxide in coastal waters. *Global Biogeochem. Cycles* **10**, 197–207.

Bange H. W., Andreae M. O., Lal S., Law C. S., Naqvi S. W. A., Patra P. K., Rixen T., and Upstill-Goddard R. C. (2001) Nitrous oxide emissions from the Arabian Sea: a synthesis. *Atmos. Chem. Phys.* **1**, 61–71.

Bates T. S., Kelly K. C., Johnson J. E., and Gammon R. H. (1995) Regional and seasonal-variations in the flux of oceanic carbon-monoxide to the atmosphere. *J. Geophys. Res.* **100**, 23093–23101.

Battle M., Bender M. L., Tans P. P., White J. W. C., Ellis J. T., Conway T., and Francey R. J. (2000) Global carbon sinks and their variability inferred from atmospheric O_2 and delta C13. *Science* **287**, 2467–2470.

Bell N., Hsu L., Jacob D. J., Schultz M. G., Blake D. R., Butler J. H., King D. B., Lobert J. M., and Maier-Reimer E. (2002) Methyl iodide: atmospheric budget and use as a tracer of marine convection in global models. *J. Geophys. Res.* **107** article no. 4340.

Berger R. and Libby W. F. (1969) Equilibration of atmospheric carbon dioxide with seawater: possible enzymatic control of the rate. *Science* **164**, 1395–1397.

Blake N. J., Blake D. R., Wingenter O. W., Sive B. C., Kang C. H., Thornton D. C., Bandy A. R., Atlas E., Flocke F., Harris J. M., and Rowland F. S. (1999) Aircraft measurements of the latitudinal, vertical, and seasonal variations of NMHCs, methyl nitrate, methyl halides, and DMS during the First Aerosol Characterization Experiment (ACE 1). *J. Geophys. Res.* **104**, 21803–21817.

Bopp L., Aumont O., Belviso S., and Monfray P. (2003) Potential impact of climate change on marine dimethyl sulfide emissions. *Tellus* **55B**, 11–22.

Boutin J. and Etcheto J. (1995) Estimating the chemical enhancement effect on the air–sea CO_2 exchange using the ERS-1 scatterometer wind speeds. In *Air–Water Gas Transfer* (eds. B. Jahne and E. C. Monahan). AEON Verlag and Studio, Hanau, pp. 827–841.

Boutin J., Etcheto J., Merlivat L., and Rangama Y. (2002) Influence of gas exchange coefficient parameterisation on seasonal and regional variability of CO_2 air–sea fluxes. *Geophys. Res. Lett.* article no. 1182.

Broadgate W. J., Liss P. S., and Penkett S. A. (1997) Seasonal emissions of isoprene and other reactive hydrocarbon gases from the ocean. *Geophys. Res. Lett.* **24**, 2675–2678.

Broecker H. C. and Siems W. (1984) The role of bubbles for gas transfer from water to air at higher wind speeds. Experiments in the wind-wave facility in Hamburg. In *Gas Transfer at Water Surfaces* (eds. W. Brutsaert and G. H. Jirka). Reidel, Dordrecht, pp. 229–236.

Broecker H. C., Petermann J., and Siems W. (1978) The influence of wind on CO_2 exchange in a wind wave tunnel, including the effects of monolayers. *J. Mar. Res.* **36**, 595–610.

Broecker W. and Peng T.-H. (1974) Gas exchange rates between air and sea. *Tellus* **26**, 21–35.

Broecker W. S., Ledwell J. R., Takahashi T., Weiss R., Merlivat L., Peng T. H., Jahne B., and Munnich K. O. (1986) Isotopic versus micrometeorologic ocean CO_2 fluxes—a serious conflict. *J. Geophys. Res.* **91**, 517–527.

Businger J. A. and Oncley S. P. (1990) Flux measurement with conditional sampling. *J. Atmos. Ocean. Technol.* **7**, 349–352.

Butler J. H. (1995) Ozone depletion—methyl bromide under scrutiny. *Nature* **376**, 469–470.

Butler J. H., Elkins J. W., Thompson T. M., Hall B. D., Swanson T. H., and Koropalov V. (1991) Oceanic consumption of CH_3CCl_3: implications for tropospheric OH. *J. Geophys. Res.* **96**, 22347–22355.

Campos M. L. A. M., Nightingale P. D., and Jickells T. D. (1996) A comparison of methyl iodide emissions from seawater and wet depositional fluxes of iodine over the southern North Sea. *Tellus* **48B**, 106–114.

Carpenter L. J. and Liss P. S. (2000) On temperate sources of bromoform and other reactive organic bromine gases. *J. Geophys. Res.* **105**, 20539–20547.

Carpenter L. J., Malin G., Liss P. S., and Kupper F. C. (2000) Novel biogenic iodine-containing trihalomethanes and other short-lived halocarbons in the coastal East Atlantic. *Global Biogeochem. Cycles* **14**, 1191–1204.

Charlson R. J., Lovelock J. E., Andreae M. O., and Warren S. G. (1987) Oceanic phytoplankton, atmospheric sulfur, cloud albedo and climate. *Nature* **326**, 655–661.

Chuck A. L., Turner S. M., and Liss P. S. (2002) Direct evidence for a marine source of C_1 and C_2 alkyl nitrates. *Science* **297**, 1151–1154.

Cicerone R. J. (1988) Has the atmospheric concentration of CO changed? In *The Changing Atmosphere: Dahlem Workshop Reports* (eds. F. S. Rowland and I. S. A. Isaksen). Wiley, Chichester, pp. 49–61.

Costa M. L. and Liss P. S. (1999) Photoreduction of mercury in sea water and its possible implications for Hg^0 air–sea fluxes. *Mar. Chem.* **68**, 87–95.

Crutzen P. J. (1970) The influence of nitrogen oxides on the atmospheric ozone content. *Q. J. Roy. Met. Soc.* **96**, 320–325.

Crutzen P. J. (1991) Methane's sinks and sources. *Nature* **350**, 308–381.

Csanady G. T. (1990) The role of breaking wavelets in air–sea gas transfer. *J. Geophys. Res.* **95**, 749–759.

Dacey J. W. H. and Wakeham S. G. (1986) Oceanic dimethylsulfide—production during zooplankton grazing on phytoplankton. *Science* **233**, 1314–1316.

Dahl E. E., Saltzman E. S., and De Bruyn W. J. (2003) The aqueous phase yield of alkyl nitrates from ROO + NO: implications for photochemical production in seawater. *Geophys. Res. Lett.* **30**, 1271–1273.

Danckwerts P. V. (1951) Significance of liquid-film coefficients in gas absorption. *Ind. Eng. Chem.* **43**, 1460–1467.

Deacon E. L. (1977) Gas transfer to and across an air–water interface. *Tellus* **29**, 363–374.

de Leeuw G., Kunz G. J., Caulliez G., Woolf D. K., Bowyer P., Leifer I., Nightingale P. D., Liddicoat M. I., Rhee T. S., Andreae M. O., Larsen S. E., Hansen F. A., and Lund S. (2002) LUMINY—an overview. In *Gas Transfer at Water Surfaces*, Geophysical Monograph (eds. M. A. Donelan, W. M. Drennan, E. S. Saltzman, and R. Wanninkhof). AGU, Washington, DC, vol. 127, pp. 291–295.

Donahue N. M. and Prinn R. G. (1990) Nonmethane hydrocarbon chemistry in the remote marine boundary-layer. *J. Geophys. Res.* **95**, 18387–18411.

Doney S. C. (1995) Comment on Phillips (1994). *J. Geophys. Res.* **100**, 14347–14350.

Dore J. E. and Karl D. M. (1996) Nitrification in the euphotic zone as a source for nitrite, nitrate, and nitrous oxide at Station ALOHA. *Limnol. Oceanogr.* **41**, 1619–1628.

Dore J. E., Popp B. N., Karl D. M., and Sansone F. J. (1998) A large source of atmospheric nitrous oxide from subtropical North Pacific surface waters. *Nature* **396**, 63–66.

Duce R. A., Liss P. S., Merrill J. T., Atlas E. L., Buat-Menard P., Hicks B. B., Miller J. M., Prospero J. M., Arimoto R., Church T. M., Ellis W., Galloway J. N., Hansen L., Jickells T. D., Knapp A. H., Reinhardt K. H., Schneider B., Soudine A., Tokos J. J., Tsunogai S., Wollast R., and Zhou M. (1991) The atmospheric input of trace species to the world ocean. *Global Biogeochem. Cycles* **5**, 193–259.

Dyrssen D. and Fogelqvist E. (1981) Bromoform concentrations of the Arctic Ocean in the Svalbard area. *Oceanolog. Acta* **4**, 313–317.

Elliott S. (1989) The effect of hydrogen peroxide on the alkaline hydrolysis of carbon disulfide. *Environ. Sci. Technol.* **24**, 264–267.

Emerson S. (1975) Chemically enhanced CO_2 gas exchange in an eutrophic lake: a general model. *Limnol. Oceanogr.* **20**, 743–753.

Emerson S., Quay P., Stump C., Wilbur D. O., and Knox M. (1991) O_2, Ar, N_2, and ^{222}Rn in surface waters of the subarctic ocean: net biological O_2 production. *Global Biogeochem. Cycles* **5**, 49–70.

Erickson D. J., Zepp R. G., and Atlas E. (2000) Ozone depletion and the air–sea exchange of greenhouse and chemically reactive gases. *Chemosph. Global Change Sci.* **2**, 137–149.

Eriksson E. (1959) The yearly circulation of chloride and sulfur in nature: meteorological, geochemical and pedological implications. *Tellus* **11**, 375–403.

Etcheto J. and Merlivat L. (1988) Satellite determination of the carbon-dioxide exchange coefficient at the ocean–atmosphere interface—a 1st step. *J. Geophys. Res.* **93**, 15669–15678.

Feely R. A., Wanninkhof R., Hansell D. A., Lamb M. F., Greeley D., and Lee K. (2002) Water column CO_2 measurements during the GasEx-98 Expedition. In *Gas Transfer at Water Surfaces*, Geophysical Monograph (eds. M. A. Donelan, W. M. Drennan, E. S. Saltzman, and R. Wanninkhof). AGU, Washington, DC, vol. 127, pp. 173–180.

Fischer R., Weller R., Jacobi H. W., and Ballschmiter K. (2002) Levels and pattern of volatile organic nitrates and halocarbons in the air at Neumayer Station (70 degrees S), Antarctic. *Chemosphere* **48**, 981–992.

Fitzgerald W. F. (1989) Atmospheric and oceanic cycling of mercury. In *Chemical Oceanography* (eds. J. P. Riley and R. Chester). Academic Press, London, vol. 10, pp. 151–186.

Flock O. R. and Andreae M. O. (1996) Photochemical and non-photochemical formation and destruction of carbonyl sulfide and methyl mercaptan in ocean waters. *Mar. Chem.* **54**, 11–26.

Fogelqvist E. (1985) Carbon-tetrachloride, tetrachloroethylene, 1,1,1-trichloroethane and bromoform in Arctic seawater. *J. Geophys. Res.* **90**, 9181–9193.

Fortescue G. E. and Pearson J. R. A. (1967) On gas absorption into a turbulent liquid. *Chem. Eng. Sci.* **22**, 1163–1176.

Frew N. M. (1997) The role of organic films in air–sea gas exchange. In *The Sea Surface and Global Change* (eds. P. S. Liss and R. A. Duce). Cambridge University Press, Cambridge, pp. 121–172.

Frew N. M., Goldman J. C., Dennett M. R., and Johnson A. S. (1990) Impact of phytoplankton-generated surfactants on air–sea gas-exchange. *J. Geophys. Res.* **95**, 3337–3352.

Gabric A. J., Ayers G. P., and Sander G. C. (1995) Independent marine and atmospheric model estimates of the sea–air flux of dimethylsulfide in the Southern Ocean. *Geophys. Res. Lett.* **22**, 3521–3524.

Garland J. A., Elzerman A. W., and Penkett S. A. (1980) The mechanism for dry deposition of ozone to seawater surfaces. *J. Geophys. Res.* **85**, 7488–7492.

Geschwend P. M., MacFarlane J. K., and Newman K. A. (1985) Volatile halogenated organic compounds released to seawater from temperate marine macroalgae. *Science* **227**, 1033–1035.

Gibb S. W., Mantoura R. F. C., Liss P. S., and Barlow R. G. (1999) Distributions and biogeochemistries of methylamines and ammonium in the Arabian Sea. *Deep-Sea Res. II* **46**, 593–615.

Gladyshev M. (1997) Biophysics of the surface film of aquatic ecosystems. In *The Sea Surface and Global Change* (eds. P. S. Liss and R. A. Duce). Cambridge University Press, Cambridge, pp. 321–338.

Glover D. M., Frew N. M., McCue S. J., and Bock E. J. (2002) A multi-year time series of global gas transfer velocity from the TOPEX dual frequency, normalised radar backscatter algorithm. In *Gas Transfer at Water Surfaces*, Geophysical Monographs (eds. M. A. Donelan, W. M. Drennan, E. S. Saltzman, and R. Wanninkhof). AGU, Washington, DC, vol. 127, pp. 325–333.

Goldman J. C. and Dennett M. R. (1983) Carbon-dioxide exchange between air and seawater—no evidence for rate catalysis. *Science* **220**, 199–201.

Goldman J. C., Dennett M. R., and Frew N. M. (1988) Surfactant effects on air sea gas-exchange under turbulent conditions. *Deep-Sea Res.* **35**, 1953–1970.

Goodwin K. D., Lidstrom M. E., and Oremland R. S. (1997) Marine bacterial degradation of brominated methanes. *Environ. Sci. Technol.* **31**, 3188–3192.

Happell J. D. and Wallace D. W. R. (1996) Methyl iodide in the Greenland/Norwegian Seas and the tropical Atlantic Ocean: evidence for photochemical production. *Geophys. Res. Lett.* **23**, 2105–2108.

Hardy J. T., Hunter K. A., Calmet D., Cleary J. J., Duce R. A., Forbes T. L., Gladyshev M., Harding G., Shenker J. M., Tratnyek P., and Zaitsev Y. (1997) Report group 2—biological effects of chemical and radiative change in the sea surface. In *The Sea Surface and Global Change* (eds. P. S. Liss and R. A. Duce). Cambridge University Press, Cambridge, pp. 35–70.

Herr F. L. and Barger W. R. (1978) Molecular hydrogen in the near-surface atmosphere and dissolved waters of the tropical North Atlantic. *J. Geophys. Res.* **83**, 6199–6205.

Hesshaimer V., Heimann M., and Levin I. (1994) Radiocarbon evidence for a smaller oceanic carbon-dioxide sink than previously believed. *Nature* **370**, 201–203.

Higbie R. (1935) The rate of absorption of a pure gas into a still liquid during short periods of exposure. *Am. Inst. Chem. Eng.* **35**, 365–389.

Ho D. T., Bliven L. F., Wanninkhof R., and Schlosser P. (1997) The effect of rain on air–water gas exchange. *Tellus* **49B**, 149–158.

Hoover T. E. and Berkshire D. C. (1969) Effects of hydration in carbon dioxide exchange across an air–water interface. *J. Geophys. Res.* **74**, 456–464.

Hunter K. A. (1997) Chemistry of the sea-surface microlayer. In *The Sea Surface and Global Change* (eds. P. S. Liss and R. A. Duce). Cambridge University Press, Cambridge.

IPCC (2001) *Climate Change 2001: The Scientific Basis. Contributions of Working Group 1 to the Third Assessment Report of the Intergovernmental Panel on Climate Change.* Cambridge University Press, Cambridge.

Jacobs C., Kjeld J. F., Nightingale P., Upstill-Goddard R., Larsen S., and Oost W. (2002) Possible errors in CO_2 air–sea transfer velocity from deliberate tracer releases and eddy covariance measurements due to near-surface concentration gradients. *J. Geophys. Res.* **107** article no. 3128.

Jacobs C. M. J., Kohsiek W., and Oost W. A. (1999) Air–sea fluxes and transfer velocity of CO_2 over the North Sea: results from ASGAMAGE. *Tellus* **51B**, 629–641.

Jahne B., Munnich K. O., Bosinger R., Dutzi A., Huber W., and Libner P. (1987) On the parameters influencing air–water gas exchange. *J. Geophys. Res.* **92**, 1937–1949.

Johnson J. E. and Bates T. S. (1996) Sources and sinks of carbon monoxide in the mixed layer of the tropical South Pacific Ocean. *Global Biogeochem. Cycles* **10**, 347–359.

Jones A. E., Weller R., Minikin A., Wolff E. W., Sturges W. T., McIntyre H. P., Leonard S. R., Schrems O., and Bauguitte S. (1999) Oxidized nitrogen chemistry and speciation in the Antarctic troposphere. *J. Geophys. Res.* **104**, 21355–21366.

Jones E. P. and Smith S. D. (1977) A first measurement of sea–air CO_2 flux by eddy correlation. *J. Geophys. Res.* **82**, 5990–5992.

Kanwisher J. (1963) On the exchange of gases between the atmosphere and the sea. *Deep-Sea Res.* **10**, 195–207.

Keeling R. F. (1993) On the role of large bubbles in air–sea gas exchange and supersaturation in the ocean. *J. Mar. Res.* **51**, 237–271.

Keeling R. F., Piper S. C., and Heimann M. (1996) Global and hemispheric CO_2 sinks deduced from changes in atmospheric O_2 concentration. *Nature* **381**, 218–221.

Keeling R. F., Stephens B. B., Najjar R. G., Doney S. C., Archer D., and Heimann M. (1998) Seasonal variations in the atmospheric O_2/N_2 ratio in relation to the kinetics of air–sea gas exchange. *Global Biogeochem. Cycles* **12**, 141–163.

Keene W. C., Sander R., Pszenny A. A. P., Vogt R., Crutzen P. J., and Galloway J. N. (1998) Aerosol pH in the marine boundary layer: a review and model evaluation. *J. Aerosol Sci.* **29**, 339–356.

Keene W. C., Khalil M. A. K., Erickson D. J., McCulloch A., Graedel T. E., Lobert J. M., Aucott M. L., Gong S. L., Harper D. B., Kleiman G., Midgley P., Moore R. M., Seuzaret C., Sturges W. T., Benkovitz C. M., Koropalov V., Barrie L. A., and Li Y. F. (1999) Composite global emissions of reactive chlorine from anthropogenic and natural sources: reactive chlorine emissions inventory. *J. Geophys. Res.* **104**, 8429–8440.

Keller M. D., Bellows W. K., and Guillard R. R. L. (1989) Dimethyl sulfide production in marine-phytoplankton. *ACS Symp. Ser.* **393**, 167–182.

Kerman B. R. (1984) A model of interfacial gas transfer for a well-roughened sea. *J. Geophys. Res.* **82**, 1439–1446.

Kesselmeier J. and Staudt M. (1999) Biogenic volatile organic compounds (VOC): an overview on emission, physiology and ecology. *J. Atmos. Chem.* **33**, 23–88.

Kettle A. J. and Andreae M. O. (2000) Flux of dimethylsulfide from the oceans: a comparison of updated data seas and flux models. *J. Geophys. Res.* **105**, 26793–26808.

Kettle A. J., Andreae M. O., Amouroux D., Andreae T. W., Bates T. S., Berresheim H., Bingemer H., Boniforti R., Curran M. A. J., DiTullio G. R., Helas G., Jones G. B., Keller M. D., Kiene R. P., Leck C., Levasseur M., Malin G., Maspero M., Matrai P., McTaggart A. R., Mihalopoulos N., Nguyen B. C., Novo A., Putaud J. P., Rapsomanikis S., Roberts G., Schebeske G., Sharma S., Simo R., Staubes R., Turner S., and Uher G. (1999) A global database of sea surface dimethylsulfide (DMS) measurements and a procedure to predict sea surface DMS as a function of latitude, longitude, and month. *Global Biogeochem. Cycles* **13**, 399–444.

Kettle A. J., Rhee T. S., von Hobe M., Poulton A., Aiken J., and Andreae M. O. (2001) Assessing the flux of different volatile sulfur gases from the ocean to the atmosphere. *J. Geophys. Res.* **106**, 12193–12209.

Kettle A. J., Kuhn U., von Hobe M., Kesselmeier J., Liss P. S., and Andreae M. O. (2002) Comparing forward and inverse models to estimate the seasonal variation of hemisphere-integrated fluxes of carbonyl sulfide. *Atmos. Chem. Phys.* **2**, 343–361.

Khalil M. A. K. and Rasmussen R. A. (1983) Atmospheric chloroform ($CHCl_3$): ocean–air exchange and global mass balance. *Tellus* **35**, 266–274.

Khalil M. A. K., Moore R. M., Harper D. B., Lobert J. M., Erickson D. J., Koropalov V., Sturges W. T., and Keene W. C. (1999) Natural emissions of chlorine-containing gases: reactive chlorine emissions inventory. *J. Geophys. Res.* **104**, 8333–8346.

Kiene R. P. (1996) Production of methanethiol from dimethylsulfoniopropionate in marine surface waters. *Mar. Chem.* **54**, 69–83.

Kiene R. P. and Bates T. S. (1990) Biological removal of dimethyl sulfide from sea-water. *Nature* **345**, 702–705.

Kiene R. P., Linn L. J., and Bruton J. A. (2000) New and important roles for DMSP in marine microbial communities. *J. Sea Res.* **43**, 209–224.

Kim G., Hussain N., and Church T. M. (2000) Excess Po-210 in the coastal atmosphere. *Tellus* **52B**, 74–80.

King D. B. and Saltzman E. S. (1997) Removal of methyl bromide in coastal seawater: chemical and biological rates. *J. Geophys. Res.* **102**, 18715–18721.

King D. B., Butler J. H., Montzka S. A., Yvon-Lewis S. A., and Elkins J. W. (2000) Implications of methyl bromide supersaturations in the temperate North Atlantic Ocean. *J. Geophys. Res.* **105**, 19763–19769.

Kirst G. O., Thiel C., Wolff H., Nothnagel J., Wanzek M., and Ulmke R. (1991) Dimethylsulfoniopropionate (DMSP) in ice-algae and its possible biological role. *Mar. Chem.* **35**, 381–388.

Klick S. and Abrahamsson K. (1992) Biogenic volatile iodated hydrocarbons in the ocean. *J. Geophys. Res.* **97**, 12683–12687.

Knutson T. R. and Tuleya R. E. (1999) Increased hurricane intensities with CO_2-induced warming as simulated using the GFDL hurricane prediction system. *Clim. Dyn.* **15**, 503–519.

Kroeze C., Mosier A., and Bouwman L. (1999) Closing the global N_2O budget: a retrospective analysis, 1500–1994. *Global Biogeochem. Cycles* **13**, 1–8.

Krysell M. (1991) Bromoform in the Nansen Basin in the Arctic-Ocean. *Mar. Chem.* **33**, 187–197.

Krysell M., Fogelqvist E., and Tanhua T. (1994) Apparent removal of the transient tracer carbon-tetrachloride from anoxic seawater. *Geophys. Res. Lett.* **21**, 2511–2514.

Lamborg C. H., Fitzgerald W. F., O'Donnell J., and Torgersen T. (2002) A non-steady state box model of global-scale mercury biogeochemistry with interhemispheric atmospheric gradients. *Abstr. Pap. Am. Chem. Soc.* **223** 072-ENVR.

Lamont J. C. and Scott D. S. (1970) An eddy cell model of mass transfer into the surface of a turbulent liquid. *AIChE J.* **16**, 513–519.

Law C. S. and Ling R. D. (2001) Nitrous oxide flux and response to increased iron availability in the Antarctic Circumpolar Current. *Deep-Sea Res. II* **48**, 2509–2527.

Law C. S. and Owens N. J. P. (1990) Significant flux of atmospheric nitrous-oxide from the northwest Indian-Ocean. *Nature* **346**, 826–828.

Law C. S., Sjoberg T. N., and Ling R. D. (2002) Atmospheric emission and cycling of carbon monoxide in the Scheldt Estuary. *Biogeochem.* **59**, 69–94.

Ledwell J. R. (1984) The variation of the gas transfer coefficient with molecular diffusivity. In *Gas Transfer at Water Surfaces* (eds. W. Brutsaert and G. H. Jirka). Reidel, Dordrecht, pp. 293–302.

Lee B. S., Bullister J. L., and Whitney F. A. (1999) Chlorofluorocarbon CFC-11 and carbon tetrachloride removal in Saanich Inlet, an intermittently anoxic basin. *Mar. Chem.* **66**, 171–185.

Lenschow D. H., Pearson R., and Stankov B. B. (1982) Measurements of ozone vertical flux to ocean and forest. *J. Geophys. Res.* **87**, 8833–8837.

Liss P. S. and Slater P. G. (1974) Flux of gases across the air–sea interface. *Nature* **247**, 181–184.

Liss P. S. (1971) Exchange of SO_2 between the atmosphere and natural waters. *Nature* **233**, 327–329.

Liss P. S. (1983) Gas transfer: experiments and geochemical implications. In *Air–Sea Exchange of Gases and Particles* (eds. P. S. Liss and W. G. N. Slinn). Reidel, Dordrecht, pp. 241–298.

Liss P. S. and Martinelli F. N. (1978) The effect of oil films on the transfer of oxygen and water vapour across an air–water interface. *Thalass. Jugoslav.* **14**, 215–220.

Liss P. S. and Merlivat L. (1986) Air–sea gas exchange rates: introduction and synthesis. In *The Role of Air–Sea Gas Exchange in Geochemical Cycling* (ed. P. Buat-Menard). Reidel, Dordrecht, pp. 113–127.

Liss P. S., Balls P. W., Martinelli F. N., and Coantic M. (1981) The effect of evaporation and condensation on gas transfer across an air–water-interface. *Oceanolog. Acta* **4**, 129–138.

Lobert J. M., Butler J. H., Montzka S. A., Geller L. S., Myers R. C., and Elkins J. W. (1995) A net sink for atmospheric CH_3Br in the East Pacific-Ocean. *Science* **267**, 1002–1005.

Lobert J. M., Yvon-Lewis S. A., Butler J. H., Montzka S. A., and Myers R. C. (1997) Undersaturation of CH_3Br in the Southern Ocean. *Geophys. Res. Lett.* **24**, 171–172.

Lovelock J. E. (1975) Natural halocarbons in the air and seawater. *Nature* **256**, 193–194.

Lovelock J. E. (1979) *Gaia: A New Look at Life on Earth.* Oxford University Press, Oxford.

Lovelock J. E., Maggs R. J., and Rasmussen R. A. (1972) Atmospheric dimethyl sulfide and the natural sulfur cycle. *Nature* **237**, 452–453.

Malin G., Wilson W. H., Bratbak G., Liss P. S., and Mann N. H. (1998) Elevated production of dimethylsulfide resulting from viral infection of cultures of *Phaeocystis pouchetii. Limnol. Oceanogr.* **43**, 1389–1393.

Manley S. L., Goodwin K., and North W. J. (1992) Laboratory production of bromoform, methylene bromide, and methyl iodide by macroalgae and distribution in nearshore southern California waters. *Limnol. Oceanogr.* **37**, 1652–1659.

Matthews B. J. H. (2000) The rate of air–sea CO_2 exchange: chemical enhancement and catalysis by marine microalgae. PhD., University of East Anglia.

McGillis W. R., Dacey J. W. H., Frew N. M., Bock E. J., and Nelson R. K. (2000) Water–air flux of dimethylsulfide. *J. Geophys. Res.* **105**, 1187–1193.

McGillis W. R., Edson J. B., Hare J. E., and Fairall C. W. (2001a) Direct covariance air–sea CO_2 fluxes. *J. Geophys. Res.* **106**, 16729–16745.

McGillis W. R., Edson J. B., Ware J. D., Dacey J. W. H., Hare J. E., Fairall C. W., and Wanninkhof R. (2001b) Carbon dioxide flux techniques performed during GasEx-98. *Mar. Chem.* **75**, 267–280.

McIntyre H. P. (2001) The measurement and implications of primary short chain alkyl nitrates in contemporary tropospheric and aged polar firn air. PhD, University of East Anglia.

Meredith M. P., VanScoy K. A., Watson A. J., and Locarnini R. A. (1996) On the use of carbon tetrachloride as a transient tracer of Weddell Sea deep and bottom waters. *Geophys. Res. Lett.* **23**, 2943–2946.

Merlivat L. and Memery L. (1983) Gas exchange across an air–water interface: experimental results and modeling of bubble contribution to transfer. *J. Geophys. Res.* **88**, 707–724.

Middelburg J. J., Nieuwenhuize J., Iversen N., Hogh N., De Wilde H., Helder W., Seifert R., and Christof O. (2002) Methane distribution in European tidal estuaries. *Biogeochemistry* **59**, 95–119.

Moncrieff J. B., Beverland I. J., O'Neill D. H., and Cropley F. D. (1998) Controls on trace gas exchange observed by a conditional sampling method. *Atmos. Environ.* **32**, 3265–3274.

Moore R. M. and Groszko W. (1999) Methyl iodide distribution in the ocean and fluxes to the atmosphere. *J. Geophys. Res.* **104**, 11163–11171.

Moore R. M. and Tokarczyk R. (1993) Volatile biogenic halocarbons in the northwest Atlantic. *Global Biogeochem. Cycles* **7**, 195–210.

Moore R. M., Groszko W., and Niven S. J. (1996a) Ocean–atmosphere exchange of methyl chloride: results from NW Atlantic Ocean and Pacific Ocean studies. *J. Geophys. Res.* **101**, 28529–28538.

Moore R. M., Webb M., Tokarczyk R., and Wever R. (1996b) Bromoperoxidase and iodoperoxidase enzymes and production of halogenated methanes in marine diatom cultures. *J. Geophys. Res.* **101**, 20899–20908.

Mosher B. W. and Duce R. A. (1987) A global atmospheric selenium budget. *J. Geophys. Res.* **92**, 13289–13298.

Mtolera M. S. P., Collen J., Pedersen M., Ekdahl A., Abrahamsson K., and Semesi A. K. (1996) Stress-induced production of volatile halogenated organic compounds in *Eucheuma denticulatum* (Rhodophyta) caused by elevated pH and high light intensities. *Euro. J. Phycol.* **31**, 89–95.

Naqvi S. W. A., Jayakumar D. A., Narvekar P. V., Naik H., Sarma V., D'Souza W., Joseph S., and George M. D. (2000) Increased marine production of N_2O due to intensifying anoxia on the Indian continental shelf. *Nature* **408**, 346–349.

Nemecek-Marshall M., Wojciechowski C., Kuzma J., Silver G. M., and Fall R. (1995) Marine Vibrio species produce the volatile organic-compound acetone. *Appl. Environ. Microbiol.* **61**, 44–47.

Nevison C. D., Weiss R. F., and Erickson D. J. (1995) Global oceanic emissions of nitrous-oxide. *J. Geophys. Res.* **100**, 15809–15820.

Nguyen B. C., Belviso S., Mihalopoulos N., Gostan J., and Nival P. (1988) Dimethyl sulfide production during natural phytoplanktonic blooms. *Mar. Chem.* **24**, 133–141.

Nightingale P. D. (1991) Low molecular weight halocarbons in seawater. PhD, University of East Anglia.

Nightingale P. D., Malin G., and Liss P. S. (1995) Production of chloroform and other low-molecular-weight halocarbons by some species of macroalgae. *Limnol. Oceanogr.* **40**, 680–689.

Nightingale P. D., Liss P. S., and Schlosser P. (2000a) Measurements of air–sea gas transfer during an open ocean algal bloom. *Geophys. Res. Lett.* **27**, 2117–2120.

Nightingale P. D., Malin G., Law C. S., Watson A. J., Liss P. S., Liddicoat M. I., Boutin J., and Upstill-Goddard R. C. (2000b) *In situ* evaluation of air–sea gas exchange parameterizations using novel conservative and volatile tracers. *Global Biogeochem. Cycles* **14**, 373–387.

Oremland R. S. (1979) Methanogenic activity in plankton samples and in fish intestines: a mechanism for *in situ* methanogenesis in oceanic waters. *Limnol. Oceanogr.* **24**, 1136–1141.

Orr J. C., Maier-Reimer E., Mikolajewicz U., Monfray P., Sarmiento J. L., Toggweiler J. R., Taylor N. K., Palmer J., Gruber N., Sabine C. L., Le Quere C., Key R. M., and Boutin J. (2001) Estimates of anthropogenic carbon uptake from four three-dimensional global ocean models. *Global Biogeochem. Cycles* **15**, 43–60.

Owens N. J. P., Law C. S., Mantoura R. F. C., Burkill P. H., and Llewellyn C. A. (1991) Methane flux to the atmosphere from the Arabian Sea. *Nature* **354**, 293–296.

Pattey E., Desjardins R. L., and Rochette P. (1993) Accuracy of the relaxed eddy-accumulation technique, evaluated using CO_2 flux measurements. *Boundary-Layer Meteorol.* **66**, 341–355.

Peng T. H., Broecker W. S., Mathieu G., Li Y. H., and Bainbridge A. E. (1979) Radon evasion rates in the Atlantic and Pacific Oceans as determined during the GEOSECS Program. *J. Geophys. Res.* **84**, 2471–2486.

Penkett S. A., Butler J. H., Reeves C. E., Singh H. B., Toohey D., and Weiss R. F. (1995) Methyl bromide. In *Scientific Assessment of Ozone Depletion: 1994, World Meteorological Organisation Global Ozone and Monitoring Project Report No. 37*, chap. 10.

Phillips L. F. (1994) Experimental demonstration of coupling of heat and matter fluxes at a gas–water interface. *J. Geophys. Res.* **99**, 18577–18584.

Phillips L. F. (1997) The physical chemistry of air–sea gas exchange. In *The Sea Surface and Global Change* (eds. P. S. Liss and R. A. Duce). Cambridge University Press, Cambridge, pp. 207–250.

Plane J. M. C. (1989) Gas-phase atmospheric oxidation of biogenic sulfur-compounds—a review. *ACS Symp. Ser.* **393**, 404–423.

Plane J. M. C., Blough N. V., Ehrhardt M. G., Waters K., Zepp R. G., and Zika R. G. (1997) Report Group 3—Photochemistry in the sea-surface microlayer. In *The Sea Surface and Global Change* (eds. P. S. Liss and R. A. Duce). Cambridge University Press, Cambridge, pp. 71–92.

Plass-Dulmer C., Koppmann R., Ratte M., and Rudolph J. (1995) Light nonmethane hydrocarbons in seawater. *Global Biogeochem. Cycles* **9**, 79–100.

Prinn R., Cunnold D., Rasmussen R., Simmonds P., Alyea F., Crawford A., Fraser P., and Rosen R. (1990) Atmospheric emissions and trends of nitrous-oxide deduced from 10 years of ALE-GAGE data. *J. Geophys. Res.* **95**, 18369–18385.

Quinn P. K., Charlson R. J., and Bates T. S. (1988) Simultaneous observations of ammonia in the atmosphere and ocean. *Nature* **335**, 336–338.

Quinn P. K., Bates T. S., Johnson J. E., Covert D. S., and Charlson R. J. (1990) Interactions between the sulfur and reduced nitrogen cycles over the central Pacific-Ocean. *J. Geophys. Res.* **95**, 16405–16416.

Ratte M., Plass-Dulmer C., Koppmann R., Rudolph J., and Denga J. (1993) Production mechanism of C2–C4 hydrocarbons in seawater—Field-measurements and experiments. *Global Biogeochem. Cycles* **7**, 369–378.

Raven J. A. (1995) Phycological reviews: 15. Photosynthetic and nonphotosynthetic roles of carbonic-anhydrase in algae and cyanobacteria. *Phycologia* **34**, 93–101.

Rayman M. P. (2000) The importance of selenium to human health. *Lancet* **356**, 233–241.

Redfield A. C. (1948) The exchange of oxygen across the sea surface. *J. Mar. Res.* **7**, 347–361.

Reifenhauser W. and Heumann K. G. (1992) Determinations of methyl-iodide in the Antarctic atmosphere and the South Polar Sea. *Atmos. Environ.* **26**, 2905–2912.

Riebesell U., Zondervan I., Rost B., Tortell P. D., Zeebe R. E., and Morel F. M. M. (2000) Reduced calcification of marine plankton in response to increased atmospheric CO_2. *Nature* **407**, 364–367.

Robertson J. E. and Watson A. J. (1992) Thermal skin effect of the surface ocean and its implications for CO_2 uptake. *Nature* **358**, 738–740.

Roether W., Klein B., and Bulsiewicz K. (2001) Apparent loss of CFC-113 in the upper ocean. *J. Geophys. Res.* **106**, 2679–2688.

Sarmiento J. L. and LeQuere C. (1996) Oceanic carbon dioxide uptake in a model of century-scale global warming. *Science* **274**, 1346–1350.

Sarmiento J. L. and Sundquist E. T. (1992) Revised budget for the oceanic uptake of anthropogenic carbon-dioxide. *Nature* **356**, 589–593.

Sarmiento J. L., Hughes T. M. C., Stouffer R. J., and Manabe S. (1998) Simulated response of the ocean carbon cycle to anthropogenic climate warming. *Nature* **393**, 245–249.

Sarmiento J. L., Monfray P., Maier-Reimer E., Aumont O., Murnane R. J., and Orr J. C. (2000) Sea–air CO_2 fluxes and carbon transport: a comparison of three ocean general circulation models. *Global Biogeochem. Cycles* **14**, 1267–1281.

Scarratt M. G. and Moore R. M. (1996) Production of methyl chloride and methyl bromide in laboratory cultures of marine phytoplankton. *Mar. Chem.* **54**, 263–272.

Scarratt M. G. and Moore R. M. (1998) Production of methyl bromide and methyl chloride in laboratory cultures of marine phytoplankton II. *Mar. Chem.* **59**, 311–320.

Schauffler S. M., Heidt L. E., Pollock W. H., Gilpin T. M., Vedder J. F., Solomon S., Lueb R. A., and Atlas E. L. (1993) Measurements of halogenated organic-compounds near the tropical tropopause. *Geophys. Res. Lett.* **20**, 2567–2570.

Schimel D. (1996) Radiative forcing of climate change. In *Climate Change 1995: The Science of Climate Change* (eds. J. T. Houghton, L. G. Meira Filho, N. Callender, N. Harris, A. Kettenberg, and K. Maskell). Cambridge University Press, Cambridge, pp. 69–131.

Schmidt U. (1974) Molecular hydrogen in the atmosphere. *Tellus* **26**, 78–90.

Shapiro S. D., Schlosser P., Smethie W. M., and Stute M. (1997) The use of ^{3}H and tritiogenic ^{3}He to determine CFC degradation and vertical mixing rates in Framvaren Fjord, Norway. *Mar. Chem.* **59**, 141–157.

Sieburth J. M. (1960) Acrylic acid, an “antibiotic” principle in Phaeocystis blooms in Antarctic waters. *Science* **132**, 676–677.

Sieburth J. M., Johnson P. W., Eberhardt M. A., Sieracki M. E., Lidstrom M., and Laux D. (1987) The 1st methane-oxidizing bacterium from the upper mixing layer of the deep ocean—methylomonas-pelagica Sp-Nov. *Current Microbiol.* **14**, 285–293.

Singh H., Chen Y., Staudt A., Jacob D., Blake D., Heikes B., and Snow J. (2001) Evidence from the Pacific troposphere for large global sources of oxygenated organic compounds. *Nature* **410**, 1078–1081.

Singh H. B., Salas L. J., and Stiles R. E. (1983) Methyl halides in and over the eastern Pacific (40-degrees-N–32-degrees-S). *J. Geophys. Res.* **88**, 3684–3690.

Singh H. B., Salas L., Herlth D., Kolyer R., Czech E., Viezee W., Li Q., Jacob D. J., Blake D., Sachse G., Harward C. N., Fuelberg H., Kiley C. M. *In situ* measurements of HCN and CH_3CN in the Pacific troposphere: sources, sinks and comparisons with spectroscopic observations. *J. Geophys. Res.* (in press).

Smethie W. M., Takahashi T., and Chipman D. W. (1985) Gas exchange and CO_2 flux in the tropical Atlantic Ocean determined from Rn-222 and pCO_2 measurements. *J. Geophys. Res.* **90**, 7005–7022.

Solomon S. (1992) Global ozone depletion—a review. *Abstr. Pap. Am. Chem. Soc.* **203**(Part 2) 308-PHYS.

Springer T. G. and Pigford R. L. (1970) Influence of surface turbulence and surfactants on gas transport through liquid interfaces. *Indust. Eng. Chem. Fundament.* **9**, 458–465.

Springer-Young M., Erickson D. J., and Carsey T. P. (1996) Carbon monoxide gradients in the marine boundary layer of the North Atlantic Ocean. *J. Geophys. Res.* **101**, 4479–4484.

Steinke M., Wolfe G. V., and Kirst G. O. (1998) Partial characterisation of dimethylsulfoniopropionate (DMSP) lyase isozymes in 6 strains of *Emiliania huxleyi*. *Mar. Ecol. Prog. Ser.* **175**, 215–225.

Steinke M., Malin G., and Liss P. S. (2002) Trophic interactions in the sea: an ecological role for climate relevant volatiles? *J. Phycol.* **38**, 630–638.

Suntharalingam P., Sarmiento J. L., and Toggweiler J. R. (2000) Global significance of nitrous-oxide production and transport from oceanic low-oxygen zones: a modeling study. *Global Biogeochem. Cycles* **14**, 1353–1370.

Takahashi T., Feely R. A., Weiss R., Wanninkhof R., Chipman D. W., Sutherland S. C., and Takahashi T. T. (1997) Global air–sea flux of CO_2: an estimate based on measurements of sea–air pCO_2 difference. In *NAS Colloquium Volume on Carbon Dioxide and Climatic Change* (ed. C. D. Keeling). *Proc. Natl. Acad. Sci. USA* **94**, 8292–8299.

Takahashi T., Sutherland S. C., Sweeney C., Poisson A., Metzl N., Tilbrook B., Bates N., Wanninkhof R., Feely R. A., Sabine C., Olafsson J., and Nojiri Y. (2002) Global sea-air CO_2 flux based on climatological surface ocean pCO_2, and seasonal biological and temperature effects. *Deep-Sea Res. II* **49**, 1601–1622.

Tanhua T., Fogelqvist E., and Basturk O. (1996) Reduction of volatile halocarbons in anoxic seawater, results from a study in the Black Sea. *Mar. Chem.* **54**, 159–170.

Tokarczyk R. and Moore R. M. (1994) Production of volatile organohalogens by phytoplankton cultures. *Geophys. Res. Lett.* **21**, 285–288.

Tsunogai S. and Tanaka N. (1980) Flux of oxygen across the air–sea interface as determined by the analysis of dissolved components in sea-water. *Geochem. J.* **14**, 227–234.

Turner S. M., Nightingale P. D., Spokes L. J., Liddicoat M. I., and Liss P. S. (1996) Increased dimethyl sulphide concentrations in sea water from *in situ* iron enrichment. *Nature* **383**, 513–517.

Uher G. and Andreae M. O. (1997) Photochemical production of carbonyl sulfide in North Sea water: a process study. *Limnol. Oceanogr.* **42**, 432–442.

Upstill-Goddard R. C., Watson A. J., Liss P. S., and Liddicoat M. I. (1990) Gas transfer in lakes measured with SF_6. *Tellus* **42B**, 364–377.

Upstill-Goddard R. C., Barnes J., and Owens N. J. P. (1999) Nitrous oxide and methane during the 1994 SW monsoon in the Arabian Sea/northwestern Indian Ocean. *J. Geophys. Res.* **104**, 30067–30084.

Upstill-Goddard R. C., Barnes J., Frost T., Punshon S., and Owens N. J. P. (2000) Methane in the southern North Sea: low-salinity inputs, estuarine removal, and atmospheric flux. *Global Biogeochem. Cycles* **14**, 1205–1217.

Valentine R. L. and Zepp R. G. (1993) Formation of carbon-monoxide from the photodegradation of terrestrial dissolved organic-carbon in natural-waters. *Environ. Sci. Technol.* **27**, 409–412.

Vogel T. M., Criddle C. S., and McCarty P. L. (1987) Transformations of Halogenated aliphatic-compounds. *Environ. Sci. Technol.* **21**, 722–736.

Wanninkhof R. (1992) Relationship between wind speed and gas exchange over the ocean. *J. Geophys. Res.* **97**, 7373–7382.

Wanninkhof R. and McGillis W. R. (1999) A cubic relationship between air–sea CO_2 exchange and wind speed. *Geophys. Res. Lett.* **26**, 1889–1892.

Wanninkhof R., Ledwell J. R., and Broecker W. S. (1985) Gas exchange wind speed relation measured with sulfur hexafluoride on a lake. *Science* **227**, 1224–1226.

Wanninkhof R., Ledwell J. R., Broecker W. S., and Hamilton M. (1987) Gas-exchange on Mono Lake and Crowley Lake, California. *J. Geophys. Res.* **92**, 14567–14580.

Wanninkhof R., Hitchcock G., Wiseman W. J., Vargo G., Ortner P. B., Asher W., Ho D. T., Schlosser P., Dickson M. L., Masserini R., Fanning K., and Zhang J. Z. (1997) Gas exchange, dispersion, and biological productivity on the west Florida shelf: results from a Lagrangian tracer study. *Geophys. Res. Lett.* **24**, 1767–1770.

Watson A. J., Upstill-Goddard R. C., and Liss P. S. (1991) Air sea gas exchange in rough and stormy seas measured by a dual tracer technique. *Nature* **349**, 145–147.

Watts S. F. (2000) The mass budgets of carbonyl sulfide, dimethyl sulfide, carbon disulfide and hydrogen sulfide. *Atmos. Environ.* **34**, 761–779.

Wever R. (1988) Ozone destruction by algae in the Arctic atmosphere. *Nature* **335**, 501.

Whitman W. G. (1923) The two-film theory of gas absorption. *Chem. Metall. Eng.* **29**, 146–148.

WHO (1996) Iodine deficiency disorders. *WHO Report Fact Sheet* **121**, http://www.who.int/inf-fs/en/fact121.html.

Woolf D. K. (1993) Bubbles and the air–sea transfer velocity of gases. *Atmos. Ocean.* **31**, 517–540.

Woolf D. K. (1997) Bubbles and their role in air–sea gas exchange. In *The Sea Surface and Global Change* (eds. P. S. Liss and R. A. Duce). Cambridge University Press, pp. 173–205.

Woolf D. K. and Thorpe S. A. (1991) Bubbles and the air–sea exchange of gases in near-saturation conditions. *J. Mar. Res.* **49**, 435–466.

Xie H. X. and Moore R. M. (1999) Carbon disulfide in the North Atlantic and Pacific Oceans. *J. Geophys. Res.* **104**, 5393–5402.

Xie H. X., Moore R. M., and Miller W. L. (1998) Photochemical production of carbon disulphide in seawater. *J. Geophys. Res.* **103**, 5635–5644.

Xie H. X., Scarratt M. G., and Moore R. M. (1999) Carbon disulphide production in laboratory cultures of marine phytoplankton. *Atmos. Environ.* **33**, 3445–3453.

Xu X., Bingemer H. G., and Schmidt U. (2002) The flux of carbonyl sulfide and carbon disulfide between the atmosphere and a spruce forest. *Atmos. Chem. Phys.* **2**, 171–181.

Yaglom A. M. and Kader B. A. (1974) Heat and mass transfer between a rough wall and turbulent flow at high Reynolds and Peclet numbers. *J. Fluid Mech.* **62**, 601–623.

Yang G. P., Watanabe S., and Tsunogai S. (2001) Distribution and cycling of dimethylsulfide in surface microlayer and subsurface seawater. *Mar. Chem.* **76**, 137–153.

Zafiriou O. C. (1975) Reaction of methyl halides with seawater and marine aerosols. *J. Mar. Res.* **33**, 75–81.

Zappa C. J., Asher W. E., and Jessup A. T. (2001) Microscale wave breaking and air–water gas transfer. *J. Geophys. Res.* **106**, 9385–9391.

Zemmelink H. J., Gieskes W. W. C., Klaassen W., de Groot H. W., de Baar H. J. W., Dacey J. W. H., Hintsa E. J., and McGillis W. R. (2002) Simultaneous use of relaxed eddy accumulation and gradient flux techniques for the measurements of sea-to-air exchange of dimethyl sulphide. *Atmos. Environ.* **36**, 5709–5717.

Zhou X. L. and Mopper K. (1997) Photochemical production of low-molecular-weight carbonyl compounds in seawater and surface microlayer and their air–sea exchange. *Mar. Chem.* **56**, 201–213.

Zika R. G., Gidel L. T., and Davis D. D. (1984) A comparison of photolysis and substitution decomposition rates of methyl iodide in the ocean. *Geophys. Res. Lett.* **11**, 353–356.

Zou Y. and Jones R. D. (1998) Reassessment of the ocean to atmosphere flux of carbon monoxide. *Chem. Ecol.* **14**, 241–257.

6.04
The Biological Pump

C. L. de la Rocha
University of Cambridge, UK

NOMENCLATURE

$H_2CO_3^*$	dissolved $CO_2 + H_2CO_3$ (μM)
$J_{C_{org}}$	flux of organic C to depth ($g\ C\ m^{-2}\ yr^{-1}$)
z	depth (m)
PP	primary production ($g\ C\ m^{-2}\ yr^{-1}$)
D	diffusivity of CO_2 in seawater ($m^2\ s^{-1}$)
r	radius of phytoplankton cell (μm)
C_e	extracellular CO_2 concentration (μM)
C_i	intracellular CO_2 concentration (μM)
k'	rate constant for $HCO_3 \rightarrow CO_2$ (s^{-1})
F_{CO_2}	flux of CO_2 to cell surface (μmol s^{-1})
τ	residence time (yr)
$\sum CO_2$	total CO_2 (μM)

6.04.1 INTRODUCTION

Despite having residence times (τ) that exceed the ~1,000 yr mixing time of the ocean (Broecker and Peng, 1982), many dissolved constituents of seawater have distributions that vary with depth and from place to place. For instance, silicic acid ($\tau = 1.5 \times 10^4$ yr), nitrate ($\tau = 3{,}000$ yr), phosphate ($\tau = (1\text{–}5) \times 10^4$ yr), and dissolved inorganic carbon (DIC; $\tau = 8.3 \times 10^4$ yr) are generally present in low concentrations in surface waters and at much higher concentrations below the thermocline (Figure 1). Additionally, their concentrations are higher in older deep waters than they are in the younger waters of the deep sea (Figure 2). This is the general distribution exhibited by elements and compounds taking part in biological processes in the ocean and is generally referred to as a "nutrient-type" distribution.

Both the lateral and vertical gradients in the concentrations of nutrients result from "the biological pump" (Figure 3). Dissolved inorganic materials (e.g., CO_2, NO_3^-, PO_4^{3-}, $Si(OH)_4$) are fixed into particulate organic matter (carbohydrates, lipids, proteins) and biominerals (silica and calcium carbonate) by phytoplankton in surface waters. Some of these particles are subsequently transported, by sinking, into the deep. The bulk of the organic material and biominerals decomposes in the upper ocean via dissolution, zooplankton grazing, and microbial hydrolysis, but a significant supply of material does survive to reach the deep sea and sediments. Thus, just as biological uptake removes certain dissolved inorganic materials in surface waters, the decomposition of sinking biogenic particles provides a source of dissolved inorganic material to deeper waters. Thus, deeper waters contain higher concentrations of biologically utilized materials than do surface waters. Older deeper waters contain higher concentrations of bums compared to newly formed deep waters or surface waters.

One side-effect of the biological pump is that CO_2 is shunted from the surface ocean and into the deep sea, thus lowering the amount in the atmosphere. For many years it has been recognized that pre-Industrial CO_2 levels in the atmosphere were about one-third of what they would be in the absence of a biological pump (Broecker, 1982). It is also known that the biological pump is not operating at its full capacity. In so-called "high-nutrient, low-chlorophyll" (HNLC) areas of the ocean, a considerable portion of the nutrients supplied to the surface waters is not utilized in support of primary production, most likely due to the limitation of phytoplankton growth by an inadequate supply of trace elements (e.g., Martin and Fitzwater, 1988). The possibility that the biological pump in HNLC areas might be stimulated by massive additions of iron both artificially as a means of removing anthropogenic CO_2 from the atmosphere and naturally as a cause for the lower glacial atmospheric CO_2 levels (Martin, 1990) is the focus of much research and debate (e.g., Martin *et al.*, 1994; Coale *et al.*, 1996; Boyd *et al.*, 2000; Watson *et al.*, 2000).

Although the biological pump is most popularly known for its impact on the cycling of carbon and major nutrients, it also has profound impacts on the geochemistry of many other elements and compounds. The biological pump heavily influences the cycling, concentrations, and residence times of trace elements—such as cadmium, germanium, zinc, nickel, iron, arsenic, selenium—through their incorporation into organic matter and biominerals (Bruland, 1980; Azam and Volcani, 1981; Elderfield and Rickaby, 2000). Scavenging by sinking biogenic particles and precipitation of materials in the microenvironment of organic aggregates and fecal pellets plays a large role in the marine geochemistry of elements such as barium, thorium, protactinium, beryllium, rare earth elements (REEs), and yttrium (Dehairs *et al.*, 1980; Anderson *et al.*, 1990; Buesseler *et al.*, 1992; Kumar *et al.*, 1993; Zhang and Nozaki, 1996). Even major elements in seawater such as Ca^{2+} and Sr^{2+} display slight surface depletions (Broecker and Peng, 1982; de Villiers, 1999) as a result of the biological pump, despite their long respective oceanic residence times of 1 Myr and 5 Myr (Broecker and Peng, 1982; Elderfield and Schultz, 1996).

6.04.2 DESCRIPTION OF THE BIOLOGICAL PUMP

The biological pump can be sectioned into several major steps: the production of organic

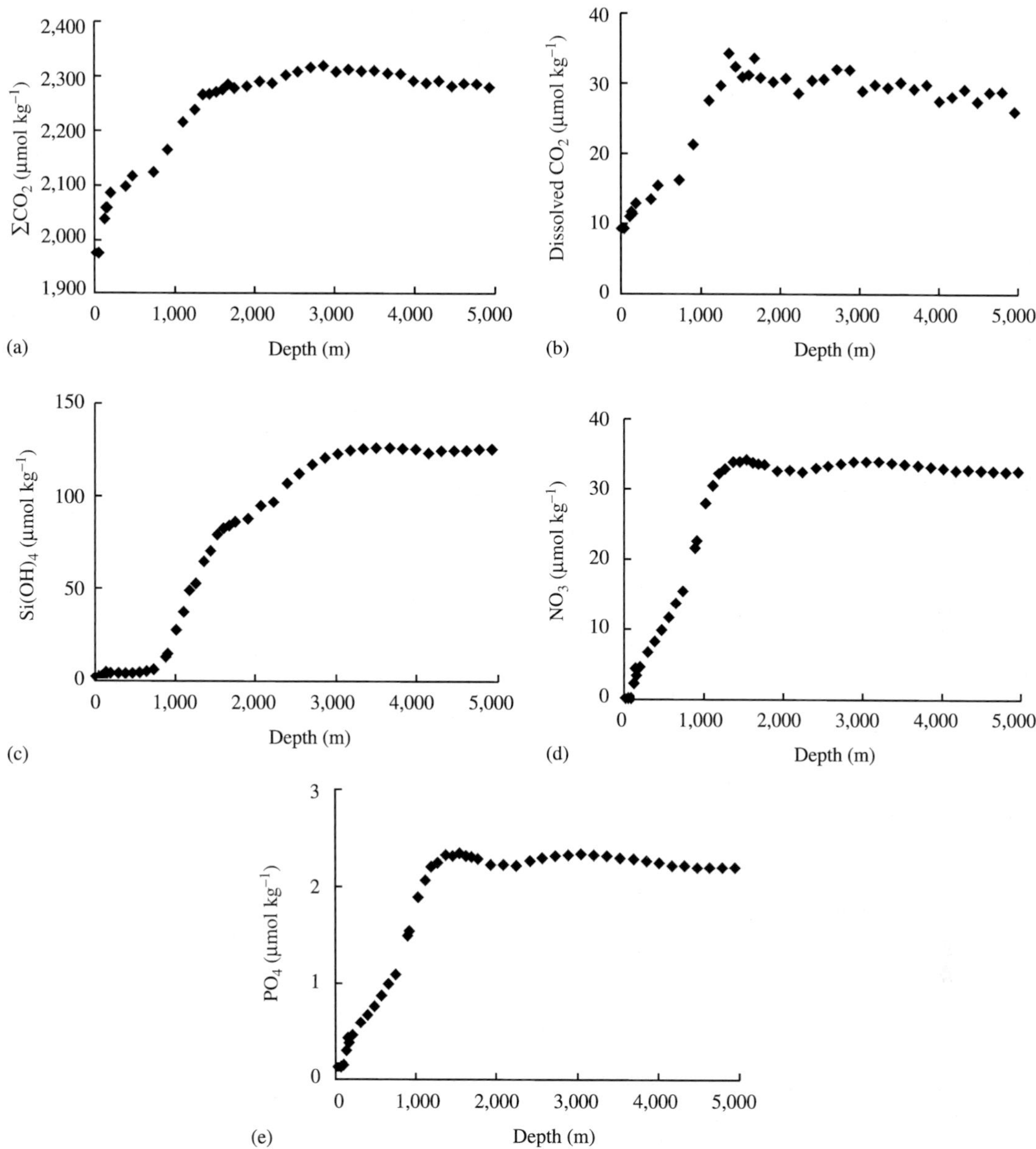

Figure 1 Depth profiles of: (a) $\sum CO_2$, (b) dissolved CO_2, (c) silicic acid, (d) nitrate, and (e) phosphate from the Indian Ocean (27° 4′ S, 56° 58′ E; GEOSECS Station 427) (source Weiss *et al.*, 1983).

matter and biominerals in surface waters, the sinking of these particles to the deep, and the decomposition of the settling (or settled) particles. In general, phytoplankton in surface waters take up DIC and nutrients. Carbon is fixed into organic material via photosynthesis and, together with nitrogen, phosphorus, and trace elements, form the carbohydrates, lipids, and proteins, which all comprise bulk organic matter. Once formed, this organic matter faces the immediate possibility of decomposition back to CO_2, phosphate, ammonia, and other dissolved nutrients through consumption by herbivorous zooplankton and degradation by bacteria. In fact, most of the primary production formed will be recycled within the upper hundred few meters of the water column (Martin *et al.*, 1987). Some portion of the primary production will, however, be exported to deeper waters or even to the sediments before decomposition and may escape remineralization entirely and remain in the sedimentary reservoir.

It is worth taking a closer look at the various steps in the biological pump (Figure 3). Rates, overall amounts, and the distribution and character of materials produced, transported, and decomposed vary wildly within the ocean.

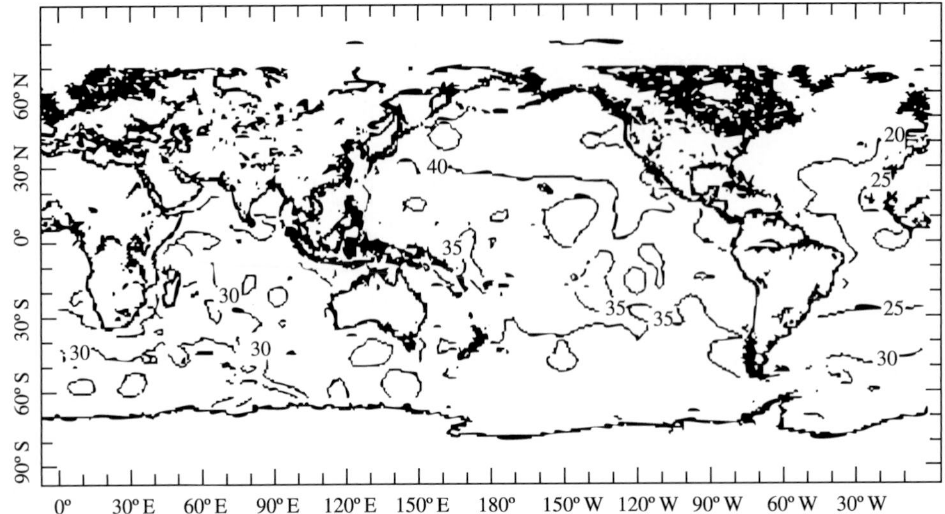

Figure 2 Nitrate concentrations along the great ocean conveyor at 2,000 m depth (source Levitus *et al.*, 1994, by way of the LDEO/IRI Data Library).

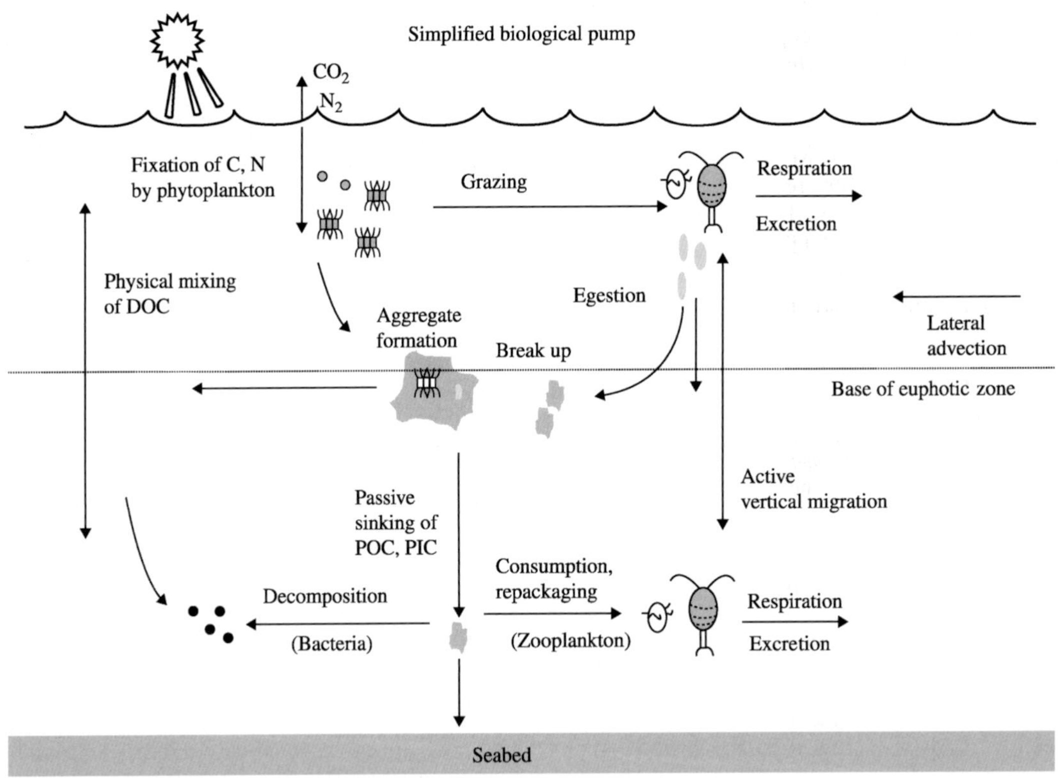

Figure 3 Diagram of the biological pump (after OCTET workshop report).

6.04.2.1 Photosynthesis and Nutrient Uptake

In the initial step of the biological pump, phytoplankton in sunlit surface waters convert CO_2 into organic matter via photosynthesis:

$$CO_2 + H_2O + \text{light} \longrightarrow CH_2O + O_2 \quad (1)$$

The first stable product of carbon fixation by the enzyme, ribulose bisphosphate carboxylase (Rubisco), is glyceraldehyde 3-phosphate, a 3-C sugar. This 3-C sugar is fed into biosynthetic pathways and forms the basis for all organic compounds produced by photosynthetic organisms. Fixed carbon and major and trace elements such as hydrogen, nitrogen, phosphorus, calcium, silicon, iron, zinc, cadmium, magnesium, iodine, selenium, and molybdenum are used for the synthesis of carbohydrates, lipids, proteins, biominerals, amino acids, enzymes, DNA, and other necessary biochemicals.

Besides carbon, the two main components of phytoplankton organic matter are nitrogen and phosphorus, in the average molar proportion of 106C:16N:1P (Redfield, 1934, 1958) known as the Redfield ratio. In addition, diatoms, by virtue of depositing opal (amorphous, hydrated silica) in their cell wall, have an average C:Si ratio of 8 (Brzezinski, 1985), although this ratio may vary from at least 3 to 40 depending on the conditions of light, temperature, and nutrient availability (Harrison *et al.*, 1977; Brzezinski, 1985). Coccolithophorids produce scales (coccoliths) made of $CaCO_3$ and contain 20–100 μmol of $CaCO_3$ per mol of organic carbon (Paasche, 1999).

Phytoplankton particulate matter (organic and biomineralized) contains many trace elements. The most abundant are magnesium, cadmium, iron, calcium, barium, copper, nickel, zinc, and aluminum (Table 1), which are important constituents of enzymes, pigments, and structural materials. Carbonic anhydrase requires zinc or cadmium (Price and Morel, 1990; Lane and Morel, 2000), nitrate reductase requires iron (Geider and LaRoche, 1994), and chlorophyll contains magnesium. Additionally, elements such as sodium, magnesium, phosphorus, chlorine, potassium, and calcium may be present as ions within cells and are important for osmoregulation and the maintenance of charge balance (e.g., Fagerbakke *et al.*, 1999).

A wide variety of ions may be adsorbed onto the surfaces of biogenic particles. The removal and deposition of particle-reactive elements such as thorium (Buesseler *et al.*, 1992) and protactinium (Kumar *et al.*, 1993) have been shown to correlate with the primary production of particles in the ocean. Additionally, thorium has been shown to complex with colloidal, surface-reactive polysaccharides (Quigley *et al.*, 2002).

Table 1 Elemental composition of marine phytoplankton from cultures and plankton tows.

Element	*Element : C ratio* (mol : mol)	*References*
N	0.15	b
Si (diatoms only)	0.13	c
P	0.009	b
Ca	0.03	d,e
Fe	2.3×10^{-6}–1.8×10^{-3}	d, e, f
Zn	6×10^{-5}	d,e
Al	1×10^{-4}	d,e
Cu	[a]3×10^{-6}–0.006	d,e
Ni	[a]2×10^{-5}–0.006	e
Cd	[a]5×10^{-7}–0.005	d,e
Mn	[a]4×10^{-6}–0.004	d,e
Ba	[a]1×10^{-5}–0.01	d,e
Mg	[a]0.02	d
Na	[a]0.1	d
Sr	[a]8×10^{-5}	d
Ti	[a]1×10^{-5}	d
Cr	[a]2×10^{-6}	d

[a] Calculated from dry weight data using an average phytoplankton C content on a dry weight basis of 50%. [b] Redfield (1958).
[c] Brzezinski (1985). [d] Martin and Knauer (1973).
[e] Collier and Edmond (1984). [f] Sunda and Huntsman (1995a).

6.04.2.1.1 *Levels of primary production*

The amount of primary production carried out in the oceans each year has been estimated from ocean color satellite data and shipboard ^{14}C incubations to be 140 g C m^{-2} for a total of 50–60 Pg C (4–5 Pmol C) fixed in the surface ocean each year (Shuskina, 1985; Martin *et al.*, 1987; Field *et al.*, 1998). This represents roughly half of the global annual 105 Pg C fixed each year (Field *et al.*, 1998), despite the fact that marine phytoplankton comprise less than 1% of the total photosynthetic biomass on Earth. Extrapolation from Redfield ratios suggests the incorporation 0.6–0.8 Pmol N, 40–50 Tmol P into biogenic particles each year in association with marine primary production. From the proportion of primary production carried out by diatoms and the average Si:C ratio of diatoms, silica production rates may be calculated to be 200–280 Tmol Si yr^{-1} (Nelson *et al.*, 1995; Tréguer *et al.*, 1995).

6.04.2.1.2 *Patterns in time and space*

Rates of primary productivity in upwelling regions of the ocean outpace those of non-upwelling coastal regions, which in turn are greater than rates in the oligotrophic open ocean (Figure 4; Ryther, 1969; Martin *et al.*, 1987). Open-ocean primary production levels are ~130 g C m^{-2} yr^{-1}, whereas in nonupwelling coastal areas and upwelling zones they are 250 g C m^{-2} yr^{-1} and 420 g C m^{-2} yr^{-1}, respectively (Martin *et al.*, 1987). However, because the open ocean constitutes 90% of the area of the ocean, the bulk (80%) of the ocean's annual carbon fixation occurs there rather than in coastal and upwelling regions.

Different types of phytoplankton dominate primary production in the different marine regimes. Diatoms perform roughly 75% of the primary production that occurs in upwelling and coastal regions of the ocean but less than 35% of that taking place in the open ocean (Nelson *et al.*, 1995). Phytoplankton biomass and primary production in the open ocean are dominated instead

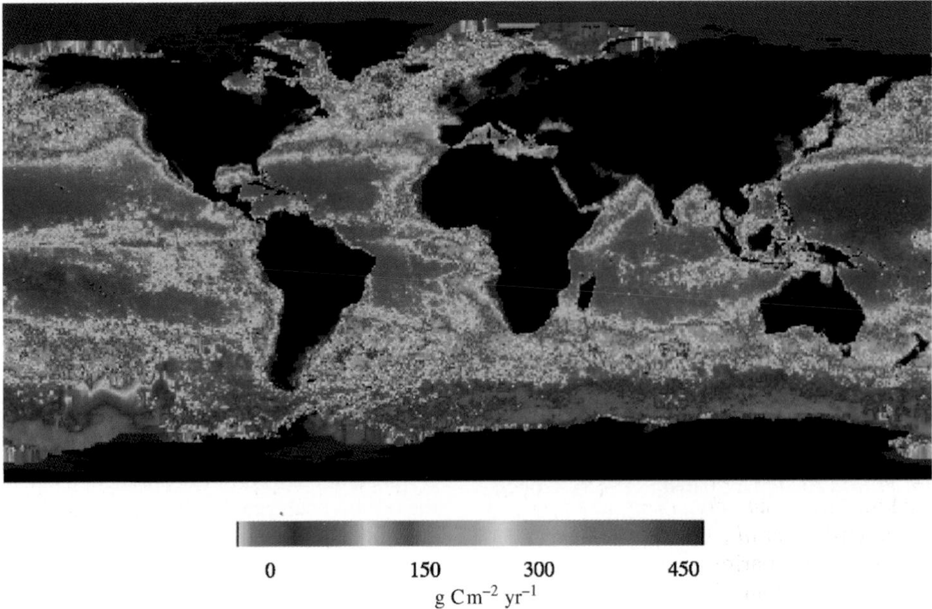

Figure 4 Distribution of primary production in the ocean (source ICMS, Rutgers University).

by prokaryotic picoplankton (Chisholm *et al.*, 1988; Liu *et al.*, 1999; Steinberg *et al.*, 2001).

Outside of the tropics, levels of marine primary productivity vary systematically throughout the year (Heinrich, 1962). Standing stocks of phytoplankton and levels of primary production peak in the spring following the onset of water column stratification and the increase in available light. Depletion of nutrients in the stratified water column in summer inhibits phytoplankton growth and grazing by zooplankton reduces standing stocks. Some areas may experience a small bloom of phytoplankton in the autumn when light levels are still adequate and the onset of winter convection and overturning injects nutrients into the euphotic zone.

6.04.2.1.3 Nutrient limitation

The upper limit of primary production is set by the supply of nutrients (nitrogen, phosphorus, silicon, iron) to the euphotic zone. Nitrogen inputs to the surface ocean may limit the primary productivity of the whole ocean over short timescales. Over timescales approaching and exceeding the $(1-5) \times 10^4$ yr residence time of phosphorus (Ruttenberg, 1993; Filippelli and Delaney, 1996), its inputs limit global ocean primary productivity (Tyrrell, 1999).

Regionally and for different types of phytoplankton, the limitation of both the rate and overall amount of primary production is more varied. Major nutrients in HNLC areas of the ocean (such as the Southern Ocean, the Equatorial Pacific, and the North Pacific subarctic) are never completely consumed in support of primary production, because low levels of iron limit phytoplankton growth (Martin and Fitzwater, 1988; Martin *et al.*, 1994; Coale *et al.*, 1996; Boyd *et al.*, 2000). Diatoms, which, unlike other dominant members of the phytoplankton, require silicon for growth, are often limited by low concentrations of silicic acid in surface waters (Brzezinski and Nelson, 1996; Nelson and Dortch, 1996). Growth of diazotrophic (N_2 fixing) phytoplankton such as the cyanobacteria, *Trichodesmium*, will be more susceptible to phosphorus and iron limitation, of course, than to nitrogen limitation. Even the concentration of dissolved CO_2 in seawater (especially in the midst of a phytoplankton bloom) may limit instantaneous rates, although not ultimate levels, of primary production (Riebesell *et al.*, 1993; Wolf-Gladrow *et al.*, 1999).

6.04.2.2 Flocculation and Sinking

6.04.2.2.1 Marine snow

The primary formation of biogenic particles in the euphotic zone represents the maximum amount of material that may be transported into the deep ocean or sediments. In practice, however, less than half of these particles survive zooplankton grazing and microbial attack long enough to be exported from the euphotic zone, and only a few percent endure to settle into the deep ocean and sediments (Martin *et al.*, 1987). Material that reaches the deep ocean and seafloor does so not as individual phytoplankton cells slowly meandering down towards the bottom, but rather arrives as larger, rapidly sinking particles (McCave, 1975; Suess, 1980; Billett *et al.*, 1983; Fowler and Knauer, 1986; Alldredge and Gotschalk, 1989) that have traversed the distance between surface and deep in a matter of days (Billett *et al.*, 1983; Asper *et al.*, 1992).

These larger particles, known collectively as "marine snow" (Alldredge and Silver, 1988), are formed either by zooplankton, which produce mucous feeding structures and fecal pellets, or by the physical coagulation of smaller particles (McCave, 1984; Alldredge *et al.*, 1993). Coagulation is the more important of the two formation pathways. The bulk of the organic material reaching the deep sea does so as aggregated phytoplankton that has not been ingested by zooplankton (Billett *et al.*, 1983; Turner, 2002).

Sinking rates of marine snow are orders of magnitude greater than those of unaggregated phytoplankton cells (Smayda, 1970; Shanks and Trent, 1980; Alldredge and Gotschalk, 1989). The shorter transit time from surface to bottom for aggregated particles results in the enhanced transport of carbon, nitrogen, phosphorus, silicon, and other materials to the deep sea and sediments despite the fact that marine snow particles are sites of elevated rates of decomposition and nutrient regeneration. Intense colonization and hydrolysis of the particles by bacteria (Smith *et al.*, 1992; Bidle and Azam, 1999, 2001) and breakup and consumption of the particles by zooplankton (Steinberg, 1995; Dilling *et al.*, 1998; Dilling and Alldredge, 2000) reduce the vertical flux of materials to the seafloor.

6.04.2.2.2 Aggregation and exopolymers

Coagulation requires the success of two activities: the collision of particles and their subsequent joining to form an aggregate. In the ocean, particles collide due to processes such as shear, Brownian motion, and differential settling (Kepkay, 1994). The probability of particles attaching following a collision is controlled by the physical and chemical properties of the particles' surfaces (Alldredge and Jackson, 1995). The probability of sticking is greatly enhanced by exopolymers produced by phytoplankton and bacteria (Alldredge and McGillivary, 1991; Alldredge *et al.*, 1993; Passow, 2000; Engel, 2000).

These exopolymers, known as transparent exopolymer particles (TEPs; Passow, 2002), turn out to be important for the transport of material to the deep. The formation of rapidly sinking aggregates is controlled more by TEP abundance than by phytoplankton concentrations (Logan *et al.*, 1995) and TEP is required for the aggregation and sedimentation of diatoms out of the water column (Passow *et al.*, 2001).

Little is known about the chemical characteristics of the polymers responsible for particle aggregation in marine systems. They are comprised of acidic polysaccharides (Alldredge *et al.*, 1993; Mopper *et al.*, 1995) and proteins (Long and Azam, 1996). The carbohydrate component of TEP contains glucose, mannose, arabinose, xylose, galactose, rhamnose, glucuronate, and O-methylated sugars (Janse *et al.*, 1996; Holloway and Cowen, 1997), and is generally rich in deoxysugars (Mopper *et al.*, 1995). Very little else is known about the specific composition of TEP, and virtually nothing is known of its structural characteristics (Holloway and Cowen, 1997; Schumann and Rentsch, 1998; Engel and Passow, 2001).

Exopolymer particles are formed from dissolved organic matter (DOM) and continue to scavenge DOM as they grow, providing a mechanism for the biological pumping of DOM into the deep sea (Engel and Passow, 2001). TEP also contains carbon and nitrogen in proportions exceeding Redfield ratios (Mari *et al.*, 2001; Engel *et al.*, 2002), providing a mechanism for pumping of carbon in excess of what would be predicted from the availability of nitrogen.

6.04.2.2.3 Sinking

Sinking rates of solitary phytoplankton cells are only about a meter per day (Smayda, 1970). Particles that sink this slowly require over a year to reach the benthos of the relatively shallow continental shelf, and ten years to reach the abyssal ocean floor. Given the rapid rates of microbial decomposition of organic material in the ocean and the abundance of zooplankton grazers, it is virtually impossible for such a slowly sinking particle to reach the seafloor.

Sinking rates of marine snow, however, are greater than $100\ m\ d^{-1}$ (Shanks and Trent, 1980; Alldredge and Gotschalk, 1989). Transit time to the deep in this case is days to weeks, which agrees with observations of a close temporal coupling between surface production and seafloor sedimentation (e.g., Billett *et al.*, 1983; Asper *et al.*, 1992).

It may easily be argued then that particle flux is controlled by rates of particle aggregation and sinking, perhaps more than it is controlled by overall levels of primary production. For example, year-to-year variability in carbon export to deep waters correlates more strongly with the size of the dominant primary producer than with year-to-year variations in levels of carbon fixation (Boyd and Newton, 1995).

6.04.2.3 Particle Decomposition and Repackaging

Organic matter in the ocean rapidly decomposes and there is intense recycling of elements even within the euphotic zone. The flux of particulate organic carbon (POC) in the ocean decreases exponentially with depth below the euphotic zone (Figure 5; Martin *et al.*, 1987). New production constitutes, on average, only 20% of the total primary production in the sea (Harrison, 1990; Laws *et al.*, 2000). Mediating this decomposition and recycling are zooplankton and heterotrophic bacteria (Cho and Azam, 1988; Smith *et al.*, 1992; Steinberg, 1995; Dilling *et al.*, 1998; Dilling and Alldredge, 2000). Bacteria and zooplankton diminish the sinking particulate flux by both consuming particles and converting them back to CO_2 and dissolved materials, and by converting large, sinking particles into smaller particles with reduced or nonexistent sinking rates.

Although the bulk of particles are broken down in the surface ocean, midwater processes are also important. Midwater decomposition of sinking particles deflates the regional variability in fluxes of POM. The POC flux range, 0.5–12 g C m^{-2} yr^{-1}, among different regions in the Atlantic at 125 m is compressed by 85% to 0.5–2.4 g C m^{-2} yr^{-1} by a depth of 3,000 m due to the biological consumption and repackaging of particles at depth (Anita *et al.*, 2001).

6.04.2.3.1 *Zooplankton grazing*

Zooplankton may reduce the sinking flux of biogenic particles in the ocean in two ways. The first is by grazing upon particles which reduces the total amount of particulate organic material (POM) in the water column and shifts its occurrence from large, fast-sinking aggregates to smaller fecal pellets which constitute only a minor portion of the sinking organic flux (Turner, 2002). The second way they reduce the particle flux is by actively breaking up aggregates into smaller particles. At stations off Southern California, for instance, the average overnight increase in the number of aggregates per liter by 15% was attributable to the fragmentation of larger particles by swimming euphausiids (Dilling and Alldredge, 2000).

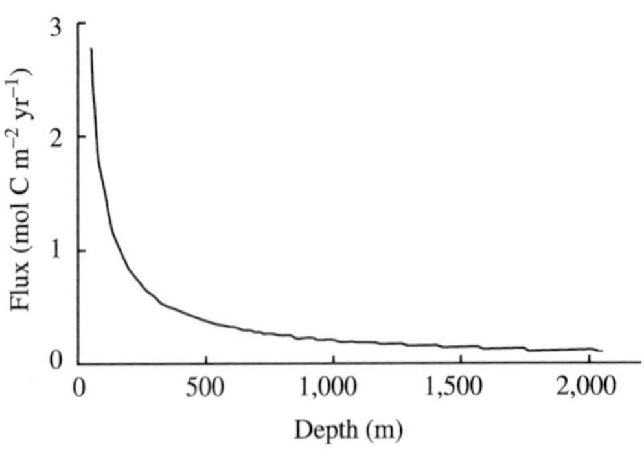

Figure 5 Sinking fluxes of C in the open ocean. Curve shown is the exponential relationship (flux = 1.53 $(z/100)^{-0.858}$) shown for carbon flux in the northeast Pacific (after Martin *et al.*, 1987).

The relative impact of zooplankton grazing on primary production decreases with increasing production levels; the proportion of primary production that is consumed by zooplankton decreases exponentially as productivity levels increase (Calbet, 2001). This supports the observation that the ratio of export production to total production is higher in areas of high productivity. Globally, ~12% of marine primary production, or 5.5 Pg C (0.5 Pmol C), is consumed by mesozooplankton each year (Calbet, 2001).

6.04.2.3.2 *Bacterial hydrolysis*

Bacterial hydrolysis plays a major role in the decomposition of sinking and suspended matter in the ocean. Bacterial organic carbon has been observed to make up over 40% of the total POC in the water column, and the proportion of the sinking flux of carbon utilized by bacteria may be equal to 40–80% of the surface primary production in near-shore areas (Cho and Azam, 1988). Marine aggregates have been shown to contain high concentrations of hydrolytic exoenzymes, such as proteases, and polysaccharidases like glucosidases (Smith *et al.*, 1992). Turnover times of organic components of marine aggregates due to hydrolysis may be short, on the order of fraction of a day to a few days (Smith *et al.*, 1992). Bacterial proteases have also been shown to enhance the dissolution rates of biogenic silica in diatom aggregates (Bidle and Azam, 2001). Much of the hydrolyzed material is not taken up by the bacteria attached to the particles but instead joins the pool of DOM present in the water (Smith *et al.*, 1992).

6.04.2.3.3 *Geochemistry of decomposition*

Ingestion of POM by zooplankton results in the respiration and excretion of a portion of that POM as CO_2, NH_4, dissolved organic nitrogen (DON), phosphate, and dissolved organic phosphorus (DOP). Assimilation efficiencies of organic matter for zooplankton grazing on phytoplankton range from 10% to 40% (Ryther, 1969; Michaels and Silver, 1988). Zooplankton do not assimilate significant quantities of silicon from diatoms consumed, leaving regeneration of silicic acid to be mediated strictly by opal dissolution rates.

6.04.2.4 Sedimentation and Burial

Of the 50–60 Pg C (4–5 Pmol C) fixed into organic material in the surface ocean each year (Shuskina, 1985; Martin *et al.*, 1987; Field *et al.*, 1998) and the 16 Pg C (1.3 Pmol C) exported to the deep sea (Falkowski *et al.*, 2000), only ~0.16 Pg C (0.13 Pmol C) reach the seafloor (Hedges and Keil, 1995). Each year roughly 0.16 Pg C (13 Tmol C) are actually preserved in ocean sediments (Hedges and Keil, 1995). Thus, the rate of accumulation of organic carbon by ocean sediments is 0.5% of that of carbon fixation in the surface ocean.

The accumulation of sedimentary organic carbon in the ocean varies greatly from place to place, with continental margin sediments accounting for the bulk of the organic carbon buildup (Hedges and Keil, 1995; de Haas *et al.*, 2002). About 94% of the sedimentary organic carbon that is preserved in the oceans is buried on continental shelves and slopes (Hedges and Keil, 1995). This leaves only 6% of the total sedimentary organic carbon to accumulate in the open ocean. Since the open ocean, due to its vast area, plays host to 80% of the annual primary production, the accumulation of only 6% of the organic carbon suggests an overall preservation efficiency of 0.02% in the open ocean. On the continental shelves and slopes, this preservation efficiency is, by comparison, large at 1.4%.

Many reasons have been suggested for the regional differences in the accumulation and preservation of organic carbon in the sediments. Differences in the flux of organic particles to the seafloor due to differences in overhead primary production levels, rates of aggregation and sinking, and depth of the water column may contribute to a higher preservation efficiency. The oxygen content of bottom waters has also been suggested as important, although a correlation between burial efficiency and bottom water oxygen concentration is not seen (Hedges and Keil, 1995) and rates of organic matter hydrolysis by bacteria may still be high, even under anoxic conditions (e.g., Arnosti *et al.*, 1994). The long-term preservation of organic material in sediments may be tied to the sorption of the organic molecules to mineral surfaces (Hedges and Keil, 1995), although the nature of the associations and the rates at which they occur have not been closely detailed.

6.04.2.5 Dissolved Organic Matter

DOM has not been as intensively studied as other aspects of the biological pump perhaps because DOM does not sink. However, DOM does play an active role in the biological pump in at least three ways. Much DOM can be utilized biologically and may directly provide phosphorus and nitrogen for primary production (Clark *et al.*, 1998; Zehr and Ward, 2002). DOM may assemble into colloidal and particulate material that can sink as well as scavenge other material to form marine snow (Alldredge *et al.*, 1993; Kepkay, 1994; Chin *et al.*, 1998). DOM is also a large reservoir of carbon in the ocean, containing at least an order of magnitude more carbon than the other organic carbon reservoirs in the ocean (Kepkay, 1994).

Although the origins of DOM have not been fully detailed, phytoplankton can serve as the dominant source of DOM to the ocean. Actively growing phytoplanktons secrete DOM (Biddanda and Benner, 1997; Soendergaard *et al.*, 2000; Teira *et al.*, 2001) and the polysaccharide composition of phytoplankton exudates resembles that of the high molecular weight fraction of DOM (Aluwihare and Repeta, 1999). Phytoplankton DOM is also released during grazing by zooplankton (Strom *et al.*, 1997). Organic matter, such as mucus, on phytoplankton cell surfaces may also be hydrolyzed by bacteria and released as DOM (Smith *et al.*, 1995).

The exact composition of marine DOM is unknown. It has, as of early 2000s, been shown to contain carbohydrates, which consist largely of polysaccharides, and amino acids, amides (such as chitin), phosphorus esters, and phosphonates (Benner *et al.*, 1992; McCarthy *et al.*, 1997; Clark *et al.*, 1998; Amon *et al.*, 2001). Microbial degradation could play a role in setting the composition of DOM in the ocean (Amon *et al.*, 2001).

One feature that has great relevance to the importance of DOM to the carbon cycle is its enrichment in carbon over the Redfield proportion. The C : N : P ratios of high molecular weight DOM are on the order of 350 : 20 : 1 (Kolowith *et al.*, 2001). Carbon enrichment is also observed for bulk DOM (Kaehler and Koeve, 2001). Part of this enrichment in carbon may be due to the enhanced remineralization of phosphorus and nitrogen from DOM (Clark *et al.*, 1998; Kolowith *et al.*, 2001), as suggested by an increase in the C : P ratio of DOM with depth (Kolowith *et al.*, 2001). DOM may also just simply be produced with high C : N and C : P ratios. TEPs which form from DOM precursors (Alldredge *et al.*, 1993; Chin *et al.*, 1998) have high C : N ratios (Engel and Passow, 2001; Mari *et al.*, 2001). The polysaccharides that eventually form TEP are exuded by phytoplankton and may represent excess photosynthate (Engel, 2002), carbohydrates, and lipids formed when nutrient limitation shuts off the supply of, for example, the nitrogen needed for the synthesis of nitrogen-containing compounds such as proteins (Morris, 1981).

The importance of the carbon-enriched DOM pool as a reservoir in the carbon cycle hinges upon both the turnover time and amount of the carbon therein. Estimates for the amount of dissolved organic carbon (DOC) in the ocean vary, although an estimate places the size of the pool at 200 Pg C (Kepkay, 1994), which is more comparable to the 750 Pg of carbon present in the atmosphere than it is to the 3.6×10^4 Pg C deep-sea reservoir of DIC (Sundquist, 1993). The average age of marine DOM is ~6,000 yr (Williams and Druffel, 1987). The bulk (~70%) of the DOM is low molecular weight (Benner *et al.*, 1992) and relatively resistant to microbial degradation (Bauer *et al.*, 1992; Amon and Benner, 1994). High molecular weight compounds that are quickly turned over by bacterial decomposition (Amon and Benner, 1994) make up the remaining 30% of the DOM pool.

6.04.2.6 New, Export, and Regenerated Production

Not all of the primary production in the ocean feeds carbon into the biological pump. The vast portion of carbon fixed globally each year in the euphotic zone is remineralized by zooplankton and bacteria in the euphotic zone and converted straight back to CO_2 and dissolved nutrients. These recycled nutrients may then be used to fuel further carbon fixation.

It has long been recognized (Dugdale and Goering, 1967) that a portion of the primary production (regenerated production) is supported by nutrients regenerated in the euphotic zone, and another portion (new production) is supported by nutrients imported into the euphotic zone through upwelling, river inputs, nitrogen fixation, or atmospheric deposition. The ratio of new to total primary production in the ocean, known as the *f*-ratio (Eppley and Peterson, 1979), is generally higher in upwelling environments than it is in oligotrophic regions of the ocean (Harrison, 1990; Laws *et al.*, 2000). On average, ~20% of the total global marine primary production is new production, although the range of values from region to region is ~0.07–0.7 (Laws *et al.*, 2000).

6.04.3 IMPACT OF THE BIOLOGICAL PUMP ON GEOCHEMICAL CYCLING

6.04.3.1 Macronutrients

6.04.3.1.1 Carbon

The influence of the biological pump on the distribution of DIC carbon in the ocean may serve as a rough model for the influence it has on the distribution of a score of other elements. Low concentrations of DIC are observed in surface waters (Figure 1) due to the uptake of dissolved CO_2 (and perhaps HCO_3^-; Raven, 1997) by phytoplankton. Concentrations of dissolved CO_2 increase most rapidly just below the euphotic zone, associated with the bulk of the decomposition of POC (Martin *et al.*, 1987; Anita *et al.*, 2001). Deep-water concentrations of dissolved CO_2 are higher than surface-water concentrations, and older deep waters contain more dissolved CO_2 than younger ones (Broecker and Peng, 1982).

Much of the current scientific interest in the biological pump revolves around the impact it has on levels of CO_2 in the atmosphere and, subsequently, on climate. This biological fixation of carbon into organic matter through photosynthesis lowers the concentration of dissolved CO_2 in surface waters and thus allows for the influx of CO_2 from the atmosphere. This fixed carbon may then be exported to deeper waters or the sediments before it decomposes back to CO_2, maintaining the observed gradient in dissolved CO_2 concentrations between waters of the surface and deep (Figure 1). Atmospheric concentrations of CO_2 are thus lower for the given size of the oceanic DIC reservoir than they would be in the absence of this biological transport of carbon to the deep. If all life in the ocean were to die off and the ocean and atmosphere came to equilibrium with respect to CO_2, concentrations of CO_2 in the atmosphere would rise by ~140 μatm (Broecker, 1982), which is a remarkable 50% of the pre-Industrial interglacial value of 280 μatm.

Details of the influence of the biological pump on the distribution and cycling of carbon in the ocean and the controlling factors are discussed later in this chapter. We briefly discuss the relationship of the biological pump to the cycling of other elements. This is by no means an exhaustive overview of the biological shuffling of elements throughout the ocean, but instead a highlight of several elements of particular biogeochemical interest.

6.04.3.1.2 Nitrogen

Of all the elements playing an important role in the regulation of the biological pump, nitrogen is the one with the most complex biologically mediated cycling. Nitrogenous species taking part in productivity range from N_2, which may be fixed into a more universally biologically available form by nitrogen fixing bacteria, to NO_3^-, which follows from the production of NO_2^- from NH_4^+ through nitrification (Figure 6). The denitrification pathway sequentially results in the transformation of NO_2^- to N_2O and N_2. In addition to inorganic nitrogen species are dissolved organic forms, such as amides, urea, free amino acids, amines (McCarthy *et al.*, 1997), which can be utilized biologically, although to varying degrees.

Most of the primary production in the ocean is supported by dissolved inorganic nitrogen (DIN)

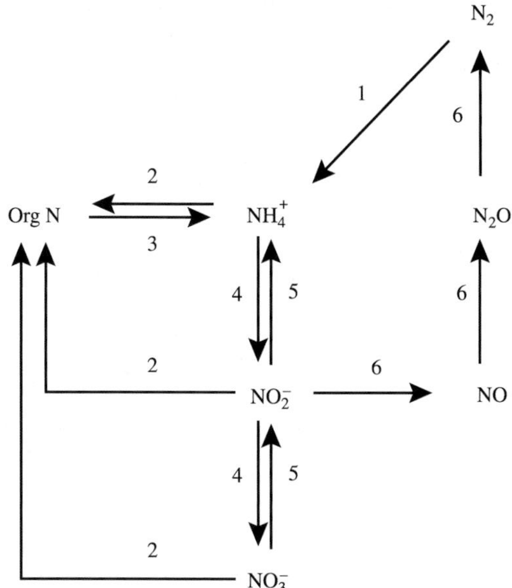

Figure 6 Microbial transformations of the nitrogen cycle. Pathways depicted are: 1—N_2 fixation; 2—DIN assimilation; 3—ammonium regeneration; 4—nitrification; 5—nitrate/nitrite reduction; and 6—denitrification.

that has recycled in the euphotic zone, as suggested by the average marine *f*-ratio of 0.2 (Laws *et al.*, 2000). This further suggests that the bulk of the primary production in the ocean relies on NH_4^+, since that is the predominant recycled form of nitrogen. Extrapolating from the Redfield C : N ratio of 6.6 and the average *f*-ratio of 0.2 and the overall estimate of primary production in the ocean each year of 4–5 Pmol C (Shuskina, 1985; Martin *et al.*, 1987; Field *et al.*, 1998) yields 0.7 Pmol of particulate organic nitrogen (PON) produced each year, 0.1 Pmol of which is exported out of the euphotic zone.

Nitrogen fixation. The two aspects of the nitrogen cycle having the greatest impact on the biological pump are nitrogen fixation and denitrification. The first provides a mechanism for drawing on the extensive atmospheric pool of N_2 gas in support of primary production. The second provides a pathway for DIN to be converted back to N_2 gas and removed from the ocean system.

In the ocean ~28 Tg N (2 Tmol N) are fixed each year (Gruber and Sarmiento, 1997). Nitrogen fixation accounts for about half of the new nitrogen used in primary production (Karl *et al.*, 1997). Only prokaryotic organisms can fix nitrogen, leaving this, at least in the ocean, in the hands of the cyanobacteria and out of the hands of eukaryotic algae such as diatoms, dinoflagellates, and coccolithophorids (unless they are hosting diazotrophic symbionts). Until recently it was believed that the filamentous, colony-forming cyanobacterium *Trichodesmium* and cyanobacterial symbionts in diatoms were responsible for the bulk of the nitrogen fixation occurring in the ocean (Capone *et al.*, 1997). However, direct measurements of the rates of nitrogen fixation by *Trichodesmium*, coupled with knowledge of their distribution and abundance, fell significantly short of nitrogen fixation rates calculated from geochemical budgets (Gruber and Sarmiento, 1997). It has now been discovered that free-living, unicellular cyanobacteria are expressing the genes for the nitrogen-fixing enzyme, nitrogenase (Zehr *et al.*, 2001), and may be contributing considerably to fixed nitrogen budgets in the ocean.

It has been suggested that rates of nitrogen fixation in the modern ocean are limited by the availability of iron. The iron requirement of *Trichodesmium*, however, turns out to be much lower than previously estimated. Instead, it is the availability of phosphate that controls the upper limit of nitrogen fixation in the modern ocean (Sañudo-Wilhelmy *et al.*, 2001).

The demonstration of the phosphorus limitation of nitrogen fixation by cyanobacteria supports the notion that over geologic time phosphorus ultimately limits productivity (Tyrrell, 1999). When cyanobacteria face a shortfall of nitrogen, they fix nitrogen to meet their demands. This influx of new nitrogen to nitrogen-limited systems allows them to draw down levels of phosphate until the system is phosphorus limited. Under phosphorus limitation, nitrogen fixation is curtailed (Sañudo-Wilhelmy *et al.*, 2001). Thus, while nitrogen may often be limiting to instantaneous rates of carbon fixation, as is frequently the case in the modern ocean, over long time periods the input of phosphorus to the oceans sets the upper attainable limit for net primary production (Tyrrell, 1999).

The control of nitrogen fixation by phosphorus availability makes the nitrogen cycle stand out against all of the other global biogeochemical cycles. In an astounding twist, the biological demand for one element relative to the other controls the *input* flux of one of the elements. Confirmation of this comes in the form of Redfield ratios, since the N : P ratio in both the seawater reservoir and the marine organic matter output is same (16 : 1) and significantly greater than unity. Due to mass balance requirements, assuming steady state over long timescales, the only way this can be the case is if the ratio of N : P inputs to the ocean is also 16 : 1. It is far more likely that the equality of N : P ratios of inputs, outputs, and reservoir is due to the grand control of nitrogen fixation by phosphate than it is likely to be due to remarkable coincidence.

6.04.3.1.3 Phosphorus

Although there is a tendency to consider dissolved inorganic phosphorus (DIP, measured as soluble reactive phosphate; Strickland and Parsons, 1968) as being simply PO_4^{3-}, DIP exists as a considerable number of species. At seawater pH, DIP is predominantly $H_2PO_4^-$ (87%) and only 12% PO_4^{3-} and 1% $H_2PO_4^-$ (Greenwood and Earnshaw, 1984). There are also numerous dissolved organic forms of phosphorus that are taken up by phytoplankton and used to fuel primary production in the ocean.

One interesting aspect of the phosphorus cycle is that, unlike the cases for the other major nutrient elements in the ocean, the phosphorus cycle faces the complexity of containing sinks that are not mediated by biological activities. For example, dissolved phosphate may scavenge onto iron or manganese oxyhydroxide particles associated with hydrothermal activity or react with basalt during the circulation of water through mid-ocean ridge hydrothermal systems (Föllmi, 1996). The exact values are poorly known, but it is estimated that the scavenging of phosphate by the hydrothermal oxyhydroxides may constitute up to 50% of the removal flux of phosphorus from the ocean (e.g., Froelich *et al.*, 1982; Berner *et al.*, 1993). Another large inorganic sink for dissolved phosphorus is the precipitation of authigenic phosphate, which account for somewhere between about 10% and 40% of the removal flux. Removal of phosphorus as sedimenting POC by comparison is thought to be 20–50% of the output of phosphorus from the ocean.

Roughly 5 Tg P (0.2 Tmol P) are removed from the ocean each year as both organic and inorganic phases. This is in reasonable balance with the roughly 5 Tg of reactive phosphorus being brought in to the system each year naturally. However, these natural sources constitute only half of the modern-day input of phosphorus to the ocean (Froelich *et al.*, 1982), anthropogenic inputs having doubled the annual phosphate flux.

A doubling of phosphorus inputs to the ocean could have a significant impact on the biological pump, although it should be noted that the ocean is already nitrogen limited and further inputs of phosphorus will only exacerbate this. Shifts in N:P ratios of surface waters may alter the structure of the phytoplankton community. For example, *Phaeocystis* grows poorly under high-phosphate conditions and may see its numbers declining.

6.04.3.1.4 Silicon

Dissolved silicic acid is required for the growth of diatoms, which deposit opal (amorphous, hydrated silica) in their cell wall and dominate the production of opal in the modern-day ocean (Lisitzin, 1972). Silica (240 Tmol) is produced by diatoms in the surface ocean each year (Nelson *et al.*, 1995; Tréguer *et al.*, 1995). Total inputs of dissolved silicic acid to the euphotic zone are, by comparison, 120 Tmol Si, mostly from rivers (5 Tmol Si) and upwelling (115 Tmol Si; Tréguer *et al.*, 1995). Half of the biogenic silica produced each year dissolves in the upper 100 m of the water column (Nelson *et al.*, 1995), and a further 47% dissolves in the deep ocean and seafloor, for a net deposition of 6 Tmol (3% of surface production) each year (Tréguer *et al.*, 1995).

Opal accumulation on the seafloor. Opal preservation efficiencies are generally highest in productive environments (e.g., 6% in the permanently open ocean zone of the Southern Ocean versus 0.4% in the oligotrophic North Atlantic). Given numbers such as these, the traditional view of opal accumulation in the sediments is that it is closely linked to opal production in overlying waters (e.g., Broecker and Peng, 1982), but many factors besides opal production govern opal preservation. Opal dissolution rates more than double for every 10 °C rise in temperature (Kamitani, 1982). Aggregation may reduce rates of silicon regeneration from diatoms (Bidle and Azam, 2001). Additionally, the fraction of produced silica that reaches the seabed correlates with high seasonality and high ratios of carbon export to production (Pondaven *et al.*, 2000), which also tend to be areas where organic matter is formed in blooms and is transported quickly to the sediments as large aggregates.

The impact of the appearance of the diatoms on the marine silica cycle. The silica cycle is an excellent example of how much the biological pump can impact concentration and distribution of elements in seawater. Presently, surface concentrations of silicic acid are low: the overall average silicic acid content of ocean waters is only 70 μM (Tréguer *et al.*, 1995). Prior to the appearance of the diatoms in the Early Tertiary (Tappan and Loeblich, 1973), oceanic concentrations of silicic acid ranged ~1,000 μM Si. The sponges and radiolarians controlling the oceanic inventory of silicic acid at that time possessed neither the numbers nor the need to draw down silicic acid concentrations. The increased output of biogenic silica from the ocean associated with the ascension of diatoms, with their high affinity for silicic acid and high cellular requirements for silicon, resulted in a precipitous drop in the silicic acid content of ocean waters over the Late Cretaceous and Paleocene (Figure 7), stabilizing in the Eocene to the <100 μM values that have held ever since (Siever, 1991).

Since sponges and radiolarians are not great players in particle flux, the rise of the diatoms must have profoundly altered the partitioning of silicic acid between surface and deep. The familiar nutrient-type distribution may have only existed for the last 50–100 million years. The approximate 14-fold drop in silicic acid concentration also suggests that the residence time of silicon in

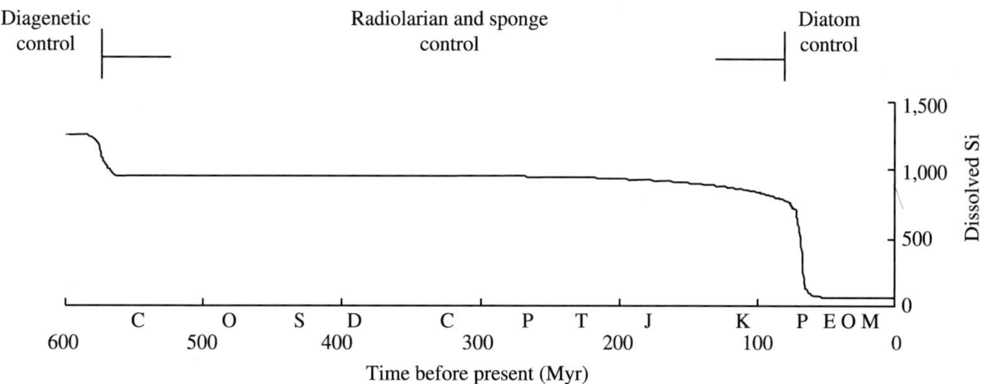

Figure 7 Estimated average marine concentrations of silicic acid over the Phanerozoic (after Siever, 1991).

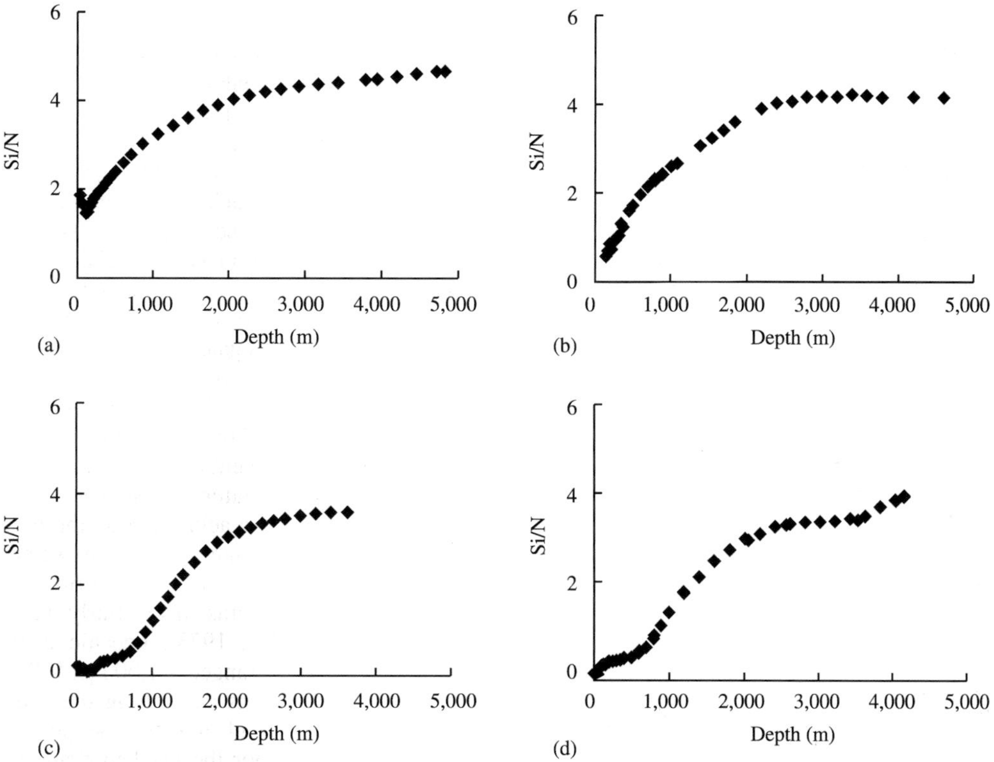

Figure 8 Ratios of silicic acid to nitrate with depth in the Pacific. Profiles shown are from: (a) the NE Pacific WOCE station 66 (47° 33′ N, 145° 33′ W); (b) HOT station Aloha (22° 45′ N, 158° 00′ W); (c) SE Pacific WOCE station 288 (38° 60′ S, 88° 00′ W); and (d) SE Pacific WOCE station 241 (53° 20′ S, 76° 36′ W).

the ocean dropped from 2×10^5 yr to today's 1.5×10^4 yr.

Excessive pumping of silicon. Silicic acid, by virtue of being regenerated from silica instead of from relatively labile POM, is regenerated more deeply than the other major nutrients (Figure 1; Dugdale *et al.*, 1995). This decoupling between silicic acid and the other nutrients, combined with the fact that not all phytoplankton utilize silicon, results in there being no Redfield relationship between silicon and carbon, nitrogen, or phosphorus. Upwelled waters contain silicic acid and nitrate close to the ~1 : 1 molar ratio of utilization by diatoms, but Si : N ratios increase with depth below the euphotic zone (Figure 8).

Impact of iron on silicon pumping. Iron limitation of diatoms increases the pumping of silicon (relative to nitrogen and carbon) to deeper waters. Iron-limited diatoms are inhibited in their utilization of nitrogen as a result of the iron requirement of nitrate reductase (Geider and LaRoche, 1994). However, iron-limited diatoms continue to take up silicic acid, although at lowered rates (De La Rocha *et al.*, 2000). As a result, the Si : N ratios of iron-limited diatoms may be as high as 2 or 3 (Takeda, 1998; Hutchins and Bruland, 1998), much higher than the 0.8 of nutrient-replete diatoms (Brzezinski, 1985).

6.04.3.2 Trace Elements

The biological pump influences, to varying degrees, the distribution of many elements in seawater besides carbon, nitrogen, phosphorus, and silicon. Barium, cadmium, germanium, zinc, nickel, iron, selenium, yttrium, and many of the REEs show depth distributions that very closely resemble profiles of the major nutrients. Additionally, beryllium, scandium, titanium, copper, zirconium, and radium have profiles where concentrations increase with depth, although the correspondence of these profiles with nutrient profiles is not as tight (Nozaki, 1997).

6.04.3.2.1 Barium

Vertical profiles of dissolved barium (Ba^{2+}) in the ocean resemble profiles of silicic acid and alkalinity (Figure 9; Lea and Boyle, 1989; Jeandel *et al.*, 1996), suggesting that biological processes strongly influence barium distributions throughout the ocean. However, the strict incorporation of barium into biogenic materials is not the dominant means of Ba^{2+} removal from ocean waters. Despite the similarity between the profiles of Ba^{2+} and $Si(OH)_4$ which suggests a common removal phase, the amount of barium incorporated into diatom opal ($<9 \times 10^{-6}$ mol Ba per mol Si; Shemesh *et al.*, 1988) cannot account for the 2×10^{-4} mol Ba^{2+} per mol $Si(OH)_4$ slope (Jeandel *et al.*, 1996) in the ocean. Ba^{2+} appears instead to be mainly removed from seawater as barite ($BaSO_4$) formed in association with opal and decaying organic material (Dehairs *et al.*, 1980; Bishop, 1988). The exact mechanism for barite precipitation is unknown, but it is thought that it forms in the SO_4^{2-}-enriched microenvironments of decaying particles that may be, thus, supersaturated with respect to barite (Dehairs *et al.*, 1980).

Although the marine budget of barium is only approximately known, it does appear to be both balanced and controlled by biogenic particle formation. Approximately 35 Gmol of Ba^{2+} are removed from surface waters every year (Dehairs *et al.*, 1980). Of this 35 Gmol, 60% settles as barite and the rest is incorporated into or adsorbed onto phases such as $CaCO_3$ or SiO_2 (Dehairs *et al.*, 1980; Dymond *et al.*, 1992). Barium (10–25 Gmol) is buried on the seafloor each year (Dehairs *et al.*, 1980).

Because barite forms in association with organic material, there is a tight correlation ($R^6 = 0.93$) between the sedimentary fluxes of carbon and barite (Dymond *et al.*, 1992). Thus, barite accumulation rates have been used to infer past levels of export production in the ocean (e.g., Dymond *et al.*, 1992; Paytan *et al.*, 1996).

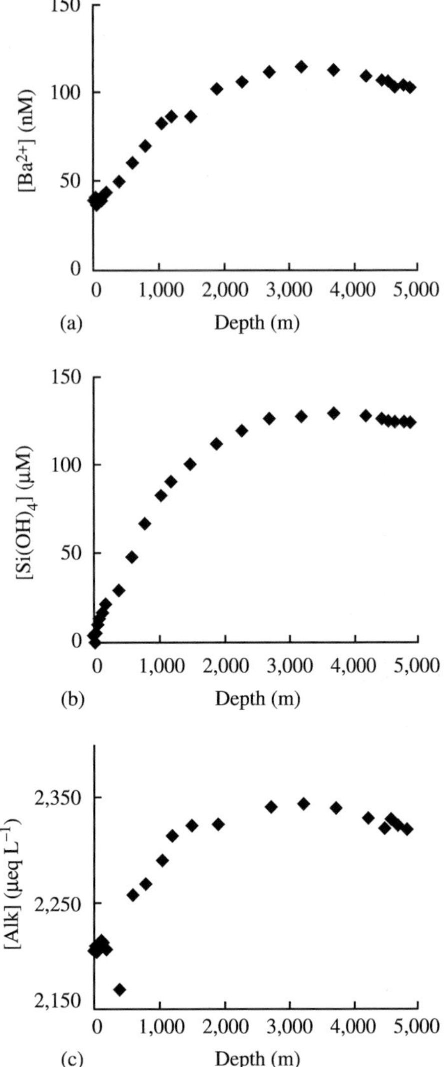

Figure 9 Profiles of Ba^{2+}, $Si(OH)_4$, and alkalinity in the Indian Ocean (06° 09′ S, 50° 55′ E) (source Jeandel *et al.*, 1996).

6.04.3.2.2 Zinc

The profiles of dissolved zinc in the ocean are also similar to those of $Si(OH)_4$ (Figure 10; Bruland, 1980), but as with barium, the main removal phase for zinc is not opal. Less than 3% of the zinc taken up by diatoms is deposited in their opaline cell wall (Ellwood and Hunter, 2000), and the Zn : Si ratio of acid-leached opal is much lower than that of dissolved Zn and $Si(OH)_4$ in the water column (Bruland, 1980; Collier and Edmond, 1984). Instead, most of

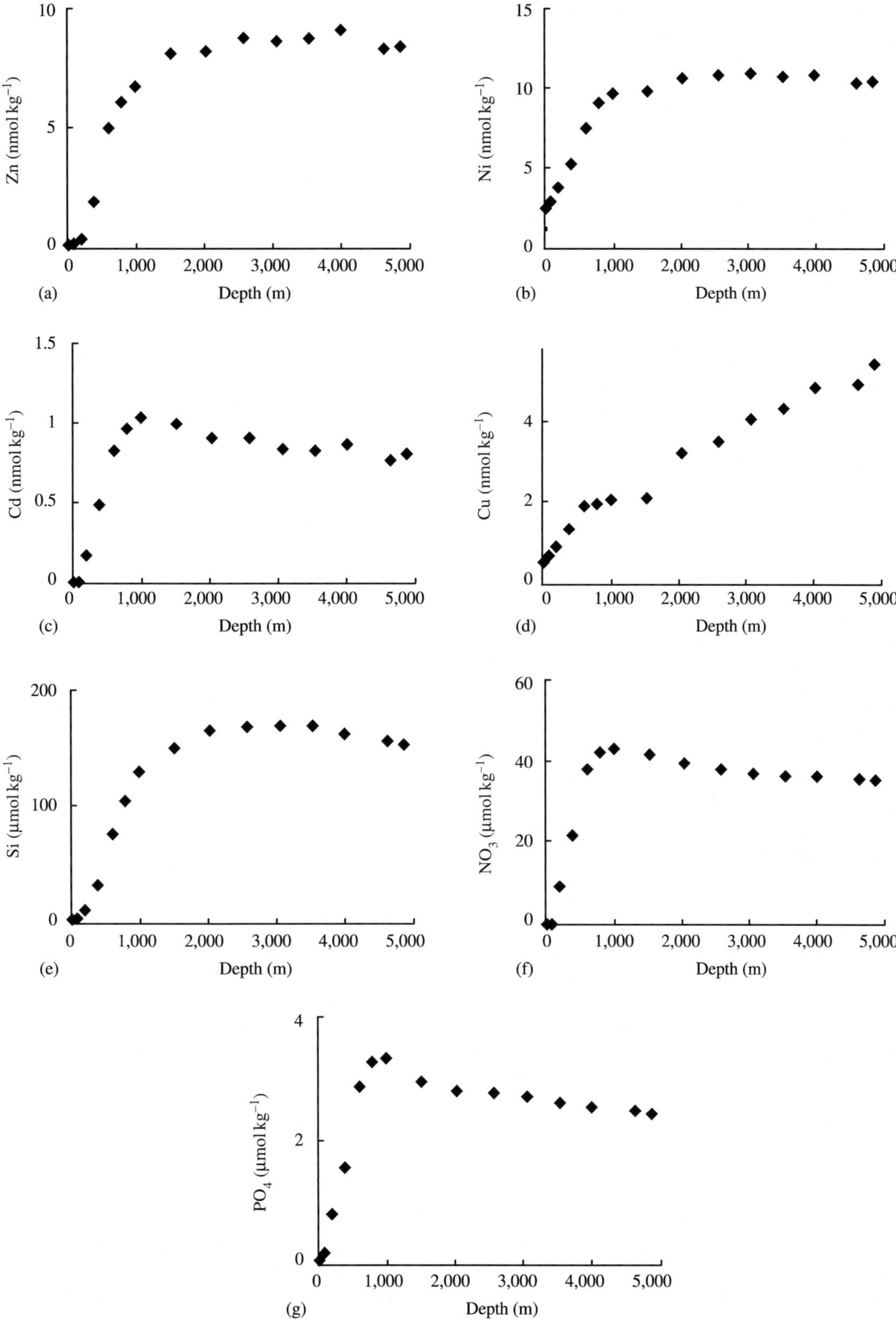

Figure 10 Profiles of Zn, Cd^{2+}, Ni, Cu, $Si(OH)_4$, NO_3^-, and PO_4^- with depth in the North Pacific (after Bruland, 1980).

the zinc removed from the surface ocean is bound up in POM.

Because zinc is required for the growth of phytoplankton, its availability affects the biological pump. Although zinc limitation of an entire phytoplankton community has never been demonstrated, levels of dissolved zinc are often low enough to limit many taxa (Morel *et al.*, 1994; Sunda and Huntsman, 1995b; Timmermans *et al.*, 2001). Zinc is an integral part of the enzyme, carbonic anhydrase (Morel *et al.*, 1994), which helps maintain an efficient supply of CO_2 to Rubisco.

The impact of low concentrations of zinc on phytoplankton growth varies. Some phytoplankton substitute cobalt for zinc in many enzymes (Price and Morel, 1990; Sunda and Huntsman, 1995b; Timmermans *et al.*, 2001) and can maintain maximal growth rates at low levels of zinc. Calcification aids in the acquisition of DIC, so calcareous phytoplankton, such as the coccolithophorids, are not as dependent on carbonic anhydrase (and therefore zinc) to maintain high rates of carbon fixation (Sunda and Huntsman, 1995b). Thus, while low levels of zinc may not curtail overall levels of primary production, they may shift the phytoplankton community structure away from the diatoms and towards the coccolithophorids (Morel *et al.*, 1994; Sunda and Huntsman, 1995b; Timmermans *et al.*, 2001), which will greatly impact the ratio of organic C to $CaCO_3$ of particles sinking to the deep sea.

6.04.3.2.3 Cadmium

Like barium and zinc, cadmium also shows a nutrient-like distribution in the ocean (Figure 10; Bruland, 1980; Löscher *et al.*, 1998), more closely mirroring those of the labile nutrients, NO_3^- and PO_4^-, than those of $Si(OH)_4$ and alkalinity. Cadmium is taken up by phytoplankton and incorporated into organic material, accounting for the similarity of its profile to that of NO_3^- and PO_4^-. Cadmium may also be adsorbed onto the surfaces of phytoplankton (Collier and Edmond, 1984).

Cadmium is taken up by phytoplankton slightly preferentially to PO_4^- (Löscher *et al.*, 1998; Elderfield and Rickaby, 2000). Waters with low PO_4^- concentrations thus have lower $Cd:PO_4^-$ ratios (0.1 nmol Cd per μmol PO_4^-) than waters with higher PO_4^- concentrations where $Cd:PO_4^-$ may approach 0.4 nmol Cd per μmol PO_4^- (Elderfield and Rickaby, 2000). Additionally, surface water $Cd:PO_4^-$ ratios drop over the development of the spring bloom in the Southern Ocean (Löscher *et al.*, 1998).

Cadmium also regenerates preferentially from decomposing particles (Knauer and Martin, 1981; Collier and Edmond, 1984). More labile components of decaying particles have higher Cd : P ratios than bulk decaying particles (Knauer and Martin, 1981). Box models of cadmium and phosphorus cycling also require enhanced regeneration of cadmium from particles for the replication of observed cadmium distributions in the ocean (Collier and Edmond, 1984).

Cadmium is generally considered to be toxic to organisms and how the marine phytoplankton utilize their cadmium is unknown. Cadmium may substitute for zinc in carbonic anhydrase at times when zinc is limiting (Price and Morel, 1990; Lane and Morel, 2000). It is possible that cadmium may play a role in polyphosphate bodies, a form of cellular storage of phosphorus that has been shown to contain significant quantities of elements such as calcium, zinc, and magnesium (Ruiz *et al.*, 2001).

The similarity between cadmium profiles and nutrient profiles has been used as a means of reconstructing past patterns of primary production and nutrient cycling. Several attempts have been made to reconstruct phosphate concentrations from the Cd : Ca ratio of foraminifera (Boyle, 1988; Elderfield and Rickaby, 2000). Foraminifera substitute Cd^{2+} for Ca^{2+} in the lattice of their calcite tests, and the ratio of Cd : Ca incorporation varies with the Cd : Ca ratio of seawater (Boyle, 1988). The Cd : Ca ratio of foraminifera in the Southern Ocean suggests that phosphorus in surface waters was not as heavily utilized by phytoplankton during the last glacial maximum (LGM), suggesting lower levels of primary production relative to nutrient flux into the euphotic zone at that time (Elderfield and Rickaby, 2000).

6.04.3.2.4 Iron

Of all the trace elements whose distributions are affected by the biological pump, iron is the one that has the most profound impact on the workings of the biological pump. Iron plays a key role in many of the crucial enzymes in biological systems, such as superoxide dismutase, ferredoxin, and nitrate reductase (Geider and LaRoche, 1994) that evolved at a time when the oceans were low in oxygen and, therefore, high in dissolved iron. As a result marine phytoplankton have a heavier demand for iron relative to the present-day availability of dissolved iron in the ocean. Growth of phytoplankton and rates of photosynthesis are frequently limited by the lack of iron in surface waters (Martin and Fitzwater, 1989).

One of the major inputs of iron to the ocean comes from the dissolution of Fe(II) from wind-borne continental dust deposited on the surface of the ocean (Zhuang *et al.*, 1990). In oxygenic environments, such as surface ocean waters, Fe(II) will quickly be oxidized to the insoluble form Fe(III) and removed from seawater. Fe(II) is also taken up by phytoplankton as well as complexed by ligands exuded into the water by marine organisms to prevent its precipitation as Fe(III).

Profiles of dissolved iron in seawater show the influence of both biotic and abiotic processes. At stations in the Northeast Pacific, dissolved iron concentrations are low in surface waters, reflecting biological uptake. Iron concentrations also show peak values at depth, corresponding to the oxygen minimum zone (Martin and Gordon, 1988), suggesting the abiotic reduction of Fe(III) back to the soluble Fe(II).

Vast tracks of the ocean, such as the equatorial Pacific, the Northeast Pacific subarctic, the Southern Ocean, and even parts of the California upwelling zone do not have sufficient supplies of iron to fully support phytoplankton growth (Martin and Fitzwater, 1988; Martin *et al.*, 1994; Coale *et al.*, 1996; Hutchins and Bruland, 1998; Boyd *et al.*, 2000). In these areas, macronutrients such as nitrogen and phosphorus are rarely depleted. Attention has turned to these HNLC areas as sites where further primary production (and the associated drawdown of atmospheric CO_2) could occur. Increased supplies of dust stimulating the biological pump in HNLC regions may be responsible for low atmospheric CO_2 concentrations during glacial times (Martin and Gordon, 1988; Watson *et al.*, 2000). Artificially stimulating the biological pump by seeding HNLC areas with chelated iron has been proposed as a means of pumping the 90 μatm of CO_2 in the atmosphere (put there by humans) into the deep sea, although there is not much consensus as to the effectiveness of such an endeavor (Chisholm *et al.*, 2001).

6.04.4 QUANTIFYING THE BIOLOGICAL PUMP

There are many different ways to quantify the biological pump. Total levels of carbon fixation in surface waters may be estimated in bottle incubations from the uptake of $^{14}CO_2$ by phytoplankton (Steemann Nielsen, 1952) or from the deviation of oxygen isotopes from their terrestrial mass fractionation line (Luz and Barkan, 2000). At the other end of the biological pump, sedimentary accumulation rates of organic carbon may be measured. In between, the impact of the feeding and vertical migration activities of mid-water organisms on the flux of particles may be investigated (e.g., Steinberg, 1995; Dilling *et al.*, 1998). However, because of the current interest in the CO_2 pumping capacity of the biological pump, we concentrate here on the methods used to estimate export production and particle flux from the euphotic zone.

Particle flux is frequently extrapolated from the measurement of new production in surface waters (Dugdale and Goering, 1967). Export of POM out of surface waters or into the deep sea may also be estimated directly through its collection in sediment traps (e.g., Martin *et al.*, 1987; Anita *et al.*, 2001). Export or sedimentation of POM may also be estimated from disequilibria between two nuclides (e.g., ^{238}U and ^{234}Th, and ^{230}Th and ^{231}Pa) that are scavenged by particles of different degrees (Buesseler *et al.*, 1992; Kumar *et al.*, 1993; François *et al.*, 1997).

Some methods for quantifying the biological pump focus the relationship between the biological pump and CO_2. POC flux measurements or estimates of nutrient removal from surface waters may be used in conjunction with various ocean models to estimate the impact of the biological pump on atmospheric concentrations of CO_2 (Sarmiento and Toggweiler, 1984; Sarmiento and Orr, 1991). Others have focused on the importance of the ratio of POC to $CaCO_3$ to the sequestering of CO_2 in the ocean (Anita *et al.*, 2001; Buitenhuis *et al.*, 2001).

6.04.4.1 Measurement of New Production

The method in most widespread use for quantifying the biological pump hinges upon the ideas that the surface ocean is at steady state on annual timescales with respect to the nitrogen budget and that nitrogen predominantly limits phytoplankton growth in the ocean. In such a system, the amount of productivity that is exported from the euphotic zone must be equal to the amount of productivity that is fuelled by the input of allocthonous or "new" nitrogen to the euphotic zone (Dugdale and Goering, 1967). According to this definition, "new production" is one that is supported from dissolved nitrogen upwelled into the euphotic zone or fixed from N_2 into PON, and export production is taken to be equal to new production. For the sake of measurement, the above definition of new production has been simplified even further. Experimentally, new production is taken to be equal to the production that uses NO_3^- (and NO_2^- also but more in theory than in practice) as its nitrogen source, as opposed to NH_4^+ or any of the organic forms of nitrogen.

It should be pointed out that measurement of new production is the measurement of the maximum amount of productivity that can be exported without running the system down with respect to the annual supply of nutrients to the euphotic zone. It may not always be appropriate to assume that new production and export production are equal; the fluxes will be equal only in systems that are not evolving. The validity of the assumption that new production equals export production depends on to what degree and over what timescale the nitrogen cycle in surface waters is at steady state.

At the moment the nitrogen cycle in the ocean is not at steady state on the decadal timescale. In the last few decades, the N : P ratio of many oceanic waters has changed (Pahlow and Riebesell, 2000; Emerson *et al.*, 2001) and near-shore waters have shifted towards silicon limitation from nitrogen limitation (Conley *et al.*, 1993) due to anthropogenic inputs of DIN to these systems and an increase in the extent of nitrogen fixation. On very long timescales, the nitrogen cycle is not at steady state either, but responds to changes in the oceanic inventory of phosphorus, which ultimately governs the rate of nitrogen fixation (Tyrrell, 1999). These imbalances may not be large enough on a yearly timescale to affect the estimate of export production from new production, but no assessment of this has been made.

Another assumption open to question is that the rate of NO_3^- uptake adequately represents the rate of uptake of new forms of nitrogen. This simplification has come about for two reasons. The first is that the uptake rate of NO_3^- by phytoplankton can be measured with reasonable ease by tracking the uptake of ^{15}N-labeled NO_3^- (Dugdale and Goering, 1967). The second is that NO_3^- is not produced in the euphotic zone to any significant degree and so its presence there can only be as a result of upwelling or atmospheric deposition.

The form of DIN released during the death, decay, and grazing of phytoplankton is NH_4^+, which is also the most easily utilizable form of DIN to phytoplankton. Oxidized forms of DIN, such as NO_3^- and NO_2^{2-}, must be reduced to NH_4^+ by nitrifying bacteria such as *Nitrosomonas*, *Nitrobacter*, *Nitrospira*, and *Proteobacteria* (Zehr and Ward, 2002) prior to assimilation by phytoplankton. The classical view is that nitrification does not occur in the euphotic zone due to the inhibition of nitrifying bacteria by light (Zehr and Ward, 2002). Eukaryotic phytoplankton are also thought to outcompete nitrifying bacteria for the supplies of NH_4^+ in surface waters. Thus, NO_3^- found in the euphotic zone must have had its origins outside of the euphotic zone, in deeper waters, in rivers or agricultural runoff, or from atmospheric deposition and is taken as the sole representative of new nitrogen in the euphotic zone (Dugdale and Goering, 1967).

Of course, NO_3^- is not likely to be the only form of allocthonous nitrogen in the euphotic zone. Concentrations of NH_4^+ in upwelled water are not zero, although they are much lower than those of NO_3^-. DON may also serve as a significant source of new nitrogen to the euphotic zone.

There is one last word of caution concerning the use of NO_3^- uptake to estimate the transport of CO_2 to depth via the biological pump. Measurement of new production divulges no information concerning the depth of decomposition of the POM formed or the ratio of POC to $CaCO_3$ of the exported particles (Anita *et al.*, 2001). For instance, CO_2 from material decomposed beneath the euphotic zone but above the maximum depth of winter mixing will be ventilated straight back out the atmosphere. $CaCO_3$ formation, a feature that is not common to all phytoplankton, diminishes the efficiency of CO_2 drawdown with primary production (see below; Buitenhuis *et al.* (2001)).

6.04.4.2 Measurement of Particle Flux

Means more direct than the measurement of new production exist for the estimation of particle fluxes into the deep. Moored or free-floating traps may be used to collect sinking particles (e.g., Martin *et al.*, 1987). Alternatively, particle flux may be estimated from particle-reactive nuclides (e.g., Buesseler *et al.*, 1992). Particle flux may also be estimated from the consumption of oxygen (associated with the decomposition of sinking POM) in waters below the surface layer (e.g., Jenkins, 1982).

6.04.4.2.1 Sediment traps

The collection of particles in sediment traps, while perhaps the most direct way of measuring the sinking flux of POM, is a method not free from a certain amount of controversy. Sediment traps both over-collect and under-collect particles. Comparison of ^{234}Th accumulating in a suite of sediment traps with ^{234}Th fluxes expected from U–Th disequilibria in the upper 300 m of the ocean suggested that the particle collection efficiency of these traps ranged from 10% to 1,000% (Buesseler, 1991). Traps deployed in the deep ocean also show a considerable variability in trapping efficiencies (e.g., Scholten *et al.*, 2001). Despite the magnitude of these biases, there is no generally applied method for correcting fluxes using particle-reactive isotopes (Anita *et al.*, 2001).

Zooplankton actively swimming into sediment traps also serve as a source of error in flux measurements. It is impossible to differentiate these "swimmers" from zooplankton that have settled passively into the cup as part of the sinking POM flux. Swimmers may constitute as much as a quarter of the POC collected by the trap (Steinberg *et al.*, 1998) and are generally removed from trap material prior to analysis. This introduces minimal error into the trap estimates of POC flux, as detrital zooplankton likely only comprise

~2% of the total organic matter sinking into the traps (Steinberg *et al.*, 1998).

6.04.4.2.2 Particle-reactive nuclides

Radionuclides in the uranium decay series serve as useful tracers of particle flux. One type of these tracers consists of a soluble parent nuclide and a particle-reactive daughter. These soluble nuclide–particle-reactive pairs include $^{238}U-^{234}Th$, $^{234}U-^{230}Th$, and $^{235}U-^{231}Pa$. The half-life of the parent exceeds the mixing time of the ocean and its distribution throughout the ocean is uniform. Once the soluble parent isotope decays to the particle-reactive daughter, the daughter is scavenged onto particulate material.

In systems with no particle scavenging, the activities of the parent and daughter nuclide will be in secular equilibrium. What is seen instead is that the activities of ^{234}Th, ^{230}Th, and ^{231}Pa are lower in surface waters than those of their parents (Figure 11). The difference in the activities of parent and daughter is a measure of the uptake of daughter onto particles (Buesseler *et al.*, 1992). With the help of a model of particle scavenging, fluxes of the particle-reactive daughter may be estimated from its vertical distribution (e.g., Coale and Bruland, 1985; Buesseler *et al.*, 1992). If the ratio of the particle-reactive nuclide to POC or PON is known, then the calculated flux of nuclide can be converted to an estimate of particle flux (Buesseler *et al.*, 1992).

Relative estimates of particle flux may also be made from the ratio of two particle-reactive nuclides, such as ^{230}Th and ^{231}Pa, which are scavenged onto particles to different degrees. The half-lives of these isotopes are much larger than their residences times in the ocean (~10^4 yr versus tens to hundreds of years), and thus there is no significant radioactive decay that occurs in the water column. The extent of the scavenging of ^{231}Pa, which is not as particle

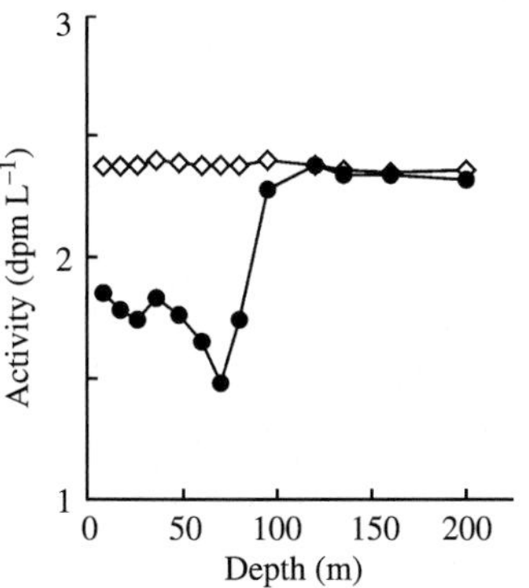

Figure 11 Profiles of ^{234}Th (circles) and ^{238}U (diamonds) in the upper ocean (after Buesseler, 1991).

reactive as ^{230}Th, is highest in areas of high particle flux (Anderson *et al.*, 1990). Thus, sediments in high flux areas exhibit $^{231}Pa/^{230}Th$ ratios in excess of the initial production ratio of 0.093, and sediments accumulating slowly exhibit ratios less than 0.093 (Anderson *et al.*, 1990). $^{231}Pa/^{230}Th$ ratios have been used to infer changes in productivity and sediment accumulation rates between the present-day interglacial and the LGM, ~2×10^4 yr ago (Kumar *et al.*, 1993; François *et al.*, 1997).

6.04.4.2.3 Oxygen utilization rates

The distribution of oxygen in ocean waters contains information about primary production. For example, the amount of excess oxygen present in the seasonal thermocline in the Pacific was long ago used to suggest that ^{14}C-based estimates of primary production were severely underestimating levels of primary production in the ocean (Shulenberger and Reid, 1981). Oxygen deficiencies in deeper waters have been used to estimate levels of export production (Jenkins, 1982).

The supplies of oxygen to waters below the euphotic zone are primarily physical: advection and mixing. Removal of oxygen from these waters takes place through the oxidation of organic matter (Equations (2) and (3)):

$$O_2 + CH_2O \rightarrow CO_2 + H_2O \quad (2)$$

$$2O_2 + NH_3 + OH^- \rightarrow NO_3^- + 2H_2O \quad (3)$$

By measuring rates of ventilation and the degree of oxygen undersaturation in deeper waters, an estimate of the rates of oxygen utilization (OUR) may be made and integrated to yield the total amount of oxygen consumed beneath the euphotic zone each year (Jenkins, 1982). From this number a flux of POC and PON may be calculated. Estimates of export production based on OURs are reasonably in line with the amount of new production that could be supported from measured fluxes of NO_3^- into surface waters (Jenkins, 1988).

One advantage of using the OUR method for estimating export fluxes is that it integrates over larger spatial and temporal scales than do estimates based on sediment traps and nuclide fluxes. Also unlike the estimates of new production based on NO_3^- uptake, OURs are directly coupled to the recycling of CO_2 via particle decomposition and thus a more direct measure of the impact of the biological pump on atmospheric CO_2. In practice, however, the measurement of NO_3^- uptake is less technically challenging and is carried out much more frequently than are estimates from nuclides and OURs.

6.04.5 THE EFFICIENCY OF THE BIOLOGICAL PUMP

6.04.5.1 Altering the Efficiency of the Biological Pump

There is much talk concerning the "efficiency" of the biological pump. Is it pumping as much carbon to the deep sea as it could be? The general consensus is that it is not operating at its full capacity, and this is generally meant to imply that globally, the nitrate flux into the euphotic zone is not fully consumed in support of marine primary production. For example, Broecker (1982) has suggested that if all of the nutrients supplied to ocean surface waters were consumed by phytoplankton, the atmospheric CO_2 content would drop by ~130 ppmv.

A simplistic estimate of how much more CO_2 the biological pump could draw out of the atmosphere gives a false impression of our understanding of the system. One hint that there is a considerable decoupling of primary production and nutrient drawdown from levels of export production is the regional variability in the ratio of export to total production. Levels of export production are higher in more highly productive areas than in low-productivity areas, as would be expected. However, a greater proportion of the total primary production is exported from mesotrophic and eutrophic regions than from oligotrophic areas of the ocean. The ratio of export production to total production may be as low as 0.12 in oligotrophic areas but as high as 0.4 in highly productive regimes (Anita *et al.*, 2001).

There is a lesson to be learnt here with regard to spurring the biological pump into action. Most of the talk regarding increasing the efficiency of the biological pump focuses only on productivity and nutrient drawdown, but there is much more to the biological pump than nutrient drawdown. The main question asked is: Can we stimulate an increase in primary production (and by extension export production) by adding iron to the ocean? From the point of view of ascertaining whether or not increased glacial dust could have stimulated the biological pump and could have been responsible for the lower levels of CO_2 then present in the atmosphere, this is an appropriate question. But from the point of view of effectively sequestering anthropogenic carbon in the deep sea via the biological pump, this is a very narrow approach. Carbon sequestration via the biological pump really cannot be considered as a practical activity until we have a better predictive understanding of not only primary production but also particle aggregation and sinking, zooplankton grazing and microbial hydrolysis, and organic carbon to $CaCO_3$ production and rain rates. These are the factors that control both the proportion and total amount of organic carbon pumped to the deep sea.

6.04.5.1.1 In HNLC areas

The biological pump in HNLC areas is not operating at full efficiency on at least two counts. Phytoplankton growth is curtailed by the lack of availability of trace metals such as iron, and so concentrations of the major nutrients, nitrogen and phosphorus, are not drawn down and carbon fixation does not occur to its maximum possible extent (Martin and Fitzwater, 1988). In addition, iron limitation heavily impacts the larger phytoplankton, such as diatoms, that are important to particle flux. When a phytoplankton community is released from iron limitation, diatom growth is stimulated more strongly than the growth of the other phytoplankton (Cavender-Bares *et al.*, 1999; Lam *et al.*, 2001).

Given all of this, the addition of iron to HNLC waters should result in higher levels of carbon fixation, increased growth of diatoms, local drawdown of CO_2, and enhanced export of carbon to the deep sea. Iron-addition experiments in bottles and *in situ* on the mesoscale unequivocally support the first two points. Addition of iron to HNLC waters results in madly blooming phytoplankton (Martin and Fitzwater, 1989; Martin *et al.*, 1994; Coale *et al.*, 1996; Boyd *et al.*, 2000; Strass, 2002). Chlorophyll concentrations may quadruple (e.g., Strass, 2002) and carbon fixation rates may triple (e.g., Coale *et al.*, 1996) over the first few days after iron addition (Figure 12).

Alternatively, support for the drawdown of CO_2 following iron addition has been mixed. In IronEx-I, the first mesoscale iron experiment, iron addition did not result in a marked drop of CO_2 concentrations in surface waters of the equatorial Pacific (Watson *et al.*, 1994). Some CO_2 was drawn down during IronEx-II, also in the equatorial Pacific, but the amount removed from surface waters was still not enough to prevent these recently upwelled, high CO_2 waters from outgassing CO_2 to the atmosphere (Coale *et al.*, 1996). The Southern Ocean, however, is a better ecosystem than the equatorial Pacific for stimulating CO_2 drawdown with iron. Both large-scale iron-addition experiments that have been carried out there (SOIREE and EisenEx-1) produced a significant lowering (25–30 μatm) of the CO_2 content of surface waters (Boyd *et al.*, 2000; Watson *et al.*, 2000; Strass, 2002).

The one key critical point that all of the iron-enrichment experiments have failed to show is an increase in the export of carbon to the deep sea. Even the Southern Ocean experiments, where a CO_2 drawdown occurred, did not result in an increase in export flux. Sediment traps set out at a depth of

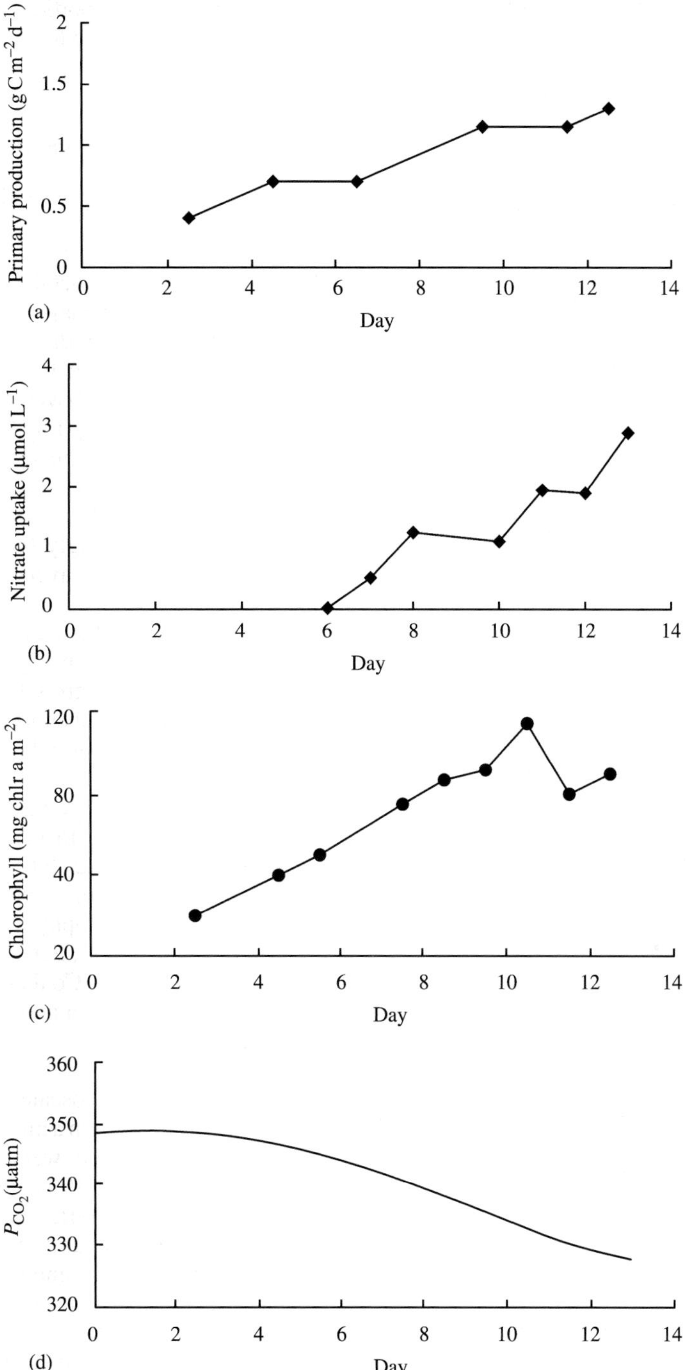

Figure 12 Responses to Fe addition to HNLC waters: chlorophyll, carbon fixation, NO_3^-, and P_{CO_2} (after Boyd *et al.*, 2000; Watson *et al.*, 2000).

100 m (below the fertilized patch) in the SOIREE experiment collected a similar amount of POC to traps stationed outside of the patch (Boyd *et al.*, 2000). In the EisenEx-1 experiment, there was no increase in sinking POC over the three-week window of the experiment (Strass, 2002).

There are many possible reasons for the lack of an observed increase in export flux in the mesoscale iron-addition experiments. The first is that there is no increase in export flux, but instead an increase in the flow of carbon through the microbial loop (microzooplankton and bacteria) which would result in the regeneration of CO_2 and nutrients in the euphotic zone. Microzooplankton biomass and bacterial activities were seen to increase in many of the mesoscale iron-addition experiments (e.g., Boyd *et al.*, 2000; Strass, 2002), suggesting that some fraction of the carbon fixed through iron addition was being shunted into the microbial loop instead of being pumped into deeper waters. Alternatively, the sediment traps used may simply have not been deployed for long enough or in the right location to catch the carbon sinking out as a result of the iron-induced bloom.

6.04.5.1.2 Through changes in community composition

Changes in community composition should profoundly affect the efficiency of the biological pump. Shifting productivity from small cells, such as photosynthetic picoplankton, which do not efficiently sink, to large cells, such as diatoms, capable of aggregating into particles that sink at hundreds of meters a day will result in the pumping of more carbon to the deep. However, there is finite amount of productivity that may be exported without running the nutrients in the system down to zero. In a steady-state system, shifting the community structure cannot increase the total amount of export production above the level of new production. However, a shift in community structure may export carbon deeper into the water column before it decomposes, thus delaying the return of that carbon to the surface ocean by a significant number of years.

A shift in community composition may also be important to the biological pump if it is from a calcareous to a noncalcareous phytoplankton, as precipitation of $CaCO_3$ diminishes the ocean's ability to hold dissolved CO_2. DIC ($\sum CO_2$) is present in seawater as several species, dissolved CO_2, carbonic acid, and the dissociated forms, bicarbonate ion, and carbonate ion:

$$\sum CO_2 = H_2CO_3^* + HCO_3^- + CO_3^{2-} \qquad (4)$$

(where $H_2CO_3^*$ represents the sum of dissolved CO_2 and carbonic acid). The ratio of the three species varies largely with pH, with the undissociated forms dominating only at low pH and HCO_3^- favored at the typical seawater pH of 8.3 (Zeebe and Wolf-Gladrow, 2001). The saturation concentration of $H_2CO_3^*$ is controlled in large part by the amount of HCO_3^- and CO_3^{2-} present in solution. Removal of HCO_3^- or CO_3^{2-} during $CaCO_3$ formation results in a lowering of the saturation concentration of dissolved CO_2, and therefore the outgassing of CO_2 from solution to the atmosphere.

Precipitation of $CaCO_3$ both produces CO_2 (Equation (5)):

$$Ca^{2+} + 2HCO_3^- \rightarrow CaCO_3 + H_2O + CO_2 \qquad (5)$$

and lowers the alkalinity (approximately equal to $2[CO_3^{2-}] + [HCO_3^-]$) of the surface waters (Zeebe and Wolf-Gladrow, 2001). Loss of alkalinity decreases the capacity of surface waters to play host to dissolved CO_2. In the surface ocean, under an atmosphere with a CO_2 partial pressure of 350 μatm (19 μatm less than the 2001 values; Keeling and Whorf, 2001), the precipitation of 1 mol of $CaCO_3$ results in the release of 0.6 mol of CO_2 concentrations (Ware *et al.*, 1992; Frankignoulle *et al.*, 1994).

The coccolithophorid, *Emiliania huxleyi*, contains, on average, 0.433 mol of $CaCO_3$ (present as scales covering the cell) for every mole of organic carbon (Buitenhuis *et al.*, 2001). Thus, for every mole of CO_2 fixed into organic matter by coccolithophorids such as *E. huxleyi*, roughly 0.26 mol of CO_2 will be released due to the formation of $CaCO_3$. A shift, therefore, from export production being carried out by coccolithophorids to a noncalcareous phytoplankton such as diatoms will result in an increased drawdown of carbon for the same amount of primary production. The fact that carbon fixation by phytoplankton that calcify is only 75% as effective at removing CO_2 as carbon fixation done by phytoplankton that do not precipitate $CaCO_3$ complicates estimates of CO_2 drawdown from the surface ocean POC export flux (Anita *et al.*, 2001).

6.04.5.1.3 By varying the C : N : P ratios of sinking material

The C : N : P ratios of sinking POM are not fixed; phosphorus and nitrogen are preferentially regenerated from sinking POM, which allows for the sequestering of more carbon in the deep ocean per unit nitrogen or phosphorus than if Redfield ratios remain unaltered during the decomposition of organic matter. It was remineralization ratios calculated from DIC and nutrient concentrations along isopycnals that initially suggested that the C : N of sinking particles might be higher than the value of 6.6 proposed by Redfield (Takahashi *et al.*, 1985). Sediment trap material also showed that the C : N and C : P ratios of sinking particles increase with depth, from near Redfield values of 6.6 and 106 at the base of the euphotic zone to ~11 and 180 by 5,000 m (Martin *et al.*, 1987).

Since the above initial observations, evidence has mounted for the preferential remineralization of nitrogen and phosphorus out of POM relative to carbon. Ocean models with continuous vertical resolution support the preferential release of nitrogen and phosphorus from sinking POM in line with the estimates of Martin *et al.* (Shaffer, 1996). Bacteria have been shown to more rapidly degrade PON than POC (Verity *et al.*, 2000). Increases in the C : N and C : P of sinking particles have also been observed at station ALOHA near Hawaii (Christian *et al.*, 1997). Differences in the C:N ratio of suspended POC (6.4) and less buoyant POC trapped at the pycnocline (8.6) in the Kattegat, just east of the North Sea, further suggest that nitrogen is remineralized more rapidly than carbon at significant levels even in the euphotic zone (Olesen and Lundsgaard, 1995).

6.04.5.1.4 By enhancing particle transport

While sediment trap evidence supports the idea that the flux of organic carbon ($J_{\mathrm{C\ org}}$) to depth (z) can be estimated from levels of primary production (PP) (Equation (6)):

$$J_{\mathrm{C\ org}} = 0.1\mathrm{PP}^{1.77} z^{-0.68} \tag{6}$$

the correlation between POC flux and overlying levels of primary production ($R^2 = 0.53$; Anita *et al.*, 2001) is not strong. The ratio of POC flux (at 125 m) to primary production also shows great variability from region to region, ranging from 0.08 to 0.38 in the Atlantic Ocean alone (Anita *et al.*, 2001).

There are many reasons that an increased downward transport of POC out of the euphotic zone does not directly follow from an increase in productivity. Export fluxes are also controlled by the packaging of smaller more numerous particles into larger less numerous aggregates and by the rate at which the resulting aggregates sink. The depth to which POC exported from the euphotic zone is transported must be determined by the balance between the carbon content of the particle, the rate of microbial hydrolysis, and the sinking rate of the particle. Carbon fixed by marine diatoms, which are large and adept at aggregating, stands a better chance at being exported to the deep than carbon fixed by very small, unsinking picoplankton.

The mode of productivity is also crucial to the sinking flux of POC. Systems exporting POC in pulses (e.g., following the periodic formation of phytoplankton blooms in temperate and high latitude areas) export a much higher proportion of their total production than do systems, such as the oligotrophic central gyres, where carbon fixation is a more static, steady process (Lampitt and Anita, 1997). This results partly from the influence of particle concentration on aggregation and partly from zooplankton population dynamics. In temperate areas, zooplankton reproduction cannot begin until after the onset of the spring phytoplankton bloom, leading to a lag in the increase in zooplankton population (and subsequently in the grazing pressure exerted by zooplankton) behind that of the increase in phytoplankton numbers (Heinrich, 1962) allowing for the aggregation and sinking of intact phytoplankton from the euphotic zone. In lower latitude, oligotrophic-area standing stocks of phytoplankton and zooplankton are not as decoupled in time and there is no clear window within which a high density of phytoplankton cells may aggregate, escape predation, and sink to the deep.

6.04.6 THE BIOLOGICAL PUMP IN THE IMMEDIATE FUTURE

What may we expect out of the biological pump in the future? We cannot currently predict how ocean biology will respond to climate warming (Sarmiento *et al.*, 1998), but there are other questions that we may begin to ask. Will, for instance, the biological pump respond to the rise in surface ocean concentrations of DIC tied to the rise in atmospheric CO_2 fuelled by anthropogenic emissions? What impact will inputs of agricultural fertilizers have on the biological pump? What role will artificial stimulation of the biological pump through deliberate ocean fertilization play in the sequestration of excess CO_2?

6.04.6.1 Response to Increased CO_2

Will the biological pump respond to the increase CO_2? Increasing concentrations of CO_2 do not appear to have the significant impact on the C:N or C:P ratios of phytoplankton (Burkhardt *et al.*, 1999) that is necessary if CO_2 is to stimulate productivity in the ocean. However, possibly the rate of carbon-rich exudation by phytoplankton and the production of TEP will rise with increasing CO_2 concentrations (Engel, 2002). It may also be possible for CO_2 levels to increase the proportion of carbon diverted into the biological pump at the expense of grazing and microbial food webs. Riebesell *et al.* (1993) suggested that carbon fixation by phytoplankton may be rate limited by the diffusive flux of CO_2 to the cell which delivers CO_2 too slowly relative to NO_3^- and PO_4^- to take up carbon, nitrogen, and phosphorus in Redfield proportions.

The diffusive flux of CO_2 to the surface of a phytoplankton cell (F_{CO_2}) is controlled not only by the diffusivity of CO_2 (D) and the radius of the cell (r), but the concentration gradient of CO_2 between the aqueous medium and the interior of the cell ($C_e - C_i$) and the rate constant for the conversion of HCO_3^- to CO_2 (k') as the carbonate system equilibrates following CO_2 removal (Equation (7)):

$$F_{CO_2} = -4\pi r D\left(1 + \frac{r}{\sqrt{D/k'}}\right)(C_e - C_i) \tag{7}$$

If the diffusive flux of CO_2 is controlling the delivery of CO_2 to the carbon-fixing enzyme, Rubisco, then an increase in the dissolved CO_2 content of seawater should stimulate phytoplankton growth rates. In particular, an increase in dissolved CO_2 concentrations should stimulate growth rates in the larger (i.e., aggregate forming) size classes and thus should also result in a shunting of carbon out of the mouths of zooplankton and into the biological pump (Riebesell *et al.*, 1993).

It is not clear that the acquisition of carbon is the rate-limiting step for photosynthesis even in the larger phytoplankton cells. Phytoplankton are not limited to the passive diffusion of CO_2 to the cell surface for carbon acquisition. HCO_3^- may be taken up directly by phytoplankton and used as a source of CO_2 for photosynthesis (Korb *et al.*, 1997; Nimer *et al.*, 1997; Raven, 1997; Tortell *et al.*, 1997). The considerably greater abundance of HCO_3^- than dissolved CO_2 in seawater would imply that carbon-fixation rates of marine phytoplankton are not CO_2 limited, and increasing concentrations of CO_2 will not have an impact on rates of primary production.

6.04.6.2 Response to Agricultural Runoff

Inputs of agricultural fertilizers are having more than one impact on the biological pump. A shift in the NO_3^-: PO_4^-: $Si(OH)_4$ of natural waters is causing a shift in the phytoplankton community structure that should impact the biological cycling of carbon in aquatic systems (Conley *et al.*, 1993). Additionally, a recent shift in the C : N : P ratios of deeper waters and an increase in export production have been observed for the northern hemisphere oceans (Pahlow and Riebesell, 2000).

6.04.6.2.1 Shift towards silicon limitation

Since the 1850s, changes in land use patterns, human population density, and extent of the use of fertilizers have resulted in an increased flux of NO_3^- and PO_4^- to rivers, lakes, and the coastal ocean (Conley *et al.*, 1993) and even to the open ocean through atmospheric deposition, for example, of anthropogenic nitrous oxides (Pahlow and Riebesell, 2000). At the same time, large freshwater systems, such as the North American Great Lakes, and coastal areas, such as the Adriatic and Baltic Seas and the Mississippi River plume, have been shifting away from nitrogen or phosphorus limitation and towards silicon limitation (Conley *et al.*, 1993; Nelson and Dortch, 1996). The extra productivity fuelled on the extra nitrogen and phosphorus has resulted in the removal of silicon from these systems and a shift in the phytoplankton community structure away from diatoms. Given that rivers are significant sources of new NO_3^-, PO_4^-, and $Si(OH)_4$ to the global ocean, this effect is expected to spread further throughout the sea.

A shift in community structure associated with silicic acid depletion may sharply reduce the amount of carbon delivered to deep waters and sediments via the biological pump. Diatoms, which are the only major phytoplankton group requiring silicic acid, are relatively large and notable for aggregating and sinking. Diatoms feed a greater portion of the organic matter they produce into the biological pump than do most other classes of phytoplankton. Currently, diatoms perform more than 75% of the primary production that occurs in high nutrient and coastal regions of the ocean (Nelson *et al.*, 1995), the exact areas that will be impacted by this input of additional NO_3^- and PO_4^- and the exact areas where carbon tends to make it down into the sediments.

One possible further impact of the decline of the diatoms lies in the identity of their probably successor. If a $CaCO_3$-producing phytoplankter, such as coccolithophorids, steps in to utilize the NO_3^- and PO_4^- the silicon-starved diatoms leave behind, the CO_2 pumping efficiency of the biological pump will decline (Robertson *et al.*, 1994) even if the same amount of organic carbon continues to be removed to the deep sea and sediments each year.

6.04.6.2.2 Shifts in export production and deep ocean C : N : P

There is evidence for anthropogenic perturbations increasing the biological pumping of carbon into deep waters. In both the North Pacific and North Atlantic, for example, deep-water N : P ratios have increased since the early 1950s (Pahlow and Riebesell, 2000) possibly due to atmospheric deposition of anthropogenic nitrogen and a subsequent shift towards phosphorus limitation in these areas. Concomitant with the rise in N : P is an increase in the apparent oxygen utilization (AOU) in both the North Atlantic and North Pacific, which suggests that levels of export production have also increased. This is estimated to have resulted in the increased oceanic sequestration of 0.2 Pg C yr^{-1} (Pahlow and Riebesell, 2000).

6.04.6.3 Carbon Sequestration via Ocean Fertilization and the Biological Pump

There have been many calls to sequester anthropogenic CO_2 in the deep ocean by stimulating primary production through the addition of literally tons of iron to HNLC surface waters. Patents have been taken out on the idea (e.g., Markels, 2001) and, counter to the recommendations of the American Society of Limnology and Oceanography (http://www.aslo.org/policy/docs/oceanfertsummary.pdf), several companies have been established to dispense carbon credits to industries willing to pay for ocean fertilization (Chisholm *et al.*, 2001). However, it remains unclear whether ocean fertilization will ever successfully transport carbon in the deep sea, how long the transported carbon might remain out of contact with the atmosphere, and what side-effect large-scale fertilization will have on marine geochemistry and ecology (Fuhrman and Capone, 1991; Peng and Broecker, 1991a,b; Sarmiento and Orr, 1991; Chisholm *et al.*, 2001; Lenes *et al.*, 2001).

A perusal of the literature suggests that carbon sequestration by iron fertilization is not a panacea for the anthropogenic carbon emissions that increased atmospheric CO_2 from 280 μatm at the start of the industrial revolution to 369 μatm by December 2000 (Keeling and Whorf, 2001). Ocean models suggest that enhanced ocean uptake of carbon with iron fertilization of the Antarctic Ocean will at best draw down atmospheric CO_2 by 70 μatm if carried out continuously for a century (Peng and Broecker, 1991a,b) and damp the annual anthropogenic CO_2 input to the atmosphere by less than 30% of current annual emission levels (Joos *et al.*, 1991).

Large-scale iron fertilization will have side-effects. If iron fertilization resulted in a complete drawdown of the nutrients available in the Southern Ocean, for example, the average O_2 content of deep waters will drop by 4–12% (Sarmiento and Orr, 1991), with areas of anoxia cropping up in the Antarctic (Peng and Broecker, 1991b) and Indian Oceans (Sarmiento and Orr, 1991). Even small-scale patches of anoxia would have a profound negative impact on the survival and distribution of metazoan fauna in the ocean and alter the balance of microbial transformations of nitrogen between reduced and oxidized phases (Fuhrman and Capone, 1991).

ACKNOWLEDGMENTS

The author thanks U. Passow for a helpful review, A. Engel, U. Passow, and A. Wischmeyer for insightful conversation, A. Alldredge, R. Collier, and L. Dilling for promptly sending reprints, A. Knoll and A. Yool for pointing out the long-term silica cycle, W. Berelson, H. Elderfield and N. McCave for helpful tips, and R. Banger and S. Bishop for finding everything the author could not.

REFERENCES

Alldredge A. L. and Gotschalk C. C. (1989) Direct observations of the mass flocculation of diatom blooms: characteristics, settling velocities and formation of diatom aggregates. *Deep-Sea Res.* **36**, 159–171.

Alldredge A. L. and Jackson G. A. (1995) Aggregation in marine systems. *Deep-Sea Res. II* **42**, 1–7.

Alldredge A. L. and McGillivary P. (1991) The attachment probabilities of marine snow and their implications for particle coagulation in the ocean. *Deep-Sea Res.* **38**, 431–443.

Alldredge A. L. and Silver M. W. (1988) Characteristics, dynamics, and significance of marine snow. *Prog. Oceanogr.* **20**, 41–82.

Alldredge A. L., Passow U., and Logan B. E. (1993) The abundance and significance of a class of large, transparent organic particles in the ocean. *Deep-Sea Res.* **40**, 1131–1140.

Aluwihare L. I. and Repeta D. J. (1999) A comparison of the chemical characteristics of oceanic DOM and extracellular DOM produced by marine algae. *Mar. Ecol. Prog. Ser.* **186**, 105–117.

Amon R. M. W. and Benner R. (1994) Rapid cycling of high-molecular-weight dissolved organic matter in the ocean. *Nature* **369**, 549–552.

Amon R. M. W., Fitznar H.-P., and Benner R. (2001) Linkages among the bioreactivity, chemical composition, and diagenetic state of marine dissolved organic matter. *Limnol. Oceanogr.* **46**, 287–297.

Anderson R. F., Lao Y., Broecker W. S., Trumbore S. E., Hofman H. J., and Wolfli W. (1990) Boundary scavenging in the Pacific Ocean: a comparison of ^{10}Be and ^{231}Pa. *Earth Planet. Sci. Lett.* **96**, 287–304.

Anita A. N., *et al.* (2001) Basin-wide particulate carbon flux in the Atlantic Ocean: regional export patterns and potential for atmospheric CO_2 sequestration. *Global Biogeochem. Cycles* **15**, 845–862.

Arnosti C., Repeta D. J., and Blough N. V. (1994) Rapid bacterial degradation of polysaccharides in anoxic marine systems. *Geochim. Cosmochim. Acta* **58**, 2639–2652.

Asper V. L., Deuser W. G., Knauer G. A., and Lohrenz S. E. (1992) Rapid coupling of sinking particles fluxes between surface and deep ocean waters. *Nature* **357**, 670–672.

Azam F. and Volcani B. E. (1981) Germanium–silicon interactions in biological systems. In *Silicon and Siliceous Structures in Biological Systems* (eds. T. L. Simpson and B. E. Volcani). Springer, pp. 69–93.

Bauer J. E., Williams P. M., and Druffel E. R. M. (1992) ^{14}C activity of dissolved organic carbon fractions in the north-central Pacific and Sargasso Sea. *Nature* **357**, 667–670.

Benner R., Pakulski J. D., McCarthy M., Hedges J. I., and Hatcher P. G. (1992) Bulk chemical characteristics of dissolved organic matter in the ocean. *Science* **255**, 1561–1564.

Berner R. A., Ruttenberg K. C., Ingall E. D., and Rao J. L. (1993) The nature of phosphorus burial in modern marine sediments. In *Interactions of C, N, P and S Biochemical Cycles and Global Change* (eds. R. Wollast, F. T. Mackenzie, and L. Chou). Springer, pp. 365–378.

Biddanda B. and Benner R. (1997) Carbon, nitrogen, and carbohydrate fluxes during the production of particulate and dissolved organic matter by marine phytoplankton. *Limnol. Oceanogr.* **42**, 506–518.

Bidle K. D. and Azam F. (1999) Accelerated dissolution of diatom silica by natural marine bacterial assemblages. *Nature* **397**, 508–512.

Bidle K. D. and Azam F. (2001) Bacterial control of silicon regeneration from diatom detritus: significance of bacterial ectohydrolases and specie identity. *Limnol. Oceanogr.* **46**, 1606–1623.

Billett D. S. M., Lampitt R. S., Rice A. L., and Mantoura R. F. C. (1983) Seasonal sedimentation of phytoplankton to the deep sea benthos. *Nature* **302**, 520–522.

Bishop J. K. B. (1988) The barite-opal-organic carbon association in oceanic particulate matter. *Nature* **332**, 341–343.

Boyd P. and Newton P. (1995) Evidence of the potential influence of planktonic community structure on the interannual variability of particulate organic carbon flux. *Deep-Sea Res. I* **42**, 619–639.

Boyd P. W., *et al.* (2000) A mesoscale phytoplankton bloom in the polar Southern Ocean stimulated by iron fertilization. *Nature* **407**, 695–702.

Boyle E. A. (1988) Cadmium: chemical tracer of deepwater paleoceanography. *Paleoceanography* **3**, 471–489.

Broecker W. S. (1982) Ocean chemistry during glacial time. *Geochim. Cosmochim. Acta* **46**, 1689–1705.

Broecker W. S. and Peng T.-H. (1982) *Tracers in the Sea*. Eldigio Press.

Bruland K. W. (1980) Oceanographic distributions of cadmium, zinc, nickel, and copper in the North Pacific. *Earth Planet. Sci. Lett.* **47**, 176–198.

Brzezinski M. A. (1985) The Si : C : N ratio of marine diatoms: interspecific variability and the effect of some environmental variables. *J. Phycol.* **21**, 347–357.

Brzezinski M. A. and Nelson D. M. (1996) Chronic substrate limitation of silicic acid uptake in the western Sargasso Sea. *Deep-Sea Res. II* **43**, 437–453.

Buesseler K. O. (1991) Do upper-ocean sediment traps provide and accurate record of particle flux? *Nature* **353**, 420–423.

Buesseler K. O., Bacon M. P., Cochran J. K., and Livingston H. D. (1992) Carbon and nitrogen export during the JGOFS North Atlantic Bloom Experiment estimated from $^{234}Th : {}^{238}U$ disequilibria. *Deep-Sea Res.* **39**, 1115–1137.

Buitenhuis E. T., van der Wal P., and de Baar H. J. W. (2001) Blooms of *Emiliania huxleyi* are sinks of atmospheric carbon dioxide: a field and mesocosm study derived simulation. *Global Biogeochem. Cycles* **15**, 577–587.

Burkhardt S., Zondervan I., and Riebesell U. (1999) Effect of CO_2 concentration on C : N : P ratio in marine phytoplankton: a species comparison.

Calbet A. (2001) Mesozooplankton grazing effect on primary production: a global comparative analysis in marine ecosystems. *Limnol. Oceanogr.* **46**, 1824–1830.

Capone D. G., Zehr J. P., Paerl H. W., Bergman B., and Carpenter E. J. (1997) *Trichodesmium*: a globally significant marine cyanobacterium. *Science* **276**, 1221–1229.

Cavender-Bares K. K., Mann E. L., Chisholm S. W., Ondrusek M. E., and Bidigare R. R. (1999) Differential response of equatorial Pacific phytoplankton to iron fertilization. *Limnol. Oceanogr.* **44**, 237–246.

Chin W. C., Orellana M. V., and Verdugo P. (1998) Spontaneous assembly of marine dissolved organic matter into polymer gels. *Nature* **391**, 568–572.

Chisholm S. W., Falkowski P. G., and Cullen J. J. (2001) Discrediting ocean fertilization. *Science* **294**, 309–310.

Chisholm S. W., Olson R. J., Zettler E. R., Goericke R., Waterbury J. B., and Welschmeyer N. A. (1988) A novel free-living prokaryote abundant in the oceanic euphotic zone. *Nature* **334**, 340–343.

Cho B. C. and Azam F. (1988) Major role of bacterial in biogeochemical fluxes in the ocean's interior. *Nature* **332**, 441–443.

Christian J. R., Lewis M. R., and Karl D. M. (1997) Vertical fluxes of carbon, nitrogen, and phosphorus in the North Pacific Subtropical Gyre near Hawaii. *J. Geophys. Res.* **102**, 15667–15677.

Clark L. L., Ingall E. D., and Benner R. (1998) Marine phosphorus is selectively remineralized. *Nature* **393**, 428.

Coale K. H. and Bruland K. W. (1985) $^{234}Th:{}^{238}U$ disequilibria within the California current. *Limnol. Oceanogr.* **30**, 22–33.

Coale K. H., *et al.* (1996) A massive phytoplankton bloom induced by an ecosystem-scale iron fertilization experiment in the equatorial Pacific Ocean. *Nature* **383**, 495–501.

Collier R. and Edmond J. (1984) The trace element geochemistry of marine biogenic particulate matter. *Prog. Oceanogr.* **13**, 113–199.

Conley D. J., Schelske C. L., and Stoermer E. F. (1993) Modification of the biogeochemical cycle of silica with eutrophication. *Mar. Ecol. Prog. Ser.* **101**, 179–192.

de Haas H., van Weering T. C. E., and de Stigter H. (2002) Organic carbon in shelf seas: sinks or sources, processes and products. *Cont. Shelf Res.* **22**, 691–717.

Dehairs F., Chesselet R., and Jedwab J. (1980) Discrete suspended particles of barite and the barium cycle in the open ocean. *Earth Planet. Sci. Lett.* **49**, 528–550.

De La Rocha C. L., Hutchins D. A., Brzezinski M. A., and Zhang Y. (2000) Effects of iron and zinc deficiency on elemental composition and silica production by diatoms. *Mar. Ecol. Prog. Ser.* **195**, 71–79.

de Villiers S. (1999) Seawater strontium and Sr/Ca variability in the Atlantic and Pacific oceans. *Earth Planet Sci. Lett.* **171**, 623–634.

Dilling L. and Alldredge A. L. (2000) Fragmentation of marine snow by swimming macrozooplankton: a new process impacting carbon cycling in the sea. *Deep-Sea Res. I* **47**, 1227–1245.

Dilling L., Wilson J., Steinberg D., and Alldredge A. (1998) Feeding by the euphausiid *Euphausia pacifica* and the copepod *Calanus pacificus* on marine snow. *Mar. Ecol. Prog. Ser.* **170**, 189–201.

Dugdale R. C. and Goering J. J. (1967) Uptake of new and regenerated forms of nitrogen in primary productivity. *Limnol. Oceanogr.* **12**, 196–206.

Dugdale R. C., Wilkerson F. P., and Minas H. J. (1995) The role of a silicate pump in driving new production. *Deep-Sea Res. I* **42**, 697–719.

Dymond J., Suess E., and Lyle M. (1992) Barium in deep-sea sediment: a geochemical proxy for paleoproductivity. *Paleoceanography* **7**, 163–181.

Elderfield H. and Rickaby R. E. M. (2000) Oceanic Cd/P ratio and nutrient utilization in the glacial Southern Ocean. *Nature* **405**, 305–310.

Elderfield H. E. and Schultz A. (1996) Mid-ocean ridge hydrothermal fluxes and the chemical composition of the ocean. *Ann. Rev. Earth Planet. Sci.* **24**, 191–224.

Ellwood M. J. and Hunter K. A. (2000) The incorporation of zinc and iron into the frustule of the marine diatom *Thalassiosira pseudonana*. *Limnol. Oceanogr.* **45**, 1517–1524.

Emerson S., Mecking S., and Abell J. (2001) The biological pump in the subtropical North Pacific Ocean: nutrient sources, Redfield ratios, and recent changes. *Global Biogeochem. Cycles* **15**, 535–554.

Engel A. (2000) The role of transparent exopolymer particles (TEP) in the increase in apparent particle stickiness (alpha) during the decline of a diatom bloom. *J. Plankton Res.* **22**, 485–497.

Engel A. (2002) Direct relationship between CO_2 uptake and transparent exopolymer particle production in natural phytoplankton. *J. Plankton Res.* **24**, 49–53.

Engel A. and Passow U. (2001) Carbon and nitrogen content of transparent exopolymer particles (TEP) in relation to their Alcian Blue absorption. *Mar. Ecol. Prog. Ser.* **219**, 1–10.

Engel A., Goldthwait S., Passow U., and Alldredge A. (2002) Temporal decoupling of carbon and nitrogen dynamics in a mesocosm diatom bloom. *Limnol. Oceanogr.* **47**, 753–761.

Eppley R. W. and Peterson B. J. (1979) Particulate organic matter lux and planktonic new production in the deep ocean. *Nature* **282**, 677–680.

Fagerbakke K. M., Norland S., and Heldal M. (1999) The inorganic ion content of native aquatic bacteria. *Can. J. Microbiol.* **45**, 304–311.

Falkowski P., *et al.* (2000) The global carbon cycle: a test of our knowledge of Earth as a system. *Science* **290**, 291–296.

Field C. B., Behrenfeld M. J., Randerson J. T., and Falkowski P. (1998) Primary production of the biosphere: integrating terrestrial and oceanic components. *Science* **281**, 237–240.

Filippelli G. M. and Delaney M. L. (1996) Phosphorus geochemistry of equatorial Pacific sediments. *Geochim. Cosmochim. Acta* **60**, 1479–1495.

Föllmi K. B. (1996) The phosphorus cycle, phosphogenesis and marine phosphate-rich deposits. *Earth Sci. Rev.* **40**, 55–124.

Fowler S. W. and Knauer G. A. (1986) Role of large particles in the transport of elements and organic compounds through the oceanic water column. *Prog. Oceanogr.* **16**, 147–194.

François R., Altabet M. A., Yu E.-F., Sigman D. M., Bacon M. P., Frank M., Bohrmann G., Bareille G., and Labeyrie L. D. (1997) Contribution of Southern Ocean surface-water stratification to low atmospheric CO_2 concentrations during the last glacial period. *Nature* **389**, 929–935.

Frankignoulle M., Canon C., and Gattuso J.-P. (1994) Marine calcification as a source of carbon dioxide: positive feedback of increasing atmospheric CO_2. *Limnol. Oceanogr.* **39**, 458–462.

Froelich P. N., Bender M. L., Luedtke N. A., Heath G. R., and De Vries T. (1982) The marine phosphorus cycle. *Am. J. Sci.* **282**, 474–511.

Fuhrman J. A. and Capone D. G. (1991) Possible biogeochemical consequences of ocean fertilization. *Limnol. Oceanogr.* **36**, 1951–1959.

Geider R. J. and LaRoche J. L. (1994) The role of iron in phytoplankton photosynthesis, and the portential for iron limitation of primary productivity in the sea. *Photosyn. Res.* **39**, 275–301.

Greenwood N. N. and Earnshaw A. (1984) *Chemistry of the Elements*. Pergamon.

Gruber N. and Sarmiento J. L. (1997) Global patterns of marine nitrogen fixation and denitrification. *Global Biogeochem. Cycles* **11**, 235–266.

Harrison P. J., Conway H. L., Holmes R. W., and Davis C. O. (1977) Marine diatoms grown in chemostats under silicate or ammonium limitation: III. Cellular chemical composition and morphology of *Chaetoceros debilis*, *Skeletonema costatum*, and *Thalassiosira gravida*. *Mar. Biol.* **43**, 19–31.

Harrison W. G. (1990) Nitrogen utilization in chlorophyll and primary productivity maximum layers: an analysis based on the *f*-ratio. *Mar. Ecol. Prog. Ser.* **60**, 85–90.

Hedges J. I. and Keil R. G. (1995) Sedimentary organic matter preservation: an assessment and speculative synthesis. *Mar. Chem.* **49**, 81–115.

Heinrich A. K. (1962) The life histories of planktonic animals and seasonal cycles of plankton communities in the ocean. *J. Cons. Int. Explor. Mer.* **27**, 15–24.

Holloway C. F. and Cowen J. P. (1997) Development of a scanning confocal laser microscopic technique to examine the structure and composition of marine snow. *Limnol. Oceanogr.* **42**, 1340–1352.

Hutchins D. A. and Bruland K. W. (1998) Iron-limited diatom growth and Si : N uptake ratios in a coastal upwelling regime. *Nature* **393**, 561–564.

Janse I., vanRijssel M., Gottschal J. C., Lancelot C., and Gieskes W. W. C. (1996) Carbonhydrates in the North Sea during spring blooms of *Phaeocystis*: a specific fingerprint. *Aquat. Microb. Ecol.* **10**, 97–103.

Jeandel C., Dupré B., Lebaron G., Monnin C., and Minster J.-F. (1996) Longitudinal distributions of dissolved barium, silica and alkalinity in the western and southern Indian Ocean. *Deep-Sea Res. I* **43**, 1–31.

Jenkins W. J. (1982) Oxygen utilization rates in North Atlantic subtropical gyre and primary production in oligotrophic systems. *Nature* **300**, 246–248.

Jenkins W. J. (1988) Nitrate flux into the euphotic zone near Bermuda. *Nature* **331**, 521–523.

Joos F. L., Sarmiento J. L., and Sigenthaler U. (1991) Estimates of the effect of Southern Ocean iron fertilization on atmospheric CO_2 concentrations. *Nature* **349**, 772–775.

Kaehler P. and Koeve W. (2001) Marine dissolved organic matter: can its C : N ratio explain carbon overconsumption? *Deep-Sea Res. I* **48**, 49–62.

Kamitani A. (1982) Dissolution rates of silica from diatoms decomposing at various temperatures. *Mar. Biol.* **68**, 91–96.

Karl D. M., Letelier R., Tupas L., Dore J., Christian J., and Hebel D. (1997) The role of nitrogen fixation in biogeochemical cycling in the subtropical North Pacific Ocean. *Nature* **388**, 533–538.

Keeling C. D. and Whorf T. P. (2001) Atmospheric CO_2 records from sites in the SiO air sampling network. In *Trends: A Compendium of Data on Global Change*. US Department of Energy, Oak Ridge National Laboratory, Carbon Dioxide Information Analysis Center, Oak Ridge, TN.

Kepkay P. E. (1994) Particle aggregation and the biological reactivity of colloids. *Mar. Ecol. Prog. Ser.* **109**, 293–304.

Knauer G. A. and Martin J. H. (1981) Phosphorus and cadmium cycling in northeast Pacific waters. *J. Mar. Res.* **39**, 65–76.

Kolowith L. C., Ingall E. D., and Benner R. (2001) Composition and cycling of marine organic phosphorus. *Limnol. Oceanogr.* **46**, 309–321.

Korb R. A., Saville P. J., Johnston A. M., and Raven J. A. (1997) Sources of inorganic carbon for photosynthesis by three species of marine diatom. *J. Phycol.* **33**, 433–440.

Kumar N., Gwiazda R., Anderson R. F., and Froelich P. N. (1993) $^{231}Pa/^{230}Th$ ratios in sediments as a proxy for past changes in Southern Ocean productivity. *Nature* **362**, 45–48.

Lam P. J., Tortell P. D., and Morel F. M. M. (2001) Differential effects of iron additions on organic and inorganic carbon production by phytoplankton. *Limnol. Oceanogr.* **46**, 1199–1202.

Lampitt R. and Anita A. N. (1997) Particle flux in deep seas: regional characteristics and temporal variability. *Deep-Sea Res. I* **44**, 1377–1403.

Lane T. W. and Morel F. M. M. (2000) A biological function for cadmium in marine diatoms. *Proc. Natl Acad. Sci. USA* **97**, 4627–4631.

Laws E. A., Falkowski P. G., Smith W. O., Jr., Ducklow H., and McCarthy J. J. (2000) Temperature effects on export production in the open ocean. *Global Biogeochem. Cycles* **14**, 1231–1246.

Lea D. W. and Boyle E. (1989) Barium content of benthic foraminifera controlled by bottom water composition. *Nature* **338**, 751–753.

Lenes J. M., *et al.* (2001) Iron fertilization and the *Trichodesmium* response on the West Florida shelf. *Limnol. Oceanogr.* **46**, 1261–1277.

Levitus S., Burgett R., and Boyer T. (1994) *World Ocean Atlas 1994 Volume 3: Nutrients*. US Department of Commerce.

Lisitzin A. P. (1972) Sedimentation in the World Ocean. *Soc. Econ. Paleontol. Mineral. Spec. Publ.* **17**, 218pp.

Liu H., Landry M. R., Vaulot D., and Campbell L. (1999) Prochlorococcus growth rates in the central equatorial Pacific: an application of the f_{max} approach. *J. Geophys. Res.* **104**, 3391–3399.

Logan B. E., Passow U., Alldredge A. L., Grossart H. P., and Simon M. (1995) Rapid formation and sedimentation of large aggregates is predictable from coagulation rates (half-lives) of transparent exopolymer particles (TEP). *Deep-Sea Res. II* **42**, 203–214.

Long R. A. and Azam F. (1996) Abundant protein-containing particles in the sea. *Aquat. Microb. Ecol.* **10**, 213–221.

Löscher B. M., de Jong J. T. M., and de Baar H. J. W. (1998) The distribution and preferential biological uptake of cadmium at 6°W in the Southern Ocean. *Mar. Chem.* **62**, 259–286.

Luz B. and Barkan E. (2000) Assessment of oceanic productivity with the triple-isotope composition of dissolved oxygen. *Science* **288**, 2028–2031.

Mari X., Beauvais S., Lemee R., and Pedrotti M. L. (2001) Non-reedfield C : N ratio of transparent exopolymeric particles in the northwestern Mediterranean Sea. *Limnol. Oceanogr.* **46**, 1831–1836.

Markels M., Jr. (2001) Method of sequestering carbon dixoide with spiral fertilization. *US Patent No. 6,200,530.*

Martin J. H. (1990) Glacial-interglacial CO_2 change: the iron hypothesis. *Paleoceanography* **5**, 1–13.

Martin J. H. and Fitzwater S. E. (1988) Iron deficiency limits phytoplankton growth in the north-east Pacific subarctic. *Nature* **331**, 341–343.

Martin J. H. and Gordon R. M. (1988) Northeast Pacific iron distributions in relation to phytoplankton productivity. *Deep-Sea Res.* **35**, 177–196.

Martin J. H. and Knauer G. A. (1973) The elemental composition of plankton. *Geochim. Cosmochim. Acta* **37**, 1639–1653.

Martin J. H., Knauer G. A., Karl D. M., and Broenkow W. W. (1987) VERTEX: carbon cycling in the northeast Pacific. *Deep-Sea Res.* **34**, 267–285.

Martin J. H., *et al.* (1994) Testing the iron hypothesis in ecosystems of the equatorial Pacific Ocean. *Nature* **371**, 123–129.

McCarthy M., Pratum T., Hedges J., and Benner R. (1997) Chemical composition of dissolved organic nitrogen in the ocean. *Nature* **390**, 150–154.

McCave I. N. (1975) Vertical flux of particles in the ocean. *Deep-Sea Res.* **22**, 491–502.

McCave I. N. (1984) Size spectra and aggregation of suspended particles in the deep ocean. *Deep-Sea Res.* **31**, 329–352.

Michaels A. F. and Silver M. W. (1988) Primary production, sinking fluxes and the microbial food web. *Deep-Sea Res.* **35**, 473–490.

Mopper K., Zhou J. A., Ramana K. S., Passow U., Dam H. G., and Drapeau D. T. (1995) The role of surface-active carbohydrates in the flocculation of a diatom bloom in a mesocosm. *Deep-Sea Res. II* **42**, 47–73.

Morel F. M. M., Reinfelder J. R., Roberts S. B., Chamberlain C. P., Lee J. G., and Yee D. (1994) Zinc and carbon co-limitation of marine phytoplankton. *Nature* **369**, 740–742.

Morris I. (1981) Photosynthetic products, physiological state, and phytoplankton growth. *Can. J. Fish. Aquat. Sci.* **210**, 83–102.

Nelson D. M. and Dortch Q. (1996) Silicic acid depletion and silicon limitation in the plume of the Mississippi River: evidence from kinetic studies in spring and summer. *Mar. Ecol. Prog. Ser.* **136**, 163–178.

Nelson D. M., Tréguer P., Brzezinski M. A., Leynaert A., and Quéguiner B. (1995) Production and dissolution of biogenic silica in the ocean: revised global estimates, comparison with regional data and relationship to biogenic sedimentation. *Global Biogeochem. Cycles* **9**, 359–372.

Nimer N. A., Iglesias-Rodriguez M. D., and Merrett M. J. (1997) Bicarbonate utilization by marine phytoplankton species. *J. Phycol.* **33**, 625–631.

Nozaki T. (1997) A fresh look at element distribution in the North Pacific. *EOS Trans., AGU Electron. Suppl* http://www.agu.org/eos_elec/97025e.html.

Olesen M. and Lundsgaard C. (1995) Seasonal sedimentation of autochthonous material from the euphotic zone of a coastal system. *Estuar. Coast. Shelf Sci.* **41**, 475–490.

Paasche E. (1999) Reduced coccolith calcite production under light-limited growth: a comparative study of three clones of *Emiliania huxleyi* (Prymnesiophyceae). *Phycologia* **38**, 508–516.

Pahlow M. and Riebesell U. (2000) Temporal trends in deep ocean Redfield ratios. *Science* **287**, 831–833.

Passow U. (2000) Formation of transparent exopolymer particles, TEP, from dissolved precursor material. *Mar. Ecol. Prog. Ser.* **192**, 1–11.

Passow U. (2002) Transparent exopolymer particles (TEP) in aquatic environments. *Prog. Oceanogr.* **55**, 287–333.

Passow U., Shipe R. F., Murray A., Pak D., Brzezinski M. A., and Alldredge A. L. (2001) The origin of transparent exopolymer particles (TEP) and their role in the sedimentation of particulate matter. *Cont. Shelf Res.* **21**, 327–346.

Paytan A., Kastner M., and Chavez F. P. (1996) Glacial to interglacial fluctuations in productivity in the Equatorial Pacific as indicated by marine barite. *Science* **274**, 1355–1357.

Peng T.-H. and Broecker W. S. (1991a) Dynamical limitations on the Antarctic iron fertilization strategy. *Nature* **349**, 227–229.

Peng T.-H. and Broecker W. S. (1991b) Factors limiting the reduction of atmospheric CO_2 by iron fertilization. *Limnol. Oceanogr.* **36**, 1919–1927.

Pondaven P., Ragueneau O., Tréguer P., Hauvespre A., Dezileau L., and Reyss J. L. (2000) Resolving the "opal paradox" in the Southern Ocean. *Nature* **405**, 168–172.

Price N. M. and Morel F. M. M. (1990) Cadmium and cobalt substitution for zinc in a marine diatom. *Nature* **344**, 658–660.

Quigley M. S., Santschi P. H., Hung C.-C., Guo L., and Honeyman B. D. (2002) Importance of acid polysaccharides for ^{234}Th complexation to marine organic matter. *Limnol. Oceanogr.* **47**, 367–377.

Raven J. A. (1997) Inorganic carbon acquisition by marine autotrophs. *Adv. Bot. Res.* **27**, 85–209.

Redfield A. C. (1934) On the proportions of organic derivatives in sea water and their relation to the composition of plankton. In *James Johnstone Memorial Volume* (ed. R. J. Daniel). University Press of Liverpool, pp. 177–192.

Redfield A. C. (1958) The biological control of chemical factors in the environment. *Am. Sci.* **46**, 205–221.

Riebesell U., Wolf-Gladrow D. A., and Smetacek V. (1993) Carbon dioxide limitation of marine phytoplankton growth rates. *Nature* **361**, 249–251.

Robertson J. E., Robinson C., Turner D. M., Holligan P., Watson A. J., Boyd P., Fernandez E., and Finsh M. (1994) The impact of a coccolithophore bloom on oceanic carbon uptake in the northeast Atlantic during summer 1991. *Deep-Sea Res. I* **41**, 297–314.

Ruiz F. A., Marchesini N., Seufferheld M., and Docampo R. (2001) The polyphosphate bodies of Chlamydomonas reinhardtii possess a proton-pumping pyrophosphatase and are similar to acidocalcisomes. *J. Biol. Chem.* **276**, 46196–46203.

Ruttenberg K. C. (1993) Reassessment of the oceanic residence time of phosphorus. *Chem. Geol.* **107**, 405–409.

Ryther J. H. (1969) Photosynthesis and fish production in the sea. *Science* **166**, 72–76.

Sañudo-Wilhelmy S. A., Kustka A. B., Gobler C. J., Hutchins D. A., Yang M., Lwiza K., Burns J., Capone D. G., Raven J. A., and Carpenter E. J. (2001) Phosphorus limitation of nitrogen fixation by *Trichodesmium* in the central Atlantic Ocean. *Nature* **411**, 66–69.

Sarmiento J. L. and Orr J. C. (1991) Three-dimensional simulations of the impact of Southern Ocean nutrient depletion on atmospheric CO_2 and ocean chemistry. *Limnol. Oceanogr.* **36**, 1928–1950.

Sarmiento J. L. and Toggweiler J. R. (1984) A new model for the role of the oceans in determining atmospheric P_{CO_2}. *Nature* **308**, 621–624.

Sarmiento J. L., Hughes T. M. C., Stouffer R. J., and Manabe S. (1998) Simulated response of the ocean carbon cycle to anthropogenic climate warming. *Nature* **393**, 245–249.

Scholten J. C., *et al.* (2001) Trapping efficiency of sediment traps from the deep eastern North Atlantic: ^{230}Th calibration. *Deep-Sea Res. II* **48**, 243–268.

Schumann R. and Rentsch D. (1998) Staining particulate organic matter with DTAF—a fluorescence dye for carbohydrates and protein: a new approach and application of a 2D image analysis system. *Mar. Ecol. Prog. Ser.* **163**, 77–88.

Shaffer G. (1996) Biogeochemical cycling in the global ocean: 2. New production, Redfield ratios, and remineralization in the organic pump. *J. Geophys. Res.* **101**, 3723–3745.

Shanks A. L. and Trent J. D. (1980) Marine snow: sinking rates and potential role in vertical flux. *Deep-Sea Res.* **27A**, 137–143.

Shemesh A., Mortlock R. A., Smith R. J., and Froelich P. N. (1988) Determination of Ge/Si in marine siliceous microfossils: separation, cleaning and dissolution of diatoms and radiolaria. *Mar. Chem.* **25**, 305–323.

Shulenberger E. and Reid J. L. (1981) The Pacific shallow oxygen maximum, deep chlorophyll maximum, and primary productivity, reconsidered. *Deep-Sea Res.* **28A**, 901–919.

Shuskina E. A. (1985) Production of principal ecological groups of plankton in the epipelagic zone of the ocean. *Oceanology* **25**, 653–658.

Siever R. (1991) Silica in the Oceans: biological–geochemical interplay. In *Scientists on Gaia* (eds. S. Schneider and P. Boston). MIT Press, pp. 287–295.

Smayda T. J. (1970) The suspension and sinking of phytoplankton in the sea. *Oceanogr. Mar. Biol. Ann. Rev.* **8**, 353–414.

Smith D. C., Simon M., Alldredge A. L., and Azam F. (1992) Intense hydrolytic enzyme activity on marine aggregates and implications for rapid particle dissolution. *Nature* **359**, 139–142.

Smith D. C., Steward G. F., Long R. A., and Azam F. (1995) Bacterial mediation of carbon fluxes during a diatom bloom in a mesocosm. *Deep-Sea Res. II* **42**, 75–97.

Soendergaard M., Williams P. J. L., Cauwet G., Riemann B., Robinson C., Terzic S., Woodward E. M. S., and Worm J. (2000) Net accumulation and flux of dissolved organic carbon and dissolved organic nitrogen in marine plankton communities. *Limnol. Oceanogr.* **45**, 1097–1111.

Steemann Nielsen E. (1952) The use of radioactive carbon (^{14}C) for measuring organic production in the sea. *J. Cons. Int. Explor. Mer.* **18**, 117–140.

Steinberg D. K. (1995) Diet of copepods (*Scolpalatum vorax*) associated with mesopelagic detritus (giant larvacean houses) in Monterey Bay, California *Mar. Biol.* **122**, 571–584.

Steinberg D. K., Pilskaln C. H., and Silver M. W. (1998) Contribution of zooplankton associated with detritus to sediment trap "swimmer" carbon in Monterey Bay, California, USA. *Mar. Ecol. Prog. Ser.* **164**, 157–166.

Steinberg D. K., Carlson C. A., Bates N. R., Johnson R. J., Michaels A. F., and Knap A. P. (2001) Overview of the US JGOFS Bermuda Atlantic Time-series Study (BATS): a decade-scale look at ocean biology and biogeochemistry. *Deep-Sea Res. II* **48**, 1405–1447.

Strass V. H. (2002) EisenEx-1: test of the iron hypothesis in a Southern Ocean eddy. *EOS: Trans., AGU Abstracts of 2002 Ocean Sciences Meeting.*

Strickland J. D. H. and Parsons T. R. (1968) *A Practical Handbook of Seawater Analysis.* Fisheries Research Board of Canada.

Strom S. L., Benner R., Ziegler S., and Dagg M. J. (1997) Planktonic grazers are a potentially important source of marine dissolved organic carbon. *Limnol. Oceanogr.* **42**, 1364–1374.

Suess E. (1980) Particulate organic carbon flux in the oceans-surface productivity and oxygen utilization. *Nature* **288**, 260–263.

Sunda W. G. and Huntsman S. A. (1995a) Iron uptake and growth limitation in oceanic and coastal phytoplankton. *Mar. Chem.* **50**, 189–206.

Sunda W. G. and Huntsman S. A. (1995b) Cobalt and zinc interreplacement in marinephytoplankton: biological and geochemical implications. *Limnol. Oceanogr.* **40**, 1404–1417.

Sundquist E. T. (1993) The global carbon dioxide budget. *Science* **259**, 934–941.

Takahashi T., Broecker W. S., and Langer S. (1985) Redfield ratio based on chemical data from isopycnal surfaces. *J. Geophys. Res.* **90**, 6907–6924.

Takeda S. (1998) Influence of iron availability on nutrient consumption ratio of diatoms in oceanic waters. *Nature* **393**, 774–777.

Tappan H. and Loeblich A. R., Jr. (1973) Evolution of the oceanic plankton. *Earth Sci. Rev.* **9**, 207–240.

Teira E., Pazó M. J., Serret P., and Fernández E. (2001) Dissolved organic carbon production by microbial populations in the Atlantic Ocean. *Limnol. Oceanogr.* **46**, 1370–1377.

Timmermans K. R., Snoek J., Gerringa L. J. A., Zondervan I., and de Baar H. J. W. (2001) Not all eukaryotic algae can replace zinc with cobalt: *Chaetoceros calcitrans* (Bacillariophyceae) versus *Emiliania huxleyi* (Prymnesiophyceae). *Limnol. Oceanogr.* **46**, 699–703.

Tortell P. D., Reinfelder J. R., and Morel F. M. M. (1997) Active uptake of bicarbonate by diatoms. *Nature* **390**, 243–244.

Tréguer P., Nelson D. M., Van Bennekom A. J., DeMaster D. J., Leynaert A., and Quéguiner B. (1995) The silica balance in the world ocean: a reestimate. *Science* **268**, 375–379.

Turner J. T. (2002) Zooplankton fecal pellets, marine snow and sinking phytoplankton blooms. *Aquat. Microbiol. Ecol.* **27**, 57–102.

Tyrrell T. (1999) The relative influences of nitrogen and phosphorus on oceanic primary production. *Nature* **400**, 525–531.

Verity P. G., Williams S. C., and Hong Y. (2000) Formation, degradation, and mass: volume ratios of detritus derived from decaying phytoplankton. *Mar. Ecol. Prog. Ser.* **207**, 53–68.

Ware J. R., Smith S. V., and Reaka-Kudla M. L. (1992) Coral reefs: sources or sinks of atmospheric CO_2. *Coral Reefs* **11**, 127–130.

Watson A. J., *et al.* (1994) Minimal effect of iron fertilization on sea-surface carbon dioxide concentrations. *Nature* **371**, 143–145.

Watson A. J., Bakker D. C. E., Ridgwell A. J., Poyd P. W., and Law C. S. (2000) Effect of iron supply on Southern Ocean CO_2 uptake and implications for glacial atmospheric CO_2. *Nature* **407**, 730–733.

Weiss R. F., Broecker W. S., Craig H., and Spencer D. (1983) *GEOSECS Indian Ocean Expedition: Volume 5. Hydrographic Data 1977–1978.* National Science Foundation.

Williams P. M. and Druffel E. R. M. (1987) Radiocarbon in dissolved organic matter in the central north Pacific Ocean. *Nature* **330**, 246–248.

Wolf-Gladrow D. A., Riebesell U., Burkhardt S., and Bijma J. (1999) Direct effects of CO_2 concentration on growth and isotopic composition of marine plankton. *Tellus (B Chem. Phys. Meteorol.)* **51B**, 176–461.

Zeebe R. E. and Wolf-Gladrow D. (2001) *CO_2 in Seawater: Equilibrium, Kinetics, Isotopes.* Elsevier.

Zehr J. P. and Ward B. B. (2002) Nitrogen cycling in the ocean: new perpsectives on processes and paradigms. *Appl. Environ. Microbiol.* **68**, 1015–1024.

Zehr J. P., Waterbury J. B., Turner P. J., Montoya J. P., Omoregie E., Steward G. F., Hansen A., and Karl D. M. (2001) Unicellular cyanobacteria fix N_2 in the subtropical North Pacific Ocean. *Nature* **412**, 635–638.

Zhang J. and Nozaki Y. (1996) Rare earth elements and yttrium in seawater: ICP_MS determinations in the East Caroline, Coral Sea, and Southern Fiji basins of the western Pacific Ocean. *Geochim. Cosmochim. Acta* **60**, 4631–4644.

Zhuang G., Duce R. A., and Kester D. R. (1990) The dissolution of atmospheric iron in surface seawater of the open ocean. *J. Geophys. Res.* **95**, 16207–16216.

6.05 Marine Bioinorganic Chemistry: The Role of Trace Metals in the Oceanic Cycles of Major Nutrients

F. M. M. Morel, A. J. Milligan, and M. A. Saito
Princeton University, NJ, USA

6.05.1 INTRODUCTION: THE SCOPE OF MARINE BIOINORGANIC CHEMISTRY

The bulk of living biomass is chiefly made up of only a dozen "major" elements—carbon, hydrogen, oxygen, nitrogen, phosphorus, sodium, potassium, chlorine, calcium, magnesium, sulfur (and silicon in diatoms)—whose proportions vary within a relatively narrow range in most organisms. A number of trace elements, particularly first

row transition metals—manganese, iron, nickel, cobalt, copper, and zinc—are also "essential" for the growth of organisms. At the molecular level, the chemical mechanisms by which such elements function as active centers or structural factors in enzymes and by which they are accumulated and stored by organisms is the central topic of bioinorganic chemistry. At the scale of ocean basins, the interplay of physical, chemical, and biological processes that govern the cycling of biologically essential elements in seawater is the subject of marine biogeochemistry. For those interested in the growth of marine organisms, particularly in the one-half of the Earth's primary production contributed by marine phytoplankton, bioinorganic chemistry and marine biogeochemistry are critically linked by the extraordinary paucity of essential trace elements in surface seawater, which results from their biological utilization and incorporation in sinking organic matter. How marine organisms acquire elements that are present at nano- or picomolar concentrations in surface seawater; how they perform critical enzymatic functions when necessary metal cofactors are almost unavailable are the central topics of "marine bioinorganic chemistry." The central aim of this field is to elucidate at the molecular level the metal-dependent biological processes involved in the major biogeochemical cycles.

By examining the solutions that emerged from the problems posed by the scarcity of essential trace elements, marine bioinorganic chemists bring to light hitherto unknown ways to take up or utilize trace elements, new molecules, and newer "essential" elements. Focusing on molecular mechanisms involved in such processes as inorganic carbon fixation, organic carbon respiration, or nitrogen transformation, they explain how the cycles of trace elements are critically linked to those of major nutrients such as carbon or nitrogen. But we have relatively little understanding of the binding molecules and the enzymes that mediate the biochemical role of trace metals in the marine environment. In this sense, this chapter is more a "preview" than a review of the field of marine bioinorganic chemistry. To exemplify the concepts and methods of this field, we have chosen to focus on one of its most important topics: the potentially limiting role of trace elements in primary marine production. As a result we center our discussion on particular subsets of organisms, biogeochemical cycles, and trace elements. Our chief actors are marine phytoplankton, particularly eukaryotes, while heterotrophic bacteria make only cameo appearances. The biogeochemical cycles that will serve as our plot are those of the elements involved in phytoplankton growth, the major algal nutrients—carbon, nitrogen, phosphorus, and silicon—leaving aside, e.g., the interesting topic of the marine sulfur cycle. Seven trace metals provide the intrigue: manganese, iron, nickel, cobalt, copper, zinc, and cadmium. But several other trace elements such as selenium, vanadium, molybdenum, and tungsten (and, probably, others not yet identified) will assuredly add further twists in future episodes.

We begin this chapter by discussing what we know of the concentrations of trace elements in marine microorganisms and of the relevant mechanisms and kinetics of trace-metal uptake. We then review the biochemical role of trace elements in the marine cycles of carbon, nitrogen, phosphorus, and silicon. Using this information, we examine the evidence, emanating from both laboratory cultures and field measurements, relevant to the mechanisms and the extent of control by trace metals of marine biogeochemical cycles. Before concluding with a wistful glimpse of the future of marine bioinorganic chemistry we discuss briefly some paleoceanographic aspects of this new field: how the chemistry of the planet "Earth"—particularly the concentrations of trace elements in the oceans—has evolved since its origin, chiefly as a result of biological processes and how the evolution of life has, in turn, been affected by the availability of essential trace elements.

6.05.2 TRACE METALS IN MARINE MICROORGANISMS

6.05.2.1 Concentrations

Bioinorganic chemists are now interested in the overall concentration and chemical speciation of trace elements in cells (O'Halloran and Culotta, 2000). In parallel with the "genome" and the "proteome" in organisms, some have begun to talk of the "metallome" (Outten and O'Halloran, 2001) to designate the suite of trace-metal concentrations, and perhaps the topic of this section could be described as "marine metallomics." What emerges from studies of trace-element concentrations in various types of cells, chiefly in unicellular organisms, is that these cellular concentrations are maintained at reasonably similar proportions among organisms from widely different taxa. This is exemplified in Figure 1, which shows the trace-metal composition of a few species of eukaryotic marine phytoplankton in cultures (Ho *et al.*, in press). Averaging the data given in Figure 1 provides an extension to Redfield formula ($C_{106}N_{16}P_1$):

$$(C_{106}N_{16}P_1)_{\times 1000}Fe_8Mn_4Zn_{0.8}Cu_{0.4}Co_{0.2}Cd_{0.2}$$

The stoichiometric coefficients of this average formula are within a factor of 3 of the elemental proportions measured for almost all individual

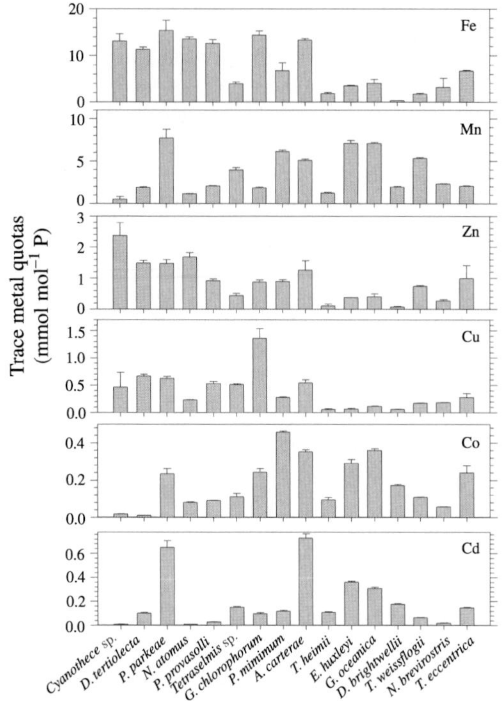

Figure 1 Cellular trace metals amounts normalized to phosphorus in a variety of phytoplankton. $Cd' = 20$ pM; $Co' = 20$ pM; $Cu' = 0.2$ pM; $Fe' = 200$ pM; $Mn' = 10$ nM; $Zn' = 20$ pM (source Ho *et al.*, in press).

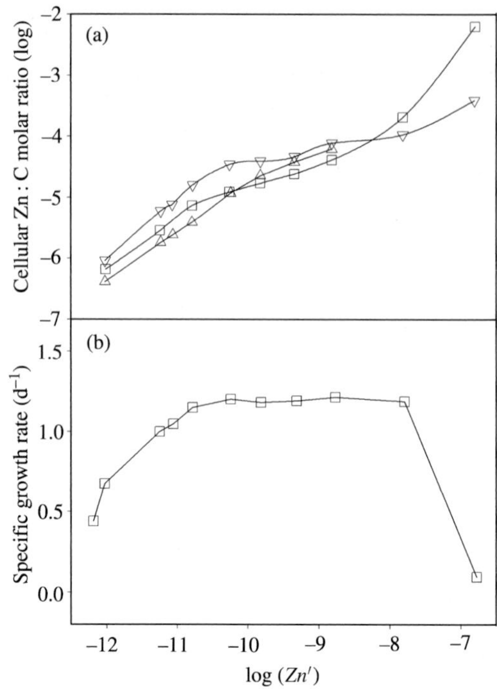

Figure 2 Cellular zinc normalized to carbon (a) and specific growth rate (b) as functions of log(Zn') (unchelated zinc concentration) in a marine diatom *Thalassiosira weissflogii* (squares) and two clones of a marine coccolithophorid *Emiliana huxleyi* (triangles) (after Sunda and Huntsman, 1992).

species. As expected, iron and manganese are quantitatively the most important trace elements in marine phytoplankton, being on average about 10 times more abundant than zinc, copper, cobalt, or cadmium. But there are obvious differences among major taxa; e.g., green algae (which are rarely dominant in the oceans) have higher iron, zinc, and copper than diatoms or coccolithophores, which contain relatively higher proportions of manganese, cobalt, and cadmium.

To what extent do the data presented in Figure 1 reflect the physiology of the organisms or the composition of their growth medium? This is obviously a key question for marine metallomics and the extant literature on marine microorganisms indeed indicates particular attention given to the relation between the composition of organisms and their growth medium (e.g., Anderson and Morel, 1982; Hudson and Morel, 1990; Saito *et al.*, 2002; Sunda and Guillard, 1976; Sunda and Huntsman, 1992). Some of the differences between the data of Figure 1 and similar data published for *S. cerevisae* and *E. coli* may owe as much to differences in growth media as to biochemical differences among organisms.

The design of culture media, in which the concentration and speciation of trace elements are tightly controlled, has been critical to physiological studies of organisms whose natural medium is trace-metal-poor seawater (Morel *et al.*, 1979; Price *et al.*, 1988/1989). In most instances, it has been found practical to control the bioavailability of trace metals by using strong chelating agents such as EDTA (designated Y). Because the rate of uptake of a metal, M, is usually proportional to its unchelated concentration (often designated as M' and sometimes referred to as the "inorganic" concentration of M), the chelated metal provides a convenient buffer that maintains M' at constant low values in the growth medium over the course of a batch culture. The value of M' can be precisely adjusted by choosing an appropriate ratio of the total concentrations of M and Y.

A general observation is that the cellular concentration of an element—its so-called cellular "quota"—varies in a sigmoidal fashion as a function of its concentration in the medium. This is illustrated in Figure 2(a) for zinc in two model species, a marine diatom and a coccolithophore (Sunda and Huntsman, 1992). Over a reasonably wide range of unchelated metal concentration, an organism regulates its metal quota by adjusting its metal uptake rate as well as, sometimes, its metal export rate. If other conditions are also optimal, this regulated metal quota normally allows the organism to grow at its maximum rate, $\mu_{\max}$ (Figure 2(b)). Below the regulated range, however, the cellular quota of the trace

metal can become too low and growth rate decreases—a straightforward case of a limitation when the organism cannot take up the metal at a rate sufficient to maintain optimal growth.

Above the regulated range, the organism accumulates the metal uncontrollably by simple diffusion of some metal species (usually electrically uncharged complexes) through the membrane or by leakage through the transport system of another metal. For example, at high concentrations, zinc enters the diatom cell in part through the manganese transport system (Sunda and Huntsman, 1998b). At some point, the intracellular metal concentration becomes toxic and the growth rate decreases. Because a decreased growth rate results in an even faster rate of metal accumulation (the metal quota being no longer "diluted" by growth), growth often stops abruptly as M' is increased above some critical threshold in the growth medium.

The unchelated concentration of a given metal in the growth medium is not the only parameter that determines the corresponding metal quotas illustrated in Figure 1. The medium concentrations of other elements can also play an important role. This occurs principally under three types of conditions:

(i) Another trace element may effectively substitute for an essential trace element M and the relative environmental concentrations of both then affect the M quota. This phenomenon is described as "biochemical substitution" and its best-documented evidence is the alternative use of zinc, cobalt, or cadmium as a metal center in the carbonic anhydrases in diatoms (Lane and Morel, 2000a; Lee *et al.*, 1995).

(ii) Quite often, another trace element may compete with M for transport or somehow interfere with its uptake. Such "competitive inhibition" has been demonstrated, e.g., for the accumulation of manganese in coastal marine phytoplankton which can be hindered by high concentrations of metals such as copper or cadmium (Sunda and Huntsman, 1996).

(iii) In most cases, the regulated cellular quota of M is affected by the biochemical need for that element and thus depends on a variety of environmental parameters. This is illustrated in several of the examples discussed in the following sections. The most dramatic of these is the case of nickel. The nickel content of phytoplankton in culture is generally quite low and not even reported in Figure 1 (because it was not added to the medium and is usually below detection in the biomass). But the nickel quota increases dramatically when the organisms are grown on urea (rather than ammonium or nitrate) and they need nickel as a metal center in urease (Price and Morel, 1991).

6.05.2.2 Uptake

The concentrations of essential trace elements are extremely low in the waters of the open ocean, typically in the nanomolar to picomolar range as illustrated in Figure 3. Zinc, copper, cadmium, and nickel all show nutrient-like vertical distributions; they are depleted in surface waters as a result of uptake by the biota and increase in concentration at depth as a result of the remineralization of sinking organic matter. Iron and cobalt, which have low deep-water concentrations (~0.6 nM and ~0.05 nM, respectively), sometimes also display nutrient-like depletion in surface waters (Figure 3). Dissolved manganese concentrations usually exhibit a surface maximum as a result of the indirect photochemical reductive dissolution of manganese oxide (Sunda and Huntsman, 1988). More complex distributions are observed near shore as the result of fluvial inputs and in some cases over anoxic sediments on the continental margin (Sundby *et al.*, 1986; Thamdrup *et al.*, 1994).

In addition to being depleted at the surface by biological uptake, most of the essential trace metals (with the apparent exception of manganese and nickel) are chelated by strong organic ligands that maintain the unchelated metal concentrations, M', at extremely low values—between 10^{-15} for Co' and 10^{-11} for Zn' (Rue and Bruland, 1995; Saito and Moffett, 2001a; Moffett, 1995; Bruland, 1989; Achterberg and van den Berg, 1997; Martin and Gordon, 1988; Sunda, 1984; Sunda and Hanson, 1987). The presence of these chelating agents has been inferred as a result of electrochemical measurements showing that the metals are not labile (i.e., that they do not react with a strong ligand or cannot be reduced at the appropriate voltage at an electrode surface) but their chemical structures are still unknown. They are certainly of biogenic origin: some may be metalloproteins or cellular ligands in various stages of remineralization; some may be specific chelators released by the biota for the purpose of dissolving, transporting, or detoxifying metals. For example, the chelated iron in seawater is probably a mixture of iron–siderophore complexes (see below) and of various species of hemes and Fe–S clusters. We note that recent data on the speciation iron in the top 500 m of the open ocean indicate a large colloidal component and a depletion of chelated iron (Wu *et al.*, 2001). In some situations, such as photodegradation of surface water metal-ligand complexes or vertical mixing of unchelated metals to the surface, the excess of metals over chelators results in a dramatic increase in unchelated metal concentrations. This has been observed for copper in the Sargasso Sea and can lead to copper toxicity to the ambient flora (Mann *et al.*, 2002).

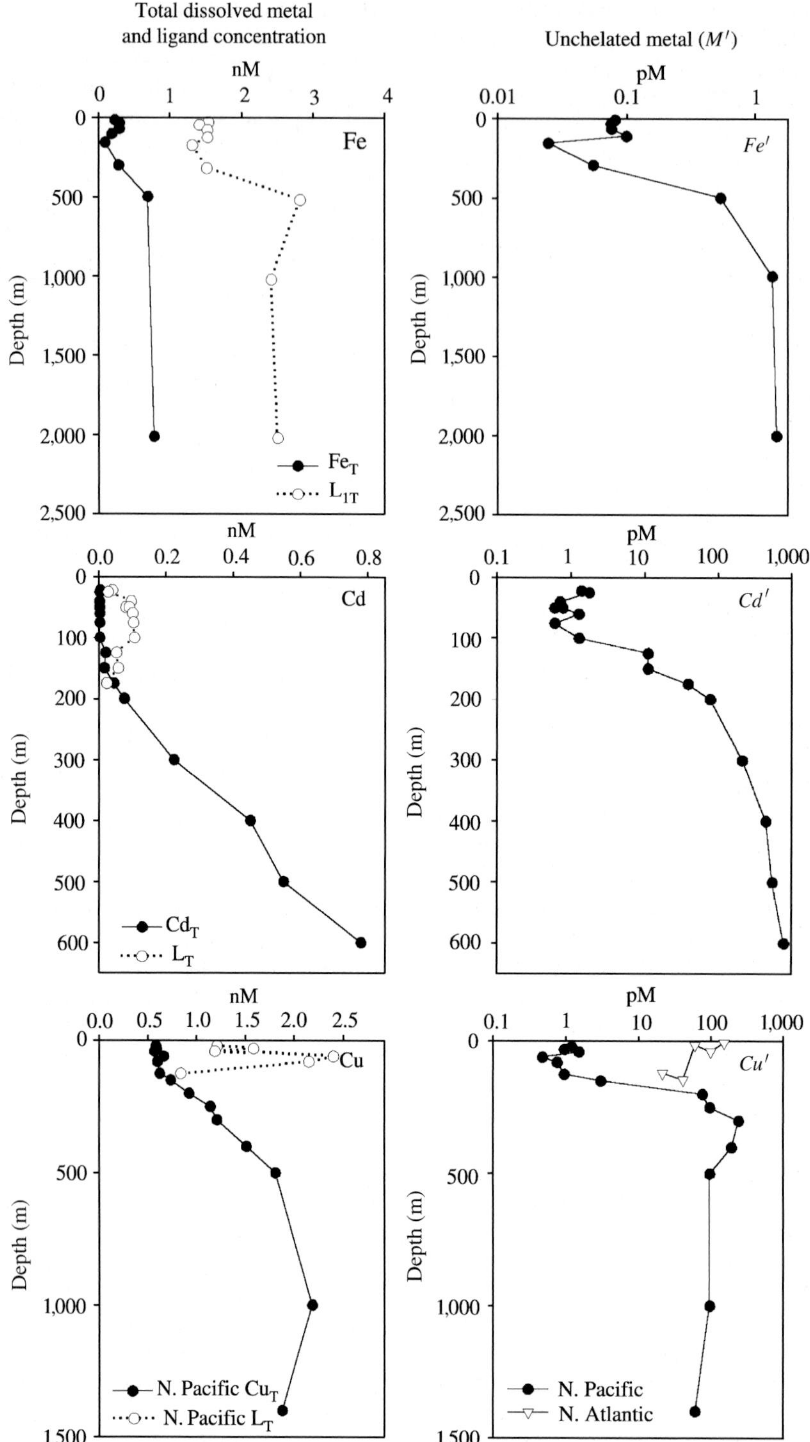

Figure 3 Concentration profiles of total dissolved metals (M_T), organic ligands specific to those metals (L_T), and uncomplexed metal as the sum of unchelated chemical species (M'). Fe data from central North Pacific (Rue and Bruland, 1995); Cd data from the central North Pacific (Bruland, 1989); Cu data from North Pacific (Coale and Bruland, 1988) and overlaid Sargasso Sea free Cu which is seasonally high (Moffett, 1995); Zn data from the central North Pacific (Bruland, 1989); Co data from the Sargasso Sea (Saito and Moffett, 2001a), Ni data from the Mediterranean Sea (Achterberg and van den Berg, 1997), total Mn data from Northeast Pacific (Martin and Gordon, 1988). Limited Mn speciation data suggests that Mn has negligible organic complexation (Sunda, 1984), thus it is assumed that $Mn' \sim Mn_T$ here. Data reported as free metal ion (e.g. M^{2+}) has been recalculated as M' (sum of all inorganic species) for the purposes of this chapter (using inorganic side reaction coefficients: $M'/M^{2+} = 35$ for Cd, 24 for Cu, 1.52 for Zn, and 1.41 for Co).

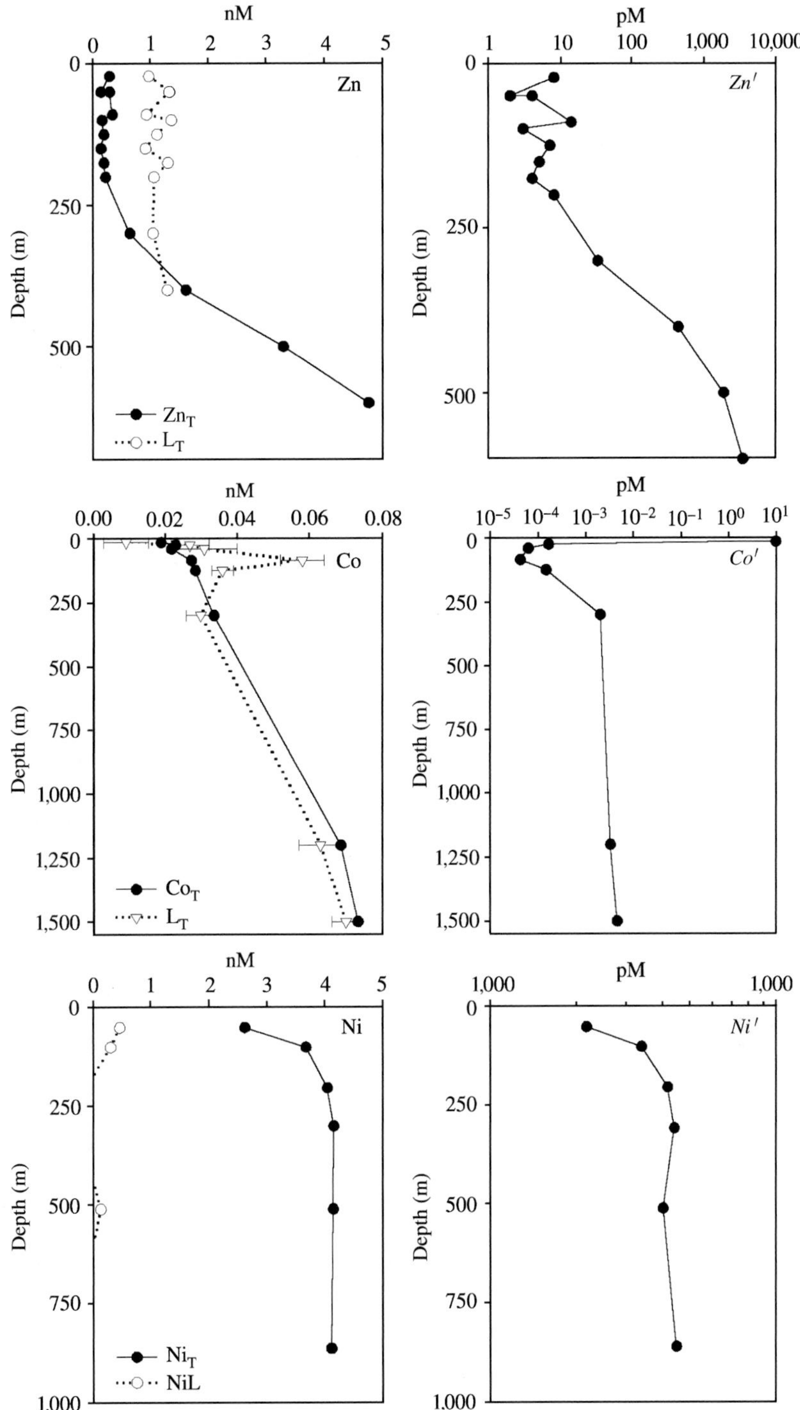

Figure 3 (continued).

To maintain sufficient concentrations of such trace elements and grow at rates of ~0.2–2.0 d^{-1}, marine microorganisms must possess particularly effective uptake systems. These systems are not generally known at the molecular level as few transport proteins have been characterized or even identified. Nonetheless, the available data, based chiefly on uptake kinetics in laboratory cultures, show that the mechanisms of uptake of trace elements of marine microorganisms are similar to those that have been characterized in model bacteria and yeasts. In most cases the uptake systems involve transmembrane transport proteins whose expressions are regulated as a function of the state of deficiency or sufficiency of the organism for the corresponding element. Figure 4 presents an example of the kinetics of (short-term) uptake of iron as a function of the

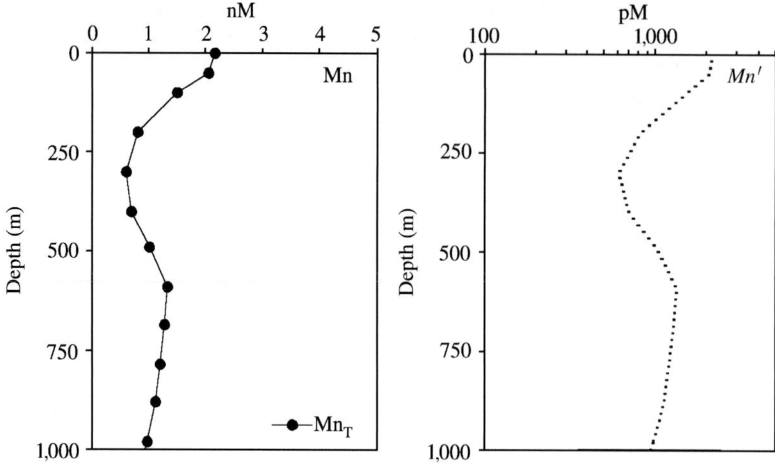

Figure 3 (continued).

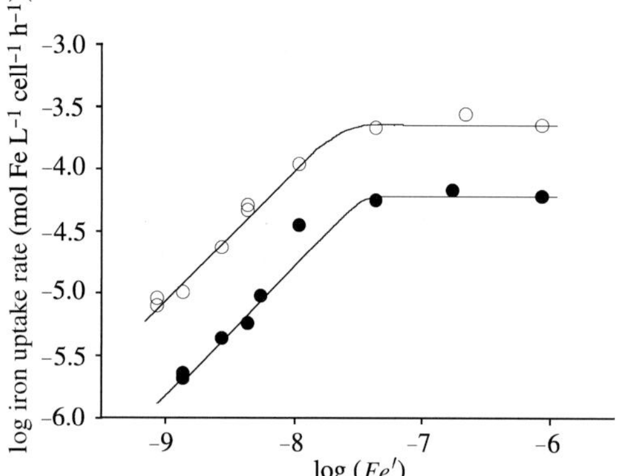

Figure 4 Short term (3 h) iron uptake rates in the diatom *Thalassiosira weissflogii* as a function of log (Fe') (unchelated iron concentration) for high iron grown (closed symbols) and low iron grown (upon symbols) (after Harrison and Morel, 1986).

unchelated iron concentration, Fe', and of the previous extent of iron sufficiency or deficiency in cultures of the diatom *T. weissflogii* (Harrison and Morel, 1986). The uptake rate follows typical Michaelis–Menten saturating kinetics and the maximum uptake rate increases several-fold when the organisms have been starved for iron. As for the growth data shown in Figure 2(b), short-term uptake kinetic data for trace metals are obtained in buffered media containing high concentrations of artificial chelators such as EDTA. By varying the nature or concentration of these chelators, one can verify that the uptake kinetics of the trace metal (i.e., its bioavailability) are determined by its unchelated concentration.

The term "bioavailability" is often used in an absolute sense when a particular form of a nutrient is either available or not available to an organism. In the case of trace-metal complexes, which dissociate over a finite timescale, the concept of bioavailability must necessarily be tied to kinetics. A particular form of a metal is more or less available to a (particular) organism depending on the rate at which it can be taken up. For example, if a metal chelate, MY, is not directly taken up, its slow dissociation will eventually make the unchelated metal, M', available to the organism (although the dissociation of MY may be too slow to allow the organism to grow).

The question arises as to whether the situation obtained in buffered laboratory cultures is applicable to the surface oceans. We know that, in surface seawater as in culture media, a major fraction of the trace metals of interest is present as complexes with strong organic ligands (Figure 3). But do these natural ligands control trace-metal availability in the same way as that EDTA does in artificial growth media? The bioavailability of chelated metals—or more precisely the mechanisms and kinetics of their uptake—must clearly depend on the nature of the chelate and the uptake system of the organism. When uptake is simply effected by transport proteins in the membrane, the kinetics of metal exchange between the chelator in the medium and the binding ligands on the proteins determine the uptake rate (the turnover rate of the transporter being rarely limiting). Unless the extracellular chelator has particular properties that allow the formation of a ternary complex chelator–metal–protein (in which case the chelator is effectively a "metallophore"; see below), this metal exchange requires release of the metal in seawater and the uptake kinetics should indeed be governed by the unchelated metal concentration in the medium (Hudson and Morel, 1993).

In some cases, however, the organism is able to accelerate the metal exchange by promoting the

release of the metal from the chelator. For example, marine diatoms possess an extracellular reductase that promotes the reduction of Fe(III) to Fe(II) in siderophore complexes, resulting in the release of Fe from the ligand (Maldonado *et al.*, 1999; Maldonado and Price, 2001). The system appears similar to that characterized in yeast (Figure 5) (Eide, 1998) which, in addition to a reductase, utilizes a multi-copper oxidase that functions in series with an Fe(III) transporter. Uptake of iron by this mechanism thus involves consecutive reduction and reoxidation of the iron. It is likely that some marine organisms have evolved ability to release iron from other types of biogenic iron compounds (Hutchins *et al.*, 1999a) or perhaps even from some iron minerals (Nodwell and Price, 2001).

Some trace-metal transport systems are even more complex than the one described in Figure 5 and involve the release of "metallophores" into the medium. The archetypes of these—and the only ones characterized so far—are the siderophores produced by various species of marine bacteria to acquire iron. In the model organisms in which they have been characterized, the mechanisms of uptake are quite varied and complex, often involving intermediate siderophores in the periplasmic space and several transport proteins (Neilands, 1981). The effect of such siderophores on iron bioavailability is clearly not the same as that of EDTA. While complexation by a siderophore makes iron directly available to the bacteria which take up the complex (and whose rate of iron uptake is proportional to FeY), it drastically reduces the bioavailability of iron to most other organisms (whose rate of iron uptake is proportional to Fe'). For organisms which are able to promote the release of iron from the siderophore, e.g., by reduction of Fe(III), the effect of complexation is a less drastic decrease in iron uptake rate. Thus the release of siderophores into the medium results in a sort of economic warfare over iron—the organisms that have the necessary FeY transporter, including those that "cheat" by synthesizing the transporter but not releasing Y into the medium, obtain almost exclusive access to the chelated iron.

A number of siderophores from marine bacteria have been isolated from cultures with their structures elucidated. Some are similar to those of terrestrial organisms but two major families—viz., the marinobactins (isolated from *Halomonas marina*) and the aquachelins (isolated from *Marinobacter* sp.)—are characterized by fatty acid tails of variable lengths (Figure 6; Martinez *et al.*, 2000). Both have a peptidic head group that contains two hydroxamates and one α-hydroxy acid to complex iron. The amphiphilic nature of these siderophores is likely to be a key to their modus operandi in the extremely dilute medium of the oceans. Of great interest also to the oceanic cycling of iron are the photochemical properties of the iron complexes of marinobactins and aquachelins. Upon illumination, the α-hydroxy acid is cleaved along with the hydrophobic tail and Fe(III) is reduced to Fe(II) (Barbeau *et al.*, 2001). While lower than that of its parent compound, the affinity of the oxidized siderophore for Fe(III) is still quite high and may be important in the marine geochemistry of iron.

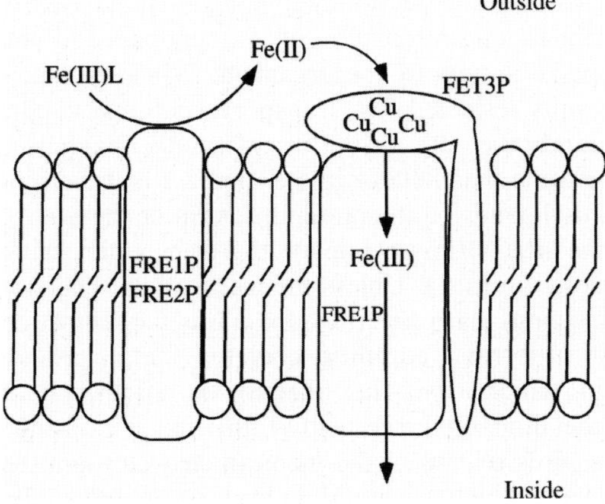

Figure 5 The iron reductase and high affinity uptake system in yeast. Fre1,2P, iron reductase proteins; Fet3P, multi-copper oxidase; FtrP, high affinity transport protein (after Eide, 1998).

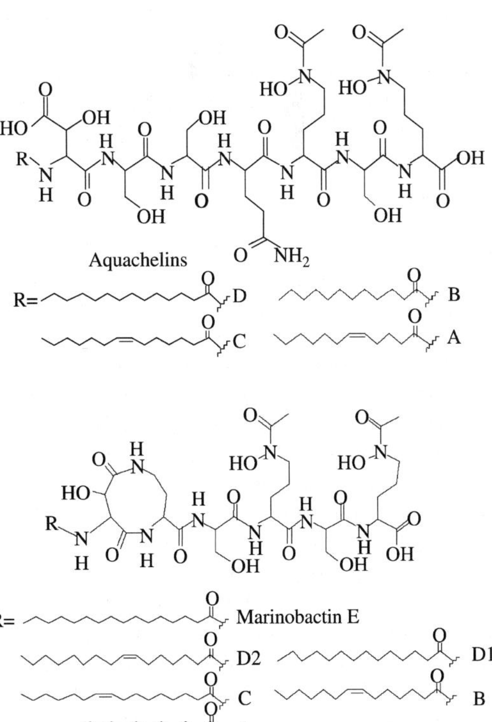

Figure 6 Two examples of ampiphilic siderophores produced by marine heterotrophic bacteria (after Martinez *et al.*, 2000).

Seawater is the most extreme environment on Earth in terms of its paucity of essential trace elements. Marine microorganisms have thus evolved some of the most efficient uptake systems possible and take up trace metals at rates near the maximum allowed by physics and chemistry. One can calculate absolute limits to the cellular uptake rates of metals by considering the simple case of uptake via transmembrane proteins (Hudson and Morel, 1990).

One limit is posed by the diffusion from the bulk medium to the cell surface. An unchelated bulk concentration of 2 pM (in the low range seen for metals such as zinc in surface seawater; see Figure 3) can supply by molecular diffusion at the most 10 amol d^{-1} to a (spherical) cell of radius $R = 4\ \mu m$. (That is, 10^{-17} mol d^{-1}; this calculation assumes the concentration of M at the surface of the cell to be 1 pM, half that of the bulk, and neglects the supply of metal from the dissociation of chelates.) Thus, for example, the diffusive supply of unchelated zinc from surface seawater to several of the organisms listed in Figure 1 (which have about the right size and contain on average ~75 amol Zn/cell) would barely allow them to grow at a rate of 0.13 d^{-1}. Clearly, the supply of trace nutrients by diffusion causes a severe limitation to the size of marine phytoplankton since the diffusive supply increases as R and the need roughly as R^3.

Another limitation to the rate of cellular uptake of a trace metal is posed by the chemical kinetics of reaction with the ligands of the transport proteins. This rate is proportional to the intrinsic second order reaction rate, *k*, the concentration of unchelated metal at the surface, and the concentration of uptake ligands on the membrane (Hudson and Morel, 1990). For a fast reacting metal with a second order rate constant of $10^7\ M^{-1}\ s^{-1}$ present at 1 pM at the cell surface, a rate of uptake of 10 amol d^{-1} requires 10 amol $cell^{-1}$ of surface ligands—as much as the metal quota itself. In a cell of 4 μm radius, this represents a surface density of uptake proteins of 3×10^4 molecules per μm^2; at such high densities, proteins with effective radii in excess of 1 nm would occupy most or all of the membrane surface area. This limit posed by chemical kinetics is less severe for fast reacting metals such as Cu^{2+} and more severe for slow reacting metals such as Ni^{2+}. It is also compounded by the need to simultaneously transport several essential elements that are all present at very low concentrations in surface seawater. The kinetic limit posed by crowding of uptake molecules also becomes increasingly severe for larger cells as the available surface area increases as R^2 and the necessary metal flux as R^3. In a competition for rare nutrients, including trace elements, size is clearly a critical factor for marine microorganisms. We note that the production of metallophores to complex trace metals in the medium may increase the bulk dissolved concentration of target metals (e.g., by dissolving some solid species) but, depending on the specifics of the metallophore uptake systems, it does not automatically resolve the problems posed by diffusion and by the kinetics of reaction with transport proteins (Völker and Wolf-Gladrow, 1999).

6.05.2.3 Trace Element Storage

Little is known of the intracellular chemistry of trace metals in marine microorganisms. In model organisms *E. coli* and *S. cerevisiae*, it has been shown that there is no cellular pool of free ions such as Zn^{2+} or Cu^{2+} (Outten and O'Halloran, 2001; Rae *et al.*, 1999). The total cellular concentrations of these metals (~mM) are either in the form of metalloproteins or bound by various chaperones or storage molecules. The same is very likely to be true for all trace metals in all type of cells, including marine microorganisms. The best known metal storage proteins are the ferritins, which are extraordinarily effective at storing iron as ferrihydrate and found in most organisms (Andrews, 1998; Grossman *et al.*, 1992). Genomic information indicates that some cyanobacteria are likely to possess certain type of bacterio-ferritin (M. Castruita, personal communication), although no such molecule has so far been isolated from marine microorganisms.

Like other plants, eukaryotic marine phytoplankton synthesize small polypeptides known as phytochelatins (Ahner *et al.*, 1994) to bind some metals intracellularly. These molecules, which have the general formula γ-(glu-cys)$_n$-gly Figure 7, bind metals through the thiol ligands of the cysteines and have thus a high affinity for soft metals such as Cd^{2+} and Cu^{2+}. Indeed these two metals have been found to be particularly effective at triggering the synthesis of phytochelatins (chiefly the dimer; see Figure 7) in marine microalgae. Phytochelatins (and perhaps other cysteine rich peptides; Ahner, personal communication) provide a rapid detoxification response in the presence of elevated concentrations of such metals. Kinetic evidence for the export of the cadmium phytochelatin complex from marine diatoms into the medium show that these organisms likely possess a metal export system similar to that characterized for similar complexes into the vacuole of yeast (Oritz *et al.*, 1995). But in diatom cultures, phytochelatins are also synthesized in the presence of the lowest concentrations of metals achievable, and their production even increases when *Zn′* is decreased to very low levels (Ahner *et al.*, 1998). Since these organisms are known to substitute cadmium for zinc when starved for zinc, it has been postulated that

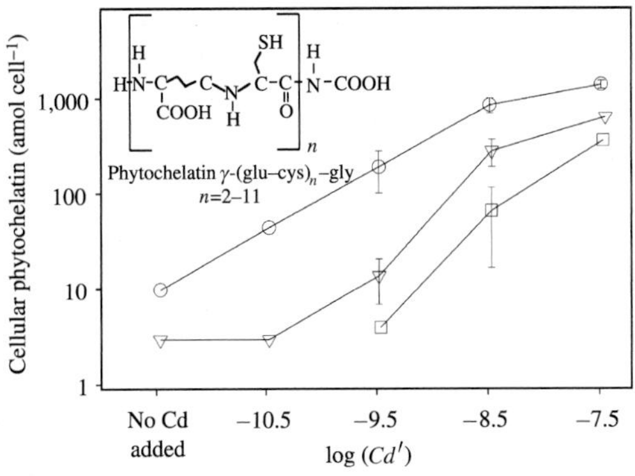

Figure 7 Phytochelatin concentrations in the marine diatom *Thalassiosira weissflogii* as a function of log (Cd') (unchelated cadmium concentration). Chain lengths for each phytochelatin dimer (circles) trimer (triangles), and tetramer (squares) (after Ahner *et al.*, 1994).

Table 1 Known biochemical functions of selected trace elements that are known to account for a sizeable fraction of cellular metals in marine microorganisms.

Metal	*Function*
Mn	Oxygen evolving complex of PS II
	Superoxide dismutase
Fe	FeS centers (e.g., aconitase, ferredoxin)
	Cytochromes
	Superoxide dismutase
	Nitrate reductase (assimilatory and respiratory)
	Nitrite reductase (assimilatory and respiratory)
Co	Carbonic anhydrase
Ni	Urease
Cu	Plastocyanin
	Ferrous oxidase
	Amine oxidase
Zn	Carbonic anhydrase
Cd	Carbonic anhydrase
	Alkaline phosphatase?

phytochelatins may serve to store cadmium as well as to detoxify metals (Ahner and Morel, 1999).

Over the course of a light–dark cycle, many metalloproteins that are involved in various aspects of nutrient acquisition, photosynthesis and respiration are in turn synthesized and degraded. This is the case, for example, of carbonic anhydrase in diatoms (Lane and Morel, 2000b), and of nitrogenase in the cyanobacterium *Trichodesmium* (Chen *et al.*, 1998)—both enzymes account for a significant fraction of the corresponding zinc and iron content of the organisms. Thus, it appears that intracellular trace metals are subject to dynamic trafficking as the cell goes through its various stages of division during the light–dark cycle. It may be so that in the metal-poor ocean, most of the storage of essential metals is in the form of functional proteins and that the metals are recycled among the metalloproteins as the need arises over the course of their growth and division.

6.05.3 THE BIOCHEMICAL FUNCTIONS OF TRACE ELEMENTS IN THE UPTAKE AND TRANSFORMATIONS OF NUTRIENTS

Not all biochemical uses of trace elements hold the same interest for oceanographers. The biochemical functions that utilize the largest fraction of a given metal in the cell are of most interest. A low metal concentration will limit the growth rate of the organism principally through impairment of these biochemical functions. Moreover, all environmental conditions that modify the need for these particular functions will also change the requirements of the organism for the metal. Although we have limited quantitative information regarding enzymatic processes in cells and the corresponding trace element requirements, Table 1 provides a list of a few biochemical functions that are thought to correspond to major trace-metal requirements in marine phytoplankton. This list is quite short, reflecting in part our limited quantitative knowledge of biochemistry, particularly in marine microorganisms. This section of the chapter and the next are devoted in large part to justify the entries in Table 1. The trace-metal requirements of various biochemical functions have been quantified on the basis of theoretical calculations and empirical data. For example, in photosynthetic organisms, metal use efficiencies—mol C assimilated per unit time and per mole catalytic metal—can be calculated on the basis of the metal content of the catalyst, its specific reaction rate and the requirement for the products of that catalyst to fix carbon at a given rate (Raven, 1988, 1990). In a few experiments, the metal concentrations associated with particular enzymes have been measured directly. In others, the predicted effect of varying environmental conditions (e.g., light or nutrient concentrations) on the requirement for a particular metal and/or the effect of varying the metal concentration on enzyme activity have been tested.

The list given in Table 1 will undoubtedly be longer as we learn more about the biochemical functions that are significant for biogeochemistry and require relatively large concentrations of particular trace elements. But this list is also notable by the absence of some "obvious" metalloproteins. For example, many well-known zinc proteins, including RNA- and DNA-polymerases, seem to represent only minute fractions of cellular zinc. In the same way, present evidence indicates that Cobalamin

(i.e., vitamin B_{12})-containing enzymes account for only a small fraction of the cobalt quota of marine phytoplankton (Sunda and Huntsman, 1995b; Wilhelm, 1995). Many such metalloenzymes are quite interesting in their own right but they probably do not represent a critical link between the geochemical cycles of trace metals and of major nutrients.

6.05.3.1 Trace Metals and the Marine Carbon Cycle

How fast marine organisms fix inorganic carbon and what controls that rate—are topics of central interest to oceanographers. So far, the only trace element that has been shown beyond doubt to limit primary production in some regions of the oceans is iron. This is perhaps not surprising in view of the extremely low concentration of iron in some surface waters (see Figure 3(a)) and of the relatively large iron requirement of marine phytoplankton (Figure 1). To better understand how various trace elements affect the marine carbon cycle we need to review briefly some of the biochemistry involved in the photosynthetic fixation of inorganic carbon and the respiration of organic compounds.

6.05.3.1.1 Light reaction of photosynthesis

The first series of steps in photosynthesis, known as the light reaction, involve the absorption of photons, the evolution of O_2, and the formation of high energy compounds (ATP) and reductants (NADPH). The absorption of photons and the transfer of electrons from H_2O to NADPH are carried out by two photosynthetic systems, PS II and PS I, working in series (Figure 8). The pigments that harvest light in each photosynthetic system and transfer the energy to the respective PS II and PS I reaction centers vary somewhat among families of algae but contain no trace elements. The formation of oxidants and reductants in the two reaction centers, as well as the transfer of electrons between them and upstream and downstream from them, involve a large number of redox intermediates such as quinones, Fe–S centers, cytochromes and ferredoxin, most of which contain Fe. An overall tally reveals 3 and 12 iron atoms per PS II and PS I reaction centers, respectively, and 8 in the electron transport chain (Raven, 1990). Depending on the relative proportions of these various components (e.g., cyanobacteria have a relatively high proportion of iron laden PS I reaction centers) and the reaction kinetics of the slowest electron transfer step, one can calculate the iron requirement for a given rate of photosynthesis (Raven, 1990). The results uniformly show that the photosynthetic apparatus contributes the largest iron requirement for oxygenic photosynthetic organisms, typically about two-third of the total. Some of the photosynthetic iron can be economized by replacement of redox intermediates: soluble cytochrome b553 can be replaced by copper-containing plastocyanin in cyanobacteria and green algae; and, in many species, ferredoxin can be replaced by flavodoxin, a flavin which contains no metal (Doucette *et al.*, 1996; La Roche *et al.*, 1993). These economies do not appear to be very large, however, they amount to ~10% of the total iron requirement.

On the oxidative side of PS II, the oxidation of water to O_2 also involves four manganese atoms,

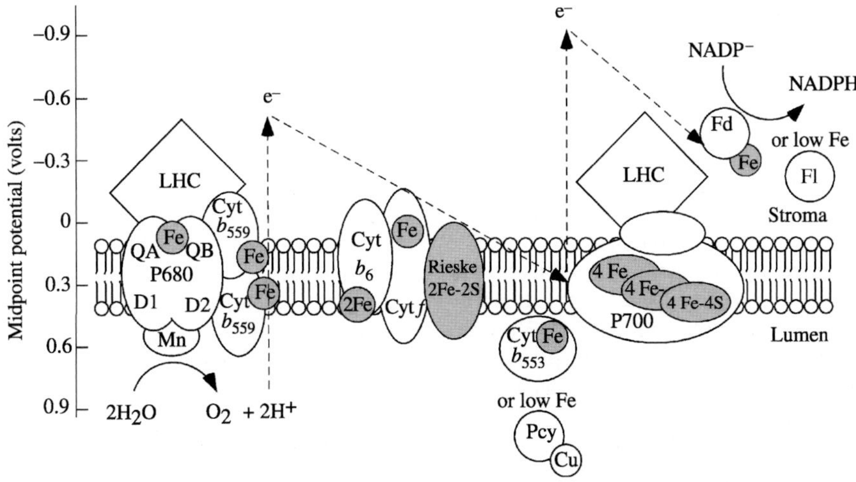

Figure 8 A diagram of the photosynthetic apparatus of oxygenic phototrophs, illustrating the proteins that contain iron (Fe) with the potential of electrons at each step superimposed. Abbreviations: LHC, light harvesting complex; Cyt, cytochrome; b, f with subscripts refer to the spectroscopic type of cytochrome; P680, the reaction center of photosystem II (PSII); P700 the reaction center of photosystem I (PSI).

each exchanging one electron. This allows four one-electron-transfer steps to yield one four-electron-transfer reaction. There is no known replacement of manganese in the oxygen evolving complex which represents a major pool of cellular manganese. All photons that are absorbed by phytoplankton do not have beneficial (photosynthetic) or benign (heat or fluorescence) consequences. Some result in the formation of extremely noxious compounds, particularly active oxygen species such as the superoxide radical, O_2^-. Organisms protect themselves from injury by these harmful species by catalyzing their reactions into less noxious ones. In particular, the dismutation of O_2^- into O_2 and H_2O_2 is catalyzed by the enzyme superoxide dismutase (SOD). In marine phytoplankton, SOD is a manganese, or perhaps, an iron enzyme (Peers and Price, personal communication), while a Cu–Zn SOD is found in other organisms.

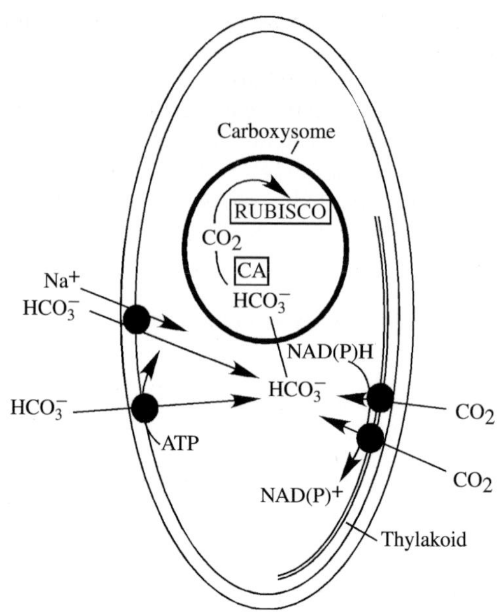

Figure 9 A model of the cyanobacterial carbon concentrating mechanism from *Synechococcus* PCC 7942, a freshwater species. Solid circles represent transporters located on the plasma membrane and interior to the cell wall. Boxes represent the catalyzing enzymes CA, Carbonic Anhydrase; RUBISCO, Ribulose 1-5 bisphosphate Carboxylase Oxygenase. The carboxysome is the site of carbon fixation (dark reactions) and the thylakoid is the site of the light reactions of photosynthesis (after Badger *et al.*, 2002).

6.05.3.1.2 Dark reaction of photosynthesis

The second series of biochemical steps in photosynthesis, known as the dark reaction, involves the use of the reductant and the energy produced in the light reaction to reduce inorganic carbon to organic carbon through a process known as the Calvin cycle. In the first step, catalyzed by the enzyme Rubisco (Ribulose 1-5-bisphosphate carboxylase oxygenase), CO_2 reacts with Rubp (Ribulose bisphosphate), a C_5 compound, to form two PGAs (phosphoglyceric acid, a C_3 compound). Subsequent phosphorylation reactions with ATP and reduction by NADPH end up providing glyceraldehyde 3P from which various compounds are synthesized including glucose. None of the nine enzymes involved in the Calvin cycle are metalloenzymes, although Rubisco is known to be activated by either manganese or magnesium (Jensen, 1990). But metals are nonetheless involved in the dark reaction of photosynthesis because of the inefficacy of Rubisco. This ancient enzyme (which presumably evolved early in Earth's life when CO_2 was plentiful and O_2 rare) requires CO_2, not HCO_3^-, as a substrate, has a low affinity for CO_2 ($K_{1/2} = 20–100$ μM) (Badger *et al.*, 1998) and, as indicated by its name, is subject to a competitive reaction with O_2. The net result is that CO_2 must somehow be concentrated near Rubisco to allow efficient carbon fixation in marine microalgae whose medium contains 2 mM HCO_3^-, and only 10 μM CO_2.

6.05.3.1.3 Carbon concentrating mechanisms

The best known carbon concentrating mechanism (CCM), is that of freshwater cyanobacteria which is depicted in Figure 9. This system consists of two HCO_3^- transporters (one powered by the inward Na^+ flux, the other by ATP), and two CO_2 pumps which appear to work by diffusion of CO_2 through the membrane down a gradient maintained by a yet to be elucidated NADPH-dependent transformation of CO_2 into HCO_3^- on the inner side of the membrane (Ohkawa *et al.*, 2002; Price *et al.*, 2002). Under normal conditions (i.e., atmospheric CO_2), the transport of HCO_3^- dominates over that of CO_2 (Price *et al.*, 1994; Yu *et al.*, 1994). The net result of the cyanobacterial CCM is the accumulation in the cytosol of a concentration of HCO_3^- far in excess of its equilibrium with CO_2. This is possible because the uncatalyzed hydration/dehydration reaction of HCO_3^-/CO_2 is relatively slow, its half-life being ~50 s, compared, for example, to a molecular diffusion time of few milliseconds in cells. The provision of CO_2 to Rusbisco in the carboxyzome is then enabled by a carbonic anhydrase (CA), a zinc enzyme that catalyzes the HCO_3^-/CO_2 reaction (Price and Badger, 1989).

With various modifications, the CCM of freshwater cyanobacteria is also generally accepted as the underlying scheme for the CCM of marine cyanobacteria and eukaryotic phytoplankters. In particular, the CCM of the model chlorophyte, *Chlamydomonas*, appears to involve similar HCO_3^- and CO_2 transporters through the plasmalemma. Often eukaryotes possess a CA in their periplasmic space. By maintaining equilibrium between HCO_3^- and CO_2, this enzyme avoids the depletion of CO_2 that would occur at the surface of these large cells as a result of CO_2 pumping and slow diffusion from the bulk medium. (In seawater, the HCO_3^- concentration is too high (2 mM) to be significantly depleted at the surface of photosynthesizing cells.)

Diatoms have been shown to accumulate inorganic carbon via a mechanism that is markedly different from the CCM of cyanobacteria and more akin to the C_4 mechanism found in some higher plants (Figure 10) (Reinfelder *et al.*, 2000). (The characterization of the inorganic carbon acquisition system of diatoms as a unicellular C_4 pathway is controversial. The central point of contention is whether a majority of the inorganic carbon—or a significant fraction of it, depending on conditions—goes through a C_4 compound before being fixed, or if the formation of C_4 compounds is chiefly an anaplerotic process and their decarboxylation an unimportant side reaction.) Inorganic carbon is transported across the plasmalemma by diffusion of CO_2 which may be supplemented by active transport of HCO_3^-, depending on concentrations in the medium. Equilibrium between CO_2 and HCO_3^- is maintained at the cell surface by a periplasmic CA whose activity is enabled by the proton buffering of the silicon frustule (Milligan and Morel, 2002). A cytoplasmic CA maintains HCO_3^-/CO_2 equilibrium on the other side of the membrane. As we shall see, the cytoplasmic CA of diatoms and, likely, their periplasmic CA, can use zinc, cobalt, or cadmium as their metal centers. Under atmospheric p_{CO_2} conditions, the dehydration of HCO_3^- to CO_2 in the periplasm and the subsequent diffusion of CO_2 into the cell appear to constitute the main source of inorganic carbon to diatoms (Tortell and Morel, 2002). Cytoplasmic HCO_3^- serves as the substrate for the enzymatic formation of a C_4 compound, oxaloacetate then malate, which is eventually decarboxylated in the chloroplast to feed CO_2 to Rubisco. The main enzymes catalyzing the carboxylation and decarboxylation reactions appear to be PEPC, which, like Rubisco can be activated by either manganese or magnesium, and PEPCK, which can be activated only by manganese.

6.05.3.1.4 Respiration

The remineralization of organic matter into CO_2, which closes the organic carbon cycle, is carried out by both the phytoplankton themselves, to produce necessary energy, and by heterotrophic bacteria which make a living from it. During the day, phytoplankton respire about a quarter of their fixed carbon and again about as much during the night (Raven, 1988). The net result is that nearly half of the carbon fixed photosynthetically is respired by the photosynthesizers themselves. Most of the rest is eventually respired by bacteria, only a small fraction of which is exported to the deep sea (Honjo, 1996).

In oxygenated waters, the respiration of sugars for the production of ATP occurs chiefly through the citric acid cycle. As in photosynthesis, the electron carrying intermediates of respiration—which include aconitase with an Fe_4S_4 center and cytochromes—are rich in iron. In phytoplankton

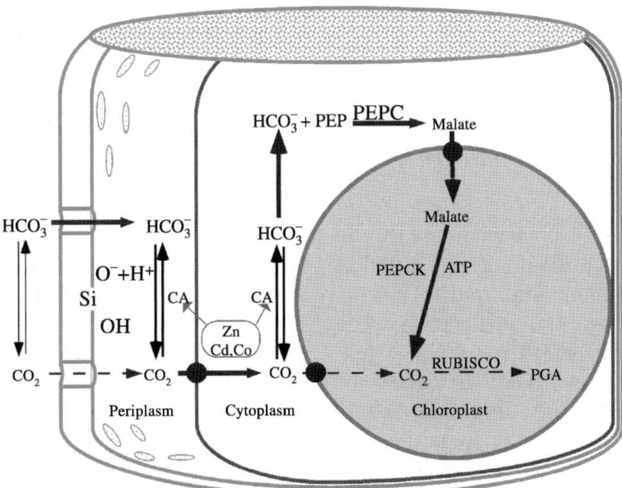

Figure 10 A hypothetical model of carbon acquisition in the marine diatom *Thalassiosira weissflogii*. Solid circles represent transporters. Catalyzing enzymes CA, carbonic anhydrase; PEPC, phosphoenol pyruvate carboxylase; PEPCK; phospoenol pyruvate carboxykinase; RUBISCO, Ribulose 1-5 bisphosphate Carboxylase Oxygenase (after Morel *et al.*, 2002).

the iron requirement for respiration is estimated to account for about one-third of the total (the bulk of the rest being involved in photosynthesis as described above; Raven, 1988).

Like phytoplankton, heterotrophic bacteria require the iron contained in the electron carriers of respiration. Various species of bacteria also express enzymes that are specialized for the degradation of particular classes of organic compounds. Many of these enzymes contain iron or zinc (a few contain copper) to carry out the additional redox reactions. The net result: most of the cellular iron in heterotrophic bacteria is directly involved in respiration and that, normalized to carbon or cellular mass, the iron requirement of these heterotrophs is predicted to be smaller by about three times than that of phytoplankton (Raven, 1988).

6.05.3.2 Trace Metals and the Nitrogen Cycle

Nitrogen, after carbon and oxygen and on par with hydrogen, constitutes a large fraction of biomass. It is generally thought to be the principal nutrient limiting the primary production of the oceans (Dugdale and Goering, 1967; Glibert and McCarthy, 1984). The question of nitrogen limitation can be considered on two scales. (i) From the point of view of a given phytoplankter in surface seawater, what are the available nitrogen species and how can the organism utilize these sources of nitrogen? (ii) At the scale of the whole ocean or ocean basins, what determines the overall mass of nitrogen available for plant growth, i.e., what controls the balance between N_2 fixation and denitrification? Most of the processes involved in the acquisition of nitrogen by phytoplankton and the overall cycling of nitrogen in seawater involve metalloenzymes and are thus of prime interest to marine bioinorganic chemists.

6.05.3.2.1 Acquisition of fixed nitrogen by phytoplankton

As depicted in Figure 11, phytoplankton can, in principle, acquire nitrogen from many different compounds, including ammonium, nitrite, nitrate, urea, aminoacids, amines, etc. In addition some species of cyanobacteria are able to fix dinitrogen, N_2, and are thus able to grow even in the absence of "fixed" nitrogen thereby increasing the oceanic pool of useable N in the process. Regardless of the original source of N, its final assimilation into aminoacids follows a unique pathway, the glutamine synthetase, glutamate oxoglutarate aminotransferase (GS-GOGAT) pathway (Zehr and Falkowski, 1988): NH_4^+ reacts with glutamate to form glutamine, followed by the transfer of an amine group from glutamine to α-ketoglutarate to form two molecules of glutamate. The enzyme GOGAT contains an Fe–S center.

Upstream from the GOGAT system, specialized enzymatic transport and transformation pathways enable the production of internal NH_4^+ from a variety of external nitrogen-containing species. Those depicted in Figure 12 are not exhaustive of all the possibilities. Vice versa, not all phytoplankton species have the enzymatic machinery to obtain nitrogen from all types of sources; on the contrary, the ability to acquire and assimilate

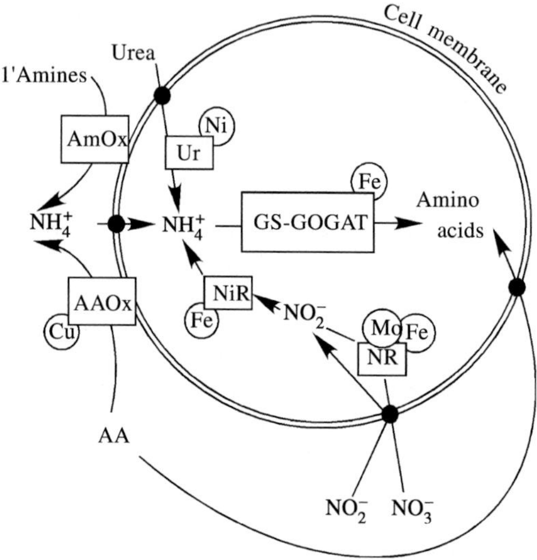

Figure 11 Nitrogen sources and metabolic pathways in marine phytoplankton. Solid circles are transporters. Boxes are the catalytic enzymes and open circles are metals associated with each enzyme. Ur, Urease; NR, Nitrate Reductase; NiR, Nitrite Reductase; AAOx, amino acid oxidase; AmOx, amine oxidase; GS-GOGAT, Glutamine Synthetase- Glutamate oxyglutarate aminotransferase (or glutamate synthase).

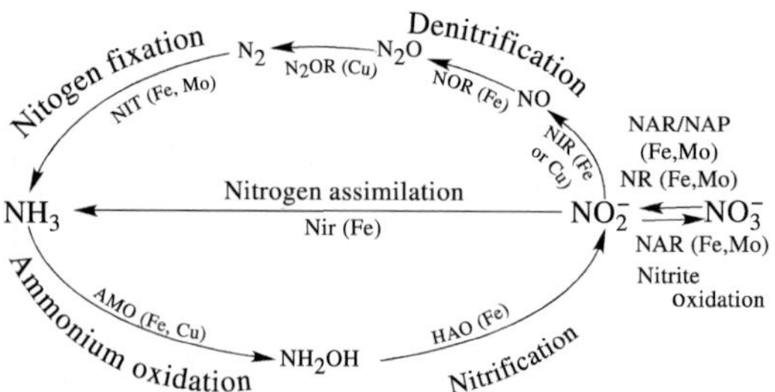

Figure 12 A diagram of the nitrogen cycle with catalyzing enzymes and metal requirements of each step. NIT, nitrogenase; AMO, ammonium mono-oxygenase; HAO, hydroxylamine oxidoreductase; NAR, membrane-bound respiratory nitrate reductase; NAP, periplasmic respiratory nitrate reductase; NR, assimilatory nitrate reductase; NIR, respiratory nitrite reductase; NiR, assimilatory nitrite reductase; NOR, nitric oxide reductase; N_2OR, nitrous oxide reductase.

nitrogen from particular sources likely represents an important ecological specialization for an organism.

Practically, all phytoplankton species are able to take up NH_4^+ actively from the external medium (Wright and Syrett, 1983). Both nitrite (NO_2^-) and nitrate (NO_3^-) are also taken up by a majority of marine phytoplankton species by what is thought to be a single transporter that does not differentiate between the two oxidation states (Cresswell and Syrett, 1982; Galvan and Fernandez, 2001). The transport mechanisms for NH_4^+, NO_2^-, and NO_3^- in marine phytoplankton are thought to be similar to those of vascular plants and involve both high and low affinity systems.

In all photoautotrophs, reduction of NO_3^- to NH_4^+ is achieved in two distinct enzymatic steps (Campbell, 2001). First, assimilatory nitrate reductase (NR) catalyzes the two electron reduction from NO_3^- to NO_2^-. NR is a large soluble cytoplasmic enzyme with FAD (flavin adinine dinucleotide), an iron-containing cytochrome and molybdopterin prosthetic groups, and requires NADH and/or NADPH as an electron donor (Guerrero *et al.*, 1981). Functional NR is in the form of a homodimer and therefore requires two atoms of iron per enzyme. Following transport into the chloroplast, NO_2^- undergoes a 6 e^- reduction to NH_4^+ via assimilatory nitrite reductase (NiR). NiR, a soluble chloroplastic enzyme, contains five iron atoms per active enzyme molecule, and requires photosynthetically reduced ferredoxin as an electron donor (Guerrero *et al.*, 1981).

For an average N : C ratio of 1 : 6, the reduction of NO_3^- to NH_4^+ necessitates eight electrons compared to the 24 for carbon reduction. It is thus not surprising that the utilization of NO_3^- as a source of nitrogen should represent a sizeable cost of both energy and iron supply and algae growing on nitrate are calculated to require 60% more iron than algae growing on ammonium (Raven *et al.*, 1992).

Many small organic molecules potentially constitute an excellent source of nitrogen for marine phytoplankton, particularly in oligotrophic waters where the concentrations of NH_4^+, NO_2^-, and NO_3^- are vanishingly low. For example, urea, $CO(NH_2)_2$, which is the most abundant form of low molecular mass organic nitrogen in the sea, has been shown to account for up to 25–50% of the nitrogen taken up by phytoplankton in various areas of the oceans (McCarthy, 1972; Varela and Harrison, 1999). In most species, urea is metabolized by hydrolysis to ammonium, catalyzed by the nickel enzyme, urease. In some classes of the Chlorophyceae (Bekheet and Syrett, 1977; Leftley and Syrett, 1973), assimilation of urea is effected by a two-step process involving the enzyme urea amidolyase.

Despite their low concentrations in seawater (<1 μM), amino acids can account for a sizeable fraction of the nitrogen uptake of phytoplankton in the sea (up to 10–50%) (Mulholland *et al.*, 1998, 2002; Pantoja and Lee, 1994). Many species of marine phytoplankton, including representative of all major classes, are able to take up actively a variety of free amino acids and to use them as nitrogen sources (Antia *et al.*, 1991). In some species of coccolithophores and dinoflagellates, a mechanism involving external cleavage of NH_4^+ has been demonstrated (Palenik and Morel, 1990a,b). The deamination reaction is catalyzed via a periplasmic L-amino acid oxidase and results in the formation of hydrogen peroxide and an α-keto acid. Under appropriate conditions, some phytoplankton species can also obtain NH_4^+ released extracellularly from amines (Palenik *et al.*, 1991). A periplasmic amine oxidase oxidizes the amine to an aldehyde, resulting in the release of H_2O_2 and NH_4^+ for uptake. As in higher plants where they have been characterized, the amine oxidases of marine phytoplankton appear to be copper metalloenzymes.

6.05.3.2.2 *N_2 fixation and the nitrogen cycle*

The overall pool of fixed nitrogen in the oceans is determined by a balance between N_2 fixation and denitrification, two of the major processes involved in the nitrogen cycle. The concentration of nitrate and nitrite that can be denitrified is maintained by nitrification of ammonium. As shown in the diagram of Figure 12, trace elements are involved in every step in the nitrogen cycle. Nitrification is carried out by two specialized classes of aerobic chemoautotrophs: first ammonia oxidizing nitrifiers use copper- and iron-containing ammonia monooxygenase (AMO) to oxidize NH_4^+ to NH_2OH and iron-containing hydroxylamine oxidoreductase (HAO) to oxidize NH_2OH to NO_2^-; then nitrite oxidizing nitrifiers oxidize NO_2^- to NO_3^- via the Fe/Mo-enzyme nitrite oxidoreductase. Denitrification is a form of anaerobic respiration utilized by a variety of heterotrophs that requires four separate types of metalloenzymes: respiratory nitrate reductase (periplasmic NAP; or membrane-bound NAR) contains Fe and Mo; respiratory nitrite reductase (NIR) contains either copper or iron; nitric oxide reductase (NOR) contains iron; and nitrous oxide reductase (N_2OR) contains copper.

In nitrogen fixation, the difficult reduction of N_2 to NH_4^+ is effected by the nitrogenase enzyme, which contains iron and molybdenum. An iron + vanadium form of nitrogenase and an iron-only form are also known, but their presence in marine phytoplankton has not yet been established (perhaps because molybdenum is the most abundant trace metal in the oceans). Diazotrophic growth is calculated to require ~7–11 times more iron than growth on ammonium (Kustka *et al.*, 2003a). It is generally thought that the colony-forming cyanobacterium *Trichodesmium* is responsible for the bulk of N_2 fixation in the oceans. Many other species of cyanobacteria also possess the ability to fix nitrogen, however, and the question of how much N_2 is fixed by what organisms is presently contentious (Zehr *et al.*, 2001).

6.05.3.3 Phosphorus Uptake

Phosphate transporters have been characterized in many model organisms, though relatively little mechanistic work has been done in marine phytoplankton. Phosphate transport is effected by high and low affinity transporters and dependent on ATP, Na^+, and Mg^{2+} in several diatoms (Cembella *et al.*, 1984). These observations are found to be consistent with the well known active transport system of yeast (Raghothama, 1999). The dependence of phosphate transport on Mg^{2+} in diatoms and yeast suggests that eukaryotes may transport an uncharged cation phosphate complex ($MeHPO_4$, where Me may be Ca^{2+}, Mg^{2+}, Co^{2+}, Mn^{2+}) as has been observed in heterotrophic bacteria (van Veen, 1997).

As is the case for nitrogen, organic compounds constitute a significant source of phosphorus for some species of marine phytoplankton, particularly in oligotrophic waters. In most organisms, the principal enzyme involved in cleaving the phosphate group from organophosphates is the zinc enzyme alkaline phosphatase. This enzyme has indeed been found in a number of species of marine phytoplankton, usually as a periplasmic enzyme.

6.05.3.4 Silicon Uptake

Because of the importance of diatoms in oceanic productivity, silicon is an important algal nutrient in seawater. A transporter of $Si(OH)_4$ has been isolated and sequenced (Hildebrand *et al.*, 1998; Hildebrand *et al.*, 1997) and the physiology of silicon uptake has been well studied (Martin-Jezequel *et al.*, 2000). Nonetheless, the molecular mechanism of $Si(OH)_4$ transport and silica frustule formation in diatoms are still largely mysterious. From indirect evidence, it appears possible that the $Si(OH)_4$ transporter may contain zinc, coordinated to cysteines, as a metal center in the portion of the protein exposed to the outside of the cell (Hildebrand, 2000; Rueter and Morel, 1981). If true, this would be an unusual example of a transport protein functioning with a metal center.

6.05.4 EFFECTS OF TRACE METALS ON MARINE BIOGEOCHEMICAL CYCLES

In the preceding section, we have seen how trace metals are involved in some of the biochemical mechanisms responsible for the uptake and transformations of carbon, nitrogen, phosphorus, and silicon by marine organisms, with a focus on phytoplankton growth and productivity. Using this information about their functions, we now examine the extent to which trace metals affect the marine biogeochemical cycles of major algal nutrients. The relevant information is sparse and often indirect. For each trace metal, we review the laboratory data that indicate the biochemical functions—and, when known, what particular metalloproteins—utilize a major fraction of the cellular quota of nutrients and identify the principal effects of metal limitation in cultures of marine microorganisms. We also review the scant field data that shed light on the role of trace metals in the acquisition and processing of nutrients by marine phytoplankton and their growth.

6.05.4.1 Iron

6.05.4.1.1 Iron and growth rates

We have seen that iron plays a key role in biochemical electron transfer processes, including the light reaction of photosynthesis and the respiration of organic carbon, and that, according to calculations, the numerous iron-containing redox intermediates involved in these processes account for the bulk of the relatively high iron requirement of both phytoplankton and heterotrophic bacteria. As a result it has been relatively easy to demonstrate iron limitation in cultures of various species of marine phytoplankton (Anderson and Morel, 1982; Hudson and Morel, 1990; Sunda and Huntsman, 1995b, 1997). For example, as illustrated in Figure 13, the growth rates of the diatom *T. weissflogii* and of the dinoflagellate *Prorocentrum minimum* become limited when the unchelated iron concentration falls below 100 pM.

It has also been relatively easy to demonstrate the iron limitation in phytoplankton growth in the field. Numerous incubation experiments and a few mesoscale experiments (involving patches of tens of square kilometers of surface ocean) have consistently shown that iron addition promotes phytoplankton growth in high nutrient–low chlorophyll (HNLC) regions of the oceans, including the northern Pacific (Coale *et al.*, 1998; Martin *et al.*, 1989), the equatorial Pacific (Coale *et al.*, 1998; Martin *et al.*, 1994; Price *et al.*, 1991, 1994), the Southern Ocean (Boyd *et al.*, 2000), and some

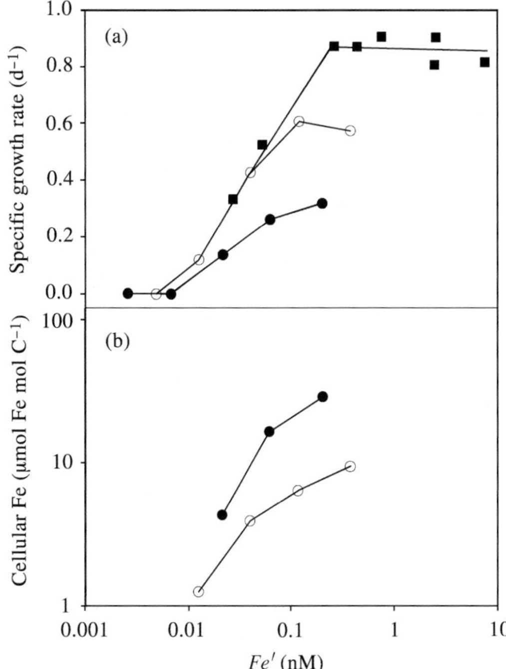

Figure 13 (a) Growth rate as a function of $\log(Fe')$ (unchelated iron concentration) for the marine diatom *Thalassiosira weissflogii* (squares) and the marine dinoflagellate *Prorocentrum minimum* (circles). (b) Intracellular Fe : C as a function of $\log(Fe')$ and irradiance in *P. minimum*. Open symbols 500 μmol quanta $m^{-2}s^{-1}$. Closed symbols 50 μmol quanta $m^{-2}s^{-1}$ (after Sunda and Huntsman, 1997).

upwelling coastal regions (Hutchins *et al.*, 1998, 2002). For example, over the seven days following the second iron addition in the IRONEX-II experiment in the equatorial Pacific, the concentration of pennate diatoms in the iron-enriched patch increased by ~100 times compared to that outside the patch (Cavender-Bares *et al.*, 1999) (Figure 14).

There are interesting differences in how major phytoplankton taxa in HLNC regions of the oceans respond to iron additions. As exemplified in Figure 14, in a majority of experiments, the most dramatic difference between the +Fe incubations and the controls is the rapid growth of relatively large diatoms which become dominant. This is not unexpected since large phytoplankters should be particularly sensitive to nutrient limitation owing to the relatively low surface area to volume ratio which, on a per mass basis, decreases the diffusion rate of nutrients from the bulk and the availability of membrane area to anchor necessary transporters (vide supra). Addition of iron promotes faster growth of diatoms which can then outpace large grazers such as copepeods, particularly under conditions where some of them may not have been representatively included in the sample being incubated (Banse, 1990).

A different response is usually seen for cyanobacteria. As seen in Figure 14, there is little change in the standing crop of *Prochlorococcus* or *Synechococcus* upon iron addition (Cavender-Bares *et al.*, 1999). But the mean fluorescence per cell of these phytoplankters increases markedly, indicating a higher chlorophyll concentration and, presumably, a faster rate of photosynthesis. An increase in the growth rate of cyanobacteria upon iron addition has indeed been confirmed by demonstrating an increase in the frequency of dividing cells in the population in the +Fe patch (Mann and Chisholm, 2000). The reason the standing crop of these faster growing cyanobacteria shows little response to iron addition is simply that their micrograzers (unlike the large zooplankters that feed on diatoms) are able to keep pace with them. In view of their small size, it is not clear why these organisms should be particularly limited under ambient iron conditions. The answer seems to be that they proportionally require more iron than eukaryotes (Brand, 1991; Raven, 1988, 1990), and that they are in fact somewhat less Fe-limited than diatoms, e.g., 60% instead of 20% of μ_{max} (Mann and Chisholm, 2000; Martin *et al.*, 1994).

The coccolithophores, which constitute the third family of phytoplankters that are most commonly found in the open oceans including some HNLC regions, do not appear to respond much to iron additions (Lam *et al.*, 2001). Because these organisms are responsible for the bulk of the precipitation of $CaCO_3$ in the open oceans, the net effect of iron addition is to increase the ratio of inorganic carbon fixed into organic biomass to that precipitated as $CaCO_3$. The absence of response of coccolithophores to iron additions presumably indicates that they are not limited by iron availability, even in very low-Fe environments. In fact, some culture studies show that oceanic coccolithophores have particularly low-Fe requirements (Brand, 1991; Brand *et al.*, 1983; Sunda and Huntsman, 1995a). In addition *Emiliana huxleyi* (the most abundant coccolithophore species) is able to take up iron at a faster rate per unit area of cell surface in heavily chelated medium than diatoms and dinoflagellates (Sunda and Huntsman, 1995a), and the maximum growth rate of coccolithophores is often low, reflecting their adaptation to low-nutrient (including iron) conditions in open-ocean environments. Thus, it is perhaps not surprising that the growth of coccolithophores should not be noticeably stimulated by iron additions.

6.05.4.1.2 *Iron uptake*

A question arises when the laboratory and field data are compared quantitatively. As is typical, the laboratory data of Figure 13(a) shows that little

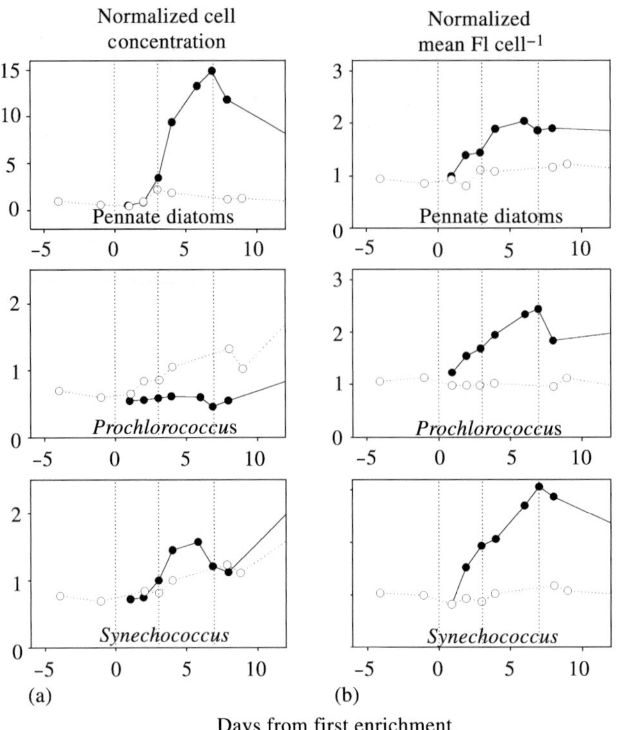

Figure 14 Response of different phytoplankton groups in the iron fertilized patch (closed symbols) relative to outside the patch (open symbols) of IronEx II in the equatorial Pacific Ocean. Normalized mean pigment fluorescence per cell is orange fluorescence for *Synechococcus* and red for *Prochlorococcus* and pennate diatoms. Fluorescence and cell abundance are normalized to values outside the patch (after Cavender-Bares *et al.*, 1999).

growth of coastal diatoms and dinoflagellates should occur below $Fe' = 50$ pM. Data for open-ocean species show growth limitation at a few pM. In the field, it has been shown by electrochemical measurements that 99% of the total dissolved iron concentration is present as strong organic complex. Thus, for example, the ambient diatom population in the equatorial Pacific, which is known to grow at about 0.3 day^{-1}, does so at ambient $Fe' \sim 0.015$ pM (Rue and Bruland, 1997). How can diatoms in the field grow so fast at such low unchelated iron concentrations that do not allow laboratory cultures to grow and the rate of diffusion to the cell surface to be sufficient for the organism? Clearly, other pools of iron must be accessible to these diatoms. Field experiments have confirmed that iron complexed to some siderophores can be taken up by the ambient flora, presumably by the mechanism illustrated in Figure 5 (Maldonado *et al.*, 1999). But additions of siderophores to field samples have also resulted in a decrease in iron uptake by phytoplankton (Hutchins *et al.*, 1999b), implying that iron–siderophore complexes may not be the only source of iron to primary producers, and that other, perhaps dominant, modes of iron acquisition are to be discovered.

6.05.4.1.3 Iron and electron transfer

In view of the dominant role of iron in the electron transport chain of photosynthesis, one can presume that a decrease in photosynthetic rate is the main effect of iron limitation on phytoplankton. This is confirmed indirectly in laboratory cultures where low iron availability can be partly compensated by high light intensity: the same specific growth rate can be achieved by the dinoflagellate *Prorocentrum minimum* at a lower ambient unchelated iron concentration when the light intensity is higher (Sunda and Huntsman, 1997) (Figure 13(a)). The photosynthetic units of cells growing at higher light intensity require less chlorophyll *a* and less iron, and this is reflected in a lower overall cellular iron quota (Figure 13(b)). In cyanobacteria, severe iron limitation causes cells to freeze, wherever they are in the cell cycle, consistent with a reduction in electron transport efficiency (Saito, 2001). The prediction that heterotrophic bacteria should require less iron than photosynthetic organisms appears also to be borne out by laboratory data. Under iron-limiting and replete conditions, the range for heterotrophs is 0.1–12 μmol Fe (mol C)$^{-1}$ (Granger and Price, 1999) and that for eukaryotic phytoplankton is 7–100 μmol Fe (mol C)$^{-1}$ (Sunda and Huntsman, 1995a). The scant available field data are also consistent with the prediction that iron stress should be reflected in the nature and the concentration of cellular electron transport carriers: in a transect from Vancouver Island to the subarctic Pacific gyre, it has been observed that the ratio of flavodoxin to ferredoxin in field populations increased as the iron concentration decreased (La Roche *et al.*, 1999).

6.05.4.1.4 Iron and nitrogen acquisition

We have seen that besides its central role in photosynthesis and respiration, iron also plays an important role in the acquisition and cycling of nitrogen. As predicted from calculations, laboratory cultures growing on NO_3^- require ~60% more iron than those growing on NH_4^+ (Maldonado and Price, 1996). However, laboratory experiments have not usually shown a difference in the growth of iron-limited cells grown on NO_3^- or NH_4^+ with the one exception of severely iron-depleted *T. oceanica* cultures (Henley and Yin, 1998; Kudo and Harrison, 1997; Maldonado and Price, 1996). In the field, it appears that the effect of low iron concentrations on NO_3^- uptake and assimilation is a significant part of diatom limitation (Price *et al.*, 1994). While small prokaryotes can acquire sufficient nitrogen from the low ambient NH_4^+ concentration, large diatoms cannot and must rely on NO_3^-, which requires additional iron for assimilation. At low iron concentrations, these organisms are thus co-limited by iron and nitrogen, as can be shown by the increase in growth rate resulting from either iron or NH_4^+ addition.

Because of the high-Fe content of nitrogenase and the additional energy requirements, the effect of iron on N_2 fixation should be even greater than that on NO_3^- uptake and assimilation. Data on relevant marine organisms have been difficult to obtain as *Trichodesmium*, the organism thought to be responsible for the bulk of N_2 fixation in the oceans, is notoriously fussy in culture. Nonetheless, laboratory studies have shown that *Trichodesmium* requires 5 times more iron when grown on N_2 than it does when grown on NH_4^+ (similar to the theoretical Fe factor of 7–11 times) (Kustka *et al.*, 2003b) (Figure 15(a)). Nitrogenase activity (measured by acetylene reduction) shows a precipitous drop below a threshold iron quota of ~35 μmol Fe (mol C)$^{-1}$ (Berman-Frank *et al.*, 2001) (Figure 15(b)). The results of such studies have been used, together with models of iron inputs to surface seawater, to predict that N_2 fixation in most of the oceans is actually limited by iron availability.

6.05.4.2 Manganese

We have seen that of all the trace metal requirements of phytoplankton, manganese is second only to iron. It is indeed also easy to limit the growth of diatoms such as *T. pseudonana* by lowering the *Mn′* concentration in the medium (Sunda and Huntsman, 1996) (Figure 16). But in the open ocean, the dissolved concentration of manganese in surface waters—in contrast to those of most other trace elements—is higher than that in deep waters (Figure 3) and usually far in excess of those that limit the growth of laboratory cultures. The high ambient Mn^{2+} concentration

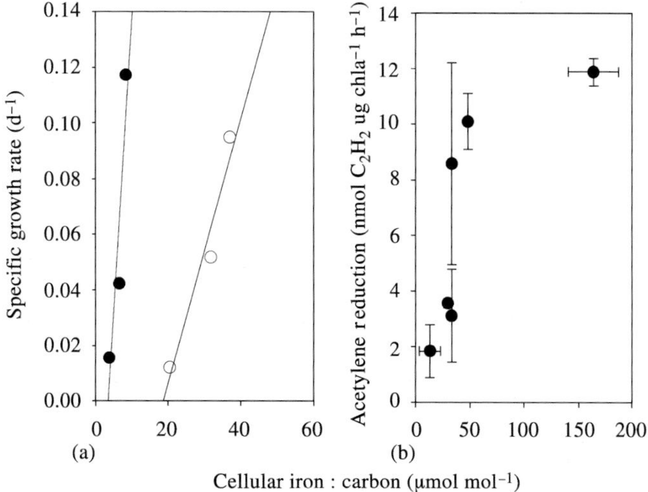

Figure 15 (a) Intracellular iron normalized to carbon in laboratory grown *Trichodesmium* sp. (a marine nitrogen-fixing cyanobacterium) with NH_4^+ (closed symbols) and N_2 (open symbols) as nitrogen sources (after Kustka *et al.*, 2003b). (b) Acetylene reduction rates (a proxy for N_2 fixation rates) as a function of cellular iron (after Berman-Frank *et al.*, 2001).

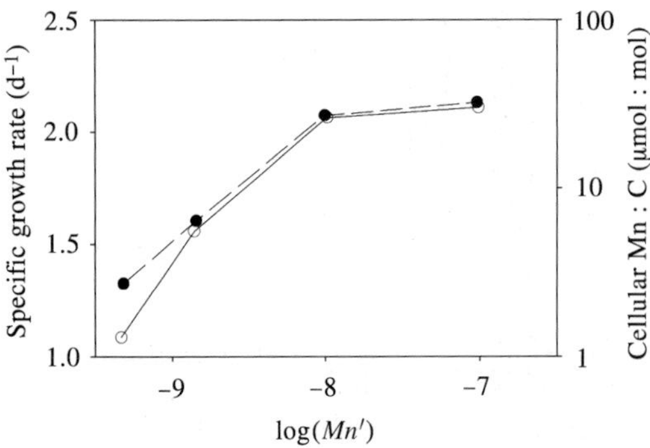

Figure 16 Specific growth rate (open symbols) and cellular Mn : C ratio (closed symbols) of the marine diatom *Thalassiosira pseudonana* as a function of log(Mn') (unchelated manganese concentration) (after Sunda and Huntsman, 1996).

is mirrored by a relatively low affinity for manganese uptake in marine phytoplankton (Sunda and Huntsman, 1998a). Zinc, copper, and cadmium have in fact a higher affinity for the manganese uptake system of model species than manganese itself. The result is that in some polluted coastal waters, a major effect of high metal concentrations is to inhibit manganese uptake by phytoplankton (Sunda and Huntsman, 1998a). Most of the manganese requirement of phytoplankton is in the water oxidizing complex of PS II. This can be seen in the slopes of specific growth rate versus Mn : C relationships obtained in laboratory cultures (Sunda and Huntsman, 1998c) which agree with those predicted from manganese use efficiency models based solely on the manganese requirement for O_2 evolution (Raven, 1990). The much greater availability of manganese than iron in surface seawater suggests that manganese might also replace iron in some other metalloenzymes in marine organisms. Indeed, manganese apparently provides a perfect replacement for iron in superoxide dismutase (SOD) in cultures of the diatom *T. pseudonana*. The cellular concentrations of manganese or iron used in SOD are sizeable fractions of the respective cellular quotas in this organism. Because the Mn- and Fe-SODs have the same molecular mass and migration patterns on non-denaturing protein gels, it has been suggested that the two metals may substitute for each other in the same enzyme (Peers and Price, personal communication).

From a bioinorganic perspective, one of the most interesting aspect of the marine geochemistry of manganese involves oxidation by bacteria. Although thermodynamically favorable, the chemical oxidation of Mn(II) to Mn(IV) by oxygen is exceedingly slow at the seawater pH. The oxidation of Mn(II) in the surface ocean is thus bacterially mediated (Sunda and Huntsman, 1988). In some cases, Mn(II) oxidation is effected extracellularly by the bacteria themselves, in others, by their spores (Francis and Tebo, 2001, 2002). The enzymes responsible for catalyzing the oxidation of Mn(II) have been partly characterized. In all cases they contain a Cu-binding motif that is typical of multi-copper oxidases (Francis and Tebo, 2001, 2002) (Figure 17). It has been verified in cultures that the bacterially mediated oxidation of Mn(II) is indeed dependent on the presence of sufficient concentrations of copper in the medium (Brouwers *et al.*, 1999; van Waasbergen *et al.*, 1996). Photoinhibition of this bacterial oxidation increases the concentration and the residence time of the photoproduced Mn(II) in surface seawater (Sunda and Huntsman, 1988).

6.05.4.3 Zinc, Cobalt, and Cadmium

It is now becoming clear that the physiological importance of zinc rivals that of iron in biology. Although the cellular concentration of zinc is typically lower than that of iron (Figure 1), the number of known Zn-metalloproteins appears to be much larger than that of Fe-metalloproteins (Maret, 2002). In cultures of marine phytoplankton, it is commonly observed that zinc, cobalt, and cadmium can partially replace each, depending on the species. Thus these three metals often play similar biochemical roles in the marine environment. The distinct geochemistries of these elements and their different utilization by various taxa of phytoplankton likely influence the community composition of marine assemblages.

It appears that cobalt plays a particularly important role in the growth of cyanobacteria (Saito *et al.*, 2002; Sunda and Huntsman, 1995b). Both *Prochlorococcus* and *Synechococcus* show an absolute cobalt requirement that zinc cannot substitute for (Figure 18(a)). The growth rate of *Synechococcus* is little affected by low zinc concentrations, except in the presence of cadmium which then becomes extremely toxic (Saito *et al.*, personal communication). The biochemical processes responsible for the major cellular utilization of zinc and cobalt in marine cyanobacteria are unknown, however. These metals may be involved in carbonic anhydrase and/or other hydrolytic enzymes. Cobalamin (vitamin B_{12}) synthesis is a function of cobalt in these organisms, yet B_{12} quotas tend to be very small (on the order of only 0.01 μmol (mol C)$^{-1}$) and hence are not likely represent a significant portion of the cellular cobalt (Wilhelm and Trick, 1995).

The importance of cobalt in the physiology and ecology of cyanobacteria is underscored by evidence showing that they produce strong, specific cobalt chelators. Production of such cobalt

(a)

Protein	Position					
MnxG	527	M	H	I	H	F
MofA	304	I	H	L	H	G
CumA	95	I	H	W	H	G
Asox	94	I	H	W	H	G
Lacc	78	V	H	W	H	G
Hcer	119	F	H	S	H	G
Fet3	8	M	H	F	H	G
CopA	99	I	H	W	H	G
			2		3	

(b)

Protein	Position						Organism
MnxG	572	F	F	H	D	H	(*Bacillus* sp. SG-1)
MofA	384	W	Y	H	D	H	(*Leptothrix discophora* SS-1)
CumA	136	W	Y	H	P	H	(*Pseudomonas putida* GB-1)
Asox	137	F	Y	H	G	H	(plants)
Lacc	121	W	Y	H	S	H	(fungi/plants)
Hcer	178	I	Y	H	S	H	(vertebrates)
Fet3	52	W	Y	H	S	H	(yeast)
CopA	140	W	Y	H	S	H	(*Pseudomonas syringae* PT23)
				3		3	

(c)

Protein	Position								
MnxG	281	H	V	F	H	Y	H	V	H
MofA	1174	H	P	V	H	F	H	L	L
CumA	391	H	P	I	H	L	H	G	M
Asox	480	H	P	W	H	L	H	G	H
Lacc	508	H	P	I	H	K	H	G	N
Hcer	994	H	T	V	H	F	H	G	H
Fet3	341	H	P	F	H	L	H	G	H
CopA	542	H	P	I	H	L	H	G	M
		1			2		3		

(d)

Protein	Position												
MnxG	334	H	C	H	L	Y	P	H	F	G	I	G	M
MofA	1279	H	C	H	I	L	G	H	E	E	N	D	F
CumA	442	H	C	H	V	I	D	H	M	E	T	G	L
Asox	542	H	C	H	I	E	P	H	L	H	M	G	M
Lacc	585	H	C	H	I	A	S	H	Q	M	G	G	M
Hcer	1039	H	C	H	V	T	D	H	I	H	A	G	M
Fet3	411	H	C	H	I	E	W	H	L	L	Q	G	L
CopA	590	H	C	H	L	L	Y	H	M	E	M	G	M
		3	1	3				1					1

Figure 17 Copper-binding sites in multi-copper oxidases. Amino acid alignment of the copper-binding sites in MnxG, MofA, CumA, and other multi-copper oxidases. The letters A–D correspond to the 4 copper binding sites. Abbreviations: Asox, ascorbate oxidase (cucumber and squash); Lacc, laccase (fungi); Hcer, human ceruloplasmin; FET3, an Fe(II)-oxidizing protein in yeast; and CopA, a copper-resistance protein. The amino acids conserved among the different proteins are shaded and the copper-binding residues are numbered according to the spectroscopic type of copper they potentially coordinate.

organic complexes has been observed during a *Synechococcus* bloom in the equatorial Pacific (Saito and Moffett, 2001b) and uptake of organically complexed cobalt has been demonstrated in *Prochlorococcus* cultures (Saito *et al.*, 2002). These results have led to the hypothesis that cobalt ligands in surface seawater are produced by cyanobacteria and that they are "cobalophores" whose function in cobalt chelation and uptake is analogous to that of siderophores for iron.

Unlike marine cyanobacteria, coccolithophores are able to substitute cobalt and zinc for each other almost indifferently (Figure 18(c)), but, in some instances, have been observed to reach maximal growth rates only at high cobalt concentrations (Sunda and Huntsman, 1995b). Addition of cadmium to low zinc cultures of the ubiquitous species *Emiliana huxleyi* grown with organic phosphate is effective in restoring high growth rate and alkaline phosphatase activity (Xu, Ho, and Morel, unpublished data). This suggests that cadmium may substitute for zinc as a metal center in alkaline phosphatase.

In diatoms, it is generally possible to demonstrate an absolute zinc requirement, but above that threshold, addition of zinc, cobalt, or cadmium to the cultures are equally effective at promoting fast growth. This is illustrated for zinc and cadmium in Figure 19 which also shows that the growth rate can be accelerated by increasing the p_{CO_2} of the medium. Native protein gels from cultures labeled with ^{65}Zn, ^{57}Co, or ^{109}Cd demonstrate that a large fraction of the cellular zinc, cobalt, or cadmium (depending on medium composition) is associated

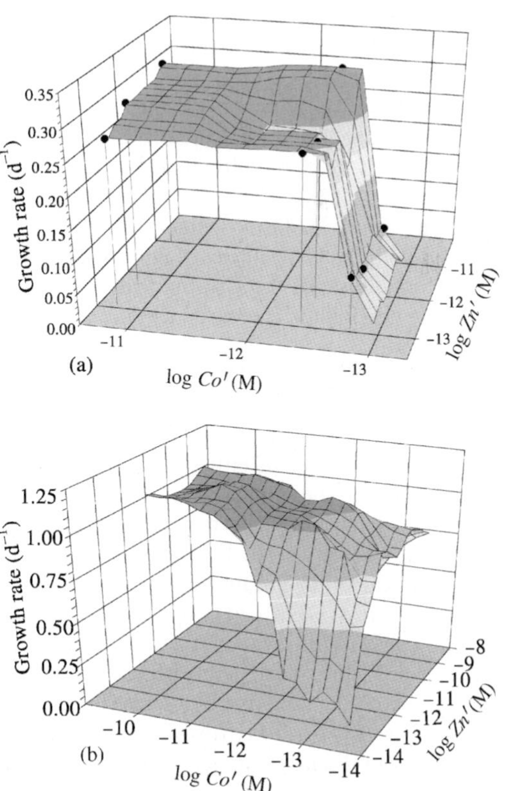

Figure 18 The effect of varying cobalt and zinc concentrations on growth rate in (a) the marine cyanobacterium *Prochlorococcus* (MED4-Ax) (source Saito *et al.*, 2002) and (b) the marine coccolithophorid *Emiliania huxleyi* (after Sunda and Huntsman, 1995b).

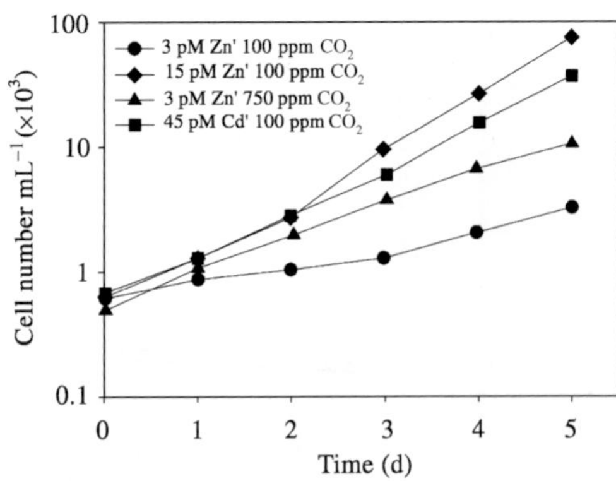

Figure 19 The effect of varying zinc, cadmium and CO_2 concentrations on growth rate of the marine diatom *Thalassiosira weissflogii*. Low Zn' (unchelated zinc concentration = 3 pM) reduces growth rate at 100 ppm CO_2. Increasing CO_2 to 750 ppm or adding Cd' (unchelated cadmium concentration = 45 pM) to the medium allows for higher growth rates (after Lane and Morel, 2000a).

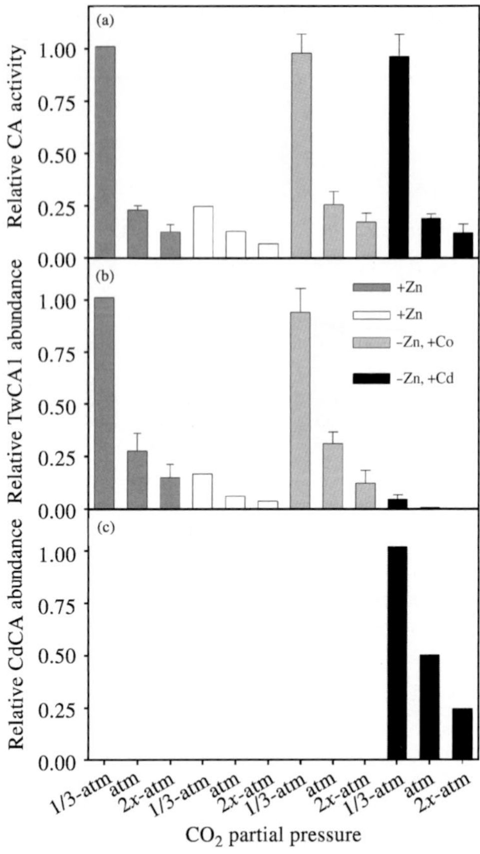

Figure 20 Relative levels of carbonic anhydrase (CA) activity (A), amounts of TWCA1 protein (a zinc containing CA) and amounts of CdCA protein (a cadmium containing CA) in the marine diatom *Thalassiosira weissflogii* as a function of metal treatment and CO_2 levels. +Zn = 15 pM Zn', −Zn = 3 pM Zn', +Co = 21 pM Co', +Cd = 45 pM Cd' (after Morel *et al.*, 2002).

with carbonic anhydrase (CA) (Morel *et al.*, 1994). There is thus little doubt that a primary role of zinc, cobalt, and cadmium in diatoms is to serve as a metal center in a CA that catalyzes inorganic carbon acquisition and that limiting these metals leads to a reduction in the rate of carbon uptake and fixation (Figure 11).

Intracellular CAs from the diatom *T. weissflogii* have been isolated and sequenced, and their activities have been monitored as a function of p_{CO_2}, Zn', Co', and Cd' in the culture medium (Lane and Morel, 2000b; Roberts *et al.*, 1997). As expected both the CA activity of cell extracts and the concentrations of the CAs increase when the p_{CO_2} of the medium is decreased (Figure 20). Low Zn' also limits CA activity and the concentration of a cytoplasmic CA, TWCA1, measured by quantitative Western analysis. The measured increase in CA activity and in TWCA1 concentration upon addition of either zinc or cobalt to low Zn' cultures show that TWCA1 can use zinc or cobalt indifferently as its metal center. In contrast, the increases in CA activity observed upon addition of cadmium to low-Zn' cultures result from the synthesis of a wholly distinct Cd enzyme, CdCA. This is the first native Cd enzyme discovered. X-ray spectroscopy (EXAFS) shows that, in TWCA1, zinc is coordinated to three imidazole nitrogens from histidines, similar to the zinc coordination in mammalian α-CAs (Cox *et al.*, 2000) (Figure 21). In contrast, XANES data indicate that, in CdCA, cadmium is likely to be coordinated to sulfur ligands from cysteins. These two CAs from diatoms have no homology to any other known CAs (or any other protein or to each other) and they thus constitute new classes, the δ-CAs and the ε-CAs (Smith and Ferry, 2000). The periplasmic CAs of diatoms have not yet been isolated or characterized. Whole cell CA activity assays under various conditions of p_{CO_2}, Zn', Co', and Cd' indicate that these periplasmic enzymes may also be able to utilize zinc, cobalt, or cadmium as their metal centers (Morel *et al.*, 2002).

The available field data show that the unchelated surface concentrations of zinc, cobalt, and cadmium are on the order of a few picomolar or even lower (Bruland, 1992, 1989; Ellwood and van den Berg, 2000, 2001; Saito and Moffett, 2001b). These are the ranges of unchelated metal concentrations over which limitation and replacement occurs in laboratory cultures of marine phytoplankton. The geographic distribution of the surface concentrations of these metals supports the idea that phytoplankton in the field take up cobalt and cadmium as a replacement for zinc. As seen in Figure 22, the disappearance of cobalt and cadmium in surface waters of the north Pacific becomes correlated with that of P (which serves as a convenient measure of algal nutrient uptake) when the concentration of zinc is depleted below approximately 1.2 nM (Martin and Gordon, 1988; Sunda and Huntsman, 1995b; Sunda and Huntsman, 2000). More directly, data on the concentration of cadmium in phytoplankton in the upwelled waters off Monterey, CA, whose flora is dominated by diatoms, show a clear increase in phytoplankton cadmium when zinc and CO_2 are low in concentrations (Cullen *et al.*, 1999) (Figure 23).

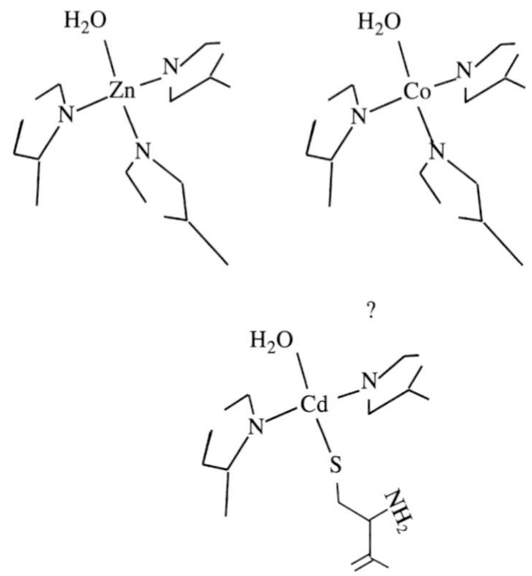

Figure 21 Active centers of carbonic anhydrase (CA) from the marine diatom *Thalassiosira weissflogii*. The zinc and cobalt centers are found in the protein TWCA1; (Cox *et al.*, 2000); Cox, unpublished data). The Cd center is found in the protein CdCA and is hypothetically based on unpublished XANES data showing sulfur binding and unpublished protein sequence data.

Despite the very low concentrations of zinc, cobalt, and cadmium in surface seawater and all the data showing that these three metals are taken up and used by phytoplankton, the evidence for limitation of primary production by any of these metals is only anecdotal (Coale, 1991; Fitzwater *et al.*, 2000). Several incubation experiments in which zinc, cobalt, or cadmium have been added have yielded negative results (although this is a difficult claim to document since most such negative results remain unpublished). One reason for these negative results may of course be the difficulty inherent in experiments where the controls must be free of zinc contamination. Another reason may simply be that in regions where both upwelling and aeolian inputs of essential trace metals are very low, iron is usually "most limiting." In this respect it is worth noting that a sizeable proportion of the few experiments that have shown increased phytoplankton growth upon zinc, cobalt, or cadmium addition were

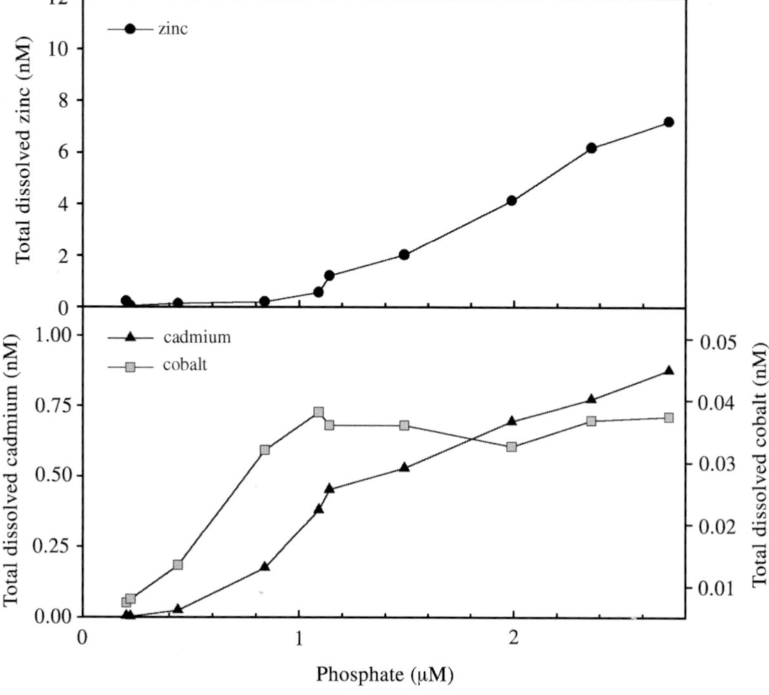

Figure 22 The sequential drawdown of zinc, cadmium, and cobalt in the north Pacific suggestive of biochemical substitution in the phytoplankton community. Metal versus phosphate concentrations are plotted from vertical profile T-5 (after Sunda and Huntsman, 1995b, 2000; Martin *et al.*, 1989).

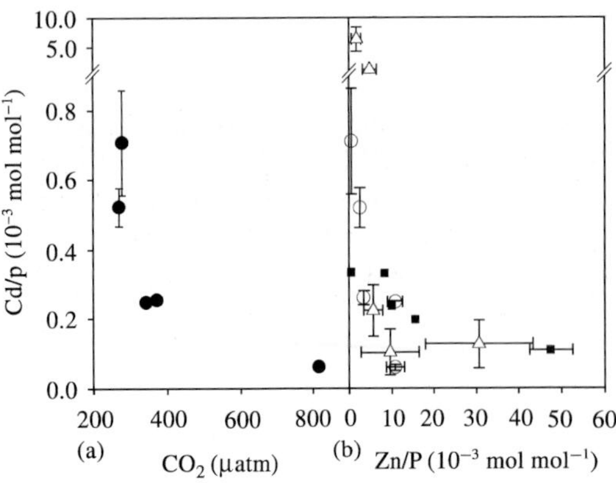

Figure 23 (a) Cd : P ratio as a function of the partial pressure of CO_2 of a natural phytoplankton assemblage from the coast of central California. (b) Cd : P versus Zn : P in different size fractions of the natural phytoplankton assemblage. Total particles >0.45 μm (circles) 5–53 μm fraction (squares) >53 μm fraction (triangles) (after Cullen *et al.*, 1999).

performed in conjunction with iron addition. For example, Saito and Moffett (2001b) reported a significant increase in chlorophyll after two days upon Fe + Co addition compared to little effect with Fe- or Co-only additions.

Another, more biochemical, explanation may be that if Zn/Cd/Co limitation manifests itself primarily as an inability to synthesize carbonic anhydrase, phytoplankton starved of these metals may acquire inorganic carbon by transporting HCO_3^- (at some cost) or by relying on CO_2 diffusion. As a result growth might only be marginally slowed down, and the major effect of Zn/Co/Cd limitation may be a shift in phytoplankton species composition rather than a decrease in the photosynthetic rate of the community as seen under iron limitation. On the basis of the differences in the inorganic carbon acquisition systems discussed above, the effects of adding these metals are expected to be different for different species of phytoplankton. Indeed it has been demonstrated in the field that diatoms take up chiefly CO_2 (derived in part from dehydration of HCO_3^- catalyzed by extracellular CA) while cyanobacteria take up mostly HCO_3^- (Tortell and Morel, 2002). It has also been shown that modulation of the ambient p_{CO_2} changes the expression of CA in diatoms in the field. Thus the need for zinc and cobalt, and their uptake by diatoms should increase at low p_{CO_2} as has been observed for cadmium (Figure 23). In cyanobacteria and coccolithophores, the link between Zn/Co and Cd supply and inorganic carbon acquisition is likely to be much looser and the variations in requirements for Zn/Co and Cd cannot be predicted since their principal biochemical roles are presently unknown.

6.05.4.4 Copper

Laboratory data demonstrating copper limitation of marine phytoplankton are rare. It has generally been difficult to limit the growth rates of cultures by lowering the copper concentration in the medium, indicating a very low absolute copper requirement and/or a very effective uptake system in most species (Sunda and Huntsman, 1995c). If, as predicted, plastocyanin normally accounts for a major fraction of the copper quota of phytoplankton, a low-copper requirement would result from the known ability to replace plastocyanin by cytochrome *c* in the electron transport chain of the light reaction in a number of taxa.

Some experiments demonstrating copper limitation in marine phytoplankton apparently involve the utilization of external copper oxidases. Under conditions where ethalonamine is the sole nitrogen source in the culture medium, the coccolithophore *Pleurochrysis carterae* requires high copper concentration for growth (Palenik and Morel, 1991). This organism possesses an extracellular amine oxidase to cleave NH_4^+ from the amine and such enzymes normally contain copper (Figure 11). Under iron-limiting conditions, cultures of several species of diatoms and a coccolithophore required high copper concentrations for rapid iron uptake and growth (Price, Maldonado and Granger, personal communication). This is consistent with the involvement of a multi-copper oxidase in the uptake system for chelated iron (Figure 5). An apparent high copper requirement, independent of iron uptake, has also been observed in oceanic species of diatoms such as *Thalassiosira oceanica* (Price, Maldonado, and Granger, personal communication). Because such species are known to have extraordinarily low iron requirements, it has been hypothesized that copper may be substituting for iron in some important biochemical pathways.

While copper limitation of marine phytoplankton has rarely been observed, many experiments have demonstrated that copper is extremely toxic to various species, often decreasing growth even at only a few pM unchelated concentrations (Anderson and Morel, 1978; Sunda and Guillard, 1976). When these data are compared to field data on copper concentration and speciation in seawater, it appears that *Cu′* concentrations may be approaching values that are toxic to some species in coastal waters. At these levels the biological effects of copper are complex and often appear to involve interference with the uptake or assimilation of other essential trace metals such as zinc or manganese (Sunda and Huntsman, 1998a). An indication that copper concentrations may be marginally toxic to the ambient flora in coastal waters is provided by the analysis of phytochelatins in the algal biomass. As seen on Figure 24, in

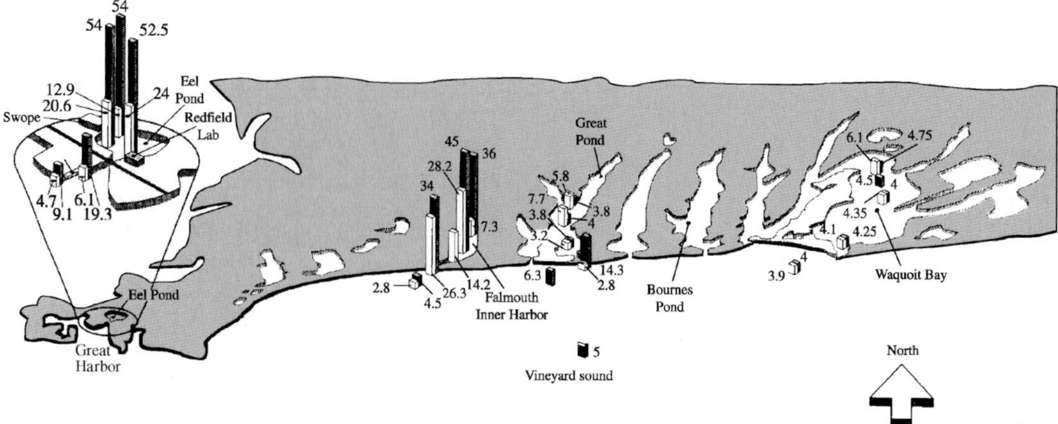

Figure 24 A map of western Cape Cod, MA, USA showing sample sites and concentrations of total dissolved copper and phytoplankton associated phytochelatin. Dark bars represent total copper (nM) and light bars are phytochelatin (μmol (g chl*a*)$^{-1}$) (after Ahner *et al.*, 1997).

embayments around Cape Cod, high phytochelatin concentrations correlate with high concentrations of unchelated copper (Ahner *et al.*, 1997). Incubation experiments confirm that copper is likely to be the major inducer of phytochelatins in these waters. In the open ocean, copper may also be toxic to some species of phytoplankton. For example, in the Sargasso Sea, it appears that the concentration of unchelated copper in the surface mixed layer is too high for the growth of the cyanobacterium *Prochlorococcus*, particularly the low-light adapted strain which thrives at great depths (Mann *et al.*, 2002). The molecular basis for the high sensitivity of *Prochlorococcus* to copper is unknown. It is worth noting that the strong copper chelators in seawater (which are still unidentified) seem to be released by another group of cyanobacteria, the *Synechococci*, which are less sensitive to copper toxicity (Moffett, 1995; Moffett *et al.*, 1990).

While most of the data on the effect of copper on marine phytoplankton has focused on copper toxicity, some data point to a potential limiting role of copper in other marine microorganisms. At very low unchelated copper concentrations, cultures of denitrifiers are unable to reduce N_2O to N_2—the last step in denitrification (Granger and Ward, 2002) (Figure 25). This inhibition results from insufficient supply of metal to the copper metalloenzyme, nitrous oxide reductase (N_2OR, see Figure 12). The denitrifiers (at least those in culture) have apparently not evolved an alternative for copper-containing N_2OR, nor are they able to reuse in N_2OR the copper contained in respiratory nitrite reductase (for many of them have the copper form of NIR). On the basis of these results, it has been hypothesized that copper availability may be an important factor, in addition to oxygen concentration, in regulating the marine production of nitrous oxide. Some

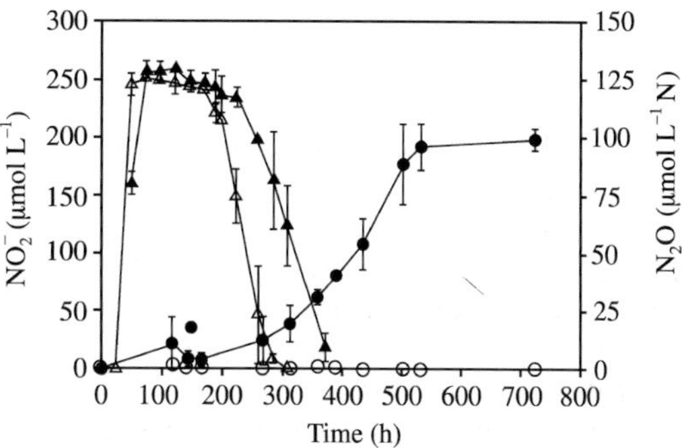

Figure 25 Effect of copper on nitrite (NO_2^-) consumption and nitrous oxide (N_2O) evolution in a culture of the denitrifying bacterium *Pseudomonas stutzeri*. Concentration of nitrite (triangles) and nitrous oxide (circles) in copper-replete (open symbols) and copper-deficient (closed symbols) medium (after Granger and Ward, 2002).

parts of the oceans such as the Arabian Sea are "hot spots" where high N_2O concentrations are accumulated at mid-depth and released to the atmosphere. As a result of advection of anoxic coastal water masses, this area is rich in sulfide which is apt to complex and precipitate copper and may result in copper limitation of nitrous oxide reductase activity.

6.05.4.5 Nickel

Of all the first row transition elements, nickel has received the least attention from oceanographers. Although the vertical profiles of nickel concentration in the oceans exhibit a surface depletion characteristic of nutrients, the surface values remain typically in the 1–5 nM range (Bruland *et al.*, 1994) (Figure 3), much in excess of the other elements discussed so far. Nonetheless, laboratory data demonstrate that

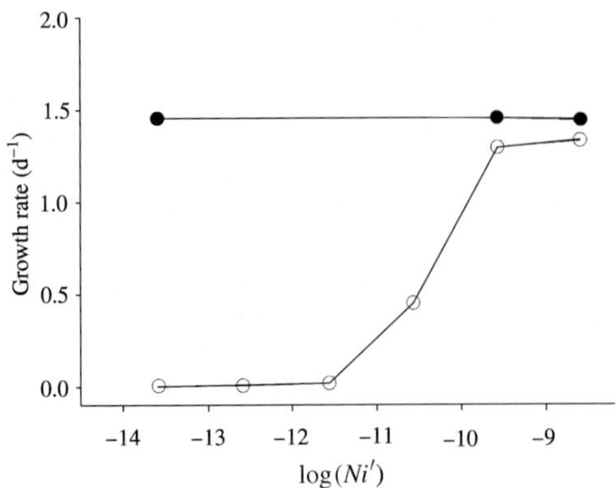

Figure 26 Growth of the marine diatom *Thalassiosira weissflogii* as a function of nickel concentration with ammonium (closed symbols) and urea (open symbols) as nitrogen sources (after Price and Morel, 1991).

nickel is necessary for the assimilation of urea. As shown in Figure 26, the diatom *T. weissflogii* grows equally well in the presence of $CO(NH_2)_2$ or NH_4^+, only when the unchelated nickel concentration is in excess of 0.2 nM (Price and Morel, 1991). Since the importance of urea as a nitrogen source for marine phytoplankton has been demonstrated in field studies, the high surface concentration of nickel in surface seawater is puzzling. As discussed above, it is possible that this high concentration reflects slow kinetics of uptake caused by the inertness of the Ni^{2+} ion. For example, the diatom *T. weissflogii* must take up roughly 15–30 amol of nickel per day to grow on urea at a specific growth rate of 1.5 day^{-1}. With a second order rate constant $\sim 1-3 \times 10^4\ M^{-1}\ s^{-1}$ for binding to an uptake ligand (reflecting the slow rate of loss of hydration water from the Ni^{2+} ion), and a nickel concentration of 1 nM, 1–10 amol of uptake proteins per cell would be necessary to achieve that uptake rate. As shown before, this is a very high density of membrane proteins. Despite their need for nickel to assimilate urea, marine phytoplankton may thus be unable to deplete the nickel concentration of surface seawater below a few nanomolars, simply because of the kinetic inertness of the Ni^{2+} ion (Hudson *et al.*, 1992). In view of the difficulty inherent in acquiring Ni^{2+} from seawater, the apparent ability of cobalt to replace nickel for urease activity in some phytoplankton species (Rees and Bekheet, 1982) may be oceanographically relevant.

6.05.5 EPILOGUE

6.05.5.1 Paleoceanographic Aspects

It is interesting to consider the variations in trace-metal availability in the oceans throughout the geologic history of the Earth relative to the trace-metal requirements of marine phytoplankton. Previous workers have postulated that trace-metal solubility was strongly influenced by the changing redox conditions of the oceans resulting from the activity of early oxygenic organisms, chiefly cyanobacteria (Williams and Frausto da Silva, 1996). It is generally agreed that Fe(II) and S(−II) at the Earth's surface have been oxidized over a period of some 2 billion years, before the build up of oxygen in the ocean and atmosphere (Holland, 1984). Thus life is thought to have originally evolved in an ocean rich in iron which would have become gradually depleted as a result of the formation of insoluble iron oxyhydroxides. Iron being in excess of sulfur in the Earth crust could not have been all precipitated as sulfides. While Fe is a trace element in the present day oceans it is the fifth most abundant element in continental rock. In parallel, trace metals, such as copper, zinc, cadmium, and molybdenum that form highly insoluble sulfides and were presumably rare in the early oceans, would have become gradually more abundant upon the oxidation of sulfide to sulfate. The concentrations of manganese and cobalt, whose sulfides are somewhat more soluble but which also form insoluble oxyhydroxides, would have, like that of iron, decreased over time. It is possible that the deep oceans may have been particularly high in sulfide for a period of time 2.5–0.5 billion years ago resulting in low iron, copper, zinc, cadmium, and molybdenum conditions (Anbar and Knoll, 2002).

As a result of selection pressure during the course of evolution, the changing concentrations of trace metals over geologic time should be somehow reflected in the trace-metal requirements of marine phytoplankton (Raven, 1988; Saito *et al.*, 2002; Scott, 1990; Sunda and Huntsman, 1995b). Indeed the trace-metal physiology of cyanobacteria is strikingly consistent with the presumed chemistry of the early oceans. Cyanobacteria have a relatively large iron quota (Raven, 1988), have an absolute cobalt requirement that zinc cannot substitute for (Saito *et al.*, 2002; Sunda and Huntsman, 1995b), and are extremely susceptible to copper and cadmium toxicity (Brand *et al.*, 1986; Mann *et al.*, 2002). In contrast, eukaryotic phytoplankton, particularly the diatoms and coccolithophores which evolved some 3.5 billion years later in an oxic and metal-poor ocean, have lower requirements for iron, lower sensitivity to copper and cadmium, and unusual enzymes with alternative metals, such as the cadmium carbonic anhydrase.

Why then haven't the cyanobacteria evolved in response to the large changes in trace-metal chemistry between an ancient reducing ocean and a modern oxidizing one? Why have they kept a biochemical apparatus that reflects early ocean chemistry instead of evolving one more similar to that of the eukaryotes? While we have no answer to such questions, we can speculate that the adaptive response of cyanobacteria has been to release into seawater specific metal chelators for the acquisition of essential metals, e.g., siderophores and cobalophores for iron and cobalt, and for the detoxification of others, e.g., strong copper chelators (Mann *et al.*, 2002; Moffett and Brand, 1996). In this view, the oceans are a *milieu externe* whose chemistry is controlled by (and for) cyanobacteria (and other prokaryotes) and has been so for billions of years. The eukaryotic phytoplankton then learned to make a living in this "inhospitable" environment by adapting their biochemistry and controlling the chemistry of their intracellular space and immediate surroundings—their *milieu interne*. As we consider the co-evolution of marine life and geochemistry which is reflected in the modern biochemistry of marine microorganisms and the chemistry of seawater, we must take into account the different timescales over which various organisms have been adapting to their marine environment (and modifying it).

6.05.5.2 A View to the Future

Over the past several decades, the importance of trace elements in the marine environment has become apparent, with major regions of world's oceans now known to be affected by iron limitation, and with the first accurate measurements of incredibly scarce quantities of other essential metals. Yet, we understand little of how metal limitation manifests itself in the biochemistry of marine microorganisms or of the role of trace elements in the geochemistry of the oceans. But we now know enough to ask relevant questions at both the molecular and global scale, and are able to use novel experimental tools to help answer these questions.

It appears that the genomic revolution and the experimental tools associated with it should also revolutionize the field of oceanography; and marine bioinorganic chemistry should be at the heart of that revolution. The sequencing of complete genomes for a variety of marine microorganisms (e.g., at the time of this writing, four marine cyanobacteria and a marine diatom have been sequenced) will provide us with the ability to search for putative metalloproteins, allow genomic microarray experiments to be conducted, and facilitate proteomic work. These genomic and proteomic tools, together with others, will allow us to address a number of difficult but important questions about the interactions of trace elements and the biota in the marine environment.

The outstanding questions in marine bioinorganic chemistry can be organized around three principal topics. First are the questions related to the trace element requirements of marine microorganisms: how do open-ocean species manage with extraordinarily low metal concentrations? What metals are they able to substitute for each other? What unusual enzymes (perhaps with "exotic" metal centers) or enzymatic pathways are they using? What differences are there between the requirements of different organisms (prokaryotes versus eukaryotes; autotrophs versus heterotrophs; diatoms versus coccolithophores; etc.)? How does the scarcity of nutrients such as carbon, nitrogen, phosphorus, and iron affect the trace element requirements of phytoplankton?

Then there are wider questions about the possible limiting roles of various trace elements in key biogeochemical processes: Do trace elements other than iron limit phytoplankton growth and primary production? Is the composition of phytoplankton assemblages controlled by trace elements? Are various processes in the cycle of nitrogen limited by trace elements (e.g., N_2 fixation by iron; N_2O reduction by copper)? What are the links between trace elements and the reduced sulfur cycle in surface seawater?

Finally, we must somehow resolve the vexing problem of the chemical speciation of trace elements in seawater: what is the chemical nature of the various metal chelators whose existence has been demonstrated by electrochemistry? Is the chemistry of several metals in surface seawater really controlled by metallophores released by prokaryotes? Or are dissolved metals chiefly present as parts of metalloproteins in the process of remineralization? How do metal chelators affect the residence times of metals (in particular, scavenged elements such as iron and cobalt) and in turn how do those chelators influence the global carbon cycle via changes in marine primary productivity?

By focusing on a molecular elucidation of key biochemical processes in the marine biogeochemical cycles of elements, marine bioinorganic chemistry should help us understand the subtle and complex interdependence of marine life and ocean geochemistry, and how they have evolved together over the history of the Earth.

ACKNOWLEDGMENTS

This work was supported in part by the Center for Environmental Bioinorganic Chemistry (NSF# CHE 0221978). Thomas Spiro, Kenneth Bruland, and, particularly, William Sunda provided helpful comments on the manuscript.

REFERENCES

Achterberg E. P. and van den Berg C. M. G. (1997) Chemical speciation of chromium and nickel in the western Mediterranean. *Deep-Sea Res. II* **44**, 693–720.

Ahner B. A. and Morel F. M. M. (1999) Phytochelatins in microalgae. *Prog. Phycol. Res.* **13**, 1–31.

Ahner B. A., Price N. M., and Morel F. M. M. (1994) Phytochelatin production by marine phytoplankton at low free metal ion concentration: laboratory studies and field data from Massachusetts Bay. *Proc. Natl. Acad. Sci.* **91**, 8433–8436.

Ahner B. A., Morel F. M. M., and Moffett J. W. (1997) Trace metal control of phytochelatin production in coastal waters. *Limnol. Oceanogr.* **42**, 601–608.

Ahner B. A., Lee J. G., Price N. M., and Morel F. M. M. (1998) Phytochelatin concentrations in the equatorial Pacific. *Deep-Sea Res. I* **45**, 1779–1796.

Anbar A. D. and Knoll A. H. (2002) Proterozoic ocean chemistry and evolution: a bioinorganic bridge. *Science* **297**, 1137–1142.

Anderson D. M. and Morel F. M. M. (1978) Copper sensitivity of *Gonyaulax-tamarensis*. *Limnol. Oceanogr.* **23**, 283–295.

Anderson M. A. and Morel F. M. M. (1982) The influence of aqueous iron chemistry on the uptake of iron by the coastal *Thalassiosira weissflogii*. *Limnol. Oceanogr.* **27**, 789–813.

Andrews S. (1998) Iron storage in bacteria. *Adv. Microb. Physiol.* **40**, 281–351.

Antia N. J., Harrison P. J., and Loiveira L. (1991) The role of dissolved organic nitrogen in phytoplankton nutrition, cell biology and ecology. *Phycologia* **30**, 1–89.

Badger M. R., Andrews T. J., Whitney S. M., Ludwig M., Yelowless D. C., Leggat W., and Price G. D. (1998) The diversity and coevolution of Rubisco, plastids, pyrenoids, and chloroplast-based CO_2-concentrating mechanisms in algae. *Can. J. Botany* **76**, 1052–1071.

Badger M. R., Hanson D., and Price G. D. (2002) Evolution and diversity of CO_2 concentrating mechanisms in cyanobacteria. *Funct. Plant Biol.* **29**, 161–173.

Banse K. (1990) Does iron really limit phytoplankton production in the offshore sub-Artic Pacific. *Limnol. Oceanogr.* **35**, 772–775.

Barbeau K., Rue E. L., Bruland K. W., and Butler A. (2001) Photochemical cycling of iron in the surface ocean mediated by microbial iron (III)-binding ligands. *Nature* **413**, 409–413.

Bekheet I. A. and Syrett P. J. (1977) Urea degrading enzymes in algae. *Br. Phycol. J.* **12**, 137–143.

Berman-Frank I., Cullen J. T., Shaked Y., Sherrell R. M., and Falkowski P. G. (2001) Iron availability, cellular iron quotas, and nitrogen fixation in *Trichodesmium*. *Limnol. Oceanogr.* **46**, 1249–1260.

Boyd P. W., Watson A. J., Law C. S., Abraham E. R., Trull T., Murdoch R., Bakker D. C. E., Bowie A. R., Buesseler K. O., Chang H., Charette M., Croot P., Downing K., Frew R., Gall M., Hadfield M., Hall J., Harvey M., Jameson G., LaRoche J., Liddicoat M., Ling R., Maldonado M. T., McKay R. M., Nodder S., Pickmere S., Pridmore R., Rintoul S., Safi K., Sutton P., Strzepek R., Tanneberger K., Turner S., Waite A., and Zeldis J. (2000) A mesoscale phytoplankton bloom in the polar Southern Ocean stimulated by iron fertilization. *Nature* **407**, 695–702.

Brand L. E. (1991) Minimum iron requirements of marine phytoplankton and the implications for biogeochemical control of new production. *Limnol. Oceanogr.* **36**, 1756–1772.

Brand L. E., Sunda W. G., and Guillard R. R. L. (1983) Limitation of marine phytoplankton reproductive rates by zinc, manganese and iron. *Limnol. Oceanogr.* **28**, 1182–1198.

Brand L. E., Sunda W. G., and Guillard R. R. L. (1986) Reduction of marine phytoplankton reproduction rates by copper and cadmium. *J. Exp. Mar. Bio. Ecol.* **96**, 225–250.

Brouwers G. J., de Vrind J. P. M., Corstjens P., Cornelis P., Baysse C., and DeJong E. (1999) cumA, a gene encoding a multi-copper oxidase, is involved in ${M_n}^{2+}$ oxidation in pseudomonas putida GB-1. *Appl. Environ. Microbiol.* **65**, 1762–1768.

Bruland K. W. (1989) Complexation of zinc by natural organic ligands in the central North Pacific. *Limnol. Oceanogr.* **34**, 269–285.

Bruland K. W. (1992) Complexation of cadmium by natural organic ligands in the central North Pacific. *Limnol. Oceanogr.* **37**, 1008–1017.

Bruland K. W., Orians K. J., and Cowen J. P. (1994) Reactive trace metals in the stratified central North Pacific. *Geochim. Cosmochim. Acta* **58**, 3171–3182.

Campbell W. H. (2001) Structure and function of eukaryotic NAD(P)H: nitrate reductase. *Cell. Mol. Life Sci.* **58**, 194–204.

Cavender-Bares K. K., Mann E. L., Chisholm S. W., Ondrusek M. E., and Bidigare R. R. (1999) Differential response of equatorial Pacific phytoplankton to iron fertilization. *Limnol. Oceanogr.* **44**, 237–246.

Cembella A. D., Antia N. J., and Harrison P. J. (1984) The utilization of inorganic and organic phosphorus-compounds as nutrients by eukaryotic microalgae—a multidisciplinary perspective: 1. *CRC CR Rev. Micro.* **10**, 317–391.

Chen Y. B., Dominic B., Mellon M. T., and Zehr J. P. (1998) Circadian rhythm of nitrogenase gene expression in the diazotrophic filamentous nonheterocystous Cyanobacterium *Trichodesmium* sp. strain IMS101. *J. Bacteriol.* **180**, 3598–3605.

Coale K. H. (1991) Effects of iron, manganese, copper, and zinc enrichments on productivity and biomass in the subarctic Pacific. *Limnol. Oceanogr.* **36**, 1851–1864.

Coale K. H. and Bruland K. W. (1988) Copper complexation in the Northeast Pacific. *Limnol. Oceanogr.* **33**, 1084–1101.

Coale K. H., Johnson K. S., Fitzwater S. E., Blain S. P. G., Stanton T. P., and Coley T. L. (1998) IronEx-I, an *in situ* iron-enrichment experiment: experimental design, implementation and results. *Deep-Sea Res. II Top. St. Ocean.* **45**, 919–945.

Cox E. H., McLendon G. L., Morel F. M. M., Lane T. W., Prince R. C., Pickering I. J., and George G. N. (2000) The active site structure of *Thalassiosira weissflogii* carbonic anhydrase 1. *Biochemistry* **39**, 12128–12130.

Cresswell R. C. and Syrett P. J. (1982) The uptake of nitrite by the diatom *Phaeodactylum*-internations between nitrite and nitrate. *J. Exp. Botany* **33**, 1111–1121.

Cullen J. T., Lane T. W., Morel F. M. M., and Sherrell R. M. (1999) Modulation of cadmium uptake in phytoplankton by seawater pCO_2. *Nature* **402**, 165–167.

Doucette G. J., Erdner D. L., Peleato M. L., Hartman J. J., and Anderson D. M. (1996) Quantitative analysis of iron-stress related proteins in *Thalassiosira weissflogii*: measurement of flavodoxin and ferredoxin using HPLC. *Mar. Ecol. Prog. Ser.* **130**, 269–276.

Dugdale R. C. and Goering J. J. (1967) Uptake of new and regenerated forms of nitrogen in primary productivity. *Limnol. Oceanogr.* **12**, 196–206.

Eide D. J. (1998) The molecular biology of metal ion transport in *Saccharomyces cerevisiae*. *Ann. Rev. Nutr.* **18**, 441–469.

Ellwood M. J. and van den Berg C. M. G. (2000) Zinc speciation in the Northeastern Atlantic Ocean. *Mar. Chem.* **68**, 295–306.

Ellwood M. J. and van den Berg C. M. G. (2001) Determination of organic complexation of cobalt in seawater by cathodic stripping voltammetry. *Mar. Chem.* **75**, 33–47.

Fitzwater S. E., Johnson K. S., Gordon R. M., Coale K. H., and Smith W. O., Jr. (2000) Trace metal concentrations in the Ross Sea and their relationship with nutrients and phytoplankton growth. *Deep-Sea Res. II* **47**, 3159–3179.

Francis C. A. and Tebo B. M. (2001) *cum*A multi-copper oxidase genes from diverse Mn(II)-oxidizing and non-Mn(II)-oxidizing *Pseudomonas* strains. *Appl. Environ. Microbiol.* **67**, 4272–4278.

Francis C. A. and Tebo B. M. (2002) Enzymatic manganese(II) oxidation by metabolically dormant spores of diverse *Bacillus* species. *Appl. Environ. Microbiol.* **68**, 874–880.

Galvan A. and Fernandez E. (2001) Eukaryotic nitrate and nitrite transporters. *Cell. Mol. Life Sci.* **58**, 225–233.

Glibert P. M. and McCarthy J. J. (1984) Uptake and assimilation of ammonium and nitrate by phytoplankton—indexes of nutritional-status for natural assemblages. *J. Plankton Res.* **6**, 677–697.

Granger J. and Price N. M. (1999) The importance of siderophores in iron nutrition of heterotrophic marine bacteria. *Limnol. Oceanogr.* **44**, 541–555.

Granger J. and Ward B. (2002) Accumulation of nitrogen oxides in copper-limited cultures of denitrifying bacteria. *Limnol. Oceanogr.* **48**, 313–318.

Grossman M. J., Hinton S. M., Minak-Bernero V., Slaughter C., and Stiefel E. I. (1992) Unification of the ferritin family of proteins. *Proc. Natl. Acad. Sci.* **89**, 2419–2423.

Guerrero M. G., Vega J. M., and Losada M. (1981) The assimilatory nitrate-reducing system and its regulation. *Ann. Rev. Plant Physiol.* **32**, 169–204.

Harrison G. I. and Morel F. M. M. (1986) Response of the marine diatom *Thalassiosira weissflogii* to iron stress. *Limnol. Oceanogr.* **31**, 989–997.

Henley W. J. and Yin Y. (1998) Growth and photosynthesis of marine *Synechococcus* (Cyanophyceae) under iron stress. *J. Phycol.* **34**, 94–103.

Hildebrand M. (2000) Silicic acid transport and its control during cell wall silicification in diatoms. In *Biomineralization: From Biology to Biotechnology and Medical Application* (ed. E. Baeuerlein). Wiley-VCH, New York, pp. 171–188.

Hildebrand M., Volcani B. E., Gassman W., and Schroeder J. I. (1997) A gene family of silicon transporters. *Nature* **385**, 688–689.

Hildebrand M., Dahlin K., and Volcani B. E. (1998) Characterization of a silicon transporter gene family in *Cylindrotheca fusiformis*: sequences, expression analysis, and identification of homologs in other diatoms. *Mol. G. Genet.* **260**, 480–486.

Ho T.-Y., Quigg A., Finkel Z. V., Milligan A. J., Wyman K., Falkowski P. J., and Morel F. M. M. The elemental composition of some marine phytoplankton. *J. Physol.* (in press).

Holland H. D. (1984) *The Chemical Evolution of the Atmosphere and Oceans*. Princeton University Press, Princeton, 582pp.

Honjo S. (1996) In *Particle Flux in the Ocean* (eds. V. Ittekkot, P. Schaefer, S. Honjo, and P. J. Depetris). Wiley, West Sussex, 372pp.

Hudson R. J. M. and Morel F. M. M. (1990) Iron transport in marine phytoplankton: kinetics of cellular and medium coordination reactions. *Limnol. Oceanogr.* **35**, 1002–1020.

Hudson R. J. M. and Morel F. M. M. (1993) Trace metal transport by marine microorganisms: implications of metal coordination kinetics. *Deep-Sea Res.* **40**, 129–150.

Hudson R. J. M., Covault D. T., and Morel F. M. M. (1992) Investigations of iron coordination and redox reactions in seawater using ^{59}Fe radiometry and ion-pair solvent extraction of amphiphilic iron complexes. *Mar. Chem.* **38**, 209–235.

Hutchins D. A., DiTullio G. R., Zhang Y., and Bruland K. W. (1998) An iron limitation mosaic in the California upwelling regime. *Limnol. Oceanogr.* **43**, 1037–1054.

Hutchins D. A., Witter A. E., Butler A., Luther I., and G. W. (1999a) Competition among marine phytoplankton for different chelated iron species. *Nature* **400**, 858–861.

Hutchins D. A., Franck M., Brezezinski M. A., and Bruland K. W. (1999b) Inducing phytoplankton iron limitation in iron-replete coastal waters with a chelating ligand. *Limnol. Oceanogr.* **44**, 1009–1018.

Hutchins D. A., Hare C. E., Weaver R. S., Zhang Y., and Firme G. F. (2002) Photoplankton iron limitation in the Humboldt current and Peru upwelling. *Limnol. Oceanogr.* **47**, 997–1011.

Jensen R. G. (1990) Ribulose 1,5-bisphospate carboxylase/oxygenase: mechanism, activation, and regulation. In *Plant Physiology, Biochemistry and Molecular Biology* (eds. D. T. Dennis and D. H. Turpin). Longmann Scientific and Technical, Essex, pp. 224–238.

Kudo I. and Harrison P. J. (1997) Effect of iron nutrition on the marine cyanobacterium *Synechococcus* grown on different N sources and irradiances. *J. Phycol.* **33**, 232–240.

Kustka A., Sanudo-Wilhelmy S., Carpenter E. J., Capone D. G., and Raven J. A. (2003a) A revised estimate of the iron use efficiency of nitrogen fixation, with special reference to the marine cyanobacterium *Trichodesmium* spp. (Cyanophyta). *J. Phycol.* **39**, 12–25.

Kustka A. B., Sañudo-Wilhelmy S., Carpenter E. J., and Sunda W. G. (2003b) Iron requirements for dinitrogen and ammonium supported growth in cultures of *Trichodesmium* (IMS 101): comparison with nitrogen fixation rates and iron: carbon ratios of field populations. *Limnol. Oceanogr.* **48**, 1869–1884.

Lam P. J., Tortell P. D., and Morel F. M. M. (2001) Differential effects of iron additions on organic and inorganic carbon production by phytoplankton. *Limnol. Oceanogr.* **46**, 1199–1202.

Lane T. W. and Morel F. M. M. (2000a) A biological function for cadmium in marine diatoms. *Proc. Natl. Acad. Sci.* **97**, 4627–4631.

Lane T. W. and Morel F. M. M. (2000b) Regulation of carbonic anhydrase expression by zinc, cobalt, and carbon dioxide in the marine diatom *Thalassiosira weissflogii*. *Plant Physiol.* **123**, 345–352.

La Roche J., Geider R. J., Graziano L. M., Murray H., and Lewis K. (1993) Induction of specific proteins in eukaryotic algae grown under iron-, phosphorus-, or nitrogen-deficient conditions. *J. Phycol.* **29**, 767–777.

La Roche J., McKay R. M. L., and Boyd P. (1999) Immunological and molecular probes to detect phytoplankton responses to environmental stress in nature. *Hydrobiologia* **401**, 177–198.

Lee J. G., Roberts S. B., and Morel F. M. M. (1995) Cadmium: a nutrient for the marine diatom *Thalassiosira weissflogii*. *Limnol. Oceanogr.* **40**, 1056–1063.

Leftley J. W. and Syrett P. J. (1973) Urease and ATP: urea amidolyase activity in unicellular algae. *J. Gen. Microbiol.* **77**, 109–115.

Maldonado M. T. and Price N. M. (1996) Influence of N substrate on Fe requirements of marine centric diatoms. *Mar. Ecol. Prog. Ser.* **141**, 161–172.

Maldonado M. T. and Price N. M. (2001) Reduction and transport of organically bound iron by *Thalassiosira oceanica* (Bacillariophyceae). *J. Phycol.* **37**, 298–309.

Maldonado M. T., Boyd P. W., Harrison P. J., and Price N. M. (1999) Co-limitation of phytoplankton growth by light and Fe during winter in the NE subarctic Pacific Ocean. *Deep-Sea Res. II* **46**, 2475–2485.

Mann E. L. and Chisholm S. W. (2000) Iron limits the cell division rate of Prochlorococcus in the eastern equatorial Pacific. *Limnol. Oceanogr.* **45**, 1067–1076.

Mann E. L., Ahlgren N., Moffett J. W., and Chisholm S. W. (2002) Copper toxicity and cyanobacteria ecology in the Sargasso Sea. *Limnol. Oceanogr.* **47**, 976–988.

Maret W. (2002) Optical methods for measuring zinc binding and release, zinc coordination environments in zinc finger proteins, and redox sensitivity and activity of zinc-bound thiols. *Meth. Enzymol.* **348**, 230–237.

Martin J. H. and Gordon R. M. (1988) Northeast Pacific iron distributions in relation to phytoplankton productivity. *Deep-Sea Res.* **35**, 177–196.

Martin J. H., Gordon R. M., Fitzwater S., and Broenkow W. W. (1989) VERTEX: phytoplankton/iron studies in the Gulf of Alaska. *Deep-Sea Res.* **36**, 649–680.

Martin J. H., Coale K. H., Johnson K. S., Fitzwater S. E., Gordon R. M., Tanner S. J., Hunter C. N., Elrod V. A., Nowicki J. L., Coley T. L., Barber R. T., Lindley S., Watson A. J., Vanscoy K., Law C. S., Liddicoat M. I., Ling R., Stanton T., Stockel J., Collins C., Anderson A., Bidigare R., Ondrusek M., Latasa M., Millero F. J., Lee K., Yao W., Zhang J. Z., Friederich G., Sakamoto C., Chavez F., Buck K., Kolber Z., Greene R., Falkowski P., Chisholm S. W., Hoge F., Swift R., Yungel J., Turner S., Nightingale P., Hatton A., Liss P., and Tindale N. W. (1994) Testing the iron hypothesis in ecosystems of the equatorial Pacific Ocean. *Nature* **371**, 123–129.

Martinez J. S., Zhang G. P., Holt P. D., Jung H.-T., Carrano C. J., Haygood M. G., and Butler A. (2000) Self-assembling amphiphilic siderophores from marine bacteria. *Science* **287**, 1245–1247.

Martin-Jezequel V., Hildebrand M., and Brzezinski M. A. (2000) Silicon metabolism in diatoms: implications for growth. *J. Phycol.* **36**, 821–840.

McCarthy J. J. (1972) Uptake of urea by natural populations of marine phytoplankton. *Limnol. Oceanogr.* **17**, 738–748.

Milligan A. J. and Morel F. M. M. (2002) A proton buffering role for silica in diatoms. *Science* **297**, 1848–1850.

Moffett J. W. (1995) The spatial and temporal variability of copper complexation by strong organic ligands in the Sargasso Sea. *Deep-Sea Res. I* **42**, 1273–1295.

Moffett J. W. and Brand L. E. (1996) Production of strong, extracellular cu chelators by marine cyanobacteria in response to Cu stress. *Limnol. Oceanogr.* **41**, 388–395.

Moffett J. W., Zika R. G., and Brand L. E. (1990) Distribution and potential sources and sinks of copper chelators in the Sargasso Sea. *Deep-Sea Res.* **37**, 27–36.

Morel F., Reuter J., Anderson D., and Guillard R. (1979) Aquil: a chemically defined phytoplankton culture medium for trace metal studies. *Limnol. Oceanogr.* **36**, 1742–1755.

Morel F. M. M., Reinfelder J. R., Roberts S. B., Chamberlain C. P., Lee J. G., and Yee D. (1994) Zinc and carbon co-limitation of marine phytoplankton. *Nature* **369**, 740–742.

Morel F. M. M., Cox E. H., Kraepiel A. M. L., Lane T. W., Milligan A. J., Schaperdoth I., Reinfelder J. R., and Tortell P. D. (2002) Acquisition of inorganic carbon by the marine diatom *Thalassiosira weissflogii*. *Funct. Plant Biol.* **29**, 301–308.

Mulholland M. R., Glibert P. M., Berg G. M., Van Heukelem L., Pantoja S., and Lee C. (1998) Extracellular amino acid oxidation by microplankton: a cross-ecosystem comparison. *Aquat. Microbiol. Ecol.* **15**, 141–152.

Mulholland M. R., Gobler C. J., and Lee C. (2002) Peptide hydrolysis, amino acid oxidation, and nitrogen uptake in communities seasonally dominated by *Aureococcus anophagefferens*. *Limnol. Oceanogr.* **47**, 1094–1108.

Neilands J. B. (1981) Iron absorption and transport in microorganisms. *Ann. Rev. Nutr.* **1**, 27–46.

Nodwell L. and Price N. (2001) Direct use of inorganic colloidal iron by marine mixotrophic phytoplankton. *Limnol. Oceanogr.* **46**, 755–777.

O'Halloran T. V. and Culotta V. C. (2000) Metallochaperones, an intracellular shuttle service for metal ions. *J. Biol. Chem.* **275**, 25057–25060.

Ohkawa H., Sonoda M., Hagino N., Shibata M., Pakrasi H. B., and Ogawa T. (2002) Functionally distinct NAD(P)H dehydrogenases and their membrane localization in Synechocystis sp. PCC6803. *Funct. Plant Biol.* **29**, 195–200.

Oritz D. F., Rusticitti T., McCue K. F., and Ow D. W. (1995) Transport of metal-binding peptides by HMT1, a fission yeast ABC-type vacuolar membrane protein. *J. Biol. Chem.* **270**, 201–205.

Outten C. E. and O'Halloran T. V. (2001) Femtomolar sensitivity of metalloregulatory proteins controlling zinc homeostasis. *Science* **292**, 2488–2492.

Palenik B. and Morel F. M. M. (1990a) Amino acid utilization by marine phytoplankton: a novel mechanism. *Limnol. Oceanogr.* **35**, 260–269.

Palenik B. and Morel F. M. M. (1990b) Comparison of cell-surface L-amino acid oxidases from several marine phytoplankton. *Mar. Ecol. Prog. Ser.* **59**, 195–201.

Palenik B. and Morel F. M. M. (1991) Amine oxidases of marine phytoplankton. *Appl. Environ. Microbiol.* **57**, 2440–2443.

Palenik B., Kieber D. J., and Morel F. M. M. (1991) Dissolved organic nitrogen use by phytoplankton: the role of cell-surface enzymes. *Biol. Oceanogr.* **6**, 347–354.

Pantoja S. and Lee C. (1994) Cell-surface oxidation of amino-acids in seawater. *Limnol. Oceanogr.* **39**, 1718–1726.

Price G. D. and Badger M. R. (1989) Isolation and characterization of high CO_2-requiring mutants of the cyanobacterium *Synechococcus* PCC7942-2 phenotypes that accumulate inorganic carbon but are apparently unable to generate CO_2 within the carboxysome. *Plant Physiol.* **91**, 514–525.

Price G. D., Maeda S., Omata T., and Badger M. R. (2002) Modes of active inorganic carbon uptake in the cyanobacterium *Synechococcus* sp. PCC7942. *Funct. Plant Biol.* **29**, 131–149.

Price N. M. and Morel F. M. M. (1991) Colimitation of phytoplankton growth by nickel and nitrogen. *Limnol. Oceanogr.* **36**, 1071–1077.

Price N. M., Harrison G. I., Hering J. G., Hudson R. J., Nirel P., Palenik B., and Morel F. M. M. (1988/1989) Preparation and chemistry of the artificial algal culture medium Aquil. *Biol. Oceanogr.* **5**, 43–46.

Price N. M., Andersen L. F., and Morel F. M. M. (1991) Iron and nitrogen nutrition of equatorial Pacific plankton. *Deep-Sea Res.* **38**, 1361–1378.

Price N. M., Ahner B. A., and Morel F. M. M. (1994) The equatorial Pacific Ocean: grazer-controlled phytoplankton populations in an iron-limited system. *Limnol. Oceanogr.* **39**, 520–534.

Rae T. D., Schmidt P. J., Pufahl R. A., Culotta V. C., and O'Halloran T. V. (1999) Undetectable intracellular free copper: the requirement of a copper chaperone for superoxide dismutase. *Science* **284**, 805–808.

Raghothama K. G. (1999) Phosphate acquisition. *Ann. Rev. Plant Physiol.* **50**, 665–693.

Raven J. A. (1988) The iron and molybdenum use efficiencies of plant growth with different energy, carbon and nitrogen sources. *New Phytol.* **109**, 279–287.

Raven J. A. (1990) Predictions of Mn and Fe use efficiencies of phototrophic growth as a function of light availability for growth and of C assimilation pathway. *New Phytol.* **116**, 1–18.

Raven J. A., Wollenweber B., and Handley L. L. (1992) A comparison of ammonium and nitrate as nitrogen sources for photolithotrophs. *New Phytol.* **121**, 19–32.

Rees T. A. V. and Bekheet I. A. (1982) The role of nickel in urea assimilation by algae. *Planta* **156**, 385–387.

Reinfelder J. R., Kraepiel A. M. L., and Morel F. M. M. (2000) Unicellular C_4 photosynthesis in a marine diatom. *Nature* **407**, 996–999.

Roberts S. B., Lane T. W., and Morel F. M. M. (1997) Carbonic anhydrase in the marine diatom *Thalassiosira weissflogii* (*Bacillariophyceae*). *J. Phycol.* **33**, 845–850.

Rue E. L. and Bruland K. W. (1995) Complexation of iron III by natural organic ligands in the Central North Pacific as determined by a new cometitive ligand equilibration/adsorptive cathodic stripping voltammetric method. *Mar. Chem.* **50**, 117–138.

Rue E. L. and Bruland K. W. (1997) The role of organic complexation on ambient iron chemistry in the equatorial

Pacific Ocean and the response of a mesoscale iron addition experiment. *Limnol. Oceanogr.* **42**, 901–910.

Rueter J. G. and Morel F. M. M. (1981) The interaction between zinc deficiency and copper toxicity as it affects the silicic acid uptake mechanisms in *Thalassiosira pseudonana*. *Limnol. Oceanogr.* **26**, 67–73.

Saito M. A. (2001) The biogeochemistry of cobalt in the Sargasso Sea. PhD Thesis, MIT-WHOI.

Saito M. A. and Moffett J. W. (2001a) Complexation of cobalt by natural organic ligands in the Sargasso Sea as determined by a new high-sensitivity electrochemical cobalt speciation method suitable for open ocean work. *Mar. Chem.* **75**, 49–68.

Saito M. A. and Moffett J. W. (2001b) Cobalt speciation in the equatorial Pacific and Peru upwelling region: sources and chemical properties of natural cobalt ligands. *Am. Soc. Limnol. Oceanogr. Meet.*

Saito M. A., Moffett J. W., Chisholm S. W., and Waterbury J. B. (2002) Cobalt limitation and uptake in *Prochlorococcus*. *Limnol. Oceanogr.* **6**, 1627–1636.

Scott I. A. (1990) Mechanistic and evolutionary aspects of vitamin B_{12} biosynthesis. *Acc. Chem. Res.* **23**, 308–317.

Smith K. S. and Ferry J. G. (2000) Prokaryotic carbonic anhydrases. *FEMS Microbiol. Rev.* **24**, 335–366.

Sunda W. G. (1984) Measurement of manganese, zinc and cadmium complexation in seawater using chelex ion-exchange equilibria. *Mar. Chem.* **14**, 365–378.

Sunda W. G. and Guillard R. R. L. (1976) The relationship between cupric ion activity and the toxicity of copper to phytoplankton. *J. Mar. Res.* **37**, 761–777.

Sunda W. G. and Hanson A. K. (1987) Measurement of free cupric ion concentration in seawater by a ligand competition technique involving copper sorption onto C_{18} SEP-PAK cartridges. *Limnol. Oceanogr.* **32**, 537–551.

Sunda W. G. and Huntsman S. A. (1988) Effect of sunlight on redox cycles of manganese in the southwestern Sargasso Sea. *Deep-Sea Res.* **35**, 1297–1317.

Sunda W. G. and Huntsman S. A. (1992) Feedback interactions between zinc and phytoplankton in seawater. *Limnol. Oceanogr.* **37**, 25–40.

Sunda W. G. and Huntsman S. A. (1995a) Iron uptake and growth limitation in oceanic and coastal phytoplankton. *Mar. Chem.* **50**, 189–206.

Sunda W. G. and Huntsman S. A. (1995b) Cobalt and zinc interreplacement in marine phytoplankton: biological and geochemical implications. *Limnol. Oceanogr.* **40**, 1404–1417.

Sunda W. G. and Huntsman S. A. (1995c) Regulation of copper concentration in the oceanic nutricline by phytoplankton uptake and regeneration cycles. *Limnol. Oceanogr.* **40**, 132–137.

Sunda W. G. and Huntsman S. A. (1996) Antagonisms between cadmium and zinc toxicity and manganese limitation in a coastal diatom. *Limnol. Oceanogr.* **41**, 373–387.

Sunda W. G. and Huntsman S. A. (1997) Interrelated influence of iron, light and cell size on marine phytoplankton growth. *Nature* **390**, 389–392.

Sunda W. G. and Huntsman S. A. (1998a) Control of Cd concentrations in a coastal diatom by interactions among free ionic Cd, Zn, and Mn in seawater. *Environ. Sci. Technol.* **32**, 2961–2968.

Sunda W. G. and Huntsman S. A. (1998b) Interactions among Cu^{2+}, Zn^{2+}, and Mn^{2+} in controlling cellular Mn, Zn, and growth rate in the coastal alga *Chlamydomonas*. *Limnol. Oceanogr.* **43**, 1055–1064.

Sunda W. G. and Huntsman S. A. (1998c) Interactive effects of external manganese, the toxic metals copper and zinc, and light in controlling cellular manganese and growth in a coastal diatom. *Limnol. Oceanogr.* **43**, 1467–1475.

Sunda W. G. and Huntsman S. A. (2000) Effect of Zn, Mn, and Fe on Cd accumulation in phytoplankton: implications for oceanic Cd cycling. *Limnol. Oceanogr.* **45**, 1501–1516.

Sundby B., Anderson L. G., Hall P. O. J., Iverfeldt A., Vanderloeff M. M. R., and Westerlund S. F. G. (1986) The effect of oxygen on release and uptake of cobalt, manganese, iron and phosphate at the sediment-water interface. *Geochim. Cosmochim. Acta* **50**, 1281–1288.

Thamdrup B., Glud R. N., and Hansen J. W. (1994) Manganese oxidation and *in-situ* manganese fluxes from a coastal sediment. *Geochim. Cosmochim. Acta* **58**, 2563–2570.

Tortell P. D. and Morel F. M. M. (2002) Sources of inorganic carbon for phytoplankton in the eastern subtropical and equatorial Pacific Ocean. *Limnol. Oceanogr.* **47**, 1012–1022.

van Veen H. W. (1997) Phosphate transport in prokaryotes: molecules, mediators and mechanisms. *Anton. Leeuwenhoek Int. J. Gen. M.* **72**, 299–315.

van Waasbergen L. G., Hildebrand M., and Tebo B. M. (1996) Identification and characterization of a gene cluster involved in manganese oxidation by spores of the marine *Bacillus* sp. strain SG-1. *J. Bacteriol.* **178**, 3517–3530.

Varela D. E. and Harrison P. J. (1999) Seasonal variability in nitrogenous nutrition of phytoplankton assemblages in the northeastern subarctic Pacific Ocean. *Deep-Sea Res. II* **46**, 2505–2538.

Völker C. and Wolf-Gladrow D. A. (1999) Physical limits on iron uptake mediated by siderophores or surface reductases. *Mar. Chem.* **65**, 227–244.

Wilhelm S. W. (1995) Ecology of iron-limited cyanobacteria: a review of physiological responses and implications for aquatic systems. *Aquat. Microbiol. Ecol.* **9**, 295–303.

Wilhelm S. W. and Trick C. G. (1995) Effects of vitamin B^{12} concentration on chemostat cultured *Synechococcus* sp. strain PCC 7002. *Can. J. Microbiol.* **41**, 145–151.

Williams R. J. P. and Frausto da Silva J. J. R. (1996) *The Natural Selection of the Chemical Elements*. Oxford University Press, New York, 646pp.

Wright S. A. and Syrett P. J. (1983) The uptake of methylammonium and dimethylammonium by the diatom, phaeodactylum-tricornutum. *New Phytol.* **95**, 189–202.

Wu J., Boyle E., Sunda W., and Wen L.-S. (2001) Soluble and colloidal iron in the oligotrophic North Atlantic and North Pacific. *Science* **293**, 847–849.

Yu J. W., Price G. D., and Badger M. R. (1994) Characterization of CO_2 and HCO_3^- uptake during steady-state photosynthesis in the cyanobacterium *Synechococcus*-pcc7942. *Austral. J. Plant Physiol.* **21**, 185–195.

Zehr J. P. and Falkowski P. G. (1988) Pathway of ammonium assimilation in a marine diatom determined with the radiotracer 13N. *J. Phycol.* **24**, 588–591.

Zehr J. P., Waterbury J. B., Turner P. J., Montoya J. P., Omoregie E., Steward G. F., Hansen A., and Karl D. M. (2001) Unicellular cyanobacteria fix N-2 in the subtropical North Pacific Ocean. *Nature* **412**, 635–638.

6.06
Organic Matter in the Contemporary Ocean

T. I. Eglinton and D. J. Repeta
Woods Hole Oceanographic Institution, MA, USA

6.06.1 INTRODUCTION

This chapter summarizes selected aspects of our current understanding of the organic carbon (OC) cycle as it pertains to the modern ocean, including underlying surficial sediments. We briefly review present estimates of the size of OC reservoirs and the fluxes between them. We then proceed to highlight advances in our understanding that have occurred since the late 1980s, especially those which have altered our perspective of the ways organic matter is cycled in the oceans. We have focused on specific areas where substantial progress has been made, although in most cases our understanding remains far from complete. These are the fate of terrigenous OC inputs in the ocean, the composition of oceanic dissolved organic matter (DOM), the mechanisms of OC preservation, and new insights into microbial inputs and processes. In each case, we discuss prevailing hypotheses concerning the composition and fate of organic matter derived from the different inputs, the reactivity and relationships between different organic matter pools, and highlight current gaps in our knowledge.

The advances in our understanding of organic matter cycling and composition has stemmed largely from refinements in existing methodologies and the emergence of new analytical capabilities. Molecular-level stable carbon and nitrogen isotopic measurements have shed new light on a range of biogeochemical processes. Natural abundance of radiocarbon data has also been increasingly applied as both a tracer and source indicator in studies of organic matter cycling. As for ^{13}C, bulk ^{14}C measurements are now complemented by measurements at the molecular level, and the combination of these different isotopic approaches has proven highly informative. The application of multinuclear solid- and liquid-state nuclear magnetic resonance (NMR) spectroscopy has provided a more holistic means to examine the complex array of macromolecules that appears to comprise both dissolved and particulate forms of organic matter. New liquid chromatography/mass spectrometry techniques provide structural information on polar macromolecules that have previously been beyond the scope of established methods. In addition to technological advances, large multidisciplinary field programs have provided important frameworks and contexts within which to interpret organic geochemical data, while novel sampling techniques have been developed that allow for the collection of more representative samples and their detailed analytical manipulation. Two particular analytical approaches are highlighted in this chapter—NMR spectroscopy as a powerful tool for structural characterization of complex macromolecules, and compound specific carbon isotope (^{13}C and ^{14}C) analysis as probes for the cycling of organic matter in the ocean through space and time.

Finally, we outline new as well as unresolved questions which provide future challenges for marine organic biogeochemists, and discuss emerging analytical approaches that may shed new light on organic matter cycling in the oceans. For example, (i) the source of "old" dissolved organic carbon (DOC) in the deep sea has yet to be resolved; (ii) the molecular-level composition of the majority of organic matter buried in marine sediments evades elucidation; and (iii) while planktonic archea have been found to be amongst the most abundant organisms in the ocean, their role in biogeochemical cycles and their legacy in the sedimentary record are only beginning to be considered. Such fundamental observations and questions continue to challenge us, and limit our understanding of the processes underpinning organic matter cycling in the oceans.

The discipline of marine organic geochemistry has expanded and evolved greatly in recent years. Hence, we have had to be selective in our coverage of new developments. This review, therefore, is by no means comprehensive and there are many important and exciting aspects of marine organic geochemistry that we have not covered. Comprehensive discussions of some of these aspects are to be found in the following review papers and chapters: soil OC (Hedges and Oades, 1997), terrestrial OC inputs to the oceans (Hedges *et al.*, 1997; Schlunz and Schneider, 2000), organic matter preservation (Tegelaar *et al.*, 1989; de Leeuw and Largeau, 1993; Hedges and Keil, 1995), lipid biomarkers (Volkman *et al.*, 1998), bacterial contributions (Sinninghe Damsté and Schouten, 1997), deep biosphere (Parkes *et al.*, 2000), eolian inputs (Prospero *et al.*, 2003), black carbon (BC) (Schmidt and Noack, 2000), gas hydrates (Kvenvolden, 1995), water column particulate organic matter (POM) (Wakeham and Lee, 1993), carbon isotopic systematics (Hayes, 1993), and use of ^{14}C and ^{13}C as tracers of OC input (Raymond and Bauer, 2001b).

In this chapter two pools of organic matter (OM) are discussed in detail. Particulate organic matter is manifestly heterogeneous, composed of all sorts of particles resulting from a wide range of inputs and a multitude of processes acting on them. In effect, sedimentary POM is chemically and spatially heterogeneous and much effort needs to be focused on sampling, fractionation, and bulk characterization rather than on detailed molecular-level studies. This situation contrasts sharply with the study of DOM, which, despite its largely macromolecular nature, appears to be remarkably uniform in composition throughout the oceans. Here, the prime need is for studies of the colloid processes involved and detailed molecular-level analysis of the composition and conformation of the refractory DOM in order to provide a basis for explaining its apparent lack of bioavailability, and to answer the question: why does DOM persist for years, even millennia, in the deep ocean?

6.06.2 RESERVOIRS AND FLUXES

6.06.2.1 Reservoirs

Figure 1 depicts the major components of the OC cycle on and in the Earth's crust. Greater than 99.9% of all carbon in the Earth's crust is stored in sedimentary rocks (Berner, 1989). About 20% of this total ($\sim 1.5 \times 10^7$ Gt) is organic, and the majority (>90%) of the OC in these consolidated

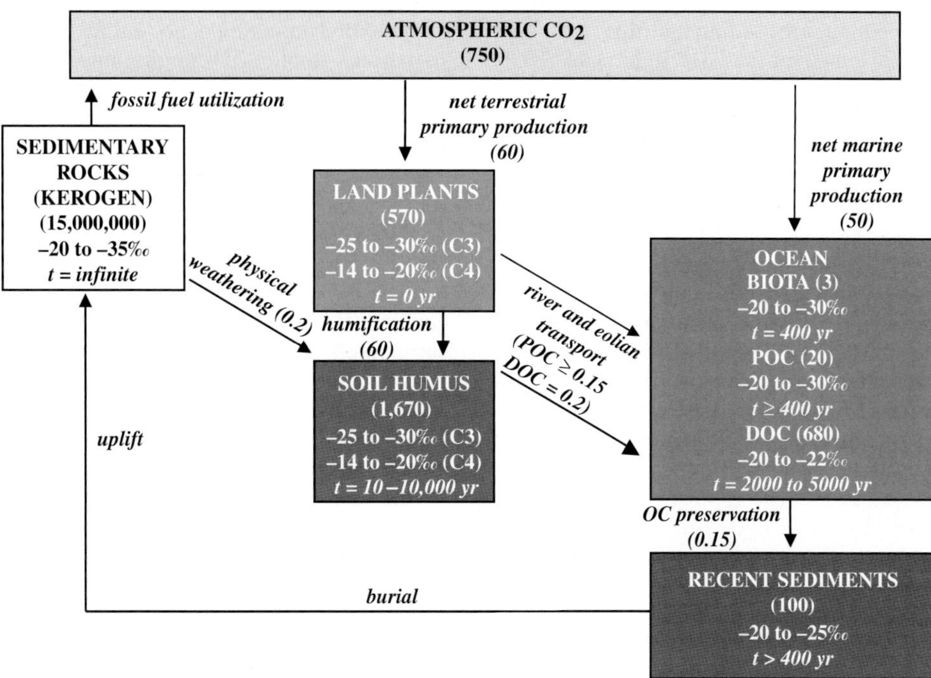

Figure 1 The global organic carbon cycle. Numbers in parentheses are approximate reservoir sizes (10^{15} g C = Gt) and italicized are approximate fluxes (10^{15} g C yr^{-1}). Nonitalicized numbers are approximate ranges for stable carbon isotopic compositions ($\delta^{13}C$, per mil) and italicized numbers are approximate radiocarbon ages (yr BP) (after Hedges, 1992).

sediments is "kerogen," operationally defined as macromolecular material that is insoluble in common organic solvents and nonoxidizing acids (Durand, 1980). Most kerogen is finely disseminated in sedimentary rocks (shales and limestones) which, on average, contain ~1% organic matter. Organic-rich deposits that include the World's fossil fuel reserves (coal, oil shales, and petroleum) account for less than 0.1% of total sedimentary OC (Hunt, 1996). Of the small fraction of organic matter that is not in the form of relict carbon sequestered in ancient sedimentary rocks, almost two-thirds resides on the continents. Approximately 25% of this "terrestrial" OC (~570 Gt) is in the form of standing biomass (plant tissue), with a further 70 Gt of plant litter on the soil surface (Post, 1993), and almost 1,600 Gt residing within the upper 1 m of soils and peat deposits (Eswaran *et al.*, 1993).

Marine biota comprise only ~3 Gt of OC, and sinking and suspended particulate OC account for a further 10–20 Gt. The majority of OC in the oceans is in the form of DOC (680 Gt) and organic matter sequestered in the upper meter of marine sediments (~100 Gt). Concentrations of marine DOC are highest in the upper ocean, and in the coastal zone. Typical open ocean DOC concentrations in surface seawater range from 60 μM to 80 μM. In the coastal zone, concentrations may climb to in excess of 200 μM, although concentrations rapidly decrease within a few kilometers of shore (Vlahos *et al.*, 2002). The inventory of marine OC is fixed by the concentration of DOC in the deep ocean, which is relatively constant at 42 μM, although small variations of a few μM C have been reported (Hansell and Carlson, 1998). These variations are intriguing, in that they suggest active cycling of DOC in the deep sea. North Atlantic Deep Water, Antarctic Bottom Water, and other deep-water masses all carry the same burden of DOC, even though they are formed at different latitudinal extremes and under very different forcing conditions. Why the concentration of DOC is so constant in the deep sea is a mystery, but the narrow range of DOC values measured in the global ocean implies a very tightly controlled feedback between production and degradation.

Approximately 100 Gt of organic matter is sequestered in the upper meter of marine sediments. In the modern ocean, ~90% of OC burial occurs under oxygenated bottom waters along continental margins (e.g., Hedges and Keil, 1995). Up to 45% of OC burial occurs in deltaic sediments, which, in spite of nature as loci of major riverine inputs, appear to include a major fraction of marine-derived organic matter (Keil *et al.*, 1997). An equivalent amount of OC is buried in nondeltaic sediments that are also proximal to land masses (Hedges and Keil, 1995), primarily on continental shelves and upper slopes (Premuzic *et al.*, 1982). Of the remainder, ~7% of OC is buried beneath highly productive regions with associated oxygen minimum zones (OMZs) and in anoxic basins. The balance (~5%) is buried pelagic abyssal sediments.

Much of the discussion in this chapter concerning sedimentary OC is centered on continental margin sediments, because this is the major locus of OC burial. In contrast, most DOM is held in the deep ocean basins, and hence discussion of the latter is largely in the context of the pelagic water column.

6.06.2.2 Fluxes

6.06.2.2.1 Terrigenous organic matter fluxes to the oceans

OC fluxes from sedimentary rock weathering on land are not well constrained but on geological timescales are believed to match OC burial in sediments (Berner, 1989). Superimposed on this background of relict OC from sedimentary rock weathering are fluxes associated with terrestrial primary production. The global rate of net terrestrial photosynthesis is estimated to be in the range of 60 Gt yr^{-1} (Post, 1993). Approximately two-thirds of the resulting total plant litter is oxidized rapidly to CO_2 (Post, 1993), while the remainder enters the soil cycle and is subject to further oxidation. Organic matter pools within soils exhibit different reactivities and turnover times that range from decades to millenia (Torn *et al.*, 1997). Over geologic timescales, however, the pervasive and continuous oxidative degradation and leaching and erosion processes on the continents result in little long-term storage of organic matter on the continents (Hedges *et al.*, 1997). However, some fraction of this terrestrial (vascular plant-derived) OC and sedimentary rock-derived (relict) OC escapes oxidation and is delivered to the oceans. The delivery of terrigenous OC to the oceans is primarily *via* riverine or atmospheric (eolian) processes.

Riverine fluxes. Approximately 0.2 Gt each of dissolved and particulate OC are carried from land to sea annually by rivers (Ludwig *et al.*, 1996). Much of this riverine organic matter appears to be soil derived based on its chemical characteristics (Meybeck, 1982; Hedges *et al.*, 1994), although autochthonous sources may be important for the dissolved fraction (Repeta *et al.*, 2002). It is now recognized that, on a global basis, riverine discharge is dominated by low-latitude tropical rivers. This not only includes major systems such as the Amazon, and Congo, but also includes the numerous smaller rivers draining mountainous tropical regions (Nittrouer *et al.*, 1995), most notably in Papua New Guinea and other parts of Oceania, which are estimated to account for nearly 50% of the global flux of river sediment to the oceans (Milliman and Syvitski, 1992). During the present-day high sea-level stand, much of the particulate OC associated with riverine discharge is trapped and buried on continental shelves (Berner, 1982; Hedges, 1992). However, some rivers discharge much of their terrestrial OC load beyond the shelf due either to turbidity flows down submarine canyons (e.g., Congo, Ganges, Brahmaputra), to the presence of a narrow shelf (e.g., on the eastern flank of Papua New Guinea), or the influence of ice-rafting as an additional mode of sediment entrainment and export on polar margins (e.g., Macdonald *et al.*, 1998).

Eolian fluxes. Eolian fluxes of organic matter from land to sea are much less well constrained than riverine inputs. They have been estimated to be <0.1 Gt yr^{-1} (Romankevich, 1984). While these flux estimates imply lesser importance of eolian inputs compared to riverine OC contributions, this mode of delivery may be significant in a regional context. In particular, marine locations downwind from major dust sources (principally in eastern Asia and western Africa) are influenced profoundly by eolian inputs of OC and other detrital components. In addition, eolian transport can deliver terrigenous materials to remote locations of the oceans, far from the influence of rivers. For such regions (e.g., central equatorial Pacific Ocean) eolian OC fluxes may be important both in terms of POM in the water column and underlying sediments (Gagosian and Peltzer, 1986; Zafiriou *et al.*, 1985; Prospero *et al.*, 2003; Eglinton *et al.*, 2002).

6.06.2.2.2 Water column fluxes and the burial of organic carbon in sediments

The turnover time and fluxes of DOC into the ocean are obtained by comparing the reservoir size and radiocarbon age. The ocean inventory of DOC is ~680 Gt, and nearly all of this carbon resides in the deep sea, where concentration profiles and radiocarbon values are constant with depth. DOC ages by ~1,000 yr as deep seawater moves from the Atlantic to the Pacific Basin, but even in the Atlantic, DOC radiocarbon values are significantly depleted relative to dissolved inorganic carbon (DIC) (Druffel *et al.*, 1992). DOC persists in seawater through several ocean mixing cycles. The average age of DOC in the deep ocean is ~5,000 yr. Assuming a steady-state ocean, and that all this carbon is synthesized with a radiocarbon age equivalent to atmospheric carbon dioxide via marine production, then the annual flux of DOC into and out of the deep-sea reservoir is ~0.1 Gt C yr^{-1}. This flux is comparable to the delivery of terrestrial DOC to the ocean by rivers, but very small compared to annual marine production (60–75 Gt C yr^{-1}). Only 0.1–0.2% of annual marine production needs to be fixed into the permanent reservoir to maintain it, making the processes that sequester and remove carbon nearly impossible to track.

Marine photosynthesis by unicellular phytoplankton produces OC at a comparable rate to land plants (Hedges, 1992). Only ~10% of the net primary production escapes the upper 100 m of the water column. This vertical export occurs in the form of sinking fecal pellets produced by zooplankton that graze upon the phytoplankton, and as aggregates of cellular debris ("marine snow") (Alldredge *et al.*, 1993; Alldredge and Silver, 1998). The rain of particulate organic carbon (POC) out of the surface ocean attenuates exponentially through the water column, and only ~10% of the OC sinking out of the euphotic zone reaches an average seafloor depth of 4,000 m (Suess, 1980). Subsequent to losses in organic matter through the water column, a further 90% or more of that deposited on the seafloor is degraded, leaving ~0.1% of organic material originally synthesized in the surface ocean to be ultimately preserved in sediments underlying most of the open ocean (Wakeham *et al.*, 1997). Global burial efficiencies exceed 0.1%, because a significant amount of organic matter is deposited on continental margins and in oxygen-deficient regions where burial efficiencies are considerably higher (Berner, 1989). Estimates for the global rate of OC burial in marine sediments range from 0.1 Gt yr^{-1} to 0.6 Gt yr^{-1} (Berner, 1989).

In addition to vertical transport, export of OC from the margins to the ocean interior is being increasingly recognized as of significance (Bauer and Druffel, 1998; Thomsen and Van Weering, 1998; Ransom *et al.*, 1998a). Some regions of the coastal ocean produce more OC than they respire (Smith and Hollibaugh, 1993), suggesting that a fraction of this nonrespired, unburied OC is available for export from margins to the deep ocean (Wollast, 1991). Lateral transport of organic matter from margins to pelagic and abyssal environments has also been invoked to help explain carbon and oxygen imbalances in the deep ocean (Smith *et al.*, 1994; Jahnke *et al.*, 1990). Radiocarbon studies also provide evidence for basin-ward export of OM from the ocean margins. For example, Bauer and co-workers (Bauer and Druffel, 1998; Bauer *et al.*, 2001) observed suspended POM (SPOM) (and DOM) in Mid-Atlantic Bight slope and rise waters that are concurrently older and higher in concentration than in the adjacent North Atlantic gyres. While there are several potential origin(s) for this old, ^{14}C-depleted carbon, sediment resuspension and advection from the shelf and upper slope (Anderson *et al.*, 1994; Churchill *et al.*, 1994) is a likely explanation.

In addition, the chemical nature of advected particulate matter may favor its preservation over biogenic debris directly produced in surface waters. Even when vertical transport of recently produced surface ocean-derived material is rapid (e.g., seasonal thermocline breakdown, rapidly sinking POM), this fresher material may be more susceptible to degradation relative to older, margin-derived material. Thus, the ^{14}C age and concentration of suspended POC in the deep ocean may be maintained by greater relative inputs from the margins than from recent surface production, and may be partly responsible for the apparent old age of POC observed in deep North Atlantic and Pacific central gyres. It remains uncertain whether or not the presence of "old" POC and DOC in slope and rise waters (Bauer and Druffel, 1998) reflects pre-aged terrigenous organic matter (from continental soils or sedimentary rocks) rather than organic matter of marine origin produced on the margin and temporarily sequestered in shelf and upper slope sediments.

6.06.3 THE NATURE AND FATE OF TERRIGENOUS ORGANIC CARBON DELIVERED TO THE OCEANS

6.06.3.1 Background

A long-standing paradox evident from global OC flux estimates (Figure 1) is that while the combined global discharge of particulate and dissolved OC from rivers is twice the OC burial rate in marine sediments, OM in both the water column and underlying sediments is apparently dominated by autochthonous inputs (e.g., Hedges and Mann, 1979; Gough *et al.*, 1993; Aluwihare *et al.*, 1997). This implies that most terrigenous OM delivered to the oceans must be efficiently mineralized, and that the ocean is operating as a net heterotrophic system, accumulating less sedimentary OC than it receives via riverine discharge alone (Smith and MacKenzie, 1987). Potential explanations for this paradox are emerging as a result of recent studies, and our improved understanding of the composition of OM in the ocean.

There are several unresolved issues underlying present estimates and assumptions on the abundance and fate of terrigenous organic matter in the oceans. The first is the composition and proportion of terrigenous OC that is buried in margin sediments proximal to the continental source, a second is the proportion of terrestrial OC that enters the deep ocean, and a third is the variability in input composition and flux, mode of delivery, and geographic distribution.

6.06.3.2 Terrestrial Organic Matter in River Systems

One particularly intriguing question is whether a significant component of deep-sea DOC is of terrestrial origin. The annual flux of DOC through rivers is of the same magnitude as the annual flux of DOC out of and into the deep ocean reservoir, and measurements of $DO^{14}C$ in rivers show the carbon to have largely modern radiocarbon values (see Figure 2). An annual flux of 0.1 Gt C with a modern radiocarbon value would support the deep-sea DOC reservoir. Lignin oxidation products, which are good biomarkers for terrestrial OC, have been measured at low concentrations in open ocean DOM, providing molecular-level confirmation of a terrestrial origin for at least some fraction of this carbon (Meyers-Schulte and Hedges, 1986; Opsahl and Benner, 1997). Terrestrial carbon could also enter the DOC reservoir by desorption from POM. As discussed later in this chapter, POM enters the ocean with a coating of terrestrial OC that is rapidly replaced by marine OC. The load of organic matter introduced by POM could be injected into seawater as DOM. Radiocarbon values for riverine POM are highly variable, but are often depleted relative to modern carbon. If new DOC is pre-aged in this manner, then the annual flux of DOC would be in excess of 0.1 Gt C yr^{-1}.

Marine DOC has stable carbon isotope values between −21‰ and −22‰ (Druffel *et al.*, 1992) consistent with a largely marine source. While these data seem to exclude a significant contribution from C3 terrestrial plants, there is increasing evidence for an important contribution from C4 plants to persistent POM in marine sediments on the continental shelf and slope (see below). Desorption of C4 plant carbon and incorporation into oceanic DOC would be difficult to detect by isotopic or molecular biomarker analyses.

The sources, abundances, and compositions of SPOM carried by rivers vary significantly, depending on the characteristics of the drainage basin (Onstad *et al.*, 2000; Raymond and Bauer, 2001a). For example, Raymond and Bauer (2001b) demonstrate significant spatial and temporal variation in $\Delta^{14}C$ and $\delta^{13}C$ values ($\Delta^{14}C$ is the measured ^{14}C concentration normalized to pre-industrial atmospheric values following Stuiver and Pollach (1977). $\Delta^{14}C = (\%\text{modern} \times e^{\lambda t} - 1) \times 1{,}000$, reported in per mil (‰), where $\lambda = {}^{14}C$ decay constant and t = calendar age) of riverine dissolved and particulate OM, reflecting varying source and ages of terrestrial and aquatic productivity. As an example of the heterogeneity in

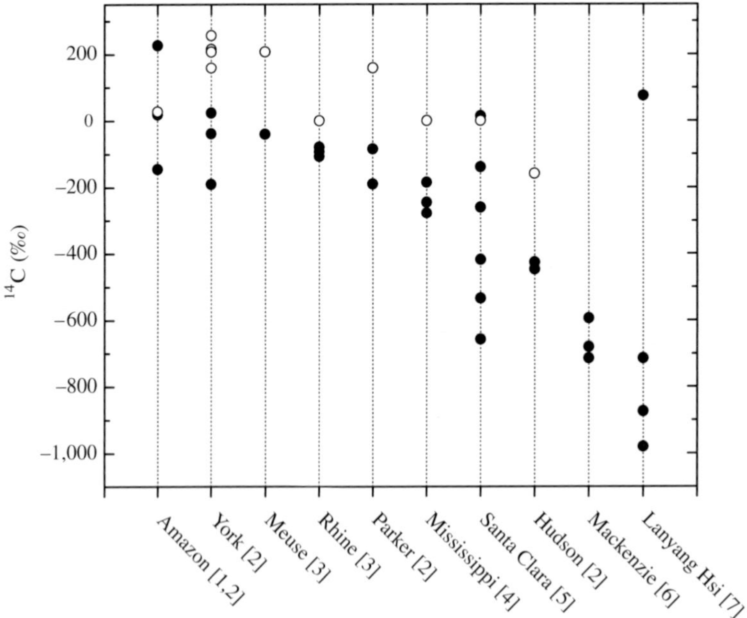

Figure 2 Variations in the ^{14}C content, expressed as $\Delta^{14}C$ (‰) of DOC (open symbols) and suspended POC (closed symbols) from a range of river systems. Data from: [1] Hedges *et al.* (1986); [2] Raymond and Bauer (2001b); [3] Megens *et al.* (2001); [4] Goni *et al.* (unpublished); [5] Masiello and Druffel (2001); [6] Eglinton *et al.* (unpublished); [7] Kao and Lui (1996).

SPOM signatures, Figure 2 shows the radiocarbon contents (expressed as $\Delta^{14}C$) of suspended OC reported for a range of river systems. Some systems transport predominantly "fresh" carbon as indicated by the presence of "bomb" ^{14}C ($\Delta^{14}C$ values greater than 0‰ reflect the atmospheric signature from aboveground nuclear weapons testing during the 1950s and early 1960s), while others carry carbon that is dominated by material older than 5×10^4 yr ($\Delta^{14}C$, −1,000‰). For some river systems, the ^{14}C contents of SPOM vary temporally, presumably reflecting variations in sediment provenance, mode of erosion or other characteristics of the drainage basin. Thus, contrary to earlier notions that rivers exclusively export ^{14}C-enriched OM to the ocean (Hedges *et al.*, 1986), many rivers export a significant fraction of old, ^{14}C-depleted DOC and POC to the oceans (Raymond and Bauer, 2001a). This old SPOM could reflect pre-aged, vascular plant-derived OC stored in an intermediate reservoir (e.g., soils), and/or contributions of relict OC from sedimentary rock weathering, and/or contributions from aquatic production utilizing old DIC (the "hardwater" effect).

The age variations highlighted above are undoubtedly coupled to differences in the chemical composition and reactivity of SPOM. Onstad *et al.* (2000) examined elemental, stable carbon isotope and lignin phenol characteristics of SPOM from rivers draining the south central US. Variations in $\delta^{13}C$ values, ranging from −18.5‰ to −26.4‰, were attributed to the contributions from C3 and C4 plants in the catchment area, and hence to temperature and hydrologic patterns in the drainage basin. Lignin–phenol compositions reflect degraded, angiosperm-rich vegetation. Results from this and other studies indicate that highly degraded soil OM is a major component of fine-grained POM transported by rivers, and that most riverine OM residing in the particulate fraction is associated with mineral phases.

In large fluvial systems, estuaries and deltas serve as the interface between the rivers and the ocean, and are sites of intensive organic matter reworking and production (Hedges and Keil, 1999). Recent work has revealed that extensive removal of terrestrial OC from suspended particles occurs at these locations. In one approach to quantify terrestrial OC losses, Keil *et al.* (1997) argued that detrital mineral surface area can serve as a conservative tracer for riverine discharged POM. Accordingly, changes in OC to specific mineral surface area (OC : SA) ratios should indicate net OC exchange between upstream and downstream locations. When applied in conjunction with $\delta^{13}C$ as a source indicator of marine and terrestrial OC, this can yield estimates of the fraction of terrestrial OC entering the ocean as riverine SPOM that is deposited in deltaic systems (Figure 3). Using this approach, Keil *et al.* (1997) calculate average OC loadings within the Amazon River (0.67 mg C m^{-2}) that are approximately twice those of the Amazon Delta (0.35 mg C m^{-2}), while $\delta^{13}C$ measurements suggest that approximately two-thirds of the TOC in deltaic sediment is terrestrial. Together, these data imply that >70% of the Amazon fluvial POM evades sequestration in deltaic sediments. Extrapolating losses of riverine SPOM for a range of river/delta systems (Columbia, USA;

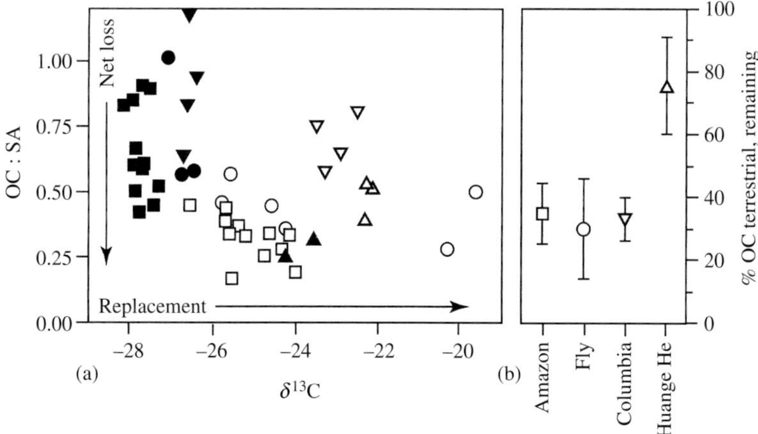

Figure 3 Loss of terrestrial OC in deltaic systems. (a) Organic carbon to mineral surface area ratio (OC : SA) plotted against bulk stable carbon isotopic compositions for riverine suspended sediments (closed symbols) and deltaic surface sediments (open symbols). A shift to lower OC : SA values indicates net loss of organic matter, and a shift to heavier (i.e., ^{13}C-enriched) isotopic compositions indicates increasing contributions from marine organic matter. (b) The average (±1 SD) total amount of terrestrial OC persisting in deltaic sediments, based on the changes in OC : SA and $\delta^{13}C$ composition between river suspended sediments and deltaic sediments for four river systems (after Keil *et al.*, 1997).

Fly, New Guinea; and Huange-He, China), Keil *et al.* (1997) calculate a global loss of fluvial POM in delta regions of ~0.1 Gt.

The magnitude of this loss is thus substantial, and comparable to flux estimates for the delivery of terrigenous OC to the oceans (Figure 1). However, while the above studies imply low burial efficiencies for fluvial POM in deltaic environments, it is uncertain whether the apparent losses of riverine POM reflect its complete mineralization or export to the ocean interior either in dissolved or particulate form (Edmond *et al.*, 1981). Moreover, the extent of terrestrial OC export and burial from river systems that do not form deltaic deposits is less well constrained.

6.06.3.3 Quantitative Importance of Terrigenous Organic Carbon in Marine Sediments

There is still much debate about the abundance and composition of terrestrial OC in both margin and pelagic sediments. Terrigenous organic matter contributions may be substantial for several reasons.

- As pointed out above, the flux of particulate OC emanating from the continents carried by rivers alone is sufficient to account for the burial of OC in marine sediments.
- Most OC burial in surface sediments occurs along continental margins, proximal to the supply of terrestrial OC.
- Much of the terrestrial OC delivered to the oceans will have been subject to extensive degradative processes in soils and rivers, implying that the fraction that survives these processes might be refractory with high potential for preservation.

Despite these considerations, terrestrial OC contributions to the marine sedimentary OM pool have generally been considered to be low. These conclusions are based on compositional studies, including bulk stable carbon isotopic (δ^{13}C) and elemental (atomic $C_{organic}/N_{total}$) and biomarker compositions. Loss of terrestrial OC in deltaic systems has already been discussed (Keil *et al.*, 1997). Hedges and Parker (1976) also observed sharp reductions in the abundance of lignin-derived phenols (as determined by yields after CuO oxidation) in Gulf of Mexico sediments with increasing distance from riverine sources (Figure 4). Compositional information stemming from the same measurements indicates that the lignin present in these samples originates from nonwoody angiosperm sources (e.g., grasses), and is highly modified. Both the terrestrial biomarker abundances and bulk stable carbon isotopic compositions (Hedges and Parker, 1976) are in accord with the replacement of terrestrial C3 vegetation with marine OC for sediments deposited progressively further offshore (Figure 4). Thus, despite proximity of a major river system (Mississippi/Atchafalaya), an important source of terrigenous sediment to the shelf, bulk sedimentary OM composition fails to indicate the presence of significant terrestrial input. Moreover, Gough *et al.* (1993) observed only trace quantities of lignin phenols in abyssal North Atlantic Ocean sediments.

However, some studies indicate that data of the type reported above may lead to underestimates of the proportion of terrigenous organic matter in marine sediments. For example, Onstad *et al.* (2000) showed that the Mississippi currently discharges isotopically "heavy" (δ^{13}C ~−20‰), lignin-poor POM that is difficult to distinguish from marine plankton remains in sediments from the Gulf of Mexico (Figure 4). Moreover, Goni *et al.* (1997, 1998) analyzed surface sediments from two offshore transects in the northwestern Gulf of Mexico using a range of techniques, including compound-specific δ^{13}C analysis of lignin-derived phenols. Bulk OC radiocarbon analyses of core top sediments yield depleted Δ^{14}C values, indicating that a significant fraction of the sedimentary carbon is "pre-aged," and most likely of allochthonous origin. Lignin phenol δ^{13}C values for inner shelf sediments are relatively depleted (average, −26.3‰), consistent with C3 vascular plant inputs, but are markedly enriched in ^{13}C at the slope sites (average, −17.5‰ for the two deepest samples) (Figure 4). Goni *et al.* (1997) interpret these molecular and isotopic compositions to indicate that a significant fraction (>50%) of the lignin, and by inference land-derived OC, in slope sediments ultimately originated from C4 plants. Consistent with Onstad *et al.* (2000) the source of this material is likely to be soil organic matter eroded from the extensive grasslands of the Mississippi River drainage basin. The mixed C3 and C4 land plant sources, the highly degraded state of this material, and the differential transport effects (i.e., winnowing, resuspension, and lateral advection through nepheloid transport) hampers its recognition and quantification in shelf and slope sediments. These findings bear upon other river-dominated margins where drainage basins may include a significant proportion of C4 vegetation, and highlight the difficulty in quantitatively assessing terrigenous OC inputs.

Prahl *et al.* (1994) also concluded that terrestrial OC contributes significantly to Washington Margin sediments. These authors determined bulk elemental and stable carbon isotopic compositions and concentrations of a range of vascular plant biomarkers (epicuticular wax-derived *n*-alkanes, lignin-derived phenols and cutin-derived hydroxyalkanoic acids) for sediments from the Columbia

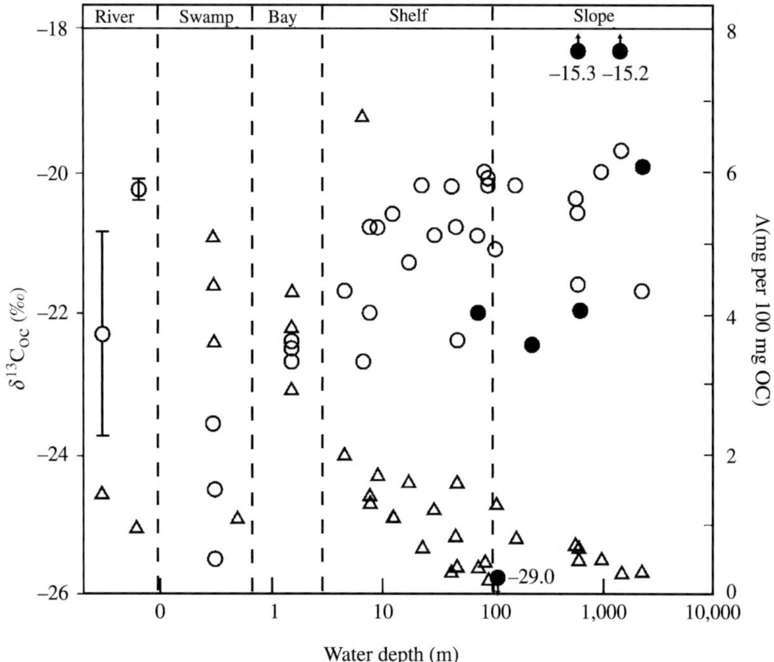

Figure 4 Variations in $\delta^{13}C$ of total OC (left axis: ‰, open circles) and the abundance of lignin-derived syringyl and vanillyl phenols (right axis: Λ, mg per 100 mg OC, triangles) in surface sediments from various water depths of the Gulf of Mexico in the vicinity of the outflow from the Mississippi River system (after Goni *et al.*, 1997). Also shown are average values for the same parameters determined on suspended particulate matter from the upper and lower reaches of the Mississippi River (extreme left column, error bars = 1 SD; data from Onstad *et al.*, 2000). In addition, $\delta^{13}C$ values are indicated for one lignin phenol (syringic acid filled circles), for selected shelf and slope samples (after Goni *et al.*, 1997).

River basin and adjacent margin. Using end-member values determined empirically by two independent means, Prahl *et al.* (1994) estimated terrestrial OC contributions of ~60%, 30%, and <15% for sediments on the shelf and slope of the Washington margin, and the adjacent Cascadia Basin, respectively.

As noted earlier, prior studies have focused on only a few major river systems (e.g., Mississippi, Amazon, Columbia) that may not be representative of fluvial inputs worldwide. While our perspective on terrigenous OC inputs has been biased towards large river systems, there is a growing body of evidence that the majority of terrestrial OC delivery to the oceans occurs via numerous small rivers draining mountain regions in the tropics (Milliman and Syvitski, 1992). The contrasting modes of OC delivery between large deltas (passive, hydrodynamic control) and small mountainous rivers (episodic, high-energy, poorly sorted particles) could have a significant influence on the proportion and location of terrestrial OC that is deposited in marine sediments (Leithold and Blair, 2001; Masiello and Druffel, 2001). In addition to terrestrial OC inputs at low latitudes, our knowledge of terrestrial OC inputs to the high-latitude oceans is limited. Here, ice-rafting serves as another mechanism for export of terrestrial OC to the ocean basins. While some data indicates that terrestrial OC may indeed comprise a substantial fraction of the OM buried in Arctic sediments (Schubert and Stein, 1996; Macdonald *et al.*, 1998), traditional tools for source apportionment are not well suited to address this problem. For example, bulk stable carbon isotopes are of limited use as source indicators because of the isotopically depleted nature of phytoplankton in the polar oceans (Rau *et al.*, 1989; Goericke and Fry, 1994). Similarly, while eolian OC inputs to the oceans are generally considered to be minor, major dust fluxes emanating from West Africa carry a substantial C4 plant signature (Huang *et al.*, 2000; Eglinton *et al.*, 2002), confounding estimates of marine and terrestrial OC in underlying sediments based on bulk stable carbon isotopic measurements.

6.06.3.3.1 Black carbon

One group of exclusively terrestrially derived organic components found in marine sediments which has been the subject of renewed interest in recent years are the carbonaceous particles from incomplete combustion processes. These products, collectively termed "black carbon" (Goldberg, 1985), are ubiquitous in the environment, and may comprise a significant fraction of OC in contemporary marine sediments (Middelburg *et al.*, 2000; Masiello and Druffel, 1998; Gustafsson and Gschwend, 1998). Schmidt and Noack (2000) reviewed the state of knowledge of BC in soils and sediments. BC can be produced via condensation of volatiles to highly graphitized particles ("soot-BC") or by formation of solid residues through direct carbonization of particulate plant material ("char-BC"). Both forms of BC are relatively inert and are distributed globally by water (fluvial) and wind (atmospheric) transport. Although BC is directly emitted and transported during biomass burning, available radiocarbon data suggest that some BC is sequestered in soils and aquatic sediments for millennia prior to subsequent export to the oceans (Masiello and Druffel, 1998; Eglinton *et al.*, 2002).

Limited progress has been made on standardizing techniques for the quantitation and characterization of BC. These techniques and definitions have primarily been developed with regard to BC in atmospheric particles. The problem of defining and quantifying BC in sediments adds a further layer of complexity, and hence reservoir sizes and fluxes are subject to considerable uncertainty. Current estimates of global BC production (0.05–0.27 Gt yr^{-1}; Kuhlbusch and Crutzen, 1995) are of the same order as those of riverine input of POC to the ocean and OC burial in marine sediments. If these flux estimates are correct, BC may represent a significant sink in the global OC cycle (Kuhlbusch, 1998), and constitute a significant fraction of the carbon buried in soils (Skjemstad *et al.*, 1996; Schmidt *et al.*, 1999) and marine sediments (Middelburg *et al.*, 2000; Masiello and Druffel, 1998).

Gustafsson and Gschwend (1998) determined concentrations and fluxes of BC in modern ocean margin sediments off northeastern USA using a thermal oxidation method (Gustafsson *et al.*, 1997). Core-top BC concentrations indicate that BC comprises a significant component of OC budgets for coastal sediments. Down-core trends were consistent with anthropogenic fossil-fuel combustion dominating BC input, while the fractional abundance of BC increased in deeper sediment sections, implying that it is resistant to degradation and may be selectively preserved (Wolbach and Anders, 1989). Suman *et al.* (1997) calculated the pre-industrial global burial of BC to be ~10 Tg yr^{-1}, which corresponds to ~11% of the estimated global marine sediment burial of TOC.

Middelburg *et al.* (1999) compared BC determined using thermal and chemical (hot nitric acid) oxidative pre-treatments in a range of marine sediments. They found that the latter significantly overestimates combustion-derived phases. Nevertheless, the lower BC estimates obtained from the thermal oxidation method account for between 15% and 30% of total OC. Examination of BC concentrations across a relict oxidation front in a Madeira Abyssal Plain turbidite provided evidence for significant BC degradation under prolonged exposure to oxygenated bottom waters.

Masiello and Druffel (1998) measured the abundance and radiocarbon content of BC (isolated by wet chemical oxidation) in sediment cores from two deep Pacific Ocean sites. They found BC comprises 12–31% of the total sedimentary OC, and was between 2,400 ^{14}C yr and 13,000 ^{14}C yr older than non-BC sedimentary OC (Figure 5). For sediment intervals deposited prior to the Industrial era (i.e., free of BC inputs from fossil fuel utilization), the authors argue that the older ages for BC must be due to storage in an intermediate reservoir before deposition. Possible intermediate pools are oceanic DOC and terrestrial soils. They conclude that if DOC is the intermediate reservoir, then BC comprises 4–22% of the DOC pool. If soils are the intermediate reservoir, then the importance of riverine OC has been underestimated.

6.06.4 A BIOPOLYMERIC ORIGIN FOR OCEANIC DISSOLVED ORGANIC MATTER

6.06.4.1 Background

As deep water upwells to the surface, some 30–40 μM carbon is added to the dissolved phase (Peltzer and Hayward, 1995). Although the mechanisms that fix carbon into DOC are not known, it has long been recognized that this new carbon represents water-soluble by-products of algal photosynthesis (Duursma, 1963; Menzel, 1974). Phytoplankton in laboratory culture and in seawater have been shown to exude OC (Hellebust, 1965; Iturriaga and Zsolnay, 1983), and the release of dissolved OC through exudation, grazing, cell lysis, or other processes fuels much of the microbial growth and respiration in the marine environment (Hellebust, 1965; Nagata and Kirchman, 1992). Stable isotope measurements of carbon standing stocks throughout the entire water column confirm an autochthonous origin for marine DOC, with values identical to marine particulate matter (−21‰ to −22‰) (Druffel *et al.*, 1992). As DOC cycles through the water column, new carbon is somehow altered and sequestered into the more permanent, deep ocean carbon reservoir.

Although DOC production is a function of primary production, DOC accumulation is not. There are a few reports of transient, localized increases in DOC inventory following algal blooms (Holmes *et al.*, 1967), and annual cycles

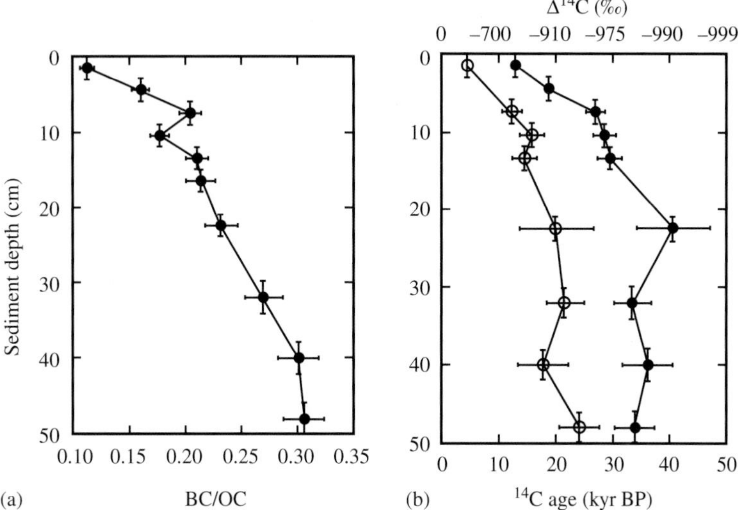

Figure 5 Abundance of radiocarbon age of black carbon in slowly accumulating (~2.5 cm kyr^{-1}) deep-sea sediments from the Southern Ocean (54 °S 176° 40′ W): (a) a plot of the ratio of black carbon to total organic carbon (BC/OC) with sediment depth and (b) Δ^{14}C (per mil) and ^{14}C age (kyr BP) of BC (solid symbols) and non-BC sedimentary OC (open symbols) as a function of depth (after Masiello and Druffel, 1998).

of DOC inventory in the upper ocean share features in common with annual cycles of primary production (Duursma, 1963; Holmes *et al.*, 1967; Carlson *et al.*, 1994). However, DOC concentrations are not well correlated with either phytoplankton biomass or primary productivity (Carlson *et al.*, 1994; Chen *et al.*, 1995). The accumulation of DOC in surface seawater results from a decoupling of production and removal processes over annual to decadal timescales, and it is convenient to subdivide DOC into at least three distinct reservoirs of differing reactivity (Carlson and Ducklow, 1995). The largest reservoir in terms of *production* is very reactive, and supports most secondary production in the ocean. This fraction of DOC consists of a few μM C of soluble biochemicals (proteins, carbohydrates, lipids, etc.), which has a turnover time of hours to days, and may equal 10–30% of total primary production (Vaccaro *et al.*, 1968; Mague *et al.*, 1980; Jørgensen *et al.*, 1993). The largest fraction of DOC in terms of total inventory is the nonreactive fraction. Nearly all the DOC in the deep ocean, and half the DOC in the surface ocean, is considered to be nonreactive based on radiocarbon measurements, which show this fraction to be 4,000–6,000 yr old (Williams *et al.*, 1969; Druffel *et al.*, 1992). This nonreactive component of DOC persists through several ocean-mixing cycles before remineralization or sequestration as particulate OC. Finally, the 30–40 μM DOC equal to the difference between deep and surface seawater DOC concentration values is considered to be reactive DOC, and cycles on seasonal to decadal timescales. Reactive DOC is produced in the surface ocean, and largely consumed as surface water is subducted into the deep ocean. The composition, cycling, and fate of reactive DOC has been the focus of much of the research on DOC completed since the 1990s.

6.06.4.2 High Molecular Weight Dissolved Organic Matter: Biopolymers or Geopolymers?

The very old radiocarbon age of DOC was originally attributed to the structural complexity of the organic constituents that make up this fraction of marine organic matter. Simple biochemicals produced by marine bacteria, algae, and animals were thought to react through abiotic or geochemically mediated reactions to form more complex humic-like substances that are metabolically unavailable to marine microorganisms (Gagosian and Stuermer, 1977). The concept of organic matter humification in seawater follows earlier models of humification reactions in soils, and early studies on marine DOC sought to characterize and compare marine humic substances with humic substances extracted from terrestrial and freshwater environments in an attempt to better understand the reactions in each environment (Hatcher *et al.*, 1980; François, 1990) These studies show marine DOM to have a low average molecular weight (<1,000 Da), to be relatively rich in nitrogen and aliphatic carbon, and poor in aromatic carbon (Stuermer and Harvey, 1974; Stuermer and Payne, 1976). Chemical differences between marine and terrestrially derived humic substances could be attributed directly to the nature of biochemicals prevalent in each environment, and the differences in reaction conditions (e.g., light, presence of catalytic surfaces) (François, 1990; Malcolm, 1990). Marine and terrestrial humic substances represent two, chemically distinct, carbon reservoirs.

Marine humic substances are isolated by adsorption onto hydrophilic resins and contribute only a small fraction (10% or less) of total marine DOC. The difficulty in sampling DOC in the presence of much more abundant inorganic salts significantly slowed progress in marine DOM research for over a decade. Application of ultrafiltration, and especially large volume ultrafiltration made a much larger fraction (up to 40%) of DOC available for study (Buesseler *et al.*, 1996). Ultrafiltration does not select strongly for the chemical characteristics of the organic matter as does resin adsorption, but selects on the basis of molecular size instead, and therefore preferentially isolates the high molecular weight (HMW) fraction of DOM. Initial studies of HMW DOM composition showed this fraction to be chemically distinct from humic substances (Benner *et al.*, 1992).

Our understanding of DOM composition and cycling has undergone a rapid change since the 1990s. Chemical studies of HMW DOM now show a composition that is rich in specific polysaccharides and proteins and remarkably uniform across diverse environments. These discoveries led Aluwihare *et al.* (1997) to propose that a major fraction of HMW DOM arises directly from biosynthesis. The concept that marine DOM has a large component of metabolically resistant bioploymers is a sharp departure from earlier ideas that described DOM as a mixture of simple biomolecules that had experienced abiotic transformation (geopolymerization) into HMW substances (fulvic and humic substances). Support for the directly formed biopolymer hypothesis comes from the chemical composition of HMW DOM itself.

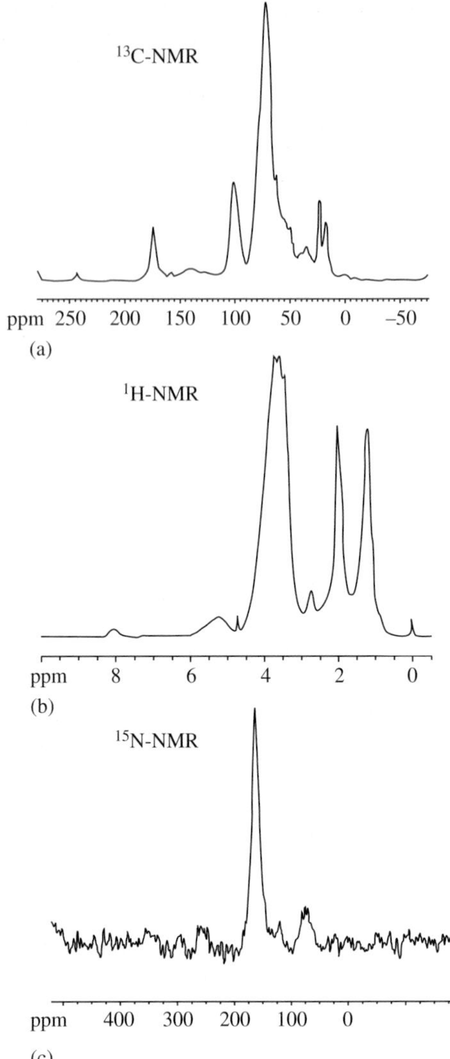

Figure 6 NMR spectra of HMW DOM from surface seawater. (a) ^{13}C-NMR spectra can be used to quantitatively determine the functional groups, and by inference, the relative importance of different biochemical classes in HMW DOM. The spectra highlight the importance of carbohydrates (100 ppm and 70 ppm), carboxylic acids (175 ppm), and alkyl carbon (10–30 ppm). (b) ^{1}H-NMR also show the importance of carbohydrates (4 ppm) and alkyl carbon (1 ppm), but additionally show that acetyl groups most likely bound to carbohydrate are an important components. ^{15}N-NMR show that 80–90% of HMW DON is amide, while 10–20% is free amine. Quantitative analyses for acetate and nitrogen suggest that most amide in surface seawater is bound as *N*-acetyl amino sugars and protein residues. In the deep ocean HMW DOM however, most amide is nonhydrolyzable, and is of unknown molecular environment.

6.06.4.2.1 Acylpolysaccharides in high molecular weight dissolved organic matter

NMR spectroscopy has proven to be the most effective technique for characterizing carbon functional groups in HMW DOM. The ^{13}C-NMR spectrum of HMW DOM collected at 15 m in the North Pacific Ocean surface is given in Figure 6(a). Nearly identical spectra have been collected from the Atlantic Ocean, as well as from some lakes and rivers (McKnight *et al.*, 1997; Repeta *et al.*, 2002). All ^{13}C-NMR spectra display a rather simple pattern of broad resonance from carboxyl (CO–(OH or NH); 175 ppm), alkene/aromatic (C=C;140 ppm), anomeric (O–C–O; 100 ppm), alcoholic (H–C–OH; 70 ppm), and alkyl (CH_x;10–40 ppm) carbon (Figure 6(a)). From these spectra we can quantify the relative amounts of carbon associated with each class of functional group, and by inference, identify the major biochemical units in DOM. ^{13}C-NMR shows HMW DOM to be especially rich in carbohydrate (100 ppm (O–C–O), and 70 ppm (H–C–OH)), which accounts for 76% of the carbon in the spectrum. The ratio of alcohol to anomeric carbon (70–100 ppm) is 4.5, within the expected range for most sugars (4–5). The importance of carbohydrate is confirmed through molecular-level analyses of monosaccharides which show neutral sugars to be abundant in HMW DOM (Sakugawa and Handa, 1985; McCarthy *et al.*, 1994; Aluwihare *et al.*, 1997; Biersmith and Benner, 1998; Borch and Kirchman, 1998). However, the amount of carbohydrate determined by NMR is much higher than that measured by molecular-level techniques (76% versus 20%, respectively for Figure 6(a)). Reconciling NMR data and molecular-level techniques is currently a major challenge in understanding the composition of HMW DOM.

Sugar distributions are dominated by six neutral sugars—fructose, galactose, glucose, mannose, rhamnose, and xylose—which occur in approximately equimolar amounts. A seventh neutral sugar (arabinose), and two amino sugars (*N*-acetyl glucosamine and *N*-acetyl galactosamine) are also abundant, but occur at only half to one quarter the concentrations of other sugars. The relative amounts of these nine sugars, and their contribution to surface water HMW DOC vary remarkably little in fresh and marine waters (Figure 7). The uniformity in molecular and spectroscopic properties of many HMW DOM samples suggests that the fraction of HMW DOM enriched in carbohydrate, acetate, and lipids is a well-defined family of biopolymers synthesized by marine algae (Aluwihare *et al.*, 1997; Aluwihare and Repeta, 1999). These polymers, referred to as acylpolysaccharide (APS) are ubiquitous in natural waters (Repeta *et al.*, 2002).

The ^{13}C-NMR spectrum of HMW DOM also includes contributions from carboxyl (CO–(OH or NH), 5% of total carbon), and alkyl (CH_x, 14% total carbon) functional groups, which may derive from proteins, lipids, or carbohydrates (deoxy- and methyl sugars). Hydrolysis of HMW DOM followed by extraction with organic solvent yields 4–8% of the total carbon in HMW DOM as acetic acid. Acetyl is easily recognized in the ^{1}H-NMR spectra of HMW DOM, where it appears as a broad singlet centered at 2 ppm (Figure 6(b)). Free acetic acid and its derivatives are not retained by ultrafiltration, and the acetyl in HMW DOM must be covalently bound to macromolecular material, most likely as an *N*-acetyl amino sugar. Acetyl contributes up to half the carboxyl carbon in the ^{13}C-NMR spectrum.

Elemental C/N and C/P ratios of HMW DOM lie between 13–16 and 300–800, respectively (McCarthy *et al.*, 1996; Karl and Bjorkman, 2002).

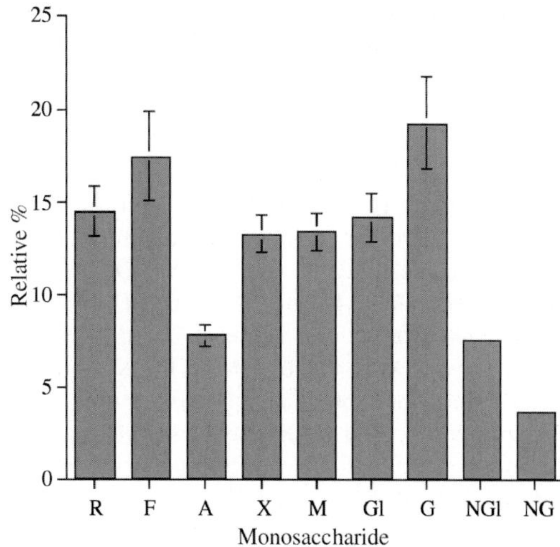

Figure 7 The distribution of neutral and basic monosaccharides in HMW DOM. Hydrolysis of HMW DOM followed by monosaccharide analysis show a remarkably uniform distribution of sugars in samples from surface and deep ocean waters in all ocean basins. Important neutral sugars include rhamnose (R), fucose (F), arabinose (A), xylose (X), mannose (M), glucose (Gi), and galactose (G). Error bars show 1 σ in relative abundance of 12 HMW DOM samples collected from the Atlantic Ocean and Pacific Ocean between 1994 and 2001. Little change in the relative abundance of sugars is observed between ocean basins. Important amino sugars include *N*-acetyl glucosamine (NGl) and *N*-acetyl galactosamine (NG), which are less abundant than the neutral sugars. If all acetate is bound as *N*-acetyl amino sugars, then monosaccharide analyses significantly under-recover the amount of amino sugars in HMW DOM.

Although C/P ratios are so high that they prohibit a major contribution from organophosphorous compounds to HMW DOC, C/N ratios are sufficiently low that N-containing biochemicals must contribute an important fraction of DOM carbon. Natural abundance ^{15}N-NMR measurements show that 80–90% of HMW dissolved organic nitrogen (DON) is amide (R-CON; Figure 6(c)) (McCarthy *et al.*, 1997). Free amine groups contribute the remaining 10–20% of HMW DON that can be observed by NMR. Small amounts of other components, such as aromatic *N*-containing compounds, may also be present but are below the limit of detection by this technique. The amount of nitrogen in HMW DON is approximately equal to the amount of carboxylic carbon. If nearly all *N* is amide, then conversely nearly all carboxyl carbon must be amide as well, and in agreement with molecular-level analyses which show only trace amounts of biomarker lipids (<0.1% total carbon), little carboxyl carbon can be components of other organic acids (fatty acids, etc.; Mannino and Harvey, 2000).

Several important biochemicals are amides, of which proteins and *N*-acetyl amino sugars are the most abundant in marine organisms, where proteins contribute up to 80%, and amino sugars up to 10% or the total nitrogen. Molecular-level analyses show both these biochemical classes to be present in HMW DON, but only at low concentrations. For most samples, amino acids account for <15–20% of HMW DON, and aminosugars <1% of the HMW DON (McCarthy *et al.*, 1996; Kaiser and Benner, 2000; Mannino and Harvey, 2000). Either HMW DON amide is derived from some other component, or the analytical protocols used in molecular-level analyses are not appropriate for measuring these biochemicals in HMW DON.

The discrepancy between molecular-level and NMR measurements of amide nitrogen can best be resolved by using the two techniques interactively. Proteins and *N*-acetyl amino sugars both yield amino compounds (free amino acids and amino sugars, respectively) on treatment with acid (Figure 8). Therefore, acid hydrolysis of HMW DON should be accompanied by a change in the ^{15}N-NMR chemical shift from amide to amine (Figure 9). *N*-acetyl amino sugars will also yield free acetic acid with hydrolysis, and the high concentrations of acetate in HMW DOM along with pyrolysis MS measurements of acetamide suggest that much of HMW DON is *N*-acetyl amino sugars (Boon *et al.*, 1998). The amount of acetate bound as amide can be quantified by comparing the amount of free acetic acid released by hydrolysis with the hydrolytic yield of free amine observed by ^{15}N-NMR. Treatment of HMW DOM with mild acid converts ~60–70% of amide to free amine. This amount equals the molar sum of free amino acids (10–20% total N, from protein hydrolysis) and acetic acid (90–50% total N, from *N*-acetyl amino sugars) measured by molecular-level techniques after HMW DON hydrolysis. The agreement between NMR and molecular-level analyses indicates that 40–50% of HMW DON is derived

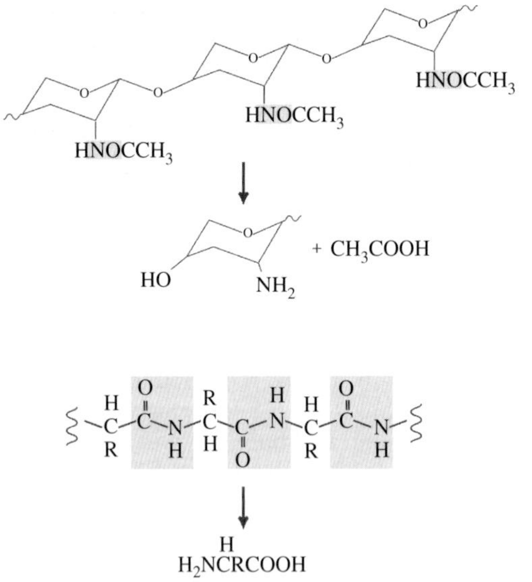

Figure 8 The effect of mild acid hydrolysis on amides in HMW DOM. Two potentially important classes of biochemicals that likely contribute to HMW DOM are (poly)-*N*-acetyl amino sugars (top) and proteins (bottom). Mild acid hydrolysis of (poly)-*N*-acetyl amino sugars will yield free acetic acid, but will not depolymerize the polysaccharide. The generation of acetic acid will be accompanied by a shift in the ^{15}N-NMR from amide to amine. In contrast, mild acid hydrolysis of proteins does not yield acetic acid, but may depolymerize the protein macromolecular segments to yield free amino acids. Free amino acids can be quantified by chromatographic techniques and compared to the shift from amide (protein) to amine (free amino acid) in ^{15}N-NMR.

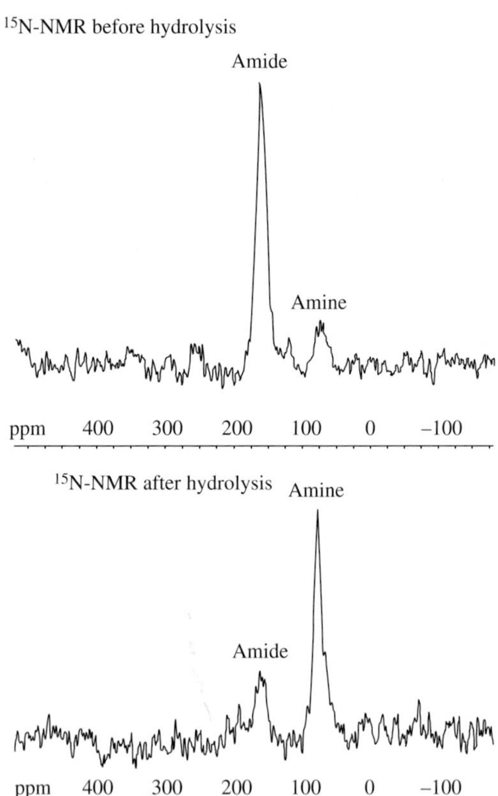

Figure 9 The effect of mild acid hydrolysis on ^{15}N-NMR of HMW DOM. Nitrogen in HMW DOM is primarily amide (180 ppm), with smaller amounts of free amine (90 ppm). Treatment of HMW DOM with dilute hydrochloric acid increases the amount of amine and decreases the amount of amide. The decrease in amide equals the amount of acetic acid and amino acids released by hydrolysis of poly-*N* acetyl amino sugars and proteins. The relative amount of protein and amino sugar can be determined by the ratio of acetic acid to amino acids in the hydrolysis product.

from *N*-acetyl amino sugars and that molecular-level analyses underestimate the amount of amino sugars in HMW DON by at least an order of magnitude.

Our knowledge of APS structure and composition is still evolving; however, Aluwihare and Repeta (1999) suggest that APS is a biopolymer that is largely, perhaps exclusively, composed of carbohydrate units. This biopolymer is approximately half neutral sugar and half *N*-acetyl amino sugar (Figure 10). Together these two components contribute 40–50% of the total carbon, 60–70% of the total carbohydrate, and an equal amount of the total new (or modern) carbon in HWM DOC (Druffel *et al.*, 1992; Guo *et al.*, 1994). Further work is needed to establish, on a molecular level, the abundance of *N*-acetyl amino sugars in APS. Are the neutral sugars and amino sugars covalently linked into a common macromolecule, and if so, what is the full structure of this polysaccharide?

6.06.4.2.2 Proteins in high molecular weight dissolved organic matter

A small fraction of the carbon, and a larger fraction of the nitrogen in HMW DOM, is protein. Tanoue *et al.* (1995) analyzed HMW DOM by gel electrophoresis and found a complex mixture of proteins with masses between 14 kDa and 66 kDa. Two major components were noted at 48 kDa and ~34–39 kDa. The 48 kDa protein was purified for *N*-terminal sequencing and shown to have significant homology with porin-P from the gram negative bacterium *Pseudomonas aeruginosa*. Tanoue's data are the most direct evidence to date that resistant biopolymers selectively survive degradation and accumulate as oceanic DOM.

Additional evidence for a bacterial contribution to HMW DOM proteins comes from molecular-level analyses of dissolved amino acids. Hydrolysis of HMW DON releases 11–29% of the nitrogen as amino acids (McCarthy *et al.*, 1996). Specific amino acids include common protein amino acids, as well as β-alanine and γ-aminobutyric acid which are nonprotein amino acid degradation products. The distribution of amino acids is similar to that of fresh plankton cells, suspended particulate matter, and total dissolved amino acids. However, stereochemical analyses show HMW DOM amino acids to be elevated in the D-enantiomer, with D/L ratios for alanine, aspartic acid, glutamic acids, and serine ranging from 0.1 to 0.5 (McCarthy *et al.*, 1998). Racemization of phytoplankton-derived L-amino acids is too slow at ocean temperatures to yield such high D/L ratios, but bacteria can synthesize D-amino acids, and it is likely that the D-amino acids in HMW DOM result from bacterial bioploymers rich in these particular amino acids. The high D/L ratios of some amino acids and the abundance of amide nitrogen in HMW DOM ^{15}N-NMR spectra led McCarthy *et al.* (1998) to

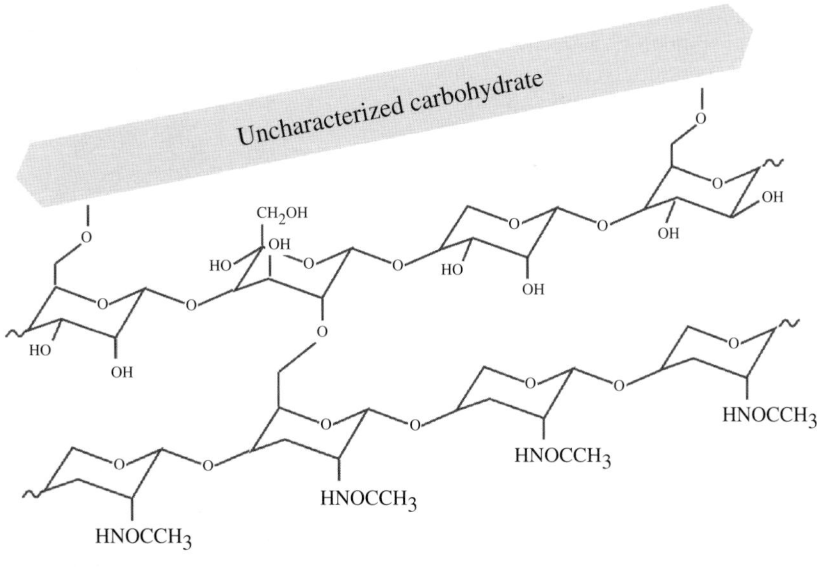

Figure 10 Hypothetical structure of APS in HMW DOM. Spectroscopic and molecular-level analyses suggest that approximately one-third of APS is neutral sugars, one-third amino sugars, and one-third unidentified carbohydrate. The consistency in monosaccharide distribution, and the fixed relative amount of acetate to total carbohydrate suggests these portions of HMW DOM are coupled into the same macromolecular structure. Further work needs to be done to establish the coupling of neutral and amino sugars directly, and to identify the unknown component of APS to bring molecular-level and spectroscopic measurements into better agreement.

postulate that peptidoglycan may be one such biopolymer that is significantly enriched in HMW DOM. Most gram-negative bacteria produce peptidoglycan as part of their cell membrane and, like porin-P discussed above, are therefore a potential source for HMW DON in the ocean.

Further chemical characterization of HMW DOM is needed to verify this hypothesis. Peptidoglycan is a polymer of *N*-acetyl glucosamine, muramic acid, and amino acids. *N*-acetyl glucosamine has already been identified by molecular-level techniques in HMW DOM. The concentration of *N*-acetyl glucosamine is low, but consistent with the amounts of D-amino acids in samples, assuming all D-amino acids are part of a peptidoglycan biopolymer. However, muramic acid has not been detected in HMW DOM samples, and the absence of muramic acid suggests pepidoglycan is not present (Repeta, unpublished). Lactic acid, a component of muramic acid, is recovered from HMW DOM in amounts of ~0.5% of total carbon, or again at levels consistent with peptidoglycan. The analytical protocols used to release lactic acid from HMW DOM are nonspecific, and it is not known whether lactic acid comes from muramic acid, or from the degradation of another polymer. Molecular-level techniques used to quantify muramic acid may suffer from the same uncertainties as those discussed previously for *N*-acetyl amino sugars, for example, current techniques may be inappropriate for the degradation resistant biopolymers in HMW DOM, and therefore they may not as yet provide accurate data on the distributions of molecular constitutents in HMW DOM. Assuming all alanine, serine, aspartic, and glutamic acid in HMW DOM is bound as peptidoglycan, then this bacterial polymer could contribute up to a maximum of 9% of the total HMW DOC. However, if these amino acids are assumed to occur only in porin-P or other bacterial proteins, then the contribution of bacterial carbon to HMW DOC would be much lower.

6.06.4.3 Gel Polymers and the Cycling of High Molecular Weight Dissolved Organic Matter

HMW biopolymers are part of the colloid-sized fraction of marine organic matter that can be visualized and enumerated with transmission electron microscopy (TEM) (Wells and Goldberg, 1991, 1993). Colloids range in size from a few nanometers up to ~1 μm in size. Very small colloids (<30 nm) are irregular in shape, while larger colloids (~30–60 nm) are more spherical assemblies of 2–5 nm sized subparticles. Concentrations of small colloids (<200 nm) range from nondetectable ($<10^4$ colloids ml^{-1}) to $>9\times10^9\ ml^{-1}$, making them the most abundant particles in seawater. Vertical profiles of colloids show a highly variable distribution throughout the water column, characteristic of a dynamic reservoir that is rapidly cycling (Wells and Goldberg, 1994). Colloids have low electron opacity by TEM, consistent with a largely organic composition, and slight enrichments of some trace metals (iron, cobalt, chromium, nickel, and vanadium) and other elements (silicon, aluminum, and calcium). Santschi *et al.* (1998) examined colloids using atomic force microscopy (AFM), which gives comparable results to TEM but has a greater resolving power for small colloids. Visualization of colloids by AFM shows fibrils, elongated particles ~1 nm in cross-section by 100–200 nm long, to be ubiquitous and a major component of colloidal material. Chemical analyses of samples artificially enriched with fibrils (through laboratory manipulation) show a parallel enrichment in carbohydrate, which may be the biochemical component of fibril material.

Colloids have been described as classic polymer gels, and their dynamics in seawater may be understood in terms of polymer gel theory (Chin *et al.*, 1998). Polymer gels are stable, three-dimensional networks of polymers and seawater (Figure 11). Gels assemble spontaneously from

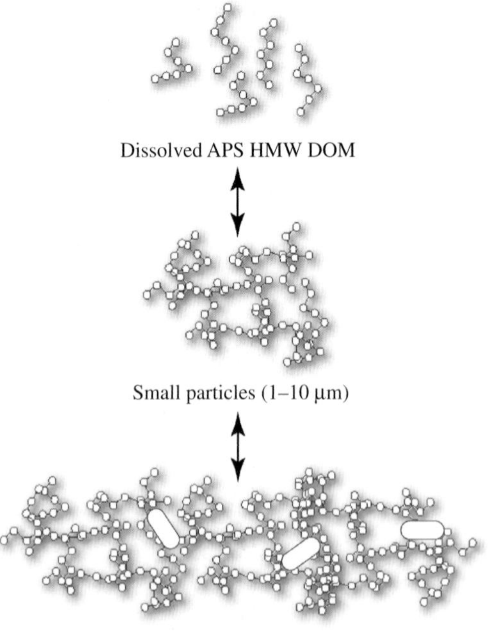

Figure 11 Formation of gel polymers by APS. The APS fraction of HMW DOM may entangle and spontaneously assemble into small particles. Further assembly will increase particle size and be accompanied by colonization by bacteria or incorporation into large, rapidly sinking particles.

DOM to form nanometer to micrometer sized particles. The kinetics of polymer gel assembly has been monitored by flow cytometry and dynamic laser scattering, which show a rapid formation of particles in filtered seawater. In less than 30 min DOM can assemble into micometer-sized particles that continue to grow as long as free polymers remain in solution (Chin *et al.*, 1998). Gels are stabilized by the presence of divalent cations, particularly Ca^{+2} and Mg^{+2}, and will disassemble in the presence of EDTA which out competes natural polymers for these cations. The composition of the gels is unknown, but the polymers that form the gels are presumed to be organic. Gels stain positively for a number of biochemicals, including carbohydrates, proteins, and lipids. The picture of colloid formation that emerges from these experiments is consistent with observations of natural colloids in seawater. Dissolved organic polymers, 1–10 nm in size, spontaneously assemble in seawater due to the relatively high concentrations of calcium ions present. Assembly is rapid and reversible, and the gel particles will grow to at least several microns in size. Particles may continue to grow until they sink out of the water column, or are removed by other processes. Polymer gel assembly couples the dissolved and particulate reservoirs of OC through the rapid exchange of colloids (Figure 12).

The polymers active in gel polymer assembly may be aged polysaccharides. Santschi *et al.* (1998) noted that fibril enriched samples of HMW DOM were rich in carbohydrate and radiocarbon. APSs are also rich in carbohydrate, and isotopic analyses of monosaccharides show this fraction to be enriched in radiocarbon relative to total DOC. APSs have a size distribution consistent with polymers that assemble into polymer gels, and are recovered from seawater as isolates rich in calcium. NMR spectra (Figure 13) and molecular-level analyses of 3 kDa, 10 kDa, and 100 kDa HMW DOM show the same major resonances and the same distribution of neutral sugars as1 kDa HMW DOM (Figure 7). It has not been established if these size fractions represent true APS polymers of increasing molecular weight, or assemblies of smaller (1–3 kDa) sized APS.

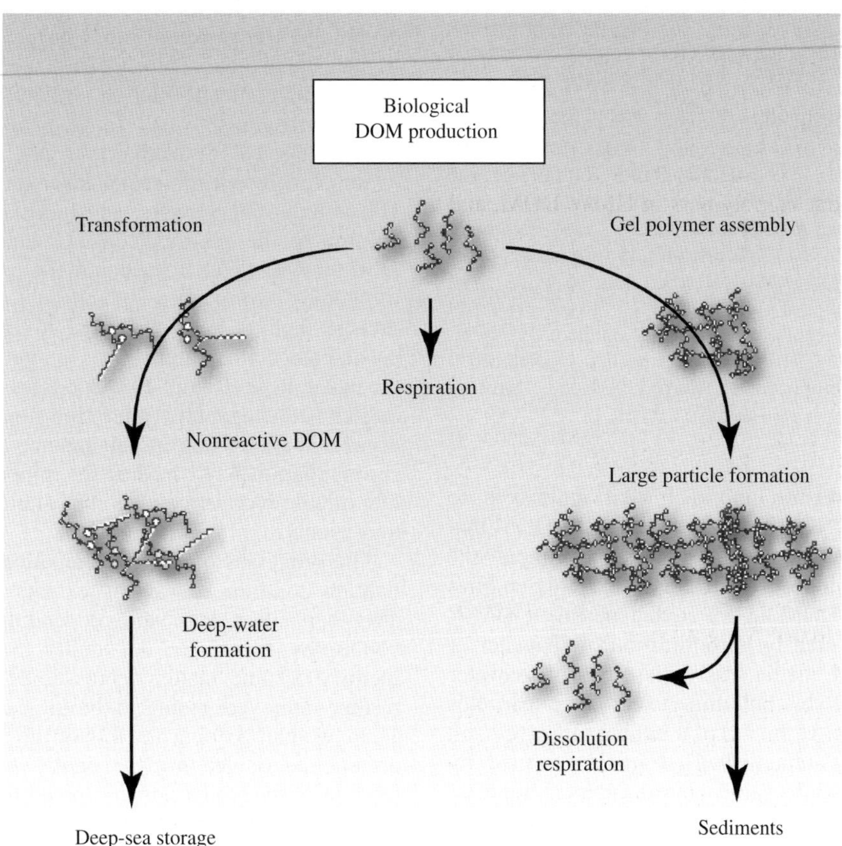

Figure 12 DOM cycling in the ocean. DOM is initially produced as a by-product of marine production in the mixed layer. Carbon stable isotope data suggests that some fraction of this organic matter becomes incorporated into the longer-term reservoir of rather refractory DOM to be subducted into the deep ocean. Newly produced DOM may also assemble into particles to enter the POM cycle. POM is oxidized, or transported into the deep ocean on large, rapidly sinking particles where biological activity may cause further oxidation, or drive re-dissolution to HMW DOM.

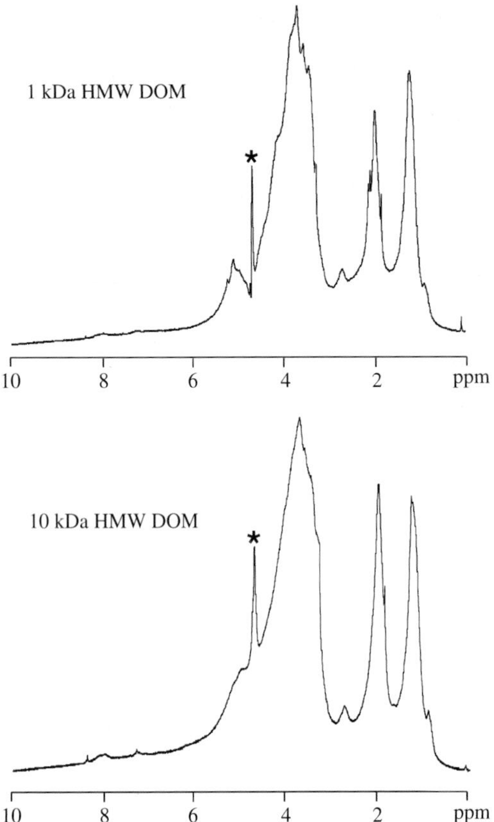

Figure 13 NMR spectra showing the similar spectral characteristics of 1 kDa and 10 kDa HMW DOM. Spectroscopic and chemical analyses of different-sized fractions of HWM DOM show each fraction to have essentially the same spectroscopic and elemental properties, and the same distribution of neutral sugars. The very HMW DOM (>10kDa) may be made up of either assemblies of 1–10 kDa APS, or true polymers of increasing molecular weight that spontaneously assemble in seawater. Further work is needed to establish if APS is the component of HMW DOM that forms gel polymer assemblies.

One promising approach that may help to establish APS as the fraction of HMW DOM that assembles into particulate matter is to compare the chemical composition of polymer gel assemblies with HMW DOM. In one such experiment, HMW DOM recovered by ultrafiltration of seawater or spent culture media was redissolved in seawater and agitated by bubbling to produce particles which collect at the top of a bubble tower (Gogou and Repeta, unpublished). Particles formed by bubbling have the same neutral sugars in approximately the same proportions as APS. NMR data likewise show particles to be rich in carbohydrate and have the same major resonances as APS, although the relative amount of major biochemicals differs between the two samples. This and other similar approaches further support the hypothesis that APS is the reactive fraction of HMW DOM that undergoes spontaneous assembly into polymer gels and larger particles.

6.06.5 EMERGING PERSPECTIVES ON ORGANIC MATTER PRESERVATION

6.06.5.1 Background

The vast majority of POM that sinks from the surface ocean is recycled during passage through the water column and burial in the upper sediment layers. The portion that escapes recycling to be sequestered in the underlying sediments serves to modulate atmospheric carbon dioxide concentrations over geological timescales, and provides a valuable archive of past ocean and climate conditions. The mechanisms by which OM survives degradation in the water column and in sediments, and the composition of the residual material are questions that have challenged marine organic chemists for many years. These processes have remained unclear, partly due to the difficulty of resolving the composition of the residual organic matter at depth using traditional wet chemical procedures and chromatographic and/or mass spectrometric techniques. In general, even for organic matter residing in the surface ocean, and in material that has recently exited the photic zone, there is a significant fraction of the OM that is no longer recognizable as biochemicals using traditional assays (Wakeham *et al.*, 1997). For example, a recent molecular-level survey of over 100 amino acids, sugars, and lipids in the water column of the central Pacific Ocean (Wakeham *et al.*, 1997) failed to account for ~15% of the molecules composing "plankton" and missed greater than 75% of the organic molecules in particulate debris raining in a matter of days to the ocean floor (Figure 14). Our assessment of the degree of diagenetic alteration undergone by organic matter using the abundance and compositional parameters based on the minor fraction of identifiable biochemicals is thus subject to major uncertainty.

The processes of signal attenuation and modification continue to varying extents on material after deposition and vary, depending on burial conditions. Moreover, our ability to structurally identify organic matter decreases as it becomes further removed from its biological source(s). Thus, we can account for less of the sedimentary organic debris leaving the photic zone derived from primary production compared to that in the original phytoplankton biomass, and organic matter in the underlying sediments contains less recognizable biological constituents than that in sinking particles. Indeed, often >80% of organic matter in surficial marine sediments remains unaccounted for in terms of readily distinguishable organic molecules (Hedges *et al.*, 2000).

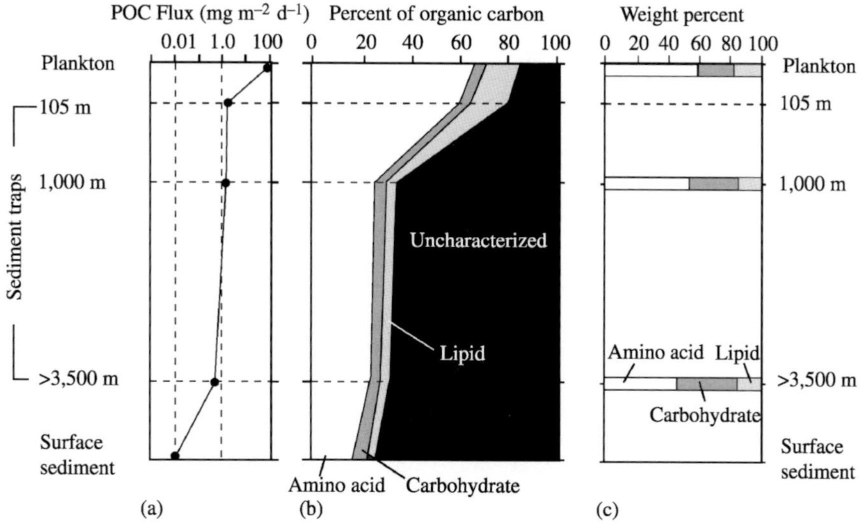

Figure 14 Fluxes and composition of particulate organic carbon in the equatorial Pacific Ocean. (a) POC fluxes ($mg\ m^{-2}\ d^{-1}$). (b) Corresponding fractions of amino acid, carbohydrate, lipid, and molecularly uncharacterized carbon (biochemical class-carbon as a percentage of total OC) in plankton, sediment traps (105 m, ~1,000m, >3,500 m) and surface sediment samples. The fraction of molecularly uncharacterized organic carbon (calculated as the difference between total OC and the sum of amino acid + carbohydrate + lipid) increases with more extensive degradation to become the major constituent in deeper POC samples (after Wakeham *et al.*, 1997). (c) Calculated weight percentages of amino acid, carbohydrate, and lipid in plankton and in sinking (sediment trap) particles in the upper and lower water column as determined by solid-state ^{13}C-NMR spectroscopy (source Hedges *et al.*, 2001).

Broad structural features of this "molecularly uncharacterized component" have been gleaned from bulk elemental and spectral analyses. However, without detailed molecular-level information the origins, reactions, and fates of this fraction are likely to remain obscure. To quote Hedges *et al.* (2000), "biogeochemists of today are playing with an extremely incomplete deck of surviving molecules, among which most of the trump cards that molecular knowledge would supply remain masked."

6.06.5.2 Compositional Transformations Associated with Sedimentation and Burial of Organic Matter

By virtue of where, when, and how the various organic matter inputs were formed and transported to the underlying sediments, it is possible to exploit specific chemical and isotopic characteristics to make inferences about the sources and composition of sedimentary organic matter. Much of this information is inaccessible at the bulk level. For example, bulk elemental compositions and stable carbon isotopic compositions are often insufficiently unique to distinguish and quantify sedimentary inputs. Abundances and distributions of source-specific organic compounds ("biomarkers") can help to identify specific inputs. However, this molecular marker approach suffers from the fact that the source diagnostic marker compounds are typically present as trace components, and extrapolation of abundances to infer overall organic matter contributions is therefore subject to considerable uncertainty. Isotopic measurements at the molecular level have provided a means to bridge the information gap between bulk and biomarker composition.

New insights into composition and transformation in sinking and SPOM have been gleaned from isotopic analyses performed in conjunction with compositional studies (e.g., Wang *et al.*, 1996, 1998; Megens *et al.*, 2001). For example, Druffel *et al.* (1992) measured ^{14}C, ^{13}C, bulk carbon, and biochemical constituents in dissolved and particulate carbon pools from the North Central Pacific Ocean and subtropical North Atlantic (Sargasso Sea). The decrease in $\Delta^{14}C$ values of suspended and sinking POC with depth has been interpreted in terms of incorporation of low-reactivity OM into the POC pool, possibly via DOC sorption or heterotrophy (Druffel *et al.*, 1992). ^{14}C and ^{13}C measurements on different classes of biochemical revealed that lipids were much "older" than corresponding amino acid and carbohydrate fractions in detrital aggregates, sediment floc, and sediments. These data indicate differences in decomposition and chemical behavior for different classes of biochemical in the deep ocean (Figure 15; Wang *et al.*, 1998).

Similar trends in organic matter reactivity and isotopic composition are evident in sediments. For example, Wang and co-workers

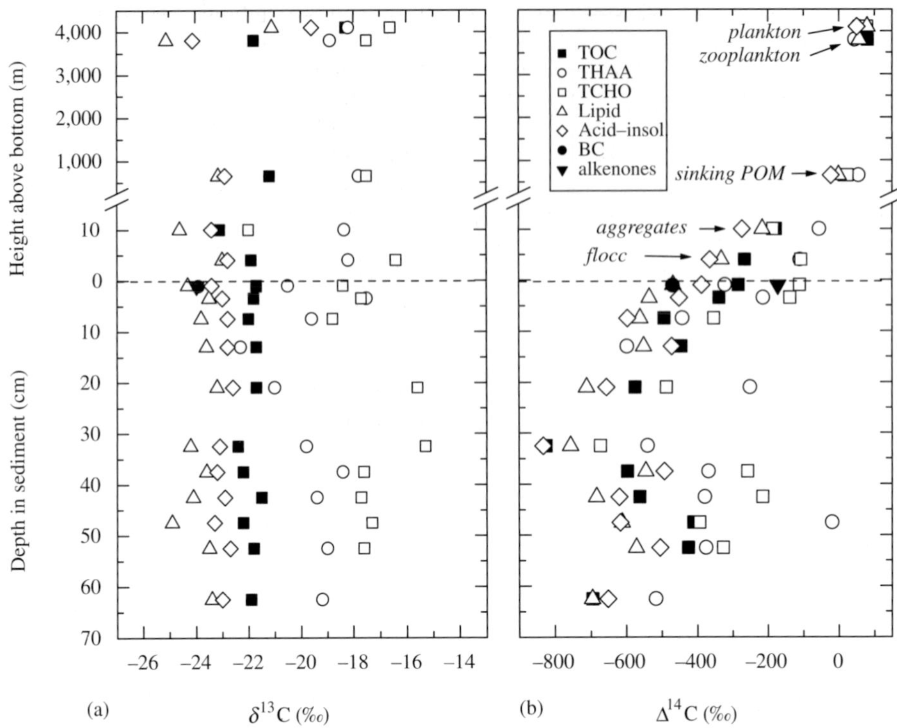

Figure 15 Variations in: (a) stable carbon isotopic composition ($\delta^{13}C$, ‰) and (b) radiocarbon content ($\Delta^{14}C$, ‰) of water column and sedimentary organic matter in the northeastern Pacific Ocean (Stn M, 34° 50′ N, 123° 00′ W; 4,100 m water depth). Samples: phytoplankton (1 m), zooplankton (100 m), sinking particulate material collected in a sediment trap 650 m above the seafloor, detrital aggregates, and surface flocc material isolated from the sediment surface and sediment core samples. Fractions measured: Total organic carbon (solid squares), total hydrolyzable amino acids (THAA, open circles), total carbohydrates (TCHO, open squares), lipids (open triangles), acid insoluble organic matter (open diamond), BC (solid circles),and alkenones (solid triangles). With the exception of black carbon (Masiello and Druffel, 1998) and alkenone (Ohkouchi and Eglinton, unpublished) data, all measurements are from Wang *et al.* (1996, 1998), or Wang and Druffel (2001).

(Wang *et al.*, 1996; Wang and Druffel, 2001) measured ^{14}C and ^{13}C compositions of total hydrolysable amino acids (THAAs), total carbohydrates (TCHO), and total lipids in deep-sea sediment profiles (Figure 15). Based on sedimentary concentration profiles, and using a "multi-G" model considering both labile and refractory organic fractions, Wang *et al.* (1998) calculated that degradation rate constants were in the order THAA ≈ TCHO > TOC ≈ TN > Total Lipid, indicating their relative reactivities in the sediment during early diagenesis. This is in good agreement with the order of average $\Delta^{14}C$ values in the sediment (THAA, −275‰; TCHO, −262‰; TOC, −371‰; lipid, −506‰), indicating that differential decomposition of organic matter may be a major process controlling the observed $\Delta^{14}C$ signatures. Alternatively, these results may indicate sorption and/or biological incorporation of "old" DOC into POC pool.

^{14}C age differences observed between biochemical classes are both expressed and magnified at the level of individual organic compounds (e.g., Eglinton *et al.*, 1997; Pearson *et al.*, 2001). This is because compound class measurements integrate molecular species derived from potentially diverse sources, whereas the full range of isotopic variability is retained in source-specific molecules. Intermolecular isotopic variability is most evident in the lipids, a biochemical class common to all organisms, but with specific compounds that may be unique to a subgroup of organisms, or even individual species.

The diversity in stable carbon isotopic compositions evident at the molecular level became apparent with the advent of gas chromatography coupled to isotope ratio mass spectrometry (IRM-GC-MS; e.g., Hayes *et al.*, 1990). More recently, ^{14}C age variations among different biomarkers have been investigated (e.g., Eglinton *et al.*, 1997; Pearson *et al.*, 2001). Substantial variability has been observed where inputs from diverse sources are encountered, even within depth horizons representing short periods of sediment accumulation. For example, surficial sediments from the Bermuda Rise in the northern Sargasso Sea deposited over a time span of less than 300 yr yield over a 3×10^4 yr spread in ^{14}C ages for different organic compounds (Figure 16). These different organic compounds that can be confidently assigned to marine photoautotrophs (alkenones), vascular plant waxes (long-chain fatty acids), and fossil (mainly thermogenic) organic matter (even carbon-numbered long chain *n*-alkanes). Their ^{14}C contents can provide valuable information on the relative importance and mode of delivery of these different inputs. In this particular example, substantial age variations are evident within a single compound class (e.g., fatty acids), highlighting the isotopic heterogeneity evident at the molecular level. Distinct ^{14}C age differences are also revealed between phytoplankton markers (alkenones) and planktonic forams—nominally both tracers of surface ocean DIC (Pearson *et al.*, 2000). These variations are potentially interpretable in terms of the provenance, modes, and timescales of delivery of different sedimentary constituents. However, even at this molecular level, it is important to recognize that ^{14}C data for a specific compound still reflect a population of otherwise identical molecules which will likely have different origins and have experienced diverse histories. For example, in the case of the alkenones from the Bermuda Rise (Figure 16), the ^{14}C age of the C_{37}–C_{39} compounds isolated from a single sediment sample undoubtedly reflects two major populations: alkenone molecules input directly from the overlying water column (and therefore presumably of similar age to the planktonic forams) and alkenone molecules which had been biosynthesized several millennia ago, stored on the margins, and subsequently transported to this location *via* lateral advection in deep currents or turbidity layers (Ohkouchi *et al.*, 2002). Although such studies are revealing the extraordinary complexity of organic matter in marine sediments, there is the exciting potential for these molecular-level age variations to provide invaluable novel insights into oceanic processes.

6.06.5.3 Controls on Organic Matter Preservation

Compositional and isotopic studies have shown that a wide range of organic materials are sequestered in marine sediments. These include direct or indirect products of marine photoautotrophy, vascular plant debris, and relict OC derived from sedimentary rock weathering.

The survival of any organic matter in the sedimentary record seems remarkable, given the

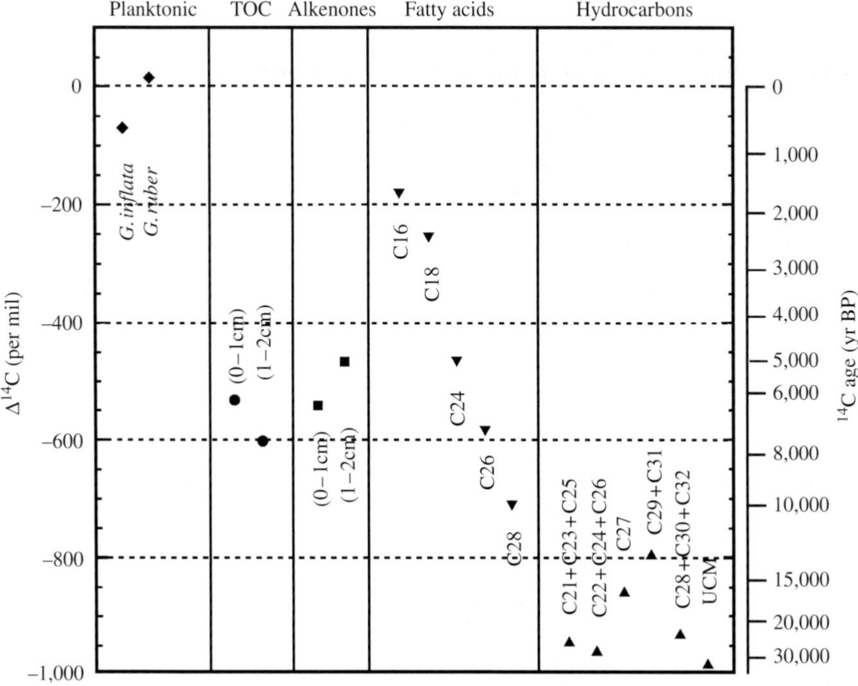

Figure 16 Bulk and molecular-level radiocarbon variations (expressed as $\Delta^{14}C$, ‰, and ^{14}C age, yr BP) in surface (0–3 cm) sediment from the northeastern Bermuda Rise in the subtropical North Atlantic (33° 41′ N, 57° 36′ W, ~4,500 m): planktonic foraminiferal calcite (diamonds), total OC (circles), C_{37}–C_{39} alkenones (squares), fatty acids (down triangles), and hydrocarbons (up triangles). Carbon numbers of fatty acids and hydrocarbons are adjacent to the symbols (UCM = unresolved complex mixture of hydrocarbons). The sedimentation rate at this site during the Holocene averages 10–20 cm kyr^{-1} (source Ohkouchi and Eglinton, unpublished).

efficiency of OM recycling in the water column and surface sediments. The means by which organic matter escapes degradation has been the subject of much debate and research. While the processes and mechanisms have remained elusive, it is evident that there are several principal factors that contribute to OM preservation in sediments, including organic matter source/provenance, the time interval over which sedimenting materials are exposed to oxic degradation, and the availability of (and physical associations with) detrital mineral phases. It seems likely that all of these factors play a role to some extent, and that their relative importance will depend on the specific depositional circumstances. Distinguishing between these factors is difficult, because they often vary in concert, and we have limited ability to recognize compositional features resulting from a given input or depositional characteristic. For example, higher burial efficiencies are encountered in many continental margin sediments yet it is unclear whether this result reflects greater inputs of recalcitrant terrigenous carbon, higher OC fluxes stemming from high primary production in coastal waters, or shorter oxygen exposure times due to the elevated fluxes of OC and other materials. We summarize prevailing concepts and supporting evidence in the following sections.

6.06.5.3.1 Physical protection

One major determinant in organic matter preservation appears to be the interaction of organic molecules with inorganic materials (Mayer, 1994, 1999; Keil *et al.*, 1994a,b; Hedges and Keil, 1995). These interactions provide a means of physically protecting labile biochemicals from degradation in the water column and underlying sediments.

Recently, Hedges *et al.* (2001) analyzed organic matter in both surface plankton and sinking particulate matter using solid-state ^{13}C-NMR spectroscopy. They observed that, despite extensive signal attenuation, minimal changes in bulk organic composition occurred, with apparently labile biochemicals such as amino-acid-like material and carbohydrates dominating organic matter content throughout the entire water column (Figure 14). These NMR measurements do not exclude the possibility of subtle modifications to these biopolymers. Nevertheless, the compositional similarity between phytoplankton biomass and the small remnant of organic matter reaching the ocean interior does imply that the formation of "unusual" macromolecules, either by chemical recombination (e.g., melanoidin formation (Tissot and Welte, 1984)) or microbial biosynthesis, is not a major process controlling the preservation of POM in pelagic waters. Instead, Hedges *et al.* (2001) suggest that OM might be shielded from degradation through association with the inorganic matrix (e.g., opal, coccoliths, and detrital aluminosilicates). This mineral matter makes up most (>80%) of the mass of sinking particles and, in addition to their protective role, also serves as ballast, expediting OM transport through the water column. Moreover, most sedimentary organic matter cannot be physically separated from its mineral matrix, indicating that the majority of OM is in intimate association with mineral phases.

Observations of organic carbon-to-mineral surface area (OC : SA) ratios by Mayer (1994) and others have revealed that sediments accumulating along continental shelves and upper slopes (excluding deltas) characteristically exhibit surface area loadings approximately equivalent to a single molecular covering on detrital mineral grains, the so-called "monolayer-equivalent" coating (Figure 17). Experiments have shown that a fraction of this organic matter is bound reversibly, and is intrinsically labile (Keil *et al.*, 1994a), but apparently escapes mineralization through association with minerals. This appears to hold true for situations where the OM passes relatively rapidly through oxygenated bottom and pore waters prior to sequestration in the underlying anoxic sediments.

In summary, the above observations and relationships indicate that the supply and availability of mineral surface area may be a primary control

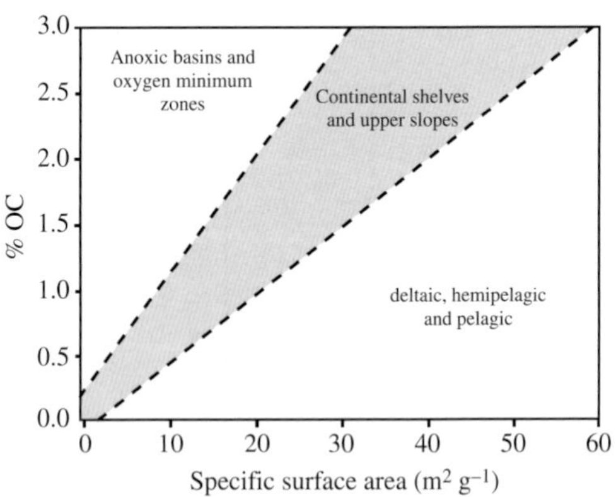

Figure 17 Generalized relationship between weight percent OC (% OC) and specific mineral surface area (SA) for marine sediments. The shaded area represents the boundaries (OC : SA ratio of 0.5–1.1 mg C m^{-2}), within which most continental shelf and upper slope sediments (outside the direct influence of rivers) fall. Sediments underlying anoxic basins and OMZs associated with high productivity (upwelling) margins tend to exhibit OC : SA ratios greater than 1.1 mg C m^{-2} whereas deltaic and abyssal sediments exhibit OC : SA ratios of less than 0.5 mg C m^{-2} (after Mayer, 1994; Hedges and Keil, 1995).

on OM preservation, particularly in open continental shelf and upper slope sediments where ~45% of all carbon burial takes place in the contemporary ocean. However, while a clear first-order relationship between mineral surface area and OC loading has been apparent for some time, the exact mode of this association remained unclear. Specifically, it was uncertain whether adsorbed organic matter is dispersed over all mineral surfaces or is more localized in occurrence. Theoretical considerations and empirical relationships have suggested the former to be the case (Keil *et al.*, 1994b), but recent work has not supported this paradigm. Transmission electron microscopic examination of sedimentary mineral grains (e.g., Ransom *et al.*, 1997, 1998b) has revealed that OM is not uniformly distributed across all mineral surfaces, but instead is preferentially associated with minerals of high surface area, particularly smectite-rich clays. Ransom *et al.* (1998a) have concluded that the association is a function of differences in the site density and chemistry of the clays, as well as differences in their flocculation behavior. Similarly, studies of gas adsorption in model systems and marine aluminosilicate sediments (Mayer, 1999) have shown that the sediments with low to moderate loading of OM (<3 mg OC m^{-2}) have generally less than 15% of their surface coated. These data imply that the abundance of non-spherical, high surface area-to-volume particles—such as clays, oxyhydroxides, and inorganic biogenic debris (e.g., diatom frustules)—controls specific surface area in most continental margin sediments. Thus, although OM is associated with mineral grains (Mayer *et al.*, 1993; Mayer, 1994; Keil *et al.*, 1994b), it is not adsorbed as a monolayer (Mayer, 1999), but must instead be locally concentrated on certain surfaces. Mayer (1999) argued that the term "monolayer equivalent" should, therefore, be removed from usage, and there appears good justification for so doing. Hence, although there appears to be a clear relationship between OC : SA and organic matter preservation, the mechanistic basis for this relationship still remains to be determined. Mayer (1999) notes that most surface area appears to be within mesopores of <10 nm diameter (Mayer, 1994), and these pores may be too small to allow attack by extracellular enzymes, hence offering protection from degradation (Mayer, 1994).

In addition to physical protection via association with detrital mineral phases, preservation of labile organic matter through association with other matrices is also possible. In particular, "encapsulation" of labile organic matter within refractory organic polymers (Knicker *et al.*, 1996; Zang *et al.*, 2001) or within biogenic mineral lattices is a possible means of stymieing organic matter degradation during burial.

6.06.5.3.2 Role of anoxia

While a protective role of minerals is inferred from the consistency of OC : SA values in open continental margin sediments, there are depositional environments that yield OC : SA values well above those predicted from sorptive controls, and where other modes of preservation must be invoked. In particular, sediments underlying anoxic or low oxygen bottom waters, or where molecular oxygen does not penetrate to a significant extent into the sediment, are characteristically rich in organic matter and tend to exhibit elevated levels of OC with OC : SA values in excess of those characteristic of margin sediments overlain by oxygenated waters (Bergamaschi *et al.*, 1997; Keil and Cowie, 1999; Figure 17). While such depositional settings account for only ~5% of total OC burial, they are of importance in relation to understanding the formation of organic-rich sedimentary rocks that are responsible for reserves of petroleum. These organic-rich sediments also represent potentially valuable, high fidelity archives of past ocean conditions, since they tend to be free from sediment mixing due to bioturbation and often accumulate relatively rapidly. They are hence conducive for preservation of labile organic compounds that carry the highest information contents. While the extent of exposure to molecular oxygen prior to entering the anoxic realm is thus frequently considered as a master variable influencing OM preservation (e.g., Demaison, 1991), much debate persists concerning the specific factors leading to enhanced organic matter burial under oxygen deficient or anoxic conditions. Indeed, is there a causal or even correlative relationship between the presence of a minimum in bottom water oxygen (BWO) concentration and the preservation of organic matter in the underlying marine sediments (Pedersen and Calvert, 1990; Cowie and Hedges, 1992).

Various approaches have been taken towards assessing the role of oxygen in OM degradation. Laboratory incubations under controlled conditions have provided detailed information on the influence of aerobic versus anaerobic microbial degradation of phytoplankton biomass (e.g., Harvey and Macko, 1997). Investigation of natural systems includes comparison of surface sediments from depocenter and periphery of anoxic or dysoxic basins (e.g., Gong and Hollander, 1997), examination of depth transects traversing OMZs that impinge on continental margins (e.g., Keil and Cowie, 1999), and comparison of turbidites overlying pelagic sediments that were subjected to oxygen "burn-down" (Cowie *et al.*, 1995; Prahl *et al.*, 1997; Hoefs *et al.*, 1998, 2002; Middelburg *et al.*, 1999).

Keil and Cowie (1999) examined OC : SA relationships in relation to BWO concentrations in sediments from the NE Arabian Sea. Sediments deposited under the oxygen minimum had OC : SA ratios in excess of 1.1 mg OC m^{-2}, while samples shallower or deeper than the oxygen-depleted water mass (BWO > 35 μM) exhibited OC : SA ratios that fall within the typically observed range (0.5–1.1 mg OC m^{-2}) (Figure 18(a)). These data indicate that organic matter preservation is enhanced within the general locale of the BWO minimum in NE Arabian Sea sediments. While OM loadings are 2–5 times the monolayer equivalent in anoxic sediments, there often remains a direct relationship between OM content and mineral surface area. Hedges and Keil (1995) speculate that organic materials sorbed in excess of a monolayer may be partially protected as result of equilibration with DOM-rich pore-waters, and brief oxygen exposure times.

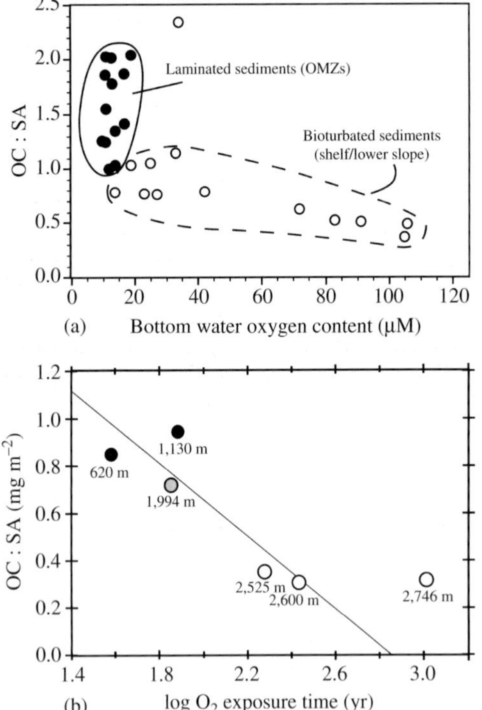

Figure 18 Oxygen controls on the degradation of OC. (a) Organic carbon to mineral surface area ratio (OC : SA) for core top sediments from the northeastern Arabian Sea plotted as a function of BWO content. Open symbols denote bioturbated sediments lying under relatively oxygenated waters on the shelf and lower slope. Solid symbols denote laminated sediments lie beneath water with the lowest BWO values (i.e., OMZ) (after Keil and Cowie, 1999). (b) Organic carbon to surface area (OC : SA) ratios for surface sediments from the Washington margin (northeastern Pacific Ocean) as a function of OET. Samples with short OETs derive from the shelf and upper slope (solid and shaded symbols). Samples with longer OETs are from the lower slope and Cascadia basin (open symbols). Water depths, in meters, are indicated (after Hedges *et al.*, 1999).

Thus, Hedges and Keil (1995) argue that organic matter preservation throughout much of the ocean may be controlled largely by competition between sorption on mineral surfaces at different protective thresholds and oxic degradation. The presence of reducing conditions slows OC degradation, leading to elevated OC contents. As an extension of this line of thinking, Hedges *et al.* (1999) hypothesize that organic matter preservation in continental margin sediments is controlled by the average residential period experienced by the accumulating organic particles in the water column and in the oxygenated pore waters immediately beneath the sediment–water interface. Trends in oxygen penetration depth, organic elemental composition, and mineral surface area for surface sediments collected along an offshore transect across the Washington continental shelf, slope, and adjacent Cascadia basin support this notion. Sediment accumulation rates decrease and dissolved molecular oxygen penetration depths increase offshore, resulting in a seaward increase in oxygen exposure times from decades (mid-shelf and upper slope) to millennia (outer Cascadia basin). Organic contents and compositions were essentially constant at each site, but varied between sites. In particular, OC : SA ratios decreased and indicators for the level of degradation increased with increasing oxygen exposure time (Figure 18(b)), indicating that sedimentary organic matter experiences a clear "oxic effect" (Hedges *et al.*, 1999). The oxygen exposure time concept helps to integrate contributing factors such as sediment accumulation rate and BWO levels that have been invoked in the past to explain OM preservation in accumulating continental margin sediments. However, we are still lacking a detailed mechanistic basis for this observed control on degradation exerted by BWO conditions.

6.06.5.3.3 Chemical protection

Intrinsically refractory biomolecules. While evidence exists for the physical protection of organic compounds from degradation (e.g., via association with mineral surfaces), and for enhanced preservation due to limited exposure to oxygenated conditions, organic matter persists in many oceanic sediments where conditions are not conducive for OM preservation based on either of the above criteria. For example, deltaic sediments often exhibit OC : SA coatings that are significantly lower than the "monolayer equivalent" (Keil *et al.*, 1997). Submonolayer equivalent OM coatings are also observed in deep-sea turbidite sediments where organic-rich margin sediments, originally sequestered under anoxic conditions on the margins, have been redeposited in abyssal locations and are exposed to oxygenated bottom waters. Submonolayer organic coatings are also observed in continental rise and abyssal plain sediments where slower sediment accumulation rates and deeper O_2 penetration depths result in increased oxygen exposure times, and little OM preservation.

The residual organic matter in such sediments is often inferred to be highly refractory, and it is assumed that the structural attributes of the organic matter may be pivotal in dictating its survival. The concept of "selective preservation" of one type of natural product over another was introduced by Tegelaar *et al.* (1989) and others. Evidence in support of selective preservation stems from analyses of insoluble macromolecules in sediments and in precursor organisms (Gelin *et al.*, 1999). Solid-state NMR spectra and pyrolytic degradation of OM in many recent and ancient sediments indicate the presence of a highly aliphatic component(s) (Van de Meent *et al.*, 1980; Eglinton, 1994). The nature and origin of this type of aliphatic component had been the subject of considerable debate until it was found that several types of organisms synthesize highly aliphatic biopolymers (Largeau *et al.*, 1984; Nip *et al.*, 1986; Goth *et al.*, 1988; Zelibor *et al.*, 1988, Gelin, 1996, Gelin *et al.*, 1997, 1999). These natural products were then proposed as the source of this sedimentary component. The recalcitrance of these aliphatic macromolecules is indicated by their relative enrichment in oxidized layers of pelagic turbidite sequences (Hoefs *et al.*, 1998).

The concept of chemical recalcitrance is not restricted to aliphatic biopolymers and one guideline for assessing reactivity and conformational relationships pertinent to the preservation of other types of biomacromolecules are structural comparisons between proteins of the same generic type occurring in hyperthermophiles and their low-temperature counterparts. X-ray-based structural studies have revealed that very minor differences evident in molecular structure and conformation brought about by a few α-amino acid substitutions can have dramatic effects on the reactivities and thermal stabilities of such molecules (Danson and Hough, 1998). The corollary for stability of marine organic matter is that similar minor changes in molecular content and architecture induced by mineralization and diagenetic processes may lead to sharp contrasts in biochemical reactivity and hence preservation (Eglinton, 1998; Hedges *et al.*, 2000).

Formation of organosulfur compounds. One factor that distinguishes organic matter accumulation under anoxic or suboxic conditions from their oxic counterparts is the presence of sulfide from microbial sulfate reduction in bottom waters and sediment pore waters. It has been established that certain organic compounds can readily react with reduced sulfur species (H_2S, polysulfides) under ambient conditions (Vairavamurthy and Mopper, 1987), providing a potential means of sequestering labile, extremely oxygen-sensitive organic molecules, such as functionalized or polyunsaturated lipids. Evidence for the reaction of organic matter with sulfides stems from laboratory studies (e.g., Krein and Aizenshtat, 1994; Schouten *et al.*, 1994), together with down-core profiles of the content and isotopic composition of organically bound sulfur (Francois, 1987; Mossmann *et al.*, 1991; Eglinton *et al.*, 1994; Putschew *et al.*, 1996; Hartgers *et al.*, 1997).

Numerous studies have investigated potential mechanisms of OM sulfurization during early

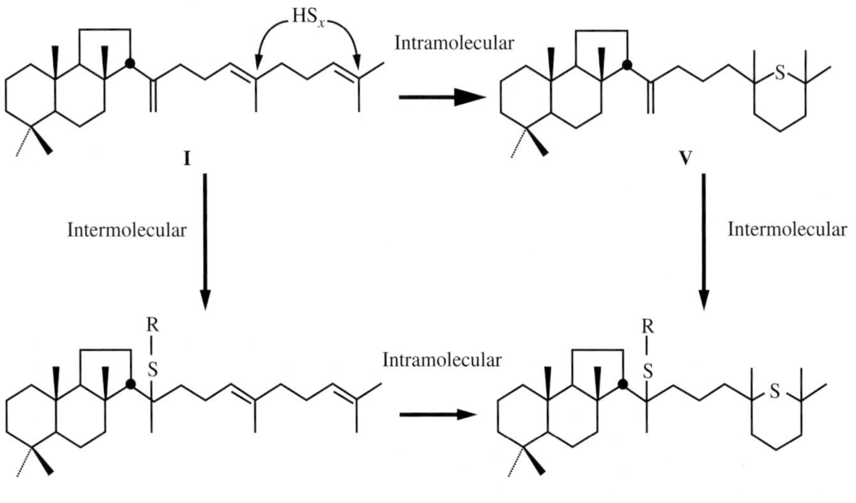

Figure 19 Model for the intramolecular and intermolecular reaction of reduced sulfur species with functionalized lipids. Proposed reaction scheme for sulfur incorporation into (17*E*)-13β(*H*)-malabarica-14(27),17,21-triene (I) (after Werne *et al.*, 2000).

diagenesis. Sulfur is considered to react in both an intramolecular and intermolecular fashion (Figure 19). Intramolecular incorporation leads to the formation of cyclic OSC (e.g., thianes, thiolanes, and thiophenes). In intermolecular reactions, the formation of sulfide or polysulfide bridges between molecules generates a wide variety of sulfur-cross-linked macromolecules (Sinninghe Damsté *et al.*, 1989; Kohnen *et al.*, 1991).

Formation of organically bound sulfur is promoted by the availability of reactive organic matter (i.e., bearing the appropriate type, position, and number of functional groups), an excess of reduced sulfur species (e.g. HS^-, polysulfides), and a limited supply of reactive iron, which would otherwise outcompete OM for sulfides (Canfield, 1989; Hartgers *et al.*, 1997). These conditions are characteristic of anoxic basins and OMZs underlying productive upwelling systems on the continental margins. The high flux of labile OM to the sediment provides an abundant source of reactive OM that both possesses the requisite functional groups, and fuels bacterial sulfate reduction, providing reduced sulfides and polysulfides. Shelf and slope sediments distal from sources of terrigenous sediment receive only limited amounts of reactive iron, except in regions of major eolian dust input. In this context, organic matter preservation through formation of organically bound sulfur is unlikely to be important on a global basis.

Most attention has been focused on the sulfurization of functionalized lipids and its influence on lipid preservation and resulting biomarker fingerprints (e.g., Kohnen *et al.*, 1991). However, lipids comprise a relatively minor fraction of the input organic matter, and hence the significance of this process as an organic matter preservation mechanism has been uncertain. Recently, it has been shown that carbohydrates can be preserved in a similar manner (Sinninghe Damsté *et al.*, 1998). Carbohydrates comprise a large fraction of the carbon fixed by photoautotrophs, especially certain diatoms which are often the dominant primary producers in upwelling systems. These biopolymers are generally considered to be highly labile and therefore poorly preserved in sediments. However, based on the premise that carbohydrates tend to be isotopically enriched relative to lipids, Sinninghe Damsté *et al.* (1998) interpret strong relationships between OC content, organic sulfur content, and $\delta^{13}C$ of bulk OC in Jurassic age sediments as a consequence of variable preservation of ^{13}C enriched carbohydrates through sulfurization reactions.

It remains unclear whether sulfurization of organic matter results in a net increase in OC preservation, for example, whether sulfurization acts to transform organic matter, but not preserve it. A key issue with respect to this question is the timing of these reactions in relation to competing diagenetic reactions that remove the organic substrates. Circumstantial evidence suggests that these reactions take place quite rapidly (Eglinton *et al.*, 1994); however, the identification of both precursor and product (Figure 19) in age-dated sediments from the anoxic Cariaco Basin provides a direct estimate of reaction rates (Werne *et al.*, 2000; Figure 20). While these results apply only for the precursor in question, they imply that sulfur incorporation is far from instantaneous.

6.06.6 MICROBIAL ORGANIC MATTER PRODUCTION AND PROCESSING: NEW INSIGHTS

6.06.6.1 Background

Compared to OM inputs from sedimentation of biogenic particles resulting from primary production in the overlying water column and transported terrigenous materials, the role of prokaryotic organisms in the production and processing of organic matter is a poorly understood component of the oceanic carbon cycle. Prokaryotes, which include the bacteria and archea, can be divided into several classes: cyanobacteria, anoxygenic photosynthetic bacteria, heterotrophic bacteria, methane oxidizing bacteria, and methanogenic archea (Balows *et al.*, 1992). All of these types of prokaryotes have a major impact on either the production or the mineralization of OC in the water column and in sediments. Indeed, viable microbial communities have now been shown to extend hundreds of meters into the sediment column (e.g., Parkes *et al.*, 1994, 2000), indicating that fresh microbial biomass may also constitute some fraction of the OM hitherto considered as "preserved" OM remnants from the overlying water column. Thus, the extent to which OM buried in sediments represents organic remnants directly inherited from the overlying surface ocean versus bacterial debris is a subject of some debate. The fact that benthic microorganisms are the last "filter" that OM passes through during burial suggests that their biomass and products might become sequestered irrespective of their reactivity (i.e., there is no one else to "eat" them). It is difficult to distinguish between these inputs, both because of the modified nature of sedimentary OM, and the fact that heterotrophic organisms utilizing multicarbon substrates will inherit the isotopic characteristics of the OM they act upon (Hayes, 1993).

De novo biosynthesis and associated reworking of organic matter by prokaryotes does not imply that all of this microbially processed organic material is transferred to the sedimentary record. Evidence of contributions from prokaryotic organisms is abundant in the form of molecular

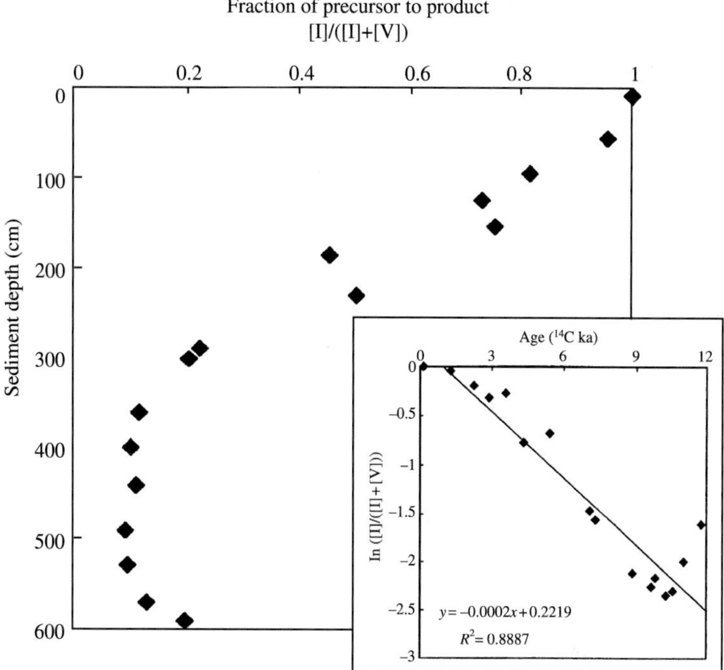

Figure 20 Progress of intramolecular sulfurization of malabaricatriene (I) with depth in anoxic Cariaco basin sediments. Plot of the ratio of the concentrations of I to the sum of the concentrations of I + V (see Figure 19 for structures) as a function of sediment depth. The progress of transformation of I–V with increasing sediment depth is indicated by the steady decrease in the relative abundance of the precursor lipid (I). Inset: plot of ln ([I]/([I] + [V])) versus sediment age (= time), used to empirically determine the first-order rate constant for sulfur incorporation (after Werne *et al.*, 2000).

fossils (e.g., hopanoids) that are virtually ubiquitous in the sedimentary record. However, while the presence of specific biomarkers may be diagnostic of specific prokaryotic inputs, their abundance is not necessarily in proportion to these inputs (the same is, of course, true for all biomarkers and incidentally POM in general). Sinninghe Damsté and Schouten (1997) reviewed evidence for prokaryotic biomass inputs to sedimentary organic matter. They argue that several lines of evidence point to limited contributions or poor preservation of bacterial biomass in the sedimentary record. Their conclusions are based on, among other things: (i) isotopic mass balance in various sedimentary settings; (ii) the effects of bacterial oxidation in organic matter-rich turbidites; and (iii) the absence of apparently recalcitrant biomacromolecules, such as aliphatic biopolymers, in prokaryotes.

In addition to the prokaryotes whose identity, physiology, and ecological role has been established, relatively recent studies using culture-independent r-RNA analyses of environmental samples have revealed genetic diversity within natural microbial communities that greatly exceeds estimates based on classical microbiological techniques. These findings underline our presently limited view of microbial activity and its biogeochemical consequences. Below, two examples are provided that highlight the potential influences of microbial processes on sedimentary organic matter composition.

6.06.6.2 Planktonic Archea

Recent culture-independent, r-RNA gene surveys have indicated the ubiquity and importance of planktonic archea in the ocean, particularly in subsurface waters. For example, Karner *et al.* (2001) found pelagic crenarcheota comprised a large fraction of total marine picoplankton, equivalent in cell numbers to bacteria at depths greater than 1,000 m (Figure 21). The fraction of crenarchaeota increased with depth, reaching 39% of the total DNA-containing picoplankton detected. Moreover, the high proportion of cells containing significant amounts of r-RNA suggests that most pelagic deep-sea microorganisms are metabolically active. The oceans are estimated to harbor $\sim 1.3 \times 10^{28}$ archeal cells and 3.1×10^{28} bacterial cells, suggesting that pelagic crenarchaeota represent one of the ocean's single most abundant cell types.

The physiologies, ecological niches, and biogeochemical roles of these organisms are yet to be fully determined. Their imprint on the sedimentary record is only now being appreciated.

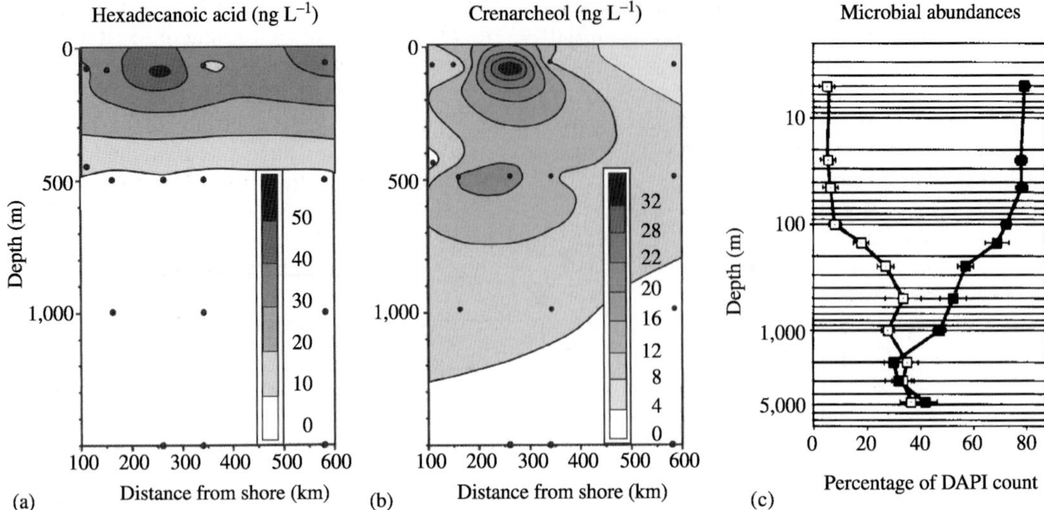

Figure 21 Comparison of vertical distribution of biomarker and microbial abundances in oceanic water columns. (a) Contour plots of concentration (ng L^{-1}) of hexadecanoic acid. (b) Crenarcheol at various depths in the water column and distances from shore on a northwest-to-southeast transect off Oman in the Arabian Sea (after Sinninghe Damsté *et al.*, 2002). Hexadecanoic acid serves as a biomarker proxy for eukaryotic and bacterial biomass and clearly shows the expected surface maximum, with concentrations dropping off steeply with increasing water depth. In contrast, crenarcheol, a molecular biomarker for planktonic crenarcheota, shows two maxima with one near 50 m and the other ~500 m. (c) Vertical distributions of microbial concentrations in the North Pacific subtropical gyre: bacteria (solid squares) and planktonic crenarcheota (open squares). Effectively, there are two microbial domains which were determined using a DAPI nucleic acid stain (after Karner *et al.*, 2001). These data show the increasing proportion of planktonic archea in deep waters, with the result that at depths greater than 2,000 m, the crenarcheota are as abundant as the bacteria.

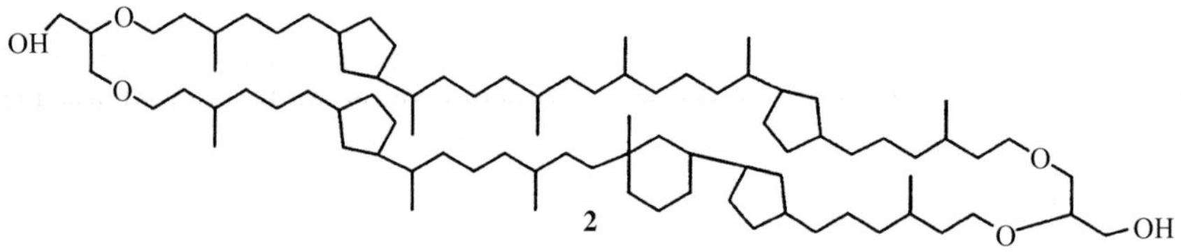

Figure 22 Molecular structure of crenarcheol (after Sinninghe Damsté *et al.*, 2002).

Sinninghe Damsté *et al.* (2002) quantified intact tetraether lipids of marine planktonic crenarchaeota in SPOM from the NE Arabian Sea. In contrast to eukaryotic and bacterial lipids (sterols and fatty acids, respectively), which were highest in surface waters, maximum concentrations of crenarcheol (Figure 22), generally occurred at 500 m, near the top of the OMZ (Figure 21). This indicates that these crenarcheota are not restricted to the photic zone of the ocean (consistent with molecular biological studies). Sinninghe Damsté *et al.* (2002) suggest that the coincidence of maximum abundances of crenarcheotal membrane lipids with the core of the OMZ indicates these organisms are probably facultative anaerobes. Moreover, calculations of cell numbers (based on membrane lipid concentrations) support other recent estimates for their significance in the World's oceans. Schouten *et al.* (1998) investigated acyclic and cyclic biphytane carbon skeletons derived from planktonic archea in a number of lacustine and marine sediments. They found these compounds to be amongst the most abundant lipids in sediments, and sometimes were present in greater proportion to those synthesized by eukaryotes and bacteria, indicating that these organisms may be an important source of sedimentary organic matter.

6.06.6.3 Anaerobic Methane Oxidation

A second newly recognized group of prokaryotes are the methane oxidizing archea. Nearly 90% of the methane produced in anoxic marine sediments is recycled through anaerobic microbial oxidation processes (Cicerone and Oremland, 1988; Reeburgh *et al.*, 1991). However, the organisms and biochemical processes responsible for the anaerobic oxidation of methane (AMO) have largely evaded elucidation until recently. Convergent lines of molecular, carbon-isotopic, and phylogenetic evidence have now implicated archea in consortia with sulfate reducing bacteria as the primary participants in this process (Hoehler *et al.*, 1994; Boetius *et al.*, 2000; Pancost *et al.*, 2000; Hinrichs *et al.*, 1999).

Lipids of these organisms are distinguished by their ether-linked nature and highly ^{13}C-depleted values. Hinrichs *et al.* (2000) investigated microbial lipids associated with AMO in gas hydrate-bearing sediments from the Eel River Basin, offshore Northern California, as well as sediments from a methane seep in the Santa Barbara Basin. In addition to archeal markers (*sn*-2-hydroxyarchaeol), these lipids are accompanied by additional ^{13}C-depleted glycerol ethers and fatty acids. Hinrichs *et al.* speculate that these ^{13}C-depleted lipids are produced by (unknown) sulfate reducing bacteria growing syntrophically with the methane utilizing archea. Interestingly, these authors note that at all of the methane seep sites examined, preservation of aquatic products is enhanced because enhanced consumption of sulfate by the methane oxidizing consortium depletes the sulfate pool that would otherwise have been available for remineralization of materials from the water column.

The identification of ^{13}C-depleted ($\delta^{13}C$ values as low as -58 per mil) archeal cyclic biphytanes in particulate matter from the Black Sea, where more than 98% of the methane released from sediments is apparently oxidized anaerobically (Reeburgh *et al.*, 1991), provides evidence for AMO in euxinic waters (Schouten *et al.*, 2001). However, the same isotopically depleted compounds were not detected in the underlying sediments, suggesting that the responsible organisms are in low abundance and/or leave no characteristic molecular fingerprint in the sedimentary record.

6.06.7 SUMMARY AND FUTURE RESEARCH DIRECTIONS

Our understanding of the composition and cycling of organic matter in the oceans has changed tremendously since the 1990s. Through the use of new analytical tools, we have a greater appreciation of the complexity of dissolved and POM while, at the same time, we can view its biomolecular building blocks with greater clarity. Despite these advances, many important unanswered questions provide impetus for continued research.

Questions of the origin and composition of marine macromolecular organic matter continue to challenge us. A key point that has emerged, especially from NMR studies, is that the formation of randomly cross-linked macromolecules ("humic substances") is no longer considered a primary pathway in the formation of either HMW DOM or sedimentary OM. Rather, these materials seem to be comprised of biochemicals that have been directly inherited from their biological precursors (primarily marine algae and bacteria) with minimal alteration. Three questions immediately arise: (i) Why are traditional wet chemical approaches unable to detect much of these apparently intact biopolymers? (ii) Given this close similarity to natural products, why is it that this material not readily biodegraded in the water column and sediments? (iii) What are the processes that render normally labile macromolecules unavailable to heterotrophic microorganisms and their hydrolytic enzymes? These issues are particularly perplexing in the case of deep-water DOM, where physical mechanisms of preservation such as sorption on mineral grains or physical encapsulation cannot be invoked, where a large fraction of the macromolecular pool is polysaccharide in nature and, paradoxically, exhibits radiocarbon ages of several millennia! It is evident that our knowledge of the fundamental factors that control organic matter degradation and transformation is far from complete. However, subtle modifications of the molecular structure of a biochemical (e.g., changes in protein conformation) are known to influence susceptibility to attack by degrading chemicals and enzymes. New analytical methods that can detect small changes in biopolymer structure are urgently required, together with new experimental approaches that can assess the impact of such molecular rearrangements on organic matter preservation at a mechanistic level.

Our limited understanding of OM composition is not restricted to the macromolecular pool. Our knowledge of the initial biological signatures carried in particles and DOM also remains highly limited. Surveys of marine microbes for their molecular signatures have largely been dictated by what can be grown axenically in culture. However, many important organisms are not available through laboratory culture, including the most abundant photoautotroph in the sea, Prochlorococus, and some of the most abundant autotrophic microbes, the planktonic crenarcheota. Clearly, given the recently recognized prevalence of these microbes in the contemporary ocean, it will be important to further study their roles in carbon and other biogeochemical cycles. The molecular signature that these organisms carry is only now being studied and much work is required to accurately interpret their corresponding molecular stratigraphic records.

A second key question for future research concerns the fate of terrestrial organic matter in the oceans. Two lines of evidence that have long argued against a significant terrestrial input are the stable carbon isotopic ($\delta^{13}C$) and elemental C_{org}/N composition of oceanic dissolved and particulate organic matter. Recent findings of extensive terrestrial organic matter degradation and exchange in deltaic systems have reinforced this notion. However, interpretation of bulk elemental and isotopic data is not necessarily straightforward. Organic matter–mineral associations, while a key determinant in OM preservation, are highly dynamic, and fluvial and eolian contributions of ^{13}C-enriched, nitrogen-replete soil OM derived from C4 vascular plant debris blur traditional distinctions based on these parameters, undermining their diagnostic capability. Several other new lines of evidence, including the recognition that terrestrial OC inputs to the oceans are largely dominated by numerous small, low latitude rivers draining mountainous regions, the recognition that black carbon in marine sediments could account for up to 30% of total OC in some continental margin deposits, and finally that relict OC eroding from sedimentary rocks on the continent represents an additional terrigenous component, all suggest that terrestrial OC may be an underestimated component of the oceanic carbon reservoir.

Multiple lines of further investigation are clearly warranted in order to address this important and complex issue. Small, low-latitude river systems have been undersampled, and our understanding of the composition and mode of terrestrial OM export from the continental shelf and slope is not well developed. At a mechanistic level, the manner in which different forms of organic are associated with particles, and move between dissolved and solid phases both during transit from the land to the oceans, and within sediment pore waters, requires detailed investigation. The timescales over which organic matter exchange and degradation take place in relation to the timescales for terrestrial OC delivery to the oceans is an important consideration in this context. Our ability to recognize and quantify terrigenous OC in the marine environment must also be improved. With respect to the preservation of POM in sediments, reactivity depends on both the intrinsic lability of the molecular structures present and on the associated material matrix. While a strong causal relationship has been established between the properties of the matrix (e.g., mineral type and surface area) and efficiency of organic matter preservation, we are lacking a mechanistic basis for it. Future research strategies need to address the nature of the associations of OM with mineral surfaces and their effects on the reactivities toward chemical and biological agents. Similarly, while recent studies have demonstrated the link between bottom and pore-water oxygen concentration and the extent of sedimentary OM degradation, we do not yet know whether the key reactant is molecular oxygen itself.

One striking aspect of marine organic geochemistry is that its material basis extends from the smallest molecules (methane, α-amino acids, short-chain fatty acids, etc.) through the ubiquitous but complex, HMW DOM to a highly diverse range of organic particles (BC, marine snow, necromass, etc.). No single set of tools is adequate to address the composition of all these materials and the ways in which they function. A variety of different analytical approaches is required. Bulk chemical, physical, and isotopic characterization, microscopy (TEM, SEM, etc.), spectroscopic techniques (NMR, etc.), fractionation and chromatographic techniques, and chemical degradation, have all been used to derive useful information. As we look ahead to the future, it is evident that many exciting opportunities lie at the interface between molecular biology and molecular organic geochemistry. For example, a better knowledge of the genetic machinery that directs the synthesis and degradation of biopolymers may help to unravel questions of OC preservation as DOC in the deep ocean and as POC in marine sediments. Such studies may also help to close the gap between organic matter characterization at the bulk and molecular level.

Finally, it is clear that human activity has and will continue to have an impact on the oceanic carbon cycle and associated biogeochemical processes. By the beginning of the 1990s, atmospheric CO_2 comprised ~750 Gt C, and this has continued to rise at a rate of ~3.4 Gt yr^{-1}. The oceans represent a major sink for anthropogenic CO_2 and it is imperative that we develop a better understanding of how organic matter cycling is impacted in order to assess the long-term consequences and health of the planet.

ACKNOWLEDGMENTS

The authors wish to dedicate this contribution to the memory of John I. Hedges. John was a source of inspiration; his contributions to marine organic geochemistry are immeasurable, but clearly evident from the extensive citations of his work throughout this chapter. The field of marine organic geochemistry has suffered a tremendous setback with his untimely departure.

The authors wish to thank Lucinda Gathercole for assistance with the preparation of this manuscript and Thomas Wagner for constructive comments on this paper. They also wish to acknowledge support from NSF Grant OCE-9907129 (T.I.E.) and the Department of Energy. This is Woods Hole Oceanographic Institution Contribution No. 10993.

REFERENCES

Alldredge A. L. and Silver M. W. (1998) Characteristics, dynamics and significance of marine snow. *Prog. Oceanogr.* **20**, 41–82.

Alldredge A. L., Passow U., and Logan B. E. (1993) The abundance and significance of a class of large, transparent organic particles in the ocean. *Deep-Sea Res. I* **40**, 1131–1140.

Aluwihare L. I. and Repeta D. J. (1999) A comparison of the chemical composition of oceanic DOM and extracellular DOM produced by marine algae. *Mar. Ecol. Prog. Ser.* **186**, 105–117.

Aluwihare L. I., Repeta D. J., and Chen R. F. (1997) A major biopolymeric component to dissolved organic carbon in seawater. *Nature* **387**, 166–167.

Anderson R., Rowe G., Kemp P., Trumbore S., and Biscaye P. (1994) Carbon budget for the mid-slope depocenter of the middle Atlantic bight. *Deep-Sea Res. Part II* **41**, 669–703.

Balows A., Trüper H. G., Dworkin M., Harder W., and Schleifer K.-H. (1992) *The Prokaryotes*. Springer, Berlin.

Barlow R. G. (1982) Phytoplankton ecology in the southern Benguela current: III. Dynamics of a bloom. *J. Exp. Mar. Bio. Ecol.* **63**, 239–248.

Bauer J. E. and Druffel E. R. M. (1998) Ocean margins as a significant source of organic matter to the deep open ocean. *Nature* **392**, 482–485.

Bauer J. E., Druffel E. R. M., Wolgast D. M., and Griffin S. (2001) Sources and cycling of dissolved and particulate organic radiocarbon in the northwest Atlantic continental margin. *Global Biogeochem. Cycles* **15**(3), 615–636.

Benner R., Pakulski J. D., McCarthy M., Hedges J. I., and Hatcher P. G. (1992) Bulk chemical characteristics of dissolved organic matter in the ocean. *Science* **255**(20 March 1992), 1561–1564.

Bergamaschi B. A., Tsamakis E., Keil R. G., Eglinton T. I., Montlucon D. B., and Hedges J. I. (1997) The effect of grain size and surface area on organic matter, lignin and carbonhydrate concentrations and molecular compositions in Peru margin sediments. *Geochim. Cosmochim. Acta* **61**, 1247–1260.

Berner R. A. (1982) Burial of organic matter and pyrite in the modern ocean: its geochemical and environmental significance. *Am. J. Sci.* **282**, 451–475.

Berner R. A. (1989) Biogeochemical cycles of carbon and sulfur and their effect on atmospheric oxygen over phanerozoic time. *Palaeogeogr. Palaeoclimatol. Palaeoecol.* **75**, 97–122.

Biersmith A. and Benner R. (1998) Carbohydrates in phytoplankton and freshly produced dissolved organic matter. *Mar. Chem.* **63**, 131–144.

Boetius A., Ravenschlag K., Schubert C. J., Rickert D., Widdel F., Gieseke A., Amann R., Jorgensen B. B., Witte U. and Pfannkuche O. (2000) A marine microbial consortium apparently mediating anaerobic oxidation of methane. *Nature* **407**(6804) 623–626.

Boon J. J., Klap V. A., and Eglinton T. I. (1998) Molecular characterization of microgram amounts of oceanic colloidal organic matter by direct temperature resolved ammonia chemical ionization mass spectrometry. *Org. Geochem.* **29**, 1051–1061.

Borch N. H. and Kirchman D. L. (1998) Concentration and composition of dissolved combined neutral sugars (polysaccharides) in seawater determined by HPLC-PAD. *Mar. Chem.* **57**, 85–95.

Buesseler K. O., Bauer J. E., Chen R. F., Eglinton T. I., Gustafsson O., Landing W., Mopper K., Moran S. B., Santschi P. H., VernonClark R., and Wells M. L. (1996) An intercomparison of cross-flow filtration techniques used for sampling marine colloids: overview and organic carbon results. *Mar. Chem.* **55**, 1–33.

Canfield D. E. (1989) Reactive iron in marine sediments. *Geochim. Cosmochim. Acta* **53**, 619–632.

Carlson C. A. and Ducklow H. W. (1995) Dissolved organic carbon in the upper ocean of the equatorial Pacific Ocean, 1992: daily and finescale vertical variations. *Deep-Sea Res. II* **42**(2–3), 639–656.

Carlson C. A., Ducklow H. W., and Michaels A. F. (1994) Annual flux of dissolved organic carbon from the euphotic zone in the northern Sargasso Sea. *Nature* **371**, 405–408.

Chen R. F., Fry B., Hopkinson C. S., Repeta D. J., and Peltzer E. T. (1995) Dissolved organic carbon on Georges Bank. *Continent. Shelf Res.* **16**, 409–420.

Chin W.-C., Orellana M. V., and Verdugo P. (1998) Spontaneous assembly of marine dissolved organic matter into polymer gels. *Nature* **391**, 568–571.

Churchill J., Wirick C., Flagg C., and Pietrafesa L. (1994) Sediments resuspension over the continental shelf east of the Delmarva Peninsula. *Deep-Sea Res. II* **41**, 341–363.

Cicerone R. J. and Oremland R. S. (1988) Biogeochemical aspects of atmospheric methane. *Global Geochem. Cycles* **2**(4), 299–327.

Cowie G. L. and Hedges J. I. (1992) The role of anoxia in organic matter preservation in coastal sediments: relative stabilities of the major biochemicals under oxic and anoxic depositional conditions. *Org. Geochem.* **19**(1–3), 229–234.

Cowie G. L., Hedges J. I., Prahl F. G., and de Lange G. J. (1995) Elemental and major biochemical changes across an oxidation front in a relict turbidite: an oxygen effect. *Geochim. Cosmochim. Acta* **59**(1), 33–46.

Danson M. J. and Hough D. W. (1998) Structure, function and stability of exoenzymes from the Archaea. *Trends Microbiol.* **6**, 307–314.

de Leeuw J. W. and Largeau C. (1993) A review of macromolecular organic compounds that comprise living organisms and their role in kerogen, coal, and petroleum formation. In *Organic Geochemistry* (eds. M. H. Engel and S. A. Macko). Plenum, New York, pp. 23–72.

Demaison G. (1991) Anoxia vs. productivity: what controls the formation of organic-carbon-rich sediments and sedimentary rocks? discussion. *Am. Assoc. Petrol. Geol. Bull.* **75**(3), 499.

Druffel E. R. M., Williams P. M., Bauer J. E., and Ertel J. R. (1992) Cycling of dissolved and particulate organic matter in the open ocean. *J. Geophys. Res.* **97**(C10), 15639–15659.

Durand B. and Nicaise G. (1980) Procedures for kerogen isolation. In *Kerogen—Insoluble Organic Matter from Sedimentary Rocks* (ed. B. Durand). Editions Technip, Paris, pp. 35–53.

Duursma E. K. (1963) The production of dissolved organic matter in the sea, as related to the primary gross production of organic matter. *Neth. J. Sea Res.* **2**, 85–94.

Edmond J. M., Boyle E. A., Grant B., and Stallard R. F. (1981) The chemical mass balance in the Amazon plume: I. The nutrients. *Deep-Sea Res.* **28A**, 1339–1374.

Eglinton G. (1998) The archaelogical and geological fate of biomolecules. In *Digging for Pathogens* (ed. C. L. Greenblatt). Balakan, Rehovot, Israel, pp. 299–327.

Eglinton T. I. (1994) Carbon isotopic evidence for the origin of macromolecular aliphatic structures in kerogen. *Org. Geochem.* **21**, 721–735.

Eglinton T. I., Minor E. C., Olson R. J., Zettler E. R., Boon J. J., Noguerola A., Eijkel G., and Pureveen J. (1994) Microscale characterization of algal and related particulate organic matter by direct temperature-resolved ("in-source") mass spectrometry. *Mar. Chem.* **52**, 27–54.

Eglinton T. I., Benitez-Nelson B. C., Pearson A., McNichol A. P., Bauer J. E., and Druffel E. R. M. (1997) Variability in radiocarbon ages of individual organic compounds from marine sediments. *Science* **277**, 796–799.

Eglinton T. I., Eglinton G., Dupont L., Sholkovitz E. R., Montluçon D., and Reddy C. M. (2002) Composition, age, and provenance of organic matter in NW African dust over the Atlantic Ocean. *Geochem. Geophys. Geosys. (G3)* **3**(8).

Eswaran H., Van den Berg E., and Reich P. (1993) Organic carbon in soils of the world. *J. Soil Sci. Soc. Am.* **57**, 192–194.

Francois R. (1987) A study of sulphur enrichment in the humic fraction of marine sediments during early diagenesis. *Geochim. Cosmochim. Acta* **51**, 17–27.

François R. (1990) Marine sedimentary humic substances: structure, genesis, and properties. *Rev. Aquat. Sci.* **3**, 41–80.

Gagosian R. B. and Peltzer E. T. (1986) The importance of atmospheric input of terrestrial organic material to deep sea sediments. *Org. Geochem.* **10**, 661–669.

Gagosian R. B. and Stuermer D. H. (1977) The cycling of biogenic compounds and their diagenetically altered transformation products in seawater. *Mar. Chem.* **5**, 605–632.

Gelin F. (1996) Origin and molecular characterization of the insoluble organic matter in marine sediments. In *Annual Report Netherlands Institute for Sea Research*, pp. 43–45.

Gelin F., Boogers I., Noordeloos A. A. M., Damsté J. S. S., Riegman R., and De Leeuw J. W. (1997) Resistant biomacromolecules in marine microalgae of the classes eustigmatophyceae and chlorophyceae: geochemical implications. *Org. Geochem.* **26**, 11–12.

Gelin F., Volkman J. K., Largeau C., Derenne S., Damsté J. S. S., and De Leeuw J. W. (1999) Distribution of aliphatic, nonhydrolyzable biopolymers in marine microalgae. *Org. Geochem.* **30**(2/3), 147–159.

Goericke R. and Fry B. (1994) Variations of marine plankton $\delta^{13}C$ with latitude, temperature and dissolved CO_2 in the world oceans. *Global Biogeochem. Cycles* **8**, 85–90.

Goldberg E. D. (1985) *Black Carbon in the Environment: Properties and Distribution*. Wiley, New York.

Gong C. and Hollander D. J. (1997) Differential contribution of bacteria to sedimentary organic matter in oxic and anoxic environments, Santa Monica Basin, California. *Org. Geochem.* **26**(9/10), 545–563.

Goni M. A., Ruttenberg K. C., and Eglinton T. I. (1997) Sources and contribution of terrigenous organic carbon to surface sediments in the Gulf of Mexico. *Nature* **389**, 275–278.

Goni M. A., Ruttenberg K. C., and Eglinton T. I. (1998) A reassessment of the sources and importance of land-derived organic matter in surface sediments from the Gulf of Mexico. *Geochim. Cosmochim. Acta* **62**, 3055–3075.

Goth K., Leeuw J. d., Puttmann W., and Tegelaar E. (1988) Origin of messel oil shale kerogen. *Nature* **336**(22/29), 759–761.

Gough M. A., Fauzi R., Mantoura C., and Preston M. (1993) Terrestrial plant biopolymers in marine sediments. *Geochim. Cosmochim. Acta* **57**(5), 945–964.

Guo L., Coleman C. H., and Santschi P. H. (1994) The distribution of colloidal and dissolved organic carbon in the Gulf of Mexico. *Mar. Chem.* **45**, 105–119.

Gustafsson O. and Gschwend P. M. (1998) The flux of black carbon to surface sediments on the New England continental shelf. *Geochim. Cosmochim. Acta* **62**, 465–472.

Gustafsson O., Haghseta F., Chan C., Macfarlane J., and Gschwend P. M. (1997) Quantification of the dilute sedimentary soot phase: implications for PAH speciation and bioavailability. *Environ. Sci. Technol.* **24**, 1687–1693.

Hansell D. A. and Carlson C. A. (1998) Deep-ocean gradients in the concentration of dissolved organic carbon. *Nature* **395**, 263–266.

Hartgers W. A., Lopez J. F., Damsté J. S. S., Reiss C., Maxwell J. R., and Grimalt J. O. (1997) Sulfur-binding in recent environments: speciation of sulfur and iron and implications for the occurrence of organo-sulfur compounds. *Geochim. Cosmochim. Acta* **61**(22), 4769–4788.

Harvey H. R. and Macko S. A. (1997) Catalysts or contributors? tracking bacterial mediation of early diagenesis in the marine water column. *Org. Geochem.* **26**(9/10), 531–544.

Hatcher P. G., Rowan R., and Mattingly M. A. (1980) 1H and 13CNMR of marine humic acids. *Org. Geochem.* **2**, 77–85.

Hayes J. (1979) Sandstone diagenesis—the hole truth. *SEPM (Spec. Publ.)* **26**(March), 127–139.

Hayes J. M. (1993) Factors controlling ^{13}C contents of sedimentary organic compounds: principles and evidence. *Mar. Geol.* **113**, 111–125.

Hayes J. M., Freeman K. H., and Popp B. N. (1990) Compound-specific isotopic analyses: a novel tool for reconstruction of ancient biogeochemical processes. *Org. Geochem.* **16**(4–6), 1115–1128.

Hedges J. I. (1992) Global biogeochemical cycles: progress and problems. *Mar. Chem.* **39**, 67–93.

Hedges J. I. and Keil R. G. (1995) Sedimentary organic matter preservation: an assessment and speculative synthesis. *Mar. Chem.* **49**, 81–115.

Hedges J. I. and Keil R. G. (1999) Organic geochemical perspectives on estuarine processes: sorption reactions and consequences. *Mar. Chem.* **65**(1–2), 55–65.

Hedges J. I. and Mann D. C. (1979) The characterization of plant tissues by their lignin oxidation products. **43**(11), 1803–1807.

Hedges J. I. and Oades J. M. (1997) Comparative organic geochemistries of soils and marine sediments. *Org. Geochem.* **27**, 319–361.

Hedges J. I. and Parker P. L. (1976) Land-derived organic matter in surface sediments from the Gulf of Mexico. *Geochim. Cosmochim. Acta* **40**, 1019–1029.

Hedges J. I., Ertel J. R., Quay P. D., Grootes P. M., Richey, J. E., Devol A. H., Farwell G. W., Schmidt F. W., and Salati E. (1986) Organic carbon-14 in the Amazon River system. *Limnol. Oceanogr.* **231**(4742), 1129–1131.

Hedges J. I., Cowie G. L., Richey J. E., Quay P. D., Benner, R., Strom M., and Forsberg B. R. (1994) Origins and processing of organic matter in the Amazon River as indicated by carbohydrates and amino acids. *Science* **39**(4), 743–761.

Hedges J. I., Keil R. G., and Benner R. (1997) What happens to terrestrial organic matter in the ocean? *Org. Geochem.* **27**(5/6), 195–212.

Hedges J. I., Hu F. S., Devol A. H., Hartnett H. E., Tsamakis E., and Keil R. G. (1999) A test for selective degradation under oxic conditions. *Am. J. Sci.* **299**, 529–555.

Hedges J. I., Eglinton G., Hatcher P. G., Kirchman D. L., Arnosti C., Derenne S., Evershed R. P., Kogel-Knabner I., Leeuw J. W. d., Littke R., Michaelis W., and Rullkotter J. (2000) The molecularly-uncharacterized component of nonliving organic matter in natural environments. *Org. Geochem.* **31**, 945–958.

Hedges J. I., Baldock J. A., Gelinas Y., Lee C., Peterson M., and Wakeham S. G. (2001) Evidence for non-selective preservation of organic matter in sinking marine particles. *Nature* **409**, 801–804.

Hellebust J. A. (1965) Excretion of some organic compounds by marine phytoplankton. *Limnol. Oceanogr.* **10**, 192–206.

Hinrichs K.-U., Hayes J. M., Sylva S. P., Brewer P. G., and DeLong E. F. (1999) Methane-consuming archaebacteria in marine sediments. *Nature* **398**, 802–805.

Hinrichs K.-U., Summons R. E., Orphan V., Sylva S. P., and Hayes J. M. (2000) Molecular and isotopic analysis of anaerobic methan-oxidizing communities in marine sediments. *Org. Geochem.* **31**, 1685–1701.

Hoefs M. J. L., Damsté J. S. S., Lange G. J. D., and Leewu J. W. d. (1998) Changes in kerogen composition across an oxidation front in Madeira abyssal plain turbidites as revealed by pyrolysis GC–MS. In *Proceedings of the Ocean Drilling Program, Scientific Results* (eds. P. P. E. Weaver, H.-U. Schmincke, J. V. Firth, and W. Duffield), vol. 157, pp. 591–607.

Hoefs M. J. L., Rijpstra W. I. C., and Damsté J. S. S. (2002) The influence of toxic degradation on the sedimentary biomarker record: I. Evidence from Madeira Abyssal plain turbidites. *Geochim. Cosmochim. Acta* **66**, 18.

Hoehler T. M., Alperin M. J., Albert D. B., and Martens C. S. (1994) Field and laboratory studies of methane oxidation in an anoxic marine sediment: evidence for a methanogen-sulfate reducer consortium. *Global Biogeochem. Cycles* **8**, 451–463.

Holmes R. W., Williams P. M., and Epply R. W. (1967) Red water in La Jolla Bay. *Limnol. Oceanogr.* **12**, 503–512.

Huang Y., Dupont L., Sarnthein M., Hayes J. M., and Eglinton G. (2000) Mapping of C4 Plant input from north west Africa into north east Atlantic sediments. *Geochim. Cosmochim. Acta* **64**, 3505–3513.

Hunt J. M. (1996) *Petroleum Geochemistry and Geology*. Freeman Press, New York.

Iturriaga R. and Zsolnay A. (1983) Heterotrophic uptake and transformation of phytoplankton extracellular products. *Botanica Marina* **26**, 375–381.

Jahnke R. A., Reimers C. E., and Craven D. B. (1990) Intensification of recycling of organic matter at the sea floor near ocean margins. *Nature* **348**(1 November 1990), 50–53.

Jørgensen N. O. G., Kroer N., Coffin R. B., Yang X.-H., and Lee C. (1993) Dissolved free amino acids and DNA as sources of carbon and nitrogen to marine bacteria. *Mar. Ecol. Prog. Ser.* **98**, 135–148.

Kaiser K. and Benner R. (2000) Determination of amino sugars in environmental samples with high salt content by high performance anion-exchange chromatography and pulsed amperometric detection. *Anal. Chem.* **72**, 2566–2572.

Kao S.-J. and Liu K. K. (1996) Particulate organic carbon export from a subtropical mountainous river (Lanyang Hsi) in Taiwan. *Limnol. Oceanogr.* **41**(8), 1749–1757.

Karl D. M. and Bjorkman K. M. (2002) Dynamics of DOP. In *Biogeochemistry of Marine Dissolved Organic Matter* (ed. D. Hansell). Elsevier, London, pp. 249–366.

Karner M. B., DeLong E. F., and Karl D. M. (2001) Archaeal dominance in the mesopelagic zone of the Pacific Ocean. *Nature* **409**, 507–510.

Keil R. G. and Cowie G. L. (1999) Organic matter preservation through the oxygen-deficient zone of the NE Arabian Sea as discerned by organic carbon: mineral surface area ratios. *Mar. Geol.* **161**, 13–22.

Keil R. G., Montlucon D. B., Prahl F. G., and Hedges J. I. (1994a) Sorptive preservation of labile organic matter in marine sediments. *Nature* **370**, 549–552.

Keil R. G., Tsamakis E., Bor Fuh C., Giddings J. C., and Hedges J. I. (1994b) Mineralogical and textural controls on the organic composition of coastal marine sediments: hydrodynamic separation using SPLITT-fractionation. *Geochim. Cosmochim. Acta* **58**(2), 879–893.

Keil R. G., Mayer L. M., Quay P. D., Richey J. E., and Hedges J. I. (1997) Loss of organic matter from riverine particles in deltas. *Geochim. Cosmochim. Acta* **61**, 1507–1511.

Knicker H., Scaroni A. W., and Hathcer P. G. (1996) 13C and 15N NMR spectroscopic investigation on the formation of fossil algal residues. *Org. Geochem.* **24**, 661–669.

Kohnen M. E. L., Sinninghe Damsté J. S., and de Leeuw J. W. (1991) Biases from natural sulphurization in palaeoenvironmental reconstruction based on hydrocarbon biomarker distributions. *Nature* **349**(28 February), 775–778.

Krein E. B. and Aizenshtat Z. (1994) The formation of isoprenoid sulfur compounds during diagenesis: simulated sulfur incorporation and thermal transformation. *Org. Geochem.* **21**(10–11), 1015–1025.

Kuhlbusch T. A. J. (1998) Research: ocean chemistry: black carbon and the carbon cycle. *Science* **280**(5371), 1903–1904.

Kuhlbusch T. A. J. and Crutzen P. J. (1995) Toward a global estimate of black carbon in residues of vegetation fires representing a sink of atmospheric CO_2 and a source of O_2. *Global Biogeochem. Cycles* **9**(4), 491–501.

Kvenvolden K. A. (1995) A review of the geochemistry of methane in natural gas hydrate. *Org. Geochem.* **23**, 997–1008.

Largeau C., Casadevall E., Kadouri A., and Metzger P. (1984) Comparative study of immature torbanite and of the extant alga *Botryococcus braunii*. *Org. Geochem.* **6**, 327–332.

Leithold E. L. and Blair N. E. (2001) Watershed control on the carbon loading of marine sedimentary particles. *Geochim. Cosmochim. Acta* **65**(14), 2231–2240.

Ludwig W., Probst J. L., and Kempe S. (1996) Predicting the oceanic input of organic carbon by continental erosion. *Global Biogeochem. Cycles* **10**, 23–41.

Macdonald R. W., Solomon S. M., Cranston R. E., Welch H. E., Yunker M. B., and Gobeil C. (1998) A sediment and organic carbon budget for the Canadian Beaufort Shelf. *Mar. Geol.* **144**, 255–273.

Mague T. H., Friberg E., Hughers D. J., and Morris I. (1980) Extracellular release of carbon by marine phytoplankton: a physiological approach. *Limnol. Oceanogr.* **25**, 262–279.

Malcolm R. L. (1990) The uniqueness of humic substances in each of soil, stream, and marine environments. *Anal. Chim. Acta* **232**, 19–30.

Mannino A. and Harvey H. R. (2000) Biochemical composition of particles and dissolved organic matter along an estuarine gradient: sources and implications for DOM reactivity. *Limnol. Oceanogr.* **45**, 775–788.

Masiello C. A. and Druffel E. R. M. (1998) Black carbon in deep-sea sediments. *Science* **280**, 1911–1913.

Masiello C. A. and Druffel E. R. M. (2001) Carbon isotope geochemistry of the Santa Clara River. *Global Biogeochem. Cycles* **15**(2), 407–416.

Mayer L. M. (1994) Surface area control of organic carbon accumulation in continental shelf sediments. *Geochim. Cosmochim. Acta* **58**.

Mayer L. M. (1999) Extent of coverage of mineral surfaces by organic matter in marine sediments. *Geochim. Cosmochim. Acta* **63**, 207–215.

Mayer L. M., Jumars P. A., Taghon G. L., Macko S. A., and Trumbore S. (1993) Low-density particles as potential nitrogenous foods for benthos. *J. Mar. Res.* **51**(2), 373–389.

McCarthy M. D., Hedges J. I., and Benner R. (1994) The chemical composition of dissolved organic matter in seawater. *Chem. Geol.* **107**, 503–507.

McCarthy M. D., Hedges J. I., and Benner R. (1996) Major biochemical composition of dissolved high molecular weight organic matter in seawater. *Mar. Chem.* **55**, 281–297.

McCarthy M., Pratum T., Hedges J. I., and Benner R. (1997) Chemical composition of dissolved organic nitrogen in the ocean. *Nature* **390**, 150–154.

McCarthy M. D., Hedges J. I., and Benner R. (1998) Major bacterial contribution to marine dissolved organic nitrogen. *Science* **281**, 231–234.

McKnight D. M., Harnish R., Wershaw R. L., Baron J. S., and Schiff S. (1997) Chemical characteristics of particulate, colloidal, and dissolved organic material in Loch Vale watershed, Rocky Mountain National Park. *Biogeochemistry* **99**, 99–124.

Megens L., Plicht J. v. d., and Leeuw J. W. d. (2001) Temporal variations in ^{13}C and ^{14}C concentrations in particulate organic matter from the southern North Sea. *Geochim. Cosmochim. Acta* **65**(17), 2899–2911.

Menzel D. W. (1974) Primary productivity, dissolved and particulate organic matter and the sites of oxidation of organic matter. In *The Sea* (ed. E. D. Goldberg). Wiley, New York, London, Toronto, pp. 659–678.

Meybeck M. (1982) Carbon, nitrogen, and phosphorus transport by world rivers. *Am. J. Sci.* **282**(4), 401–450.

Meyers-Schulte K. J. and Hedges J. I. (1986) Molecular evidence for a terrestrial component of organic matter dissolved in ocean water. *Nature* **321**, 61–63.

Middelburg J. J., Nieuwenhuize J., and van Breugel P. (1999) Black carbon in marine sediments. *Mar. Chem.* **65**, 245–252.

Middelburg J. J., Barranguet C., Boschker H. T. S., Herman P. M. J., Moens T., and Heip C. H. R. (2000) The fate of intertidal microphytobenthos carbon: an *in situ* super(13)C-labeling study. *Limnol. Oceanogr.* **45**(6), 1224–1234.

Milliman J. D. and Syvitski J. P. M. (1992) Geomorphic/tectonic control of sediment discharge to the ocean: the importance of small mountainous rivers. *J. Geol.* **100**, 525–544.

Mossmann J.-R., Aplin A. C., Curtis C. D., and Coleman M. L. (1991) Geochemistry of inorganic and organic sulphur in organic-rich sediments from the Peru margin. *Geochim. Cosmochim. Acta* **55**, 3581–3595.

Nagata T. and Kirchman D. L. (1992) Release of dissolved organic matter by heterotrophic protozoa, implications for microbial food webs. *Arch. Hydrobiol.* **35**, 99–109.

Nip M., Tegelaar E. W., Leeuw J. W. D., Schenck P. A., and Holloway P. J. (1986) A new non-saponifiable highly aliphatic and resistant biopolymer in plant cuticles: evidence from pyrolysis and ^{13}C NMR analysis of present day and fossil plants. *Naturwissenschaften* **73**, 579–585.

Nittrouer C. A., Kuehl S. A., Sternberg R. W., Jr., Figueiredo A. G., Jr., and Faria L. E. C. (1995) An introduction to the geological significance of sediment transport and accumulation on the Amazon continental shelf. *Mar. Geol.* **125**, 177–192.

Ohkouchi N., Eglinton Timothy I., Keigwin Lloyd D., and Hayes John M. (2002) Spatial and temporal offsets between proxy records in a sediment drift. *Science* **298**(5596), 1224–1227.

Onstad G. D., Canfield D. E., Quay P. D., and Hedges J. I. (2000) Sources of particulate organic matter in rivers from the continental USA: lignin phenol and stable carbon isotope compositions. *Geochim. Cosmochim. Acta* **64**, 3539–3546.

Opsahl S. and Benner R. (1997) Distribution and cycling of terrigenous dissolved organic matter in the ocean. *Nature* **386**, 480–482.

Pancost R., Damsté J. S. S., de Lint S., vanderMaarel M. J. E. C., Gottschal J. C., and Party M. S. (2000) Widespread anaerobic methane oxidation by methanogens in mediterranean sediments. *Appl. Environ. Microbiol.* **66**, 1126–1136.

Parkes R. J., Cragg B. A., Bale S. J., Getliff J. M., Goodman K., Rochelle P. A., Fry J. C., Weightman A. J., and Harvey S. M. (1994) Deep bacterial biosphere in Pacific Ocean sediments. *Nature* **371**, 410–412.

Parkes R. J., Cragg B. A., and Wellsbury P. (2000) Recent studies on bacterial populations and processes in subseafloor sediments: a review. *Hydrogeol. J.* **8**(1), 11–28.

Pearson A. and Eglinton T. I. (2000) The origin of n-alkanes in Santa Monica Basin surface sediment: a model based on compound-specific ^{14}C and ^{13}C data. *Org. Geochem.* **31**, 1103–1116.

Pearson A., Eglinton T. I., and McNichol A. P. (2000) An organic tracer for surface ocean radiocarbon. *Paleoceanography* **15**, 541–550.

Pearson A., McNichol A. P., Benitez-Nelson B. C., Hayes J. M., and Eglinton T. I. (2001) Origins of lipid biomarkers in Santa Monica Basin surface sediment: a case study using compound-specific D14C analysis. *Geochim. Cosmochim. Acta* **65**(18), 3123–3137.

Pedersen T. F. and Calvert S. E. (1990) Anoxia vs. productivity: what controls the formation of organic-carbon-rich sediments and sedimentary rocks? *Am. Assoc. Petrol. Geol. Bull.* **74**, 454–466.

Peltzer E. T. and Hayward N. A. (1995) Spatial and temporal variability of total organic carbon along 140° W in the equatorial Pacific Ocean in 1992. *Deep-Sea Res. Spec. EqPac (II)* **43**, 1155–1180.

Post W. M. (1993) Organic carbon in soil and the global carbon cycle. In *The Global Carbon Cycle* (ed. M. Heimann). Springer, New York, pp. 277–302.

Prahl F. G., Ertel J. R., Goni M. A., Sparrow M. A., and Eversmeyer B. (1994) Terrestrial OC on the Washington margin. *Geochim. Cosmochim. Acta* **58**, 3035–3048.

Prahl F. G., Lange G. J. D., Scholten S., and Cowie G. L. (1997) A case of post-depositional aerobic degradation of terrestrial organic matter in trubidite deposits from the Madeira Abyssal plain. *Org. Geochem.* **27**(3/4), 141–152.

Premuzic E. T., Benkovitz C. M., Gaffney J. S., and Walsh J. J. (1982) The nature and distribution of organic matter in the surface sediments of the world oceans and seas. *Org. Geochem.* **4**, 63–77.

Prospero J. M., Ginoux P., Torres O., Nicholson S. E., and Gill T. E. (2003) Environmental characterization of global sources of atmospheric soil dust identified with the NIMBUS-7 TOMS absorbing aerosol product. *Rev. Geophys.* (in press).

Putschew A., Scholz-Bottcher B. M., and Rullkotter J. (1996) Early diagenesis of organic matter and related sulfur incorporation in surface sediments of meromictic Lake Cadagno in the Swiss Alps. *Org. Geochem.* **25**, 379–390.

Ransom B., Bennett R. H., and Baerwald R. (1997) TEM study of *in situ* organic matter on continental margins: occurrence and the 'monolayer' hypothesis. *Mar. Geol.* **138**, 1–9.

Ransom B., Kim D., Kastner M., and Wainwright S. (1998a) Organic matter preservation on continental slopes: importance of mineralogy and surface area. *Geochim. Cosmochim. Acta* **62**(8), 1329–1345.

Ransom B., Shea K. F., Burkett P. J., Bennett R. H., and Baerwald R. (1998b) Comparison of pelagic and nepheloid layer marine snow: implications for carbon cycling. *Mar. Geol.* **150**, 39–50.

Rau G. H., Takahashi T., and Des Marais D. J. (1989) Latitudinal variations in plankton δ^{13}C: implications for CO_2 and productivity in past oceans. *Nature* **341**(12 October 1989), 516–518.

Raymond P. A. and Bauer J. E. (2001a) Riverine export of aged terrestrial organic matter to the North Atlantic Ocean. *Nature* **409**, 497–500.

Raymond P. A. and Bauer J. E. (2001b) Use of super(14)C and super(13)C natural abundances for evaluating riverine, estuarine, and coastal DOC and POC sources and cycling: a review and synthesis. *Org. Geochem.* **32**(4), 469–485.

Reeburgh W. S., Ward B. B., Whalen S. C., Sandbeck K. A., Kilpatrick K. A., and Kerkhof L. J. (1991) Black Sea methane geochemistry. *Deep-Sea Res.* **38**(suppl. 2), S1189–S1210.

Repeta D. J., Quan T. M., Aluwihare L. I., and Accardi A. M. (2002) Chemical characterization of high molecular weight dissolved organic matter from fresh and marine waters. *Geochim. Cosmochim. Acta* **66**, 955–962.

Romankevich E. A. (1984) *Geochemistry of Organic Matter in the Ocean*, Springer, Berlin, 351pp.

Sakugawa H. and Handa N. (1985) Isolation and chemical characterization of dissolved and particulate polysaccharides in Mikawa Bay. *Geochim. Cosmochim. Acta* **49**, 1185–1193.

Santschi P. H., Balnois E., Wilkinson K. J., Zhang J., and Buffle J. (1998) Fibrillar polysaccharides in marine macromolecular organic matter as imaged by atomic force microscopy and transmission electron microscopy, **43**, 896–908.

Schlunz B. and Schneider R. R. (2000) Transport of terrestrial organic carbon to the oceans by rivers: re-estimating flux-and burial rates. *Int. J. Earth Sci.* **88**, 599–606.

Schmidt M. W. I. and Noack A. G. (2000) Black carbon in soils and sediments: analysis, distribution, implications, and current challenges. *Global Geochem. Cycles* **14**(3), 777–793.

Schmidt M. W. I., Skjemstad J. O., Gehrt E., and Kogel-Knabner I. (1999) Charred organic carbon in German chernozemic soils. *Euro. J. Soil. Sci.* **50**, 351–365.

Schouten S., de Graaf W., Sinninghe Damsté J. S., van Driel G. B., and de Leeuw J. W. (1994) Laboratory simulation of natural sulfurization: II. Reaction of multifunctionalized lipids with inorganic polysulfides at low temperatures. *Org. Geochem.* **22**, 825–834.

Schouten S., Hoefs M. J. L., Koopmans M. P., Bosch H.-J., and Damsté J. S. S. (1998) Structural identification, occurrence and fate of archaeal ether-bound acyclic and cyclic biphytanes and corresponding diols in sediments. *Org. Geochem.* **29**, 1305–1319.

Schouten S., Wakeham S. G., and Sinninghe Damsté J. S. (2001) Evidence for anaerobic methane oxidation by archaea in euxinic waters of the Black Sea. *Org. Geochem.* **32**, 1277–1281.

Schubert C. J. and Stein R. (1996) Deposition of organic carbon in Arctic Ocean sediments: terrigenous supply vs. marine productivity. *Org. Geochem.* **24**(4), 421–436.

Siegenthaler U. and Sarmiento J. L. (1993) Atmospheric carbon dioxide and the ocean, **365**, 119–125.

Sinninghe Damsté J. S. and Schouten S. (1997) Is there evidence for a substantial contribution of prokaryotic biomass to organic carbon in phanerozoic carbonaceous sediments? *Org. Geochem.* **26**(9/10), 517–530.

Sinninghe Damsté J. W., Eglinton T. I., de Leeuw J. W., and Schenck P. A. (1989) Organic sulfur in macromolecular sedimentary organic matter: I. Structure and origin of sulfur-containing moieties in kerogen, asphaltenes and coal as revealed by flash pyrolysis. *Geochim. Cosmochim. Acta* **53**, 873–889.

Sinninghe Damsté J. S., Kok M. D., Koster J., and Schouten S. (1998) Sulfurized carbohydrates: an important sedimentary sink for organic carbon? *Earth Planet. Sci. Lett.* **164**, 7–13.

Sinninghe Damsté J. S., Rijpstra W. I. C., Hopmans E. C., Prahl F. G., Wakeham S. G., and Schouten S. (2002) Distribution of membrane lipids of planktonic Crenarchaeota in the Arabian Sea. *Appl. Environ. Microbiol.* **68**(6), 2997–3002.

Skjemstad J. O., Clarke P., Taylor J. A., Oades J. M., and McClure S. G. (1996) The chemistry and nature of protected carbon in soil. *Austral. J. Soil Res.* **34**(2), 251–271.

Smith K., Kaufman R., and Baldwin R. (1994) Coupling of near-bottom pelagic and benthic processes at abyssal depths. *Limnol. Oceanol.* **39**, 1101–1118.

Smith S. V. and Hollibaugh J. T. (1993) Coastal metabolism and the oceanic organic carbon balance. *Rev. Geophys.* **31**, 75–89.

Smith S. V. and MacKenzie F. T. (1987) The ocean as a net heterotrophic system: implications from the carbon biogeochemical cycle. *Global Biogeochem. Cycles* **1**, 187–198.

Stuermer D. and Harvey G. R. (1974) Humic substances from seawater. *Nature* **250**, 480–481.

Stuermer D. and Payne J. R. (1976) Investigations of seawater and terrestrial humic substances with carbon-13 and proton magnetic resonance. *Geochim. Cosmochim. Acta* **40**, 1109–1114.

Stuiver M. and Pollach H. A. (1977) On the reporting of ^{14}C ages. *Radiocarbon* **35**, 355–365.

Suess E. (1980) Particulate organic carbon flux in the oceans; surface productivity and oxygen utilization. *Nature (London)* **288**(5788), 260–263.

Suman D. O., Khulbusch T. A. J., and Lim B. (1997) Marine sediments: a reservoir for black carbon and their use as spatial and temporal records of combustion. In *Sedimental Records of Biomass Burning and Global Change* (ed. J. S. Clark). Springer, New York, pp. 271–293.

Tanoue E., Nishiyama S., Kamo M., and Tsugita A. (1995) Bacterial membranes: possible source of a major dissolved protein in seawater. *Geochim. Cosmochim. Acta* **59**, 2643–2648.

Tegelaar E. W., de Leeuw J. W., Derenne S., and Largeau C. (1989) A reappraisal of kerogen formation. *Geochim. Cosmochim. Acta* **53**, 3103–3106.

Thomsen L. and Van Weering T. J. (1998) Spatial and temporal variability of particulate matter in the benthic boundary layer at the NW European continental margin (Goban Spur). *Prog. Oceanogr.* **42**, 61–76.

Tissot B. P. and Welte D. H. (1984) *Petroleum Formation and Occurrence*. Springer, New York.

Torn M. S., Trumbore S. E., Chadwick O. A., Vitousek P. M., and Hendricks D. M. (1997) Mineral control on carbon storage and turnover. *Nature* **389**, 170–173.

Vaccaro R. F., Hicks S. E., Jannasch H. W., and Carey F. G. (1968) The occurrence and role of glucose in seawater. *Limnol. Oceanogr.* **13**, 356–360.

Vairavamurthy A. and Mopper K. (1987) Geochemical formation of organosulphur compounds (thiols) by addition of H_2S to sedimentary organic matter. *Nature* **329**(6140), 623–625.

van de Meent D., Brown S. C., Philp R. P., and Simoneit B. R. T. (1980) Pyrolysis-high resolution gas chromatography and pyrolysis-gas chromatography-mass spectrometry of kerogen and kerogen precursors. *Geochim. Cosmochim. Acta* **44**, 999–1013.

Vlahos P., Chen R. F., and Repeta D. J. (2002) Dissolved organic carbon in the mid-Atlantic bight. *Deep-Sea Res.* **49**, 4369–4385.

Volkman J. K., Barrett S. M., Blackburn S. I., Mansour M. P., Sikes E. L., and Gelin F. (1998) Microalgal biomarkers: a review of recent research developments. *Org. Geochem.* **29**, 1163–1176.

Wakeham S. G. and Lee C. (1993) Production, transport, and alteration of particulate organic matter in the marine water column. In *Organic Geochemistry* (eds. M. H. Engel and S. A. Macko). Plenum, New York, pp. 145–169.

Wakeham S. G., Lee C., Hedges J. I., Hernes P. J., and Peterson M. L. (1997) Molecular indicators of diagenetic status in marine organic matter. *Geochim. Cosmochim. Acta* **61**, 5363–5369.

Wang X.-C. and Druffel E. R. M. (2001) Radiocarbon and stable carbon isotope compositions of organic compound classes in sediments from the NE Pacific and Southern Oceans. *Mar. Chem.* **73**, 65–81.

Wang X.-C., Druffel E. R. M. and Lee C. (1996) Radiocarbon in organic compound classes in particulate organic matter and sediment in the deep northeast Pacific Ocean. *Geophys. Res. Lett.* **23**, 3583–3586.

Wang X. C., Druffel E. R. M., Griffin S., Lee C., and Kashgarian M. (1998) Radiocarbon studies of organic compound classes in plankton and sediment of the northeastern Pacific Ocean. *Geochim. Cosmochim. Acta* **62**, 1365–1378.

Wells M. L. and Goldberg E. D. (1991) Occurrence of small colloids in sea water. *Nature* **353**, 342–344.

Wells M. L. and Goldberg E. D. (1993) Colloid aggregation in seawater. *Mar. Chem.* **41**, 353–358.

Wells M. L. and Goldberg E. D. (1994) The distribution of colloids in the North Atlantic and Southern Oceans. *Limnol. Oceanogr.* **39**, 286–302.

Werne J. P., Hollander D. J., Behrens A., Schaeffer P., Albrecht P., and Damsté J. S. S. (2000) Timing of early diagenetic sulfurization of organic matter: a precursor-product relationship in holocene sediments of the anoxic Cariaco Basin, Venezuela. *Geochim. Cosmochim. Acta* **64**(10), 1741–1751.

Williams P. M., Oeschger H., and Kinney P. (1969) Natural radiocarbon activity of dissolved organic carbon in the north-east Pacific Ocean. *Nature* **224**, 256–259.

Wolbach W. S. and Anders E. (1989) Elemental carbon in sediments: determination and isotopic analysis in the presence of kerogen. *Geochim. Cosmochim. Acta* **53**, 1637–1647.

Wollast R. (1991) The coastal organic carbon cycle: fluxes, sources, and sinks. In *Physical, Chemical, and Earth Sciences Research Report*, vol. 9, pp. 365–381.

Zafiriou O. C., Gagosian R. B., Peltzer E. T., Alford J. B., and Loder T. (1985) Air-to-sea fluxes of lipids at Enewetak Atoll. *J. Geophys. Res.* **90**(D1), 2409–2423.

Zang X., Nguyen R. T., Harvey H. R., Knicker H., and Hatcher P. G. (2001) Preservation of proteinaceous material during the degradation of the green alga *Botryococcus braunii*: a solid-state 2D 15N 13C NMR spectroscopy study. *Geochim. Cosmochim. Acta* **65**(19), 3299–3305.

Zelibor J. L., Romankiw L., Hatcher P. G., and Colwell R. R. (1988) Comparative analysis of the chemical composition of mixed and pure cultures of green algae and their decomposed residues by ^{13}C nuclear resonance spectroscopy. *Appl. Environ. Microbiol.* **54**, 1051–1060.

6.07
Hydrothermal Processes

C. R. German
Southampton Oceanography Centre, Southampton, UK
and
K. L. Von Damm
University of New Hampshire, Durham, NH, USA

6.07.1 INTRODUCTION

6.07.1.1 What is Hydrothermal Circulation?

Hydrothermal circulation occurs when seawater percolates downward through fractured ocean crust along the volcanic mid-ocean ridge (MOR) system. The seawater is first heated and then undergoes chemical modification through reaction with the host rock as it continues downward, reaching maximum temperatures that can exceed 400 °C. At these temperatures the fluids become extremely buoyant and rise rapidly back to the seafloor where they are expelled into the overlying water column. Seafloor hydrothermal circulation plays a significant role in the cycling of energy and mass between the solid earth and the oceans; the first identification of submarine hydrothermal venting and their accompanying chemosynthetically based communities in the late 1970s remains one of the most exciting discoveries in modern science. The existence of some form of hydrothermal circulation had been predicted almost as soon as the significance of ridges themselves was first recognized, with the emergence of plate tectonic theory. Magma wells up from the Earth's interior along "spreading centers" or "MORs" to produce fresh ocean crust at a rate of $\sim 20\ km^3\ yr^{-1}$, forming new seafloor at a rate of $\sim 3.3\ km^2\ yr^{-1}$ (Parsons, 1981; White *et al.*, 1992). The young oceanic lithosphere formed in this way cools as it moves away from the ridge crest. Although much of this cooling occurs by upward conduction of heat through the lithosphere, early heat-flow studies quickly established that a significant proportion of the total heat flux must also occur via some additional *convective* process (Figure 1), i.e., through circulation of cold seawater within the upper ocean crust (Anderson and Silbeck, 1981).

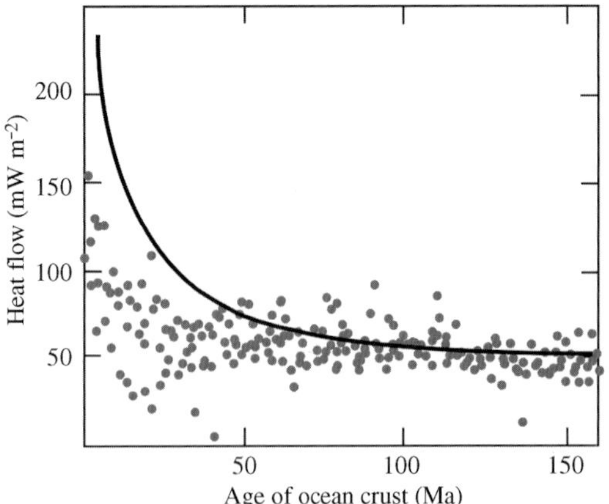

Figure 1 Oceanic heat flow versus age of ocean crust. Data from the Pacific, Atlantic, and Indian oceans, averaged over 2 Ma intervals (circles) depart from the theoretical cooling curve (solid line) indicating convective cooling of young ocean crust by circulating seawater (after C. A. Stein and S. Stein, 1994).

The first *geochemical* evidence for the existence of hydrothermal vents on the ocean floor came in the mid-1960s when investigations in the Red Sea revealed deep basins filled with hot, salty water (40–60 °C) and underlain by thick layers of metal-rich sediment (Degens and Ross, 1969). Because the Red Sea represents a young, rifting, ocean basin it was speculated that the phenomena observed there might also prevail along other young MOR spreading centers. An analysis of core-top sediments from throughout the world's oceans (Figure 2) revealed that such metalliferous sediments did, indeed, appear to be concentrated along the newly recognized global ridge crest (Boström *et al.*, 1969). Another early indication of hydrothermal activity came from the detection of plumes of excess ^3He in the Pacific Ocean Basin (Clarke *et al.*, 1969)—notably the >2,000 km wide section in the South Pacific (Lupton and Craig, 1981)—because ^3He present in the deep ocean could only be sourced through some form of active degassing of the Earth's interior, at the seafloor.

One area where early heat-flow studies suggested hydrothermal activity was likely to occur was along the Galapagos Spreading Center in the eastern equatorial Pacific Ocean (Anderson and Hobart, 1976). In 1977, scientists diving at this location found hydrothermal fluids discharging chemically altered seawater from young volcanic seafloor at elevated temperatures up to 17 °C (Edmond *et al.*, 1979). Two years later, the first high-temperature (380 ± 30 °C) vent fluids were found at 21° N on the East Pacific Rise (EPR) (Spiess *et al.*, 1980)—with fluid compositions remarkably close to those predicted from the lower-temperature Galapagos findings (Edmond *et al.*, 1979). Since that time, hydrothermal activity has been found at more than 40 locations throughout the Pacific, North Atlantic, and Indian Oceans (e.g., Van Dover *et al.*, 2002) with further evidence—from characteristic chemical anomalies in the ocean water column—of its occurrence in even the most remote and slowly spreading ocean basins (Figure 3), from the polar seas of the Southern Ocean (German *et al.*, 2000; Klinkhammer *et al.*, 2001) to the extremes of the ice-covered Arctic (Edmonds *et al.*, 2003).

The most spectacular manifestation of seafloor hydrothermal circulation is, without doubt, the high-temperature (>400 °C) "black smokers" that expel fluids from the seafloor along all parts of the global ocean ridge crest. In addition to being visually compelling, vent fluids also exhibit important enrichments and depletions when compared to ambient seawater. Many of the dissolved chemicals released from the Earth's interior during venting precipitate upon mixing

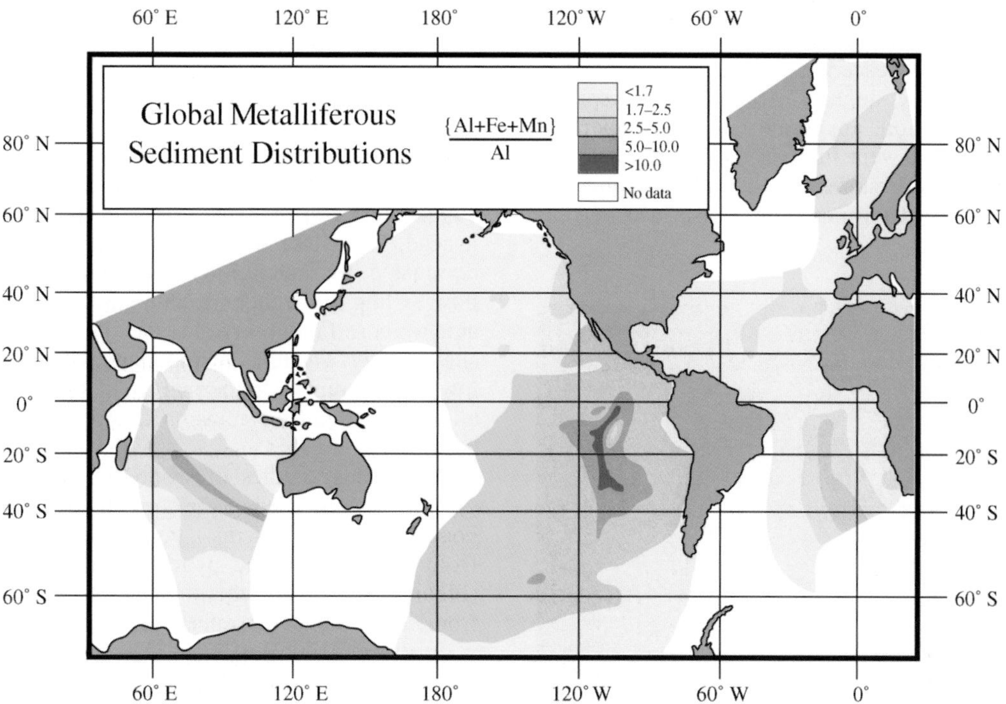

Figure 2 Global map of the (Al + Fe + Mn):Al ratio for surficial marine sediments. Highest ratios mimic the trend of the global MOR axis (after Boström *et al.*, 1969).

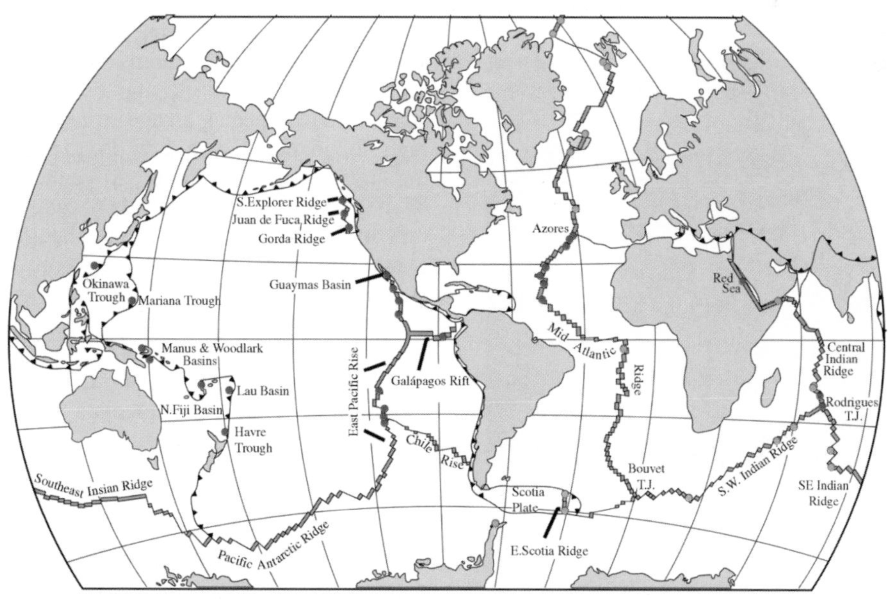

Figure 3 Schematic map of the global ridge crest showing the major ridge sections along which active hydrothermal vents have already been found (red circles) or are known to exist from the detection of characteristic chemical signals in the overlying water column (orange circles). Full details of all known hydrothermally active sites and plume signals are maintained at the InterRidge web-site: http://triton.ori.u-tokyo.ac.jp/~intridge/wg-gdha.htm

with the cold, overlying seawater, generating thick columns of black metal-sulfide and oxide mineral-rich smoke—hence the colloquial name for these vents: "black smokers" (Figure 4). In spite of their common appearance, high-temperature hydrothermal vent fluids actually exhibit a wide range of temperatures and chemical compositions, which are determined by subsurface reaction conditions. Despite their spectacular appearance, however, high-temperature vents may only represent a small fraction—perhaps as little as 10%—of the total hydrothermal heat flux close to ridge axes. A range of studies—most notably along the Juan de Fuca Ridge (JdFR) in the NE Pacific Ocean (Rona and Trivett, 1992; Schultz *et al.*, 1992; Ginster *et al.*, 1994) have suggested that, instead, axial hydrothermal circulation may be dominated by much lower-temperature diffuse flow exiting the seafloor at temperatures comparable to those first observed at the Galapagos vent sites in 1977. The relative importance of high- and low-temperature hydrothermal circulation to overall ocean chemistry remains a topic of active debate.

(a)

(b)

Figure 4 (a) Photograph of a "black smoker" hydrothermal vent emitting hot (>400 °C) fluid at a depth of 2,834 m into the base of the oceanic water column at the Brandon vent site, southern EPR. The vent is instrumented with a recording temperature probe. (b) Diffuse flow hydrothermal fluids have temperatures that are generally <35 °C and, therefore, may host animal communities. This diffuse flow site at a depth of 2,500 m on the EPR at 9° 50′ N is populated by *Riftia* tubeworms, mussels, crabs, and other organisms.

While most studies of seafloor hydrothermal systems have focused on the currently active plate boundary (~0–1 Ma crust), pooled heat-flow data from throughout the world's ocean basins (Figure 1) indicate that convective heat loss from the oceanic lithosphere actually continues in crust from 0–65 Ma in age (Stein *et al.*, 1995). Indeed, most recent estimates would indicate that hydrothermal circulation through this older (1–65 Ma) section, termed "flank fluxes," may be responsible for some 70% or more of the total hydrothermal heat loss associated with spreading-plate boundaries—either in the form of warm (20–65 °C) altered seawater, or as cooler water, which is only much more subtly chemically altered (Mottl, 2003).

When considering the impact of hydrothermal circulation upon the chemical composition of the oceans and their underlying sediments, however, attention returns—for many elements—to the high-temperature "black smoker" systems. Only here do many species escape from the seafloor in high abundance. When they do, the buoyancy of the high-temperature fluids carries them hundreds of meters up into the overlying water column as they mix and eventually form nonbuoyant plumes containing a wide variety of both dissolved chemicals and freshly precipitated mineral phases. The processes active within these dispersing hydrothermal plumes play a major role in determining the net impact of hydrothermal circulation upon the oceans and marine geochemistry.

6.07.1.2 Where Does Hydrothermal Circulation Occur?

Hydrothermal circulation occurs predominantly along the global MOR crest, a near-continuous volcanic chain that extends over ~6×10^4 km (Figure 3). Starting in the Arctic basin this ridge system extends south through the Norwegian-Greenland Sea as far as Iceland and then continues southward as the Mid-Atlantic Ridge (MAR), passing through the Azores and onward into the far South Atlantic, where it reaches the Bouvet Triple Junction, near 50° S. To the west, a major transform fault connects this triple junction to the Sandwich and Scotia plates that are separated by the East Scotia Ridge (an isolated back-arc spreading center). These plates are also bound to north and south by two major transform faults that extend further west between South America and the Antarctic Peninsula before connecting to the South Chile Trench. To the east of the Bouvet Triple Junction lies the SW Indian Ridge, which runs east and north as far as the Rodrigues Triple Junction (~25° S, 70° E), where the ridge crest splits in two. One branch, the Central Indian Ridge, extends north through the western Indian Ocean and Gulf of Aden ending as the incipient ocean basin that is the Red Sea (Section 6.07.1.1). The other branch of the global ridge crest branches south east from the Rodrigues Triple Junction to form the SE Indian and Pacific-Antarctic Ridges which extend across the entire southern Indian Ocean past Australasia and on across the southern Pacific Ocean as far as ~120° W, where the ridge again strikes north. The ridge here, the EPR, extends from ~55° S to ~30° N but is intersected near 30° S by the Chile Rise, which connects to the South Chile Trench. Further north, near the equator, the Galapagos Spreading Center meets the EPR at another triple junction. The EPR (and, hence, the truly continuous portion of the global ridge crest, extending back through the Indian and Atlantic Oceans) finally ends where it runs "on-land" at the northern end of the Gulf of California. There, the ridge crest is offset to the NW by a transform zone, more commonly known as the San Andreas Fault, which continues offshore once more, off northern California at ~40° N, to form the Gorda, Juan de Fuca, and Explorer Ridges—all of which hug the NE Pacific/N. American margin up to ~55° N. Submarine hydrothermal activity is also known to be associated with the back-arc spreading centers formed behind ocean–ocean subduction zones which occur predominantly around the northern and western margins of the Pacific Ocean, from the Aleutians via the Japanese archipelago and Indonesia all the way south to New Zealand. In addition to ridge-crest hydrothermal venting, similar circulation also occurs associated with hot-spot related intraplate volcanism—most prominently in the central and western Pacific Ocean (e.g., Hawaii, Samoa, Society Islands), but these sites are much less extensive, laterally, than ridge crests and back-arc spreading centers, combined. A continuously updated map of reported hydrothermal vent sites is maintained by the InterRidge community as a Vents Database (http://triton.ori.u-tokyo.ac.jp/~intridge/wg-gdha.htm).

As described earlier, the first sites of hydrothermal venting to be discovered were located along the intermediate to fast spreading Galapagos Spreading Center (6 cm yr^{-1}) and northern EPR (6–15 cm yr^{-1}). A hypothesis, not an unreasonable one, influenced heavily by these early observations but only formalized nearly 20 years later (Baker *et al.*, 1996) proposed that the incidence of hydrothermal venting along any unit length of ridge crest should correlate positively with spreading-rate because the latter is intrinsically linked to the magmatic heat flux at that location. Thus, the faster the spreading rate the more abundant the hydrothermal activity, with the most abundant venting expected (and found: Charlou *et al.*, 1996; Feely *et al.*, 1996; Ishibashi *et al.*, 1997) along the superfast spreading southern EPR (17–19° S), where ridge-spreading rate is among the fastest known (>14 cm yr^{-1}). Evidence for reasonably widespread venting has also been found most recently along some of the slowest-spreading sections of the global ridge crest, both in the SW Indian Ocean (German *et al.*, 1998a; Bach *et al.*, 2002) and in the Greenland/Arctic Basins (Connelly *et al.*, 2002; Edmonds *et al.*, 2003). Most explorations so far, however, have focused upon ridge crests closest to nations with major oceanographic research fleets and in the low- to mid-latitudes, where weather conditions are most favorable toward use of key research tools such as submersibles and deep-tow vehicles. Consequently, numerous active vent sites are known along the NE Pacific ridge crests, in the western Pacific back-arc basins and along the northern MAR (Figure 3), while other parts of the global MOR system remain largely unexplored (e.g., southern MAR, Central Indian Ridge, SE Indian Ridge, Pacific–Antarctic Ridge).

Reinforcing how little of the seafloor is well explored, as recently as December 2000 an entirely new form of seafloor hydrothermal activity, in a previously unexplored geologic setting was discovered (Kelley *et al.*, 2001). Geologists diving at the Atlantis fracture zone, which offsets part of the MAR near 30° N, found moderate-temperature fluids (40–75 °C) exiting from tall (up to 20 m) chimneys, formed predominantly from calcite [$CaCO_3$], aragonite [$CaCO_3$], and brucite [$Mg(OH)_2$]. These compositions are quite unlike previously documented hydrothermal vent fluids (Section 6.07.2), yet their geologic setting is one that may recur frequently along slow and very slow spreading ridges (e.g., Gracia *et al.*, 1999, 2000, Parson *et al.*, 2000; Sauter *et al.*, 2002). The Lost City vent site may, therefore, represent a new and important form of hydrothermal vent input to the oceans, which has hitherto been overlooked.

6.07.1.3 Why Should Hydrothermal Fluxes Be Considered Important?

Since hydrothermal systems were first discovered on the seafloor, determining the magnitude of their flux to the ocean and, hence, their importance in controlling ocean chemistry has been the overriding question that numerous authors have tried to assess (Edmond *et al.*, 1979, 1982; Staudigel and Hart, 1983; Von Damm *et al.*, 1985a; C. A. Stein and S. Stein, 1994; Elderfield and Schultz, 1996; Schultz and Elderfield, 1997; Mottl, 2003). Of the total heat flux from the interior of the Earth (~43 TW) ~32 TW is associated with cooling through oceanic crust and, of this, some 34% is estimated to occur in the form of hydrothermal circulation through ocean crust up to 65 Ma in age (C. A. Stein and S. Stein, 1994). The heat supply that drives this circulation is of two parts: magmatic heat, which is actively *emplaced* close to the ridge axis during crustal formation, and heat that is *conducted* into the crust from cooling lithospheric mantle, which extends out beneath the ridge flanks.

At the ridge axis, the magmatic heat available from crustal formation can be summarized as (i) heat released from the crystallization of basaltic magma at emplacement temperatures (latent heat), and (ii) heat mined from the solidified crust during cooling from emplacement temperatures to hydrothermal temperatures, assumed by Mottl (2003) to be 1175 ± 25 °C and 350 ± 25 °C, respectively. For an average crustal thickness of ~6 km (White *et al.*, 1992) the mass of magma emplaced per annum is estimated at 6×10^{16} g yr^{-1} and the maximum heat available from crystallization of this basaltic magma and cooling to hydrothermal temperatures is 2.8 ± 0.3 TW (Elderfield and Schultz, 1996; Mottl, 2003). If all this heat were transported as high-temperature hydrothermal fluids expelled from the seafloor at 350 °C and 350 bar, this heat flux would equate to a volume flux of $5-7 \times 10^{16}$ g yr^{-1}. It should be noted, however, that the heat capacity (c_p) of a 3.2% NaCl solution becomes extremely sensitive to increasing temperature under hydrothermal conditions of temperature and pressure, as the critical point is approached. Thus, for example, at 350 bar, a moderate increase in temperature near 400 °C could cause an increase in c_p approaching an order of magnitude resulting in a concomitant drop in the water flux required to transport this much heat (Bischoff and Rosenbauer, 1985; Elderfield and Schultz, 1996).

Of course, high-temperature hydrothermal fluids may not be entirely responsible for the transport of all the axial hydrothermal heat flux. Elderfield and Schultz (1996) considered a uniform distribution, on the global scale, in which only 10% of the total axial hydrothermal flux occurred as "focused" flow (heat flux = 0.2–0.4 TW; volume flux = $0.3-0.6 \times 10^{16}$ g yr^{-1}). In those calculations, the remainder of the axial heat flux was assumed to be transported by a much larger volume flux of lower-temperature fluid ($280-560 \times 10^{16}$ g yr^{-1} at ~5 °C). But how might such diffuse flow manifest itself? Should diffuse fluid be considered as diluted high-temperature vent fluid, conductively heated seawater, or some combination of the above? Where might such diffuse fluxes occur? Even if the axial hydrothermal heat flux were only restricted to 0–0.1 Ma crust, the associated fluid flow might still extend over the range of kilometers from the axis on medium-fast ridges—i.e., out onto young ridge flanks. For slow and ultraslow spreading ridges (e.g., the MAR) by contrast, all 0–0.1 Ma and, indeed 0–1 Ma crustal circulation would occur within the confines of the axial rift valley (order 10 km wide). The partitioning of "axial" and "near-axial" hydrothermal flow, on fast and slow ridges and between "focused" and "diffuse" flow, remains very poorly constrained in the majority of MOR settings and is an area of active debate.

On older oceanic crust (1–65 Ma) hydrothermal circulation is driven by upward conduction of heat from cooling of the underlying lithospheric mantle. Heat fluxes associated with this process are estimated at 7 ± 2 TW (Mottl, 2003). These values are significantly greater than the total heat fluxes associated with axial and near-axis circulation combined, and represent as much as 75–80% of Earth's total *hydrothermal* heat flux, >20% of the total *oceanic* heat flux and >15% of the Earth's *entire* heat flux. Mottl and Wheat (1994) chose to subdivide the fluid circulation associated with this heat into two components, warm (>20 °C) and cool (<20 °C) fluids, which exhibit large and small changes in the composition of the circulating seawater, respectively. Constraints from the magnesium mass balance of the oceans suggest that the cool (less altered) fluids carry some 88% of the total *flank* heat flux, representing a cool-fluid water flux (for 5–20 °C fluid temperatures) of $1-4 \times 10^{19}$ g yr^{-1} (Mottl, 2003).

To put these volume fluxes in context, the maximum flux of cool (<20 °C) hydrothermal fluids, calculated above is almost identical to the global riverine flux estimate of $3.7-4.2 \times 10^{19}$ g yr^{-1} (Palmer and Edmond, 1989). The flux of high-temperature fluids close to the ridge axis, by contrast, is ~1,000-fold lower. Nevertheless, for an ocean volume of ~1.4×10^{24} g, this still yields a (geologically short) oceanic residence time, with respect to high-temperature circulation, of ~20–30 Ma—and the hydrothermal fluxes will be important for those elements which exhibit high-temperature fluid concentrations more than 1,000-fold greater than river waters. Furthermore, high-temperature fluids emitted from "black smoker" hydrothermal systems typically entrain large volumes of ambient seawater during the formation of buoyant and neutrally buoyant plumes (Section 6.07.5) with typical dilution ratios of $\sim 10^4$:1 (e.g., Helfrich and Speer, 1995). If 50% of the fluids circulating at high temperature through young ocean crust are entrained into hydrothermal plumes then the total water flux through hydrothermal plumes would be approximately one order of magnitude greater than all other hydrothermal fluxes *and* the global riverine flux to the oceans (Table 1). The associated residence time of the global ocean, with respect to cycling through hydrothermal plume entrainment, would be just 4–8 kyr, i.e., directly comparable to the mixing time of the global deep-ocean conveyor (~1.5 kyr; Broecker and Peng, 1982). From that perspective, therefore, we can anticipate that hydrothermal circulation should play an important role in the marine geochemistry of any tracer which exhibits a residence time greater than ~1–10 kyr in the open ocean (see Chapter 6.02).

The rest of the chapter is organized as follows. In Section 6.07.2 we discuss the chemical composition of hydrothermal fluids, why they are important, what factors control their compositions, and how these compositions vary, both in space, from one location to another, and in time. Next (Section 6.07.3) we identify that the fluxes established thus far represent gross fluxes into and out of the ocean crust associated with high-temperature venting. We then examine the other source and sink terms associated with hydrothermal circulation, including alteration of the oceanic crust, formation of hydrothermal mineral deposits, interactions/uptake within hydrothermal plumes and settling into deep-sea sediments. Each of these "fates" for hydrothermal material is then considered in more detail. Section 6.07.4 provides a detailed discussion of near-vent deposits, including the formation of polymetallic sulfides and other minerals, as well as near-vent sediments. In Section 6.07.5 we present a detailed description of the processes associated with hydrothermal plumes, including a brief explanation of basic plume dynamics, a discussion of how plume processes modify the gross flux from high-temperature venting and further discussions of how plume chemistry can be both determined by, and influence, physical oceanographic, and biological interactions. Section 6.07.6 discusses the fate of hydrothermal products and concentrates on ridge-flank metalliferous sediments, including their potential for paleoceanographic investigations and role in "boundary scavenging" processes. We conclude (Section 6.07.7) by identifying some of the unresolved questions associated with hydrothermal circulation that are most in need of further investigation.

6.07.2 VENT-FLUID GEOCHEMISTRY

6.07.2.1 Why are Vent-fluid Compositions of Interest?

The compositions of vent fluids found on the global MOR system are of interest for several reasons; how and why those compositions vary has important implications. The overarching question, as mentioned in Section 6.07.1.3, is to determine how the fluids emitted from these systems influence and control ocean chemistry, on both short and long timescales. This question is very difficult to address in a quantitative manner because, in addition to all the heat flux and related water flux uncertainties discussed in Section 6.07.1, it also requires an understanding of the range of chemical variation in these systems and an understanding of the mechanisms and variables that control vent-fluid chemistries and temperatures. Essentially every hydrothermal vent that is discovered has a different composition (e.g., Von Damm, 1995) and we now know that these compositions often vary profoundly on short

Table 1 Overview of hydrothermal fluxes: heat and water volume: data from Elderfield and Schultz (1996) and Mottl (2003).

(I) Summary of global heat fluxes		
Heat flux from the Earth's interior	43 TW	
Heat flux associated with ocean crust	32 TW	
Seafloor hydrothermal heat flux	11 TW	
(II) Global hydrothermal fluxes: heat and water		
	Heat flux (TW)	Water flux (10^{16} g yr^{-1})
Axial flow (0–1 Ma)		
All flow at 350 °C	2.8 ± 0.3	5.6 ± 0.6
10%@350 °C/90%@5 °C	2.8 ± 0.3	420 ± 140
Hydrothermal plumes (50%)		28,000 ± 3,000
Off-axis flow (1–65 Ma)	7 ± 2	1,000–4,000
Global riverine flux		3,700–4,200

(minutes to years) timescales. Hence, the flux question remains a difficult one to answer. Vent fluid compositions also act as sensitive and unique indicators of processes occurring within young oceanic crust and at present, this same information cannot be obtained from any other source. The "window" that vent fluids provide into subsurface crustal processes is especially important because we can not yet drill young oceanic crust, due to its unconsolidated nature, unless it is sediment covered. The chemical compositions of the fluids exiting at the seafloor provide an integrated record of the reactions and the pressure and temperature ($P-T$) conditions these fluids have experienced during their transit through the crust. Vent fluids can provide information on the depth of fluid circulation (hence, information on the depth to the heat source), as well as information on the residence time of fluids within the oceanic crust at certain temperatures. Because the dissolved chemicals in hydrothermal fluids provide energy sources for microbial communities living within the oceanic crust, vent-fluid chemistries can also provide information on whether such communities are active at a given location. Vent fluids may also lead to the formation of metal-rich sulfide and sulfate deposits at the seafloor. Although the mineral deposits found are not economic themselves, they have provided important insights into how metals and sulfide can be transported in the same fluids and, thus, how economically viable mineral deposits are formed. Seafloor deposits also have the potential to provide an integrated history of hydrothermal activity at sites where actively venting fluids have ceased to flow.

6.07.2.2 Processes Affecting Vent-fluid Compositions

In all known cases the starting fluid for a submarine hydrothermal system is predominantly, if not entirely, seawater, which is then modified by processes occurring within the oceanic crust. Four factors have been identified: the two most important are (i) phase separation and (ii) water–rock interaction; the importance of (iii) biological processes and (iv) magmatic degassing has yet to be established.

Water–rock interaction and phase separation are processes that are inextricably linked. As water passes through the hydrothermal system it will react with the rock and/or sediment substrate that is present (Figure 5). These reactions begin in the downflow zone, and continue throughout. When

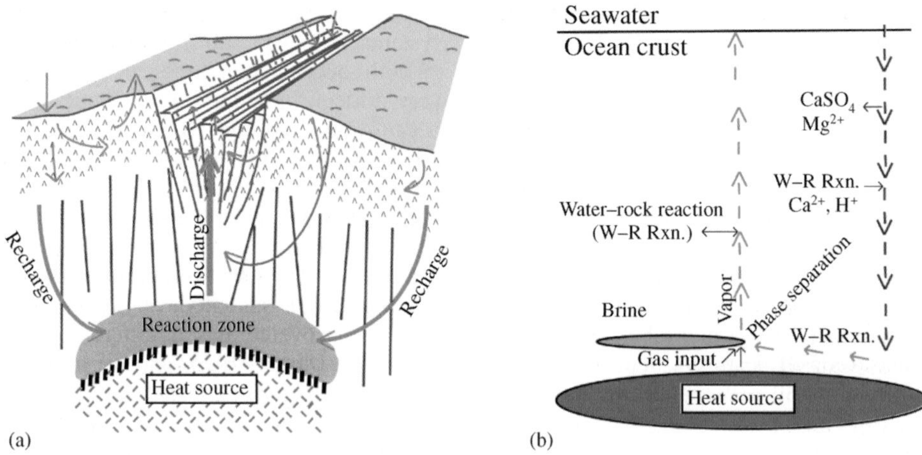

Figure 5 (a) Schematic illustration of the three key stages of submarine hydrothermal circulation through young ocean crust (after Alt, 1995). Seawater enters the crust in widespread "recharge" zones and reacts under increasing conditions of temperature and pressure as it penetrates downward. Maximum temperatures and pressures are experienced in the "reaction zone," close to the (magmatic or hot-rock) "heat source" before buoyant plumes rise rapidly back toward the seafloor—the "discharge" zone. (b) Schematic of processes controlling the composition of hydrothermal vent fluid, as it is modified from starting seawater (after Von Damm, 1995). During recharge, fluids are heated progressively. Above ~130 °C anhydrite ($CaSO_4$) precipitates and, as a result of water–rock reaction, additional calcium (Ca^{2+}) is leached from the rock in order to precipitate most of the sulfate (SO_4^{2-}) derived from seawater. Magnesium (Mg^{2+}) is also lost to the rock and protons (H^+) are added. As the fluid continues downward and up the temperature gradient, water–rock interactions continue and phase separation may occur. At at least two sites on the global MOR system, direct degassing of the magma must be occurring, because very high levels of gas (especially CO_2, and helium) are observed in the hydrothermal fluids. The buoyant fluids then rise to the seafloor. In most cases the fluids have undergone phase separation, and in at least some cases storage of the liquid or brine phase has occurred which has been observed to vent in later years from the same sulfide structure (Von Damm *et al.*, 1997). See Figure 6 for additional discussion of phase separation.

vent fluids exit at the seafloor, what we observe represents the net result of all the reactions that have occurred along the entire hydrothermal flow path. Because the kinetics of most reactions are faster at higher temperatures, it is assumed that much of the reaction occurs in the "reaction zone." Phase separation may also occur at more than one location during the fluid's passage through the crust, and may continue as the $P-T$ conditions acting on the fluid change as it rises through the oceanic crust, back toward the seafloor. However, without a direct view into any of the active seafloor hydrothermal systems, for simplicity of discussion, and because we lack better constraints, we often view the system as one of: (i) water–rock reaction on the downflow leg; (ii) phase separation and water–rock reaction in the "reaction zone"; (iii) additional water–rock reaction after the phase separation "event" (Figure 5). Unless confronted with clear inconsistencies in the (chemical) data that invalidate this approach, we usually employ this simple "flow-through" as our working conceptual model. Even though we are unable to rigorously constrain the complexities for any given system, it is always important to remember that the true system is likely far more complex than any model we employ.

In water–rock reactions, chemical species are both gained and lost from the fluids. In terms of differences from the major-element chemistry of seawater, magnesium and SO_4 are lost, and the pH is lowered so substantially that all the alkalinity is titrated. The large quantities of silicon, iron, and manganese that are frequently gained may be sufficient for these to become "major elements" in hydrothermal fluids. For example, silicon and iron can exceed the concentrations of calcium and potassium, two major elements in seawater. Much of the dissolved SO_4 in seawater is lost on the downflow leg of the hydrothermal system as $CaSO_4$ (anhydrite) precipitates at temperatures of ~130 °C—just by heating seawater. Because there is more dissolved SO_4 than calcium in seawater, on a molar basis, additional calcium would have to be leached from the host rock if more than ~33% of all the available seawater sulfate were to be precipitated in this way. In fact, it is now recognized that at least some dissolved SO_4 must persist down into the reaction zone, based on the inferred redox state at depth (see later discussion). Some seawater SO_4 is also reduced to H_2S, substantial quantities of which may be found in hydrothermal fluids at any temperature, based on information from sulfur isotopes (Shanks, 2001). The magnesium is lost by the formation of Mg–OH silicates. This results in the generation of H^+, which accounts for the low pH and titration of the alkalinity. Sodium can also be lost from the fluids due to Na–Ca replacement reactions in plagioclase feldspars, known as albitization. Potassium (and the other alkalis) are also involved in similar types of reaction that can also generate acidity. Large quantities of iron, manganese, and silicon are also leached out of the rocks and into the fluids.

An element that is relatively conservative through water–rock reaction is chlorine in the form of the anion chloride. Chloride is key in hydrothermal fluids, because with the precipitation and/or reduction of SO_4 and the titration of HCO_3^-/CO_3^{2-}, chloride becomes the overwhelming and almost only anion (Br is usually present in the seawater proportion to chloride). Chloride becomes a key component, therefore, because almost all of the cations in hydrothermal fluids are present as chloro-complexes; thus, the levels of chloride in a fluid effectively determine the total concentration of cationic species that can be present. A fundamental aspect of seawater is that the major ions are present in relatively constant ratios—this forms the basis of the definition of salinity (see Volume Editor's Introduction). Because these constant proportions are not maintained in vent fluids and because chloride is the predominant anion, discussions of vent fluids are best discussed in terms of their *chlorinity*, not their *salinity*.

Although small variations in chloride may be caused by rock hydration/dehydration, there are almost no mineralogic sinks for chloride in these systems. Therefore, the main process that effects changes in the chloride concentrations in the vent fluids is phase separation (Figure 6). Phase separation is a ubiquitous process in seafloor hydrothermal systems. Essentially no hydrothermal fluids are found with chlorinities equal to the local ambient seawater value. To phase separate seawater at typical intermediate-to-fast spreading MOR depths of ~2,500 m requires temperatures ≧389 °C (Bischoff, 1991). This sets a minimum temperature that fluids must have reached, therefore, during their transit through the oceanic crust. The greater the depth, the higher the temperature required for phase separation to occur. Known vent systems occur at depths of 800–3,600 m, requiring maximum temperatures in the range 297–433 °C to phase separate seawater. Seawater, being a two-component system, $H_2O + NaCl$ (to a first approximation) exhibits different phase separation behavior from pure water. The critical point for seawater is 407 °C and 298 bar (Bischoff and Rosenbauer, 1985) compared to 374 °C and 220 bar for pure water. For the salt solution, the two-phase curve does not stop at the critical point but, instead, continues beyond it. As a solution crosses the two-phase curve, it will separate into two phases, one with chlorinities greater than starting seawater, and the other with chlorinities less than starting seawater. If the fluid reaches the two-phase curve at temperature and pressure

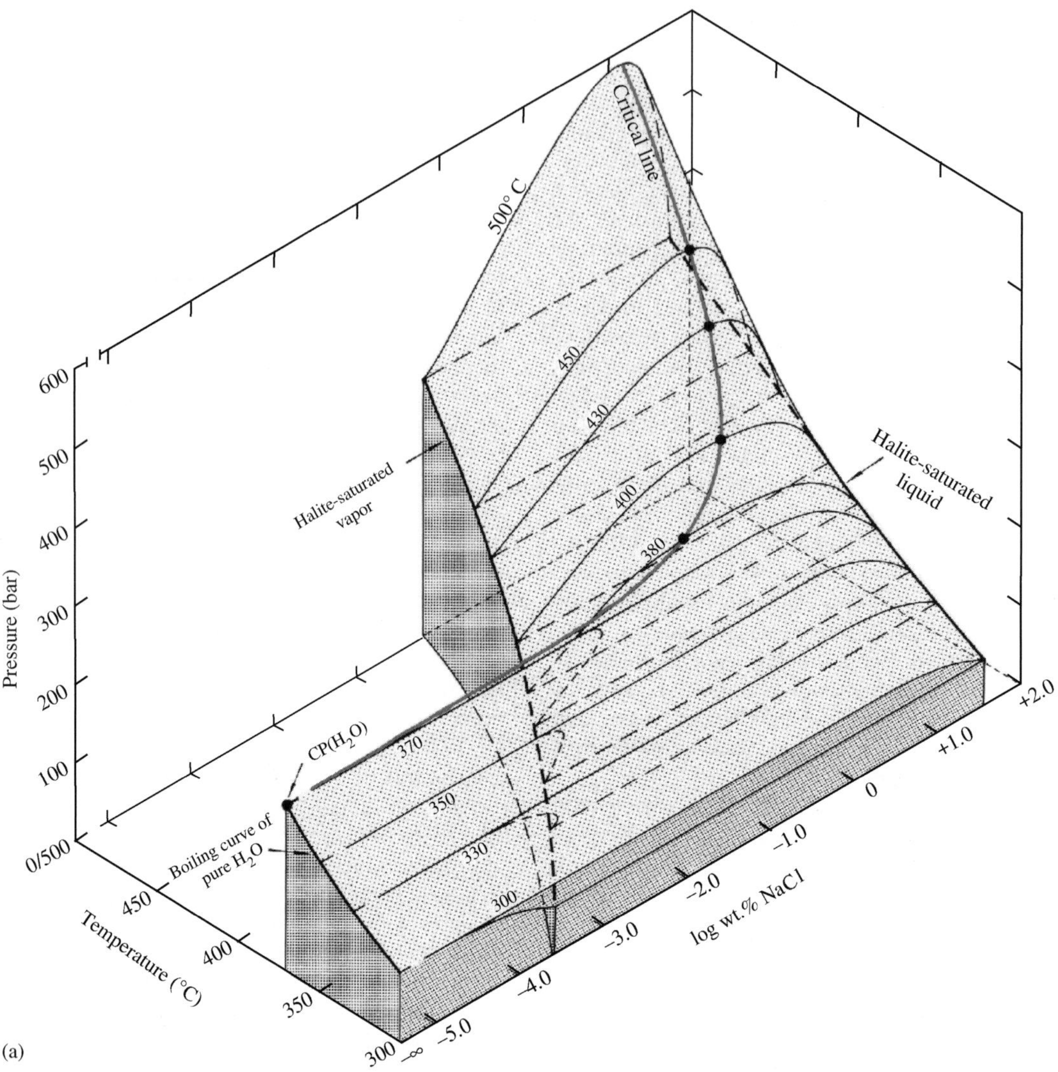

Figure 6 Phase relations in the NaCl–H_2O system. (a) The amount of salt (NaCl) in the NaCl–H_2O system varies as both a function of temperature and pressure. Bischoff and Pitzer (1989) constructed this figure of the three dimensional relationships between pressure (P), temperature (T), and composition (x) in the system based on previous literature data and new experiments, in order to better determine the phase relationships for seafloor hydrothermal systems. The P–T–x relationships define a 3D space, but more commonly various projections are shown. (b) The P–T properties for seawater including the two phase curve (solid-line) separating the liquid stability field from the liquid + vapor field, and indicating the location of the critical point (CP) at 407 °C and 298 bar. Halite can also be stable in this system and the region where halite + vapor is stable is shown, separated from the liquid + vapor stability field by the dotted line. This figure is essentially a "slice" from (a) and is a commonly used figure to show the phase relations for seawater (after Von Damm *et al.*, 1995). (c) A "slice" of (a) can also be made to better demonstrate the relationships in the system in T–x space. Here isobars show the composition of the conjugate vapor and brine (liquid) phases formed by the phase separation of seawater. This figure can be used to not only determine salt contents of the conjugate phases, but also their relative amounts. On this figure the compositions of the vapor and liquid phases sampled from "F" vent in 1991 and 1994, respectively, are shown (after Von Damm *et al.*, 1997).

conditions *less* than the critical point, subcritical phase separation (also called boiling) will occur, with the generation of a low chlorinity "vapor" phase. This phase contains salt, the amount of which will vary depending on where the two-phase curve was intersected (Bischoff and Rosenbauer, 1987). What is conceptually more difficult to grasp, is that when a fluid intersects the two-phase curve at P–T conditions *greater* than the critical point, the process is called supercritical phase separation (or condensation). In this case a small amount of a relatively high chlorinity liquid phase is condensed out of the fluid. Both sub- and super-critical phase separation occur in seafloor hydrothermal systems. To complete the phase relations in this system, halite may also precipitate

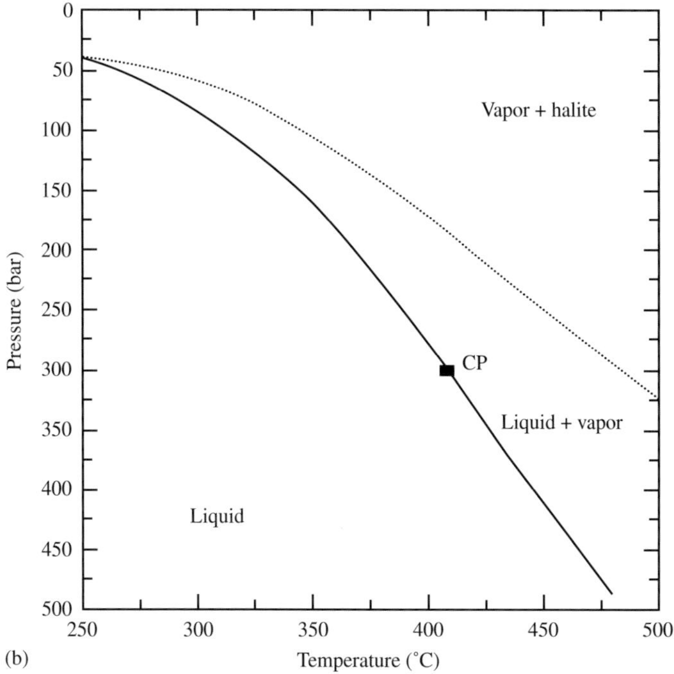

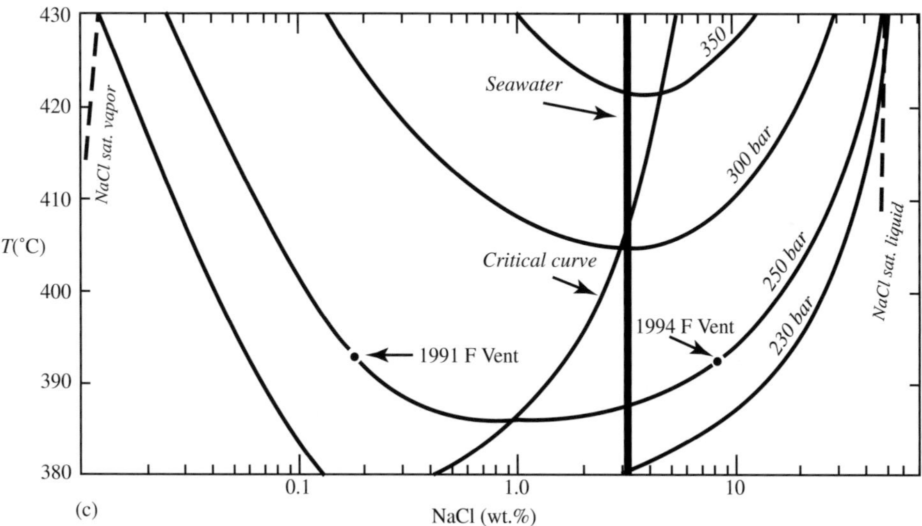

Figure 6 (continued).

(Figure 6). There is evidence that halite forms, and subsequently redissolves, in some seafloor hydrothermal systems (Oosting and Von Damm, 1996; Berndt and Seyfried, 1997; Butterfield *et al.*, 1997; Von Damm, 2000). The $P-T$ conditions at which the fluid intersects the two-phase curve, will determine the relative compositions of the two phases, as well as their relative amounts. Throughout this discussion, we have assumed the starting fluid undergoing phase separation is seawater (or, rather, an NaCl equivalent, because the initial magnesium and SO_4 will already be lost by this stage). If the NaCl content is different, the phase relations in this system change, forming a family of curves or surfaces that are a function of the NaCl content, as well as pressure and temperature. The critical point is also a function of the salt content, and hence is really a critical curve in $P-T-x$ (x referring to composition) space.

As phase separation occurs, substantially changing the chloride content of vent fluids (values from <6% to ~200% of the seawater concentration have been observed), other chemical species will change in concert. It has been shown, both experimentally as well as in the

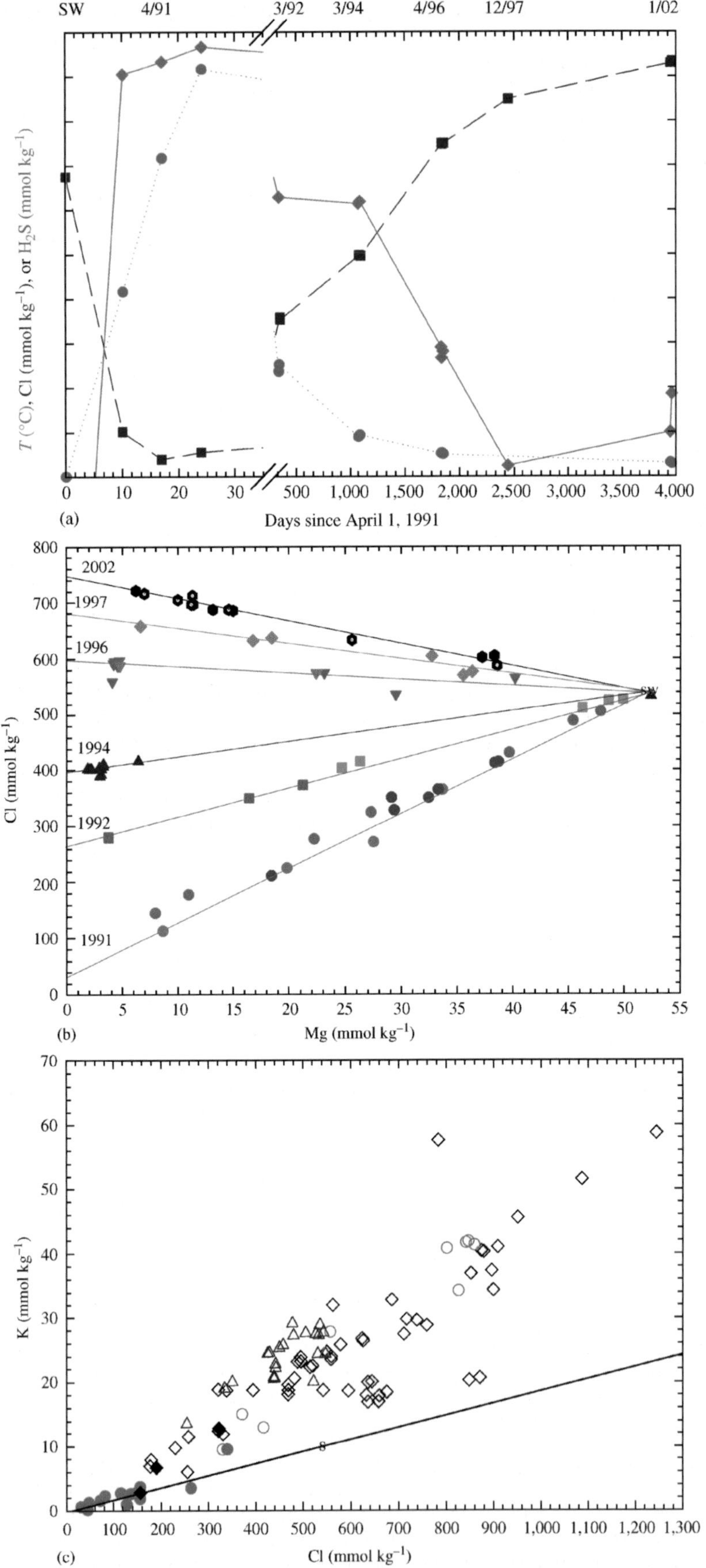
SW
4/91
3/92
3/94
4/96
12/97
1/02
T (°C), Cl (mmol kg⁻¹), or H₂S (mmol kg⁻¹)
0
10
20
30
500
1,000
1,500
2,000
2,500
3,000
3,500
4,000
(a)
Days since April 1, 1991
2002
1997
1996
1994
1992
1991
SW
Cl (mmol kg⁻¹)
Mg (mmol kg⁻¹)
(b)
K (mmol kg⁻¹)
Cl (mmol kg⁻¹)
(c)

field, that most of the cations (and usually bromide) maintain their element-to-Cl ratios during the phase separation process (Berndt and Seyfried, 1990; Von Damm, 2000; Von Damm *et al.*, 2003), i.e., most elements are conservative with respect to chloride. Exceptions do occur, however—primarily for those chemical species not present as chlorocomplexes. Dissolved gases (e.g., CO_2, CH_4, He, H_2, H_2S) are preferentially retained in the low chlorinity or vapor phase, and boron, which is present as a hydroxyl complex, is relatively unaffected by phase separation (Bray and Von Damm, 2003a). Bromide, as viewed through the Br/Cl ratio, is sometimes seen to be fractionated from chloride; this occurs whenever halite is formed or dissolves because bromide is preferentially excluded from the halite structure (Oosting and Von Damm, 1996; Von Damm, 2000). Fluids that have deposited halite, therefore, will have a high Br/Cl ratio, while fluids that have dissolved halite will have a low Br/Cl ratio, relative to seawater. It is because of the ubiquity of phase separation that vent-fluid compositions are often now viewed or expressed as ratios with respect to chloride, rather than as absolute concentrations. This normalization to chloride *must* be used when trying to evaluate net gains and losses of chemical species as seawater traverses the hydrothermal circulation cell, to correct for the fractionation caused by phase separation.

Aside from early eruptive fluids (discussed later), the chemical composition of most high-temperature fluids (Figure 7) appears to be controlled by equilibrium or steady state with the rock assemblage. (Equilibrium requires the assemblage to be at its lowest energy state, but the actual phases present may be metastable, in which case they are not at true thermodynamic equilibrium but, rather, have achieved a steady-state condition.) When vent-fluid data are modeled with geochemical modeling codes using modern thermodynamic databases, the results suggest equilibrium, or at least steady state, has been achieved. The models cannot be rigorously applied to many of the data, however, because the fluids are often close to the critical point and in that region the thermodynamic data are not as well constrained. Based on results from both geochemical modeling codes and elemental ratios, current data indicate that not only the major elements, but also many minor elements (e.g., rubidium, caesium, lithium, and strontium) are controlled by equilibrium, or steady-state, conditions between the fluids and their host-rocks (Bray and Von Damm, 2003b). The rare earth elements (REE) in vent fluids present one such example. REE distributions in hydrothermal fluids are light-REE enriched and exhibit strong positive europium anomalies, apparently quite unrelated to host-rock MORB compositions (Figure 8). However, Klinkhammer *et al.* (1994) have shown that when these same REE concentrations are plotted versus their ionic radii, the fluid trends not only become linear but also show the same fractionation trend exhibited by plagioclase during magma segregation, indicating that vent-fluid

Figure 7 Compositional data for vent fluids. (a) Time series data from "A" vent for chloride and H_2S concentrations and measured temperature (*T*). When time-series data are available, this type of figure, demonstrating the change in fluid composition in a single vent over time, is becoming more common. The data plotted are referred to as the "end-member" data (data from Von Damm *et al.*, 1995 and unpublished). Points on the *y*-axis are values for ambient seawater. Note the low chlorinity (vapor phase) fluids venting initially from A vent; over time the chloride content has increased and the fluids sampled more recently, in 2002, are the liquid (brine) phase. As is expected, the concentration of H_2S, a gas that will be partitioned preferentially into the vapor phase, is anticorrelated with the chloride concentration. The vertical axis has 10 divisions with the following ranges: *T* (°C) 200–405 (ambient seawater is 2 °C); Cl (mmol kg^{-1}) 0–800 (ambient is 540 mmol kg^{-1}); H_2S (mmol kg^{-1}) 0–120 (ambient is 0 mmol kg^{-1}). (b) Whenever vent fluids are sampled, varying amounts of ambient seawater are entrained into the sampling devices. Vent fluids contain 0 mmol kg^{-1}, while ambient seawater contains 52.2 mmol kg^{-1}. Therefore, if actual sample data are plotted as properties versus magnesium, least squares linear regression fits can be made to the data. The calculated "end-member" concentration for a given species, which represents the "pure" hydrothermal fluid is then taken as the point where that line intercepts the *y*-axis (i.e., the calculated value at Mg = 0 mmol kg^{-1}). These plots versus magnesium are therefore mostly sampling artifacts and are referred to as "mixing" diagrams. While these types of plots were originally used to illustrate vent-fluid data, they have been largely superceded by figures such as (a) or (c). This figure (b) shows the data used to construct the time series represented in (a). Note the different lines for the different years. In some years samples were collected on more than one date. All samples for a given year are shown by the same shape, the different colors within a year grouping indicate different sample dates. In some years the chemical composition varied from day-to-day, but for simplicity a single line is shown for each year in which samples were collected (Von Damm *et al.*, 1995 and unpublished data). (c) As the chloride content of a vent fluid is a major control on the overall composition of the vent composition, most of the cations vary as a function of the chloride-content. Variations in the chloride content are a result of phase separation. This shows the relationship between the potassium (K) and chloride content in vent fluids in the global database, as of 2000. The line is the ratio of K/Cl in ambient seawater. Closed circles are from 9–10° N EPR following the 1991 eruption, open circles are other 9–10° N data not affected by the eruptive events; triangles are from sites where vents occur on enriched oceanic crust, diamonds are from bare-basalt (MORB) hosted sites, filled diamonds are other sites impacted by volcanic eruptions. Data sources and additional discussion in Von Damm (2000).

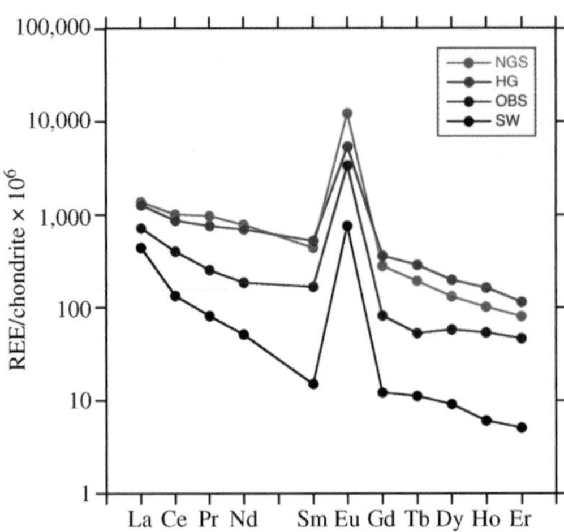

Figure 8 End-member REE concentrations in vent fluids from four different "black smokers" at the 21° N vent site, EPR, normalized to chondrite (REE data from Klinkhammer *et al.*, 1994). NGS = *National Geographic Smoker*; HG = *Hanging Gardens*; OBS = *Ocean Bottom Seismometer*; SW = *South West* vent.

REE concentrations may be intrinsically linked to the high-temperature alteration of this particular mineral.

Two other processes are known to influence the chemistry of seafloor vent fluids: biological processes and magmatic degassing. Evidence for "magmatic degassing" has been identified at two sites along the global MOR system—at 9° 50′ N and at 32° S on the EPR (M. D. Lilley, personal communication; Lupton *et al.*, 1999a). These sites have very high levels of CO_2, and very high He/heat ratios. The interpretation is that we are seeing areas with recent magma resupply within the crust and degassing of the lavas, resulting in very high gas levels in the hydrothermal fluids found at these sites. We do not know the spatial–temporal variation of this process, hence, we cannot yet evaluate its overall importance. Presumably, every site on the global MOR system undergoes similar processes episodically. What is not known, however, is the frequency of recurrence at any one site. Consequently, the importance of fluxes due to this degassing process, versus more "steady-state" venting, cannot currently be assessed. At 9° 50′ N, high gas contents have now been observed for almost a decade; no signature of volatile-metal enrichment has been observed in conjunction with these high gas contents (Von Damm, 2003).

The fourth process influencing vent-fluid compositions is biological change, which can take the form of either consumption or production of various chemical species. As the current known limit to life on Earth is ~120 °C (e.g., Holland and Baross, 2003) this process can only affect fluids at temperatures lower than this threshold. This implies that high-temperature vents should not be subject to these effects whereas they may occur in both lower-temperature axial diffuse flow and beneath ridge flanks. From observations at the times of seafloor eruptions and/or diking events, it is known that there are microbial communities living within the oceanic crust (e.g., Haymon *et al.*, 1993). Their signatures can be seen clearly in at least some low-temperature fluids, as noted in particular by changes in the H_2, CH_4, and H_2S contents of those fluids (Von Damm and Lilley, 2003). Hence, biological influences have been observed; how widespread this is, which elements are affected, and what the overall impact on chemical fluxes may be all remain to be resolved.

6.07.2.3 Compositions of Hydrothermal Vent Fluids

6.07.2.3.1 Major-element Chemistry

The known compositional ranges of vent fluids are summarized in Figure 9 and Table 2. Because no two vents yet discovered have exactly the same composition, these ranges often change with each new site. As discussed in Section 6.07.2.2, vent fluids are modified seawater characterized by the loss of magnesium, SO_4, and alkalinity and the gain of many metals, especially on a chloride normalized basis.

Vent fluids are acidic, but not as acid as may first appear from pH values measured at 25 °C and 1 atm (i.e., in shipboard laboratories). The cation H^+ in vent fluids is also present as a chloro-complex with the extent of complexation increasing as *P* and *T* increase. At the higher *in situ* conditions of *P* and *T* experienced at the seafloor, therefore, much of the H^+ is incorporated into the HCl–aqueous complex; hence, the activity of H^+ is reduced and the *in situ* pH is substantially higher than what is measured at laboratory temperatures. The K_w for water also changes as a function of *P* and *T* such that neutral pH is not necessarily 7 at other *P*–*T* conditions. For most vent fluids, the *in situ* pH is 1–2 pH units more acid than neutral, not the ~4 units of acidity that the measured (25 °C, 1 atm) data appear to imply. Most high-temperature vent fluids have (25 °C measured) pH values of 3.3 ± 0.5 but a few are more acidic whilst several are less acid. If fluids are more acid than pH 3.3 ± 0.5, it is often an indicator that metal sulfides have precipitated below the seafloor because such reactions produce protons. Two mechanisms are known that can cause fluids to be less acidic than the norm: (i) cases where the rock substrate appears to be more highly altered—the rock cannot buffer the solutions to as low a pH; (ii) when organic matter is present, ammonium is often present and

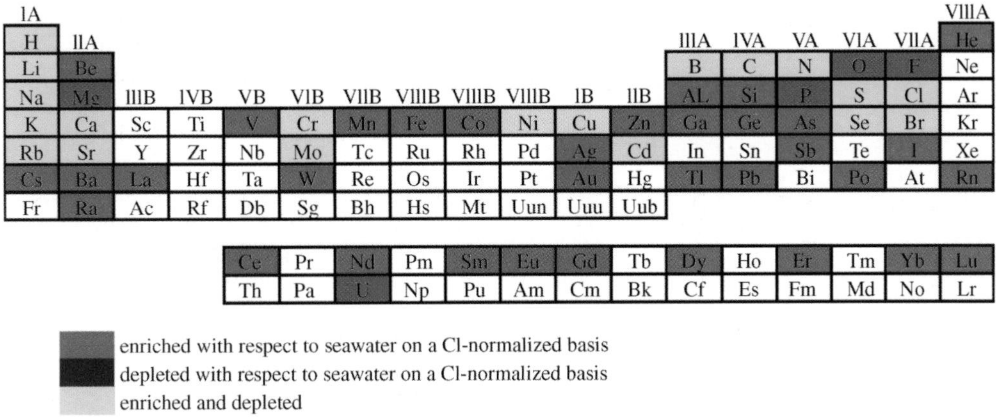

1A																	VIIIA
H	IIA											IIIA	IVA	VA	VIA	VIIA	He
Li	Be											B	C	N	O	F	Ne
Na	Mg	IIIB	IVB	VB	VIB	VIIB	VIIIB	VIIIB	VIIIB	IB	IIB	AL	Si	P	S	Cl	Ar
K	Ca	Sc	Ti	V	Cr	Mn	Fe	Co	Ni	Cu	Zn	Ga	Ge	As	Se	Br	Kr
Rb	Sr	Y	Zr	Nb	Mo	Tc	Ru	Rh	Pd	Ag	Cd	In	Sn	Sb	Te	I	Xe
Cs	Ba	La	Hf	Ta	W	Re	Os	Ir	Pt	Au	Hg	Tl	Pb	Bi	Po	At	Rn
Fr	Ra	Ac	Rf	Db	Sg	Bh	Hs	Mt	Uun	Uuu	Uub						
				Ce	Pr	Nd	Pm	Sm	Eu	Gd	Tb	Dy	Ho	Er	Tm	Yb	Lu
				Th	Pa	U	Np	Pu	Am	Cm	Bk	Cf	Es	Fm	Md	No	Lr

enriched with respect to seawater on a Cl-normalized basis
depleted with respect to seawater on a Cl-normalized basis
enriched and depleted

Figure 9 Periodic table of the elements showing the elements that are enriched in hydrothermal vent fluids relative to seawater (red), depleted (blue) and those which have been shown to exhibit both depletions and enrichments in different hydrothermal fluids (yellow) relative to seawater. All data are normalized to the chloride content of seawater in order to evaluate true gains and losses relative to the starting seawater concentrations.

the NH_3/NH_4^+ couple serves to buffer the pH to a higher level.

Vent fluids are very reducing, as evidenced by the presence of H_2S rather than SO_4, as well as H_2, CH_4 and copious amounts of Fe^{2+} and Mn^{2+}. In rare cases, there can be more H_2S and/or H_2 than chloride on a molar basis and it is the prevailing high acidity that dictates that H_2S rather than HS^- or S^{2-} is the predominant form in high-temperature vent fluids. Free H_2 is derived as a result of water–rock reaction and there is substantially more H_2 than O_2 in these fluids. Therefore, although redox calculations are typically given in terms of the $\log f_{O_2}$, the redox state is best calculated based on the H_2/H_2O couple for seafloor vent fluids. This can then be expressed in terms of the $\log f_{O_2}$. The K for the reaction:

$$H_2 + \tfrac{1}{2}O_2 = H_2O$$

also changes as a function of temperature and pressure. Another way to determine how reducing vent fluids are is by comparing them to various mineralogic buffers such as pyrite–pyrrhotite–magnetite (PPM) or hematite–magnetite–pyrite (HMP). Most vent fluids lie between these two extremes, but there is some systematic variation (Seyfried and Ding, 1995). The observation that vent fluids are more oxic than would be expected based on the PPM buffer, provides one line of evidence that the reaction zone is not as reducing as initially predicted, consistent with at least some dissolved seawater SO_4 penetrating into the deeper parts of the system rather than being quantitatively removed by anhydrite precipitation within shallower levels of the downflow limb.

Lower-temperature (<100 °C) vent fluids found right at the axis are in most known cases a dilution of some amount of high-temperature fluids with seawater, or a low-temperature fluid with a composition close to seawater. There is some evidence for an "intermediate" fluid, perhaps most analogous to a crustal "ground water," with temperatures of ~150 °C within the oceanic crust. Evidence for the latter is found in some Ocean Drilling Program data (Magenheim *et al.*, 1992), some high-temperature vent fluids from 9° 50′ N on the EPR (Ravizza *et al.*, 2001), and some very unusual ~90 °C fluids from the southern EPR (O'Grady, 2001). To conclude, the major-element composition of high-temperature vent fluids can be described as acidic, reducing, metal-rich NaCl solutions whilst lower-temperature fluids are typically a dilution of this same material with seawater. The few exceptions to this will be discussed below.

6.07.2.3.2 Trace-metal chemistry

Compared to the number of vent fluids sampled and analyzed for their major-element data, relatively little trace-metal data exist. This is because when hot, acidic vent fluids mix with seawater, or even just cool within submersible- or ROV-deployed sampling bottles, they become supersaturated with respect to many solid phases and, thus, precipitate. Once this occurs, everything in the sampling apparatus must be treated as one sample: a budget can only be constructed by integrating these different fractions back together. In the difficult sampling environment found at high-temperature vent sites, pieces of chimney structure are also sometimes entrained into the sampling apparatus. It is necessary, therefore, to be able to discriminate between particles that have precipitated from solution in the sampling bottle and "contaminating" particles that are extraneous to the sample. In addition, water samples are often

Table 2 Ranges in chemical composition for all known vent fluids.

Chemical species	*Units*	*Seawater*	*Overall range*	*Slow*[a] *(0–2.5 cm yr^{-1})*	*Intermediate*[a] *>2.5–6*	*Fast*[a] *>6–12*	*Ultrafast*[a] *>12*	*Sediment covered*[b]	*Ultramafic hosted*[c]	*Arc, back-arc*[d]
T	°C	2	>2–405	40–369	13–382	8–403	16–405	100–315	40–364	278–334
pH	25 °C, 1 atm	7.8	2.0–9.8	2.5–4.85	2.8–4.5	2.45– > 6.63	2.96–5.53	5.1–5.9	2.7–9.8	2.0–5.0
Alkalinity	meq kg^{-1}	2.4	− 3.75–10.6	−3.4–0.31	−3.75–0.66	−2.69– < 2.27	−1.36–0.915	1.45–10.6		−0.20–3.51
Cl	mmol kg^{-1}	540	30.5–1,245	357–675	176–1,245	30.5–902	113–1,090	412–668	515–756	255–790
SO_4	mmol kg^{-1}	28	<0– < 28	−3.5–1.9	−25–0.763	>− 8.76	−0.502–9.53	0	<12.9	0
H_2S	mmol kg^{-1}	0	0–110	0.5–5.9	0–19.5	0–110	0–35	1.10–5.98	0.064–1.0	2.0–13.1
Si	mmol kg^{-1}	0.03–0.18	<24	7.7–22	11–24	2.73–22.0	8.69–21.3	5.60–13.8	6.4–8.2	10.8–14.5
Li	umol kg^{-1}	26	4.04–5,800	238–1,035	160–2,350	4.04–1,620	248–1,200	370–1,290	245–345	200–5,800
Na	mmol kg^{-1}	464	10.6–983	312–584	148–924	10.6–983	109–886	315–560	479–553	210–590
K	mmol kg^{-1}	10.1	−1.17–79.0	17–28.8	6.98–58.7	−1.17–51	2.2–44.8	13.5–49.2	20.2–22	10.5–79.0
Rb	umol kg^{-1}	1.3	0.156–360	9.4–40.4	22.9–59	0.156–31.1	0.39–6.8	22.5–105	28–37.1	8.8–360
Cs	nmol kg^{-1}	2	2.3–7,700	100–285	168–364	2.3–264		1,000–7,700	331–385	
Be	nmol kg^{-1}	0	10–91		10–37			12–91		
Mg	mmol kg^{-1}	52.2	0	0	0	0	0	0	<19	0
Ca	mmol kg^{-1}	10.2	−1.31–109	9.9–43	9.75–109	−1.31–106	4.02–65.5	26.6–81.0	21.0–67	6.5–89.0
Sr	umol kg^{-1}	87	−29–387	42.9–133	0.0–348	− 29–387	10.7–190	160–257	138–203	20–300
Ba	umol kg^{-1}	0.14	1.64–100	<52.2	>8– >46		1.64–18.6	>12	>45–79	5.9–100
Mn	umol kg^{-1}	<0.001	10–7,100	59–1,000	140–4,480	62.7–3,300	20.6–2,750	10–236	330–2,350	12–7,100
Fe	mmol kg^{-1}	<0.001	0.007–25.0	0.0241–5.590	0.009–18.7	0.007–12.1	0.038–14.7	0–0.18	2.5–25.0	13–2,500
Cu	umol kg^{-1}	0.007	0–162	0–150	0.1–142	0.18–97.3	2.6–150	<0.02–1.1	27–162	0.003–34
Zn	umol kg^{-1}	0.012	0–3,000	0–400	2.2–600	13–411	1.9–740	0.1–40.0	29–185	7.6–3,000
Co	umol kg^{-1}	0.00003	<0.005–14.1	0.130–0.422	0.022–0.227			<0.005	11.8–14.1	
Ni	umol kg^{-1}	0.012							2.2–3.6	
Ag	nmol kg^{-1}	0.02	<1–230		<1–38			<1–230	11–47	
Cd	nmol kg^{-1}	1.0	0–180	75–146	0–180			<10–46	63–178	
Pb	nmol kg^{-1}	0.01	<20–3,900	221–376	183–360			<20–652	86–169	36–3,900
B	umol kg^{-1}	415	356–3,410	356–480	465–1,874	430–617	400–499	<2,160		470–3,410
Al	umol kg^{-1}	0.02	0.1–18.7	1.03–13.9	4.0–5.2	0.1–18	9.3–18.7	0.9–7.9	1.9–4	4.9–17.0
Br	umol kg^{-1}	840	29.0–1,910	666–1,066	250–1789	29.0–1370	216–1910	770–1,180		306–1,045
F	umol kg^{-1}	68	<38.8	16.1–38.8	"0"					
CO_2	mmol kg^{-1}			3.56–39.9	<5.7	<200	8.4–22			14.4–200
CH_4	umol kg^{-1}	0.0003		150–2,150	<52		7–133		130–2,200	
NH_4	mmol kg^{-1}	0	<15.6	<0.06	<0.65			5.6–15.6		
H_2	umol kg^{-1}	0.0003	<38,000	1.1–727	<0.45	<38,000	40–1300		250–13,000	

[a] These omit sedimented covered and um hosted. [b] Includes: Guaymas, Escanaba, Middle Valley. [c] Includes Rainbow, Lost City, kvd unpublished data for Logatchev. [d] Compilation from Ishibashi and Urabe (1995).

subdivided into different fractions, aboard ship, making accurate budget reconstructions difficult if not impossible to complete. It is because of these difficulties that there are few robust analyses of many trace metals, especially those that precipitate as, or co-precipitate with, metal sulfide phases. Some general statements can, however, be made. In high-temperature vent fluids, most metals are enriched relative to seawater, sometimes by 7–8 orders of magnitude (as is sometimes true for iron). At least some data exist demonstrating the enrichment of all of vanadium, cobalt, nickel, copper, zinc, arsenic, selenium, aluminum, silver, cadmium, antimony, caesium, barium, tungsten, gold, thallium, lead, and REE relative to seawater. Data also exist showing that molybdenum and uranium are often lower than their seawater concentrations. These trace-metal data have been shown to vary with substrate and the relative enrichments of many of these trace metals varies significantly between MOR hydrothermal systems, those located in back arcs, and those with a significant sedimentary component. Even fewer trace-metal data exist for low-temperature "diffuse" fluids. The original work on the <20 °C GSC fluids (Edmond *et al.*, 1979) showed these fluids to be a mix of high-temperature fluids with seawater, with many of the transition metals present at *less than* their seawater concentrations due to precipitation and removal below the seafloor. Essentially the same results were obtained by James and Elderfield (1996) using the MEDUSA system to sample diffuse-flow fluids at TAG (26° N, MAR).

6.07.2.3.3 Gas chemistry of hydrothermal fluids

In general, concentrations of dissolved gases are highest in the lowest-chlorinity fluids, which represent the vapor phase. However, there are exceptions to this rule, and gas concentrations vary significantly between vents, even at a single location. In the lowest chlorinity and hottest fluids, H_2S may well be the dominant gas. However, because H_2S levels are controlled by metal-sulfide mineral solubility, this H_2S is often lost via precipitation. While the first vent sites discovered contained less than twice the CO_2 present in seawater (Welhan and Craig, 1983), more vents have higher levels of CO_2 than is commonly realized. Few MOR vent fluids have CO_2 levels less than or equal to the total CO_2 levels present in seawater ($\sim$2.5 mmol kg^{-1}). Instead, many fluids have concentrations approaching an order of magnitude more CO_2 than seawater; the highest approach two orders of magnitude more CO_2 than seawater, but these highest levels are uncommon (M. D. Lilley, personal communication). Back-arc systems more commonly have higher levels of CO_2 in their vent fluids, but concentrations two orders of magnitude greater than seawater are, again, close to the upper maximum of what has been sampled so far (Ishibashi and Urabe, 1995). CH_4 is much less abundant than CO_2 in most systems. Vent-fluid CH_4 concentrations are typically higher in sedimented systems and in systems hosted in ultramafic rocks, when compared to bare basaltic vent sites. CH_4 is also enriched in low-temperature vent fluids when compared to concentrations predicted from simple seawater/vent-fluid mixing (Von Damm and Lilley, 2003). Longer-chain organic molecules have also been reported from some sites, usually at even lower abundances than CO_2 and/or CH_4 (Evans *et al.*, 1988). The concentrations of H_2 gas in vent fluids vary substantially, over two orders of magnitude (M. D. Lilley, personal communication). Again the highest levels are usually observed in vapor phase fluids, especially those sampled immediately after volcanic eruptions or diking events. Relatively high values (several mmol kg^{-1}) have also been reported from sites hosted by ultramafic rocks. Of the noble gases, helium, especially ^{3}He, is most enriched in vent fluids. ^{3}He can be used as a conservative tracer in vent fluids, because its entire source in vent fluids is primordial, from within the Earth (see Section 6.07.5). Radon, a product of radioactive decay in the uranium series, is also greatly enriched in vent fluids (e.g., Kadko and Moore, 1988; Rudnicki and Elderfield, 1992). Less data are available for the other noble gases, at least some of which appear to be relatively conservative compared to their concentration in starting seawater (Kennedy, 1988).

6.07.2.3.4 Nutrient chemistry

The concentrations of nutrients available in seawater control biological productivity. Consequently, the dissolved nutrient concentrations in natural waters are always of great interest. Compared to deep-ocean seawater, the PO_4 contents of vent fluids are significantly lower, but are not zero. Much work remains to be done on the distribution of nitrogen species, and the nitrogen cycle in general, in vent fluids. Generally, in basalt-hosted systems, the nitrate + nitrite content is also lower than local deepwaters but, again, is not zero. Ammonium in these systems typically measures less than 10 μmol kg^{-1} but, in some systems in which no sediment cover is present, values of 10s to even 100s of μmol kg^{-1} are sometimes observed (cf. Von Damm, 1995). In Guaymas Basin, in contrast, ammonium concentrations as high as 15 mmol kg^{-1} have been measured (Von Damm *et al.*, 1985b). N_2 is also present. Silica concentrations are extremely high, due to interaction with host rocks at high temperature, at depth. Of course, in these systems, it could be debated what actually constitutes a "nutrient." For example, both dissolved H_2 and H_2S (as well as numerous other reduced species), represent important primary energy sources for the chemosynthetic communities invariably found at hydrothermal vent sites.

6.07.2.3.5 Organic geochemistry of hydrothermal vent fluids

Studies of the organic chemistry of vent fluids are truly in their infancy and little field data exist (Holm and Charlou, 2001). There are predictions of what should be present based on experimental work (e.g., Berndt *et al.*, 1996; Cruse and Seewald, 2001) and thermodynamic modeling (e.g., McCollom and Shock, 1997, 1998). These results await confirmation "in the field." Significant data on the organic geochemistry are uniquely available for the Guaymas Basin hydrothermal system (e.g., Simoneit, 1991), which underlies a very highly productive area of the ocean, and is hosted in an organic-rich sediment-filled basin.

6.07.2.4 Geographic Variations in Vent-fluid Compositions

6.07.2.4.1 The role of the substrate

There are systematic reasons for some of the variations observed in vent-fluid compositions. One of the most profound is the involvement of sedimentary material in the hydrothermal circulation cell, as seen at sediment-covered ridges (Von Damm *et al.*, 1985b; Campbell *et al.*, 1994; Butterfield *et al.*, 1994). The exact differences this imposes depend upon the nature of the sedimentary material involved: the source/nature of the aluminosilicate material, the proportions and type of organic matter it contains and the proportion and type of animal tests present, calcareous and/or siliceous. In the known sediment-hosted systems (Guaymas Basin, EPR; Escanaba Trough, Gorda Ridge; Middle Valley, JdFR; and perhaps the Red Sea) basalts are intercalated with the sediments or else underlie them. Hence, in these systems, reactions with basalt are overprinted by those with the sediments. In most cases, depending on the exact nature as well as thickness of the sedimentary cover, this causes a rise in the pH, which results in the precipitation of metal sulfides before the fluids reach the seafloor. The presence of carbonate and/or organic matter also buffers the pH to significantly higher levels (at least pH 5 at 25 °C and 1 atm).

The chemical composition of most seafloor vent fluids can be explained by reaction of unaltered basalt with seawater. However, in some cases the best explanation for the fluid chemistry is that the fluids have reacted with basalt that has already been highly altered (Von Damm *et al.*, 1998). Two indicators for this are higher pH values (pH ~4 versus pH ~3.3 at 25 °C), as well as lower K/Na molar ratios and lower concentrations of the REEs. In the last several years, several vent sites have been sampled that cannot be explained by these mechanisms. At some locations, vent fluids must be generated by reaction of seawater with ultramafic rocks (Douville *et al.*, 2002). These fluids can also have major variations from each other, depending on the temperature regime. In high-temperature fluids that have reacted with an ultramafic substrate, silicon contents are generally lower than in basalt-hosted fluids; H_2, calcium, and iron contents are generally higher, but these fluids remain acidic (Douville *et al.*, 2002). Not much is yet known about the seafloor fluids that are generated from lower-temperature ultramafic hydrothermal circulation. In the one example studied thus far (Lost City) the measured pH is greater than that in seawater, and fluid compositions are clearly controlled by a quite distinct set of serpentinization-related reactions (Kelley *et al.*, 2001). An illustration of this fundamental difference is given by magnesium which is quantitatively stripped from "black smoker" hydrothermal fluids but exhibits ~20–40% of seawater concentrations (9–19 mM) in the Lost City vents, leading to the unusual magnesium-rich mineralization observed at this site (see later). The seafloor fluids from Lost City are remarkably similar to those found in continental hydrothermal systems hosted in ultramafic environments (Barnes *et al.*, 1972). In back-arc spreading centers such as those found in the western Pacific, andesitic rock types are common and profound differences in vent-fluid compositions arise (Fouquet *et al.*, 1991; Ishibashi and Urabe, 1995, Gamo *et al.*, 1997). These fluids can be both more acidic and more oxidizing than is typical and the relative enrichments of transition metals and volatile species in these fluids are quite distinct from what is observed in basalt-hosted systems.

Major differences in substrate are, therefore, reflected in the compositions of vent fluids. Insufficient trace-metal data for vent fluids exist, however, to discern more subtle substrate differences, e.g., between EMORB and NMORB on non-hotspot influenced ridges. Where the ridge axis is influenced by hot-spot volcanism, some differences may be seen in the fluid compositions, as, for example, the high barium in the Lucky Strike vent fluids, but in this case most of the fluid characteristics (e.g., potassium concentrations) do not show evidence for an enriched substrate (Langmuir *et al.*, 1997; Von Damm *et al.*, 1998).

6.07.2.4.2 The role of temperature and pressure

Temperature, of course, plays a major role in determining vent-fluid compositions. Pressure is often thought of as less important than temperature, but the relative importance of the two depends on the exact $P-T$ conditions of the fluids. Because of the controls that pressure and temperature conditions exert on the thermodynamics as well as the physical properties of the fluids, the two effects cannot be discussed completely independently from each other. Not only do $P-T$ conditions govern phase separation, as discussed above, they control transport in the fluids and mineral dissolution and precipitation reactions. Temperature, especially, plays a role in the quantities of elements that are leached from the host rocks. When temperature decreases, as it often does due to conductive cooling as fluids rise through the oceanic crust, most minerals become less soluble. Due to these decreasing mineral solubilities, transition metals and sulfide, in particular, may be lost from the ascending fluids. $P-T$ conditions in the fluids also control the strength of the aqueous complexes. In general, as P and T rise, aqueous species become more associated. Because of the properties of water at the critical point (the dielectric constant goes to zero), all the species must become associated, as there can be no charged species in solution at the critical point. Therefore, transport of species can increase markedly as the critical point is approached because there will be smaller amounts of the (charged) species present in solution which are needed if mineral solubility products are to be exceeded (Von Damm *et al.*, 2003). It is in this critical point region that small changes in pressure can be particularly significant—for example, as a fluid is rising in the upflow ("discharge") limb of a hydrothermal cell (Figure 5). Because most vent-fluid compositions are controlled by equilibrium or steady state, and because the equilibrium constants for these reactions change as a function of pressure and temperature, $P-T$ conditions will ultimately control all vent-fluid compositions. One problem associated with modeling vent fluids and trying to understand the controls on their compositions is that we really do not know the temperature in the "reaction zone." Basaltic lavas are emplaced at temperatures of 1,100–1,200 °C, but rocks must be brittle to retain fractures that allow fluid flow, and this brittle–ductile transition lies in the range 500–600 °C. A commonplace statement is that the reaction zone temperature is ~450 °C, but we do not really know this value with any accuracy, nor how variable it may be from one location to another. At the seafloor, we have sampled fluids with exit temperatures as high as ~405 °C. Hence, in addition to the constraints provided by the recognition that at least subcritical phase separation is pervasive (see above) we can further determine that (i) reaction zone temperatures must exceed 405 °C, at least in some cases, and (ii) that in cases where evidence for supercritical phase separation has been determined (e.g., Butterfield *et al.*, 1994; Von Damm *et al.*, 1998) temperatures must exceed 407 °C within the oceanic crust.

The pressure conditions at any hydrothermal field are largely controlled by the depth of the overlying water column. Pressure is most critical in terms of phase separation and vent fluids are particularly sensitive to small changes in pressure when close to the critical point. It is in this region, close to the critical point, when fluids are very expanded (i.e., at very low density) that small changes in pressure can cause significant changes in vent-fluid composition.

6.07.2.4.3 The role of spreading rate

When one considers tables of vent-fluid chemical data, one cannot separate vents from ultrafast- versus slow-spreading ridges (Table 2); the range of chemical compositions from each of these two end-member types of spreading regime overlap. There has been much debate in the marine geological literature whether rates of magma supply, rather than spreading rate, should more correctly be applied when defining ridge types (e.g., "magma-starved" versus "magma-rich" sections of ridge crest). While any one individual segment of ridge crest undoubtedly passes through different stages of a volcanic–tectonic cycle, regardless of spreading rate (e.g., Parson *et al.*, 1993; Haymon, 1996), it is generally the case that slow-spreading ridges are relatively magma-starved whilst fast spreading ridges are more typically magma-rich. Consequently, we continue to rely upon (readily quantified) spreading rate (De Mets *et al.*, 1994) as a proxy for magma supply. To a first approximation, ridge systems in the Atlantic Ocean are slow-spreading, while fast-spreading ridges are only found in the Pacific. The Pacific contains ridges that spread at rates from ~15 cm yr^{-1} full opening rate to a minimum of ~2.4 cm yr^{-1} on parts of the Gorda Ridge (comparable to the northern MAR). In the Indian Ocean, ridge spreading varies between intermediate (~6 cm yr^{-1}, CIR and SEIR) and very slow rates (<2.0 cm yr^{-1}, SWIR). In the Arctic Ocean the spreading rate is the slowest known, decreasing to <1.0 cm yr^{-1} from west to east as the Siberian shelf is approached. Discussion of vent-fluid compositions from different oceans, therefore, often approximates closely to variations in vent-fluid chemistries at different spreading rates. Although tables of the ranges of vent-fluid chemistries do not show distinct differences between ocean basins, some important differences do become apparent when those data are modeled. (NB: Although there is now firm evidence for hydrothermal activity in the Arctic, those systems have not yet been sampled for vent fluids; similarly, in the Indian Ocean, only two sites have recently been discovered). Consequently, meaningful comparisons can only readily be made, at present, between Atlantic and Pacific vent-fluid compositions. In systems on the slow spreading MAR, the calculated f_{O_2} of hydrothermal fluids is higher, suggesting that these systems are more oxidizing. Also, for example, both TAG and Lucky Strike vent fluids contain relatively little potassium compared to sodium (Edmond *et al.*, 1995; Von Damm *et al.*, 1998). Boron is also low in some of the Atlantic sites, especially at TAG and Logatchev (You *et al.*, 1994; Bray and Von Damm, 2003a). The explanation for these observations is that on the slow-spreading MAR, hydrothermal activity is active for a much longer period of time at any given site (also reflected in the relative sizes of the hydrothermal deposits formed: Humphris *et al.*, 1995; Fouquet *et al.*, submitted). Consequently, MAR vent fluids become more oxic because more dissolved seawater SO_4 has penetrated as deep as the reaction zone; the rocks within the hydrothermal flow cell have been more completely leached and altered. Because of the more pronounced tectonic (rather than volcanic) activity that is associated with slow spreading ridges, rock types that are normally found at greater depths within the oceanic crust can be exposed at or near the surface. Thus, hydrothermal sites have been located along the MARs that are hosted in ultramafic rocks: the Rainbow, Logatchev, and Lost City sites. No ultramafic-hosted systems are expected to occur, by contrast, along the fast-spreading ridges of the EPR.

6.07.2.4.4 *The role of the plumbing system*

Another fundamental difference observed in the nature of hydrothermal systems at fast- and slow-spreading ridges concerns intra-field differences in vent-fluid compositions. (Note that the terms "vent field" and "vent area" are often used interchangeably, have no specific size classifications, and may be used differently by different authors. In our usage, "vent area" is smaller, referring to a cluster of vents within 100 m or so, while a "vent field" may stretch for a kilometer or more along the ridge axis—but this is by no means a standard definition.) On a slow spreading ridge, such as the MAR, all of the fluids venting, for example, at the TAG site, can be shown to have a common source fluid that may have undergone some change in composition due to near surface processes such as mixing and/or conductive cooling. Many of the fluids can also be related to each other at the Lucky Strike site. In contrast, on fast spreading ridges, vents that may be within a few tens of meters of each other, clearly have different source fluids. A plausible explanation for these differences would be that vents on slow spreading ridges are fed from greater depths than those on fast-spreading ridges, with emitted fluids channeled upward from the subsurface along fault planes or other major tectonic fractures. Hence, in at least some cases, hydrothermal activity found on slow spreading ridges may be located wherever fluids have been preferentially channeled. Active vent sites on slow spreading ridges also appear to achieve greater longevity, based on the size of the sulfide structures and mounds they have produced. Fluids on fast spreading ridges, by contrast, are fed by much shallower heat sources and the conduits for these fluids appear to be much more localized, resulting in the very pronounced chemical differences often observed between immediately adjacent vent structures. Clearly, the plumbing systems at fast and slow spreading ridges must be characterized by significant, fundamental differences.

6.07.2.5 Temporal Variability in Vent-fluid Compositions

The MOR is, in effect, one extremely long, continuous submarine volcano. While volcanoes are commonly held to be very dynamic features, however, little temporal variability was observed for more than the first decade of work on hydrothermal systems. Indeed, a tendency arose not to view the MOR as an active volcano, at least on the timescales that were being worked on. This perspective was changed dramatically in the early 1990s. Together with evidence for recent volcanic eruptions at several sites, profound temporal variability in vent-fluid chemistries, temperatures, and styles of venting were also observed (Figure 7). In one case, the changes observed at a single vent almost span the full range of known compositions reported from throughout the globe. These temporal variations in hydrothermal venting reflect changes in the nature of the underlying heat source. The intrusion of a basaltic dike into the upper ocean crust, which may or may not be accompanied by volcanic extrusion at the seafloor, has been colloquially termed "the quantum unit of ocean accretion." These dikes are of the order 1 m wide, 10 km long, and can extend hundreds of meters upward through the upper crust toward the ocean floor. These shallow-emplaced and relatively small, transient, heat sources provide most, if not all of the heat that supports venting immediately following magmatic emplacement. Over timescales of as little as a year, however, an individual dike will have largely cooled and the heat source driving any continuing vent activity deepens. An immediate result is a decrease in measured exit temperatures for the vent fluids, because more heat is now lost, conductively, as the fluids rise from deeper within the oceanic crust. Vent-fluid compositions change, too, because the conditions of phase separation change; so, too, do the subsurface path length and residence time, such that the likelihood that circulating fluids reach equilibrium or steady state with the ocean crust also vary. Detailed time-series studies at sites perturbed by magmatic emplacements have shown that it is the vapor phase which vents first, in the earliest stages after a magmatic/volcanic event, while the high-chlorinity liquid phase is expelled somewhat later. In the best documented case study available, "brine" fluids were actively venting at a location some three years after the vapor phase had been expelled from the same individual hydrothermal chimney; at other event-affected vent sites, similar evolutions in vent-fluid composition have been observed over somewhat longer timescales.

The temporal variability that has now been observed has revolutionized our ideas about the functioning of hydrothermal systems and the timescales over which processes occur on the deep ocean floor. It is no exaggeration to state that processes we thought would take 100–1,000 yr, have been seen to occur in <10 yr. The majority of magmatic intrusions/eruptions that have been detected have been along the JdFR (Cleft Segment, Co-axial Segment, and Axial Volcano) where acoustic monitoring of the T-phase signal that accompanies magma migration in the upper crust has provided real-time data for events in progress and allowed "rapid response" cruises to be organized at these sites, within days to weeks. We also have good evidence for two volcanic events on the ultrafast spreading southern EPR, but the best-studied eruption site, to date, has been at 9°45–52′ N on the EPR. Serendipitously, submersible studies began at 9–10° N EPR less than one month after volcanic eruption at this site (Haymon *et al.*, 1993; Rubin *et al.*, 1994). Profound chemical changes (more than a factor of two in some cases) were noted at some of these vents during a period of less than a month (Von Damm *et al.*, 1995; Von Damm, 2000). Subsequently, it has become clear that very rapid changes occur within an initial one-year period which are related to changes in the conditions of phase separation and water rock reaction. These, in turn, are presumed to reflect responses to the mining of heat from the dike-intrusion, including lengthening of the reaction path and increases in the residence time of the fluids within the crust. At none of the other eruptive sites has it been possible to complete comparable direct sampling of vent fluids within this earliest "post-event" time period. It is now clear that the first fluid to be expelled is the vapor phase (whether formed as a result of sub- or supercritical phase separation), probably because of its lower density. What happens next, however, is less clear. In several cases, the "brine" (liquid) phase has been emitted next. In some vents this has occurred as a gradual progression to higher-chlorinity fluids; in other vents, the transition appears to occur more as a step function—although those observations may be aliased by the episodic nature of the sampling programs involved. What is most certainly the case at 9° N EPR, however, is that following initial vapor-phase expulsion, some vents have progressed to venting fluids with chlorinities greater than seawater much faster (≤3 yr), than others (~10 yr), and several have never made the transition. Further, in some parts of the eruptive area, vapor phase fluids are still the predominant type of fluid being emitted more than a decade after the eruption event. Fluids exiting from several of the vents have begun to decrease in chlorinity again, without ever having reached seawater concentrations. Conversely, other systems (most notably those from the Cleft segment) have been emitting vent fluids with chlorinities approximately twice that of seawater for more than a decade. If one wanted to sustain an argument that hydrothermal systems followed a vapor-to-brine phase evolution as a system ages (e.g., Butterfield *et al.*, 1997), it would be difficult to reconcile the observation that systems that are presumed to be relatively "older" (e.g., those on the northern Gorda Ridge) vent vapor-phase fluids, only (Von Damm *et al.*, 2001). Finally, one vent on the southern EPR, in an area with no evidence for a recent magmatic event, is emitting fluids which are phase separating "real time," with vapor exiting from the top of the structure, and brine from the bottom, simultaneously (Von Damm *et al.*, 2003). Fundamentally, there is a chloride-mass balance problem at many known hydrothermal sites. For example, at 21° N EPR, where high-temperature venting was first discovered and an active system is now known to have persisted for at least 23 years (based on sampling expeditions from 1979 and 2002) *only* low-chlorinity fluids are now being emitted (Von Damm *et al.*, 2002). Clearly there must be some additional storage and/or transport of higher-chlorinity fluids within the underlying crust. Our understanding of such systems is, at best, poor. What *is* clear is that pronounced temporal variability occurs at many vent sites, most notably at those that have been affected by magmatic events. There is also evidence for changes accompanying seismic events that are not related to magma-migration, but most likely related to cracking within the upper ocean crust (Sohn *et al.*, 1998; Von Damm and Lilley, 2003).

In marked contrast to those sites where volcanic eruptions and/or intrusions (diking events) have been detected, there are several other sites that have been sampled repeatedly over timescales of about two decades where no magmatic activity is known to have occurred. At some of these sites, chemical variations observed over the entire sampling period fall within the analytical error of our measurement techniques. The longest such time-series is for hydrothermal venting at 21° N on the EPR, where black smokers were first discovered in 1979, and where there has been remarkably little change in the composition of at least some of the vent fluids. Similarly, the Guaymas Basin vent site was first sampled in January 1982 (Von Damm *et al.*, 1985b), South Cleft on the JdF ridge in 1984 (Von Damm and Bischoff, 1987), and TAG on the MAR in 1986 (Campbell *et al.*, 1988). All these sites have exhibited very stable vent-fluid chemistries, although only TAG would be considered as a "slow spreading" vent site. Nevertheless, it is the TAG vent site that has shown perhaps the most remarkable stability in its vent-fluid compositions; these have remained invariant for more than a decade, even after perturbation from the drilling of a series of 5 ODP holes direct into the top of the active sulfide mound (Humphris *et al.*, 1995; Edmonds *et al.*, 1996).

Accounting for temporal variability (or lack thereof) when calculating hydrothermal fluxes is, clearly, problematic. It is very difficult to calculate the volume of fluid exiting from a hydrothermal system accurately. Many of the differences from seawater are most pronounced in the early eruptive period (~1 yr), which is also the time when fluid temperatures are hottest (Von Damm, 2000). Visual observations suggest that this is a time of voluminous fluid flow, which is not unexpected given that an enhanced heat source will have recently been emplaced directly at the seafloor in the case of an eruption, or, in the case of a dike intrusion, at shallow depths beneath. The upper oceanic crust exhibits high porosity, filled with ambient seawater. At eruption/intrusion, this seawater will be heated rapidly, its density will decrease, the resultant fluid will quickly rise, and large volumes of unreacted, cooler seawater will be drawn in and quickly expelled. It is not unreasonable to assume, therefore, that the water flux through a hydrothermal system may be at its largest during this initial period, just when chemical compositions are most extreme (Von Damm *et al.*, 1995; Von Damm, 2000). The key to the problem, therefore, probably lies in determining how much time a hydrothermal system spends in its "waxing" (immediate post-eruptive) stage when compared to the time spent at "steady state" (e.g., as observed at 21° N EPR) and in a "waning" period, together with an evaluation of the relative heat, water, and chemical fluxes associated with each of these different stages. If fluid fluxes and chemical anomalies are greatest in the immediate post-eruptive period, for example, the initial 12 months of any vent-field eruption may be geochemically more significant than a further 20 years of "steady-state" emission. At fast-spreading ridges, new eruptions might even occur faster than such a vent "lifecycle" can be completed. Alternately, the converse may hold true: early-stage eruptions may prove relatively insignificant over the full lifecycle of a prolonged, unperturbed hydrothermal site.

To advance our understanding of the chemical variability of vent fluids, it will be important to continue to find new sites that may be at evolutionary stages not previously observed. Equally, it will be important to continue studies of temporal variability at known sites: both those that have varied in the past and those that have appeared to be stable over the time intervals at which they have been sampled. Understanding the mechanisms and physical processes that control these vent-fluid compositions are key to calculating hydrothermal fluxes.

6.07.3 THE *NET* IMPACT OF HYDROTHERMAL ACTIVITY

It is important to remember that the gross chemical flux associated with expelled vent fluids (Section 6.07.2) is not identical to the net flux from hydrothermal systems. Subsurface hydrothermal circulation can have a net negative flux for some chemicals, the most obvious being magnesium which is almost quantitatively removed from the starting seawater and is added instead to the oceanic crust. Another example of such removal is uranium, which is also completely removed from seawater during hydrothermal circulation and then recycled into the upper mantle through subduction of altered oceanic crust. Even where, compared to starting seawater, there is no concentration gain or loss, an element may nevertheless undergo almost complete isotopic exchange within the oceanic crust indicating that none of the substance originally present in the starting seawater has passed conservatively through the hydrothermal system—the most obvious example being that of strontium. We present a brief summary of ocean crust mineralization in Section 6.07.4.1, but a more detailed treatment of ocean crust alteration is given elsewhere (see Chapter 3.15).

For the remainder of this chapter we concentrate, instead, upon the fate of hydrothermal discharge once it reaches the seafloor. Much of this material, transported in dissolved or gaseous form in warm or hot fluids, does not remain in solution but forms solid phases as fluids cool and/or mix with colder, more alkaline seawater. These products, whose formation may be mediated as well as modified by a range of biogeochemical processes, occur from the ridge axis out into the deep ocean basins. Massive sulfide as well as silicate, oxide, and sometimes carbonate deposits formed from high-temperature fluids are progressively altered by high-temperature metasomatism, as well as low-temperature oxidation and mass wasting—much of which may be biologically mediated. Various low-temperature deposits may also be formed, again often catalyzed by biological activity. In addition to these near-vent hydrothermal products, abundant fine-grained particles are formed in hydrothermal plumes, which subsequently settle to the seafloor to form metalliferous sediments, both close to vent sites and across ridge flanks into adjacent ocean basins. The post-depositional fates for these near- and far-field deposits remain poorly understood. Sulfide deposits, for example, may undergo extensive diagenesis and dissolution, leading to further release of dissolved chemicals into the deep ocean. Conversely, oxidized hydrothermal products may remain well-preserved in the sedimentary record and only be recycled via subduction back into the Earth's interior. On ridges where volcanic eruptions are frequent, both relatively fresh and more oxidized deposits may be covered over by subsequent lava flows (on the timescale of a decade) and, thus, become assimilated into the oceanic crust, isolated from the overlying water- and sediment-columns. In the following sections we discuss the fates of various hydrothermal "products" in order of their distance from the vent-source: near-vent deposits (Section 6.07.4); hydrothermal plumes (Section 6.07.5); and hydrothermal sediments (Section 6.07.6).

6.07.4 NEAR-VENT DEPOSITS

6.07.4.1 Alteration and Mineralization of the Upper Ocean Crust

Hydrothermal circulation causes extensive alteration of the upper ocean crust, reflected both in mineralization of the crust and in changes to physical properties of the basement (Alt, 1995). The direction and extent of chemical and isotopic exchange between seawater and oceanic crust depends on variations in temperature and fluid penetration and, thus, vary strongly as a function of depth. Extensive mineralization of the upper ocean crust can occur where metals leached from large volumes of altered crust become concentrated at, or close beneath, the sediment–water interface (Hannington *et al.*, 1995, Herzig and Hannington, 2000). As we have seen (Section 6.07.2), hydrothermal circulation within the ocean crust can be subdivided into three major components—the *recharge*, *reaction,* and *discharge* zones (Figure 5).

Recharge zones, which are broad and diffuse, represent areas where seawater is heated and undergoes reactions with the crust as it penetrates, generally downwards. (It is important to remember, however, that except where hydrothermal systems are sediment covered, the location of the recharge zone has never been established; debate continues, for example, whether recharge occurs *along* or *across* axis.) Important reactions in the recharge zone, at progressively increasing temperatures, include: low-temperature oxidation, whereby iron oxyhydroxides replace olivines and primary sulfides; fixation of alkalis (potassium, rubidium, and caesium) in celadonite and nontronite (ferric mica and smectite, respectively) and fixation of magnesium, as smectite (<200 °C) and chlorite (>200 °C). At temperatures exceeding ~130–150 °C, two other key reactions occur: formation of anhydrite ($CaSO_4$) and mobilization of the alkalis (potassium, rubidium, lithium) (Alt, 1995). Upon subduction, the altered mineralogy and composition of the ocean crust can lead to the development of chemical and isotopic heterogeneities, both in the mantle and in the composition of island arc volcanic rocks. This subject is discussed in greater detail by Staudigel (see Chapter 3.15).

The *reaction* zone represents the highest pressure and temperature (likely >400 °C) conditions reached by hydrothermal fluids during their subsurface circulation; it is here that hydrothermal vent fluids are believed to acquire much of their chemical signatures (Section 6.07.2). As they become buoyant, these fluids then rise rapidly back to the seafloor through *discharge* zones. The deep roots of hydrothermal discharge zones have only ever been observed at the base of ophiolite sequences (e.g., Nehlig *et al.*, 1994). Here, fluid inclusions and losses of metals and sulfur from the rocks indicate alteration temperatures of 350–440 °C (Alt and Bach, 2003) in reasonable agreement with vent-fluid observations (Section 6.07.2). Submersible investigations and towed camera surveys of the modern seafloor have allowed surficial hydrothermal deposits to be observed in some detail (see next section). By contrast, the alteration pipes and "stockwork" zones that are believed to form the "roots" underlying all seafloor hydrothermal deposits (Figure 10) and which are considered to be quantitatively important in global geochemical

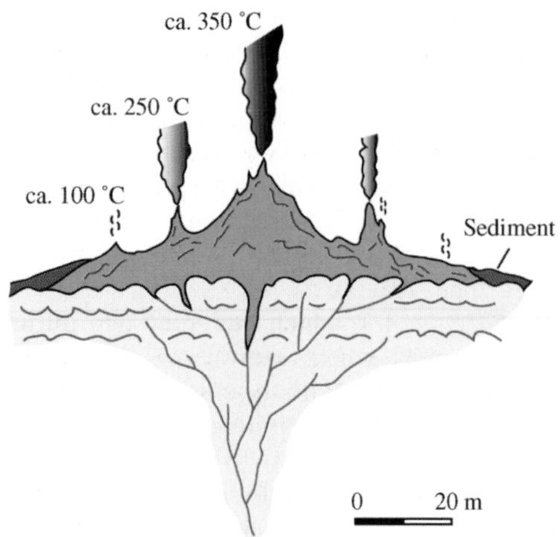

Figure 10 Schematic illustration of an idealized hydrothermal sulfide mound including: branching stockwork zone beneath the mound; emission of both high-temperature (350 °C) and lower-temperature (100–250 °C) fluids from the top of the mound together with more ephemeral diffuse flow; and deposition by mass-wasting of an apron of sulfidic sediments around the edges of the mound.

cycles (e.g., Peucker-Ehrenbrink *et al.*, 1994) remain relatively inaccessible. Direct sampling of the stockwork beneath active seafloor hydrothermal systems has only been achieved on very rare occasions, e.g., through direct sampling from ODP drilling (Humphris *et al.*, 1995) or through fault-face exposure at the seabed (e.g., Fouquet *et al.*, 2003). Because of their relative inaccessibility, even compared to all other aspects of deep-sea hydrothermal circulation, study of sub-seafloor crustal mineralization remains best studied in ancient sulfide deposits preserved in the geologic record on-land (Hannington *et al.*, 1995).

6.07.4.2 Near-vent Hydrothermal Deposits

The first discoveries of hydrothermal vent fields (e.g., Galapagos; EPR, 21° N) revealed three distinctive types of mineralization: (i) massive sulfide mounds deposited from focused high-temperature fluid flow, (ii) accumulations of Fe–Mn oxyhydroxides and silicates from low-temperature diffuse discharge, and (iii) fine-grained particles precipitated from hydrothermal plumes. Subsequently, a wide range of mineral deposits have been identified that are the result of hydrothermal discharge, both along the global ridge crest and in other tectonic settings (Koski *et al.*, 2003). Of course, massive sulfides only contain a fraction of the total dissolved load released from the seafloor. Much of this flux is delivered to ridge flanks via dispersion in buoyant and nonbuoyant hydrothermal plumes (Section 6.07.5). In addition, discoveries such as the carbonate-rich "Lost City" deposits (Kelley *et al.*, 2001), silica-rich deposits formed in the Blanco Fracture Zone (Hein *et al.*, 1999), and metal-bearing fluids on the flanks of the JdFR (Mottl *et al.*, 1998) remind us that there is still much to learn about the formation of hydrothermal mineral deposits.

Haymon (1983) proposed the first model for how a black smoker chimney forms (Figure 11). The first step is the formation of an anhydrite ($CaSO_4$) framework due to the heating of seawater, and mixing of vent fluids with that seawater. The anhydrite walls then protect subsequent hydrothermal fluids from being mixed so extensively with seawater, as well as providing a template onto which sulfide minerals can precipitate as those fluids cool within the anhydrite structure. As the temperature and chemical compositions within the chimney walls evolve, a zonation of metal sulfide minerals develops, with more copper-rich phases being formed towards the interior, zinc-rich phases towards the exterior, and iron-rich phases ubiquitous. This model is directly analogous to the concept of an "intensifying hydrothermal system" developed by Eldridge *et al.* (1983), in which initial deposition of a fine-grained mineral carapace restricts mixing of hydrothermal fluid and seawater at the site of discharge. Subsequently, less-dilute, higher-temperature (copper-rich) fluids interact with the sulfides within this carapace to precipitate chalcopyrite and mobilize more soluble, lower-temperature metals such as lead and zinc toward the outer, cooler parts of the deposit. Thus, it is the steep temperature and chemical gradients, caused by both mixing and diffusion, which account for the variations in wall mineralogy and Cu–Zn zonation observed in both chimneys and larger deposits. These processes, initially proposed as part of a conceptual model, have subsequently been demonstrated more rigorously by quantitative geochemical modeling of hydrothermal fluids and deposits (Tivey, 1995).

Drilling during Ocean Drilling Program Leg 158 revealed that similar internal variations can also occur on much larger scales—e.g., across the entire TAG mound (Humphris *et al.*, 1995; Petersen *et al.*, 2000). That work revealed the core of the mound to be dominated by chalcopyrite-bearing massive pyrite, pyrite–anhydrite and pyrite–silica breccias whilst the mound top and margins contained little or no chalcopyrite but more sphalerite and higher concentrations of metals soluble at lower temperatures (e.g., zinc, gold). The geochemical modeling results of Tivey (1995) point to a mechanism of entrainment of seawater into the

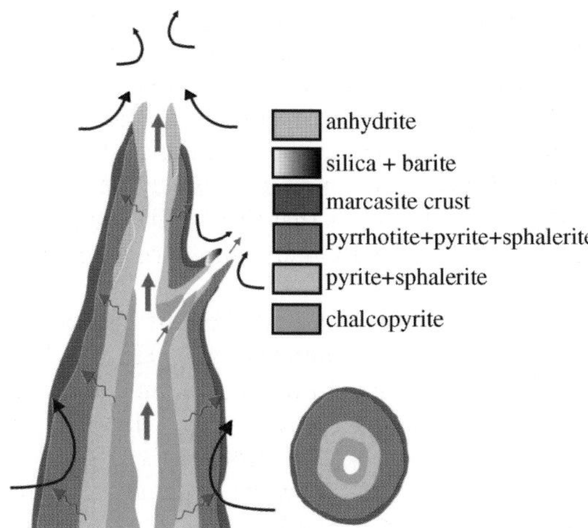

Figure 11 Schematic diagram showing mineral zonation in cross-section and in plan view for a typical black smoker chimney. Arrows indicate direction of inferred fluid flow (after Haymon, 1983).

focused upflow zone within the mound which would, almost simultaneously: (i) induce the precipitation of anhydrite, chalcopyrite, pyrite, and quartz; (ii) decrease the pH of the fluid; and (iii) mobilize zinc and other metals. When combined, the processes outlined above—zone refining and the entrainment of seawater into active sulfide deposits—appear to credibly explain mineralogical and chemical features observed both in modern hydrothermal systems (e.g., the TAG mound) and in "Cyprus-type" deposits found in many ophiolites of orogenic belts (Hannington *et al.*, 1998).

A quite different form of hydrothermal deposit has also been located, on the slow-spreading MAR. The Lost City vent site (Kelley *et al.*, 2001) occurs near 30° N on the MAR away from the more recently erupted volcanic ridge axis. Instead, it is situated high up on a tectonic massif where faulting has exposed variably altered peridotite and gabbro (Blackman *et al.*, 1998). The Lost City field hosts at least 30 active and inactive spires, extending up to 60 m in height, on a terrace that is underlain by diverse mafic and ultramafic lithologies. Cliffs adjacent to this terrace also host abundant white hydrothermal alteration both as flanges and peridotite mineralization, which is directly akin to deposits reported from Alpine ophiolites (Früh-Green *et al.*, 1990). The Lost City chimneys emit fluids up to 75 °C which have a very high pH (9.0–9.8) and compositions which are rich in H_2S, CH_4, and H_2—consistent with serpentinization reactions (Section 6.07.2)—but low in dissolved silica and metal contents (Kelley *et al.*, 2001). Consistent with this, the chimneys of the Lost City field are composed predominantly of magnesium and calcium-rich carbonate and hydroxide minerals, notably calcite, brucite, and aragonite.

In addition to the sulfide- and carbonate-dominated deposits described above, mounds and chimneys composed of Fe- and Mn-oxyhydroxides and silicate minerals also occur at tectonically diverse rift zones, from MORs such as the Galapagos Spreading Center to back-arc systems such as the Woodlark Basin (Corliss *et al.*, 1978; Binns *et al.*, 1993). Unlike polymetallic sulfides, Fe–Mn oxide-rich, low-temperature deposits should be chemically stable on the ocean floor. Certainly, metalliferous sediments in ophiolites—often referred to as "umbers"—have long been identified as submarine hydrothermal deposits formed in ancient ocean ridge settings. These types of ophiolite deposit may be intimately linked to the Fe–Mn–Si oxide "mound" deposits formed on pelagic ooze near the Galapagos Spreading Center (Maris and Bender, 1982). It has proven difficult, however, to determine the precise temporal and genetic relationship of umbers to massive sulfides, not least because no gradation of Fe–Mn–Si oxide to sulfide mineralization has yet been reported from ophiolitic terranes. The genetic relationship between sulfide and oxide facies deposits formed at modern hydrothermal sites also remains enigmatic. Fe–Mn–Si oxide deposits may simply represent "failed" massive sulfides. Alternately, it may well be that there are important aspects of, for example, axial versus off-axis plumbing systems (e.g., porosity, permeability, chemical variations caused by phase separation) or controls on the sulfur budget of hydrothermal systems that remain inadequately understood. What seems certain is that the three-dimensional problem of hydrothermal deposit formation (indeed, 4D if one includes temporal evolution) cannot be solved from seafloor observations alone. Instead, what is required is a continuing program of seafloor drilling coupled with analogue studies of hydrothermal deposits preserved on land.

6.07.5 HYDROTHERMAL PLUME PROCESSES

6.07.5.1 Dynamics of Hydrothermal Plumes

Hydrothermal plumes form wherever buoyant hydrothermal fluids enter the ocean. They represent an important dispersal mechanism for the thermal and chemical fluxes delivered to the oceans while the processes active within these plumes serve to modify the gross fluxes from venting, significantly. Plumes are of further interest to geochemists because they can be exploited in the detection and location of new hydrothermal fields *and* for the calculation of total integrated fluxes from any particular vent field. To biologists, hydrothermal plumes represent an effective transport mechanism for dispersing vent fauna, aiding gene-flow between adjacent vent sites along the global ridge crest (e.g., Mullineaux and France, 1995). In certain circumstances the heat and energy released into hydrothermal plumes could act as a driving force for mid-depth oceanic circulation (Helfrich and Speer, 1995).

Present day understanding of the dynamics of hydrothermal plumes is heavily influenced by the theoretical work of Morton *et al.* (1956) and Turner (1973). When high-temperature vent fluids are expelled into the base of the much colder, stratified, oceanic water column they are buoyant and begin to rise. Shear flow between the rising fluid and the ambient water column produces turbulence and vortices or eddies are formed which are readily visible in both still and video-imaging of active hydrothermal vents. These eddies or vortices act to entrain material from the ambient water column, resulting in a continuous dilution of the original vent fluid as the plume rises. Because the oceans exhibit stable density-stratification, this mixing causes progressive dilution of the buoyant plume with water which is denser than both the initial vent fluid and the overlying water column into which the plume is rising. Thus, the plume becomes progressively less buoyant as it rises and it eventually reaches some finite maximum height above the seafloor, beyond which it cannot rise (Figure 12). The first, rising stage of hydrothermal plume evolution is termed the *buoyant plume*. The later stage, where plume rise has ceased and hydrothermal effluent begins to be dispersed laterally, is termed the *nonbuoyant plume* also referred to in earlier literature as the *neutrally buoyant plume*.

The exact height reached by any hydrothermal plume is a complex function involving key properties of both the source vent fluids and the water column into which they are injected—notably the initial buoyancy of the former and the degree of stratification of the latter. A theoretical approach to calculating the maximum height-of-rise that can be attained by any hydrothermal plume is given by Turner (1973) with the equation:

$$z_{max} = 3.76 F_0^{1/4} N^{-3/4} \tag{1}$$

where F_0 and N represent parameters termed the buoyancy flux and the Brunt–Väisälä frequency, respectively. The concept of the buoyancy flux, F_0 (units $cm^4\ s^{-3}$) can best be understood from an explanation that the product $F_0\rho_0$ represents the total weight deficiency produced at the vent source per unit time (units $g\ cm\ s^{-3}$). The Brunt–Väisälä frequency, also termed the buoyancy frequency, N (units s^{-1}) is defined as:

$$N^2 = -(g/\rho_0)\mathrm{d}\rho/\mathrm{d}z \tag{2}$$

where g is the acceleration due to gravity, ρ_0 is the background density at the seafloor and $\mathrm{d}\rho/\mathrm{d}z$ is the ambient vertical density gradient. In practice, buoyant hydrothermal plumes always exceed this theoretical maximum because, as they reach the level z_{max} the plume retains some finite positive vertical velocity. This leads to "momentum overshoot" (Turner, 1973) and "doming" directly above the plume source before this (now negatively buoyant) dome collapses back to the level of zero buoyancy (Figure 12).

Note also the very weak dependence of emplacement height (z_{max}) upon the buoyancy flux or heat flux of any given vent source. A doubling of z_{max} for any plume, for example, could only be achieved by a 16-fold increase in the heat flux provided by its vent source. By contrast, the ambient water column with which the buoyant plume becomes progressively more diluted exerts a significant influence because the volumes entrained are nontrivial. For a typical plume with $F_0 = 10^{-2}\ m^4\ s^{-3}$, $N = 10^{-3}\ s^{-1}$ the entrainment flux is of the order $10^2\ m^3\ s^{-1}$ (e.g., Helfrich and Speer, 1995) resulting in very rapid dilution of the primary vent fluid (10^2–10^3:1) within the first 5–10 m of plume rise and even greater dilution ($\sim 10^4$:1) by the time of emplacement within the nonbuoyant, spreading hydrothermal plume (Feely *et al.*, 1994). Similarly, the *time* of rise for a buoyant hydrothermal plume, is entirely dependent on the background buoyancy frequency, N (Middleton, 1979):

$$\tau = \pi N^{-1} \tag{3}$$

which, for a typical value of $N = 10^{-3}\ s^{-1}$, yields a plume rise-time of ≤ 1 h.

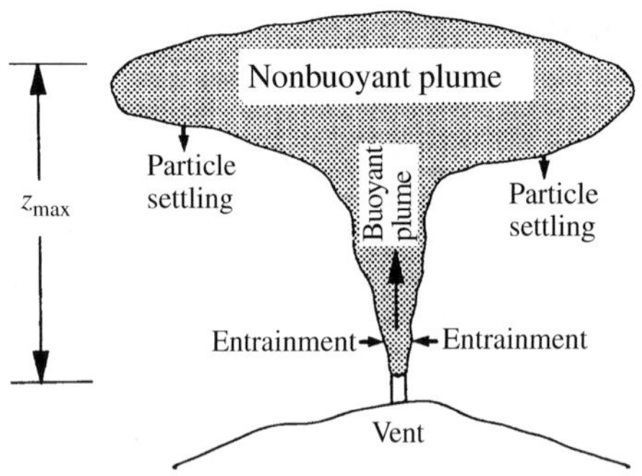

Figure 12 Sketch of the hydrothermal plume rising above an active hydrothermal vent, illustrating entrainment of ambient seawater into the buoyant hydrothermal plume, establishment of a nonbuoyant plume at height z_{max} (deeper than the maximum height of rise actually attained due to momentum overshoot) and particle settling from beneath the dispersing nonbuoyant plume (after Helfrich and Speer, 1995).

6.07.5.2 Modification of Gross Geochemical Fluxes

Hydrothermal plumes represent a significant dispersal mechanism for chemicals released from seafloor venting to the oceans. Consequently, it is important to understand the physical processes that control this dispersion (Section 6.07.5.1). It is also important to recognize that hydrothermal plumes represent non-steady-state fluids whose chemical compositions evolve with age (Figure 13). Processes active in hydrothermal plumes can lead to significant modification of *gross* hydrothermal fluxes (cf. Edmond *et al.*, 1979; German *et al.*, 1991b) and, in the extreme, can even reverse the sign of *net* flux to/from the ocean (e.g., German *et al.*, 1990, 1991a).

6.07.5.2.1 Dissolved noble gases

For one group of tracers, however, inert marine geochemical behavior dictates that they do undergo conservative dilution and dispersion within hydrothermal plumes. Perhaps the simplest example of such behavior is primordial dissolved ^{3}He, which is trapped in the Earth's interior and only released to the deep ocean through processes linked to volcanic activity—notably, submarine hydrothermal vents. As we have seen, previously, end-member vent fluids typically undergo ~10,000-fold dilution prior to emplacement in a nonbuoyant hydrothermal plume. Nevertheless, because of the large enrichments of dissolved ^{3}He in hydrothermal fluids when compared to the low background levels in seawater, pronounced

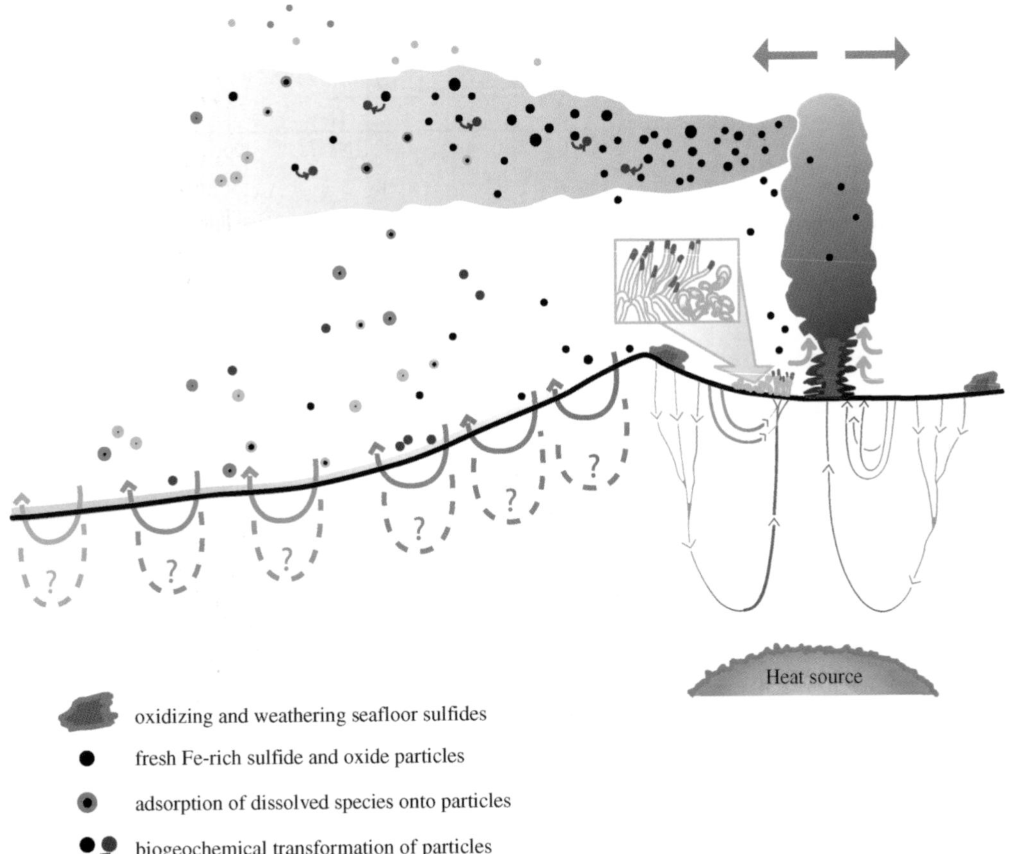

Figure 13 Schematic representation of an MOR hydrothermal system and its effects on the overlying water column. Circulation of seawater occurs within the oceanic crust, and so far three types of fluids have been identified and are illustrated here: high-temperature vent fluids that have likely reacted at >400 °C; high-temperature fluids that have then mixed with seawater close to the seafloor; fluids that have reacted at "intermediate" temperatures, perhaps ~150 °C. When the fluids exit the seafloor, either as diffuse flow (where animal communities may live) or as "black smokers," the water they emit rises and the hydrothermal plume then spreads out at its appropriate density level. Within the plume, sorption of aqueous oxyanions may occur onto the vent-derived particles (e.g., phosphate, vanadium, arsenic) making the plumes a sink for these elements; biogeochemical transformations also occur. These particles eventually rain-out, forming metalliferous sediments on the seafloor. While hydrothermal circulation is known to occur far out onto the flanks of the ridges, little is known about the depth to which it extends or its overall chemical composition because few sites of active ridge-flank venting have yet been identified and sampled (Von Damm, unpublished).

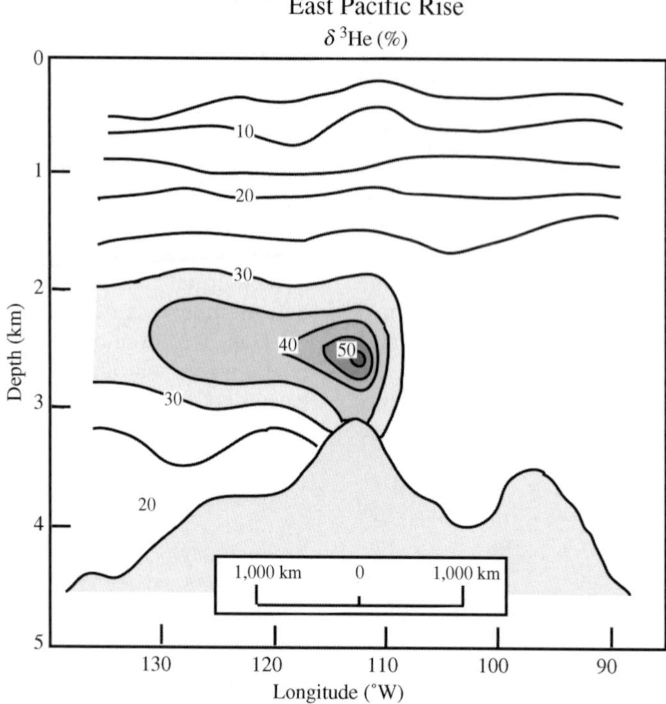

Figure 14 Distribution of δ^3He across the Pacific Ocean at 15° S. This plume corresponds to the lobe of metalliferous sediment observed at the same latitude extending westward from the crest of the EPR (Figure 2) (after Lupton and Craig, 1981).

enrichments of dissolved ^{3}He relative to ^{4}He can be traced over great distances in the deep ocean. Perhaps the most famous example of such behavior is the pronounced ^{3}He plume identified by Lupton and Craig (1981) dispersing over ~2,000 km across the southern Pacific Ocean, west of the southern EPR (Figure 14). A more recent example of the same phenomenon, however, is the large-scale ^{3}He anomaly reported by Rüth *et al.* (2000) providing the first firm evidence for high-temperature hydrothermal venting anywhere along the southern MAR. Rn-222, a radioactive isotope of the noble element radon, is also enriched in hydrothermal fluids; while it shares the advantages of being a conservative tracer with ^{3}He, it also provides the additional advantage of acting as a "clock" for hydrothermal plume processes because it decays with a half-life of 3.83 d. Kadko *et al.* (1990) used the fractionation of ^{222}Rn/^{3}He ratios in a dispersing nonbuoyant hydrothermal plume above the JdFR to estimate rates of dispersion or "ages" at different locations "down-plume" and, thus, deduce rates of uptake and/or removal of various nonconservative plume components (e.g., H_2, CH_4, manganese, particles). A complication to that approach arises, however, with the recognition that—in at least some localities—maximum plume-height ^{222}Rn/^{3}He ratios exceed pristine high-temperature vent-fluid values; clearly, entrainment from near-vent diffuse flow can act as an important additional source of dissolved ^{222}Rn entering ascending hydrothermal plumes (Rudnicki and Elderfield, 1992).

6.07.5.2.2 Dissolved reduced gases (H_2S, H_2, CH_4)

The next group of tracers that are important in hydrothermal plumes are the reduced dissolved gases, H_2S, H_2, and CH_4. As we have already seen, dissolved H_2S is commonly the most abundant dissolved reduced gas in high-temperature vent fluids (Section 6.07.2). Typically, however, any dissolved H_2S released to the oceans undergoes rapid reaction, either through precipitation of polymetallic sulfide phases close to the seafloor and in buoyant plumes, or through oxidation in the water column. As of the early 2000s, there has only been one report of measurable dissolved H_2S at nonbuoyant plume height anywhere in the deep ocean (Radford-Knöery *et al.*, 2001). That study revealed maximum dissolved H_2S concentrations of ≤ 2 nM, representing a 5×10^5-fold decrease from vent-fluid concentrations (Douville *et al.*, 2002) with complete oxidative removal occurring in just 4–5 h, within ~1–10 km of the vent site. Dissolved H_2, although not commonly abundant in such high concentrations in vent fluids (Section 6.07.2), exhibits similarly short-lived behavior within hydrothermal plumes. Kadko *et al.* (1990) and German *et al.* (1994) have reported maximum plume-height dissolved H_2 concentrations of 12 nM and 32 nM above the Main Endeavour vent site, JdFR and above the Steinahóll vent site, Reykjanes Ridge. From use of the $^{222}Rn/^{3}He$ "clock," Kadko *et al.* (1990) estimated an apparent "oxidative-removal" half-life for dissolved H_2 of ~10 h.

The most abundant and widely reported dissolved reduced gas in hydrothermal plumes is methane which is released into the oceans from both high- and low-temperature venting and the serpentinization of ultramafic rocks (e.g., Charlou *et al.*, 1998). Vent-fluid concentrations are significantly enriched over seawater values (10–2,000 $\mu mol\ kg^{-1}$ versus $<5\ nmol\ kg^{-1}$) but the behavior of dissolved methane, once released, appears variable from one location to another: e.g., rapid removal of dissolved CH_4 was observed in the Main Endeavour plume (half-life = 10 d; Kadko *et al.*, 1990) yet near-conservative behavior for the same tracer has been reported for the Rainbow hydrothermal plume, MAR, over distances up to 50 km from the vent site (Charlou *et al.*, 1998; German *et al.*, 1998b). Possible reasons for these significant variations are discussed later (Section 6.07.5.4).

6.07.5.2.3 Iron and manganese geochemistry in hydrothermal plumes

The two metals most enriched in hydrothermal vent fluids are iron and manganese. These elements are present in a reduced dissolved form (Fe^{2+}, Mn^{2+}) in end-member vent fluids yet are most stable as oxidized Fe(III) and Mn(IV) precipitates in the open ocean. Consequently, the dissolved concentrations of iron and manganese in vent fluids are enriched ~10^6:1 over open-ocean values (e.g., Landing and Bruland, 1987; Von Damm, 1995; Statham *et al.*, 1998). When these metal-laden fluids first enter the ocean, two important processes occur. First, the fluids are instantaneously cooled from >350 °C to ≤30 °C. This quenching of a hot saturated solution leads to precipitation of a range of metal sulfide phases that are rich in iron but not manganese because the latter does not readily form sulfide minerals. In addition, turbulent mixing between the sulfide-bearing vent fluid and the entrained, oxidizing, water column leads to a range of redox reactions resulting in the rapid precipitation of high concentrations of suspended iron oxyhydroxide particulate material. The dissolved manganese within the hydrothermal plume, by contrast, typically exhibits much more sluggish oxidation kinetics and remains predominantly in dissolved form at the time of emplacement in the nonbuoyant plume. Because iron and manganese are so enriched in primary vent fluids, nonbuoyant plumes typically exhibit total (dissolved and particulate) iron and manganese concentrations, which are ~100-fold higher than ambient water column values immediately following nonbuoyant plume emplacement. Consequently, iron and manganese, together with CH_4 and ^3He (above), act as extremely powerful tracers with which to identify the presence of hydrothermal activity from water-column investigations. The fate of iron in hydrothermal plumes is of particular interest because it is the geochemical cycling of this element, more than any other, which controls the fate of much of the hydrothermal flux from seafloor venting to the oceans (e.g., Lilley *et al.*, 1995).

Because of their turbulent nature, buoyant hydrothermal plumes have continued to elude detailed geochemical investigations. One approach has been to conduct direct sampling using manned submersibles or ROVs (e.g., Rudnicki and Elderfield, 1993; Feely *et al.*, 1994). An alternate, and indirect, method is to investigate the geochemistry of precipitates collected both rising in, and sinking from, buoyant hydrothermal plumes using near-vent sediment traps (e.g., Cowen *et al.*, 2001; German *et al.*, 2002). From direct observations it is apparent that up to 50% of the total dissolved iron emitted from hydrothermal vents is precipitated rapidly from buoyant hydrothermal plumes (e.g., Mottl and McConachy, 1990; Rudnicki and Elderfield, 1993) forming polymetallic sulfide phases which dominate (>90%) the iron flux to the near-vent seabed (Cowen *et al.*, 2001; German *et al.*, 2002). The remaining dissolved iron within the buoyant and nonbuoyant hydrothermal plume undergoes oxidative precipitation. In the well-ventilated N. Atlantic Ocean, very rapid Fe(II) oxidation is observed with a half-life for oxidative removal from solution of just 2–3 min (Rudnicki and Elderfield, 1993). In the NE Pacific, by contrast, corresponding half-times of up to 32 h have been reported from JdFR hydrothermal plumes (Chin *et al.*, 1994; Massoth *et al.*, 1994). Field and Sherrell (2000) have predicted that the oxidation rate for dissolved Fe^{2+} in hydrothermal plumes should decrease along the path of the thermohaline circulation, reflecting the progressively decreasing pH and dissolved oxygen content of the deep ocean (Millero *et al.*, 1987):

$$-d[Fe(II)]/dt = k[OH^-]^2[O_2][Fe(II)] \quad (4)$$

The first Fe(II) incubation experiments conducted within the Kairei hydrothermal plume, Central Indian Ridge, are consistent with that prediction, yielding an Fe(II) oxidation half-time of ~2 h (Statham *et al.*, 2003, submitted).

6.07.5.2.4 Co-precipitation and uptake with iron in buoyant and nonbuoyant plumes

There is significant co-precipitation of other metals enriched in hydrothermal fluids, along with iron, to form buoyant plume polymetallic sulfides. Notable among these are copper, zinc, and lead. Common accompanying phases, which also sink rapidly from buoyant plumes, are barite and anhydrite (barium and calcium sulfates) and amorphous silica (e.g., Lilley *et al.*, 1995). In nonbuoyant hydrothermal plumes, where Fe- and (to a lesser extent) Mn-oxyhydroxides predominate, even closer relationships are observed between particulate iron concentrations and numerous other tracers. To a first approximation, three differing iron-related behaviors have been identified (German *et al.*, 1991b; Rudnicki and Elderfield, 1993): (i) co-precipitation; (ii) fixed molar ratios to iron; and (iii) oxyhydroxide scavenging (Figure 15). The first is that already alluded to above and loosely termed "chalcophile" behavior—namely preferential co-precipitation with iron as sulfide phases followed by preferential settling from the nonbuoyant plume. Such elements exhibit a generally positive correlation with iron for plume particle concentrations but with highest X:Fe ratios closest to the vent site (X = Cu, Zn, Pb) and much lower values farther afield.

The second group are particularly interesting. These are elements that establish fixed X:Fe ratios in nonbuoyant hydrothermal plumes which do not vary with dilution or dispersal distance "down-plume" (Figure 15). Elements that have been shown to exhibit such "linear" behavior include potassium, vanadium, arsenic, chromium, and uranium (e.g., Trocine and Trefry, 1988; Feely *et al.*, 1990, 1991; German *et al.*, 1991a,b). Hydrothermal vent fluids are not particularly enriched in any of these elements, which typically occur as rather stable "oxyanion" dissolved species in seawater. Consequently, this uptake must represent a significant removal flux, for at least some of these elements, from the deep ocean. The P:Fe ratios observed throughout all Pacific hydrothermal plumes are rather similar (P:Fe = 0.17–0.21; Feely *et al.*, 1998) and distinctly higher than the value observed in Atlantic hydrothermal plumes (P:Fe = 0.06–0.12). This has led to speculation that plume P:Fe ratios may reflect the ambient dissolved PO_4^{3-} concentration of the host water column and, thus, may offer potential as a long-term tracer of past ocean circulation, if preserved faithfully in metalliferous marine sediments (Feely *et al.*, 1998).

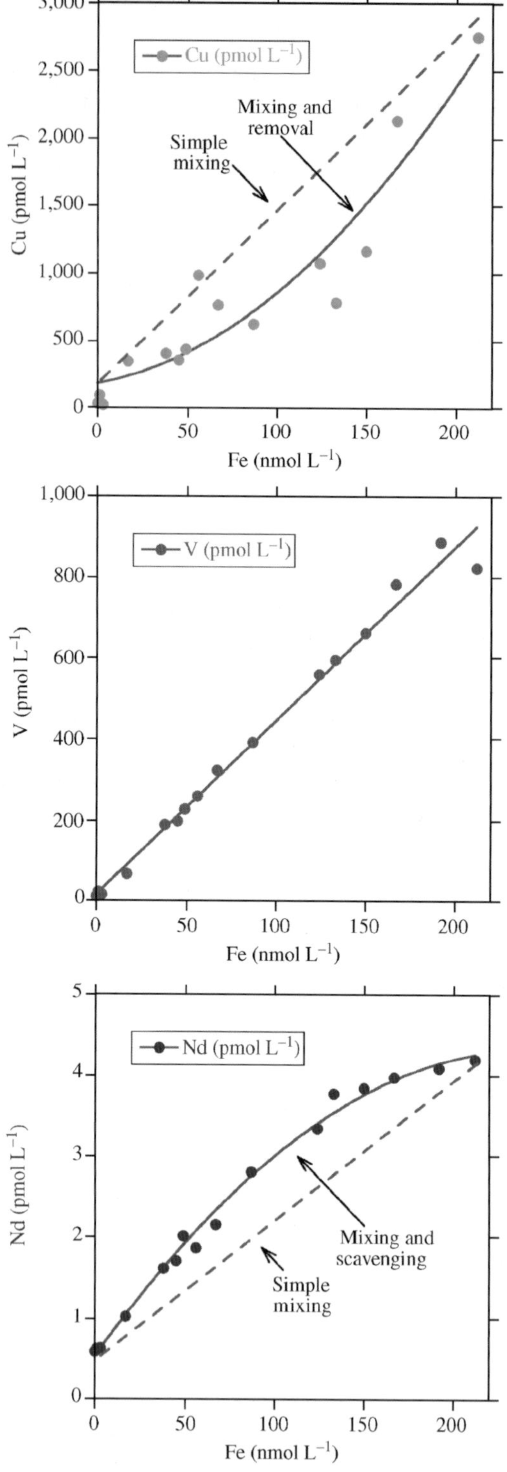

Figure 15 Plots of particulate copper, vanadium, and neodymium concentrations versus particulate iron for suspended particulate material filtered *in situ* from the TAG hydrothermal mound, MAR, 26° N (data from German *et al.*, 1990, 1991b). Note generally positive correlations with particulate Fe concentration for all three tracers but with additional negative (Cu) or positive (Nd) departure for sulfide-forming and scavenged elements, respectively.

6.07.5.2.5 *Hydrothermal scavenging by Fe-oxyhydroxides*

A final group of elements identified from hydrothermal plume particle investigations are of particular importance: "particle-reactive" tracers. Perhaps the best examples are the REE although other tracers that show similar behavior include beryllium, yttrium, thorium, and protactinium (German *et al.*, 1990, 1991a,b, 1993; Sherrell *et al.*, 1999). These tracers, like the two preceding groups, exhibit generally positive correlations with particulate iron concentrations in hydrothermal plumes (Figure 15). Unlike the "oxyanion" group, however, these tracers do not show constant X:Fe ratios within the nonbuoyant plume. Instead, highest values (e.g., for REE:Fe) are observed at increasing distances away from the vent site, rather than immediately above the active source (German *et al.*, 1990; Sherrell *et al.*, 1999). One possible interpretation of this phenomenon is that the Fe-oxyhydroxide particles within young nonbuoyant plumes are undersaturated with respect to surface adsorption and that continuous "scavenging" of dissolved, particle-reactive species occurs as the particles disperse through the water column (German *et al.*, 1990, 1991a,b). An alternate hypothesis (Sherrell *et al.*, 1999) argues, instead, for two-stage equilibration within a nonbuoyant hydrothermal plume: close to the vent source, a maximum in (e.g.,) particulate REE concentrations is reached, limited by equilibration at fixed distribution constants between the high particulate iron concentrations present and the finite dissolved tracer (e.g., REE) concentrations present in plume-water. As the plume disperses and undergoes dilution, however, particulate iron concentrations also decrease; re-equilibration between these particles and the diluting "pristine" ambient seawater, at the same fixed distribution constants, would then account for the higher REE/Fe ratios observed at lower particulate iron concentrations. More work is required to resolve which of these interpretations (kinetic versus equilibrium) more accurately reflects the processes active within hydrothermal plumes. What is beyond dispute concerning these particle-reactive tracers is that their uptake fluxes, in association with hydrothermal Fe-oxyhydroxide precipitates, far exceeds their dissolved fluxes entering the oceans from hydrothermal vents. Thus, hydrothermal plumes act as *sinks* rather than *sources* for these elements, even causing local depletions relative to the ambient water column concentrations (e.g., Klinkhammer *et al.*, 1983). Thus, even for those "particle-reactive" elements which are greatly enriched in vent fluids over seawater concentrations (e.g., REE), the processes active within hydrothermal plumes dictate that hydrothermal circulation causes a net *removal* of these tracers not just relative to the vent fluids themselves, but also from the oceanic water column (German *et al.*, 1990; Rudnicki and Elderfield, 1993). In the extreme, these processes can be sufficient to cause geochemical fractionations as pronounced as those caused by "boundary scavenging" in high-productivity ocean margin environments (cf.Anderson *et al.*, 1990; German *et al.*, 1997).

Thus far, we have treated the processes active in hydrothermal plumes as inorganic geochemical processes. However we know this is not strictly the case: microbial processes are well known to mediate key chemical reactions in hydrothermal plumes (Winn *et al.*, 1995; Cowen and German, 2003) and more recently the role of larger organisms such as zooplankton has also been noted (e.g., Burd *et al.*, 1992; Herring and Dixon, 1998). The biological modification of plume processes is discussed more fully below (Section 6.07.5.4).

6.07.5.3 Physical Controls on Dispersing Plumes

Physical processes associated with hydrothermal plumes may affect their impact upon ocean geochemistry; because of the fundamentally different hydrographic controls in the Pacific versus Atlantic Oceans, plume dispersion varies between these two oceans. In the Pacific Ocean, where deep waters are warmer and saltier than overlying water masses, nonbuoyant hydrothermal plumes which have entrained local deep water are typically warmer and more saline at the point of emplacement than that part of the water column into which they intrude (e.g., Lupton *et al.*, 1985). The opposite has been observed in the Atlantic Ocean where deep waters tend to be colder and less saline than the overlying water column. Consequently, for example, the TAG nonbuoyant plume is anomalously cold and fresh when compared to the background waters into which it intrudes, 300–400 m above the seafloor (Speer and Rona, 1989).

Of perhaps more significance, geochemically, are the physical processes which affect plume-dispersion *after* emplacement at nonbuoyant plume height. Here, topography at the ridge crest exerts particular influence. Along slow and ultraslow spreading ridges (e.g., MAR, SW Indian Ridge, Central Indian Ridge) nonbuoyant plumes are typically emplaced within the confining bathymetry of the rift-valley walls (order 1,000 m). Along faster spreading ridges such as the EPR, by contrast, buoyant plumes typically rise clear of the confining topography (order 100 m, only). Within rift-valley "corridors" plume dispersion is highly dependent upon along-axis current flow. At the TAG hydrothermal field (MAR 26° N), for example, residual currents are dominated by tidal excursions (Rudnicki *et al.*, 1994). A net effect of these relatively "stagnant" conditions is that plume material trapped within the vicinity of the vent site appears to be recycled multiple times through the TAG buoyant and nonbuoyant plume, enhancing the scavenging effect upon "particle-reactive" tracers (e.g., thorium isotopes) within the local water column (German and Sparks, 1993). At the Rainbow vent site (MAR 36° N) by contrast, much stronger prevailing currents ($\sim$10 cm s^{-1}) are observed and a more unidirectional, topographically controlled flow is observed (German *et al.*, 1998b). Failure to appreciate the potential complexities of such dispersion precludes the "informed" sampling required to resolve processes of geochemical evolution within a dispersing hydrothermal plume. Nor should it be assumed that such topographic steering is entirely a local effect confined to slow spreading ridges' rift-valleys. In recent work, Speer *et al.* (2003) used a numerical simulation of ocean circulation to estimate dispersion along and away from the global ridge crest. A series of starting points were considered along the entire ridge system, 200 m above the seafloor and at spacings of 30–100 km along-axis; trajectories were then calculated over a 10 yr integrated period. With few exceptions (e.g., major fracture zones) the net effect reported was that these dispersal trajectories tend to be constrained by the overall form of the ridge and flow parallel to the ridge axis over great distances (Speer *et al.*, 2003).

The processes described above are relevant to established "chronic" hydrothermal plumes. Important exceptions (only identified rarely–so far) are Event (or "*Mega-*") Plumes. One interpretation of these features is as follows: however a hydrothermal system may evolve, it must first displace a large volume of cold seawater from pore spaces within the upper ocean crust. Initial flushing of this system must be rapid, especially on fast ridges that are extending at rates in excess of ten centimeters per year. In circumstances where there is frequent recurrence, either of intrusions of magma close beneath the seafloor or dike-fed eruptions at the seabed, seafloor venting may commence with a rapid exhalation of a large volume of hydrothermal fluid to generate an "event" plume high up in the overlying water column (e.g., Baker *et al.*, 1995; Lupton *et al.*, 1999b). Alternately, Palmer and Ernst (1998, 2000) have argued that cooling of pillow basalts, erupted at $\sim$1,200 °C on the seafloor and the most common form of submarine volcanic extrusion, is responsible for the formation of these same "event" plume features. Whichever eruption-related process causes their formation, an important question that follows is: how much hydrothermal flux is overlooked if we fail to intercept "event" plumes at the onset of venting at any given location? To address this problem, Baker *et al.* (1998) estimated that the event plume triggered by dike-intrusion at the co-axial vent field (JdFR) contributed less than 10% of the total flux of heat and chemicals released during the $\sim$3 yr life span of that vent. If widely applicable, those deliberations suggest that event plumes can safely be ignored when calculating global geochemical fluxes (Hein *et al.*, 2003). They remain of great interest to microbiologists, however, as a potential "window" into the deep, hot biosphere (e.g., Summit and Baross, 1998).

6.07.5.4 Biogeochemical Interactions in Dispersing Hydrothermal Plumes

Recognition of the predominantly along-axis flow of water masses above MORs as a result of topographic steering has renewed speculation that hydrothermal plumes may represent important vectors for gene-flow along the global ridge crest, transporting both chemicals and vent larvae alike, from one adjacent vent site to another (e.g., Van Dover *et al.*, 2002). If such is the case, however, it is also to be expected that a range of biogeochemical processes should also be active within nonbuoyant hydrothermal plumes. One particularly good example of such a process is the microbial oxidation of dissolved manganese. In the restricted circulation regime of the Guaymas Basin, formation of particulate manganese is dominated by bacteria and the dissolved manganese residence time is less than a week (Campbell *et al.*, 1988). Similarly, uptake of dissolved manganese in the Cleft Segment plume (JdFR) is bacterially dominated, albeit with much longer residence times, estimated at 0.5 yr to >2 yr (Winn *et al.*, 1995). Distributions of dissolved CH_4 and H_2 in hydrothermal plumes have also been shown to be controlled by bacterially mediated oxidation (de Angelis *et al.*, 1993) with apparent residence times which vary widely for CH_4 (7–177 d) but are much shorter for dissolved H_2 (<1 d).

The release of dissolved H_2 and CH_4 into hydrothermal plumes provides suitable substrates for both primary (chemolithoautotrophic) and secondary (heterotrophic) production within dispersing nonbuoyant hydrothermal plumes. Although the sinking organic carbon flux from hydrothermal plumes (Roth and Dymond, 1989; German *et al.*, 2002) may be less than 1% of the total oceanic photosynthetic production (Winn *et al.*, 1995) hydrothermal production of organic carbon is probably restricted to a corridor extending no more than $\sim$10 km to either side of the ridge axis. Consequently, microbial activity within hydrothermal plumes may have a

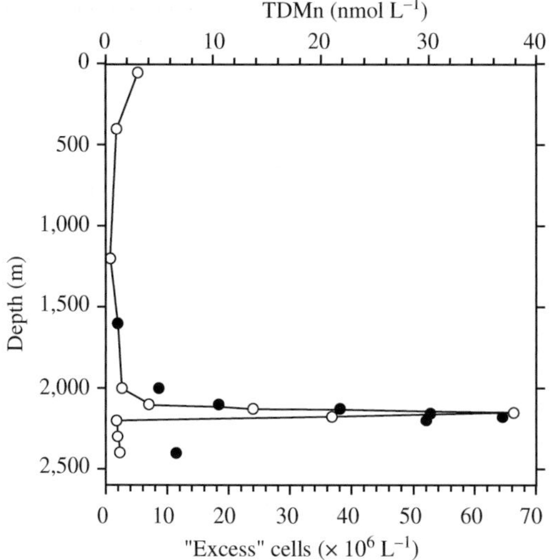

Figure 16 Plot showing co-registered enrichments of TDMn concentrations and "excess" cell counts (after subtraction of a "background" N. Atlantic cell-count profile) in the Rainbow nonbuoyant hydrothermal plume near 36°14' N, MAR (source German *et al.*, 2003).

pronounced local effect—perhaps 5–10-fold greater than the photosynthetic-flux driven from the overlying upper ocean (Cowen and German, 2003). Although the detailed microbiology of hydrothermal plumes remains poorly studied, as of the early 2000s, bacterial counts from the Rainbow plume have identified both an increase in total cell concentrations at plume-height compared to background (Figure 16) and, further, that 50–75% of the microbial cells identified in that work were particle-attached compared to typical values of just 15% for the open ocean. Detailed molecular biological analysis of those particle-associated microbes have also revealed that the majority (66%) are archeal in nature rather than bacterial (German *et al.*, 2003); in the open ocean, by contrast, it is the bacteria which typically dominate (Fuhrmann *et al.*, 1993; Mullins *et al.*, 1995). It is tempting to speculate that these preliminary data may provide testament to a long-established (even on geological timescales) chemical-microbial symbiosis at hydrothermal vents.

6.07.5.5 Impact of Hydrothermal Plumes Upon Ocean Geochemical Cycles

Hydrothermal plumes form by the entrainment of large volumes of ambient ocean water into rising buoyant plumes driven by the release of vent fluids at the seafloor. The effect of this dilution is such that the entire volume of the oceans is cycled through buoyant and nonbuoyant hydrothermal plumes, on average, every few thousand years—a timescale comparable to that for mixing of the entire deep ocean.

Close to the vent source, rapid precipitation of a range of polymetallic sulfide, sulfate, and oxide phases leads to a strong modification of the gross dissolved metal flux from the seafloor. Independent estimates by Mottl and McConachy (1990) and German *et al.* (1991b), from buoyant and nonbuoyant plume investigations in the Pacific and Atlantic Oceans respectively, concluded that perhaps only manganese and calcium achieved a significant dissolved metal flux from hydrothermal venting to the oceans. (For comparison, the following 27 elements exhibited no evidence for a significant dissolved hydrothermal flux: iron, beryllium, aluminum, magnesium, chromium, vanadium, cobalt, copper, zinc, arsenic, yttrium, molybdenum, silver, cadmium, tin, lanthanum, cerium, praseodymium, neodymium, samarium, europium, gadolinium, terbium, holmium, erbium, lead, and uranium.) In addition to rapid co-precipitation and deposition of vent-sourced metals to the local sediments (e.g., Dymond and Roth, 1988; German *et al.*, 2002), hydrothermal plumes are also the site of additional uptake of other dissolved tracers from the water column—most notably large dissolved "oxyanion" species (phosphorus, vanadium, chromium, arsenic, uranium) and particle-reactive species (beryllium, yttrium, REE, thorium, and protactinium). For many of these tracers, hydrothermal plume removal fluxes are as great as, or at least significant when compared to, riverine input fluxes to the oceans (Table 3). What remains less certain, however, is the extent to which these particle-associated species are subsequently retained within the hydrothermal sediment record.

6.07.6 HYDROTHERMAL SEDIMENTS

Deep-sea metalliferous sediments were first documented in the reports of the Challenger Expedition, 1873–1876 (Murray and Renard, 1891), but it took almost a century to recognize that such metalliferous material was concentrated along all the world's ridge crests (Figure 2). Boström *et al.* (1969) attributed these distributions to some form of "volcanic emanations"; the accuracy of those predictions was confirmed some ten years later with the discovery of ridge-crest venting (Corliss *et al.*, 1978; Spiess *et al.*, 1980) although metalliferous sediments had already been found in association with hot brine pools in the Red Sea (Degens and Ross, 1969). Following the discovery of active venting, it has become recognized that hydrothermal sediments can be classified into two types: those derived

Table 3 Global removal fluxes from the deep ocean into hydrothermal plumes (after Elderfield and Schultz, 1996).

Element	*Hydrothermal input* (mol yr^{-1})	*Plume removal* (mol yr^{-1})	*Riverine input* (mol yr^{-1})
Cr	0	4.8×10^7	6.3×10^8
V	0	4.3×10^8	5.9×10^8
As	$(0.01-3.6) \times 10^7$	1.8×10^8	8.8×10^8
P	-4.5×10^7	1.1×10^{10}	3.3×10^{10}
U	-3.8×10^5	4.3×10^7	3.6×10^7
Mo	0	1.9×10^6	2.0×10^8
Be		1.7×10^6	3.7×10^7
Ce	9.1×10^5	1.0×10^6	1.9×10^7
Nd	5.3×10^5	6.3×10^6	9.2×10^6
Lu	2.1×10^3	0.6×10^5	1.9×10^5

from plume fall-out (including the majority of metalliferous sediments reported from ridge flanks) and those derived from mass wasting close to active vent sites (see, e.g., Mills and Elderfield, 1995).

6.07.6.1 Near-vent Sediments

Near-vent metalliferous sediments form from the physical degradation of hydrothermal deposits themselves, a process which begins as soon as deposition has occurred. Whilst there is ample evidence for extensive mass wasting in ancient volcanogenic sulfide deposits, only limited attention has been paid, to date, to this aspect of modern hydrothermal systems. Indeed, much of our understanding comes from a series of detailed investigations from a single site, the TAG hydrothermal field at 26° N on the MAR. It has been shown, for example, that at least some of the weathered sulfide debris at TAG is produced from collapse of the mound itself. This collapse is believed to arise from waxing and waning of hydrothermal circulation which, in turn, leads to episodic dissolution of large volumes of anhydrite within the mound (e.g., Humphris *et al.*, 1995; James and Elderfield, 1996). The mass-wasting process at TAG generates an apron of hydrothermal detritus with oxidized sulfides deposited up to 60 m out, away from the flanks of the hydrothermal mound.

Similar ponds of metalliferous sediment are observed close to other inactive sulfide structures throughout the TAG area (Rona *et al.*, 1993). Metz *et al.* (1988) characterized the metalliferous sediment in a core raised from a sediment pond close to one such deposit, ~2 km NNE of the active TAG mound. That core consisted of alternating dark red-brown layers of weathered sulfide debris and lighter calcareous ooze. Traces of pyrite, chalcopyrite, and sphalerite, together with elevated transition-metal concentrations were found in the dark red-brown layers, confirming the presence of clastic sulfide debris. Subsequently, German *et al.* (1993) investigated a short-core raised from the outer limit of the apron of "stained" hydrothermal sediment surrounding the TAG mound itself. That core penetrated through 7 cm of metal-rich degraded sulfide material into pelagic carbonate ooze. The upper "mass-wasting" layer was characterized by high transition-metal contents, just as observed by Metz *et al.* (1988), but also exhibited REE patterns similar to vent fluids (see earlier) and high uranium contents attributed to uptake from seawater during oxidation of sulfides (German *et al.*, 1993). Lead isotopic compositions in sulfidic sediments from both sites were indistinguishable from local MORB, vent fluids and chimneys (German *et al.*, 1993, Mills *et al.*, 1993). By contrast, the underlying/intercalated carbonate/calcareous ooze layers from each core exhibited lead isotope, REE, and U–Th compositions which much more closely reflected input of Fe-oxyhydroxide particulate material from nonbuoyant hydrothermal plumes (see below).

6.07.6.2 Deposition from Hydrothermal Plumes

Speer *et al.* (2003) have modeled deep-water circulation above the global ridge crest and concluded that this circulation is dominated by topographically steered flow along-axis. Escape of dispersed material into adjacent deep basins is predicted to be minimal, except in key areas where pronounced across-axis circulation occurs. If this model proves to be generally valid, the majority of hydrothermal material released into nonbuoyant hydrothermal plumes should not be dispersed more than ~100 km off-axis. Instead, most hydrothermal material should settle out in a near-continuous rain of metalliferous sediment along the length of the global ridge crest. Significant off-axis dispersion is only predicted (i) close to the equator (~5° N to 5° S), (ii) where the ridge intersects boundary currents or regions of deep-water formation, and (iii) in the Antarctic Circumpolar Current (Speer *et al.*, 2003). One good example of strong across-axis flow is at the equatorial MAR where pronounced eastward flow of both Antarctic Bottom Water and lower North Atlantic Deep Water has been reported, passing through the Romanche and Chain Fracture Zones (Mercier and Speer, 1998).

Another location where the large-scale off-axis dispersion modeled by Speer *et al.* (2003) has already been well documented is on the southern EPR (Figure 14). There, metalliferous sediment enrichments underlie the pronounced dissolved ^{3}He plume which extends westward across the southern Pacific Ocean at ~15° S (cf. Boström *et al.*, 1969; Lupton and Craig, 1981). Much of our understanding of ridge-flank metalliferous sediments comes from a large-scale study carried out at this latitude (19° S) by Leg 92 of the Deep Sea Drilling Project (DSDP). That work targeted sediments underlying the westward-trending plume to investigate both temporal and spatial variability in hydrothermal output at this latitude (Lyle *et al.*, 1987). A series of holes were drilled extending westward from the ridge axis into 5–28 Ma crust; the recovered cores comprised mixtures of biogenic carbonate and Fe–Mn oxyhydroxides. One important result of that work was the demonstration, based on lead isotopic analyses, that even the most distal sediments, collected at a range of >1,000 km from the ridge axis, contained 20–30% mantle-derived lead (Barrett *et al.*, 1987). In contrast, analysis of the same samples indicated that REE distributions in the metalliferous sediments were dominated by a seawater source (Ruhlin and Owen, 1986). This is entirely consistent with what has subsequently been demonstrated for hydrothermal plumes (see Section 6.07.5, above) with the caveat that REE/Fe ratios in DSDP Leg 92 sediments are everywhere higher than the highest REE/Fe ratios yet measured in modern nonbuoyant hydrothermal plume particles (German *et al.*, 1990; Sherrell *et al.*, 1999).

6.07.6.3 Hydrothermal Sediments in Paleoceanography

Phosphorus and vanadium, which are typically present in seawater as dissolved oxyanion species, have been shown to exhibit systematic plume-particle P : Fe and V : Fe variations which differ from one ocean basin to another (e.g., Trefry and Metz, 1989; Feely *et al.*, 1990). This has led to the hypothesis (Feely *et al.*, 1998) that (i) plume P : Fe and V : Fe ratios may be directly linked to local deep-ocean dissolved phosphate concentrations and (ii) ridge-flank metalliferous sediments, preserved under oxic diagenesis, might faithfully record temporal variations in plume-particle P : Fe and/or V : Fe ratios. Encouragingly, a study of slowly accumulating (~0.5 cm kyr^{-1}) sediments from the west flank of the JdFR has revealed that V : Fe ratios in the hydrothermal component from that core appear faithfully to record local plume-particle V : Fe ratios for the past ~200 kyr (German *et al.*, 1997; Feely *et al.*, 1998). More recently, however, Schaller *et al.* (2000) have shown that while cores from the flanks of the southern EPR (10° S) also exhibit V : Fe ratios that mimic modern plume-values, in sediments dating back to 60–70 kyr, the complementary P : Fe and As : Fe ratios in these samples are quite different from contemporaneous nonbuoyant plume values. These variations have been attributed to differences in the intensity of hydrothermal iron oxide formation between different hydrothermal plumes and/or significant uptake/release of phosphorus and arsenic, following deposition (Schaller *et al.*, 2000).

Unlike vanadium, REE/Fe ratios recorded in even the most recent metalliferous sediments are much higher than those in suspended hydrothermal plume particles (German *et al.*, 1990, 1997; Sherrell *et al.*, 1999). Further, hydrothermal sediments' REE/Fe ratios increase systematically with distance away from the paleo-ridge crest (Ruhlin and Owen, 1986; Olivarez and Owen, 1989). This indicates that the REE may continue to be taken up from seawater, at and near the sediment–water interface, long after the particles settle from the plume to the seabed. Because increased uptake of dissolved REE from seawater should also be accompanied by continuing fractionation across the REE series (e.g., Rudnicki and Elderfield, 1993) reconstruction of deep-water REE patterns from preserved metalliferous sediment records remain problematic. Much more tractable, however, is the exploitation of these same sample types for isotopic reconstructions.

Because seawater uptake dominates the REE content of metalliferous sediment, neodymium isotopic analysis of metalliferous carbonate can provide a reliable proxy for contemporaneous seawater, away from input of near-vent sulfide detritus (Mills *et al.*, 1993). Osmium also exhibits a similar behavior and seawater dominates the isotopic composition of metalliferous sediments even close to active vent sites (Ravizza *et al.*, 1996). Consequently, analysis of preserved metalliferous carbonate sediments has proven extremely useful in determining the past osmium isotopic composition of the oceans, both from modern marine sediments (e.g., Ravizza, 1993; Peucker-Ehrenbrink *et al.*, 1995) and those preserved in ophiolites (e.g., Ravizza *et al.*, 2001). Only in sediments close to an ultramafic-hosted hydrothermal system, have perturbations from a purely seawater osmium isotopic composition been observed (Cave *et al.*, 2003, in press).

6.07.6.4 Hydrothermal Sediments and Boundary Scavenging

It has been known for sometime that sediments underlying areas of high particle settling flux exhibit pronounced fractionations between particle-reactive tracers. Both ^{231}Pa and ^{10}Be, for example, exhibit pronounced enrichments relative to ^{230}Th, in ocean margin environments, when compared to sediments underlying mid-ocean gyres (e.g., Bacon, 1988; Anderson *et al.*, 1990; Lao *et al.*, 1992). Comparable fractionations between these three radiotracers (^{230}Th, ^{231}Pa, and ^{10}Be) have also been identified in sediments underlying hydrothermal plumes (German *et al.*, 1993; Bourlès *et al.*, 1994; German *et al.*, 1997). For example, a metalliferous sediment core raised from the flanks of the JdFR exhibited characteristic hydrothermal lead-isotopic and REE/Fe compositions, together with high $^{10}Be/^{230}Th$ ratios indicative of net focusing relative to the open ocean (German *et al.*, 1997). The degree of fractionation observed was high, even compared to high-productivity ocean-margin environments (Anderson *et al.*, 1990; Lao *et al.*, 1992), presumably due to intense scavenging onto hydrothermal Fe-oxyhydroxides. Of course, the observation that REE and thorium are scavenged into ridge-crest metalliferous sediments is not new; sediments from the EPR near 17° S, with mantle lead, excess ^{230}Th and seawater-derived REE compositions were reported more than thirty years ago by Bender *et al.* (1971). More recently, however, examination of ridge crest sediments and near-vent sediment-traps has revealed that the settling flux of scavenged tracers (e.g., ^{230}Th) from hydrothermal plumes is higher than can be sustained by *in situ* production in the overlying water column alone (German *et al.*, 2002). Thus, uptake onto Fe-oxyhydroxide material in hydrothermal plumes and sediments may act as a special form of deep-ocean "boundary scavenging" leading to the net focusing and deposition of these dissolved tracers in ridge-flank metalliferous sediments.

6.07.7 CONCLUSION

The field of deep-sea hydrothermal research is young; it was only in the mid-1970s when it was first discovered, anywhere in the oceans. To synthesize current understanding of its impact on marine geochemistry, therefore, could be considered akin to explaining the significance of rivers to ocean chemistry in the early part of the last century. This chapter has aimed to provide a brief synposis of the current state of the art, but much more surely remains to be learnt. There are three key questions that will continue to focus efforts within this vigorous research field:

(i) What are the geological processes that control submarine hydrothermal venting? How might these have varied during the course of Earth's history?

(ii) To what extent do geochemical and biological processes interact to regulate hydrothermal fluxes to the ocean? How might past-ocean processes have differed from the present-day ones?

(iii) What are the timescales relevant to hydrothermal processes? Whilst some long-term proxies do exist (sufide deposits, metalliferous sediments) for active processes, we do not have any time-series records longer than 25 years!

REFERENCES

Alt J. C. (1995) Subseafloor processes in mid-ocean ridge hydrothermal systems. *Geophys. Monogr. (AGU)* **91**, 85–114.

Alt J. C. and Bach W. (2003) Alteration of oceanic crust: subsurface rock–water interactions. In *Energy and Mass Transfer in Marine Hydrothermal Systems* (eds. P. Halbach, V. Tunnicliffe, and J. Hein). DUP, Berlin, pp. 7–28.

Anderson R. N. and Hobart M. A. (1976) The relationship between heat flow, sediment thickness and age in the eastern Pacific. *J. Geophys. Res.* **81**, 2968–2989.

Anderson R. N. and Silbeck J. N. (1981) Oceanic heat flow. In *The Oceanic Lithosphere, The Seas* (ed. C. Emiliani), Wiley, New York, vol. 7, pp. 489–523.

Anderson R. F., Lao Y., Broecker W. S., Trumbore S. E., Hofmann H. J., and Wolfli W. (1990) Boundary scavenging in the Pacific Ocean: a comparison of ^{10}Be and ^{231}Pa. *Earth Planet. Sci. Lett.* **96**, 287–304.

Bach W., Banerjee N. R., Dick H. J. B., and Baker E. T. (2002) Discovery of ancient and active hydrothermal systems along the ultra-slow spreading southwest Indian ridge 10 degrees–16 degrees E. *Geophys. Geochem. Geosys.* **3**, paper 10.1029/2001GC000279.

Bacon M. P. (1988) Tracers of chemical scavenging in the ocean: boundary effects and largescale chemical fractionation. *Phil. Trans. Roy. Soc. London Ser. A* **325**, 147–160.

Baker E. T. (1998) Patterns of event and chronic hydrothermal venting following a magmatic intrusion: new perspectives from the 1996 Gorda ridge eruption. *Deep-Sea Res. II* **45**, 2599–2618.

Baker E. T., Massoth G. J., Feely R. A., Embley R. W., Thomson R. E., and Burd B. J. (1995) Hydrothermal event plumes from the CoAxial seafloor eruption site, Juan de Fuca ridge. *Geophys. Res. Lett.* **22**, 147–150.

Baker E. T., Chen Y. J., and Phipps Morgan J. (1996) The relationship between near-axis hydrothermal cooling and the spreading rate of mid-ocean ridges. *Earth Planet. Sci. Lett.* **142**, 137–145.

Baker E. T., Massoth G. J., Feely R. A., Cannon G. A., and Thomson R. E. (1998) The rise and fall of the CoAxial hydrothermal site, 1993–1996. *J. Geophys. Res.* **103**, 9791–9806.

Barnes I., Rapp J. T., O'Neill J. R., Sheppard R. A., and Gude A. J. (1972) Metamorphic assemblages and the flow of metamorphic fluid in four instances of serpentenization. *Contrib. Mineral. Petrol.* **35**, 263–276.

Barrett T. J., Taylor P. N., and Lugowski J. (1987) Metalliferous sediments from DSDP leg 92, the East Pacific rise transect. *Geochim. Cosmochim. Acta* **51**, 2241–2253.

Bender M., Broecker W., Gornitz V., Middel U., Kay R., Sun S., and Biscaye P. (1971) Geochemistry of three cores from the East Pacific Rise. *Earth Planet. Sci. Lett.* **12**, 425–433.

Berndt M. E. and Seyfried W. E. (1990) Boron, bromine, and other trace elements as clues to the fate of chlorine in mid-ocean ridge vent fluids. *Geochim. Cosmochim. Acta* **54**, 2235–2245.

Berndt M. E. and Seyfried W. E. (1997) Calibration of Br/Cl fractionation during sub-critical phase separation of seawater: possible halite at 9 to 10°N East Pacific Rise. *Geochim. Cosmochim. Acta* **61**, 2849–2854.

Berndt M. E., Allen D. E., and Seyfried W. E. (1996) Reduction of CO_2 during serpentinization of olivine at 300 degrees C and 500 bar. *Geology* **24**, 351–354.

Binns R. A., Scott S. D., Bogdanov Y. A., Lisitzin A. P., Gordeev V. V., Gurvcich E. G., Finlayson E. J., Boyd T., Dotter L. E., Wheller G. E., and Muravyev K. G. (1993) Hydrothermal oxide and gold-rich sulfate deposits of Franklin seamount, Western Woodlark Basin, Papua New Guinea. *Econ. Geol.* **88**, 2122–2153.

Bischoff J. L. (1991) Densities of liquids and vapors in boiling $NaCl–H_2O$ solutions: a PVTX summary from 300° to 500°C. *Am. J. Sci.* **291**, 309–338.

Bischoff J. L. and Pitzer K. S. (1989) Liquid-vapor relations for the system $NaCl–H_2O$: summary of the P–T–x surface from 300deg to 500deg C. *Am. J. Sci.* **289**, 217–248.

Bischoff J. L. and Rosenbauer R. J. (1985) An empirical equation of state for hydrothermal seawater (3.2% NaCl). *Am. J. Sci.* **285**, 725–763.

Bischoff J. L. and Rosenbauer R. J. (1987) Phase separation in seafloor systems—an experimental study of the effects on metal transport. *Am. J. Sci.* **287**, 953–978.

Blackman D. K., Cann J. R., Janssen B., and Smith D. K. (1998) Origin of extensional core complexes: evidence from the Mid-Atlantic Ridge at Atlantis fracture zone. *J. Geophys. Res.* **103**, 21315–21333.

Boström K., Peterson M. N. A., Joensuu O., and Fisher D. E. (1969) Aluminium-poor ferromanganoan sediments on active ocean ridges. *J. Geophys. Res.* **74**, 3261–3270.

Bourlès D. L., Brown E. T., German C. R., Measures C. I., Edmond J. M., Raisbeck G. M., and Yiou F. (1994) Hydrothermal influence on oceanic beryllium. *Earth Planet. Sci. Lett.* **122**, 143–157.

Bray A. M. and Von Damm K. L. (2003a) The role of phase separation and water–rock reactions in controlling the boron content of mid-ocean ridge hydrothermal vent fluids. *Geochim. Cosmochim. Acta* (in revision).

Bray A. M. and Von Damm K. L. (2003b) Controls on the alkali metal composition of mid-ocean ridge hydrothermal fluids: constraints from the 9–10°N East Pacific Rise time series. *Geochim. Cosmochim. Acta* (in revision).

Broecker W. S. and Peng T. H. (1982) *Tracers in the Sea.* Eldigio Press, Columbia University, New York, 690pp.

Burd B. J., Thomson R. E., and Jamieson G. S. (1992) Composition of a deep scattering layer overlying a mid-ocean ridge hydrothermal plume. *Mar. Biol.* **113**, 517–526.

Butterfield D. A., McDuff R. E., Mottl M. J., Lilley M. D., Lupton J. E., and Massoth G. J. (1994) Gradients in the composition of hydrothermal fluids from the endeavour segment vent field: phase separation and brine loss. *J. Geophys. Res.* **99**, 9561–9583.

Butterfield D. A., Jonasson I. R., Massoth G. J., Feely R. A., Roe K. K., Embley R. E., Holden J. F., McDuff R. E., Lilley M. D., and Delaney J. R. (1997) Seafloor eruptions and evolution of hydrothermal fluid chemistry. *Phil. Trans. Roy. Soc. London* **A355**, 369–386.

Campbell A. C., Palmer M. R., Klinkhammer G. P., Bowers T. S., Edmond J. M., Lawrence J. R., Casey J. F., Thomson G., Humphris S., Rona P., and Karson J. A. (1988) Chemistry of hot springs on the Mid-Atlantic Ridge. *Nature* **335**, 514–519.

Campbell A. C., German C. R., Palmer M. R., Gamo T., and Edmond J. M. (1994) Chemistry of hydrothermal fluids from the Escanaba Trough, Gorda Ridge. In *Geologic, Hydrothermal and Biologic Studies at Escanaba Trough, Gorda Ridge*, US Geol. Surv. Bull. 2022 (eds. J. L. Morton, R. A. Zierenberg, and C. A. Reiss). Offshore Northern California, pp. 201–221.

Cave R. R., Ravizza G. E., German C. R., Thomson J., and Nesbitt R. W. (2003) Deposition of osmium and other platinum-group elements beneath the ultramafic-hosted rainbow hydrothermal plume. *Earth Planet. Sci. Lett.* **210**, 65–79.

Charlou J.-L., Fouquet Y., Donval J. P., Auzende J. M., Jean Baptiste P., Stievenard M., and Michel S. (1996) Mineral and gas chemistry of hydrothermal fluids on an ultrafast spreading ridge: East Pacific Rise, 17° to 19°S (NAUDUR cruise, 1993) phase separation processes controlled by volcanic and tectonic activity. *J. Geophys. Res.* **101**, 15899–15919.

Charlou J. L., Fouquet Y., Bougault H., Donval J. P., Etoubleau J., Jean-Baptiste P., Dapoigny A., Appriou P., and Rona P. A. (1998) Intense CH_4 plumes generated by serpentinization of ultramafic rocks at the intersection of the 15 degrees 20' N. *Geochim. Cosmochim. Acta* **62**, 2323–2333.

Chin C. S., Coale K. H., Elrod V. A., Johnson K. S., Johnson G. J., and Baker E. T. (1994) *In situ* observations of dissolved iron and manganese in hydrothermal vent plumes, Juan de Fuca Ridge. *J. Geophys. Res.* **99**, 4969–4984.

Clarke W. B., Beg M. A., and Craig H. (1969) Excess ^{3}He in the sea: evidence for terrestrial primordial helium. *Earth Planet. Sci. Lett.* **6**, 213–220.

Connelly D. P., German C. R., Egorov A., Pimenov N. V., and Dohsik H. (2002) Hydrothermal plumes overlying the ultraslow spreading Knipovich Ridge, 72–78°N. *EOS Trans. AGU (abstr.)* **83**, T11A-1229.

Corliss J. B., Lyle M., and Dymond J. (1978) The chemistry of hydrothermal mounds near the Galapagos rift. *Earth Planet. Sci. Lett.* **40**, 12–24.

Cowen J. P. and German C. R. (2003) Biogeochemical cycling in hydrothermal plumes. In *Energy and Mass Transfer in Marine Hydrothermal Systems* (eds. P. Halbach, V. Tunnicliffe, and J. Hein). DUP, Berlin, pp. 303–316.

Cowen J. P., Bertram M. A., Wakeham S. G., Thomson R. E., Lavelle J. W., Baker E. T., and Feely R. A. (2001) Ascending and descending particle flux from hydrothermal plumes at endeavour segment, Juan de Fuca Ridge. *Deep-Sea Res. I* **48**, 1093–1120.

Cruse A. M. and Seewald J. S. (2001) Metal mobility in sediment-covered ridge-crest hydrothermal systems: experimental and theoretical constraints. *Geochim. Cosmochim. Acta* **65**, 3233–3247.

de Angelis M. A., Lilley M. D., Olson E. J., and Baross J. A. (1993) Methane oxidation in deep-sea hydrothermal plumes of the endeavour segment of the Juan de Fuca Ridge. *Deep-Sea Res.* **40**, 1169–1186.

Degens E. T. and Ross D. A. (eds.) (1969) *Hot Brines and Heavy Metal Deposits in the Red Sea.* Springer, New York.

De Mets C., Gordon R. G., Argus D. F., and Stein S. (1994) Effect of recent revisions to the geomagnetic reversal timescale on estimates of current plate motions. *Geophys. Res. Lett.* **21**, 2191–2194.

Douville E., Charlou J. L., Oelkers E. H., Bienvenu P., Colon C. F. J., Donval J. P., Fouquet Y., Prieur D., and Appriou P. (2002) The rainbow vent fluids (36 degrees 14' N, MAR): the influence of ultramafic rocks and phase separation on trace metal content in Mid-Atlantic Ridge hydrothermal fluids. *Chem. Geol.* **184**, 37–48.

Dymond J. and Roth S. (1988) Plume dispersed hydrothermal particles—a time-series record of settling flux from the endeavour ridge using moored sensors. *Geochim. Cosmochim. Acta* **52**, 2525–2536.

Edmond J. M., Measures C. I., McDuff R. E., Chan L. H., Collier R., Grant B., Gordon L. I., and Corliss J. B. (1979) Ridge crest hydrothermal activity and the balances of the major and minor elements in the ocean: the Galapagos data. *Earth Planet. Sci. Lett.* **46**, 1–18.

Edmond J. M., Von Damm K. L., McDuff R. E., and Measures C. I. (1982) Chemistry of hot springs on the East Pacific Rise and their effluent dispersal. *Nature* **297**, 187–191.

Edmond J. M., Campbell A. C., Palmer M. R., Klinkhammer G. P., German C. R., Edmonds H. N., Elderfield H., Thompson G., and Rona P. (1995) Time-series studies of vent-fluids from the TAG and MARK sites (1986, 1990) Mid-Atlantic Ridge: a new solution chemistry model and a mechanism for Cu/Zn zonation in massive sulphide ore-bodies. In *Hydrothermal Vents and Processes*, Geol. Soc. Spec. Publ. 87 (eds. L. M. Parson, C. L. Walker, and D. R. Dixon), The Geological Society Publishing House, Bath, UK, pp. 77–86.

Edmonds H. N., German C. R., Green D. R. H., Huh Y., Gamo T., and Edmond J. M. (1996) Continuation of the hydrothermal fluid chemistry time series at TAG and the effects of ODP drilling. *Geophys. Res. Lett.* **23**, 3487–3489.

Edmonds H. N., Michael P. J., Baker E. T., Connelly D. P., Snow J. E., Langmuir C. H., Dick H. J. B., German C. R., and Graham D. W. (2003) Discovery of abundant hydrothermal venting on the ultraslow-spreading Gakkel Ridge in the Arctic Ocean. *Nature* **421**, 252–256.

Elderfield H. and Schultz A. (1996) Mid-ocean ridge hydrothermal fluxes and the chemical composition of the ocean. *Ann. Rev. Earth Planet. Sci.* **24**, 191–224.

Eldridge C. S., Barton P. B., and Ohmoto H. (1983) Mineral textures and their bearing on formation of the Kuroko orebodies. *Econ. Monogr.* **5**, 241–281.

Evans W. C., White L. D., and Rapp J. B. (1988) Geochemistry of some gases in hydrothermal fluids from the southern Juan de Fuca Ridge. *J. Geophys. Res.* **93**, 15305–15313.

Feely R. A., Massoth G. J., Baker E. T., Cowen J. P., Lamb M. F., and Krogslund K. A. (1990) The effect of hydrothermal processes on mid-water phosphorous distributions in the northeast Pacific. *Earth Planet. Sci. Lett.* **96**, 305–318.

Feely R. A., Trefry J. H., Massoth G. J., and Metz S. (1991) A comparison of the scavenging of phosphorous and arsenic from seawater by hydrothermal iron hydroxides in the Atlantic and Pacific oceans. *Deep-Sea Res.* **38**, 617–623, 1991.

Feely R. A., Massoth G. J., Trefry J. H., Baker E. T., Paulson A. J., and Lebon G. T. (1994) Composition and sedimentation of hydrothermal plume particles from North Cleft Segment, Juan de Fuca Ridge. *J. Geophys. Res.* **99**, 4985–5006.

Feely R. A., Baker E. T., Marumo K., Urabe T., Ishibashi J., Gendron J., Lebon G. T., and Okamura K. (1996) Hydrothermal plume particles and dissolved phosphate over the superfast-spreading southern East Pacific Rise. *Geochim. Cosmochim. Acta* **60**, 2297–2323.

Feely R. A., Trefry J. H., Lebon G. T., and German C. R. (1998) The relationship between P/Fe and V/Fe in hydrothermal precipitates and dissolved phosphate in seawater. *Geophys. Res. Lett.* **25**, 2253–2256.

Field M. P. and Sherrell R. M. (2000) Dissolved and particulate Fe in a hydrothermal plume at 9°45' N, East Pacific Rise: slow Fe(II) oxidation kinetics in Pacific plumes. *Geochim. Cosmochim. Acta* **64**, 619–628.

Fouquet Y., Von Stackelberg U., Charlou J. L., Donval J. P., Erzinger J., Foucher J. P., Herzig P., Mühe R., Soakai S., Wiedicke M., and Whitechurch H. (1991) Hydrothermal activity and metallogenesis in the Lau back-arc basin. *Nature* **349**, 778–781.

Fouquet Y., Henry K., Bayon G., Cambon P., Barriga F., Costa I., Ondreas H., Parson L., Ribeiro A., and Relvas G. (2003) The Rainbow hydrothermal field: geological setting, mineralogical and chemical composition of sulfide deposits (MAR 36°14' N). *Earth Planet. Sci. Lett.* (submitted).

Früh-Green G. L., Weissert H., and Bernoulli D. (1990) A multiple fluid history recorded in alpine ophiolites. *J. Geol. Soc. London* **147**, 959–970.

Fuhrmann J. A., McCallum K., and Davis A. A. (1993) Phylogenetic diversity of subsurface marine microbial communities from the Atlantic and Pacific oceans. *Appl. Environ. Microbiol.* **59**, 1294–1302.

Gamo T., Okamura K., Charlou J. L., Urabe T., Auzende J. M., Ishibashi J., Shitashima K., and Chiba H. (1997) Acidic and sulfate-rich hydrothermal fluids from the Manus back-arc basin, Papua New Guinea. *Geology* **25**, 139–142.

German C. R. and Sparks R. S. J. (1993) Particle recycling in the TAG hydrothermal plume. *Earth Planet. Sci. Lett.* **116**, 129–134.

German C. R., Klinkhammer G. P., Edmond J. M., Mitra A., and Elderfield H. (1990) Hydrothermal scavenging of rare earth elements in the ocean. *Nature* **316**, 516–518.

German C. R., Fleer A. P., Bacon M. P., and Edmond J. M. (1991a) Hydrothermal scavenging at the Mid-Atlantic Ridge: radionuclide distributions. *Earth Planet. Sci. Lett.* **105**, 170–181.

German C. R., Campbell A. C., and Edmond J. M. (1991b) Hydrothermal scavenging at the Mid-Atlantic Ridge: modification of trace element dissolved fluxes. *Earth Planet. Sci. Lett.* **107**, 101–114.

German C. R., Higgs N. C., Thomson J., Mills R., Elderfield H., Blusztajn J., Fleer A. P., and Bacon M. P. (1993) A geochemical study of metalliferous sediment from the TAG hydrothermal mound, 26°08' N, Mid-Atlantic Ridge. *J. Geophys. Res.* **98**, 9683–9692.

German C. R., Briem J., Chin C., Danielsen M., Holland S., James R., Jónsdottir A., Ludford E., Moser C., Ólafsson J., Palmer M. R., and Rudnicki M. D. (1994) Hydrothermal activity on the Reykjanes ridge: the steinahóll vent-field at 63°06' N. *Earth Planet. Sci. Lett.* **121**, 647–654.

German C. R., Bourlès D. L., Brown E. T., Hergt J., Colley S., Higgs N. C., Ludford E. M., Nelsen T. A., Feely R. A., Raisbeck G., and Yiou F. (1997) Hydrothermal scavenging on the Juan de Fuca Ridge: Th-230(xs), Be-10 and REE in ridge-flank sediments. *Geochim. Cosmochim. Acta* **61**, 4067–4078.

German C. R., Baker E. T., Mevel C. A., Tamaki K., and the FUJI scientific team (1998a) Hydrothermal activity along the south-west Indian ridge. *Nature* **395**, 490–493.

German C. R., Richards K. J., Rudnicki M. D., Lam M. M., Charlou J. L., and FLAME scientific party (1998b) Topographic control of a dispersing hydrothermal plume. *Earth Planet. Sci. Lett.* **156**, 267–273.

German C. R., Livermore R. A., Baker E. T., Bruguier N. I., Connelly D. P., Cunningham A. P., Morris P., Rouse I. P., Statham P. J., and Tyler P. A. (2000) Hydrothermal plumes above the East Scotia Ridge: an isolated high-latitude back-arc spreading centre. *Earth Planet. Sci. Lett.* **184**, 241–250.

German C. R., Colley S., Palmer M. R., Khripounoff A., and Klinkhammer G. P. (2002) Hydrothermal sediment trap fluxes: 13°N, East Pacific Rise. *Deep Sea Res. I* **49**, 1921–1940.

German C. R., Thursherr A. M., Radford-Kröery J., Charlou J.-L., Jean-Baptiste P., Edmonds H. N., Patching J. W., and the FLAME 1 & II science teams (2003) Hydrothermal fluxes from the Rainbow vent-site, Mid-Atlantic Ridge: new constraints on global ocean vent-fluxes. *Nature* (submitted).

Ginster U., Mottl M. J., and VonHerzen R. P. (1994) Heat-flux from black smokers on the endeavor and Cleft segments, Juan de Fuca Ridge. *J. Geophys. Res.* **99**, 4937–4950.

Gracia E., Bideau D., Hekinian R., and Lagabrielle Y. (1999) Detailed geological mapping of two contrasting second-order segments of the Mid-Atlantic Ridge between oceanographer and Hayes fracture zones (33 degrees 30' N-35 degrees N). *J. Geophys. Res.* **104**, 22903–22921.

Gracia E., Charlou J. L., Radford-Knoery J., and Parson L. M. (2000) Non-transform offsets along the Mid-Atlantic Ridge south of the Azores (38 degrees N-34 degrees N): ultramafic exposures and hosting of hydrothermal vents. *Earth Planet. Sci. Lett.* **177**, 89–103.

Hannington M. D., Jonasson I., Herzig P., and Petersen S. (1995) Physical and chemical processes of seafloor

mineralisation at mid-ocean ridges. *Geophys. Monogr. (AGU)* **91**, 115–157.
Hannington M. D., Galley A. G., Herzig P. M., and Petersen S. (1998) Comparison of the TAG mound and stock work complex with Cyprus-type massive sulfide deposits. In *Proceedings of the Ocean Drilling Program, Scientific Results*. **158**, pp. 389–415.
Haymon R. M. (1983) Growth history of hydrothermal black smoker chimneys. *Nature* **301**, 695–698.
Haymon R. M. (1996) The response of ridge-crest hydrothermal systems to segmented, episodic magma supply. In *Tectonic, Magmatic, Hydrothermal, and Biological Segmentation of Mid-ocean Ridges*, Geol. Soc. Spec. Publ. 118 (eds. C. J. McLeod, P. A. Tyler, and C. L. Walker), Geological Society Publishing House, Bath, UK, pp. 157–168.
Haymon R., Fornari D., Von Damm K., Lilley M., Perfit M., Edmond J., Shanks W., Lutz R. A., Grebmeier J. M., Carbotte S., Wright D., McLaughlin E., Smith M., Beedle N., and Olson E. (1993) Volcanic eruption of the mid-ocean ridge along the EPR crest at 9°45–52' N: I. Direct submersible observations of seafloor phenomena associated with an eruption event in April 1991. *Earth Planet. Sci. Lett.* **119**, 85–101.
Hein J. R., Koski R. A., Embley R. W., Reid J., and Chang S.-W. (1999) Diffuse-flow hydrothermal field in an oceanic fracture zone setting, northeast Pacific: deposit composition. *Explor. Mining Geol.* **8**, 299–322.
Hein J. R., Baker E. T., Cowen J. P., German C. R., Holzbecher E., Koski R. A., Mottl M. J., Pimenov N. V., Scott S. D., and Thurnherr A. M. (2003) How important are material and chemical fluxes from hydrothermal circulation to the ocean? In *Energy and Mass Transfer in Marine Hydrothermal Systems* (eds. P. Halbach, V. Tunnicliffe, and J. Hein). DUP, Berlin, pp. 337–355.
Helfrich K. R. and Speer K. G. (1995) Ocean hydrothermal circulation: mesoscale and basin-scale flow. *Geophys. Monogr. (AGU)* **91**, 347–356.
Herring P. J. and Dixon D. R. (1998) Extensive deep-sea dispersal of postlarval shrimp from a hydrothermal vent. *Deep-Sea Res.* **45**, 2105–2118.
Herzig P. M. and Hannington M. D. (2000) Input from the deep: hot vents and cold seeps. In *Marine Geochemistry* (eds. H. D. Schultz and M. Zabel). Springer, Heidelberg, pp. 397–416.
Holland M. E. and Baross J. A. (2003) Limits to life in hydrothermal systems. In *Energy and Mass Transfer in Marine Hydrothermal Systems* (eds. P. Halbach, V. Tunnicliffe, and J. Hein). DUP, Berlin, pp. 235–248.
Holm N. G. and Charlou J. L. (2001) Initial indications of abiotic formation of hydrocarbons in the rainbow ultramafic hydrothermal system, Mid-Atlantic Ridge. *Earth Planet. Sci. Lett.* **191**, 1–8.
Humphris S. E., Herzig P. M., Miller D. J., Alt J. C., Becker K., Brown D., Brügmann G., Chiba H., Fouquet Y., Gemmel J. B., Guerin G., Hannington M. D., Holm N. G., Honnorez J. J., Iturrino G. J., Knott R., Ludwig R., Nakamura K., Petersen S., Reysenbach A.-L., Rona P. A., Smith S., Sturz A. A., Tivey M. K., and Zhao X. (1995) The internal structure of an active sea-floor massive sulphide deposit. *Nature* **377**, 713–716.
Ishibashi J.-I. and Urabe T. (1995) Hydrothermal activity related to arc-backarc magmatism in the western Pacific. In *Backarc Basins: Tectonics and Magmatism* (ed. B. Taylor). Plenum, NY, pp. 451–495.
Ishibashi J., Wakita H., Okamura K., Nakayama E., Feely R. A., Lebon G. T., Baker E. T., and Marumo K. (1997) Hydrothermal methane and manganese variation in the plume over the superfast-spreading southern East Pacific Rise. *Geochim. Cosmochim. Acta* **61**, 485–500.
James R. H. and Elderfield H. (1996) Chemistry of ore-forming fluids and mineral formation rates in an active hydrothermal sulfide deposit on the Mid-Atlantic Ridge. *Geology* **24**, 1147–1150.
Kadko D. and Moore W. (1988) Radiochemical constraints on the crustal residence time of submarine hydrothermal fluids—endeavour ridge. *Geochim. Cosmochim. Acta* **52**, 659–668.
Kadko D. C., Rosenberg N. D., Lupton J. E., Collier R. W., and Lilley M. D. (1990) Chemical reaction rates and entrainment within the endeavour ridge hydrothermal plume. *Earth Planet. Sci. Lett.* **99**, 315–335.
Kelley D. S., Karson J. A., Blackman D. K., Früh-Green G. L., Butterfield D. A., Lilley M. D., Olson E. J., Schrenk M. O., Roe K. K., Lebon G. T., Rivizzigno P., and the AT3-60 shipboard party, (2001) An off-axis hydrothermal vent field near the Mid-Atlantic Ridge at 30°N. *Nature* **412**, 145–149.
Kennedy B. M. (1988) Noble gases in vent water from the Juan de Fuca Ridge. *Geochim. Cosmochim. Acta* **52**, 1929–1935.
Klinkhammer G., Elderfield H., and Hudson A. (1983) Rare earth elements in seawater near hydrothermal vents. *Nature* **305**, 185–188.
Klinkhammer G. P., Elderfield H., Edmond J. M., and Mitra A. (1994) Geochemical implications of rare earth element patterns in hydrothermal fluids from mid-ocean ridges. *Geochim. Cosmochim. Acta* **58**, 5105–5113.
Klinkhammer G. P., Chin C. S., Keller R. A., Dahlmann A., Sahling H., Sarthou G., Petersen S., and Smith F. (2001) Discovery of new hydrothermal vent sites in Bransfield strait, Antarctica. *Earth Planet. Sci. Lett.* **193**, 395–407.
Koski R. A., German C. R., and Hein J. R. (2003) Fate of hydrothermal products from mid-ocean ridge hydrothermal systems: near-field to global perspectives. In *Energy and Mass Transfer in Marine Hydrothermal Systems* (eds. P. Halbach, V. Tunnicliffe, and J. Hein). DUP, Berlin, pp. 317–335.
Landing W. M. and Bruland K. W. (1987) The contrasting biogeochemistry of iron and manganese in the Pacific Ocean. *Geochim. Cosmochim. Acta* **51**, 29–43.
Langmuir C., Humphris S., Fornari D., Van Dover C., Von Damm K., Tivey M. K., Colodner D., Charlou J.-L., Desonie D., Wilson C., Fouquet Y., Klinkhammer G., and Bougault H. (1997) Description and significance of hydrothermal vents near a mantle hot spot: the lucky strike vent field at 37°N on the Mid-Atlantic Ridge. *Earth Planet. Sci. Lett.* **148**, 69–92.
Lao Y., Anderson R. F., Broecker W. S., Trumbore S. E., Hoffman H. J., and Wölfli W. (1992) Transport and burial rates of ^{10}Be and ^{231}Pa in the Pacific ocean during the Holocene period. *Earth Planet. Sci. Lett.* **113**, 173–189.
Lilley M. D., Feely R. A., and Trefry J. H. (1995) Chemical and biochemical transformation in hydrothermal plumes. *Geophys. Monogr. (AGU)* **91**, 369–391.
Lupton J. E. and Craig H. (1981) A major ^{3}He source on the East Pacific Rise. *Science* **214**, 13–18.
Lupton J. E., Delaney J. R., Johnson H. P., and Tivey M. K. (1985) Entrainment and vertical transport of deep ocean water by buoyant hydrothermal plumes. *Nature* **316**, 621–623.
Lupton J. E., Butterfield D., Lilley M., Ishibashi J., Hey D., and Evans L. (1999a) Gas chemistry of hydrothermal fluids along the East Pacific Rise, 5°S to 32°S. *EOS Trans. AGU (abstr.)* **80**, F1099.
Lupton J. E., Baker E. T., and Massoth G. J. (1999b) Helium, heat and the generation of hydrothermal event plumes at mid-ocean ridges. *Earth Planet. Sci. Lett.* **171**, 343–350.
Lyle M., Leinen M., Owen R. M., and Rea D. K. (1987) Late tertiary history of hydrothermal deposition at the East Pacific Rise, 19°S—correlation to volcano-tectonic events. *Geophys. Res. Lett.* **14**, 595–598.
Magenheim A. J., Bayhurst G., Alt J. C., and Gieskes J. M. (1992) ODP leg 137, borehole fluid chemistry in hole 504B. *Geophys. Res. Lett.* **19**, 521–524.
Maris C. R. P. and Bender M. L. (1982) Upwelling of hydrothermal solutions through ridge flank sediments shown by pore-water profiles. *Science* **216**, 623–626.
Massoth G. J., Baker E. T., Lupton J. E., Feely R. A., Butterfield D. A., VonDamm K. L., Roe K. K., and

LeBon G. T. (1994) Temporal and spatial variability of hydrothermal manganese and iron at Cleft segment, Juan de Fuca Ridge. *J. Geophys. Res.* **99**, 4905–4923.

McCollom T. M. and Shock E. L. (1997) Geochemical constraints on chemolithoautotrophic metabolism by microorganisms in seafloor hydrothermal systems. *Geochim. Cosmochim. Acta* **61**, 4375–4391.

McCollom T. M. and Shock E. L. (1998) Fluid-rock interactions in the lower oceanic crust: thermodynamic models of hydrothermal alteration. *J. Geophys. Res.* **103**, 547–575.

Mercier H. and Speer K. G. (1998) Transport of bottom water in the Romanche Fracture Zone and the Chain Fracture Zone. *J. Phys. Oc.* **28**, 779–790.

Metz S., Trefry J. H., and Nelsen T. A. (1988) History and geochemistry of a metalliferous sediment core from the Mid-Atlantic Ridge at 26°N. *Geochim. Cosmochim. Acta* **52**, 2369–2378.

Middleton J. H. (1979) Times of rise for turbulent forced plumes. *Tellus* **31**, 82–88.

Millero F. J., Sotolongo S., and Izaguirre M. (1987) The oxidation kinetics of Fe(II) in seawater. *Geochim. Cosmochim. Acta* **51**, 793–801.

Mills R. A. and Elderfield H. (1995) Hydrothermal activity and the geochemistry of metalliferous sediment. *Geophys. Monogr. (AGU)* **91**, 392–407.

Mills R. A., Elderfield H., and Thomson J. (1993) A dual origin for the hydrothermal component in a metalliferous sediment core from the Mid-Atlantic Ridge. *J. Geophys. Res.* **98**, 9671–9678.

Mottl M. J. (2003) Partitioning of energy and mass fluxes between mid-Ocean ridge axes and flanks at high and low temperature. In *Energy and Mass Transfer in Marine Hydrothermal Systems* (eds. P. Halbach, V. Tunnicliffe, and J. Hein). DUP, Berlin, pp. 271–286.

Mottl M. J. and McConachy T. F. (1990) Chemical processes in buoyant hydrothermal plumes on the East Pacific Rise near 21°N. *Geochim. Cosmochim. Acta* **54**, 1911–1927.

Mottl M. J. and Wheat C. G. (1994) Hydrothermal circulation through mid-ocean ridge flanks: fluxes of heat and magnesium. *Geochim. Cosmochim. Acta* **58**, 2225–2237.

Mottl M. J., Wheat G., Baker E., Becker N., Davis E., Feely R., Grehan A., Kadko D., Lilley M., Massoth G., Moyer C., and Sansome F. (1998) Warm springs discovered on 3.5 Ma oceanic crust, eastern flank of the Juan de Fuca Ridge. *Geology* **26**, 51–54.

Morton B. R., Taylor G. I., and Turner J. S. (1956) Turbulent gravitational convection from maintained and instantaneous sources. *Proc. Roy. Soc. London Ser. A.* **234**, 1–23.

Mullineaux L. S., and France S. C. (1995) Disposal mechanisms of deep-sea hydrothermal vent fauna. *Geophys. Monogr. (AGU)* **91**, 408–424.

Mullins T. D., Britschgi T. B., Krest R. L., and Giovannoni S. J. (1995) Genetic comparisons reveal the same unknown bacterial lineages in Atlantic and Pacific bacterioplankton communities. *Limnol. Oceanogr.* **40**, 148–158.

Murray J. and Renard A. F. (1891) *Deep-sea Deposits*. Report "Challenger" Expedition (1873–1876), London.

Nehlig P., Juteau T., Bendel V., and Cotten J. (1994) The root zones of oceanic hydrothermal systems—constraints from the Samail ophiolite (Oman). *J. Geophys. Res.* **99**, 4703–4713.

O'Grady K. M. (2001) The geochemical controls on hydrothermal vent fluid chemistry from two areas on the ultrafast spreading southern East Pacific Rise. MSc Thesis, University of New Hampshire, 134pp.

Olivarez A. M. and Owen R. M. (1989) REE/Fe variation in hydrothermal sediments: implications for the REE content of seawater. *Geochim. Cosmochim. Acta* **53**, 757–762.

Oosting S. E. and Von Damm K. L. (1996) Bromide/chloride fractionation in seafloor hydrothermal fluids from 9–10°N East Pacific Rise. *Earth Planet. Sci. Lett.* **144**, 133–145.

Palmer M. R. and Edmond J. M. (1989) The strontium isotope budget of the modern ocean. *Earth Planet. Sci. Lett.* **92**, 11–26.

Palmer M. R. and Ernst G. G. J. (1998) Generation of hydrothermal megaplumes by cooling of pillow basalts at mid-ocean ridges. *Nature* **393**, 643–647.

Palmer M. R. and Ernst G. G. J. (2000) Comment on Lupton *et al.* (1999b). *Earth Planet. Sci. Lett.* **180**, 215–218.

Parson L. M., Murton B. J., Searle R. C., Booth D., Evans J., Field P., Keetin J., Laughton A., McAllister E., Millard N., Redbourne L., Rouse I., Shor A., Smith D., Spencer S., Summerhayes C., and Walker C. (1993) En echelon axial volcanic ridges at the Reykjanes ridge: a life cycle of volcanism and tectonics. *Earth Planet. Sci. Lett.* **117**, 73–87.

Parson L., Gracia E., Coller D., German C. R., and Needham H. D. (2000) Second order segmentation—the relationship between volcanism and tectonism at the MAR, 38°N–35°40' N. *Earth Planet. Sci. Lett.* **178**, 231–251.

Parsons B. (1981) The rates of plate creation and consumption. *Geophys. J. Roy. Astron. Soc.* **67**, 437–448.

Petersen S., Herzig P. M., and Hannington M. D. (2000) Third dimension of a presently forming VMS deposit: TAG hydrothermal mound, Mid-Atlantic Ridge, 26°N. *Mineralium Deposita* **35**, 233–259.

Peucker-Ehrenbrink B., Hofmann A. W., and Hart S. R. (1994) Hydrothermal lead transfer from mantle to continental crust—the role of metalliferous sediments. *Earth Planet. Sci. Lett.* **125**, 129–142.

Peucker-Ehrenbrink B., Ravizza G., and Hofmann A. W. (1995) The marine Os-187/Os-186 record of the past 180 million years. *Earth Planet. Sci. Lett.* **130**, 155–167.

Radford-Knoery J., German C. R., Charlou J.-L., Donval J.-P., and Fouquet Y. (2001) Distribution and behaviour of dissolved hydrogen sulfide in hydrothermal plumes. *Limnol. Oceanogr.* **46**, 461–464.

Ravizza G. (1993) Variations of the 187Os/186Os ratio of seawater over the past 28 million years as inferred from metalliferous carbonates. *Earth Planet. Sci. Lett.* **118**, 335–348.

Ravizza G., Martin C. E., German C. R., and Thompson G. (1996) Os isotopes as tracers in seafloor hydrothermal systems: a survey of metalliferous deposits from the TAG hydrothermal area, 26°N Mid-Atlantic Ridge. *Earth Planet. Sci. Lett.* **138**, 105–119.

Ravizza G., Blusztajn J., Von Damm K. L., Bray A. M., Bach W., and Hart S. R. (2001) Sr isotope variations in vent fluids from 9°46–54' N EPR: evidence of a non-zero-Mg fluid component at Biovent. *Geochim. Cosmochim. Acta* **65**, 729–739.

Rona P. A. and Trivett D. A. (1992) Discrete and diffuse heat transfer at ASHES vent field, axial volcano, Juan de Fuca Ridge. *Earth Planet. Sci. Lett.* **109**, 57–71.

Rona P. A., Bogdanov Y. A., Gurvich E. G., Rimskikorsakov N. A., Sagalevitch A. M., Hannington M. D., and Thompson G. (1993) Relict hydrothermal zones in the TAG hydrothermal field, Mid-Atlantic Ridge 26°N 45°W. *J. Geophys. Res.* **98**, 9715–9730.

Roth S. E. and Dymond J. (1989) Transport and settling of organic material in a deep-sea hydrothermal plume—evidence from particle-flux measurements. *Deep Sea Res.* **36**, 1237–1254.

Rubin K. H., MacDougall J. D., and Perfit M. R. (1994) $^{210}Po/^{210}Pb$ dating of recent volcanic eruptions on the seafloor. *Nature* **468**, 841–844.

Rudnicki M. D. and Elderfield H. (1992) Helium, radon and manganese at the TAG and SnakePit hydrothermal vent fields 26° and 23°N, Mid-Atlantic Ridge. *Earth Planet. Sci. Lett.* **113**, 307–321.

Rudnicki M. D. and Elderfield H. (1993) A chemical model of the buoyant and neutrally buoyant plume above the TAG vent field, 26 degrees N, Mid-Atlantic Ridge. *Geochim. Cosmochim. Acta* **57**, 2939–2957.

Rudnicki M. D., James R. H., and Elderfield H. (1994) Near-field variability of the TAG nonbuoyant plume 26°N Mid-Atlantic Ridge. *Earth Planet. Sci. Lett.* **127**, 1–10.

Ruhlin D. E. and Owen R. M. (1986) The rare earth element geochemistry of hydrothermal sediments from the East Pacific Rise: examination of a seawater scavenging mechanism. *Geochim. Cosmochim. Acta* **50**, 393–400.

Rüth C., Well R., and Roether W. (2000) Primordial ^{3}He in South Atlantic deep waters from sources on the Mid-Atlantic Ridge. *Deep-Sea Res.* **47**, 1059–1075.

Sauter D., Parson L., Mendel V., Rommevaux-Jestin C., Gomez O., Briais A., Mevel C., and Tamaki K. (2002) TOBI sidescan sonar imagery of the very slow-spreading southwest Indian Ridge: evidence for along-axis magma distribution. *Earth Planet. Sci. Lett.* **199**, 81–95.

Schaller T., Morford J., Emerson S. R., and Feely R. A. (2000) Oxyanions in metalliferous sediments: tracers for paleoseawater metal concentrations? *Geochim. Cosmochim. Acta* **64**, 2243–2254.

Schultz A. and Elderfield H. (1997) Controls on the physics and chemistry of seafloor hydrothermal circulation. *Phil. Trans. Roy. Soc. London A* **355**, 387–425.

Schultz A., Delaney J. R., and McDuff R. E. (1992) On the partitioning of heat-flux between diffuse and point-source sea-floor venting. *J. Geophys. Res.* **97**, 12299–12314.

Seyfried W. E. and Ding K. (1995) Phase equilibria in subseafloor hydrothermal systems: a review of the role of redox, temperature, pH and dissolved Cl on the chemistry of hot spring fluids at mid-ocean ridges. *Geophys. Monogr. (AGU)* **91**, 248–272.

Shanks W. C., III (2001) Stable isotopes in seafloor hydrothermal systems: vent fluids, hydrothermal deposits, hydrothermal alteration, and microbial processes. In *Stable Isotope Geochemistry,* Rev. Mineral. Geochem. 43 (eds. J. W. Valley and D. R. Cole). Mineralogical Society of America, pp. 469–525.

Sherrell R. M., Field M. P., and Ravizza G. (1999) Uptake and fractionation of rare earth elements on hydrothermal plume particles at 9°45' N, East Pacific Rise. *Geochim. Cosmochim. Acta* **63**, 1709–1722.

Simoneit B. R. T. (1991) Hydrothermal effects on recent diatomaceous sediments in Guaymas Basin—generation, migration, and deposition of petroleum. In *AAPG Memoir 47: The Gulf and Peninsular Province of the Californias*, American Association of Petroleum Geologists, Tulsa, OK, chap. 38, pp. 793–825.

Sohn R. A., Fornari D. J., Von Damm K. L., Hildebrand J. A., and Webb S. C. (1998) Seismic and hydrothermal evidence for a cracking event on the East Pacific Rise at 9°50' N. *Nature* **396**, 159–161.

Speer K. G. and Rona P. A. (1989) A model of an Atlantic and Pacific hydrothermal plume. *J. Geophys. Res.* **94**, 6213–6220.

Speer K. G., Maltrud M., and Thurnherr A. (2003) A global view of dispersion on the mid-oceanic ridge. In *Energy and Mass Transfer in Marine Hydrothermal Systems* (eds. P. Halbach, V. Tunnicliffe, and J. Hein). DUP, Berlin, pp. 287–302.

Spiess F. N., Ken C. M., Atwater T., Ballard R., Carranza A., Cordoba D., Cox C., Diaz Garcia V. M., Francheteau J., Guerrero J., Hawkins J., Haymon R., Hessler R., Juteau T., Kastner M., Larson R., Luyendyk B., Macdongall J. D., Miller S., Normark W., Orcutt J., and Rangin C. (1980) East Pacific Rise: hot springs and geophysical experiments. *Science* **207**, 1421–1433.

Statham P. J., Yeats P. A., and Landing W. M. (1998) Manganese in the eastern Atlantic Ocean: processes influencing deep and surface water distributions. *Mar. Chem.* **61**, 55–68.

Statham P. J., Connelly D. P., and German C. R. (2003) Fe(II) oxidation in Indian Ocean hydrothermal plumes. *Nature* (submitted).

Staudigel H. and Hart S. R. (1983) Alteration of basaltic glass—mechanisms and significance for oceanic-crust seawater budget. *Geochim. Cosmochim. Acta* **47**, 337–350.

Stein C. A. and Stein S. (1994) Constraints on hydrothermal heat flux through the oceanic lithosphere from global heat flow. *J. Geophys. Res.* **99**, 3081–3095.

Stein C. A., Stein S., and Pelayo A. M. (1995) Heat flow and hydrothermal circulation. Geophys. *Monogr. (AGU)* **91**, 425–445.

Summit M. and Baross J. A. (1998) Thermophilic subseafloor microorganisms from the 1996 north Gorda ridge eruption. *Deep-Sea Res.* **45**, 2751–2766.

Tivey M. K. (1995) The influence of hydrothermal fluid composition and advection rates on black smoker chimney mineralogy—insights from modelling transport and reaction. *Geochim. Cosmochim. Acta* **59**, 1933–1949.

Trefry J. H. and Metz S. (1989) Role of hydrothermal precipitates in the geochemical cycling of vanadium. *Nature* **342**, 531–533.

Trocine R. P. and Trefry J. H. (1988) Distribution and chemistry of suspended particles from an active hydrothermal vent site on the Mid-Atlantic Ridge at 26°N. *Earth Planet. Sci. Lett.* **88**, 1–15.

Turner J. S. (1973) *Buoyancy Effects in Fluids*. Cambridge University Press, 368pp.

Van Dover C. L., German C. R., Speer K. G., Parson L. M., and Vrijenhoek R. C. (2002) Evolution and biogeography of deep-sea vent and seep invertebrates. *Science* **295**, 1253–1257.

Von Damm K. L. (1995) Controls on the chemistry and temporal variability of seafloor hydrothermal fluids. *Geophys. Monogr. (AGU)* **91**, 222–247.

Von Damm K. L. (2000) Chemistry of hydrothermal vent fluids from 9–10°N, East Pacific Rise: time zero the immediate post-eruptive period. *J. Geophys. Res.* **105**, 11203–11222.

Von Damm K. L. (2003) Evolution of the hydrothermal system at East Pacific Rise 9°50' N: geochemical evidence for changes in the upper oceanic crust. *Geophys. Monogr. (AGU)* (submitted).

Von Damm K. L. and Bischoff J. L. (1987) Chemistry of hydrothermal solutions from the southern Juan de Fuca Ridge. *J. Geophys. Res.* **92**, 11334–11346.

Von Damm K. L. and Lilley M. D. (2003) Diffuse flow hydrothermal fluids from 9°50' N East Pacific Rise: origin, evolution and biogeochemical controls. *Geophys. Monogr. (AGU)* (in press).

Von Damm K. L., Edmond J. M., Grant B., Measures C. I., Walden B., and Weiss R. F. (1985a) Chemistry of submarine hydrothermal solutions at 21°N, East Pacific Rise. *Geochim. Cosmochim. Acta* **49**, 2197–2220.

Von Damm K. L., Edmond J. M., Measures C. I., and Grant B. (1985b) Chemistry of submarine hydrothermal solutions at Guaymas Basin, Gulf of California. *Geochim. Cosmochim. Acta* **49**, 2221–2237.

Von Damm K. L., Oosting S. E., Kozlowski R., Buttermore L. G., Colodner D. C., Edmonds H. N., Edmond J. M., and Grebmeier J. M. (1995) Evolution of East Pacific Rise hydrothermal vent fluids following a volcanic eruption. *Nature* **375**, 47–50.

Von Damm K. L., Buttermore L. G., Oosting S. E., Bray A. M., Fornari D. J., Lilley M. D., and Shanks W. C., III (1997) Direct observation of the evolution of a seafloor black smoker from vapor to brine. *Earth Planet. Sci. Lett.* **149**, 101–112.

Von Damm K. L., Bray A. M., Buttermore L. G., and Oosting S. E. (1998) The geochemical relationships between vent fluids from the lucky strike vent field, Mid-Atlantic Ridge. *Earth Planet. Sci. Lett.* **160**, 521–536.

Von Damm K. L., Gallant R. M., Hall J. M., Loveless J., Merchant E., and Scientific party of R/V Knorr KN162-13, (2001) The Edmond hydrothermal field: pushing the envelope on MOR brines. *EOS Trans. AGU (abstr.)* **82**, F646.

Von Damm K. L., Parker C. M., Gallant R. M., Loveless J. P., and the AdVenture 9 Science Party, (2002) Chemical evolution of hydrothermal fluids from EPR 21°N: 23 years later in a phase separating world. *EOS Trans. AGU (abstr.)* **83**, V61B-1365.

Von Damm K. L., Lilley M. D., Shanks W. C., III, Brockington M., Bray A. M., O'Grady K. M., Olson E., Graham A., Proskurowski G., and the SouEPR Science Party (2003) Extraordinary phase separation and segregation in vent fluids from the southern East Pacific Rise. *Earth Planet. Sci. Lett.* **206**, 365–378.

Welhan J. and Craig H. (1983) Methane, hydrogen, and helium in hydrothermal fluids at 21°N on the East Pacific Rise. In *Hydrothermal Processes at Seafloor Spreading Centres* (ed. P. A. Rona, K. Boström, L. Laubier, L. Laubier, and K. L. Smith, Jr.). NATO Conference Series IV: 12, Plenum, New York, pp. 391–409.

White R. S., McKenzie D., and O'Nions R. K. (1992) Oceanic crustal thickness from seismic measurements and rare earth element inversions. *J. Geophys. Res.* **97**, 19683–19715.

Winn C. D., Cowen J. P., and Karl D. M. (1995) Microbiology of hydrothermal plumes. In *Microbiology of Deep-sea Hydrothermal Vent Habitats* (ed. D. M. Karl), CRC, Boca Raton.

You C.-F., Butterfield D. A., Spivack A. J., Gieskes J. M., Gamo T., and Campbell A. J. (1994) Boron and halide systematics in submarine hydrothermal systems: effects of phase separation and sedimentary contributions. *Earth Planet. Sci. Lett.* **123**, 227–238.

6.08 Tracers of Ocean Mixing

W. J. Jenkins

Woods Hole Oceanographic Institution, MA, USA

6.08.1 INTRODUCTION

The distributions of chemicals within the ocean are governed by a variety of biological, chemical, physical, and geological processes. The relative importance of each process in controlling a geochemical element's distribution varies from substance to substance, depending on its chemical reactivity and role in various biogeochemical cycles. Often, these processes are poorly understood and inextricably intertwined. Consider, for example, the distribution of dissolved oxygen in the Atlantic (Figure 1).

Oxygen is driven toward saturation on contact with the atmosphere, produced by biological fixation in the upper ocean, and consumed by bacterial degradation of organic material in the water column and sediments below. These processes are clearly variable in both space and time, and are not well quantified. In addition to these biogeochemical processes, one sees the effects of large-scale circulation, ventilation, and mixing on its distribution. Note, for example, the penetration of high-oxygen water into the abyss from the north, and into intermediate depths from the south. Since the large-scale dissolved oxygen distribution appears to be in approximate steady state, it must represent a balance between both physical and biogeochemical processes.

Physical processes, however, are a common determinant for the distributions of all geochemical distributions, and their quantification may lead to insight into the magnitude aznd nature of the other processes. For example, determination of tracer ventilation rates has been used to quantify oxygen utilization rates, which in turn can be used to estimate export production in the ocean (Jenkins, 1984; Jenkins and Wallace, 1992).

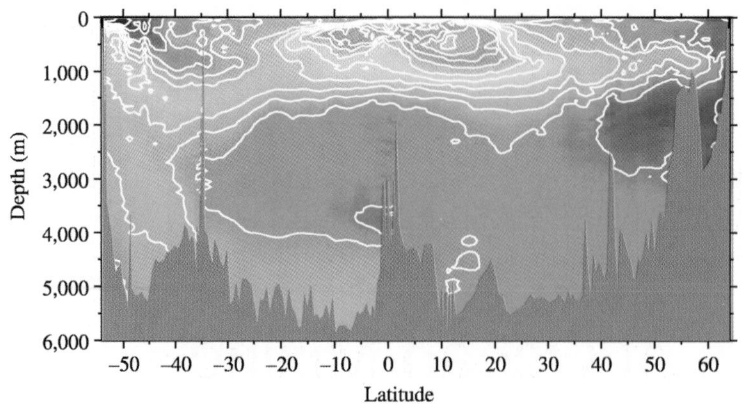

Figure 1 The meridional distribution of dissolved oxygen in the central Atlantic. Blue corresponds to high concentrations and red to low.

Conversely, the distributions of some geochemicals whose biogeochemical behaviors are well understood or sufficiently simple may be used to obtain quantitative information regarding the magnitude of physical processes in the ocean.

Fluid flow in the oceans is generally turbulent. This turbulence occurs on space scales ranging from millimeters to hundreds of kilometers, and in concert with convection and molecular diffusion, works to homogenize properties within the ocean. Nevertheless, the oceans are *not* homogeneous because large-scale processes that change seawater properties are at work both in the ocean interior (e.g., particle sinking, see Chapter 6.09) and at the oceanic boundary layers, where they contact the atmosphere and the solid earth. The resultant distributions arise from an approximate balance between this "biogeochemical and boundary forcing" and ocean ventilation, circulation, and mixing. The nature and magnitude of these physical processes are crucial in determining the oceanic physical and biogeochemical states, and how they will evolve in the face of global change.

Since the ocean is a key element in the climate system, global-scale ocean models play a central role in climate change forecasting, from the viewpoint of both heat transport and carbon dioxide uptake and sequestration. These models are used to predict the shift in the biogeochemical state of the ocean due to direct anthropogenic impact and changes in climate forcing. Although becoming increasingly sophisticated and powerful, such models are necessarily idealized. This arises from the fact that physical processes occur on space scales spanning more than 18 orders of magnitude (from molecular to planetary scale), whereas we are presently computationally limited to a mere 3 or 4 orders of magnitude. Subgridscale processes, i.e., processes occurring on spatial and temporal scales not actually resolved by the models, must be represented by some integrated quasi-statistical parametrization. Mixing is a key example of such processes. The magnitude and, indeed, the nature of the parametrization can be (and have been) estimated heuristically, but ultimately, the magnitude and validity of this parametrization is rooted in oceanic observations.

Tracers represent a potent tool in this regard. Tracers constitute a general class of materials in the ocean that occur in sufficiently small quantities (they are *trace*) that they do not directly influence the fluid's density and hence behavior. (Temperature and salinity have been referred to as "active" tracers (as opposed to the other "passive" tracers), and technically do not fall within this strict definition.) More importantly, however, is the fact that they *trace* fluid flow in some fashion. An example of this is the penetration of bomb-produced tritium into the North Atlantic (Figure 2), which traces the entry point of newly ventilated water into the global overturning circulation. In a qualitative sense, such distributions provide compelling visualizations of ventilation pathways and timescales of the circulation simply by showing where the tracer *is* and where the tracer *is not*. Presence of a bomb-produced tracer at a location in the abyssal Atlantic in the 1980s, for example, implies a connection to the ocean surface within the past few decades. Comparing Figures 1 and 2, for example, leads one to infer the importance and rates of ventilation pathways in determining the distribution of oxygen in the deep North Atlantic.

Tracers integrate over space- and timescales processes that occur in a patchy and often episodic fashion. Thus, they provide space- and time-averaged information on these processes, often on scales that are relevant to important oceanographic and climate problems. These attributes, however, prove to be significant challenges as well. The manner in which a tracer integrates these processes is intrinsically colored by the

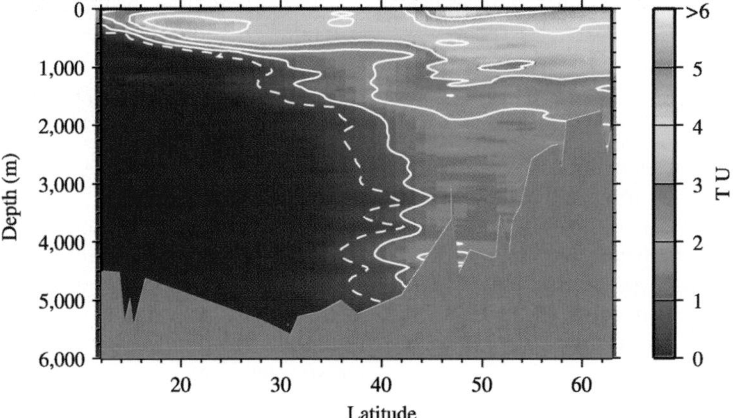

Figure 2 The meridional distribution of tritium in the central North Atlantic in the early 1980s.

nature of the tracer, its reactivity (if any), boundary conditions, and time history (if transient) (e.g., see Doney and Jenkins, 1988). These aspects may not be perfectly known, and the actual observational database tends to be more sparse or noisy than desired. Moreover, the information extracted from the observed tracer distributions is in general model dependent. Different tracers, however, tend to illuminate different aspects of biogeochemical and physical processes, so that combining these different observations provides stronger constraints than from individual tracers. An analogy might be observing a scene through a single color filter: the basic features of the scene are revealed, but adding color components builds a clearer, more complete picture, because each component adds new and unique features to the scene.

The objective of this chapter is to provide an overview of recent advances in the use of tracers to measure ocean mixing, circulation, and ventilation. This is an extremely broad and active area of research, so the discussion will be by no means comprehensive. Rather than providing a litany encompassing everything that has been published in the past few decades on tracers and ocean mixing, I will attempt to discuss a small number of recent studies within a simplified theoretical framework. The hope is that this will provide the reader with a better synthesis of where the field stands as of early 2000s, and where significant advances are likely to occur. The discussion will not include tracers that are dominated by processes such as sorption/desorption in the marine environment (e.g., particle reactive tracers; see Chapter 6.09), and will not dwell on large-scale modeling efforts (whether prognostic or inverse), although each are of considerable importance in marine geochemistry.

To set the stage for a discussion of current work on using tracers to study mixing, I begin in the next section by discussing the basic theory behind advection–diffusion modeling. This is followed in Section 6.08.3 with a brief characterization of mixing in the ocean. The following two sections (Sections 6.08.4 and 6.08.5) are an outline some of the recently developed framework theory for discussing tracer results. These will be followed with a brief overview of steady-state tracers (Section 6.08.6), and a more in-depth discussion of transient tracers (Section 6.08.7). Section 6.08.8 is a discussion of age tracer techniques, and Section 6.08.9 treats purposeful tracer release experiments. A final section includes some concluding remarks and some speculation about the future.

6.08.2 THEORETICAL FRAMEWORK 1: THE ADVECTION–DIFFUSION EQUATION

We shall consider the distribution of a scalar property C in one dimension for simplicity. It can be described by an advection–diffusion relationship

$$\frac{\partial C}{\partial t} + \frac{\partial}{\partial x}(uC) = D\frac{\partial^2 C}{\partial x^2} + J \qquad (1)$$

where D is the molecular diffusivity, u is the fluid velocity, and J is an *in situ* source/sink term. In a turbulent-fluid environment, there exist random fluctuations in fluid velocity and scalar concentrations that can be described in terms of mean and fluctuating components:

$$u = \bar{u} + u', \qquad C = \bar{C} + C' \qquad (2)$$

where

$$\bar{u} = \frac{1}{T}\int_T u\,\mathrm{d}t, \qquad \bar{C} = \frac{1}{T}\int_T C\,\mathrm{d}t \qquad (3)$$

where T is an appropriate averaging timescale. Thus, by definition

$$\int_T u' \, \mathrm{d}t = 0, \qquad \int_T C' \, \mathrm{d}t = 0 \tag{4}$$

The assumption is that there exists some timescale that allows us to separate the random fluctuations from the mean (e.g., see Tennekes and Lumley, 1972). That is, there exists some meaningful average for these properties. Substituting these definitions into Equation (1), integrating over T, and utilizing Equations (4) to eliminate terms, we obtain

$$\frac{\partial \bar{C}}{\partial t} + \frac{\partial}{\partial x}(\bar{u}\bar{C} + \overline{(u'C')}) = D\frac{\partial^2 \bar{C}}{\partial x^2} + J \tag{5}$$

where the overbar represents integration over the interval T. The resultant equation resembles the original Equation (1) except for the averaged product of the fluctuations, which is called the *Reynolds flux*. If the fluid and concentration fluctuations were uncorrelated, then this term would integrate to zero, but in practice there is a correlation, because there is likely a causal relationship between fluid motion and concentration fluctuations. By assuming that concentration fluctuations are caused by random fluid parcel displacements of length l' in a macroscopic concentration gradient (e.g., see Garrett, 1989), we can reformulate the Reynolds flux term

$$\frac{\partial \bar{C}}{\partial t} + \frac{\partial}{\partial x}\left(\bar{u}\bar{C} - \overline{u'l'}\frac{\partial \bar{C}}{\partial x}\right) = D\frac{\partial^2 \bar{C}}{\partial x^2} + J \tag{6}$$

Here, the putative length scale represents some kind of mean displacement, or some Lagrangian decorrelation scale that is evidenced by a granularity in the tracer distribution (see Armi and Stommel, 1983; Jenkins, 1987; Joyce and Jenkins, 1993). The $\overline{u'l'}$ term is usually characterized as a turbulent diffusion coefficient κ, because of its functional similarity to the molecular diffusion coefficient:

$$\frac{\partial C}{\partial t} + \frac{\partial}{\partial x}(uC) = D\frac{\partial^2 C}{\partial x^2} + \frac{\partial}{\partial x}\left(\kappa\frac{\partial C}{\partial x}\right) + J \tag{7}$$

At this point, the variables C and u are now used to connote the average (macroscopic) tracer concentration and velocity, respectively, and the overbar is dropped for convenience. Implicit in this formulation is the belief that fundamentally Lagrangian (particle) dispersion can be modeled as a continuum Eulerian phenomenon in a fashion analogous to the Fickian formulation of molecular transport by Brownian motion. This is a useful fiction for simple modeling exercises, but must be used with caution (see the next section on isopycnal diffusion).

Expanding the turbulent diffusion term, one obtains an additional pseudo-advective term:

$$\frac{\partial C}{\partial t} + \frac{\partial}{\partial x}(uC) = D\frac{\partial^2 C}{\partial x^2} + \frac{\partial \kappa}{\partial x}\left(\frac{\partial C}{\partial x}\right) + \kappa\frac{\partial^2 C}{\partial x^2} + J \tag{8}$$

which should be accounted for in regions of strong eddy kinetic energy gradients (and hence changing mixing rates) since it behaves like a velocity (e.g., see Armi, 1979). If κ is assumed to be spatially invariant Equation (7) can be reduced to a simpler form

$$\frac{\partial C}{\partial t} + \frac{\partial}{\partial x}(uC) = (D + \kappa)\frac{\partial^2 C}{\partial x^2} + J \tag{9}$$

Now it will become evident that for most oceanic circumstances $\kappa \gg D$, usually by several orders of magnitude, so that the molecular diffusivity can be safely ignored, giving the familiar time-dependent advection–diffusion equation

$$\frac{\partial C}{\partial t} + \frac{\partial}{\partial x}(uC) = \kappa\frac{\partial^2 C}{\partial x^2} + J \tag{10}$$

An important characteristic of a property distribution is encapsulated in the *Peclet number*, $Pe = UL/\kappa$, which is the ratio of diffusive timescale to advective timescale of the system. In this definition, U and L are the characteristic velocity and length scales of the flow. The Peclet number is a measure of the relative importance of advection *versus* diffusion, where a large number indicates an advectively dominated distribution, and a small number indicates a diffuse flow. Numerical modeling indicates that certain tracer distributions, in particular tracer–tracer relationships, are significantly affected by the Peclet number, and consequently can be used to determine the nature of the fluid flow (Jenkins, 1988; Musgrave, 1985, 1990).

Equation (10) can be expressed in three dimensions as

$$\frac{\partial C}{\partial t} = \kappa\nabla^2 C - \vec{u}\cdot\nabla C + J \tag{11}$$

where κ now becomes an anisotropic eddy diffusivity tensor. Since buoyancy forces suppress fluid motion in a direction orthogonal to isopycnal surfaces (or more properly, neutral surfaces McDougall, 1987), which tends to be approximately vertical, terms associated with that direction are much smaller than horizontal. To the extent that such surfaces exhibit slope in response to horizontal pressure gradients (largely but not solely geostrophic in nature), there will be off-diagonal terms in the diffusivity tensor if expressed in "geodesic" coordinates (Redi, 1982). Expressing the tracer balance in a more "natural" density (isopycnal) coordinate system (where the x- and y-directions are aligned with the isopycnal, and the "z"-direction is orthogonal) effectively diagonalizes the diffusivity tensor

$$\frac{\partial C}{\partial t} = \kappa_\rho \nabla_\rho^2 C - \vec{u} \cdot \nabla_\rho C + J \qquad (12)$$

where the subscript ρ indicates operation in the density coordinate space.

Equation (12), when used in large-scale coarse resolution (non-eddy resolving) models, is incomplete in one other respect. The effect of mesoscale eddies on tracer transport includes an additional, advective-like mechanism sometimes referred to as "bolus transport" (see Gent and McWilliams, 1990; Gent *et al.*, 1995). This can be included in Equation (12) as

$$\frac{\partial C}{\partial t} = \kappa_\rho \nabla_\rho^2 C - \left(\vec{u} + \frac{\overline{h'_\rho u'}}{\overline{h_\rho}} \right) \cdot \nabla_\rho C + J \qquad (13)$$

where h_ρ is the isopycnal layer thickness. This underlines an important limitation to the turbulent diffusion concept: it is an Eulerian continuum approximation to a fundamentally Lagrangian process. In subsequent discussion, this factor will be ignored, but it is of fundamental importance in large-scale numerical modeling.

A final modification to these equations is necessary if the substance under consideration is radioactive, or subject to some first-order concentration-dependent consumption or removal process. Thus in general (11) can be rewritten as

$$\frac{\partial C}{\partial t} = \kappa \nabla^2 C - \vec{u} \cdot \nabla C - \lambda C + J \qquad (14)$$

Within the framework of this equation, tracers may be broadly categorized as

- *conservative* ($J = 0$) or *nonconservative*, counterexamples being dissolved atmospheric argon *versus* dissolved oxygen;
- *steady state* ($\partial C/\partial t = 0$) or *transient*, counterexamples being silica *versus* bomb produced tritium;
- *stable* ($\lambda = 0$) or *radioactive*, counterexamples being dissolved atmospheric helium *versus* radon; and
- *passive* (nondensity affecting) or *active*, counterexamples being CFC-11 *versus* temperature or salinity.

This classification has been discussed extensively within the context of a one-dimensional advection–diffusion model, along with simple solutions to the relevant equations (Craig, 1969). It should be noted, however, that specific tracers may fall into different categories depending on the nature of the specific application. For example, radiocarbon is a transient tracer in the surface waters of the ocean because its natural inventory (due to cosmic ray production) has been affected both by dilution with fossil fuel carbon and by massive production from atmospheric nuclear weapons testing. In deep Pacific ocean waters, however, it is more a steady-state tracer (see Section 6.08.6). Moreover, there are regions of the ocean where both components are present.

6.08.3 THE NATURE OF OCEANIC MIXING

Think of the ocean in its true geometry: the ratio of its vertical to horizontal dimensions is comparable to the page of paper on which you read these words. The truly remarkable aspect of ocean mixing is that water masses can travel horizontally for tens of thousands of kilometers before being mixed away through a vertical distance of only a kilometer or so. Because the ocean is largely a stratified fluid, mixing does not occur in a spatially isotropic fashion. Fluid motion is generally suppressed by buoyancy forces in the vertical, or more properly the diapycnal direction, i.e., the direction orthogonal to the surface of constant density, so that diapycnal (approximately vertical) mixing is generally many orders of magnitude smaller than isopycnal (largely horizontal) mixing. (A more exact terminology would be "dianeutral," which is orthogonal to the local "neutral surface" (McDougall, 1987).) Despite this, each are thought to be of roughly equal importance in affecting oceanic distributions, and in fact diapycnal mixing may have a dominant impact on the behavior of large-scale ocean models (Bryan, 1987; McWilliams, 1996).

A way to compare the relative importance of diapycnal and isopycnal mixing, especially in relation to the planetary-scale circulation is to compare their effective timescales. The shorter the timescale, the greater is the influence on property distributions. The dominant timescale for global-scale ocean circulation is ~500–1,000 yr (as determined by radiocarbon measurements; see Section 6.08.6). In order to estimate the diffusive timescale, one characterizes the strength of turbulent mixing in terms of a Fickian diffusion coefficient, by analogy to molecular diffusion (see discussion in Section 6.08.3). For diapycnal mixing, the dominant length scale ranges from 300 m to 3,000 m, corresponding to a typical main thermocline thickness to nearly the entire ocean depth:

$$T_D \sim \frac{L_D^2}{\kappa_D} = \frac{10^5 - 10^7 \text{ m}^2}{10^{-5} - 10^{-4} \text{ m}^2 \text{ s}^{-1}} = 10^9 - 10^{12} \text{ s} = 30\text{–}30{,}000 \text{ yr} \qquad (15)$$

The actual diffusive timescale is likely in the middle of this range, since mixing appears less vigorous in the thermocline than in more weakly stratified waters (see below). For isopycnal mixing, the dominant horizontal scales range from a few hundred kilometers to ocean basin scale (10^4 km):

$$T_{\mathrm{I}} \sim \frac{L_{\mathrm{I}}^2}{\kappa_{\mathrm{I}}} = \frac{10^{11}-10^{14}\ \mathrm{m}^2}{10^2-10^3\ \mathrm{m}^2\ \mathrm{s}^{-1}} = 10^9-10^{12}\ \mathrm{s} = 30\text{–}30{,}000\ \mathrm{yr} \quad (16)$$

Thus, despite their large numeric disparity, their net affects are comparable, and demonstrably important with respect to the global circulation timescale. A detailed discussion of the mechanics of mixing is beyond the scope of this chapter, but it is worthwhile to outline some of their major aspects. The mechanisms of mixing differ fundamentally between the two modes, so the remaining discussion occurs in two parts: diapycnal and isopycnal mixing.

6.08.3.1 Diapycnal Mixing in the Ocean

In a stratified fluid, buoyancy resists any vertical displacement of a water parcel. For a small displacement, a fluid parcel experiences an acceleration in the opposite direction linearly dependent on the displacement *l*:

$$B = -\frac{g}{\rho}\frac{\partial \rho}{\partial z} l \quad (17)$$

which resembles a spring constant in a simple harmonic oscillator, whose fundamental frequency is given by

$$N = \left(-\frac{g}{\rho}\frac{\partial \rho}{\partial z}\right)^{1/2} \quad (18)$$

where ρ is the fluid density and g is the gravitational acceleration. This is often referred to as the "buoyancy frequency." If one compares the work needed to overcome the stability gradient to the energy available to turbulence from the vertical shear in the horizontal velocity field, $\partial u/\partial z$, one obtains the gradient Richardson number

$$Ri = \frac{N^2}{(\partial u/\partial z)^2} \quad (19)$$

which is a dimensionless number that is inversely related to the potential for mixing. Laboratory experiments, field observations, and numerical experimentation suggest that when Ri is sufficiently small (~ 0.25), vertical mixing readily occurs.

Thus, one would expect in general that diapycnal or vertical mixing would on average be suppressed by stratification. It has been suggested, partly from quasi-theoretical arguments (Gargett and Holloway, 1984), and partly on the basis of observational evidence (Gargett, 1984; Quay *et al.*, 1980; Sarmiento *et al.*, 1976), that there is an inverse relationship between N and diapycnal mixing rates generally with

$$\kappa_{\mathrm{D}} \propto N^{-a} \quad (20)$$

where $1 \leq a \leq 2$. These suppositions are likely correct, but there probably also exists several other contributing factors that may be independent of stratification (see Gargett, 1984; Garrett, 1979).

Diapycnal mixing occurs in a number of ways within the ocean. The mechanisms include (in no particular order):

- breaking inertial and internal waves (due to the superposition of waves with different frequencies);
- shear instability (when Ri becomes small, particularly in strong currents);
- double diffusion (salt fingering caused by the difference in molecular diffusivities of heat and salt);
- tidal mixing (caused by the interaction of tidal motion with the bottom or sides; important in certain geographic areas, such as over rough terrain);
- cabbeling (due to the nonlinearity of the seawater equation of state); and
- convection (when buoyancy is decreased at the ocean surface by cooling).

Measurement of temperature variance on very small scales (often referred to as microstructure) has been used to quantify the amount of mixing that is occurring by assuming a balance between its creation by turbulence and destruction by mixing (e.g., see Kunze and Sanford, 1996; Polzin *et al.*, 1995). Such measurements typically give small mixing rates ($\sim 10^{-5}\ \mathrm{m}^2\ \mathrm{s}^{-1}$) in the open ocean. Double diffusion becomes important when warm salty water overlies cold fresher water, and results in characteristic temperature–salinity structures, and the so-called "density ratio" signatures (e.g., see Schmitt, 1990, 1994, 1998). This form of mixing counterintuitively results in an enhanced transport of more slowly (molecularly) diffusing salt over temperature, and appears to set the shape of the T–S relationship for the main thermocline water masses in the subtropical North Atlantic (Schmitt, 1981).

Overall, diapycnal mixing appears to be small in the ocean interior, i.e., $\sim 10^{-5}\ \mathrm{m}^2\ \mathrm{s}^{-1}$ (Kunze and Sanford, 1996; Ledwell *et al.*, 1993) but dramatically larger over rough topography being $\sim 10^{-4}\ \mathrm{m}^2\ \mathrm{s}^{-1}$ (Ledwell *et al.*, 2000; Polzin *et al.*, 1997) and is even larger near the ocean surface (due to wind induced mixing) and the bottom boundary layer (due to friction).

6.08.3.2 Isopycnal Mixing in the Ocean

Lateral mixing may be regarded as the result of two processes (Eckart, 1948; Garrett, 1983), stirring, which streaks out properties, thereby sharpening gradients, and mixing, which operates on these enhanced gradients (in a "diffusive" manner) thereby homogenizing properties on smaller scales. It should be noted that lateral turbulent mixing exhibits a space-scale dependence, with κ increasing with length scale. A simple theory predicts that κ varies as $L^{4/3}$, where L is the scale of the dye patch (Stommel, 1949) although some later work indicates a somewhat weaker dependence approximating $L^{1.1}$ (Okubo, 1971). Part of the challenge is the precise definition of L in the determination. Figure 3 is an illustrative plot of some horizontal mixing estimates, showing the general scale length dependence.

This "non-Fickian behavior" arises from the fact that the range of fluid motions responsible for dispersion of a tracer depends on the size of the tracer patch in relation to the spectrum of fluid motions occurring, and the distinction between stirring and mixing (e.g., see Csanady, 1972; Rhines and Young, 1983; Young *et al.*, 1982).

Mixing on the meter (subkilometer) scale is dominated by stirring due to shear dispersion by internal and inertial waves and subsequent diapycnal mixing (Young *et al.*, 1982), which gives

$$\kappa_{\mathrm{IS}} \approx \left(\frac{N}{f}\right)^2 \kappa_{\mathrm{D}} \qquad (21)$$

where the subscripts I, S, and D correspond to "isopycnal," "small scale," and "diapycnal," respectively. Here N and f are the buoyancy and the Coriolis frequencies respectively. Under typical subtropical, mid-thermocline conditions (i.e., at a few hundred meters depth at mid-latitudes), this leads to mixing rates of $\sim 0.1\ \mathrm{m^2\,s^{-1}}$ (Ledwell *et al.*, 1998; Sundermeyer and Price, 1998).

At 1–30 km scales, properties begin to feel the straining effect of mesoscale eddies, which tease out properties into long streaks. The r.m.s. tracer displacement grows exponentially with time, largely due to rapid stretching of tracer along mesoscale eddy streamlines, and cross-streak mixing, both by the small scale mixing described above, and possibly vortical motions (Ledwell *et al.*, 1998; Polzin *et al.*, 1995). Isopycnal diffusivity on this scale approaches a few $\mathrm{m^2\,s^{-1}}$.

For space scales exceeding a hundred kilometers, mesoscale eddies prove effective in transporting, stirring and mixing tracer. Traditionally, such mixing rates are estimated from the large-scale distributions of tracers, yielding isopycnal diffusivities of order 100–1,000 $\mathrm{m^2\,s^{-1}}$ (e.g., see Armi and Stommel, 1983; Arons and Stommel, 1967; Jenkins, 1987, 1991, 1998; Ledwell *et al.*, 1998; Olbers *et al.*, 1985).

6.08.4 THEORETICAL FRAMEWORK 2: TRACER AGES

In some instances, it is possible to use specific tracer concentrations, in particular radioactive or transient tracers to define a "tracer age." The underlying premise is that the tracer age is set to zero at some starting point (usually the ocean surface) and progressively increases after contact

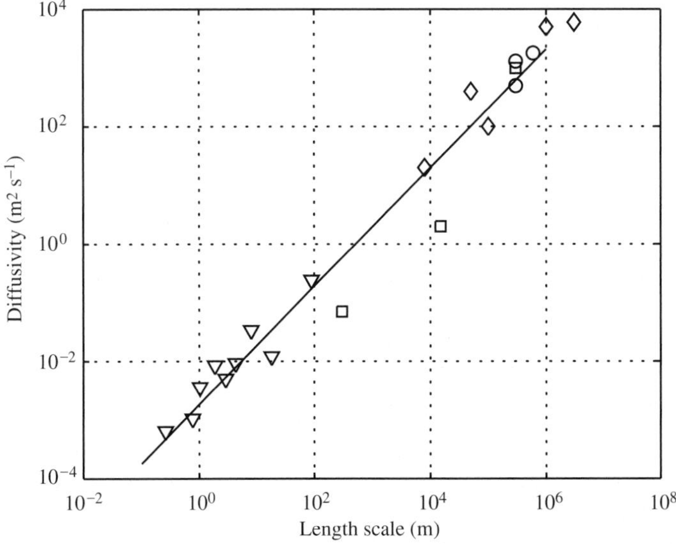

Figure 3 Some horizontal mixing estimates as a function of space scale. Results are from a small-scale float deployment (triangles, Stommel, 1949), the tracer release experiment (squares, Ledwell *et al.*, 1998), and some larger-scale advection–diffusion balances (circles, see later sections) and estimates from radium isotopes (diamonds, see Section 6.08.6). The solid line represents $\kappa \propto L^1$.

is lost with its starting point. This is an attractive approach in that it provides a "direct" visualization of the pathways of ventilation and circulation. Although a useful concept in a qualitative sense, this approach is not without its complexity.

We begin by developing the concept of an ideal ventilation age (after England, 1995a; Theile and Sarmiento, 1990)

$$\frac{\partial A}{\partial t} = \kappa\nabla^2 A - \vec{u}\cdot\nabla A + 1 \tag{22}$$

If the age A of an individual water parcel is in steady state (i.e., $\partial A/\partial t = 0$), then the age tendency term (the last term in Equation (22), the tendency to increase at a rate of 1 s s^{-1} or 1 yr yr^{-1}), is balanced by advection and diffusion of age. Such an age is different from a truly Lagrangian parcel age, which in steady state would be in the absence of mixing:

$$\vec{u}\cdot\nabla A_L = 1 \tag{23}$$

In this case, the velocity field would seem to be obtainable directly from observation of the gradient of A_L but this is not entirely the case.

It is instructive to consider how a "real" tracer age may differ from the idealized behavior evinced by Equation (22). There are three basic tracer age techniques that have been used extensively in the literature: "radiometric dating," "transient concentration dating," and "transient concentration ratio dating." We analyze the first two approaches here. The third is too complicated to yield useful insight from simple analysis. The common thread of this discussion is that tracer ages do not necessarily represent the true ventilation age, as given by Equation (22), and in general, because of their "nonlinear" behavior, tend to *underestimate* the true age of the water (Deleersnijder *et al.*, 2001; Delhez *et al.*, 2003). Thus considerable caution should be used when interpreting such age distributions.

6.08.4.1 Radiometric Dating

Consider the transient tracer pair tritium (^{3}H) and ^{3}He:

$$^3\text{H} \xrightarrow{12.5\ \text{yr}} {}^3\text{He} \tag{24}$$

where tritium decays to ^{3}He with a half-life of 12.45 yr (Unterweger *et al.*, 1980). In a geochemical sense, these may be regarded as ideal tracers, since tritium, being an isotope of hydrogen, exists virtually solely as part of the water molecule, and ^{3}He is a stable, inert gas. The concept model is simple: a water parcel near the ocean surface containing tritium will not accrue any excess ^{3}He since this gas will be lost to the atmosphere. (There is, in fact, a background of atmospheric helium, containing both ^{3}He and ^{4}He. For simplicity, we will assume perfect equilibrium with the atmosphere, and will deal only with *excess* ^{3}He, ignoring the atmospheric background. Further, we will simply refer to the *excess* ^{3}He as simply "^{3}He.") Once the parcel sinks below the surface, ^{3}He begins to accumulate, and at some arbitrary time after subduction, we can compute the age of the water using

$$\tau = \frac{1}{\lambda}\ln\left(\frac{[^3\text{H}] + [^3\text{He}]}{[^3\text{H}]}\right) \tag{25}$$

where τ is the age, λ is the radioactive decay constant (in inverse time units), and the square brackets indicate concentration of the substances in appropriate units. The above equation can be derived in a straightforward way from the radioactive decay equation.

It is therefore possible to construct maps of this age property in the oceans, and gain an immediate qualitative grasp of the ventilation pathways and "stagnation points" in the circulation. Figure 4 is an example of this, taken from the WOCE Pacific expeditions during the early to mid-1990s. One sees the equatorward penetration of newly ventilated waters into the gyres, and the pool of old water emanating from the eastern tropics. A strong temptation arises to relate the observed age gradients to the flow field in a quantitative manner. The difficulty with this simple view arises when mixing occurs.

It is evident from inspection of Equation (25) that the mixing of water parcels with differing tritium concentrations will yield a disproportionate weighting toward the water with higher tritium. This can be quantified by considering the coupled tritium and ^{3}He advection–diffusion equations:

$$\frac{\partial\vartheta}{\partial t} = \kappa\nabla^2\vartheta - \vec{u}\cdot\nabla\vartheta - \lambda\vartheta \tag{26}$$

and

$$\frac{\partial\varphi}{\partial t} = \kappa\nabla^2\varphi - \vec{u}\cdot\nabla\varphi + \lambda\vartheta \tag{27}$$

where ϑ and φ are the tritium and ^{3}He concentrations, respectively, and both tracers are transient (see Section 6.08.7). Note the coupling imposed by the last term in each equation.

Scale analysis of the above equations reveals an interesting formulation of the Peclet number (a measure of the relative importance of advection and mixing) giving

$$Pe_R = \frac{U^2}{\kappa\lambda} \tag{28}$$

(see Jenkins, 1988) where the subscript R distinguishes this from the traditional definition. Here the intrinsic scale length L is determined by

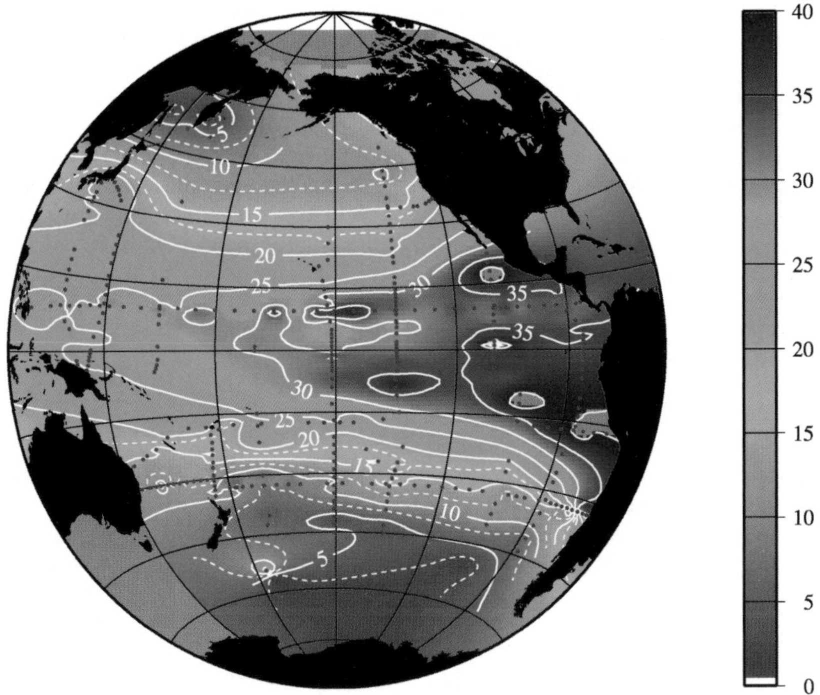

Figure 4 The distribution of tritium–^{3}He age on the $\sigma_0 = 26.5$ kg m^{-3} isopycnal during the 1990s.

the advective length over which the radiotracer decays. Consider the relationship between tritium and ^{3}He in the subtropical North Atlantic (Figure 5). The data fall in a hook-shaped pattern, corresponding to the transition between young, high tritium, low ^{3}He waters and old low tritium waters. The shape of this distribution is most comparable to a low Peclet number (~1 or less) flow in a one-dimensional stream tube simulation.

This result seems at first counterintuitive, since an estimate of the traditional Peclet number can be constructed for horizontal flow using "reasonable" mid-gyre numbers:

$$Pe = \frac{UL}{\kappa} \approx \frac{0.01\,\mathrm{m\,s^{-1}} \times 3{,}000\,\mathrm{km}}{1{,}000\,\mathrm{m^2\,s^{-1}}} = 30 \qquad (29)$$

where we have used typical large-scale isopycnal mixing rates, a typical mid-gyre velocity, and the characteristic gyre scale. The reason for the disparity can be seen by comparing the radiotracer length scale

$$L_\mathrm{R} = \frac{U}{\lambda} \approx \frac{0.01\,\mathrm{m\,s^{-1}}}{1.7 \times 10^{-9}\,\mathrm{s^{-1}}} \approx 6{,}000 \text{ km} \qquad (30)$$

That is, the radiotracer length scale exceeds the gyre scale, which means that the putative stream tube folds back on itself within the gyre circulation, short-circuiting the advective–diffusive balance, and lowering the effective Peclet number of the flow.

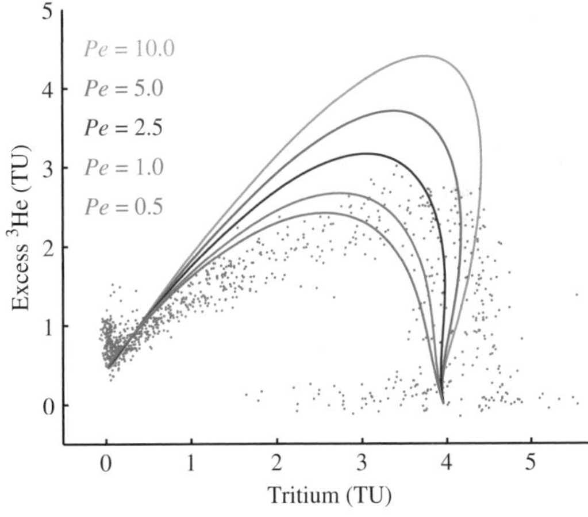

Figure 5 The tritium–^{3}He relationship for the subtropical North Atlantic in the early 1980s (solid dots). The different colored lines correspond to stream tube simulations with various values of Pe_R.

Continuing with the derivation, the two tracers may be combined by addition to provide a stable, conservative dye-tracer:

$$\zeta = \vartheta + \varphi, \qquad \frac{\partial \zeta}{\partial t} = \kappa \nabla^2 \zeta - \vec{u} \cdot \nabla \zeta \qquad (31)$$

This is a useful way of combining two coupled tracers, both of which respond to mixing and decay in a specific way, to fabricate two new composite tracers, one which responds predominantly to decay and hence is a time tracer (the age tracer τ), and a dye tracer ζ that responds largely to dilution.

Combining the above Equations (25)–(27), and (31) with some algebraic manipulation (Jenkins, 1987, 1998), one can derive an advection–diffusion equation for the tritium–^{3}He age as

$$\frac{\partial \tau}{\partial t} = \kappa \nabla^2 \tau - \vec{u} \cdot \nabla \tau + 1 + \kappa \left(\frac{\nabla \zeta}{\zeta} + \frac{\nabla \vartheta}{\vartheta} \right) \cdot \nabla \tau \quad (32)$$

or, alternatively,

$$\frac{\partial \tau}{\partial t} = \kappa \nabla^2 \tau - \vec{u} \cdot \nabla \tau + 1 + \kappa [\nabla \ln(\zeta \vartheta)] \cdot \nabla \tau \quad (33)$$

which resembles Equation (22), except for the peculiar mixing term on the far right-hand-side. Inasmuch as it operates on the age gradient, the term appears as a pseudo-advective divergence term, and is generally negative. Consider the early days of the bomb tritium transient, when tritium was advecting and mixing into an essentially tritium-free ocean. Downstream gradients in tritium (specifically ζ and tritium) will be *negative*, which will make the mixing term the same sign as, and hence augmenting the advective term. This means that the age "signal" will propagate more rapidly into the interior, resulting in a tracer age "depression" relative to the ideal ventilation age. This is consistent with the intuitive argument made earlier about the age mixing nonlinearly because mixtures tend to favor the higher tritium end-member.

Although complicated, this relationship quantifies the effect of mixing, and can be used to assess mixing rates within the ocean (see Section 6.08.8, and Jenkins, 1998; Robbins and Jenkins, 1998; Robbins *et al.*, 2000).

6.08.4.2 Transient Concentration Dating

Where atmospheric concentrations have been changing with time (e.g., as in the case of anthropogenicaly released gases), it could be argued that in the absence of mixing, water parcels could be "time stamped" with equilibrium concentrations. Thus, a fluid parcel could be assigned a vintage "V" by using the time-dependent concentration history

$$C = f(V) \quad (34)$$

providing that there is a unique (or determinable) relationship such as a monotonically increasing atmospheric abundance due to anthropogenic release. An example would be CFC-12 concentrations in the atmosphere prior to the Montreal Protocol.

Such temporal signatures would only be preserved in the absence of mixing, for dilution of recently ventilated waters with tracer-free or differently aged waters would alter the concentration and hence the inferred age in a nonlinear fashion. That is, the age of the mixture would not be equal to the mixture of the ages of the component parts. Following the work of Doney *et al.* (1997) and ignoring the second-order effects associated with the nonlinearity of the solubility function for the gas, one can derive a relationship for the age equivalent to Equation (32) by substituting (34) into (11) and expanding the derivatives using

$$\begin{aligned} \frac{\partial C}{\partial t} &= \frac{\partial f}{\partial V} \frac{\partial V}{\partial t} \\ \nabla C &= \frac{\partial f}{\partial V} \nabla V \\ \nabla^2 C &= \frac{\partial f}{\partial V} \nabla^2 V + \frac{\partial^2 f}{\partial V^2} (\nabla V)^2 \end{aligned} \quad (35)$$

and using $f' = \partial f / \partial V$, $f'' = \partial^2 f / \partial V^2$, and $\tau_C = t - V$ to obtain

$$\frac{\partial \tau_C}{\partial t} = \kappa \nabla^2 \tau_C - \vec{u} \cdot \nabla \tau_C + 1 - \kappa \left(\frac{f''}{f'} \nabla \tau_C \right) \cdot \nabla \tau_C$$

or (36)

$$\frac{\partial \tau_C}{\partial t} = \kappa \nabla^2 \tau_C - \vec{u} \cdot \nabla \tau_C + 1 + \kappa \left(\frac{f''}{(f')^2} \nabla C \right) \cdot \nabla \tau_C$$

The relationship thus derived is similar to that for the radiometric age Equation (33), with an additional pseudo-advective term related to the intensity of mixing and material gradients. An important corollary, however, is that the effect of mixing is related to the curvature of the time history of the atmospheric concentration (see Doney *et al.*, 1997). Thus, if the growth rate of atmospheric concentrations were perfectly linear, we would have $f'' = 0$, so that the additional mixing term becomes zero and the age tracer obeys the ideal transient tracer age relationship:

$$\frac{\partial \tau_C}{\partial t} = \kappa \nabla^2 \tau_C - \vec{u} \cdot \nabla \tau_C + 1 \quad (37)$$

Examination of the atmospheric concentration histories for CFCs, it would appear that this would be a reasonable approximation for CFC-11 in the 1970s through early 1980s, and for CFC-12 for the 1980s, but the abrupt turnover in release curves in the early 1990s would lead to significant departures from the ideal. Further analysis of (36), however, might prove useful in studying CFC age distributions and analyzing model simulation results.

6.08.5 THEORETICAL FRAMEWORK 3: OPTIMUM MULTIPARAMETER ANALYSIS AND TRACER AGE SPECTRA

It can be argued that the concentration of a conservative, steady-state tracer C at a particular location in the ocean arises from the mixture of a number of water masses with their own characteristic concentrations C_j, which in turn were set at their "origins" (usually, but not exclusively thought of as the sea surface in a specific geographic region). In principle, if enough tracers are measured at this given location, and if the tracers are "independent," i.e., they have different water mass concentration "fingerprints," it should be possible to estimate the relative contributions of the various source water masses to the mixture. Hence for a mixture of N water masses with M distinct tracers, one has

$$\begin{aligned} X_1C_{11} + X_2C_{12} + X_3C_{13} + \cdots + X_NC_{1N} &= C_1 \\ X_1C_{21} + X_2C_{22} + X_3C_{23} + \cdots + X_NC_{2N} &= C_2 \\ &\vdots \\ X_1C_{M1} + X_2C_{M2} + X_3C_{M3} + \cdots + X_NC_{MN} &= C_M \end{aligned} \tag{38a}$$

and

$$X_1 + X_2 + X_3 + \cdots + X_N = 1 \tag{38b}$$

and for all

$$0 \le X_i \le 1 \tag{38c}$$

where C_{ij} is the ith tracer concentration of the jth source water mass, C_i is the ith tracer concentration of the resultant mixture, and X_j is the fractional contribution of the jth source water mass. Implicit within this formulation is that the properties mix conservatively (i.e., linearly). Equations (38) are a $(M+1)$ set of linear simultaneous equations in N unknowns that can be written as a matrix equation

$$S\chi = c \tag{39}$$

where χ is the vector of water mass fractions

$$\chi = \begin{bmatrix} X_1 \\ X_2 \\ \vdots \\ X_N \end{bmatrix} \tag{40}$$

c is the vector of tracer concentrations at the location in question,

$$c = \begin{bmatrix} C_1 \\ C_2 \\ \vdots \\ C_M \\ 1 \end{bmatrix} \tag{41}$$

and S is the water mass source values

$$S = \begin{bmatrix} C_{11} & C_{12} & \cdots & C_{1N} \\ C_{21} & C_{22} & \cdots & C_{2N} \\ \vdots & \vdots & \vdots & \vdots \\ C_{M1} & C_{M2} & \cdots & C_{MN} \\ 1 & 1 & \cdots & 1 \end{bmatrix} \tag{42}$$

which can be solved in a non-negative weighted least-squares sense (e.g., see Mackas *et al.*, 1987) provided that $M > N$. This achieved by minimization of the quantity

$$D^2 = (S\chi - c)^{\mathrm{T}} W^{-1} (S\chi - c) \tag{43}$$

where W is the covariance matrix for the tracers. Two important issues in this calculation are

- The choice of appropriate tracers for this constraint. The tracers must be conserved in mixing, and they should be linearly independent (i.e., they should provide unique information).
- The tracer concentrations should be normalized (i.e., weighted) to account for their relative measurement uncertainties. This is included in W.

From this, maps of the relative contribution of water masses can be made. For further discussion on the finer points of this approach, the reader is referred to numerous works by Tomczak and co-workers (e.g., Hamann and Swift, 1991; Poole and Tomczak, 1999; Tomczak, 1999; Tomczak and Large, 1989).

The approach described above is usually applied to a modest number of source water masses, typically 3–6, a number practically limited by the number of independent conservative tracers available to constrain the mixture. With a subtle shift of emphasis, it is possible to conceptually extend this approach to a continuum. Rewriting Equation (11) slightly,

$$\frac{\partial C(\vec{x},t)}{\partial t} - \kappa \nabla^2 C(\vec{x},t) + \vec{u} \cdot \nabla C(\vec{x},t) = J(\vec{x},t) \tag{44}$$

where we have made explicit the space and time dependence of both the concentration and the source/sink terms, we can then seek a Green function solution where

$$\frac{\partial G}{\partial t} - \kappa\nabla^2 G + \vec{u}\cdot\nabla G = \delta(\vec{x} - \vec{x}')\delta(t - t') \qquad (45)$$

(the right-hand side is the Dirac-delta function, or an impulse), which allows us to construct the tracer concentration at a given place and time as a superposition by the convolution

$$C(\vec{x}, t) = \int \mathrm{d}^3\vec{x}' \int_{t_0}^{t} \mathrm{d}t'\, J(\vec{x}', t) G(\vec{x}, t|\vec{x}', t') \qquad (46)$$

That is, the Green function is a solution of the time-dependent advection–diffusion equation for a point source impulse at a location $\vec{x}'$ and time t' which has unit area:

$$\int \mathrm{d}^3\vec{x}' \int_{t_0}^{t} \mathrm{d}t'\, G(\vec{x}, t|\vec{x}', t') = 1 \qquad (47)$$

and hence represents a normalized weighting function for the contribution of water masses at all points in space and time to the observation point $\vec{x}$. Note that it is dependent only on the fluid flow field, and not on specific tracers: it is *tracer independent*. In a crude way of thinking, the Green function plays the role of the χ vector in Equation (39). Although it is not possible to uniquely define the structure of G over all space and time with available tracer measurements (the problem becomes underdetermined given a finite number of tracer constraints), one may seek to use tracer observations to learn about some *aspect* of the space and time structure of G.

Tracer age spectrum analysis is an important extension of this technique. (We have closely followed the discussion of Haine and Hall (2002) with a change in nomenclature to fit with usage within this chapter.) It is predicated on the recognition that a given fluid parcel represents a mixture of fluid "particles" that originated from a variety of locations at the sea surface (where the age was zero) with different transit times to reach this location. Hence, there must exist an "age spectrum" that is solely a characteristic of the fluid advective–diffusive regime. Individual tracer age measurements may represent some aspect of this age spectrum, but precisely what aspect depends on the nature of the age tracer, its history, and its boundary conditions. For a tracer with no internal sources or sinks, but which is set to some value at the ocean surface, here designated by Ω, one can construct a Green function type solution that satisfies

$$\frac{\partial G_\Omega}{\partial t} - \kappa\nabla^2 G_\Omega + \vec{u}\cdot\nabla G_\Omega = \delta(\vec{x} - \vec{x}')\delta(t - t')$$

$$\int \mathrm{d}^3\vec{x} \int_{t_0}^{t} \mathrm{d}t'\, G_\Omega(\vec{x}, t|\vec{x}', t') = 1 \qquad (48)$$

$$\vec{x}' \in \Omega$$

Here, G_Ω may be called the "multiple source boundary propagator," which describes the water mixture at a location along with the origin/transit times $\tau = (t - t')$ of those water masses at the sea surface. It is, by definition, dependent solely on fluid transport properties, and independent of the particular tracer. Since there are no internal sinks or sources of C, we can construct the tracer distribution in a fashion analogous to (46) with

$$C(\vec{x}, t) = \int_\Omega \mathrm{d}^2\vec{x}' \int_{t_0}^{t} \mathrm{d}t' C(\vec{x}', t')\, G_\Omega(\vec{x}, t|\vec{x}', t')$$

$$\vec{x}' \in \Omega \qquad (49)$$

where $C(\vec{x}', t)$ gives the variation in time and space of the tracer concentration at the sea surface as a function of space and time. The attractive feature of this formulation is that it separates the dependence of the tracer distribution into the fluid transport G and the boundary condition $C(\Omega, t)$.

This approach can be connected with the tracer-age concept in the following manner. Consider the transient concentration ("vintage age") approach used for CFC distributions (see previous section). We define the tracer concentration age as

$$\tau_C(\vec{x}, t) = t - V(C(\vec{x}, t)) \qquad (50)$$

where V is the vintage age as a function of surface concentration (i.e., the inverse of the surface concentration *versus* age function). Following the concept above we have

$$\tau_C(\vec{x}, t) = t - V\left(\int_\Omega \mathrm{d}^2\vec{x}' \int_{-\infty}^{t} \mathrm{d}t' C(\vec{x}', t') \times G_\Omega(\vec{x}, t|\vec{x}', t')\right) \qquad (51)$$

In the special case where the surface concentration is a linear function of time (i.e., monotonically increasing), then we have $C(\Omega, t) \propto t$, so the above reduces to

$$\tau_C(\vec{x}, t) = \int_\Omega \mathrm{d}^2\vec{x}' \int_0^\infty \mathrm{d}\tau\, \tau G_\Omega(\vec{x}, t|\vec{x}', t - \tau)$$

$$= \int_\Omega \mathrm{d}^2 x\, A(\vec{x}, t) \qquad (52)$$

where A is a solution of the equation

$$\frac{\partial A}{\partial t} - \kappa\nabla^2 A + \vec{u}\cdot\nabla A = 1 \qquad (53)$$

That is, the tracer concentration age of a transient tracer whose concentration in the ocean surface water is increasing in an exactly linear fashion is equal to the first moment of the age spectrum, and thus equates to an ideal age tracer. This is precisely the conclusion reached in Section 6.08.4 (Equations (36) and (37)).

This new approach is currently receiving considerable attention in oceanography, and is a useful adjunct to the traditional strategies for analysis of tracer distributions and numerical models. A number of illustrative examples have been discussed in the literature (Deleersnijder *et al.*, 2001, 2002; Delhez *et al.*, 2003; Haine and Hall, 2002; Hall and Haine, 2002).

6.08.6 STEADY-STATE TRACERS

The distributions of steady-state tracers have traditionally been used to infer the qualitative nature of ocean circulation, tracing the origins and pathways of water masses. There is a rich history that need not be repeated here. In this discussion, the distributions of dissolved oxygen and major inorganic nutrients (nitrate, phosphate, and silicate) are not utilized. The reasoning is that although they boast a long history and large database, their involvement in the carbon system and biogeochemical processes makes them suspect in drawing inferences about physical quantities. The one exception to this is the use of quasi-conservative constructs that utilize the Redfield stoichiometry to "see through" the biogeochemical "interference" in order to reconstruct water-mass histories or origins (e.g., see the discussion of optimum multiparameter analysis in the previous section) (Broecker and Peng, 1982; Broecker *et al.*, 1976, 1980b).

6.08.6.1 Radiocarbon

Despite its nonconservative behavior, however, it is important to consider radiocarbon amongst the number of examples whereby inferences have been made regarding the magnitude of mixing in the ocean using radioactive steady-state tracers. Munk (1966) and later Craig (1969) made a simple scaling calculation to estimate the net vertical mixing rate for deep Pacific by assuming that the vertical distributions of temperature and salinity were in one-dimensional advective–diffusive balance, and adding the vertical profile of natural radiocarbon to solve for mixing and vertical advection. The argument was that there existed a one-dimensional advective–diffusive subrange between the incoming common water at ~3,000 m depth and the intermediate water above at 1,000 m. They assumed that the water between these end-members was dominated by vertical mixing and advection. Here, the vertical advection is an assumed generalized upwelling of deep waters formed by sinking in polar regions. This, in turn, must be compensated by downward diffusion of heat (and other tracers) to establish steady state.

Whereas Munk ignored biogeochemical effects on the radiocarbon profiles, Craig improved the original estimate by accounting for the biological regeneration of near-surface "modern" carbon using dissolved oxygen profiles and using Redfield stoichiometry (Craig, 1969). By iteratively solving a set of simultaneous partial differential equations, they could obtain the individual mixing and advection terms. Starting with the vertical profiles of salinity and temperature (treating them as passive tracers), one has

$$\begin{aligned}\frac{\partial T}{\partial t} &= 0 = \kappa\frac{\partial^2 T}{\partial z^2} - w\frac{\partial T}{\partial z}\\ \frac{\partial S}{\partial t} &= 0 = \kappa\frac{\partial^2 S}{\partial z^2} - w\frac{\partial S}{\partial z}\end{aligned} \tag{54}$$

whose solutions are exponential curves between the end-member values which have a characteristic scale length defined by

$$z^* = \frac{\kappa}{w} \tag{55}$$

The second constraint is given by the observed oxygen profile, which satisfies

$$\frac{\partial O_2}{\partial t} = 0 = \kappa\frac{\partial^2 O_2}{\partial z^2} - w\frac{\partial O_2}{\partial z} - J_{O_2} \tag{56}$$

which provide (given z^*) an additional relationship:

$$J^* = \frac{J_{O_2}}{w} \tag{57}$$

now given a radiocarbon concentration profile, where (Note that the radiocarbon *concentration* must be used in this relationship, rather than the traditionally reported $\Delta^{14}C$, which is roughly an isotopic ratio anomaly.)

$$\frac{\partial C_{14}}{\partial t} = 0 = \kappa\frac{\partial^2 C_{14}}{\partial z^2} - w\frac{\partial C_{14}}{\partial z} - \lambda_{14}C_{14} - RJ \tag{58}$$

Here R is the known stoichiometric ratio of carbon oxygen in organic remineralization (based on observations of AOU and $\sum CO_2$ in the oceans). From the profile, one obtains

$$\lambda^* = \frac{\lambda_{14}}{w} \tag{59}$$

Since λ_{14} is a known constant, it is possible to solve Equations (55), (57), and (59) to obtain κ, w, and J. From this they obtained a net vertical mixing rate of $\sim 10^{-4}\,m^2\,s^{-1}$ which became a canonical measure of vertical mixing in the abyssal ocean.

The distribution of natural radiocarbon, with its half-life of ~5,000 yr is unique among steady-state tracers as a diagnostic of ocean mixing and circulation. Aside from its long (but not too long) half-life, it also holds the distinction of being a very slow exchanger with the atmosphere due to the large inorganic carbonate buffer system. This latter feature will prove an interesting diagnostic of large-scale ocean models, as it is a sensitive diagnostic of the residence time of upwelled waters at the ocean surface (see Broecker and Peng, 1982; Broecker *et al.*, 1978). Moreover, this isotope is dutifully recorded in corals, providing a backwards perspective of surface ocean radiocarbon records over many centuries (e.g., Druffel, 1981). There are complications, however, to its use. First, its distribution is affected in the deep waters by remineralization of organic material raining down from above, which significantly affects its deep distributions (Craig, 1969). Second, the atmospheric inventories have been severely perturbed first by the Suess Effect (dilution with old, fossil fuel carbon (Druffel, 1981), and then subsequently by nuclear weapons produced radiocarbon (see next section). Thus, in order to effectively utilize this tracer, the different components must be sorted out (Broecker *et al.*, 1985, 1980a).

6.08.6.2 Radon-222

Within ocean sediments, the decay of uranium and thorium isotopes leads to the creation of ^{222}Rn, which is released to sedimentary pore waters and subsequently diffuses into the overlying seawater. Near the seafloor, excess ^{222}Rn can be seen against the background of a natural standing stock of this isotope in the water column, which is produced by *in situ* decay of ^{226}Ra, a long-lived and relatively uniformly distributed isotope. Because of its short half-life, the existence of this excess isotope some several hundred meters above the seafloor implies a significant flux into the bottom waters, and the shape of the profiles has been modeled as a vertical diffusive balance with radioactive decay of radon and *in situ* production from the decay of ^{226}Ra,

$$\kappa\frac{\partial^2 C_{222}}{\partial z^2} - \lambda_{222}C_{222} + \lambda_{226}C_{226} = 0 \quad (60)$$

leading to estimates of order 10^{-4}–10^{-2} m^2 s^{-1} (Broecker *et al.*, 1968; Chung and Craig, 1972; Sarmiento and Biscaye, 1986; Sarmiento and Broecker, 1980; Sarmiento *et al.*, 1978), and have been argued to anticorrelate with the vertical stability (Sarmiento *et al.*, 1976). Although such estimates support the existence of a strong effectively vertical mixing rate near the seafloor, it is not possible to distinguish a truly local vertical mixing from shearing and advection of smaller scale bottom mixed layers from nearby topographic features (e.g., see Armi, 1977; Armi and D'Asaro, 1980; Armi and Millard, 1976).

6.08.6.3 Radium

No stable isotopes of radium exist. Radium is produced by the decay of thorium isotopes in the natural decay chain sequences, and contrary to its parent thorium, is relatively soluble and conservative in seawater. There are four isotopes, with half-lives ranging from a few days (3.7 d for ^{224}Ra, and 11.4 d for ^{223}Ra) through years (5.7 yr for ^{228}Ra) to millennia (1,600 yr for ^{226}Ra). The longest-lived isotope is relatively homogeneously distributed in the global ocean, because its decay timescale is longer than the planetary circulation timescale, but exhibits enhanced abundance near coastal regions. Moreover, the shorter-lived isotopes have the rather useful boundary condition of being dominated by production in marine sediments, particularly in continental shelf/slope environments. Here, radium levels are elevated by a combination of desorption from particles (either in river water or sediments) and direct groundwater discharge. (Radium isotopes have been used recently to quantify the magnitude of tidal groundwater pumping and direct groundwater discharge to the coastal environment, a potentially significant component for geochemical budgets of some elements for the ocean (Moore, 1997, 1999).)

Because of its short half-life, the distribution of ^{224}Ra can be sensitive indicator of small-scale horizontal mixing processes. Its distribution in Long Island Sound, a narrow embayment a few tens of kilometers wide, has been used to measure mixing rates of order 5–50 m^2 s^{-1} (Torgersen *et al.*, 1996), a value compatible with the spatial scale (see Section 6.08.3, and Ledwell *et al.*, 1998). Moore (2000) has made measurements of all four isotopes in coastal waters of the Mid-Atlantic Bight. He assumed a steady-state horizontal diffusion–decay balance:

$$\kappa\frac{\partial^2 C}{\partial x^2} = \lambda C \quad (61)$$

where x is the off-shore distance, to compute mixing coefficients in the range 300–400 m^2 s^{-1} within 50 km of shore. Sarmiento *et al.*, 1982) modeled the distribution of ^{228}Ra on isopycnal surfaces in the North Atlantic assuming diffusion–decay balance, and estimated basin-scale horizontal mixing rates of ~6,000 m^2 s^{-1}. Huh and Ku (1998) have similarly modeled the zonal distribution of ^{228}Ra and ^{226}Ra in the northeast Pacific, yielding $\kappa \sim 100$ m^{2}s^{-1} within ~100 km of the coast, but substantially larger κ (~5,000 m^2 s^{-1}) up to 1,000 km off-shore. This latter value appears somewhat larger than conventional estimates on these space scales (e.g., see Jenkins, 1991, 1998; Ledwell *et al.*, 1998; Robbins *et al.*, 2000), and is likely a reflection of the author's choice of the simple diffusion–decay model to model the radium distributions, ignoring the role of gyre-scale circulation and advection.

6.08.6.4 Argon-39

This extremely rare isotope of argon is produced by cosmic ray interactions in the atmosphere, and decays with a half-life of 269 yr. It offers some intriguing advantages as a steady-state tracer, most notably the ideal length of its half-life (making it useful on decade to millennial timescales) and its biogeochemical inertness (making it conservative, and hence simple). Cosmic ray production rate of this isotope have been roughly constant ($\pm 7\%$) over the past 1,000 years, and its gas exchange characteristics are straightforward. The biggest problem with this tracer is the challenge of obtaining and processing large seawater samples ($\sim 1\ m^3$ water is required per sample) and the extreme difficulty making the very low background radiometric measurements themselves. These challenges conspire to limit the amount of data available, but results in the deep eastern North Atlantic show consistency between radiocarbon and ^{39}Ar (Schlitzer *et al.*, 1985) estimates of deep water renewal rates, mixing and circulation in that area.

6.08.6.5 Dissolved Atmospheric Argon

Curiously, it appears that an inert, nonradioactive gas can provide a measure of vertical mixing rates in the ocean. This arises from the fact that seasonal heating in the subtropics leads to a supersaturation of this gas in the upper 50–100 m. The exhalation of this gas during summer months is incomplete, being restricted by stratification and slow mixing across the seasonal thermocline. Combining observations with an upper ocean seasonal model (the Price–Weller–Pinkel model modified for gas exchange: Musgrave *et al.*, 1988) reveals that the shallow argon evolution over summer months constrains vertical mixing rates near the sea surface (Spitzer and Jenkins, 1989). The mixing rate obtained, $\sim 10^{-4} m^2\ s^{-1}$ is somewhat higher than thermocline mixing rates (e.g., see Ledwell *et al.*, 1993), but smaller than mixed layer rates.

6.08.7 TRANSIENT TRACERS

Embellishing the categorization made in Section 6.08.2, transient tracers can be defined as substances whose concentration distributions are changing on interannual timescales and in a noncyclical nature. Thus, dissolved oxygen or argon in the seasonal layer would be excluded from this discussion, despite the fact that their concentrations change with time due to seasonal forcing and the passage of events such as phytoplankton blooms or storms. I refer instead to tracers whose distributions have been affected on regional or global scales by humankind's activities.

Of the tracers discussed here, there have been three generic release styles. They are:

- *Nuclear weapons testing fallout*, which is characterized by a pulse-like injection into the atmosphere predominantly in the 1950s and 1960s, and includes the radioisotopes ^3H (tritium), ^{14}C (radiocarbon), ^{90}Sr, ^{129}I, ^{137}Cs, etc.
- *Industrial and domestic release* of substances associated with societal or military activities, such as fossil fuel CO_2, chlorofluorocarbons (CFCs) from a variety of sources, ^{85}Kr from nuclear fuel reprocessing.
- *Point source and episodic releases* of primarily radioisotopes from nuclear fuel reprocessing (e.g., Windscale effluent) and from accidents (e.g., the Chernobyl disaster).

Figure 6 is a schematic of the time history of representative tracers characterized by the first two types of release styles for the northern hemisphere. Three of the tracers (radiocarbon and the CFCs) are gaseous phase tracers, and are consequently relatively homogeneously distributed between the hemispheres, with the northern hemisphere leading the southern hemisphere by 1–2 yr: both CFCs and radiocarbon were released preferentially in the northern hemisphere. Radiocarbon was released from atmospheric nuclear weapons tests that were almost solely in the northern hemisphere, and CFC releases were dominated by North American, European, and Soviet production. The time lag

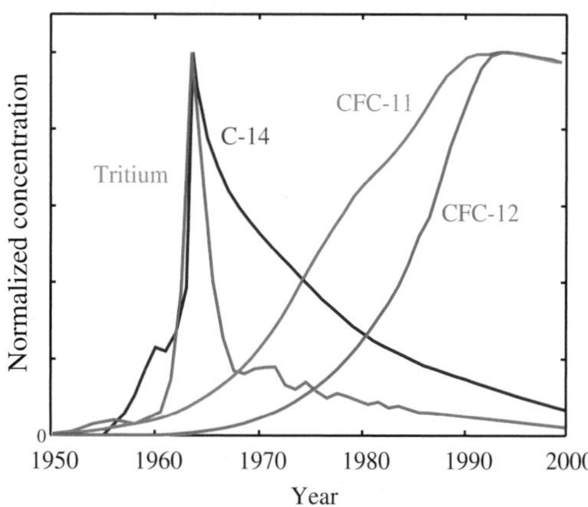

Figure 6 Normalized selected transient tracer histories for the northern hemisphere.

between northern- and southern-hemispheric abundances is characteristic of the exchange timescale between the two hemispheres in the troposphere.

Tritium, however, was largely released directly to the stratosphere, where it is oxidized to tritiated water vapor. It is subseqently mixed back into the troposphere, predominantly at mid-latitudes and rapidly rained out as tritiated rainfall (e.g., see Weiss and Roether, 1980). Thus, there is a far greater asymmetry between the two hemispheres, and characteristically different "deposition" histories. Figure 7 shows the first two EOFs of the tritium in precipitation as analyzed by Doney (Doney *et al.*, 1992) from IAEA data. The pattern decomposes into two dominant factors: a northern impulse (the largest component) and a smaller, more diffuse southern component.

Another difference arises from the way in which the tracers are introduced into the ocean. CFCs have the simplest boundary conditions, as they dissolve as inert gases, following gas exchange and solubility rules (Warner *et al.*, 1996; Warner and Weiss, 1985). A typical gas exchange timescale for CFCs is of order 1–2 months, depending on wind speed and mixed-layer depth. Radiocarbon also enters the ocean via gas exchange (as CO_2), but its gas exchange timescale is amplified by the large isotopic inertia associated with the carbonate system in seawater. Thus, the gas exchange timescale for radiocarbon is much longer, i.e., of order 10–12 yr (Broecker and Peng, 1982). Tritium, however, is deposited by both direct precipitation and water vapor exchange:

$$D_{\rm atm} = PC_{\rm P} + E\frac{h}{1-h}C_{\rm V} - E\frac{1}{\alpha(1-h)}C_{\rm S} \quad (62)$$

where P and E are the precipitation and evaporation rates, $C_{\rm P}$, $C_{\rm V}$, and $C_{\rm S}$ are the tritium concentrations in precipitation, atmospheric water vapor, and the sea surface, respectively, and where h and α are the relative humidity and isotopic fractionation factor (Weiss and Roether, 1980).

The impact of the difference in the boundary conditions between tracers can be seen in Figure 8, which compares the meridional distributions of tritium and CFC-11 in the central Pacific. The common element of both distributions is the tongue of elevated tracer concentrations being subducted in the northern subtropics, advected equatorward, and upwelled in the tropics. This feature was first noted in tritium by Fine *et al.* (1981, 1983, 1987) and places important constraints on the exchange timescales for subtropical–tropical overturning in the Pacific. This, in turn is potentially an important regulatory element

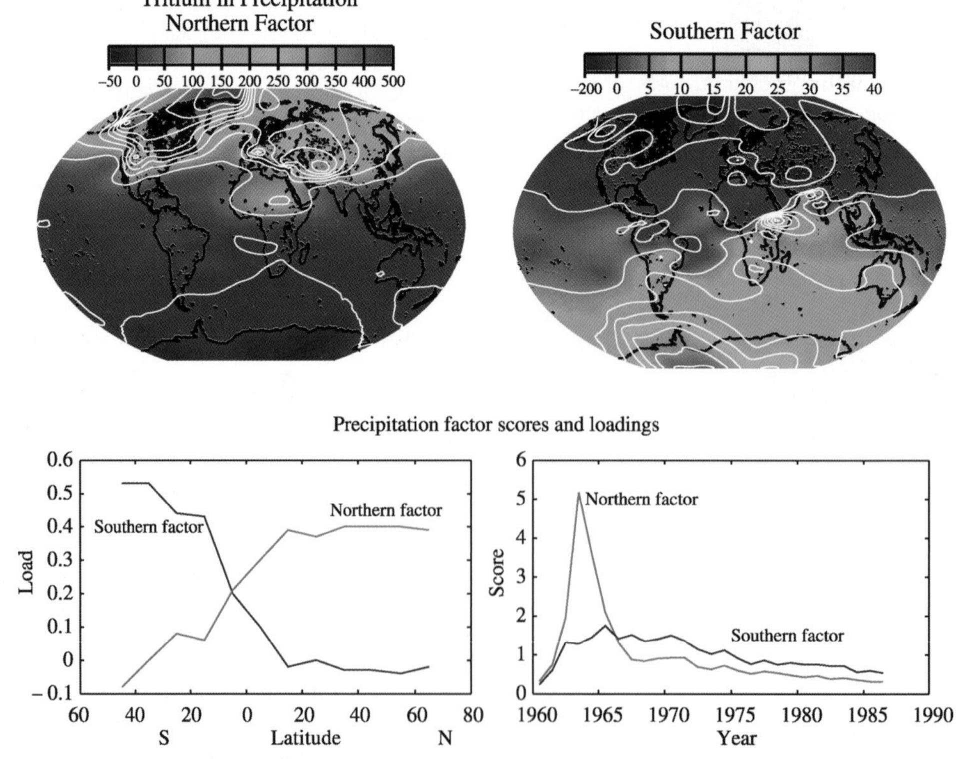

Figure 7 The northern and southern EOFs for tritium concentrations in precipitation (after Doney *et al.*, 1992).

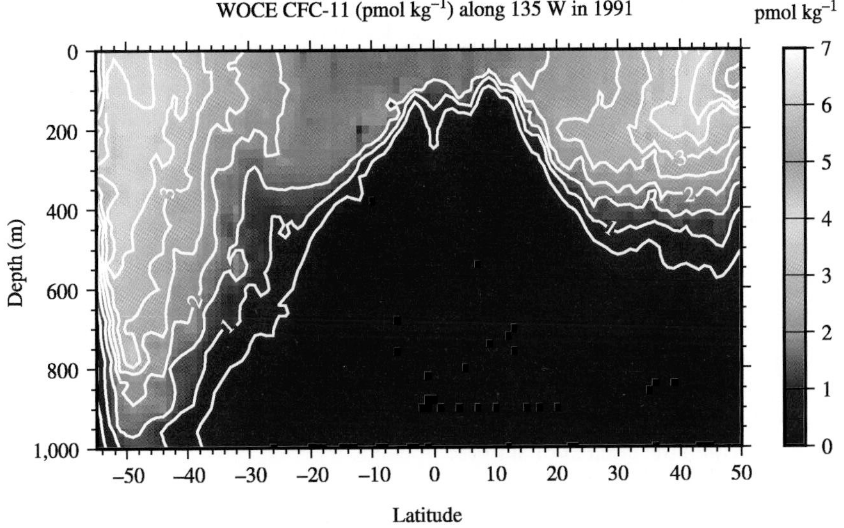

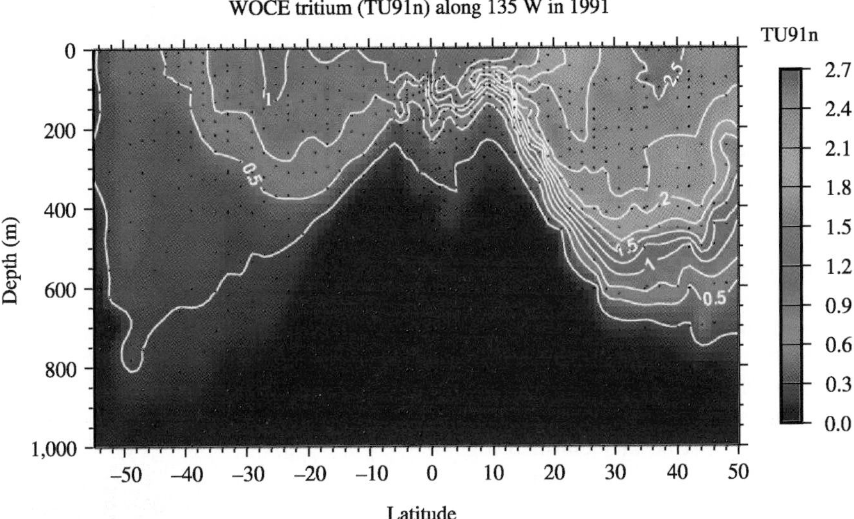

Figure 8 The meridional distributions of CFC-11 (upper panel) and tritium (lower panel) along 135° W in the Pacific during the WOCE expedition. CFC data were obtained from the WHPO data release no. 3 (see also Bullister *et al.*, 2000; Fine *et al.*, 2001; Schlosser, *et al.*, 2001).

in decadal variations in El Nino/ENSO strength (An and Jin, 2001; Deser *et al.*, 1996; Gu and Philander, 1997; Guilderson and Schrag, 1997; Zhang *et al.*, 1998).

The contrast between these tracers is rooted in the pronounced meridional asymmetry in the tritium pattern (lower-southern-hemispheric values), compared to the more balanced CFC distribution. Another aspect is the northern mid-latitude maximum in tritium compared to a poleward enhancement of CFC-11. The latter is a result of the temperature dependence of CFC solubility, but the former arises from the progressive sequestration of tritium in the subtropical gyres, a phenomenon similar to that of radiocarbon (see Broecker *et al.*, 1985), and driven by the unique temporal history and depositional mode of the tracer. Such contrasting behavior is an important attribute that can be exploited in mixing calculations and model evaluation.

Multitracer approaches offer the most powerful constraints on mixing and ventilation. For example, Schlosser and co-workers have exploited the differing time histories and boundary conditions of a suite of tracers (notably tritium, ^{3}He, CFCs, radiocarbon, and ^{39}Ar) to study Arctic ventilation and mixing (Schlosser *et al.*, 1990, 1994, 1995b), and to demonstrate climatic variations in deep water formation in the Greenland/Norwegian Seas (Schlosser *et al.*, 1991, 1995a; Schlosser and Smethie, 1995).

There are four basic, but overlapping approaches to using transient tracers. They include:

- *Flow visualization,* where the tracer is, and where it is not, what pathways does it trace on entering the ocean, and what scales of dilution are occurring.
- *Direct "age" computation*, based on simple aging or vintage models (see Sections 6.08.4 and 6.08.8). The argument is that the tracer age distribution resembles the *ideal* age distribution to some approximation.
- *Diagnostic calculations,* using observed tracer distributions (e.g., tracer or age gradients, or relationships with other tracers) it may be possible to calculate mixing, velocity or ventilation rates directly within the context of simple advective–diffusive or box models.
- *Comparison with prognostic models,* by comparing observed tracer fields with model simulations, it may be possible to improve choice of model parametrizations, identify regions or processes where the model does/ does not perform well

To some extent, all of these approaches have been used with some success over recent years. In the remainder of this section, I discuss some illustrative examples, with emphasis on the third of these approaches.

There are many examples of how transient tracers have been used to visualize ventilation pathways, but perhaps the most striking are the sections across the deep western boundary current (DWBC) in the North Atlantic. An early section of tritium across the DWBC (Jenkins and Rhines, 1980) revealed a strikingly outlined high tritium core which indicated the rapid southward flow of recently ventilated water. There were a number of significant attributes of this observation, however: (i) the dislocation of the tracer maximum from the high velocity core, (ii) the scale of dilution (about 10 : 1) relative to the source waters, and (iii) the remarkable absence of Labrador Sea Water tritium. The first two features hinted at the scale of the recirculation impact on entrainment and detrainment in the DWBC and the resultant downstream tracer propagation speed (see Doney and Jenkins, 1994), and the last underlined the importance of climatic variations in Labrador Sea Water formation. This was shown clearly in subsequent DWBC transections by Smethie, Fine, and others (Fine and Molinari, 1988; Pickart *et al.*, 1989, 1996; Smethie, 1993).

The penetration of tritium into the thermocline has been used to estimate vertical and diapycnal mixing rates. The first quantitative attempt was by Rooth and Ostlund (1972), who utilized an empirical relationship between bomb-tritium and temperature in the main thermocline to construct a pair of coupled advection–diffusion equations, and use the empirical relationship to eliminate advective terms in the equations, and solve for mixing rates. Starting with the advection–diffusion equations for tritium and temperature:

$$\frac{\partial C}{\partial t} = \kappa_\rho \nabla_\rho^2 C - \vec{u}\cdot\nabla_\rho C - \lambda C$$
$$0 = \kappa_\rho \nabla_\rho^2 T - \vec{u}\cdot\nabla_\rho T \quad (63)$$

they recognized a log–log relationship between the temperature and tritium within the thermocline:

$$\phi = \mu\theta + \phi_0 \quad (64)$$

where

$$\phi = \ln C \quad \text{and} \quad \theta = \ln(T - 2.3) \quad (65)$$

The choice of the 2.3 °C offset in the temperature transformation was predicated on the assumption that the vertical temperature distribution was pinned at the lower end by North Atlantic Deep Water flow. Transforming and rearranging Equations (63) with the definitions in (65) gives

$$\frac{\partial \phi}{\partial t} + \vec{u}\cdot\nabla_\rho\phi + \lambda = \kappa_\rho((\nabla_\rho\phi)^2 + \nabla_\rho\kappa\cdot\nabla_\rho\phi + \nabla_\rho^2\phi)$$
$$\vec{u}\cdot\nabla_\rho\theta = \kappa_\rho((\nabla_\rho\theta)^2 + \nabla_\rho\kappa\cdot\nabla_\rho\theta + \nabla_\rho^2\theta) \quad (66)$$

and coupling the equations with the empirical relationship expressed in (64), they obtained

$$\frac{\partial \phi}{\partial t} + \lambda = \left(\mu^2 - \mu\right)\kappa_\rho(\nabla_\rho\theta)^2$$
$$\approx \left(\mu^2 - \mu\right)\left(\frac{\kappa_D}{H^2} + \frac{\kappa_I}{L^2}\right) \quad (67)$$

which relates the time evolution of the tritium distribution in the main thermocline to the large scale diffusive temperature fluxes, and ultimately to the spatial scales (H for vertical or diapycnal length, and L for horizontal or isopycnal) of temperature gradients. Using the empirical slope $\mu \approx 5$ and the vertical thermal scale height for the main thermocline $H \approx 440$ m, they obtained

$$1.8\times10^{-5} \le \left(\kappa_D + \frac{H^2}{L^2}\kappa_I\right)$$
$$\le 2.3\times10^{-5}\,\text{m}^2\,\text{s}^{-1} \quad (68)$$

and argued that since the second term is positive definite, sets an upper bound on diapycnal diffusion in the thermocline of ~2×10^{-5} m^2 s^{-1}.

The significance of this analysis was that it was the first tracer-based estimate of diapycnal diffusivity that pointed to relatively low mixing rates in the thermocline (2×10^{-5} m^2 s^{-1}), smaller than the canonical values derived from the earlier "abyssal recipes" papers, and much smaller than average *vertical* mixing rates estimated by simple one-dimensional calculations (e.g., see Li *et al.*, 1984). These lower values were supported by consideration of the relationship between bomb-tritium and its daughter isotope ^{3}He in the thermocline (Jenkins, 1980), by a subsequent analysis of tritium profiles in the North Pacific (Kelley and Van Scoy, 1999), and by tracer release experiments in the North Atlantic (see Section 6.08.9). Resolution of this apparent discrepancy lies in the recognition that the one-dimensional balance calculation incorporates an *effective* vertical mixing rate that is a composite between true diapycnal mixing and isopycnal mixing along sloping isopycnal surfaces. Put another way, the vertical mixing rate of Li *et al.* (Li and Garrett, 1997) is dominated by off-diagonal elements in the eddy diffusion tensor in Equation (12).

Considerable progress has been made in comparing models to tracer distributions, both for radiocarbon (Rodgers *et al.*, 1999) and for CFCs (England, 1995b; England *et al.*, 1994; England and Maier-Reimer, 2001). The ability of models to replicate tracer observations is a strong function of mixing parametrization. The success and utility of future comparisons will depend strongly on the existence of a uniformly high-quality data set of as many tracers as possible, and an improved understanding of the boundary conditions and histories of these tracers.

6.08.8 TRACER AGE DATING

Although a compelling intuitive concept for yielding insights into ocean ventilation (see Figure 4) tracer age dating is affected by the nonideal behavior in the presence of mixing. This significantly complicates interpretation(e.g., see Doney *et al.*, 1997; Robbins and Jenkins, 1998), but this complication can be mastered and exploited in a constructive way. An example of this is a study carried out in the "subduction area," a well-surveyed region in the eastern subtropical North Atlantic. The distributions of tritium and ^{3}He were determined over an $\sim$1,500 km by 1,500 km area and analyzed on isopycnal surfaces. The approach was to combine information provided by geostrophic balance (relative to some deeper reference level) with advective–diffusive balance for tritium–helium age and salinity. The subsurface distribution of the tracers is demonstrably dominated by isoycnal processes, so the balance on a given isopycnal can be expressed as

$$\frac{\partial \tau}{\partial x}u_{\mathrm{r}} + \frac{\partial \tau}{\partial y}v_{\mathrm{r}} - (\nabla_\rho^2 \tau + \nabla \ln(\zeta\vartheta)\cdot\nabla\tau)\kappa_\rho = 1 - \frac{\partial \tau}{\partial t} - \frac{\partial \tau}{\partial x}u_{\mathrm{g}} - \frac{\partial \tau}{\partial y}v_{\mathrm{g}} \qquad (69)$$

and

$$\frac{\partial S}{\partial x}u_{\mathrm{r}} + \frac{\partial S}{\partial y}v_{\mathrm{r}} - (\nabla_\rho^2 S)\kappa_\rho = 1 - \frac{\partial S}{\partial x}u_{\mathrm{g}} - \frac{\partial S}{\partial y}v_{\mathrm{g}} \qquad (70)$$

where x and y refer to the along isopycnal east and northward directions, and the subscripts r and g refer to the reference-level and geostrophic velocity components. The terms on the right-hand side of Equations (69) and (70) are known (measured) for each surface, along with the gradients and Laplacians on the left-hand side. The unknowns are the reference level velocity components u_{r} and v_{r}, and the isopycnal diffusivity κ_ρ, the latter of which is allowed to be a function of isopycnal surface. Thus, for a selection of N isopycnal surfaces, one obtains a set of $2N$ equations with $N+2$ unknowns (N isopycnal diffusivities, and 2 reference-level velocity components). This may be solved in a least squares sense to obtain both absolute velocities, and isopycnal mixing rates as a function of depth (see Figure 9), along with their statistical uncertainties.

Note the general structure of the isopycnal mixing rate as a function of depth: the gradual decrease into the deeper thermocline is consistent with expectations, but the lower values near the surface appear problematic. This surface decrease is likely a result of ignoring possible unsteady terms in the salinity Equation (70) on the shallowest surfaces.

An additional interpretive strategy is to analyze the temporal evolution of the tracer age field. Although the magnitude of the nonlinear terms in Equation (33) are now small (Jenkins, 1987, 1998), the early evolution of the tritium transient was characterized by large nonlinearities, so that the present age distribution is still relaxing on deeper density horizons, where adjustment times are large (Robbins and Jenkins, 1998). This can be used to advantage, as the temporal evolution, and the spatial structure of the age fields become sensitive to mixing rates and ventilation pathways. This has been used to quantify the mixing of material across the Azores front, and its role in ventilation of the deep thermocline (Robbins *et al.*, 2000).

6.08.9 TRACER RELEASE EXPERIMENTS

Dye release experiments have been performed in the marine environment for many decades, but

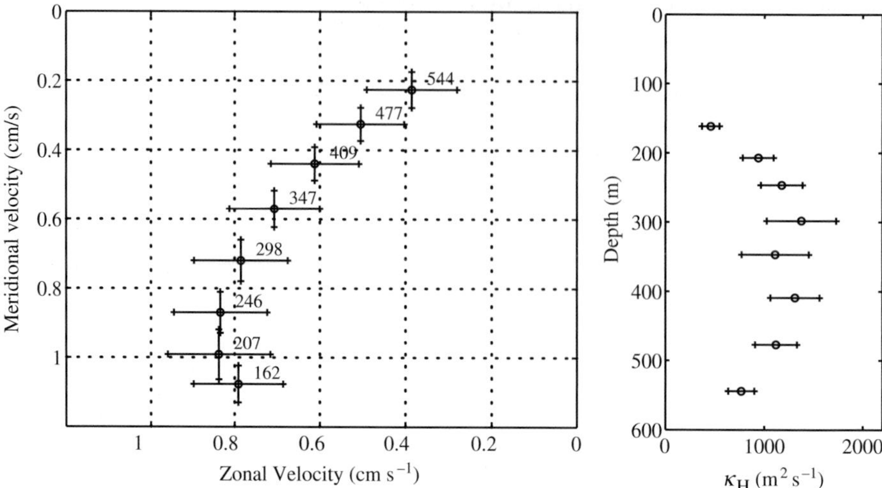

Figure 9 Tracer age estimated absolute velocities (left panel, with numbers indicating depth) and isopycnal mixing coefficients (right panel).

suffered from limitations in the scale of the release, and the complication of nonconservative behavior in some of the dyes used. However, recent technical improvements in deployment and measurement technology has resulted in a series of definitive experiments on a large enough scale to compare with traditional tracer measurements. Early experiments releasing SF_6 were performed in the Santa Monica Basin (Ledwell and Watson, 1991) which revealed a relatively weak interior diapycnal mixing rate ($\sim 2.5 \times 10^{-5}$ m^2 s^{-1}). Once the tracer had reached the basin walls, however, boundary mixing contributed to its vertical spread, leading to a substantially larger effective mixing rate (Ledwell and Bratkovich, 1995; Ledwell and Hickey, 1995).

The SF_6 release experiment was scaled up and performed at a depth of ~300 m in the North Atlantic main thermocline of the subduction area in 1992 (Ledwell *et al.*, 1993), where the dye patch was monitored for 2.5 yr. During this time, the distribution of dye in the vertical evolved as a spreading Gaussian profile, whose vertical scale increased with time consistent with a diapycnal mixing rate of initially $(1.2 \pm 0.2) \times 10^{-5}$ m^2 s^{-1} and subsequently a somewhat larger rate of $(1.7 \pm 0.2) \times 10^{-5}$ m^2 s^{-1}, confirming the earlier tracer work, and being consistent with microstructure based estimates made locally (St. Laurent and Schmitt, 1999).

Analysis of the early spread of the SF_6 dye patch on isopycnals revealed isopycnal mixing rates of order 0.07 m^2 s^{-1} for spatial scales of 100–1,000 m, which increased to 2 m^2 s^{-1} on 1–10 km scales. At this intermediate stage, the character of the tracer patch was dominated by straining and streaking by the mesoscale eddy field, and the streaks grew at an exponential rate. At the terminal stages of the experiment, mesoscale eddies dominated mixing, with effective mixing rates approaching 1,000 m^2 s^{-1} (Ledwell *et al.*, 1998), values consistent with float statistics obtained during the experiment (Sundermeyer and Price, 1998).

The presence of relatively weak diapycnal mixing below the main thermocline, however, appears problematic in terms of interpreting the large-scale one-dimensional balances studied by Munk and Craig. Turbulent dissipation measurements reveal that diapycnal mixing is weak over smooth ocean floor, but highly enhanced over rough topography (St. Laurent *et al.*, 2001; Toole *et al.*, 1994) due to tidal interaction with terrain. This was dramatically demonstrated by another tracer release experiment, this time in the abyssal Brazil Basin (Ledwell *et al.*, 2000; Polzin *et al.*, 1997) where mixing rates of $\sim(2-4) \times 10^{-4}$ m^2 s^{-1} were observed in the water column over rough topography, and values approaching 10^{-3} m^2 s^{-1} were reached near the seafloor. Such values, when regionally averaged, lead to sufficiently large mean mixing rates to close buoyancy budgets for the abyssal Brazil Basin based on large-scale hydrography (Morris *et al.*, 2001).

6.08.10 CONCLUDING REMARKS

Thinking about the use of tracers to study ocean mixing has evolved since the halcyon days of GEOSECS. The simple questions of "what is the vertical mixing rate?" and "what is the age of the deep water?" have become much more complex with the realization that the tools we are using require considerable skill and have intrinsic limitations. The oceans and the practice of oceanography have become much more complicated with time. However, the pictures of evolving transient tracer fields that are emerging from WOCE and CLIVAR efforts are compelling and evocative. The large-scale structures are both reassuringly consistent with what we know about ocean ventilation, circulation and mixing, and yet also show a wealth of detail and structure that promises more insight into physical processes in the ocean. The challenge is utilizing this information in an effective and quantitative way.

The ocean modeling community is beginning to provide us with simulated tracer fields that are beginning to more realistically resemble the observations, but with informative differences. Modelers are increasingly turning to tracers (beyond temperature and salinity) for model evaluation, testing and validation (England and Maier-Reimer, 2001). The areas of agreement and disparity are important diagnostics of model performance and parametrization. Such tests are critical if reliable global change predictions are to be produced by such simulations. The ability of such models to "get the right answer for the wrong reasons" is a serious concern, but the risk becomes increasingly remote as additional independent tracer fields are added as constraints.

A limitation of tracer constraints on oceanic processes is the sparseness of data, and imperfectly known boundary conditions such as air–sea behavior and history. We should consider improving global coverage of tracer measurements, as well as targeted process-oriented experiments to improve on the former. We also ought to encourage activities that make inroads in the latter. The primary advantage of transient tracers lies in the time evolution of their distributions. This, coupled with the clear indication that the physical environment is changing as well, leads us to the conclusion that repeat observations are not just a luxury, but a necessity.

REFERENCES

An S.-I. and Jin F.-F. (2001) Collective role of thermocline and advective feedbacks in the ENSO mode. *J. Climatol.* **14**, 3421–3432.

Armi L. (1977) Dynamics of the bottom boundary layer of the deep ocean. In *Bottom Turbulence, Proceedings of the 8th International Liege Colloqium on Ocean Hydrodynamics* (ed. J. C. J. Nihoul). Elsevier, New York, pp. 153–164.

Armi L. (1979) The effects of variations in eddy diffusivity on property distributions in the ocean. *J. Mar. Res.* **37**, 515–530.

Armi L. and D'Asaro E. A. (1980) Flow structures of the Benthic ocean. *J. Geophys. Res.* **85**(C1), 469–484.

Armi L. and Millard R. C. (1976) The bottom boundary layer of the deep ocean. *J. Geophys. Res.* **81**(27), 4983–4990.

Armi L. and Stommel H. (1983) Four views of a portion of the North Atlantic subtropical gyre. *J. Phys. Oceanogr.* **13**, 828–857.

Arons A. B. and Stommel H. (1967) On the abyssal circulation of the world ocean: III. An advection-lateral mixing model of the distribution of a tracer property in an ocean basin. *Deep-Sea Res.* **14**, 441–457.

Broecker W. S. and Peng T. H. (1982) *Tracers in the Sea.* Eldigio Press, Palisades, NY.

Broecker W. S., Cromwell J., and Li Y.-H. (1968) Rates of vertical eddy diffusion near the ocean floor based on measurements of the distribution of excess ^{222}Rn. *Earth Planet. Sci. Lett.* **5**, 101–105.

Broecker W. S., Takahashi T., and Li Y.-H. (1976) Hydrography of the central Atlantic: I. The two-degree discontinuity. *Deep-Sea Res.* **23**, 1083–1104.

Broecker W. S., Peng T. H., and Stuiver M. (1978) An estimate of the upwelling rate in the equatorial Atlantic based on the distribution of bomb radiocarbon. *J. Geophys. Res.* **83**(C12), 6179–6186.

Broecker W. S., Peng T. H., and Takahashi T. (1980a) A strategy for the use of bomb-produced radiocarbon as a tracer for the transport of fossil fuel CO_2 into the deep-sea source regions. *Earth Planet. Sci. Lett.* **49**, 438–463.

Broecker W. S., Takahashi T., and Stuiver M. (1980b) Hydrography of the central Atlantic: II. Waters beneath the two-degree discontinuity. *Deep-Sea Res.* **27A**, 397–419.

Broecker W. S., Peng T. H., Ostlund H. G., and Stuiver M. (1985) The distribution of bomb radiocarbon in the ocean. *J. Geophys. Res.* **90**(C4), 6953–6970.

Bryan F. (1987) Parameter sensitivity of primitive equation ocean general circulation models. *J. Phys. Oceanogr.* **17**(7), 970–985.

Bullister J. L., Fine R. A., Smethie W. M., Warner M. J., and Weiss R. F. (2000) Global integration and interpretation of WOCE CFC data. *USWOCE Report* **12**, 21–25.

Chung J.-Y. and Craig H. (1972) Excess radon and temperature profiles from the eastern equatorial Pacific. *Earth Planet. Sci. Lett.* **14**, 55–64.

Craig H. (1969) Abyssal carbon and radiocarbon in the Pacific. *J. Geophys. Res.* **74**(23), 5491–5506.

Csanady G. T. (1972) *Turbulent Diffusion in the Environment.* Reidel, Boston, MA.

Deleersnijder E., Campin J.-M., and Delhez E. J. M. (2001) The concept of age in marine modelling: I. Theory and preliminary model results. *J. Mar. Syst.* **28**, 229–267.

Deleersnijder E., Mouchet A., Delhez E. J. M., and Beckers J.-M. (2002) Transient behavior of water ages in the world ocean. *Math. Comp. Model.* **36**, 121–127.

Delhez E. J. M., Deleersnijder E., Mouchet A., and Beckers J.-M. (2003) A note on the age of radioactive tracers. *J. Mar. Syst.* **38**, 277–286.

Deser C., Alexander M. A., and Timlin M. S. (1996) Upper-ocean thermal variations in the North Pacific during 1970–1991. *J. Clim.* **9**, 1840–1855.

Doney S. C. and Jenkins W. J. (1988) The effect of boundary conditions on tracer estimates of thermocline ventilation rates. *J. Mar. Res.* **46**, 947–965.

Doney S. C. and Jenkins W. J. (1994) Ventilation of the deep western boundary current and abyssal western North Atlantic: estimates from tritium and ^{3}He distributions. *J. Phys. Oceanogr.* **24**, 638–659.

Doney S. C., Glover D. M., and Jenkins W. J. (1992) A model function of the global bomb-tritium distribution in precipitation, 1960–1986. *J. Geophys. Res.* **97**, 5481–5492.

Doney S. C., Jenkins W. J., and Bullister J. L. (1997) A comparison of ocean tracer dating techniques on a meridional section in the eastern North Atlantic. *Deep-Sea Res.* **44**(4), 603–626.

Druffel E. R. M. (1981) Radiocarbon in annual coral rings from the eastern tropical Pacific Ocean. *Geophys. Res. Lett.* **8**(1), 59–61.

Eckart C. (1948) An analysis of the stirring and mixing processes in incompressible fluids. *J. Mar. Res.* **7**(3), 265–275.

England M. H. (1995a) The age of water and ventilation timescales in a global ocean model. *J. Phys. Oceanogr.* **25**(11), 2756–2777.

England M. H. (1995b) Using chlorofluorcarbons to assess ocean climate models. *Geophys. Res. Lett.* **22**(22), 3051–3054.

England M. H. and Maier-Reimer E. (2001) Using chemical tracers to assess ocean models. *Rev. Geophys.* **39**(1), 29–70.

England M. H., Garcon V., and Minster J.-F. (1994) CFC uptake in a world ocean model: 1. Sensitivity to the surface gas forcing. *J. Geophys. Res.* **99**(C12), 25215–25233.

Fine R. A. and Molinari R. L. (1988) A continuous deep western boundary current between Abaco (26.5N) and Barbados (13N). *Deep-Sea Res.* **35**(9), 1441–1450.

Fine R. A., Reid J. L., and Ostlund H. G. (1981) Circulation of tritium in the Pacific Ocean. *J. Phys. Oceanogr.* **11**, 3–14.

Fine R. A., Peterson W. H., Rooth C. G., and Ostlund H. G. (1983) Cross-equatorial tracer transport in the upper waters of the Pacific Ocean. *J. Geophys. Res.* **88**, 763–769.

Fine R. A., Peterson W. H., and Ostlund H. G. (1987) The penetration of tritium into the tropical Pacific. *J. Phys. Oceanogr.* **17**, 553–564.

Fine R. A., Maillet K. A., Sullivan K. F., and Willey D. (2001) Circulation and ventilation flux of the Pacific Ocean. *J. Geophys. Res.* **106**(C10), 22159–22178.

Gargett A. E. (1984) Vertical eddy diffusivity in the ocean interior. *J. Mar. Res.* **42**, 359–393.

Gargett A. E. and Holloway G. (1984) Dissipation and diffusion by internal wave breaking. *J. Mar. Res.* **42**(1), 15–27.

Garrett C. (1979) Mixing in the ocean interior. *Dyn. Atmos. Oceans* **3**, 239–265.

Garrett C. (1983) On the initial streakiness of a dispersing tracer in two- and three-dimensional turbulence. *Dyn. Atmos. Oceans* **7**, 265–277.

Garrett C. (1989) A mixing length interpretation of fluctuations in passive scalar concentration in homogeneous turbulence. *J. Geophys. Res.* **94**(C7), 9710–9712.

Gent P. R. and McWilliams J. C. (1990) Isopycnal mixing in ocean circulation models. *J. Phys. Oceanogr.* **20**, 150–155.

Gent P. R., Willebrand J., McDougall T. J., and McWilliams J. C. (1995) Parameterizing eddy-induced tracer transports in ocean circulation models. *J. Phys. Oceanogr.* **25**, 463–474.

Gu D. and Philander S. G. (1997) Interdecadal climate fluctuations that depend on exchanges between the tropics and extratropics. *Science* **275**, 805–807.

Guilderson T. and Schrag D. (1997) Abrupt shift in subsurface temperatures in the tropical Pacific associated with changes in El Niño. *Science* **281**, 240–243.

Haine T. W. N. and Hall T. M. (2002) A generalized transport theory: water-mass composition and age. *J. Phys. Oceanogr.* **32**, 1932–1946.

Hall T. M. and Haine T. W. N. (2002) On ocean transport diagnostics: the idealized age tracer and the age spectrum. *J. Phys. Oceanogr.* **32**(6), 1987–1991.

Hamann I. M. and Swift J. H. (1991) A consistent inventory of water mass factors in the intermediate and deep Pacific Ocean derived from conservative tracers. *Deep-Sea Res.* **38**(suppl.1), S129–S169.

Huh C.-A. and Ku T.-L. (1998) A 2-D section of ^{228}Ra and ^{226}Ra in the northeast Pacific. *Oceanolog. Acta* **21**(4), 533–542.

Jenkins W. J. (1980) Tritium and ^{3}He in the Sargasso Sea. *J. Mar. Res.* **38**, 533–569.

Jenkins W. J. (1984) The use of tracers and water masses to estimate rates of respiration. In *Heterotrophic Activity in the Sea* (eds. J. E. Hobbie and P. L. Williams). Plenum, New York, pp. 391–403.

Jenkins W. J. (1987) ^{3}H and ^{3}He in the beta triangle: observations of gyre ventilation and oxygen utilization rates. *J. Phys. Oceanogr.* **17**, 763–783.

Jenkins W. J. (1988) Using anthropogenic tritium and 3He to study subtropical gyre ventilation and circulation. *Phil. Trans. Roy. Soc. London* **A325**, 43–61.

Jenkins W. J. (1991) Determination of isopycnal diffusivity in the Sargasso Sea. *J. Phys. Oceanogr.* **21**, 1058–1061.

Jenkins W. J. (1998) Studying thermocline ventilation and circulation using tritium and ^{3}He. *J. Geophys. Res.* **103**, 15817–15831.

Jenkins W. J. and Rhines P. B. (1980) Tritium in the deep North Atlantic Ocean. *Nature* **286**, 877–880.

Jenkins W. J. and Wallace D. W. R. (1992) Tracer based inferences of new primary production in the sea. In *Primary Production and Biogeochemical Cycles in the Sea* (eds. P. G. Falkowski and A. D. Woodhead). Plenum, New York, pp. 299–316.

Joyce T. M. and Jenkins W. J. (1993) Spatial variability of subducted water in the North Atlantic. *J. Geophys. Res.* **98**, 10111–10124.

Kelley D. and Van Scoy K. A. (1999) A basin-wide estimate of vertical mixing in the upper pycnocline: spreading of bomb tritium in the North Pacific Ocean. *J. Phys. Oceanogr.* **19**, 1759–1771.

Kunze E. and Sanford T. B. (1996) Abyssal mixing: where it is not. *J. Phys. Oceanogr.* **26**(10), 2286–2296.

Ledwell J. R. and Bratkovich A. (1995) A tracer study of mixing in the Santa Cruz basin. *J. Geophys. Res.* **100**(C10), 20681–20704.

Ledwell J. R. and Hickey B. M. (1995) Evidence for enhanced boundary mixing in the Santa Monica basin. *J. Geophys. Res.* **100**(C10), 20665–20679.

Ledwell J. R. and Watson A. (1991) The Santa Monica basin tracer experiment: a study of diapycnal and isopycnal mixing. *J. Geophys. Res.* **96**(C5), 8695–8718.

Ledwell J. R., Watson A. J., and Law C. S. (1993) Evidence for slow mixing rates across the pycnocline from an open-ocean tracer-release experiment. *Nature* **364**, 701–703.

Ledwell J. R., Watson A., and Law C. S. (1998) Mixing of a tracer in the pycnocline. *J. Geophys. Res.* **103**(C10), 21499–21529.

Ledwell J. R., Montgomery E. T., Polzin K. L., St. Laurent L. C., Schmitt R. W., and Toole J. M. (2000) Evidence for enhanced mixing over rough topography in the abyssal ocean. *Nature* **403**, 179–182.

Li M. and Garrett C. (1997) Mixed layer deepening due to Langmuir circulation. *J. Phys. Oceanogr.* **27**(1), 121–132.

Li Y.-H., Peng T. H., Broecker W. S., and Ostlund H. G. (1984) The average vertical mixing coefficient for the oceanic thermocline. *Tellus* **36B**, 212–217.

Mackas D. L., Denman K. L., and Bennett A. F. (1987) Least squares multiple tracer analysis of water mass composition. *J. Geophys. Res.* **92**(C3), 2907–2918.

McDougall T. J. (1987) Neutral surfaces in the ocean: implications for modelling. *Geophys. Res. Lett.* **14**(8), 797–800.

McWilliams J. C. (1996) Modelling of the general ocean circulation. *Ann. Rev. Fluid Mech.* **28**, 215–248.

Moore W. S. (1997) High fluxes of radium and barium from the mouth of the Ganges–Brahmaputra rivers during low river discharge suggest a large groundwater source. *Earth Planet. Sci. Lett.* **150**, 141–150.

Moore W. S. (1999) The subterranean estuary: a reaction zone of groundwater and seawater. *Mar. Chem.* **65**, 111–125.

Moore W. S. (2000) Determining coastal mixing rates using radium isotopes. *Continent. Shelf Res.* **20**, 1993–2007.

Morris M. Y., Hall M. M., St. Laurent L. C., and Hogg N. G. (2001) Abyssal mixing in the Brazil basin. *J. Phys. Oceanogr.* **31**, 3331–3348.

Munk W. H. (1966) Abyssal recipes. *Deep-Sea Res.* **13**, 707–730.

Musgrave D. L. (1985) A numerical study of the roles of subgyre-scale mixing and the western boundary current on homgenization of a passive tracer. *J. Geophys. Res.* **90**(C4), 7037–7043.

Musgrave D. L. (1990) Numerical studies of tritium and ^{3}He in the thermocline. *J. Phys. Oceanogr.* **20**, 344–373.

Musgrave D. L., Chou J., and Jenkins W. J. (1988) Application of a model of upper-ocean physics for studying seasonal cycles of oxygen. *J. Geophys. Res.* **93**, 15679–15700.

Okubo A. (1971) Oceanic diffusion diagrams. *Deep-Sea Res.* **18**, 789–802.

Olbers D., Wenzel M., and Willebrand J. (1985) The inference of North Atlantic circulation patterns from climatological hydrographic data. *Rev. Geophys.* **23**, 313–356.

Pickart R. S., Hogg N. G., and Smethie W. M. (1989) Determining the strength of the deep western boundary current using the chlorfluoromethane ratio. *J. Phys. Oceanogr.* **19**(7), 940–951.

Pickart R. S., Smethie W. M., Lazier J. R. N., Jones E. P., and Jenkins W. J. (1996) Eddies of newly formed upper Labrador Sea Water. *J. Geophys. Res.* **101**, 20711–20726.

Polzin K. L., Toole J. M., and Schmitt R. W. (1995) Finescale parametrizations of turbulent dissipation. *J. Phys. Oceanogr.* **25**, 306–328.

Polzin K. L., Toole J. M., Ledwell J. R., and Schmitt R. W. (1997) Spatial variability of turbulent mixing in the abyssal ocean. *Science* **276**, 93–96.

Poole R. and Tomczak M. (1999) Optimum multiparameter analysis of the water mass structure in the Atlantic Ocean thermocline. *Deep-Sea Res.* **46**, 1895–1921.

Quay P., Broecker W. S., Hesslein R. H., and Schindler D. W. (1980) Vertical diffusion rates determined by tritium tracer experiments in the thermocline and hypolimnion of two lakes. *Limnol. Oceanogr.* **25**(2), 201–218.

Redi M. H. (1982) Oceanic isopycnal mixing by coordinate rotation. *J. Phys. Oceanogr.* **12**, 1154–1158.

Rhines P. B. and Young W. R. (1983) How rapidly is a passive scalar mixed within closed streamlines. *J. Fluid Mech.* **133**, 133–145.

Robbins P. E. and Jenkins W. J. (1998) Observations of temporal changes of tritium-^{3}He age in the eastern North Atlantic thermocline: evidence for changes in ventilation? *J. Mar. Res.* **56**, 1125–1161.

Robbins P. E., Price J. F., Owens W. B., and Jenkins W. J. (2000) On the importance of lateral diffusion for the ventilation of the lower thermocline in the subtropical North Atlantic. *J. Phys. Oceanogr.* **30**, 67–89.

Rodgers K. B., Schrag D., Cane M. A., and Naik N. H. (1999) The bomb ^{14}C transient in the Pacific Ocean. *J. Geophys. Res.* **105**(C4), 8489–8512.

Rooth C. G. and Ostlund H. G. (1972) Penetration of tritium into the North Atlantic thermocline. *Deep-Sea Res.* **19**, 481–492.

Sarmiento J. L. and Broecker W. S. (1980) Ocean floor ^{222}Rn standing crops in the Atlantic and Pacific Oceans. *Earth Planet. Sci. Lett.* **49**, 341–350.

Sarmiento J. L. and Biscaye P. (1986) Radon 222 in the Benthic boundary layer. *J. Geophys. Res.* **91**(C1), 833–844.

Sarmiento J. L., Feely H. W., Moore W. S., Bainbridge A. E., and Broecker W. S. (1976) The relationship between vertical eddy diffusion and buoyancy gradient in the deep sea. *Earth Planet. Sci. Lett.* **32**, 357–370.

Sarmiento J. L., Broecker W. S., and Biscaye P. (1978) Excess bottom radon 222 distribution in deep ocean passages. *J. Geophys. Res.* **83**(C10), 5068–5076.

Sarmiento J. L., Rooth C. G., and Broecker W. S. (1982) Radium 228 as a tracer of basin wide processes in the abyssal ocean. *J. Geophys. Res.* **87**(C12), 9694–9698.

Schlitzer R., Roether W., Weidmann U., Kalt P., and Loosli H. H. (1985) A meridional ^{14}C and ^{39}Ar section in the northeast Atlantic deep water. *J. Geophys. Res.* **90**(C4), 6945–6952.

Schlosser P., Bonisch G., Kromer B., Muennich K. O., and Koltermann K. P. (1990) Ventilation rates of waters in the Nansen Basin of the Arctic Ocean derived from a multitracer approach. *J. Geophys. Res.* **C95**, 3265–3272.

Schlosser P., Bonisch G., Rhein M., and Bayer R. (1991) Reduction of deepwater formation in the Greenland Sea during the 1980s: evidence from tracer data. *Science* **251**, 1054–1056.

Schlosser P., Kromer B., Ostlund H. G., Ekwurzel B., Bonisch G., Loosli H. H., and Purtschert R. (1994) On the ^{14}C and ^{39}Ar distribution in the central Arctic Ocean: implications for deep water formation. *Radiocarbon* **36**, 327–343.

Schlosser P. and Smethie W. M. (1995) Transient tracers as a tool to study variability of ocean circulation. In *Natural Variability on Decade-to-Century Time Scales* (ed. N. R. Council). National Research Council, pp. 274–289.

Schlosser P., Bonisch G., Kromer B., Loosli H. H., Buhler R., Bayer R., Bonani G., and Koltermann K. P. (1995a) Mid-1980s distribution of tritium, ^{3}He, ^{14}C, and ^{39}Ar in the Greenland/Norwegian seas and the Nansen basin of the Arctic Ocean. *Prog. Oceanogr.* **35**, 1–28.

Schlosser P., Swift J. M., Lewis D., and Pfirman S. (1995b) The role of large-scale Arctic Ocean circulation on the transport of contaminants. *Deep-Sea Res.* **42**(6), 1341–1367.

Schlosser P., Bullister J. L., Fine R. A., Jenkins W. J., Key R. M., Lupton J. E., Roether W., and Smethie W. M. (2001) Transformation and age of water masses. In *Ocean Circulation and Climate: Observing and Modelling the Global Ocean* (eds. G. Siedler, J. Church, and J. Gould). Academic Press, London, vol. 77, pp. 431–452.

Schmitt R. W. (1981) Form of the temperature-salinity relationship in the central water: evidence for double-diffusive mixing. *J. Phys. Oceanogr.* **11**, 1015–1026.

Schmitt R. W. (1990) On the density ratio balance in the central water. *J. Phys. Oceanogr.* **20**(6), 900–906.

Schmitt R. W. (1994) Double diffusion in oceanography. *Ann. Rev. Fluid Mech.* **26**, 255–285.

Schmitt R. W. (1998) Double-diffusive convection. In *Ocean Modelling and Parametrization* (eds. E. P. Chassignet and J. Verron). Kluwer, Amsterdam, pp. 215–234.

Smethie W. M. (1993) Tracing the thermohaline circulation in the western North Atlantic using chlorofluorocarbons. *Prog. Oceanogr.* **31**, 51–99.

Spitzer W. S. and Jenkins W. J. (1989) Rates of vertical mixing, gas exchange and new production: estimates from seasonal gas cycles in the upper ocean near Bermuda. *J. Mar. Res.* **47**, 169–196.

St. Laurent L. C. and Schmitt R. W. (1999) The contribution of salt fingers to vertical mixing in the North Atlantic tracer release experiment. *J. Geophys. Res.* **29**, 1404–1424.

St. Laurent L. C., Toole J. M., and Schmitt R. W. (2001) Buoyancy forcing by turbulence above rough topography in the abyssal Brazil basin. *J. Phys. Oceanogr.* **31**, 3476–3495.

Stommel H. (1949) Horizontal diffusion due to oceanic turbulence. *J. Mar. Res.* **8**(3), 199–225.

Sundermeyer M. A. and Price J. F. (1998) Lateral mixing and the North Atlantic tracer release experiment: observations and numerical simulations of Lagrangian particles and a passive tracer. *J. Geophys. Res.* **103**(C10), 21481–21497.

Tennekes H. and Lumley J. L. (1972) *A First Course in Turbulence*. MIT Press, Cambridge, MA.

Theile G. and Sarmiento J. L. (1990) Tracer dating and ocean ventilation. *J. Geophys. Res.* **95**, 9377–9391.

Tomczak M. (1999) Some historical, theoretical and applied aspects of quantitative water mass analysis. *J. Mar. Res.* **57**(3), 275–303.

Tomczak M. and Large D. G. B. (1989) Optimum multi-parameter analysis of mixing in the thermocline of the eastern Indian Ocean. *J. Geophys. Res.* **94**(C11), 16141–16149.

Toole J. M., Polzin K. L., and Schmitt R. W. (1994) Estimates of diapycnal mixing in the abyssal ocean. *Science* **264**, 1120–1123.

Torgersen T., Turekian K. K., Turekian V. C., Tanaka N., DeAngelo E., and O'Donnell J. (1996) ^{224}Ra distribution in surface and deep water of long island sound: sauces and horizontal transport rates. *Continent. Shelf Res.* **16**(12), 1545–1559.

Unterweger M. P., Coursey B. M., Schima F. J., and Mann W. B. (1980) Preparation and calibration of the 1978 national bureau of standards tritiated-water standards. *Int. J. Appl. Radiat. Isotopes* **31**, 611–614.

Warner M. J. and Weiss R. F. (1985) Solubilities of chlorofluorocarbons 11 and 12 in water and seawater. *Deep-Sea Res.* **32**(12), 1485–1497.

Warner M. J., Bullister J. L., Wisegarver D. P., Gammon R. H., and Weiss R. F. (1996) Basin-wide distributions of chlorofluorcarbons CFC-11 and CFC-12 in the North Pacific, 1985–1989. *J. Geophys. Res.* **101**, 20525–20542.

Weiss W. M. and Roether W. (1980) The rates of tritium input to the world oceans. *Earth Planet. Sci. Lett.* **49**, 435–446.

Young W. R., Rhines P. B., and Garrett C. (1982) Shear-flow dispersion, internal waves and horizontal mixing in the ocean. *J. Phys. Oceanogr.* **12**(6), 515–527.

Zhang R.-H., Rothstein L. M., and Busalacchi A. J. (1998) Origin of upper-ocean warming and El Niño change in decadal scales in the tropical Pacific Ocean. *Nature* **391**, 879–883.

6.09
Chemical Tracers of Particle Transport

R. F. Anderson

Lamont-Doherty Earth Observatory of Columbia University, Palisades, NY, USA

NOMENCLATURE

A concentration in activity units (decays/time/volume) (superscripts d, c, s, and L refer to dissolved, colloidal, small particles, and large particles, respectively)

k_1 first-order rate constant for adsorption of dissolved tracer to particles

k_{-1} first-order rate constant for desorption of particulate tracer

k_2 rate constant for loss of particulate substance from the water column assuming that a single constant applies to all particles

k_{ads} first-order rate constant for adsorption of dissolved tracer to colloids

k_s rate constant for loss of particulate substance from the water column when applied to large, rapidly sinking particles

K_z vertical eddy diffusivity

S sinking rate of particles; may be applied either to small or to large particle classes
w vertical advection velocity
β_{-1} rate constant for remineralization of labile particles
β_2 rate constant for aggregation of small particles into large, rapidly sinking aggregates
β_{-2} rate constant for disaggregation of large aggregates into small suspended particles
λ radioactive decay constant
ρ *in situ* density of sediment (g dry wt. cm^{-3} wet sediment)
τ residence time
ψ sediment focusing factor, equivalent to the ratio of the rate of sediment burial at a specific location divided by the regional average rain rate of particles to the seabed

6.09.1 PARTICLE TRANSPORT AND OCEAN BIOGEOCHEMISTRY

Particles represent important agents of transport in global ocean cycles of many trace elements, of carbon, and of other substances. Once introduced into the oceans, many trace elements are removed from seawater by scavenging (sorption, complexation, and other forms of surface reactions) to particles (Goldberg, 1954; Turekian, 1977). Scavenging and burial in marine sediments represents the principal loss process influencing the biogeochemical cycle of many trace elements in the ocean (Li, 1981).

Carbon dioxide is effectively transferred from atmosphere to deep sea by sinking biogenic particles. The magnitude of this transfer depends on the depth scale of organic matter regeneration by dissolution, oxidation, and other processes that convert organic phases back into dissolved inorganic constituents. The greater the depth of organic matter regeneration, the larger will be the transfer of CO_2 from the atmosphere into the deep sea. In part this reflects the longer timescale for deep waters to return to the surface and exchange their gases with the atmosphere. Furthermore, there is a positive feedback created by the reaction of respiratory CO_2 with $CaCO_3$ that amplifies the ocean's uptake of atmosphere CO_2 (Archer *et al.*, 2000). The depth of organic matter regeneration depends on the competing rates of regeneration and of sinking of particulate organic matter. Sinking rates, in turn, are strongly dependent on particle dynamics (aggregation and disaggregation) as well as on particle composition, which influences particle density. Accurate representations of particle dynamics and regeneration length scales are vital elements of ocean carbon cycle models.

Lateral transport of particles is an important component of modern biogeochemical cycles. Furthermore, lateral redistribution of sediments modifies the sedimentary record of past changes in environmental conditions. Therefore, paleoceanographic studies that rely on reconstructing accumulation rates of sedimentary constituents must account for lateral redistribution of sediments within the ocean.

The examples above illustrate the importance of understanding the transport and transformation of particulate matter in the ocean. Vertical transport processes are primarily considered when examining the fate of biogenic particles formed in surface waters. Lateral transport processes influence the dispersion into the ocean interior of particulate material eroded from continents. Much of this dispersion occurs by eddy mixing, but advection by ocean currents plays an important role in transport of marine particles as well. A complete understanding of the cycling of material within the oceans relies on an accurate knowledge of particle transport.

6.09.2 TRACERS OF PARTICLE TRANSPORT

Several naturally occurring radionuclides provide useful tracers of particle transport. A favorable characteristic of these tracers is that each has a well-defined source. Furthermore, they often have only a single significant source in seawater, through the radioactive decay of a dissolved parent nuclide. These radionuclide tracers form as dissolved chemical species, but then become attached to particles at a rate that is characteristic of the chemical properties of the tracer. Consequently, information about particle dynamics (e.g., aggregation and disaggregation) and particle transport (both vertical and horizontal) can be obtained by combining measured concentrations of dissolved and particulate radionuclides with mass budget approaches and simple models.

Several useful particle-reactive tracers are found within the natural uranium (^{238}U, ^{235}U) and thorium (^{232}Th) decay series:

$$^{238}U \xrightarrow{\alpha} {}^{234}Th \xrightarrow{\beta} {}^{234}Pa \xrightarrow{\beta} {}^{234}U \xrightarrow{\alpha} {}^{230}Th \xrightarrow{\alpha} \cdots$$

$$^{226}Ra \xrightarrow{\alpha} {}^{210}Pb \xrightarrow{\beta} {}^{210}Bi \xrightarrow{\beta} {}^{210}Po \xrightarrow{\alpha} {}^{206}Pb$$

(short-lived intermediates between ^{226}Ra and ^{210}Pb have been omitted)

$$^{235}U \xrightarrow{\alpha} {}^{231}Th \xrightarrow{\beta} {}^{231}Pa \xrightarrow{\alpha} {}^{227}Ac \rightarrow \cdots$$

$$^{232}Th \xrightarrow{\alpha} {}^{228}Ra \xrightarrow{\beta} {}^{228}Ac \xrightarrow{\beta} {}^{228}Th \xrightarrow{\alpha} {}^{224}Ra \xrightarrow{\alpha} \cdots$$

Uranium-238, ^{235}U, and ^{232}Th are long lived and have been present since the formation of the Earth. Uranium forms soluble carbonate complexes, so it is readily released into surface waters during weathering of continental crust. Its stable soluble chemical form gives uranium a long residence time (~4×10^5 yr; Henderson, 2002) and, consequently, a uniform distribution in the ocean (Chen *et al.*, 1986). Thorium, by contrast, is quite insoluble, so most of the ^{232}Th mobilized by continental weathering remains within particulate phases which, after being transported to the sea by rivers, are buried in nearshore sediments. Thorium isotopes produced by decay of uranium are similarly insoluble, and it is this property that makes them useful tracers of particle transport. The nearly uniform concentration of uranium throughout the ocean provides a well-defined source of the radiogenic thorium isotopes, and it is this known source that permits mass budget approaches to be exploited in evaluating rates of particle transport. Among the thorium isotopes produced by uranium decay, ^{234}Th (half-life 24.1 d) and ^{230}Th (half-life 75 kyr) are widely used as particle tracers. The half-life of ^{231}Th (25 h) is too short to permit useful exploitation as a particle tracer. However, its decay product, ^{231}Pa (half-life 32.5 kyr), is highly particle reactive and it provides the most useful particle tracer in the ^{235}U decay series. Radioactive decay of ^{232}Th in sediments releases soluble ^{228}Ra, which diffuses into the water column where it is mixed into the ocean interior. Decay of ^{228}Ra (half-life 5.75 yr) produces ^{228}Th (half-life 1.91 yr) which, like other thorium isotopes, is rapidly scavenged by particles.

Analogous to the process releasing ^{228}Ra to seawater, decay of ^{230}Th in sediments releases dissolved ^{226}Ra which is then mixed into the ocean interior. Radium-226 decays through a series of short-lived nuclides to ^{210}Pb (half-life 22.3 yr) which, like thorium and protactinium, is insoluble and readily sorbs to particles. Radioactive decay of gaseous ^{222}Rn in the atmosphere also produces ^{210}Pb, which is then deposited on the sea surface with aerosols and in precipitation. Although ^{210}Pb and, to a lesser extent, ^{231}Pa have found many applications as tracers of particle transport, by far the greatest use has been made of thorium isotopes, which form the focus of this review.

6.09.3 TRANSFER FROM SOLUTION TO PARTICLES (SCAVENGING)

Fundamental to the use of natural radionuclides as tracers of particle transport is an understanding of the rates and mechanisms of radionuclide transfer from solution into the particulate phase. In a closed system, where no mechanism exists for the physical separation of a daughter nuclide from its parent, the decay series reach steady state where the decay rate of each daughter equals its rate of production. Because the production rate of a daughter equals the decay rate of its parent, at steady state in a closed system the decay rate (activity) of parent equals decay rate (activity) of the daughter. In an open system, where physical, chemical, and biological processes may separate a daughter from its parent, radioactive disequilibrium exists where the activity of the daughter differs from that of its parent. The extent of disequilibrium can be modeled to derive a rate constant for the process creating the disequilibrium. This principle was exploited in the 1960s to evaluate the scavenging rate of ^{230}Th in the deep sea (Moore and Sackett, 1964) and the scavenging rate of ^{234}Th in surface waters (Bhat *et al.*, 1969).

Assuming horizontal concentration gradients to be negligible, and further assuming uniform parameters in the vertical dimension, a mass balance for dissolved ^{234}Th or ^{230}Th (Figure 1) can be written as

$$\frac{\partial A_{\mathrm{Th}}^{\mathrm{d}}}{\partial t} = K_z \frac{\partial^2 A_{\mathrm{Th}}^{\mathrm{d}}}{\partial z^2} - w \frac{\partial A_{\mathrm{Th}}^{\mathrm{d}}}{\partial z} + \lambda_{\mathrm{Th}} A_{\mathrm{U}} - (\lambda_{\mathrm{Th}} + k_1) A_{\mathrm{Th}}^{\mathrm{d}} \qquad (1)$$

Here the mass budget is constructed in terms of thorium activity ($A_{\mathrm{Th}} = \lambda_{\mathrm{Th}} N_{\mathrm{Th}}$, where N is the number of atoms of thorium and λ_{Th} is the decay constant for the appropriate thorium isotope). K_z is

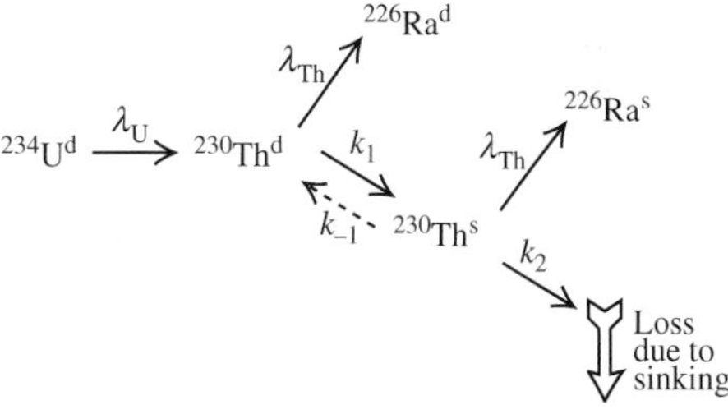

Figure 1 Schematic illustration of the production and fate of ^{230}Th in seawater. Radioactive decay of dissolved ^{234}U produces ^{230}Th that initially exists as a dissolved species. Dissolved ^{230}Th may either undergo radioactive decay to ^{236}Ra, or it may be adsorbed to particles. Radioactive decay is represented by a decay constant, λ (λ = ln(2)/radioactive half-life)), and uptake by particles (scavenging) is represented by a first-order rate constant, k_1. ^{230}Th is initially sorbed by small slowly settling particles (supersript "s"), which form the vast majority of particle mass in the ocean. ^{230}Th sorbed to small particles may undergo radioactive decay to ^{226}Ra; it may desorb (return to solution, represented by the first-order rate constant k_{-1}); or it may sink from the water column, where the loss of particulate ^{230}Th is represented by the first-order rate constant k_2. Similar processes influence ^{234}Th, which is produced by radioactive decay of ^{238}U and which decays to ^{234}Pa.

the vertical eddy diffusivity, k_1 is the rate constant for uptake of dissolved thorium by particles, w is the vertical advection velocity, $\lambda_{Th}A_U$ is the production rate of the appropriate thorium isotope, and z is depth. Scavenging is represented here as a first-order irreversible process, although this assumption is often not valid (see below). Because the concentration of uranium is essentially uniform throughout the ocean (Chen *et al.*, 1986), the production rate of ^{234}Th and of ^{230}Th is independent of depth. In the deep ocean the vertical advection and diffusion terms are negligible (Nozaki *et al.*, 1981), although this is not true in surface waters where upwelling may supply a significant source of dissolved thorium (see Section 6.09.5).

For steady-state conditions ($\partial/\partial t = 0$) and neglecting vertical advection and diffusion, Equation (1) can be rearranged to solve for the residence of dissolved thorium with respect to uptake by particles:

$$\tau^{d}_{Th} = \frac{1}{k_1} = \frac{A^{d}_{Th}}{\lambda_{Th}(A_U - A^{d}_{Th})} \tag{2}$$

where τ^{d}_{Th} is the residence time of dissolved thorium with respect to its sorption by particles. An equivalent residence time for total thorium (dissolved plus particulate) with respect to removal from the water column by scavenging processes can be derived simply by replacing the concentration of dissolved thorium by the concentration of total thorium in Equation (2).

Neglecting vertical advection and diffusion, a mass balance for particulate thorium (A^{p}_{Th}; Figure 1) can be expressed as

$$\frac{\partial A^{p}_{Th}}{\partial t} = -S\frac{\partial A^{p}_{Th}}{\partial z} + k_1 A^{d}_{Th} - (\lambda_{Th} + k_2)A^{p}_{Th} \tag{3}$$

where S is the mean sinking rate of a uniform population of particles and k_2 is a rate constant for loss of particulate thorium from the water column, represented here to be first order with respect to the concentration of particulate thorium.

If the vertical distribution of particulate thorium is uniform, as would be the case in the surface mixed layer of the ocean for example, and if the system is at steady state, then Equation (3) reduces to

$$0 = k_1 A^{d}_{Th} - (\lambda_{Th} + k_2)A^{p}_{Th} \tag{4}$$

from which the residence time of particulate thorium can be derived as

$$\tau^{p}_{Th} = \frac{1}{k_2} = \frac{A^{p}_{Th}}{k_1 A^{d}_{Th} - \lambda_{Th}A^{p}_{Th}} \tag{5}$$

The solution to Equation (1) for conditions of steady state and negligible advection and diffusion predicts a uniform concentration of dissolved thorium in the ocean:

$$A^{d}_{Th} = \frac{\lambda_{Th}A_U}{\lambda_{Th} + k_1} \tag{6}$$

It was evident in the earliest profiles obtained for dissolved ^{230}Th that this prediction does not hold, in that concentrations of dissolved ^{230}Th were observed to increase with depth in a nearly linear fashion (Bacon and Anderson, 1982; Nozaki *et al.*, 1981). These results were interpreted to indicate that sorption processes are a reversible reaction in which particulate thorium may be released back into solution, as represented by k_{-1} in Figure 1 (Bacon and Anderson, 1982; Nozaki *et al.*, 1981). Neglecting advection and diffusion, mass balance equations for dissolved and particulate thorium can be written, respectively, as

$$\frac{\partial A^{d}_{Th}}{\partial t} = \lambda_{Th}A_U + k_{-1}A^{p}_{Th} - (\lambda_{Th} + k_1)A^{d}_{Th} \tag{7}$$

and

$$\frac{\partial A^{p}_{Th}}{\partial t} = -S\frac{\partial A^{p}_{Th}}{\partial z} + k_1 A^{d}_{Th} - (\lambda_{Th} + k_{-1} + k_2)A^{p}_{Th} \tag{8}$$

At a fixed depth and for steady state, the partitioning of thorium between particulate and dissolved phases can be represented as

$$\frac{A^{p}_{Th}}{A^{d}_{Th}} = \frac{k_1}{\lambda_{Th} + k_{-1} + k_2} \tag{9}$$

Neglecting radioactive decay for ^{230}Th, which is reasonable given that its 75 kyr half-life is much greater than its residence time in the water column (10–40 yr; Anderson *et al.*, 1983a,b), the steady-state distributions for dissolved and particulate concentrations can be represented as (Bacon and Anderson, 1982)

$$A^{d}_{Th} = \frac{\lambda_{Th}A_U + k_{-1}A^{p}_{Th}}{k_1} \tag{10}$$

and

$$A^{p}_{Th} = \frac{\lambda_{Th}A_U}{k_2}\left[1 - \exp\left(\frac{-k_2}{S}\right)z\right] \tag{11}$$

Here it is seen that the concentration of particulate ^{230}Th increases with increasing depth, as does the concentration of dissolved ^{230}Th, which has a constant relationship with the concentration of particulate ^{230}Th according to the reversible scavenging model.

In this model S represents the average sinking rate of the population of small, slowly settling particles that contain most of the inventory of particulate thorium in the ocean. According to this model, particulate ^{230}Th lost from the system ($k_2A^{p}_{Th}$) is removed permanently. Subsequent studies have shown this latter condition not to be strictly true (Section 6.09.6). Nevertheless, the principles of reversible scavenging have found support in many subsequent studies (e.g., Moore and Hunter, 1985; Moore and Millward, 1988; Niven and Moore, 1993). Furthermore, reversible scavenging is consistent with the principles of equilibrium reactions involving balanced forward and reverse sorption processes that have been applied to the exchange of trace metals between solution and solid surfaces (e.g., Balistrieri *et al.*, 1981; Schindler, 1975).

Reversible scavenging models similar, in principle, to that described above have now been applied successfully to profiles of ^{230}Th in many regions of the ocean, including cases where dissolved ^{230}Th profiles are influenced by rapid ventilation of deep waters (as in the North Atlantic Ocean; Moran *et al.*, 2001; Vogler *et al.*, 1998). Average sinking rates for the population of fine-grained particles, derived by fitting equations such as above to the measured ^{230}Th profiles, vary little from one region to another and fall within the range of 400–1,000 m yr^{-1} (Bacon and Anderson, 1982; Moran *et al.*, 2001; Nozaki *et al.*, 1981). Similar mean particle settling rates have been derived by applying an irreversible scavenging model to profiles of particulate ^{230}Th in the Pacific Ocean and Indian Ocean (Krishnaswami *et al.*, 1976, 1981). The consistency among particle-settling rates derived by these various studies suggests that this settling rate is a robust parameter that can be applied throughout the world ocean.

Likewise, the remarkable similarity among ^{230}Th profiles (Figure 2) obtained throughout the open Pacific Ocean (i.e., away from continental margins or the upwelling regions of the Southern Ocean; e.g., Bacon and Anderson, 1982; Chase *et al.*, 2003b) suggests that the rates of scavenging and removal of reactive elements vary little from one region to another, despite regional differences in the composition and flux of particulate material to which ^{230}Th is scavenged. If so, then the rate constants derived by Bacon and Anderson (1982) may be generally applicable as well. By fitting simultaneously the equations for the equilibrium partitioning of thorium between particulate and dissolved phases to measured

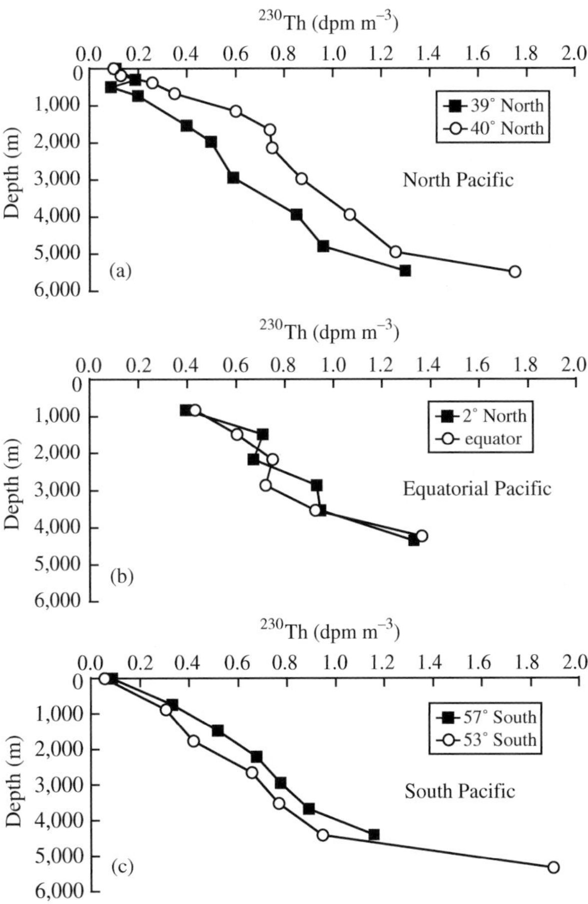

Figure 2 Profiles of total (dissolved plus particulate) concentrations of ^{230}Th in the: (a) North (Nozaki *et al.*, 1981), (b) equatorial (Anderson, unpublished), and (c) South (Chase *et al.*, 2003b) Pacific Ocean. Throughout the Pacific Ocean, the concentration profiles of ^{230}Th are similar, both in terms of their linearity and the absolute concentrations.

distributions of ^{230}Th and of ^{234}Th, Bacon and Anderson were able to derive explicit values for sorption, desorption, and removal rate constants. They reported rate constants for the uptake of thorium from solution (k_1) that increased with increasing particle concentration, rising approximately from 0.2 yr^{-1} to 1.0 yr^{-1} over a particle concentration of 5–25 μg L^{-1}. Their rate constants for desorption (k_{-1}) were independent of particle concentration, and ranged between 1.5 yr^{-1} and 4 yr^{-1}. Reasonable fits to ^{230}Th profiles could only be obtained with rate constants for removal of particulate Th (k_2) that were less than 0.1 yr^{-1}. The fact that k_2 is much smaller than k_1 and k_{-1} implies that ^{230}Th atoms, on average, cycle between dissolved and particulate phases many times before eventually being buried in sediments.

Recognizing that scavenging of dissolved radionuclide tracers is a reversible process was a milestone in understanding the removal of particle-reactive substances from the ocean. This principle must be taken into account in using natural radionuclides as tracers of particle transport, and when using natural radionuclides to infer the rates of processes responsible for scavenging particle-reactive substances (e.g., trace metals) from the ocean.

6.09.4 COLLOIDAL INTERMEDIARIES

6.09.4.1 Development of the Colloidal Pumping Hypothesis

Rates of sorption reactions inferred from U–Th disequilibria in natural waters (see above) as well as sorption kinetics measured in laboratory studies are much slower than predicted from physicochemical theory of surface coordination reactions. This finding led to the suggestion that colloidal intermediates are responsible for the slow transfer of dissolved species onto particulate matter (Morel and Gschwend, 1987; Santschi *et al.*, 1986). Specifically, it was suggested that dissolved species (e.g., thorium) equilibrate rapidly with sorption sites on colloids. Colloids then coagulate to form larger (filterable) particles, and the rate-limiting step in the scavenging and removal of thorium (and other reactive species) from solution is the coagulation process.

Honeyman and Santschi (1989, 1992) developed these ideas more formally into the "colloidal pumping" hypothesis. Building upon existing coagulation theory, they were able to model adsorption and coagulation processes to produce rates of scavenging by large (filterable) particles that were consistent with observations. Furthermore, using the colloidal pumping hypothesis, Honeyman and Santschi were also able to explain the "particle concentration effect"—a characteristic tendency for the partition coefficient for sorbed trace elements ((atoms/g particles)/(atoms/g water)) to decrease with increasing particle concentration. This feature of metal uptake by particles in natural waters, such as the slow sorption kinetics, is inconsistent with classical theories of adsorption.

At about the time that Honeyman and Santschi published their colloidal pumping hypothesis, lab and field studies showed that thorium and other metals in seawater are, indeed, associated with colloids (Baskaran *et al.*, 1992; Moran and Buesseler, 1992; Moran and Moore, 1989; Niven and Moore, 1988). Subsequently, many more studies have demonstrated that colloids are abundant in seawater and that they play an important role in the marine biogeochemical cycles of organic carbon and of many trace elements (Guo and Santschi, 1997).

6.09.4.2 Important Features of Colloids

Although inorganic forms of colloidal material occur in seawater, the vast majority of marine colloids are composed of organic matter (Benner *et al.*, 1992; Santschi *et al.*, 1995; Wells and Goldberg, 1994). Colloidal organic matter (COM; $>\sim$1,000 nominal molecular weight, i.e., $>$1 kDa) constitutes a large fraction (roughly 20–40%) of the operationally defined dissolved organic carbon (DOC) in seawater (Benner *et al.*, 1992; Guo and Santschi, 1997; Guo *et al.*, 1997; Santschi *et al.*, 1995). Furthermore, high molecular weight COM ($>$10 kDa) constitutes a reactive component of operationally defined DOC (Benner *et al.*, 1992; Santschi *et al.*, 1995; Wells and Goldberg, 1994). Colloid aggregation may be a significant abiotic sink for DOC in the ocean. Consequently, the objective of better understanding the cycling and fate of DOC (COC) in the ocean has been one factor motivating research on marine colloids.

The mass of COM is generally greater than that of suspended particulate matter in the ocean, and it can be as much as an order of magnitude greater (Guo *et al.*, 1997). The small size of colloids offers a large surface area for sorption, potentially providing an important vector for scavenging and removal of particle-reactive trace elements. Indeed, a large fraction of certain trace metals in seawater have been found to occur in association with colloids (Guo and Santschi, 1997), primarily in estuarine and coastal waters (e.g., Guo *et al.*, 2000; Wells *et al.*, 1998; Wen *et al.*, 1999). However, the importance of colloidal metals has also been demonstrated in open-ocean waters where, for example, the majority of the operationally defined dissolved (passing through a filter of 0.2 μm pore diameter) iron is found to be in colloidal size fractions (Nishioka *et al.*, 2001; Wu *et al.*, 2001). The presence of iron in surface waters as colloidal rather than soluble forms reduces its bioavailability, and influences mechanisms of iron removal (Wu *et al.*, 2001). Consequently, coagulation of colloidal iron may represent a significant loss of iron from the ocean.

Binding of trace metals to colloids has been shown to affect their mechanism of uptake by plankton as well as their bioavailability (e.g., Chen and Wang, 2001; Wang and Guo, 2000). Decomposition of plankton, in turn, sometimes releases metals to seawater in colloidal forms (Wang and Guo, 2001). Consequently, the objective of better understanding the biogeochemical cycling of trace elements, as well as the bioavailability of those elements, has also been motivating research on marine colloids.

6.09.4.3 Evidence for Coagulation of Colloids and of Colloidal Thorium (and Other Metals)

With many aspects of the marine biogeochemical cycles of organic matter and of trace elements linked to the transport and fate of colloids, exploring the tenets of the colloidal pumping hypothesis has become a high priority. Colloids exist in seawater and they do not sink, so their removal must involve either dissolution or coagulation into larger particles capable of sinking. Transmission electron micrographs of marine colloids do, indeed, show that they consist of aggregates of smaller granules (Wells and Goldberg, 1994), providing *prima facie* evidence for colloid aggregation.

Laboratory studies have demonstrated that colloids of uniform composition (hematite) do coagulate as predicted by the colloidal pumping hypothesis (Honeyman and Santschi, 1991). Quigley *et al.* (1996) later showed that thorium sorbed to the hematite surfaces reliably and quantitatively traces the coagulation process. Thorium was suggested to be an ideal monitor of the coagulation process (coagulometer), because it adsorbs irreversibly to the hematite, and because its affinity for sorption is so great that it is sorbed quantitatively for all practical purposes. Quigley *et al.* (1996) noted that the irreversible sorption of thorium to hematite is inconsistent with the reversible scavenging of thorium in seawater inferred by previous studies (see Section 6.09.3). Later, Quigley *et al.* (2001) concluded that sorption of thorium to COM is also irreversible. As a result of these studies, Quigley and coworkers suggested that the apparent desorption of thorium is actually an artifact of the breakup of aggregated particles into colloids that appear in the operationally defined dissolved fraction. If thorium is truly sorbed irreversibly to colloids, then this would be beneficial for using thorium as a coagulometer because, once adsorbed, the transfer of thorium among size fractions is entirely due to the aggregation and disaggregation properties of the colloids, and the behavior of thorium no longer depends on its own chemistry. This aspect of thorium sorption to marine particles requires further investigation. Meanwhile, laboratory studies using marine particles and marine COM have demonstrated that colloid aggregation is capable of moving thorium, mercury, and other metals from colloidal forms into larger, filterable particles (Quigley *et al.*, 2001; Stordal *et al.*, 1996; Wen *et al.*, 1997).

6.09.4.4 Rate Constants for Colloid Coagulation

Exploiting the well-known tendency for thorium to sorb to marine particles, several investigators have sought to determine rate constants for the coagulation of COM in seawater (Baskaran *et al.*, 1992; Dai and Benitez-Nelson, 2001; Guo and Santschi, 1997; Guo *et al.*, 1997; Moran and Buesseler, 1992, 1993; Niven *et al.*, 1995; Santschi *et al.*, 1995). In each case a simple model illustrated in Figure 3(a) was applied to measured distributions of dissolved, colloidal, and particulate ^{234}Th. Dissolved thorium concentrations were defined using cross-flow filtration (CFF) and membranes with a nominal size cutoff of either <1 kDa or <10 kDa, depending on the study. Particulate thorium concentrations were measured by filtration through membranes with pore sizes that ranged from 0.2 μm to 1.0 μm, again depending on the study. The colloidal size fraction was defined as that which passed through the filter used to retain particulate thorium, but which was retained by the CFF membrane.

Measured distributions of ^{234}Th were modeled assuming irreversible scavenging of ^{234}Th to colloids, and serial coagulation of colloidal material into larger size fractions which are eventually removed from the system by sinking. Neither desorption of thorium from colloids nor disaggregation of particles into smaller size classes was considered in evaluating coagulation rates, even though these processes are included in the fully developed colloidal pumping model (Figure 3(b)), because the system is underdetermined when reverse reactions are allowed. In evaluating coagulation rates the system is generally assumed to be at steady state, although Moran and Buesseler (1993) applied a non-steady-state model to time-series data. Further assuming that advection and diffusion terms are negligible compared to production and removal fluxes, the mass balance equations for ^{234}Th, as represented in Figure 3(a), can be written as

$$\frac{\partial A^{\mathrm{d}}_{\mathrm{Th}}}{\partial t} = \lambda_{\mathrm{Th}} A_{\mathrm{U}} - \lambda_{\mathrm{Th}} A^{\mathrm{d}}_{\mathrm{Th}} - k_{\mathrm{ads}} A^{\mathrm{d}}_{\mathrm{Th}} = 0 \quad (12)$$

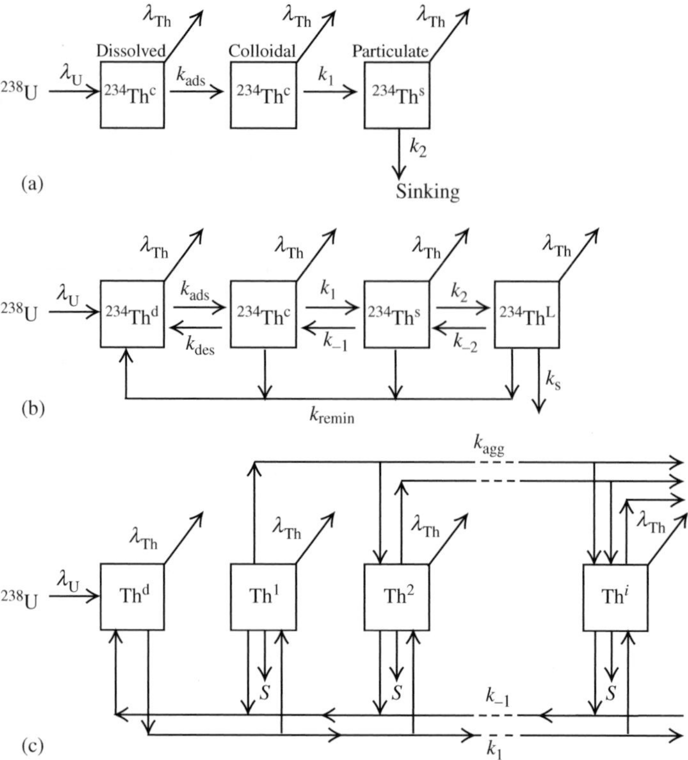

Figure 3 Three conceptual models of increasing complexity simulate the uptake of dissolved ^{234}Th onto colloids, and the coagulation of colloidal ^{234}Th into larger particles. (a) Simple model in which dissolved ^{234}Th is initially sorbed by colloids which, in turn, coagulate into larger particles that may sink from the water column. Each step is considered to be irreversible and rate constants are treated as first order. Adsorption is represented by the rate constant, k_{ads}. Other terms are as defined in Figure 1. (b) A more complex model in which adsorption of dissolved ^{234}Th to colloids (k_{ads}), coagulation of colloids into small particles (k_1), and further coagulation of particulate Th into larger, rapidly sinking particles (k_2) are each represented as a reversible process (k_{des} is a first-order rate constant for desorption of colloidal Th; k_{-1} and k_{-2} represent rate constants for disaggregation of small and large particles, respectively). Here, k_s, the rate constant for loss of particulate Th by sinking, is equivalent to k_2 in panel (a). Remineralization of particles in all size classes, including colloids, is considered to occur at a rate represented by the first-order rate constant k_{remin}. (c) Schematic of the coagulation model developed by Burd *et al.* (2000) that contains a broad spectrum of particle size classes (1, 2, ..., *i*) ranging in diameter from 1 nm to 100 μm. Adsorption of Th to (k_1), and desorption from (k_{-1}), each size class is simulated. Aggregation (k_{agg}) occurs between all size classes, although disaggregation is not simulated. Radioactive decay and loss by sinking occur for all size classes (after Burd *et al.*, 2000).

$$\frac{\partial A^{c}_{Th}}{\partial t} = k_{ads}A^{d}_{Th} - \lambda_{Th}A^{d}_{Th} - k_1 A^{c}_{Th} = 0 \quad (13)$$

$$\frac{\partial A^{p}_{Th}}{\partial t} = k_1 A^{c}_{Th} - \lambda_{Th}A^{p}_{Th} - k_2 A^{p}_{Th} = 0 \quad (14)$$

Here A^d, A^c, and A^p represent the concentration (in units of activity per unit volume) of ^{234}Th in the dissolved, colloidal, and particulate fractions, respectively; k_{ads} is the rate constant for sorption of dissolved thorium onto colloids, k_1 is the rate constant for coagulation of colloids into filterable particles, and k_2 is the rate constant for loss of filterable particles by sinking. Note that the combined effects here of adsorption and coagulation are equivalent to k_1 in Equations (1)–(11) above, whereas k_2 is essentially the same in both cases. In each case the loss term of one pool represents a source for the next pool in the coagulation sequence, and eventually all ^{234}Th that does not decay is lost from the system by sinking of particles.

Regardless of the size cutoff used to define the colloid fraction (1 kDa or 10 kDa), turnover times of colloidal material were found to be short (ranging from about a day to a few weeks). In studies where COM was further divided into size fractions of >1 kDa and >10 kDa, the larger size fraction was found to have a shorter turnover time (about a week in the Mid-Atlantic Bight and in the Gulf of Mexico) than the smaller size fraction (about a month; Guo *et al.*, 1997). These residence times represent lower limits, because desorption of thorium and disaggregation of particles was not considered. Nevertheless, even allowing for substantial errors due to these limitations, the results imply a high degree of reactivity for natural marine COM.

6.09.4.5 Limitations and Questions

Beyond the limitations and assumptions mentioned above, one must ask if the rate constants derived by modeling the size distribution of ^{234}Th can be applied equally to all marine colloids, or even to all forms of COM. Guo *et al.* (1997) found that ^{234}Th and organic carbon have similar distributions among size fractions, allowing for the possibility that thorium is distributed uniformly among various fractions of COM. They also found that the partition coefficient for uptake of thorium ($\sim 10^6$ mL g^{-1}) is similar among operationally defined size classes of colloidal (1–10 kDa; and 10 kDa to 0.2 μm) as well as particulate (>0.2 μm) organic carbon, suggesting relatively similar binding of thorium among size classes. However, in contrast to the earlier studies of Guo *et al.* (1997), Dai and Benitez-Nelson (2001) found that thorium partitions among size classes differently than does organic carbon. Their results indicated that thorium is bound preferentially to the larger and more reactive portion of COM. Hence, coagulation rates derived from ^{234}Th may not apply equally to all colloidal species.

Some studies provide both indirect and direct evidence that thorium binds preferentially to certain forms of COM. Niven *et al.* (1995) found that ^{234}Th is incorporated into the colloidal pool to a greater extent than average following phytoplankton blooms, leading them to suggest that phytoplankton exudates have a greater affinity for thorium than do other forms of COM. The rapid turnover time of colloidal thorium is inconsistent with the slow turnover of COM inferred from ^{14}C ages, especially when applied to the smaller (>1 kDa) size fraction (Guo and Santschi, 1997; Santschi *et al.*, 1995). Laboratory studies have shown that thorium does not bind to all COM fractions with equal affinity (Quigley *et al.*, 2001), and that thorium reacts preferentially with the acid polysaccharide fraction of COM (Quigley *et al.*, 2002), consistent with the suggestion by Niven *et al.* (1995) that its uptake by COM increased with increasing production of phytoplankton exudates. Quigley *et al.* (2002) found that >76% of the ^{234}Th added to isolates of marine COM appeared in the acid polysaccharide fraction following separation of organic compound classes by gradient gel electrophoresis. Based on the evidence above, it appears that ^{234}Th serves as a tracer primarily for the more reactive forms of marine COM, which include acid polysaccharide compounds produced by marine phytoplankton.

Models of colloid dynamics are far ahead of the ability to test the models using *in situ* data. As noted above, the models applied to measured ^{234}Th distributions do not include desorption or disaggregation. Neither do they include parallel sorption of dissolved ^{234}Th to multiple size classes of particles, although some studies have found evidence for parallel sorption of thorium to multiple size classes of colloids and particles (Quigley *et al.*, 2001). State-of-the-art models do include these features, as well as a well-resolved spectrum of particle size classes ranging from 1 nm to 100 μm (Figure 3(c); Burd *et al.*, 2000). The next step in this area of research will be to find clever methods to test these more elaborate models using results from laboratory and field studies.

Finally, it is worth noting that studies relevant to the colloidal pumping hypothesis have been applied largely to surface waters. It remains to be determined if colloids are as important for uptake of trace metals in the deep sea, where concentrations of DOM are lower than in the surface ocean and, presumably, concentrations of the more-reactive high molecular weight COM are much lower. Chase *et al.* (2002) found that the partitioning of ^{230}Th and ^{231}Pa between seawater and particles collected by sediment traps is determined primarily by the macroscopic composition of the particles, specifically by the $CaCO_3$ and opal contents of the particles. First-order control on the sorption of thorium and protactinium by the composition of large particles suggests that colloidal intermediates may play much less of a role in the scavenging of particle-reactive trace elements in the deep sea than they do in surface waters. Further studies are needed to test this hypothesis.

6.09.5 EXPORT OF PARTICLES FROM SURFACE OCEAN WATERS

6.09.5.1 Scavenging Rates and Particle Flux

Evaluating the flux of particulate matter sinking out of surface waters, often referred to as the "export flux," represents the most widespread use of natural radionuclides as tracers of particle transport. Basic methods for this application were developed in the late 1980s. The approach was subsequently applied at many locations throughout the ocean during the 1990s as part of the Joint Global Ocean Flux Study (JGOFS) and related programs seeking to understand the processes and conditions that regulate the flux of particulate organic matter from the surface ocean into the deep sea. Some reviews have described results obtained by applying this approach to natural (unperturbed) systems (Buesseler, 1998; Cochran and Masqué, in press; Moran *et al.*, 2003). In addition, export fluxes of particulate organic carbon (POC) have been evaluated using ^{234}Th during fertilization experiments in which iron has been added to nutrient-rich regions of the ocean (e.g., Bidigare *et al.*, 1999; Charette and Buesseler, 2000; Nodder *et al.*, 2001).

Radioactive disequilibrium between ^{238}U and ^{234}Th in surface waters was first described by Bhat *et al.* (1969), who recognized that scavenging to sinking particles removed ^{234}Th from the water column. The rate of ^{234}Th scavenging and removal from surface waters was later shown to be positively correlated with primary productivity (Coale and Bruland, 1985) as well as with the sinking flux of POC (Bruland and Coale, 1986). These relationships demonstrated that biogenic particles formed by marine phytoplankton represented the principal vector removing scavenged thorium from the upper water column.

Bruland and Coale (1986) and Coale and Bruland (1985) applied a steady-state irreversible scavenging model (Equations (1)–(5) above) to measured profiles of dissolved and particulate ^{234}Th in order to evaluate the rate constant for sorption (k_1) and the mean residence time of particulate thorium (Equation (5)). They reasoned that desorption could be neglected because the calculated residence time of particulate thorium (2–18 d) was much less than the residence time of particulate thorium with respect to desorption (3–4 months; Bacon and Anderson, 1982). Therefore, an irreversible scavenging model could be applied, at least to surface waters of the productive California Current, which served as their study region. Similar reasoning has been applied to the Southern Ocean (Rutgers van der Loeff *et al.*, 1997).

Rate constants for uptake of dissolved thorium (k_1) in surface waters of the California Current (Coale and Bruland, 1985) were found to be much greater (7–66 yr^{-1}) than in the deep sea (0.2–1.2 yr^{-1}; Bacon and Anderson, 1982), reflecting the much greater concentrations of particulate matter in surface waters of the California Current (200–2,000 $\mu g\ L^{-1}$) compared to the deep sea (5–25 $\mu g\ L^{-1}$; Bacon and Anderson, 1982).

In a later study of the oligotrophic North Pacific gyre, Coale and Bruland (1987) found a layered structure to the scavenging of ^{234}Th from the upper water column. There, scavenging of thorium was found to be more intense within the euphotic zone below the mixed layer than within the mixed layer, despite the fact that rates of primary production were greater within the mixed layer. These findings supported the view that the POC produced by phytoplankton in the mixed layer of oligotrophic gyres is recycled efficiently, and that very little is exported to depth. In contrast, a greater flux of POC was found to be exported from that portion of the euphotic zone lying below the mixed layer, despite lower rates of primary productivity within the deeper layer, because recycling there is much less efficient than within the mixed layer. Coale and Bruland (1987) concluded that the rate of scavenging and removal of ^{234}Th from surface waters is regulated more by the flux of POC exported from surface waters than by the rate of primary production. This view was supported by the strong correlation between the scavenging rate constant for thorium (k_1) and the flux of POC, as established using results from environments ranging from oligotropic gyres to productive estuaries (Bruland and Coale, 1986).

Coale and Bruland (1985) evaluated residence times of particulate thorium (Equation (5)) and concluded that residence times were regulated, in part, by the rates of zooplankton grazing and by the types of zooplankton present. Eppley (1989) suggested that if the residence time of particulate thorium could be applied to the inventory of POC in the upper water column, then one could derive the export flux of POC from measured inventories of POC and residence times of particulate thorium. Unfortunately, differential recycling of POC and particulate thorium precludes the application of this idea. Labile POC is regenerated more rapidly than refractory particulate thorium, causing the residence time of POC with respect to loss by sinking to be greater than the corresponding residence time of particulate thorium by a factor of 2–3 (Coale and Bruland, 1985; Murray *et al.*, 1989).

6.09.5.2 Export Flux of POC

Conceptual problems with the residence time of particulate thorium, as defined by Equation (5), have been noted by Kim *et al.* (1999) and by Alleau *et al.* (2001). However, Buesseler and Charette (2000) argued that the overall uncertainties in the evaluation of particle residence times associated with the propagated errors in the measurement of dissolved and particulate ^{234}Th are much greater than the errors in the conceptual model. More important, Buesseler and Charette (2000) noted that export fluxes of POC can be evaluated without relying on a determination of particle residence time. Their recommended approach is to integrate the deficit of ^{234}Th throughout the upper water column (Figure 4), and apply the mass balance constraint that the integrated deficit of ^{234}Th is equivalent to the flux of particulate ^{234}Th (F_{Th}) exported below the depth of integration:

$$F_{Th} = \lambda_{Th} \int_0^z (A_U - A_{Th}^t)\, dz \qquad (15)$$

Here, the total activity of ^{234}Th is used to evaluate the exported flux of particulate ^{234}Th.

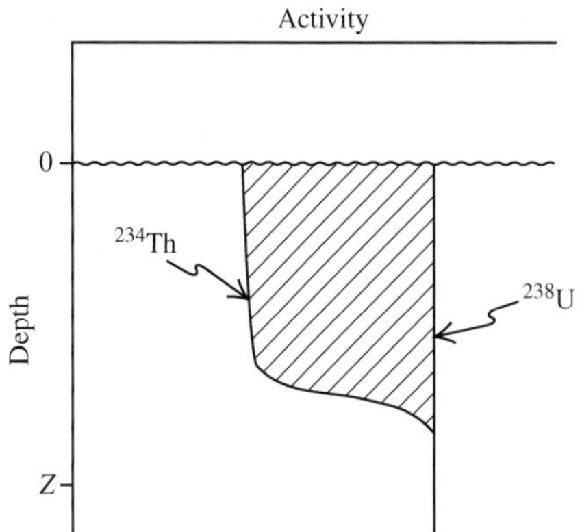

Figure 4 Schematic illustration of concentration profiles of ^{234}Th and ^{238}U in the upper water column. The integrated deficit of ^{234}Th (hatched area) is used to compute the flux of particulate ^{234}Th exported from surface waters.

The export flux of POC (F_{POC}) at depth z can then be evaluated as

$$F_{POC} = F_{Th}\left(\frac{POC}{A^{p}_{Th}}\right) \quad (16)$$

where POC/A^{p}_{Th} is the average $POC/^{234}Th$ ratio of exported particles. This approach, equivalent to that recommended by Buesseler and Charette (2000), is the one that has generally been employed since ^{234}Th was first used to evaluate export fluxes of POC during the JGOFS North Atlantic Bloom Experiment (NABE) in 1989 (Buesseler *et al.*, 1992). This approach makes no assumptions about the similar behaviors of POC and of particulate thorium. It relies only on two principles: (i) the export flux of ^{234}Th can be evaluated from measured $^{234}Th/^{238}U$ disequilibria, and (ii) the particles analyzed to evaluate the POC/A^{p}_{Th} ratio are actually characteristic of those carrying the exported POC and ^{234}Th. Confidence in this approach was provided by the good agreement between ^{234}Th-derived POC fluxes and independent estimates of POC export derived from nutrient budgets and by modeling CO_2 fluxes. Note, however, that while this approach has experienced much success, some substantial uncertainties remain, as described below.

6.09.5.3 Non-steady-state Conditions and Advected Fluxes

Simple mass budget relationships developed by Coale and Bruland (1985) to evaluate scavenging rates and residence times hold inherent assumptions, including steady-state conditions and negligible fluxes of ^{234}Th by advection or diffusion. Even though these conditions may apply at certain times and in certain places, there are many situations when they do not. For example, vertical advection and diffusion represented negligible ^{234}Th fluxes during the NABE (Buesseler *et al.*, 1992). However, the strong seasonality of phytoplankton growth and of POC fluxes associated with the spring bloom that is typical of this region precluded any meaningful application of a steady-state model. Buesseler *et al.* (1992) applied a non-steady-state model to a time series of four profiles of dissolved and particulate ^{234}Th, and compared the results to those that would have been obtained using a steady-state model. During times of decreasing ^{234}Th inventory, in the early stages of the bloom when scavenging intensity is increasing, the steady-state model underestimated the scavenging rate constant (k_1), and vice versa. The error in the inferred particle flux that would have resulted from the inappropriate application of a steady-state model would have been as large as a factor of 4. Similarly, during the early phase of the spring bloom (i.e., scavenging intensity increasing) near the Antarctic Polar Front in the Atlantic Sector of the Southern Ocean, Rutgers van der Loeff *et al.* (1997) found that a steady-state model underestimated the export flux of particles by more than a factor of 2.

An advantage of ^{234}Th is that its radioactive half-life (24.1 d) is comparable to the timescales of phytoplankton blooms. Consequently, time-series measurements of thorium deficits can be used to assess the changes in scavenging intensity and particle export throughout different phases of a bloom (e.g., growth and senescence). They can also be used during other transient events involving changes in biological productivity or other changes in particle flux (e.g., deposition of dust from the atmosphere or resuspension of sediments).

In regions of strong upwelling or lateral transport, steady-state models may be valid, whereas advection and diffusion cannot be neglected. For example, in the central equatorial Pacific Ocean, upwelling at the equator represents a significant source of dissolved ^{234}Th to surface waters (Bacon *et al.*, 1996; Buesseler *et al.*, 1995; Murray *et al.*, 1996). Although dramatic changes in upwelling rate in this region are associated with transitions between El Nino and La Nina conditions, the time rate of change in upwelling is sufficiently slower than the processes influencing the supply and removal of ^{234}Th that steady-state models can be applied (Bacon *et al.*, 1996; Buesseler *et al.*, 1995). Buesseler *et al.* (1995) applied horizontal and vertical transport velocities from an ocean General Circulation Model to measured distributions of ^{234}Th. Zonal and meridional fluxes of ^{234}Th were found to be small terms in the overall mass balance for ^{234}Th; however, the flux of ^{234}Th associated with upwelling at the equator could not be neglected. Including an upwelling term in the mass balance increased the ^{234}Th-derived export flux by $\sim$50% at 140° W, and by 25–30% in regions further to the east (Buesseler *et al.*, 1995). Within the upwelling system of the Arabian Sea, where a strong seasonal cycle in primary production and export flux are associated with the summer monsoon, Buesseler *et al.* (1998) found it necessary to include both upwelling and non-steady-state conditions in a model of ^{234}Th-derived export fluxes.

6.09.5.4 Limitations and Prospects

A number of uncertainties are inherent in the evaluation of export fluxes using the ^{234}Th method (Buesseler, 1998; Cochran and Masqué, 2003; Moran *et al.*, 2003). By far the greatest of these is associated with the determination of the POC/(particulate ^{234}Th) ratio used in Equation (16). Measured POC/(particulate ^{234}Th) ratios vary with particle size in a nonsystematic fashion, i.e., the ratio may either increase or decrease with increasing particle size, depending on the location (Buesseler, 1998). Measured POC/(particulate ^{234}Th) ratios generally decrease with increasing water depth (Bacon *et al.*, 1996; Buesseler, 1998), reflecting the greater lability of POC compared to particulate thorium. Measured POC/(particulate ^{234}Th) ratios also differ depending on whether particles are collected by sediment traps or by *in situ* filtration (e.g., Bacon *et al.*, 1996; Buesseler, 1998; Buesseler *et al.*, 1995; Murray *et al.*, 1996). Part of the disagreement among methods likely reflects the fact that the determination of POC concentrations is itself method dependent, with differences among methods exceeding an order of magnitude in extreme cases (Gardner *et al.*, 2003).

Despite these uncertainties, much has been learned about the global pattern of the export flux of POC, and its relationship to primary production, using the ^{234}Th technique. Throughout most of the ocean, the export flux of POC is a small fraction (2–10%) of primary production (Buesseler, 1998; Cochran and Masqué, 2003). Export ratios this low indicate tight coupling between production of POC by phytoplankton and its regeneration by zooplankton and bacteria. Exceptions occur during seasonal blooms at mid- to high latitudes, where 25–50% of POC produced by primary production may be exported (Buesseler *et al.*, 2003; Buesseler, 1998). High export ratios are a consequence of the decoupling that occurs when large phytoplankton, particularly diatoms, are able to grow rapidly in the absence of intense grazing by zooplankton.

In addition to the wealth of new information about POC fluxes gained by application of the ^{234}Th technique, fluxes of other particulate constituents, such as $CaCO_3$ (Bacon *et al.*, 1996) and biogenic opal (e.g., Buesseler *et al.*, 2001; Nelson *et al.*, 2002), have been determined as well. In principle, the flux out of surface waters of any particulate phase can be evaluated using the ^{234}Th method, provided that the average ratio of the constituent to ^{234}Th in exported particulate matter can be determined accurately.

Polonium-210 (half-life 138 d) has been examined as a possible tracer of export flux (e.g., Friedrich and Rutgers van der Loeff, 2002; Murray *et al.*, submitted). The following summary and prospectus is taken from these references. Supply of ^{210}Po occurs almost entirely by decay of ^{210}Pb which, in turn, is either produced *in situ* by decay of ^{226}Ra or is delivered to the sea surface in precipitation or with dry fallout. Lead-210 is itself scavenged by particles, but the rate constant for scavenging of ^{210}Po is greater than that of ^{210}Pb, allowing radioactive disequilibrium between ^{210}Pb and ^{210}Po to be used to evaluate the rate of scavenging and export of ^{210}Po following principles much like those described above for the ^{238}U/^{234}Th system.

Polonium is suggested to have an advantage over thorium as a tracer for POC export because polonium is assimilated into cells, where it seems to be associated with organic phases (possibly proteins), whereas lead, like thorium, is sorbed to the outer surfaces of particles, both living and detritus (Fisher *et al.*, 1983). Therefore, it has been argued that polonium may be a more specific tracer for POC flux rather than for the flux of total particulate matter. This general prediction is supported by water column data showing that particulate Po/Pb ratios correlate positively with POC/bSi ratios (bSi = biogenic silica) in the Southern Ocean (Friedrich and Rutgers van der Loeff, 2002). Furthermore, particulate ^{210}Po concentrations are positively correlated with POC concentrations in the equatorial Pacific Ocean, and the POC/(particulate ^{210}Po) ratio of suspended particles is indistinguishable from that of particles collected by sediment traps (Murray *et al.*, submitted), unlike the situation for the POC/(particulate ^{234}Th) ratio (see above).

The two principal disadvantages of using ^{210}Po as an export tracer are its half-life and the labor-intensive nature of its measurements. The half-life of ^{210}Po (138 d) is substantially greater than the timescale for seasonal phytoplankton blooms. Consequently, the amplitude of the seasonal cycle of scavenging and export will be damped by the time constant for ^{210}Po production and decay. Application of the ^{210}Po method requires independent determination of ^{210}Pb and ^{210}Po concentrations, both of which are labor intensive. Typically, ^{210}Po is determined by isotope dilution alpha spectrometry, and ^{210}Pb isolated from the sample is set aside for several months to permit ingrowth of ^{210}Po, which is measured a second time to evaluate the concentration of ^{210}Pb. Despite these limitations, if further studies confirm that ^{210}Po is a more specific tracer than thorium for POC, and if it can be demonstrated that uncertainties in the POC/(particulate ^{210}Po) ratio are much less than for ^{234}Th, then ^{210}Po may yet find widespread application as a tracer of POC export from surface waters.

6.09.6 PARTICLE DYNAMICS AND REGENERATION OF LABILE PARTICLES

6.09.6.1 Particle Dynamics, Fluxes, and Regeneration

Most of the downward flux of particulate material is carried by aggregates that settle at rates of ~50–150 m d^{-1}, orders of magnitude greater than would be expected for individual cells or other particles of similar size (Jackson and Burd, 1998). The transport and fate of material carried by aggregates is regulated by the interplay between the sinking rate of the aggregates and the factors that control their interaction with dissolved and suspended phases. Adsorption, desorption, aggregation, disaggregation, dissolution, and ingestion are factors that influence the size, composition, density, and sinking rate of particles, as well as the fate of material carried by rapidly settling aggregates.

Downward transport and regeneration of POC in the ocean represents a case of special interest. The depth scale of POC regeneration influences the timescale over which respired CO_2 is isolated from contact with the atmosphere. In general, the greater the depth of POC regeneration, the longer the time before respired CO_2 will reach the sea surface where it can exchange with the atmosphere. Consequently, the overall partitioning of CO_2 between the atmosphere and the deep sea is dependent, in part, on the depth scale of POC regeneration. Early efforts to characterize the depth scale of POC regeneration relied on empirical fits to POC fluxes collected by sediment traps (e.g., Martin *et al.*, 1987). These early efforts demonstrated that most of the POC exported from surface waters is regenerated within the upper few hundred meters of the water column. However, they offered little insight into the mechanisms regulating the depth scale of regeneration.

A mechanistic understanding of the processes regulating particle dynamics and regeneration depth scales is essential to the development of ocean carbon cycle models that simulate accurately the partitioning of CO_2 between the atmosphere and the ocean. Consequently, the oceanographic community is investing considerable effort into studies of these parameters, and into the development of models to simulate the relevant processes (e.g., Athias *et al.*, 2000; Boehm and Grant, 2002; Boyd *et al.*, 1999; Boyd and Stevens, 2002; Burd *et al.*, 2000; Dadou *et al.*, 2001). Armstrong *et al.* (2002) and Berelson (2002) have demonstrated that particle sinking rates and regeneration depth scales vary with time, location, and particle composition. However, further research is needed to identify the processes regulating this variability, and to determine their sensitivity to climate change.

6.09.6.2 Conceptual Models of Aggregation and Disaggregation

In a study of sediment trap samples collected at a depth of 3,200 m in the Sargasso Sea, Bacon *et al.* (1985) found that the fluxes of several radionuclides exhibited a seasonal cycle in phase with the total mass flux of particles. Mass flux, in turn, was found to be closely coupled to the seasonal cycle of primary production in surface waters of this region (Deuser *et al.*, 1981). Seasonally varying fluxes of radionuclides produced in the deep sea (e.g., ^{230}Th, ^{231}Pa, and ^{210}Pb) are inconsistent with the view that these nuclides are removed by scavenging to small particles which constitute the bulk of particle mass in the deep sea and which are inferred to sink at an average rate of several hundred meters per year (Section 6.09.3). A seasonal cycle in the flux of these nuclides implies that scavenging in the deep sea responds rapidly to changes in the export of particles from surface waters.

Bacon *et al.* (1985) found that the radionuclide content of the large rapidly settling particles collected by the deep sediment trap remained relatively constant throughout the year, despite large seasonal changes in the flux of large particles. This finding led to the view that sinking particles in the deep sea (3,200 m in this case) are drawn from a reservoir of relatively uniform composition, but at seasonally varying rates. Furthermore, these rates are proportional to the flux of biogenic particles exported from surface waters.

To describe these observations, Bacon *et al.* (1985) developed a simple conceptual model involving two size classes of particles (Figure 5(a)): (i) small suspended particles which have negligible sinking rates constitute most of the mass of particulate matter in the deep sea and are responsible for most scavenging of radionuclides and other dissolved particle-reactive species; and (ii) large aggregates which represent a small fraction of total particulate material in the deep sea but which, because of their large setting velocities (~100 m d^{-1}), carry most of the flux of sinking particulate material. A similar conceptual model was invoked previously by Bishop *et al.* (1977) based on the analysis of particles collected by *in situ* filtration.

Downward transport of substances such as ^{230}Th, which are scavenged initially to small particles, is induced by aggregation processes that

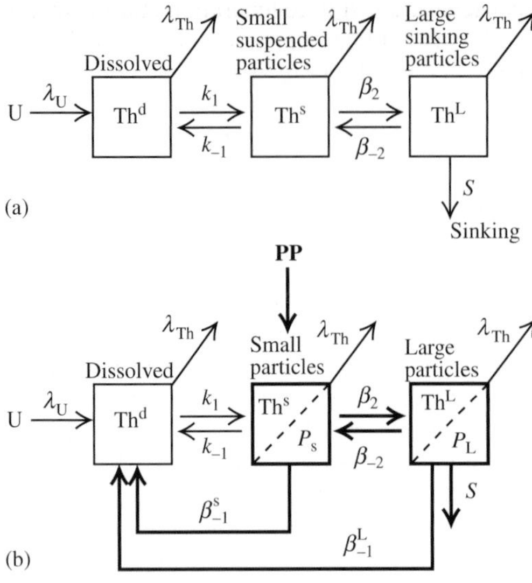

Figure 5 (a) Scavenging model of Bacon *et al.* (1985) illustrating a two-step process for removal of dissolved ^{230}Th, and other particle-reactive substances, from seawater. Adsorption of dissolved ^{230}Th by small suspended particles is treated as a reversible process, with rate constants for adsorption and desorption represented by k_1 and k_{-1}. Small particles constitute most of the particle mass in the oceans, and are considered to sink at a negligible rate. Small particles are assembled into large aggregates, also by a reversible process (β_2 and β_{-2}), which carry most of the sinking flux of particulate material in the ocean. (b) Schematic illustration of the model used by Clegg and co-workers (Clegg *et al.*, 1991; Clegg and Whitfield, 1993) and similar to that used by Cochran and co-workers (Cochran *et al.*, 1993, 2000) to simulate scavenging and particle dynamics. Three compartments of the model consist of the dissolved phase, small particles (P_s) and large particles (P_L). Elements in bold represent the behavior of particle mass, which is produced by primary production (PP) and lost to remineralization (both size classes; β_{-1}) and to sinking (large size class only; S). First-order rate constants for aggregation of small particles and disaggregation of large particles are represented by β_2 and β_{-2}, respectively. Elements in lighter type, and above the diagonal dashed lines, represent the behavior of Th, which undergoes reversible sorption to small particles (k_1, k_{-1}) and tracks particles through aggregation, disaggregation, remineralization, and sinking.

incorporate small suspended particles into large rapidly settling aggregates according to the model of Bacon *et al.* (1985). The rate of aggregation, in turn, is proportional to the flux of large particles exported from the surface (i.e., a function of primary production). Disaggregation processes were invoked, as well, to maintain a steady-state distribution of small suspended particles in the deep sea. This general conceptual model has served as the basis for numerous subsequent studies of the dynamics of marine particles.

6.09.6.3 Strategies to Evaluate Rate Constants

Several strategies have been employed to evaluate rate constants inherent in the conceptual model of Bacon *et al.* (1985). Nozaki *et al.* (1987) derived aggregation and disaggregation rates from measured vertical distributions of dissolved and particulate ^{230}Th. Clegg and co-workers (e.g., Clegg *et al.*, 1991; Clegg and Whitfield, 1990, 1991, 1993) added to Bacon's conceptual model an explicit source of particles produced by primary production as well as regeneration of labile particles (Figure 5(b)). Cochran and co-workers (Cochran *et al.*, 1993, 2000) solved simultaneously the equations expressing mass balances for ^{234}Th and ^{228}Th to derive rate constants for aggregation and disaggregation. Murnane and co-workers (Murnane, 1994; Murnane *et al.*, 1996, 1994, 1990) applied inverse techniques to measured distributions and fluxes of thorium isotopes and of particulate matter to derive rate constants for aggregation, disaggregation, and particle regeneration. In each of these studies, adsorption and desorption were treated as reversible processes. Rate constants for adsorption and desorption evaluated during these studies were in all cases found to be consistent with those determined previously (Bacon and Anderson, 1982; Nozaki *et al.*, 1981).

Efforts to derive rate constants for aggregation and disaggregation build upon the work of Nozaki *et al.* (1987), who incorporated these constants into mass balance expressions for thorium isotopes contained in three phases: dissolved thorium, small nonsinking particles, and large rapidly settling particles. Mass balances, expressed in terms of thorium activity (A_{Th}), are

$$\frac{\partial A^d_{Th}}{\partial t} = \lambda_{Th}A_U + k_{-1}A^p_{Th} - (\lambda_{Th} + k_1)A^d_{Th} \quad (17)$$

$$\frac{\partial A^p_{Th}}{\partial t} = k_1A^d_{Th} + \beta_{-2}A^L_{Th} - (\lambda_{Th} + k_{-1} + \beta_2)A^p_{Th} \quad (18)$$

$$\frac{\partial A^L_{Th}}{\partial t} = \beta_2A^p_{Th} - (\lambda_{Th} + \beta_{-2})A^L_{Th} - S\frac{\partial A^L_{Th}}{\partial z} \quad (19)$$

where superscripts d, p, and L refer, respectively, to the concentration (activity, dpm m^{-3}) of thorium (^{234}Th or ^{230}Th) in dissolved, small particulate, and large particulate forms. Rate constants for adsorption (k_1), desorption (k_{-1}), aggregation (β_2), and disaggregation (β_{-2}) are as depicted in Figure 5(a). S is the sinking rate of large particles.

Nozaki *et al.* (1987) presented solutions to the above equations assuming a steady state ($\partial/\partial t = 0$) and assuming that vertical diffusion and advection, as well as horizontal transport, are negligible. Further assuming no depth dependence of the rate constants, the solutions to the equations above for ^{230}Th, for which radioactive decay can be neglected due to its long half-life, reduce to the following expressions:

$$A_{\mathrm{Th}}^{\mathrm{L}} = \frac{\lambda_{\mathrm{Th}} A_{\mathrm{U}}}{S} z \tag{20}$$

$$A_{\mathrm{Th}}^{\mathrm{p}} = \frac{\lambda_{\mathrm{Th}} A_{\mathrm{U}}}{S} \frac{\beta_{-2}}{\beta_2} z + \frac{\lambda_{\mathrm{Th}} A_{\mathrm{U}}}{\beta_2} \tag{21}$$

$$A_{\mathrm{Th}}^{\mathrm{d}} = \frac{\lambda_{\mathrm{Th}} A_{\mathrm{U}}}{S} \frac{\beta_{-2}}{\beta_2} \frac{k_{-1}}{k_1} z + \frac{\beta_2 + k_{-1}}{k_1 \beta_2} \lambda_{\mathrm{Th}} A_{\mathrm{U}} \tag{22}$$

As shown by these expressions, the concentration of ^{230}Th in each phase, as well as the total concentration of ^{230}Th in the water column, is predicted to increase linearly with depth. This is consistent with observations (Nozaki *et al.*, 1987; Figure 2).

The average settling rate of large particles can be estimated from the depth dependence of the activity of ^{230}Th (dpm m^{-3}) associated with large particles (rearrange Equation (20) to solve for S). Alternatively, sinking rates can be estimated from the rate of downward propagation of a transient signal generated in surface waters and captured in a series of sediment traps placed at different depths (Berelson, 2002). With S determined, the ratio β_{-2}/β_2 can be derived from the measured depth dependence of $A_{\mathrm{Th}}^{\mathrm{p}}$ (Equation (21)), whereas the value of β_2 can be estimated from the surface intercept of the relationship between $A_{\mathrm{Th}}^{\mathrm{p}}$ and depth. Adsorption and desorption rate constants can be estimated from the depth dependence of dissolved ^{230}Th, as described by Nozaki *et al.* (1987). Adsorption and desorption rate constants estimated by this approach are in good agreement with those derived using dual thorium isotopes (Section 6.09.3).

The approach of Nozaki *et al.* (1987) is limited by its assumptions, in particular, by the assumed uniformity of the rate constants throughout the water column. This is likely not the case in the upper ocean, where most regeneration of particulate matter occurs. Nevertheless, to the extent that this approach is valid, it suggests that the timescale for aggregation of small particles ranges from a few weeks to a year ($\beta_2 = 0.006-0.03$ d^{-1}), whereas the corresponding timescale for disaggregation of large particles is on the order of a day ($\beta_{-2} = 0.4-65$ d^{-1}), indicating that individual large particles have a short lifetime in the deep sea.

Clegg *et al.* (1991) constructed a particle cycle model in which the upper water column was divided into a series of boxes, and the mass balance for thorium isotopes and for total particle mass was constrained for each box. Biogenic particles are introduced to the surface box using measured or estimated rates of primary production, while particles are lost to regeneration and to sinking in all boxes. Equations for particle mass balance can be expressed (Figure 5(b)) as

$$\frac{\partial P_{\mathrm{S}}}{\partial t} = \mathrm{PP} - (\beta_{-1} + \beta_2) P_{\mathrm{S}} + \beta_{-2} P_{\mathrm{L}} \tag{23}$$

$$\frac{\partial P_{\mathrm{L}}}{\partial t} = \beta_2 P_{\mathrm{S}} - (\beta_{-1} + \beta_{-2}) P_{\mathrm{L}} - S \frac{\partial P_{\mathrm{L}}}{\partial z} \tag{24}$$

where P_{S} and P_{L} are the concentrations of small and large particles, respectively, PP represents the rate of particle production in surface waters, β_2 and β_{-2} are defined above, and β_{-1} is the rate constant for regeneration, which is applied to all particles. Assuming steady state and negligible transport by advection and diffusion, Clegg *et al.* (1991) derived rate constants for adsorption, desorption, aggregation, disaggregation, and regeneration using data from Station P in the northeastern Pacific, and from three sites in the equatorial Pacific Ocean. In each case, a particle settling rate of 150 m d^{-1} was assumed.

Different approaches were used by Clegg *et al.* (1991) depending on the nature of the data available from each site. For the equatorial Pacific sites, mass budget information for total particle mass was combined with a mass balance for ^{234}Th to derive the rate parameters. The system at Station P was underdetermined, so the disaggregation rate constant from the equatorial Pacific was applied to allow calculation of aggregation and regeneration rates. At Station P the regeneration rate (β_{-1}) was found to decrease exponentially from 0.05 d^{-1} at 150 m to 0.001 d^{-1} in the deep sea. At the equatorial Pacific sites, β_{-1} decreased from ~0.1 d^{-1} in the upper 100 m to ~0.004 d^{-1} below 200 m. Aggregation rate constants (β_2) at the equatorial Pacific sites decreased from 0.02–0.1 d^{-1} in the upper 100 m to 0.0001–0.001 d^{-1} below 200 m. Rate constants for disaggregation (β_{-2}) decreased from 1–10 d^{-1} in the upper 100 m to ~0.2 d^{-1} below 200 m. Applying this value of β_{-2} to dissolved and particulate thorium distributions at Station P, Clegg *et al.* (1991) estimated β_2 values to decrease from ~0.1 d^{-1} at the surface to ~0.002 d^{-1} in the deep sea.

Clegg and Whitfield (1993) applied their particle cycle model to the northeastern Atlantic site studied during the JGOFS NABE. Primary production data, together with concentrations and fluxes of particulate matter measured during the experiment, were used to model particle dynamics following the procedure of Clegg and Whitfield (1990). The particle cycle model was developed independently of the thorium data that were available from that site. Assuming an average settling rate of large particles to be 150 m d^{-1}, and a particle disaggregation rate constant equivalent to that derived in the earlier equatorial Pacific study (Clegg *et al.*, 1991), Clegg and Whitfield (1993) found that the measured concentrations and fluxes of ^{234}Th (Buesseler *et al.*, 1992) could be predicted well with the particle cycle model, indicating that the ^{234}Th/^{238}U disequilibria were consistent with the model simulation of the particle cycle.

Cochran *et al.* (1993, 2000) evaluated rate constants for aggregation and disaggregation by solving simultaneously separate expressions of Equation (19) for ^{234}Th and for ^{228}Th. Assuming steady state, negligible transport of thorium by advection and diffusion, and a settling rate of 150 m d^{-1} for large particles, β_2 and β_{-2} were derived using measured concentrations of ^{234}Th and ^{228}Th in large and small particles. During the JGOFS NABE, Cochran *et al.* (1993) analyzed sediment trap samples to obtain thorium concentrations in large particles. During the JGOFS study in the Ross Sea, large particles were collected for analysis by *in situ* filtration onto Teflon screen having a mesh size of 70 μm (Cochran *et al.*, 2000). Rate constants determined in the Ross Sea (0.04–0.2 d^{-1} for β_2 and 2.4–13.8 d^{-1} for β_{-2}) were greater than the corresponding values in the North Atlantic study (0.003–0.09 d^{-1} for β_2 and 0.34–1.1 d^{-1} for β_{-2}). Results from the North Atlantic were derived for the depth interval between 150 m and 300 m, whereas those from the Ross Sea apply to the upper 100 m, and this difference may account, in part, for the larger rate constants derived for the Ross Sea.

Aggregation rates in the Ross Sea increased with increasing biomass from spring to summer, whereas disaggregation rates decreased over the same time period (Cochran *et al.*, 2000). Aggregation rates decreased with increasing water depth in the Ross Sea, whereas disaggregation rates exhibited the opposite trend. In contrast to the Ross Sea, both aggregation and disaggregation rates increased with increasing biomass during the spring bloom in the North Atlantic study (Cochran *et al.*, 1993).

Large uncertainties exist, both in the absolute values of the rate constants and in the trends described above, due to the propagation of errors through the large number of measured parameters required to derive the rate constants (Cochran *et al.*, 1993, 2000). Consequently, alternative strategies have been explored using larger data sets to better constrain rate constants for aggregation and disaggregation. Murnane and co-workers (Murnane, 1994; Murnane *et al.*, 1996, 1994, 1990) pioneered the use of inverse techniques to derive rate constants for particle dynamics. Murnane *et al.* (1990) noted that nine independent equations can be derived by writing mass balance expressions for three thorium isotopes (^{228}Th, ^{230}Th, and ^{234}Th) in each of the three phases (dissolved, small particle, and large particle). In principle, then, one could solve for adsorption, desorption, aggregation, disaggregation, sinking, and regeneration rates (Figure 5(b)) if the concentration of each of the three thorium isotopes were measured in the dissolved phase, as well as in small and large particulate phases, throughout the water column.

Murnane (1994) compared different regression techniques for evaluating rate constants using thorium data collected at Station P in the northeastern Pacific Ocean. Murnane *et al.* (1994) extended this approach to the northwestern Atlantic Ocean, where previous studies provided dissolved and particulate concentrations of ^{228}Th and ^{230}Th, dissolved concentrations of ^{228}Ra, total small particle concentration, and total large particle flux. Despite the effort to include a large amount of information from previous studies, both in the northeastern Pacific and in the northwestern Atlantic, the rate constants for aggregation, disaggregation, and remineralization were poorly constrained. Murnane and co-workers suggested that the large uncertainties may have resulted from non-negligible advection and diffusion terms in the mass balance equations, or from significant temporal variability that violated the assumption of steady state. Although the model results contained large uncertainties, the estimated median values for aggregation (0.006–0.01 d^{-1}) and disaggregation (0.4–0.5 d^{-1}) rate constants are consistent with values obtained by previous studies (see above). The median estimate of the rate constant for regeneration (β_{-1}) decreased from 0.2 d^{-1} in surface waters to 0.0005 d^{-1} in the deep North Atlantic Ocean.

Murnane *et al.* (1996) took advantage of the larger suite of data collected during the JGOFS NABE to derive rate constants for aggregation, disaggregation, and regeneration that had uncertainties much smaller than in other studies. In addition to dissolved and particulate concentrations of ^{234}Th and ^{228}Th, they included the concentrations and fluxes of POC and nitrogen in the inverse model, as well as fluxes of thorium isotopes and of total mass collected by sediment traps. Murnane *et al.* (1996) found rate constants for aggregation and disaggregation to increase over the course of the bloom, ranging from 0.005 ± 0.0005 d^{-1} to 0.21 ± 0.024 d^{-1} for β_2 and 0.4 ± 0.05 d^{-1} to 1.4 ± 0.2 d^{-1} for β_{-2}. Rate constants for regeneration in the upper water column ranged from 0.04 d^{-1} to 0.2 d^{-1}.

Each of the models described above is highly idealized in order to estimate rate constants for particle dynamics using a limited amount of available data. For example, the models above assume only two classes of particles: only one sinking rate for large particles, and only one regeneration rate where regeneration is allowed at all. Models containing a spectrum of particle sizes, particle sinking rates, and regeneration rates have been developed (e.g., Boyd and Stevens, 2002; Burd *et al.*, 2000; Jackson, 2001). A challenge for the future will be to design an experimental strategy under which the particle-tracing characteristics of thorium isotopes, together with the rate information provided by their radioactive decay constants, can be exploited to constrain parameters representing particle dynamics in models simulating the behavior of many classes of particles.

6.09.7 LATERAL REDISTRIBUTION OF SEDIMENTS

6.09.7.1 Focusing and Winnowing of Sediments

Efforts to construct geochemical mass budgets and many areas of paleoceanographic research rely on an accurate evaluation and interpretation of particle fluxes and of sediment accumulation rates. Ocean currents transport particles laterally throughout the ocean. Therefore, geochemical mass budgets and paleoceanographic interpretations may both suffer significant errors if lateral redistribution of particles by ocean currents is not taken into account.

Sediment focusing is a term used to describe conditions when ocean currents produce a net lateral transport of particles to a reference site. That is, the rate of particle deposition on the seabed at a site experiencing sediment focusing is greater than the regional average rain rate of particles settling through the water column. For example, preferential deposition of particles may occur at sites where the interaction between deep currents and bottom topography creates a decrease in turbulent energy that maintains particles in suspension. Winnowing is the term that is applied when currents lead to the net lateral transport of particles away from a reference site. In this case, the rate of sediment deposition at the reference site is less than the regional average rain rate of settling particles.

Particle-reactive radionuclides provide a tracer with which to evaluate the net lateral redistribution of sediments. In some cases, inventories or accumulation rates of these nuclides can be compared to their known rate of supply to estimate the net extent of lateral redistribution of sediments. In other cases, inventories (or accumulation rates) of nuclides can be compared among several sites to assess the relative rate of sediment deposition among the sites, even if the rate of supply of the tracer is not known accurately. Inherent in this approach is the assumption that sediments deposited at all of the sites have a uniform initial radionuclide content. Deposition of sediments must be roughly at steady state over the mean life of the nuclide, although short-term fluctuations in the deposition of sediment do not invalidate the approach.

In principle, this approach can be applied to any environment. However, application to shallow nearshore environments is complicated by conditions that lead to net lateral fluxes of dissolved nuclides. For example, ocean currents may carry dissolved nuclides from a region of low particle flux to a region of high particle flux, where intensified scavenging strips the dissolved nuclide from the water column. Under these circumstances, the inventory of the nuclide in the sediments at the high particle flux site would be much greater than at the low particle flux site. Substantial lateral fluxes of dissolved ^{210}Pb, ^{234}Th, and ^{7}Be to coastal sites have been documented (e.g., Carpenter *et al.*, 1987; Gustafsson *et al.*, 1998). Failure to account for the lateral supply of dissolved nuclides under such conditions would lead to the misinterpretation of sediment nuclide inventories to indicate focusing of sediment to the high particle flux site. For this reason, natural radionuclides have been applied more successfully to the study of sediment transport in the deep sea, where lateral gradients in particle flux and in dissolved nuclide concentrations are much smaller than in coastal surface waters.

6.09.7.2 Lead-210

An advantage of using ^{210}Pb (half-life 22.3 yr) in the study of sediment transport in the deep sea is that its average rate of supply to the sediments can be determined simply by measuring its inventory. Assuming steady-state depositional conditions over the mean life of ^{210}Pb, its flux to the sediments (F_{Pb}) is related to its inventory (I_{Pb}) as $F_{Pb} = \lambda_{Pb} I_{Pb}$. In typical deep-sea sediments, the entire inventory of ^{210}Pb supplied by scavenging from the water column, often referred to as excess ^{210}Pb to distinguish it from ^{210}Pb supported by its uranium-series progenitors in the sediments, is contained within the upper 10–20 cm of sediment. Consequently, the inventory of ^{210}Pb can be estimated with relatively few measurements, and the average flux of ^{210}Pb can be determined without the need to evaluate sediment accumulation rates. Inventories of ^{210}Pb can then be used, for example, to characterize regional patterns of sediment deposition, or to discriminate between sediment supplied to a study site by sinking from above versus that supplied by lateral transport.

Sediment bedforms, referred to as mudwaves, often form in regions of strong abyssal currents. A model for the formation of mudwaves suggested that, on average, there should be enhanced sediment deposition on the upstream face of a wave, and reduced deposition or erosion on the downstream side (Flood, 1988; Flood *et al.*, 1993). In a study of mudwaves in the Argentine Basin, inventories of ^{210}Pb were found to be greater on the upstream face of mudwaves than on the downstream side (Anderson *et al.*, 1993), supporting the proposed mechanism of formation.

Biological productivity in coastal waters is generally much greater than in the open ocean, which raises questions about the role of coastal systems in the global ocean carbon cycle (Liu *et al.*, in preparation). The Shelf-Edge Exchange Processes (SEEP) program was designed to investigate the production, transport, transformation, and fate of POC in an ocean-margin region of the northwestern Atlantic Ocean (Biscaye *et al.*, 1994). Mean annual fluxes of POC collected by sediment traps deployed on the continental slope increased with depth by about a factor of 5 (from ~100 m to the bottom, at ~1,000 m; Biscaye and Anderson, 1994). This increase with depth in POC flux could have resulted either from the collection of POC exported laterally from the adjacent continental shelf, or from the collection of locally resuspended bottom sediments. In the latter case, short-term fluxes collected by sediment traps would have been much greater than long-term average rates of sediment accumulation.

Annual fluxes of ^{210}Pb collected by sediment traps increased with depth in much the same manner as the increase in POC flux, suggesting that similar processes were affecting ^{210}Pb and POC. Furthermore, annual fluxes of ^{210}Pb collected by near-bottom sediment traps were found to be in good agreement with the mean annual fluxes of ^{210}Pb required to maintain the measured inventories of ^{210}Pb in slope sediments of the SEEP study area (Anderson *et al.*, 1994). This agreement indicated that the higher fluxes of ^{210}Pb collected by the deeper traps were not created by collection of locally resuspended sediments. Rather, particulate ^{210}Pb was supplied by lateral transport at mid-depth to the study area. By analogy, it was inferred that similar lateral transport processes were supplying POC at mid-depth, and it was concluded that the flux of POC collected by deep sediment traps was an accurate representation of the flux of POC to sediments. Applying ^{210}Pb in this manner helped SEEP investigators constrain the flux of POC exported from the continental shelf to be less than 2% of the POC produced by primary production in shelf waters (Anderson *et al.*, 1994). Future programs seeking to construct mass budgets in regions of dynamic sediment transport would similarly benefit from using ^{210}Pb to discriminate between lateral supply of particles and local resuspension of bottom sediments as contributors to fluxes collected by sediment traps.

6.09.7.3 Thorium-230

Focusing and winnowing of particles by deep-sea currents create potential artifacts in paleoceanographic research. Many studies rely on an accurate determination of the accumulation rate of marine sediments, as well as of biogenic and geochemical proxies contained within the sediments, to infer past changes in ocean processes and environmental conditions. One area of paleoceanographic research using this approach involves the assessment of past changes in biological productivity of the ocean. Another is the evaluation of the transport of continental dust via the atmosphere to remote regions of the ocean. Paleoceanographers working on these topics require a means with which to determine lateral fluxes of particulate matter. Furthermore, an accurate assessment of past changes in the lateral transport of sediments by deep-sea currents offers valuable insights into climate-related changes in ocean circulation. The well-defined source of ^{230}Th, together with its tendency to be removed rapidly from the water column by scavenging to settling particles, has enabled paleoceanographers to evaluate, and correct for, sediment redistribution by deep-sea currents (for reviews see Frank *et al.*, 1999; Henderson and Anderson, 2003).

In using ^{230}Th to evaluate sediment transport, it is assumed that ^{230}Th produced by uranium decay in the water column is removed quantitatively by scavenging to settling particles. That is, the flux of particulate ^{230}Th to the sediments (F_{Th}) is assumed to be equal to the production rate of ^{230}Th in the water column ($P_{Th} = \lambda_{230}A_{234}z$), where λ_{230} is the decay constant of ^{230}Th, A_{234} is the activity of ^{234}U, and z is water depth. P_{Th} is a linear function of water depth, amounting to a production rate of 2.63×10^{-3} dpm ^{230}Th cm^{-2} kyr^{-1} per meter of water depth.

The strong affinity of thorium for sorption to particles gives it a residence time with respect to scavenging and removal from the deep sea of between 10 yr and 40 yr (Anderson *et al.*, 1983a,b). This residence time is sufficiently short that there is little net lateral transport of dissolved ^{230}Th between its production and removal to sediments. Numerous studies of particles collected by sediment traps have confirmed that F_{Th} is approximately equal to P_{Th} (to within about ±30%) throughout most of the open ocean (Scholten *et al.*, 2001; Yu *et al.*, 2001a,b and references therein). Sensitivity tests using an ocean general circulation model also confirmed that F_{Th} is within about ±30% of P_{Th} throughout most of the open ocean (Henderson *et al.*, 1999). Because its flux is constrained so well, ^{230}Th has come to be used widely as a "constant flux proxy" (CFP) for paleoceanographic research, allowing investigators to evaluate, and correct for, lateral redistribution of particles.

Measured concentrations of ^{230}Th in marine sediments consist of three components: one scavenged from seawater, another supported by uranium contained within lithogenic minerals, and the third produced by radioactive decay of authigenic uranium. In order to use the scavenged component of ^{230}Th as a CFP, the measured ^{230}Th concentration must be corrected for the other components, and the scavenged component (generally referred to as "excess") must be corrected for radioactive decay since the time of sediment deposition. This decay correction requires an independent chronology for the sediments, which is generally provided by ^{14}C or ^{18}O stratigraphy. Details concerning these corrections are provided by Henderson and Anderson (2003).

Sediment focusing can be evaluated by comparing the accumulation rate of excess ^{230}Th with its rate of production in the overlying water column. Using measured concentrations of excess ^{230}Th in a sediment core, a sediment focusing factor (ψ) can be defined as (Francois *et al.*, 1990; Suman and Bacon, 1989)

$$\psi = \frac{\int_{z_2}^{z_1}(\mathrm{xs}^{230}\mathrm{Th}^0)\rho_\mathrm{b}\,\mathrm{d}z}{P_\mathrm{Th}(t_1 - t_2)} \qquad (25)$$

where z_i is the depth at interval i in the core, t_i is the age at depth i, ($\mathrm{xs}^{230}\mathrm{Th}^0$) is the decay-corrected concentration of excess ^{230}Th in the sediment, P_Th is as defined above, and ρ_b is the bulk dry density of the sediment. Sediment focusing factors greater than 1 imply a flux of excess ^{230}Th to the sediments greater than that expected by production in the water column (i.e., deep-ocean currents laterally transport particles containing ^{230}Th to the site of interest). Focusing factors less than 1 indicate that sediment winnowing has taken place (i.e., net export of particles that contain excess ^{230}Th).

Focusing factors in excess of 10 occur in the Southern Ocean (Frank *et al.*, 1999), where the Antarctic circumpolar current contains regions of high velocity throughout the water column. In other regions focusing factors are generally less than the extreme cases that occur in the Southern Ocean. However, sediment focusing factors may change systematically with climate (e.g., Higgins *et al.*, 2002; Marcantonio *et al.*, 2001, 1996), and this must be taken into account when interpreting accumulation rates of sedimentary material (as illustrated below with barite accumulation).

With the constraint that F_Th is constant and approximately equal to P_Th, accumulation rates of sedimentary constituents can be evaluated using the ^{230}Th-profiling method developed by Bacon and co-workers (Bacon, 1984; Francois *et al.*, 1990; Suman and Bacon, 1989). The flux of any sedimentary constituent "i" (F_i) can be estimated as

$$F_i = \frac{C_i P_\mathrm{Th}}{\mathrm{xs}^{230}\mathrm{Th}^0} \qquad (26)$$

where F_i is the flux of constituent "i," C_i is the measured concentration of "i" in the sediments, and other terms are defined above. Three principal advantages are offered by this approach: (i) fluxes of sedimentary constituents can be corrected for sediment focusing; (ii) fluxes can be evaluated at a higher temporal resolution than is normally achieved using traditional stratigraphic methods, providing temporal resolution that is limited roughly by the ratio of bioturbation depth to sediment accumulation rate; and (iii) fluxes are relatively insensitive to errors in sediment chronology of up to several thousand years. More important, fluxes derived by this method are insensitive to random errors in sediment chronology that may cause inferred sediment accumulation rates between two age control points to be much greater or less than the true value. Uncertainty in the absolute flux of sedimentary constituents is limited by the ~30% uncertainty in the assumption that $F_\mathrm{Th} = P_\mathrm{Th}$. However, within a single core, where local processes influencing the scavenging and removal of ^{230}Th are unlikely to experience extreme changes through time, the uncertainty in the relative fluxes of sedimentary constituents at different depths in a single core is much less than this.

Since its introduction by Bacon (1984), the ^{230}Th-profiling method has been used to evaluate accumulation rates of biogenic and lithogenic sediments at many locations throughout the ocean. Bacon (1984) determined that the glacial–interglacial cycles in the $CaCO_3$ content of North Atlantic sediments were caused by dilution of a relatively constant flux of $CaCO_3$ by a variable supply of lithogenic material. Francois *et al.* (1990) discovered a short-lived increase in $CaCO_3$ burial during the last deglaciation in tropical Atlantic sediments. There, the duration of increased $CaCO_3$ deposition was too short to have been evident using traditional stratigraphic techniques.

Paleoproductivity studies in the Southern Ocean, where the deep-reaching Antarctic circumpolar current causes substantial sediment redistribution, have been especially dependent on the ^{230}Th-profiling method to obtain accumulation rates of opal, excess barium, organic carbon, and other paleoproductivity proxies (Anderson *et al.*, 2002, 1998; Chase *et al.*, 2003a; Francois *et al.*, 1997, 1993; Frank, 1996; Frank *et al.*, 2000; Kumar *et al.*, 1995). Studies of the Atlantic and Indian sectors of the Southern Ocean concluded that productivity in regions south of the Antarctic polar front was lower during the last glacial maximum (LGM) than during the Holocene, whereas productivity north of the APF was greater than today during the LGM. Whether or not there was a net change in productivity of the Southern Ocean from the LGM to the Holocene remains a topic of debate. One view is that there was little net change between glacial and Holocene conditions (e.g., Francois *et al.*, 1997; Frank, 1996; Frank *et al.*, 2000). Anderson *et al.* (2002, 1998) and Kumar *et al.* (1995) concluded that productivity was greater overall during the LGM. These differences reflect the greater emphasis given to accumulation rates of opal and of excess barium by those advocating little net change, and the greater emphasis given to organic carbon and authigenic uranium accumulation rates by those inferring greater overall productivity during the LGM. These patterns are confined to the Atlantic and western Indian sectors of the Southern Ocean. In the Pacific sector, Chase *et al.* (2003a) found evidence for lower overall productivity during the LGM. In all cases, internally consistent results were obtained, despite extensive sediment redistribution in the Southern Ocean, because accumulation rates of paleoproductivity proxies were evaluated using the ^{230}Th-profiling method. Where disagreements remain, they pertain to the interpretation of specific proxy records rather than to uncertainties in the estimated proxy accumulation rates.

Barite accumulation in equatorial Pacific sediments provides a good illustration of the pronounced differences that obtain depending on whether the ^{230}Th-profiling technique or traditional stratigraphic methods, such as ^{18}O, are used. Barite accumulation rates are often used to infer past changes in biological productivity, and barite accumulation rates in equatorial Pacific sediments evaluated using ^{18}O stratigraphy have been interpreted to indicated substantially greater productivity of that region during glacial periods compared to interglacials (Paytan *et al.*, 1996). In contrast to the results obtained using ^{18}O stratigraphy, where glacial–interglacial changes in barite accumulation exhibit a peak-to-trough amplitude of about a factor of 5, barite accumulation rates evaluated using the ^{230}Th-profiling method show little variability and no systematic pattern of glacial–interglacial change (Figure 6; Marcantonio *et al.*, 2001). Upon correcting the barite results for sediment focusing, the record shows little evidence for glacial–interglacial changes in productivity of the central equatorial Pacific Ocean.

Thomas *et al.* (2000) suggested that variable scavenging intensity might have biased the ^{230}Th-normalized barite fluxes in equatorial Pacific sediments. Marcantonio *et al.* (2001) refuted this argument by showing that ^{10}Be/^{230}Th ratios, which are sensitive to changes in scavenging intensity, were relatively constant throughout the period studied, further supporting the view that there has been little glacial–interglacial change in particle flux.

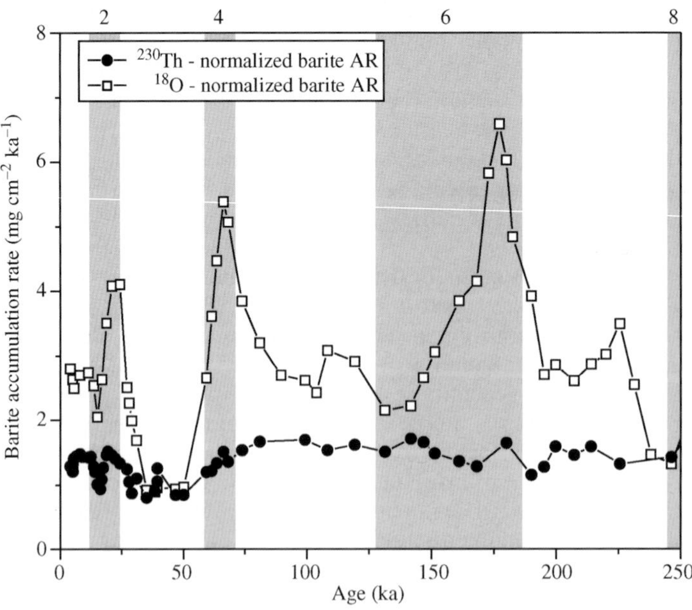

Figure 6 Accumulation rate of barite in equatorial Pacific core TT013-PC72 (0°, 140° W) determined by traditional ^{18}O stratigraphy (open squares) and by normalizing to ^{230}Th (solid circles). Shaded regions represent glacial periods (marine isotope stages 2, 4, 6, and 8 shown along the top of the figure). Correcting for sediment focusing by normalizing to ^{230}Th eliminates the glacial increase in barite accumulation apparent in the ^{18}O-derived record, and thus substantially alters the reconstructed changes in biological productivity inferred from the barite record (sources Paytan, 1995; Paytan *et al.*, 1996) (after Marcantonio *et al.*, 2001).

6.09.7.4 Helium-3

Although ^{230}Th has found widespread use as a CFP, applications are limited by its 75 ka half-life to sediments deposited during the past ~300 ka. Paleoceanographic research would clearly benefit from a CFP that can be applied further back in time. Calibration work is underway to determine the suitability of extraterrestrial ^{3}He, contained in interplanetary dust particles (IDPs), as a CFP that may be applied to much older sediments.

Extraterrestrial ^{3}He is implanted in IDPs, or micrometeorites, by the solar wind. The measured $^{3}He/^{4}He$ ratio in these IDPs is $\sim 2 \times 10^{-4}$ (Nier and Schlutter, 1990, 1992), roughly two orders of magnitude above the atmospheric background ratio ($^{3}He/^{4}He = 1.384 \times 10^{-6}$) and as much as four orders of magnitude above the range of $^{3}He/^{4}He$ ratios found in average continental crust (e.g., 2×10^{-8}; Mamyrin and Tolstikhin, 1984). As the cosmic dust is enriched in ^{3}He by ca. eight orders of magnitude compared to terrigenous matter, IDPs are the principal contributor to the total ^{3}He concentration in pelagic marine sediments (Farley, 1995; Farley and Patterson, 1995; Marcantonio *et al.*, 1996, 1995; Takayanagi and Ozima, 1987).

If the accretion of extraterrestrial ^{3}He-containing IDPs can be determined to have been constant through time, or if variability in the accretion rate can be constrained within known limits, then normalization to the concentration of extraterrestrial ^{3}He in marine sediments could be used to evaluate the preserved fluxes of other sedimentary constituents (similar to the approach used with ^{230}Th; Equation (26)). Because ^{3}He is a stable isotope, and it is retained within IDPs over timescales of at least the past 10^{8} yr (Farley, 1995; Patterson *et al.*, 1998), it holds potential for application as a CFP to paleoceanographic studies well back into the Mesozoic.

Fluxes of extraterrestrial ^{3}He have been evaluated at a number of sites around the world by normalizing to ^{230}Th (i.e., by exploiting ^{230}Th as a CFP). Records of temporal variability covering periods of up to 300 ka have been constructed at 14 sites worldwide (Higgins *et al.*, in preparation). Two situations have been discovered in which the flux of ^{3}He falls outside the modern range of values. First, low ^{3}He fluxes coincide with extreme sediment focusing ($\psi > 10$). Second, high ^{3}He fluxes were found during glacial periods in regions of extensive iceberg melting (as identified through increased deposition of terrigenous material, including ice-rafted debris; Higgins *et al.*, in preparation). Eliminating three high-latitude sites in the North Atlantic Ocean and Southern Ocean bearing evidence for glacial input by icebergs, and a site in the Cape Basin with extreme sediment focusing, 10 records of ^{230}Th-normalized ^{3}He accumulation during the Late Pleistocene have been constructed, and none of these exhibit any systematic pattern of temporal variability. Temporal variability in the flux of ^{3}He is dominated by single-point extreme values, a phenomenon that may be related to the statistics of sampling a small population of IDPs (Farley *et al.*, 1997; Patterson and Farley, 1998). These anomalies are relatively rare, and can be identified through replicated analyses. Excluding the single-point extreme values, the ^{230}Th-normalized flux of extraterrestrial ^{3}He has varied by no more than ~30% during the past ~300 ka (Higgins *et al.*, in preparation). If the flux of extraterrestrial ^{3}He has varied by no more than this amount over longer periods of time, then ^{3}He will find many applications in paleoceanography as a constant flux proxy.

Although calibration of ^{3}He as a CFP continues, an example from the North Pacific Ocean is provided here to illustrate the dramatic differences that sometimes exist between accumulation rates derived using normal stratigraphic techniques and those obtained using ^{3}He as a CFP. Rea *et al.* (1998) described one the most comprehensive studies of the accumulation of Asian dust, as well as biogenic opal, in North Pacific sediments (ODP Site 885/886; Figure 7(a)). Accumulation rates reported by Rea *et al.* were derived using paleomagnetic and biostratigraphic techniques.

Higgins *et al.* (submitted) measured ^{3}He in sediments from Sites 885/886 at depths corresponding to samples analyzed by Rea *et al.* (1998). Assuming that the flux of ^{3}He has been constant over the past 7 Ma, and equal to the average flux determined by normalizing to ^{230}Th in Late Pleistocene sediments, Higgins *et al.* derived fluxes for opal and dust by normalizing to measured concentrations of extraterrestrial ^{3}He, following the approach used previously with ^{230}Th (Equation (26)). Accumulation rates derived by normalizing to ^{3}He were found to be substantially different from those estimated using traditional stratigraphic methods. The modest opal accumulation between 4 Ma and 5 Ma reported by Rea *et al.* (1998) became a pronounced maximum when recomputed by normalizing to ^{3}He (Figure 7(b)), consistent with opal records from other sites in the North Pacific Ocean (Haug *et al.*, 1999; Rea *et al.*, 1998; Snoeckx *et al.*, 1995). Instead of an order of magnitude rise in dust accumulation beginning at ~3.6 Ma, the ^{3}He-normalized accumulation rates of eolian material exhibit a more gradual fourfold rise, beginning at ~6 Ma. The ^{3}He-normalized dust fluxes at Sites 885/886 are consistent with recent evidence from the loess plateau in China, showing the onset of dust deposition at ~7 Ma (e.g., An *et al.*, 2001; Ding *et al.*, 1999; Qiang *et al.*, 2001; Sun *et al.*, 1998a,b). The improved consistency with other records of the revised (^{3}He-normalized) fluxes of opal and dust at Sites 885/886 illustrates the

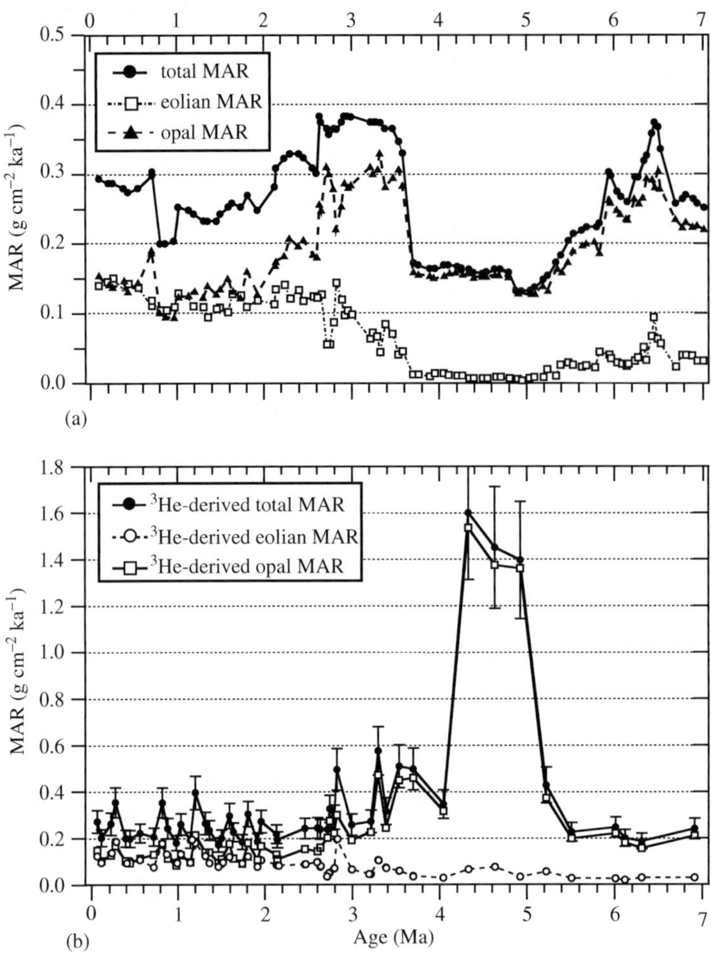

Figure 7 Accumulation rates of total mass, opal, and the eolian component (continental dust) of North Pacific sediments recovered at ODP Sites 885/886. Accumulation rates derived by: (a) Rea *et al.* (1998) using an age model based on biostratigraphic and paleomagnetic age control points and (b) Higgins *et al.* (submitted) by normalizing to ^{3}He.

potential value added in applying ^{3}He as a long-term CFP to paleoceanographic records.

6.09.8 SUMMARY

Naturally occurring radionuclides have provided valuable insights into diverse ocean processes. Scavenging and removal of insoluble trace elements have been shown to have characteristics of reversible adsorption–desorption equilibrium. High molecular weight COM, which constitutes a substantial fraction of the operationally defined dissolved organic matter in surface ocean water, has been shown to be highly reactive, with turnover times of days to weeks. Scavenging and removal from surface waters of particle-reactive trace elements occur at a rate that is proportional to the flux of POC exported to depth. The export flux of POC, as well as its spatial and temporal variability, have been evaluated using ^{234}Th–^{238}U disequilibria. Combining information from multiple thorium isotopes offers promise of evaluating rate constants for particle aggregation, disaggregation, and regeneration. Lateral redistribution of sediments by deep-sea currents has been evaluated by comparing burial rates of ^{230}Th with its known production rate, and this approach has led to significant reinterpretation of certain paleoceanographic records.

Much of this progress has been achieved using basic techniques that were developed in the early 1980s. New techniques and improved interpretations await discovery and development. A coordinated study at the basin to global scale of the distribution of natural radionuclides, as well as their partitioning among dissolved, colloidal, and particulate phases, would improve our ability to exploit these tracers.

REFERENCES

Alleau Y., Colbert D., Covert P., Haley B., Qiu X., Collier R., Falkner K., Hales B., Prahl F., and Gordon L. (2001) ^{234}Th applied to particle removal rates from the surface ocean: a mathematical treatment revisited. *Geophys. Res. Lett.* **28**(14), 2855–2857.

An Z. S., Kutzbach J. E., Prell W. L., and Porter S. C. (2001) Evolution of Asian monsoons and phased uplift of the Himalayan Tibetan plateau since Late Miocene times. *Nature* **411**, 62–66.

Anderson R. F., Bacon M. P., and Brewer P. G. (1983a) Removal of ^{230}Th and ^{231} Pa at ocean margins. *Earth Planet. Sci. Lett.* **66**(1–3), 73–90.

Anderson R. F., Bacon M. P., and Brewer P. G. (1983b) Removal of ^{230}Th and ^{231}Pa from the open ocean. *Earth Planet. Sci. Lett.* **62**(1), 7–23.

Anderson R. F., Fleisher M. Q., and Manley P. L. (1993) Uranium-series tracers of mudwave migration in the Argentine Basin. *Deep-Sea Res. II: Top. Stud. Oceanogr.* **40**(4–5), 889–909.

Anderson R. F., Rowe G. T., Kemp P. F., Trumbore S., and Biscaye P. E. (1994) Carbon budget for the mid-slope depocenter of the Middle Atlantic Bight. *Deep-Sea Res. II: Top. Stud. Oceanogr.* **41**(2–3), 669–703.

Anderson R. F., Kumar N., Mortlock R. A., Froelich P. N., Kubik P., Dittrich-Hannen B., and Suter M. (1998) Late-Quaternary changes in productivity of the Southern Ocean. *J. Mar. Sys.* **17**(1–4), 497–514.

Anderson R. F., Chase Z., Fleisher M. Q., and Sachs J. (2002) The Southern Ocean's biological pump during the Last Glacial Maximum. *Deep-Sea Res II* **49**, 1909–1938.

Archer D., Winguth A., Lea D., and Mahowald N. (2000) What caused the glacial/interglacial atmospheric p_{CO_2} cycles? *Rev. Geophys.* **38**(2), 159–189.

Armstrong R. A., Lee C., Hedges J. I., Honjo S., and Wakeham S. G. (2002) A new, mechanistic model for organic carbon fluxes in the ocean: a quantitative role for the association of POC with ballast minerals. *Deep-Sea Res. II* **49**, 219–236.

Athias V., Mazzega P., and Jeandel C. (2000) Selecting a Global optimization method to estimate the oceanic particle cycling rate constants. *J. Mar. Res.* **58**(5), 675–707.

Bacon M. P. (1984) Glacial to interglacial changes in carbonate and clay sedimentation in the Atlantic Ocean estimated from ^{230}Th measurements. *Isotope Geosci.* **2**(2), 97–111.

Bacon M. P. and Anderson R. F. (1982) Distribution of thorium isotopes between dissolved and particulate forms in the deep-sea. *J. Geophys. Res. Oceans Atmos.* **87**(C3), 2045–2056.

Bacon M. P., Huh C. A., Fleer A. P., and Deuser W. G. (1985) Seasonality in the flux of natural radionuclides and plutonium in the deep Sargasso Sea. *Deep-Sea Res. I: Oceanogr. Res. Pap.* **32**(3), 273–286.

Bacon M. P., Cochran J. K., Hirschberg D., Hammar T. R., and Fleer A. P. (1996) Export flux of carbon at the equator during the EqPac time-series cruises estimated from ^{234}Th measurements. *Deep-Sea Res. II: Top. Stud. Oceanogr.* **43**(4–6), 1133–1153.

Balistrieri L. S., Brewer P. G., and Murray J. W. (1981) Scavenging residence times of trace metals and surface chemistry of sinking particles in the deep ocean. *Deep-Sea Res.* **28A**, 101–121.

Baskaran M., Santschi P. H., Benoit G., and Honeyman B. D. (1992) Scavenging of thorium isotopes by colloids in seawater of the Gulf of Mexico. *Geochim. Cosmochim. Acta* **56**(9), 3375–3388.

Benner R., Pakulski J. D., McCarthy M., Hedges J. I., and Hatcher P. G. (1992) Bulk chemical characteristics of dissolved organic matter in the ocean. *Science* **255**, 1561–1564.

Berelson W. (2002) Particle settling rates increase with depth in the ocean. *Deep-Sea Res. II* **49**, 237–251.

Bhat S. G., Krishnaswami S., Lal D., Rama, and Moore W. S. (1969) Th^{234}/U^{238} ratio in the ocean. *Earth Planet. Sci. Lett.* **5**, 483–491.

Bidigare R. R., Hanson K. L., Buesseler K. O., Wakeham S. G., Freeman K. H., Pancost R. D., Millero F. J., Steinberg P., Popp B. N., Latasa M., Landry M. R., and Laws E. A. (1999) Iron-stimulated changes in C-13 fractionation and export by equatorial Pacific phytoplankton: toward a paleogrowth rate proxy. *Paleoceanography* **14**(5), 589–595.

Biscaye P. E. and Anderson R. F. (1994) Fluxes of particulate matter on the slope of the southern Middle Atlantic Bight—Seep-II. *Deep-Sea Res. II: Top. Stud. Oceanogr.* **41**(2–3), 459–509.

Biscaye P. E., Flagg C. N., and Falkowski P. G. (1994) The shelf edge exchange processes experiment, SEEP II: an introduction to hypotheses, results and conclusions. *Deep-Sea Res. II: Top. Stud. Oceanogr.* **41**(2–3), 231–252.

Bishop J. K. B., Edmond J. M., Ketten D. R., Bacon M. P., and Silker W. B. (1977) Chemistry, biology, and vertical flux of particulate matter from upper 400 m of Equatorial Atlantic Ocean. *Deep-Sea Res.* **24**(6), 511–548.

Boehm A. B. and Grant S. B. (2002) A steady state model of particulate organic carbon flux below the mixed layer and application to the Joint Global Ocean Flux Study. *J. Geophys. Res. Oceans* **106**(C12), 31227–31237.

Boyd P. W. and Stevens C. L. (2002) Modelling particle transformations and the downward organic carbon flux in the NE Atlantic Ocean. *Prog. Oceanogr.* **52**(1), 1–29.

Boyd P. W., Sherry N. D., Berges J. A., Bishop J. K. B., Calvert S. E., Charette M. A., Giovannoni S. J., Goldblatt R., Harrison P. J., Moran S. B., Roy S., Soon M., and Strom S. (1999) Transformations of biogenic particulates from the pelagic to the deep ocean realm. *Deep-Sea Res. II: Top. Stud. Oceanogr.* **46**(11–12), 2761–2792.

Bruland K. W. and Coale K. H. (1986) Surface water ^{234}Th/^{238}U disequilibria: spatial and temporal variations of scavenging rates within the Pacific Ocean. In *Dynamic Processes in the Chemistry of the Upper Ocean* (eds. J. D. Burton, P. G. Brewer, and R. Chesselet). Plenum, New York, pp. 159–172.

Buesseler K., Ball L., Andrews J., Benitez-Nelson C., Belastock R., Chai F., and Chao Y. (1998) Upper ocean export of particulate organic carbon in the Arabian Sea derived from thorium-234. *Deep-Sea Res. II: Top. Stud. Oceanogr.* **45**(10–11), 2461–2487.

Buesseler K., Ball L., Andrews J., Cochran J. K., Hirschberg D., Bacon M. P., Fleer A., and Brzezinski M. A. (2003) Upper ocean export of particulate organic carbon and biogenic silica in the Southern Ocean along 170° W. *Deep-Sea Res. II* **50**, 579–603.

Buesseler K. O. (1998) The decoupling of production and particulate export in the surface ocean. *Global Biogeochem. Cycles* **12**(2), 297–310.

Buesseler K. O. and Charette M. A. (2000) Commentary on Kim *et al.* (1999). *Geophys. Res. Lett.* **27**(13), 1939–1940.

Buesseler K. O., Bacon M. P., Cochran J. K., and Livingston H. D. (1992) Carbon and nitrogen export during the JGOFS North-Atlantic Bloom Experiment estimated from ^{234}Th : ^{238}U disequilibria. *Deep-Sea Res. I: Oceanogr. Res. Pap.* **39**(7–8A), 1115–1137.

Buesseler K. O., Andrews J. A., Hartman M. C., Belastock R., and Chai F. (1995) Regional estimates of the export flux of particulate organic-carbon derived from ^{234}Th during the JGOFS EqPac program. *Deep-Sea Res. II: Top. Stud. Oceanogr.* **42**(2–3), 777–804.

Buesseler K. O., Ball L., Andrews J., Cochran J. K., Hirschberg D. J., Bacon M. P., Fleer A., and Brzezinski M. (2001) Upper ocean export of particulate organic carbon and biogenic silica in the Southern Ocean along 170 degrees W. *Deep-Sea Res. II: Top. Stud. Oceanogr.* **48**(19–20), 4275–4297.

Burd A. B., Moran S. B., and Jackson G. A. (2000) A coupled adsorption-aggregation model of the POC/^{234}Th ratio of marine particles. *Deep-Sea Res. I: Oceanogr. Res. Pap.* **47**(1), 103–120.

Carpenter R., Bennett J. T., and Peterson M. L. (1981) Pb-210 activities in and fluxes to sediments of the Washington continental-slope and shelf. *Geochim. Cosmochim. Acta* **45**(7), 1155–1172.

Charette M. and Buesseler K. (2000) Does iron fertilization lead to rapid carbon export in the Southern Ocean? *Geochem. Geophys. Geosys.* **1**, 2000GC000069.

Chase Z., Anderson R. F., Fleisher M. Q., and Kubik P. (2002) The influence of particle composition on scavenging of Th, Pa and Be in the ocean. *Earth Planet. Sci. Lett.* **204**, 215–229.

Chase Z., Anderson R. F., Fleisher M. Q., and Kubik P. (2003a) Accumulation of biogenic and lithogenic material in the Pacific sector of the Southern Ocean during the past 40,000 years. *Deep-Sea Res. II* **50**, 799–832.

Chase Z., Anderson R. F., Fleisher M. Q., and Kubik P. (2003b) Scavenging of ^{230}Th, ^{231}Pa and ^{10}Be in the Southern Ocean (SW Pacific sector): the importance of particle flux and advection. *Deep-Sea Res. II* **50**, 739–768.

Chen J. H., Edwards R. L., and Wasserburg G. J. (1986) ^{238}U, ^{234}U and ^{232}Th in seawater. *Earth Planet. Sci. Lett.* **80**, 241–251.

Chen M. and Wang W. X. (2001) Bioavailability of natural colloid-bound iron to marine plankton: influences of colloidal size and aging. *Limnol. Oceanogr.* **46**(8), 1956–1967.

Clegg S. L. and Whitfield M. (1990) A generalized-model for the scavenging of trace-metals in the open ocean: 1. Particle cycling. *Deep-Sea Res. I: Oceanogr. Res. Pap.* **37**(5), 809–832.

Clegg S. L. and Whitfield M. (1991) A generalized-model for the scavenging of trace-metals in the open ocean: 2. Thorium scavenging. *Deep-Sea Res. I: Oceanogr. Res. Pap.* **38**(1), 91–120.

Clegg S. L. and Whitfield M. (1993) Application of a generalized scavenging model to time-series ^{234}Th and particle data obtained during the JGOFS North-Atlantic Bloom Experiment. *Deep-Sea Res. I: Oceanogr. Res. Pap.* **40**(8), 1529–1545.

Clegg S. L., Bacon M. P., and Whitfield M. (1991) Application of a generalized scavenging model to thorium isotope and particle data at equatorial high-latitude sites in the Pacific Ocean. *J. Geophys. Res.: Oceans* **96**(C11), 20655–20670.

Coale K. H. and Bruland K. W. (1985) ^{234}Th–^{238}U disequilibria within the California Current. *Limnol. Oceanogr.* **30**(1), 22–33.

Coale K. H. and Bruland K. W. (1987) Oceanic stratified euphotic zone as elucidated by ^{234}Th–^{238}U disequilibria. *Limnol. Oceanogr.* **32**(1), 189–200.

Cochran J. K. and Masqué P. (2003) ^{234}Th/^{238}U disequilibria in the ocean: scavenging rates, export fluxes and particle dynamics. *Rev. Mineral. Geochem.* **52**, 461–492.

Cochran J. K., Buesseler K. O., Bacon M. P., and Livingston H. D. (1993) Thorium isotopes as indicators of particle dynamics in the upper ocean—results from the JGOFS North-Atlantic Bloom Experiment. *Deep-Sea Res. I: Oceanogr. Res. Pap.* **40**(8), 1569–1595.

Cochran J. K., Buesseler K. O., Bacon M. P., Wang H. W., Hirschberg D. J., Ball L., Andrews J., Crossin G., and Fleer A. (2000) Short-lived thorium isotopes (^{234}Th, ^{228}Th) as indicators of POC export and particle cycling in the Ross Sea, Southern Ocean. *Deep-Sea Res. II: Top. Stud. Oceanogr.* **47**(15–16), 3451–3490.

Dadou I., Lamy F., Rabouille C., Ruiz-Pino D., Andersen V., Bianchi M., and Garcon V. (2001) An integrated biological pump model from the euphotic zone to the sediment: a 1-D application in the Northeast tropical Atlantic. *Deep-Sea Res. II: Top. Stud. Oceanogr.* **48**(10), 2345–2381.

Dai M. H. and Benitez-Nelson C. R. (2001) Colloidal organic carbon and ^{234}Th in the Gulf of Maine. *Mar. Chem.* **74**(2–3), 181–196.

Deuser W. G., Ross E. H., and Anderson R. F. (1981) Seasonality in the supply of sediment to the deep Sargasso Sea and implications for the rapid transfer of matter to the deep ocean. *Deep-Sea Res. I: Oceanogr. Res. Pap.* **28**(5), 495–505.

Ding Z. L., Xiong S. F., Sun J. M., Yang S. L., Gu Z. Y., and Liu T. S. (1999) Pedostratigraphy and paleomagnetism of a ~7.0 Ma eolian loess-red clay sequence at Lingtai, Loess Plateau, north-central China and the implications for paleomonsoon evolution. *Paleogeogr. Paleoclimatol. Paleoecol.* **152**, 49–66.

Eppley R. W. (1989) New production: history, methods, problems. In *Productivity of the Ocean: Present and Past* (eds. W. H. Berger, V. S. Smetacek, and G. Wefer). Wiley, New York, pp. 85–97.

Farley K. A. (1995) Cenozoic variations in the flux of interplanetary dust recorded by ^{3}He in a deep-sea sediment. *Nature* **376**, 153–156.

Farley K. A. and Patterson D. B. (1995) A 100-kyr periodicity in the flux of extraterrestrial ^{3}He to the seafloor. *Nature* **378**, 600–603.

Farley K. A., Love S. G., and Patterson D. B. (1997) Atmospheric entry heating and helium retentivity of interplanetary dust particles. *Geochim. Cosmochim. Acta* **61**, 2309–2316.

Fisher N. S., Burns K. A., Cherry R. D., and Heyraud M. (1983) Accumulation and cellular-distribution of ^{241}Am, ^{210}Po, and ^{210}Pb in 2 marine-algae. *Mar. Ecol.: Prog. Ser.* **11**(3), 233–237.

Flood R. D. (1988) A lee wave model for deep-sea mudwave activity. *Deep-Sea Res. I: Oceanogr. Res. Pap.* **35**(6), 973–983.

Flood R. D., Shor A. N., and Manley P. L. (1993) Morphology of abyssal mudwaves at Project MUDWAVES sites in the Argentine Basin. *Deep-Sea Res. II: Top. Stud. Oceanogr.* **40**(4–5), 859–888.

Francois R., Bacon M., and Suman D. O. (1990) Thorium 230 profiling in deep-sea sediments: high-resolution records of flux and dissolution of carbonate in the equatorial Atlantic during the last 24,000 years. *Paleoceanography* **5**(5), 761–787.

Francois R., Bacon M. P., Altabet M. A., and Laberyrie L. D. (1993) Glacial–interglacial changes in sediment rain rate in the SW Indian sector of sub-Antarctic waters as recorded by ^{230}Th, ^{231}Pa, ^{238}U, and ∂^{15}N. *Paleoceanography* **8**(5), 611–629.

Francois R., Altabet M. A., Yu E. F., Sigman D. M., Bacon M. P., Frank M., Bohrmann G., Bareille G., and Labeyrie L. D. (1997) Contribution of Southern Ocean surface-water stratification to low atmospheric CO_2 concentrations during the last glacial period. *Nature* **389**(6654), 929–935.

Frank M. (1996) Reconstructions of Late Quaternary environmental conditions applying the natural radionuclides ^{230}Th, ^{10}Be, ^{231}Pa and ^{238}U: a study of deep-sea sediments from the eastern sector of the Antarctic Circumpolar Current System. *Rep. Polar Res.* **186**, 1–136.

Frank M., Gersonde R., and Mangini A. (1999) Sediment redistribution, $^{230}Th_{ex}$-normalization and implications for reconstruction of particle flux and export productivity. In *Use of Proxies in Paleoceanography: Examples from the South Atlantic* (eds. G. Fischer and G. Wefer). Springer, Berlin, pp. 409–426.

Frank M., Gersonde R., Rutgers van der Loeff M. M., Bohrmann G., Nürnberg C. C., Kubik P. W., Suter M., and Mangini A. (2000) Similar glacial and interglacial export bioproductivity in the Atlantic sector of the Southern Ocean: multiproxy evidence and implications for glacial atmospheric CO_2. *Paleoceanography* **15**(6), 642–658.

Friedrich J. and Rutgers van der Loeff M. M. (2002) A two-tracer (^{210}Po–^{234}Th) approach to distinguish organic carbon and biogenic silica export flux in the Antarctic Circumpolar Current. *Deep-Sea Res. I: Oceanogr. Res. Pap.* **49**(1), 101–120.

Gardner W. D., Richardson M. J., Carlson C. A., Hansell D. A., and Mishonov A. V. (2003) Determining true particulate organic carbon: bottles, pumps and methodologies. *Deep-Sea Res. II* **50**, 655–674.

Goldberg E. D. (1954) Marine geochemistry: 1. Chemical scavengers of the sea. *J. Geol.* **62**, 249–265.

Guo L. D. and Santschi P. H. (1997) Composition and cycling of colloids in marine environments. *Rev. Geophys.* **35**(1), 17–40.

Guo L. D., Santschi P. H., and Baskaran M. (1997) Interactions of thorium isotopes with colloidal organic matter in oceanic environments. *Coll. Surf. I: Physicochem. Eng. Aspects* **120**(1–3), 255–271.

Guo L. D., Santschi P. H., and Warnken K. W. (2000) Trace metal composition of colloidal organic material in marine environments. *Mar. Chem.* **70**(4), 257–275.

Gustafsson O., Buesseler K. O., Geyer W. R., Moran S. B., and Gschwend P. M. (1998) An assessment of the relative importance of horizontal and vertical transport of particle-reactive chemicals in the coastal ocean. *Continent. Shelf Res.* **18**(7), 805–829.

Haug G. H., Sigman D. M., Tiedemann R., Pedersen T. F., and Sarnthein M. (1999) Onset of permanent stratification in the subarctic Pacific Ocean. *Nature* **401**, 779–782.

Henderson G. M. (2002) Seawater (^{234}U/^{238}U) during the last 800 thousand years. *Earth Planet. Sci. Lett.* **199**, 97–110.

Henderson G. M. and Anderson R. F. (2003) The U-series toolbox for paleoceanography. *Rev. Mineral. Geochem.* **52**, 493–531.

Henderson G. M., Heinze C., Anderson R. F., and Winguth A. M. E. (1999) Global distribution of the ^{230}Th flux to ocean sediments constrained by GCM modelling. *Deep-Sea Res. I: Oceanogr. Res. Pap.* **46**(11), 1861–1893.

Higgins S. M., Anderson R. F., Marcantonio F., Schlosser P., and Stute M. (2002) Sediment focusing creates 100 ka cycles in Interplanetary Dust accumulation on the Ontong-Java Plateau. *Earth Planet. Sci. Lett.* **203**, 383–397.

Higgins S. M., Anderson R. F., Marcantonio F., Schlosser P., Stute M., Schwarz B., Frank M., and Strobl C. Fluxes of Extraterrestrial ^{3}He, continental dust, and opal in North Pacific sediments over the past 7 Ma. *Paleoceanography* (submitted for publication).

Higgins S. M., Anderson R. F., Marcantonio F., Schlosser P., Stute M., and Fleisher M. Q. A global estimate of the interplanetary dust flux over the last 250 ka based on extraterrestrial ^{3}He in marine sediments (in preparation).

Honeyman B. D. and Santschi P. H. (1989) A Brownian-pumping model for oceanic trace metal scavenging: evidence from Th isotopes. *J. Mar. Res.* **47**, 951–992.

Honeyman B. D. and Santschi P. H. (1991) Coupling adsorption and particle aggregation—laboratory studies of colloidal pumping using ^{59}Fe-labeled hematite. *Environ. Sci. Technol.* **25**(10), 1739–1747.

Honeyman B. D. and Santschi P. H. (1992) The role of particles and colloids in the transport of radionuclides and trace metals in the ocean. In *Environmental Particles* (eds. J. Buffle and H. P. van Leeuwen). Lewis Publishers, Boca Raton, vol. 1, pp. 379–423.

Jackson G. A. (2001) Effect of coagulation on a model planktonic food web. *Deep-Sea Res. I: Oceanogr. Res. Pap.* **48**(1), 95–123.

Jackson G. A. and Burd A. B. (1998) Aggregation in the marine environment. *Environ. Sci. Technol.* **32**(19), 2805–2814.

Kim G. B., Hussain N., and Church T. M. (1999) How accurate are the ^{234}Th based particulate residence times in the ocean? *Geophys. Res. Lett.* **26**(5), 619–622.

Krishnaswami S., Lal D., Somayajulu B. L. K., Weiss R. F., and Craig H. (1976) Large volume *in situ* filtration of deep Pacific waters: mineralogical and radioisotope studies. *Earth Planet. Sci. Lett.* **32**, 420–429.

Krishnaswami S., Sarin M. M., and Somayajulu B. L. K. (1981) Chemical and radiochemical investigations of surface and deep particles of the Indian Ocean. *Earth Planet. Sci. Lett.* **54**, 81–96.

Kumar N., Anderson R. F., Mortlock R. A., Froelich P. N., Kubik P., Dittrich-Hannen B., and Suter M. (1995) Increased biological productivity and export production in the glacial Southern Ocean. *Nature* **378**(6558), 675–680.

Li Y.-H. (1981) Ultimate removal mechanisms of elements from the ocean. *Geochim. Cosmochim. Acta* **45**, 1159–1164.

Liu K. K., Atkinson L., Quiñones R., and Talaue-McManus L. *Carbon and Nutrient Fluxes in Continental Margins: A Global Synthesis*. Springer (in preparation).

Mamyrin B. A. and Tolstikhin I. N. (1984) *Helium Isotopes in Nature*. Elsevier, Amsterdam.

Marcantonio F., Kumar N., Stute M., Anderson R. F., Seidl M. A., Schlosser P., and Mix A. (1995) A comparative study of accumulation rates derived by He and Th isotope analysis of marine sediments. *Earth Planet. Sci. Lett.* **133**, 549–555.

Marcantonio F., Anderson R. F., Stute M., Kumar N., Schlosser P., and Mix A. (1996) Extraterrestrial ^{3}He as a tracer of marine sediment transport and accumulation. *Nature* **383**(6602), 705–707.

Marcantonio F., Anderson R. F., Higgins S., Stute M., Schlosser P., and Kubik P. (2001) Sediment focusing in the central equatorial Pacific Ocean. *Paleoceanography* **16**(3), 260–267.

Martin J. H., Knauer G. A., Karl D. M., and Broenkow W. W. (1987) Vertex-carbon cycling in the northeast Pacific. *Deep-Sea Res. I: Oceanogr. Res. Pap.* **34**(2), 267–285.

Moore R. M. and Hunter K. A. (1985) Thorium adsorption in the ocean: reversibility and distribution amongst particle sizes. *Geochim. Cosmochim. Acta* **49**, 2253–2257.

Moore R. M. and Millward G. E. (1988) The kinetics of reversible Th reactions with marine particles. *Geochim. Cosmochim. Acta* **52**, 113–118.

Moore W. S. and Sackett W. M. (1964) Uranium and thorium series inequilibrium in seawater. *J. Geophys. Res.* **69**, 5401–5405.

Moran S. B. and Buesseler K. O. (1992) Short residence time of colloids in the upper ocean estimated from ^{238}U–^{234}Th disequilibria. *Nature* **359**(6392), 221–223.

Moran S. B. and Buesseler K. O. (1993) Size-fractionated ^{234}Th in continental-shelf waters off New-England—implications for the role of colloids in oceanic trace-metal scavenging. *J. Mar. Res.* **51**(4), 893–922.

Moran S. B. and Moore R. M. (1989) The distribution of colloidal aluminum and organic carbon in coastal and open ocean waters off Nova Scotia. *Geochim. Cosmochim. Acta* **53**, 2519–2527.

Moran S. B., Shen C.-C., Weinstein S. E., Hettinger L. H., Hoff J. H., Edmonds H. N., and Edwards R. L. (2001) Constraints on deep water age and particle flux in the equatorial and south Atlantic Ocean based on seawater ^{231}Pa and ^{230}Th. *Geophys. Res. Lett.* **28**(18), 3437–3440.

Moran S. B., Weinstein S. E., Edmonds H. N., Smith J. N., Kelly R. P., Pilson M. E. Q., and Harrison W. G. (2003) Does ^{234}Th/^{238}U disequilibria provide an accurate record of the export flux of particulate organic carbon from the upper ocean? *Limnol. Oceanogr.* **48**, 1018–1029.

Morel F. M. M. and Gschwend P. M. (1987) The role of colloids in the partitioning of solutes in natural waters. In *Aquatic Surface Chemistry: Chemical Processes at the Particle–Water Interface* (ed. W. Stumm). Wiley, New York, pp. 405–422.

Murnane R. J. (1994) Determination of thorium and particulate matter cycling parameters at Station-P—a reanalysis and comparison of least-squares techniques. *J. Geophys. Res.: Oceans* **99**(C2), 3393–3405.

Murnane R. J., Sarmiento J. L., and Bacon M. P. (1990) Thorium isotopes, particle cycling models, and inverse calculations of model rate constants. *J. Geophys. Res.: Oceans* **95**(C9), 16195–16206.

Murnane R. J., Cochran J. K., and Sarmiento J. L. (1994) Estimates of particle-cycling and thorium-cycling rates in the northwest Atlantic-Ocean. *J. Geophys. Res.: Oceans* **99**(C2), 3373–3392.

Murnane R. J., Cochran J. K., Buesseler K. O., and Bacon M. P. (1996) Least-squares estimates of thorium, particle, and nutrient cycling rate constants from the JGOFS North Atlantic Bloom Experiment. *Deep-Sea Res. I: Oceanogr. Res. Pap.* **43**(2), 239–258.

Murray J. W., Downs J. N., Strom S., Wei C. L., and Jannasch H. W. (1989) Nutrient assimilation, export production and ^{234}Th scavenging in the eastern equatorial Pacific. *Deep-Sea Res. I: Oceanogr. Res. Pap.* **36**(10), 1471–1489.

Murray J. W., Young J., Newton J., Dunne J., Chapin T., Paul B., and McCarthy J. J. (1996) Export flux of particulate organic carbon from the central equatorial Pacific determined using a combined drifting trap ^{234}Th approach. *Deep-Sea Res. II: Top. Stud. Oceanogr.* **43**(4–6), 1095–1132.

Murray J. W., Paul B., Dunne J., and Chapin T. ^{234}Th, ^{210}Pb, ^{210}Po and stable Pb in the central equatorial Pacific: tracers for particle cycling. *Deep-Sea Res. I* (in press).

Nelson D. E., Anderson R. F., Barber R. T., Brzezinski M. A., Buesseler K., Chase Z., Collier R., Dickson M. L., Francois R., Hiscock M. R., Honjo S., Marra J., Martin W. R., Sambrotto R. N., Sayles F. L., and Sigmon D. E. (2002) Vertical budgets for organic carbon and biogenic silica in the Pacific sector of the Southern Ocean, 1996–1998. *Deep-Sea Res. II* **49**, 1645–1673.

Nier A. O. and Schlutter D. J. (1990) Helium and neon in stratospheric particles. *Meteoritics* **25**, 263–267.

Nier A. O. and Schlutter D. J. (1992) Extraction of helium from individual interplanetary dust particles by step-heating. *Meteoritics* **27**, 166–173.

Nishioka J., Takeda S., Wong C. S., and Johnson W. K. (2001) Size-fractionated iron concentrations in the northeast Pacific Ocean: distribution of soluble and small colloidal iron. *Mar. Chem.* **74**(2–3), 157–179.

Niven S. E. H. and Moore R. M. (1988) Effect of natural colloidal matter on the equilibrium adsorption of thorium in seawater. In *Radionuclides: A Tool for Oceanography* (eds. J. C. Guary, P. Guiegueniat, and R. J. Pentreath). Elsevier, London.

Niven S. E. H. and Moore R. M. (1993) Thorium sorption in seawater suspensions of aluminum oxide particles. *Geochim. Cosmochim. Acta* **57**, 2169–2179.

Niven S. E. H., Kepkay P. E., and Boraie A. (1995) Colloidal organic-carbon and colloidal ^{234}Th dynamics during a coastal phytoplankton bloom. *Deep-Sea Res. II: Top. Stud. Oceanogr.* **42**(1), 257–273.

Nodder S. D., Charette M. A., Waite A. M., Trull T. W., Boyd P. W., Zeldis J., and Buesseler K. O. (2001) Particle transformations and export flux during an *in situ* iron-stimulated algal bloom in the Southern Ocean. *Geophys. Res. Lett.* **28**(12), 2409–2412.

Nozaki Y., Horibe Y., and Tsubota H. (1981) The water column distribution of thorium isotopes in the western North Pacific. *Earth Planet. Sci. Lett.* **54**, 203–216.

Nozaki Y., Yang H. S., and Yamada M. (1987) Scavenging of thorium in the ocean. *J. Geophys. Res.: Oceans* **92**(C1), 772–778.

Patterson D. B. and Farley K. A. (1998) Extraterrestrial ^{3}He in seafloor sediments: evidence for correlated 100 kyr periodicity in the accretion rate of interplanetary dust, orbital parameters, and Quaternary climate. *Geochim. Cosmochim. Acta* **62**, 3669–3682.

Patterson D. B., Farley K. A., and Schmitz B. (1998) Preservation of extraterrestrial ^{3}He in 480-Ma-old marine limestones. *Earth Planet. Sci. Lett.* **163**, 315–325.

Paytan A. (1995) Marine barite, a recorder of ocean chemistry, productivity, and circulation. PhD thesis, University of California, San Diego.

Paytan A., Kastner M., and Chavez F. P. (1996) Glacial to interglacial fluctuations in productivity in the equatorial Pacific as indicated by marine barite. *Science* **274**(5291), 1355–1357.

Qiang X. K., Li Z. X., Powell C. M., and Zheng H. B. (2001) Magnetostratigraphic record of the Late Miocene onset of the East Asian monsoon, and Pliocene uplift of northern Tibet. *Earth Planet. Sci. Lett.* **187**, 83–93.

Quigley M. S., Honeyman B. D., and Santschi P. (1996) Thorium sorption in the marine environment: equilibrium partitioning at the hematite/water interface, sorption/desorption kinetics and particle tracing. *Aquat. Geochem.* **1**, 277–301.

Quigley M. S., Santschi P. H., Guo L. D., and Honeyman B. D. (2001) Sorption irreversibility and coagulation behavior of ^{234}Th with marine organic matter. *Mar. Chem.* **76**(1–2), 27–45.

Quigley M. S., Santschi P. H., Hung C. C., Guo L. D., and Honeyman B. D. (2002) Importance of acid polysaccharides for ^{234}Th complexation to marine organic matter. *Limnol. Oceanogr.* **47**(2), 367–377.

Rea D. K., Snoeckx H., and Joseph L. H. (1998) Late Cenozoic eolian deposition in the North Pacific: Asian drying, Tibetan uplift, and cooling of the Northern Hemisphere. *Paleoceanography* **13**, 215–224.

Rutgers van der Loeff M. M., Friedrich M., and Bathmann U. V. (1997) Carbon export during the Spring Bloom at the Antarctic Polar Front, determined with the natural tracer ^{234}Th. *Deep-Sea Res. II: Top. Stud. Oceanogr.* **44**(1–2), 457–478.

Santschi P. H., Nyffeler U. P., Li Y.-H, and O'Hara P. (1986) Radionuclide cycling in natural waters: relevance of scavenging kinetics. In *Sediments and Water Interactions* (ed. P. G. Sly). Springer, New York, pp. 437–449.

Santschi P. H., Guo L. D., Baskaran M., Trumbore S., Southon J., Bianchi T. S., Honeyman B., and Cifuentes L. (1995) Isotopic evidence for the contemporary origin of high-molecular-weight organic-matter in oceanic environments. *Geochim. Cosmochim. Acta* **59**(3), 625–631.

Schindler P. W. (1975) Removal of trace metals from the oceans: a zero order model. *Thalassia Jugoslavica* **11**, 101–111.

Scholten J. C., Fietzke J., Vogler S., Rutgers van der Loeff M. M., Mangini A., Koeve W., Waniek J., Stoffers P., Antia A., and Kuss J. (2001) Trapping efficiencies of sediment traps from the deep eastern North Atlantic: the ^{230}Th calibration. *Deep-Sea Res. II* **48**, 2383–2408.

Snoeckx H., Rea D. K., Jones C. E., and Ingram L. (1995) Eolian and silica deposition in the central North Pacific: results from sites 885/886. In *Proceedings of the Ocean Drilling Program, Scientific Results* (eds. D. K. Rea, I. A. Basov, D. W. Scholl, and J. F. Allan). ODP, College Station, TX, pp. 219–230.

Stordal M. C., Santschi P. H., and Gill G. A. (1996) Colloidal pumping: evidence for the coagulation process using natural colloids tagged with ^{203}Hg. *Environ. Sci. Technol.* **30**(11), 3335–3340.

Suman D. O. and Bacon M. P. (1989) Variations in Holocene sedimentation in the North-American Basin determined from ^{230}Th measurements. *Deep-Sea Res. I: Oceanogr. Res. Pap.* **36**(6), 869–878.

Sun D. G., An Z. S., Shaw J., Bloemendal J., and Sun Y. B. (1998a) Magnetostratigraphy and palaeoclimatic significance of late tertiary aeolian sequences in the Chinese Loess Plateau. *Geophys. J. Int.* **134**, 207–212.

Sun D. H., Shaw J., An Z. S., Cheng M. Y., and Yue L. P. (1998b) Magnetostratigraphy and paleoclimatic interpretation of a continuous 7.2 Ma Late Cenozoic eolian sediments from the Chinese Loess Plateau. *Geophys. Res. Lett.* **25**, 85–88.

Takayanagi M. and Ozima M. (1987) Temporal variation of ^{3}He/^{4}He ratio recorded in deep-sea sediment cores. *J. Geophys. Res.* **92**, 12531–12538.

Thomas E., Turekian K. K., and Wei K.-Y. (2000) Productivity control of fine particle transport to equatorial Pacific sediment. *Global Biogeochem. Cycles* **14**, 945–955.

Turekian K. K. (1977) The fate of metals in the ocean. *Geochim. Cosmochim. Acta.* **41**, 1139–1144.

Vogler S., Rutgers van der Loeff M. M., and Mangini A. (1998) ^{230}Th in the eastern North Atlantic: the importance of water mass ventilation in the balance of ^{230}Th. *Earth. Planet. Sci. Lett.* **156**, 61–74.

Wang W. X. and Guo L. D. (2000) Bioavailability of colloid-bound Cd, Cr, and Zn to marine plankton. *Mar. Ecol.: Prog. Ser.* **202**, 41–49.

Wang W. X. and Guo L. D. (2001) Production of colloidal organic carbon and trace metals by phytoplankton decomposition. *Limnol. Oceanogr.* **46**(2), 278–286.

Wells M. L. and Goldberg E. D. (1994) The distribution of colloids in the North-Atlantic and Southern Oceans. *Limnol. Oceanogr.* **39**(2), 286–302.

Wells M. L., Kozelka P. B., and Bruland K. W. (1998) The complexation of "dissolved" Cu, Zn, Cd and Pb by soluble and colloidal organic matter in Narragansett Bay, RI. *Mar. Chem.* **62**(3–4), 203–217.

Wen L. S., Santschi P. H., and Tang D. G. (1997) Interactions between radioactively labeled colloids and natural particles: evidence for colloidal pumping. *Geochim. Cosmochim. Acta* **61**(14), 2867–2878.

Wen L. S., Santschi P., Gill G., and Paternostro C. (1999) Estuarine trace metal distributions in Galveston Bay: importance of colloidal forms in the speciation of the dissolved phase. *Mar. Chem.* **63**(3–4), 185–212.

Wu J. F., Boyle E., Sunda W., and Wen L. S. (2001) Soluble and colloidal iron in the oligotrophic North Atlantic and North Pacific. *Science* **293**(5531), 847–849.

Yu E. F., Francois R., Bacon M. P., and Fleer A. P. (2001a) Fluxes of ^{230}Th and ^{231}Pa to the deep sea: implications for the interpretation of excess ^{230}Th and ^{231}Pa/^{230}Th profiles in sediments. *Earth Planet. Sci. Lett.* **191**, 219–230.

Yu E. F., Francois R., Bacon M. P., Honjo S., Fleer A. P., Manganini S. J., van der Loeff M. M. R., and Ittekot V. (2001b) Trapping efficiency of bottom-tethered sediment traps estimated from the intercepted fluxes of ^{230}Th and ^{231}Pa. *Deep-Sea Res. I: Oceanogr. Res. Pap.* **48**(3), 865–889.

6.10
Biological Fluxes in the Ocean and Atmospheric p_{CO_2}

D. Archer
University of Chicago, IL, USA

6.10.1 INTRODUCTION

The basic outlines for the carbon cycle in the ocean, as it is represented in ocean carbon cycle models, were summarized by Broecker and Peng (1982). Since then, ongoing field research has revised that picture, unearthing new degrees of freedom and sensitivities in the ocean carbon cycle, which may provide clues to changes in the behavior of the carbon cycle over the glacial cycles and deeper into geological time. These sensitivities include new ideas about the chemistry and cycling of phytoplankton nutrients such as nitrate and iron, the physics of sinking organic matter, and the production and redissolution of phytoplankton companion minerals $CaCO_3$ and SiO_2, in the water column and in the sediment.

The carbon cycle is typically broken down into two components, both of which will be summarized here. The first is the biological pump, effecting the redistributing of biologically active elements like carbon, nitrogen, and silicon within the circulating waters of the ocean. The second is the ultimate removal of these elements by burial in sediments. These two components of the carbon cycle together control the mean concentrations of many chemicals in the ocean, including ocean pH and the p_{CO_2} of the atmosphere.

This chapter overlaps considerably with the only two others from this Treatise I have had the pleasure to read: Chapters 6.04 and 6.19. My chapter is distinguished, I suppose, by the perspective of a geochemical ocean modeler, attempting to integrate new field observations into the context of the ocean control of the p_{CO_2} of the atmosphere.

6.10.2 CHEMICAL REARRANGEMENT OF THE WATERS OF THE OCEAN

6.10.2.1 The Organic Carbon Biological Pump

The cycling of dissolved carbon and associated chemicals (such as oxygen, nitrogen, and phosphorus) in the ocean is driven by the lives and deaths of phytoplankton. Sunlight is required as an energy source for photosynthesis, and is available in sufficient intensity only into the top 100 m or so of the ocean; the rest of the ocean is dark, or at least too dark to make a living harvesting energy from light. The net geochemical effect of photosynthesis in the ocean is to convert dissolved constituents into particles, pieces of solid material that can take leave of their parent fluid by sinking. As they do, they strip the surface waters of whichever component locally limits the growth of phytoplankton (typically nitrogen or iron), and carry the entire suite of biologically useful elements to depth.

As the particles sink, they degrade in the water column, or they reach the seafloor, to degrade within the sediments or to exit the ocean by burial. This combination of removal at the sea surface and addition at depth maintains higher concentrations of bioactive elements and compounds at depth in the ocean than at the surface. The shape of a depth profile of dissolved nitrogen in the form of nitrate, a limiting "fertilizer" or nutrient to phytoplankton, resembles those of other bioactive compounds, including C, PO_4, Si, Fe, and Cd. This process, fundamental to the ocean carbon cycle, has been called the biological pump (Broecker and Peng, 1982; see Chapter 6.04).

6.10.2.1.1 Nutrient limitation and new production

Phytoplankton in surface waters, upon completion of their earthly sojourns, suffer one of two fates. Either they are degraded and consumed within the surface waters of the ocean, or they may sink to the deep sea. On average, the removal of dissolved nutrients from the surface zone to depth must be balanced by mixing up of dissolved nutrients from below, or other sources such as deposition from the atmosphere (as in the case of iron), or local production (as in nitrogen fixation). Traditionally, oceanographers have distinguished "new" versus "recycled" primary production using the different forms of biologically available nitrogen (Eppley and Peterson, 1979). When phytoplankton degrade, their nitrogen is released in the reduced form, as ammonia, which corresponds to nitrogen's chemical oxidation state in proteins and DNA. Ammonia released within the euphotic zone is quickly assimilated by the next generation of phytoplankton. Ammonia released below the euphotic zone escapes this fate, because the darkness precludes photosynthesis. Instead, this ammonia gets oxidized by bacteria to nitrate. Therefore, most of the biologically available nitrogen in deep waters of the oxic ocean is in the form of nitrate. Using this dichotomy, oceanographers can distinguish between new production, supported by upwelling from below, as the uptake of isotopically labeled nitrate, versus recycled production as the uptake of isotopically labeled ammonia. The ratio of new production to production has been called the *f*-ratio by biological oceanographers (Dugdale and Goering, 1967). Values of the *f*-ratio range from 10% to 20% in warm, oligotrophic, low-nutrient open-ocean conditions, to 50% in blooms and near-shore conditions (Eppley and Peterson, 1979). However, the apparent correlation between *f*-ratio and primary production may be an artifact of differences in sea-surface temperature (Laws *et al.*, 2000).

6.10.2.1.2 Iron as a limiting micronutrient

Martin and Fitzwater (1988) originally proposed that the availability of the micronutrient iron may be limiting phytoplankton production in remote parts of the world ocean. This proposal was convincingly proven by a series of open-ocean fertilization experiments, in the equatorial Pacific (Coale *et al.*, 1996; Martin *et al.*, 1994) and in the Southern Ocean (Boyd *et al.*, 2000). Although photosynthesis rates and chlorophyll concentrations increased significantly in all cases, the iron fertilization was less effective at stimulating carbon sinking to depth, especially in the Southern Ocean, where the added iron recycled in the euphotic zone until it dispersed by lateral mixing some months later (Abraham *et al.*, 2000). Several attempts have been made to model the iron cycle in the ocean (Archer and Johnson, 2000; Fung *et al.*, 2000; Lefevre and Watson, 1999). Although dust deposition delivers more iron to the surface of the ocean (Fung *et al.*, 2000), uncertainty remains about the availability of that iron to the phytoplankton. It has been argued that dust deposition may play only a minor role in regulating phytoplankton abundance (Archer and Johnson, 2000; Lefevre and Watson, 1999). The North Pacific Ocean, in particular, appears to be a problem area, with high deposition of dust (Duce and Tindale, 1991) failing to drive depletion of sea-surface nutrients.

6.10.2.1.3 Measuring carbon export

In the steady state (that is to say, on a long enough time average), the uptake of new nutrients, nitrate for example, must be balanced by the removal of nutrients in exported organic matter. Oceanographers have defined a quotient called the *e*-ratio, describing the export efficiency of euphotic zone organic matter, export production/gross production. On long enough timescales, the *e*-ratio must equal the *f*-ratio. In this way, the biological pump acts to limit the productivity of the ocean, pacing it according to the overturning timescale of the ocean circulation.

Sediment traps and sinking particles. The measurement of organic carbon export from the surface ocean has remained problematic over the decades. The workhorse of the oceanographic community for determining sinking particulate fluxes is the moored sediment trap. Sediment traps are plagued by biases from hydrodynamics and from "swimmers," which actively enter the open collection cups of the traps to feed, either consuming our signal or perhaps perishing in the preservative and adding to it (Hedges *et al.*, 1993). Swimmers are operationally excluded by picking them out individually. Hydrodynamic over- or under-trapping can be detected by comparing the delivery rate of radiogenic thorium with its known source rate in the water column above (Buesseler, 1991). In general, deeper traps (below 1,500 m) are found to be unbiased by hydrodynamic trapping errors, whereas shallower traps may undersample the true sinking flux by a factor of 2 or more, on average (Yu *et al.*, 2001). Thus, historical estimates of euphotic zone carbon export based on sediment traps must be evaluated with caution (Berger, 1989; Martin *et al.*, 1987).

Export of DOC. Another recent revision of our understanding of carbon export has been the generation of dissolved organic matter (DOM). Dissolved matter is unable to physically separate itself from its parent fluid, the way particulate organic matter (POM) does by sinking. However, significant gradients in the concentration of DOC exist in the ocean, so fluid transport through the steady-state DOC concentration field represents the export of significant quantities of organic matter in the dissolved form. Concentrations of DOC in the deep ocean range fall in a relatively low and narrow range, 37–42 μM. DOC reaches higher concentrations in the surface ocean, up to 80–100 μM in oligotrophic surface waters. Vertical convection thus drives export of sea-surface DOC to depth. Upwelling brings low-DOC water to the sea surface, where photosynthesis begins the production of DOC. This DOC can be exported laterally or back to the ocean interior. On a global mean, DOC export accounts for less than half of the export of organic matter from the surface ocean, and only a tiny fraction to the deep sea (Yamanaka and Tujika, 1997).

Nutrient supply. In spite of the developments of sediment trap calibration by radiogenic isotopes and export of organic matter in the suspended or dissolved states, problems remain in balancing carbon and nutrient export against supply by physical fluid transport. This problem is most acute in the subtropical gyres, intensely sampled in two time-series JGOFS stations: one near Bermuda (BATS) and one near Hawaii (HOT). Net vertical motions in the upper ocean are determined by the divergence of surface currents, driven by friction with the wind. The direction and intensity of the resulting vertical motion is determined by the curl of the wind stress, which leads to net upwelling in the subpolar gyres, carrying nutrients vertically into the euphotic zone, but downward in the subtropical gyres, supplying no nutrients for photosynthesis at all (Peixoto and Oort, 1992). On a smaller spatial scale, eddies and fronts lead to vertical motion and offsets in the depth of the constant-density (isopycnal) surfaces (Mahadevan and Archer, 2000; McGillicuddy and Robinson, 1997; Oschlies and Koeve, 2000). These mesoscale motions may carry nutrients vertically into the euphotic zone, particularly in Bermuda where the proximity to the Gulf Stream generates high levels of mesoscale motions. Sea-surface helium concentrations (Jenkins, 1988) seem to support the idea of vertical transport by physical fluid motions. At Hawaii in particular, nitrogen fixation, the production of biologically available nitrogen in the form of ammonia from the generally inert but ubiquitous N_2 gas, is thought to supply a large fraction of the required nitrogen (Karl *et al.*, 1997). Nutrients may be mined from deeper waters and actively transported into the euphotic zone by diatoms (Villareal *et al.*, 1999).

6.10.2.2 $CaCO_3$ Production and Export

6.10.2.2.1 Calcite and aragonite

Several mineral phases are produced by phytoplankton and exported to depth in the ocean, associated with and analogous to the production and export of organic matter. First among these is $CaCO_3$, comprised of two mineral phases: calcite and the more soluble aragonite (Milliman, 1974; Mucci, 1983). The solubility of calcite depends also on the concentration of magnesium; higher magnesium leads to higher solubility (Plath *et al.*, 1980). The dominant mineral in deep-sea sediments is low-magnesium calcite, the least soluble form. Calcite is thermodynamically stable in the surface ocean in most parts of the world, and down to several kilometers depth (Broecker and Peng, 1982; Millero, 1982).

6.10.2.2.2 *Temperature and [CO_3^{2-}] regulation of $CaCO_3$ production*

In general, $CaCO_3$ sinking fluxes in the deep sea tend to correlate with fluxes of organic matter (Dymond and Lyle, 1993; Klaas and Archer, 2002; Milliman, 1993; Tsunogai and Noriki, 1991). However, $CaCO_3$ production decreases dramatically when sea-surface temperatures drop below ~10 °C (Honjo *et al.*, 2000; Wefer *et al.*, 1990). This could be because CO_2 gas solubility increases with decreasing temperature, driving CO_3^{2-} to a lower concentration by pH equilibrium chemistry. Laboratory and controlled system growth experiments have observed a correlation between water [CO_3^{2-}] and calcification rates in coccolithophorids (Riebesell *et al.*, 2000) and in corals (Langdon *et al.*, 2000). Shells of foraminifera tend to be thicker from higher [CO_3^{2-}] (warmer) source waters (Barker and Elderfield, 2002). However, in deep traps there is no significant correlation between the rain ratio and sea-surface temperature, once SST rises above the critical 10 °C level (Klaas and Archer, 2002).

6.10.2.2.3 *Latitudinal distribution discrepancy*

A discrepancy exists between satellite observations of calcification by coccolithophorids and geochemical estimates of the spatial distribution of calcification rate. Massive blooms of coccolithophorids are seen from space as the organisms shed their $CaCO_3$ plates (coccoliths), which scatter light in the water and, therefore, appear light from space. These blooms are found in high latitudes, especially in the North Atlantic (Brown and Yoder, 1994). In contrast, geochemical estimates of calcification, such as the sources of alkalinity in the water column of the ocean (Feely *et al.*, submitted; Sarmiento *et al.*, submitted), and the distribution of $CaCO_3$ on the seafloor (Archer, 1996a) tend to reveal a world where total calcification is most intense in lower latitudes.

6.10.2.2.4 *Organic C/$CaCO_3$ production ratio*

The classical estimate, 4 mol of organic matter per mole of $CaCO_3$, was derived from the relative deep-surface gradients in alkalinity (mostly from dissolved $CaCO_3$) and some proxy for organic carbon degradation, such as nitrate or total CO_2 corrected for $CaCO_3$ (Li *et al.*, 1969). However, this estimate is sensitive to the depth scale of particle degradation in the water column. Organic matter degrades more quickly than $CaCO_3$ dissolves, and so much of the organic matter produced may degrade in the top few hundred meters of the water column, where its impact on the abyss/surface chemical gradients will be smaller than if it degraded in the abyss itself. A recent analysis of the 100–200 m gradients in phosphate and alkalinity from WOCE data seems to indicate a production ratio of 10 or more (Sarmiento *et al.*, submitted). Reconciling the shallow- and whole-ocean gradients of soft-tissue and $CaCO_3$ pump products will require some knowledge of the depth to which these components both sink.

6.10.2.3 SiO_2 Production and Export

The second most important mineral phase of the biological pump is biogenic opal, SiO_2. Opal shells are produced by diatoms, a marine alga, and by radiolarians, a heterotrophic protist. On a large scale, production of SiO_2 is limited by the availability of dissolved silica, in the form of silicic acid, H_4SiO_4.

In general, diatoms appear to be adapted to turbulent, unstable conditions (Margalef, 1978), and changes in water column turbulence associated with climate may be responsible for decadal-timescale changes in the relative production rates of diatoms and coccolithophorids in the North Atlantic (Antia *et al.*, 2001). The dynamics of diatom growth are documented most clearly in the North Atlantic, where deep wintertime mixing limits the growth of phytoplankton until springtime stratification and the spring bloom. The sequence of events that follows is that diatoms bloom most quickly, stripping the surface ocean of dissolved silica, and partially depleting it of the soft-tissue nutrients nitrate and phosphate (Lochte *et al.*, 1993). The same sequence is seen in the Southern Ocean, where diatoms apparently deplete surface water of H_4SiO_4 before the elimination of dissolved nitrate and phosphate (Levitus, 1982; Levitus *et al.*, 1993). Diatoms are found in coastal waters where nutrients are replete (Nelson *et al.*, 1995), and in the blooming plume of iron fertilization experiments (Coale *et al.*, 1996). Diatom export from the surface ocean has been correlated with organic carbon export, and it has been proposed that a few large diatoms might, in fact, dominate the export of organic carbon (Goldman, 1987). (Iron may play another role in the silica cycle, as it has been shown that the diatoms that grow in iron-limited Southern Ocean conditions carry more silica per mole of organic carbon or nitrogen (Franck *et al.*, 2000; Hutchins and Bruland, 1998; Takeda, 1998).)

These considerations have led to a model paradigm wherein abundant supply of silica might stimulate the diatoms, depleting available sea-surface nitrate and thus starving out the coccolithophorids who would produce $CaCO_3$. Honjo refers to the Atlantic as a "carbonate ocean" and the Pacific as a "silicate" ocean based on a comparison of Atlantic and Pacific $CaCO_3$ and SiO_2 sinking fluxes (Honjo, 1996). Models of ocean biogeochemistry (Archer *et al.*, 2000b; Heinze *et al.*, 1999; Matsumoto *et al.*, in press) tend to include some silica dependence for $CaCO_3$ production rates, based on the idea that diatoms outcompete coccolithophorids given sufficient dissolved silica. In deep sediment traps, however, the sinking flux of silica tends to correlate positively with $CaCO_3$ flux, rather than negatively as one might expect if silica inhibits $CaCO_3$ production (Figure 1; Klaas and Archer, 2002).

6.10.2.4 Geochemical Signature of the Biological Pump

The intensity of the biological pump signature on the chemistry of the ocean depends, of course, on the total rate at which material is exported from the surface euphotic zone, but it also depends on how deep it sinks before it degrades back to dissolved constituents. This is because shallower

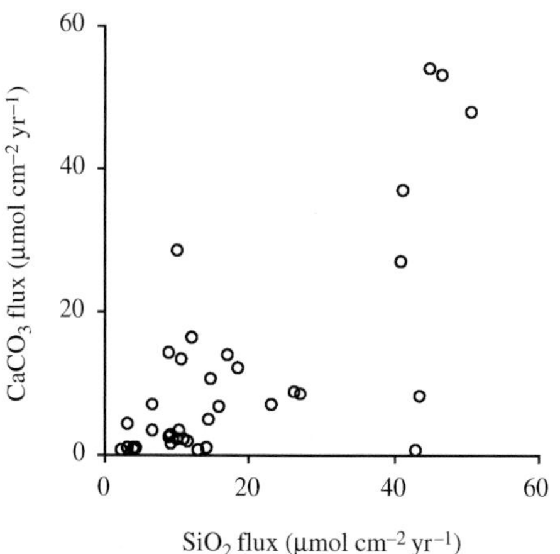

Figure 1 Sediment trap sinking fluxes of $CaCO_3$ plotted against SiO_2. The point of this plot is that high fluxes of SiO_2 tend to correlate with high fluxes of $CaCO_3$, except in regions of cold SST (sources Brewer *et al.*, 1980; Deuser, 1986; Dymond and Collier, 1988; Dymond and Lyle, 1993; Honjo, 1990; Honjo *et al.*, 1982, 1998; Ittekkot *et al.*, 1991; Martin *et al.*, 1987; Milliman, 1993; Noriki and Tsunogai, 1986; Pilskaln and Honjo, 1987; Tsunogai and Noriki, 1991; Wefer *et al.*, 1982, 1988).

subsurface water masses return to the sea surface more quickly than the deepest waters of the ocean. Carbon and nutrients deposited in the shallow thermocline will, therefore, have a smaller impact on the surface/deep fractionation of the ocean than abyssal redissolution. The mineral phases calcite and opal, and to a lesser extent aragonite, tend to dissolve more deeply in the water column.

6.10.2.4.1 Organic carbon redissolution depth

Martin *et al.* (1987) published an initial formulation of a redissolution curve for organic carbon, which has extremely been influential in the ocean modeling community. Their formulation was based on sediment trap data from the North East Pacific Ocean (the VERTEX program) and described the sinking flux of organic matter as a power law:

$$\mathrm{Flux}(z) = \mathrm{Flux}(100\ \mathrm{m})(z/100)^{-0.858}$$

Subsequent analysis has revealed spatial heterogeneity in the organic carbon sinking flux (Lutz *et al.*, in press). In general, the fluxes in the interior of the ocean are less variable than export fluxes determined from the surface ocean; in other words, more productive regions tend to be relatively less efficient at sinking to depth than less productive regions. Other analyses have focused on the role of heavier materials, such as $CaCO_3$, SiO_2, and lithogenic clays, as ballast, in transporting organic matter to depth (Armstrong *et al.*, 2002; Klaas and Archer, 2002) (see below).

6.10.2.4.2 $CaCO_3$ water column redissolution

A significant fraction of the $CaCO_3$ produced in the surface ocean reaches the seafloor, to dissolve or accumulate in the sediments. The traditional view has always been that water column $CaCO_3$ dissolution is relatively insignificant, consistent with its thermodynamic stability. Sediment trap results suggest that $CaCO_3$ survives relatively intact its trip through the water column of the ocean. Measured fluxes of $CaCO_3$ on multiple sediment traps at different depths on a single mooring show no clear trend toward decreasing $CaCO_3$ flux with depth (Figure 2). In addition, visual examination of $CaCO_3$ shells captured in sediment traps does not reveal any overwhelmingly obvious signs of dissolution within the water column (Honjo, 1976). However, three lines of evidence suggest that, in spite of sediment trap data, a considerable fraction of the $CaCO_3$ produced in the upper ocean dissolves as it sinks through the water column.

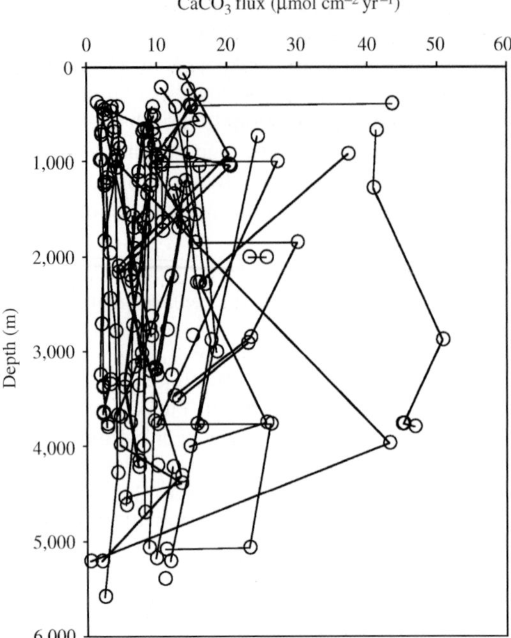

Figure 2 Sediment trap sinking fluxes of $CaCO_3$. Multiple traps which share a common mooring are connected with lines. The point of this messy plot is that the sinking flux of $CaCO_3$ does not appear to decrease with depth in response to water column dissolution (sources Brewer *et al.*, 1980; Deuser, 1986; Dymond and Collier, 1988; Dymond and Lyle, 1993; Honjo, 1990; Honjo *et al.*, 1982, 1998; Ittekkot *et al.*, 1991; Martin *et al.*, 1987; Milliman, 1993; Noriki and Tsunogai, 1986; Pilskaln and Honjo, 1987; Tsunogai and Noriki, 1991; Wefer *et al.*, 1982, 1988).

Chemical signature. One such line of evidence arises from the distribution of alkalinity in the water column. Using Geosecs data, Fiadiero (1980) concluded that the alkalinity distribution in the deep Pacific Ocean could only be explained by water column dissolution of $CaCO_3$. More recently, the WOCE program global ocean survey has provided us with a new, much more extensive and higher-quality global survey of ocean chemistry. Feely *et al.* (submitted) have used these data to essentially confirm the Fiadiero findings. In the Pacific Ocean, they conclude that up to 75% of the $CaCO_3$ production flux could be degrading in the water column of the ocean. They find two hot zones of $CaCO_3$ dissolution, one correlating with the aragonite saturation horizon in the depth range of 1–2 km, comprising approximately half of the total dissolution flux, and another source in the depth range of 2–5 km. The deep alkalinity source is comparable to calculated rates of sedimentary dissolution (Archer, 1996b), which may be difficult to distinguish from dissolution in the water column. Addressing a different question, that of the location of diapycnal mixing in the ocean, Munk and Wunsch (1998) showed that boundary mixing, advected into the ocean interior along isopycnal surfaces, is indistinguishable from more diffuse mixing throughout the interior. We may presume that the same reasoning applies to sources of alkalinity.

$CaCO_3$ budgets. Another argument comes from comparison of measured calcification rates in the surface ocean with rates of $CaCO_3$ accumulation in sediments. Milliman *et al.* (2000) make a strong case for dissolution in the water column above the saturation horizon for $CaCO_3$, based on a comparison of euphotic zone calcification rate estimates (Balch *et al.*, 2000; Balch and Kilpatrick, 1996) and sediment accumulation rates in regions where the overlying water is supersaturated with respect to calcite. Ocean carbon cycle models typically impose a significant water column dissolution fraction of $CaCO_3$ rain (Archer and Maier-Reimer, 1994; Orr *et al.*, 2001). The mechanism for this dissolution is unclear. The Feely results seem to indicate that the flux of aragonite in the open ocean may be higher than had previously been assumed. However, they also find a significant dissolution flux uncorrelated with the aragonite saturation horizon, which is presumably calcite. It has been proposed that calcite might be dissolving in the guts of zooplankton (Jansen and Wolf-Gladrow, 2001), or within acidic microenvironments inside marine snow.

Planktonic shell thickness. A third indicator of $CaCO_3$ dissolution within the water column comes from a more subtle examination of the shells of foraminifera, the thickness of their walls in fact. This technique was pioneered by Lohmann (1995) and further pursued by Broecker and Clark (2001). What they find is that the shell thickness of forams increases throughout the first kilometer of the water column, and decreases below ~3 km, representative of dissolution within the water column.

6.10.2.4.3 SiO_2 sinking and redissolution

The dissolution of SiO_2 as it sinks through the water column is another ongoing topic of research and discussion. The Southern Ocean is a crucial depocenter for SiO_2 in deep-sea sediments, in the form of a band of high-opal sediments known as the opal belt. In the view of Nelson *et al.* (1995), the intensity of opal burial in this location is not due to high rates of opal production, but rather a peculiar high efficiency of opal transfer to the seafloor in this region. Gnanadesikan *et al.* (1999) point to extreme temperature sensitivity of opal dissolution kinetics, and find that an ocean circulation/silica cycle model is better able to reproduce the observed distribution of opal when this dependence is included. Opal dissolves faster in warmer water, in part explaining the seafloor transport efficiency within the cold Southern Ocean waters. An additional factor may be the episodic nature of primary production, which precludes a standing crop of grazers.

6.10.2.4.4 *The ballast model for sinking particles*

One of the more robust features of the sediment trap data set is the constancy of the "organic carbon": $CaCO_3$ rain ratio in the deep sea. Although the absolute fluxes vary widely by area and time, the ratio of their fluxes (mol : mol) seems quite stable in the range of 0.5–1.0. One explanation for this observation, which has significant implications for the ability of the ocean to change its pH by changing the rain ratio, is called the ballast model (Armstrong *et al.*, 2002; Klaas and Archer, 2002). The density of organic matter is similar enough to that of water that organic matter, by itself, would not sink at all. Armstrong *et al.* (2002) did a regression of sediment trap from the equatorial Pacific and found that organic carbon sinking fluxes could be described remarkably well by combining a ballasted flux (a simple function of the ballast flux at depth times some uniform "carrying coefficient") with a "Free POC" component, which degrades over a 600 m depth scale in the surface ocean. Klaas and Archer (2002) extended this analysis to the global sediment trap dataset, and distinguished three forms of ballast: $CaCO_3$, biogenic silica, and lithogenic material, mainly terrestrial clays. They found that the carrying coefficient (grams organic carbon per gram ballast) for $CaCO_3$ was twice as high as for the other phases, a result which is qualitatively consistent with the greater density of $CaCO_3$. When the global fluxes of the three ballast phases are multiplied by their carrying coefficients, they found that 80% of the organic matter flux to the seafloor was carried by $CaCO_3$. This explains the relative constancy of the rain ratio in the deep sea, and it also seems to limit the ability of the ocean carbon cycle to vary this rain ratio by changing, e.g., the relative productivities of diatoms and coccolithophorids. Some ecological change in the surface ocean, impacting the sinking flux of $CaCO_3$, would merely change the flux of organic matter in concert with the changing $CaCO_3$ flux.

This idea is somewhat at odds with the common observation from near the sea surface, either from shallow sediment traps or from phytoplankton growth studies, that diatoms dominate the export of carbon from the surface ocean (Goldman, 1987). However, there is room in the "Free POC" component of Armstrong *et al.* (2002) for diatoms and carbon export. Perhaps the particles, produced in the organic carbon to $CaCO_3$ ratio of 4–10, are unable to sink very quickly until organic matter degradation decreases the organic ratio to the 0.5–1.0 range seen in the deep sea, at which point the express elevator takes them to the seafloor.

6.10.2.5 Direct Atmospheric p_{CO_2} Signature of the Biological Pump

6.10.2.5.1 *Organic carbon pump*

The biological pump has a first-order impact on the CO_2 concentration of the atmosphere, and therefore on the climate of the earth. The removal of dissolved CO_2 from surface water during photosynthesis decreases the equilibrium partial pressure of CO_2 in the overlying atmosphere (this quantity is defined as the p_{CO_2} of the surface water). Changes in the biological pump in some form may be responsible for the glacial/interglacial p_{CO_2} cycles (Petit *et al.*, 1999). The strongest direct impact of the biological pump on p_{CO_2} is driven by the depletion of surface waters of dissolved CO_2 by the production of organic carbon (the "soft-tissue" pump).

Timescale. The timescale for sea-surface depletion of CO_2 to affect atmospheric p_{CO_2} is that of ocean rearrangement and atmosphere/ocean gas exchange equilibration: on the order of hundreds of years. Looking into the past for biological pump-driven changes in the p_{CO_2} of the atmosphere, we would look for changes that occur on this timescale (unless the efficiency of the biological pump itself is changing more slowly than that).

Model sensitivity. One of the more recent surprise was the discovery that the sensitivity of atmospheric p_{CO_2} to the biological pump is quite model specific (Archer *et al.*, 2000a; Bacastow, 1996; Broecker *et al.*, 1999). In particular, box models of ocean chemistry are more sensitive to the efficiency of the biological pump than are ocean models based on a continuum representation of circulation and chemistry, most notably the general circulation models (GCMs) (Figure 3). The horizontal axis of this plot is the inventory of the dissolved nutrient phosphorus, in the form of PO_4^{3-}, within the top 50 m of the world ocean. The sea-surface phosphorus inventory depends on a balance between delivery by upwelling and mixing of phosphorus from below, and removal as sinking biogenic particles.

The reason for the discrepancy is that most of the available sea-surface nutrients (which could be depleted were the phytoplankton more efficient or the circulation less energetic) are to be found in the high latitudes, notably in the Southern Ocean. A startling and seminal finding in the 1980s was

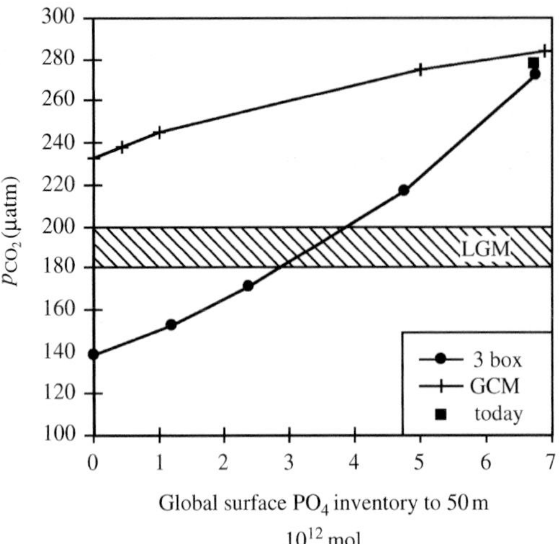

Figure 3 Steady-state model atmospheric p_{CO_2} response to the efficiency of the biological pump, as indicated by the inventory of PO_4^{3-} in the top 50 m of the ocean (source Archer *et al.*, 2000a).

that changes to the nutrient concentration in the high-latitude surface ocean in a simple type of ocean chemistry models called box models have a disproportionately large impact on the steady-state CO_2 concentration of the atmosphere (Knox and McElroy, 1984; Sarmiento and Toggweiler, 1984; Siegenthaler and Wenk, 1984). The low-latitude surface ocean is conversely less effective at determining the p_{CO_2} of the atmosphere.

A box model is comprised of some small number of ocean reservoirs, typically three to a few tens. The chemical characteristics of the seawater with each box is taken to be homogeneous (well mixed). Water flow in the ocean is described as fluxes between the boxes; given in units of Sv (Sverdrups, $10^6\,m\,s^{-1}$), these typically range to values less than 100 Sv. A flow between two boxes carries with it the chemical signature of the source box; this numerical technique is called "upstream" differencing. Gas exchange acts to pull sea-surface gas concentrations, such as CO_2 and $^{14}CO_2$, toward equilibrium values with the atmosphere. The fluxes imposed between boxes are determined by tuning various tracers, usually ^{14}C, toward observed values.

More recently, however, it was found that GCMs have a greater sensitivity to low-latitude forcing (Bacastow, 1996; Broecker *et al.*, 1999), and therefore lower sensitivity to high-latitude forcing (Archer *et al.*, 2000a). A GCM aims to represent the chemical and physical fields of the ocean in a continuous way, sampled on a finite difference grid (discretized). Flows across the grid are determined by balancing the forces of inertia, gravity, stress, friction, and the rotation of the Earth. Because atmospheric p_{CO_2} values above GCMs are less sensitive to the high-latitude sea surface than it is above box models, atmospheric p_{CO_2} is also less sensitive to the soft-tissue pump, variations in which are focused in the high latitudes by the availability of nutrients. Conversely, the effects of the $CaCO_3$ pump seem to be stronger in GCMs than in box models (Matsumoto *et al.*, in press), perhaps because $CaCO_3$ production occurs predominatly in warm surface waters. The reasons for the discrepancy are still unclear. Archer *et al.* (2000a) pointed to vertical diffusion, which dominates the residence time of water in the surface layers of the ocean. Toggweiler (2003) points to the extent of equilibration of surface waters, in particular in the high latitudes. Until the discrepancy between box models and GCMs is resolved, it is difficult to predict the sensitivity of the real ocean to changes in the soft-tissue biological pump.

$CaCO_3$ pump. The $CaCO_3$ component of the biological pump also affects the p_{CO_2} of surface waters and therefore of the atmosphere. $CaCO_3$ production depletes the parent water of dissolved carbonate ion, CO_3^{2-}, which actually drives the dissolved CO_2 concentration up by the reaction

$$CO_2 + CO_3^{2-} + H_2O \leftrightarrow 2HCO_3^- \qquad (1)$$

The $CaCO_3$ pump therefore acts to increase the p_{CO_2} of the atmosphere, counteracting, to a small extent, the effect of the soft-tissue pump. This seeming paradoxical behavior is a result of the pH chemistry of the carbonate buffer system. Dissolved CO_2, in its hydrated form H_2CO_3, is an acid, and it exists in buffer chemistry equilibrium with carbonate ion, CO_3^{2-}, the base salt form of H_2CO_3. The removal of CO_3^{2-} by calcification acidifies the parent seawater, shifting the carbonate buffer system equilibrium toward higher CO_2 concentration. A converse of this process is the dissolution of $CaCO_3$ on the seafloor or in coral reefs in response to the invasion of fossil fuel CO_2 into the ocean (Archer *et al.*, 1989b, 1997; Walker and Kasting, 1992). One might naively expect coral reefs to grow larger with the increased availability of carbon to build them from, but the pH equilibrium chemistry overrides the absolute abundance of dissolved carbon.

6.10.3 CONTROLS OF MEAN OCEAN CHEMISTRY

The mean concentrations of constituents of seawater are determined not by simple distillation of river water but by their various mechanisms of removal from the ocean. The dominant cation in river water, e.g., is calcium from weathering of carbonate and silicate rocks, whereas the dominant cation in the ocean is sodium, because there are no efficient loss mechanisms for sodium analogous to the formation of $CaCO_3$ as a loss mechanism for calcium (Holland, 1978). The cycle of $CaCO_3$ in the ocean is of particular interest, as it determines the pH of the ocean and the p_{CO_2} of the atmosphere, on timescales of thousands to hundreds of thousands of years. Organic matter burial can be considered a net source of oxygen to the atmosphere, and mechanisms for regulating atmospheric O_2 focus on the mechanics of burying organic matter in sediments (Cappellen and Ingall, 1996; Walker, 1977).

The redissolution or burial of organic matter in sediments is a decision that is made jointly by the physics of diffusion, chemistry of organic matter oxidation, and the biology which mediates the chemical reactions. The concentration profile of a solute in sediment pore water is governed by the diffusion equation, which can be written most simply as

$$\frac{\partial C}{\partial t} = 0 = \frac{\partial}{\partial z} D_M \frac{\partial C}{\partial z} + j_C$$

where C is a solute concentration, D_M is a molecular diffusion coefficient, and j_C is a production or comsumption rate. When applied to the distribution of dissolved oxygen, this expression results in a generic oxygen profile within the sediments (e.g., Figure 4(a)). At any depth in the sediments, the downward diffusive flux of oxygen is balanced by the integrated consumption below that depth. When applied to the concentration of a dissolution product such as H_4SiO_4 of CO_3^{2-}, the diffusion equation results in a generic profile as illustrated in Figure 4(b). With depth, the pore waters become increasingly isolated from diffusive contact with the overlying water, and the concentration of the solute approaches equilibrium with the solid phase. The chemistry of the overlying water serves as an upper boundary condition for this equation.

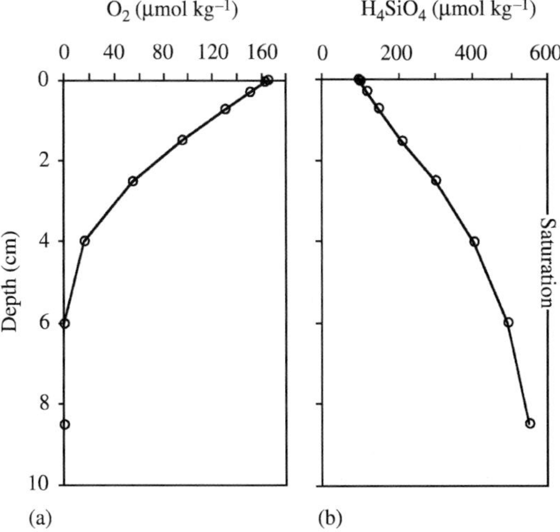

Figure 4 Model profiles of (a) O_2 and (b) H_4SiO_4 in surface sediments. From the Muds model (Archer *et al.*, 2001).

6.10.3.1.1 Controls on organic matter burial

Hypsometry. Globally, the burial of organic matter in sediments is accomplished largely within a relatively small area of the seafloor, in coastal waters of the continental shelf and slope. There are several reasons for this. First, rates of primary production tend to be higher in coastal waters, in part because of coastal upwelling, and in part because of riverine and recycling sources of nutrients such as iron (Johnson *et al.*, 2001). Accumulation rates of abiotic material such as clay minerals are orders of magnitude higher in near-shore sediments than in the abyssal ocean (Middelburg *et al.*, 1997; Tromp *et al.*, 1995). This tends to promote organic carbon burial. Second, the efficiency of organic matter transfer to the sediments is higher in coastal waters. The water column itself is thinner, minimizing degradation of sinking material. Also, coastal phytoplankton themselves tend to be larger than their pelagic cousins, increasing their export efficiency (see Chapter 6.04). The highest rates of primary production tend to be found on continental shelves or just along the shelf/slope break, where coastal upwelling is manifested most strongly. Results of the SEEP program indicate that most of the organic matter produced on the slope respires there (Biscaye *et al.*, 1994). However, organic carbon concentrations in sandy shelf sediments are often rather low. Measured sediment respiration rates often reach a maximum somewhat further offshore, in finer-grained continental slope sediments that deposit in this lower-energy setting (Jahnke *et al.*, 1990). Although carbon export from the shelf to the slope may be a negligible term in the shelf carbon budget, it is by no means trivial to the budget of the slope, analogous to crumbs from the high table.

Degradation rate kinetics. Because most of the chemical and physical processes that control organic matter degradation are determined by biology, most of them are determined for modeling purposes empirically. The first order of business is a rate constant for the degradation reaction itself:

$$\text{OrgC} + \text{electron acceptor} \rightarrow CO_2 + \text{electron donor}$$

$$\text{Rate} = k(\text{OrgC})$$

where k is the rate constant, in units of inverse time. It has been found that degradation rates are independent of the concentration of the electron acceptor (e.g., O_2) until the concentration of O_2 reaches some critical low level ($\sim\mu$M).

This psuedo-zero-order dependence is called Monod kinetics (Devol, 1978). These kinetic rate formulations become important if the rate constants for organic degradation differ for different electron acceptors. Oxic degradation, in particular, appears to be faster than sulfate degradation, but it is confined to the upper oxic zone of the sediment (Figure 5).

Rate constants for organic degradation can be estimated in the field from pore-water data (e.g., high-resolution microelectrode oxygen profiles (Hales *et al.* 1993)) or from laboratory incubations (for sulfate respiration in particular, Weber *et al.* (2001)). These rate constants vary widely, with values orders of magnitude higher in shelf and slope than in abyssal sediments. The rate constants have been parametrized as a function of total sediment accumulation rate (Middelburg *et al.*, 1997; Tromp *et al.*, 1995), organic carbon rain (Archer *et al.*, 2001), or the mean age of the sedimentary organic matter (Emerson, 1985). In fact, there is not simply one homogeneous kind of organic matter in sediments, but rather modeling and field studies are able to resolve at least two types of organic matter with degradation kinetics differing by several orders of magnitude (Berner, 1980).

Bioturbation. Other biologically mediated relevant processes include bioturbation, the physical mixing of solid sediment, and pore-water irrigation, an enhancement of solute transport by benthic fauna. Bioturbation is so ubiquitous that its absence in sedimentary deposits (indicated by fine laminae) is taken as strong evidence of anoxia during deposition. In general, the rate of bioturbation scales similarly to that of organic carbon respiration, increasing in shallower water. The depth of bioturbation appears to be much less variable (Boudreau, 1994, 1998). The bioturbation rate appears to decrease smoothly with depth (Archer *et al.*, 2001; Martin and Sayles, 1990), but can also be treated as a constant within the bioturbated layer. To make things even more complicated, bioturbation rates derived from radionuclides seem to depend on the lifetime of the radionuclide, with faster mixing for the shorter-lived tracer (Fornes *et al.*, 1999; Smith *et al.*, 1993).

Pore-water irrigation. Pore-water irrigation is the exchange of solute chemistry with overlying water resulting from the action of benthic macrofauna, who pump overlying water through their burrows within the sediment. Rates of irrigation can be determined by comparing total oxygen uptake rates, from benthic flux chambers, with diffusive uptake rates, from microelectrodes, which are assumed to be independent of any irrigation influence. In general, pore-water irrigation becomes negligible, or at least unmeasurable, in abyssal sediments (Archer and Devol, 1992; Archer *et al.*, 2001; Reimers and Smith, 1986).

Oxygen. A traditional view of oxygen dependence on organic carbon concentrations of sediments was challenged in the 1980s by laboratory measurements of the degradation of fresh phytoplankton using either O_2 or SO_4^{2-} as an electron acceptor (Henrichs and Reeburgh, 1987). However, degradation rate constants derived from field data appear to diverge as they slow down in the deep (Tromp *et al.*, 1995), to values that differ by five or six orders of magnitude (Figure 5). Calvert and Pederson (1993) compared organic carbon concentrations between oxic and anoxic (Black Sea) settings and concluded that oxygen plays little part in organic carbon preservation. However, Archer *et al.* (2001) summarized data from a wide variety of locations, and found dependence both on oxygen concentration and on organic carbon rain to the seafloor.

Protection by physical adsorption. A third viewpoint comes from the observation that the organic matter concentration of sediment appears

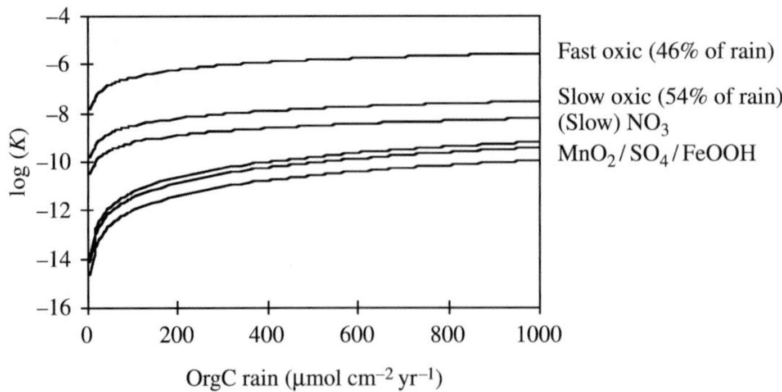

Figure 5 Model-tuned respiration rate constants for various electron acceptors (O_2, NO_2, Mn, Fe, and S), and for two fractions of organic matter rain to the seafloor (faster fraction shown for O_2 only). Dependence on organic carbon rain was based on data from Toth and Lehrman (1977), Tromp *et al.* (1995), Westrich and Berner (1984) tuned to reproduce sediment pore-water and solid phase field measurements (Archer *et al.*, 2001).

to correlate inversely with the grain size of the sediment, in such a way as to maintain a constant ratio of organic matter concentration to the surface area of the sediment grains (Hedges and Keil, 1995). Keil *et al.* (1994) were able to chemically extract organic matter from the surfaces of clays, and found that its reactivity was increased significantly by this extraction. We note also the surprising success of the Muds sediment model at predicting organic carbon concentration in sediments, with no consideration of organic adsorption effects at all.

Global sedimentary budget. Finally, on a global scale, we point to an interesting model by Berner and Canfield (1989), which took as a given an organic carbon and reduced sulfur concentration of sediments of different type (nearshore sands, offshore clays, abyssal clays, etc.). In this view, a net source or sink of O_2 to the atmosphere could be driven by rearrangement of the sedimentary rock mass of the Earth, from abyssal clays to nearshore clays, for example.

6.10.3.1.2 $CaCO_3$

Dissolution kinetics. In contrast to the degradation of organic matter, the dissolution of $CaCO_3$ in sediments relies on abiotic dissolution which ought, in principle, to be more conservative across regions of the ocean. A complication is that $CaCO_3$ dissolution into seawater is comprised of multiple chemical reactions such as

$$>CO_3^{2-} + H^+ \rightarrow Ca^{2+} + HCO_3^-$$

and

$$>Ca^+ + H_2CO_3 \rightarrow >CO_3HO + CaHCO_3^+$$

(Arakaki and Mucci, 1995; Berner and Morse, 1974). The total loss of solid-phase $CaCO_3$ is determined as the sum of these reactions, and therefore exhibits a complicated dependence on the saturation state of the fluid. This behavior has been parametrized as high-order kinetic rate law of the form

$$R = k(1 - \Omega)^n$$

where k is a rate constant, Ω is the saturation state $(= [Ca^{2+}][CO_3^{2-}]/K'_{sp})$, and n is a potentially nonunity dissolution rate order (Keir, 1980; Morse, 1979). However, the numerical value of the rate order (the exponent n above) is difficult, if not impossible, to extract from field data (Archer *et al.*, 1989a; Hales and Emerson, 1996, 1997a), and sensitive to the imprecision of the saturation state for $CaCO_3$ in laboratory measurements (Hales and Emerson, 1997b). In general, it has been found that the rate constant (the pre-exponential factor k above) varies by orders of magnitude from place to place in the ocean (Hales and Emerson, 1996, 1997a), but less so if linear dissolution kinetics are used. We should also note that laboratory determinations of the rate constant tend to exceed field determinations by two to three orders of magnitude.

Pore-water pH. The dissolution of $CaCO_3$ in sediments is also complicated by interaction with organic matter diagenesis via the pH equilibrium chemistry of seawater (reaction (1), above). The addition of dissolved CO_2 from organic carbon degradation shifts the equilibrium to the right, decreasing the concentration of CO_3^{2-} in the pore waters. Figure 6 shows a model-generated concentration profile of CO_2, HCO_3^-, and CO_3^{2-} in sediment pore waters above the saturation horizon for $CaCO_3$ (i.e., where the $[CO_3^{2-}]$ of the overlying water is higher than the equilibrium

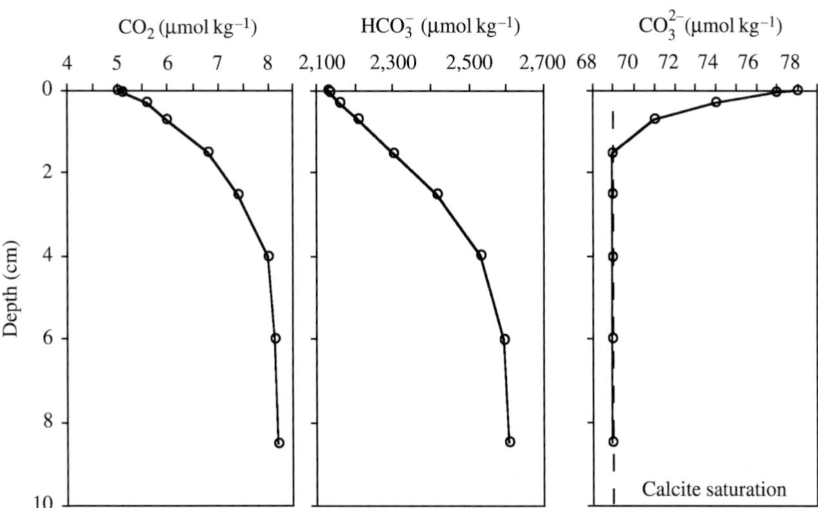

Figure 6 Model profiles of CO_2, HCO_3^-, and CO_3^{2-} in surface sediments. From the Muds model (Archer *et al.*, 2001).

value with respect to calcite). Sediment chemistry models predict that respiration accounts for a significant dissolution flux from sediments, e.g., 20–40% of the calcite rain to the seafloor may be dissolving, even where the overlying water is saturated with respect to calcite (Archer, 1991; Emerson and Archer, 1990).

The implication of a role for respiration as a driver for $CaCO_3$ dissolution is that any hypothetical change in the ratio of organic carbon to calcite raining to the seafloor may impact the global burial rate of calcite, and thus (by a mechanism to be explained below) the steady-state pH of the ocean (Archer and Maier-Reimer, 1994).

6.10.3.1.3 SiO_2

On the face of it, the dissolution of SiO_2 from sediments ought to be simpler than that of organic matter, which is driven biologically, or $CaCO_3$, with its coupling to pore-water pH and organic carbon respiration. The dissolution reaction for SiO_2 is simply

$$SiO_2 + H_2O \rightarrow H_4SiO_4\ (aq)$$

Although H_4SiO_4 (aq) could be affected by the dissociation reaction

$$H_4SiO_4 \leftrightarrow H^+ + H_3SiO_4^-$$

the equilibrium constant for this reaction makes it insignificant in seawater.

However, opal diagenesis in sediments, despite its apparent simplicity, is not without mystery. The asymptotic concentration of silicic acid in pore waters in contact with measurable solid-phase opal ought to reflect the solubility of opal. Instead, what is observed are generally lower concentrations than the laboratory-measured equilibrium values, with a tendency toward correlation with the productivity of the surface ocean. Asymptotic silicon in Southern Ocean sediments reaches 600 μM and up, while in the subtropical oligotrophic ocean asymptotic silicon is only 150–300 μM (Schink *et al.*, 1974). Archer *et al.* (1993) argued that dynamically, with the framework of the diffusive pore-water system, the equilibrium silicon has a greater impact on opal preservation than the rate constant for dissolution. Recent work has found a correlation between the asymptotic (equilibrium) silicon concentration and the concentration of aluminum in the sediment solid phase (Cappellen and Qiu, 1997; Dixit *et al.*, 2001), rationalizing the decreasing solubility as a function of the formation of aluminosilicates on the opal surface. Similar behavior has been documented in the laboratory (Lewin, 1961).

6.10.3.1.4 *Indirect atmospheric p_{CO_2} signature of the biological pump*

The biological pump also has an indirect mechanism of affecting the p_{CO_2} of the atmosphere, called $CaCO_3$ compensation. The mechanism is based on the constraint that on long enough timescales, the inputs and outputs of alkalinity to the ocean must balance. The input arises from weathering of carbonate and silicate rocks on land, and removal is by the production and sedimentation of $CaCO_3$. The crucial component of this mechanism is that the fraction of $CaCO_3$ produced that is buried depends on the acidity of the ocean. If input of alkalinity exceeds output, e.g., the buildup of alkalinity in the ocean pulls the pH to higher values, allowing a greater fraction of the $CaCO_3$ produced to be buried (Broecker and Peng, 1982, 1987). The system approaches equilibrium on a timescale of roughly 5,000 years (Archer *et al.*, 1997). Because the mechanism relies on changing the mean chemistry of the ocean by interaction with the sediments, this mechanism has also been called the "open system" response (and the direct effect described above has correspondingly been called the closed system response (Sigman *et al.*, 1998)). The link to atmospheric p_{CO_2} arises from the dependence of the dissolved CO_2 concentration on seawater pH. As pH rises, less and less of the dissolved carbon is found in the acid form CO_2.

Weathering. $CaCO_3$ compensation provides several ways in which the p_{CO_2} of the atmosphere could be affected by the biological pump. Most obviously, if the delivery rate of alkalinity to the ocean from weathering were to increase, then a greater fraction of $CaCO_3$ production would have to be buried to balance weathering. This would drive ocean pH up, decreasing atmospheric p_{CO_2} (Berger, 1982; Opdyke and Walker, 1992). One mechanism for increasing weathering during glacial time is the exposure of shallow-water reefs and carbonate banks, which therefore transform from significant depocenters of $CaCO_3$ from the ocean (Kleypas, 1997; Milliman, 1993) to significant weathering sources (Gibbs and Kump, 1994; Munhoven and Francois, 1993). This mechanism must have played some role in glacial p_{CO_2} variations, although the timing of events at the deglacialtion precludes a primary sea-level driver of p_{CO_2} (Broecker and Henderson, 1998).

Rain ratio. Somewhat less obviously, a decrease in the production of $CaCO_3$ would mandate an increase in the burial fraction that must be maintained in order to achieve throughput balance. Third, $CaCO_3$ dissolution in sediments is promoted by the oxic degradation of organic matter (Emerson and Bender, 1981). Therefore, an increase in organic matter rain to the seafloor could ultimately drive pH toward the basic to compensate, decreasing atmospheric p_{CO_2}. Both the direct and the indirect effects of organic carbon and $CaCO_3$ rain depend on the ratio of organic carbon to $CaCO_3$ produced. This ratio therefore has a name, the "rain ratio." The effects of organic carbon and $CaCO_3$ on $CaCO_3$ compensation are collectively called the rain-ratio hypothesis (Archer and Maier-Reimer, 1994).

6.10.4 SUMMARY

In summary, some advances in the biogeochemical cycles in the ocean have suggested several new sensitivities, ways in which the past or future climate change could impact the p_{CO_2} concentration of the atmosphere. Some of these mechanisms act on the 10^3-year timescale of ocean/atmosphere p_{CO_2} equilibration, such as the effect of iron on the biological pump. This effect is minimized if (Section 6.10.2.1.2) iron is predominantly recycled within the ocean rather than supplied by deposition of continental dust, and (Section 6.10.2.5.1) if the real ocean behaves as a general circulation model rather than as a box model, in the way that the high-latitude surface ocean interacts with the p_{CO_2} of the atmosphere. Other potential sensitivities include a decrease in calcification with acidity of surface waters (Section 6.10.2.2.2) and an increase in water column redissolution of $CaCO_3$ (Section 6.10.2.4.2). Alternatively, $CaCO_3$ production may increase with an increase in the stratification of the water column in a future global warming world (Section 6.10.2.3; Antia *et al.*, 2001), or perturbations in dust deposition due to changes in land use (Sections 6.10.2.1.2 and 6.10.2.3).

On timescales of $CaCO_3$ compensation (Section 6.10.3.1.4), any changes in calcification rate, or the ratio of calcite to organic carbon reaching the seafloor, have an amplified effect on the p_{CO_2} of the atmosphere. This effect may be minimized if organic matter requires ballasting, predominately from $CaCO_3$, to reach the seafloor (Section 6.10.2.4.4). The impact of iron deposition on Si : C ratios of diatoms may affect atmospheric p_{CO_2} (Matsumoto *et al.*, in press), if an increase in silicon availability depresses calcification, an effect which is not clearly seen in sediment trap data (Klaas and Archer, 2002).

REFERENCES

Abraham E. R., Law C. L., Boyd P. W., Lavender S. J., Maldonado M. T., and Bowle W. R. (2000) Importance of stirring in the development of an iron-fertilized phytoplankton bloom. *Nature* **407**, 727–703.

Antia A. N., Koeve W., Fischer G., Blanz T., Schulz-Bull D., Scholten J., Neuer S., Kremling K., Kuss J., Peinert R., Hebbeln D., Bathmann U., Conte M., Fehner U., and Zeitzschel B. (2001) Basin-wide particulate carbon flux in the Atlantic Ocean: regional export patterns and potential for atmospheric CO_2 sequestration. *Global Biogeochem. Cycles* **15**, 845–862.

Arakaki T. and Mucci A. (1995) A continuous and mechanistic representation of calcite reaction-controlled kinetics in dilute solutions at 25 °C at 1 atm total pressure. *Aquat. Chem.* **1**, 105–130.

Archer D. E. (1991) Modeling the calcite lysocline. *J. Geophys. Res.* **96**(C9), 17037–17050.

Archer D. E. (1996a) An atlas of the distribution of calcium carbonate in sediments of the deep sea. *Global Biogeochem. Cycles* **10**, 159–174.

Archer D. E. (1996b) A data-driven model of the global calcite lysocline. *Global Biogeochem. Cycles* **10**, 511–526.

Archer D. E. and Devol A. (1992) Benthic oxygen fluxes on the Washington shelf and slope: a comparison of *in situ* microelectrode and chamber flux measurements. *Limnol. Oceanogr.* **37**, 614–629.

Archer D. E. and Johnson K. (2000) A model of the iron cycle in the ocean. *Global Biogeochem. Cycles* **14**, 269–279.

Archer D. E. and Maier-Reimer E. (1994) Effect of deep-sea sedimentary calcite preservation on atmospheric CO_2 concentration. *Nature* **367**, 260–264.

Archer D., Emerson S., and Reimers C. (1989a) Dissolution of calcite in deep-sea sediments: pH and O_2 microelectrode results. *Geochim. Cosmochim. Acta* **53**, 2831–2846.

Archer D., Emerson S., and Smith C. R. (1989b) Direct measurement of the diffusive sublayer at the deep sea floor using oxygen microelectrodes. *Nature* **340**, 623–626.

Archer D., Lyle M., Rodgers K., and Froelich P. (1993) What controls opal preservation in tropical deep-sea sediments. *Paleoceanography* **8**, 7–21.

Archer D., Kheshgi H., and Maier-Riemer E. (1997) Multiple timescales for neutralization of fossil fuel CO_2. *Geophys. Res. Lett.* **24**, 405–408.

Archer D., Eshel G., Winguth A., Broecker W. S., Pierrehumbert R. T., Tobis M., and Jacob R. (2000a) Atmospheric p_{CO_2} sensitivity to the biological pump in the ocean. *Global Biogeochem. Cycles* **14**, 1219–1230.

Archer D. E., Winguth A., Lea D., and Mahowald N. (2000b) What caused the glacial/interglacial atmospheric p_{CO_2} cycles? *Rev. Geophys.* **38**, 159–189.

Archer D. E., Morford J. L., and Emerson S. (2001) A model of suboxic sedimentary diagenesis suitable for automatic tuning and gridded global domains. *Global Biogeochem. Cycles* (in press).

Armstrong R. A., Lee C., Hedges J. I., Honjo S., and Wakeham S. G. (2002) A new mechanistic model for organic carbon fluxes in the ocean based on the quantitative association of POC with ballast minerals. *Deep-Sea Res. II* **49**, 219–236.

Bacastow R. B. (1996) The effect of temperature change of the warm surface waters of the oceans on atmospheric CO_2. *Global Biogeochem. Cycles* **10**, 319–334.

Balch W. M., Drapeau D. T., and Fritz J. J. (2000) Monsoonal forcing of calcification in the Arabian Sea. *Deep-Sea Res. II* **47**, 1301–1337.

Barker S. and Elderfield H. (2002) Foraminiferal calcification response to glacial–interglacial changes in atmospheric CO_2. *Science* **297**, 833–836.

Balch W. M. and Kilpatrick K. (1996) Calcification rates in the equatorial Pacific along 140° W. *Deep-Sea Res. II* **43**(4–6), 971–994.

Berger W. H. (1982) Deglacial CO_2 buildup: constraints on the coral reef model. *Palaeogeogr. Palaeoclimatol. Palaeoecol.* **40**, 235–253.

Berger W. H. (1989) Global maps of ocean productivity. In *Productivity of the Ocean: Present and Past* (eds. S. V. S. W. H. Berger, and G. Wefer). Wiley, pp. 429–453.

Berner R. A. (1980) A rate model for organic matter decomposition during bacterial sulfate reduction in marine sediments. In *Biogeochemistry of Organic Matter at the Sediment–Water Interface*. Comm. Natl. Recherche Scientifique, France.

Berner R. A. and Canfield D. E. (1989) A new model for atmospheric oxygen over Phanerozoic time. *Am. J. Sci.* **289**, 333–361.

Berner R. A. and Morse J. W. (1974) Dissolution kinetics of calcium carbonate in sea water IV. Theory of calcite dissolution. *Am. J. Sci.* **274**, 107–134.

Biscaye P. E., Flagg C. N., and Falkowski P. G. (1994) The shelf edge exchange processes experiment, SEEP-II: an introduction to hypotheses, results and conclusions. *Deep-Sea Res. II* **41**, 231–252.

Boudreau B. P. (1994) Is burial velocity a master parameter for bioturbation? *Geochim. Cosmochim. Acta* **58**, 1243–1249.

Boudreau B. P. (1998) Mean mixed depth of sediments: the wherefore and the why. *Limnol. Oceanogr.* **43**, 524–526.

Boyd P. W., Watson A. J., Law C. S., Abraham E. R., Trull T., Murdoch R., Bakker D. C. E., Bowie A. R., Buesseler K. O., Chang H., Charette M., Croot P., Downing K., Frew R., Gall M., Hadfield M., Hall J., Harvey M., Jameson G., LaRoche J., Liddicoat M., Ling R., Maldonado M. T., McKay R. M., Nodder S., Pickmere S., Pridmore R., Rintous S., Safi K., Sutton P., Strzepek R., Tannaberger K., Turner S., Waite A., and Zeldis J. (2000) A mesoscale phytoplankton bloom in the polar Southern Ocean stimulated by iron fertilization. *Nature* **407**, 695–702.

Brewer P. G., Nozaki Y., Spencer D. W., and Fleer A. P. (1980) Sediment trap experiments in the deep North Atlantic: isotopic and elemental fluxes. **38**(4), 703–728.

Broecker W. S. and Clark E. (2001) An evaluation of Lohmann's foraminifera weight dissolution index. *Paleoceanography* **16**, 531–534.

Broecker W. S. and Henderson G. (1998) The sequence of event surrounding termination II and their implications for the cause of glacial–interglacial CO_2 changes. *Paleoceanography* **13**, 352–364.

Broecker W. S. and Peng T. H. (1982) *Tracers in the Sea*. Eldigio Press, Palisades, NY.

Broecker W. S. and Peng T. H. (1987) The role of $CaCO_3$ compensation in the glacial to interglacial atmospheric CO_2 change. *Global Biogeochem. Cycles* **1**, 15–29.

Broecker W., Lynch-Steiglitz J., Archer D., Hofmann M., Maier-Reimer E., Marchal O., Stocker T., and Gruber N. (1999) How strong is the Harvardton–Bear constraint? *Global Biogeochem. Cycles* **13**, 817–821.

Brown C. W. and Yoder J. A. (1994) Coccolithophorid blooms in the global ocean. *J. Geophys. Res.* **99**, 7467–7482.

Buesseler K. O. (1991) Do upper-ocean sediment traps provide an accurate record of particle flux? *Nature* **353**, 420–423.

Calvert S. E. and Padersen T. F. (1993) Geochemistry of recent oxic and anoxic marine sediments: implications for the geological record. *Mar. Geol.* **113**, 67–88.

Cappellen P. V. and Ingall E. (1996) Redox stabilization of the atmosphere and oceans by phosphorus-limited marine productivity. *Science* **271**, 493–495.

Cappellen P. V. and Qiu L. (1997) Biogenic silica dissolution in sediments of the Southern Ocean: I. Solubility. *Deep-Sea Res. II* **44**, 1109–1128.

Coale K. H., Johnson K. S., Fitzwater S. E., Gordon R. M., Tanner S., Chavez F. P., Ferioli L., Sakamoto C., Rogers P., Millero F., Steinberg P., Nightingale P., Cooper D., Cochlan W. P., Landry M. R., Constantinou J., Rollwagen G., Trasvina A., and Kudela R. (1996) A massive phytoplankton bloom induced by an ecosystem-scale iron fertilization experiment in the equatorial Pacific Ocean. *Nature* **383**, 495–501.

Deuser W. G. (1986) Seasonal and interannual variations in deep-water particle fluxes in the Sargasso Sea and their relation to surface hydrography. *Deep-Sea Res.* **33**, 225–246.

Devol A. H. (1978) Bacterial oxygen uptake kinetics as related to biological processes in oxygen deficient zones of the oceans. *Deep-Sea Res.* **25**, 137–146.

Dixit S., Cappellen P. V., and Bennekom A. J. V. (2001) Processes controlling solubility of biogenic silica and pore water build-up of silicic acid in marine sediments. *Mar. Chem.* **73**, 333–352.

Duce R. A. and Tindale N. W. (1991) Atmospheric transport of iron and its deposition in the ocean. *Limnol. Oceanogr.* **36**, 1715–1726.

Dugdale R. C. and Goering J. J. (1967) Uptake of new and regenerated forms of nitrogen in primary productivity. *Limnol. Oceanogr.* **12**, 196–207.

Dymond J. and Collier R. (1988) Biogenic particle fluxes in the Equatorial Pacific: evidence for both high and low productivity during the 1982–1983 El Nino. *Global Biogeochem. Cycles* **2**, 129–137.

Dymond J. and Lyle M. (1993) Particle fluxes in the ocean and implications for sources and preservation of ocean sediments. In *Geomaterial Fluxes, Glacial to Recent* (eds. W. Hey and T. Usselman). National Academy of Sciences, Washington, DC.

Emerson S. (1985) Organic carbon preservation in marine sediments. In *The Carbon Cycle and Atmospheric CO_2: Natural Variations Archean to Present* (ed. B. Sundquist). American Geophysical Union, Washington, DC.

Emerson S. and Bender M. L. (1981) Carbon fluxes at the sediment–water interface of the deep sea: calcium carbonate preservation. *J. Mar. Res.* **39**, 139–162.

Emerson S. and Archer D. (1990) Calcium carbonate preservation in the ocean. *Phil. Trans. Roy. Soc. London A* **331**, 29–41.

Eppley R. W. and Peterson B. J. (1979) Particulate organic matter flux and planktonic new production in the deep ocean. *Nature* **282**, 677–680.

Feely R. A., Sabine C. L., Lee K., Millero F. J., Lamb M. F., Greeley D., Bullister J. L., Key R. M., Peng T.-H., Kozyr A., Ono T., and Wong C. S. *In situ* calcium carbonate dissolution in the Pacific Ocean. *Global Biogeochem. Cycles* (submitted for publication).

Fiadeiro M. (1980) The alkalinity of the deep Pacific. *Earth Planet. Sci. Lett.* **49**, 499–505.

Fornes W. L., DeMaster D. J., Levin L. A., and Blair N. E. (1999) Bioturbation and particle transport in California slope sediments: a radiochemical approach. *J. Mar. Res.* **57**, 335–355.

Franck V. M., Brzezinski M. A., Coale K. H., and Nelson D. M. (2000) Iron and silicic acid concentrations regulate Si uptake north and south of the Polar Frontal Zone in the Pacific Sector of the Southern Ocean. *Deep-Sea Res. II* **47**, 3315–3338.

Fung I. Y., Mayn S. K., Tegen I., Doney S. C., John J. G., and Bishop J. K. B. (2000) Iron supply and demand in the upper ocean. *Global Biogeochem. Cycles* **14**, 281–295.

Gibbs M. and Kump L. R. (1994) Global chemical erosion during the last glacial maximum and the present: sensitivity or changes in lithology and hydrology. *Paleoceanography* **9**, 529–543.

Gnanadesikan A. (1999) A global model of silica cycling: sensitivity to eddy parameterization and remineralization. *Global Biogeochem. Cycles* **13**, 199–220.

Goldman J. C. (1987) Spatial and temporal discontinuities of biological processes in pelagic surface waters. In *Towards a Theory on Biological and Physical Processes in the World Ocean* (ed. R. B. J.). Reidal, Dordrecht, The Netherlands.

Hales B. and Emerson S. (1996) Calcite dissolution in sediment of the Ontong-Java Plateau: *in situ* measurements of pore water O_2 and pH. *Global Biogeochem. Cycles* **10**, 527–541.

Hales B. and Emerson S. (1997a) Calcite dissolution in sediments of the Ceara Rise: *in situ* measurements of porewater O_2, pH, and CO_2 (aq). *Geochim. Cosmochim. Acta* **61**, 501–514.

Hales B. and Emerson S. (1997b) Evidence in support of first-order dissolution kinetics of calcite in seawater. *Earth Planet. Sci. Lett.* **148**, 317–327.

Hales B., Emerson S., and Archer D. (1993) Respiration and dissolution in the sediments of the western North Atlantic: estimates from models of *in situ* microelectrode measurements of porewater oxygen and pH. *Deep-Sea Res.* **41**, 695–719.

Hedges J. I. and Keil R. G. (1995) Sedimentary organic matter preservation—an assessment and speculative synthesis. *Mar. Chem.* **49**, 81–115.

Hedges J. I., Lee C., Wakeham S. G., Hernes P. J., and Peterson M. L. (1993) Effects of poisons and preservatives on the fluxes and elemental composition of sediment trap materials. *J. Mar. Res.* **51**, 651–668.

Heinze C., Maier-Reimer E., Winguth A. M. E., and Archer D. (1999) A global oceanic sediment model for long term climate studies. *Global Biogeochem. Cycles* **13**, 221–250.

Henrichs S. M. and Reeburgh W. S. (1987) Anaerobic mineralization of marine sediment organic matter: rates and the role of anaerobic processes in the oceanic carbon economy. *Geomicrobiology J.* **5**, 191–237.

Holland H. D. (1978) *The Chemistry of the Atmosphere and Oceans*, 351pp. Wiley, New York.

Honjo S. (1976) Coccoliths: production, transportation, and sedimentation. *Mar. Micropaleo.* **1**, 65–79.

Honjo S. (1990) Particle fluxes and modern sedimentation in the polar oceans. In *Polar Oceanography: Part B. Chemistry, Biology, and Geology*. Academic Press, pp. 687–739.

Honjo S. (1996) Fluxes of particles to the interior of the open oceans. In *Particle Flux in the Ocean* (eds. V. Ittekkot, P. Aschauffer, S. Honjo, and P. Depetris). Wiley, New York.

Honjo S., Manganini S. J., and Cole J. J. (1982) Sedimentation of biogenic matter in the deep ocean. *Deep-Sea Res.* **29**(5A), 609–625.

Honjo S., Manganini S. J., and Wefer G. (1998) Annual particle flux and a winter outburst of sedimentation in the northern Norwegian Sea. *Deep-Sea Res.* **8**, 1223–1234.

Honjo S., Francois R., Manganini S., Dymond J., and Collier R. (2000) Particle fluxes to the interior of the Southern Ocean in the Western Pacific sector along 170 W. *Deep-Sea Res. II* **47**, 3521–3548.

Hutchins D. A. and Bruland K. W. (1998) Iron-limited diatom growth and Si : N uptake ratios in coastal upwelling regime. *Nature* **393**, 561–564.

Ittekkot V., Nair R. R., and Honjo S. (1991) Enhanced particle fluxes in Bay of Bengal induced by injection of fresh water. *Nature* **351**, 385–387.

Jahnke R. A., Reimers C. E., and Craven D. B. (1990) Intensification of recycling of organic matter at the seafloor near ocean margins. *Nature* **348**, 50–54.

Jansen H. and Wolf-Gladrow D. A. (2001) Carbonate dissolution in copepod guts: a numerical model. *Mar. Ecol. Prog. Ser.* **221**, 199–207.

Jenkins W. J. (1988) Nitrate flux into the euphotic zone near Bermuda. *Nature* **331**, 521–524.

Johnson K. S., Chavez F. P., Elrod V. A., Fitzwater S. E., Pennington J. T., Buck K. R., and Walz P. M. (2001) The annual cycle of iron and the biological response in central California coastal waters. *Geophys. Res. Lett.* **28**, 1247–1250.

Karl D., Letelier R., Tupas L., Dore J., Christian J., and Hebel D. (1997) The role of nitrogen fixation in geoceochemical cycling in the subtropical North Pacific Ocean. *Nature* **388**, 533–538.

Keil R. G., Montlucon D. B., Prahl F. G., and Hedges J. I. (1994) Sorptive preservation of labile organic matter in marine sediments. *Nature* **370**, 549–552.

Keir R. S. (1980) The dissolution kinetics of biogenic calcium carbonates in seawater. *Geochim. Cosmochim. Acta* **44**, 241–252.

Klaas C. and Archer D. E. (2002) Association of sinking organic matter with various types of mineral ballast in the deep sea: implications for the rain ratio. *Global Biogeochem. Cycles* **16**, doi:10.1029/2001GB001765.

Kleypas J. A. (1997) Modeled estimates of global reef habitat and carbonate production since the last glacial maximum. *Paleoceanography* **12**, 533–545.

Knox F. and McElroy M. (1984) Change in atmospheric CO_2: influence of the marine biota at high latitude. *J. Geophys. Res.* **89**, 4629–4637.

Langdon C., Takahashi T., Sweeney C., Chipman D., Goddard J., Marubini F., Aceves H., Barnett H., and Atkinson M. J. (2000) Effect of calcium carbonate saturation state on the calcification rate of an experimental reef. *Global Biogeochem. Cycles* **14**, 639–654.

Laws E. A., Falkowski P. G., W. O. S, Jr., Ducklow H., and McCarthy J. J. (2000) Temperature effects on export production in the open ocean. *Global Biogeochem. Cycles* **14**, 1231–1246.

Lefevre N. and Watson A. (1999) Modeling the geochemical cycle of iron in the oceans and its impact on atmospheric CO_2 concentrations. *Global Biogeochem. Cycles* **13**, 727–736.

Levitus S. (1982) *Climatological Atlas of the World Ocean*. US Government Printing Office, Washington, DC, 173pp.

Levitus S., Conkright M. E., Reid J. L., Najjar R. G., and Mantyla A. (1993) Distribution of nitrate, phosphate, and silicate in the world's oceans. *Prog. Oceanogr.* **31**, 245–273.

Lewin J. C. (1961) The dissolution of silica from diatom walls. *Geochim. Cosmochim. Acta* **21**, 182–198.

Li Y.-H., Takahashi T., and Broecker W. S. (1969) Degree of saturation of $CaCO_3$ in the Oceans. *J. Geophys. Res.* **74**, 5507.

Lochte K., Ducklow H. W., Fasham M. J. R., and Stienen C. (1993) Plankton succession and carbon cycling at 47 °N 20 °W during the JGOFS North Atlantic Bloom Experiment. *Deep-Sea Res. II* **40**, 91–114.

Lohmann G. P. (1995) A model for variation in the chemistry of planktonic foraminifera due to secondary calcification and selective dissolution. *Paleoceanography* **10**, 445–458.

Lutz M., Dunbar R., and Caldeira K. Regional variability in the vertical flux of particulate organic carbon in the ocean interior. *Global Biogeochem. Cycles* (in press).

Mahadevan A. and Archer D. (2000) Effect of mesoscale circulation on nutrient supply to the upper ocean. *J. Geophys. Res.* **105**, 1209–1225.

Margalef R. (1978) Life-forms of phytoplankton as survival alternatives in an unstable environment. *Oceanol. Acta* **1**, 493–509.

Martin J. H. and Fitzwater S. E. (1988) Iron deficiency limits photoplankton growth in the north-east Pacific subarctic. *Nature* **331**, 341–343.

Martin J. H., Knauer G. A., Karl D. M., and Broenkow W. M. (1987) VERTEX: carbon cycling in the northeast Pacific. *Deep-Sea Res.* **34**(2), 267–285.

Martin W. R. and Sayles F. L. (1990) Seafloor diagenetic fluxes, Woods Hole Oceanographic Institution.

Martin J., Coale K. H., Johnson K. S., Fitzwater S. E., Gordon R. M., Tanner S. J., Hunter C. N., Elrod V. E., Nowicki J. L., Coley T. L., Barber R. T., Lindley S., Watson A. J., Scoy K. V., Law C. S., Liddicoat M. I., Ling R., Stanton T., Stockel J., Collins C., Anderson A., Bidigare R., Ondrusek M., Latasa M., Millero F. J., Lee K., Yao W., Zhang J. Z., Friederich G., Sakamoto C., Chavez F., Buck K., Kolber Z., Greene R., Falkowski P., Chisolm S. W., Hoge F., Swift R., Yungel J., Turner S., Nightingale P., Hatton A., Liss P., and Tindale N. W. (1994) Testing the iron hypothesis in ecosystems of the equatorial Pacific Ocean. *Nature* **371**, 123–129.

Matsumoto K., Sarmiento J. L., and Brzezinski M. A. Silicic acid "leakage" from the Southern Ocean as a possible mechanism for explaining glacial atmospheric p_{CO_2}. *Global Biogeochem. Cycles* (in press).

McGillicuddy D. J. and Robinson A. R. (1997) Eddy induced nutrient supply and new production in the Sargasso Sea. *Deep-Sea Res. I* **44**, 1427–1450.

Middelburg J. J., Soetaert K., and Herman P. M. J. (1997) Empirical relations for use in global diagenetic models. *Deep-Sea Res. I* **44**, 327–344.

Millero F. J. (1982) The effect of pressure on the solubility of minerals in water and seawater. *Geochim. Cosmochim. Acta* **46**, 11–22.

Milliman J. D. (1974) *Marine Carbonates*. Springer, Heidelberg, 375pp.

Milliman J. D. (1993) Production and accumulation of calcium carbonate in the ocean: budget of a non-steady state. *Global Biogeochem. Cycles* **7**, 927–957.

Milliman J. D., Troy P. J., Balch W. M., Adams A. K., Li Y.-H., and Mackenzie F. T. (2000) Biologically mediated dissolution of calcium carbonate above the chemical lysocline? *Deep-Sea Res. I* **46**, 1653–1699.

Morse J. W. (1979) The kinetics of calcium carbonate dissolution and precipitation, 227.

Mucci A. (1983) The solubility of calcite and aragonite in seawater at various salinities, temperatures, and one atmosphere total pressure. *Am. J. Sci.* **283**, 780–799.

Munhoven G. and Francois L. M. (1993) Glacial–interglacial changes in continental weathering: possible implications for atmospheric CO_2. In *Carbon Cycling in the Glacial Ocean: Constraints on the Ocean's Role in Global Change* (ed. R. Zahn). Springer, Berlin.

Munk W. and Wunsch C. (1998) Abyssal recipes: II. Energetics of tidal and wind mixing. *Deep-Sea Res. I* **45**, 1977–2010.

Nelson D. M., Treguer P., Brzezinski M. A., Laynaert A., and Queguiner B. (1995) Production and dissolution of biogenic silica in the ocean: revised global estimates, comparison with regional data and relationship to biogenic sedimentation. *Global Biogeochem. Cycles* **9**, 359–372.

Noriki S. and Tsunogai S. (1986) Particulate fluxes and major components of settling particles from sediment trap experiments in the Pacific Ocean. **33**(7), 903–912.

Opdyke B. N. and Walker J. C. G. (1992) Return of the coral reef hypothesis: basin to shelf partitioning of $CaCO_3$ and its effect on atmospheric p_{CO_2}. *Geology* **20**, 733–736.

Orr J. C., Maier-Reimer E., and Mikolajewicz U. (2001) Estimates of anthropogenic carbon uptake from four three-dimensional global ocean models. *Global Biogeochem. Cycles* **15**, 43–60.

Oschlies A. and Koeve W. (2000) An eddy-premitting coupled physical–biological model of the North Atlantic 2. Ecosystem dynamics and comparison with satellite and JGOFS local studies data. *Global Biogeochem. Cycles* **14**, 499–523.

Peixoto J. P. and Oort A. H. (1992) *Physics of Climate*. American Institute of Physics, New York, 520pp.

Petit J. R., Jouzel J., Raynaud D., Barkov N. I., Barnola J.-M., Basile I., Bender M., Chappellaz J., Davis M., Delaygue G., Delmotte M., Kotlyakov V. M., Legrand M., Lipenkov V. Y., Lorius C., Pepin L., Ritz C., Saltzman E., and Stievenard M. (1999) Climate and atmospheric history of the past 420,000 years from the Vostok ice core, Antarctica. *Nature* **399**, 429–436.

Pilskaln C. A. and Honjo S. (1987) The fecal pellet fraction of biogeochemical particle fluxes to the deep sea. *Global Biogeochem. Cycles* **1**(1), 31–48.

Plath D. C., Johnson K. S., and Pytkowicz R. M. (1980) The solubility of calcite—probably containing magnesium—in seawater. *Mar. Chem.* **10**, 9–29.

Reimers C. E. and Smith K. L. J. (1986) Reconciling measured and predicted fluxes of oxygen across the deep sea sediment–water interface. *Limnol. Oceanogr.* **31**, 305–318.

Riebesell U., Zondervan I., Rost B., Tortell P. D., Zeebe R. E., and Morel F. M. M. (2000) Reduced calcification of marine plankton in response to increased atmospheric CO_2. *Nature* **407**, 364–367.

Sarmiento J. L. and Toggweiler R. (1984) A new model for the role of the oceans in determining atmospheric p_{CO_2}. *Nature* **308**, 621–624.

Sarmiento J. L., Dunne J., Gnanadesikan A., Key R. M., Matsumoto K., and Slater R. A new estimate of the $CaCO_3$ to organic carbon export ratio. *Global Biogeochem. Cycles* (submitted for publication).

Schink D. R., Fanning K. A., and Pilson M. E. Q. (1974) Dissolved silica in the upper pore waters of the Atlantic Ocean floor. *JGR* **79**(15), 2243–2250.

Siegenthaler U. and Wenk T. (1984) Rapid atmospheric CO_2 variations and ocean circulation. *Nature* **308**, 624–626.

Sigman D. M., McCorkle D. C., and Martin W. R. (1998) The calcite lysocline as a constraint on glacial/interglacial low latitude production changes. *Global Biogeochem. Cycles* **12**, 409–428.

Smith C. R., Pope R. H., DeMaster D. J., and Magaard L. (1993) Age-dependent mixing of deep sea sediments. *Geochim. Cosmochim. Acta* **57**, 1473–1488.

Takeda S. (1998) Influence of iron availability on nutrient consumption ratio of diatoms in oceanic waters. *Nature* **393**, 774–777.

Toggweiler J. R., Gnanadesikan A., Carson S., Murnane R., and Sarmiento J. L. (2003) Representation of the carbon cycle in box models and GCMs:1. Solubility pump. *Global Biogeochem. Cycles* **17**(1), 1026, doi: 10.1029/2001GB001401.

Toth D. J. and Lehrman A. (1977) Organic matter reactivity and sedimentation rates in the ocean. *Am. J. Sci.* **277**, 465–485.

Tromp T. K., Cappellen P. V., and Key R. M. (1995) A global model for the early diagenesis of organic carbon and organic phosphorus in marine sediments. *Geochim. Cosmochim. Acta* **59**, 1259–1284.

Tsunogai S. and Noriki S. (1991) Particulate fluxes of carbonate and organic carbon in the ocean. Is the marine biological activity working as a sink of the atmospheric carbon? *Tellus* **43B**, 256–266.

Villareal T. V., Pilskain C., Brzezinski M., Lipschultz F., Dennet M., and Gardner G. B. (1999) Upward transport of oceanic nitrate by migrating diatom mats. *Nature* **397**, 423–425.

Walker J. C. G. (1977) *Evolution of the Atmosphere*. Macmillan, New York, 318pp.

Walker J. C. G. and Kasting J. F. (1992) Effects of fuel and forest conservation on future levels of atmospheric carbon dioxide. *Palaeogeogr. Palaeoclimat. Palaeoecol. (Global and Planetary Change Section)* **97**, 151–189.

Weber A., Riess W., Wenzhoefer F., and Jorgensen B. B. (2001) Sulfate reduction in Black Sea sediments: *in situ* and laboratory radiotracer measurements from the shelf to 2,000 m depth. *Deep-Sea Res. I* **48**, 2073–2096.

Wefer G., Suess E., Balzer W., Liebezeit G., Muller P. J., Ungerer C. A., and Zenk W. (1982) Fluxes of biogenic components from sediment trap deployment incircumpolar waters of the Drake Passage. *Nature* **299**, 145–147.

Wefer G., Fischer G., Fuetterer D., and Gersonde R. (1988) Seasonal particle flux in the Bransfield Strait, Antarctica. *Deep-Sea Res.* **35**, 891–898.

Wefer G., Fischer R., Futterer D. K., Gersonde R., Honjo S., and Ostermann D. (1990) Particle sedimentation and productivity in Antarctic waters of the Atlantic sector. 363–379.

Westrich J. T. and Berner R. A. (1984) The role of sedimentary organic matter in bacterial sulfate reduction: the G model tested. *Limnol. Oceanogr.* **29**, 236–249.

Yamanaka Y. and Tujika E. (1997) Role of dissolved organic matter in the marine biogeochemical cycle: studies using an ocean biogeochemical general circulation model. *Global Biogeochem. Cycles* **11**, 599–612.

Yu E.-F., Francois R., Bacon M. P., Honjo S., Fleer A. P., Manganini S. J., Loeff M. M. R. v. d., and Ittekot V. (2001) Trapping efficiency of bottom-tethered sediment traps estimated from the intercepted fluxes of 230 Th and 231 Pa. *Deep-Sea Res. I* **48**, 865–8891.

6.11 Sediment Diagenesis and Benthic Flux

S. Emerson and J. Hedges
University of Washington, Seattle, WA, USA

6.11.1 INTRODUCTION

Chemical reactions in marine sediments and the resulting fluxes across the sediment–water interface influence the global carbon cycle and the pH of the sea and affect the abundance of $CaCO_3$ and opal-forming plankton in the ocean. On very long timescales these diagenetic reactions control carbon burial in sedimentary rocks and the oxygen content of the atmosphere. Sedimentary deposits that remain after diagenesis are the geochemical artifacts used for interpreting past changes in ocean circulation, biogeochemical cycles, and climate. This chapter is about the processes of diagenesis and burial of the chemical elements that make up the bulk of the particulate matter that reaches the seafloor (organic matter, $CaCO_3$, SiO_2, Fe, Mn, and aluminosilicates).

Understanding of sediment diagenesis and benthic fluxes has evolved with advances in both experimental methods and modeling. Measurements of chemical concentrations in sediments, their associated pore waters and fluxes at the sediment–water interface have been used to identify the most important reactions. Because transport in pore waters is usually by molecular diffusion, this medium is conducive to interpretation by models of heterogeneous chemical equilibrium and kinetics. Large chemical changes and manageable transport mechanisms have led to elegant models of sediment diagenesis and great advances in understanding of diagenetic processes.

We shall see, though, that the environment does not yield totally to simple models of chemical equilibrium and chemical kinetics, and laboratory determined constants often cannot explain the field observations. For example, organic matter degradation rate constants determined from modeling are so variable that there are essentially no constraints on these values from laboratory experiments. In addition, reaction rates of $CaCO_3$ and opal dissolution determined from modeling pore waters usually cannot be reproduced in laboratory experiments of these reactions. The inability to mechanistically understand reaction kinetics calculated from diagenesis models is an important uncertainty in the field today.

Processes believed to be most important in controlling the preservation of organic matter have evolved from a focus on the lability of the substrate to the protective mechanisms of mineral–organic matter interactions. The specific electron acceptor is not particularly important during very early diagenesis, but the importance of oxygen to the degradation of organic matter during later stages of diagenesis has been clarified by the study of diagenesis in turbidites deposited on the ocean floor during glacial periods.

Evolution of thinking about the importance of reactions between seawater and detrital clay minerals has come full circle since the mid-1960s. "Reverse weathering" reactions were hypothesized in very early chemical equilibrium (Sillen, 1961) and mass-balance (Mackenzie and Garrels, 1966) models of the oceans. Subsequent observations that marine clay minerals generally resemble those weathered from adjacent land and the discovery of hydrothermal circulation put these ideas on the back burner. Recent studies of silicate and aluminum diagenesis, however, have rekindled awareness of this process, and it is back in the minds of geochemists as a potentially important process for closing the marine mass balance of some elements.

6.11.2 DIAGENESIS AND PRESERVATION OF ORGANIC MATTER

Roughly 90% of the organic matter that exits the euphotic zone of the ocean is degraded in the water column. Of the ~10% of the organic carbon flux that reaches the seafloor, only about one-tenth escapes oxidation and is buried. Degradation of the organic matter that reaches the ocean sediments drives the reactions that control sediment diagenesis and benthic flux.

We begin our discussion with what we call the "pillars" of knowledge in the field—those concepts that are basic to understanding the mechanisms of organic matter diagenesis and on which future developments rested. This is followed by a description of the dominant mechanisms of organic-matter diagenesis as one progresses from oxic through the anoxic conditions. Finally, we will discuss factors controlling the reactivity of organic matter and the mechanisms of organic matter preservation.

6.11.2.1 The Pillars of Organic Matter Diagenesis

The basic concepts of organic matter diagenesis are described here as (i) the *thermodynamic* sequence of reactions of electron acceptors and their *stoichiometry*, and (ii) the *kinetics* of organic matter degradation as described by the diagenesis equations and observations of degradation rates. These ideas derived mainly from studies of ocean sediments in which pore-water transport is controlled by molecular diffusion (deepsea oxic and anoxic-SO_4 reducing), but represent the intellectual points of departure for studying near-shore systems where transport is more complicated, but where the bulk of marine organic matter is degraded.

6.11.2.1.1 Thermodynamics and sequential use of electron acceptors

The large highly structured molecules of organic matter are formed by energy from the sun and exist at atmospheric temperature and pressure in a reduced, thermodynamically unstable state. These compounds subsequently undergo reactions with oxidants to decrease the free energy of the system. The oxidants accept electrons from the organic matter during oxidation reactions (Stumm and Morgan, 1981). The electron acceptors that are in major abundance in the environment include O_2, NO_3^-, Mn(IV), Fe(III), SO_4^{2-}, and organic matter itself during fermentation (described here as methane production). These reactions are listed in the order of the free energy gained in Table 1. Half reactions for both the organic matter oxidation and the electron-acceptor reductions are represented. The changes in free energy for the reactions depend on the free energy of formation of the solids involved (organic matter, iron, and manganese oxides), and thus vary slightly among compilations in the literature. Note that the amounts of free energy gain for the whole reactions involving oxygen, nitrate, and manganese are similar. Values drop-off dramatically for iron and sulfate reactions and then again for methane production. The sequence

Table 1 The standard free energy of reaction, ΔG_r^o, for the main environmental redox reactions.

Reaction	ΔG_r^o (kJ mol^{-1}) *(half reaction)*	ΔG_r^o (kJ mol^{-1}) *(whole reaction)*
Oxidation		
$CH_2O^a + H_2O \rightarrow CO_2 + 4H^+ + 4e^-$	−27.4	
Reduction		
$4e^- + 4H^+ + O_2 \rightarrow 2H_2O$	−491.0	−518.4
$4e^- + 4.8H^+ + 0.8NO_3^- \rightarrow 0.4N_2 + 2.4H_2O$	−480.2	−507.6
$4e^- + 8H^+ + 2MnO_2(s) \rightarrow 2Mn^{2+} + 4H_2O$	−474.5	−501.9
$4e^- + 12H^+ + 2Fe_2O_3(s) \rightarrow 4Fe^{2+} + 6H_2O$	−253.2	−280.6
$4e^- + 5H^+ + \frac{1}{2}SO_4^{2-} \rightarrow \frac{1}{2}H_2S + 2H_2O$	−116.0	−143.4
$4e^- + 4H^+ + CH_2O \rightarrow CH_4 + H_2O$	−7.0	−34.4

Standard free energies of formation from Stumm and Morgan, 1981.
[a] CH_2O represents organic matter ($\Delta G_f = -129$ kJ mol^{-1}).

Table 2 Stoichiometric organic matter oxidation reactions. Redfield ratios for x, y, and z are 106, 16, 1.

Redox process	*Reaction*
Aerobic respiration	$(CH_2O)_x(NH_3)_y(H_3PO_4)_z + (x + 2y)O_2$ $\rightarrow xCO_2 + (x + y)H_2O + yHNO_3 + zH_3PO_4$
Nitrate reduction	$5(CH_2O)_x(NH_3)_y(H_3PO_4)_z + 4xNO_3^- 3$ $\rightarrow xCO_2 + 3x H2O + 4xHCO_3^- + 2xN_2 + 5yNH_3 + 5zH_3PO_4$
Manganese reduction	$(CH_2O)_x(NH_3)_y(H_3PO_4)_z + 2xMnO_2(s) + 3xCO_2 + xH_2O$ $\rightarrow 2xMn^{2+} + 4xHCO_3^- + yNH_3 + zH_3PO_4$
Iron reduction	$(CH_2O)_x(NH_3)_y(H_3PO_4)_z + 4xFe(OH)_3 + 3xCO_2 + xH_2O$ $\rightarrow 4xFe^{2+} + 8xHCO_3^- + 3xH_2O + yNH_3 + zH_3PO_4$
Sulfate reduction	$2(CH_2O)_x(NH_3)_y(H_3PO_4)_z + xSO_4^{2-}$ $\rightarrow xH_2S + 2xHCO_3^- + 2yNH_3 + 2zH_3PO_4$
Methane production	$(CH_2O)_x(NH_3)_y(H_3PO_4)_z$ $\rightarrow xCH_4 + xCO_2 + 2yNH_3 + 2zH_3PO_4$

Source: Tromp *et al.* (1995).

of electron acceptor use is sometimes categorized into oxic diagenesis (O_2 reduction), suboxic diagenesis (NO_3^- and Mn(IV) reduction), and anoxic diagenesis (Fe(III) and SO_4^{2-} reduction and methane formation). This terminology is not used here because the definition of suboxic is vague and ambiguous. We recommend referring to these reactions using the true meaning of the terms—oxic for O_2 reduction and anoxic for the rest (anoxic-NO_3^- reduction, anoxic-Mn(IV) reduction and so forth).

While all of these reactions are favored thermodynamically, they are almost always enzymatically catalyzed by bacteria. It has been observed from the study of pore waters in deepsea sediments (e.g., Froelich *et al.*, 1979) and anoxic basins (e.g., Reeburgh, 1980) that there is an ordered sequence of redox reactions in which the most energetically favorable reactions occur first and the active electron acceptors do not overlap significantly. Bacteria are energy opportunists. Using estimates of the stoichiometry of the diagenesis reactions (Table 2) one can sketch the order and shape of reactant profiles actually observed in sediment pore-water chemistry (Figure 1). The schematic figure shows all electron acceptors in a single sequence. This is rarely observed in the environment because regions with abundant bottom-water oxygen and moderate organic matter flux to the sediments (i.e., the deep ocean) run out of reactive organic carbon before sulfate reduction becomes important. In near-shore environments, where there is sufficient organic matter flux to the sediments to activate sulfate reduction and deplete sulfate in pore waters, zones of oxygen, nitrate, and Mn(IV) reduction are very thin or obscured by benthic animal irrigation and bioturbation.

Recent global models of the importance of the different electron acceptors (Figure 2; Archer *et al.*, 2002) indicate that oxic respiration accounts for ~95% of the organic matter oxidation below 1,000 m in the ocean. However, between 80% and 90% of organic matter is buried in sediments above 1,000 m in river deltas and on continental margins (Archer *et al.*, 2002; Hedges and Keil, 1995). Anoxic diagenesis is more important in these regions, and when they are included, oxic diagenesis accounts for ~70% of the total organic matter oxidation in marine sediments.

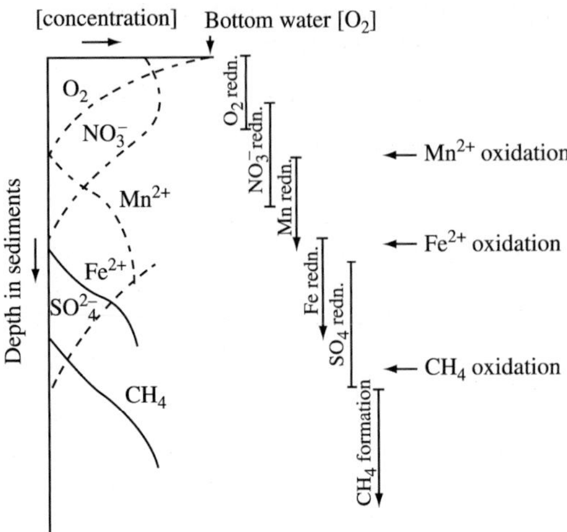

Figure 1 A schematic representation of the pore-water result of organic matter degradation by sequential use of electron acceptors.

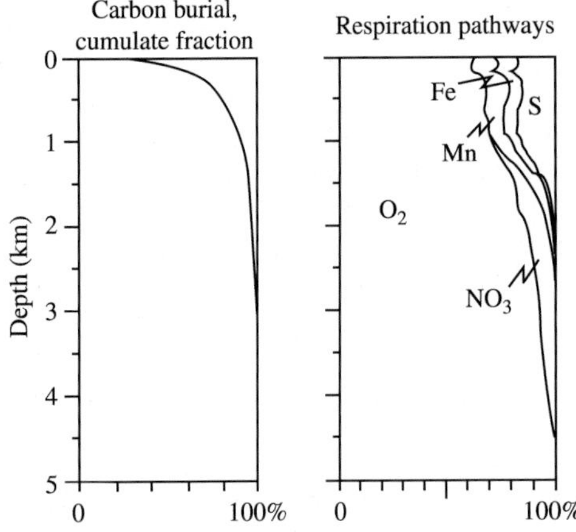

Figure 2 The cumulative fraction of carbon burial and the respiration pathways as a function of depth in the ocean derived from the global diagenesis model of Archer *et al.* (2002) (after Archer *et al.*, 2002).

6.11.2.1.2 Kinetics of organic matter degradation

The second "pillar" in our understanding of organic matter diagenesis and benthic flux consists of advances in quantifying the rates of organic matter degradation and burial. The kinetics of organic matter degradation have been determined by modeling environmental pore-water data and in laboratory studies. Most models of organic matter degradation have derived in some way from the early studies by Berner on this subject (e.g., Berner, 1980). In its simplest form, one-dimensional diagenesis, the change in organic carbon concentration (C_s, g C g_s^{-1}; s = dry sediment) with respect to time and depth is

$$\frac{\partial}{\partial t}(\rho(1-\phi)C_s) = \frac{\partial}{\partial z}\left(D_b\frac{\partial(\rho(1-\phi)C_s)}{\partial z}\right) - \frac{\partial}{\partial z}(\omega\rho(1-\phi)C_s) + MR \quad (1)$$

where z is depth below the sediment–water interface (positive downward), ϕ is porosity (cm_{pw}^3/cm_b^3; where pw is pore water and b is bulk), ρ is the dry sediment density (g cm_s^{-3}), M (g mol^{-1}) is the molecular weight of carbon, D_b ($cm_b^2\ s^{-1}$; s = second) is the sediment bioturbation rate, ω is the sedimentation rate ($cm_b\ s^{-1}$), and R (mol $cm_b^{-3}\ s^{-1}$) is the reaction term. Organic matter degradation is usually considered to be first order with respect to substrate concentration so

$$R = kC_s(1-\phi)\rho/M \quad (2)$$

where k is the first-order degradation rate constant. Probably the largest simplification here is that stirring of sediments by animals is modeled as a random process analogous to molecular diffusion. This is a gross simplification to reality. It has been shown that different tracers of bioturbation yield different results and animal activity varies with organic matter flux to the sediment–water interface (Aller, 1982; Smith *et al.*, 1993; Smith and Rabouille, 2002).

The reaction–diffusion equation for the concentration of the pore-water constituent, C_d (mol cm_{pw}^{-3}) is

$$\frac{\partial}{\partial t}(\phi C_d) = \frac{\partial}{\partial z}\left(D\frac{\partial(\phi)C_d}{\partial z}\right) - \frac{\partial}{\partial z}(v\phi C_d) + \gamma R \quad (3)$$

where D now represents the molecular diffusion coefficient and v (cm s^{-1}) is the velocity of water (which is only the same as the sediment burial, ω, when porosity is constant and there is no outside-induced flow; Imboden (1975)) and γ is a stoichiometric ratio of the pore-water element to organic carbon. At steady state with respect to compaction and the boundary conditions, the left side of Equation (3) is zero and v and ω are equal below the depth of porosity change. A very detailed treatment of many different cases is presented in Berner (1980) and Boudreau (1997).

After application of the diagenesis equations to a variety of marine environments, it became clear that the organic matter degradation rate constant, k, derived to fit the pore water and sediment profiles, was highly variable. The organic fraction of such sediments is thus often modeled as a mixture of a small number of discrete components (G_i), each of which has a finite initial amount and first-order decay constant (e.g., Jørgensen, 1979; Westrich and Berner, 1984). An alternative to such discrete models is to treat sedimentary organic matter as containing either one component whose reactivity decreases continuously over time (Middelburg, 1989), or as a continuum of multiple components whose distribution changes over time (Boudreau and Ruddick, 1991). All these discrete and continuum models capture the fundamental feature that bulk organic matter breaks down at an increasingly slower rate as it degrades.

A consequence of this broad continuum in reactivity is that sedimentary organic matter can be observed to degrade on essentially all timescales of observation. Although slightly different degradation rates may be measured for various components of a sedimentary mixture, such as different elements or biochemicals, the range of absolute values of the measured rate constants closely correspond to the time span represented by the experimental data (Emerson and Hedges, 1988; Middelburg, 1989). This direct correspondence in observation period (in units of time) and measured degradation rate constant (in units of inverse time) extends over eight orders of magnitude from days to millions of years (Figure 3). A result of the high order kinetics is that components of sedimentary mixtures that react more slowly will become a much greater fraction of unreacted material while the reaction rate constant (the curvature) is mainly described by the more labile components.

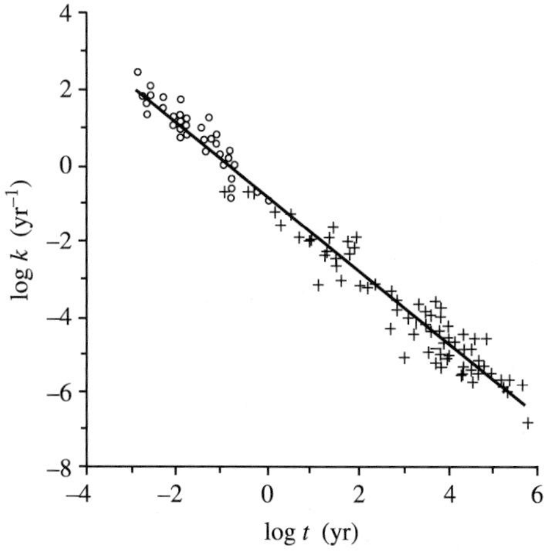

Figure 3 The rate constant, k, for organic matter degradation versus the age of the organic matter undergoing degradation determined from models of pore waters and laboratory experiments (after Middelburg, 1989) (reproduced by permission of Elsevier from *Geochim. Cosmochim. Acta* **1989**, *53*, 1577–1581). Circles are data from laboratory experiments. Crosses were determined by models of organic carbon versus depth. The line is drawn through the points.

6.11.2.2 Organic Matter Diagenesis Down the Redox Progression

6.11.2.2.1 O_2, NO_3^-, and Mn(IV) diagenesis

The relationships among the flux of organic carbon to the sediment–water interface and its diagenesis and burial in deep-ocean sediments where oxygen is the primary electron acceptor is depicted in Figure 4(a) (Emerson *et al.*, 1985). In this case one identifies two cases—one is carbon limited, where measurable oxygen persists in the pore waters at a depth of about one meter and one that is oxygen limited in which oxygen goes to zero at some depth within the bioturbated zone. Emerson *et al.* (1985) demonstrated that the measurements of carbon flux in near-bottom sediment traps was consistent with pore-water metabolite fluxes and the organic carbon content of the sediments assuming reasonable bioturbation rates at three sites in the eastern and central equatorial Pacific. Examples of carbon limited diagenesis (Figure 4(b)) are in locations of relatively low particulate organic carbon flux to the sediments such as the pelagic North Pacific and Atlantic and some carbonate-rich locations in the western equatorial Pacific (Grundmanis and Murray, 1982; Wilson *et al.*, 1985; Murray and Kuivila, 1990; Rutgers van der Loeff, 1990). Diagenesis in most of the rest of marine sediments >1,000 m is oxygen limited. Emerson *et al.* (1985) used an oxygen-only model for the oxygen-limited case to suggest that the two dominant factors controlling the organic carbon content in deep-ocean sediments are the carbon rain rate to the sediments and the concentration of oxygen in the bottom waters. This suggestion was challenged by others who believed anoxic diagenesis is efficient enough to consume most of the carbon left after the pore waters become oxygen depleted (e.g., Pedersen and Calvert, 1990; Calvert and Pedersen, 1992). We return to this subject in Section 6.11.2.4.

Nitrate plays an important role as a tracer of both oxic and anoxic (NO_3^--reducing) diagenesis because it is both produced and consumed during these processes (Table 2). Bender *et al.* (1977) and Froelich *et al.* (1979) used these relationships to infer the depth of the zone of oxic diagenesis. Middelburg *et al.* (1996) estimated that the contribution of denitrification to total sediment organic matter degradation is 7–11%, and that the global denitrification rate in sediments is $\sim 18 \times 10^{12}$ mol N yr^{-1}. The latter value is at least a factor of 2 greater than water-column denitrification, indicating the importance of marine sediments as sink for fixed nitrogen.

Manganese and iron oxides are two solid phase electron acceptors that play important roles in organic matter degradation. Their effect is limited by the liability of the solid to

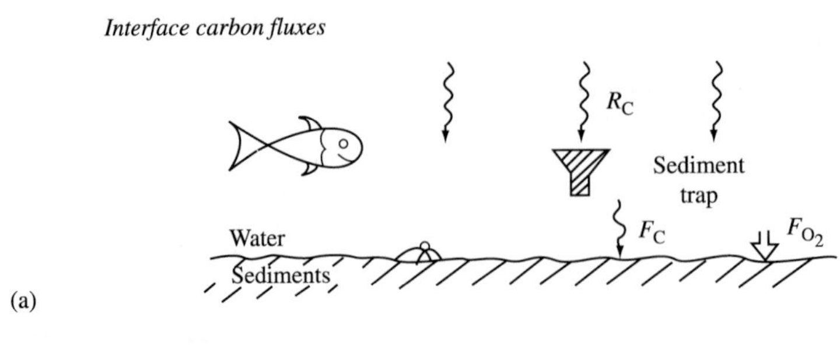

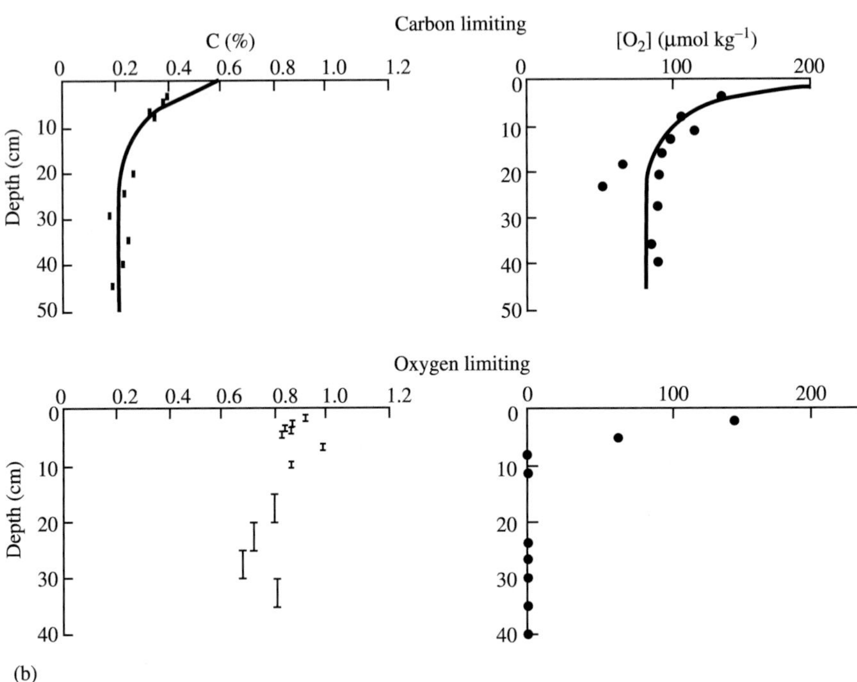

Figure 4 (a) A schematic representation of the fluxes of organic matter and oxygen at the sediment–water interface of the deep sea. R_C is the rain rate of particulate organic carbon and F_C and F_{O_2} are the fluxes of organic carbon and oxygen at the sediment–water interface (source Emerson *et al.*, 1985). (b) Data indicating the carbon- and oxygen-limiting cases. The carbon-limiting case is redrawn from Grundmanis and Murray (1982), and the oxygen-limited data are from Murray and Kuivila (1990).

dissolution, which is not easily quantified experimentally (e.g., Raiswell *et al.*, 1994), creating another unknown in models of these reactions. Manganese is reduced nearly simultaneously with nitrate, which is consistent with the comparable amounts of free energy available (Table 1). The produced Mn^{2+} is then either transported to the overlying water or reoxidized by oxygen. This relocation process is ubiquitous in oxygen-limited areas of marine sediments and creates manganese enrichment in surface sediments of the deep ocean. Although manganese cycling has the most important impact on organic matter diagenesis in near-shore sediments (Figures 2 and 5), it also has been shown to play an extremely important role in deepsea locations of relatively high organic carbon rain and labile, hydrothermally derived manganese. Aller (1990) demonstrated that nearly all organic matter degradation in sediments of the Panama Basin (eastern equatorial Pacific) is oxidized by the reduction of Mn(IV). The flux of oxygen determined by microelectrode O_2 profiles nearly balanced the flux of Mn(II) from below. The net redox reaction in this location is dominated by organic matter oxidation and O_2 reduction, but with the intermediate oxidation and reduction of manganese. The unusual availability of oxidized Mn(IV) (~3 wt.%) creates a very different redox environment than in deepsea locations where solid manganese concentrations are near shale values (~0.08 wt.%).

Sediment diagenesis reactions cause sources and sinks that are of global importance to the geochemical mass balance of manganese and some other metals sensitive to redox changes.

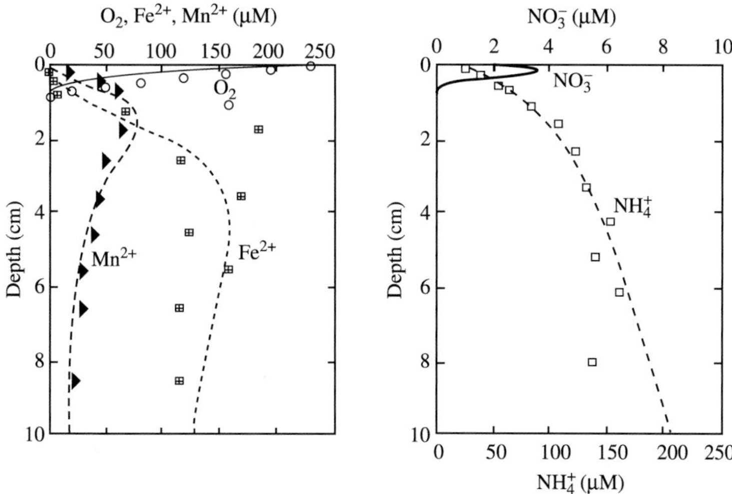

Figure 5 Pore-water profiles of O_2, Fe(II), Mn(II), NO_3^-, and NH_4^+ from sediments of the near-shore waters of Denmark. Symbols represent data from Canfield *et al.* (1993) and lines are model results from Wang and Van Cappellen (1996) (after Wang and Van Cappellen, 1996) (reproduced by permission of Elsevier from *Geochim. Cosmochim. Acta* **1996**, *60*, 2993–3014).

Morford and Emerson (1999) were able to quantify this effect using experimental evidence that indicated manganese and vanadium are remobilized to the ocean when oxygen penetrates pore waters 1 cm or less. Other redox-sensitive elements (rhenium, cadmium, and uranium) are taken up in the sediments under these redox conditions. They then applied the oxygen- and carbon-only models discussed in Section 6.11.2.2 and distributions of bottom-water O_2 and particulate organic carbon rain to the sediments below 1,000 m (Jahnke, 1996) to determine that the global extent of this redox condition is ~3% of the sediments. For manganese, vanadium, and rhenium the fluxes from and to the more reducing areas are very large even though the calculation must be considered a minimum because the sediments shallower than 1,000 m were not considered. For example, the manganese flux from the reducing sediments on the ocean margins is 1.4–2.6 times that from rivers indicating a massive redistribution of this element within the ocean basins.

6.11.2.2.2 Fe, SO_4^{2-} reduction, and CH_4 production

As one approaches continents from the deep ocean, overlying productivity becomes greater and the water depth shoals so that particles are degraded less while sinking. Both factors increase the particulate organic matter flux to the sediment–water interface. This creates more extensive anoxia in the sediments, which is sometimes compounded on continental slopes by low bottom-water oxygen conditions. A natural result of the greater supply of organic matter to the sediments is that benthic animals become bigger and more diverse. The consequence to organic matter diagenesis is that bioturbation is deeper and more intense and that animal irrigation activities are rapid enough to compete with molecular diffusion as the mechanism of pore-water transport.

The relative roles of diffusion and animal-induced advection across the sediment–water interface has been quantified by comparing oxygen fluxes determined by benthic lander measurements with those calculated from pore-water micro-electrode oxygen profiles (Figure 6; Archer and Devol, 1992). As one progresses up the continental slope and onto the shelf of the northwest US the fluxes determined by these two methods diverge. Those determined from the benthic lander become greater at depths $<\sim$100 m indicating the local importance of animal irrigation activity. This process complicates the diagenetic redox balance in coastal marine sediments, where 80–90% of marine organic matter is buried, because it is much more difficult to generalize on the mechanism and magnitude of animal irrigation than molecular diffusion.

Aller (1984) created a mechanistic model for the multi-dimensional transport of dissolved pore-water species by animals. He observed that ammonia profiles caused by sulfate reduction in the top-ten-centimeter layer of Long Island Sound sediments could not be interpreted by one-dimensional diffusion (Equation (3)). The multi-dimensional effects of irrigation were reproduced mathematically by characterizing the top layer of

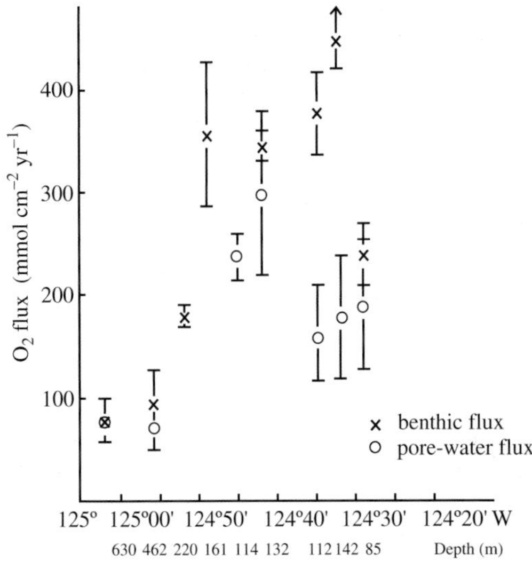

Figure 6 Oxygen fluxes determined from benthic lander measurements (×) and calculated from microelectrode oxygen gradients measured in the top few centimeters of pore waters (○) as a function of latitude and depth on the northwest US continental margin. Error bars are the standard deviation of replicate measurements (after Archer and Devol, 1992).

the sediments as a system of closely packed hollow cylinders. Integration of the equations for radial diffusion using geometries for the inner- and outer-cylinder radii determined from field measurements resulted in pore-water ammonia profiles that reproduced observations. Many subsequent "nonlocal" models of irrigation (Christensen *et al.*, 1984; Emerson *et al.*, 1984; Boudreau, 1984) have been able to reproduce the essential characteristics of animal irrigation without the complexity or elegance of the multidimensional model, but require a nonlocal transport rate. This approach has been adopted in most of the general models of sediment diagenesis (e.g., Wang and Van Cappellen, 1996; Archer *et al.*, 2002).

Wang and Van Cappellen (1996) demonstrated the intense interplay between the redox coupling of iron and manganese and transport by animal activity in the sediments of the eastern Skagerrak between Denmark and Norway (Canfield *et al.*, 1993). Sediment pore-water profiles from this area (Figure 5) indicate that most Mn(IV) reduction is coupled to oxidation of Fe(II) which was formed during organic matter and H_2S oxidation. Again the manganese and iron redox cycles shuttle electrons between more oxidized and reduced species. Adsorption of the reduced dissolved form of these metals to sediment surfaces plays an important role in their reactivity and transport by bioturbation and irrigation back to the surface sediments where they are reoxidized or transported to the overlying waters. In general, most of the iron redox cycling occurs within the sediments because of the relatively rapid oxidation kinetics of Fe(II) while some of the recycled Mn(II) escapes to the bottom waters because it is reduced nearer the sediment–water interface and has slower oxidation kinetics.

Shallow environments at the mouths of tropical rivers are the deposition sites of ~60% of the sediment delivered to the ocean (Nittrouer *et al.*, 1991). In some of these locations seasonal resuspension of the sediments occurs to a depth of 1–2 m to form fluid muds. This setting creates a very different type of sediment diagenesis that is characterized by intense iron and manganese reactions but little sulfate reduction or methane formation. Because organic matter is abundant at these shallow river-mouth locations, oxygen is depleted relatively rapidly after deposition. Abundant oxidized iron and manganese in these highly weathered sediments are reduced, creating massive, time-dependent increases in pore-water iron and manganese (Aller *et al.*, 1986). The period of diagenesis; however, is not long enough between resuspension events for sulfate reduction to become established. This situation is one of extreme non-steady-state diagenesis in which pore-water transport is dominated by physical mechanisms rather than animal irrigation or molecular diffusion.

In near-shore regions where organic matter flux to the sediments is high or bottom-water oxygen concentrations are low and horizontal sediment transport does not dominate, sulfate reduction and subsequent methane formation are important processes (Martins *et al.*, 1980; Reeburgh, 1980). Some of the earliest studies of steady-state diagenesis were applied to measurements of pore waters in Long Island Sound and Santa Barbara Basin sediments to interpret organic matter degradation by anoxic, SO_4^{2-}-reduction reactions (Berner, 1974). Early measurements of SO_4^{2-} and CH_4 in marine pore waters indicated that methane appears only after most of the SO_4^{2-} has been reduced (Figure 7), creating profiles that do not overlap significantly much like those of O_2 and Mn(IV) and Fe(II) and NO_3^-. Locations of methane formation are restricted in the marine environment because of the high sulfate concentrations in seawater. This is not true in freshwater systems where abundant CH_4 production occurs because organic matter is abundant and SO_4^{2-} concentrations are low.

Reeburgh (1980) suggested that the pore-water distributions of SO_4^{2-} and CH_4 indicate that CH_4 is being oxidized anaerobically with SO_4 being the electron acceptor. This suggestion, which is virtually unavoidable based on the metabolite distributions and interpretation by diffusion equations (see also Murray *et al.*, 1978), was not

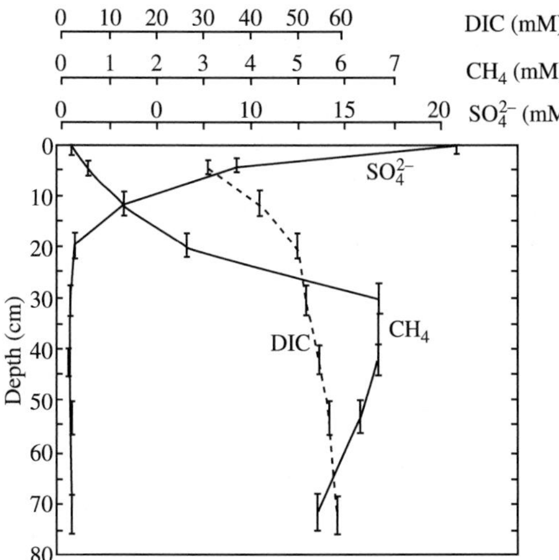

Figure 7 Pore-water profiles of SO_4^{2-}, CH_4, and DIC from the sediments of Scan Bay Alaska an anoxic fjord (after Reeburgh, 1980) (reproduced by permission of Elsevier from *Earth Planet. Sci.* **1980**, *47*, 345–352).

accepted initially by many microbiologists because it has been difficult to culture the SO_4^{2-} reducing/CH_4 oxidizing bacteria.

6.11.2.3 Benthic Respiration

Jahnke (1996) extrapolated available benthic-flux measurements from landers and pore-water determinations into a global map of benthic oxygen flux for sediments deeper than 1,000 m. When compared with global primary production rates and sediment-trap particle fluxes, these data indicate that ~1% of the primary production reaches deep-sea sediments and is oxidized there (Table 3). Also ~45% of respiration in the ocean below 1,000 m occurs within sediments.

Because particulate carbon export from surface waters is seasonal, and it is known that the majority of the particle flux is by large rapidly falling particles, one might expect that the benthic flux would also vary seasonally. This was shown to be true in the North Pacific by a series of benthic respiration deployments (Smith and Baldwin, 1984), but not in the North Atlantic near Bermuda (Sayles *et al.*, 1994). The reason for these differences is not presently understood.

The isotopes of oxygen and nitrogen gas should ideally be tracers of the relative amount of O_2 and NO_3^- reduction that takes place in the water column and sediments because the fractionation factor in a diffusive medium is theoretically equal to the square root of that in a completely open system (Bender, 1990). Brandes and Devol (1997) tested this model by measuring the fractionation of oxygen and nitrogen gas in sediments using a benthic lander. They found that the situation was more complicated than that predicted from the diffusion model. The observed fractionation factors for O_2 and NO_3^- reduction in sediments were much less than the square root of the open-system isotope fractionation factor. The most logical explanation is that both oxygen reduction and denitrification take place in sediment micro-environments where most of the O_2 and NO_3^- are consumed before they have time to communicate with the overlying waters. While there have been only a few benthic-flux isotope fractionation studies, it is clear that broad generalizations about the distribution of respiration between the water column and the sediments is not possible based on isotope measurements until the isotope fractionation in sediments is better understood. This caveat should hold for all isotope systems undergoing diagenesis in marine sediments.

Table 3 Comparison of benthic oxygen fluxes at the sediment–water interface and primary production (PP) in the ocean's euphotic zone.

Latitude	*PP* (10^{14} mol C yr^{-1})	*Benthic flux* (10^{14} mol C yr^{-1})	*% of PP*
10° N–10° S	2.08	0.020	1.0
11° N–37° N	2.67	0.026	1.0
38° N–50° N	0.83	0.008	1.0
50° N–60° N	0.41	0.005	1.1

Source: Jahnke (1996).

6.11.2.4 Factors Controlling Organic Matter Degradation

There are many factors that contribute to the seemingly universal slowing of organic matter decomposition with time. One of these is that the physical form and distribution of organic matter within sediments is not uniform. A second is that the rate and extent of organic matter degradation can vary with the different inorganic electron acceptors available at different stages of degradation. Finally, the structural features of the residual organic matter mixture may vary over time as more readily utilized components are oxidized or converted into less-reactive products.

6.11.2.4.1 Mineral association

It has long been recognized that organic matter tends to concentrate in fine-grained continental margin sediments, as opposed to coarser silts and sands (Premuzic *et al.*, 1982). Since the early 1990s, it has become clear that organic matter and fine-grained minerals in marine sediments are physically associated. One line of evidence is that only a small fraction (~10%) of the bulk organic matter in unconsolidated marine sediments can be separated as discrete particles by flotation in heavy liquids or hydrodynamic sorting (Mayer, 1994a,b; Keil *et al.*, 1994a). In addition, the concentrations of organic carbon in bulk sediments (Suess, 1973; Mayer, 1994a; Ransom *et al.*, 1998) and their size fractions (Keil *et al.*, 1994a; Bergamaschi *et al.*, 1997) increase directly with external mineral surface area as measured by N_2 adsorption (Figure 8).

Most sediments collected under oxic waters along continental margins exhibit organic carbon (OC) concentrations ~0.5–1.0 mg OC m^{-2} (e.g., Figure 8), a "loading" that is similar to that expected for a single layer of protein spread uniformly across the surfaces of mineral grains. Such organic matter concentrations were initially referred to as being "monolayer-equivalent" a term that was introduced with the caveat that the actual distribution pattern of organic matter over mineral-grain surfaces was then unknown (Mayer, 1994a). The notion that organic matter might be spread one-molecule deep on essentially all mineral grains implies sorption of previously dissolved organic substances that are physically shielded on the mineral surface from direct degradation by bacteria and their exoenzymes. Evidence in support of the protective function came from the demonstration that over 75% of dissolved organic matter desorbed from sedimentary minerals deposited for hundreds of years could be respired within five days once removed from this matrix (Keil *et al.*, 1994b).

Although the concept that sedimentary organic matter is strongly associated with mineral surfaces has stood the test of time, the monolayer-equivalent hypothesis has not. Transmission electron micrographs of organic matter in continental margin sediments indicate that most organic matter is discontinuously distributed in discrete "blebs," mucus networks and smears that are associated with domain junctions in clay-rich flocks (Ransom *et al.*, 1997). Mayer (1999) deduced from the energetics of gas adsorption onto minerals from continental-margin sediments that generally less than 15% of the surfaces of typical sedimentary minerals are coated with organic matter.

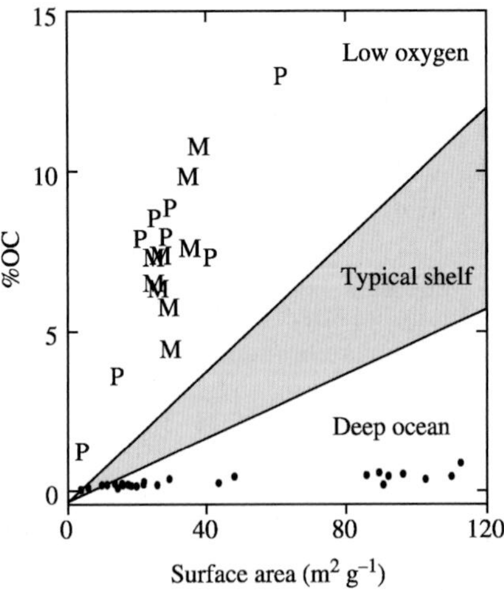

Figure 8 Weight percentages of organic carbon (%OC) plotted versus mineral surface area for surficial sediments from a range of depositional regimes. M and P represent data for samples from the Mexican and Peruvian margins, respectively (source Hedges and Keil, 1995).

Physical protection is insufficient alone to explain the distribution of organic matter in marine sediments. For example, marine sediments deposited under bottom waters with little or no dissolved O_2 usually have surface-normalized organic carbon concentrations substantially greater than 0.5–1.0 mg OC m^{-2}, whereas fine-grained deepsea clays typically contain a tenth or less of the organic concentration exhibited by continental-margin sediments of equivalent surface area (Figure 8). In particular, additional processes must account for that fact that open-ocean sediments that cover ~80% of total seafloor account for less than 5% of global organic carbon burial (Berner, 1989; Hedges and Keil, 1995).

6.11.2.4.2 The importance of oxygen

A commonly made assumption in descriptions of sedimentary diagenesis is that degradation rate and extent are largely controlled by the "quality" of available organic substrate(s), as opposed to the relative supply of different electron acceptors (Berner, 1980). This perspective is supported by a variety of field and laboratory studies (Calvert and Pedersen, 1992). In particular, freshly dissolved organic substrates (Lee, 1992; Kristensen and Holmer, 2001) and polysaccharide- and protein-rich materials (Westrich and Berner, 1984) are often degraded at similar rates in the presence or absence of molecular oxygen. However, some laboratory experiments show much slower and less efficient anoxic degradation of aged organic matter (Kristensen and Holmer, 2001) and carbon-rich substrates such as lipids (Atlas *et al.*, 1981) and pigments (Sun *et al.*, 1993a,b). Harvey *et al.* (1995) observed that the total carbon, total nitrogen, protein, lipid, and carbohydrate fractions of a diatom and coccolith were all more rapidly degraded in oxic versus anoxic laboratory incubations. Lignin, a biomacromolecule that is carbon-rich, insoluble in water and difficult to hydrolyze, is very sparingly degraded in the absence of O_2 (Benner *et al.*, 1984; Hedges *et al.*, 1985).

This apparent contradiction may be partially explained by selective initial use of easily degraded proteins and polysaccharides and the resulting concentration of carbon-rich, hydrolysis-resistant substrates such as lipids and lignin whose effective degradation requires O_2 (Emerson and Hedges, 1988; Canfield, 1994). The rate-determining step for both aerobic and anaerobic microbial degradation of polysaccharides and proteins is hydrolysis by extracellular enzymes, after which the released oligosaccharides and peptides less than ~600 amu are taken into cells for further alteration (Weiss *et al.*, 1991). Given this commonality and the fact that molecular oxygen is not required in the initial depolymerization phase, it is not surprising that these two major biochemical types often are both degraded effectively, although not necessarily at the same rates, under both oxic and anoxic conditions (Harvey *et al.*, 1995). In contrast, effective degradation of carbon-rich substrates and hydrolysis-resistant materials such as lignin, hydrocarbons, and pollen requires molecular oxygen, as opposed to simply water addition. Such degradation is often accomplished by O_2-requiring enzymes that catalyze electron (or hydrogen) removal or directly insert one or two oxygen atoms into organic molecules (Sawyer, 1991).

The most direct field evidence that the extent of sedimentary organic matter preservation is affected by exposure to bottom-water oxygen comes from oxidation fronts in deepsea turbidites of various ages and depositional settings (Wilson *et al.*, 1985; Weaver and Rothwell, 1987). One of these deposits in which the timing of the exposure to oxic and anoxic conditions is well documented, is the relict f-turbidite from the Madeira abyssal plain (MAP) ~700 km offshore of northwest Africa (Prahl *et al.*, 1989, 1997; Cowie *et al.*, 1995). This ~4 m-thick deposit was emplaced ~140 ka at a water depth of ~5,400 m when fine-grained carbonate-rich sediments slumped off the African continental slope and flowed down to cover the entire MAP with a texturally and compositionally uniform layer. This deposit was subsequently exposed to oxygenated bottom water for thousands of years during which time an oxidation front slowly penetrated approximately one-half meter into the turbidite before diffusive O_2 input was halted by accumulating sediment and the entire turbidite relaxed back to anoxic conditions (Buckley and Cranston, 1988). Pore-water sulfate concentrations measured within the sediments indicate little or no *in situ* sulfate reduction (De Lange *et al.*, 1987).

Comparative elemental analyses of the upper and lower sections of two sediment cores collected on the MAP show that organic concentrations decreased at both locations from values of 0.93–1.02 wt.% OC below the oxidation front to values 0.16–0.21 wt.% within the surface-oxidized layer (Figure 9). Pollen abundances decreased in the same samples from ~1,600 grains g^{-1} below the oxidation front to zero above it. Overall, 80% of the organic matter and essentially all of the pollen that has been stable for 140 kyr in the presence of pore-water sulfate was degraded in the upper section of the MAP cores as a result of long-term exposure to dissolved O_2.

The broad implication of these observations is that somewhere between upper continental margins and the deep ocean, depositional conditions lead to greatly increased exposure times of sedimentary organic matter to O_2 that are

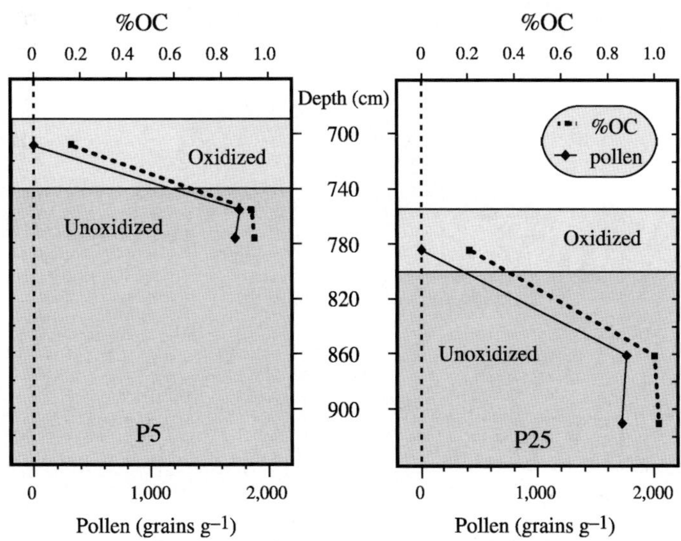

Figure 9 Profiles of the %OC and pollen abundances down two sequences of the f-turbidite from the MAP (source Cowie *et al.*, 1995) (reproduced by permission of Elsevier from *Geochim. Cosmochim. Acta* **1995**, *59*, 33–46).

sufficient to create the greatly reduced organic matter concentrations typical of modern-pelagic sediments (e.g., Figure 8). A test for this hypothesis was carried out for modern sediments depositing at varying distances and rates off the Washington State coast (see Hartnett *et al.*, 1998; Hedges *et al.*, 1999). Oxygen penetration depths were found to increase offshore from fractions of a centimeter in continental shelf and upper-slope sediments to over three centimeters in deeper (~3,000 m) offshore deposits. Average sediment accumulation rates decreased offshore from ~15 cm to 3 cm per thousand years. The combined result of these two trends is that oxygen exposure time (OET, the depth of oxygen penetration divided by the sedimentation rate) increases consistently offshore from decades on the continental shelf and upper slope to hundreds of years on the lower slope to over 1,000 yr at the most offshore study site. Corresponding concentrations of organic matter per unit sediment-surface area decreased consistently offshore from maximal values typical of upper-continental margin sediments depositing under oxygenated bottom water (~1 mg OC m^{-2}) to substantially lower concentrations (~0.3 mg OC m^{-2}) in offshore deep sediments. Mole percentages of nonprotein amino acids and physically corroded pollen increased consistently offshore, indicating that the remnant organic matter in deeper deposits is appreciably more degraded. Thus, over time spans of hundreds to thousands of years, exposure to molecular oxygen appears to affect both the amount and composition of organic matter preserved in continental margin sediments.

6.11.3 DIAGENESIS AND PRESERVATION OF CALCIUM CARBONATE

Between 20% and 30% of the carbonate produced in the surface ocean is preserved in marine sediments. The fraction of $CaCO_3$ produced that is buried dramatically affects the alkalinity and dissolved inorganic carbon (DIC) of seawater, and is thus important for understanding the processes that control the partial pressure of carbon dioxide in the atmosphere. Paleoceanographers have observed that the $CaCO_3$ content of marine sediments has changed with time in concert with glacial–interglacial periods. By studying the mechanisms that presently control $CaCO_3$ preservation, one seeks to understand what past changes imply about the chemistry of the ocean through time.

Sedimentary calcium carbonates are formed as the shells of marine plants and animals. Biologically produced $CaCO_3$ consists primarily of two minerals—aragonite and calcite. Shallow-water carbonates, primarily corals and shells of benthic algae (e.g., *Halimeda*) are heterogeneous in their mineralogy and chemical composition but are mainly composed of aragonite and magnesium-rich calcite (see Morse and Mackenzie (1990) for a discussion). Carbonate tests of microscopic plants and animals that live in the surface ocean (there are also benthic animals that produce carbonate shells) are primarily made of the mineral calcite, which composes the bulk of the $CaCO_3$ in deep-ocean sediments. A large fraction of the ocean floor consists of $CaCO_3$ from these tests (Figure 10). Note that the topographic rises on the ocean floor are $CaCO_3$-rich,

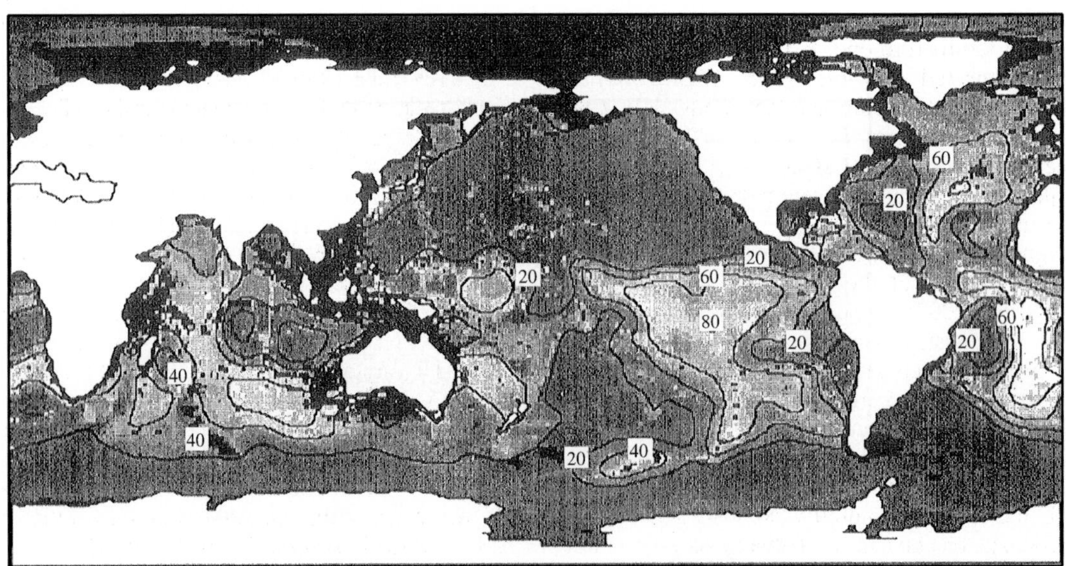

Figure 10 Global distribution of the wt.% $CaCO_3$ in surface sediments of the deep ocean (>1,000 m) (source Archer, 1996).

while the abyssal planes are barren of this mineral. The other noticeable major trend is that there is relatively little $CaCO_3$ in the sediments of the North Pacific. We focus next on the processes that control these distributions.

6.11.3.1 Mechanisms Controlling $CaCO_3$ Burial: Thermodynamics

The solubility of $CaCO_3$ in seawater has been studied extensively because of its great abundance in sedimentary rocks and the ocean. The equation for dissolution of pure calcium carbonate:

$$CaCO_{3(s)} \Leftrightarrow Ca^{2+} + CO_3^{2-} \quad (4)$$

has the simple "apparent" solubility product in seawater:

$$K'_{sp} = [Ca^{2+}][CO_3^{2-}] \quad (5)$$

The apparent constant, K'_{sp}, is related to thermodynamic constants, K_{sp}, via the total activity coefficients of Ca^{2+} and CO_3^{2-}. Apparent constants are usually used in seawater because laboratory determinations of the constants are determined in this medium.

The saturation state of seawater with respect to the solid is sometimes denoted by the Greek letter omega, Ω.

$$\Omega = [Ca^{2+}][CO_3^{2-}]/K'_{sp} \quad (6)$$

The numerator of the right side is the product of measured total concentrations of calcium and carbonate in the water—the ion concentration product (ICP). If $\Omega = 1$ then the system is in equilibrium and should be stable. If $\Omega > 1$, the waters are supersaturated, and the laws of thermodynamics would predict that the mineral should precipitate removing ions from solution until Ω returned to one. If $\Omega < 1$, the waters are undersaturated and the solid $CaCO_3$ should dissolve until the solution concentrations increase to the point where $\Omega = 1$. In practice it has been observed that $CaCO_3$ precipitation from supersaturated waters is rare probably because of the presence of the high concentrations of magnesium in seawater blocks nucleation sites on the surface of the mineral (e.g., Morse and Arvidson, 2002). Supersaturated conditions thus tend to persist. Dissolution of $CaCO_3$, however, does occur when $\Omega < 1$ and the rate is readily measurable in laboratory experiments and inferred from pore-water studies of marine sediments. Since calcium concentrations are nearly conservative in the ocean, varying by only a few percent, it is the apparent solubility product, K'_{sp}, and the carbonate ion concentration that largely determine the saturation state of the carbonate minerals.

The apparent solubility products of calcite and aragonite have been determined repeatedly in seawater solutions. We adopt the values of Mucci (1983) ($4.35(\pm0.20) \times 10^{-7}$ and $6.65 (\pm0.12) \times 10^{-7}$ $mol^2\ kg^{-2}$) for calcite and aragonite, respectively, at 25 °C, S = 35, and 1 atm pressure. These data agree within error with previous measurements of Morse *et al.* (1980) and represent many repetitions to give a clear estimate of the reproducibility ($\sim\pm5\%$).

Because of the great depth of the ocean, the most important physical property determining the solubility of carbonate minerals in the sea is pressure. The pressure dependence of the equilibrium constants is related to the difference in volume, ΔV, occupied by the ions of Ca^{2+} and CO_3^{2-} in solution versus in the solid phase. The volume difference between the dissolved and solid phases is called the partial molal volume change, ΔV:

$$\Delta V = V_{Ca} + V_{CO_3} - V_{CaCO_3} \quad (7)$$

The change in partial molal volume for calcite dissolution is negative, meaning that the volume occupied by solid $CaCO_3$ is greater than the combined volume of the component Ca^{2+} and CO_3^{2-} in solution. Since with increasing pressure Ca^{2+} and CO_3^{2-} prefer the phase occupying the least volume, calcite becomes more soluble with pressure (depth) by a factor of ~2 for a depth increase of 4 km. Values of the partial molal volume change determined by laboratory experiments and *in situ* measurements result in a range of 35–45 $cm^3\ mol^{-1}$ (see Sayles, 1980; Millero and Berner, 1972; and Chapter 6.19). The uncertainty in this value is thus $\sim\pm10\%$.

The final important factor affecting the solubility of $CaCO_3$ in the ocean is the concentration of carbonate ion. The high ratio of organic carbon to carbonate carbon in the particulate material degrading and dissolving in the deepsea causes the deep waters to become more acidic and carbonate poor as they progress along the conveyer belt circulation network from the North Atlantic to deep Indian and northern Pacific oceans. Carbonate ion concentrations change from ~250 $\mu mol\ kg^{-1}$ in surfaces waters to mean values in the deep waters of 113 $\mu mol\ kg^{-1}$ in the Atlantic, 83 in the Indian and South Pacific, and 70 $\mu mol\ kg^{-1}$ in the deep North Pacific oceans. There is little vertical difference in these values below 1,500 m (see Chapter 6.19). Thus the tendency for $CaCO_3$ minerals to be preserved is greatest in surface waters of the world's oceans and decreases "downstream" in deep waters from the Atlantic to Indian and Pacific oceans. The mean saturation horizon for calcite shoals from a depth of ~4.5 km in the equatorial Atlantic to 3.0 km in the Indian Ocean and South Pacific to less than 1.0 km in the North Pacific (Broecker and Peng, 1981; Feely *et al.*, 2003).

There have been many attempts to correlate the presence of calcite in marine sediments with

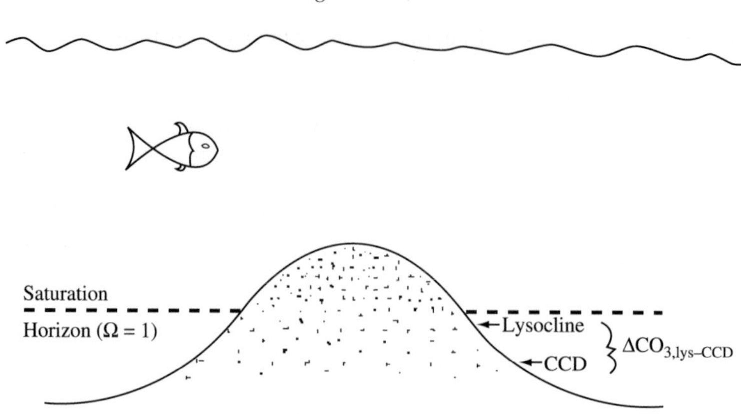

Figure 11 A sketch of the theoretical relationships among the depths of the lysocline, the CCD, and the saturation horizon ($\Omega = 1$). Dot density represents relative $CaCO_3$ content in the sediments.

the degree of saturation in the overlying water. The sketch in Figure 11 demonstrates the ideal relationship between the "saturation horizon" in the water (where $\Omega = 1$) and the presence of $CaCO_3$ in the sediments. The terminology for the presence of $CaCO_3$ in sediments is a little esoteric with the word "lysocline" coined to be the depth at which the first indication of dissolution of carbonates occurs and "carbonate compensation depth" (CCD) being the depth where the rain rate of calcium carbonate to the seafloor is exactly compensated by the rate of dissolution of $CaCO_3$ (i.e., where there is no longer burial of $CaCO_3$). It would seem that one could determine the importance of thermodynamics in determining calcite preservation by letting the ocean do the work and simply comparing lysocline and saturation-horizon relationships. There are two main problems with these attempts at direct observation. The first is the poor accuracy with which we know the degree of saturation in the ocean and the second is our inability to precisely determine the onset of $CaCO_3$ dissolution within sediments.

How well can we presently determine the saturation-horizon depth (where $\Omega = 1$) for calcite in the sea? If we assume that we know the calcium concentration exactly, then the error in Ω is determined by the errors in K'_{sp} and the measured carbonate ion concentration, $[CO_3^{2-}]$. Mucci (1983) was able to determine repeated laboratory measurements of the apparent solubility product, K'_{sp}, at 1 atm pressure to $\sim\pm5\%$, and the pressure dependence at 4 km is known to $\sim\pm10\%$. These errors compound to $\pm11\%$ in the value of K'_{sp} (4 km). Carbonate ion concentrations in the sea are almost always calculated from A_T and DIC. Being slightly conservative about accuracy of these values in ocean surveys (±4 μeq kg^{-1} for A_T and ±2 μmol kg^{-1} for DIC; they can be determined with errors about half these values if conditions are perfect), and assuming we know exactly the value of the equilibrium constants in the carbonate system, the error in $[CO_3^{2-}]$ is $\approx\pm4\%$. This uncertainty plus that for K'_{sp} compound the error in Ω to $\pm12\%$. Because $[CO_3^{2-}]$ values for the individual ocean basins are nearly vertical below 1.5 km, most of the change in Ω with depth is due to the pressure effect. The slope of K'_{sp} with depth at 4 km is equivalent to the change in saturation carbonate ion concentration, $[CO_3^{2-}]_{sat}$, with depth ($\sim$16 μmol km^{-1}; see Chapter 6.19). For the Atlantic Ocean where the $[CO_3^{2-}]$ below 1.5 km is $\sim$110 μmol kg^{-1}, this represents a change of 14% km^{-1}. Thus, the uncertainty in the value of Ω is such that one does not know the saturation horizon in the ocean to better than ±0.90 km depth.

Determining the depth of the onset of dissolution from measurements of the $CaCO_3$ content of marine sediments involves nearly as much error. The reason is that the content of $CaCO_3$ in the sediments is insensitive to the fraction of $CaCO_3$ dissolved when the sediments are highly concentrated in $CaCO_3$. This is because the fraction dissolved is determined by flux balance, and measurements determine the percent carbonate in the sediment. For example, assume that the raining particulate material is 90 wt.% $CaCO_3$, with the remaining 10% being unreactive solids such as clay minerals. If there is no dissolution in the sediments then they also will be 90% $CaCO_3$. If half of the $CaCO_3$ dissolves, the sediments are still $4.5/(4.5 + 1) = 0.82$ or 82 wt.% $CaCO_3$. Thus, if we assume an error of $\pm5\%$ in the $CaCO_3$ measurement, we would not be able to detect the onset of dissolution till about one-quarter of the raining material dissolved. There is often a very gradual change in the $CaCO_3$ concentration with depth at the top of the sedimentary transition zone or enough scatter in the data to cause determination of the depth of the onset of dissolution to have an error of at least ±0.5 km.

While errors in evaluating the depths of both the saturation horizon and the onset of $CaCO_3$ dissolution complicate "field" tests of the importance of chemical equilibrium, the difference in carbonate ion concentration over the depth range of the transition between calcite-rich and calcite-poor sediments is more clear because it is easier to know the difference in both $[CO_3^{2-}]$ and $CaCO_3$ wt.% than the absolute value. The difference, $\Delta CO_{3,\ lys-CCD}$ ($= [CO_3^{2-}]_{lys} - [CO_3^{2-}]_{CCD}$; Figure 11), has been mapped by Archer (1996) in all areas of the ocean where both $[CO_3^{2-}]$ in the bottom water and $CaCO_3$ wt.% in sediments have been determined. The ocean mean is $19 \pm 12\ \mu mol\ kg^{-1}$ ($n = 30$). The simple fact that the transition from $CaCO_3$-rich to $CaCO_3$-poor sediments occurs over a broad range of $[CO_3^{2-}]$ values indicates that the pattern of $CaCO_3$ preservation cannot be based on thermodynamics alone. Kinetic processes must be important.

6.11.3.2 Mechanism of $CaCO_3$ Dissolution: Kinetics

6.11.3.2.1 The dissolution rate constant

The dissolution rates of the minerals of calcium carbonate have been shown in laboratory experiments to follow the rate law:

$$R = k\{K'_{sp} - \mathrm{ICP}\}^n \qquad (8)$$

where k is the dissolution rate constant which has units necessary to match those of the rate. The exponent n is one for diffusion-controlled reactions and usually some higher number for surface-controlled reaction rates (see Morse and Arvidson, 2002). The above equation can be recast for $CaCO_3$ dissolution in two ways:

$$R = k\{[Ca^{2+}]_s[CO_3^{2-}]_s - [Ca^{2+}][CO_3^{2-}]\}^n \qquad (9)$$

and, if $[Ca]_s = [Ca]$,

$$R = k[Ca^{2+}]^n\{[CO_3^{2-}] - [CO_3^{2-}]\}^n$$

where $k^* = k[Ca]^n$, alternatively,

$$R = k\{K'_{sp}(1 - \Omega)\}^n = kK'^n_{sp}(1 - \Omega)^n \qquad (10)$$

where $k^* = kK'^n_{sp}$.

These equations are indistinguishable for rate measurements at a single temperature and pressure, but predict different results for the variation of the rate constant, k^*, with environmental variables (i.e., T, P, and $[Ca^{2+}]$). Because $[Ca^{2+}]$ is not constant in laboratory experiments during the course of dissolution, Equation (10) is normally used to interpret these results (e.g., Morse and Arvidson, 2002; Keir, 1980). In the ocean, where $[Ca^{2+}]$ is nearly a constant but K'_{sp} varies dramatically with pressure, Equation (9) is more convenient.

There are several studies that have been successful in determining the dissolution rate at conditions near seawater saturation. Acker *et al.* (1987) was able to employ very precise determinations of pH to measure the rate of dissolution of a single pteropod shell at different pressures from 15 atm to 300 atm. Because his measurements were at different pressures and K'_{sp} is a function of pressure, he was able to determine whether the rate constant is indeed a function of K'_{sp}. He found that Equation (9) fit his data better than (10), suggesting that the constant is not pressure dependent and the former is a more accurate universal rate law. An exponent of $n = 1.9$ was obtained for this surface-controlled dissolution reaction and a partial molal volume, ΔV, of $-39\ cm^3\ mol^{-1}$ (very close to the mean of the values determined in laboratory experiments for calcite) best fit the data.

The most extensive laboratory measurements of the dissolution of carbonates (Keir, 1980) employed a steady-state "chemostat" reactor to measure the dissolution rate of reagent grade calcite, coccoliths, foraminifera, synthetic aragonite, and pteropods. In these experiments the rate constant varied by a factor of ~100 between the different forms of calcite (after making the correction for surface area) and the data were interpreted with $n = 4.5$ order kinetics. While these measurements are still the standard for calcium carbonate dissolution rate kinetics, the high order kinetics have been reinterpreted (Hales and Emerson, 1997a) using more defendable K'_{sp} values to have a rate law that has an order of $n = 1$–2 (Figure 12). This result agrees much

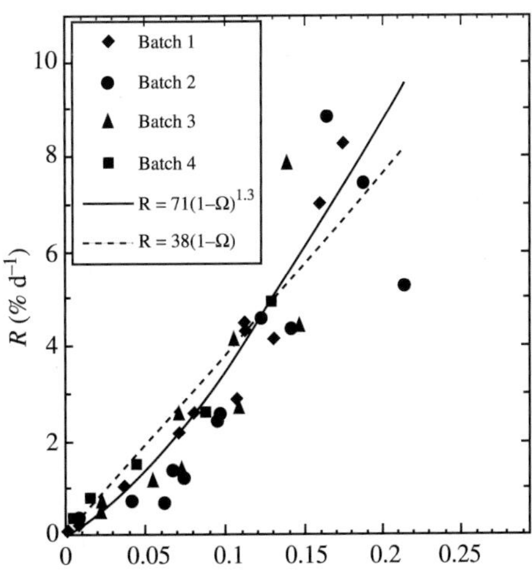

Figure 12 The dissolution rate (R in % d^{-1}) for calcite from laboratory experiments of Keir (1980) as a function of the degree of undersaturation, $(1 - \Omega)$. Lines represent the rate laws with exponents of $n = 1.3$ and 1.0 (source Hales and Emerson, 1997a) (reproduced by permission of Elsevier from *Earth Planet. Sci. Lett.* **1997**, *148*, 317–327).

more closely with the aragonite experiments and dissolution rate laws determined for other minerals, and we adopt it as more likely than $n = 4.5$. Note that the units of the rate constant (Figure 12; $k = 0.38\ d^{-1}$ or 100 times this value, 38% d^{-1}) are normalized to the concentration of solid in the experimental reactor. A convenient way to view this rate constant is that the units represent moles $CO_3^{2-}\ cm^{-3}\ d^{-1}$ released to the water per mole of $CaCO_3\ cm^{-3}$ in the solution, thus mol $cm^{-3}\ d^{-1}$/mol $cm^{-3} = d^{-1}$. At the time of using the rate constant to calculate dissolution in the environment the units must be "transformed."

One of the great uncertainties in our understanding of the kinetics of $CaCO_3$ dissolution at this time is that the dissolution rate constants required to interpret ocean observations are much smaller than those measured in the laboratory. This can be illustrated rather simply by applying the observed laboratory kinetic rate constant to determine the ΔCO_3 necessary to produce the transition in $CaCO_3$ concentrations observed in marine sediments (Appendix A). To make the calculation we have to assume dissolution begins when the bottom waters become undersaturated with respect to calcite and then determine the flux of calcium from the sediments as the degree of undersaturation increases. It is assumed that the dissolution flux of $CaCO_3$ is equal to the flux of CO_3^{2-} from the sediments and that CO_3^{2-} does not react with other carbonate system ions in solution. These assumptions are, of course, incorrect but acceptable for a first order of approximation. (The 2 : 1 change in A_T and DIC, that accompanies $CaCO_3$ dissolution, increases the $[CO_3^{2-}]$ by ~35% of the DIC increase at pH = 7.5. Thus, the gradient of $[Ca^{2+}]$ would be ~3 times the gradient in $[CO_3^{2-}]$ at the pH of pore waters. This is considered an acceptable error for this demonstration.)

The result of the calculation (Appendix A) indicates that the rate constant necessary to create the $\Delta CO_{3,lys-CCD} = 19\ \mu mol\ kg^{-1}$, has to be at least 100 times less than the laboratory-determined values. This has been confirmed by *in situ* measurements (see later). Perhaps the most important lesson here is that models and laboratory experiments consider primarily pure phases whereas impurities and surface coatings may greatly influence dissolution rates in nature. Until there is better agreement between the laboratory and *in situ* model-determined values, the latter will have to be used in model reconstructions of the relationship between ocean chemistry and sedimentary carbonate content.

6.11.3.2.2 *The effect of organic matter degradation*

The relatively slow dissolution of calcite can explain the gradual change between carbonate-rich and carbonate-poor sediments. There is, however, another important issue that we have not considered—the role of organic matter degradation in sediments in promoting *in situ* calcite dissolution. It was long suspected that organic matter degradation would promote dissolution, but this was not quantified until the 1980s and 1990s. Two factors lead to the realization that the inorganic kinetic interpretation of $CaCO_3$ burial in the ocean was incomplete. First, observations of the carbon content of particles that rain to the ocean floor collected in sediment traps 1 km or less above the bottom suggest the molar organic carbon to $CaCO_3$ carbon rain ratio is about one. A few centimeters below the sediment surface this ratio is more like 0.1, indicating that something like 90% of the organic carbon that reaches the surface is degraded rather than buried. Second, sediment pore-water studies in the same areas as the sediment trap deployments show strong oxygen depletions in the top few centimeters. Simple-flux calculations require that the bulk of the organic matter degradation between the sinking particles and that buried takes place within the sediments (Emerson *et al.*, 1985; Figure 4). Thus, most particles that reach the seafloor are stirred into the sediments before they have a chance to degrade while sitting on the surface. If this were not the case, and the particles degraded on the surface, there would be little oxygen depletion within the sediments.

Organic matter degradation within the sediments creates a microenvironment that is corrosive to $CaCO_3$ even if the bottom waters are not, because addition of DIC and no A_T to the pore water causes it to have a lower pH and smaller $[CO_3^{2-}]$. Using a simple analytical model and first-order dissolution rate kinetics, Emerson and Bender (1981) predicted that this effect should result in up to 50% of the $CaCO_3$ that rains to the seafloor being dissolved even at the saturation horizon, where the bottom waters are saturated with respect to calcite. Because the percent $CaCO_3$ in sediments is so insensitive to dissolution and the saturation-horizon depth so uncertain, this suggestion was well within the constraints of environmental observations.

The effect of organic matter driven dissolution is to raise the carbonate transition in sediments relative to the saturation horizon in the water column (Emerson and Archer, 1990; Figure 11), but there should be little change in the $\Delta CO_{3,lys-CCD}$ necessary to create the transition in %$CaCO_3$. Thus, the argument about the relationship between the dissolution rate constant and observed transition of %$CaCO_3$ in sediments (Appendix A) should not be affected.

The suggestion of "organic $CaCO_3$ dissolution" in sediments has been tested by determining the gradient of oxygen and $[CO_3^{2-}]$ in the sediment pore waters. This had to be done on a very fine (millimeter) scale because the important region for the reaction is near the sediment–water interface. The test required *in situ* measurements because it has been shown the pH values of pore waters change when they are depressurized. To do this Archer *et al.* (1989) and later Hales and Emerson (1996) built an instrument capable of traveling to the deepsea sediment surface, slowly inserting oxygen and pH microelectrodes, one millimeter at a time, into the sediments and recording the data *in situ*. The results of these experiments, some of which are reproduced in Figure 13, confirmed the prediction that a significant amount of $CaCO_3$ dissolves because of organic matter degradation. The pH of the pore waters cannot be interpreted without assuming dissolution of $CaCO_3$ in response to organic matter degradation measured by the pore-water oxygen profiles. This process makes the burial of $CaCO_3$ in the ocean dependent not only on thermodynamics and kinetics but also on the particulate rain ratio of organic carbon to calcium carbonate carbon. The kinetic rate constant required to model the measured pH profiles (Figure 13) is more than 2 orders of magnitude smaller than that measured in laboratory experiments which is consistent with the simple argument presented earlier in Appendix A.

There has always been some reluctance to assume that the pH in a porous medium is controlled solely by the carbonate buffer system in the pore waters. There are arguments that H^+ ions on particle surfaces can affect pH measurements (Stumm and Morgan, 1981) and, even more importantly, comprehensive models of pore-water chemistry are beginning to demonstrate that H^+ adsorption on mineral surfaces may play an important role in controlling the pH of pore waters. Conclusions of the pH measurements, however, have been confirmed by millimeter-scale measurements of both p_{CO_2} and calcium concentration in the pore waters (Hales *et al.*, 1997; Cai *et al.*, 1995; Wenzhofer *et al.*, 2001.)

Observations from benthic-flux experiments in which A_T and calcium fluxes have been measured both below and above the calcite saturation horizon confirm the effect of organic matter degradation on the dissolution of $CaCO_3$ in most (e.g., Berelson *et al.*, 1990; Jahnke *et al.*, 1997) but not all situations. R. J. Jahnke and D. B. Jahnke (2002) interpret their benthic-flux measurements at five locations in the world's ocean to show that in regions with sediments of very high $CaCO_3$ content there appears to be no evidence of metabolic dissolution of $CaCO_3$ based on A_T and calcium benthic fluxes. They suggest that the reason for this observation is that the pH of pore waters is controlled by adsorption of hydrogen ion and carbonate species on the surface of $CaCO_3$. Transport of these species to and from the sediment surface by bioturbation creates a flux that is more important than that caused by molecular diffusion in the pore waters.

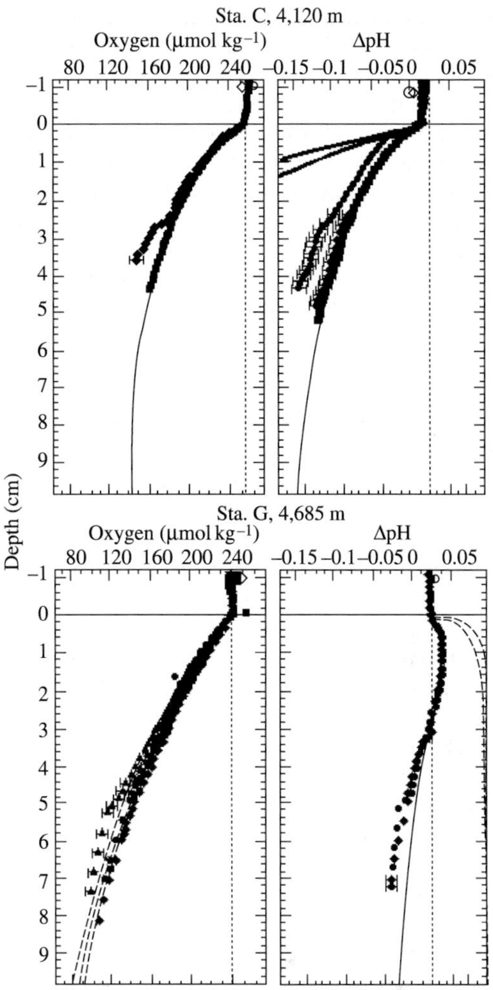

Figure 13 Pore-water profiles of oxygen concentration and ΔpH (The pH difference between the value in the pore water and the value in bottom water) in the top ~10 cm of sediments from two locations on the Ceara Rise. Points are individual measurements, sometimes from different electrodes on the same deployment. Open symbols in the overlying water are measurements in the bottom water after the pore-water profile. Solid and dashed curves are model solutions. In the top figure (Sta. C) the bottom waters are saturated with respect to calcite. The thick solid lines on the top right figure indicate the predicted ΔpH if there were no $CaCO_3$ dissolution caused by organic matter degradation in the sediments. In the bottom figure (Sta. G) bottom waters are undersaturated with respect to calcite. The dashed lines in the right figure indicate the trend if there were only dissolution promoted by bottom-water undersaturation (after Hales and Emerson, 1997b) (reproduced by permission of Elsevier from *Geochim. Cosmochim. Acta* **1997**, *61*, 501–514).

Potential problems with this interpretation are that it stems from very small changes in the chambers of the benthic landers, because open ocean areas high in $CaCO_3$ are locations with relatively low organic matter degradation rates. In addition, the surface properties of $CaCO_3$ in seawater have not been measured sufficiently well to incorporate adsorption effects into diagenesis models. The observations, however, demand interpretation, and, if adsorption effects turn out to be important, it will have ramifications for interpreting mechanisms of $CaCO_3$ preservation because of the wide-spread distribution of $CaCO_3$-rich sediments.

Recent measurements of calcium and alkalinity in the ocean above the calcite saturation horizon (Milliman *et al.*, 1999; Chen, 2002) suggest dissolution in supersaturated waters. The proposed mechanisms are variations of the organic matter driven $CaCO_3$ dissolution mechanism. In these cases the authors suggest that microenvironments in falling particulate material (Milliman *et al.*, 1999) or anerobic dissolution in sediments of the continental shelves and marginal seas (Chen, 2002) are locations of $CaCO_3$ dissolution. As the details and accuracy of measurements improve, thermodynamic and kinetic mechanisms required to interpret the results become more and more complex.

6.11.3.2.3 The interpretation of the ^{14}C age of surface sediments

Determining the ^{14}C accumulation rate as a function of the degree of saturation should be a sensitive way of determining the extent and mechanism of $CaCO_3$ dissolution in sediments. The ^{14}C age of the bioturbated layer (the top 4–8 cm of sediments based on the constancy of ^{14}C measurements) and the accumulation rate below depend on the rain rate and dissolution flux of $CaCO_3$. Two simple end-member cases illustrate the complexity of interpreting core top ^{14}C data. If dissolution takes place within the sediments, as it must for organic matter-driven dissolution to be effective, and all $CaCO_3$ particles dissolve at the same rate, the radiocarbon age should decrease with the extent of dissolution—the reservoir size decreases but the influx remains the same. In this case the probability of survival of all $CaCO_3$ particles is the same so there would be fewer older particles in the sediment mixed layer with more intense dissolution. If, however, dissolution takes place primarily at the sediment–water interface before the particles are mixed into the sediment, or for some reason younger particles dissolve faster, the mixed layer ages would increase with progressive dissolution because the input flux decreases faster than the reservoir size.

Broecker *et al.* (1999 and references therein) argued, based on bioturbated layer sediment ^{14}C-$CaCO_3$ measurements, that increase in age with depth in the western equatorial Pacific is either due to interface or non-steady-state dissolution. This would argue against the importance of organic matter driven dissolution. DuBois and Prell (1988), however, present data that indicate the ^{14}C age of the sediment bioturbated layer on the Sierra Leone Rise in the eastern equatorial Atlantic decreases as dissolution increases with greater depth (Figure 14) suggesting that homogeneous dissolution dominates. Keir and Michel (1993) concluded that these simple end-member calculations are complicated by a non-steady-state increase in dissolution during the Holocene.

While there appear to be a variety of interpretations of the origin of core-top ^{14}C data, recent measurements of pore-water ^{14}C (Martin *et al.*, 2000) have helped to identify the age of the $CaCO_3$ particles undergoing dissolution. Pore-water $DI^{14}C$ measurements are sensitive to the mechanism of dissolution because dissolved $DI^{14}C$ changes due to calcite dissolution directly reflect the age of the dissolving particles. An age for the added ^{14}C that is similar to the age of the mixed layer (4,000 yr or so) indicates homogeneous dissolution is most important; a much younger age indicates interface dissolution is more important. The other advantage of this method is that it traces processes that have occurred relatively recently. The timescale for replacement of the top-ten centimeters of

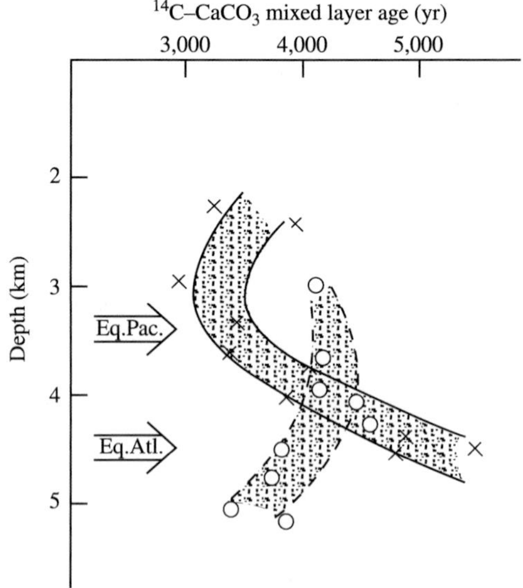

Figure 14 The ^{14}C-$CaCO_3$ age of the sediment bioturbated layer as a function of water depth on the Ontong–Java Plateau (X) in the western equatorial Pacific (Broecker *et al.*, 1999) and the Sierra Leone Rise (O) in the eastern equatorial Atlantic (DuBois and Prell, 1988). The arrows indicate the depth of the first signs of $CaCO_3$ dissolution in the sediments as based on $\%CaCO_3$ (for the Pacific) and $\%CaCO_3$ fragments (for the Atlantic).

pore-water DIC by diffusion is years rather than thousands of years for the processes that control sediment ages. Martin *et al.* (2000) measured pore-water $DI^{14}C$ at a stations above and below the saturation horizon in the western equatorial Pacific. Below the saturation horizon, where $CaCO_3$ dissolves spontaneously by inorganic processes, $DI^{14}C$ values indicate younger particles are preferentially dissolved, in accordance with the sediment ^{14}C ages that increase with greater undersaturation and interface dissolution. Pore-water DIC from the sediments above the saturation horizon, however, had ages that are more consistent with a mixture of interface particles undergoing homogeneous dissolution.

It appears at the time of writing this chapter that the mechanisms determining the ^{14}C age of surface sediments are variable and may be different for different saturation states even at the same ocean location. This observation, compounded by the non-steady-state possibility (Keir and Michel, 1993) indicates that the mechanisms controlling core top ^{14}C ages are probably too complicated to distinguish the importance of organic matter driven dissolution.

6.11.4 DIAGENESIS AND PRESERVATION OF SILICA

Biogenic silica, in the form of opal, makes up an important part of marine sediments, particularly in the southern and eastern equatorial oceans (Figure 15). These deposits are formed primarily from tests of diatoms that lived in the surface oceans. More than half of the opal formed in the surface ocean is dissolved within the upper 100 m and only a small percentage of the production is ultimately buried in marine sediments (Nelson *et al.*, 1995). Rewards to be gained by understanding the mechanisms that control opal diagenesis in sediments are evaluating the utility of the SiO_2 concentration changes in sediments as a tracer for past diatom production and understanding the role of authigenic silicates as a sink for major ions in marine geochemical mass balances.

The main tool for studying diagenesis and preservation of SiO_2 in marine sediments has been the measurement of H_4SiO_4 in pore waters which has occurred since the 1970s (e.g., the early work of Hurd (1973) and Fanning and Pilson (1974); Figure 16). The difficulty in interpreting these results has been that the asymptotic values in pore-water profiles, the concentration that is achieved by 10 cm below the sediment–water interface, is highly variable geographically where solid opal is preserved in the sediments. Possible explanations for these observations fall into three general categories (McManus *et al.*, 1995). First, asymptotic values may be different because the solubility of opal formed in surface waters varies geographically. Second, pore waters may never achieve equilibrium but opal formed in surface waters has a number of phases of different reactivity which, in concert with sediment bioturbation, create different steady-state asymptotic values. Finally, diagenesis reactions within the sediments may create authigenic phases other than opal that control the pore-water solubility and chemical kinetics of H_4SiO_4.

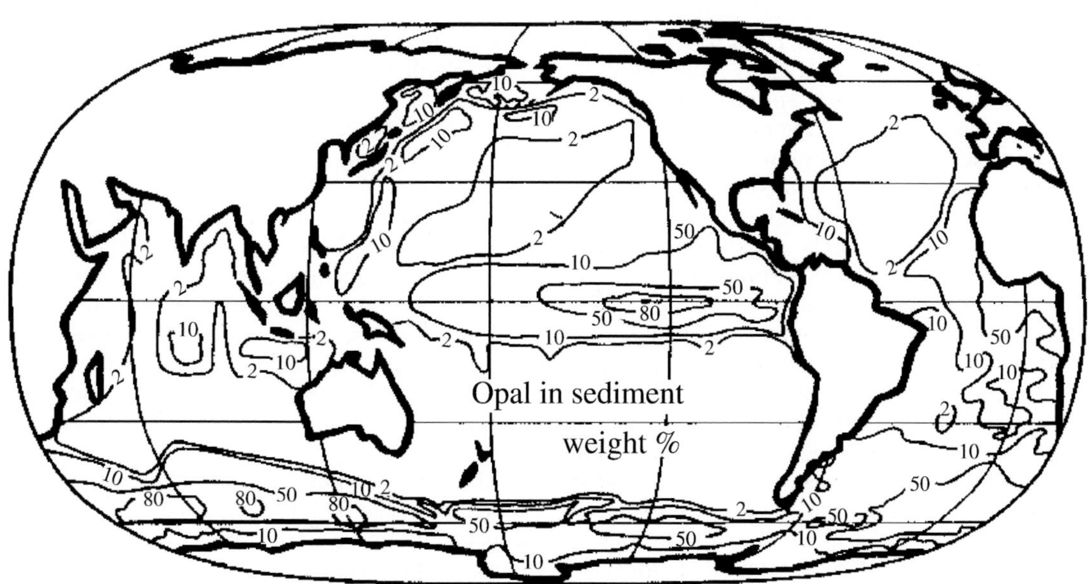

Figure 15 The global distribution of SiO_2 in marine sediments in weight percent (after Broecker and Peng, 1982).

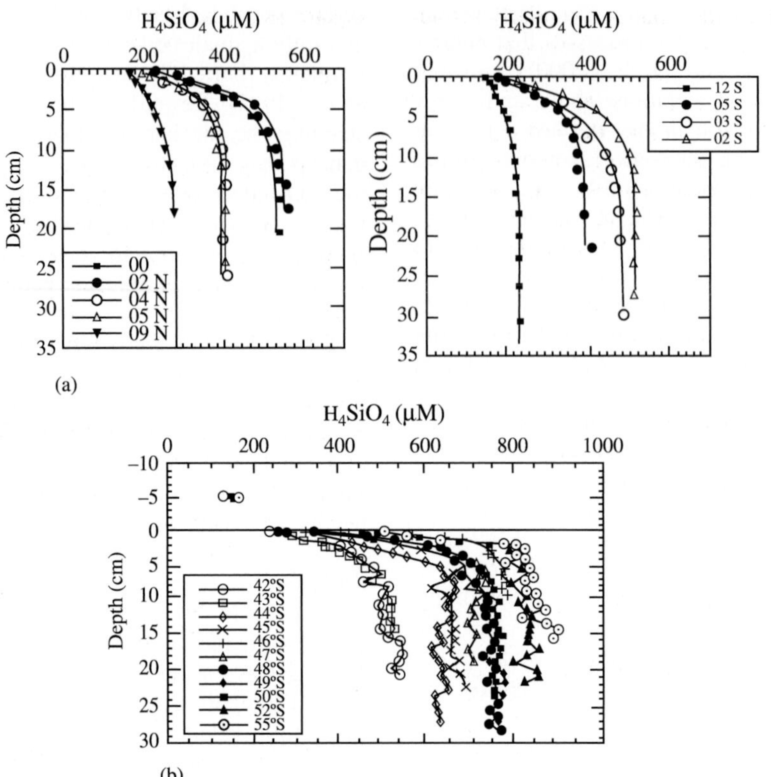

Figure 16 Pore-water H_4SiO_4 concentrations as a function of depth in sediment on (a) a North–South transect along 140° W in the equatorial Pacific (after McManus *et al.*, 1995) (reproduced by permission of Elsevier from *Deep-Sea Res. II* **1995**, *42*, 871–902) and (b) a North–South transect through the Indian sector of the Polar Front (after Rabouille *et al.*, 1997) (reproduced by permission of Elsevier from *Deep-Sea Res. II* **1997**, *44*, 1151–1176).

6.11.4.1 Controls on the H_4SiO_4 Concentration in Sediment Pore Waters: Thermodynamics

Even before the wide variety of asymptotic pore-water values in marine sediments were observed, it was shown (e.g., Hurd, 1973; Lawson *et al.*, 1978) that the H_4SiO_4 concentration obtained by incubating diatom frustrules obtained from surface waters was greater than that observed in pore waters. The equilibrium solubility is much too high to explain the results, and equilibrium values for other SiO_2 minerals are even higher. More recent experiments in sediments from the Southern Ocean using stirred flow-through reactors indicate saturation concentrations that range between 1,000 mol L^{-1} H_4SiO_4 and 1,600 mol L^{-1} H_4SiO_4 (Van Cappellen and Qiu, 1997a), values that are much greater than those determined in the pore-water profiles from the same location (Figure 16). These results lend little support to the hypothesis that the asymptotic values are equilibrium concentrations of different opals produced in the surface waters. However, we shall see later that thermodynamics does play a role in explaining the observations.

6.11.4.2 Controls on the H_4SiO_4 Concentration in Sediment Pore Waters: Kinetics

Early studies of the rate of opal dissolution in the laboratory indicated that the dissolution rate of acid-cleaned planktonic diatoms varied as a linear function of the degree of undersaturation (e.g., Hurd, 1973; Lawson *et al.*, 1978).

$$\mathrm{d}[H_4SiO_4]/\mathrm{d}t = k_{Si}S\{[H_4SiO_4]_{sat} - [H_4SiO_4]\} \quad (11)$$

where k_{Si} is the rate constant for SiO_2 dissolution (cm s^{-1}) and S is the solid surface area (cm^2 cm^{-3}). Generally, the dissolution rate constants determined by these methods are much greater than those needed to model the pore-water profiles, indicating some important differences between the laboratory experiments and field results, just as in the case for $CaCO_3$ dissolution kinetics.

The first attempt to use the dynamics of dissolution and burial in marine sediments to explain the pore-water observations was the elegant interpretation by Schink *et al.* (1975). They assumed opal fractions of different reactivity—one that dissolves rapidly and completely and another that is essentially refractory. The idea was that the first fraction, in concert with sediment bioturbation, sets the pore-water asymptotic concentration of H_4SiO_4 and the refractory portion determines the sediment concentration of opal. The reason that bioturbation is important in defining the pore-water concentration of H_4SiO_4 is that it stirs the opal deeper into the sediments where dissolution is more effective in creating a strong concentration gradient.

One can illustrate the relationships among bioturbation, dissolution kinetics and the asymptotic concentration in sediment pore waters rather simply in the following way. If we assume that the reactive portion of the opal rain to the sediment–water interface dissolves in the top h cm of the sediments, then the residence times for the reactive opal concentration, $[SiO_2]_{sed}$ (mol cm_s^{-3}), with respect to *in situ* dissolution is:

$$\tau_{diss} = h(1 - \varphi)f[SiO_2]_{sed}/F_{Si} \qquad (12)$$

where φ is the porosity ($cm_{pw}^3\,cm_b^{-3}$), f is the fraction of the sediment that is opal and F_{Si} (mol $cm^{-2}\,s^{-1}$) is the flux of H_4SiO_4 to the overlying water. Since all the reactive opal dissolves, this residence time is roughly equal to the mean time required to stir opal, by bioturbation, to the depth, h:

$$\tau_{trub} = h^2/2D_b \qquad (13)$$

where D_b is the bioturbation coefficient ($cm^2\,s^{-1}$). If we assume that a simple steady-state balance between diffusion and dissolution controls the distribution of H_4SiO_4 (mol cm_{pw}^{-3}) in sediment pore waters, then:

$$D_{Si}\frac{d^2[H_4SiO_4]}{dz^2} = -kS\{[H_4SiO_4]_{asy} - [H_4SiO_4]\} \qquad (14)$$

where D_{Si} is the molecular diffusion coefficient of silicic acid ($cm^2\,s^{-1}$), and kS is the kinetic rate constant (s^{-1}). The boundary condition at $z = 0$ constrains the surface concentration to be that of the bottom water, $[H_4SiO_4]_{BW}$, and the value at depth when $z \rightarrow \infty$, approaches $[H_4SiO_4]_{asy}$, the asymptotic concentration. This relationship is exactly analogous to the diffusion–reaction equation for CO_3^{2-} (Appendix A; Equation (20)) and the flux at the sediment–water interface is

$$F_{Si} = (kSD_{Si})^{0.5}\Delta H_4SiO_4 \qquad (15)$$

where $\Delta H_4SiO_4 = \{[H_4SiO_4]_{asy} - [H_4SiO_4]_{BW}\}$. Substituting this value for the flux in Equation (12) and equating the two residence-time equations gives:

$$\Delta H_4SiO_4 = \{2(1 - \varphi)f[SiO_2]/h(kSD_{Si})^{0.5}\}D_b \qquad (16)$$

The asymptotic concentration is proportional to the value of the bioturbation coefficient. Observed values for the terms on the right of the above equation indicate that these approximations reproduce measured pore-water values adequately given the crudeness of the model (Appendix B).

While the dynamics of this interpretation are instructive for realizing the importance of bioturbation, explaining the field observations by this mechanism requires changes in the bioturbation coefficient over short distances that have not been measured and provides no explanation for the refractory portion of the opal flux.

Van Cappellen and Qiu (1997b) used flow-through reactor studies to demonstrate that the dissolution rate of unaltered silicon-rich sediment from the Southern Ocean follows a rate law that is exponential rather than linear with respect to the degree of undersaturation. The implication is that the rate of dissolution is much greater near the sediment–water interface than below. In fact they find that, when the laboratory kinetic studies are applied to the sediments, SiO_2 dissolution is pretty much finished below a few centimeters. Rabouille *et al.* (1997) and McManus *et al.* (1995) interpreted pore-water H_4SiO_4 profiles (Figure 16) using a model that incorporated a depth-dependent rate constant.

The question remains as to what processes cause the kinetics of opal dissolution to change as the mineral ages in marine sediments. A striking clue to the answer was the observation that the asymptotic H_4SiO_4 pore-water values in Southern Ocean sediments are strongly dependent on the amount of detrital material present in the opal-rich sediments (Figure 17). Since earlier studies had established that aluminum is reactive in marine sediments (Mackin and Aller, 1984), this correlation implies that detrital aluminosilicates supply the aluminum necessary for reactions that take place very soon after deposition.

6.11.4.3 The Importance of Aluminum and the Rebirth of "Reverse Weathering"

Field observations implicating the importance of aluminosilicates to opal diagenesis were followed by laboratory experiments to determine the effect of Al(III) derived from detrital aluminum silicate on the solubility and dissolution kinetics of opal. Dixit *et al.* (2001) mixed opal-rich (~90% SiO_2) sediments from the Southern Ocean with different amounts of either kaolinite or ground basalt in long-term (21 months) batch

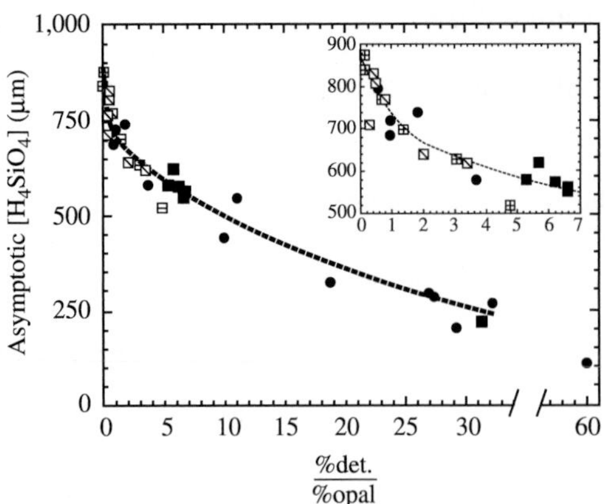

Figure 17 The relationship between the asymptotic H_4SiO_4 concentration in pore waters and the relative detrital and opal content of the sediments (Van Cappellen, personal communication). (Data from (●) Rickert (2000); (⊠) King *et al.* (2000); (■) Koning *et al.* (1997); (⊞) Van Cappellen and Qiu (1997a)).

experiments. The observed concentration of H_4SiO_4 at the end of these experiments was strongly influenced by the presence of the aluminosilicate phase. Values ranged from ~1,000 μmol kg^{-1} H_4SiO_4 for nearly pure opal to ~400 μmol kg^{-1} for a 1 : 4 aluminosilicate : opal mixture. An authigenic phase forms in the presence of dissolved silicon and aluminum that is less soluble than pure opal. They found that the pore-water aluminum concentration and the Al : Si ratio of diatom frustrules in surface sediments were also proportional to the amount of detrital material in the sediments. The substitution of Al(III) for one in 70 of the silicon atoms in opal decreased its solubility by ~25%.

The depression in opal solubility with incorporation of aluminum clarifies many aspects of the field observations but not all. Equilibrium arguments alone cannot explain why pore waters in sediments with high concentrations of detrital material are observed to remain undersaturated with respect to the experimentally determined opal solubility. Kinetic experiments using flow-through reactors (Dixit *et al.*, 2001) revealed that active precipitation of authigenic aluminosilicates prevented the pore waters from reaching equilibrium with the opal phase present in the sediments. Also, studies of the surface chemistry of biogenic silicates (Dixit and Van Cappellen, 2002) indicate that changes in the surface chemical structure during diagenesis contributes to slower kinetics. Thus, the laboratory studies indicate the mechanisms that explain model interpretations of pore-water H_4SiO_4 concentrations are surface chemical changes and authigenic aluminosilicate formation. Precipitation of authigenic aluminosilicate minerals must be one of the most widespread diagenesis reactions occurring in marine sediments.

While these reactions presently imply only the substitution of aluminum for silicon, this mechanism, and field observations of the geochemistry of aluminum in tropical near-shore sediments (e.g., Mackin and Aller, 1984), bring us back to authigenic process suggested by Mackenzie and Garrels (1966) to close the marine geochemical imbalance for Mg^{2+}, K^+, and HCO_3^- created during weathering on land. The generalized scheme for the proposed "reverse weathering" reaction was

$$
\begin{aligned}
&H_4SiO_4 + \text{degraded clay from land} + HCO_3^- \\
&\quad + Mg^{2+},\ K^+ \rightarrow \text{ion-rich authigenic clay} \\
&\quad + CO_2 + H_2O
\end{aligned}
$$

The popularity of this proposal waned owing to lack of strong evidence to prove that it occurred, and fluxes suggested in the early studies of hydrothermal processes (Edmond *et al.*, 1979) obviated the need for low-temperature reactions to balance the river inflow of Mg^{2+} and HCO_3^-. This has changed since we now know that the flow of water through high-temperature zones of hydrothermal areas is less than previously suggested, and the importance of low-temperature reactions and flows are uncertain.

The quantitative importance of "reverse weathering" reactions were demonstrated by the rapid formation of authigenic aluminosilicates in the sediments of the Amazon delta (Michalopoulos and Aller, 1995). These authors placed seed materials (glass beads, quartz grains, and quartz grains coated with iron oxide) into anoxic Amazon delta sediments. After 12–36 months they observed the formation of K–Fe–Mg-rich clay minerals on the seed materials and suggest that formation of these materials in Amazon sediments alone could account for removal of 10% of the global riverine input of K^+. Since environments like the Amazon delta account for ~60% of the flux of detrital material to the oceans, the impact of these reactions globally might be much greater than in this delta alone.

Both the tropical "reverse weathering" studies and the recent discovery of the process controlling opal diagenesis in surface sediments demonstrate the importance of rapid authigenic aluminosilicate formation in marine sediments. The focus is now back on determining the importance of these reactions in marine geochemical mass balances. The role of detrital material in the preservation of opal complicates the utility of SiO_2 as a paleoceanographic tracer. This role will depend on understanding whether there is a proportionality between opal flux to sediments and the preservation rate.

APPENDIX A

Approximating the role of calcite dissolution kinetics in determining the width of the $CaCO_3$ transition from 90% to 10% in marine sediments

(i) *The carbon-flux balance at the sediment–water interface.* The $CaCO_3$ rain rate to the sediment–water interface, R_{Ca} is assigned to be 1.5 g_{CaCO_3} m^{-2} yr^{-1} and the burial rate, B_{Ca}, is calculated from Equation (17) for different values of the dissolution flux, F_{Ca}.

$$R_{Ca} = B_{Ca} + F_{Ca}(\text{mol cm}^{-2}\text{s}^{-1}) \quad (17)$$

(ii) *Derivation of the $CaCO_3$ dissolution flux, F_{Ca}, as a function of the degree of carbonate saturation from CO_3 dynamics at the sediment–water interface.*

Assume $F_{Ca} = F_{CO_3}$, and as a first order of approximation, ignore all carbonate equilibrium reactions in the solution (see text). In this case the steady-state diffusion–reaction equation for CO_3 concentration, $[CO_3^{2-}]$ (mol cm^{-3}), in the pore waters is

$$0 = D\{d^2\varphi[CO_3]/dz^2\} - \varphi J'(\text{mol cm}^{-3}\text{s}^{-1}) \quad (18)$$

where z is depth in the pore waters, φ is porosity, and D (cm^2 s^{-1}) is the molecular diffusion coefficient for CO_3^{2-}. Assuming first-order kinetics, the dissolution rate is given by

$$J' = k'\{[CO_3] - [CO_3]_s\} \quad (19)$$

where $[CO_3]_s$ is the carbonate concentration in equilibrium with calcite, and, k' (s^{-1}) is the *in situ* dissolution rate constant. If porosity is not a function of z, $[CO_3] = [CO_3]^0$ at $z=0$, and $[CO_3]$ approaches $[CO_3]_{sat}$ as $z \to \infty$, the flux at the sediment–water interface, $z=0$ is

$$F_{CO_3} = -\varphi D\{d[CO_3]/dz\}_{z=0}$$
$$= \phi(Dk')^{1/2}\{[CO_3]_0 - [CO_3]_s\} \quad (20)$$

Table 4 shows the relationship between the degree of undersaturation, ΔCO_3, and the fraction of $CaCO_3$ in sediments, f, for different values of the dissolution rate constant, k. The results are based on the mass balance model described below.

(iii) *The relationship between $CaCO_3$ rain, burial, and the fraction of $CaCO_3$ in sediments* (R_D and B_D = detrital rain rate and burial rate, mol cm^{-2} s^{-1}).

Above the lysocline, where there is no $CaCO_3$ dissolution:

$$R_{Ca}/(R_{Ca} + R_D) = f_0 \quad (21)$$

Below the lysocline:

$$B_{Ca}/(B_{Ca} + B_D) = f \quad (22)$$

Since $R_D = B_D$:

$$f = B_{Ca}/\{(R_{Ca}/f_0) - R_{Ca} + B_{Ca}\} \quad (23)$$

(iv) The values of the fraction of $CaCO_3$, f, preserved in the sediments as a function of the degree of undersaturation, ΔCO_3, calculated from Equations (17), (20), and (23) for three different rate constants are shown in Table 4.

Laboratory dissolution rate constants, k, were transferred to values used in Equation (19), k', by multiplying the lab rate equation, $j = k(1 - \Omega)$, by the concentration of $CaCO_3$ in sediments, $[CaCO_3]$, (J' in Equation (19) = $J[CaCO_3]$)

$$J[CaCO_3] = k[CaCO_3]\ (1 - \Omega)$$

To change from $(1 - \Omega)$ to $\{[CO_3^{2-}]_s - [CO_3^{2-}]\}$ multiply by $K'_{sp}/[Ca^{2+}]$:

$$K'_{sp}/[Ca^{2+}](1 - [Ca^{2+}][CO_3^{2-}]/K'_{sp})$$
$$= ([CO_3^{2-}]_s - [CO_3^{2-}])$$

Thus,

$$J[CaCO_3] = -k[CaCO_3]\{[Ca^{2+}]/K'_{sp}\}([CO_3^{2-}] - [CO_3^{2-}]_s)$$

and

$$k' = k[CaCO_3]([Ca^{2+}]/K'_{sp})$$

Table 4

ΔCO_3 (mol cm^{-3} $\times 10^{-9}$)	F_{Ca}[a] (mol m^{-2} s^{-2} $\times 10^{13}$)			B_{Ca}[b] (mol m^{-2} s^{-2} $\times 10^{13}$)			f[c]		
	k = 0.38 d^{-1}	k = 0.038 d^{-1}	k = 0.0038 d^{-1}	k = 0.38 d^{-1}	k = 0.038 d^{-1}	k = 0.0038 d^{-1}	k = 0.38 d^{-1}	k = 0.038 d^{-1}	k = 0.0038 d^{-1}
0	0	0	0	4.7	4.7	4.7	0.90	0.90	0.90
5	10.6	3.3	1.0		1.4	3.7	0	0.73	0.88
10	21.6	6.7	2.1			2.6	0	0	0.83
20	42.2	13.3	4.2			0.3	0	0	0.36

[a] Equation (20) with $D_{CO_3} = 1 \times 10^{-6}$ cm^2 s^{-1}, $\varphi = 0.9$ and k the three cases in $k = 0.38$ d^{-1}, 0.038 d^{-1}, and 0.0038 d^{-1}. [b] Equation (17) with $R_{Ca} = 4.7 \times 10^{-13}$ mol m^{-2} s^{-1} (1.5 g_{CaCO_3} cm^{-2} kyr^{-1}). [c] Equation (23) with $f_0 = 0.9$.

Evaluating $[CaCO_3]$ and $[Ca^{2+}]/K'_{sp}$

$$[CaCO_3] = f(1-\varphi)\rho/M = 0.5(0.1)2.5/100$$
$$= 1.25\times10^{-3}\ \mathrm{mol_{CaCO_3}\ cm_{bulk}^{-1}}$$

$$[Ca^{2+}]/K'_{sp} = 0.01\ \mathrm{mol\ cm^{-3}}/10^{-12}\ \mathrm{mol^2\ cm^{-6}}\ \text{(at 4 km)}$$
$$= 10^7(\mathrm{mol^{-1}\ cm^3})$$

APPENDIX B

Importance of bioturbation to the asymptotic concentration of H_4SiO_4 in marine sediment pore waters

We consider the simple balance presented in Equation (16):

$$\Delta H_4SiO_4 = \{2(1-\varphi)f[SiO_2]/h(kSD)^{0.5}\}D_b$$

where:

- $f = 0.5$ is the fraction of opal in the sediments;
- $(1-\varphi) = 0.3\ (\mathrm{cm_s^3\ cm_b^{-3}})$, porosity, $\varphi = 0.7$;
- $[SiO_2] = \rho/M = 2.0\ \mathrm{g\ cm_s^{-3}}/65\ \mathrm{g_{SiO_2}\ mol^{-1}} = 0.031\ \mathrm{mol\ cm_s^{-3}}$;
- $h = 5$ cm, the $1/e$ depth of the pore-water profiles (Figure 16);
- $kS = 1\times10^{-7}\ \mathrm{s^{-1}}$ (2 °C) from pore-water models (Broecker and Peng, 1982; Hurd, 1973; Vanderborght *et al.*, 1977); $D_{H_4SiO_4}$ (2 °C) $= 2\times10^{-6}$.

Values of D_B in the equatorial Pacific Ocean are $(7-40)\times10^{-9}\ \mathrm{cm^2\ s^{-1}}$ from ^{210}Pb and ^{234}Th profiles, respectively (McManus *et al.*, 1995):

D_b ($\mathrm{cm^2\ s^{-1}}$)	$[H_4SiO_4]_{asy} - [H_4SiO_4]_{BW}$ (μmol L^{-1})
10^{-9}	4
10^{-8}	42
10^{-7}	416

REFERENCES

Acker J. G., Byre R. H., Ben-Yaakov S., Feely R. A., and Betzer P. R. (1987) The effect of pressure on aragonite dissolution rates in seawater. *Geochim. Cosmochim. Acta* **51**, 2171–2175.

Aller R. C. (1982) The effects of macrobenthos on chemical properties of marine sediment and overlying water. In *Animal-sediment Relations* (eds. D. L. McCall and M. J. S. Tevesz). Plenum, New York, pp. 53–102.

Aller R. C. (1984) The importance of relict burrow structures and burrow irrigation in controlling sedimentary solute distributions. *Geochim. Cosmochim. Acta* **48**, 1929–1934.

Aller R. C. (1990) Bioturbation and manganese cycling in hemipelagic sediments. *Phil. Trans. Roy. Soc. London A* **331**, 51–68.

Aller R. C., Mackin J. E., and Cox R. T. (1986) Diagenesis of Fe and S in Amazon inner shelf muds: apparent dominance of Fe reduction and implications for the genesis of ironstones. *Continent. Shelf Res.* **6**, 263–389.

Archer D. (1996) An atlas of the distribution of calcium carbonate in sediments of the deep sea. *Global Biogeochem. Cycles* **10**, 159–174.

Archer D. and Devol A. (1992) Benthic oxygen fluxes on the Washington shelf and slope: a comparison of *in situ* microelectrode and chamber flux measurements. *Limnol. Oceanogr.* **37**, 614–629.

Archer D., Emerson S. R., and Reimers C. E. (1989) Dissolution of calcite in deep-sea sediments: pH and O_2 microelectrode results. *Geochim. Cosmochim. Acta* **53**, 2831–2845.

Archer D. E., Morford J. L., and Emerson S. (2002) A model of suboxic sedimentary diagenesis suitable for automatic tuning and gridded global domains. *Global Biogeochem. Cycles* **16** (1), doi 10.1029/2000G001288.

Atlas R. M., Boehm P. D., and Calder J. A. (1981) Chemical and biological weathering of oil from the Amoco Cadiz spillage within the littoral zone. *Estuar. Coast. Shelf Sci.* **12**, 589–608.

Bender M. L. (1990) The $\delta^{18}O$ of dissolved O_2 in seawater: a unique tracer of circulation and respiration in the deep sea. *J. Geophys. Res.* **95**, 22243–22252.

Bender M. L., Fanning K., Froelich P. N., Heath G. R., and Maynard V. (1977) Interstitial nitrate profiles and oxidation of sedimentary organic matter in the eastern equatorial Atlantic. *Science* **198**, 605–609.

Benner R., Maccubbin A. E., and Hodson R. E. (1984) Anaerobic biodegradation of the lignin and polysaccharide components of lignocellulose and synthetic lignin by sediment microflora. *Appl. Environ. Microbiol.* **47**, 998–1004.

Berelson W. M., Hammond D., and Cutter G. A. (1990) *In situ* measurements of calcium carbonate dissolution rates in deep-sea sediments. *Geochim. Cosmochim. Acta* **54**, 3013–3020.

Bergamaschi B. A., Tsamakis E., Keil R. G., Eglinton T. I., Montluçon D. B., and Hedges J. I. (1997) The effect of grain size and surface area on organic matter, lignin and carbohydrate concentration, and molecular compositions in Peru margin sediments. *Geochim. Cosmochim. Acta* **61**, 1247–1260.

Berner R. A. (1974) Kinetic models for the early diagenesis of nitrogen, sulfur, phosphorus, and silicon in anoxic marine sediments. In *The Sea* (ed. E. D. Goldberg). Wiley, New York, vol. 5, pp. 427–450.

Berner R. A. (1980) *Early Diagenesis: A Theoretical Approach*. Princeton University Press, Princeton, NJ.

Berner R. A. (1989) Biogeochemical cycles of carbon and sulfur and their effect on atmospheric oxygen over phanerozoic time. *Palaeogeogr. Palaeoclimatol. Palaeoecol.* **73**, 97–122.

Boudreau B. P. (1984) On the equivalence of non-local and radial diffusion models for pore water irrigation. *J. Mar. Res.* **42**, 731–735.

Boudreau B. P. (1997) *Diagenetic Models and their Implication: Modeling Transport and Reactions in Aquatic Sediments*. Springer, Berlin.

Boudreau B. P. and Ruddick B. R. (1991) On a reactive continuum representation of organic matter diagenesis. *Am. J. Sci.* **291**, 507–538.

Brandes J. A. and Devol A. H. (1997) Isotopic fractionation of oxygen and nitrogen in coastal marine sediments. *Geochim. Cosmochim. Acta* **61**, 1793–1801.

Broecker W. S. and Peng T.-H. (1982) *Tracers in the Sea*. Lamont-Doherty Geological Observatory, Palisades, NY.

Broecker W. S., Clark E., McCorkle D. C., Hajdas I., and Bonani G. (1999) Core top ^{14}C ages as a function of latitude and water depth on the Ontong-Java plateau. *Paleoceanography* **14**, 13–22.

Buckley D. E. and Cranston R. E. (1988) Early diagenesis in deep sea turbidites: the imprint of paleo-oxidation zones. *Geochim. Cosmochim. Acta* **52**, 2925–2939.

Cai W.-J., Reimers C. E., and Shaw T. (1995) Microelectrode studies of organic carbon degradation and calcite dissolution at a California continental rise site. *Geochim. Cosmochim. Acta* **59**, 497–511.

Calvert S. E. and Pedersen T. F. (1992) Organic carbon accumulation and preservation in marine sediments: how important is anoxia? In *Organic Matter* (eds. J. Whelan and J. W. Farrington). University Press, New York, pp. 231–263.

Canfield D. E. (1994) Factors influencing organic carbon preservation in marine sediments. *Chem. Geol.* **114**, 315–329.

Canfield D. E., Thamdrup B., and Hansen J. W. (1993) The anaerobic degradation of organic matter in Danish coastal sediments: iron reduction, manganese reduction and sulfate reduction. *Geochim. Cosmochim. Acta* **57**, 3867–3883.

Chen C.-T. A. (2002) Shelf vs. dissolution-generated alkalinity above the chemical lysocline. *Deep-Sea Res. II* **49**, 5365–5375.

Christensen J. P., Devol A. H., and Smethie W. M. (1984) Biological enhancement of solute exchange between sediments and bottom water on the Washington continental shelf. *Continent. Shelf Res.* **3**, 9–23.

Cowie G. L., Hedges J. I., Prahl F. G., and De Lange G. J. (1995) Elemental and major biochemical changes across an oxidation front in a relict turbidite: a clear-cut oxygen effect. *Geochim. Cosmochim. Acta* **59**, 33–46.

De Lange G. J., Jarvis I., and Kuijpers A. (1987) Geochemical characteristics and provenance of late quaternary sediments from the Madeira abyssal plains N. Atlantic. In *Geology and Geochemistry of Abyssal Plains* (eds. P. P. E. Weaver and J. Thomson). Blackwell, London, pp. 147–165.

Dixit S. and Van Cappellen P. (2002) Surface chemistry and reactivity of biogenic silica. *Geochim. Cosmochim. Acta* **66**, 2559–2568.

Dixit S., Van Cappellen P., and van Bennekom A. J. (2001) Process controlling solubility of biogenic silica and pore water build-up of silicic acid in marine sediments. *Mar. Chem.* **73**, 333–352.

DuBois L. G. and Prell W. L. (1988) Effects of carbonate dissolution on the radiocarbon age structure of sediment mixed layers. *Deep-Sea Res.* **35**, 1875–1885.

Edmond J. M., *et al.* (1979) Ridge crest hydrothermal activity and the balances of the major and minor elements in the ocean: the galapagos data. *Earth Planet. Sci. Lett.* **46**, 1–18.

Emerson S. R. and Archer D. (1990) Calcium carbonate preservation in the ocean. *Phil. Trans. Roy. Soc. London A* **331**, 29–40.

Emerson S. R. and Bender M. I. (1981) Carbon fluxes at the sediment–water interface of the deep-sea: calcium carbonate preservation. *J. Mar. Res.* **39**, 139–162.

Emerson S. and Hedges J. I. (1988) Processes controlling the organic carbon content of open ocean sediments. *Paleoceanography* **3**, 621–634.

Emerson S., Fisher K., Reimers C., and Heggie D. (1985) Organic carbon dynamics and preservation in deep-sea sediments. *Deep-Sea Res.* **32**, 1–21.

Emerson S., Jahnke R., and Heggie D. (1984) Sediment–water exchange in shallow water estuarine sediments. *J. Mar. Res.* **42**, 709–730.

Fanning K. A. and Pilson M. E. Q. (1974) Diffusion of dissolved silica out of deep-sea sediments. *J. Geophys. Res.* **79**, 1293–1297.

Feely R. A., Sabine C. L., Lee K., Millero F. J., Lamb M. F., Greeley D., Bullister J. L., Key R. M., Peng T.-H., Kozyr A., Ono T., and Wong C. S. (2002) *In-situ* calcium carbonate dissolution in the Pacific Ocean. *Global Biogeochem. Cycles* **16** (4), 1144, doi 10.1029/2002GB001866.

Froelich P. N., Klinkhammer G. P., Bender M. L., Ludtke N. A., Heath G. R., Cullin D., Dauphin P., Hammond D., Hartman B., and Maynard V. (1979) Early oxidation of organic matter in pelagic sediments of the eastern equatorial Atlantic: suboxic diagenesis. *Geochim. Cosmochim. Acta*, 1075–1090.

Grundmanis V. and Murray J. W. (1982) Aerobic respiration in pelagic marine sediments. *Geochim. Cosmochim. Acta* **46**, 1101–1120.

Hales B. and Emerson S. R. (1996) Calcite dissolution in sediments of the Ontong-Java plateau: *in situ* measurements of pore water O_2 and pH. *Global Biogeochem. Cycles* **10**, 527–541.

Hales B. and Emerson S. R. (1997a) Evidence in support of first-order dissolution kinetics of calcite in seawater. *Earth Planet. Sci. Lett.* **148**, 317–327.

Hales B. and Emerson S. R. (1997b) Calcite dissolution in sediments of the Ceara rise: *in situ* measurements of pore water O_2, pH, and CO_2(aq). *Geochim. Cosmochim. Acta* **61**, 501–514.

Hales B., Burgess L., and Emerson S. (1997) An absorbance-based fiber-optic sensor for CO_2(aq) measurements in pore waters of seafloor sediments. *Mar. Chem.* **59**, 51–62.

Hartnett H. E., Keil R. G., Hedges J. I., and Devol A. H. (1998) Influence of oxygen exposure time on organic carbon preservation in continental margin sediments. *Nature* **391**, 572–574.

Harvey H. R., Tuttl J. H., and Bell J. T. (1995) Kinetics of phytoplankton decay during simulated sedimentation: changes in biochemical composition and microbial activity under oxic and anoxic conditions. *Geochim. Cosmochim. Acta* **59**, 3367–3377.

Hedges J. I. and Keil R. G. (1995) Sedimentary organic matter preservation: an assessment and speculative synthesis. *Mar. Chem.* **49**, 81–115.

Hedges J. I., Cowie G. L., Ertel J. R., Barbour R. J., and Hatcher P. G. (1985) Degradation of carbohydrates and lignins in buried woods. *Geochim. Cosmochim. Acta.* **49**, 701–711.

Hedges J. I., Hu F. S., Devol A. H., Hartnett H. E., and Keil R. G. (1999) Sedimentary organic matter preservation: a test for selective oxic degradation. *Am. J. Sci.* **299**, 529–555.

Hurd D. C. (1973) Interactions of biogenic opal sediment and seawater in the central equatorial Pacific. *Geochim. Cosmochim. Acta* **37**, 2257–2282.

Imboden D. M. (1975) Interstitial transport of solutes in non-steady state accumulating and compacting sediments. *Earth Planet. Sci. Lett.* **27**, 221–228.

Jahnke R. A. (1996) The global ocean flux of particulate organic carbon: a real distribution and magnitude. *Global Biogeochem. Cycles* **10**, 71–88.

Jahnke R. J. and Jahnke D. B. (2002) Calcium carbonate dissolution in deep-sea sediments: implications of bottom water saturation state and sediment composition. *Geochim. Cosmochim. Acta* (in press).

Jahnke R. J., Craven D. B., McCorkle D. C., and Reimers C. E. (1997) $CaCO_3$ dissolution in California continental margin sediments: the influence of organic matter remineralization. *Geochim. Cosmochim. Acta* **61**, 3587–3604.

Jørgensen B. B. (1979) A comparison of methods for the quantification of bacterial sulfate reduction in coastal marine sediments: II. Calculation from mathematical models. *Geomicrobiol. J.* **1**, 29–47.

Keil R. G., Tsamakis E., Fuh C. B., Giddings J. C., and Hedges J. I. (1994a) Mineralogical and textural controls on organic composition of coastal marine sediments: hydrodynamic separation using SPLITT fractionation. *Geochim. Cosmochim. Acta.* **57**, 879–893.

Keil R. G., Montluçon D. B., Prahl F. G., and Hedges J. I. (1994b) Sorptive preservation of labile organic matter in marine sediments. *Nature* **370**, 549–552.

Keir R. S. (1980) The dissolution kinetics of biogenic calcium carbonates in seawater. *Geochim. Cosmochim. Acta* **44**, 241–252.

Keir R. S. and Michel R. L. (1993) Interface dissolution control of the ^{14}C profile in marine sediment. *Geochim. Cosmochim. Acta* **57**, 3563–3573.

King S. L., Froelich P. N., and Jahnke R. A. (2000) Early diagenesis of germanium in sediments of the Antarctic South Atlantic: in Search of the missing Ge-sink. *Geochim. Cosmochim. Acta* **64**, 1375–1390.

Koning E., Brummer G.-J., van Raaphorst W., van Bennekom J., Helder W., and van Iperen J. (1997) Settling dissolution and burial of biogenic silica in the sediments off Somalia (northwestern Indian Ocean). *Deep-Sea Res. II* **44**, 1341–1360.

Kristensen E. and Holmer M. (2001) Decomposition of plant materials in marine sediment exposed to different electron acceptors (O_2, NO_3^-, and SO_4^{2-}), with emphasis on substrate origin, degradation kinetics, and the role of bioturbation. *Geochim. Cosmochim. Acta* **65**, 419–433.

Lawson D. S., Hurd D. C., and Pankratz H. S. (1978) Silica dissolution rates of decomposing assemblages at various temperatures. *Am. J. Sci.* **278**, 1373–1393.

Lee C. (1992) Controls on organic carbon preservation: the use of stratified water bodies to compare intrinsic rates of decomposition in oxic and anoxic systems. *Geochim. Cosmochim. Acta* **56**, 3323–3335.

Mackenzie F. T. and Garrels R. M. (1966) Chemical mass balance between rivers and oceans. *Am. J. Sci.* **264**, 507–525.

Mackin J. E. and Aller R. C. (1984) Dissolved Al in sediments and waters of the East China Sea: implications for authigenic mineral formation. *Geochim. Cosmochim. Acta* **48**, 281–297.

Martin W. R., McNichol A. P., and McCorkle D. C. (2000) The radiocarbon age of calcite dissolving at the sea floor: estimates from pore water data. *Geochim. Cosmochim. Acta* **64**, 1391–1404.

Martins C., Kipphut G. W., and Klump J. V. (1980) Sediment–water chemical exchange in the coastal zone traced by *in situ* radon-222 flux measurements. *Science* **208**, 285–288.

Mayer L. M. (1994a) Surface area control of organic carbon accumulation in continental shelf sediments. *Geochim. Cosmochim. Acta* **58**, 1271–1284.

Mayer L. M. (1994b) Relationships between mineral surfaces and organic carbon concentrations in soils and sediments. *Chem. Geol.* **114**, 347–363.

Mayer L. M. (1999) Extent of coverage of mineral surfaces by organic matter in marine sediments. *Geochim. Cosmochim. Acta* **63**, 207–215.

McManus J., Hammond D. E., Berelson W. M., Kilgore T. E., DeMaster D. J., Ragueneau O. G., and Collier R. W. (1995) Early diagenesis of biogenic opal: dissolution rates, kinetics, and paleoceanographic implications. *Deep-Sea Res. II* **42**, 871–902.

Michalopoulos P. and Aller R. C. (1995) Rapid clay mineral formation in the Amazon delta sediments: reverse weathering and ocean elemental cycles. *Science* **270**, 614–617.

Middelburg J. J. (1989) A simple rate model for organic matter decomposition in marine sediments. *Geochim. Cosmochim. Acta* **53**, 1577–1581.

Middelburg J. J., Soetaert K., Herman P. M. J., and Heip C. H. R. (1996) Denitrification in marine sediments: a model study. *Global Biogeochem. Cycles* **10**, 661–673.

Millero F. J. and Berner R. A. (1972) Effect of pressure on carbonate equilibria in seawater. *Geochim. Cosmochim. Acta* **36**, 92–98.

Milliman J. D., Troy P. J., Balch W. M., Adams A. K., Li Y. H., and Mackenzie F. T. (1999) Biologically mediated dissolution of calcium carbonate above the chemical lysocline? *Deep-Sea Res. I* **46**, 1653–1669.

Morford J. and Emerson S. (1999) The geochemistry of redox sensitive trace metals in sediments. *Geochim. Cosmochim. Acta* **63**, 1735–1750.

Morse J. W. and Arvidson R. S. (2002) The dissolution kinetics of major sedimentary carbonate minerals. *Earth Sci. Rev.* **58**, 51–84.

Morse J. W. and Mackenzie F. T. (1990) *Geochemistry of Sedimentary Carbonates*. Elsevier, New York, NY.

Morse J. W., Mucci A., and Millero F. J. (1980) The solubility of calcite and aragonite in seawater at 35 ppt salinity at 25°C and atmospheric pressure. *Geochim. Cosmochim. Acta* **44**, 85–94.

Mucci A. (1983) The solubility of calcite and aragonite in seawater at various salinities, temperatures, and one atmosphere total pressure. *Am. J. Sci.* **283**, 780–799.

Murray J. W. and Kuivila K. M. (1990) Organic matter diagenesis in the northeast Pacific: transition from aerobic red clay to suboxic hemipelagic sediments. *Deep-Sea Res.* **37**, 59–80.

Murray J. W., Grundmanis V., and Smethie W. (1978) Interstitial water chemistry in the sediments of Saanich Inlet. *Geochim. Cosmochim. Acta* **42**, 1011–1026.

Nelson D. M., Treguer P., Brzezinski M. A., Leynaert A., and Queguiner B. (1995) Production and dissolution of biogenic silica in the ocean: revised global estimates, comparison with regional data and relationship to biogenic sedimentation. *Global Biogeochem. Cycles* **9**, 359–372.

Nittrouer C. A., *et al.* (1991) Sedimentology and stratigraphy of the Amazon continental shelf. *Oceanography* **4**, 33–38.

Pedersen T. F. and Calvert S. E. (1990) Anoxia vs. productivity: what controls the formation of organic-carbon-rich sediments and sedimentary rocks? *Am. Assoc. Petrol. Geol. Bull.* **74**, 454–466.

Prahl F. G., De Lange G. J., Lyle M., and Sparrow M. A. (1989) Post-depositional stability of long-chain alkenones under contrasting redox conditions. *Nature* **341**, 434–437.

Prahl F. P., De Lang G. J., Scholten S., and Cowie G. L. (1997) A case of post-depositional aerobic degradation of terrestrial organic matter in turbidite deposits from the Madeira abyssal plain. *Org. Geochem.* **27**, 141–152.

Premuzic E. T., Benkovitz C. M., Gaffney J. S., and Walsh J. J. (1982) The nature and distribution of organic matter in the surface sediments of world oceans and seas. *Org. Geochem.* **4**, 63–77.

Rabouille C., Gaillard J.-F., Treguer P., and Vincendeau M.-A. (1997) Biogenic silica recycling in surficial sediments across the Polar front of the Southern Ocean (Indian sector). *Deep-Sea Res. II* **44**, 1151–1176.

Raiswell R., Canfield D. E., and Berner R. A. (1994) A comparison of iron extraction methods for the determination of the degree of pyritization and the recognition of iron-limited pyrite formation. *Chem. Geol.* **111**, 101–110.

Ransom B., Bennett R. H., Baerwald R., and Shea K. (1997) TEM study of "*in situ*" organic matter on continental margins: occurrence and the "monolayer" hypothesis. *Mar. Geol.* **138**, 1–9.

Ransom B., Ki D., Kastner M., and Wainright S. (1998) Organic matter preservation on continental slopes: importance of mineralogy and surface area. *Geochim. Cosmochim. Acta* **62**, 1329–1345.

Reeburgh W. S. (1980) Anaerobic methane oxidation: rate depth distribution in Skan Bay sediments. *Earth Planet. Sci. Lett.* **47**, 345–352.

Rickert D. (2000) Dissolution kinetics of biogenic silica in marine environments. PhD Thesis, Ber. Polarforsch., 351.

Rutgers van der Loeff M. M. (1990) Oxygen in pore waters of deep-sea sediments. *Phil. Trans. Roy. Soc. London A* **331**, 69–84.

Sawyer D. T. (1991) *Oxygen Chemistry*. Oxford Press, New York.

Sayles F. L. (1980) The solubility of $CaCO_3$ in seawater at 2°C based upon *in-situ* sampled pore water composition. *Mar. Chem.* **9**, 223–235.

Sayles F. L., Martin W. R., and Deuser W. G. (1994) Response of benthic oxygen demand to particulate organic carbon supply in the deep sea near Bermuda. *Nature* **371**, 686–689.

Schink D. R., Guinasso N. L., and Fanning K. A. (1975) Processes affecting the concentration of silica at the sediment–water interface of the Atlantic Ocean. *J. Geophys. Res.* **80**, 2013–3031.

Sillen L. G. (1961) The physical chemistry of sea water. In *Oceanography*, Am. Assoc. Adv. Sci. Pub. 67 (ed. M. Sears). Washington, DC, pp. 549–581.

Smith C. R. and Rabouille C. (2002) What controls the mixed-layer depth in deep-sea sediments? The importance of POC flux. *Limnol. Oceanogr.* **47**, 418–426.

Smith C. R., Pope R. H., DeMaster D. J., and Magaard L. (1993) Age-dependent mixing of deep-sea sediments. *Geochim. Cosmochim. Acta* **57**, 1473–1488.

Smith K. L. and Baldwin R. J. (1984) Seasonal fluctuations in deep-sea sediment community oxygen consumption: central and eastern North Pacific. *Nature* **307**, 624–626.

Stumm W. and Morgan J. J. (1981) *Aquatic Chemistry*. Wiley, New York, NY.

Suess E. (1973) Interaction of organic compounds with calcium carbonate: II. Organo-carbonate association in recent sediments. *Geochim. Cosmochim. Acta* **37**, 2435–2447.

Sun M.-Y., Lee C., and Aller R. C. (1993a) Laboratory studies of oxic and anoxic degradation of chlorophyll-a in long island sound sediments. *Geochim. Cosmochim. Acta* **57**, 147–157.

Sun M.-Y., Lee C., and Aller R. C. (1993b) Anoxic and oxic degradation of ^{14}C-labeled chloropigments and a ^{14}C-labeled diatom in long island sound sediments. *Limnol. Oceanogr.* **38**, 1438–1451.

Tromp T. K., Van Cappellen P., and Key R. M. (1995) A global model for the early diagenesis of organic carbon and organic phosphorus in marine sediments. *Geochim. Cosmochim. Acta* **59**, 1259–1284.

Van Cappellen P. and Qiu L. (1997a) Biogenic silica dissolution in sediments of the Southern Ocean: I. Solubility. *Deep-Sea Res. II* **44**, 1109–1128.

Van Cappellen P. and Qiu L. (1997b) Biogenic silica dissolution in sediments of the Southern Ocean: I. Kinetics. *Deep-Sea Res. II* **44**, 1129–1149.

Vanderborght J. P., Wollast R., and Billen B. (1977) Kinetic model of diagenesis in disturbed sediments: mass transfer properties and silica diagenesis. *Limnol. Oceanogr.* **33**, 787–793.

Wang Y. and Van Cappellen P. (1996) A multicomponent reactive transport model of early diagenesis: application to redox cycling in coastal marine sediments. *Geochim Cosmochim. Acta* **60**, 2993-3014 .

Weaver P. P. E. and Rothwell R. G. (1987) Sedimentation on the Madeira abyssal plain over the last 300,000 years. In *Geology and Geochemistry of Abyssal Plains* (eds. P. P. E. Weaver and J. Thomson). Blackwell, London, pp. 71–86.

Weiss M. S., Abele U., Weckesser J., Welte W., Schultz E., and Schultz G. E. (1991) Molecular architecture and electrostatic properties of a bacterial porin. *Science* **254**, 1627–1630.

Wenzhofer F., Adler M., Kohls O., Hensen C., Strotmann B., Boehme S., and Schulz H. D. (2001) Calcite dissolution driven by benthic mineralization in the deep sea: *in situ* measurements of Ca^{2+}, pH, pCO_2 and O_2. *Geochim. Cosmochim. Acta* **65**, 2677–2690.

Westrich J. T. and Berner R. A. (1984) The role of sedimentary organic matter in bacterial sulfate reduction: the G model tested. *Limnol. Oceanogr.* **29**, 236–249.

Wilson T. R. S., Thomson J., Colley S., Hydes D. J., Higgs N. C., and Sørensen J. (1985) Early organic diagenesis: the significance of progressive subsurface oxidation fronts in pelagic sediments. *Geochim. Cosmochim. Acta* **49**, 811–822.

6.12
Geochronometry of Marine Deposits

K. K. Turekian

Yale University, New Haven, CT, USA

and

M. P. Bacon

Woods Hole Oceanographic Institution, MA, USA

6.12.1 INTRODUCTION

Marine deposits in the form of sediment accumulations, ferromanganese nodules, and crusts and corals are important recorders not only of marine events but also of global environmental changes. Until accurate methods of establishing chronometries of these deposits were developed, however, they were of limited value for coupling global events with the environmental proxies in the deposits. The chronometry of marine deposits ultimately depends on accurate calibration by radioactive methods. The chronometric potential of tracking the change in abundance of a radioactive species or its products lies at the heart of all successful dating methods. After radioactive geochronometry has been established, the system can be further exploited with stratigraphic tools such as periodic changes in properties over time or comparative chronometries with deposits on land.

6.12.2 PRINCIPLES

6.12.2.1 Radioactive Geochronometry

All absolute dating methods that have proven dependable are based on radioactive decay. Virtually all of the methods depend on the methodical decrease in the amount of the radioactive nuclide and the growth of the corresponding daughter product. The one exception is in the area of radiation-induced damage in solids, which is the basis of thermoluminescence or ESR dating. This latter scheme will be dealt with later in the chapter. Here the basic principles of canonical radioactive geochronometry as applied to marine deposits is reviewed.

The radioactive decay equation is

$$N = N_0\exp(-\lambda t) \qquad (1)$$

Where N_0 is the number of atoms of the radioactive species at the time of formation of the system under study, N is the amount left after a time t, and λ is the decay constant. Where the ratio of the daughter to parent nuclide is measured after a time t has elapsed, then the equation becomes

$$(N_0 - N)/N = \exp(\lambda t) - 1 \qquad (2)$$

Here $N_0 - N$ represents the amount of daughter that has been produced after time t has elapsed.

Where the daughter is also radioactive, the equations become more complex. The full discussion of the complexities can be found in Friedlander *et al.* (1981) or Ivanovich and Harmon (1992).

There are several fundamental requirements for the use of the radioactive systems for geochronometry: (i) there must not be any initial daughter nuclides in the system or, if there are, there must be an adequate method of identifying and accommodating their presence; (ii) there must not be loss or addition of parent or daughter during the life of the system but, if there is, a method for assessing and accommodating the loss or gain of parent or daughter must be devised; and (iii) the half-life of the radioactive species being studied must be compatible with the time range of the system.

6.12.2.2 Secondary Stratigraphic Procedures

The classical geologic methods for arriving at relative ages based on sequencing and correlation via physical or paleontological markers do not provide chronometry. The stratigraphic arguments can, however, be utilized to extend information on time once suitable radiometric dating has been established in some part of the system. There are several such examples which will be addressed in this chapter:

(i) The $\delta^{18}O$ record (i.e., the $^{18}O/^{16}O$ value normalized to a standard) in foraminifera from deep-sea cores can be used to track changes in climate. The deconvolution of these records into the orbitally driven Milankovitch cycles allows extension of the record beyond the radiometrically calibrated sequence for the past 3×10^5 yr based on the uranium decay chain nuclides.

(ii) The changes, during the Cenozoic, of the global ocean temperatures as recorded in foraminifera from deep-sea sediments can be dated using the K–Ar technique on volcanic-ash layers.

(iii) The calibration of the magnetic-reversal timescale in various geologic settings can be used to date deep-sea deposits that are continuously accumulated.

6.12.3 RADIOACTIVE SYSTEMS USED IN MARINE GEOCHRONOMETRY

There are four radionuclide systems used in the geochronometry of marine deposits: ^{14}C, the uranium and thorium decay chains, cosmogenic nuclides other than ^{14}C and, for detrital volcanic deposits, $^{40}K-^{40}Ar$.

6.12.3.1 Radiocarbon

Natural ^{14}C (half-life = 5,730 yr) is produced by cosmic-ray interaction with the atmosphere. It enters the oceans by exchange of CO_2 between the ocean and atmosphere. The ratio of $^{14}C/^{12}C$ in the dissolved inorganic carbon species in seawater is determined by isotopic fractionation, as tracked by following the stable carbon isotopes: ^{12}C

and ^{13}C. The $^{14}C/^{12}C$ ratio is diminished in the deep ocean as the water ages during thermohaline circulation. Upwelling low $^{14}C/^{12}C$ water equilibrates with the atmospheric CO_2 but this is not instantaneous. This delay causes the $^{14}C/^{12}C$ value of surface seawater to lag the value in the atmosphere.

Moreover, there are fluctuations in the ^{14}C production rate so that the initial value of $^{14}C/^{12}C$ ratio in the atmosphere and surface ocean also fluctuates proportionately. This variation has been observed in independently dated tree rings on land and in coral accumulations in the sea (Bard *et al.*, 1990). All these factors plus the intrinsic properties of accumulation, bioturbation, and preservation must be considered to translate a $^{14}C/^{12}C$ ratio to an age expressed in calendar years.

6.12.3.2 Uranium and Thorium Decay Chain Nuclides

Table 1 shows the decay schemes for ^{238}U, ^{235}U, and ^{232}Th. Some of the daughters of each of these nuclides have been used in the study of marine deposits, either for rates of sediment accumulation or rates of bioturbation. The fundamental procedure for the use of these nuclides in determining rates depends on the separation of the various members of the uranium and thorium decay series in the aqueous medium as they are produced so that the ratio of daughter to parent nuclide is far from the secular equilibrium expected after the proper lapse of time. For example, the thorium isotopes produced in the uranium and thorium decay chains are quickly removed, whereas radium and uranium remain in solution until included in biological tests.

6.12.3.3 Cosmogenic Nuclides

Other radionuclides, in addition to ^{14}C, are produced in the atmosphere by the bombardment of the gases in the atmosphere by cosmic rays. The production rate for the Earth as a whole is determined by how well the Earth is magnetically shielded. This shield is provided intrinsically by the Earth's magnetic moment and externally by the growth and contraction of the solar magnetic envelope. As in the case of ^{14}C, the production rate of the cosmogenic nuclides will be highest when Earth's magnetic moment is low or when solar activity is small. These fluctuations must be accommodated if the cosmogenic nuclides are to be useful in geochronometry. The short-lived

Table 1 Uranium and thorium decay chains.

	U-238 SERIES							Th-232 SERIES					U-235 SERIES				
Np																	
U	U-238 4.51×10^9 yr		U-234 2.48×10^5 yr										U-235 7.13×10^8 yr				
Pa	↓	Pa-234 1.18 m	↓										↓	Pa-231 3.2×10^4 yr			
Th	Th-234 24.1 d		Th-230 7.52×10^4 yr					Th-232 1.39×10^{10} yr		Th-228 1.90 yr			Th-231 25.6 h	↓	Th-227 18.6 d		
Ac			↓					↓	Ac-228 6.13 h	↓				Ac-227 22.0 yr	↓		
Ra			Ra-226 1622 yr					Ra-228 5.75 yr		Ra-224 3.64 d					Ra-223 11.4 d		
Fr			↓							↓					↓		
Rn			Rn-222 3.825 d							Rn-220 54.5 s					Rn-219 3.92 s		
At			↓							↓					↓		
Po			Po-218 3.05 m		Po-214 1.6×10^{-4} s		Po-210 138.4 d			Po-216 0.158 s	65%	Po-212 30×10^{-7} s			Po-215 1.83×10^{-3} s		
Bi			↓	Bi-214 19.7 m	↓	Bi-210 5.0 d	↓			↓	Bi-212 60.5 m	↓			↓	Bi-211 2.16 m	
Pb			Pb-214 26.8 m		Pb-210 22.3 yr		Pb-206			Po-212 10.6 h	35% ↓	Pb-208			Pb-211 36.1 m	↓	Pb-207
Tl											Tl-208 3.1 m					Tl-207 4.79 m	

cosmogenic nuclides such as ^{7}Be (half-life = 53 d) are used for the study of bioturbation. The longer-lived radionuclides such as ^{32}Si (half-life = 140 yr) can be used to study both mixing rates and rates of accumulation. The long-lived ^{10}Be (half-life = 1.5 Myr) has been used for dating back to ~10 Myr.

6.12.3.4 Potassium–Argon

The decay of ^{40}K (half-life = 1.250×10^{9} yr with 10.5% of the decays going to ^{40}Ar) is used in deep-sea deposits in association with volcanic-ash layers. Dating using the ratio of ^{40}Ar to ^{39}Ar (a surrogate for ^{40}K resulting from irradiation of the sample with neutrons) has been widely used in volcanic-ash layers associated with biostratigraphic units and is the basis of the Cenozoic chronology presented by Berggren *et al.* (1995). There have also been efforts to use the scheme for dating low-temperature minerals such as glauconite in continental margin sediments, but these will not be discussed in this chapter.

6.12.4 COASTAL DEPOSITS

6.12.4.1 Applicable Methods and Requirements

The rates of accumulation in coastal deposits are commonly greater than deep-sea deposits. This disparity is not true everywhere since coastal areas can be deficient in sediment supply or may be subject to efficient erosive processes. Accumulations can occur in estuaries, coastal depressions, and salt marshes. Attempts at geochronometry with a number of nuclides have been made in all of these areas. Confounding the record for both coastal and deep-sea sediments is the effect of bioturbation. Sediments deposited under anoxic conditions are free of this effect but all other sediments are subject to a variety of scales of bioturbation.

6.12.4.2 Unbioturbated Deposits

The requirement for obtaining unbioturbated deposits is either the deposition under reducing conditions or deposition in a rigid structure not capable of biological mechanical activity. The former includes anoxic or suboxic basins such as parts of the Gulf of California, Santa Barbara basin, the Black Sea, or Cariaco Trench. The latter is mainly restricted to "high" salt marshes (cores of middle salt marshes show significant bioturbation according to Saffert and Thomas (1998). Both types of deposits have been dated by radioactive methods.

Anoxic basins in the Gulf of California have been dated by the excess ^{210}Pb decrease with depth (DeMaster and Turekian, 1987). Because of the 22 yr half-life of ^{210}Pb, the dating is restricted to the latest 100 yr of sediment accumulation. DeMaster and Turekian (1987) also used ^{32}Si as a chronometric tool although their main goal initially was to determine the half-life of ^{32}Si by using the ^{210}Pb-dated layers and extension to deeper parts of the sediment pile. Actually the results were flawed as the method for evaluating the half-life of ^{32}Si by independent laboratory determination showed that the half-life was ~140 yr and not as high as the value inferred from the sediment-based measurements. The assumption of constant-sediment accumulation beyond the highly varved ^{210}Pb dated section to greater depths in the sediment was wrong. The use of the correct ^{32}Si half-life shows that the layers accumulated at different rates beyond the 100 yr are datable by ^{210}Pb.

Santa Barbara basin sediments back to 100 yr have been dated by ^{210}Pb (Koide *et al.*, 1972) and back to $\sim 2 \times 10^{4}$ yr by ^{14}C (Kennett and Ingram, 1995). Santa Barbara basin shows sedimentation changing between a suboxic unbioturbated regime during interglacials and an aerated bioturbated regime during glacials. The pattern tracks other climate-controlled features in ice cores and North Atlantic sediments.

Certain smaller suboxic environments within larger estuarine systems also show unbioturbated sections of sediments. Because of their generally rapid rate of accumulation, the cosmogenic nuclide ^{7}Be (53 d half-life) may be used in addition to ^{210}Pb for sediment accumulation rate assessment for the more recently deposited sediments. Similarly the pattern of bomb-produced ^{137}Cs in sediments can serve as a chronometer. In addition, radiocarbon ages commonly can be obtained from calcareous fractions in the accumulating sediment. All these approaches were used at the FOAM site in Long Island Sound by Krishnaswami *et al.* (1984).

Aside from anoxic or suboxic basins, the other marine environment suitable for radioactive geochronometry is salt-marsh deposits. As sea level has risen over the past 100 years, salt marshes have kept up by vertical growth of a vegetated framework that supports sediment accumulation. In addition, since high salt marshes are inundated by seawater only ~5% of the year, the surface becomes an accumulator of atmospherically derived species including ^{210}Pb. The radioactive decay of ^{210}Pb can then be used to determine the age of levels in the salt marsh and thereby the accumulation rate of the salt marsh and its components. Since the salt-marsh vertical growth depends on the rise in sea level, the ^{210}Pb chronometer becomes a proxy for the rate of rise

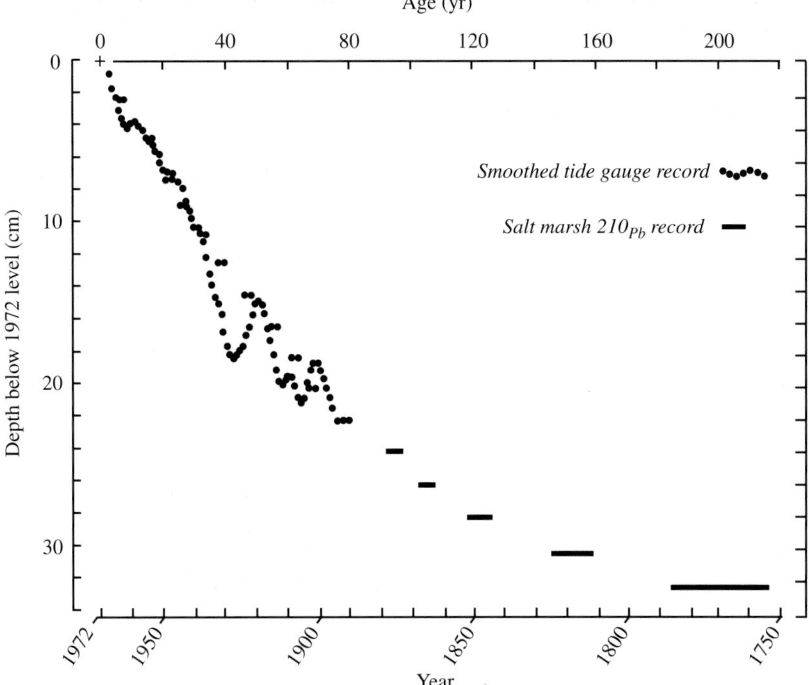

Figure 1 Comparison of the change in sea level measured in a tide gauge in New York versus the excess ^{210}Pb derived from the atmosphere in a core from the Farm River Salt Marsh, Branford, Connecticut (source McCaffrey and Thomson, 1980).

of sea level recorded along coasts if the tectonic or isostatic upward movement of the land can be corrected for. The procedures have been described by McCaffrey and Thomson (1980) and a plot for the Farm River salt marsh in Connecticut is shown in Figure 1. Other studies of salt-marsh vertical growth and its relation to atmospheric and sediment fluxes have since been published (Varekamp and Thomas, 1998).

6.12.4.3 Bioturbated Deposits

Under oxic conditions all sediments are mixed as the result of the actions of a variety of types of biota. Depending on the depth of biological activity, the sedimentary record will reflect this mixing process in the distribution of radionuclides. The full equation describing the distribution of a radioactive species in a sediment pile is

$$\frac{\partial A}{\partial t} = D_B \frac{\partial^2 A}{\partial z^2} - S\frac{\partial A}{\partial z} - \lambda A \tag{3}$$

where A is the radioactivity per unit mass of the nuclide of interest, z is depth in a core, D_B is the particle mixing coefficient treated as a diffusion phenomenon, and S is the sedimentation rate. The solution to the above equation is

$$A(z) = A_0 \exp\left(\frac{S - \sqrt{S^2 + 4\lambda D_B}}{2D_B}\right)z \tag{4}$$

Where S is slow compared to the value of λ for the nuclide used the equation is approximated by

$$A(z) = A_0 \exp\left(\frac{-z\sqrt{4\lambda D_B}}{2D_B}\right) \tag{5}$$

A plot of ln $A(z)$ against z yields the value of D_B for the system under consideration.

Benninger *et al.* (1979) measured the distribution of ^{234}Th (half-life = 24 d), ^{210}Pb (half-life = 22 yr) and considered the distribution of ^{14}C (half-life = 5,730 yr) in their discussion of the different mixing modes of a core from Long Island Sound. The ^{234}Th measures the mixing D_B of clams, the ^{210}Pb D_B primarily of worms, and the D_B of burrowing crustaceans (*Squilla*). This is shown in Figure 2. The D_B of Squilla bioturbation is small. This fact can explain the decrease in ^{14}C specific activity of the organic component with depth if the sediment accumulation rate was very slow. The ^{14}C was interpreted by Benoit *et al.* (1979) as representing sediment accumulation rate. This interpretation is compromised by the bioturbation effect. Other ^{14}C profiles published

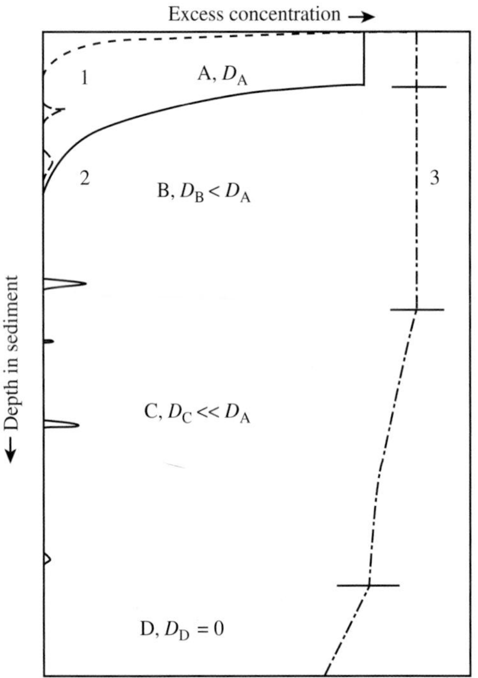

Figure 2 Schematic representation of depth profiles of excess radionuclides in a sediment column undergoing biological mixing. Curves 1, 2, and 3 represent patterns for different scales of mixing and involving different radionuclides. Decay constants are assumed to decrease with the depths represented and the mixing ("particle diffusion") coefficients (D_A, D_B, D_C, D_D) decrease with depth (zones A, B, C, D). The concentration profiles are continuous over depth intervals where mixing is rapid on the timescale of radioactive decay and discontinuous below except that discontinuities may occur at depths where the mixing regime changes (e.g., curve 2). Mixing and sediment accumulation both influence the shapes of the profiles in the continuous segments, except in zone D which is unmixed.

for coastal sediments also appear to be indicators of bioturbation rather than accumulation (Tanaka *et al.*, 1991).

6.12.5 DEEP-SEA SEDIMENTS

6.12.5.1 Radiocarbon

The use of radiocabon as a dating tool depends on the deposition of carbon in the form either of calcareous tests or of organic carbon. The $^{14}C/^{12}C$ ocean ratio is not homogeneous. The deeper waters generally are lower in $^{14}C/^{12}C$ values because the circulation time of the oceans is ~1,000 yr. The results from the GEOSECS (1987) program show the oceanic profiles in all the oceans. The surface oceans are most directly impacted by the $^{14}C/^{12}C$ ratio of the atmosphere, so much so that surface waters have the imprint of the bomb ^{14}C, thus track the atmospheric burden. The GEOSECS profiles clearly show this effect and the distribution in the oceans has been used to determine the rates of water-mass formation in the North Atlantic.

Prior to the bomb effect, which became important in 1950 and grew to a maximum in 1962, surface seawater had a $^{14}C/^{12}C$ imprint that was generally lower than that expected for equilibrium with the atmosphere. Indeed the average age of surface, inorganic carbon species dissolved in surface seawater appeared to be ~400 yr virtually everywhere in the oceans. The oldest surface ages are found at the sites of upwelling such as the equatorial oceans and the eastern boundaries of the oceans. The "reservoir" age ~400 yr is the consequence of the supply of aged upwelled water to the surface where subsequent exchange with the atmosphere results in the nonzero age surface-carbon value. All dating based on measurements of tests or organic material derived from the surface oceans will have this initial bias in age that must be accommodated independent of other concerns about the radiocarbon dating scheme.

The first uses of radiocarbon in deep-sea core dating were based on few data points and depended on extrapolation assuming the constant rate of titanium deposition (Arrhenius *et al.*, 1951) or interpolation (Suess, 1956) for determination of rates of accumulation and chronology. The first systematic study of radiocarbon incorporating possible changes in accumulation rates with depth in a core was performed by Broecker *et al.* (1958). They showed that accumulation rates of both the carbonate fraction and the detrital fraction varied with time in the equatorial Atlantic and those variations were linked to paleoclimatic indicators inferred from paleontologic data (Figure 3).

It has been observed that the surface deposits in marine cores do not have the expected age of zero. This disparity has been ascribed to loss of core tops at time of core recovery and the reservoir effect discussed above. It has also been discovered that the dominant control of the apparently constant dates in the top 8 cm of a core is bioturbation. Nozaki *et al.* (1977) studied a core obtained by a submersible research vessel and therefore exempt from the artifact of mixing or loss of the top of the sediment pile, which commonly occurs during piston coring and recovery. They measured ^{14}C and ^{210}Pb, using the latter to determine that bioturbation has indeed occurred and establish its rate constant (Figures 4 and 5). Clearly bioturbation has occurred to a depth of ~8 cm, below that depth the absence of bioturbation permits the use of ^{14}C to determine an accumulation rate and establish a chronology. The relationship between depth of bioturbation

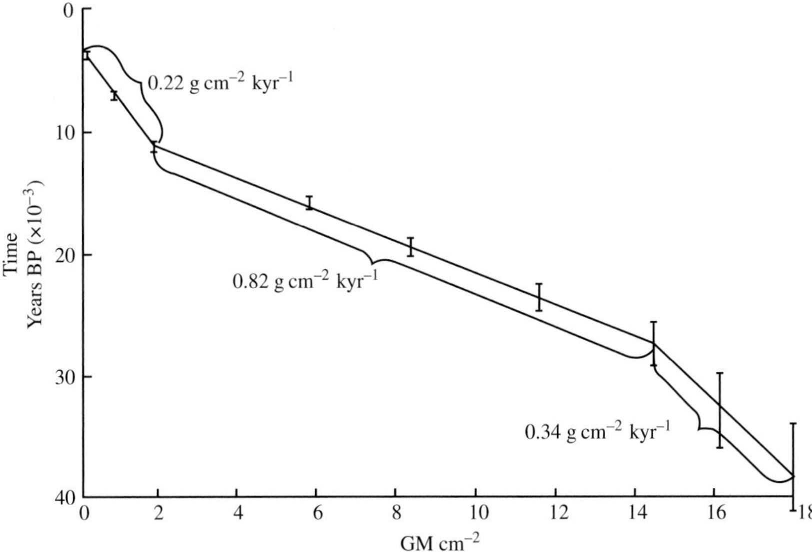

Figure 3 Cumulative curve of weight of acid-insoluble fraction ("clay") as a function of time (BP) in an Atlantic equatorial core from the Mid-Atlantic Ridge (LDEO core A 180-74) (source Broecker *et al.*, 1958).

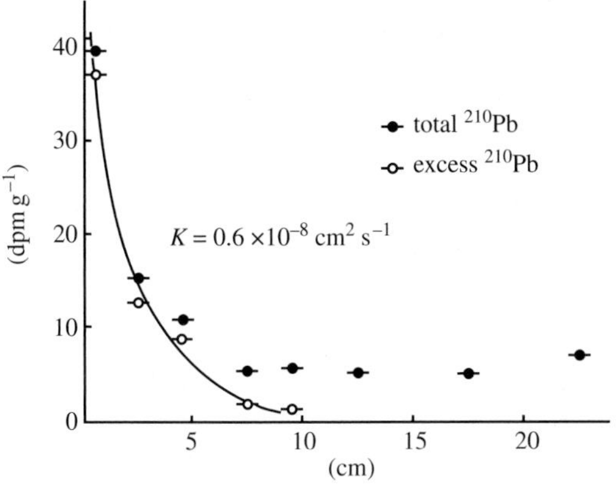

Figure 4 ^{210}Pb distribution in core 527-3 of the FAMOUS expedition at the Mid-Atlantic Ridge west of the Azores. Excess ^{210}Pb was calculated by subtracting ^{226}Ra activity from the total ^{210}Pb activity (source Nozaki *et al.*, 1977).

and sediment accumulation rate is

$$\text{Age} = \frac{1}{\lambda_{^{14}\text{C}}} \ln\left[1 + \left(\frac{\lambda_{^{14}\text{C}}}{S}\right) X_{\text{M}}\right] \quad (6)$$

where the age is the radiocarbon age of the mixed layer, X_{M} is the depth of the mixed layer, S is the accumulation rate of sediment and λ is the decay constant for ^{14}C. In principle, for the length of the core that is accumulating at the present rate, a radiocarbon age of the surface-mixed layer and the assumption that the mixed layer is 8 cm thick allows the determination of S and therefore a chronometry for the portion of the core below 8 cm.

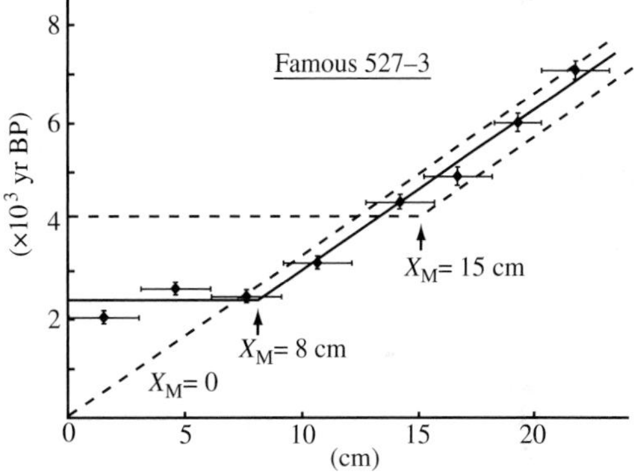

Figure 5 Model fit of ^{14}C distribution in FAMOUS core 527-3. The two dashed lines show the cases of $X_{\text{M}} = 0$ cm and $X_{\text{M}} = 15$ cm and the solid line is the case of $X_{\text{M}} = 8$ cm (source Nozaki *et al.*, 1977).

The organic fraction of deep-sea sediments may also be dated so long as the contribution of older detritus can be excluded. This is a problem in coastal and continental margin sediments primarily (Tanaka *et al.*, 1991). Dating of anoxic or suboxic continental margin sediments by ^{14}C has been discussed above. Generally, radiocarbon dating of organic carbon is applied to noncalcareous deposits where organic carbon is of sufficient abundance for dating. Turekian and Stuiver (1964) determined the rates of accumulation in the Argentine basin where sediments contained 0.8% C and virtually no calcium carbonate. The accumulation rate of detrital sediment was the highest observed for the deep ocean. The source of sediment, based on the quartz-surface texture

(Krinsley *et al.*, 1973), was the circum-Antarctic area by bottom transport rather than stream transport from South America.

(Goldberg and Bruland, 1974; Ku, 1976; Turekian and Cochran, 1978; Ivanovich *et al.*, 1992; Huh and Kadko, 1992).

6.12.5.2 ^{230}Th and ^{231}Pa

6.12.5.2.1 *The basic theory*

The use of ^{230}Th and ^{231}Pa in determining the chronology of deep-sea sediments is based on their production in the oceanic water column from the decay of uranium isotopes (Table 1) dissolved in seawater and their incorporation in the bottom deposit by strong adsorption on particle surfaces followed by the sedimentation of the particles. This combination of the processes leading to removal of particle-reactive radionuclides such as these from the water column is referred to as chemical scavenging (see Chapter 6.09). At the time of deposition, sediment contains an initial quantity N_0 of excess (unsupported) ^{230}Th or ^{231}Pa activity along with small quantities supported by decay of the parent uranium isotopes that are present. The decay of the excess nuclides with time and their burial is governed by Equation (1) and leads to an exponential decrease of the unsupported component as a function of the depth in the sediment.

The ^{230}Th and ^{231}Pa methods are not used directly for absolute dating of individual sedimentary horizons, because the assumption that N_0 is constant over time does not hold exactly but can be upset by fluctuations due to changes in sediment-deposition rate and other factors. Instead, the common practice is to use the decreasing activity with depth to derive an average rate of sediment accumulation over the length of a core or some other long interval. If t is the age (time since deposition) of a sediment horizon at depth z, and if S is the average sediment accumulation rate (thickness per unit time), then from Equation (1)

$$\ln N = -\frac{\lambda}{S}z + \ln N_0 \tag{7}$$

The customary practice is to plot $\ln N$ against z and by regression analysis determine the slope, from which S is calculated. From this result ages can be assigned to events recorded in a core, and uncertainties can be estimated from the regression statistics.

The ^{230}Th and ^{231}Pa methods can each be used independently, and concordance between the two improves confidence in the result. Because of the relative difficulty of analysis, the ^{230}Th method is more often applied alone. Sediment ages of up to 3.5×10^5 yr by the ^{230}Th method and 1.5×10^5 yr by the ^{231}Pa method can be determined. There are several good reviews containing more information on these methods (Goldberg and Bruland, 1974; Ku, 1976; Turekian and Cochran, 1978; Ivanovich *et al.*, 1992; Huh and Kadko, 1992).

6.12.5.2.2 *The underlying assumptions*

Use of the ^{230}Th and ^{231}Pa methods assumes closed-system behavior, i.e., no diffusional mobility that would change the slope of the $\ln N$ versus z plot. Because of the very strong adsorption of thorium and protactinium on solid phases, significant mobility is unlikely, and no evidence for it has been reported. The concordance between sediment accumulation rates from ^{230}Th and from ^{231}Pa (Ku, 1976) also argues against significant mobility.

Equation (7) assumes that N_0, the initial amount of ^{230}Th or ^{231}Pa per unit of sediment, remains constant over time. In evaluating this assumption, it is helpful to think of N_0 as the ratio between the flux of the radionuclide and the flux of the particles making up the sediment. For N_0 to remain constant, either the individual fluxes must both remain constant, or they must covary exactly. Because of the long-residence time of uranium in the oceanic water column (4.5×10^5 yr; Cochran, 1992), it is expected that its concentration in seawater should not change very much over the applicable time periods of the ^{230}Th and ^{231}Pa methods, and this is supported by the constancy of the uranium content of fossil corals (Broecker, 1971). Thus, the supply of the two decay products is unlikely to have changed significantly over these periods. Because of the strongly particle-reactive nature of both thorium and protactinium, they are removed from the water column by scavenging on very short timescales (~10–100 yr) compared to the radioactive half-lives of ^{230}Th (75,200 yr) and ^{231}Pa (32,700 yr), so that there is negligible loss by decay in the water column. Thus, the deposition rates of ^{230}Th and ^{231}Pa are, within a small fraction of 1%, equal to their supply rates and are very unlikely to have changed significantly over time. However, although this simple balance between supply and deposition must hold over the whole ocean, because of horizontal redistributions of the supply, it does not necessarily hold at any particular location, and this is one potential source of this variability in N_0.

A more significant source of variation in N_0 is change in particle flux, which has unquestionably occurred. Indeed, it is the variations in flux that are of greatest interest in paleoclimatic and paleoceanographic studies; the traditional ^{230}Th and ^{231}Pa methods do not resolve them but instead average over them. This point is taken up further in the next section.

6.12.5.2.3 Applications

In spite of their limitations, the ^{230}Th and ^{231}Pa methods have made an important contribution to the establishment of the Late Pleistocene chronology of deep-sea sediments. They provided the timescale upon which the deep-sea δ^{18}O record of global ice volume could be correlated with solar insolation, thus providing strong support for the astronomical theory of climate change (Broecker and Van Donk, 1970).

The more recent work with ^{230}Th and ^{231}Pa has been concerned less with their use to establish absolute chronology and more with the interpretation of their profiles in sediment cores to determine shorter-term variability in particle flux. Particle-flux measurements with sediment traps and other studies of the behavior of ^{230}Th and ^{231}Pa in the oceanic water column have resulted in a better understanding of the extent to which they can be laterally redistributed following their production. It has been shown that, over much of the ocean, the redistribution of ^{230}Th is minimal, so that the flux of particulate ^{230}Th to the seafloor is nearly in balance with the fixed rate of supply from ^{234}U integrated over the water column above (Yu *et al.*, 2001). This is in contrast to the behavior of ^{231}Pa, which shows a stronger tendency toward lateral redistribution and preferential deposition in areas of higher particle flux and higher rates of scavenging around ocean margins and other high-productivity areas (Yang *et al.*, 1986; Bacon, 1988; Walter *et al.*, 1999; Yu *et al.*, 2001).

If it is assumed, because of its constant rate of supply and the minimal potential for lateral redistribution, that the flux of particulate ^{230}Th to any point on the seafloor has remained constant with time, then any variability in N_0 must be due to changes in particle flux, and a simple inverse relationship should hold. This consideration has led to the development of ^{230}Th as a constant-flux reference tracer against which variations in the mass flux or the fluxes of individual sediment components can be measured (Bacon, 1984; Suman and Bacon, 1989; Francois *et al.*, 1990). Downcore profiles of excess ^{230}Th are measured and then converted to profiles of N_0 by removing the primary exponential trend or by decay corrections based on ages determined independently from radiocarbon dating or oxygen-isotope stratigraphy. The normalized flux (rain rate) F_i of any sediment component is then given by

$$F_i = \frac{\beta \cdot Z \cdot f_i}{^{230}\mathrm{Th}^0_{\mathrm{ex}}} \tag{8}$$

where f_i is the weight fraction of component i, $^{230}\mathrm{Th}^0_{\mathrm{ex}}$ is the activity of excess ^{230}Th decay corrected to the time of deposition, β is the constant rate of production of ^{230}Th from ^{234}U in the water column (2.63×10^{-5} dpm cm^{-3} kyr^{-1}), and Z is the water depth. The ^{230}Th profiling method has been applied to the studies of carbonate, opal, and clay sedimentation (Yang *et al.*, 1990; Francois and Bacon, 1991), pulsed inputs of ice-rafted debris known as Heinrich events (Francois and Bacon, 1994; Thomson *et al.*, 1995, 1999; McManus *et al.*, 1998), the flux of cosmogenic nuclides such as ^{10}Be (Anderson *et al.*, 1990; Frank *et al.*, 1995), and the flux of interplanetary dust particles to Earth as recorded by the ^{3}He content of deep-sea sediments (Marcantonio *et al.*, 1995, 1996, 1998, 1999, 2001).

The approach just described allows records of paleoflux to be inferred, but only to the extent that a component is preserved in the sediment. For example, a carbonate paleoflux determined by this method is the net flux, i.e., carbonate rain, to the seafloor minus dissolution after deposition. Because organic matter is so poorly preserved in deep-sea sediments, there is little possibility of using the method to arrive at an unambiguous record of organic productivity. However, the variations in ^{231}Pa flux and their correlation with variations in particle flux, which is the main factor causing the lateral redistribution of ^{231}Pa, has led to the proposed use of the ^{231}Pa/^{230}Th ratio in sediments as an indicator of past changes in biological productivity or particle export flux of surface waters. The theory is such that the ratio is preserved in the sediment even if the biogenic phases (organic matter, carbonate, opal) are remineralized. However, the ^{231}Pa/^{230}Th ratio in the particle flux depends on other variables such as horizontal transport and particle composition, and these complications may limit the usefulness of the approach. A review of this method and its limitations is given by Walter *et al.* (1999). It has been suggested that the ^{10}Be/^{26}Al ratio may prove to be a more reliable paleoproductivity indicator (Luo *et al.*, 2001), though many of the same considerations and limitations apply to its use.

6.12.5.2.4 Problems of erosion and focusing

Deep-sea sediments are not ideal accumulators of the pelagic rain of particles from above. Instead, the action of bottom currents can redistribute the arriving particles so that there is preferential winnowing from topographic highs and accumulation in lows. The accumulation of sediment at a rate that is greater than the local pelagic rain is often called "sediment focusing." Changes in the degree of focusing can cause changes in the rate of sediment accumulation at a given point on the seafloor that are not related to changes in supply by the pelagic rain.

If, as argued above, ^{230}Th can serve as a constant-flux reference, i.e., if its pelagic rain rate remains constant over time and is equal to the integrated production over the water column, then it is possible to quantify the degree of focusing in a core. A focusing factor ψ can be defined as follows:

$$\psi = \frac{\int_{z_2}^{z_1} [^{230}\mathrm{Th}_{\mathrm{ex}}^0] \cdot \rho_{\mathrm{b}} \mathrm{d}z}{\beta \cdot Z \cdot (t_2 - t_1)} \quad (9)$$

where ρ_b is the dry bulk density of the sediment, and t_1 and t_2 are the ages of horizons z_1 and z_2, which can be approximated from the depth profile of $^{230}Th_{ex}$ in the core or can be obtained by independent means such as radiocarbon dating or oxygen-isotope stratigraphy, and the other symbols are as defined before. The numerator on the right-hand side of Equation (9) is then the amount of $^{230}Th_{ex}$ that accumulated between t_2 and t_1, and the denominator is the amount expected from the known rate of supply. Ideally $\psi = 1$. Focusing is indicated if $\psi > 1$ and erosion if $\psi < 1$.

Suman and Bacon (1989) used this method to determine focusing factors over the past 10^4 years ranging from 4 to 13 on the Bermuda rise, a drift deposit where sediment-accumulation rates are unusually high because of lateral input of sediments by bottom currents of the Gulf Stream return flow. Marcantonio *et al.* (2001) used ^{230}Th to quantify sediment focusing in the equatorial Pacific and to show in this and earlier papers (e.g., Marcantonio *et al.*, 1996) that climate-related variations in the burial rate of extra-terrestrial ^{3}He are due mainly to the variations in sediment focusing and that the supply rate of the ^{3}He-bearing extraterrestrial dust is relatively constant. Thomas *et al.* (2000) have argued that the variations in sediment focusing in the equatorial Pacific may be caused not so much by variable bottom currents but more by variations in biological productivity of the upper ocean, which would have caused variation in the scavenging of fine particles and also in the flux of ^{230}Th. More work is needed to resolve the questions that they have raised, including an examination of the ^{231}Pa record in equatorial Pacific cores.

6.12.5.3 ^{10}Be

The cosmogenic nuclide ^{10}Be (half-life = 1.5 Myr) is a logical candidate for dating deep-sea deposits. The production in the atmosphere is primarily in the stratosphere. Its entry into the troposphere from the stratosphere occurs primarily around 40–50° latitude where tropopausal folding occurs. It differs in that regard from the short-lived ^{7}Be (half-life = 53 d), whose major flux to the ocean surface is from the troposphere and subject to latitudinal production variations due to the Earth's magnetic lines of force (greater production at the poles, least production at the equator). The deposition of ^{10}Be in deep-sea sediments is primarily controlled by efficiency of scavenging by particles as ocean circulation blurs the ^{10}Be concentration variations in ocean water to make it effectively independent of its locus of delivery (Turekian and Cochran, 1978). The mean-residence time of beryllium in the oceans is ~100 yr.

Early measurements (Tanaka *et al.*, 1977; Tanaka and Inoue, 1979) were made by radioactive counting but the use of accelerator mass spectrometry (Raisbeck *et al.*, 1979; Turekian *et al.*, 1979) has improved the quality and quantity of measurements. Tanaka and Inoue (1979) summarized their research on Pacific cores using the radioactive counting technique. Generally the results were compatible, in each core where it could be tested, with magnetic-reversal chronometry. The data up to 1978 have been reviewed by Turekian and Cochran (1978). They indicated that several factors control the deposition of ^{10}Be with time and location in the oceans. Tanaka and Inoue (1979) specifically showed that biological productivity strongly controlled the flux of ^{10}Be to the ocean bottom (Figure 6). Mangini *et al.* (1984), using accelerator mass spectrometry, studied a deep-sea core from the Pacific and showed that the rate of accumulation of sediments at the site of the core (GPC-3) varied as follows: 2 mm kyr^{-1} from 1.1 Myr to the present, 1.1 mm kyr^{-1} from 3.3 Myr to 1.1 Myr, and 0.5 mm kyr^{-1} from 3.3 Myr to ~14.5 Myr or longer. This range shows that the rate of deposition need not be constant over long time periods even where the viscissitudes of the glacial cycles were not operative. Mangini *et al.* (1984) also showed that the ^{10}Be flux to sediments was higher in the high-productivity upwelling areas of ocean margins as Tanaka and Inoue (1979) had shown for the equatorial Pacific.

The refinement of ^{10}Be measurements and the fact that its concentrations can also be measured in ice cores for the last several hundred thousand years allows the evaluation of its production variations over time as shown by Frank *et al.* (1995). They normalized the ^{10}Be concentration with respect to ^{230}Th to accommodate regional focusing of the two nuclides as the result of particle scavenging in much the same way as ^{3}He accumulation rate was calibrated (see discussion in next section). Sharma (2002) used this approach to show that the ^{10}Be flux changes, once corrected for the Earth's intrinsic magnetic moment changes,

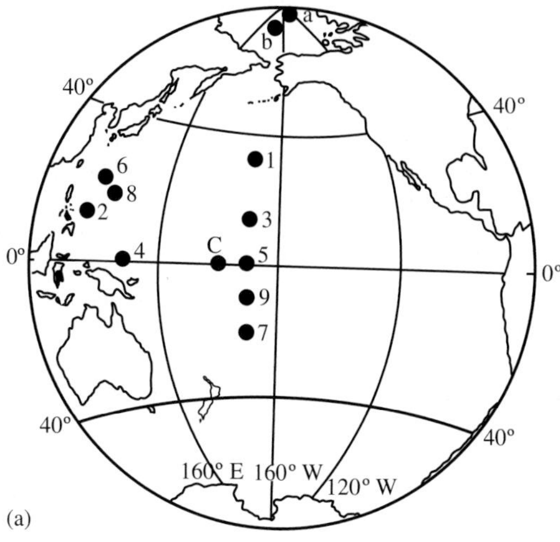

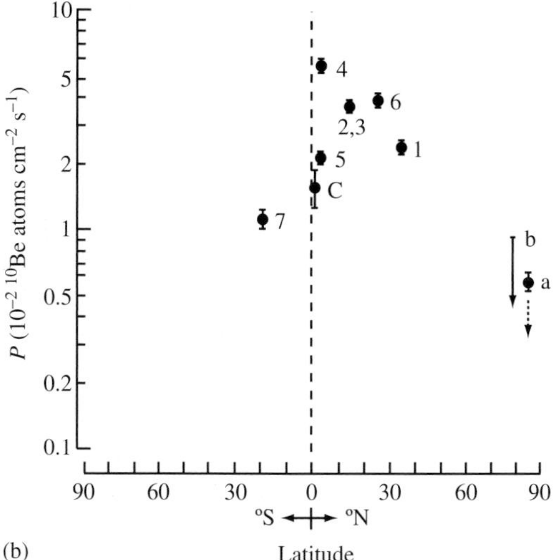

Figure 6 (a) Sampling locations for ^{10}Be measurements in Pacific deep-sea cores (source Tanaka and Inoue, 1979). (b) Latitudinal effect on the rates of ^{10}Be deposition (P) in the Pacific. The sites are those in Figure 6(a). Cores 8 and 9 are excluded because of large scatter of ^{10}Be concentrations with depth. Note the increased flux of ^{10}Be at the equator relative to other sites because of high biological productivity and the low values in the Arctic because of low productivity (source Tanaka and Inoue, 1979).

were related to solar-magnetic activity. As solar-magnetic activity is coupled to the photon flux, there should be a climate signal linked to the ^{10}Be flux signal, and Sharma (2002) showed that indeed there is a strong correlation in deep-sea sediments between δ^{18}O record and the ^{10}Be flux.

For long timescales the fluctuations in ^{10}Be flux are damped out by bioturbation in slowly accumulating sediments or averaged out in very slowly growing manganese nodules and crusts as discussed below.

6.12.5.4 ^{3}He

The accumulation of cosmic dust on Earth is generally assumed to be constant as it is not a function of any terrestrial process of enrichment or depletion as is known to be the case for cosmic rays (and therefore cosmic-ray produced nuclides such as ^{10}Be discussed above).

There are characteristic chemical properties of cosmic dust that have been involved in the study of sediment accumulation rates. The platinum group elements, such as iridium and osmium, offer good examples. Attempts to use iridium in this way have had the important result of indicating a giant meteorite impact at the Cretaceous–Tertiary boundary (Alvarez *et al.*, 1980) but it has not been proven important in determining chronometry.

The nuclide that has shown some promise is ^{3}He as mentioned above. Measurements of ^{3}He concentration in deep-sea sediments were first made by Ozima *et al.* (1984). Farley and Patterson (1995) measured ^{3}He concentrations in a deep-sea core and found that the concentration showed an apparent periodicity of 10^5 yr. Marcantonio *et al.* (1996) showed that monitoring the ^{3}He concentration with the ^{230}Th concentration produced independently at a constant rate in the ocean-water column (see above) indicated that indeed the ^{3}He flux was constant at least over the past 4.5×10^5 yr. The concentration changes were the result of both sediment accumulation and sediment focusing. The use of ^{230}Th helped to separate these two processes. On this basis, in areas not subject to focusing over long periods the ^{3}He concentration of the sediment would be a direct measure of the flux of cosmic dust and terrigenous sediment. If the cosmic-dust flux has been calibrated with ^{230}Th and if it has remained constant over the Cenozoic, then it can be used as a chronometer.

One such extended record has been obtained for the Cenozoic for a North Pacific clay core (GPC3) by Farley (1995). There is some indication that the flux of ^{3}He may have varied, perhaps due to meteorite impacts or possibly focusing, but the uncertainties in calibration of the ^{3}He flux by independent methods of sediment-accumulation rate determination are far from perfect and more study is called for. Nevertheless, the prospect of using a constant ^{3}He flux as a chronometric tool remains tantalizing. Farley (2001) provides a summary of results and concepts to the present time.

6.12.5.5 Volcanic Layers

Volcanic debris from explosive volcanism occurring at convergent plate boundaries can be

Table 2 Comparison of fission track ages and potassium–argon ages of volcanic material in deep-sea sediments.

Sample	*Location*	*K–Ar age* (10^6 yr)	*Magnetic reversal age* (10^6 yr)	*Fission track age* (10^6 yr)
V21-145 815 cm	34° 03′ N, 164° 50′ E	1.45 (±0.08)	1.4	1.47 (±0.16)
V21-173 725 cm	44° 22′ N, 163° 33′ W	1.62 (±0.08)	1.6	1.62 (±0.17)
EM 8-13	28° 59′ N, 117° 30′ W	11.4 (±0.6)		10.50 (±0.6)

Source: MacDougall (1971).

deposited in deep-sea sediments. As volcanic-ash layers provide the opportunity of dating strata by a number of radiometric methods.

Dymond (1969) obtained four sediment cores from the Pacific that showed the presence of volcanic layers and dated them by the ^{230}Th method, calibrated paleomagnetic normal/reverse stratigraphy and potassium–argon dating of volcanic fragments. In one core (V19-153, 8° 51′ S, 102° 07′ E) he obtained ages progressing down the length of the core from 65 cm to 690 cm that spanned K/Ar datable time intervals from 6×10^4 yr to 1.84 Myr.

Macdougall (1971) used fission-track dating of glass shards to determine the ages of volcanic layers in deep-sea sediments. He compared his results to K/Ar dates and showed that both methods gave identical results (Table 2).

The volcanic layer dating has been extended by stratigraphic correlation of diagnostic chemical imprints of volcanic ash dated on land adjacent to the deep-sea sediments of eastern Africa (Brown *et al.*, 1992).

6.12.5.6 Extension of Dating Techniques

6.12.5.6.1 *The Milankovitch cycles and chronology*

The periodicities of the Milankovitch orbital forcing on environmental parameters has now been established through deep-sea sediment records. The primary proxy has been δ^{18}O variations in foraminifera, but relative abundance of marine species has also been shown to have the periodicities ascribable to the Milankovitch pattern. Since the periods of the three major components of the Milankovitch cycle, precession, obliquity and eccentricity, have fixed values of $\sim 2 \times 10^4$ yr, 4×10^4 yr, and 10^5 yr, respectively, this pattern can be tracked through a sedimentary record and provide a precise chronology once the periods have been well established and certified in the datable parts of cores. Bassinot *et al.* (1994) showed the value of the procedure by dating the Brunhes/Matuyama magnetic-reversal boundary. At first the Milankovitch reconstruction seemed to give a higher value (7.7×10^5 yr BP) than permissible by the then available K/Ar dates, but subsequent, more precise dating using the ^{40}Ar/^{39}Ar method showed that the boundary was actually 7.7×10^5 yr BP as inferred from the Milankovitch reconstruction. Using this technique, dating has been extended through the Oligocene (Shackleton *et al.*, 2000).

6.12.5.6.2 *Oxygen and carbon isotopes in carbonate tests*

Ocean-drilling campaigns have provided deep-sea sediment cores ranging through the Cenozoic era. These cores often provide continuous records of the changing oxygen and carbon isotope signature of the oceans and the temperature of deposition of the carbonate test. Prior to ~33.5 Myr BP there were no large ice caps so that all the oxygen isotopic variations are presumed to be due to the temperature of the ocean in which the foraminiferan grew. Generally, if the bottom waters of the oceans had a high-latitude source the oxygen-isotope signature would reflect water colder than surface waters. At ~33.5 Myr BP, the formation of the Antarctic ice cap sequestered water with a light oxygen-isotope signature. This accumulation of water of low δ^{18}O drove the δ^{18}O of the oceans heavier. The climatic cooling associated with this event also cooled the oceans and resulted in a heavier signature in the foraminifera from the temperture effect as well. This situation was further enhanced as the result of the development of the northern hemisphere ice caps since ~2.7 Myr BP. These changes in the δ^{18}O recorded by foraminifera provide a unique and diagnostic pattern that can be used as a chronostratigraphic tool.

In the case of carbon isotopes the δ^{13}C is the complex result of the relative importance of organic carbon sequestration or release by weathering compared to calcium carbonate deposition since the former has a δ^{13}C of about −25 and the latter a δ^{13}C of ~0. A further effect can be the episodic release of extremely light carbon (δ^{13}C ~ −75) in the form of methane. Methane is known to be held in clathrate structures in sediments under high-productivity areas at depths and temperatures characteristic of their formation (Dickens *et al.*, 1995). The carbon isotopic record through the Cenozoic combined with the oxygen-isotope record provides a chronostratigraphic tool for dating core sequences (Figure 7).

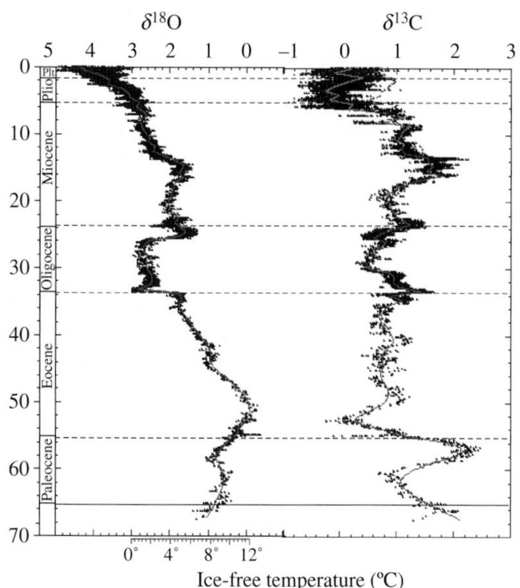

Figure 7 Global deep-sea oxygen and carbon isotope records based on data from more than 40 DSDP and ODP sites (source Zachos *et al.*, 2001).

There have been attempts to use the oxygen-isotope pattern in fish debris to date cores from below the carbonate compensation depth, which are devoid of calcium carbonate tests because of dissolution. Fish teeth are composed of apatite and the phosphate retains its initial oxygen signature since that can be affected seriously only by enzymatic recycling of the phosphate. The dating is established using the pattern of oxygen-isotope variation with time shown in Figure 7. A preliminary attempt to do just that was made on a red-clay core from the north Pacific (GPC3 discussed above in another context) by Blake and Turekian (2000). The analysis was able to discern the characteristic pattern of oxygen isotopes imprinted in a biologically derived test that is in equilibrium with seawater at the ambient temperature of growth. The method has the benefit of providing not only a chronostratigraphic tool but also independent records of surface-ocean temperatures of the times prior to 35 Myr BP (assuming that the fish spent most of their lives in the mixed layer).

6.12.5.6.3 *Magnetic-reversal stratigraphy*

Cande and Kent (1992) have compiled a magnetic stratigraphy for the Earth as recorded from deep-sea sediments and ocean-floor magnetic intensity patterns. As the sequence is accurately dated by independent means, it can provide a chronostratigraphic tool in the absence of other chronometers. Berggren *et al.* (1995) published an integration of bio-magnetostratigraphy with $^{40}Ar/^{39}Ar$ radiometric dates. There has been a joining of the orbital forcing, volcanic dating, and magnetic-reversal stratigraphy to date the Brunhes-Matuyama reversal boundary precisely (Shackleton *et al.*, 1993).

The Fe/Ca ratio in an ODP core from the western North Atlantic has been used as an index of orbitally driven climate change allowing deconvolution of the orbital forcing pattern and extension back to the Late Eocene. In this time period the timing of orbital pattern was used for fine-tuning of the magnetic-reversal record (Palike *et al.*, 2001).

6.12.5.6.4 *Element accumulation: Titanium and Cobalt*

There have been two attempts to use elemental measurements to determine accumulation rates of marine deposits independent of the cosmic dust ^3He approach or the tacitly assumed constant-flux model used in the ^{230}Th and ^{231}Pa approaches. One depends on a trace metal that tracks the fine-grained (clay) fraction of deep-sea sediments, the other on the addition of a hydrogenous element to the accumulating marine deposit.

The so-called titanium method used by Arrhenius *et al.* (1951) assumes a so-called "lutite veil" depositing at a constant rate throughout the deep oceans. Titanium associated with this lutite fraction then is also assumed to accumulate at a constant rate. Variations in the accumulation rates of biogenic components can then be assessed. The initial titanium method was calibrated by a single radiocarbon date as discussed above. The method has not been used since it was discovered that accumulation rates of all detrital and biogenic components of deep-sea sediments are subject to change as a function of climatic history and focusing processes.

Another element used on the assumption of constant accumulation is cobalt. This element is used on the assumption that a component of the sediment contains this element by addition from a hydrogenous source. That is, it is scavenged from seawater as sedimentation occurs but the removal is at a constant rate. The element cobalt was first suggested for ferromanganese oxide crusts (Halbach *et al.*, 1983). It was used for a deep-sea sediment core by Kyte *et al.* (1993). They calibrated the imported constant cobalt addition to the sediments with time by stratigraphically dated horizons such as the 65 Myr old iridium anomaly, due to a large meteorite impact, identified as marking the Cretaceous–Tertiary boundary. They also used epoch boundaries from fish-debris stratigraphy. The assumption of constant cobalt addition to deep-sea sediments assumes that the concentration of cobalt in the oceans has been invariant with time and that the mechanism of sequestration is time independent. These assumptions are difficult to assess. In the absence of any other dating tool, however, this approximation may sometimes be useful.

6.12.6 FERROMANGANESE DEPOSITS

6.12.6.1 Applicable Methods

Iron- and manganese-rich deposits on the seafloor are widespread and occur as nodules (the commercially important type), crusts, or thin coatings on rocks. They form authigenically by precipitation of metals from seawater. As they accrete, they incorporate from seawater significant quantities of radionuclides, including ^{230}Th, ^{231}Pa, and ^{10}Be, which allow the determination of their age or rate of growth (accretion). In all cases, the age determination is based on the decay of unsupported radioactivity and is governed by Equation (1). As with sediments, the usual practice is to measure the activity as a function of depth and apply Equation (3) to obtain an average growth rate over depth intervals corresponding to ~1–4 half-lives of the radionuclide. Each of the three radionuclides can be applied individually as an independent method, and variants based on the ratios ^{230}Th/^{232}Th, ^{231}Pa/^{230}Th, or ^{10}Be/^{9}Be have often been employed. The long half-life of ^{10}Be, allowing ages of several million years to be determined, makes this nuclide especially attractive for dating slowly growing deep-sea ferromanganese deposits.

Ferromanganese nodules and crusts often grow around a nucleus or upon a substrate consisting of volcanic minerals or glass, which are well suited for dating by the K–Ar method, and provide an alternate approach to determine their age (Barnes and Dymond, 1967). If it is assumed that accretion of the deposit began just after the formation of the nucleus or substrate, then the average rate of accretion over its lifetime can be estimated. This method can give only a maximum age of the deposit, or a minimum rate of accretion, because of the unknown time by which the onset of accretion might have been delayed or dissolution might have occurred (Ku, 1977). Fission-track dating has also been used in a similar way (Aumento, 1969).

A limited amount of work has been done on the dating of shallow-water nodules, which, because of diagenetic remobilization of manganese within the sediment column, generally form more rapidly than deep-ocean deposits. Unlike deep-water nodules, the rapidly growing shallow-water nodules do not contain excess ^{230}Th or ^{231}Pa but instead contain, initially, quantities of ^{230}Th and ^{231}Pa that are less than their equilibrium activities. The ingrowth over time of ^{230}Th and ^{231}Pa toward equilibrium with ^{234}U and ^{235}U provides the basis for dating these deposits (Ku and Glasby, 1972), as it does for corals (see below). An important limitation of the uranium-series methods for dating shallow-water nodules is that not all of the ^{230}Th or ^{231}Pa are necessarily produced by decay *in situ*. Significant amounts may be present initially, so that the derived ages must be interpreted as maximum values. Phosphorites are another type of deposit that can be dated by the ingrowth method (Burnett and Veeh, 1977; Burnett *et al.*, 1982, 2000; Kress and Veeh, 1980; Roe *et al.*, 1983; Kim and Burnett, 1986).

Ku (1977) published an excellent review of the earlier literature on growth rates of ferromanganese deposits.

6.12.6.2 The Underlying Assumptions

As is the case with deep-sea sediments, all of the dating methods based on decay of unsupported radionuclides assume that N_0—the initial amount of ^{230}Th, ^{231}Pa, or ^{10}Be per unit of deposit—remains constant over time. In general, this requires that both the accretion rate of the deposit and the uptake rate of the radionuclide from seawater have remained constant. The various ratio methods are used primarily so that the latter assumption can be relaxed, the theory being that the variations in the uptake rate of two isotopes, ^{230}Th and ^{232}Th, e.g., would tend to cancel each other out. Because of variations in seawater chemistry over time, it is not likely that the assumption of constant N_0, or of the initial ratios, is strictly true, but the linearity generally observed in $\ln N$ versus z plots indicates that it is a sufficiently good approximation that valid estimates of average accretion rates over long time intervals can be obtained.

The ^{230}Th, ^{231}Pa, and ^{10}Be methods are all based on a concentration gradient with depth below the outer surface of the deposit, which, in principle, would drive a diffusion that would reduce the gradient and lead to apparent growth rates that are too high. This was a controversial point in some of the earlier literature, but recent work has shown quite convincingly that ^{230}Th and ^{10}Be have very low effective diffusivities in ferromanganese crusts and may be regarded as essentially immobile (Mangini *et al.*, 1986; Chabaux *et al.*, 1997; Henderson and Burton, 1999). Because of the inferred large distribution coefficient for uptake of protactinium from seawater by manganese oxides (Anderson *et al.*, 1983), it is likely that a similar immobility (closed-system behavior) applies for ^{231}Pa as well.

6.12.6.3 Applications

Probably the most important contribution of radiometric dating to the study of deep-sea ferromanganese deposits was the establishment of their very slow growth rates, measured in millimeters per *million* years (Ku, 1977; Burnett and Morgenstein, 1976; Guichard *et al.*, 1978; Krishnaswami and Cochran, 1978; Ku *et al.*, 1979; Moore *et al.*, 1981; Krishnaswami *et al.*, 1982; Huh and Ku, 1984), though manganese crusts from sites near seafloor spreading centers, close to hydrothermal sources of manganese, can grow considerably faster (Moore and Vogt, 1976). The generally slow growth is in contrast to the more rapid accumulation of deep-sea sediments, which is measured in millimeters (or more) per *thousand* years, and it raises the problem of explaining how manganese nodules avoid burial and remain at the sediment surface. Possibilities include the action of bottom currents or episodic nudging by benthic animals, but the exact mechanism is still not understood.

More detailed sampling has revealed discontinuities in the radionuclide-depth profiles indicating episodic growth (Krishnaswami and Cochran, 1978; Eisenhauer *et al.*, 1992) or relatively sudden changes in rate of growth (Krishnaswami *et al.*, 1982; Mangini *et al.*, 1986). Variations in the rate of growth over time may explain many of the discordant results that have been obtained when ^{230}Th, ^{231}Pa, and ^{10}Be profiles have been compared on the same samples (Krishnaswami *et al.*, 1982), because each radionuclide averages over a different length of time. Comparison of radionuclide distributions, including those of ^{226}Ra, between the top and bottom surfaces of nodules has given evidence they have rolled over at times in the past (Krishnaswami and Cochran, 1978; Huh and Ku, 1984; Moore, 1984). The inferred times between rollover "events" range from 10^3 yr to 10^5 yr.

The more recent work with ferromanganese deposits has focused on further resolving shorter-term variations in their rate of growth (Mangini *et al.*, 1990; Eisenhauer *et al.*, 1992), and there is growing interest in their use as recorders of past changes in seawater chemistry (Huh and Ku, 1990). These studies have focused on crusts more than nodules. The nodules are suspect because of their close contact with the sediments and possible diagenetic supply of metals, and it is believed that crusts, which form, e.g., on the flanks of seamounts away from the sediments, provide a more purely hydrogenous deposit and thus a more accurate record of variations in seawater composition.

Mangini *et al.* (1990) and Eisenhauer *et al.* (1992) have proposed that a constant ^{230}Th flux model, analogous to the one described above for deep-sea sediments, be applied to ferromanganese deposits to determine short-term variations in growth rate. Depth profiles of ^{230}Th are measured and converted to profiles of N_0 by removing the primary exponential trend, and Equation (4) is applied to obtain point-by-point profiles of growth rate. Eisenhauer *et al.* (1992) also showed that very high resolution sampling can be achieved by selecting crusts that have a very flat, laminar structure, and they obtained a resolution of 0.02 mm, corresponding to a time ~5,000 yr, in two crusts from the Pacific Ocean. Both crusts showed systematic variations in ^{230}Th concentrations, which they correlated with the Late Pleistocene glaciation cycle and interpreted as higher growth rates during the post-glacial and the last interglacial and lower growth rates (or growth hiatuses) during glacials. It is clear from this study that such high-resolution records of varying composition contain valuable information, but it remains to be seen whether the constant ^{230}Th flux assumption can be justified. In contrast to the ^{230}Th flux to the sediments, which is limited to a value equal to its rate of supply, the ^{230}Th flux to ferromanganese deposits is typically only 10–20% of the total supply from the water column, so it is far less certain that it would remain constant over time. Chabaux *et al.* (1997) have shown evidence for significant variations in the Th/U and ^{230}Th/^{232}Th ratios recorded in ferromanganese crusts over the past 1.5×10^5 yr.

6.12.7 CORALS

One of the most important applications of uranium-series methods of age determination has been the dating of fossil corals and other carbonate materials. In contrast to deep-sea sediments, which accumulate excess ^{230}Th and ^{231}Pa that decay over time, carbonates accumulate uranium by co-precipitation from seawater that is essentially free of ^{230}Th and ^{231}Pa. The radioactive ingrowth of ^{230}Th and ^{231}Pa over time toward secular equilibrium with ^{238}U and ^{235}U is the basis of the two methods.

With current measurement techniques, ages as great as 550 ka (^{230}Th) and 200 ka (^{231}Pa) can be determined accurately. The equation for the ingrowth of ^{230}Th is

$$\frac{^{230}\text{Th}}{^{238}\text{U}} = 1 - e^{-\lambda_{230}t} + \left(\frac{^{234}\text{U}}{^{238}\text{U}} - 1\right)\left(\frac{\lambda_{230}}{\lambda_{230} - \lambda_{234}}\right) \times (1 - e^{-(\lambda_{230}-\lambda_{234})t}) \tag{10}$$

where λ_{230} and λ_{234} are the decay constants of ^{230}Th and ^{234}U and the ratios are activity ratios measured in a sample of age t. For ingrowth of

^{231}Pa, the equation is

$$\frac{^{231}\text{Pa}}{^{235}\text{U}} = 1 - e^{-\lambda_{231}t} \quad (11)$$

where λ_{231} is the decay constant of ^{231}Pa. The terms in Equation (10) involving the ^{238}U/^{234}U ratio are necessary because of the ~15% excess of ^{234}U in seawater caused by the preferential mobility of ^{234}U in natural waters due to alpha recoil during the decay of ^{238}U. The decay of excess ^{234}U provides another independent chronometer if it can be assumed that the initial excess (the seawater value) remains constant over time

$$\frac{^{234}\text{U}}{^{238}\text{U}} = 1 + \left(\left[\frac{^{234}\text{U}}{^{238}\text{U}}\right]_0 - 1\right)e^{-\lambda_{234}t} \quad (12)$$

(decay of ^{238}U and ^{235}U over the times of interest are negligible). Figure 8, based on Equations (10)–(12), shows how the ^{234}U excess and the ^{230}Th/^{238}U and ^{231}Pa/^{235}Pa activity ratios evolve with time in a closed system.

Until the early 1980s, the development and application of uranium-series methods were based on measurements by alpha counting, which is limited in its precision and sensitivity by the slow disintegration rates (low count rates) of the radionuclides of interest. Beginning in the late 1980s, measurement techniques based on thermal-ionization mass spectrometry (TIMS) were developed (Edwards *et al.*, 1986–1987), markedly improving both precision and sensitivity. Because of the higher precision, dating of older materials which requires measurement of small departures from equilibrium ratios (Figure 8) is possible, and because of the higher sensitivity smaller and younger samples can be dated. The earlier work based on alpha counting is well reviewed by Ku (1976), and a later review by Burnett and Veeh (1992) includes an extensive discussion of marine phosphorite dating, whose problems are analogous to those of carbonate dating. An account of the development of mass-spectrometric methods is given by Chen *et al.* (1992).

Equations (10) and (11) assume that initial ^{230}Th and ^{231}Pa concentrations are zero or can be corrected for. The ^{232}Th/^{238}U ratio in surface corals is similar to that in seawater, suggesting that thorium does not fractionate from uranium significantly during coral growth (Edwards *et al.*, 1986–1987). Because of the extremely low ^{230}Th/^{238}U ratio in seawater (activity ratio ~10^{-5}), initial ^{230}Th must be negligible. Concordance between ^{230}Th and ^{231}Pa ages indicates that initial ^{231}Pa must also be negligible (Edwards *et al.*, 1997).

As with all radiometric dating methods in applying Equations (10)–(12), a critical requirement, and often the one most difficult to satisfy, is that closed-system behavior has been maintained over the time interval of interest. The ^{234}U/^{238}U ratio is often used as a means of testing the closed-system assumption and assessing the reliability of dates based on the ^{230}Th/^{238}U ratio. If the initial ratio, as inferred from the ^{230}Th/^{238}U age, departs from the seawater value, it can be concluded that the system has been disrupted and that the measured age is unreliable. Use of the ^{234}U/^{238}U ratio for this purpose assumes that the initial ratio (the seawater value) has remained constant, and this has been borne out in recent studies using high-precision mass-spectrometric measurements, at least for the past 360 ka (Henderson *et al.*, 1993; Gallup *et al.*, 1994; Henderson, 2002), though the possibility of small glacial-to-interglacial

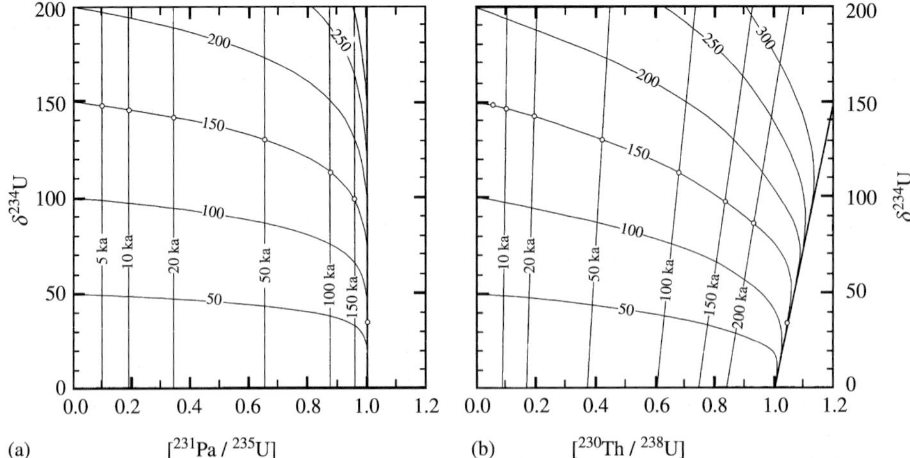

Figure 8 Plots of: (a) δ^{234}U versus ^{231}Pa/^{235}U activity ratio and (b) δ^{234}U versus ^{230}Th/^{238}U activity ratio. The δ expression is defined from the uranium isotope activity ratio as follows: $\delta^{234}\text{U} = (^{234}\text{U}/^{238}\text{U} - 1) \times 1{,}000$, modern seawater having a value of ~145 (Henderson, 2002). The nearly vertical lines are contours of constant age, and the curves running from left to right are contours of constant-initial δ^{234}U (source Cheng *et al.*, 1998).

variations in the ratio cannot be ruled out completely (Henderson, 2002). Checks for concordancy between $^{230}Th/^{238}U$ and $^{231}Pa/^{235}U$ ages are another means of detecting open-system behavior.

Unaltered fossil corals that are free of contaminating materials have proven in general to approximate closed-system behavior closely enough to yield reliable uranium-series ages. This has not, unfortunately, proven to be the case for fossil marine-mollusk shells, which typically contain secondary uranium from the environment that was not present when the animal was alive and also show frequent evidence of extraneous ^{230}Th (Kaufman *et al.*, 1971). The availability of high-precision mass-spectrometric measurements of ^{231}Pa as well as the uranium and thorium isotopes now makes it possible to detect more subtle departures from closed-system behavior in coral samples and other materials, thus providing more stringent tests for validity of $^{230}Th/^{238}U$ and $^{231}Pa/^{235}U$ dates and possibly allowing corrections to be derived for certain types of open-system behavior. Cheng *et al.* (1998) have derived in detail the systematics of U–Th–Pa dating for several types of open-system behavior as well as for closed systems.

One of the early triumphs of uranium-series dating was the use of ^{230}Th to establish the chronology of raised coral terraces on Barbados (Broecker *et al.*, 1968; Mesolella *et al.*, 1969) and New Guinea (Veeh and Chappell, 1970). The dates obtained for high sea-level stands (interglacials) provided strong support for the Milankovitch astronomical theory of climate relating ice volume with solar insolation. The more precise measurements and the more stringent concordancy checks obtained by TIMS have given confirmation in several locations and have also provided for more exact correlation with astronomical calculations and for resolution of earlier disputes (Edwards *et al.*, 1987; Gallup *et al.*, 1994; Muhs and Szabo, 1994; Stirling *et al.*, 1995; Szabo *et al.*, 1994; Bard *et al.*, 1996). Precise ^{230}Th dating of submerged corals collected by drilling have provided the best record yet of sea level since the last glacial maximum (Bard *et al.*, 1996; Edwards *et al.*, 1993).

The availability of precise ^{230}Th ages for younger corals allows a comparison with ^{14}C ages. Bard *et al.* (1990) used coral cores raised off Barbados to calibrate the ^{14}C timescale over the past 3×10^4 yr, extending the calibration beyond the range of dendrochronology, which is limited to the past ~9,000 yr. The ^{14}C ages were systematically younger than the ^{230}Th ages by as much as 3,500 yr at 20 kyr BP, indicating a significantly higher $^{14}C/^{12}C$ ratio in the atmosphere at the time of the last glacial maximum, which they attributed to changes in cosmogenic-nuclide production rate linked to changes in the strength of Earth's magnetic field. Dating of younger corals has also provided a chronology for other paleoclimatic reconstructions such as sea–surface temperature records based on oxygen isotope or Sr/Ca paleothermometry (Beck *et al.*, 1992; Guilderson *et al.*, 1994; McCulloch *et al.*, 1996).

Most of the work with corals has been with reef-building corals, which grow very near the sea surface. An important development, however, is the demonstration that reliable ^{230}Th ages (Adkins *et al.*, 1998; Cheng *et al.*, 2000) and ^{231}Pa ages (Goldstein *et al.*, 2001) can also be obtained on deep-sea corals, and the new field of deep-water coral paleoceanography has begun to grow from this. Apparent ^{14}C ages, coupled with ^{230}Th ages, can be used to infer past variations in the ocean's circulation and rate of convective overturning (Adkins *et al.*, 1998; Goldstein *et al.*, 2001). A difficult problem with deep-sea corals not usually encountered with surface corals is the often serious contamination with extraneous ^{230}Th and ^{231}Pa from manganese oxide coatings or other adsorbing surfaces. It can be overcome by a combination of careful cleaning of samples to remove noncarbonate phases and the use of appropriate corrections based on: (i) ^{232}Th measurements and an assumed $^{230}Th/^{232}Th$ ratio in the contaminating component (Adkins *et al.*, 1998) or (ii) on a whole-rock isochron method that assumes a two-component mixing of carbonate and contaminant (Lomitschka and Mangini, 1999; Cheng *et al.*, 2000; Goldstein *et al.*, 2001).

6.12.8 METHODS NOT DEPENDING ON RADIOACTIVE DECAY

6.12.8.1 Amino Acid Racemization

Chemical production of optically active amino acids yields equal numbers of left-handed and right-handed molecules. The mixture is called racemic. Biological amino acids are all left handed or L-enantiomers. With time the initially L-enantiomers are reconfigured to approach a racemic mixture. The kinetics are primarily determined by temperature but chemical environment can also play a part. In a constant-temperature, constant-chemistry environment of the top 5 m of a deep-sea sediment, the gradual racemization of the amino acid being studied provides a chronometer. Bada *et al.* (1970) analyzed a core from the Atlantic Ocean using the transformation of isoleucine to alloisoleucine, its racemization product, to establish a chronometry (Figure 9). Amino acid racemization dating depends on knowing the temperature history of the sediment pile as sediment accumulates and is subject to errors due to terrestrial heat flow.

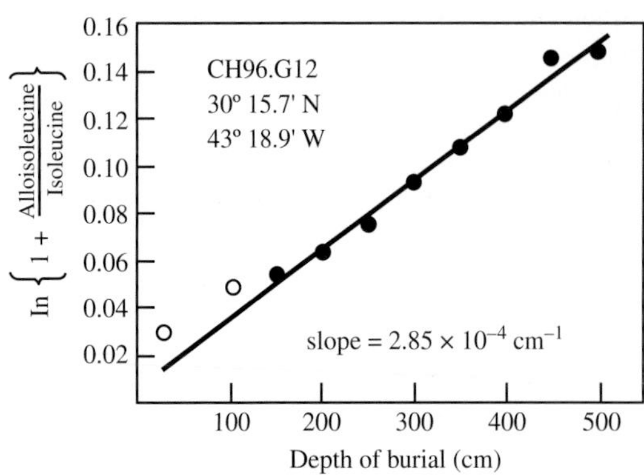

Figure 9 Plot of the extent of racemization of isoleucene against depth in an Atlantic deep-sea sediment core (source Bada *et al.*, 1970).

Further studies using this tool have not been pursued.

6.12.8.2 Thermoluminescence

When quartz or other suitable mineral detector is deposited into a matrix of minerals containing the radioactive nuclides of the ^{238}U, ^{235}U, and ^{232}Th decay series, and ^{40}K, the detector mineral is subject to radiation damage. If the grain had been cleared of all memory of radiation damage prior to deposition, then the extent of damage is a function of the time the detector mineral has been immersed in the radiation-producing matrix.

The method of assessing is commonly by thermoluminescence, wherein the number of photons released during the annealing process or heating are detected by a phototube. The procedure was developed for pottery where the kilning process annealed the detector mineral and therefore provided information on the time since annealing. An alternative method is to use ESR.

In the case of marine deposition, the transport of particles through the air results in the annealing by ultraviolet radiation from the Sun. The reset minerals then act as accumulators of lattice damage once they are buried. The technique has been established for wind-blown deposits such as dunes and loess deposits. Wintle and Huntley (1979) applied it to deep-sea sediments, building on initial studies by Huntley and Johnson (1976). Subsequently, there has not been a great deal of work on deep-sea sediments, and most of the interest has shifted to sand dunes and loess deposits where the method provides a unique dating tool.

ACKNOWLEDGMENTS

Ellen Thomas and Gwyneth Williams provided thoughtful reviews of this chapter. We thank them, with the caveat that they are not responsible for our persistent shortcoming.

MPB is grateful for generous financial support provided by the US Department of Energy (most recently through its Ocean Carbon Sequestration Research Program, Biological and Environmental Reasearch grant #DE-FG02-00ER63020), and the US National Science Foundation (most recently through its Chemical Oceanography Program, grant #OCE-0117922).

REFERENCES

Adkins J. F., Cheng H., Boyle E. A., Druffel E. R. M. and Edwards R. L. (1998) Deep-sea coral evidence for rapid change in ventilation of the deep North Atlantic 15,400 years ago. *Science* **280**, 725–728.

Alvarez L. W., Alvarez W., Asaro F., and Michel H. V. (1980) Extraterrestrial cause for the Cretaceous–Tertiary extinction. *Science* **208**, 1095–1108.

Anderson R. F., Bacon M. P., and Brewer P. G. (1983) Removal of ^{230}Th and ^{231}Pa at ocean margins. *Earth Planet. Sci. Lett.* **66**, 73–90.

Anderson R. F., Lao Y., Broecker W. S., Trumbore S. E., Hofmann H. J., and Wolfli W. (1990) Boundary scavenging in the Pacific Ocean: a comparison of ^{10}Be and ^{231}Pa. *Earth Planet. Sci. Lett.* **96**, 287–304.

Arrhenius G., Kjellberg G., and Libby W. F. (1951) Age determination of Pacific chalk ooze by radiocarbon and titanium content. *Tellus* **3**, 222–229.

Aumento F. (1969) The Mid-Atlantic Ridge near 45° N. V. Fission track and ferro-manganese chronology. *Can. J. Earth Sci.* **6**, 1431–1440.

Bacon M. P. (1984) Glacial to interglacial changes in carbonate and clay sedimentation in the Atlantic Ocean estimated from ^{230}Th measurements. *Isotope Geosci.* **2**, 97–111.

Bacon M. P. (1988) Tracers of chemical scavenging in the ocean: boundary effects and large-scale chemical fractionation. *Phil. Trans. Roy. Soc. London A* **325**, 147–160.

Bada J. L., Luyendyk B. P., and Maynard J. B. (1970) Marine sediments: dating by the racemization of amino acids. *Science* **170**, 730–732.

Bard E., Hamelin B., Fairbanks R. G., and Zindler A. (1990) Calibration of the ^{14}C timescale over the past 30,000 years using mass spectrometric U–Th ages from Barbados corals. *Nature* **345**, 405–410.

Bard E., Jouannic C., Hamelin B., Pirazzoli P., Arnold M., Faure G., Sumosusastro P., and Syaefudin (1996) Pleistocene sea levels and tectonic uplift based on dating of corals from Sumba Island, Indonesia. *Geophys. Res. Lett.* **23**, 1473–1476.

Barnes S. S. and Dymond J. R. (1967) Rates of accumulation of ferromanganese nodules. *Nature* **213**, 1218–1219.

Bassinot F. C., Labeyrie L. D., Vincent E., Quidelleur X., Shackleton N. J., and Lancelot Y. (1994) The astronomical theory of climate and the age of the Brunhes-Matuyama magnetic reversal. *Earth Planet. Sci. Lett.* **126**, 91–108.

Beck J. W., Edwards R. L., Ito E., Taylor F. W., Recy J., Rougerie F., Joannot P., and Henin C. (1992) Sea-surface temperature from coral skeletal strontium/calcium ratios. *Science* **257**, 644–647.

Benninger L. K., Aller R. C., Cochran J. K., and Turekian K. K. (1979) Effects of biological sediment mixing on the ^{210}Pb chronology and trace metal distribution in a Long Island Sound sediment core. *Earth Planet. Sci. Lett.* **43**, 241–259.

Benoit G. J., Turekian K. K., and Benninger L. K. (1979) Radiocarbon dating of a core from Long Island Sound. *Estuar. Coast. Mar. Sci.* **9**, 171–180.

Berggren W. A., Kent D. V., Swisher C. C., III, and Aubry M.-P. (1995) A revised Cenozoic geochronology and chronostratigraphy. *Soc. Sed. Geol. Spec. Publ.* **54**, 129–212.

Blake R. E. and Turekian K. K. (2000) Phosphate oxygen isotope composition of fish debris as a chronostratigraphic tool: results from LL44-GPC3 a Pacific red clay core. *EOS* **81**(suppl. 2), F707.

Broecker W. S. (1971) A kinetic model for the chemical composition of sea water. *Quat. Res.* **1**, 188–207.

Broecker W. S. and Van Donk J. (1970) Insolation changes, ice volumes, and the O^{18} record in deep-sea cores. *Rev. Geophys. Space Phys.* **8**, 169–198.

Broecker W. S., Turekian K. K., and Heezen B. C. (1958) The relation of deep-sea sedimentation rates to variations in climate. *Am. J. Sci.* **256**, 503–517.

Broecker W. S., Thurber D. L., Goddard J., Ku T.-L., Matthews R. K., and Mesolella K. J. (1968) Milankovitch hypothesis supported by precise dating of coral reefs and deep-sea sediments. *Science* **159**, 297–300.

Brown F. H., Sarna-Wojcicki A. M., Meyer C. E., and Haileab B. (1992) Correlation of Pliocene and Pleistocene tephra layers between the Turkana Basin of East Africa and the Gulf of Aden. *Quat. Int.* **13/14**, 55–67.

Burnett W. C. and Morgenstein M. (1976) Growth rates of Pacific manganese nodules as deduced by uranium-series and hydration-rind dating techniques. *Earth Planet. Sci. Lett.* **33**, 208–218.

Burnett W. C. and Veeh H. H. (1977) Uranium-series disequilibrium studies in phosphorite nodules from the west coast of South America. *Geochim. Cosmochim. Acta* **41**, 755–764.

Burnett W. C. and Veeh H. H. (1992) Uranium-series studies of marine phosphates and carbonates. In *Uranium-series Disequilibrium: Applications to Earth, Marine, and Environmental Sciences* (eds. M. Ivanovich and R. S. Harmon). Clarendon Press, Oxford, pp. 487–512.

Burnett W. C., Beers M. J., and Roe K. K. (1982) Growth rates of phosphate nodules from the continental margin off Peru. *Science* **215**, 1616–1618.

Burnett W. C., Glenn C. R., Yeh C. C., Schultz M., Chanton J., and Kashgarian M. (2000) U-series ^{14}C, and stable isotope studies of recent phosphatic "protocrusts" from the Peru margin. In *Marine Authigenesis: from Global to Microbia* (eds. C. R. Glenn, L. Prévôt-Lucas, and J. Lucas). SEPM (Society for Sedimentary Geology), Tulsa, OK, pp. 163–183.

Cande S. C. and Kent D. V. (1992) A new geomagnetic polarity time scale for the Late Cretaceous and Cenozoic. *J. Geophys. Res.* **97**, 13917–13951.

Chabaux F., O'Nions R. K., Cohen A. S., and Hein J. R. (1997) $^{238}U-^{234}U-^{230}Th$ disequilibrium in hydrogenous oceanic Fe–Mn crusts: palaeoceanographic record or diagenetic alteration? *Geochim. Cosmochim. Acta* **61**, 3619–3632.

Chen J. H., Edwards R. L., and Wasserburg G. J. (1992) Mass spectrometry and applications to uranium-series disequilibrium. In *Uranium-series Disequilibrium: Applications to Earth, Marine, and Environmental Sciences* (eds. M. Ivanovich and R. S. Harmon). Clarendon Press, Oxford, pp. 174–206.

Cheng H., Edwards R. L., Murrell M. T., and Benjamin T. M. (1998) Uranium—thorium–protactinium dating systematics. *Geochim. Cosmochim. Acta* **62**, 3437–3452.

Cheng H., Adkins J., Edwards R. L., and Boyle E. A. (2000) U–Th dating of deep-sea corals. *Geochim. Cosmochim. Acta* **64**, 2401–2416.

Cochran J. K. (1992) The oceanic chemistry of the uranium- and thorium-series nuclides. In *Uranium-series Disequilibrium: Applications to Earth, Marine, and Environmental Sciences* (eds. M. Ivanovich and R. S. Harmon). Clarendon Press, Oxford, pp. 334–395.

DeMaster D. J. and Turekian K. K. (1987) The radiocarbon record in varved sediments of Carmen Basin, Gulf of California: a measure of upwelling intensity variation during the past several hundred years. *Paleoceanography* **2**, 249–254.

Dickens G. R., O'Neil J. R., Rea D. K., and Owen R. M. (1995) Dissociation of oceanic methane hydrate as a cause of the carbon isotope excursion at the end of the Palaeocene. *Paleoceanography* **10**, 965–971.

Dymond J. (1969) Age determinations of deep-sea sediments: a comparison of three methods. *Earth Planet. Sci. Lett.* **6**, 9–14.

Edwards R. L., Chen J. H., and Wasserburg G. J. (1986–1987) $^{238}U-^{234}U-^{230}Th-^{232}Th$ systematics and the precise measurement of time over the past 500,000 years. *Earth Planet. Sci. Lett.* **81**, 175–192.

Edwards R. L., Chen J. H., Ku T.-L., and Wasserburg G. J. (1987) Precise timing of the last interglacial period from mass spectrometric determination of thorium-230 in corals. *Science* **236**, 1547–1553.

Edwards R. L., Beck J. W., Burr G. S., Donahue D. J., Chappell J. M. A., Bloom A. L., Druffel E. R. M., and Taylor F. W. (1993) A large drop in atmospheric $^{14}C/^{12}C$ and reduced melting in the Younger Dryas, documented with ^{230}Th ages of corals. *Science* **260**, 962–968.

Edwards R. L., Cheng H., Murrell M. T., and Goldstein S. J. (1997) Protactinium-231 dating of carbonates by thermal ionization mass spectrometry: implications for quaternary climate change. *Science* **276**, 782–786.

Eisenhauer A., Gögen K., Pernicka E., and Mangini A. (1992) Climatic influences on the growth rates of Mn crusts during the Late Quaternary. *Earth Planet. Sci. Lett.* **109**, 25–36.

Farley K. A. (1995) Cenozoic variations in the flux of interplanetary dust recorded by ^{3}He in a deep-sea sediment. *Nature* **376**, 153–156.

Farley K. A. (2001) Extraterrestrial helium in seafloor sediments: identification, characteristics and accretion rates over geologic time. In *Accretion of Extraterrestrial Matter throughout Earth's History* (eds. B. Peucker-Ehrenbrinck and B. Schmitz). Kluwer, New York, pp. 179–204.

Farley K. A. and Patterson D. B. (1995) A 100-kyr periodicity in the flux of extraterrestrial ^{3}He to the seafloor. *Nature* **376**, 600–603.

Francois R. and Bacon M. P. (1991) Variations in terrigenous input into the deep equatorial Atlantic during the past 24,000 years. *Science* **251**, 1473–1476.

Francois R. and Bacon M. P. (1994) Heinrich events in the North Atlantic: radiochemical evidence. *Deep-Sea Res. I* **41**, 315–334.

Francois R., Bacon M. P., and Suman D. O. (1990) Thorium 230 profiling in deep-sea sediments: high-resolution records of flux and dissolution of carbonate in the equatorial Atlantic during the last 24,000 years. *Paleoceanography* **5**, 761–787.

Frank M., Eisenhauer A., Bonn W. J., Walter P., Grobe H., Kubik P. W., Dittrich-Hannen B., and Mangini A. (1995) Sediment redistribution versus paleoproductivity change: Weddell Sea margin sediment stratigraphy and biogenic particle flux of the last 250,000 years deduced from $^{230}Th_{ex}$, ^{10}Be and biogenic barium profiles. *Earth Planet. Sci. Lett.* **136**, 559–573.

Friedlander G., Kennedy J. W., Macias E. S., and Miller J. M. (1981) *Nuclear and Radiochemistry*, 3rd edn. Wiley-Interscience, New York.

Gallup C. D., Edwards R. L., and Johnson R. G. (1994) The timing of high sea levels over the past 200,000 years. *Science* **263**, 796–800.

GEOSECS (1987) *Atlantic, Pacific, and Indian Ocean Expeditions, 7: Shorebased Data and Graphics*. National Science Foundation, Washington.

Goldberg E. D. and Bruland K. W. (1974) Radioactive geochronologies. In *The Sea, 5* (ed. E. D. Goldberg). Wiley-Interscience, New York, pp. 451–489.

Goldstein S. J., Lea D. W., Chakraborty S., Kashgarian M., and Murell M. T. (2001) Uranium-series and radiocarbon geochronology of deep-sea corals: implications for Southern

Ocean ventilation rates and the oceanic carbon cycle. *Earth Planet. Sci. Lett.* **193**, 167–182.

Guichard F., Reyss J.-L., and Yokoyama Y. (1978) Growth rate of manganese nodule measured with ^{10}Be and ^{26}Al. *Nature* **272**, 155–156.

Guilderson T. P., Fairbanks R. G., and Rubenstone J. L. (1994) Tropical temperature variations since 20,000 years ago: modeling interhemispheric climate change. *Science* **263**, 663–665.

Halbach P., Segl M., Puteanus D., and Mangini A. (1983) Co-fluxes and growth rates in ferromanganese deposits from Central Pacific seamount areas. *Nature* **304**, 716–719.

Henderson G. M. (2002) Seawater (^{234}U/^{238}U) during the last 800 thousand years. *Earth Planet. Sci. Lett.* **199**, 97–110.

Henderson G. M. and Burton K. W. (1999) Using (^{234}U/^{238}U) to assess diffusion rates of isotope tracers in ferromanganese crusts. *Earth Planet. Sci. Lett.* **170**, 169–179.

Henderson G. M., Cohen A. S., and O'Nions R. K. (1993) ^{234}U/^{238}U ratios and ^{230}Th ages for Hateruma Atoll corals: implications for coral diagenesis and seawater ^{234}U/^{238}U ratios. *Earth Planet. Sci. Lett.* **115**, 65–73.

Huh C.-A. and Kadko D. C. (1992) Marine sediments and sedimentation processes. In *Uranium-series Disequilibrium: Applications to Earth, Marine, and Environmental Sciences* (eds. M. Ivanovich and R. S. Harmon). Clarendon Press, Oxford, pp. 460–486.

Huh C.-A. and Ku T.-L. (1984) Radiochemical observations on manganese nodules from three sedimentary environments in the north Pacific. *Geochim. Cosmochim. Acta* **48**, 951–963.

Huh C.-A. and Ku T.-L. (1990) Distribution of thorium 232 in manganese nodules and crusts: paleoceanographic implications. *Paleoceanography* **5**, 187–195.

Huntley D. J. and Johnson H. P. (1976) Thermoluminescence as a potential means of dating siliceous ocean sediments. *Can. J. Earth Sci.* **13**, 593–596.

Ivanovich M. and Harmon R. S. (ed.) (1992) *Uranium Series Disequilibrium: Applications to Earth, Marine, and Environmental Problems*. Clarendon Press, Oxford.

Ivanovich M., Latham A. G., and Ku T.-L. (1992) Uranium-series disequilibrium applications in geochronology. In *Uranium-series Disequilibrium: Applications to Earth, Marine, and Environmental Sciences* (eds. M. Ivanovich and R. S. Harmon). Clarendon Press, Oxford, pp. 62–94.

Kaufman A., Broecker W. S., Ku T.-L., and Thurber D. L. (1971) The status of U-series methods of mollusk dating. *Geochim. Cosmochim. Acta* **35**, 1155–1183.

Kennett J. P. and Ingram B. L. (1995) A 20,000-year record of ocean circulation and climate change from Santa Barbara basin. *Nature* **377**, 510–514.

Kim K. H. and Burnett W. C. (1986) Uranium-series growth history of a quaternary phosphatic crust from the Peruvian continental margin. *Chem. Geol.* **58**, 227–244.

Koide M., Soutar A., and Goldberg E. D. (1972) Marine geochronology with ^{210}Pb. *Earth Planet. Sci. Lett.* **14**, 442–446.

Kress A. G. and Veeh H. H. (1980) Geochemistry and radiometric ages of phosphatic nodules from the continental margin of northern New South Wales. *Australia. Mar. Geol.* **36**, 143–157.

Krinsley D., Biscaye P. E., and Turekian K. K. (1973) Argentine Basin sediment sources as indicated by quartz surface textures. *J. Sedim. Petrol.* **43**, 251–257.

Krishnaswami S. and Cochran J. K. (1978) Uranium and thorium series nuclides in oriented ferromanganese nodules: growth rates, turnover times and nuclide behavior. *Earth Planet. Sci. Lett.* **40**, 45–62.

Krishnaswami S., Mangini A., Thomas J. H., Sharma P., Cochran J. K., Turekian K. K., and Parker P. D. (1982) ^{10}Be and Th isotopes in manganese nodules and adjacent sediments: nodule growth histories and nuclide behavior. *Earth Planet. Sci. Lett.* **59**, 217–234.

Krishnaswami S., Monaghan M. C., Westrich J. T., Bennett J. T., and Turekian K. K. (1984) Chronologies of sedimentary processes in sediments of the FOAM site, Long Island Sound, Connecticut. *Am. J. Sci.* **234**, 706–733.

Ku T.-L. (1976) The uranium-series methods of age determination. *Ann. Rev. Earth Planet. Sci.* **4**, 347–379.

Ku T. L. (1977) Rates of accretion. In *Marine Manganese Deposits* (ed. G. P. Glasby). Elsevier, New York, pp. 249–267.

Ku T. L. and Glasby G. P. (1972) Radiometric evidence for the rapid growth rate of shallow-water, continental margin manganese nodules. *Geochim. Cosmochim. Acta* **36**, 699–703.

Ku T. L., Omura A., and Chen P. S. (1979) ^{10}Be and U-series isotopes in manganese nodules from the central North Pacific. In *Marine Geology and Oceanography of the Pacific Manganese Nodule Province* (eds. J. L. Bischoff and D. Z. Piper). Plenum, pp. 791–814.

Kyte F. T., Leinen M., Heath G. R., and Zhou L. (1993) Cenozoic sedimentation history of the central North Pacific: inferences from the elemental geochemistry of core LL44-GPC3. *Geochim. Cosmochim. Acta* **57**, 1719–1740.

Lomitschka M. and Mangini A. (1999) Precise Th/U-dating of small and heavily coated samples of deep sea corals. *Earth Planet. Sci. Lett.* **170**, 391–401.

Luo S., Ku T.-L., Wang L., Southon J. R., Lund S. P., and Schwartz M. (2001) ^{26}Al, ^{10}Be and U–Th isotopes in Blake Outer Ridge sediments: implications for past changes in boundary scavenging. *Earth Planet. Sci. Lett.* **185**, 135–147.

Macdougall D. (1971) Fission track dating of volcanic glass shards in marine sediments. *Earth Planet. Sci. Lett.* **10**, 403–406.

Mangini A., Segl M., Bonani G., Hofmann H. J., Morenzoni E., Nessi M., Suter M., Wölfli W., and Turekian K. K. (1984) Mass-spectrometric ^{10}Be dating of deep-sea sediments applying the Zurich tandem accelerator. *Nucl. Instr. Meth. Phys. Res. B* **5**, 353–358.

Mangini A., Segl M., Kudrass H., Wiedicke M., Bonani G., Hofmann H. J., Morenzoni E., Nessi M., Suter M., and Wölfli W. (1986) Diffusion and supply rates of ^{10}Be and ^{230}Th radioisotopes in two manganese encrustations from the South China Sea. *Geochim. Cosmochim. Acta* **50**, 149–156.

Mangini A., Eisenhauer A., and Walter P. (1990) Response of manganese in the ocean to the climatic cycles in the Quaternary. *Paleoceanography* **5**, 811–821.

Marcantonio F., Kumar N., Stute M., Anderson R. F., Seidl M. A., Schlosser P., and Mix A. (1995) A comparative study of accumulation rates derived by He and Th isotope analysis of marine sediments. *Earth Planet. Sci. Lett.* **133**, 549–555.

Marcantonio F., Anderson R. F., Stute M., Kumar N., Schlosser P., and Mix A. (1996) Extraterrestrial ^{3}He as a tracer of marine sediment transport and accumulation. *Nature* **383**, 705–707.

Marcantonio F., Higgins S., Anderson R. F., Stute M., Schlosser P., and Rasbury E. T. (1998) Terrigenous helium in deep-sea sediments. *Geochim. Cosmochim. Acta* **62**, 1535–1543.

Marcantonio F., Turekian K. K., Higgins S., Anderson R. F., Stute M., and Schlosser P. (1999) The accretion rate of extraterrestrial ^{3}He based on oceanic ^{230}Th flux and the relation to Os isotope variation over the past 200,000 years in an Indian Ocean core. *Earth Planet. Sci. Lett.* **170**, 157–168.

Marcantonio F., Anderson R. F., Higgins S., Stute M., and Schlosser P. (2001) Sediment focusing in the central equatorial Pacific Ocean. *Paleoceanography* **16**, 260–267.

McCaffrey R. J. and Thomson J. (1980) A record of the accumulation of sediment and trace elements in a Connecticut salt marsh. *Adv. Geophys.* **22**, 165–236.

McCulloch M., Mortimer G., Esat T., Xianhua L., Pillans B., and Chappell J. (1996) High resolution windows into early Holocene climate: Sr/Ca coral records from the Huon Peninsula. *Earth Planet. Sci. Lett.* **138**, 169–178.

McManus J. F., Anderson R. F., Broecker W. S., Fleisher M. Q., and Higgins S. M. (1998) Radiometrically determined sedimentary fluxes in the sub-polar North Atlantic during the last 140,000 years. *Earth Planet. Sci. Lett.* **155**, 29–43.

Mesolella K. J., Matthews R. K., Broecker W. S., and Thurber D. L. (1969) The astronomical theory of climate change: Barbados data. *J. Geol.* **77**, 250–274.

Moore W. S. (1984) Thorium and radium isotopic relationships in manganese nodules and sediments at MANOP Site S. *Geochim. Cosmochim. Acta* **48**, 987–992.

Moore W. S. and Vogt P. R. (1976) Hydrothermal manganese crusts from two sites near the Galapagos spreading axis. *Earth Planet. Sci. Lett.* **29**, 349–356.

Moore W. S., Ku T.-L., MacDougall J. D., Burns V. M., Burns R., Dymond J., Lyle M. W., and Piper D. Z. (1981) Fluxes of metals to a manganese nodule: radiochemical, chemical, structural, and mineralogical studies. *Earth Planet. Sci. Lett.* **52**, 151–171.

Muhs D. R. and Szabo B. J. (1994) New uranium-series ages of the Waimanalo Limestone, Oahu, Hawaii: implications for sea level during the last interglacial period. *Mar. Geol.* **118**, 315–326.

Nozaki Y., Cochran J. K., Turekian K. K., and Keller G. (1977) Radiocarbon and ^{210}Pb distribution in submersible-taken deep-sea cores from Project FAMOUS. *Earth Planet. Sci. Lett.* **34**, 167–173.

Ozima M., Takayanagi M., Zashu S., and Amari S. (1984) High ^{3}He/^{4}He ratio in ocean sediments. *Nature* **311**, 448–450.

Palike H., Shackleton N. J., and Rohl U. (2001) Astronomical forcing in Late Eocene marine sediments. *Earth Planet. Sci. Lett.* **193**, 589–602.

Raisbeck G. M., Yiou F., Fruneau M., Loiseau J. M., and Lieuvin M. (1979) ^{10}Be concentration and residence time in the oceans. *Earth Planet. Sci. Lett.* **43**, 237–240.

Roe K. K., Burnett W. C., and Lee A. I. N. (1983) Uranium disequilibrium dating of phosphate deposits from the Lau Group, Fiji. *Nature* **302**, 603–606.

Saffert H. L. and Thomas E. (1998) Living foraminifera in salt marsh peat cores: Kelsey marsh (Clinton, CT) and the Great Marshes (Barnstable, MA). *Mar. Micropaleontol.* **33**, 175–202.

Shackleton N. J., Hagelberg T. K., and Crowhurst S. J. (1993) Evaluating the success of astronomical tuning: Pitfalls of using coherence as a criterion for assessing Pre-Pleistocene timescales. *Paleoceanography* **10**, 693–697.

Shackleton N. J., Hall M. A., Raffi I., Tauxe L., and Zachos J. (2000) Astronomical calibration age for the Oligocene–Miocene boundary. *Geology* **28**, 447–450.

Sharma M. (2002) Variations in solar magnetic activity during the lst 200,000 years: is there a Sun-climate connection? *Earth Planet. Sci. Lett.* **199**, 459–472.

Stirling C. H., Esat T. M., McCulloch M. T., and Lambeck K. (1995) High-precision dating of corals from Western Australia and implications for the timing and duration of the Last Interglacial. *Earth Planet. Sci. Lett.* **135**, 115–130.

Suess H. E. (1956) Absolute chronology of the last glaciation. *Science* **123**, 355–357.

Suman D. O. and Bacon M. P. (1989) Variations in Holocene sedimentation in the North American Basin determined from Th-230 measurements. *Deep-Sea Res.* **36**, 869–878.

Szabo B. J., Ludwig K. R., Muhs D. R., and Simmons K. R. (1994) Th-230 ages of corals and duration of the Last Interglacial sea level high stand on Oahu Hawaii. *Science* **266**, 93–96.

Tanaka N., Turekian K. K., and Rye D. M. (1991) The radiocarbon, δ^{13}C, ^{210}Pb, and ^{137}Cs record in box cores from the continental margin of the Middle Atlantic Bight. *Am. J. Sci.* **291**, 90–105.

Tanaka S. and Inoue T. (1979) ^{10}Be dating of north Pacific sediment cores up to 2.5 million years B.P. *Earth Planet. Sci. Lett.* **45**, 181–187.

Tanaka S., Inoue T., and Imamura M. (1977) The ^{10}Be method of dating marine sediments—comparison with the paleomagnetic method. *Earth Planet. Sci. Lett.* **37**, 55–60.

Thomas E., Turekian K. K., and Wei K.-Y. (2000) Productivity control of fine particle transport to equatorial Pacific sediment. *Global Biogeochem. Cycles* **14**, 945–955.

Thomson J., Higgs N. C., and Clayton T. (1995) A geochemical criterion for the recognition of Heinrich events and estimation of their depositional fluxes by the $(^{230}Th_{excess})_0$ profiling method. *Earth Planet. Sci. Lett.* **135**, 41–56.

Thomson J., Nixon S., Summerhayes C. P., Schönfeld J., Zahn R., and Grootes P. (1999) Implications for sedimentation changes on the Iberian margin over the last two glacial/interglacial transitions from $(^{230}Th_{excess})_0$ systematics. *Earth Planet. Sci. Lett.* **165**, 255–270.

Turekian K. K. and Cochran J. K. (1978) Determination of marine chronologies with natural radionuclides. In *Chemical Oceanography, 7* (eds. J. P. Riley and R. Chester). Academic Press, London, pp. 313–360.

Turekian K. K. and Stuiver M. (1964) Clay and carbonate accumulation rates in three South Atlantic deep-sea cores. *Science* **146**, 55–56.

Turekian K. K., Cochran J. K., Krishnaswami S., Lanford W. A., Parker P. D., and Bauer K. A. (1979) The measurement of ^{10}Be in manganese nodules using a tandem Van de Graaff accelerator. *Geophys. Res. Lett.* **6**, 417–420.

Varekamp J. C. and Thomas E. (1998) Sea level rise and climate change over the last 1000 years. *EOS* **79**, 69–75.

Veeh H. H. and Chappell J. (1970) Astronomical theory of climatic change: support from New Guinea. *Science* **167**, 862–865.

Walter H.-J., Rutgers van der Loeff M. M., and François R. (1999) Reliability of the ^{231}Pa/^{230}Th activity ratio as a tracer for bioproductivity of the ocean. In *Use of Proxies in Paleoceanography: Examples from the South Atlantic* (eds. G. Fischer and G. Wefer). Springer, pp. 393–408.

Wintle A. G. and Huntley D. J. (1979) Themoluminescence dating of a deep-sea sediment core. *Nature* **279**, 710–712.

Yang H.-S., Nozaki Y., Sakai H., and Masuda A. (1986) The distribution of ^{230}Th and ^{231}Pa in the deep-sea surface sediments of the Pacific Ocean. *Geochim. Cosmochim. Acta* **50**, 81–99.

Yang Y.-L., Elderfield H., and Ivanovich M. (1990) Glacial to Holocene changes in carbonate and clay sedimentation in the equatorial Pacific Ocean estimated from thorium 230 profiles. *Paleoceanography* **5**, 789–809.

Yu E.-F., Francois R., Bacon M. P., and Fleer A. P. (2001) Fluxes of ^{230}Th and ^{231}Pa to the deep sea: implications for the interpretation of excess ^{230}Th and ^{231}Pa/^{230}Th profiles in sediments. *Earth Planet. Sci. Lett.* **191**, 219–230.

Zachos J., Pagani M., Sloan L., Thomas E., and Billups K. (2001) Trends, rhythms, and aberrations in global climate 65 Ma to present. *Science* **292**, 686–692

6.13 Geochemical Evidence for Quaternary Sea-level Changes

R. L. Edwards, K. B. Cutler, and H. Cheng
University of Minnesota, Minneapolis, MN, USA

and

C. D. Gallup
University of Minnesota, Duluth, MN, USA

6.13.1 INTRODUCTION

Throughout the Quaternary, sea level has risen and fallen as continental ice sheets have waned and waxed. The main cause of sea-level change has been variation in the total volume of continental ice and resulting change in the fraction of the Earth's surface H_2O contained in the ocean. Today more than 97% of the Earth's surface H_2O is in the ocean and less than 2% is stored as ice in continental glaciers, with groundwater making up the bulk of the remainder. Of the total continental ice (ice above sea level), ~80% is contained in the east Antarctic ice sheet, 10% in the west Antarctic ice sheet, and the final 10% in the Greenland ice sheet. If all continental ice were to melt, sea level would rise by ~70 m. During the last glacial maximum (LGM), sea level was ~125 m lower than present, equivalent to 3% more surface H_2O stored as continental ice. Because of its relationship to continental ice volume, an accurate Quaternary sea-level curve has been a long-term goal of scientists interested in ice-age cycles and their causes.

Although sea level is closely related to continental ice volume, it is also a function of other variables, most notably isostatic effects (see Clark *et al.* (1978) and review of Peltier (1998) and references therein). Sea-level change can be divided into a eustatic component, which depends only on shifts in the mass of H_2O between continental ice and the ocean, and an isostatic component, which is site specific and is a function of the history of the geographic distribution of H_2O on the planet. As H_2O is redistributed geographically (during glacial melting or growth), the shape of the geoid changes. Thus, the shape of the ocean surface changes immediately. The redistribution of H_2O also generates flow in the mantle from newly loaded areas to newly unloaded areas. The flow has a timescale of a few to several thousand years. The flow changes the shape and elevation of the seafloor and further changes the shape of the geoid, both of which change sea level as monitored at a particular locality. Understanding and quantifying isostatic effects is complex requiring some knowledge of the three-dimensional (3D) viscosity structure of the mantle, the 3D elastic properties of the lithosphere, and the geographic history of H_2O redistribution on the Earth's surface. Much of the vast literature in this area is beyond the scope of this chapter. We will, however, address this issue and present sea-level data from so-called "far-field" sites, localities far from glaciated and formerly glaciated areas. Sea level at such sites is dominated by the eustatic component. We will compare sea-level curves from different far-field sites as a means of determining empirically the degree to which far-field sites are affected differentially by isostatic effects. Finally, some of the geochemical methods for determining ice volume are independent of isostatic effects.

6.13.2 METHODS OF SEA-LEVEL RECONSTRUCTION

Two methods of sea-level reconstruction have been successful. The first involves direct reconstruction of sea level at individual localities by dating features that record past sea level. The second involves determining the history of variations in the oxygen isotopic composition of the ocean.

6.13.2.1 Methods of Direct Sea-level Reconstruction

Direct sea-level reconstruction has been accomplished using a number of different strategies. The main approach has been ^{230}Th dating of carbonates that mark or constrain sea level, supplemented in some cases with ^{231}Pa dating. Protactinium-231 dating provides a critical check on the accuracy of ^{230}Th dates. ^{230}Th dating has a range from 3 yr to 6×10^5 yr. Therefore, in principle, this method can be used to reconstruct sea level over several glacial–interglacial cycles. In the late 1960s and early 1970s, this approach was used to establish aspects of the timing of sea-level change over the last interglacial–glacial cycle (Veeh, 1966; Broecker *et al.*, 1968; Mesolella *et al.*, 1969; Veeh and Chappell, 1970; Chappell, 1974; Bloom *et al.*, 1974). This approach has seen renewed activity since late 1980s, spurred by the development of high-precision, high-sensitivity mass spectrometric techniques for ^{230}Th dating (Edwards *et al.*, 1987a) and ^{231}Pa dating (Pickett *et al.*, 1994; Edwards *et al.*, 1997). This strategy has been applied to aragonite from fossil skeletons of coral species that grow near the sea surface. A date and an elevation yield a point on a sea-level curve. It has also been applied to speleothems (cave calcite). As speleothems only grow subaerially, a date on a subsample of speleothem represents a time when sea level was below the elevation of the speleothem. The ^{230}Th and ^{231}Pa dating approaches are powerful as they yield absolute ages for materials that directly record sea level, over timescales of several interglacial–glacial cycles. Challenges include the discontinuous nature of such records, difficulties in assessing whether diagenesis has affected age accuracy, the accuracy of corrections for tectonic uplift or subsidence, and difficulties in recovering submerged samples. As we live in an interglacial period, most recorders of past sea levels are submerged. Many of these difficulties have been overcome in reaching our current state of knowledge on sea-level history.

An alternate approach for times in the last 4×10^4 yr involves ^{14}C dating of materials that record sea level. Compared to the ^{230}Th method, the ^{14}C method is restricted to a smaller time range and requires an accurate ^{14}C calibration curve; however, it can be applied to a wider range of materials, including other carbonates and organic materials. A number of coral studies have used ^{14}C and ^{230}Th dating (e.g., Bard *et al.*, 1990a; Edwards *et al.*, 1993), producing information on ^{14}C calibration as well as sea level. Such ^{230}Th-based calibrations can then be applied to sea-level records based solely on ^{14}C dating, providing calendar-age timescales for such sea-level curves.

6.13.2.2 Methods of Sea-level Reconstruction from Oxygen Isotope Measurements

A completely different strategy involves reconstruction of past ocean $^{18}O/^{16}O$ ratios as a proxy for past ice volume. Glacial ice has a lower $^{18}O/^{16}O$ ratio than seawater. Therefore, when ice volume is high, seawater has relatively high $^{18}O/^{16}O$ ratio. When ice volume is low, seawater has relatively low $^{18}O/^{16}O$ ratio. If the average $^{18}O/^{16}O$ ratio of glacial ice is constant with time, then the average $^{18}O/^{16}O$ ratio of seawater approximates a linear function of ice volume. As opposed to direct sea-level measurements, the seawater $^{18}O/^{16}O$ ratio is independent of isostatic effects. The $^{18}O/^{16}O$ method also integrates all glacial ice, whereas the direct sea-level method measures that portion of glacial ice which lies above the sea level. However, the precision with which sea level can be determined is limited, as analytical error is significant compared to the glacial–interglacial amplitude in seawater $^{18}O/^{16}O$ values. Furthermore, there are difficulties in separating the seawater $^{18}O/^{16}O$ signal from the $^{18}O/^{16}O$ record of interest, and in general such records cannot be dated by radiometric means.

Methods for reconstruction have focused on estimating the ocean-water component of $^{18}O/^{16}O$ ratio from records of the $^{18}O/^{16}O$ values of foraminiferal calcite isolated from marine sediments (Emiliani, 1955). The $^{18}O/^{16}O$ ratio of foraminiferal calcite is primarily a function of the $^{18}O/^{16}O$ ratio of the water and the temperature of the water. A number of strategies have been used to separate the water and temperature components of such records. One method utilizes comparisons between discontinuous direct sea-level records with continuous deep-sea foram records to develop empirical relationships between the two. These relationships can then be applied to the full foram record as a means of separating out the two components (Chappell and Shackleton, 1986).

Direct measurements of the $^{18}O/^{16}O$ ratio of LGM pore fluids and comparison with LGM foram calcite $^{18}O/^{16}O$ ratio has also been used to directly resolve the two components for the LGM. Applications of the Chappell and Shackleton (1986) method using the constraints of Schrag *et al.* (1996) have further refined our ability to resolve the ice volume and temperature components (Cutler *et al.*, 2003). The Schrag (1996) pore-fluid work is important because it establishes, through direct measurement, a key point on the seawater $^{18}O/^{16}O$ curve. Analogous measurements on the chloride content of LGM pore fluids test and complement the $^{18}O/^{16}O$ data (Adkins and Schrag, 2001). Chloride is proportional to salinity, a direct measure of ice volume. The average oceanic $^{18}O/^{16}O$ value can only be recast as ice volume if the average $^{18}O/^{16}O$ ratio of the ice is known; however, average oceanic salinity is a direct measure of ice volume as ice contains essentially no salt. Because both sea and glacial ice have low-salt contents, the salinity measurement integrates both types of ice. Because glacial ice is tens of per mil different (lower) in $^{18}O/^{16}O$ ratio than seawater, but sea ice is similar to seawater in $^{18}O/^{16}O$ ratio ($\sim$2 per mil higher), the seawater $^{18}O/^{16}O$ measurement is sensitive only to glacial ice. Both methods establish an important benchmark. Unfortunately, these measurements cannot be extended back in time because the smoothing effect of pore-fluid advection has erased earlier variations.

A second method for resolving the components involves using the Mg/Ca ratio of the forams to determine temperature (see Chapter 6.14; Nurnberg, 1995; Nurnberg *et al.*, 1996; Lea *et al.*, 1999; Elderfield and Ganssen, 2000), then subtracting the temperature component from the foram $^{18}O/^{16}O$ record. This method has been applied to planktonic records to reconstruct seawater $^{18}O/^{16}O$ ratio at specific sites in the surface ocean (see Elderfield and Ganssen, 2000; Lea *et al.*, 2000, 2002). Analogous benthic records are yet to be determined; however, efforts are underway to refine the Mg/Ca thermometer to the point where similar benthic records can be established (Martin *et al.*, 2002). A benthic record would be very important because the $^{18}O/^{16}O$ ratio of surface waters may deviate significantly from the oceanic average because of hydrologic effects (evaporation/precipitation/runoff/riverine input). Whereas deep waters may also deviate from the oceanic average, (i) the deviations are smaller, (ii) the $^{18}O/^{16}O$ values for a particular site are representative of large volumes of ocean water, and (iii) extrapolation to average ocean values may be possible with a very limited number of sites. Given our current knowledge of present and past ocean circulation, deep Pacific waters are likely to have values close to the average ocean, whereas deep Atlantic waters are likely to have values that are also sensitive to the particular mixture of southern and northern source waters present at the site as a function of time.

Traditionally foram records have not been datable by radiometric means (Delaney and Boyle, 1983; Henderson and O'Nions, 1995). Generally ages are established by orbital tuning (Imbrie *et al.*, 1984). However, in specific instances where forams have been separated from uranium-rich carbonate bank sediments, the bank sediments have been dated by ^{230}Th methods, yielding radiometric chronologies for the foram $^{18}O/^{16}O$ record (Slowey *et al.*, 1996; Henderson and Slowey, 2000; Robinson *et al.*, 2002).

The foram approach has the great advantage that it is continuous and can be potentially used to reconstruct sea levels over long-time intervals (millions of years). Challenges include limits to temporal resolution imposed by sedimentation rate and bioturbation depth, dating problems, limits in ability to resolve sea level because the analytical error in $^{18}O/^{16}O$ estimate is significant compared to the glacial to interglacial amplitude, and issues related to separating the ice volume and temperature signals. Whereas this approach is independent of isostatic effects, it is not independent of changes in the average isotopic composition of glacial ice, nor is it independent of differences between the average $^{18}O/^{16}O$ value of seawater and its value at the site in question.

A second method for reconstructing ocean water $^{18}O/^{16}O$ ratio utilizes records of $^{18}O/^{16}O$ estimates in atmospheric O_2 (Bender *et al.*, 1985, 1994 and references therein; Sowers *et al.*, 1993; Petit *et al.*, 1999; Shackleton, 2000). Such records, established by extracting and analyzing gas from bubbles in glacial ice, extend back to 4×10^5 yr (Petit *et al.*, 1999). As with foram records, atmospheric O_2 records have a component that results from changes in the $^{18}O/^{16}O$ ratio of seawater. The basic link is photosynthesis. Water in plants is ultimately derived from the ocean and therefore has an oxygen isotopic composition that reflects the oxygen isotopic composition of seawater. The oxygen isotopic composition of plant waters is then reflected in the oxygen isotopic composition of the oxygen produced during photosynthesis. The timescale over which changes in seawater isotopic values are transmitted to the atmosphere is ~1 kyr (Bender *et al.*, 1994). The modern atmospheric $^{18}O/^{16}O$ value is shifted from the $^{18}O/^{16}O$ value in seawater by a large amount (~23.5 per mil), an offset called the "Dole effect" (see Dole *et al.*, 1954). The two major contributors to changes in atmospheric $^{18}O/^{16}O$ ratio are changes in the Dole effect and changes in seawater $^{18}O/^{16}O$ ratio. The major hurdle in extracting the ocean-water $^{18}O/^{16}O$ signal from atmospheric $^{18}O/^{16}O$ record is the lack of a clear rationale for separating the Dole and ocean-water components. Shackleton (2000) separated the components by comparing power spectra of the deep-sea oxygen isotope record (which has temperature and seawater $^{18}O/^{16}O$ components) and the atmospheric oxygen isotope record (which has Dole effect and seawater $^{18}O/^{16}O$ components), and making reasonable presumptions about how the power in each orbital band is partitioned between the components. For example, in the precession band, the atmospheric record has twice the power of the deep-sea record. Thus, half or more of the atmospheric record's power in the precession band must result from Dole effect changes and half or less must result from seawater $^{18}O/^{16}O$ ratio changes (Shackleton, 2000). In this fashion, estimates of the relative contributions of the two components can be made in the frequency domain, and reconstructed as a seawater $^{18}O/^{16}O$ record and a Dole effect record in the time domain. As the deep-sea oxygen isotope record is also part of this analysis, a deep-sea temperature record also results from this exercise (see Chapter 6.14).

As with most foram records, the atmospheric record cannot be dated with radiometric techniques. Dating of the record has been accomplished by correlation with the orbitally tuned deep-sea record (Sowers *et al.*, 1993), using estimates of ice accumulation rates, ice flow models, and estimates of offsets between snowfall and gas closure ages (Petit *et al.*, 1999), and by direct orbital tuning of the atmospheric oxygen record itself (Shackleton, 2000). Separate tuning of the marine and atmospheric records establishes independent chronologies for the two.

The atmospheric $^{18}O/^{16}O$ approach has many of the advantages and disadvantages of the deep-sea $^{18}O/^{16}O$ approach. The strengths are the long, continuous nature of the record. The atmospheric record is independent of isostatic effects, but not from changes in the average $^{18}O/^{16}O$ ratio of the ocean. Similar challenges are faced in the dating of the two types of $^{18}O/^{16}O$ records. There are issues with separating the Dole effect and seawater components of the record. Finally, as with the foram record analytical error in $^{18}O/^{16}O$ measurement is significant compared to the glacial–interglacial amplitude in seawater $^{18}O/^{16}O$ records. In sum, the strengths of the direct sea-level reconstruction are the capability to date material with radiometric methods and the precision with which past sea level can be determined. The strengths of the seawater $^{18}O/^{16}O$ methods are continuity and length. In the final analysis, the best sea-level reconstruction may be the most accurate direct sea-level points, combined with $^{18}O/^{16}O$ records as a means of interpolating between and extrapolating from these points.

6.13.3 HISTORY AND CURRENT STATE OF DIRECT SEA-LEVEL RECONSTRUCTION

The history of direct sea-level reconstruction is closely tied to the development of radiometric dating techniques. Because of the discontinuous nature of deposits upon which these reconstructions are based, the ability to date the materials is critical. Thorium-230 dating and improvements in ^{230}Th dating are central to the history of direct sea-level reconstruction, with ^{231}Pa dating playing an increasingly important role.

6.13.3.1 ^{230}Th and ^{231}Pa Dating: Current Status and Historical Overview

Thorium-230 dating, also referred to as U/Th dating or $^{238}U-^{234}U-^{230}Th$ dating, involves calculating ages from radioactive decay and ingrowth relationships among ^{238}U, ^{234}U, and ^{230}Th. Thorium-232 is also typically measured as a long-lived, essentially stable index isotope (over the timescales relevant to ^{230}Th dating). Thorium-230 dating can, in principle, be used to date materials as young as 3 yr and as old as 6×10^5 yr (Edwards *et al.*, 1987a; Edwards, 1988). Protactinium-231 dating, also referred to as U/Pa dating, involves calculating ages from the ingrowth of ^{231}Pa from its grandparent ^{235}U. At present ^{231}Pa dating can be used to date materials as young as 10 yr and as old as 2.5×10^5 yr (Edwards *et al.*, 1997). Thorium-230 dating covers all of the ^{231}Pa time range and more, with somewhat higher precision, and is therefore the method of choice if a single method is applied. However, the combination of ^{231}Pa and ^{230}Th dating is of great importance in assessing possible diagenetic mobilization of the pertinent nuclides, and thereby, the accuracy of the ages (Allegre, 1964; Ku, 1968). Even if the primary age exceeds the 2.5×10^5 yr limit of ^{231}Pa dating, the combined methods can be used to assess the degree to which the samples have remained closed over the past 2.5×10^5 yr (e.g., Edwards *et al.*, 1997). Thus, ^{231}Pa analysis can play an important role in assessing age accuracy. Taken together, ^{230}Th and ^{231}Pa dating cover the last several glacial–interglacial cycles, a period that typically cannot be accessed with other radiometric dating techniques.

Dating of the following three materials has been critical for sea-level studies: coralline aragonite, aragonitic carbonate bank sediments, and speleothem calcite. All three have relatively high primary uranium concentrations (~100 ppb to ~5 ppm) and low primary thorium and protactinium concentrations. As isotopes of uranium are the parents of both ^{230}Th and ^{231}Pa dating schemes, the relatively high uranium concentrations make them good candidates for ^{230}Th and ^{231}Pa dating. The high uranium concentrations of these materials distinguish them from most biogenic carbonates (e.g., mollusk shells; Broecker, 1963), which generally have much lower uranium concentrations, as much as five orders of magnitude lower (Edwards *et al.*, 1987b).

Historically, our ability to apply ^{230}Th and ^{231}Pa dating techniques has been limited by technical capabilities. Limitations result from the small concentrations of the key intermediate daughter nuclides. A relatively high-concentration material (a 5×10^4 yr old coral) typically has a ^{230}Th concentration of ~100 fm g^{-1} (60 billion atoms g^{-1}) and a ^{231}Pa concentration of ~3 fm g^{-1} (2 billion atoms g^{-1}). At the extreme low end of the concentration range, surface seawater contains ~4×10^{-24} mol of ^{230}Th/g (3,000 ^{230}Th atoms g^{-1}) and ~80×10^{-21} mol of ^{231}Pa/g (50 ^{231}Pa atoms g^{-1}). We are still limited by analytical capabilities for some applications and are actively pursuing analytical improvements; however, measurements of ^{230}Th and ^{231}Pa in all of the above materials can now be made with relatively small samples and relatively high precision.

Historically this was not always the case. Almost 50 years after ^{230}Th was first identified by Boltwood (1907). Barnes *et al.* (1956) were the first to quantify ^{230}Th levels in a coral, using alpha-counting techniques. This work spawned a series of sea-level studies that established the main characteristics of sea level in the last interglacial–glacial cycle (Veeh, 1966; Broecker *et al.*, 1968; Mesolella *et al.*, 1969; Bloom *et al.*, 1974; Chappell, 1974; Ku *et al.*, 1974). The field ultimately languished because of the technical limitations of alpha-counting techniques, in terms of sample size and precision (Edwards, 1988, 2000; Wasserburg, 2000). The basic problem was a limit on the fraction of atoms that can be detected by decay-counting techniques given the large difference between the half-lives of the pertinent nuclides (>7.5×10^4 yr) and reasonable laboratory counting times (weeks). This problem was solved with the development of mass spectrometric methods for the measurement of ^{234}U (Chen *et al.*, 1986) and ^{230}Th (Edwards *et al.*, 1987a) in natural materials. Mass spectrometric measurements obviate the need to wait for the nuclides of interest to decay as mass spectrometers detect the ions/atoms of interest directly. In this regard, the development of mass-spectrometric techniques for ^{230}Th and ^{234}U measurement is analogous to the development of accelerator mass-spectrometer techniques for ^{14}C measurement, which improved upon traditional beta-counting techniques. Mass spectrometric methods for measuring ^{230}Th and ^{234}U greatly reduced sample-size requirements and improved analytical precision. These technical improvements invigorated the sea-level work by improving the precision of ^{230}Th ages, extending the range of ^{230}Th dating, and improving our ability to detect diagenetic alteration of nuclides used in ^{230}Th dating.

Protactinium-231 measurements follow a similar history. Despite its discovery early in the twentieth century, the first systematic sea-level study applying ^{231}Pa dating was carried out only much later (Ku, 1968), using decay-counting techniques. Analogous to the development of mass-spectrometric techniques for ^{230}Th and ^{234}U, similar techniques were developed for the measurement of ^{231}Pa (Pickett *et al.*, 1994), with the first application to coral dating and sea-level studies coming soon thereafter (Edwards *et al.*, 1997). As with ^{230}Th, the use of mass-spectrometric ^{231}Pa measurement techniques (as opposed to traditional decay-counting techniques) improves the precision of ages, reduces sample-size requirements, and extends the range of ^{231}Pa dating to both older and younger times. As discussed below, ^{231}Pa dating has played and will continue to play a critical role in combination with ^{230}Th dating in testing for diagenesis and age accuracy.

6.13.3.2 ^{230}Th and ^{231}Pa Dating: Theory

Thorium-230 dating is based on the initial portion of the ^{238}U decay chain. Uranium-238 decays by alpha emission with a half-life of $(4.4683 \pm 0.0048) \times 10^9$ yr (Jaffey *et al.*, 1971) to ^{234}Th, which in turn decays (half-life = 24.1 d) by beta emission to ^{234}Pa, which decays (half-life = 6.7 h) by beta emission to ^{234}U, which decays (half-life = 245,250 ± 490 yr; Cheng *et al.*, 2000) by alpha emission to ^{230}Th (half-life = 75,690 ± 230 yr; Cheng *et al.*, 2000). Because of their short half-lives, ^{234}Th and ^{234}Pa can be ignored in carbonate dating applications. ^{230}Th dating is based on the radioactive ingrowth of ^{230}Th from initially very low levels.

Protactinium-231 dating is based on the initial portion of the ^{235}U decay chain. Uranium-235 decays by alpha emission (half-life = $(7.0381 \pm 0.0096) \times 10^8$ yr; Jaffey *et al.*, 1971) to ^{231}Th, which decays (half-life = 1.06 d) to ^{231}Pa (half-life = 32,760 ± 220 yr; Robert *et al.*, 1969). Because of its short half-life, ^{231}Th can be ignored for ^{231}Pa dating applications.

Protactinium-231 dating is based on the radioactive ingrowth of ^{231}Pa from initially very low levels. Low initial ^{230}Th and ^{231}Pa concentrations result from the extreme fractionation of uranium from both thorium and protactinium during the weathering process. All are actinides, but have different valences under oxidizing conditions. Uranium is typically +6 under oxidizing conditions, protactinium, +5, and thorium, +4. Uranium is soluble as uranyl ion and in various uranyl carbonate forms. Thorium has extremely low solubility in virtually all natural waters. Protactinium also has low solubility in natural waters, although its solubility is generally higher than for thorium. The initial fractionation of uranium from thorium and protactinium takes place during weathering and soil formation where a significant proportion of the uranium tends to dissolve in the aqueous phase and both thorium and protactinium tend to remain associated with solid phases. Low concentrations of thorium and protactinium are maintained in seawater as thorium and protactinium produced by radioactive decay are removed by adsorption onto solid particles and complexation with organic molecules associated with solid particles. When coralline aragonite forms from surface seawater, there is little fractionation among uranium, protactinium, and thorium so that corals incorporate the three elements in approximately their proportions in seawater.

The ^{230}Th age equation, calculated assuming (i) initial ^{230}Th/^{238}U = 0 and (ii) all changes in isotope ratios are the result of radioactive decay and ingrowth (no chemical/diagenetic shifts in isotope ratios), is

$$[^{230}\mathrm{Th}/^{238}\mathrm{U}] - 1 = -e^{-\lambda_{230}t} + (\delta^{234}\mathrm{U_m}/1{,}000) \times (\lambda_{230}/(\lambda_{230} - \lambda_{234}))(1 - e^{-(\lambda_{230}-\lambda_{234})t}) \qquad (1)$$

The brackets around ^{230}Th/^{238}U indicate that this is an activity ratio. λs are decay constants; t is age; and $\delta^{234}\mathrm{U_m}$ is the measured deviation in parts per thousand (per mil) of the ^{234}U/^{238}U ratio from secular equilibrium: $\delta^{234}\mathrm{U} = ([^{234}\mathrm{U}/^{238}\mathrm{U}] - 1)1{,}000$. Given measured ^{230}Th/^{238}U and ^{234}U/^{238}U, the only unknown is age, which can be calculated from Equation (1). Because age appears twice, the equation must be solved by iteration. The general form of Equation (1) was first solved by Bateman (1910). Broecker (1963) presented the specific equation (Equation (1)).

^{234}U/^{238}U ratios vary in nature and in surface waters, in what turns out to be a happy complication. As is clear from Equation (1), this phenomenon requires measurements of both ^{234}U/^{238}U and ^{230}Th/^{238}U ratios to solve for age. However, given both values, we can solve for age uniquely. Furthermore, a second equation that relates measured and initial ^{234}U/^{238}U can be calculated from the equations of radioactive production and decay, subject only to the assumption that chemical reactions (diagenesis) involving uranium have not occurred since precipitation of the aragonite:

$$\delta^{234}\mathrm{U_m} = (\delta^{234}\mathrm{U_i})e^{-\lambda_{234}t} \qquad (2)$$

where the subscript "i" refers to the initial value and the decay constant is that of ^{234}U. Thus, Equations (1) and (2) can be solved for two unknowns (age and initial δ^{234}U). For marine samples, knowledge of initial δ^{234}U is of great importance in assessing dating accuracy. A number of arguments suggest that marine δ^{234}U has been constant within fairly tight bounds (Edwards, 1988; Hamelin *et al.*, 1991; Richter and Turekian, 1993). If so, deviations from marine values indicate diagenetic shifts in uranium and potential inaccuracy in ^{230}Th age.

The ^{231}Pa age equation, calculated assuming no chemical shifts in protactinium or uranium and an initial ^{231}Pa/^{235}U = 0, is analogous to the ^{230}Th age equation (Equation (1)), but simpler in comparison. There is no term analogous to δ^{234}U, because there is no long-lived intermediate daughter isotope between ^{235}U and ^{231}Pa:

$$[^{231}\mathrm{Pa}/^{235}\mathrm{U}] - 1 = -e^{-\lambda_{231}t} \qquad (3)$$

Perhaps the greatest strength of $^{231}Pa/^{235}U$ data is their use in concert with $^{230}Th-^{234}U-^{238}U$ data in testing age accuracy (Cheng *et al.*, 1998). A good method for interpreting these data is through a $^{231}Pa/^{235}U$ versus $^{230}Th/^{234}U$ concordia diagram (Figure 1; Allegre, 1964; Ku, 1968; see Cheng *et al.*, 1998). The ordinate is the key isotopic ratio for ^{231}Pa dating and the abscissa is the key isotopic ratio for ^{230}Th dating. Plotted parametrically is age along the locus of isotopic compositions for which ^{231}Pa and ^{230}Th age are identical: concordia. This plot is analogous to the U–Pb concordia plot (Wetherill, 1956a,b) except that the $^{231}Pa/^{235}U$ versus $^{230}Th/^{234}U$ diagram has different concordia for different initial $\delta^{234}U$ values. If a data point plots off of concordia, the sample must have been altered chemically and one or both of the ages are not accurate. If a data point plots on concordia, then the sample's isotopic composition is consistent with closed-system behavior of the pertinent nuclides. We consider the most reliable sea-level data cited below to be those based upon both ^{230}Th and ^{231}Pa dating, for which the ages are concordant.

6.13.3.3 Tests of Dating Assumptions

One of the keys in ^{230}Th and ^{231}Pa dating of corals is establishing negligible initial $^{230}Th/^{238}U$ and $^{231}Pa/^{235}U$ values. This has been established by analyzing very young coral subsamples of known age. Initial ^{230}Th (Edwards *et al.*, 1988) and initial ^{231}Pa (Edwards *et al.*, 1997) are both negligible, at levels below the equivalent of several years of ingrowth for both daughter isotopes. Similarly, initial ^{230}Th is not significant in the speleothem sea-level studies discussed below. However, dating of aragonitic bank sediments requires correction for initial ^{230}Th, which can be accomplished with isochron techniques (Ludwig and Titterington, 1994; Bischoff and Fitzpatrick, 1991; Luo and Ku, 1991) and correction for pore-water processes associated with alpha-recoil (Henderson *et al.*, 2001).

The second assumption, used in deriving Equations (1)–(3), is that the system has remained closed to chemical exchange. Testing this assumption is more difficult. A number of approaches have been used. Perhaps the most powerful is combined ^{231}Pa and ^{230}Th dating, but most of the methods discussed below play a role. As primary coral skeletons are aragonitic, a standard test for diagenesis is mineralogy. Most fossil corals have at least partially recrystallized to calcite, a diagenetic phenomenon that can easily be detected by X-ray diffraction. Samples that show no evidence for recrystallization can be further tested using petrographic criteria (e.g., Chappell and Polach, 1972; Stein *et al.*, 1993; Bar-Mathews *et al.*, 1993).

Two additional tests specifically screen for open-system behavior of the nuclides of interest.

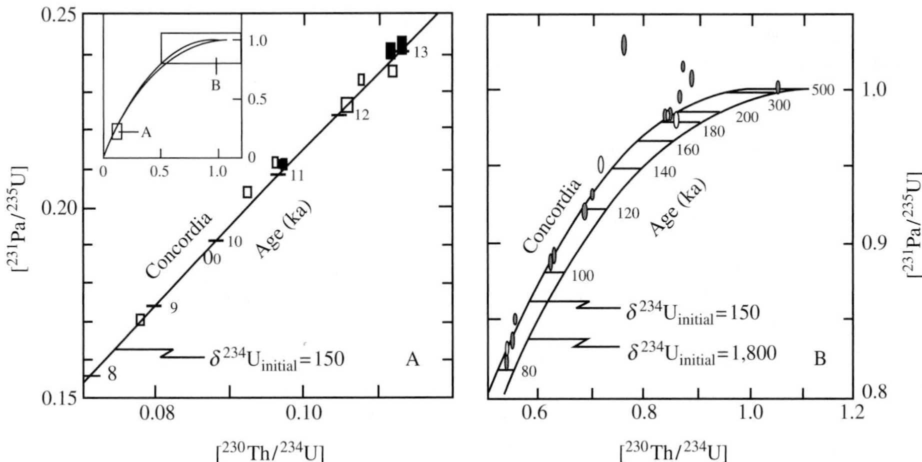

Figure 1 $^{231}Pa-^{230}Th$ concordia plot. The full concordia plot is shown in the inset of the left diagram. The curves extending from the lower left to upper right of each diagram are concordia curves. Concordia represents the locus of points for which the isotopic composition yields identical ^{231}Pa and ^{230}Th ages. Separate concordia curves correspond to different initial uranium isotopic compositions (initial $\delta^{234}U$). Except as indicated, illustrated concordia curves correspond to initial $\delta^{234}U = 150$, the approximate marine value, and the appropriate curve for corals. Coral data (<15 kyr) from the Huon Peninsula, Papua New Guinea (left diagram) show no evidence for diagenesis, with all points plot on or close to concordia. Barbados corals (>80 kyr, right diagram) either plot on concordia (no evidence for diagenesis) or above concordia (consistent with addition of ^{231}Pa and ^{230}Th or loss of uranium). Older Papua New Guinea corals (not illustrated) plot on (no evidence for diagenesis), above (consistent with ^{230}Th and ^{231}Pa gain or uranium loss), and below (consistent with ^{230}Th and ^{231}Pa loss or uranium gain) concordia. The concordia test is one of our most rigorous tests for diagenesis and age accuracy. The most reliable sea-level points have been established with this test (after Edwards *et al.*, 1997).

The first is a comparison of the initial δ^{234}U value of a coral with the modern marine value. Corals incorporate marine uranium without significant isotopic fractionation. The modern marine δ^{234}U value as measured in modern corals is 145.8 ± 1.7‰ (using updated half-lives; Cheng *et al.*, 2000). The marine δ^{234}U value does not vary more than the analytical error with depth or geographic location (Chen *et al.*, 1986, consistent with the long residence time of uranium, (2–4) × 10^5 yr; Ku *et al.*, 1977). The wide range in riverine δ^{234}U values (0–2,000; Cochran, 1982) suggests that the elevation of the marine δ^{234}U value above secular equilibrium results from a complex combination of weathering and alpha-recoil processes. At some level the oceanic δ^{234}U value must change with time as a result of some combination of changes in the average δ^{234}U of riverine input, changes in the riverine flux of uranium to the ocean, or changes in the removal rate of uranium from the ocean. However, a number of models suggest that it is unlikely that this value has changed by more than ~10 per mil over timescales of hundreds of thousands of years (Edwards, 1988; Hamelin *et al.*, 1991; Richter and Turekian, 1993). Thus, an initial δ^{234}U value that differs from the modern marine value by more than ~10 per mil indicates open-system behavior.

Figures 2 and 3 show coral data in plots of δ^{234}U versus ^{230}Th/^{238}U. These plots illustrate both the power of the δ^{234}U test for diagenesis and the magnitude of the diagenesis "problem" in dating corals significantly older than 2 × 10^4 yr. Figure 2 shows a series of corals from the Huon Peninsula, Papua New Guinea (Edwards *et al.*, 1993), with ages younger than 1.4 × 10^4 yr. All samples plot within error of the modern marine initial δ^{234}U contour, showing no evidence for open-system behavior, on the basis of this test. Figure 3 shows data from a series of samples from Barbados in a similar plot. These samples are all older than 8 × 10^4 yr. Most samples plot significantly off of the modern marine initial δ^{234}U contour, indicating that most have not behaved as closed systems. This is a common problem affecting samples older than ~2 × 10^4 yr: most show evidence for diagenesis. This point emphasizes the biggest hurdle to obtain an accurate sea-level curve: finding and identifying those corals that have behaved as closed systems. In this regard, the δ^{234}U test is a powerful ally. In applying this test, a common threshold for the lack of alteration is whether a sample's initial δ^{234}U value lies within 8 per mil of the marine value (Gallup *et al.*, 1994; Cutler *et al.*, 2003). Initial marine uranium isotopic composition is a necessary, but not sufficient condition to ensure closed-system behavior. For example, there are coral samples that have marine uranium isotopic composition, but have discordant ^{230}Th and ^{231}Pa ages (Edwards *et al.*, 1997; Gallup *et al.*, 2002; Cutler *et al.*, 2003); hence, the importance of a second diagenesis test can be realized.

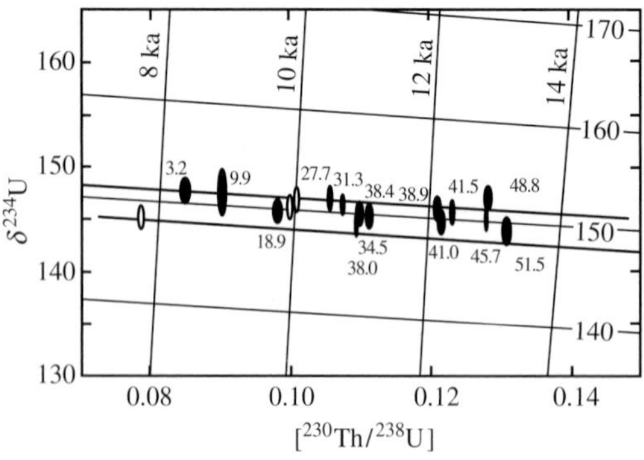

Figure 2 Plot of measured δ^{234}U versus measured ^{230}Th/^{238}U activity ratio with data from Huon Peninsula, Papua New Guinea corals (<14 kyr). Sub-vertical contours are ^{230}Th age (calculated from Equation (1)) and sub-horizontal contours are initial δ^{234}U (calculated from Equation (2)). Modern seawater δ^{234}U lies between the contours that bound the 150 per mil contour. All points plot within error of these bounds, indicating that these young corals show no evidence for diagenetic addition or exchange of uranium (after Edwards *et al.*, 1993).

Combined ^{231}Pa and ^{230}Th dating of corals provides perhaps the most rigorous test for closed-system behavior. Because mass spectrometric ^{231}Pa techniques are still fairly new and since the measurements themselves are not easy, there is still limited data of this sort. In coming years, ^{231}Pa measurements will play a major role in assessing the accuracy of ^{230}Th-based chronologies. The few such data sets on reef-building corals include data reported by Edwards *et al.* (1997), Gallup *et al.* (2002), Cutler *et al.* (2003), and Koetsier *et al.* (1999).

These data have helped to establish with high probability the accuracy of ages of key corals that record sea level (Gallup *et al.*, 2002; Cutler *et al.*, 2003). They have also demonstrated that coral samples that might otherwise be considered pristine are, in fact, altered. Data from these samples, which plot off of Concordia, include those that plot above concordia and those that plot below concordia. Figure 1 (Edwards *et al.*, 1997) shows a set of samples from the Huon Peninsula, Papua New Guinea, and Barbados. The younger (around 10 kyr old) Huon Peninsula samples (Figure 1, left graph) are all on cordia or are very close to it suggesting that these ages are affected little by diagenetic processes. Older Huon Peninsula samples (10 s to more than 100 kyr old) generally plot on, below, and above concordia (not illustrated). This indicates

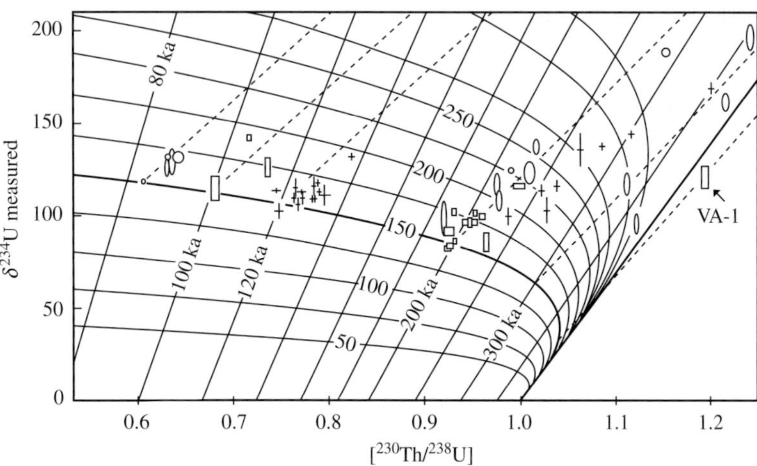

Figure 3 Analogous to Figure 2, but for older corals (>80 kyr) from Barbados. These data are typical of older corals, which either have uranium isotopic composition close to the marine value (close to the 150 per mil contour), or elevated values. Samples with values below the marine value have also been documented, but are less common. The samples that plot close to the 150 contour show no evidence for diagenesis on the basis of this test. This is an important diagenesis/age accuracy test for corals. Gallup *et al.* (1994) showed that one could model the data illustrated above with a ^{230}Th–^{234}U addition model (dashed lines show equal age samples). The model suggests that diagenetically elevated uranium isotopic composition is associated with diagenetically elevated ^{230}Th age; however, precise corrections for this effect cannot be made (after Gallup *et al.*, 1994).

that a variety of diagenetic conditions on the Huon Peninsula can result in daughter loss/parent gain as well as daughter gain/parent loss (Cheng *et al.*, 1998). In general, this contrasts with Barbados samples of about the same age, which have isotopic compositions that generally plot on or above concordia (Figure 1, right graph). Thus, Barbados diagenetic processes tend to be characterized by parent loss/daughter gain (Cheng *et al.*, 1998).

Combined ^{231}Pa–^{230}Th studies will be important in understanding diagenetic processes, and perhaps using this knowledge to constrain age even in cases where the data are discordant. However, given our state of knowledge regarding these issues, the most immediate benefit is the identification, along with other tests, of material that has remained closed to chemical exchange and therefore records accurate ^{230}Th and ^{231}Pa ages. The relatively few analyses that have been made so far indicate that concordant material can be identified and can be used to establish solid chronologies.

6.13.3.4 Sources of Error in Age

A quick glance at Equations (1)–(3) shows sources of error that contribute to error in the age, presuming that the assumptions (about initial condition and the closed system) used in calculating the equations hold. These include errors in the decay constants/half-lives and errors in the measurement of the pertinent isotope ratios. Relationships among error in half-lives, laboratory standardization procedures, and ^{230}Th age are discussed in detail by Cheng *et al.* (2000). Cheng *et al.* (2000) have also redetermined the half-lives of ^{234}U and ^{230}Th. Our recommended half-life values are shown above, including the two new values determined by Cheng *et al.* (2000). Data published prior to Cheng *et al.* (2000) do not use the revised values, whereas data published subsequently may or may not use the new values. The revised half-lives do have a small, but significant effect on calculated ^{230}Th ages, particularly ages older than ~100 ka. Furthermore, the new value for the ^{234}U half-life changes δ^{234}U values as these are calculated from measured ^{234}U/^{238}U atomic ratios using the secular equilibrium ^{234}U/^{238}U value: $\lambda_{238}/\lambda_{234}$. δ^{234}U values calculated with the new λ_{234} are ~3 per mil lower than those calculated with commonly used λ_{234} values, hence the revised modern seawater δ^{234}U of 145.8 ± 1.7 per mil (Cheng *et al.*, 2000), compared to earlier values ~3 per mil higher. In general, half-lives are now known precisely enough, so that their contribution to error in age is comparable to or smaller than typical errors in isotope ratios (determined with mass-spectrometric techniques).

As noted above, mass-spectrometric techniques supercede earlier decay-counting techniques because of their ability to detect a much larger fraction of the nuclides of interest, thereby reducing sample size and counting statistics error by large amounts (see Wasserburg, 2000; Edwards, 2000). Mass-spectrometric measurements on corals typically result in errors in ^{238}U, ^{234}U, ^{235}U, and ^{230}Th of 2 per mil or better (2σ), with the exception that fractional error in ^{230}Th typically increases progressively from this value for samples progressively younger than several ka. This results from the low concentrations of ^{230}Th in very young corals. Errors in ^{231}Pa are typically somewhat larger than those of the other isotopes, with errors of ± several per mil, except for corals younger than a few thousand years before present.

Errors in age resulting from the analytical methods are given in Figure 4. Considering ^{230}Th age, a few hundred year old samples have errors in age of ~±3 yr, 10^4 year old samples have errors in age of ~±30 yr, and 10^5 year old samples have errors of ±1 ka. Three-year old samples are distinguishable from their zero-age counterparts and the ones which are as old as 700 ka are distinguishable from infinite age samples. Considering ^{231}Pa ages, samples younger than 1,000 yr have errors of several to 10 yr in age, 10^4 yr old samples have errors of several tens to 100 yr in age, and 10^5 yr old samples have errors of a few ka in age. Samples as young as 7 yr are distinguishable from their zero-age counterparts and the ones which are as old as 250 ka are distinguishable from infinite age samples. Hence, the age of ^{230}Th samples can be measured precisely for materials younger than ~6×10^5 yr, and ^{231}Pa analyses can be used to test the closed-system assumption and age accuracy over the past 2.5×10^5 yr. Thus, in terms of analytical capabilities, age of ^{230}Th samples can provide for direct reconstruction of sea level over the past several glacial–interglacial cycles.

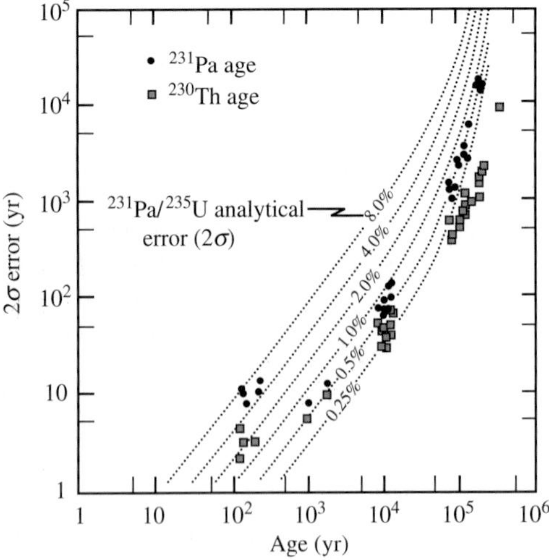

Figure 4 Plot of 2σ error in age versus age, both on log scales. ^{231}Pa dates are illustrated with solid circles; ^{230}Th dates are illustrated with gray squares. Contours indicate analytical error in ^{231}Pa/^{235}U ratio and pertain only to the solid circles. All data were collected using mass spectrometric techniques, and illustrate the high precision of this technique despite extremely low natural concentrations of the isotopes of interest (after Edwards *et al.*, 1997).

6.13.3.5 Current Status of Direct Sea-level Reconstruction: The Past 500 kyr

Figures 5–8 summarize our knowledge of sea-level history over the past 500 kyr, as determined using direct sea-level measurements. Note that we have more information about high sea levels than about low sea levels. In tectonically stable areas, materials that record low sea levels are submerged. In areas of tectonic uplift, materials that record low sea levels tend to lie stratigraphically below younger deposits from subsequent sea-level high stands. For this reason, much of our limited information about low sea levels comes from drilling or coring offshore (e.g., Fairbanks, 1989; Hanebuth *et al.*, 2000; Yokoyama *et al.*, 2000) or along shorelines in stable areas (Bard *et al.*, 1996a) and areas of tectonic uplift (e.g., Chappell and Polach, 1991; Edwards *et al.*, 1993; Cutler *et al.*, 2003).

In addition to this difference in coverage between low and high sea-level data, there is more data for more recent times than earlier times. We have virtually continuous data since the LGM (Figure 8), data on most of the important features of sea level since the end of Termination II (Figure 7), a number of key points during the penultimate glacial and interglacial periods (Figures 6 and 7), and scattered data beyond (Figure 5). This temporal trend is basically a function of more limited preservation further back in time.

Following the sea-level data in Figures 5–8 from oldest to youngest, the highest position of sea level in this interval is ~20 m above present sea level, at some time previous to 420 ± 30 kyr (Hearty *et al.*, 1999). The actual Hearty *et al.* data do not give an upper bound on the age, but other arguments suggest a marine isotope stage 11 (or MIS 11) correlation, which places it within the last 500 kyr. Confirmation of this data from another tectonically stable locality would be a major contribution. This sea-level elevation implies that the east Antarctic ice sheet must have been smaller than its present size, as the combined sea-level rise expected from complete melting of the Greenland and west Antarctic ice sheets is less than 20 m. This datum may have important implications for climate in the immediate future as orbital conditions during MIS 11 were similar to today's. Sea level

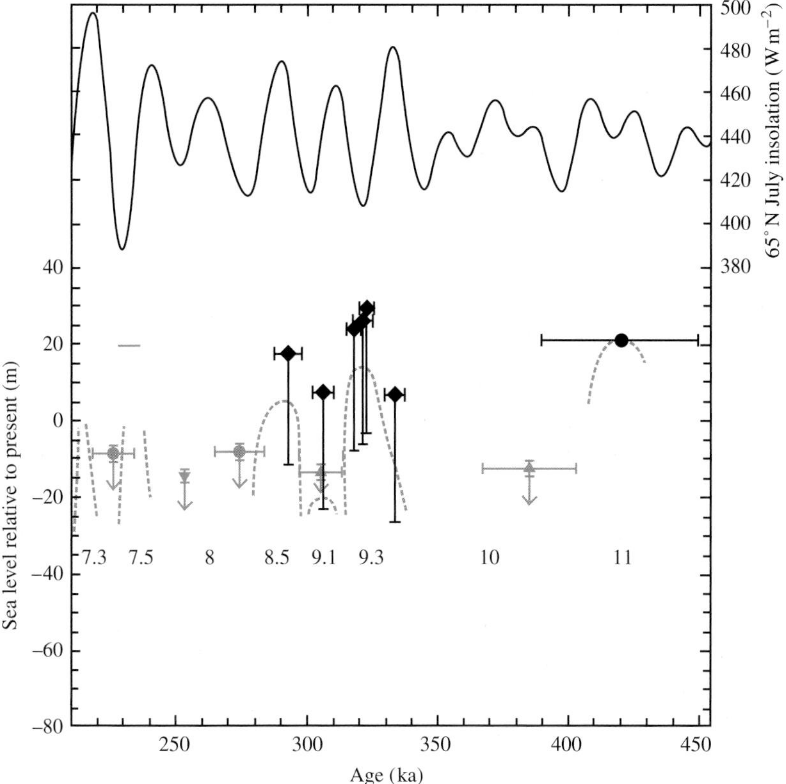

Figure 5 Direct sea-level data (450–210 kyr) from corals (diamonds, Stirling *et al.*, 2001), speleothems (solid circle, Hearty *et al.*, 1999; triangles pointing up, Richards, 1995; triangle pointing down, Borton *et al.*, 1996; gray circles, Li *et al.*, 1989 and Lundberg and Ford, 1994). The gray symbols represent maximum possible sea level based on speleothem growth. MISs are indicated by numbers. The line segment above MIS 7.5 represents the timing and duration of peak MIS 7.5 on the basis of dating of the marine oxygen isotope record (Robinson *et al.*, 2002). The placement of the line segment has no significance in terms of sea-level elevation. The MIS 11 point is actually a minimum age (Hearty *et al.*, 1999). July summer solar insolation at 65 degrees north latitude is depicted in the top panel (Berger, 1978).

fell from MIS 11 levels, below −12 m by 385 ± 15 kyr, signaling the start of MIS 10 (Richards, 1995).

On the basis of coral dating (Stirling *et al.*, 2001), sea level reached peak MIS 9 levels (likely MIS 9.3) by 324 ± 3 kyr, dropped to ~25 m below peak levels by 318 ± 3 kyr (likely the beginning of MIS 9.2), then reached subsequent maxima at 306 ± 4 kyr (plausibly MIS 9.1) and 293 ± 5 kyr (plausibly MIS 8.5). The Stirling *et al.* (2001) corals are remarkably well preserved, on the basis of their δ^{234}U values; however, as yet, there is no confirmation of closed-system behavior from ^{231}Pa dating. The Stirling *et al.* (2001) data place maximum sea level during MIS 9 between −4 m and 29 m. Stirling *et al.* (2001) present current elevations for the dated corals (highest elevation is 29 m) and stipulate an uplift rate of <0.1 m kyr^{-1}. The highest possible uplift rate implies a correction of 33 m, yielding a peak sea-level value of >−4 m. Presuming the corals grew at the sea level, the maximum possible sea level is the uncorrected value: <29 m. Speleothem growth stipulates that sea level must have dropped below −13 m by 305 ± 8 kyr, remained below −13 m until at least 276 ± 8 kyr, constraining the MIS 9/8 boundary (Richards, 1995). The timing of the MIS 9 sea levels is broadly consistent with orbital forcing. Stirling *et al.* (2001) also determine a date of 632 ± 21 kyr for the age of MIS 15 (not shown in Figure 5).

MIS 7 sea levels are known in somewhat more detail, from a number of different studies (Figures 5–7). At the beginning of MIS 7.5, sea level rose above −14 m subsequent to 253 ± 4 kyr (Borton *et al.*, 1996). Sea level dropped below −8.5 m from peak MIS 7.5 levels by 225 +6/−7 kyr and rose above −8.5 m to peak MIS 7.3 levels subsequent to this date (Li *et al.*, 1989; Lundberg and Ford, 1994; Toscano and Lundberg, 1999). The length of MIS 7.5 (from 237 kyr to 228 kyr), based on direct dating of the marine oxygen isotope record (Robinson *et al.*, 2002), is

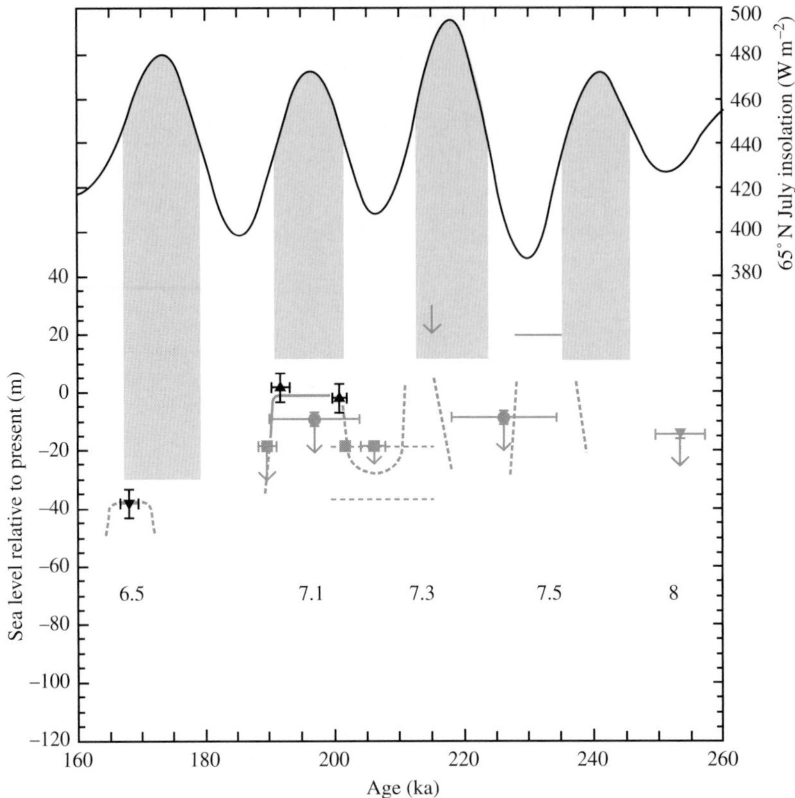

Figure 6 Direct sea-level data (260–160 kyr) from corals (solid triangles, Gallup *et al.*, 1994, 2002; Edwards *et al.*, 1997) and speleothems (gray symbols: triangle, Borton *et al.*, 1996; circles, Li *et al.*, 1989 and Lundberg and Ford, 1994; squares, Bard *et al.*, 2002). The gray symbols represent maximum possible sea level based on speleothem growth. MISs are indicated by numbers. The line segment above MIS 7.5 represents the timing and duration of peak MIS 7.5 on the basis of dating of the marine oxygen isotope record (Robinson *et al.*, 2002). The arrow above the MIS 7.3 peak indicates the beginning of MIS 7.3, determined by direct dating of the marine oxygen isotope record (Robinson *et al.*, 2002). The placements of the line segment and arrow have no significance in terms of sea-level elevation. The pair of dashed lines indicate constraints on MIS 7.2 sea-level elevation. This constraint comes from a coral with significant uncertainty in age, but which nevertheless constrains MIS 7.2 sea-level elevation (Gallup *et al.*, 1994). July summer solar insolation at 65 degrees north latitude is depicted in the top panel (Berger, 1978). The vertical gray bars indicate times of high insolation and are intended as aids in comparing the curves.

consistent with the direct sea-level data. The Robinson *et al.* data also place the beginning of MIS 7.3 at 215 kyr, consistent with the 225 +6/−7 kyr maximum MIS 7.3 age from the speleothem data. Coral data indicate that sea level fell to elevations as low as −28 ± 9 m during MIS 7.2 (Gallup *et al.*, 1994). Consistent with this value, speleothem data indicate that sea level during MIS 7.2 was lower than −18 m from at least 206.1 ± 1.9 kyr to 201.6 ± 1.8 kyr, and rose above −18 m at some time subsequent to the latter date (Bard *et al.*, 2002). The 201.6 kyr age for the end of MIS 7.2 is consistent with previous coral data that place the beginning of MIS 7.1 at 200.8 ± 1.0 kyr (Gallup *et al.*, 1994). Coral data indicate that MIS 7.1 lasted until at least 191.7 ± 1.6 kyr (Edwards *et al.*, 1997). The coral data place peak MIS 7.1 sea level between −6 m and 9 m. Speleothem data indicate that sea level fell below −15 m by 190 ± 5 kyr (Richards, 1995), and below −18 m by 189.7 ± 1.5 kyr (Bard *et al.*, 2002), signaling the start of MIS 6. The speleothem data for the beginning of MIS 6 are consistent with the coral data for the end of MIS 7, illustrating the complementary nature of the two types of information. The coral data give a minimum duration for MIS 7.1 of 9.1 ± 1.9 kyr, at an elevation between −6 m and 9 m and the speleothem data give a maximum duration of 11.9 ± 2.6 kyr at an elevation above −18 m.

Note that the coral MIS 7.1 elevation data (−6 m to 9 m; Gallup *et al.*, 1994) and some Bahamas speleothem data (Li *et al.*, 1989; Lundberg and Ford, 1994; Toscano and Lundberg, 1999) are nominally inconsistent. However, considering 2σ dating errors, it is possible that the critical speleothem growth phase actually correlates to MIS 6. If so, the speleothem was

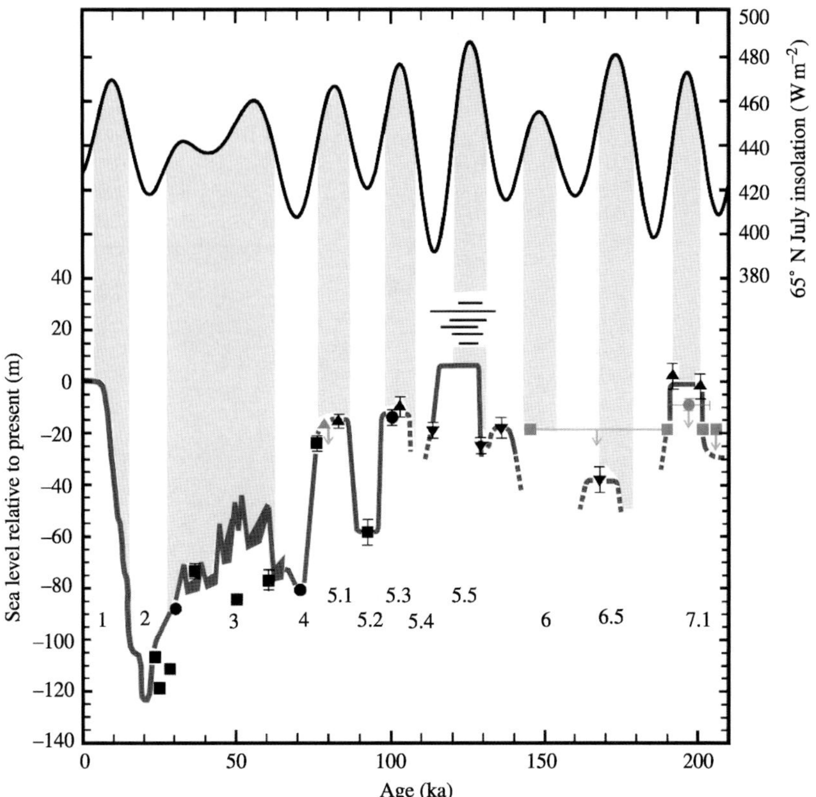

Figure 7 Direct sea-level data (210 kyr to present) from corals (solid symbols: triangles pointing up, Edwards *et al.*, 1997; triangles pointing down, Gallup *et al.*, 2002; squares, Cutler *et al.*, 2003; circles, Bard *et al.*, 1990b) and speleothems (gray symbols: squares, Bard *et al.*, 2002; circle, Li *et al.*, 1989 and Lundberg and Ford, 1994; triangle, Richards *et al.*, 1994). The gray symbols represent maximum possible sea level based on speleothem growth. MISs are indicated by numbers. The line segments above MIS 5.5 are different estimates of the timing and duration of peak MIS 5.5 sea level from corals and from direct dating of the marine oxygen isotope record. From top to bottom the line segments are from the following studies: Edwards *et al.* (1987b), Muhs *et al.* (2002), Chen *et al.* (1991), Stirling *et al.* (1998), Slowey *et al.* (1996), and Stirling *et al.* (1995). The placements of the line segments have no significance in terms of sea-level elevation. The spline between 65 kyr and 35 kyr is from Chappell (2002). Note that the spline disagrees with a ^{231}Pa–^{230}Th concordant point at ~50 kyr. This indicates that either the spline or the point are in error, or that the spline does not capture the full range of sea-level variability in this time-range. July summer solar insolation at 65 degrees north latitude is depicted in the top panel (Berger, 1978). The vertical gray bars indicate times of high insolation and are intended as aids in comparing the curves.

not active during MIS 7.1, thus resolving the discrepancy.

The date of the beginning of MIS 6 is best constrained by the speleothem data outlined above (Richards, 1995; Bard *et al.*, 2002). Within MIS 6, the most stringent speleothem constraints place sea level below −18 m from 189 ± 1.5 kyr to 145.2 ± 1.1 kyr (Bard *et al.*, 2002). Coral data place MIS 6.5 sea level at −38 ± 5 m at ~168.0 ± 1.3 kyr (Gallup *et al.*, 2002).

The transition between MIS 6 and MIS 5 (Termination II) has been a subject of intense study and debate. Much of this debate concerns whether or not a significant fraction of sea-level rise preceded northern-hemisphere summer insolation rise. If so, northern summer insolation could not have forced Termination II directly. Corals that give concordant ^{231}Pa and ^{230}Th ages place sea level at −18 m at 135.8 ± 0.8 kyr (Gallup *et al.*, 2002). Presuming minimum MIS 6 sea levels were at least as low as LGM values, sea level rise, prior to the main insolation rise, must be ~100 m to this level. There is some suggestion from coral work that sea level may have then fallen to levels of −25 m at 129.5 ± 0.5 kyr (Gallup *et al.*, 2002) or perhaps as low as −90 m at 130 ± 2 kyr (Esat *et al.*, 1999). However, there are some questions about the stratigraphy of the former site and the latter data have not been confirmed with ^{231}Pa analyses. Direct dating of the marine oxygen isotope record places the mid-point of Termination II at 135 ± 2.5 kyr, broadly consistent with the Gallup *et al.* (2002) data. Thus, at present, the evidence suggests that a good

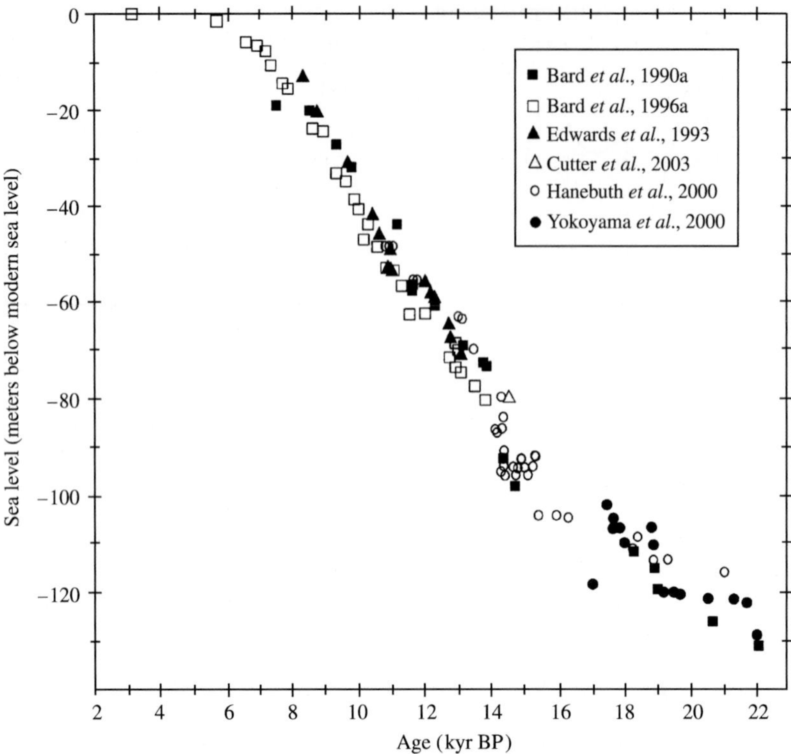

Figure 8 Direct sea-level data (23 kyr to present) from the indicated references. The data from different sites generally agree, indicating that different far-field sites have not been affected differentially by isostatic effects. Note also the apparent rapid rises (melt-water pulses) in sea level at ~19 kyr, 14.6 kyr, and 11 kyr. The latter two correspond to times of abrupt rise in Greenland temperature and east Asian monsoon intensity.

portion, but not all of Termination II sea-level rise, took place before the bulk of Termination II insolation rise.

A large number of studies have determined the timing and height of sea level during the last interglacial period (MIS 5.5). Most studies place sea level at 5 ± 3 m during this interval (see Ku *et al.* (1974) for early work and Gallup *et al.* (1994) and references therein), with some indication of a minor regression and subsequent transgression in the middle of MIS 5.5 (at ~125 kyr; Chen *et al.*, 1991). Despite the large number of ^{230}Th dates, only one ^{230}Th–^{231}Pa concordant date exists for MIS 5.5 corals (Edwards *et al.*, 1997), illustrating the early status of ^{231}Pa analyses. There is a large spread in the hundreds of MIS 5.5 ^{230}Th ages (considered reliable on the basis of uranium isotopic composition (Edwards *et al.*, 1987a,b, 1997; Bard *et al.*, 1990b, 1996b; Chen *et al.*, 1991; Hamelin *et al.*, 1991; Collins *et al.*, 1993a,b; Stein *et al.*, 1993; Zhu *et al.*, 1993; Gallup *et al.*, 1994; McCulloch and Esat, 2000; Muhs *et al.*, 1994, 2002; Szabo *et al.*, 1994; Stirling *et al.*, 1995, 1998; Eisenhauer *et al.*, 1996; Esat *et al.*, 1999)). For example, Muhs *et al.* (2002) give a range of 134 kyr to 113 kyr for corals from Oahu. However, dates tend to be concentrated toward the middle of this range. For example, Stirling *et al.* (1998) suggest that the main period of MIS 5.5 reef growth took place between 128 kyr and 121 kyr, with the possibility that less robust reef growth continued as late as 116 ± 1 kyr. Furthermore, ^{230}Th–^{231}Pa concordant dates place sea level below MIS 5e levels at 129.1 ± 0.5 kyr (Gallup *et al.*, 2002) and at 113.1 ± 0.7 kyr (Cutler *et al.*, 2003). Thus, it would appear that at least some of the ^{230}Th data on the old end of the range and perhaps the young end of the range are diagenetic outliers. Based on our arguments, our spline in Figure 7 places the final rise to peak MIS 5.5 sea levels between 129.1 kyr and 128 kyr and the drop from peak MIS 5.5 sea level at 116 ± 1 kyr. If our interpretation of the vast body of data for Termination II and MIS 5e is correct, then much of the rise in sea level during Termination II could not have been forced by orbital geometry, but the final rise to peak MIS 5.5 sea level could well have been forced by orbital changes.

In our discussion of sea level between MIS 5.5 and MIS 2, we follow the results of Cutler *et al.* (2003), who have summarized most of the latest data in this range and Chappell (2002), who presents some of our best constraints on sea level during MIS 3. Surprisingly, we have no data for MIS 5.4, nor are there data for MIS 5.3 that are reliable on the basis of both uranium isotopic composition and ^{230}Th–^{231}Pa concordance. On the basis of both criteria, Cutler *et al.* (2003) place MIS 5.2 sea level at −58 m at 92.6 ± 0.5 kyr. Sea level then rose to levels of −13 m to −18 m at 83.3 ± 0.4 kyr (^{231}Pa–^{230}Th concordant data from Barbados; Gallup *et al.*, 1994; Edwards *et al.*, 1997) during MIS 5.1. There is clear evidence that MIS 5.1 sea level reached higher values than the Gallup *et al.* (1994) Barbados point. The strongest evidence is found at localities north of Barbados. Evidence from Florida indicates that MIS 5.1 sea level reached values as high as −7 m to −9 m (Ludwig *et al.*, 1996; Toscano and Lundberg, 1999), with the possibility that MIS 5.1 sea levels reached values as high as 1–2 m above present as recorded on Bermuda (Ludwig *et al.*, 1996) and along the US Atlantic Coastal Plain (Szabo, 1985). The temporal relationship between the Gallup *et al.* (1994) data and the "high sea-level" data is not clear, although much of the "high sea-level" data are nominally younger than the Gallup *et al.* (1994) point. Two possibilities seem plausible. (i) MIS 5.1 sea level rose from the level reported by Gallup *et al.* (1994) to the higher values reported in Ludwig *et al.* (1996), Toscano and Lundberg (1999), and Szabo (1985). If so, evidence for these higher values are yet to be identified on Barbados. (ii) Isostatic effects cause an increase in apparent MIS 5.1 sea level to the north. Such an explanation is plausible, considering that the northernmost sites (Virginia) are not far from the Laurentide ice sheet. Regardless of the explanation, speleothem data from the Bahamas (Richards *et al.*, 1994) constrain sea level to below −17 m by 79.6 ± 1.8 kyr. Coral data from Barbados (^{230}Th–^{231}Pa concordant; Cutler *et al.*, 2003) indicate a sea level of −21 m at 76.2 ± 0.3 kyr. This point is critical as it constrains the age of the end of MIS 5.1.

A coral data point places early MIS 4 sea level at −81 m at 70.8 kyr (Bard *et al.*, 1990b). The Cutler *et al.* point for the end of MIS 5.1 and the Bard *et al.* point for early MIS 4 delineate the magnitude, rate, and timing of the largest sea-level drop in the last interglacial–glacial cycle. Sea level dropped 60 m, half the full glacial–interglacial amplitude, in 5.4 kyr, an average rate of 10.4 m kyr^{-1} (Cutler *et al.*, 2003). This rate corresponds to a minimum average accumulation rate of ice on the continents of 18 cm yr^{-1}. We calculate a minimum value by assuming ice accumulation over the maximum area covered by ice sheets on North America and Eurasia during the LGM. Ice buildup of this magnitude is a major challenge, as ice buildup at much lower rates has been difficult to simulate. It is likely that this interval was characterized by sharp latitudinal changes in temperature such that warm-moist air originating from low latitudes is cooled rapidly at mid- to high latitudes. In such a scenario, latitudinal moisture transport from the tropics to northern glaciers would be efficient. Supporting this scenario is the peak in northern-winter latitudinal insolation gradient coincident with this interval (Cutler *et al.*, 2003).

MIS 3 sea levels are poorly known. There are only three ^{230}Th–^{231}Pa concordant data points between 65 kyr and 35 kyr (recording sea levels between −85 m and −74 m; Cutler *et al.*, 2003; Figure 7). Perhaps our best estimates of sea levels during this interval come from a study by Chappell (2002) from the Huon Peninsula, Papua New Guinea. Chappell uses a series of ^{230}Th-dated corals, combined with a geomorphic analysis and model of coral terrace structure. We show his data and a spline derived from his geomorphic analysis in Figure 7. Although the ^{230}Th dates have not been confirmed with ^{231}Pa analysis, the points and spline agree broadly with the limited number of concordant points. The Chappell analysis suggests that MIS 3 sea level was characterized by ~6,000 yr cycles of gradual sea-level fall followed by rapid sea-level rise, with an amplitude of 10–15 m. Chappell correlates the cycles with Bond cycles, with sea-level rise corresponding broadly to the timing of Heinrich events. Presuming these correlations are correct, a particular sea-level rise could either be associated with the Heinrich event itself, or the subsequent interstadial event. In either case, the Chappell contribution makes a strong case for millennial-scale sea-level variations during MIS 3.

Two coral data constrain sea level during the MIS 3–MIS 2 transition, one from Barbados (−88 m at 30.3 kyr, ^{230}Th only; Bard *et al.*, 1990b) and one from the Huon Peninsula (−107 m at 23.7 kyr, ^{231}Pa–^{230}Th concordant; Cutler *et al.*, 2003). Two other concordant data are consistent with the latter point, but are from corals that likely grew several meters below sea level (Cutler *et al.*, 2003).

The sea-level curve from the LGM through Termination I is now known in some detail, largely from studies at five localities (Figure 8): Barbados (^{230}Th and ^{14}C dating of corals; Fairbanks, 1989; Bard *et al.*, 1990a), the Huon Peninsula Papua New Guinea (^{230}Th and ^{14}C dating of corals; Chappell and Polach, 1991; Edwards *et al.*, 1993), Tahiti (^{230}Th and ^{14}C dating of corals; Bard *et al.*, 1996a), the Sunda Shelf, southeast Asia (^{14}C dating of sea-level indicators; Hanebuth *et al.*, 2000), and the Bonaparte Gulf, northwest Australia (^{14}C dating of sea-level indicators; Yokoyama *et al.*, 2000). The data from these far-field sites agree remarkably well (Figure 8) despite potential differences from isostatic effects. The total amount of sea-level change during this interval is large, the full glacial–interglacial amplitude. Furthermore, rates of sea-level change are extremely high during much of this interval. The Earth should, therefore, be far from isostatic equilibrium during portions of this interval. Because the sites record the same deglacial sea-level history (Figure 8), isostatic adjustments must not have affected these far-field sites differentially by more than a several meters. The one clearly documented instance of different sea-level history at different far-field sites comes from mid-Holocene sea level, where tropical Pacific sites record an "overshoot" to values slightly higher than the modern level, whereas Caribbean sites record an asymptotic approach to modern values (not shown in Figure 8). The difference is ~2–3 m. Thus, the empirical evidence suggests that these far-field sites are not differentially affected by isostatic adjustments at a resolution of several meters. This observation in itself may place constraints on the melting history of the ice sheets. For example, based on the similarity of sea-level histories, Clark *et al.* (2002) demonstrated that the sea-level rise at ~14.5 kyr (Figure 8) could not result solely from melting of the Laurentide, but required contributions from other ice sheets. In any case, the empirical evidence suggests relatively small differential isostatic effects for far-field sites.

The combined LGM–deglacial data show sea-level rise taking place between 19 kyr and 6 kyr, starting from an LGM value of about −121 m (Fairbanks, 1989) and ending close to the modern level. Sea levels during the several previous glacial periods in this interval are within 20 m of the LGM value (Rohling *et al.*, 1998). The deglacial sequence is punctuated by rapid jumps (melt-water pulses; Figure 8) at ~19 kyr, 14.5 kyr, and 11 kyr to 10 kyr. Sea-level rise rates reached values ~20 m kyr^{-1} during these intervals. The latter two pulses coincide with or just postdate times of jumps in Greenland temperature (GISP, 1997; Dansgaard *et al.*, 1993) and east Asian monsoon intensity (Wang *et al.*, 2001). The jumps correspond to the beginning of the Bolling-Allerod (14.66 kyr; Wang *et al.*, 2001) and the end of the Younger Dryas (11.55 kyr; Wang *et al.*, 2001), further supporting the idea that these events along with other millennial-scale events affect a large geographic area. If the correlations between sea level, Greenland temperature, and east Asian monsoon intensity can be generalized back in time, this would suggest that the Chappell (2002) sea-level jumps (which are about the same magnitude as the deglacial pulses) correlate with the beginning of the larger Greenland interstadials (immediately after Heinrich events).

6.13.4 HISTORY AND CURRENT STATE OF SEA-LEVEL DETERMINATIONS FROM OXYGEN ISOTOPE MEASUREMENTS

6.13.4.1 History of Marine Oxygen Isotope Measurements and Sea Level

The first deep-sea oxygen isotope record was put forth by Emiliani (1955), who interpreted the $^{18}O/^{16}O$ variations largely in terms of temperature change. Shackleton (1967) challenged this interpretation, suggesting instead that the changes in isotopic composition were largely the result of ice-volume changes. Shackleton based his arguments upon: (i) calculations based on estimates of the average isotopic composition and volume of the ice sheets; (ii) comparisons between benthic and planktonic records; and (iii) benthic records from localities (deep Pacific) where water temperatures were within 1–2 °C of the freezing point (and therefore could not have cooled by more than this difference). Quantitative separations of the two components were not attempted until Chappell and Shackleton (1986) compared direct sea-level measurements from the Huon Peninsula with a benthic deep-sea oxygen isotope record. They suggested that Pacific benthic oxygen isotope records had a significant temperature component, with 1.5 °C increases in temperature at terminations I and II and a similar decrease at the end of MIS 5.5. This placed Pacific deep waters at close to the freezing point between MIS 5.4 and MIS 2 (see Chapter 6.14). Cutler *et al.* (2003) have performed a similar analysis, but with the key additional constraint of the direct LGM seawater $^{18}O/^{16}O$ measurement (Schrag *et al.*, 1996). Recent years have seen the development of foram Mg/Ca thermometry (Nurnberg, 1995; Nurnberg *et al.*, 1996; Lea *et al.*, 1999; Elderfield and Ganssen, 2000; Chapter 6.14) and its use to separate the seawater and temperature components of the marine oxygen isotope record (Elderfield and Ganssen, 2000; Lea *et al.*, 2000, 2002).

6.13.4.2 Comparison of Direct Sea-level and Benthic Foram Records

The most recent analysis of comparisons between benthic foram and direct sea-level records has been presented by Cutler *et al.* (2003). These comparisons, coupled with LGM pore-fluid data (Schrag *et al.*, 1996), demonstrate that deep-sea temperatures changed significantly. The Cutler *et al.* (2003) separations of ice volume and temperature components of benthic Atlantic (EW9209-1; Curry and Oppo, 1997) and Pacific (V19-30; Shackleton *et al.*, 1983) records are shown in Figure 9. Temperatures reach values near the freezing point at the LGM in both the

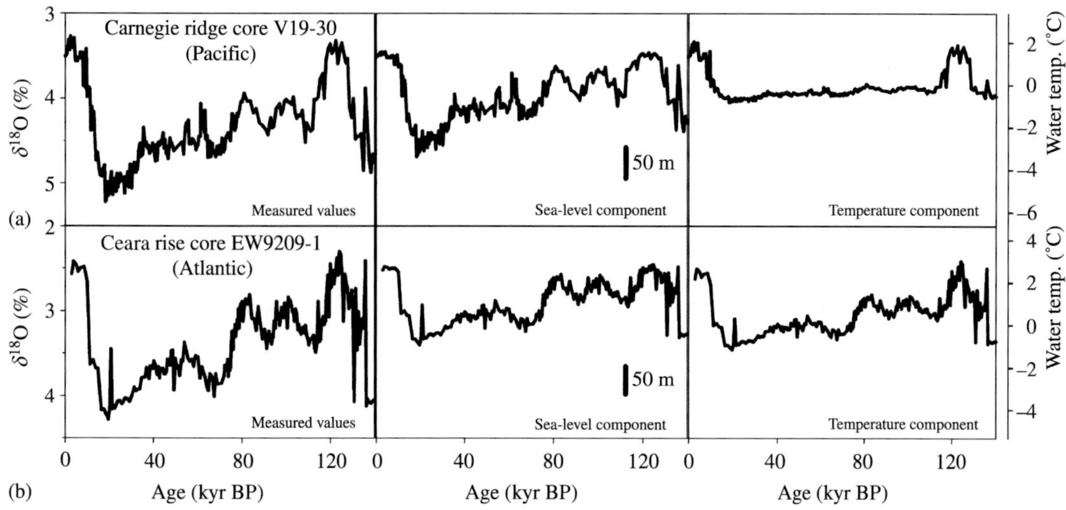

Figure 9 Benthic marine oxygen isotope records (left diagrams) from the Carnegie Ridge (a) (Pacific: Shackleton *et al.*, 1983) and the Ceara Rise (b) (Atlantic: Curry and Oppo, 1997) and their separation into sea-level components (middle diagrams) and temperature components (right diagrams) (after Cutler *et al.*, 2003).

Atlantic and Pacific. The largest shifts in temperature take place at the terminations and at the end of MIS 5.5 (confirming the earlier Chappell and Shackleton (1986) results), with smaller temperature changes during the noninterglacial stages. The calculated sea-level curves are broadly consistent with the direct sea-level curve; however, the calculated curves are continuous. One of the main strengths of this sort of reconstruction is the incorporation of direct sea-level data into the analysis. The accuracy of the separation of temperature and sea-level components is highest at relatively high sea levels and relatively low sea levels where the LGM pore-fluid value adds a key constraint. However, even for intermediate sea levels, the temperature component must be significant, considering the largest possible errors.

6.13.4.3 $^{18}O/^{16}O$-based Sea-level Records

Two recent records, based solely on $^{18}O/^{16}O$ measurements, are shown in Figure 10, along with the direct sea-level spline (Figures 5–8) and the Cutler *et al.* Pacific record (Figure 9, top center). The thin dark gray trace is the Shackleton (2000) record based on the $^{18}O/^{16}O$ history of atmospheric O_2. The thin solid black trace is the Lea *et al.* (2002) east Pacific planktonic foram record, for which the temperature component has been subtracted using Mg/Ca thermometry. If one allows for possible errors in timescales, the Shackleton (2000, Figure 10), Lea *et al.* (2002, Figure 10), Cutler *et al.* (2003) foram-based sea-level curve (dashed thin black trace, Figure 10), and the direct sea-level curves (thick gray trace, Figure 10) are remarkably consistent, broadly

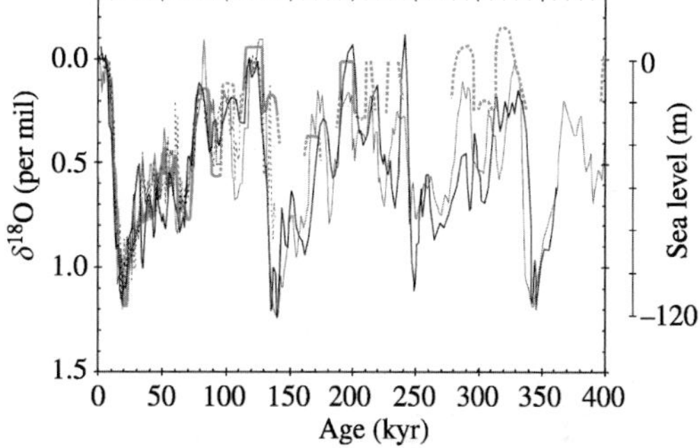

Figure 10 Comparisons of the direct sea-level record (thick gray trace from Figures 5–8), the sea-level record from the top-center panel of Figure 9 (finely dashed line between 130 kyr and present; Cutler *et al.*, 2003), a sea-level record based on subtraction of the temperature component (using Mg/Ca thermometry) from a planktonic marine oxygen isotope record (solid black trace; Lea *et al.*, 2002), and a sea-level record based on the separation of oxygen isotope components from the record of $^{18}O/^{16}O$ variations in atmospheric O_2 (coarsely dashed trace, Shackleton, 2000).

justifying the various methods employed. Despite the general agreement among the different types of records, there are conspicuous discrepancies. For example, the direct sea-level record places MIS 7.1 sea level close to modern levels and MIS 6.5 sea level at −38 m. The Lea *et al.* record agrees with both of these values; however, the Shackleton record places MIS 7.1 and MIS 6.5 sea levels at about the same elevation, below modern levels, but above −38 m. The direct record places peak MIS 9 at about present sea level or higher; however, the Lea *et al.* record has MIS 9 sea levels significantly below modern values. For MIS 8.5, the Lea *et al.* and Shackleton records disagree, with the Shackleton value agreeing with the direct sea-level constraints. Broadly speaking, the oxygen-isotope-based records are less precise in "sea-level elevation" and generally have chronologies based on orbital tuning; however, the records are continuous over long time intervals. The direct sea-level records constrain sea-level elevation more precisely and accurately and have chronologies based on radiometric dating; however, with the exception of the last deglacial sequence, these records are discontinuous. In general, the oxygen-isotope-based records are most useful where direct sea-level constraints are lacking or where dating errors are large. Broadly speaking, this is in the older portion of time interval in question and the lower portion of the sea-level range.

6.13.5 CAUSES OF SEA-LEVEL CHANGE AND FUTURE WORK

A large body of evidence indicates that orbital geometry is an important control on sea-level change. The modern version of this theory was put forth by Milankovitch (1941). The simplest form of this theory maintains that northern high-latitude summer insolation forces the glacial cycles: cool summers favoring ice buildup and warm summers favoring loss of ice. Early evidence in support of this idea was provided by direct sea-level measurements (Broecker *et al.*, 1968; Mesolella *et al.*, 1969) and by spectral analysis of the marine oxygen isotope record (Hays *et al.*, 1976). The early evidence also showed that sea level did not respond in a linear fashion to insolation changes. For example, the marine oxygen isotope record is dominated by the power in the 100 kyr band; however, the northern high-latitude summer insolation curve has negligible power in this band (Hays *et al.*, 1976; Imbrie *et al.*, 1984). The so-called 100 kyr problem has attracted much attention over the years, but as yet has no generally accepted solution. Another manifestation of the 100 kyr problem is the general sawtooth pattern of the sea-level curve (Figures 5–10), with rapid deglaciation (terminations) leading to interglacial periods followed by more gradual ice buildup leading to glacial maximum conditions (Broecker and Van Donk, 1970). The period for each cycle is ~100 kyr. The insolation curve, alternatively, is smoothly varying, with no features analogous to the terminations (Figures 5–7).

Our understanding of sea-level history supports the general concept of orbital forcing of sea level, but also points to other controls. Some relationship between sea level and orbital forcing is clear from a simple comparison between the two curves (Figures 5–7). To the extent that orbital forcing does control sea level, it is likely that northern summer insolation is not the only important aspect of such forcing. For example, periods of rapid glaciation (e.g., the MIS 5–MIS 4 transition) take place at times of high northern low-latitude winter insolation and high northern winter latitudinal-insolation gradient (Cutler *et al.*, 2003), supporting the idea that these conditions are important controls on the efficiency of moisture transport (Ruddiman *et al.*, 1980; Cutler *et al.*, 2003), and thereby affect glacial buildup and decay. It has also been proposed that high-latitude southern insolation has a bearing on sea level, perhaps by triggering atmospheric CO_2 rise (Henderson and Slowey, 2000).

Beyond orbital forcing, evidence is mounting for sea-level changes related to millennial-scale climate variations. For example, deglacial sea level rises rapidly at times that correspond to or immediately postdate abrupt increases in Greenland temperature and increases in east Asian monsoon intensity. During MIS 3, sea-level shifts have "periodicities" similar to Bond cycles. The causes of millennial-scale events are themselves a matter of some debate; however, it is likely that massive atmospheric and oceanic circulation changes are involved. Although largely confined to the Holocene, there is some evidence that solar changes may be involved (Flint, 1971; Denton and Karlen, 1973; Bond *et al.*, 2000; Neff *et al.*, 2001). Thus, circulation changes likely affect sea level, and perhaps solar changes as well.

For years, studies of the timing of Termination II have been a focal point of sea-level work. Recent studies suggest that the bulk of Termination II sea-level rise did, in fact, take place before the bulk of northern hemisphere summer insolation rise (Henderson and Slowey, 2000; Gallup *et al.*, 2002). Possibilities for the cause of early Termination II sea-level rise include northern hemisphere insolation forcing modulated by isostatic effects (e.g., Birchfield *et al.*, 1981; Peltier and Hyde, 1986), southern hemisphere insolation forcing tied to the atmospheric CO_2 rise (Henderson and Slowey, 2000), sea-level rise associated with millennial-scale climate change and circulation changes, and CO_2 rise due to other factors such as changing dust contents affecting ocean productivity (Broecker and Henderson, 1998). It does appear that the final increment of sea-level rise during Termination II took place at about the time of the northern hemisphere summer insolation high. Thus, orbital geometry may well have caused the rise to full interglacial sea-level elevation and the onset of full interglacial conditions worldwide.

Future challenges in sea-level work include increasing the resolution of the direct sea-level curve for virtually all times, increasing the reliability of direct sea-level data using ^{231}Pa dating to test the reliability of ^{230}Th dates, improving our ability to resolve the seawater $^{18}O/^{16}O$ record from foram and O_2 records, working with new sea-level proxies such as the pore-fluid chloride measurements. Our ability to improve the resolution and reliability of the direct sea-level curve will depend on our ability to identify and collect unaltered material from key time intervals. Such efforts are of great importance in understanding the basic controls on the Earth's climate.

REFERENCES

Adkins J. F. and Schrag D. P. (2001) Pore fluid constraints on past deep-ocean temperature and salinity. *Geophys. Res. Lett.* **28**, 771–774.

Allegre M. C. (1964) De l'extension de la methode de calcul graphique Concordia aux mesures d'ages absolus effectues a l'aide du desequilibre radioactif. Cas des mineralisations secondaires d'uranium. *Note (*) de. C. R. Acad. Sci. Paris t.* **259**, 4086–4089.

Bard E., Hamelin B., Fairbanks R. G., and Zindler A. (1990a) Calibration of ^{14}C time scaleover the last 30,000 years using mass spectrometric U–Th ages from Barbados corals. *Nature* **345**, 461–468.

Bard E., Hamelin B., and Fairbanks R. G. (1990b) U–Th ages obtained by mass spectrometry in corals from Barbados: sea level during the past 130,000 years. *Nature* **346**, 456–458.

Bard E., Hamelin B., Arnold M., Montaggioni L., Cabioch G., Faure G., and Rougerie F. (1996a) Deglacial sea-level record from Tahiti corals and the timing of global meltwater discharge. *Nature* **382**, 241–244.

Bard E., Jouannic C., Hamelin B., Pirazzoli P., Arnold M., Faure G., Sumosusastro P., and Syaefudin (1996b) Pleistocene sea levels and tectonic uplift based on dating of corals from Sumba Island, Indonesia. *Geophys. Res. Lett.* **23**, 1473–1476.

Bard E., Antonioli F., and Silenzi S. (2002) Sea-level during the penultimate interglacial period based on a submerged stalagmite from Argentarola Cave (Italy). *Earth Planet. Sci. Lett.* **196**, 135–146.

Bar-Matthews M., Wasserburg G. J., and Chen J. H. (1993) Diagenesis of fossil coral skeletons—correlation between trace-elements, textures, and U-234/U-238. *Geochim. Cosmochim. Acta* **57**, 257–276.

Barnes J. W., Lang E. J., and Potratz H. A. (1956) The ratio of ionium to uranium in coral limestone. *Science* **124**, 175–176.

Bateman H. (1910) The solution of a system of differential questions occurring in the theory of radioactive transformations. *Proc. Cambridge Phil. Soc.* **15**, 423–427.

Bender M., Labeyrie L. D., Raynaud D., and Lorius C. (1985) Isotopic composition of atmospheric O_2 in ice linked with deglaciation and global primary productivity. *Nature* **318**, 349–352.

Bender M., Sowers T., Dickson M. L., Orchardo J., Grootes P., Mayewski P. A., and Meese D. A. (1994) Climate correlations between Greenland and Antarctica during the past 100,000 years. *Nature* **372**, 663–666.

Berger A. L. (1978) Long-term vatiations of caloric insolation resulting from the earth's orbital elements. *Quat. Res.* **9**, 139–167.

Birchfield G. E., Weertman J., and Lunde A. T. (1981) A paleoclimatic model of Northern Hemisphere ice sheet. *Quat. Res.* **15**, 126–142.

Bischoff J. L. and Fitzpatrick J. A. (1991) U-series dating of impure carbonates: an isochron technique using total-sample dissolution. *Geochim. Cosmochim. Acta* **55**, 543–554.

Bloom A. L., Broecker W. S., Chappel J. M. A., Matthews R. K., and Mesolella K. J. (1974) Quaternary sea-level fluctuations on a tectonic coast: new $^{230}Th/^{234}U$ dates from the Huon Peninsula, New Guinea. *Quat. Res.* **4**, 185–205.

Boltwood B. B. (1907) Note on a new radio-active element. *Am. J. Sci.* **24**, 370–372.

Bond G., Kromer B., Beer J., Muscheler R., Evans M. N., Showers W., Hoffmann S., Lotti-Bond R., Hajdas I., and Bonani G. (2001) Persistent solar influence on North Atlantic climate during the Holocene. *Science* **294**, 3130–3136.

Borton C. J., Edwards R. L., and Richards D. A. (1996) Constraints on sea level from Bahamas. *AGU 1996 Spring Meeting, Trans. Am. Geophys. Union* **77**, S168.

Broecker W. S. (1963) A preliminary evaluation of uranium series inequilibrium as a tool for absolute age measurements on marine carbonates. *J. Geophys. Res.* **68**, 2817–2834.

Broecker W. S. and Henderson G. M. (1998) The sequence of events surrounding Termination II and their implications for the cause of glacial–interglacial CO_2 change. *Paleoceanography* **13**, 352–364.

Broecker W. S. and Van Donk J. (1970) Insolation changes, ice volume and the ^{18}O record in deep-sea cores. *Rev. Geophys. Space Phys.* **8**, 169–198.

Broecker W. S., Thurber D. L., Goddard J., Ku T. K., Matthews R. K., and Mesolella K. J. (1968) Milankovich hypothesis supported by precise dating of coral reefs and deep sea sediments. *Science* **159**, 297–300.

Chappell J. (1974) Geology of coral terraces, Huon Peninsula, New Guinea: a study of Quaternary tectonic movements and sea-level changes. *Geol. Soc. Am. Bull.* **85**, 553–570.

Chappell J. (2002) Sea level changes forced ice breakouts in the Last Glacial cycle: new results from coral terraces. *Quat. Sci. Rev.* **21**, 1229–1240.

Chappell J. C. and Polach H. (1972) Some effects of partial recrystallization on ^{14}C dating Late Pleistocene corals and mollusks. *Quat. Res.* **2**, 244–252.

Chappell J. and Polach H. (1991) Post-glacial sea-level rise from a coral record at Huon Peninsula, Papua New Guinea. *Nature* **349**, 147–149.

Chappell J. and Shackleton N. J. (1986) Oxygen isotopes and sea level. *Nature* **324**, 137–140.

Chen J. H., Edwards R. L., and Wasserburg G. J. (1986) ^{238}U, ^{234}U and ^{232}Th in seawater. *Earth Planet. Sci. Lett.* **80**, 241–251.

Chen J. H., Curran H. A., White B., and Wasserburg G. J. (1991) Precise chronology of the last interglacial period: $^{238}U/^{230}Th$ data from fossil coral reefs in the Bahamas. *Geol. Soc. Am. Bull.* **103**, 82–97.

Cheng H., Edwards R. L., Murrell M. T., and Goldstein S. (1998) The systematics of uranium–thorium–protactinium dating. *Geochim. Cosmochim. Acta* **62**, 3437–3452.

Cheng H., Edwards R. L., Hoff J., Gallup C. D., Richards D. A., and Asmerom Y. (2000) The half-lives of uranium-234 and thorium-230. *Chem. Geol.* **169**, 17–33.

Clark J. A., Farrell W. E., and Peltier W. R. (1978) Global changes in postglacial sea level: a numerical calculation. *Quat. Res.* **9**, 265–287.

Clark U., Mitrovica J. X., Milne G. A., and Tamisiea M. E. (2002) Sea-level fingerprinting as a direct test for the source of global meltwater pulse IA. *Science* **295**, 2438–2441.

Cochran J. K. (1982) The oceanic chemistry of U and Th series nuclides. In *Uranium Series Disequilibrium: Application to Environmental Problems* (eds. M. Ivanovich and R. S. Harmon). Clarendon press, Oxford, pp. 384–430.

Collins L. B., Zhu Z. R., Wyrwoll K. H., Hatcher B. G., Playford P. E., Chen J. H., Eisenhauer A., and Wasserburg G. J. (1993a) Late Quaternary evolution of coral reefs on a cool-water carbonate margin: the Abrolhos carbonate platforms, Southwest Australia. *Mar. Geol.* **110**, 203–212.

Collins L. B., Zhu Z. R., Wyrwoll K. H., Hatcher B. G., Playford P. E., Eisenhauer A., Chen J. H., Wasserburg G. J., and Bonani G. (1993b) Holocene growth history of a reef complex on a cool-water carbonate margin: Easter Group of the Houtman Abrolhos, eastern Indian Ocean. *Mar. Geol.* **115**, 29–46.

Curry W. B. and Oppo D. W. (1997) Synchronous, high-frequency oscillations in tropical sea surface temperatures and North Atlantic deep water production during the last glacial cycle. *Paleoceanography* **12**, 1–14.

Cutler K. B., Edwards R. L., Taylor F. W., Cheng H., Adkins J., Gallup C. D., Cutler P. M., Burr G. S., and Bloom A. L. (2003) Rapid sea-level fall and deep-ocean temperature change since the last interglacial. *Earth Planet. Sci. Lett.* **206**, 253–271.

Dansgaard W., Johnson S. J., Clausen H. B., Dahl-Jensen D., Gundestrup N. S., Hammer C. U., Hvidberg C. S., Steffensen J. P., Sveinbjornsdottir A. E., Jouzel J., and Bond G. (1993) Evidence for general instability of past climate from a 250-kyr ice-core record. *Nature* **364**, 218–220.

Delaney M. L. and Boyle E. A. (1983) Uranium and thorium isotope concentrations in foraminiferal calcite. *Earth Planet. Sci. Lett.* **62**, 258–262.

Denton G. H. and Karlen W. (1973) Holocene climatic variations: their pattern and possible cause. *Quat. Res.* **3**, 155–205.

Dole M., Lane G. A., Rudd D. P., and Zaukelies D. A. (1954) Isotopic composition of atmospheric oxygen and nitrogen. *Geochim. Cosmochim. Acta* **6**, 65–78.

Edwards R. L. (1988) High-precision Th-230 ages of corals and the timing of sea level fluctuations in the late Quaternary. PhD Dissertation, California Institute of Technology.

Edwards R. L. (2000) C. C. Patterson Award acceptance speech. *Geochim. Cosmochim. Acta* **64**, 759–761.

Edwards R. L., Chen J. H., and Wasserburg G. J. (1987a) ^{238}U, ^{234}U, ^{230}Th, ^{232}Th systematics and the precise measurement of time over the past 500,000 years. *Earth Planet. Sci. Lett.* **81**, 175–192.

Edwards R. L., Chen J. H., Ku T. L., and Wasserburg G. J. (1987b) Precise timing of the last interglacial period from mass spectrometric determination of ^{230}Th in corals. *Science* **236**, 1547–1553.

Edwards R. L., Taylor F. W., and Wasserburg G. J. (1988) Dating earthquatkes with high-precision thorium-230 ages of very young corals. *Earth Planet. Sci. Lett.* **90**, 371–381.

Edwards R. L., Beck J. W., Burr G. S., Donahue D. J., Chappell J. M. A., Blom A. L., Druffel E. R. M., and Taylor F. W. (1993) A large drop in atmospheric ^{14}C/^{12}C and reduced melting in the Younger Dryas, documented with ^{230}Th ages of corals. *Science* **260**, 962–968.

Edwards R. L., Cheng H., Murrell M. T., and Goldstein S. J. (1997) High resolution protactinium-231 dating of carbonates: implications for the causes of Quaternary climate change. *Science* **276**, 782–786.

Eisenhauer A., Zhu Z. R., Collins L. B., Wyrwoll K., and Eichstätter R. (1996) The last interglacial sea level change: new evidence from the Abrolhos islands, West Australia. *Geol. Rundsch.* **85**, 606–614.

Elderfield H. and Ganssen G. (2000) Past temperature and ^{18}O of surface ocean waters inferred from foraminiferal Mg/Ca ratios. *Nature* **405**, 442–445.

Emiliani C. (1955) Pleistocene temperatures. *J. Geol.* **63**, 537–578.

Esat T. M., McCulloch M. T., Chappell J., Pillans B., and Omura A. (1999) Rapid fluctuations in sea level recorded at Huon Peninsula during the Penultimate Deglaciation. *Science* **283**, 197–201.

Fairbanks R. G. (1989) A 17,000-year glacio-eustatic sea-level record: influence of glacial melting rates on the Younger Dryas event and deep-ocean circulation. *Nature* **342**, 637–642.

Flint R. F. (1971) *Glacial and Quaternary Geology*. Wiley, New York.

Gallup C. D., Edwards R. L., and Johnson R. G. (1994) The timing of high sea levels in the past 200,000 years. *Science* **263**, 796–800.

Gallup C. D., Cheng H., Taylor F. W., and Edwards R. L. (2002) Direct determination of the timing of sea level change during Termination II. *Science* **295**, 310–313.

Greenland Summit Ice Cores [CD-ROM] (1997). Available from the National Snow and Ice Data Center, University of Colorado at Boulder, and the World Data Center-A for Paleoclimatology, National Geophysical Data Center, Boulder, CO, www.ngdc.noaa.gov/paleo/icecore/greenland/summit/index.html

Hamelin B., Bard E., Zindler A., and Fairbanks R. G. (1991) ^{234}U/^{238}U mass spectrometry of corals: how accurate is the U–Th age of the last interglacial period. *Earth Planet. Sci. Lett.* **106**, 169–180.

Hanebuth T., Stattegger K., and Grootes P. M. (2000) Rapid flooding of the Sunda Shelf: a late-glacial sea-level record. *Science* **288**, 1033–1035.

Hays J. D., Imbrie J., and Shackelton N. J. (1976) Variations in the Earth's orbit: pacemaker of the ice ages. *Science* **194**, 1121–1132.

Hearty P., Kindler P., Cheng H., and Edwards R. L. (1999) Evidence for a +20 m sea level in the Mid-Pleistocene. *Geology* **27**, 375–378.

Henderson G. M. and O'Nions R. K. (1995) U-234/U-238 ratios in Quaternary planktonic foraminifera. *Geochim. Cosmochim. Acta* **59**(22), 4685–4694.

Henderson G. and Slowey N. (2000) Evidence from U–Th dating against Northern Hemisphere forcing of the penultimate deglaciation. *Nature* **404**, 61–66.

Henderson G. M., Slowey N. C., and Fleisher M. Q. (2001) U–Th dating of carbonate platform and slope sediments. *Geochim. Cosmochim. Acta* **65**, 2757–2770.

Imbrie J., Hays J. D., Martinson D. G., McIntyre A., Mix A. C., Morley J. J., Pisias N. G., Prell W. L., and Shackleton N. J. (1984) The orbital theory of Pleistocene climate: support from a revised chronology of the marine δ^{18}O record. In *Milankovitch and Climate* (eds. A. L. Berger, G. Hays, G. Kukla, and B. Salzman). D. Reidel, Dordrecht, vol. 1, pp. 269–305.

Jaffey A. H., Flynn K. F., Glendenin L. E., Bentley W. C., and Essling A. M. (1971) Precision measurement of Half-lives and specific activities of ^{235}U and ^{238}U. *Phys. Rev. C.* **4**, 1889–1906.

Koetsier G., Elliott T., and Fruijtier C. (1999) Constrants on diagenetic age disturbance: combined uranium–protactinium and uranium–thorium ages of the key Largo Formation, Florida Keys, USA. *9th Annual V. M. Goldschmidt Conf.*, pp. 157–158.

Ku T.-L. (1968) Protactinium-231 method of dating coral from Barbados Island. *J. Geophys. Res.* **73**, 2271–2276.

Ku T. L., Kimmel M. A., Easton W. H., and O'Neil T. J. (1974) Eustatic sea level 120,000 years ago on Oahu, Hawaii. *Science* **183**, 959–962.

Ku T. L., Knauss K. G., and Mathieu G. G. (1977) Uranium in the open ocean: concentration and isotopic composition. *Deep-Sea Res.* **24**, 1005–1017.

Lea D. W., Mashiotta T. A., and Spero H. J. (1999) Controls on magnesium and strontium uptake in planktonic foraminifera determined by live culturing. *Geochim. Cosmochim. Acta* **63**, 2369–2379.

Lea D. W., Pak D. K., and Spero H. J. (2000) Climate impact of Late Quaternary equatorial Pacific sea surface temperature variations. *Science* **289**, 1719–1724.

Lea D. W., Martin P. A., Pak D. K., and Spero H. J. (2002) Reconstructing a 350 kyr history of sea level using planktonic Mg/Ca and oxygen isotope records from a Cocos Ridge core. *Quat. Sci. Rev.* **21**, 283–293.

Li W. X., Lundberg J., Dickin A. P., Ford D. C., Schwarcz H. P., McNutt R., and Williams D. (1989) High precision mass-spectrometric uranium-series dating of cave deposits and implications for paleoclimate studies. *Nature* **339**, 534–536.

Ludwig K. R. and Titterington D. M. (1994) Calculation of ^{230}Th / U isochrons, ages, and errors. *Geochim. Cosmochim. Acta* **58**, 5031–5042.

Ludwig K. R., Muhs D. R., Simmons K. R., Halley R. B., and Shinn E. A. (1996) Sea-level records at ~80 ka from tectonically stable platforms: Florida and Bermuda. *Geology* **24**(3), 211–214.

Lundberg J. and Ford D. C. (1994) Late Pleistocene sea level change in the Bahamas from mass spectrometric U-series dating of submerged speleothem. *Quat. Sci. Rev.* **13**, 1–14.

Luo S. and Ku T.-L. (1991) U-series isochron dating: a generalized method employing total-sample dissolution. *Geochim. Cosmochim. Acta* **55**, 555–564.

Martin P. A., Lea D. W., Rosenthal Y., Shackleton N. J., Sarnthein M., and Papenfuss T. (2002) Quaternary deep sea temperature histories derived from benthic foraminiferal Mg/Ca. *Earth Planet. Sci. Lett.* **198**, 193–209.

McCulloch M. T. and Esat T. (2000) The coral record of last interglacial sea levels and sea surface temperatures. *Chem. Geol.* **169**, 107–129.

Mesolella K. J., Matthews R. K., Broecker W. S., and Thurber D. L. (1969) The astronomical theory of climatic change: Barbados data. *J. Geol.* **77**, 250–274.

Milankovitch M. (1941) *The Canon of the Ice Ages*. Royal Serb. Acad, Beograd.

Muhs D. R., Kennedy G. L., and Rockwell T. K. (1994) Uranium-series ages of marine terrace corals from the Pacific coast of North America and implications for last-interglacial sea level history. *Quat. Res.* **42**, 72–87.

Muhs D. R., Kathleen Simmons R., and Steinke B. (2002) Timing and warmth of the Last Interglacial period: new U-series evidence from Hawaii and Bermuda and a new fossil compilation for North America. *Quat. Sci. Rev.* **21**, 1355–1383.

Neff U., Burns S. J., Mangini A., Mudelsee M., Fleitmann D., and Matter A. (2001) Strong coherence between solar variability and the monsoon in Oman between 9 and 6 kyr ago. *Nature* **411**, 290–293.

Nurnberg D. (1995) Magnesium in tests of *Neogloboquadrina pachyderma* sinistral from high northern and southern latitudes. *J. Foraminiferal Res.* **25**, 350–368.

Nurnberg D., Bijma J., and Hemleben C. (1996) Assessing the reliability of magnesium in foraminiferal calcite as a proxy for water mass temperatures. *Geochim. Cosmochim. Acta* **60**, 803–814.

Peltier W. R. (1998) Postglacial variations in the level of the sea: implications for climate dynamics and solid earth geophysics. *Rev. Geophys.* **36**, 603–689.

Peltier W. R. and Hyde W. T. (1986) Sensitivity experiments with a model of the ice age cycle: the response to Milankovitch forcing. *J. Atmos. Sci.* **44**, 1351–1375.

Petit J. R., Jouzel J., Raynaud D., Barkov N. I., Barnola J. M., Basile I., Bender M., Chappellaz J., Davis J., Delaygue G., Delmotte M., Kotlyakov V. M., Legrand M., Lipenkov V. M., Lorius C., Pépin L., Ritz C., Saltzman E., and Stievenard M. (1999) Climate and atmospheric history of the past 420,000 years from the Vostok Ice Core, Antarctica. *Nature* **399**, 429–436.

Pickett D. A., Murrell M. T., and Williams R. W. (1994) Determination of femtogram quantities of protactinium in geologic samples by thermal ionization mass spectrometry. *Anal. Chem.* **66**, 1044–1049.

Richards D. A. (1995) Pleistocene sea levels and palaeoclimate of the Bahamas based on Th-230 ages of speleothems. PhD Thesis, University of Bristol.

Richards D. A., Smart P. L., and Edwards R. L. (1994) Maximum sea levels for the last glacial period from U-series ages of submerged speleothems. *Nature* **367**, 357–360.

Richter F. M. and Turekian K. K. (1993) Simple models for the geochemical response of the ocean to climatic and tectonic forcing. *Earth Planet. Sci. Lett.* **119**, 121–131.

Robert J., Miranda C. F., and Muxart R. (1969) Mesure de la periode du protactinium-231 par microcalorietrie. *Radiochim. Acta* **11**, 104–108.

Robinson L. F., Henderson G. M., and Slowey N. C. (2002) U–Th dating of marine isotope stage 7 in Bahamas slope sediments. *Earth Planet. Sci. Lett.* **196**, 175–187.

Rohling E. J., Fenton M., Jorissen F. J., Bertrand P., Ganssen G., and Caulet J. P. (1998) Magnitudes of sea-level lowstands of the past 500,000 years. *Nature* **394**, 162–165.

Ruddiman W. F., McIntyre A., Niebler-Hunt V., and Durazzi J. T. (1980) Oceanic evidence for the mechanism of rapid Northern Hemisphere glaciation. *Quat. Res.* **13**, 33–64.

Schrag D. P., Hampt G., and Murray D. W. (1996) Pore fluid constraints on the temperature and oxygen isotopic composition of the glacial ocean. *Science* **272**, 1930–1932.

Shackleton N. J. (1967) Oxygen isotope analyses and Pleistocene temperatures reassessed. *Nature* **215**, 15–17.

Shackleton N. J. (2000) The 100,000-year ice-age cycle identified and found to lag temperature, carbon dioxide, and orbital eccentricity. *Science* **289**, 1897–1902.

Shackleton N. J., Imbrie J., and Hall M. A. (1983) Oxygen and carbon isotope record of east Pacific core V19-30: implications for the formation of deep water in the late Pleistocene North Atlantic. *Earth Planet. Sci. Lett.* **65**, 233–244.

Slowey N. C., Henderson G. M., and Curry W. B. (1996) Direct U–Th dating of marine sediments from the two most recent interglacial periods. *Nature* **383**, 242–244.

Sowers T., Bender M., Labeyrie L., Martinson D., Jouzel J., Raynaud D., Pichon J. J., and Korotkevich Y. S. (1993) A 135,000 year Vostok-SPECMAP common temporal framework. *Paleoceanography* **8**, 737–766.

Stein M., Wasserburg G. J., Aharon P., Chen J. H., Zhu Z. R., Bloom A. L., and Chapell J. (1993) TIMS U series dating and stable isotopes of the last interglacial event at Papua New Guinea. *Geochim. Cosmochim. Acta* **57**, 2541–2554.

Stirling C. H., Esat T. M., McCulloch M. T., and Lambeck K. (1995) High-precision U-series dating of corals from Western Australia and implications for the timing and duration of the last interglacial. *Earth Planet. Sci. Lett.* **135**, 115–130.

Stirling C. H., Esat T. M., Lambeck K., and McCulloch M. T. (1998) Timing and duration of the last interglacial: evidence for a restricted interval of widespread coral reef growth. *Earth Planet. Sci. Lett.* **160**, 745–762.

Stirling C., Esat T., Lambeck K., McCulloch M., Blake G., Lee D.-C., and Halliday A. (2001) Orbital forcing of the marine isotope stage 9 interglacial. *Science* **291**, 290–293.

Szabo B. J. (1985) Uranium-series dating of fossil corals from marine sediments of southeastern United States of America coastal plain. *Geol. Soc. Am. Bull.* **96**, 398–406.

Szabo B. J., Ludwig K. R., Muhs D. R., and Simmons K. R. (1994) Thorium-230 ages of corals and duration of the last interglacial sea-level high stand on Oahu, Hawaii. *Science* **266**, 93–96.

Toscano M. A. and Lundberg J. (1999) Submerged Late Pleistocene reefs on the tectonically-stable S. E. Florida margin: high-precision geochronology, stratigraphy, resolution of substage 5a sea-level elevation, and orbital forcing. *Quat. Sci. Rev.* **18**, 753–767.

Veeh H. H. (1966) ^{230}Th/^{238}U and ^{234}U/^{238}U ages of Pleistocene high sea level stand. *J. Geophys. Res.* **71**, 3379–3386.

Veeh H. H. and Chappell J. (1970) Astronomical theory of climatic changes: support from New Guinea. *Science* **167**, 862–865.

Wang Y. J., Cheng H., Edwards R. L., An Z. S., Wu J. Y., Shen C.-C., and Dorale J. A. (2001) A high-resolution absolute-dated Late Pleistocene monsoon record from Hulu Cave, China. *Science* **294**, 2345–2348.

Wasserburg G. J. (2000) Citation for presentation of the 1999 C. C. Patterson Award to R. Lawrence Edwards. *Geochim. Cosmochim. Acta* **64**, 755–757.

Wetherill G. S. (1956a) An interpretation of the Rhodesia and Witwatersrand age patterns. *Geochim. Cosmochim. Acta* **9**, 290–292.

Wetherill G. S. (1956b) Discordant uranium–lead ages. *Trans. Am. Geophys. Union* **37**, 320–326.

Yokoyama Y., Lambeck K., De Deckker P., Johnston P., and Fifield L. K. (2000) Timing of the last glacial maximum from observed sea-level minima. *Nature* **406**, 713–716.

Zhu Z. R., Wyrwoll K. H., Chen L. B., Chen J. H., Wasserburg G. J., and Eisenhauer A. (1993) High-precision U-series dating of last Interglacial events by mass spectrometry: Houtman Abrolhos Islands, Western Australia. *Earth Planet. Sci. Lett.* **118**, 281–293.

6.14
Elemental and Isotopic Proxies of Past Ocean Temperatures

D. W. Lea

University of California, Santa Barbara, CA, USA

6.14.1 INTRODUCTION

Determining the temperature evolution of the oceans is a fundamental problem in the geosciences. Temperature is the most primary representation of the state of the climate system, and the temperature of the oceans is critical because the oceans are the most important single component of the Earth's climate system. A suite of isotopic and elemental proxies, mostly preserved in marine carbonates, are the essential method by which earth scientists determine past ocean temperatures (see Table 1). This is a field with both a long history and a great deal of recent progress. Paleotemperature research has been at the forefront of geoscience research since 1950s, and, with our need to understand the global climate system heightened by the threat of global warming, it promises to remain a vibrant and important area well into the future.

In this chapter I begin by reviewing the history of the elemental and isotopic proxies and how that history shapes research priorities today. I then review the state of our knowledge at present (as of early 2000s), including areas that are well developed (i.e., oxygen isotopes in coral aragonite), areas that are experiencing phenomenal growth (i.e., Mg/Ca in foraminifer shells), and areas that are just starting to develop (i.e., Ca isotopes in carbonates). In these reviews I include an estimation of the uncertainty in each of these techniques and areas, including aspects that particularly need to be addressed. I also address the question of overlap and confirmation between proxies, and in particular how information from one proxy can augment a second proxy.

6.14.2 A BRIEF HISTORY OF EARLY RESEARCH ON GEOCHEMICAL PROXIES OF TEMPERATURE

Geologists have been interested in establishing the temperature history of the oceans for as long as they have documented historical variations in marine sediments. Probably the first realization that geochemical variations might reflect temperatures can be traced to the great American geochemist Frank Wigglesworth Clarke (1847–1931), the namesake of the Geochemical Society's Clarke award. Aside from his voluminous contributions to establishing precise atomic weights, the composition of the Earth's crust, and natural waters, Clarke found time to document a provocative relationship between the magnesium content of biogenic carbonates and their growth temperature (Clarke and Wheeler, 1922). The authors speculated that this relationship had a definite cause and could

Table 1 Summary of major paleotemperature techniques.

	Phases	*Sensitivity* (per °C)	*Estimated SE*	*Major secondary effects*	*Time scale*[a]
Oxygen Isotopes	Foraminifera	0.18–0.27‰	0.5 °C if $\delta^{18}O$-sw is known	Effect of $\delta^{18}O$-sw	0–100 Ma
	Corals	~0.2‰	0.5 °C if $\delta^{18}O$-sw is known	Kinetic effects Effect of $\delta^{18}O$-sw	0–130 ka
	Opal			Effect of $\delta^{18}O$-sw	0–30 ka
Mg/Ca	Foraminifera	9 ± 1%	~1 °C	Dissolution Secular Mg/Ca variations (>10 Ma)	0–40 Ma
	Ostracodes	~9%	~1 °C	Dissolution? Calibration	0–3.2 Ma
Sr/Ca	Corals	−0.4 to −1.0%	0.5 °C?	Growth effects Secular Sr/Ca changes (>5 ka)	0–130 ka
Ca isotopes	Foraminifera	0.02–0.24‰	unknown	Species effects, calcification	0–125 ka
Alkenone unsaturation index[b]	Sediment organics	0.033 (0.023–0.037) in $U_{37}^{K'd}$	~1.5 °C (global calib.)	Transport, species variation	0–3 Ma
Faunal transfer functions[c]	Foraminifera, Radiolaria, Dinoflagellates	NA	1.5 °C	Ecological shifts	0–?

[a] Timescale over which the technique has been applied. [b] Chapter 6.15. [c] (Imbrie and Kipp, 1971) [d] (Müller *et al.*, 1998 and Pelejero and Calvo, 2003).

possibly be useful: "This rule, or rather tendency, we are inclined to believe is general, although we must admit that there are probably exceptions to it." Clarke and Wheeler recognized that the magnesium to calcium ratio of the oceans was nearly constant, and that therefore the trend had to have some cause other than compositional variations: "That warmth favors the assimilation of magnesia by marine invertebrates seems to be reasonably certain, but why it should be so is not clear. The relation is definite but as yet unexplained. *We hope it is not inexplicable*" (quotations added for emphasis). Further along in this same paragraph Clarke and Wheeler presaged the field of geochemical paleoceanography and paleoclimatology: "Attempts will likely be made to use our data in studies of climatology, but are such attempts likely to be fruitful?" The authors envisioned researchers using the bulk magnesium content of ancient rocks to determine past temperatures, an approach they deemed doubtful because it would depend on the ratios of particular organisms in rocks. Of course, Clarke and Wheeler did not envision the powerful analytical techniques available to the present-day analyst, where the trace element content of individual chambers of plankton shells can be readily analyzed. Such single species analysis is what eventually enabled the useful application of Clarke and Wheeler's original insight to paleothermometry (see below). However, it was not until the late 1990s that the observed magnesium relationship became both explicable and fruitful for climatology.

The next great step forward came after World War II, when Harold Urey (1893–1981), a 1934 Nobel Laureate for his discovery of deuterium, the heavy isotope of hydrogen, took up a professorship at the University of Chicago. There, he became interested in the utilization of natural fractionations in stable isotope systems for geological purposes (Urey, 1947). He theorized that the effect of temperature on the partitioning of oxygen isotopes between water and carbonate might become a useful geological tool: "Accurate determinations of the O^{18} content of carbonate rocks could be used to determine the temperature at which they were formed." Interestingly, Urey did not envision that compositional variations in the ocean would complicate such determinations: "First, there is the large reservoir of oxygen in the oceans which *cannot have changed* in isotopic compositions during geological time" (italics added for emphasis). He, of course, recognized that if such variations occurred, they would complicate paleotemperature determinations, and he pointed out in the same paragraph that variations in the isotopic composition of calcium would hinder its potential use for paleotemperature analysis, a point relevant to present research (see below). Of course, Urey would not have known about or envisioned the considerable effects continental glaciation and crustal exchange would have on the oxygen isotopic composition of the oceans (Sturchio, 1999).

It was left to Urey's students and post-doctoral scholars to exploit his original insights. Major advances came from the establishment of a so-called "paleotemperature equation" by Samuel Epstein (1919–2001), mainly based on calcite precipitated by mollusks in either controlled experiments or field-collected samples (Epstein *et al.*, 1953). Analysis of this calcite yielded a paleotemperature equation that demonstrated a sensitivity of $\sim -0.2‰$ change in $\delta^{18}O$ per °C. It is important to note that during this time period, and, as shown above, from the conception of the original idea, the emphasis in using oxygen isotopes was to reconstructing paleotemperatures. A few years later, Cesare Emiliani (1922–1995), a student and later a post-doctoral scholar with Urey, exploited these advances when he documented regular cyclic variations in the oxygen isotopic composition of planktonic foraminifera taken from eight sediment cores in the Caribbean (Emiliani, 1955). Although Emiliani did allow for small variations in the isotopic composition of seawater, he largely interpreted the observed $\delta^{18}O$ variations as a reflection of recurring cold intervals in the past during which tropical surface waters cooled by 6–8 °C. Many of the questions raised by Emiliani's classic 1955 study are still relevant today and are addressed in detail below.

During this same period scholars at Chicago examined the potential temperature dependence of trace elements in carbonates, focusing mainly on magnesium and strontium incorporation (Chave, 1954). Although these studies provided some additional insights beyond Clarke and Wheeler's (1922) original findings, they failed to yield advances of the kind that were spurring research on isotopic variations. Samuel Epstein (personal communication, 1992) felt that the general sense in the Chicago group was that elemental substitution was likely to be less regular than isotopic substitution, presumably because the individual activity coefficients of each element would introduce additional complexity beyond a simple temperature dependence.

Following Emiliani's (1955) discovery, other laboratories established the capability to apply oxygen isotope variations to oceanic temperature history. It is worth a brief mention of two further major advances relevant to Urey's original conception. In 1967, Nicholas Shackleton of Cambridge University reported the first systematic down-core variations in benthic foraminifera (Shackleton, 1967). He argued that benthic fauna, because they lived in the near-freezing bottomwaters of the ocean, would mainly record the change in isotopic variation of seawater. By demonstrating that the observed benthic variations were of similar magnitude to those in planktics, he was able to demonstrate that the major portion of the isotopic signal recorded in marine sediments reflected oscillations in the oxygen isotopic composition of the ocean, which in turn occurred in response to the periodic transfer of isotopically depleted water onto continental ice sheets. Once this paradigm shift was in place, Shackleton and others were able to use oxygen isotopic variations as a stratigraphic and chronometric tool for marine sediments (Shackleton and Opdyke, 1973), an advance which lead to the establishment of a precise timescale for the Late Quaternary marine record and ultimately the verification of orbital variations as the pacemakers of the Pleistocene Ice Ages (Hays *et al.*, 1976). So from Urey's conception to Shackleton's insight, a tool originally envisioned as a paleothermometer found its most profound use as a recorder of the compositional variations in seawater that Urey considered to be an unlikely influence. Parallel advances in the utilization of elemental variations did not occur until the 1990s (see below), but it is worth mentioning as a close to this historical summary that recent research indicates that the Mg/Ca content of foraminifera is the long-sought solution to the "Urey dilemma," because it provides, in combination with oxygen isotopes, a simultaneous temperature and isotopic composition history for seawater (Lea *et al.*, 2000; Lear *et al.*, 2000).

6.14.3 OXYGEN ISOTOPES AS A PALEOTEMPERATURE PROXY IN FORAMINIFERA

6.14.3.1 Background

The use of oxygen isotope ratios as a paleotemperature indicator in carbonate minerals is based on the thermodynamic fractionation between ^{16}O and ^{18}O that occurs during precipitation (Urey, 1947). This fractionation, which offsets the $\delta^{18}O$ ($\delta^{18}O = [(^{18/16}O_{sample}/^{18/16}O_{standard}) - 1]$) of carbonate minerals relative to seawater by $\sim +30‰$, is a logarithmic function of temperature with a slope, over the oceanic temperature range of -2 °C to 30 °C, of between $-0.20‰$ and $-0.27‰$ per °C, in agreement with thermodynamic predictions (O'Neil *et al.*, 1969; Shackleton, 1974; Kim and O'Neil, 1997; Zhou and Zheng, 2003). Because the oxygen isotope proxy is based on a thermodynamic principle, it is expected to be robust and relatively unaffected by secondary kinetic factors. For foraminifera, unicellular zooplankton and benthos that precipitate calcite and, less commonly, aragonite shells (sometimes called tests), oxygen isotopic ratios do appear to be quite robust, although there are clear indications of a secondary effect from factors such as ontogenetic variations and seawater carbonate ion (see below).

The most significant complication in using oxygen isotopes to determine both absolute temperatures as well as relative temperature changes is that the $\delta^{18}O$ of carbonate solids reflects both temperature fractionation and the $\delta^{18}O$ of the seawater from which the carbonate precipitated. The $\delta^{18}O$ of seawater in turn reflects two major factors: (i) the mean $\delta^{18}O$ of seawater (Schrag *et al.*, 1996), which is determined by the amount of continental ice, which varies on timescales of 10^4-10^5 yr, and by interaction between seafloor basalts and seawater, which varies on timescales of 10^7-10^8 yr and (ii) the evaporation–precipitation (rainfall) balance ($E-P$) for that part of the ocean or, for subsurface waters, the balance that applied to the source waters of those deep waters. The second factor is often described as a "salinity effect" because $\delta^{18}O$ tends to track with salinity variations because both respond to $E-P$. The relationship between $\delta^{18}O$ and salinity, however, varies considerably over the ocean because of the varying isotopic composition of freshwater (Schmidt *et al.*, 2001). It is important to emphasize that both effects cast considerable uncertainty into the use of oxygen isotopic ratios for absolute and relative paleothermometry on essentially all timescales.

Despite these caveats, oxygen isotopic ratios are probably the most widely used climate proxy in ocean history research. Reasons for this widespread use relate to the history of oxygen isotopes in geological research (see Section 6.14.2), the fact that they can be measured quite precisely by mass spectrometry and are relatively immune, at least in younger deposits, to secondary effects, the fact that $\delta^{18}O$ records tend to be quite reproducible and clearly record climate variability, and finally, because $\delta^{18}O$ records have proven so useful for stratigraphic and chronological purposes.

An important aspect of the application of oxygen isotopic ratios in foraminifera is that the planktonic foraminifera occupy several ecological niches, including surface waters, shallow subsurface waters, and deeper thermocline waters. Along with their benthic counterparts, this makes it possible to recover oxygen-isotopic records representative of different part of the water column. A complication, however, is that many of the planktonic species migrate vertically, potentially compounding the signals they record.

6.14.3.2 Paleotemperature Equations

There are a number of calibrations in use for oxygen isotopes in foraminifera, some derived from other organisms (Epstein *et al.*, 1953), some derived from culturing (Erez and Luz, 1983; Bemis *et al.*, 1998), and some derived from core-top calibrations (Shackleton, 1974). These calibrations take the form of a polynomial paleotemperature equation:

$$T = a + b * (\delta^{18}O_{calcite} - \delta^{18}O_{water}) + c * (\delta^{18}O_{calcite} - \delta^{18}O_{water})^2 \quad (1)$$

where T is temperature (°C), a is temperature when $\delta^{18}O_{calcite} - \delta^{18}O_{water}$ (both on V-PDB scale) is 0, b is the slope, and c is the second-order term for curvature (not always included). The inverse of the slope b represents the change in $\delta^{18}O$ (in ‰) for a 1 °C change in temperature. If the second-order term c is included, then the slope is not constant. The value of the slope is predicted to increase with decreasing temperature, because isotopic fractionation increases with decreasing temperature (Urey, 1947). Experimental evidence from inorganic calcite precipitation studies indicates that slope ranges from 0.27‰ per °C at 0 °C to 0.20‰ per °C at 25 °C (O'Neil *et al.*, 1969; Kim and O'Neil, 1997). Observational evidence from calibration of foraminiferal $\delta^{18}O$ indicates a similar or slightly larger range of values (Bemis *et al.*, 1998). The status of oxygen isotope calibrations is extensively reviewed in Bemis *et al.* (1998). For warm-water studies, the low light *Orbulina universa* calibration of Bemis *et al.* (1998) appears to work well:

$$T = 16.5 - 4.80 * (\delta^{18}O_{calcite} - \delta^{18}O_{water}) \quad (2)$$

For cold waters and certain benthics (e.g., *Uvigerina*), the Shackleton (1974) expression, which is a polynomial expansion of O'Neil (1969), appears to work well:

$$T = 16.9 - 4.38 * (\delta^{18}O_{calcite} - \delta^{18}O_{water}) + 0.1 * (\delta^{18}O_{calcite} - \delta^{18}O_{water})^2 \quad (3)$$

The quantitative applicability of these equations is becoming more important with the growing interest in combining independent foraminiferal temperature estimates from Mg/Ca with $\delta^{18}O$ to calculate $\delta^{18}O_{water}$ (Mashiotta *et al.*, 1999; Elderfield and Ganssen, 2000; Lea *et al.*, 2000; Lear *et al.*, 2000).

6.14.3.3 Secondary Effects and Diagenesis

There has been a long discussion in the literature as to the extent to which foraminifera shells are in oxygen-isotopic equilibrium. The precise state of equilibrium cannot be defined with sufficient precision from theory, so the general practice is to compare observed foraminiferal calibrations to inorganic experiments (Shackleton, 1974; Kim and O'Neil, 1997; Bemis *et al.*, 1998). Such comparisons suggest that foraminifera shells can be offset from equilibrium for a number of reasons (Figure 1), ranging from unknown vital effects that offset benthic species (Duplessy *et al.*, 1970), offsets due to the effect of light on symbiotic algae (Spero and Lea, 1993; Bemis *et al.*, 1998), offsets due to ontogeny (growth) (Spero and Lea, 1996), offsets due to carbonate ion concentration of seawater (Spero *et al.*, 1997), and offsets due to the addition of gametogenic calcite at depth (Duplessy and Blanc, 1981). All of these effects can complicate the use of oxygen isotopes for paleotemperature or paleo-$\delta^{18}O_{water}$ studies. Some of these effects, such as the effect of gametogenic calcite, also apply for Mg/Ca paleothermometry (Rosenthal *et al.*, 2000; Dekens *et al.*, 2002), so it is likely that complimentary studies will improve our understanding of the limitations these effects impose.

On Quaternary timescales, the main diagenetic concern is partial shell dissolution that takes place on the seafloor or within the sedimentary mixed layer. This effect has been demonstrated to increase the $\delta^{18}O$ of shells by ~0.2‰ per km in deeper, more dissolved sediments (Wu and Berger, 1989; Dekens *et al.*, 2002), presumably through the loss of individual shells and/or shell material with more negative $\delta^{18}O$. Because this

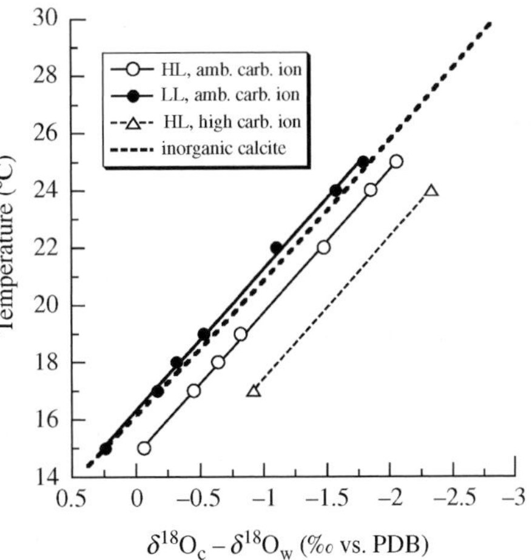

Figure 1 Comparison of oxygen-isotope paleotemperature equations for *O. universa*, a symbiont-bearing planktonic foraminifera, with values for inorganic calcite precipitation (Kim and O'Neil, 1997). These data demonstrate that the low light (LL) oxygen isotope equation at ambient carbonate ion concentration for *O. universa* is essentially indistinguishable from the inorganic equation. For high light (HL) conditions, in which symbiont photosynthetic activity is maximized, $\delta^{18}O$ shifts to more negative values. For high carbonate ion conditions, $\delta^{18}O$ shifts to even more negative values. These trends demonstrate the range of potential biological influences on foraminiferal $\delta^{18}O$ (after Bemis *et al.*, 1998).

effect appears to be coincident for both oxygen isotopes and Mg/Ca (Brown and Elderfield, 1996; Rosenthal *et al.*, 2000; Dekens *et al.*, 2002), it is likely that complimentary studies will allow a better assessment of the potential biases imposed by dissolution. On longer timescales, diagenetic effects multiply to include gradual replacement of the primary calcite (Schrag *et al.*, 1995; Schrag, 1999a). Recent studies suggest that shells with unusually good levels of preservation, such as those preserved in impermeable clay-rich sediments, record much more negative $\delta^{18}O$ (and hence warmer temperatures) than shells from deeply buried open ocean sequences (Pearson *et al.*, 2001).

6.14.3.4 Results on Quaternary Timescales

The many important results achieved in paleoceanography and paleoclimatology using oxygen isotope ratios in foraminifera shells are well known and have been reviewed in many other places: e.g., Shackleton (1987), Imbrie *et al.* (1992), and Mix (1992). Because of the outstanding geological importance of oxygen isotopic results, they take a central place in the history of proxy development, a subject discussed in Sections 6.14.2 and 6.14.3.1. As was emphasized in Section 6.14.2, the pioneers in the utilization of oxygen isotopes envisioned them as a paleotemperature tool. With the realization that change in the isotopic composition of the ocean exceeded the temperature influence on foraminiferal calcite (Shackleton, 1967), emphasis shifted to the use of Quaternary oxygen isotopic variations as a tool for stratigraphy and chronology (Shackleton and Opdyke, 1973; Hays *et al.*, 1976) and for calibration of the magnitude of past sea-level and ice volume change (Chappell and Shackleton, 1986; Shackleton, 1987). With the advent of independent geochemical paleotemperature proxies, it became possible to deconvolve the isotopic signal into its temperature and compositional components (Rostek *et al.*, 1993; Mashiotta *et al.*, 1999; Elderfield and Ganssen, 2000). Current research (see Sections 6.14.6.4 and 6.14.6.5) is focused on the veracity of this approach and the separation of ice volume and hydrological influences in the extracted $\delta^{18}O$-water records (Lea *et al.*, 2000, 2002; Lear *et al.*, 2000; Martin *et al.*, 2002).

6.14.3.5 Results for the Neogene, Paleogene, and Earlier Periods

The Cenozoic benthic foraminiferal $\delta^{18}O$ record is one of the major success of the geochemical approach to paleoclimate research (Zachos *et al.*, 2001). Because it is a record of the combined influences of temperature and ice volume influence, which are evolving semi-independently over the course of the Cenozoic, it is more a record of earth system processes than of temperature (see Chapter 6.20). There has been a great deal of interest and controversy concerning the utilization of oxygen isotopes in low-latitude planktonic foraminifera to determine tropical ocean temperatures on longer timescales (Barron, 1987). Recent research suggests that diagenetic overprinting of the primary foraminiferal $\delta^{18}O$ is a major influence (Schrag, 1999a; Pearson *et al.*, 2001). Results from the Pearson *et al.* (2001) study suggest that low-latitude sea surface temperatures (SSTs) during the Late Cretaceous and Eocene epochs were at least 28–32 °C compared to previous estimates, based on less well preserved material, of 15–23 °C. Obviously such a large shift requires a reevaluation of Paleogene and Cretaceous climates, but it might also point the way towards a means of correcting less well preserved samples for diagenesis, perhaps in combination with the Mg/Ca approach.

6.14.3.6 Summary of Outstanding Research Issues

Of the primary paleothermometric techniques reviewed in this chapter, oxygen isotopes in foraminifera have the longest history of development and application. It is probably safe to say that oxygen isotopes are on the most firm ground in terms of known influences and inherent accuracy. New results, such as a primary seawater carbonate ion influence (Spero *et al.*, 1997) and secondary diagenetic overprints (Schrag, 1999a; Pearson *et al.*, 2001), suggest that there is still major progress to made in this area. The most outstanding research issues today certainly must include progress and prospects for combining Mg/Ca paleothermometry with oxygen isotopes on both Quaternary, Neogene, and Paleogene timescales, and the need for reevaluation of the integrity and interpretation of Paleogene oxygen isotope ratios in foraminifera.

6.14.4 OXYGEN ISOTOPES AS A CLIMATE PROXY IN REEF CORALS

6.14.4.1 Background

Many of the factors described in Section 6.14.3.1 apply equally to oxygen isotopes in corals, which are dominantly composed of aragonite. Corals, however, have many unique aspects which require separate consideration. First, their oxygen isotopic composition is invariably depleted relative to equilibrium by

~1–6‰, presumably because of their different biochemical mechanisms of precipitation as well as the influence of symbiotic zooxanthellae (McConnaughey, 1989a; McConnaughey, 1989b). This offset from equilibrium actually discouraged early researchers, who assumed that corals would not be reliable temperature recorders (S. Epstein, Caltech, personal communication, 1998). Eventually, however, researchers began to investigate the prospect of attaining climate records from coral skeletons, and despite the offset from equilibrium research revealed that the oxygen-isotopic composition recorded subseasonal variations in seawater temperature and salinity (Weber and Woodhead, 1972; Emiliani *et al.*, 1978; Fairbanks and Dodge, 1979; Druffel, 1985; Dunbar and Wellington, 1985; McConnaughey, 1989a; Cole and Fairbanks, 1990; Shen *et al.*, 1992). These early discoveries have been followed by a fantastic array of results from longer coral time series that have enabled researchers to elevate coral climate records to the same level of importance as tree ring records (Gagan *et al.*, 2000). In general, coral $\delta^{18}O$ records are not interpreted directly as temperature records but rather as climate records whose variability reflects some combination of temperature and salinity effects.

Because aragonite is more susceptible to dissolution than calcite, especially under the influence of meteoric waters, and because most fossil corals are recovered from uplifted terrestrial deposits, diagenesis is an especially important limiting factor in recovering older coral records. This problem can be circumvented by drilling into submerged fossil deposits, but because of logistical difficulties, so far this has been accomplished in only a few key spots such as Barbados and Tahiti (Fairbanks, 1989; Bard *et al.*, 1996).

6.14.4.2 Paleotemperature Equations

Weber and Woodhead (1972) first demonstrated that oxygen isotopes in corals respond to temperature but are offset to negative $\delta^{18}O$ values relative to equilibrium. The oxygen isotope paleotemperature equation is well calibrated for corals (Gagan *et al.*, 2000). The systematic offset from the equilibrium or inorganic aragonite value is attributed to a biological or vital effect (McConnaughey, 1989a,b). This offset, however, appears to be stable over time in many different settings (Gagan *et al.*, 2000), although researchers recognize that vital effects can offset coral $\delta^{18}O$ to varying degrees (McConnaughey, 1989a,b; Spero *et al.*, 1997). The slope of the calibrations, however, appears to be nearly constant at ~0.2‰ per °C, in general agreement with the slope from inorganic experiments (Zhou and Zheng, 2003); the constancy of the slope suggests that historical changes in temperature will be accurately recorded.

Of course, the oxygen-isotopic composition of carbonate is also a function of the $\delta^{18}O$ of seawater, which varies with the local $E-P$, and hence salinity. Because temperature and salinity often vary together in the tropics, researchers tend to use coral $\delta^{18}O$ variations as climate proxies rather than temperature proxies. This approach has been very successful (see Section 6.14.4.4), but it still leaves the problem of attributing the observed changes in $\delta^{18}O$ to some specific combination of temperature and $\delta^{18}O_{water}$ changes. One solution is to use an independent temperature proxy such as Sr/Ca (see Section 6.14.8) at the same time to separate the coral $\delta^{18}O$ signal into its components (McCulloch *et al.*, 1994; Gagan *et al.*, 1998). It is also possible that comparison of records of coral $\delta^{18}O$ from different areas with contrasting climatology could be used to separate temperature and $\delta^{18}O_{water}$ influence.

6.14.4.3 Secondary Effects and Diagenesis

The major secondary effect for oxygen isotopes in hermatypic reef corals is the negative offset from equilibrium (Emiliani *et al.*, 1978). The degree to which this offset is stable in space and time (McConnaughey, 1989a,b) is critical to the interpretation of observed $\delta^{18}O$ variations in terms of absolute temperature and salinity changes. Measurements of $\delta^{18}O$ in coral heads from a single reef do reveal up to 1‰ variability in absolute values from specimen to specimen. The biological factors that cause these differences obviously have the potential to affect the use of coral $\delta^{18}O$ for paleoclimate research.

There is some evidence that the buildup of aragonite cements in living coral skeletons can affect the fidelity of coral $\delta^{18}O$ and Sr/Ca (see Section 6.14.8.3) records (Muller *et al.*, 2001). This occurs as pore spaces in the older part of the coral heads fill with inorganic aragonite, which has a more enriched $\delta^{18}O$ relative to the original coral material. As a result, the recorded climate signal in the oldest part of the coral is shifted to systematically colder and/or more salty values. Müller *et al.* (2001) observed that both $\delta^{18}O$ and Sr/Ca were biased similarly by inorganic aragonite precipitation, so cross-checks between proxies within the same coral does not alleviate the problem. Fortunately, however, it is possible to observe the precipitation of the secondary aragonite using petrography.

Because aragonite reverts to calcite when it interacts with meteoric water, subaerial exposure of fossil corals has the potential to change the $\delta^{18}O$ of the coral. Generally, diagenetically altered corals can be avoided by using X-ray crystallography to screen for the presence of calcite. A recent study suggests that small, restricted levels of aragonite alteration have minimal effects on coral $\delta^{18}O$ (McGregor and Gagan, 2003).

6.14.4.4 Results on Historical Timescales

Measurement of oxygen isotopic variations in coral heads up to 500 yr old has to be counted among the great success of the geochemical approach to paleoclimate research (Figure 2). These records have become important paleoclimatic archives of tropical climate change, and they have been incorporated into historical climate records used to assess global warming in the past century (Mann *et al.*, 1998; Crowley, 2000a). Some of the outstanding findings from the coral records include: (i) most of the longer records show a secular shift to more negative $\delta^{18}O$ values starting in the nineteenth century; (ii) a series of decadal or longer coherent shifts in the nineteenth century that might reflect regional to global cooling patterns; and (iii) shifts in the magnitude and frequency of the El Niño/Southern Oscillation (ENSO) phenomenon and Indian Ocean monsoon over the past two centuries (Cole *et al.*, 2000; Urban *et al.*, 2000; Cobb *et al.*, 2001).

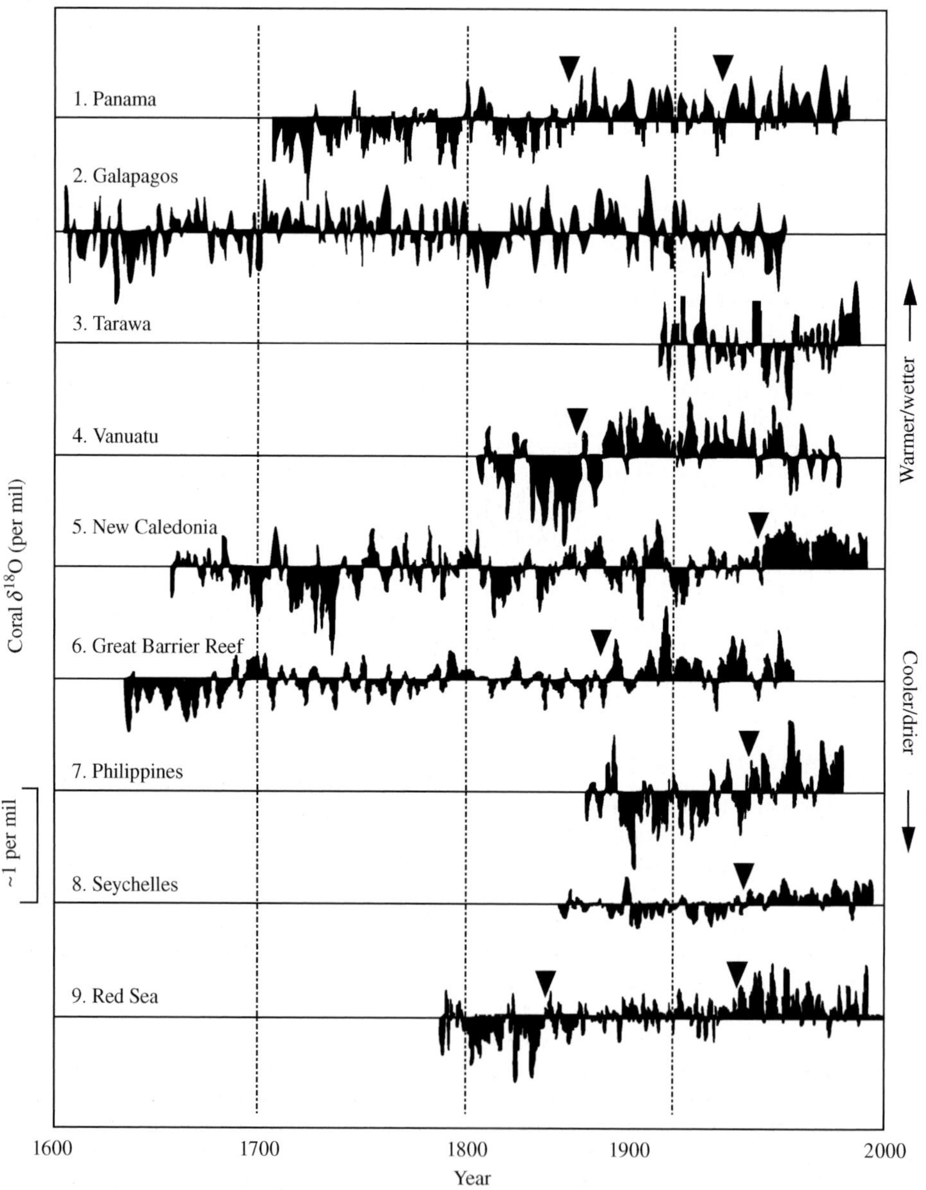

Figure 2 Comparison of annual mean coral oxygen isotope records in the Pacific and Indian Ocean region extending back more than 100 yr (Gagan *et al.*, 2000) (reproduced by permission of Elsevier from *Quat. Sci. Rev.* **2000**, *19*, 45–64). These records demonstrate the considerable potential of this approach for documenting historical climate variation in the tropical oceans.

These coral records have great value as "generic" proxy climate records, in the same sense that tree ring records have been used without an explicit attribution of the observed variations to temperature, precipitation, etc. Because these records are based on a geochemical parameter that follows thermodynamic rules, it should be possible to eventually extract true temperature and/or salinity records from the coral time series. Another potential complication, however, is that the coral records might reflect biological effects such as gradual growth into shallower waters as coral heads grow (Gagan *et al.*, 2000). Many coral heads show a secular shift to more negative $\delta^{18}O$ values in the most modern period of growth, a result generally attributed to warming of surface waters in response to anthropogenic factors (Gagan *et al.*, 2000). But in at least some cases, this shift appears to be larger than can explained by SST shifts recorded by instrumental records (the records in Urban *et al.*, 2000; Cobb *et al.*, 2001 are good examples), so this trend either reflects a coincident decline in surface salinity or the aforementioned biological factors or perhaps undetected secondary aragonite precipitation in the oldest parts of the coral (Muller *et al.*, 2001).

6.14.4.5 Results on Late Quaternary Timescales

With the realization that oxygen isotopes in coral heads record subannual ocean climate variations came the idea of using such records from fossil corals to reconstruct both absolute and relative climate change for past geological periods (Fairbanks and Matthews, 1978). Because fossil corals from emerging coastlines have been exposed to meteoric fluids and weathering, this approach requires consideration of the potential for diagenetic changes (see Section 6.14.4.3). Although the complications of using fossil corals as paleoclimate tools are greater, the information to be gained is of great importance because it applies to climate systems under different boundary conditions (Tudhope *et al.*, 2001).

The first generation of studies on oxygen isotopes in fossil corals attempted to use absolute differences as a gauge of changes in mean ocean $\delta^{18}O$ in response to continental glaciation and mean SST changes. Early studies used coral $\delta^{18}O$ to better define the relationship between sea-level change and mean ocean $\delta^{18}O$ changes (see Section 6.14.3.4) (Fairbanks and Matthews, 1978). A number of studies published during the 1990s use coral $\delta^{18}O$ to try to establish SST history for the tropics (Guilderson *et al.*, 1994, 2001; McCulloch *et al.*, 1999). These studies suggest quite large glacial cooling of 4–6 °C for both tropical Atlantic and Pacific SST. Such large cooling is generally not supported by other approaches (Bard *et al.*, 1997; Lea *et al.*, 2000; Nürnberg *et al.*, 2000). Obviously the coral $\delta^{18}O$ approach depends heavily on knowledge of the $\delta^{18}O$ of local seawater, which will shift with both ice volume and local changes in *E* versus *P* (see Section 6.14.3). Sr/Ca paleothermometry in corals provides a way around this problem, but this approach appears to have its own set of limitations (see Section 6.14.8.3).

More recent studies are focusing on the climate variability encoded in annual and subannual fossil coral $\delta^{18}O$ (Hughen *et al.*, 1999; Tudhope *et al.*, 2001). This approach does not require separating the oxygen-isotope signal into its components, but rather uses the coral $\delta^{18}O$ signal as a climate proxy, with particular attention to the spectral characteristics of the time series. This approach is also less subject to diagenetic constraints, because corals that maintain distinct seasonal signatures are likely to be relatively unaltered. The results of these studies have been quite impressive in demonstrating that the nature of ENSO variability has been different under varying geological boundary conditions. The Tudhope *et al.* (2001) study documented coral climate proxy variability for seven different time slices. These records demonstrate that the amplitude of ENSO variability (2–7 yr band) has generally been weaker in the geological past relative to the twentieth century. The amplitudes appear to have been weakest in the mid-Holocene (~6.5 ka) and during most of the cold glacial episodes (Tudhope *et al.*, 2001). These records are by necessity fragmentary and only comprise a short window into ENSO variability in the past. These records also do not address the question of changes in ENSO frequency in the past, as has been suggested by studies of other climate proxies (e.g., Rodbell *et al.*, 1999).

6.14.4.6 Summary of Outstanding Research Issues

High-resolution coral oxygen isotope records have to be counted among the great successes of the geochemical approach to paleoclimate research. This is ironic given that early researchers were highly skeptical about the fidelity of coral $\delta^{18}O$ because of the clear lack of equilibrium (Emiliani *et al.*, 1978). Although coral $\delta^{18}O$ is clearly a valuable indicator of climate history, many challenges remain in direct assignment of the observed trends to an absolute history of temperature and salinity. High priorities for future research include establishing a coincident temperature proxy such as Sr/Ca (see Section 6.14.8) and determining the degree to which factors associated with the growth of large coral heads might influence longer-term records.

6.14.5 OXYGEN ISOTOPES AS A CLIMATE PROXY IN OTHER MARINE BIOGENIC PHASES

Oxygen isotopes have been used as temperature or climate proxies in a number of other marine biogenic phases, although far less widely than in foraminifera or reef corals. Probably the most important work has been on oxygen isotopes in diatom opal (Shemesh *et al.*, 1992, 1994, 1995). Because many sites in the Southern Ocean contain virtually no carbonate, opal $\delta^{18}O$ becomes critical for both stratigraphic and paleoclimatological purposes. Unfortunately, the systematics of oxygen isotopes in diatoms appears to be considerably more complex than for carbonates (Labeyrie and Juillet, 1982; Juillet-Leclerc and Labeyrie, 1986). Oxygen isotopes have also been measured as temperature or climate proxies in ahermatypic solitary corals (Smith *et al.*, 1997), in coralline sponges (Böhm *et al.*, 2000), in fish otoliths (Andrus *et al.*, 2002), as well as in pteropods (Grossman *et al.*, 1986) and other mollusks.

6.14.6 MAGNESIUM AS A PALEOTEMPERATURE PROXY IN FORAMINIFERA

6.14.6.1 Background and History

At the time of this writing (Spring 2002), research in the use of Mg/Ca ratios in foraminifer shells is probably advancing as fast as any area in climate proxy research. As a result of these advances since the late 1990s, researchers now have a good idea of the main advantages and limitations of this approach. It is fair to say that this new paleothermometry approach, perhaps more than any other, is revolutionizing the means by which paleoceanographers and paleoclimatologists unravel ocean and climate history. For this reason, I review the history of this development more closely than for the other proxies.

As discussed in Section 6.14.2, the observation that magnesium was higher in marine carbonates precipitated in water warmers dates to the early part of the twentieth century. Several studies confirmed these early observations for neritic foraminifer shells, which are composed of high magnesium calcite (>5% $MgCO_3$) (Chave, 1954; Chilingar, 1962). Another study suggested that pelagic foraminifera, which are composed of low-magnesium calcite (<1% $MgCO_3$), might also follow this pattern (Savin and Douglas, 1973). Several studies also demonstrated that inorganic carbonates followed the same pattern (Chilingar, 1962; Katz, 1973). This latter point is important, because it indicates that the temperature influence could not be entirely biological.

Another important milestone in research in this area came with the recognition that dissolution on the seafloor or within the sediments could significantly alter the Mg/Ca ratio of foraminifer shells (Bender *et al.*, 1975; Hecht *et al.*, 1975; Lorens *et al.*, 1977). Lorens *et al.* (1977) went so far as to state that "*diagenesis rules out* using Mg/Ca ratios of whole tests as growth temperature indicators (italics added for emphasis)." Despite this clear hindrance, studies documenting systematic down-core variations (Cronblad and Malmgren, 1981) and possible links to growth temperature (Delaney *et al.*, 1985) kept interest in the possibility of this proxy's usefulness alive. Several studies in the 1990s confirmed the early observations of a dissolution effect and species rankings (Rosenthal and Boyle, 1993; Russell *et al.*, 1994; Brown and Elderfield, 1996). It was not until Dirk Nürnberg, a doctoral student at Bremen University, Germany, used electron microprobe determinations on shell surfaces to documented more convincing Mg/Ca–temperature relationships in cultured, core-top and down-core planktonic foraminifera (Nürnberg, 1995; Nürnberg *et al.*, 1996a,b), that the international community recognized the potential importance of this new tool. Progress on documenting a potential response of magnesium in benthic foraminifera to bottomwater temperatures, a result presaged by Scot Izuka's (University of Hawaii) pioneering study of magnesium in *Cassidulina* (Izuka, 1988), occurred at about the same time (Russell *et al.*, 1994; Rathburn and Deckker, 1997; Rosenthal *et al.*, 1997). The Rosenthal *et al.* (1997) paper is notable for its broad calibration and for being the first to point out that the magnesium relationship to temperature is predicted, albeit with a smaller slope, by thermodynamic calculations.

Progress has been very rapid since these initial findings, in part because of improvements in analytical instrumentation and methods. Milestones include the first attempt to deduce glacial tropical SSTs using Mg/Ca (Hastings *et al.*, 1998), the first culturing calibrations made on whole shells (Lea *et al.*, 1999), the first attempt to combine Mg/Ca paleotemperatures with oxygen isotopic ratios to deduce variations in $\delta^{18}O$-seawater (Mashiotta *et al.*, 1999; Elderfield and Ganssen, 2000), the first long tropical SST and $\delta^{18}O_{water}$ records (Lea *et al.*, 2000), the first application of benthic magnesium to Cenozoic climate evolution (Lear *et al.*, 2000), and the first detailed Late Quaternary benthic magnesium records (Martin *et al.*, 2002). The following sections detail the most important of these findings and research priorities for the future.

6.14.6.2 Calibration and Paleotemperature Equations

The underlying basis for magnesium paleothermometry is that the substitution of magnesium in calcite is endothermic and therefore is favored at higher temperatures. The enthalpy change for the reaction based on the more recent thermodynamic data is 21 kJ mol^{-1} (Koziol and Newton, 1995), which can be shown, using the van't Hoff equation, to equate to an exponential increase in Mg/Ca of 3% per °C (Lea *et al.*, 1999). The thermodynamic prediction of an exponential response is one of the reasons that magnesium paleotemperature calibrations are generally parametrized this way. Available inorganic precipitation data generally follows the thermodynamic prediction (Chilingar, 1962; Katz, 1973; Burton and Walter, 1987; Mucci, 1987), with the most extensive data set (Oomori *et al.*, 1987) yielding a 3.1 ± 0.4% per °C increase in D_{Mg} for calcites precipitated in seawater over 10–50 °C (all responses given as percentages are calculated as exponentials, with 95% CI).

Foraminifera shells differ from the thermodynamic prediction in two fundamental ways. First, foraminifera contain 5–10 times lower magnesium than predicted from thermodynamic calculations (Bender *et al.*, 1975). Second, the response of shell magnesium to temperature is ~3 times larger than the thermodynamic prediction and inorganic observation, averaging 9 ± 1% per °C (Lea *et al.*, 1999). Why the latter effect is so is not known, but it has several important implications for magnesium paleothermometry. First, it increases the sensitivity of this approach, which is critical in determining its real error, which depends on large part on the relative magnitude of the temperature response versus that of all the combined sources of error, including measurement error, population variability, and secondary effects. Second, it raises the question of why the response is so much greater in foraminifera and if the augmentation of the response depends on secondary factor(s) that might change over geological time. One possibility is that the much smaller magnesium content of foraminifer shells increases the thermodynamic response (Figure 3). Data from a recent study (Toyofuku *et al.*, 2000), which calibrated two neritic high-magnesium benthic species in culturing

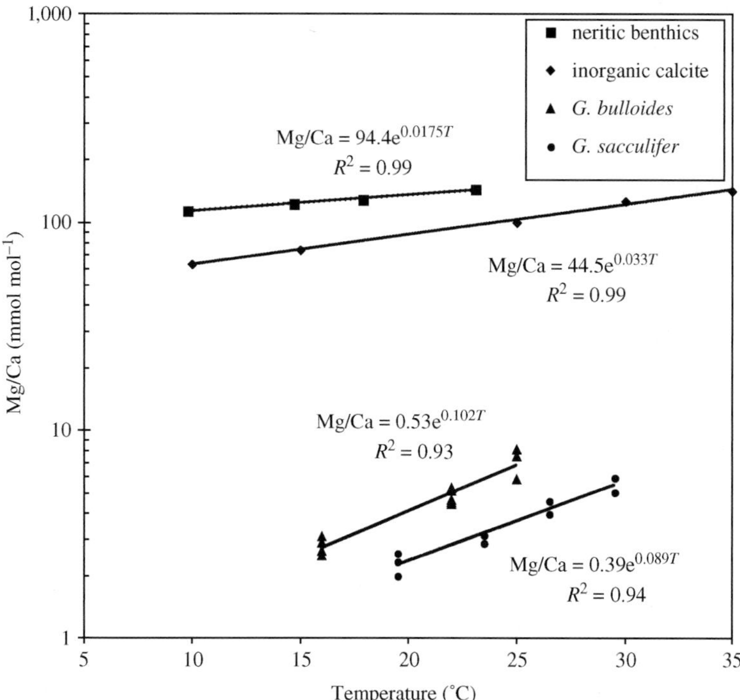

Figure 3 Comparison of Mg/Ca–temperature relationships for inorganic calcite precipitation (Oomori *et al.*, 1987), neritic benthic foraminifera (Toyofuku *et al.*, 2000), a tropical spinose symbiont bearing planktonic foraminifera, *G. sacculifer* (Nürnberg *et al.*, 1996a,b), and a subpolar spinose symbiont barren planktonic foraminifera, *G. bulloides* (Lea *et al.*, 1999). All of the foraminifera results are from culturing. Mg/Ca is plotted on a log scale because of the wide range of values. All of the relationships are fit with an exponential. Note that high Mg inorganic and benthic calcite has a shallower slope and much smaller exponential constant (2–3%); the low Mg foraminiferal calcite has a steeper slope and higher exponential constant (9–10%). Low Mg benthic foraminifera have exponential constants of ~10% (source Rosenthal *et al.*, 1997).

experiments, suggest that Mg/Ca in these species increases by between 1.8% per °C and 2.6% per °C, a far smaller increase than is observed for low magnesium foraminifera. The magnesium response to temperature found by Toyofuku *et al.* (2000) is actually much closer to the ~3% per °C observed for inorganic calcite (Oomori *et al.*, 1987), which contain magnesium contents similar to neritic benthics. This correspondence suggests the magnesium response to temperature might scale with the magnesium contents of calcite

At present, three planktonic species, *Globigerinoides sacculifer, Globigerina bulloides,* and *O. universa*, have been calibrated by culturing and fit with equations of the form

$$\text{Mg/Ca (mmol mol}^{-1}) = be^{mT} \qquad (4)$$

where b is the pre-exponential constant, m the exponential constant, and T the temperature (Lea *et al.*, 1999). Fitting Mg/Ca–temperature data with an equation of this form has the dual advantage of allowing for an exponential response while also parametrizing, by the use of the natural logarithm e, the exponential constant as the change in Mg/Ca per °C. It should be noted that it is the exponential constant that determines the magnitude of temperature change calculated from down-core variations in Mg/Ca and the pre-exponential constant that determines the absolute temperature. Calibration results for these three species indicates exponential constants between 0.085 and 0.102, equivalent to 8.5% to 10.2% increase in Mg/Ca per °C (Nürnberg *et al.*, 1996a,b; Lea *et al.*, 1999). A recent study utilizing planktonic foraminifera from a sediment trap time series off Bermuda extends calibration to seven other species, which in aggregate have a temperature response of 9.0 ± 0.3% (Anand *et al.*, 2003). The pre-exponential constant b ranges between 0.3 and 0.5, with the exception of higher values for *O. universa*, which appears to be unique in many aspects of its shell geochemistry (Nürnberg *et al.*, 1996a,b; Lea *et al.*, 1999; Anand *et al.*, 2003). Core-top calibrations are in general agreement with the culturing results, and include calibrations of eight planktonic species (Elderfield and Ganssen, 2000; Lea *et al.*, 2000; Dekens *et al.*, 2002; Rosenthal and Lohmann, 2002). The Dekens *et al.* (2002) calibrations, which include a second term to account for dissolution in the form of a water depth or saturation effect, are discussed in Section 6.14.6.3.

Calibration for benthic species is somewhat more uncertain. The first comprehensive calibration was carried out by Yair Rosenthal, then at MIT, who used *Cibicidoides* spp. from shallow sediments on the Bahamas outer bank to calibrate benthic magnesium between 5 °C and 18 °C (Rosenthal *et al.*, 1997). When this calibration is augmented with *C. wuellerstorfi* data from deeper sites and adjusted for an analytical offset between atomic absorption spectrophotometry and ICP-MS (Martin *et al.*, 2002), the calibration yields

$$\text{Mg/Ca} = 0.85e^{0.11T} \qquad (5)$$

The value of the exponential constant, 0.109 ± 0.007 (95% CI), overlaps with estimates from planktonic species, suggesting that a magnesium response to temperature of this magnitude is a common factor among foraminifera. It should be noted that the Martin *et al.* (2002) data set suggests that the magnesium response might be steeper at bottomwater temperatures, <4 °C. Resolving the calibration for benthic magnesium at the coldest temperatures is an important research priority because there are is a great deal of research interest in establishing the temperature evolution of cold bottomwaters. Establishing calibrations for other species is also a high priority, in part because of the insight this provides into the basis for magnesium paleothermometry.

6.14.6.3 Effect of Dissolution

It has been known since the 1970s that the magnesium content of foraminifer shells, as well as other carbonates, is susceptible to change via dissolution (Hecht *et al.*, 1975; Lorens *et al.*, 1977). As mentioned previously, this factor was one of the main reasons that little hope was held out for the usefulness of foraminiferal magnesium as a paleotemperature proxy. At present, researchers accept that dissolution alters the Mg/Ca content of foraminifera shells and instead are investigating the degree to which such changes occur, how dissolution can be assessed and whether correction factors are possible, and the degree to which dissolution affects Mg/Ca and oxygen isotopes similarly or dissimilarly.

In the mid-1990s, a number of groups measured Mg/Ca in planktonic foraminifera from oceanic depth transects, mostly as support for studies of other metals (F, U, and V) in the shells (Rosenthal and Boyle, 1993; Russell *et al.*, 1994; Brown and Elderfield, 1996; Hastings *et al.*, 1996). The advantage of the depth transect approach is that one can assume that shells with similar compositions rain down from overlying surface waters to all the sites, and that observed differences must be due to post-depositional processes. These studies demonstrated, to a greater or lesser degree, that Mg/Ca in the shells decreased with water depth and inferred increasing dissolution. The Rosenthal and Boyle (1993) study in particular documented both the general relationship between Mg/Ca and δ^{18}O as well as the drop in Mg/Ca with water depth in both spinose and nonspinose species. In general, these studies indicated that the drop in Mg/Ca was more pronounced for nonspinose species such as *Globorotalia tumida*, a result that was interpreted to reflect preferential dissolution of magnesium-rich chamber calcite over magnesium-poor keel calcite (Brown and Elderfield, 1996). One of these studies also revived the idea, first suggested by Savin and Douglas (1973), that the magnesium content of the shells influenced their solubility (Brown and Elderfield, 1996). Calculations suggest that the saturation horizon for ontogenetic calcite with Mg/Ca of 10 mmol mol^{-1}, about twice the value found in typical tropical shells, could be 300 m shallower. Magnesium loss presumably occurs when shells are on the seafloor and/or when they pass through the sediment mixed layer where metabolic CO_2 is available for dissolution. The fact that surface dwelling *Globigerinoides ruber* indicates decreasing Mg/Ca with water depth in the western equatorial Pacific (Lea *et al.*, 2000; Dekens *et al.*, 2002), an area with minimal temporal and spatial variation in mixed layer temperatures, suggests that magnesium loss might occur via preferential dissolution of magnesium-rich portions of the shell (Lohmann, 1995; Brown and Elderfield, 1996). It is also quote possible, however, that the progressive loss of the less robust individuals, which might have preferentially calcified in the warmest waters, shifts the mean Mg/Ca to lower values in deeper sediments.

A clear complexity in utilizing shell Mg/Ca for paleotemperature is that many species migrate vertically and/or add gametogenic calcite at depths significantly deeper than their principal habitat depth (Bé, 1980). This complicates the dissolution question, because these different shell portions are likely to have slightly different solubilities and Mg/Ca ratios. An innovative approach to this problem was suggested by Rosenthal *et al.* (2000), who argued that the relationship between size-normalized shell mass and dissolution loss could be used to correct shell Mg/Ca. This approach, which relies on a constant relationship between shell mass changes and Mg/Ca changes, has yet to be validated in downcore studies, although a new study (Rosenthal and Lohmann, 2002), demonstrates that this approach can yield consistent glacial–interglacial SST changes from cores both above and below the lysocline.

A somewhat different approach has been taken by others (Lea *et al.*, 2000; Dekens *et al.*, 2002). They quantified the Mg/Ca loss in depth transects as a percentage loss per kilometer water depth, thus allowing direct comparison of the magnitude of potential dissolution loss versus the magnitude of the temperature effect. If independent estimates of past shifts in lysocline depth are available, it is then possible to estimate the magnitude and direction of dissolution bias down-core. Dekens *et al.* (2002) found, based on core tops from the tropical Atlantic and Pacific, that Mg/Ca loss ranged from 3% per km water depth for *G. sacculifer*, 5% for *G. ruber*, and 22% for *N. dutertrei*, a nonspinose thermocline dweller. These terms equate to a bias on magnesium paleothermometry of 0.4 °C per km, 0.6 °C per km, and 2.8 °C per km effective shift, respectively, in foraminiferal lysocline, or depth of effective dissolution. Given that evidence for late Quaternary lysocline shifts is generally between 0.2 km and 0.8 km (Farrell and Prell, 1989), this approach suggests that down-core dissolution biases on magnesium paleothermometry will be less than 0.5 °C for spinose surface dwellers. Calibration equations derived from the Dekens *et al.* (2002) calibration set are also parametrized using Δcarbonate ion (the difference between *in situ* and saturation values) to account for differences in dissolved carbonate ion between basins.

The evidence for dissolution effects on magnesium in benthic foraminifera is less certain. For one, it is more difficult to discern a dissolution trend because benthic Mg/Ca is decreasing with increasing water depth and decreasing bottom-water temperature. Data from a depth transect on the Ontong Java Plateau for Sr/Ca, Ba/Ca, and Cd/Ca have been used to infer a dissolution effect on these elements (McCorkle *et al.*, 1995), although alternative interpretations such as carbonate ion or pressure effects on biomineralization have also been suggested (Elderfield *et al.*, 1996). Martin *et al.* (2002) suggested that the steeper trend of benthic Mg/Ca in the coldest waters, estimated at ~20% per °C versus 11% per °C, might reflect dissolution and magnesium loss in the deepest, most undersaturated waters. Alternatively, it might reflect the influence of other factors, such as carbonate ion saturation. Regardless, this will be a critical issue in validating benthic Mg/Ca for use on the coldest bottomwaters.

6.14.6.4 Other Secondary Effects: Salinity, pH, Gametogenesis, and Changes in Seawater Mg/Ca

Factors other than temperature and dissolution also appear to influence Mg/Ca in planktonic shells. Based on culturing, there is clear evidence for differences in uptake between species (Lea *et al.*, 1999), with as much as a factor of two variations. For this reason, species-specific calibrations are necessary, although it is difficult to do this by any means other than culturing because of the complication of habitat depth. Salinity appears to exert a small effect on shell Mg/Ca, with an observed increase of between $6 \pm 4\%$ for *O. universa* (Lea *et al.*, 1999) and $8 \pm 3\%$ for *G. sacculifer* (Nürnberg *et al.*, 1996a) per salinity unit (SU) increase. (Note: this and all other relationships of this kind are quoted as the exponential constant, with 95% confidence intervals, for an exponential fit to the observational data; the original published data was not always fit this way. An exponential fit has the advantage of giving the response in terms of a constant percentage, which then can be easily related to the exponential constant in the temperature response equation.) Assuming an Mg/Ca temperature response of 10% per °C (see Section 6.14.6.2), the salinity influence is equivalent to a positive bias of between 0.6–0.8 °C per SU increase. More extensive culturing data is needed; however, before such an influence can be accepted as likely to apply for salinity differences of <3.

Investigation of the effect of seawater pH indicates that pH has a significant effect on magnesium uptake, with an observed decrease of $-6 \pm 3\%$ per 0.1 pH unit increase for *G. bulloides* and *O. universa*. Again assuming an Mg/Ca temperature response of 10% per °C (Section 6.14.6.2), pH influence is equivalent to a bias −0.6 °C per 0.1 pH unit increase. Past variability in oceanic pH (Sanyal *et al.*, 1996) and water-column variability in pH could therefore both exert significant biases on Mg paleothermometry.

A bias that applies equally to any foraminifera-based proxy is the problem of gametogenic calcite addition in the subsurface, as well as other vertical migration effects. For magnesium, one early study claimed that gametogenic calcite from *G. sacculifer* cultures is highly enriched in magnesium (Nürnberg *et al.*, 1996a), but this observation has not confirmed by studies of shells from sediments (Elderfield and Ganssen, 2000; Nürnberg *et al.*, 2000; Rosenthal *et al.*, 2000; Dekens *et al.*, 2002). The observation that Mg/Ca in *G. sacculifer*, a species known to add ~30% gametogenic calcite (Bé, 1980), is generally lower (Elderfield and Ganssen, 2000; Lea *et al.*, 2000; Dekens *et al.*, 2002) than in *G. ruber*, a species that adds little or no gametogenic calcite (Caron *et al.*, 1990), suggests that the addition of gametogenic calcite takes place in cold, subsurface waters and reduces Mg/Ca and inferred temperatures for those species that add significant shell calcite this way.

6.14.6.5 Results on Quaternary Timescales

Although magnesium paleothermometry has only been used for ~5 yr, it has already led to a number of important and unprecedented findings for paleoceanographic and paleoclimatic research. These include documenting the history of sub-polar Antarctic SST variations (Mashiotta *et al.*, 1999; Rickaby and Elderfield, 1999), tropical Atlantic and Pacific SST changes (Hastings *et al.*, 1998; Elderfield and Ganssen, 2000; Lea *et al.*, 2000; Nürnberg *et al.*, 2000), and changes in bottomwater temperature in the Atlantic and Pacific (Martin *et al.*, 2002). Secondary products include δ^{18}O-seawater records for the sub-polar Antarctic (Mashiotta *et al.*, 1999), equatorial Pacific (Lea *et al.*, 2000, 2002), and, for five different planktonic species in one core, the tropical Atlantic (Elderfield and Ganssen, 2000). Several other high-resolution records from the tropical Pacific have been published (Koutavas *et al.*, 2002; Stott *et al.*, 2002; Rosenthal *et al.*, 2003).

Among these Mg/Ca results, perhaps the most important are those that are available for the tropics. Past SST changes in the tropics have been a contentious issue (Crowley, 2000b), mostly because the actual glacial–interglacial changes are relatively small (<5 °C) and therefore more difficult to detect unambiguously using either faunal or geochemical methods. The faunal approach, in particular, is hampered by the fact that glacial tropical assemblages in the warm pools are not very different from their interglacial counterparts (Crowley, 2000b). Even with re-examination and major refinements, the faunal approach does not yield significant cooling in the tropical warm pools (Mix *et al.*, 1999; Trend-Staid and Prell, 2002). The Mg/Ca approach works especially well in the tropics, because the calibration curve at warm temperatures shows the largest absolute change in Mg/Ca per °C (Figures 3 and 4). Oligotrophic tropical sites, which are poor candidates for the alkenone unsaturation paleotemperature approach, generally contain abundant specimens of *G. ruber* and *G. sacculifer*, which are well calibrated for Mg/Ca. These low productivity sites also have minimal potential for diagenetic changes, which removes one confounding factor for trace element work.

From this vantage point, it appears that Mg/Ca paleothermometry has cracked the problem of glacial cooling of the tropical warm pools (Hastings *et al.*, 1998; Lea *et al.*, 2000), although it must be said that the modest but systematic degree of cooling recorded by Mg/Ca was presaged by earlier results from the alkenone unsaturation technique (Lyle *et al.*, 1992; Bard *et al.*, 1997; Pelejero *et al.*, 1999). But the Mg/Ca results put the ~3 °C level of glacial cooling, relative to modern conditions, on a very firm footing, especially for the western Pacific warm pool (Lea *et al.*, 2000; Stott *et al.*, 2002; Visser *et al.*, 2003; Rosenthal *et al.*, 2003), which is the largest and warmest tropical water mass in the oceans. Results from a core on the Ontong Java Plateau, which lies on the equator in the center of the western Pacific warm pool, span the last 500 kyr and indicate that glacial cooling was systematically ~3 °C cooler than modern conditions and that this cooling occurred during each

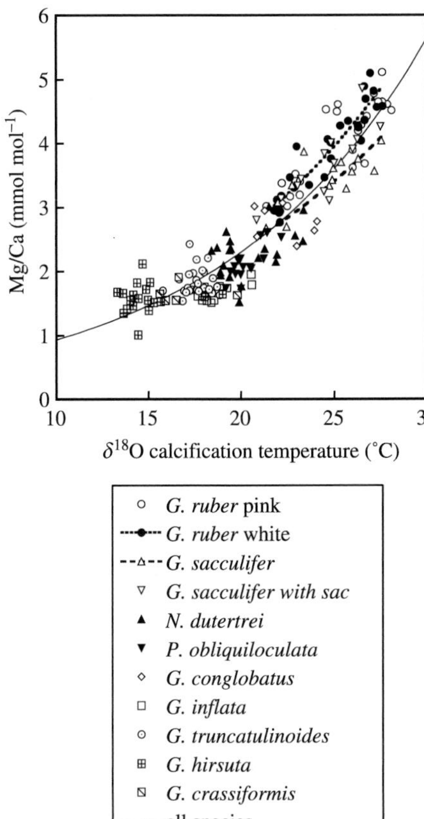

Figure 4 Mg/Ca of different planktonic foraminifera from a Bermuda sediment trap time series, plotted versus calcification temperatures calculated from the oxygen isotopic composition of the shells (Anand *et al.*, 2003) (reproduced by permission of American Geophysical Union from *Paleoceanography* **2003**, *18*, 1050). The aggregate fit to all the data in the plot is: Mg/Ca = 0.38 exp(0.09*T*), very similar to relationships derived from culturing and core-top studies.

of the last five major glacial episodes (Figure 5). Of great interest is the fact that glacial warming appears to lead ice sheet demise by ~3 kyr (Lea *et al.*, 2000). This unanticipated SST lead, which suggests a prominent role for the tropics in pacing ice age cycles, has now also been observed in high-resolution records of the last deglaciation (Stott *et al.*, 2002; Rosenthal *et al.*, 2003).

The Mg/Ca approach has also led to a number of other important findings. One of the strengths of this approach is that the recorded paleotemperature is recorded simultaneously with the δ^{18}O composition of the shell. Combining these factors using an oxygen-isotope paleotemperature equation yields the δ^{18}O-water at the time of shell precipitation. The limited number of studies using this approach already suggests that it is likely to yield important results on shifts in global ice volume as well as regional salinity shifts (Mashiotta *et al.*, 1999; Elderfield and Ganssen, 2000; Lea *et al.*, 2000, 2002; Stott *et al.*, 2002; Rosenthal *et al.*, 2003). A paleosalinity proxy has always been a difficult prospect, but it appears now that comparison of extracted δ^{18}O-water will make it possible to reconstruct patterns of salinity change in the past (Lea, 2002). Such reconstructions will be invaluable in understanding paleoclimate shifts in the tropics.

Another strength of the magnesium paleothermometry is that it can be applied to benthic fauna, including ostracode shells (see Section 6.14.7). There are no other techniques that provide direct estimates of bottomwater temperatures. There is only one published study which contains detailed records of benthic foraminiferal Mg/Ca variations, from the eastern tropical Atlantic and Pacific, in the Quaternary (Martin *et al.*, 2002). These two records indicate the great promise of this approach in elucidating deep-water temperature variations, which appear to have been ~2–3 °C. However, there are also considerable challenges. The absolute magnitude of the Mg/Ca change is much smaller in the cold-temperature region, and therefore other influences, such as vital effects, dissolution or calcification effects (Elderfield *et al.*, 1996), can exert significant biases. Separating these effects will undoubtedly be a major research area in the future.

6.14.6.6 Results for the Neogene

One of the most exciting prospects for magnesium paleothermometry is combining this approach with the benthic oxygen isotope curve for the Cenozoic (Zachos *et al.*, 2001) to separate the influence of temperature and ice volume. Two studies already suggest the great potential of this approach (Lear *et al.*, 2000; Billups and Schrag, 2002). The Lear *et al.* (2000) study, which is based on a data set extending back to the Eocene, reveals that benthic Mg/Ca records the gradual ~12 °C cooling of bottomwaters that occurred during the Cenozoic and that had been inferred from oxygen isotopes. Combining the Mg/Ca-based temperatures with measured δ^{18}O allows calculation of the δ^{18}O evolution of seawater, which can be traced to the expansion and contraction of ice sheets. Comparison of magnesium temperature trends with δ^{18}O over the Eocene–Oligocene boundary reveals that the δ^{18}O shifts are dominated by global ice volume shifts. There are significant uncertainties, such as species offsets, diagenesis, and changes in seawater Mg/Ca, in extending magnesium paleothermometry to longer timescales, but there are also great prospects for major discoveries. One can only await with anticipation the revelations yet to come when high-resolution benthic and planktonic Mg/Ca Neogene and Paleogene records are available from a number of sites!

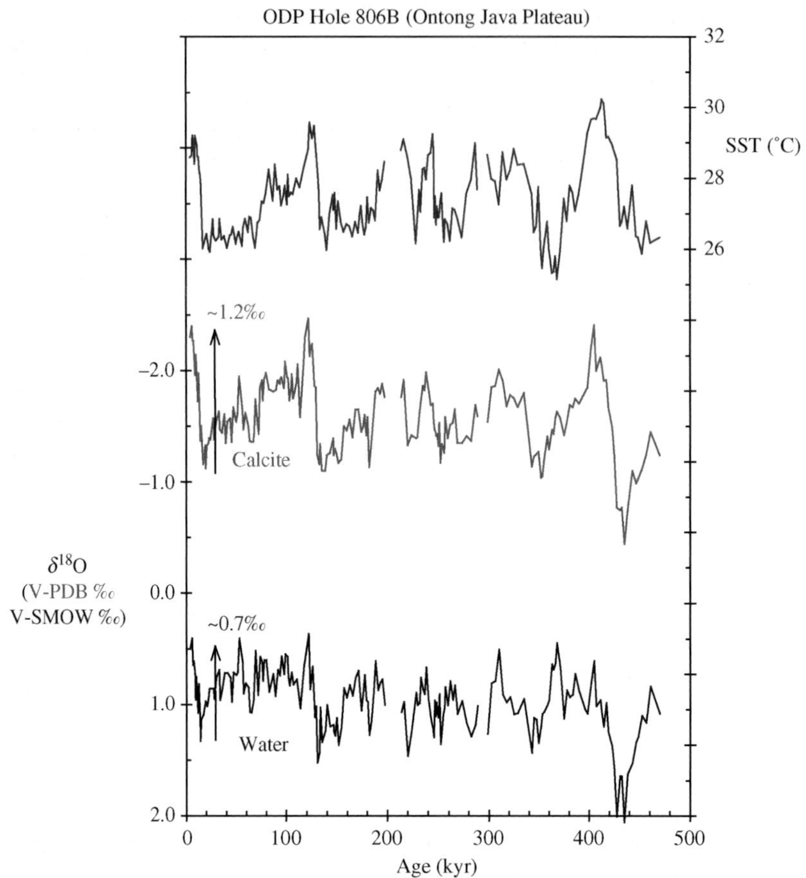

Figure 5 Down-core records of Mg/Ca-based SSTs, $\delta^{18}O$, and $\delta^{18}O_{water}$ derived from the surface dwelling planktonic foraminifera *G. ruber* in ODP Hole 806B on the Ontong Java Plateau in the western equatorial Pacific (Lea *et al.*, 2000). Glacial marine isotope stages (MIS) are indicated. Note the reduction, from 1.2‰ to 0.7‰, in glacial–interglacial $\delta^{18}O$ amplitude when the temperature portion of the signal is removed. The fact that the $\delta^{18}O_{water}$ amplitude in the WEP is smaller than the oceanic mean change is attributed to a hydrological shift to less saline surface waters during glacial episodes (Lea *et al.*, 2000; Rosenthal *et al.*, 2003; Oppo *et al.*, 2003). Note the lead of Mg/Ca over $\delta^{18}O$, especially prominent on the MIS 12 to 11 transition.

6.14.6.7 Summary of Outstanding Research Issues

Magnesium paleothermometry is in the midst of a period of phenomenal growth, and it has quickly taken its place as one of the most useful means paleoceanographers have at their disposal to study past climates. Many questions, such as ecological bias, species offsets, environmental influences other than temperature, dissolution and diagenetic overprinting, must be addressed before the ultimate reliability of magnesium paleothermometry is known. One inherent advantage is the enormous amount already known about foraminiferal ecology and geochemistry, much of which applies equally to oxygen isotopes and Mg/Ca. At this stage, the most fundamental issues are: (i) establishing the spatial and temporal stability of magnesium temperature calibrations for the important paleoceanographic species; (ii) establishing the extent to which dissolution biases down-core Mg/Ca records; and (iii) establishing the degree to which benthic magnesium variations record temperature variations in the coldest part of the bottomwater temperature range (<4 °C).

6.14.7 MAGNESIUM AS PALEOTEMPERATURE PROXIES IN OSTRACODA

Magnesium paleothermometry applied to ostracode shells has proved to be an important means of discerning past variations in bottomwater temperatures (Dwyer *et al.*, 1995; Cronin *et al.*, 1996, 2000; Correge and Deckker, 1997). This approach is based on the same principle as magnesium paleothermometry in foraminifera, although ostracode calibrations have been fit with linear calibrations. These calibrations suggest that the increase in Mg/Ca in ostracodes is ~9% per °C, and therefore similar to the foraminiferal calibrations (see Section 6.14.6.2). The results in Dwyer *et al.* (1995) indicate that ostracode Mg/Ca can be used quite effectively to separate bottomwater temperature and $\delta^{18}O_{water}$ influences in both the Quaternary and Pliocene. In light of subsequent discoveries, it is interesting to note that Dwyer *et al.* (1995) saw a lead of temperature over benthic $\delta^{18}O$ in their Late Quaternary ostracode Mg/Ca record of 3,500 yr. Similarly, a study of magnesium in benthic foraminifera in a tropical Atlantic core saw a lead of ~4,000 yr in benthic magnesium of temperature over benthic $\delta^{18}O$ over the last 200 kyr (Martin *et al.*, 2002). These results highlight the importance of independent paleothermometers.

6.14.8 STRONTIUM AS A CLIMATE PROXY IN CORALS

6.14.8.1 Background

The idea of using Sr/Ca in corals as a paleothermometer goes back to the early 1970s, but it is one that that did not come to fruition until the early 1990s with the application of more precise analytical techniques. Early studies indicated that there was an inverse relationship between seawater temperature and strontium content of both inorganically precipitated and coral aragonite, with a relatively small inverse temperature dependence of just under 1% per °C (Kinsman and Holland, 1969; Weber, 1973; Smith *et al.*, 1979; Lea *et al.*, 1989). The breakthrough study for coral Sr/Ca, led by Warren Beck, then at University of Minnesota (Beck *et al.*, 1992), utilized extremely precise isotope dilution thermal ionization mass spectrometric (ID-TIMS) determinations to establish the relationship between Sr/Ca and temperature. Their calibration data indicated a 0.6% decrease in Sr/Ca per °C, and with determinations of ±0.03% (2 SD) possible by ID-TIMS, the Beck *et al.* (1992) approach indicated a possible paleotemperature determination of a remarkable ±0.05 °C! Along with their calibration data, Beck *et al.* (1992) presented Sr/Ca data from a fossil coral from Vanuatu that had been dated to the late Younger Dryas/Early Holocene period. These data indicated a 5.5 °C cooling of SST in this region, and it must be counted among the first strong evidence challenging the CLIMAP (1981) view of relatively unchanged tropical SST during glacial episodes.

Following the Beck *et al.* (1992) publication, a number of laboratories undertook more detailed studies of the calibration and also investigated the application of this approach to paleoceanography and paleoclimatology (de Villiers *et al.*, 1994, 1995; Guilderson *et al.*, 1994; McCulloch *et al.*, 1994; Gagan *et al.*, 1998). Generally these studies have supported the Beck *et al.* (1992) original insights, although two major problems with the Sr/Ca approach have been identified: (i) it appears that growth rate and symbiont activity have a marked influence on coral Sr/Ca (de Villiers *et al.*, 1994; Cohen *et al.*, 2001, 2002) and (ii) there is evidence for a secular shift in seawater Sr/Ca on glacial–interglacial timescales, with generally higher values during glacial episodes (Stoll and Schrag, 1998; Martin *et al.*, 1999; Stoll *et al.*, 1999). Another major step forward in the application of the Sr/Ca paleothermometer came with the development of a very rapid but still precise atomic absorption spectrophotometry method (Schrag, 1999b); this technique enables researchers to generate the large data sets required for long, high-resolution climate records (Linsley *et al.*, 2000).

6.14.8.2 Paleotemperature Equations

The relationship between coral Sr/Ca and seawater temperature is parametrized as a linear function of the form

$$\text{Sr/Ca}_{\text{coral}}\ (\text{mmol mol}^{-1}) = b + m(\text{SST}) \quad (6)$$

The thermodynamic prediction for strontium substitution in aragonite is actually an exponential response with an inverse temperature dependence, a consequence of the negative enthalpy (exothermic) nature of the reaction in which strontium substitutes for calcium in aragonite. The observed exponential constant for inorganic aragonite precipitation is quite small: −0.45% per °C (Kinsman and Holland, 1969). Therefore, over the small range of coralline strontium paleothermometry, the relationship can be quite adequately expressed as a linear relationship. The single inorganic aragonite precipitation study indicates a slope m of 0.039 and an intercept b of 10.66 mmol mol^{-1} (Kinsman and Holland, 1969). Calibrations are available for a number of coral species, but mostly for species of *Porites* (Smith *et al.*, 1979; Beck *et al.*, 1992; de Villiers *et al.*, 1994; Mitsuguchi *et al.*, 1996; Shen, 1996; Gagan *et al.*, 1998; Sinclair *et al.*, 1998; Cohen *et al.*, 2001, 2002). Values of the intercept b, which determines the absolute Sr/Ca for a particular temperature, range from 10.3 to 11.3; values of the slope m, which determines the temperature sensitivity, range from 0.036 to 0.086 (Figure 6). The variability in the slope is a critical problem for coral strontium paleothermometry, because the cited slopes equate to a variability in temperature dependence of −0.4% per °C to −1.0% per °C. Therefore, a recorded change of Sr/Ca in corals of 1% can imply between a 1 °C and 2.5 °C shift in paleotemperature. Of course, in practice, it is

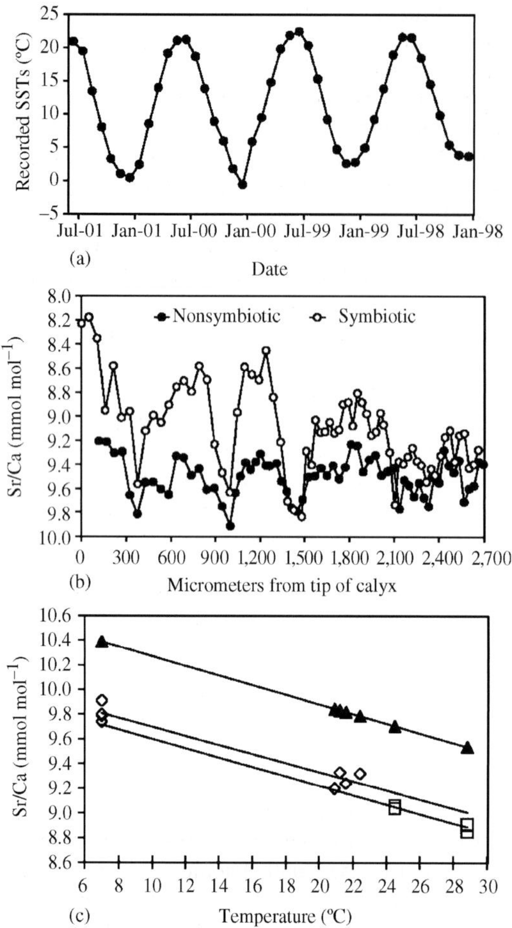

Figure 6 Comparison of Sr/Ca records in symbiont-bearing and symbiont barren ahermatypic corals (Cohen *et al.*, 2002). (a) Average monthly SSTs at 12 ft (4 m) depth in the Woods Hole harbor, the coral collection site, between January 1998 and July 2001. (b) Life history Sr/Ca profiles from symbiotic (open circles) and asymbiotic (solid circles) skeleton of *A. poculata* colonies collected in Woods Hole in July 2001. Skeletal Sr/Ca in the first year of life is the same in both samples, but the similarities decrease as the corallites mature, a divergence caused by a progressive decrease in summertime Sr/Ca in the symbiotic skeleton. (c) The Sr/Ca–SST relationship in asymbiotic *Astrangia* skeleton ($=-0.036x+10.065$) is compared with night-time skeleton of the tropical reef coral *Porites* ($=-0.038x+9.9806$) and inorganic aragonite precipitated at equilibrium ($=-0.039x+10.66$). The slope of the regression equations, indicative of the temperature sensitivity of Sr/Ca uptake into the coral skeleton, are similar for all three precipitates (−0.036, −0.038, and −0.039, respectively). This agreement establishes temperature as the primary control of Sr/Ca in the asymbiotic skeleton (source Cohen *et al.*, 2002) (reproduced by permission of American Association for the Advancement of Science from *Science*, **2002**, *296*, 331–333).

possible to narrow this uncertainty by conducting local calibrations (e.g., Correge *et al.*, 2000).

The key question is the degree to which the relationship between coral Sr/Ca and SST stays constant in time and space. There have been a number of investigations of this question (de Villiers *et al.*, 1994; Alibert and McCulloch, 1997; Cohen *et al.*, 2001, 2002), and the answer seems to be that growth and environmental factors clearly do affect Sr uptake. The more recent of these studies, based on ahermatypic solitary corals, indicates that algal symbionts might be the main influence on the slope, with an enhancement of 65% associated with enhance calcification during periods of strong symbiotic photosynthesis (Cohen *et al.*, 2002). If this result applies generally, it will have important implications for the use of corals in paleoclimate research, because most of the long records are based on the symbiont-bearing reef coral *Porites* (Schrag and Linsley, 2002). However, there is a lot of evidence that coral strontium paleothermometry works remarkably well, and the key question appears to be ascertaining the meaningfulness of long-term secular shifts on both historical and prehistoric timescales (see Sections 6.14.8.4 and 6.14.8.5) Figure 6.

6.14.8.3 Secondary Effects and Diagenesis

The main secondary effects on Sr/Ca paleothermometry appear to be related to growth rate and symbiosis (see Section 6.14.8.2). It is not yet known how these effects might influence the use of coral Sr/Ca for paleoclimate studies, but one imagines that the effects could be important if the growth conditions are changing over the lifetime of the coral. One of the most impressive long-term Sr/Ca records, from Rarotonga in the South Pacific (Linsley *et al.*, 2000), implies a large cooling in ~1760, early in the coral head's life history. This cooling does not appear in some other climate records (Cane and Evans, 2000) and could therefore reflect secondary effects.

The effect of diagenesis has not been widely studied for the Sr/Ca paleothermometer. For young corals growing during historical times, there is clear evidence that precipitation of inorganic aragonite in the pores of the oldest parts of the coral heads can affect the bulk coral Sr/Ca (Enmar *et al.*, 2000; Muller *et al.*, 2001). This secondary aragonite has a higher Sr/Ca ratio than the original coral material, and it therefore biases Sr/Ca paleotemperatures to colder values.

As subaerially exposed corals interact with meteoric waters, they are altered to calcite. This alteration results in a lost of strontium from the skeleton, and will obviously have strong effects on Sr/Ca paleothermometry. To avoid this problem, researchers routinely screen for the presence of calcite in fossil corals. A new study (McGregor and Gagan, 2003) demonstrates that local diagenesis can have marked affects on coral Sr/Ca, with a very large positive bias on reconstructed SST. This occurs because of the large drop in Sr/Ca that accompanies conversion of aragonite to calcite. Interestingly, the bias on oxygen isotopes is much smaller, mainly because the absolute difference in aragonite and calcite end-member compositions is much smaller for $\delta^{18}O$ (see Section 6.14.4.3). McGregor and Gagan (2003) suggest that such localized low-level diagenesis can be detected through a combination of X-ray diffraction techniques, thin section analysis, and high-resolution spatial sampling of the coral skeleton.

A unique secondary complication for Sr/Ca is the potential influence of small changes in seawater Sr/Ca. Although spatial variability in seawater Sr/Ca in the present ocean is very small ($\leq$2%) (Brass and Turekian, 1974; de Villiers *et al.*, 1994), the small sensitivity of the coral Sr/Ca paleothermometer makes it sensitive to these variations. For example, an observed 2% variation in coralline Sr/Ca, which equates to the maximum seawater variation, is equivalent to between a 2 °C and 5 °C temperature change, depending on the slope of the calibration (see Section 6.14.8.2). In practice, it is likely that most locations do not experience variations of more than 0.5% in seawater Sr/Ca (de Villiers *et al.*, 1994), but on historical timescales it is at least possible that larger shifts might have taken place.

On geological timescales, there is growing evidence that shifts in seawater Sr/Ca large enough to affect Sr/Ca paleothermometry have taken place. Such shifts were first hypothesized by a group at Harvard, who recognized that changes in sea level associated with changing continental ice volume had the potential to change seawater Sr/Ca, because calcium carbonate deposition on the continental shelves is dominated by aragonite, which contains ~5 times more strontium than calcite, which dominates deep-sea carbonate deposition (Stoll and Schrag, 1998), Lowered sea level during continental glaciation favors deposition in the deep sea, which therefore raises the Sr/Ca of seawater. Stoll and Schrag (1998) calculated that this change could result in a 1–2% enrichment of seawater Sr/Ca during of just after sea-level lowstands, which, depending on which calibration is used, would result in a −1 °C to −5 °C bias in Sr/Ca paleothermometry.

Subsequently, Pamela Martin (UCSB) and co-workers demonstrated that systematic glacial–interglacial variations in foraminiferal Sr/Ca similar to and even somewhat larger than those predicted by the Stoll and Schrag (1998) study are indeed preserved in deep-sea records (Martin *et al.*, 1999). Several studies have confirmed this observation (Stoll *et al.*, 1999; Elderfield *et al.*, 2000; Shen *et al.*, 2001). These studies indicate foraminifer shells record variations of up to 6% on glacial–interglacial timescales. Some of this variation is undoubtedly due to secondary (kinetic) effects on foraminiferal Sr/Ca, such as temperature, pH and salinity, all of which are known to have small influences on shell Sr/Ca (Lea *et al.*, 1999). But comparison of benthic records from different ocean basins suggests that there is a strong common signal in these records, with a common glacial–interglacial amplitude of ~3% (Martin *et al.*, 1999). Comparison of this stacked benthic record with coral Sr/Ca values of fossil corals suggests that up to half of the observed coral Sr/Ca variation might be attributable to seawater Sr/Ca variation (Martin *et al.*, 1999). This might explain why tropical SST drops during glacial episodes based on fossil coral Sr/Ca (Beck *et al.*, 1992; Guilderson *et al.*, 1994; McCulloch *et al.*, 1996, 1999) are typically twice that suggested by other geochemical proxies (Lea *et al.*, 2000). Regardless of the exact details, it is clear that secular changes in seawater Sr/Ca have the potential to influence the Sr/Ca paleothermometer on longer timescales.

6.14.8.4 Results on Historical Timescales

A few long time series of Sr/Ca have been published, from Rorotonga and the Great Barrier Reef (GBR) in the South Pacific (Linsley *et al.*, 2000; Hendy *et al.*, 2002). These records show clear interannual variability as well as distinct secular trends. For example, the GBR (Hendy *et al.*, 2002) records, which are an average of eight different coral cores, indicate a secular shift to warmer SST in the youngest part of the records (after ~1950). This shift is corroborated by $\delta^{18}O$ and U/Ca measurements in the same corals and appears to track well with instrumental records. This is quite important because of the question of attribution for the prominent trend towards more negative $\delta^{18}O$ observed in many large corals (see Section 6.14.4.4). Combining the metal paleotemperature records with the $\delta^{18}O$ record yields a residual $\delta^{18}O$-water record that suggests that GBR waters have become progressively less salty since the mid-nineteenth century. Results from the Rorotonga site (Linsley *et al.*, 2000) are somewhat different and suggest a prominent cooling in ~1750 followed by a series of decadal oscillations that correlate in the twentieth century with the Pacific Decadal Oscillation (Mantua *et al.*, 1997). The warm period recorded in the mid-eighteenth century at the Rorotonga site appears to be corroborated in the GBR sites, although with a reduced magnitude. Because this time interval was cold in much of the northern hemisphere, the warm South Pacific SSTs and high salinities might be an important clue to the source of what is know as the Little Ice age in the NH (Hendy *et al.*, 2002).

6.14.8.5 Results on Geological Timescales

The first detailed Sr/Ca results published for well-dated fossil corals (Beck *et al.*, 1992) were interpreted as indicating a 6 °C cooling of tropical SST in Vanuatu region of the South Pacific during the latest Younger Dryas/earliest Holocene. Although the degree of cooling has not been supported by other data, presumably because of a secular increase in seawater Sr/Ca (Stoll and Schrag, 1998; Martin *et al.*, 1999; Lea *et al.*, 2000), the Beck *et al.* (1992) result was one of the first to seriously challenge the prevailing view of warm tropical oceans during glacial episodes. Subsequently, Sr/Ca data from sites in the Caribbean (Guilderson *et al.*, 1994) and western Pacific (McCulloch *et al.*, 1999), apparently supported by coincident shifts in $\delta^{18}O$, were published to indicate 5–6 °C cooling of tropical SSTs during glacial episodes. Holocene changes in SST have also been reconstructed using Sr/Ca (Beck *et al.*, 1997; Gagan *et al.*, 1998). In retrospect, it appears that the glacial estimates of cooling were too large, in part because they would have rendered large parts of the tropical sea inhospitable to massive reef corals (Crowley, 2000b). In addition, terrestrial shifts such as the well-known drop in snowlines during the last glacial maximum are compatible with tropical SST drops of ~3 °C (Pierrehumbert, 1999). Pinpointing the exact cause of why coral Sr/Ca appears to give an excess cooling signature of glacial episodes is obviously an important research problem, but regardless of the exact causes of that offset, the tropical cooling results from coral Sr/Ca were a very important initial part of the motivation that lead to a growing focus on the paleoclimatic role of the tropics.

The spectral characteristics of Sr/Ca records in fossil corals can, like oxygen isotopes, provide insight into changes in interannual climate change such as ENSO without requiring conversion into absolute SSTs (Hughen *et al.*, 1999; Correge *et al.*, 2000). The Hughen *et al.* (1999) results are notable for providing evidence, in the form of interannual variations in Sr/Ca and $\delta^{18}O$, of ENSO-like variability in a fossil coral from Sulawesi, Indonesia dated to the last interglacial sea-level highstand, 124 ka.

6.14.8.6 Summary of Outstanding Research Issues

During the 1990s, the coral Sr/Ca paleothermometer has grown to be a fundamental tool for paleoclimate research on historical and geological timescales. Optimal use of this tool requires a better understanding of how the coral Sr/Ca temperature calibration is in both space and time. The most critical question is the degree to which symbiosis (and other kinetic factors) influence the sensitivity of the paleothermometer (Cohen *et al.*, 2002). If the Sr/Ca of coral material precipitated by hermatypic corals during the day is strongly biased by photosynthesis, as suggested by Cohen *et al.* (2002), it will place a severe limitation on both the usefulness and the accuracy of the Sr/Ca paleothermometer.

On geological timescales, the question of secular change in seawater Sr/Ca as well as diagenetic influence require further investigation. Foraminiferal Sr "stacks" of seawater Sr/Ca change (Martin *et al.*, 1999) could be improved by adding more cores and improving the precision of the analyses (Shen *et al.*, 2001). It might also be possible to correct for offsets between sites by taking into account environmental factors that also influence foraminiferal Sr/Ca. In this way it should be possible to eventually generate definite secular records of seawater Sr/Ca that can be suitably applied to coral Sr/Ca paleothermometry.

6.14.9 MAGNESIUM AND URANIUM IN CORALS AS PALEOTEMPERATURE PROXIES

The ratios of Mg/Ca and U/Ca in corals appear to serve as paleothermometers, although with apparently less fidelity than is found for Sr/Ca. The results for U/Ca (Min *et al.*, 1995; Shen and Dunbar, 1995) and Mg/Ca (Mitsuguchi *et al.*, 1996; Sinclair *et al.*, 1998; Fallon *et al.*, 1999) show convincing annual cycles, but with some complications; for example, Sinclair *et al.* (1998) observe, using laser ICP-MS analyses of coral surfaces, that there are high-frequency oscillations in Mg/Ca that range up to 50% of the total signal. In their comparative study, Sinclair *et al.* (1998) also observed differences in the seasonal profile of U/Ca and Sr/Ca, suggesting that other factors might be at play for uranium incorporation. Fallon *et al.* (1999) observed that magnesium incorporation tracked with SST but also evinced variability not related to temperature, suggesting that Mg/Ca paleothermometry in corals is not going to be as simple as it appeared in the initial study (Mitsuguchi *et al.*, 1996).

Two of these studies also looked at boron incorporation into the coral skeleton and observed that it also appears to be, at least in part, related to temperature (Sinclair *et al.*, 1998; Fallon *et al.*, 1999). The fact that at least four elements follow a seasonal pattern related to temperature suggests that elemental incorporation in coral skeletons is linked to calcification and is not simply driven by a thermodynamic temperature effect. If this applies generally, than all of the coral metal paleothermometers will have to be applied with attention to the possibility of distortions caused by growth factors.

6.14.10 CALCIUM ISOTOPES AS A PALEOTEMPERATURE PROXY

The possibility of using calcium isotopes for paleothermometry is a very new idea that is based on the empirical observation of temperature-related fractionation between the isotopes ^{40}Ca and ^{44}Ca (reported as δ^{44}Ca). Measurements of precise calcium-isotopic variations are quite challenging (Russell *et al.*, 1978), and this has limited investigations of this isotopic system to relatively small data sets. This potential paleothermometer has been calibrated in neritic benthic foraminifera (De La Rocha and DePaolo, 2000), a spinose tropical planktonic foraminifera (Nägler *et al.*, 2000), and a subtropical planktonic foraminifera (Gussone *et al.*, 2003). In addition, Zhu and Macdougall (1998) compared three pairs of warm and cold foraminifera and demonstrated a systemic difference. The most convincing evidence of the potential utility of this new paleothermometer comes from the study of Nägler *et al.* (2000), which demonstrates an increase of 0.24‰ in shell δ^{44}Ca per °C, based on three cultured points of *G. sacculifer*. Down-core measurements from a core in the tropical Atlantic indicate that shells from glacial intervals are ~0.5–1.0‰ depleted in δ^{44}Ca, consistent with colder glacial temperatures. Although the results in Nägler *et al.* (2000) indicate that calcium isotope paleothermometry has great promise, they are somewhat confounded by the fact a new study shows a much weaker response of δ^{44}Ca to temperature (~0.02‰ per °C) in a subtropical foraminifera, *O. universa*, calibrated over a wide temperature range by culturing (Gussone *et al.*, 2003). The weaker response in *O. universa* is mirrored by a similar response in inorganically precipitated aragonite (Gussone *et al.*, 2003). Given a measurement precision of ~0.12‰ for replicate samples (Gussone *et al.*, 2003), the slope for *G. sacculifer* allows for a resolution of ~0.5 °C in paleothermometry, whereas the slope for *O. universa* allows for a resolution of only ~5 °C. Obviously, the utility of this approach will rely heavily on which slope is more generally representative, and if that slope is stable in space and time.

6.14.11 CONCLUSIONS

Geochemists have, since the 1950s, already come up with a remarkable array of paleotemperature proxies in marine carbonates. These proxies work in diverse oceanic settings, in different organisms, in different parts of the water column, and on varied timescales. Each proxy has different strengths and weaknesses, and some of the proxies, such as Mg/Ca and oxygen isotopes in foraminifera, reinforce each other when applied together.

Perhaps most remarkable is the amount of progress that has been made since the mid-1980s on three new or revived paleothermometric approaches, each of which work particularly well in the tropics: oxygen isotopes in corals, Mg/Ca in foraminifera, and Sr/Ca in corals. This progress, in conjunction with advances in alkenone unsaturation paleothermometry (see Chapter 6.15), has not only expanded the importance of geochemistry in paleoclimate research, but changed its focus from mainly a chronostratigraphic and sea-level tool (i.e., oxygen isotopes in foraminifera) to a series of proxies that can be used to gauge the temporal and spatial history of oceanic temperatures. This shift, and, for example, the general level of agreement between geochemical paleothermometers for such fundamental problems as the cooling of the glacial tropics, suggests that major breakthroughs to long-standing paleoclimatological issues are now within reach. Although many problems and challenges remain, and although none of the available proxies work perfectly, it is clear that recent research progress has elevated geochemical paleothermometers to an even more fundamental role in quantitative paleoclimate research.

ACKNOWLEDGMENTS

The author thanks A. Cohen for sharing her compilation of coral strontium calibrations, and J. Cole, J. Clark, D. Schrag, and H. Spero for discussions. Some of this material is based upon work supported by the US National Science Foundation, more recently under grant OCE-0117886.

REFERENCES

Alibert C. and McCulloch M. T. (1997) Strontium/calcium ratios in modern porites corals from the Great Barrier Reef as a proxy for sea surface temperature: calibration of the thermometer and monitoring of ENSO. *Paleoceanography* **12**(3), 345–363.

Anand P., Elderfield H., and Conte M. H. (2003) Calibration of Mg/Ca thermometry in planktonic foraminifera from a sediment trap time series. *Paleoceanography* **18**(2), 1050, doi: 10.1029/2002PA000846.

Andrus C. F. T., Crowe D. E., Sandweiss D. H., Reitz E. J., and Romanek C. S. (2002) Otolith δ^{18}O record of mid-Holocene sea surface temperatures in Peru. *Science* **295**(5559), 1508–1511.

Bard E., Hamelin B., Arnold M., Montaggioni L., Cabioch G., Faure G., and Rougerie F. (1996) Deglacial sea-level record from Tahiti corals and the timing of global meltwater discharge. *Nature* **382**, 241–244.

Bard E., Rostek F., and Sonzogni C. (1997) Interhemispheric synchrony of the last deglaciation inferred from alkenone palaeothermometry. *Nature* **385**, 707–710.

Barron E. J. (1987) Eocene equator-to-pole surface ocean temperatures: a significant climate problem? *Paleoceanography* **2**, 729–739.

Bé A. W. H. (1980) Gametogenic calcification in a spinose planktonic foraminifer, *Globigerinoides sacculifer* (Brady). *Mar. Micropaleontol.* **5**, 283–310.

Beck J. W., Edwards R. L., Ito E., Taylor F. W., Recy J., Rougerie F., Joannot P., and Henin C. (1992) Sea-surface temperature from coral skeletal strontium/calcium ratios. *Science* **257**(5070), 644–647.

Beck J. W., Recy J., Taylor F., Edwards R. L., and Cabloh G. (1997) Abrupt changes in early Holocene Tropical sea surface temperature derived from coral records. *Nature* **385**, 705–707.

Bemis B. E., Spero H. J., Bijma J., and Lea D. W. (1998) Reevaluation of the oxygen isotopic composition of planktonic foraminifera: experimental results and revised paleotemperature equations. *Paleoceanography* **13**(2), 150–160.

Bender M. L., Lorens R. B., and Williams D. F. (1975) Sodium, magnesium, and strontium in the tests of planktonic foraminifera. *Micropaleontology* **21**, 448–459.

Billups K. and Schrag D. P. (2002) Paleotemperatures and ice volume of the past 27 Myr revisited with paired Mg/Ca and $^{18}O/^{16}O$ measurements on benthic foraminifera. *Paleoceanography* 10.1029/2000PA000567.

Böhm F., Joachimski M. M., Dullo W. C., Eisenhauer A., Lehnert H., Reitner J., and Wörheide G. (2000) Oxygen isotope fractionation in marine aragonite of coralline sponges. *Geochim. Cosmochim. Acta* **64**(10), 1695–1703.

Brass G. W. and Turekian K. K. (1974) Strontium distribution in GEOSECS oceanic profiles. *Earth Planet. Sci. Lett.* **23**, 141–148.

Brown S. and Elderfield H. (1996) Variations in Mg/Ca and Sr/Ca ratios of planktonic foraminifera caused by postdepositional dissolution—evidence of shallow Mg-dependent dissolution. *Paleoceanography* **11**(5), 543–551.

Burton E. A. and Walter L. M. (1987) Relative precipitation rates of aragonite and Mg calcite from seawater: temperature or carbonate ion control? *Geology* **15**, 111–114.

Cane M. A. and Evans M. (2000) Do the tropics rule? *Science* **290**, 1107–1108.

Caron D. A., Anderson O. R., Lindsey J. L., Faber J. W. W., and Lin Lim E. (1990) Effects of gametogenesis on test structure and dissolution of some spinose planktonic foraminifera and implications for test preservation. *Mar. Micropaleontol.* **16**, 93–116.

Chappell J. and Shackleton N. J. (1986) Oxygen isotopes and sea level. *Nature* **324**, 137–140.

Chave K. E. (1954) Aspects of the biogeochemistry of magnesium: 1. Calcareous marine organisms. *J. Geol.* **62**, 266–283.

Chilingar G. V. (1962) Dependence on temperature of Ca/Mg ratio of skeletal structures of organisms and direct chemical precipitates out of sea water. *Bull. Soc. CA Acad. Sci.* **61**(part 1), 45–61.

Clarke F. W. and Wheeler W. C. (1922) The inorganic constituents of marine invertebrates. *USGS Prof. Pap.* **124**, 55.

CLIMAP. (1981) Seasonal reconstructions of the Earth's surface at the last glacial maximum. In *The Geological Society of America Map and Chart Series* **MC-36**.

Cobb K. M., Charles C. D., and Hunter D. E. (2001) A central tropical Pacific coral demonstrates Pacific, Indian, and Atlantic decadal climate connections. *Geophys. Res. Lett.* **V28**(N11), 2209–2212.

Cohen A., Layne G., Hart S., and Lobel P. (2001) Kinetic control of skeletal Sr/Ca in a symbiotic coral: implications for the paleotemperature proxy. *Paleoceanography* **V16**(N1), 20–26.

Cohen A. L., Owens K. E., Layne G. D., and Shimizu N. (2002) The effect of algal symbionts on the accuracy of Sr/Ca paleotemperatures from coral. *Science* **296**(5566), 331–333.

Cole J. E. and Fairbanks R. G. (1990) The southern oscillation recorded in the $\delta^{18}O$ of corals from Tarawa Atoll. *Paleoceanography* **5**(5), 669–683.

Cole J. E., Dunbar R. B., McClanahan T. R., and Muthiga N. A. (2000) Tropical Pacific forcing of decadal SST variability in the western Indian Ocean over the past two centuries. *Science* **287**, 617–619.

Correge T. and Deckker P. D. (1997) Faunal and geochemical evidence for changes in intermediate water temperature and salinity in the western Coral Sea (northeast Australia) during the late Quaternary. *Palaeogeogr. Palaeoclimat. Palaeoecol.* **313**, 183–205.

Correge T., Delcroix T., Recy J., Beck W., Cabioch G., and Le Cornec F. (2000) Evidence for stronger El Nino-southern oscillation (ENSO) events in a Mid-Holocene massive coral. *Paleoceanography* **15**(4), 465–470.

Cronblad H. G. and Malmgren B. A. (1981) Climatically controlled variation of Sr and Mg in quaternary planktonic foraminifera. *Nature* **291**, 61–64.

Cronin T. M., Raymo M. E., and Kyle K. P. (1996) Pliocene (3.2–2.4 Ma) ostracode faunal cycles and deep ocean circulation, North Atlantic Ocean. *Geology* **24**(8), 695–698.

Cronin T. M., Dwyer G. S., Baker P. A., Rodriguez-Lazaro J., and DeMartino D. M. (2000) Orbital and suborbital variability in North Atlantic bottom water temperature obtained from deep-sea ostracod Mg/Ca ratios. *Palaeogeogr. Palaeoclimat.* **162**(1–2), 45–57.

Crowley T. J. (2000a) Causes of climate change over the past 1000 years. *Science* **289**(5477), 270–277.

Crowley T. J. (2000b) CLIMAP SSTs re-revisited. *Clim. Dyn.* **16**(4), 241–255.

De La Rocha C. L. and DePaolo D. J. (2000) Isotopic evidence for variations in the marine calcium cycle over the Cenozoic. *Science* **289**(5482), 1176–1178.

de Villiers S., Shen G. T., and Nelson B. K. (1994) The Sr/Ca-temperature relationship in coralline aragonite: influence of variability in $(Sr/Ca)_{seawater}$ and skeletal growth parameters. *Geochim. Cosmochim. Acta* **58**(1), 197–208.

de Villiers S., Nelson B. K., and Chivas A. R. (1995) Biological controls on coral Sr/Ca and $\delta^{18}O$ reconstructions of sea surface temperatures. *Science* **269**, 1247–1249.

Dekens P. S., Lea D. W., Pak D. K., and Spero H. J. (2002) Core top calibration of Mg/Ca in tropical foraminifera: refining paleo-temperature estimation. *Geochem. Geophys. Geosys.* **3**, 1022, 10.1029/2001GC000200.

Delaney M. L., Bé A. W. H., and Boyle E. A. (1985) Li, Sr, Mg, and Na in foraminiferal calcite shells from laboratory culture, sediment traps, and sediment cores. *Geochim. Cosmochim. Acta* **49**, 1327–1341.

Druffel E. R. M. (1985) Detection of El Nino and decade timescale variations of sea surface temperature from banded coral records: implications for the carbon dioxide cycle. In *The Carbon Cycle and Atmospheric CO_2: Natural Variations Archean to Present*. American Geophysical Union, Washington, DC, pp. 111–122.

Dunbar R. B. and Wellington G. M. (1985) Stable isotopes in a branching coral monitor seasonal temperature variation. *Nature* **293**, 453–455.

Duplessy J.-C. and Blanc P.-L. (1981) Oxygen-18 enrichment of planktonic foraminifera due to gametogenic calicification below the euphotic zone. *Science* **213**, 1247–1250.

Duplessy J. C., Lalou C., and Vinot A. C. (1970) Differential isotopic fractionation in benthic foraminifera and paleotemperatures reassessed. *Science* **168**, 250–251.

Dwyer G. S., Cronin T. M., Baker P. A., Raymo M. E., Buzas J. S., and Correge T. (1995) North Atlantic deepwater temperature change during Late Pliocene and Late Quaternary climatic cycles. *Science* **270**, 1347–1351.

Elderfield H. and Ganssen G. (2000) Past temperature and $\delta^{18}O$ of surface ocean waters inferred from foraminiferal Mg/Ca ratios. *Nature* **405**, 442–445.

Elderfield H., Bertram C., and Erez J. (1996) Biomineralization model for the incorporation of trace elements into foraminiferal calcium carbonate. *Earth Planet. Sci. Lett.* **142**(3–4), 409–423.

Elderfield H., Cooper M., and Ganssen G. (2000) Sr/Ca in multiple species of planktonic foraminifera: implications for reconstructions of seawater Sr/Ca. *Geochem. Geophys. Geosys.* **1**, paper no. 1999GC00031.

Emiliani C. (1955) Pleistocene temperatures. *J. Geol.* **63**, 538–578.

Emiliani C., Hudson J., Shinn E., and George R. (1978) Oxygen and carbon isotopic growth record in a reef coral from the Florida keys and a deep-sea coral from Blake Plateau. *Science* **202**, 627–628.

Enmar R., Stein M., Bar-Matthews M., Sass E., Katz A., and Lazar B. (2000) Diagenesis in live corals from the Gulf of Aqaba: I. The effect on paleo-oceanography tracers. *Geochim. Cosmochim. Acta* **64**(18), 3123–3132.

Epstein S., Buchsbaum R., Lowenstam H. A., and Urey H. C. (1953) Revised carbonate-water isotopic temperature scale. *Geol. Soc. Am. Bull.* **64**, 1315–1325.

Erez J. and Luz B. (1983) Experimental paleotemperature equation for planktonic foraminifera. *Geochim. Cosmochim. Acta* **47**, 1025–1031.

Fairbanks R. G. (1989) A 17,000-year glacio-eustatic sea level record: influence of glacial melting rates on the Younger Dryas event and deep-ocean circulation. *Nature* **342**, 637–642.

Fairbanks R. G. and Dodge R. E. (1979) Annual periodicities in $^{16}O/^{16}O$ and $^{13}C/^{12}C$ ratios in the coral *Montrastrea annularis*. *Geochim. Cosmochim. Acta* **43**, 1009–1020.

Fairbanks R. G. and Matthews R. K. (1978) The marine oxygen isotope record in Pleistocene coral, Barbados, West Indies. *Quat. Res.* **10**, 181–196.

Fallon S. J., McCulloch M. T., Woesik R. v., and Sinclair D. J. (1999) Corals at their latitudinal limits: laser ablation trace element systematics in *Porites* from Shirigai Bay, Japan. *Earth Planet. Sci. Lett.* **172**, 221–238.

Farrell J. W. and Prell W. L. (1989) Climatic change and $CaCO_3$ preservation: an 800,000 year bathymetric reconstruction from the central equatorial Pacific Ocean. *Paleoceanography* **4**(4), 447–466.

Gagan M. K., Ayliffe L. K., Hopley D., Cali J. A., Mortimer G. E., Chappell J., McCulloch M. T., and Head M. J. (1998) Temperature and surface-ocean water balance of the Mid-Holocene tropical western pacific. *Science* **279**, 1014–1018.

Gagan M. K., Ayliffe L. K., Beck J. W., Cole J. E., Druffel E. R. M., Dunbar R. B., and Schrag D. P. (2000) New views of tropical paleoclimates from corals. *Quat. Sci. Rev.* **19**(1–5), 45–64.

Grossman E., Betzer P. R., Walter C. D., and Dunbar R. B. (1986) Stable isotopic variation in pteropods and atlantids from North Pacific sediment traps. *Mar. Micropaleontol.* **10**, 9–22.

Guilderson T. P., Fairbanks R. G., and Rubenstone J. L. (1994) Tropical temperature variations since 20,000 years ago: modulating interhemispheric climate changes. *Science* **263**, 663–665.

Guilderson T., Fairbanks R., and Rubenstone J. (2001) Tropical Atlantic coral oxygen isotopes: glacial-interglacial sea surface temperatures and climate change. *Mar. Geol.* **V172**(N1–2), 75–89.

Gussone N., Eisenhauer A., Heuser A., Dietzel M., Bock B., Böhm F., Spero H. J., Lea D. W., Bijma J., and Nägler T. F. (2003) Model for kinetic effects on calcium isotope fractionation ($\delta^{44}Ca$) in inorganic aragonite and cultured planktonic foraminifera. *Geochim. Cosmochim. Acta* **67**(7), 1375–1382.

Gussone N., Eisenhauer A., Heuser A., Dietzel M., Bock B., Böhm F., Spero H., Lea D. W., Bijma J., Zeebe R., and Nägler T. F. (2003) Model for kinetic effects on calcium isotope fractionation ($\delta^{44}Ca$) in inorganic aragonite and cultured planktonic foraminifera. *Geochim. Cosmochim. Acta* **67**(7), 1375–1382.

Hastings D. W., Emerson S. R., and Nelson B. K. (1996) Vanadium in foraminifera calcite: evaluation of a method to determine paleo-seawater concentrations. *Geochim. Cosmochim. Acta* **60**(19), 3701–3715.

Hastings D. W., Russell A. D., and Emerson S. R. (1998) Foraminiferal magnesium in *Globeriginoides sacculiferas* a paleotemperature proxy. *Paleoceanography* **13**(2), 161–169.

Hays J. D., Imbrie J., and Shackleton N. J. (1976) Variations in the Earth's orbit: pacemaker of the Ice ages. *Science* **194**, 1121–1132.

Hecht A. D., Eslinger E. V., and Garmon L. B. (1975) Experimental studies on the dissolution of planktonic foraminifera. In *Dissolution of Deep-sea Carbonates* (eds. W. V. Sitter, A. W. H. Bé, and W. H. Berger). Cushman Foundation for Foraminiferal Research, pp. 59–69.

Hendy E. J., Gagan M. K., Alibert C. A., McCulloch M. T., Lough J. M., and Isdale P. J. (2002) Abrupt decrease in tropical Pacific Sea surface salinity at end of little Ice age. *Science* **295**(5559), 1511–1514.

Hughen K. A., Schrag D. P., Jacobsen S. B., and Hantoro W. (1999) El Nino during the last interglacial period recorded by a fossil coral from Indonesia. *Geophys. Res. Lett.* **26**(20), 3129–3132.

Imbrie J. and Kipp N. G. (1971) A new micropaleontological method quantitative paleoclimatology: application to a late pleistocene Caribbean core. In *The Late Cenozoic Ice Ages* (ed. K. Turekian). Yale University Press, New Haven and London, pp. 71–181.

Imbrie J., Boyle E. A., Clemens S. C., Duffy A., Howard W. R., Kukla G., Kutzbach J., Martinson D. G., McIntyre A., Mix A. C., Molfino B., Morley J. J., Peterson L. C., Pisias N. G., Prell W. L., Raymo M. E., Shackleton N. J., and Toggweiler J. R. (1992) On the structure and origin of major glaciation cycles: 1. Linear responses to Milankovitch forcing. *Paleoceanography* **7**(6), 701–738.

Izuka S. K. (1988) Relationships of magnesium and other minor elements in tests of *Cassidulina Subglobosa* and *C. Oriangulata* to physical oceanic properties. *J. Foraminiferal Res.* **18**(2), 151–157.

Juillet-Leclerc A. D. and Labeyrie L. D. (1986) Temperature dependence of the oxygen isotopic fractionation between diatom silica and water. *Earth Planet. Sci. Lett.* **84**, 69–74.

Katz A. (1973) The interaction of magnesium with calcite during crystal growth at 25–90 °C and one atmosphere. *Geochim. Cosmochim. Acta* **37**, 1563–1586.

Kim S.-T. and O'Neil J. R. (1997) Equilibrium and nonequilibrium oxygen isotope effects in synthetic calcites. *Geochim. Cosmochim. Acta* **61**(16), 3461–3475.

Kinsman D. J. J. and Holland H. D. (1969) The co-precipitation of cations with $CaCO_3$: IV. The co-precipitation of Sr^2+ with aragonite between 16° and 96 °C. *Geochim. Cosmochim. Acta* **33**, 1–17.

Koutavas A., Lynch-Stieglitz J., Marchitto T. M. J., and Sachs J. P. (2002) Deglacial and Holocene climate record from the Galapagos Islands: El Niño linked to Ice Age climate. *Science* **297**(5579), 226–230.

Koziol A. M. and Newton R. C. (1995) Experimental determination of the reactions magnesite + quartz = enstatite + CO_2 and magnesite = periclase + CO_2 and the enthalpies of formation of enstatite and magnesite. *Am. Mineral.* **80**, 1252–1260.

Labeyrie L. D. and Juillet A. (1982) Oxygen isotopic exchangeability of diatom valve silica: interpretation and consequences for paleoclimatic studies. *Geochim. Cosmochim. Acta* **46**, 967–975.

Lea D. W. (2002) The glacial tropical Pacific: not just a west side story. *Science* **297**, 202–203.

Lea D. W., Shen G. T., and Boyle E. A. (1989) Coralline barium records temporal variability in equatorial Pacific upwelling. *Nature* **340**, 373–376.

Lea D. W., Mashiotta T. A., and Spero H. J. (1999) Controls on magnesium and strontium uptake in planktonic foraminifera determined by live culturing. *Geochim. Cosmochim. Acta* **63**(16), 2369–2379.

Lea D. W., Pak D. K., and Spero H. J. (2000) Climate impact of Late Quaternary equatorial Pacific sea surface temperature variations. *Science* **289**(5486), 1719–1724.

Lea D. W., Martin P. A., Pak D. K., and Spero H. J. (2002) Reconstructing a 350 ky history of sea level using planktonic Mg/Ca and oxygen isotopic records from a Cocos Ridge core. *Quat. Sci. Rev.* **21**(1–3), 283–293.

Lear C. H., Elderfield H., and Wilson P. A. (2000) Cenozoic deep-sea temperatures and global ice volumes from Mg/Ca in benthic foraminiferal calcite. *Science* **287**, 269–272.

Linsley B. K., Wellington G. M., and Schrag D. P. (2000) Decadal sea surface temperature variability in the subtropical South Pacific from 1726 to 1997 AD. *Science* **290**, 1145–1148.

Lohmann G. P. (1995) A model for variation in the chemistry of planktonic foraminifera due to secondary calcification and selective dissolution. *Paleoceanography* **10**(3), 445–457.

Lorens R. B., Williams D. F., and Bender M. L. (1977) The early nonstructural chemical diagenesis of foraminiferal calcite. *J. Sed. Petrol.* **47**(4), 1602–1609.

Lyle M. W., Prahl F. G., and Sparrow M. A. (1992) Upwelling and productivity changes inferred from a temperature record in the central equatorial Pacific. *Nature* **355**, 812–815.

Mann M. E., Bradley R. S., and Hughes M. K. (1998) Global-scale temperature patterns and climate forcing over the past six centuries. *Nature* **392**(23), 779–787.

Mantua N. J., Hare S. R., Zhang Y., Wallace J. M., and Francis R. C. (1997) A Pacific interdecadal climate oscillation with impacts on salmon production. *Bull Am. Meteorol. Soc.* **78**(6), 1069–1079.

Martin P. A., Lea D. W., Mashiotta T. A., Papenfuss T., and Sarnthein M. (1999) Variation of foraminiferal Sr/Ca over Quaternary glacial-interglacial cycles: evidence for changes in mean ocean Sr/Ca? *Geochem. Geophys. Geosys.* **1**, Paper Number 1999GC000006.

Martin P. A., Lea D. W., Rosenthal Y., Shackleton N. J., Sarnthein M., and Papenfuss T. (2002) Quaternary deep sea temperature histories derived from benthic foraminiferal Mg/Ca. *Earth Planet. Sci. Lett.* **198**, 193–209.

Mashiotta T. A., Lea D. W., and Spero H. J. (1999) Glacial-interglacial changes in Subantarctic sea surface temperature and δ^{18}O-water using foraminiferal Mg. *Earth Planet. Sci. Lett.* **170**(4), 417–432.

McConnaughey T. (1989a) ^{13}C and ^{18}O isotopic disequilibrium in biological carbonates: I. Patterns. *Geochim. Cosmochim. Acta* **53**(1), 151–162.

McConnaughey T. A. (1989b) ^{13}C and ^{18}O isotopic disequilibrium in biological carbonates: II. *In vitro* simulation of kinetic isotopic effects. *Geochim. Cosmochim. Acta* **53**, 163–171.

McCorkle D. C., Martin P. A., Lea D. W., and Klinkhammer G. P. (1995) Evidence of a dissolution effect on benthic shell chemistry: δ^{13}C, Cd/Ca, Ba/Ca, and Sr/Ca from the Ontong Java Plateau. *Paleoceanography* **10**(4), 699–714.

McCulloch M. T., Gagan M. K., Mortimer G. E., Chivas A. R., and Isdale P. J. (1994) A high-resolution Sr/Ca and δ^{18}O coral record from the great barrier reef, Australia, and the 1982–1983 El Nino. *Geochim. Cosmochim. Acta* **58**(12), 2747–2754.

McCulloch M., Mortimer G., Esat T., Xianhua L., Pillans B., and Chappell J. (1996) High resolution windows into early Holocene climate: Sr/Ca coral records from the Huon Peninsula. *Earth Planet. Sci. Lett.* **138**, 169–178.

McCulloch M. T., Tudhope A. W., Esat T. M., Mortimer G. E., Chappell J., Pillans B., Chivas A. R., and Omura A. (1999) Coral record of equatorial sea-surface temperatures during the penultimate deglaciation at Huon peninsula. *Science* **283**, 202–204.

McGregor H. V. and Gagan M. K. (2003) Diagenesis and geochemistry of Porites corals from Papua New Guinea: implications for paleoclimate reconstruction. *Geochim. Cosmochim. Acta* (in review).

Min G., Edwards R., Taylor F., Gallup C., and Beck J. (1995) Annual cycles of U/Ca in coral skeletons and U/Ca thermometry. *Geochim. Cosmochim. Acta* **59**(10), 2025–2042.

Mitsuguchi T., Matsumoto E., Abe O., Uchida T., and Isdale P. J. (1996) Mg/Ca thermometry in coral skeletons. *Science* **274**, 961–963.

Mix A. C. (1992) The marine oxygen isotope record: constraints on timing and extent of ice-growth events (120–65 ka). In *The Last Interglacial Transition in North America*, Geological Society of America Special Paper (eds. P. U. Clark and P. D. Lea). GSA, Boulder, Co. vol. 270, pp. 19–30.

Mix A. C., Morey A. E., Pisias N. G., and Hostetler S. W. (1999) Foraminiferal faunal estimates of paleotemperature: circumventing the no-analog problem yields cool ice age tropics. *Paleoceanography* **14**(3), 350–359.

Mucci A. (1987) Influence of temperature on the composition of magnesian calcite overgrowths precipitated from seawater. *Geochim. Cosmochim. Acta* **51**, 1977–1984.

Müller P. J., Kirst G., Ruhland G., von Storch I., and Rosell-Melé A. (1998) Calibration of the alkenone paleotemperature index UK'37 based on core-tops from the eastern South Atlantic and the global ocean (60°N–60°S). *Geochim. Cosmochim. Acta* **62**(10), 1757–1772.

Müller A., Gagan M. K., and McCulloch M. T. (2001) Early marine diagenesis in corals and geochemical consequences for paleoceanographic reconstructions. *Geophys. Res. Lett.* **28**(23), 4471–4474.

Nägler T. F., Eisenhauer A., Müller A., Hemleben C., and Kramers J. (2000) The δ^{44}Ca-temperature calibration on fossil and cultured *Globigerinoides sacculifer*: new tool for reconstruction of past sea surface temperatures. *Geochem. Geophys. Geosys.* **1**, (Paper number: 2000GC000091).

Nürnberg D. (1995) Magnesium in tests of *Neogloboquadrina Pachyderma* sinistral from high Northern and Southern latitudes. *J. Foraminiferal Res.* **25**(4), 350–368.

Nürnberg D., Bijma J., and Hemleben C. (1996a) Assessing the reliability of magnesium in foraminiferal calcite as a proxy for water mass temperatures. *Geochim. Cosmochim. Acta* **60**(5), 803–814.

Nürnberg D., Bijma J., and Hemleben C. (1996b) Erratum: assessing the reliability of magnesium in foraminiferal calcite as a proxy for water mass temperatures. *Geochim. Cosmochim. Acta* **60**(13), 2483–2484.

Nürnberg D., Müller A., and Schneider R. R. (2000) Paleo-sea surface temperature calculations in the equatorial east Atlantic from Mg/Ca ratios in planktic foraminifera: a comparison to sea surface temperature estimates from $U^{K'}37'$, oxygen isotopes, and foraminiferal transfer function. *Paleoceanography* **15**(1), 124–134.

O'Neil J. R., Clayton R. N., and Mayeda T. K. (1969) Oxygen isotope fractionation in divalent metal carbonates. *J. Chem. Phys.* **51**(12), 5547–5558.

Oomori T., Kaneshima H., and Maezato Y. (1987) Distribution coefficient of Mg^{2+} ions between calcite and solution at 10–50 °C. *Mar. Chem.* **20**, 327–336.

Oppo D. W., Linsley B. K., Rosenthal Y., Dannenmann S., and Beaufort L. (2003) Orbital and suborbital climate variability in the Sulu Sea, western tropical Pacific. *Geochem. Geophys. Geosys.* **4**(1), 1003, doi: 10.1029/2001GC000260.

Pearson P. N., Ditchfield P. W., Singano J., Harcourt-Brown K. G., Nicholas C. J., Olsson R. K., Shackleton N. J., and Hall M. A. (2001) Warm tropical sea surface temperatures in the late Cretaceous and Eocene epochs. *Nature* **413**, 481–487.

Pelejero C. and Calvo E. (2003) The upper end of the $U_{37}^{K'}$ temperature calibration revisited. *Geochem. Geophys. Geosys.* **4**, 1014. doi:10.1029/2002GC000431.

Pelejero C., Grimalt J. O., Heilig S., Kienast M., and Wang L. (1999) High resolution U_{37}^{K} temperature reconstructions in the South China Sea over the past 220 kyr. *Paleoceanography* **14**(2), 224–231.

Pierrehumbert R. (1999) Huascaran delta O-18 as an indicator of tropical climate during the last glacial maximum. *Geophys. Res. Lett.* **26**(9), 1345–1348.

Rathburn A. E. and Deckker P. D. (1997) Magnesium and strontium compositions of recent benthic foraminifera from the Coral Sea, Australia and Prydz Bay, Antarctica. *Mar. Micropaleontol.* **32**, 231–248.

Rickaby R. E. M. and Elderfield H. (1999) Planktonic foraminiferal Cd/Ca: paleonutrients or paleotemperature? *Paleoceanography* **14**(3), 293–303.

Rodbell D. T., Seltzer G. O., Anderson D. M., Abbot M. B., Enfield D. B., and Newman J. H. (1999) An ~15,000-year record of El Niño-driven alluviation in southwestern Ecuador. *Science* **283**, 516–520.

Rosenthal Y. and Boyle E. A. (1993) Factors controlling the fluoride content of planktonic foraminifera: an evaluation of its paleoceanographic utility. *Geochim. Cosmochim. Acta* **57**(2), 335–346.

Rosenthal Y. and Lohmann G. P. (2002) Accurate estimation of sea surface temperatures using dissolution corrected calibrations for Mg/Ca paleothermometry. *Paleoceanography* **17**, 1044. doi:10.1029/2001PA000749.

Rosenthal Y., Boyle E. A., and Slowey N. (1997) Temperature control on the incorporation of Mg, Sr, F, and Cd into benthic foraminiferal shells from Little Bahama Bank: prospects for thermocline paleoceanography. *Geochim. Cosmochim. Acta* **61**(17), 3633–3643.

Rosenthal Y., Lohmann G. P., Lohmann K. C., and Sherrell R. M. (2000) Incorporation and preservation of Mg in *Globigerinoides sacculifer* implications for reconstructing the temperature and $^{18}O/^{16}O$ of seawater. *Paleoceanography* **15**(1), 135–145.

Rosenthal Y., Dannenmann S., Oppo D. W., and Linsley B. K. (2003) The amplitude and phasing of climate change during the last deglaciation in the Sulu Sea, western equatorial Pacific. *Geophys. Res. Lett.* **30**(8), 1428. doi:10.1029/2002GL016612.

Rostek F., Ruhland G., Bassinot F. C., Müller P. J., Labeyrie L. D., Lancelot Y., and Bard E. (1993) Reconstructing sea surface temperature and salinity using $\delta^{18}O$ and alkenone records. *Nature* **364**, 319–321.

Russell A. D., Emerson S., Nelson B. K., Erez J., and Lea D. W. (1994) Uranium in foraminiferal calcite as a recorder of seawater uranium concentrations. *Geochim. Cosmochim. Acta* **58**(2), 671–681.

Russell W. A., Papanastassiou D. A., and Tombrello T. A. (1978) Ca isotope fractionation on the Earth and other solar system materials. *Geochim. Cosmochim. Acta* **42**, 1075–1090.

Sanyal A., Hemming N. G., Broecker W. S., Lea D. W., Spero H. J., and Hanson G. N. (1996) Oceanic pH control on the boron isotopic composition of foraminifera: evidence from culture experiments. *Paleoceanography* **11**(5), 513–517.

Savin S. M. and Douglas R. G. (1973) Stable isotope and magnesium geochemistry of recent planktonic foraminifera from the South Pacific. *Geol. Soc. Am. Bull.* **84**, 2327–2342.

Schmidt G. A., Hoffmann G., and Thresher D. (2001) Isotopic tracers in coupled models: a new paleo-tool. *PAGES News* **9**(1), 10–11.

Schrag D. P. (1999a) Effects of diagenesis on the isotopic record of late paleogene tropical sea surface temperatures. *Chem. Geol.* **161**(1–3), 215–224.

Schrag D. P. (1999b) Rapid analysis of high-precision Sr/Ca ratios in corals and other marine carbonates. *Paleoceanography* **14**(2), 97–102.

Schrag D. P. and Linsley B. K. (2002) Paleoclimate: corals, chemistry, and climate. *Science* **296**(5566), 277–278.

Schrag D. P., Depaolo D. J., and Richter F. M. (1995) Reconstructing past sea surface temperatures—correcting for diagenesis of bulk marine carbonate. *Geochim. Cosmochim. Acta* **59**(11), 2265–2278.

Schrag D. P., Hampt G., and Murray D. W. (1996) Pore fluid constraints on the temperature and oxygen isotopic composition of the glacial ocean. *Science* **272**, 1930–1932.

Shackleton N. (1967) Oxygen isotope analyses and Pleistocene temperatures re-assessed. *Nature* **215**, 15–17.

Shackleton N. J. (1974) Attainment of isotopic equilibrium between ocean water and the benthonic foraminifera genus *Uvigerina*: isotopic changes in the ocean during the last glacial. *Cent. Nat. Rech. Sci. Colloq. Int.* **219**, 203–209.

Shackleton N. J. (1987) Oxygen isotopes, ice volume and sea level. *Quat. Sci. Rev.* **6**, 183–190.

Shackleton N. J. and Opdyke N. D. (1973) Oxygen isotope and paleomagnetic stratigraphy of equatorial Pacific core V28-238: oxygen isotope temperatures and ice volumes on a 10^5 and 10^6 year scale. *Quat. Res.* **3**, 39–55.

Shemesh A., Charles C. D., and Fairbanks R. G. (1992) Oxygen isotopes in biogenic silica—global changes in ocean temperature and isotopic composition. *Science* **256**(5062), 1434–1436.

Shemesh A., Burckle L. H., and Hays J. D. (1994) Meltwater input to the Southern Ocean during the last glacial maximum. *Science* **266**(5190), 1542–1544.

Shemesh A., Burckle L. H., and Hays J. D. (1995) Late Pleistocene oxygen isotope records of biogenic silica from the Atlantic sector of the Southern Ocean. *Paleoceanography* **10**(2), 179–196.

Shen C.-C., Hastings D. W., Lee T., Chiu C.-H., Lee M.-Y., Wei K.-Y., and Edwards R. L. (2001) High precision glacial-interglacial benthic foraminiferal Sr/Ca records from the eastern equatorial Atlantic Ocean and Caribbean Sea. *Earth Planet. Sci. Lett.* **190**(3–4), 197–209.

Shen G. T. (1996) Rapid changes in the tropical ocean and the use of corals as monitoring systems. In *Geoindicators: Assessing Rapid Environmental Changes in Earth Systems* (eds. A. Berger and W. Iams). A. A. Balkema, Rotterdam, pp. 155–169.

Shen G. T. and Dunbar R. B. (1995) Environmental controls on uranium in reef corals. *Geochim. Cosmochim. Acta* **59**(10), 2009–2024.

Shen G. T., Cole J. C., Lea D. W., Linn L. J., McConnaughey T. A., and Fairbanks R. G. (1992) Surface ocean variability at Galapagos from 1936 to 1982: calibration of geochemical tracers in corals. *Paleoceanography* **7**(5), 563–588.

Sinclair D. J., Kinsley L. P. J., and McCulloch M. T. (1998) High resolution analysis of trace elements in corals by laser ablation ICP-MS. *Geochim. Cosmochim. Acta* **62**(11), 1889–1901.

Smith J. E., Risk M. J., Schwarcz H. P., and McConnaughey T. A. (1997) Rapid climate change in the North Atlantic during the Younger Dryas recorded by deep-sea corals. *Nature* **386**, 818–820.

Smith S. V., Buddemeier R. W., Redalje R. C., and Houck J. E. (1979) Strontium-calcium thermometry in coral skeletons. *Science* **204**, 404–407.

Spero H. J. and Lea D. W. (1993) Intraspecific stable isotope variability in the planktonic foraminifera *Globigerinoides sacculifer*: results from laboratory experiments. *Mar. Micropaleontol.* **22**, 221–234.

Spero H. J. and Lea D. W. (1996) Experimental determination of stable isotopic variability in *Globigerina bulloides*: implications for paleoceanographic reconstructions. *Mar. Micropaleontol.* **28**, 231–246.

Spero H. J., Bijma J., Lea D. W., and Bemis B. (1997) Effect of seawater carbonate chemistry on planktonic foraminiferal carbon and oxygen isotope values. *Nature* **390**, 497–500.

Stoll H. M. and Schrag D. P. (1998) Effects of Quaternary sea level cycles on strontium in seawater. *Geochim. Cosmochim. Acta* **62**(7), 1107–1118.

Stoll H. M., Schrag D. P., and Clemens S. C. (1999) Are seawater Sr/Ca variations preserved in Quaternary foraminifera. *Geochim. Cosmochim. Acta* **63**(21), 3535–3547.

Stott L., Poulsen C., Lund S., and Thunnell R. (2002) Super ENSO and global climate oscillations at millennial time-scales. *Science* **279**(5579), 222–226.

Sturchio N. C. (1999) A conversation with Harmon Craig. *Geochem. News* **98**, 12–20.

Toyofuku T., Kitazato H., Kawahata H., Tsuchiya M., and Nohara M. (2000) Evaluation of Mg/Ca thermometry in foraminifera: comparison of experimental results and measurements in nature. *Paleoceanography* **15**(4), 456–464.

Trend-Staid M. and Prell W. L. (2002) Sea surface temperature at the Last Glacial Maximum: a reconstruction using the modern analog technique. *Paleoceanography* **17**(4) 10.1029/2000PA000506.

Tudhope A. W., Chilcott C. P., McCulloch M. T., Cook E. R., Chappell J., Ellam R. M., Lea D. W., Lough J. M., and Shimmield G. B. (2001) Variability in the El Niño southern oscillation through a glacial-interglacial cycle. *Science* **291**(5508), 1511–1517.

Urban F., Cole J., and Overpeck J. (2000) Influence of mean climate change on climate variability from a 155-year Tropical Pacific coral record. *Nature* **407**(6807), 989–993.

Urey H. C. (1947) The thermodynamic properties of isotopic substances. *J. Chem. Soc.*, 562–581.

Visser K., Thunell R., and Stott L. (2003) Magnitude and timing of temperature change in the Indo-Pacific warm pool during deglaciation. *Nature* **421**, 152–155.

Weber J. N. (1973) Incorporation of strontium into reef coral skeletal carbonates. *Geochim. Cosmochim. Acta* **37**, 2173–2190.

Weber J. N. and Woodhead P. M. J. (1972) Temperature dependence of oxygen-18 concentration in reef coral carbonates. *J. Geophys. Res.* **77**, 463–473.

Wu G. and Berger W. H. (1989) Planktonic foraminifera: differential dissolution and the Quaternary stable isotope record in the west equatorial Pacific. *Paleoceanography* **4**(2), 181–198.

Zachos J., Pagani M., Sloan L., Thomas E., and Billups K. (2001) Trends, rhythms, and aberrations in global climate 65 Ma to present. *Science* **292**(5517), 686–693.

Zhou G.-T. and Zheng Y.-F. (2003) An experimental study of oxygen isotope fractionation between inorganically precipitated aragonite and water at low temperatures. *Geochim. Cosmochim. Acta* **67**(3), 387–399.

Zhu P. and Macdougall J. D. (1998) Ca isotopes in the marine environment and the oceanic Ca cycle. *Geochim. Cosmochim. Acta* **62**(10), 1691–1698.

6.15
Alkenone Paleotemperature Determinations

T. D. Herbert
Brown University, Providence, RI, USA

6.15.1 INTRODUCTION

The organic biomarker proxy for past sea surface temperatures ("U^{k}_{37}") came to paleoceanography from an unexpected direction. Nearly all paleoceanographic tools rely on some aspect of the fossilized hard parts of marine organisms. Thus, assemblages of calcareous microplankton such as foraminifera and coccoliths, or of siliceous plankton such as radiolaria and diatoms, provided the basis for the CLIMAP reconstruction of the Ice Age ocean (CLIMAP, 1976, 1981). Additionally, a host of chemical methods relies on the same hard parts to furnish isotopic and trace element signatures, and generally requires that skeletal material be well preserved. The alkenone method differs in several important ways. Individual molecules, extracted and separated from a matrix of hundreds to thousands of other organic compounds, are the targets. In most cases, the remnant alkenones and alkenoates that are the subject of this review constitute no more, and often considerably less, than a few percent of their initial flux that left the surface layer of the ocean and fell toward the sediments. Good preservation is thus not a major issue for use of the proxy. In addition, while many geochemical techniques assume that skeletal material is a passive recorder of isotopic and trace element composition of seawater, and that incorporation of paleoenvironmental signals follows thermodynamic laws that can be modeled using nonbiogenic phases in the laboratory, the alkenone method assumes that the ratios of biomarkers measured were actively regulated by the producing organisms in life according to the temperature of the water in which they grew.

Alkenone paleothermometry promises a direct estimate of near-surface ocean temperatures. Alkenones and the related alkenoates come exclusively from a few species of haptophyte algae. These organisms require sunlight, and they generally prefer the upper photic zone. The environmental information contained in their molecular fossils therefore is quite specific, although, as will be discussed at length in a later section, ambiguities still exist on the depth and seasonal variations of alkenone-producing species in the ocean. In contrast, many assemblages of planktonic organisms such as foraminifera and radiolaria contain many species known to live well below the surface mixed layer. The link between microfossil assemblages and sea surface temperature and salinity is therefore indirect and statistical, rather than mechanistic.

As originally defined by the Bristol organic geochemistry group (Brassell, 1986a,b), the U^{k}_{37} index reflected the proportions of the di-($C_{37:2}$), tri-($C_{37:3}$), and tetra-($C_{37:4}$) unsaturated ketones. Subsequent work showed that there was no empirical benefit to including the $C_{37:4}$ ketone in a paleotemperature equation. The currently accepted $U^{k'}_{37}$ index (Prahl and Wakeham, 1987) varies positively with temperature, and is defined as $C_{37:2}/(C_{37:2} + C_{37:3})$, where $C_{37:2}$ represents the quantity of the di-unsaturated ketone and $C_{37:3}$ the quantity of the tri-unsaturated form. The alkenone paleotemperature proxy thus depends only on the relative proportions of the common C_{37} ketones and not on their absolute amounts. Furthermore, although the alkenones are produced by calcareous algae, they survive in sediments where carbonate has dissolved, as first recognized by Marlowe *et al.* (1984a,b) and Brassell *et al.* (1986a). The above expression for the index shows that it can vary between 0 and 1.0; thus, it may saturate at either extremely cold or warm temperatures.

Alkenones appear recalcitrant to diagenesis in the water column and within sediments relative to other large macromolecules. Indeed, the first reported occurrence of alkenones came not from recent material, but from Miocene age sediments of the Walvis Ridge (Boon *et al.*, 1978). Shortly thereafter, these compounds were linked to modern haptophyte algae, principally *Emiliania huxleyi* (de Leeuw *et al.*, 1980; Volkman *et al.*, 1980a,b; Marlowe *et al.*, 1984a,b). Reviews of lipid analyses of Deep Sea Drilling Project sediments revealed that most sediments of Pleistocene through mid-Eocene age appeared to contain measurable quantities of alkenones and alkenoates (Marlowe *et al.*, 1984a, 1990; Brassell, 1993). Brassell *et al.* (1986a) provided the seminal study linking alkenone unsaturation to paleotemperature fluctuations in the Late Pleistocene. After noting that modern surface sediments differed in their unsaturation ratios depending on latitude, Brassell *et al.* (1986a,b) reconstructed alkenone unsaturation in conjunction with benthic and planktonic foraminiferal $\delta^{18}O$ over the last 8×10^5 yr in a core from the subtropical North Atlantic. The unsaturation index declined during glacial periods, suggesting cooler surface ocean temperatures during ice age conditions. The authors further demonstrated that the alkenone index gave a continuous paleoclimatic curve, even in intervals barren of foraminifera due to dissolution. Prahl and Wakeham (1987) and Prahl *et al.* (1988) proposed the first quantitative calibration of alkenone unsaturation to growth temperature. Unsaturation parameters measured on a strain of *E. huxleyi* grown in the laboratory at known temperatures were compared to the unsaturation index on particulate material collected from the near-surface ocean in the northeast Pacific. Prahl and Wakeham (1987) showed that the laboratory calibration appeared to apply well to the field observations of unsaturation and the water temperature in which the alkenones apparently were synthesized. The calibration of alkenone unsaturation to temperatures expanded with the first systematic study of core-top sediments by Sikes *et al.* (1991). That study produced two important results: (i) the unsaturation index in recent sediments followed a relation to overlying sea surface temperatures (SSTs) very similar to the Prahl *et al.* (1988) calibration, and (ii) there appeared to be no ill effects on the unsaturation index over the time of core storage. Pristine or frozen samples were therefore not needed to produce good estimates of the $U^{k'}_{37}$ index for paleoceanographic studies.

As with any paleoceanographic proxy, a number of uncertainties must be evaluated that could affect the accuracy measurement as an estimate of past SSTs. The principal caveats raised can be broadly categorized as ecological, physiological, genetic, and diagenetic. All describe factors, which could cause the $U^{k'}_{37}$ index to deviate from a unique relation to SSTs. Ecological concerns come from observations that alkenone-producing species do not inhabit precisely the same depth throughout the ocean, and that they vary in abundance seasonally. The alkenone unsaturation parameter recorded by sediments could therefore measure past temperatures very precisely, but at which depths, and with what seasonal bias? It is also possible that the proportions of alkenones synthesized by haptophyte algae vary with growth rate, independent of temperature. Our present state of ignorance dictates that we do not know the growth phase of haptophyte material exported out of the photic zone—whether the products represent the initial exponential growth phase observed in culture or stationary growth. Natural populations also differ in their genetic composition. Alkenone-producing species are notable for their wide range of environmental tolerances. The consequences for the $U_{37}^{k'}$ index of genetic variations within strains of the same producing species and between the different alkenone-synthesizing species are still debated. In addition, alkenones measured in sediments represent the surviving molecules of a series of degradational pathways that begin in the water column, proceed to the sediment/water interface, and may continue into the sediment. Should there be a bias in the relative lability of the $C_{37:2}$ and $C_{37:3}$ ketones, this would be imparted to paleoceanographic reconstructions of temperature.

As should become clear, the $U_{37}^{k'}$ index appears nevertheless to provide a remarkably faithful estimate of paleotemperatures near the sea surface. At the same time, difficulties in matching the space and timescales of modern process studies to the information contained in sediments mean that the caveats raised above remain significant. Field studies provide only snapshots of haptophyte abundance and alkenone unsaturation parameters, sediment traps provide only a few years of data at only a few locations in the global ocean, and it is unclear how well laboratory cultures replicate the natural environment. I have endeavored to treat different lines of evidence systematically, but I have found it difficult to discuss each aspect in a purely serial way. The reader will therefore be asked to digest a review in which very diverse measurements and paradigms are woven together to answer the central question of how to reconstruct past ocean surface temperatures with the $U_{37}^{k'}$ proxy.

6.15.2 SYSTEMATICS AND DETECTION

Alkenones occur as a typical suite of 37-, 38-, and 39-carbon chained (C_{37}, C_{38}, C_{39}) ketones in marine particles and sediments. This set of compounds (Figure 1) constitutes a "fingerprint" for alkenones extracted from sediments. Their existence went undetected until chromatographic columns capable of sustaining the high temperatures at which long-chained alkenones elute came into existence in the late 1970s (Volkman *et al.*, 1980b). Although all alkenones are straight-chain hydrocarbons, they may differ in the number of double bonds (unsaturation) in the chain, and in the structure of the terminal ketone group (terminal carbon in the chain bonded to either a methyl or ethyl group). The C_{37} alkenones used in the $U_{37}^{k'}$ index are methyl ketones. 38-carbon chained molecules include not only tri- and di-unsaturated methyl ketone forms, but also $C_{38:3}$ and $C_{38:2}$ ethyl ketones (de Leeuw *et al.*, 1980; Volkman *et al.*, 1980a). A $C_{38:4}$ ethyl ketone has been reported from sediments underlying cold waters (Marlowe *et al.*, 1984b). As discussed at greater length in a later section, the total concentration of the four common C_{38} ketones is generally nearly the same as the sum of the two common C_{37} ketones in marine particulates and sediments (Conte *et al.*, 2001). Di- and tri-unsaturated C_{39} ethyl ketones (Figure 1) are also commonly observed at ~10–20% of the concentration of C_{37} ketones in analyses of alkenone-containing materials (Prahl *et al.*, 2001). Novel C_{35} and C_{36} methyl ketones have been reported recently from Black Sea sediments (Xu *et al.*, 2001); these apparently come from precursor organisms not commonly found in normal marine waters.

C_{36} and C_{37} fatty acid methyl esters (alkenoates) and C_{37} and C_{38} alkenes also accompany alkenones in alkenone-producing species (Volkman *et al.*, 1980b; Marlowe *et al.*, 1984b; Conte and Eglinton, 1993; Rosell-Mele *et al.*, 1994; Grossi *et al.*, 2000; Mouzdahir *et al.*, 2001). The relative proportions of alkenones, alkenoates, and alkenes vary greatly between different strains of alkenone-containing species grown in culture (Conte *et al.*, 1994a, 1995). The chain length and degree of unsaturation of alkenoates and alkenes follows the pattern of the long-chained alkenones (Marlowe *et al.*, 1984b; Conte and Eglinton, 1993; Grossi 2000). Unfortunately, alkenoates and alkenes rarely reach more than 10% of the concentrations of the C_{37} ketones in sediment extracts (the author has never observed significant quantities of C_{37} or C_{38} alkenes), and thus have limited utility for paleoceanographic reconstructions.

Our laboratory has found few marine sediment locations where alkenones cannot be extracted in enough quantity for quantification. Exceptions to the rule include very oligotrophic oceanic gyre locations, such as the Ontong-Java Plateau, and the red clay province of the North Pacific, where alkenone determinations are exceedingly difficult. Typical quantities of C_{37} ketones in marine sediments range from 100 ppb to 10 ppm of total sediment (dry weight). Alkenones can be extracted from sediments and particulates as part of a total lipid extract. From 1 g to 5 g of dry sediment is extracted with organic solvent (typically 9 : 1 methanol : methylene chloride) by Soxhlet apparatus, by repeated sonication at room temperature, or by an Accelerated Solvent

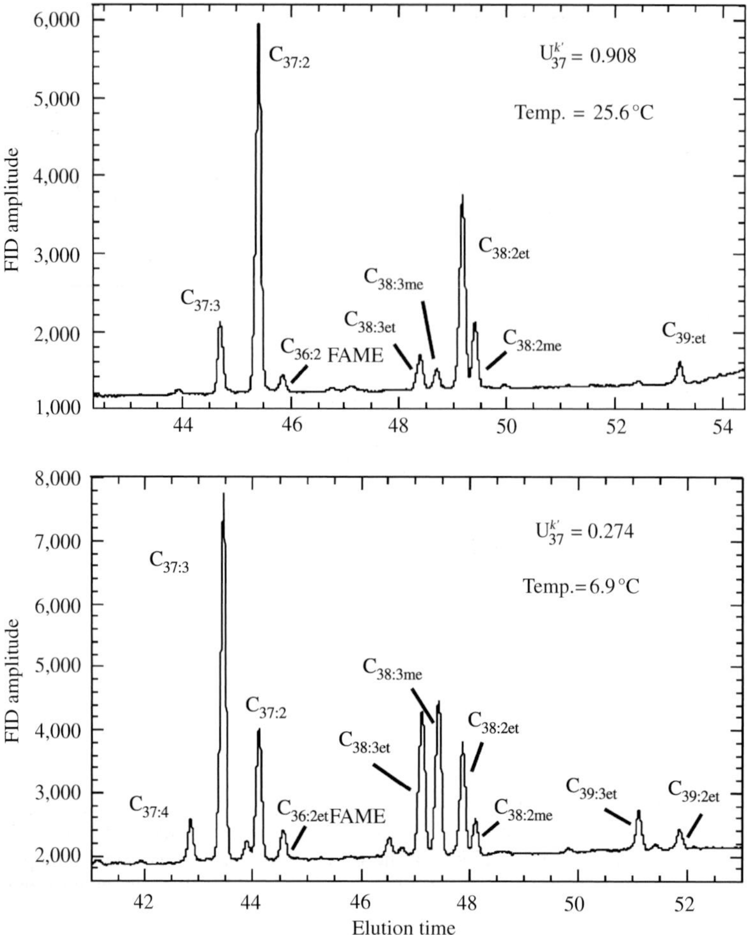

Figure 1 Typical GC-FID chromatograms of alkenone-containing sediment extracts.

Extraction system such as that offered by Dionex Corp. Some laboratories analyze alkenones as part of the total lipid extract; others prefer to run "cleaner" fractions following one of a number of schemes of fraction purification using silica gel columns or thin-layer chromatography (Villanueva *et al.*, 1997). Most open-ocean sediment extracts do not contain appreciable amounts of interfering lipids; however, some samples with more complex matrices may benefit from cleanup procedures.

Gas chromatography coupled to one of several detectors separates alkenones from their lipid matrix and permits their quantification. Relative to most lipid compounds in sediments, alkenones have high molecular weights and high boiling points. Elution through a nonpolar chromatographic column separates alkenones largely through boiling point, and subordinately by chemical interactions with the column film. Boiling point is largely determined by the molecular weight of the lipid. Because the mass differences between $C_{37:4}$, $C_{37:3}$, and $C_{37:2}$ ketones are small (molecular weights of 526 amu, 528 amu, and 530 amu, respectively), long chromatographic columns and/or slow temperature programs are required to completely resolve the component alkenones. In analogy with the $U^{k'}_{37}$ index, unsaturation ratios can be defined by the proportions of di- and tri-unsaturated C_{38} ketones (Conte *et al.*, 1998a). These relate linearly to the $U^{k'}_{37}$ index in culture (Conte *et al.*, 1998a; Yamamoto *et al.*, 2000), water-column (Conte and Eglinton, 1993; Conte *et al.*, 2001) and sediment studies (Figure 2). However, analytical challenges are greater for the four C_{38} ketones, whose pairs of ethyl and methyl di- and tri-unsaturated ketones prove very difficult to separate completely (Figure 1). Unsaturation ratios based on the C_{38} ketones therefore rarely appear in the alkenone literature. C_{39} ketones make up only a minor component of an alkenone-containing extract, and are rarely quantified.

Most alkenone determinations are made with a flame ionization detector (FID). This detector is simple, reliable, and highly sensitive. However, the FID functions essentially as a carbon detector, and does not give diagnostic information on the structure of the compounds detected. Alkenones are identified by FID by their elution times and by

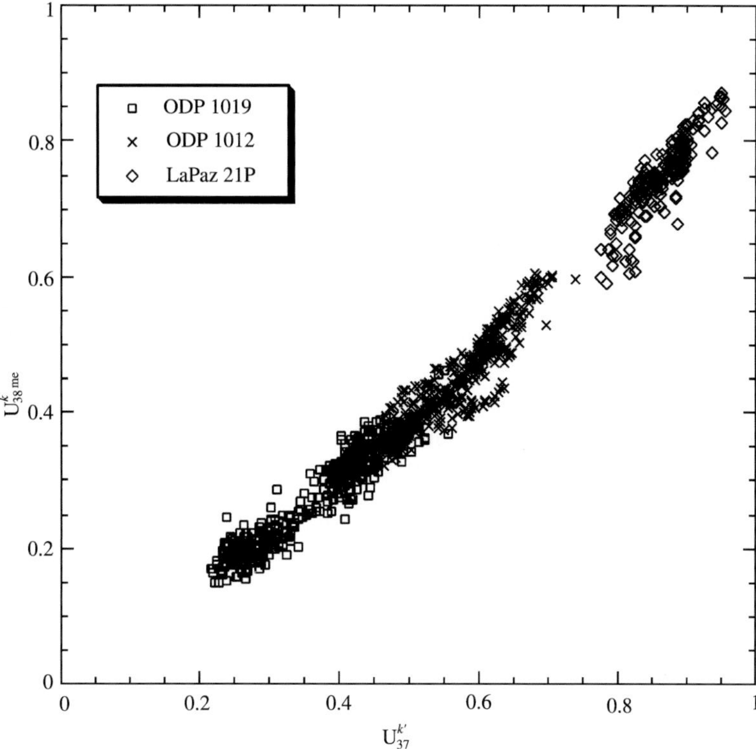

Figure 2 Relationship observed between the $U^{k'}_{37}$ index and the U^{k}_{38me} index $[C_{38:2me}/(C_{38:2me} + C_{38:3me})]$ in a set of cores along the California margin (Herbert, unpublished). See Figure 1 for identification of the alkenone peaks. Note that the two indices are closely related, although the U^{k}_{38me} index has a slope and intercept offset from the $U^{k'}_{37}$ index. Scatter in the relationship primarily comes from analytical difficulties in quantifying the C_{38me} peaks accurately.

reference to external standards added to the injected solvent. Should other compounds share a common elution time (co-elution) with the alkenone chromatographic peaks, they cannot be separated from the signals of interest. The accuracy of gas chromatography (GC)-FID analysis therefore depends on the quality of chromatographic separations in the GC column, and on the presence or absence of interfering compounds. Fortunately, experience shows that interfering high molecular weight compounds are very rare in extracts of marine particulates in sediments. The stability of alkenones during microbial degradation, relative to other high molecular weight lipids, apparently enhances the alkenone signal in sediment extracts. Alkenones stand out not only as some of the most abundant compounds detected in typical lipid extracts, but also as one of the few sets of chromatographic peaks to elute late in the separation. Indeed, the virtual isolation of the alkenone "fingerprint" from the majority of chromatographic peaks almost certainly led to the initial recognition of their paleotemperature significance.

Other detection techniques offer more specificity, but require more analytical time and effort. Alkenones can be determined by GC-mass spectrometry (GC-MS) (Boon *et al.*, 1978; de Leeuw *et al.*, 1980), which identifies each compound by the molecular ion and characteristic fragmentation patterns. However, because the ionization efficiency for alkenones is low, GC-MS generally has lower sensitivity than GC-FID. Chemical ionization coupled to mass spectrometry can improve detection limits for alkenones (Rosell-Mele *et al.*, 1995a; Chaler *et al.*, 2000), but few laboratories employ this technique. GC-MS is therefore used primarily to confirm compound identification by GC-FID, and to detect the presence of co-eluting peaks in the region of alkenone elution if such contamination is suspected. Recently developed variants of GC-FID, such as multidimensional gas chromatography (Thomsen *et al.*, 1998) and comprehensive GC × GC (Xu *et al.*, 2001), promise to improve the specificity of alkenone detection, while maintaining the high throughput and sensitivity of GC-FID. This GC × GC method introduces "slices" of a separation by a first chromatographic column via thermal modulation into a second column of different polarity. The two-dimensional separations achieved appear to enhance the discrimination between potentially interfering compounds (Xu *et al.*, 2001).

In the absence of agreed-upon alkenone reference standards, the results of a recent interlaboratory comparison (Rosell-Mele *et al.*, 2001) offered a largely pleasant surprise. The intercomparison produced a greater consistency in the alkenone unsaturation parameter between laboratories than might have been anticipated. Splits of homogenized unidentified sediment samples were sent to each participating laboratory. Bristol University also prepared mixtures of synthetic $C_{37:3}$ and $C_{37:2}$ ketones. Laboratories were asked to perform replicate extractions and chromatographic determinations following their standard procedures. Interlaboratory variability for the $U^{k'}_{37}$ index, the most commonly measured alkenone unsaturation parameter, translated to a maximum of 2.1 °C between laboratories (95% confidence limit). Individual laboratories routinely obtain precision on replicate measurements on the order of 0.005 $U^{k'}_{37}$ units, or ~0.15 °C. The interlaboratory comparison also found no difference between the mean values determined by laboratories that put lipid extracts through cleanup procedures and those that analyzed total lipid extracts. This result indicates that reliable alkenone determinations can be made for most marine sediments with minimal sample preparation. Care should, however, be used to make sure that compounds are correctly identified and resolved in samples with unusually large amounts of labile compounds, such as found in shallow continental margin sediments. Early diagenesis generally removes interfering compounds in more slowly accumulating pelagic sediments.

Analytical challenges most likely occur at the extreme ends of the unsaturation index, where either the $C_{37:3}$ or $C_{37:2}$ alkenone abundances become very small, and in coastal sediments, where one encounters a more complex matrix of lipids than true for open-ocean sediments. Ternois *et al.* (1997) showed that a short (30 m) column on Norwegian Fjord samples ($U^{k'}_{37}$ ~0.2) did not resolve $C_{37:2}$ ketones from $C_{36:3}$ FAMEs, and hence produced erroneous estimates of sea-surface temperature. Using a 50 m column on the same samples obviated these problems. A more subtle error may come from irreversible adsorbtion of alkenones on the chromatographic column (Villanueva and Grimalt, 1996; Grimalt *et al.*, 2001). The effect is negligible for most samples, but can become significant when one is analyzing either the $C_{37:2}$ or $C_{37:3}$ alkenone near its limit of detection. Grimalt *et al.* (2001) recommend determining the irreversible adsorbtion for a GC system by varying the concentrations of extracts and determining the intercepts of a plot of $C_{37:3}$ and $C_{37:2}$ peak area versus concentration. Their modeling demonstrates that temperature errors can arise at both the high and low ends of the $U^{k'}_{37}$ scale, but will more likely affect results at the warm end of the $U^{k'}_{37}$ range (Grimalt *et al.*, 2001). Chromatographic biases will be toward warmer apparent temperatures at high $U^{k'}_{37}$ values and lower apparent temperatures at low values. Villanueva and Grimalt (1996) recommend introducing at least 5–10 ng of alkenones into the GC to avoid the bias of irreversible adsorbtion, corresponding to a threshold of 50 ng g^{-1} alkenone concentration in the sediments if 1 g is extracted. Our laboratory has also found that buildup of nonvolatile compounds on either a guard or chromatographic column will lead to preferential absorbtion of the $C_{37:3}$ alkenone, or a bias toward warmer temperature estimates, even in samples with large quantities of both ketones (see also Villanueva and Grimalt, 1996). The solution is to trim the column at the injection point by 20–30 cm daily, and to monitor the deterioration of the column by running known material frequently to detect drifts.

6.15.3 OCCURRENCE OF ALKENONES IN MARINE WATERS AND SEDIMENTS

Long-chained alkenones were first identified in Miocene through Pleistocene age marine sediments on the Walvis Ridge (Boon *et al.*, 1978). A group led by de Leeuw (de Leeuw *et al.*, 1980) in the Netherlands subsequently identified C_{37} and C_{38} alkenones in recent sediments of the Black Sea, where *E. huxleyi* dominates the coccolithophorid flora. Simultaneously, work at Bristol University established the connection between alkenone synthesis and certain species of extant haptophyte (coccolithophorid) algae (Volkman *et al.*, 1980a,b; Marlowe *et al.*, 1984a,b 1990). Researchers at Bristol began to culture the producing species in the lab and to examine modern sediments for the presence of long-chained alkenones. It is now clear that two widely distributed species of coccolithophorid algae, *E. huxleyi* and *Gephyrocapsa oceanica*, are the principal alkenone synthesizers in the modern ocean (Conte *et al.*, 1994a; Volkman *et al.*, 1995). Other coccolithophorid species tested to date do not yield alkenones. In addition, alkenone production has been reported from some, but not all, species of the noncalcifying haptophyte genera *Isochrysis* and *Chrysotila* (Conte *et al.*, 1994a; Marlowe *et al.*, 1984b; Versteegh *et al.*, 2001; Volkman *et al.*, 1989). Cruises along the California margin and the equatorial Pacific suggest that noncalcifying haptophyte algae often dominate over coccolithophorid forms (Thomsen *et al.*, 1994). The possibility therefore exists that other noncalcifying haptophyte species that synthesize long-chained alkenones may well have gone undetected (Brassell *et al.*, 1987).

As far as is known, no marine algal groups other than the haptophytes synthesize long-chained alkenones.

The coccolithophorid algae, of which the principal alkenone producers represent a subset, play an important role in the cycles of both organic and inorganic carbon in the ocean (Balch *et al.*, 1992; Sikes and Fabry, 1994; Thomsen *et al.*, 1994). Coccolithophorids surround their cell with a number (in the case of *E. huxleyi*, ~23) of minute (micron-sized) calcite platelets. The intact aggregate is referred to as a coccosphere; after death the coccosphere may disintegrate into its component coccoliths. The biogeography of coccolithophorids is quite well understood, thanks to the distinctive liths produced by different species (Winter *et al.*, 1994). *E. huxleyi* is the most abundant and ubiquitous extant coccolithophore (Okada and Honjo, 1973; Winter *et al.*, 1994). In many plankton studies, it may compose 60–80% of the coccolithophorid assemblage. *E. huxleyi* appears to tolerate a large range of temperatures, salinities, nutrient levels, and light availability (Winter *et al.*, 1994). It therefore occurs in waters of nearly all temperatures, excluding those of the truy polar oceans, and in waters ranging from the highly saline (41‰) Red Sea to the brackish Sea of Azov and Black Sea (Bukry, 1974). Unlike many coccolithophorids, which thrive in the oligotrophic central gyre regions of the ocean, *E. huxleyi* can flourish in more eutrophic environments. *E. huxleyi* blooms appear to be triggered under conditions of a highly stratified upper water column, a shallow (10–20 m deep) mixed layer, and high light intensities (Nanninga and Tyrell, 1996). *E. huxleyi* may have a competitive advantage over other phytoplankton under conditions of high light intensity and low inorganic phosphate ability, as concluded by modeling work on mesocosm experiments in Norwegian fjords (Aksnes *et al.*, 1994). Doubling rates can be as fast as 2.8 d^{-1} (Brand and Guillard, 1981). Under bloom conditions, *E. huxleyi* occurs in large enough quantities in the surface ocean that scattering of light by the platelets can modify the optical properties of the water observed by satellite (Balch *et al.*, 1992). Indeed, dense milky patches of *E. huxleyi* blooms as large as 10^5 km^2 have been observed from space in the high latitude (40–60° N) North Atlantic and North Pacific oceans, off the Falklands in the South Atlantic, and on the north Australian margin (Brown and Yoder, 1994).

G. oceanica, the other known alkenone-synthesizing species of importance, has a more limited oceanic distribution. *G. oceanica* apparently does not occur in waters colder than ~12 °C (Okada and McIntyre, 1979). It commonly occurs in tropical and subtropical waters, in particular, the high fertility regions of the eastern Pacific and the Arabian Sea (Houghton and Guptha, 1991; Roth, 1994). This preference is consistent with a pulse of *G. oceanica* observed along the California margin during spring upwelling conditions (Ziveri *et al.*, 1995). However, the abundance of *G. oceanica* in the western equatorial Pacific warm pool demonstrates that the species does not require high nutrient concentrations. It reportedly tolerates very high salinities (45–51‰) (Winter, 1982), but cannot grow in water of low (15‰) salinity (Brand, 1984). As with *E. huxleyi*, *G. oceanica* can often dominate coccolithophorid assemblages, where it thrives (Roth, 1994).

The alkenone-producing species are typically reduced in abundance or excluded in oceanic provinces that favor diatom growth (Brand, 1994). Thus, they do not occur in the truly polar Arctic waters, and in the much broader siliceous province in the Southern Ocean, where they taper to zero abundance poleward of ~60° S (Nishida, 1986; Sikes and Volkman, 1993). The abundance of *E. huxleyi* and *G. oceanica* is also reduced in regions of high silicate availability, such as in many coastal zones and upwelling regions. Where nutrients such as nitrate and phosphate, or trace metals, such as iron, are low, the competitive advantage reverses to favor coccolithophores over the siliceous phytoplankton (Brand, 1991).

A series of reports of alkenone occurrences in sediments from brackish and freshwaters indicates that haptophytes other than the recognized marine producers may generate inputs of long-chained ketones in these environments. In some cases, the alkenone unsaturation ratio is consistent with open-marine settings, and may point to the extreme tolerance of *E. huxleyi* to low salinities. For example, Ficken and Farrimond (1995) studied a Norwegian Fjord recently opened to the open ocean by human activity. The sediments recorded the transition from fresh to brackish (10‰) conditions by an abrupt increase in C_{37} alkenones in the marine section. The $U_{37}^{k'}$ temperatures recorded in the recent interval lie between 7 °C and 10 °C using the standard marine temperature calibration, suggesting to the authors that *E. huxleyi* was likely the producing organism. The earlier lacustrine sediment sequence also contained alkenones, although at low levels. These differed from the marine sequence in their high abundance of the $C_{37:4}$ ketone (Ficken and Farrimond, 1995). Indeed, unusually high concentrations of the tetraunsaturated ketone seem to signal a "brackish" component to the haptophyte flora in environments such as the Baltic Sea (Schulz *et al.*, 2000) and Chesapeake Bay (Mercer *et al.*, 1999), and also to characterize lacustrine sediments (Cranwell, 1985; Volkman *et al.*, 1988; Li *et al.*, 1996; Thiel *et al.*, 1997; Mayer and Schwark, 1999; Zink *et al.*, 2001). No study has yet confidently connected particular producing organisms to these "freshwater" alkenones.

6.15.3.1 Genetic and Evolutionary Aspects of Alkenone Production

Given that the synthesis of various alkenones and alkenoates is evidently regulated closely by the haptophyte producers, the question arises as to whether a universal response of parameters such as the $U_{37}^{k'}$ unsaturation index would be expected in the face of genetic diversity. Genetic variations could affect the use of alkenone paleothermometry in at least three different scales. In wide-ranging organisms such as *E. huxleyi* and *Gephyrocapsa*, strains evolve with physiologies attuned to the local environment (Paasche, 2002). The extent of genetic exchange between strains is unknown. If as a result of genetic isolation, the relations between alkenone unsaturation and growth temperature differ from strain to strain, then one would somehow have to account for this variation in paleoceanographic studies, or at least attempt to estimate how much uncertainty should be added to paleotemperature determinations due to strain effects. At a larger biogeographic scale, where the relative abundance of *E. huxleyi* and *G. oceanica* varies widely, one needs to decide whether each species has a significantly different $U_{37}^{k'}$–temperature relation, and, if so, then how to proceed to quantify the proportions of alkenone flux due to the two species. A temporal scale problem exists as well. The stratigraphic range of alkenones in sediments far exceeds the paleontological range of the extant alkenone-producing species (Marlowe *et al.*, 1990; Brassell, 1993). Does one assume that the $U_{37}^{k'}$ measured before the appearance of *E. huxleyi* or *G. oceanica* denotes the same temperature as a calibration derived from the modern organisms?

Genetic differences between strains of *E. huxleyi* and *G. oceanica* undoubtedly exist. Growing isolates in cultures is one way to establish which features might show genetic variation. Among the characters whose variation appears genetic are growth rate, temperature preference, salinity tolerance, calcite production rate, chloroplast pigment composition, and long-chained lipid composition (Fisher and Honjo, 1989; Paasche, 2002). As examples, Brand (1982) found that clones isolated from the cold water Gulf of Maine are genetically adapted to lower temperatures than those from the warm Sargasso Sea. Brand also determined that the oceanic genotypes from the Sargasso Sea cannot grow in 15‰ salinity, while coastal genotypes can (Brand, 1984). The relative success of the strains reversed when the strains were grown in saline conditions. Reproductive rates of *G. oceanica* clones from open ocean and neritic strains also differed in cultures (Brand, 1982). However, genetic variation does not always correspond to geographical separation. Young and Westbroek (1991) studied several genetically different morphological forms of *E. huxleyi,* whose populations were only partially separated by temperature preferences. Brand (1982) discovered that while strains of *G. oceanica* from the coastal waters might differ genetically, strains of *E. huxleyi* from the same regions did not.

At the same time, molecular and paleontological evidence is growing to support the strong genetic similarity of all marine alkenone producers. Two dominant morphotypes (A and B) of *E. huxleyi* are recognized. Yet Medlin *et al.* (1996) found little genetic variation when they compared ssu rRNA from three geographically isolated type A clones and a type B clone. In their judgment, the degree of separation did not warrant separating the A and B morphotypes at the species level. Two studies that sequenced portions of the genome of a number of coccolithophorid species were also revealing. Fujiwara *et al.* (2001) recognized four clades within the coccolithophores based on the gene for the subunit of the Rubisco enzyme *rbc*L. They recognized *Emiliania–Gephyrocapsa* as one clade, with identical nucleotide sequences for the *rbc*L subunit. By sequencing the spacer region of the Rubisco operon, Fujiwara *et al.* (2001) found that the noncalcifying alkenone-producing genus *Isochrysis* is an ingroup of the *Emiliania–Gephyrocapsa* clade. The authors suggest that *Isochrysis* may have secondarily lost the ability to form coccoliths. Ultrastructural similarities in the endoplasmic reticulum of *Emiliania, Isochrysis, Gephyrocapsa*, and *Chrysotila* also link the known alkenone-producing species (Fujiwara *et al.*, 2001). An independent study of the sequence of 18S rDNA by Edvardsen *et al.* (2000) confirmed the genetic clustering of the *Emiliania–Gephyrocapsa* clade within the coccolithophores.

A recent morphometric study of the distribution of *Gephyrocapsid* coccoliths hints that the previously recognized species may all belong to one biological species with strong morphological variations along environmental gradients. Bollman (1997) looked at a set of 70 globally distributed Holocene sediment assemblages to make quantitative determinations of *Gephyrocapsid* coccolith morphometric parameters such as size, bridge angle, ellipticity, and the width of the central collar. Bollman found strong environmental controls on the occurrence of morphotypes. Temperature and chlorophyll abundance (inferred from satellite data) proved to be the strongest predictors, although there was also a weak influence of salinity on the occurrence of some morphotypes. The largest morphological differences between samples corresponded to the greatest environmental differences. Furthermore, within assemblages, Bollman found gradations in morphotypes between the end-members he recognized. Bollman proposed that six major morphotypes would be conventionally associated with different *Gephyrocapsid* species (his *Gephyrocapsa* Equatorial = *G. oceanica*; *Gephyrocapsa* Oligotrophic = *G. carribbeanica*; *Gephyrocapsa* Transitional = *G. margerli*; *Gephyrocapsa* Cold = *G. muellerae*; *Gephyrocapsa* Large = *G. oceanica rodela*; *Gephyrocapsa* Minute = *G. aperta*). Although Bollman admitted that paleontological examination cannot answer the question of whether his six morphotypes represent six biological species or only one species with strong morphological plasticity, he clearly leans toward the latter. This result is in accord with what many nannofossil authors see as difficulty in *Gephyrocapsid* species concepts (Ziveri *et al.*, 1995; W. Wei, personal communication). For example, Wei's (1993) stratigraphic study of the evolution of Plio-Pleistocene nannofossils distinguishes the first appearance of several "species" of *Gephyrocapsids* that would include most definitions of *G. oceanica*, at ~0.9 Ma. However, Wei labels these as *Gephyrocapsa* sp. C–D to denote the state of taxonomic uncertainty.

Results suggesting the close evolutionary relationship between alkenone-producing species are encouraging as paleoceanographers apply the alkenone method to sediments that predate the appearance of modern taxa. *E. huxleyi* first appeared during marine oxygen isotope stage 8 (Thierstein *et al.*, 1977), at ~280 ka. Its appearance can be shown to be globally synchronous to within 5–10 kyr by comparing its first appearance datum to oxygen isotope data measured in foraminifera in the same sediments. How sharply the evolutionary first occurrence of *E. huxleyi* appears depends on sediment location and sample interval; in some cases its coccoliths rapidly become abundant, while in others its arrival is less dramatic (Thierstein *et al.*, 1977). *Emiliania* almost certainly descended from a *Gephyrocapsid* ancestor (McIntyre, 1970). The species apparently reached its modern levels of abundance earlier in the tropics (ca. 85 ka) than at higher latitudes (Jordan *et al.*, 1996). Before the advent of *E. huxleyi*, *G. oceanica* was globally dominant during some intervals of the Pleistocene (Thierstein *et al.*, 1977; Bollman, 1997). The genus *Gephyrocapsa* has been reported in sediments as old as Middle Miocene (South Atlantic: Jiang and Gartner, 1984; Equatorial Pacific: Pujos, 1987), although its first frequent occurrence is in the mid-Pliocene (~3.5 Ma, Bollman, 1997). All morphological variations of *Gephyrocapsa* have existed since at least 620 ka, with the exception of Bollman's (1997) *Gephyrocapsa Cold* and *Gephyrocapsa Oligotrophic*.

Marlowe *et al.* (1990) produced an important review that compared the occurrence of alkenones in sediments with micropaleontological data on the coccoliths of the same material. Their synthesis indicated that a number of morphologically related species in the family *Gephyrocapsaceae* were potential sources of alkenones and alkenoates (genera *Crenalithus*, *Dictyococcites*, *Emiliania*, *Gephyrocapsa*, *Pseudoemiliania*, and *Reticulofenestrata*). The *Gephyrocapsaceae* were the only nannofossil family uniformly associated with alkenone-bearing sediments. The extinct genus *Reticulofenestra* evolved early in the Eocene and continued through the Pliocene, where it most likely left the genus *Gephyrocapsa* as its descendant (Rio, 1982). *Reticulofenestra* is the only genus of *Gephyrocapsids* to accompany alkenones in Miocene, Oligocene, and Eocene samples (Marlowe *et al.*, 1990). In the Pleistocene, *Gephyrocapsa* coccoliths always appear where alkenones have been detected (Marlowe *et al.*, 1990).

Cretaceous precursor alkenones have been reported; their evolutionary relationship with the modern clade of alkenone-producing organisms is unknown. Cretaceous alkenones (Farrimond *et al.*, 1986; Yamamoto *et al.*, 1996) differ from the modern suite in that C_{41} and C_{42} ketones dominate, and these consist (to date) only of di-unsaturated forms.

In sum, genetic and micropaleontological data suggest very close evolutionary relationships between the alkenone-producing species. Alkenone production occurs (and occurred) in species or variants already known to be closely related by micropaleontologists. Critical questions such as the degree of genetic exchange between species variants, and the degree of genetic variability within living populations, remain to be determined. Ignorance of the role of alkenones in the cells of haptophyte algae, and of genetic versus environmental controls on alkenone parameters, means that genetic effects in both space and time remain important unknowns in alkenone paleotemperature research.

6.15.4 FUNCTION

Alkenones occur in both motile and coccolith-bearing forms of *E. huxleyi* (Volkman *et al.*, 1980b; Conte *et al.*, 1995; Bell and Pond, 1996). The role that alkenones play in the producing organisms is not understood. That they play some critical role in *E. huxleyi* and related species is suggested by the large fraction of cellular investment accounted for by alkenones and alkenoates (Prahl *et al.*, 1988; Conte *et al.*, 1994b; Epstein *et al.*, 2001; Versteegh *et al.*, 2001)—generally

5–10% of cell carbon. Marlowe *et al.* (1984a) and Brassell *et al.* (1986b) proposed that alkenone unsaturation helps to regulate membrane fluidity at different temperatures, in analogy with the known function of membrane lipids in many plants. This model clearly associates the unsaturation index with a physiological response to growth temperature. However, alkenones have not been conclusively associated with the membranes of haptophyte algae. Conte and Eglinton (1993) did not detect alkenones in fragmented membranes of *E. huxleyi*. Mouzdahir *et al.* (2001) interpreted the rapid light-dependent degradation of C_{31} and C_{33} alkenones in contrast to the recalcitrance of C_{37} and C_{38} alkenones as an indication that the shorter-chained alkenes resided in membranes, while the longer-chained ketones resided elsewhere in the cell. Epstein *et al.* (2001) noted that cell quotas of alkenones increase with decreasing growth rate in cultures grown from an initial stock of nutrients. In analogy with triacylglycerides in cultures of other marine plankton, they proposed that alkenones may serve as storage molecules for the producing organisms. If this view is correct, then we have no clear reason why the unsaturation should relate to growth temperature.

6.15.5 ECOLOGICAL CONTROLS ON ALKENONE PRODUCTION AND DOWNWARD FLUX

Since alkenone-producing haptophytes can live in a range of depths in the photic zone, and vary greatly in their productivity over the course of an annual cycle, there is much to be learned about how the ecology of these organisms can affect the $U^{k'}_{37}$ signal eventually encoded in the sediments. Depending on the depth of maximum production relative to the mixed layer, alkenone producers may synthesize biolipids in temperatures representative of SST, or offset to colder temperatures by several degrees. Phytoplankton production is also quite seasonal in most ocean locations. If alkenone production follows a strong annual cycle, then again the $U^{k'}_{37}$ temperature could contain a bias relative to mean annual conditions (in the latter case, the bias could be toward either colder or warmer than average temperatures, depending on the season of maximum production). Profiles of coccolith and alkenone abundance through the photic zone, and over the yearly cycle, give indications of how ecological biases may vary between different oceanic provinces. Sediment traps intercept falling coccoliths and lipids to give windows into how the annual cycle in the near-surface is exported downward to the seafloor. A number of such studies will be reviewed below for their implications for the $U^{k'}_{37}$ thermometer. Many are micropaleontological, assessing both the absolute vertical fluxes of the alkenone producers *E. huxleyi* and *G. oceanica,* and the proportions of these species relative to the entire coccolithophorid assemblage. Lipid analyses also are available to measure the abundance of alkenones in the upper water column, and their flux into sediment traps. Unfortunately, very few studies combine both micropaleontological and geochemical approaches.

Nearly all micropaleontological time-series data collected on near-surface samples and sediment traps found either *E. huxleyi* or *G. oceanica* to dominate the coccolith flora on an annual basis. The density of living coccolithophorids is best measured by counting the number of intact coccospheres per unit volume of water sampled (e.g., Haidar and Thierstein, 2001). *E. huxleyi* accounts for the majority of the coccolithophorids surveyed in time series acquired off Bermuda (Haidar and Thierstein, 2001; 64% on an annual basis), the San Pedro Basin off Southern California (Ziveri *et al.*, 1995; 30–80% during the annual cycle), the Northeast Atlantic (Broerse *et al.*, 2000b; 69% and 72% at two sites surveyed), and off the coast of Northwest Africa (Sprengel *et al.*, 2000). *G. oceanica* was the most important taxon in year-long sediment trap studies of the Arabian Sea upwelling area, followed by *E. huxleyi* (Andruleit *et al.*, 2000; Broerse *et al.*, 2000c).

Depth profiles of cell densities in the photic zone generally show *E. huxleyi* to live within the mixed layer. Cortés *et al.* (2001) studied the seasonal depth distribution of coccolithophorid species off Hawaii. Sampling showed that the main production occurred in the middle photic zone (50–100 m), which lay within the mixed layer for most of the year. While the depth of maximum *E. huxleyi* density varied during the annual cycle, it generally lay between the shallowest sampling level (10 m) and 100 m. Depth profiles off Bermuda (Haidar and Thierstein, 2001) found that maximum densities of *E. huxleyi* were nearly always shallower than 100 m, and more commonly within the upper 50 m. The highest cell densities for *E. huxleyi* recorded were at 1 m depth in March, after the seasonal advection of nitrate into the mixed layer. Seven years of water-column particulate data off Bermuda confirm that alkenone concentrations in the surface mixed layer are 2–4 times higher than in the deep fluorescence maximum at 75–110 m (Conte *et al.*, 2001).

Hamanaka *et al.* (2000) conducted the only published study that directly estimated alkenone production as a function of depth habit. These authors incubated phytoplankton at 0 m, 5 m, 10 m, 25 m, 40 m, and 60 m depth, and measured growth rates by spiking the water with $\delta^{13}C$. Maximum incorporation into alkenones occurred at 5 m depth, which coincided with the peak in alkenone concentration in particulates. Production rates at the surface and below 5 m were very low (from 269 ng l^{-1} d^{-1} at 5 m to 1.4 ng l^{-1} d^{-1} at 25 m). The depth of maximum alkenone synthesis lay well above the depth of the chlorophyll maximum at 25 m, indicating that the alkenone-producing species were offset vertically from the majority of the phytoplankton community. An array of shallow sediment traps intercepted the largest alkenone flux at 15 m, just below the layer most productive of alkenones.

There is, however, indirect evidence to suggest that alkenone synthesis may occur at depths below the mixed layer in some environments. $U_{37}^{k'}$ values in particulates often correspond to temperatures colder than the mixed layer. For example, Prahl *et al.* (2001) found that the alkenone flux to sediment traps in the Wilkinson Basin (coastal northeastern United States) peaked during summer time when surface waters stratified and a subsurface chlorophyll maximum was established in the upper seasonal thermocline. The $U_{37}^{k'}$ temperatures of the summer particulates corresponded to temperatures at the base of the upper thermocline, 6–7 °C colder than the ~5 m thick summer mixed layer. Similarly, Ternois *et al.* (1996) used the $U_{37}^{k'}$ of sediment trap material to time series of upper water-column temperatures in the western Mediterranean to deduce that the depth of maximum alkenone production varies over the annual cycle. The $U_{37}^{k'}$ temperature during the spring season corresponded to a maximum production depth of 50 m, while that depth apparently shallowed to 30 m during the second phase of high production in the fall months. Both Ohkouchi *et al.* (1999) and Prahl *et al.* (1993) have further argued that $U_{37}^{k'}$ values should be systematically offset to colder temperatures than the mixed layer in gyre regions, where the deep nutricline tends to produce subsurface chlorophyll maxima. It should be noted that in order to use the $U_{37}^{k'}$ value to estimate the depth of alkenone synthesis, one has to assume a temperature calibration. The suggestion of subsurface production is not purely circular, however, as one can compare the qualitative features of the $U_{37}^{k'}$ values over the annual cycle with the temperature time series. Thus, the $U_{37}^{k'}$ of falling particles actually decreased from spring into the onset of summer water-column stratification in the Wilkinson Basin study (Prahl *et al.*, 2001), in direct contrast to the warming of the surface layer.

Alkenone production also occurs in the context of the annual cycle of upper water-column temperatures. Standing stocks of alkenone-producing species and their falling products display strong changes over the course of the year in all studies to date. In most cases, the time of maximum abundance of *E. huxleyi* and/or *G. oceanica*, and the maximum flux of alkenones into sediment traps, coincides with the dominant period for phytoplankton blooming. Thus, production peaks in spring months in most subtropical and mid-latitude locations (Prahl *et al.*, 1993; Sprengel *et al.*, 2000, 2002; Broerse *et al.*, 2000b; Antia *et al.*, 2001; Cortés *et al.*, 2001; Harada *et al.*, 2001; Haidar and Thierstein, 2001). In the typical cycle, cell densities of *E. huxleyi* increase in the upper photic zone after the seasonal advection of nitrate that occurs with winter and early spring mixing (Haidar and Thierstein, 2001). The abundance of *E. huxleyi* may increase by an order of magnitude (Cortés *et al.*, 2001) or more (Haidar and Thierstein, 2001) in the upper water column during the spring bloom, as does its flux to sediment traps in locations such as the Northeast Atlantic (Broerse *et al.*, 2000b). A sediment trap sample obtained in May in the Norwegian Sea intercepted a nearly monospecific bloom of *E. huxleyi* (Cadee, 1985). Flux maxima are, however, more commonly diffuse, occupying 2–3 months of time (Broerse *et al.*, 2000b). Alkenone fluxes in sediment traps generally show a spring peak, indicating that surface ecological signals are exported to depth (Ternois *et al.*, 1997; Sicre *et al.*, 1999).

Maximum alkenone production may be shifted to the summer months in some regions. For example, Prahl *et al.* found that alkenone flux to sediment traps peaked during the early summer months at the Wilkinson Basin location (43° N), after the water column had stratified. At higher-latitude locations such as the Norwegian Sea and the Barents Sea in the northern hemisphere (Samtleben and Bickert, 1990; Thomsen *et al.*, 1998), and in the Indian Ocean sector of the Southern Ocean (Ternois *et al.*, 1998), the flux of *E. huxleyi* and alkenones is phased even more strictly to the summer months. Winter production can reach vanishingly small amounts during winter months in such harsh environments (Ternois *et al.*, 1998; Broerse *et al.*, 2000a). While such a pattern follows the classic seasonal progression of maximum production with latitude, exceptions to the rule occur. Goni *et al.* (2001) recorded a two- to threefold increase in the organic-carbon normalized fluxes of alkenones in the Gulf of California from June to October, well past the expected spring bloom. The latter authors speculated that maximal alkenone production could be separated from that of other phytoplankton groups by competition. Groups such as diatoms that often dominate peak bloom periods (Giraudeau *et al.*, 1993; Broerse *et al.*, 2000a,c) also generally show higher seasonal variability than does *E. huxleyi* (Beaufort and Heussner, 1999, 2001; Harada *et al.*, 2001).

A competitive interaction apparently was observed by Ziveri *et al.* (1995), who found that the highest coccolith and coccosphere flux occurred in winter in the San Pedro basin off Southern California. This peak preceded the spring upwelling season favored by diatoms. Further evidence of *E. huxleyi* and/or alkenone production offset from that of other phytoplankton groups can be found in the studies of Ziveri *et al.* (2000), Harada *et al.* (2001), and Muller and Fischer (2001).

A number of oceanic regimes also produce twice-yearly flux maxima of alkenone production. In the Mediterranean, a fall bloom of alkenone production occurs (Ternois *et al.*, 1996; Sicre *et al.*, 1999). This is also true off Hawaii (Cortés *et al.*, 2001), in the central equatorial Pacific (Harada *et al.*, 2001), in the Sea of Okhotsk (Broerse *et al.*, 2000a), and in the Norwegian Sea (Thomsen *et al.*, 1998). A lack of dissolved silica may inhibit diatom growth and promote haptophyte production during the fall months in such locations (Broerse *et al.*, 2000a).

The monsoonally driven upwelling system of the Arabian Sea offers a special occasion to study the seasonality of alkenone production. In this environment, *G. oceanica* dominates, although *E. huxleyi* still occupies the second position in the coccolith flora (Andruleit *et al.*, 2000; Broerse *et al.*, 2000c). Winds favorable to upwelling occur twice per year. The Southwest monsoon in late spring/early summer sees the peak in upwelling conditions (Prahl *et al.*, 2000), while a secondary maximum occurs when the winds shift direction during the northeast monsoon (Andruleit *et al.*, 2000; Prahl *et al.*, 2000). Coccolithophore fluxes, comprised of 60–70% *G. oceanica* and *E. huxleyi*, increase during the southwest monsoon, at the same time as the maximum in silica flux (Broerse *et al.*, 2000c). Flux maxima lasted for nearly three months in this sediment trap study. In another sediment trap experiment conducted in the Arabian Sea, Prahl *et al.* (2000) determined that distinct alkenone flux maxima occur at the start and stop of the northeast and southwest monsoons. A lag in alkenone flux relative to other phytoplankton biomarkers suggested a successional delay in haptophyte production during the more dramatic southwest monsoon (Prahl *et al.*, 2000). No observable offset in production occurred during the less distinctive northeast monsoon, or at the close of the southwest monsoon. During peak monsoonal upwelling, alkenone fluxes reached ~25 times those of the unproductive months of the year.

Because only a few sediment trap experiments report data for much more than one year's duration, we have only a glimpse at the importance of interannual variability in the productivity of alkenone-synthesizing species. Reports do suggest large changes in the downward flux of either *E. huxleyi* or alkenones between years. Three years of sediment trap data from the northwestern Mediterranean Sea revealed, in addition to the bi-annual flux maxima mentioned earlier, considerable interannual variability (Sicre *et al.*, 1999). In general, Sicre *et al.* (1999) found that maximum alkenone fluxes coincided with maximum total organic carbon (TOC) fluxes. In the year 1994, however, there were comparable TOC fluxes to other bloom periods without correspondingly large C_{37} ketone fluxes. This study also reported large variability in the amplitudes of the spring and fall blooms in alkenones. For example, the spring and fall blooms produced fluxes of C_{37} ketones of $8\ \mu g^{-1}\ m^{-2}\ d^{-1}$and $16\ \mu g^{-1}\ m^{-2}\ d^{-1}$, respectively, in 1989–1990, but in 1994, the fall bloom amounted to a flux of only $0.1\ \mu g^{-1}\ m^{-2}\ d^{-1}$ to the 200 m trap. Muller and Fischer's (2001) four-year sediment trap study in the upwelling region of North Africa also documented approximately sevenfold variations in the annual flux of alkenones to sediment traps. These authors found no repeatable seasonal cycle in alkenone fluxes at their study location. Strong year-to-year variations in coccolith and coccosphere fluxes have also been identified in sediment trap data from the North Atlantic (Ziveri *et al.*, 2000) and in water-column censuses off Bermuda (Haidar and Thierstein, 2001). Conte *et al.* (1998b) report a particularly interesting short-lived alkenone flux peak in a Sargasso Sea sediment trap experiment. This region produces a classic spring bloom following the winter deepening of the mixed layer. The pulses of alkenone production detected by Conte *et al.* (1998b) preceded this predictable part of the annual cycle. Several short-lived flux events occurred, usually during December to January, but not for every year studied. During the flux events, falling organic matter was enriched in labile, phytoplankton-derived debris, including alkenones. For unknown reasons, a transient event in the surface water was inefficiently degraded and sent rapidly to depth (Conte *et al.*, 1998b).

Tiered sediment trap arrays present a picture of how seasonal and episodic production works its way toward the seafloor. Nearly all such arrays show that the near-surface temporal variability is attenuated with depth. The seasonal variability, so evident in many shallow sediment trap time series, is reduced by factors of 2 (Broerse *et al.* (2000a), Sea of Okhotsk; Ziveri *et al.* (2000), Northwest Atlantic; Muller and Fischer (2001), northwest African margin), to 3 (Harada *et al.* (2001), central equatorial Pacific; Thomsen *et al.* (1998), Norwegian Sea). This attenuation presumably comes both from the selective biological degradation of more labile lipids at shallow depths, and a diffuse supply of fine-grained sediment particles at depth that partially masks the variability of surface production (see more below).

These observations raise the question of exactly what factors control the downward transport of alkenones. Clearly, as molecules associated with organisms only 10–20 μm size (before decomposition and disaggregation of coccoliths), they must have vanishingly low settling rates without the aid of processes that cause aggregation. The majority of sediment trap studies find that the coccolith and/or alkenone flux is highly correlated with the total mass flux over the yearly cycle (Beaufort and Heussner, 1999; Andruleit *et al.*, 2000; Broerse *et al.*, 2000a,b). This may signal the general synchronization of *E. huxleyi* and/or *G. oceanica* production with the total biogenic flux, which then sweeps smaller particles out of the photic zone (Thomsen *et al.*, 1998; Ziveri and Thunell, 2000). To support this idea, shallower sediment trap collections find that the intact coccosphere flux, which must represent the most recently produced and least remineralized component of the haptophyte flux, coincides with the time of highest detached coccolith flux, and that these peak the same time as the highest total mass flux (Broerse *et al.*, 2000a). Broerse *et al.* (2000a) note the role of marine snow in entangling coccospheres in the autumn bloom in the Sea of Okhotsk, and a complementary role played by large diatoms in the spring. Coccospheres represent only a tiny portion of the coccolith carbonate flux (Broerse *et al.*, 2000a; Ziveri *et al.*, 2000; Ziveri and Thunell, 2000); most of the coccolith flux must undergo multiple cycles of release and reaggregation (Ziveri *et al.*, 2000).

Studies that report alkenone fluxes relative to total organic carbon (normalized to TOC in typical units of 100–1,000 $\mu g\ g^{-1}$ TOC) also provide evidence that much of the alkenone flux from the photic zone comes from bloom episodes. Such normalized alkenone concentrations in sediment traps ranged from lows of 29 $\mu g\ g^{-1}$ C during unproductive winter months in the Wilkinson Basin (northeast margin of the US) to highs of 1,054 $\mu g\ g^{-1}$ C during the early summer bloom period (Conte *et al.*, 1998b; Prahl *et al.*, 2001). Similar results were reported in the Southern Ocean (Ternois *et al.*, 1997), Mediterranean Sea (Sicre *et al.*, 1999), and Norwegian Sea (Thomsen *et al.*, 1998). Alkenone accumulations during times of very low flux may indeed represent the sedimentation of "relict" material synthesized during more productive seasons (Conte *et al.*, 1998b; Sicre *et al.*, 1999; Prahl *et al.*, 2001). There are, however, several sediment trap studies that do not report significant correlations between the alkenone flux and the total organic flux (Ternois *et al.*, 1996; Muller and Fischer, 2001).

Evidence has recently emerged that some benthic environments may also receive a diffuse supply of fine-grained particles with associated alkenones. Processes such as resuspension of continental margin and slope sediments during major storms, and benthic currents may thus contribute lateral supplies of alkenones to the sediments. Fluxes of coccoliths in a deep-sea canyon setting in the Bay of Biscay increased significantly during the fall and winter stormy months (Beaufort and Heussner, 1999), as did coccolith and alkenone fluxes during summer resuspension months in the Norwegian Sea (Andruleit, 1997; Thomsen *et al.*, 1998). Core-top $U^{k'}_{37}$ estimates can apparently be affected by the input of fossil alkenones in the Norwegian and Barents Sea. Thomsen *et al.* (1998) determined $U^{k'}_{37}$ values in shallow sediment traps consistent with production in (cold) surface temperatures, but showed that deeper traps and surface sediments had $U^{k'}_{37}$ indices 5–10 °C too warm for the locations. The presence of pre-Quaternary coccoliths in floral assemblages unequivocally indicates that some portion of the flux comes from the erosion of older slope and shelf sediments (Beaufort and Heussner, 1999, 2001; Weaver *et al.*, 1999). A number of other sediment traps have found higher coccolith fluxes at depth than in shallow traps, indicating a lateral source of material (Sprengel *et al.*, 2000, 2002; Ziveri *et al.*, 2000; Antia *et al.*, 2001).

The effect of lateral transport on $U^{k'}_{37}$ temperature estimates will depend on the age of the transported material (e.g., contemporary or fossil), the location from which it comes (e.g., from similar latitude and surface temperature, or from long distance), and the ratio of the advected flux to the alkenones sedimented in the vertical sense. Deep equatorward benthic currents are apparently responsible for transporting alkenones, along with the fine fraction, a great distance from their source in at least two instances. In the southwestern Atlantic, recent age sediments under the Brazil–Malvinas Confluence and Malvinas Current produce anomalously cold (by 2–6 °C) $U^{k'}_{37}$ temperature values (Benthien and Müller, 2000). The remainder of the 87 core tops that these authors analyzed in a large region from 5° N to 50° S had $U^{k'}_{37}$ values in good agreement with global core-top calibrations to local mean annual SST. Benthien and Müller (2000) argue that sediments in these specific regions were transported northward and offshore by benthic currents, and hence contain $U^{k'}_{37}$ signals of their origin in cold waters. Benthic boundary current advection of alkenones has been documented in a different way by Ohkouchi *et al.* (2002) in Bermuda Rise drift sediments. Compound-specific AMS ^{14}C dating of alkenones shows that they may be significantly older than foraminifera in the same layers. The authors attribute this to southward advection of the fossil fine fraction material from the Nova Scotia margin. However, there may also be environments with significant lateral transport in which the effect of allocthonous material does not swamp primary coccolithophorid ecological signals. For example, Beaufort and Heussner (1999) found that the seasonal cycle of succession in coccolithophorid species was preserved in the Bay of Biscay sediment trap time series, despite the input from resuspended margin sediments. The primary $U^{k'}_{37}$ signal would presumably be preserved as well in such a case.

6.15.5.1 Effects of Water-column Recycling and Sediment Diagenesis on the Alkenone Unsaturation Index

As they descend through the water column and are incorporated into sediments, alkenone and other lipid biomarkers encounter different degradational conditions (Prahl *et al.*, 1989a; McCaffrey *et al.*, 1990; Sun and Wakeham, 1994; Koopmans *et al.*, 1997; Rontani *et al.*, 1997; Teece *et al.*, 1998; Sinninghe-Damsté *et al.*, 2002). The amount of time that lipids spend exposed to these metabolic pathways varies substantially, as does the degradational efficiency of each pathway. Rapid transit under oxic conditions generally occurs through the water column. Slow passage, often under suboxic to anoxic conditions, characterizes the entrance of alkenones into the sedimentary record. The key question for paleoceanographers is whether these steps produce measurable changes in the alkenone unsaturation index.

The fraction of alkenones buried in sediments represents less than 1% of the initial flux from the photic zone in most cases. This attenuation can be measured by comparing tiered sediment trap fluxes to surficial sediment fluxes. The synchronization of alkenone flux peaks between shallow and deep traps shows that the time required for alkenone-containing particles to reach the seafloor is only one to two weeks (Conte *et al.*, 1998b; Muller and Fischer, 2001). Somewhat slower average sinking velocities pertain to periods of low alkenone and total mass flux (Muller and Fischer, 2001). During their descent, alkenones are strongly recycled, both relative to their initial flux, and relative to bulk organic matter (Sicre *et al.*, 1999; Muller and Fischer, 2001). Alkenones do appear more resistant to degradation than most other lipids of planktonic origin (Prahl *et al.*, 2000; Volkman *et al.*, 1980a). Recycling in the sediments adds further attenuation to the roughly one order of magnitude loss in the water column. In the slow deposition rate of most marine environments, alkenones will be exposed to degradation under oxic conditions for thousands of years. However, where sediment accumulates more rapidly, as along continental margins, and where oxygen levels in either the bottom water or sediment pore waters plummet, alkenones will be exposed to nonoxic bacterial metabolic pathways. Comparison of sediment trap fluxes to surface sediments in pelagic regions suggests alkenone preservation factors of 0.2% (relative to the deep trap flux) (Muller and Fischer, 2001), 0.6% (Prahl *et al.*, 1989b), and 1% (Prahl *et al.*, 2000). In pelagic environments, degradation of alkenones in surface sediments appears approximately an order of magnitude more efficient than the reduction in bulk organic carbon (Prahl and Muelhausen, 1989; Prahl *et al.*, 2000; Muller and Fischer, 2001). However, much better preservation occurs in high sediment flux settings. The highest preservation efficiency of alkenones reported is 44% in the shallow (~300 m) and high deposition rate setting of the Wilkinson Basin (Prahl *et al.*, 2001) and nearly 100% in the anoxic bottom sediments of the Guaymas Basin (Goni *et al.*, 2001). Prahl *et al.* (1993) used an onshore–offshore sediment trap and sediment transect to demonstrate that the preservation efficiency of alkenones and total organic carbon was high, and similar (~25%) at their near-shore site (water depth 2,717 m), but fell systematically in the direction of the open ocean to only 0.25% in the most distal site. The deepest water site clearly displayed preferential loss of alkenones relative to bulk organic carbon (Prahl *et al.*, 1993). Preservation efficiency in sediments thus seems to depend strongly on exposure time to oxic degradation (Madureira *et al.*, 1995; Prahl *et al.*, 2001).

None of the sediment trap studies reports significant shifts in the alkenone unsaturation index as a consequence of degradation. A sediment trap deployment in the northwestern Mediterranean Sea found that despite a fivefold loss in alkenones between shallow and deep traps, there was no apparent offset in the $U^{k'}_{37}$ ratio of the biomarkers. Sawada *et al.* (1998) similarly found no shift in the unsaturation index through the enormous vertical path length of the northwest Pacific (traps arrayed at 1,674 m, 4,180 m, 5,687 m, and 8,688 m). Furthermore, the $U^{k'}_{37}$ composition of underlying sediments agreed with sediment trap estimates, indicating that early diagenesis in the sediments did not modify the unsaturation index relative to the incoming composition. Other comparisons of the $U^{k'}_{37}$ index between sediment traps and core tops find very good agreement as well (Muller and Fischer, 2001; Prahl *et al.*, 1989b, 2001).

Several investigators have conducted laboratory studies of the effects of degradation on the alkenone unsaturation index. In the early history of the development of the $U^{k'}_{37}$ index,

Volkman *et al.* (1980a) compared the unsaturation index in fecal pellets of the copepod *Calanus helgolandicus* with that of the *E. huxleyi* used to feed them. No change in the index was observed after passage through the guts of these zooplankton. Grice *et al.* (1998) obtained similar results by feeding the alkenone-synthesizing haptophyte *I. Galbana* to the copepod *Temora*. The best study of sedimentary processes comes from Teece *et al.* (1998), who exposed alkenones to microbial degradation under different conditions for almost 800 days. Oxic, sulfate-reducing, and methanogenic experiments produced time series of the degradation rates of alkenones and other lipids. After rapid initial degradation of lipids, the fate of alkenones varied significantly. About 85% of the initial alkenone inventory had been degraded under oxic conditions by the end of the experiment. The two anoxic pathways yielded different results. Under sulfate-reducing conditions, degradation essentially ceased with ~60% of the alkenones remaining. Methanogenic conditions led to preservation not much better than for oxic conditions (~80% degradation). Teece *et al.* (1998) conclude that the different apparent alkenone degradation rate constants under different anoxic conditions lead to subtleties in understanding the preservation of alkenones under varying redox states. A very important finding, however, was that the $U_{37}^{k'}$ index did not shift beyond a very modest increase (0.03 units, equivalent to less than 1 °C apparent temperature change) reported for the oxic experiment (Teece *et al.*, 1998).

Two sediment studies present the dissenting view that the alkenone unsaturation index may shift to higher $U_{37}^{k'}$ values (higher apparent temperature) during diagenesis. Gong and Hollander (1999) compared near $U_{37}^{k'}$ sediment data acquired down-core at two sites in the Santa Monica Basin that differed in bottom-water oxygenation. They attribute a positive $U_{37}^{k'}$ offset at the oxic site to indicated preferential degradation of the $C_{37:3}$ ketone. In order to estimate the offset, Gong and Hollander had to match samples of the same age between the cores. Thus, some of the apparent temperature offsets of up to 2.5 °C depend on the quality of the age models. Nevertheless, Gong and Hollander (1999) present evidence over the last two centuries of deposition for an average $U_{37}^{k'}$ offset equal to 0.9 °C at the oxic site. Another way to study the possibility of diagenetic alteration of the $U_{37}^{k'}$ index is to look for gradients in the index with preservation of initially homogeneous material. Fine-grained turbidites, the sedimentary product of near-instantaneous emplacement of well-mixed sediment on the seafloor, offer this opportunity. After the emplacement of the turbidite layer (often 10–100 cm thick), a well-developed redox front moves downward from the sediment–water interface. The top of the turbidite layer will experience significant oxidation, while the base may suffer none at all. Hoefs *et al.* (1998) studied $U_{37}^{k'}$ indices through oxidation fronts in a number of fossil turbidite layers cored on the Madeira abyssal plain. They reversed the conclusion of Prahl *et al.* (1989b), who had earlier found negligible shift in the $U_{37}^{k'}$ index in a relatively young turbidite sequence in the same region. Hoefs *et al.* reported that alkenones were degraded to a far greater degree (factors of 50–1,000) in the oxidized zones of the turbidites relative to total organic carbon. The samples with the lowest alkenone concentrations produced $U_{37}^{k'}$ estimates 2.5–3.5 °C warmer than samples in the unaltered bases of the turbidites. Hoefs *et al.* (1998) conclude that diagenesis may indeed selectively degrade the $C_{37:3}$ ketone and produce significant artifacts for paleoceanography.

Grimalt *et al.* (2000) mounted a serious criticism of the evidence for differential diagenesis based on the turbidite studies. They noted that the amount of apparent temperature change depended strongly on the reported alkenone concentrations. The strongest evidence for differential diagenesis (Hoefs *et al.*, 1998) comes from samples with extraordinarily low alkenone concentrations. Grimalt *et al.* conclude that the shift to higher $U_{37}^{k'}$ values could well be an analytical artifact of attempting to analyze the $C_{37:3}$ ketone at its limit of detection. As noted earlier, irreversible chromatographic column adsorbtions become significant at very low concentrations of either ketone. In Madeira Abyssal Plain sediments, the initial (unoxidized) $C_{37:3}$ ketone concentration is quite low relative to the di-unsaturated ketone, and it could well reach its limit of detection under the concentrations analyzed by Hoefs *et al.* (1998).

What is the long-term fate of the alkenone unsaturation index in sediments? The time available for alteration in sediments dwarfs the few thousand years of near-surface exposure. It is difficult to completely answer the question, since without knowing *a priori* what correct SST estimates would be for ancient sediments, we cannot get a direct estimate of diagenetic offsets or lack thereof. Nevertheless, some qualitative lines of reasoning suggest that the alkenone index probably remains quite stable on geological timescales. The signature of differential diagenesis on the $U_{37}^{k'}$ index would include long-term trends toward reduced concentrations of alkenones with greater time and burial depth. Most authors would expect that a diagenetic overprint, if there is one, would be at the expense of the $C_{37:3}$ ketone. If one models the degradation as first-order with concentration of the C_{37} ketones, then one can imagine that slightly different rate constants could be involved. The available evidence suggests that, once the alkenones have survived the relatively high metabolic activity of the upper few centimeters to tens of centimeters of the sediment column, these rate constants must be very small. That is because the time alkenones and other lipids spend below the surficial layer is one order of magnitude longer at 10^4 yr, and three orders of magnitude greater at 1 Myr than the time spent during early diagenesis. Over these long timescales, decay rate constants of perceptible size should produce nearly monotonic trends in alkenone loss as a function of sediment age, even if there were primary variations in the initial alkenone concentration of the sediment, because the sediment age would be many e-folding times of the inverse rate constant. Our laboratory has found that, to the contrary, alkenone concentrations are frequently higher in sediments hundreds of thousands to millions of years old than they are in the core-top material of the same sites (Figure 3). The implication we draw is that variations in initial near-surface alkenone concentration persist for very long durations in the deeper sediment column without diagenetic attenuation. Furthermore, alkenones appear stable when normalized to other products of photosynthesis. Figure 4 displays the alkenone content of sediments from the Oman Margin (ODP Site 723) normalized to total organic nitrogen and chlorins, an early diagenetic transformation product of chlorophylls, as a function of age. Considered individually, all three data sets show high-amplitude variations with time that are related to changes in their production and preservation in this dynamic upwelling zone. This variability disappears into near-monotonic

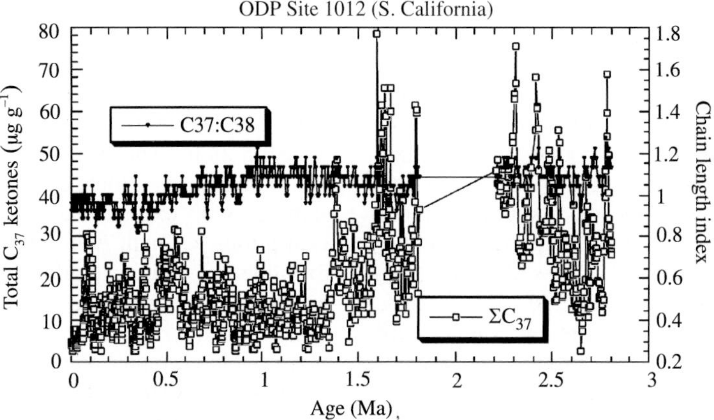

Figure 3 Alkenone concentrations (μg g^{-1} dry sediment weight) determined over a 2.8 Ma record from ODP Site 1012 off the southern California coast. Note the absence of a down-core decrease in alkenone concentrations that would suggest progressive diagenetic loss of alkenones in older sediments. Evolutionary conservatism in alkenone synthesis is suggested by the stability of the chain length index $(\sum C_{37}/(\sum C_{37} + \sum C_{38}))$ over the record.

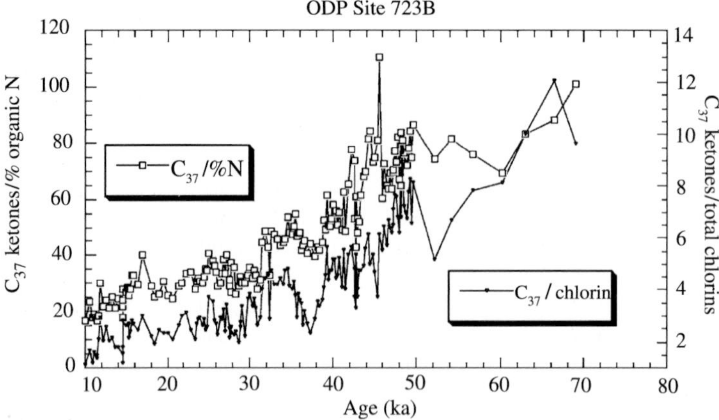

Figure 4 Down core trends in alkenone-normalized organic Nitrogen and total chlorins from ODP Site 723 (Oman Margin). The records extend from the late Holocene at the core top to ~70 ka. Note the monotonic decrease of the two organic classes relative to alkenones. This behavior is consistent with the progressive degradation of these more labile components and the diagenetic stability of the C_{37} alkenones.

trends of increasing normalized alkenone concentration with greater age (Figure 4), which are best explained as preferential degradation of organic nitrogen and chlorophyll relative to alkenones.

Sediments with elevated pore-water temperatures provide at least one exception to this general rule of the diagenetic stability of alkenones. Simoneit *et al.* (1994) measured alkenone abundances in sediments of the Middle Valley of the Juan de Fuca spreading center. Hydrothermal alteration of the sediments resulted in the loss of alkenones at temperatures greater than 200 °C. The Simoneit *et al.* (1994) study thus suggests caution in using the alkenone unsaturation index in other regions of unusually high geothermal gradients, or in very deeply buried sediments.

6.15.6 CALIBRATION OF $U_{37}^{k'}$ INDEX TO TEMPERATURE

The calibration of the alkenone unsaturation index actually resolves into two questions for paleoceanographers to address. The first concerns defining an equation, or sets of equations if one should not prove globally applicable, which relate the unsaturation index to the water temperature in which the producing organisms grew. The second involves understanding how the unsaturation index recorded in sediments, which represents an enormous integration of growth–temperature histories of individual organisms, and relates to a consistent measure of sea-surface temperature (e.g., SST versus subsurface depth, and annual versus seasonal temperature). These are not one and the same question, because *E. huxleyi* and related species may not always live in the mixed layer, and because their production may vary seasonally. For this reason, I evaluate evidence in the following section for how the influences of water-column habitat, seasonality, and particle transport through the ocean may affect the interpretation of the sedimentary $U_{37}^{k'}$ index.

One can approach the question of the quantitative relation of the $U_{37}^{k'}$ index to growth temperature at considerably different scales of time, genetic, and environmental variability. Culture studies allow the experimentalist to eliminate many confounding variables in the natural environment (genetically mixed populations, varying nutrient availability, different depth habitats, etc.) to isolate the influence of factors such as growth temperature, physiological state (exponential versus late-log, versus stationary growth), and nutrient availability on alkenone and alkenoate systematics. Despite the elegance of the experimental approach, results from such studies must be viewed as models for what may exist under natural conditions, not necessarily as calibrations. By collecting particulate material in the water column, either by filtering material in the euphotic zone, or collecting falling particles in sediment traps and relating their alkenone composition to time series of near-ocean temperatures, we move closer to relating the growth environment of the alkenone-producing algae to the signals sent to the sediment. This comes at the cost of unknown genetic variability in the natural populations, and some ambiguity in the actual time and depth of alkenone synthesis and, hence, the appropriate growth temperature versus alkenone/alkenoate relations. Near-surface sediments now provide a global database to examine the paleo-environmental information contained in the preserved record of alkenones and alkenoates. The time averaging inherent in sedimentation means that one is integrating temporal, physiological, and genetic influences on scales not approached in the laboratory or in field studies. However, correlations from the sediments to environmental variables in the water column rely on statistical relations, since one cannot directly link the sediment $U_{37}^{k'}$ index to particular ecological controls on the unsaturation–temperature relation.

6.15.6.1 Culture Calibrations

Prahl and co-workers (Prahl and Wakeham, 1987; Prahl *et al.*, 1988) conducted seminal studies that still provide the widely accepted calibration of the $U_{37}^{k'}$ index to growth temperature. These papers report the $U_{37}^{k'}$ values of one strain of *E. huxleyi* grown at different temperatures. Prahl and his collaborators argued that the good comparison of their laboratory results to the results of $U_{37}^{k'}$ analyses of water-column particulates collected at known temperatures in the eastern Pacific suggested that they had arrived at a valid calibration between 5 °C and 25 °C. It may well be that linking laboratory to field data helped ensure the robustness of their calibration. The relationship derived by Prahl and Wakeham (1987 and amended slightly by Prahl *et al.* (1988)) followed a linear relationship with temperature. Extrapolated to the limits of the $U_{37}^{k'}$ index of 0 and 1, it suggests a lower temperature limit of ~1 °C, and an upper limit of ~28 °C. Both the very coldest and warmest surface temperatures of the ocean would therefore lie outside the range of the $U_{37}^{k'}$ index.

Numerous culture studies of alkenone-producing species and strains have followed Prahl's initial studies, under the theory that cultures uniquely allow the experimentalist to isolate factors that may influence alkenone and alkenoate distributions. Experience shows that differences can exist between results obtained from using batch or continuous culture methods on the same strain of alkenone-producing algae (Popp *et al.*, 1998), from the phase of growth from which alkenones are harvested (Conte *et al.*, 1998a; Epstein *et al.*, 1998, unpublished; Yamamoto *et al.*, 2000), and from different laboratories culturing the same strain (see results of culturing *E. huxleyi* strain VAN556 by Conte *et al.* (1998), compared to Prahl *et al.* (1988), or to data presented by Sawada *et al.* (1996) showing differences between two laboratories). Furthermore, replicate cultures grown in the same laboratory under ostensibly similar conditions can yield a spread of $U_{37}^{k'}$ values (Conte *et al.*, 1995; Versteegh *et al.*, 2001). In batch culture, a nutrient medium is provided to a strain inoculate. After a period of rapid (exponential) growth, the cell density approaches a limit, and may even decline. The investigator can harvest cells at various times during the sequence to determine alkenone and alkenoate concentrations. Continuous cultures maintain the organisms in the exponential growth phase by supplying nutrients. Growth of alkenone-producing haptophyte algae in chemostats (Popp *et al.*, 1998) represents a particularly sophisticated manipulation, as these cultures grow in a medium of constant (low) nutrient availability. It is not clear whether batch or continuous growth models better represent natural conditions, or whether the sinking flux of alkenones and alkenoates in the ocean comes from populations in exponential, late logarithmic, or stationary growth state.

All culture studies confirm the first-order dependence of $U_{37}^{k'}$ and other alkenone (Conte *et al.*, 1998a) unsaturation parameters on growth temperature, but produce results that conflict in many ways. The study of Prahl *et al.* (1988) demonstrated that haptophytes grown in the laboratory adjust their unsaturation to temperature changes on a timescale of days; culture work by Conte *et al.* (1998a) suggested rapid adjustment of alkenoate/alkenone ratios to changes in growth temperature as well. However, culture calibration studies suggest very large variations in the relation of unsaturation to growth temperature that may depend on genetic and physiological factors (Conte *et al.*, 1995, 1998a; Epstein *et al.*, 1998). Twenty-four strains of alkenone-producing species cultured by Conte *et al.* (1995) at 15 °C gave $U_{37}^{k'}$ values that ranged from 0.3 to 0.55. Only one of these approached the value of ~0.56 appropriate for the Prahl *et al.* (1988) temperature calibration and, indeed, Conte *et al.* (1995) obtained a $U_{37}^{k'}$ value of ~0.4 at 15 °C for VAN55, the strain used by Prahl and Wakeham (1987) and Prahl *et al.* (1988). Volkman *et al.* (1995) suggested that cultures of *G. oceanica* produce a significantly different relation of unsaturation to growth temperature; however, their experimental results do not agree with *G. oceanica* cultures grown by Sawada *et al.* (1996) or Conte *et al.* (1998a).

It also seems clear that factors such as growth phase, light, and nutrient levels can significantly influence the unsaturation index of haptophytes grown in the laboratory (see summary in tables 4 and 5 of Versteegh *et al.*, 2001). Both Conte *et al.* (1998a) and Epstein *et al.* (1998) found changes in the unsaturation index between log, late-log, and stationary phases of growth. Epstein *et al.* (1998) proposed that nutrient availability, which would control growth rates of cultured and natural populations, could significantly affect the calibration of unsaturation to growth temperature. Both investigations found increasing alkenone concentrations (pg cell^{-1}) in late logarithmic and stationary phase growth. Conte *et al.* (1998a) also documented very large ranges in the ratios of alkenoates to C_{37} and C_{38} ketones (0–2.8), and in the $\sum C_{37}:\sum C_{38}$ ketone ratio depending on growth phase. Comparison of these parameters to field data led Conte *et al.* to conclude that natural populations most closely resemble late-log of stationary populations grown in the laboratory. In contrast to the batch culture experiments discussed above, Popp *et al.* (1998) used chemostats to control steady-state growth rates, which they argue may be a better model for natural systems. The latter study found no significant dependence of $U_{37}^{k'}$ on growth rate at constant temperature.

A number of studies have investigated whether genetic and/or physiological differences create "fingerprints" in other aspects of alkenone/alkenoate systematics that might allow investigators to distinguish past variations in species/strain production in sediments. Volkman *et al.* (1995) and Sawada *et al.* (1996) suggested that the proportions of C_{37} to C_{38} ketones ("chain length index"), or alkenoate/alkenone ratios, might relate to the proportions of *E. huxleyi* to *G. oceanica* at the time of production (see also Yamamoto *et al.*, 2000). Further culture work by Conte *et al.* (1998a) does not support either suggestion (see also Section 6.15.4).

Nearly all culture calibration studies predict higher growth temperatures for the same unsaturation index than postulated by the Prahl *et al.* (1988) regression, although several (Conte *et al.* (1998a) and *G. oceanica* cultures of Sawada *et al.* (1996)) fall very close to the Prahl relation. If this ensemble of culture data is correct, field and sediment studies applying the Prahl *et al.* (1988) calibration might frequently overestimate temperatures. As we evaluate water-column, sediment-trap, and core-top data, we should assess whether the large range of possible $U_{37}^{k'}$–temperature relations suggested by culture studies, whether of genetic or physiological origin, demonstrably affect the accuracy of a unified calibration relation. One would expect to find that different haptophyte biogeographic regions produce distinct calibrations of unsaturation to temperature, and to see the influence of nutrient availability in offsets between upwelling and nonupwelling regions.

6.15.6.2 Particulates

By studying alkenone parameters in particulate matter collected in the photic zone, we lose the ability to manipulate potential genetic or physiological influences, but we gain the ability to compare alkenone systematics to temperatures in the natural setting. Calibration equations can be generated and tested by comparing the alkenone unsaturation index in suspended particles with ambient water temperatures. One generally assumes that the measured water temperature is the same as the temperature in which the alkenones and alkenoates were synthesized. This may not always be correct for particles sinking or mixing through a temperature-stratified water column. In fact, given a general tendency for particles to sink, the temperature of alkenone synthesis might be higher than the temperature in which the particles are collected, but almost certainly not be lower (Sicre *et al.*, 2002). A potential temporal offset also exists between the time of alkenone synthesis and the measurement. Thus, the alkenone temperatures could be set to the temperature of previous "bloom" conditions, rather than the currently measured water-column temperature.

In contrast to culture studies, relationships between $U^{k'}_{37}$ and temperature in suspended particulate organic carbon show much more agreement with the Prahl *et al.* (1988) calibration equation. Several large-scale compilations of water-column unsaturation ratios have been presented (Brassell, 1993; Sikes *et al.*, 1997; Conte *et al.*, in press), as well as regional studies in the North Atlantic and Mediterranean (Conte *et al.*, 1992; Conte and Eglinton, 1993; Sikes and Volkman, 1993; Ternois *et al.*, 1997; Sicre *et al.*, 2002). Data have been variously interpreted as requiring regional calibrations of growth temperature (Conte and Eglinton, 1993; Ternois *et al.*, 1997), or as requiring modifications of the original Prahl *et al.* (1988) culture-based $U^{k'}_{37}$ equation to a more appropriate relation based on water-column particulates equation (Brassell, 1993; Sikes and Volkman, 1993; Conte *et al.*, in press). The arguments favoring regional calibrations are based on regression estimates of a small number of samples, and are not yet compelling. In the case of the Black Sea, however, Freeman and Wakeham (1992) determined a water-column $U^{k'}_{37}$ that would underestimate SST by more than 5 °C. It now appears that the Black Sea represents a special case of mixing of brackish water haptophyte alkenone-producing algae with open-ocean varieties. In other cases, a regional particulate $U^{k'}_{37}$ calibration does not accurately predict surficial sediment values in the western Mediterranean Sea (Ternois *et al.*, 1998; Cacho *et al.*, 1999). This discrepancy points out the dangers of using calibrations based on limited temperature ranges and short space- and timescales to derive accurate $U^{k'}_{37}$ paleotemperature equations.

The most extensive synthesis of water-column $U^{k'}_{37}$–temperature relations argues for the global applicability of a single calibration equation (Conte *et al.*, in press; Figure 5). This study examines 392 samples from all major ocean basins, gathered in the mixed layer (0–30 m) to prevent including samples acquired in the seasonal thermocline, which might be falling from the warmer layer above. Although the data set is weighted heavily to the North Atlantic, it includes provinces dominated by *G. oceanica* as well as *E. huxleyi* (see also the recent study by Bentaleb *et al.* (2002) for additional water-column data in the *G. oceanica* province of the western equatorial Pacific), upwelling zones and gyres. Regional data sets fall nicely along a global relation, contradicting earlier interpretations (e.g., Conte and Eglinton, 1993; Ternois *et al.*, 1998), which proposed that regional $U^{k'}_{37}$–temperature equations were needed for accurate temperature estimates. The similarity of the regional data sets is striking enough for Conte *et al.* (in press) to conclude that differences in the genetic makeup of natural alkenone-synthesizing populations, and

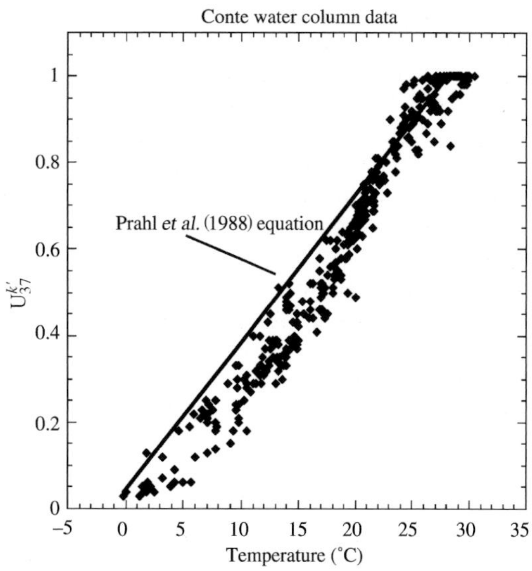

Figure 5 Compilation of the $U^{k'}_{37}$ index of mixed-layer particulates in relation to *in situ* temperature (Conte *et al.*, in press). The heavy solid line indicates the linear Prahl *et al.* (1988) paleotemperature relation used for sediment estimates.

differences in their growth environment (differing nutrient fluxes and/or water-column stability) do not significantly detract from the use of a global calibration equation for paleotemperature estimation.

In my judgment, however, Conte *et al.* (in press) overstep their data by arguing that the correct form of a global equation must be nonlinear, and that water-column calibrations conflict with sediment-based $U_{37}^{k'}$–temperature regressions. Conte *et al.* propose a third-order polynomial equation to describe the flattening of the $U_{37}^{k'}$–temperature relation near the cold and warm ends of the data set. As they note, the data suggest more variability of the index in relation to temperature at the warm and cold extremes of the surface ocean. The scatter may reflect some combination of the importance of nonthermal factors on alkenone synthesis near the extremes of temperature encountered by haptophyte algae (Conte *et al.*, in press), but it may also include analytical errors, as $C_{37:3}$ and $C_{37:2}$ alkenones approach their detection limits in warm and cold waters, respectively (cf. Grimalt *et al.*, 2001; Pelejero and Calvo, in press). In any event, the third-order polynomial fit of water-column $U_{37}^{k'}$ to *in situ* temperatures improves the r^2 value to 0.97 as compared to the r^2 value of 0.96 for a linear fit, and reduces the standard error of estimate from 1.4 °C to 1.2 °C. The slope (0.038) and intercept (−0.104) of a linear fit of water-column particulate $U_{37}^{k'}$ to *in situ* temperature may not be statistically different from the canonical Prahl *et al.* (1988) equation or its nearly identical core-top version (Muller *et al.*, 1998). If the Conte *et al.* (in press) calibration is accurate, then the authors find offsets between the temperatures derived from $U_{37}^{k'}$ analysis of core-top sediments and the mean annual temperature of the surface waters overlying the core sites. $U_{37}^{k'}$ sediment values at mid-latitudes would be systematically high compared to mean annual growth temperatures. Whether the sediment bias of 2–3 °C at mid-latitudes inferred by Conte *et al.* (in press) exists depends critically on the choice of the third-order polynomial description of the $U_{37}^{k'}$–temperature relationship, and may be premature. As discussed below, sediment-up calibrations do not favor the polynomial formulation preferred by Conte and colleagues.

6.15.6.3 Sediment Traps

Sediment trap material provides a valuable view of the $U_{37}^{k'}$ and quantity of alkenones transiting to the seafloor. Few time-series experiments have been reported to date, although data from the Gulf of California (Goni *et al.*, 2001) and off the coast of Angola (Muller and Fischer, 2001) offer reasonable resolution for 1.5 yr and 4 yr periods, respectively. Goni *et al.* (2001) compared the $U_{37}^{k'}$ of monthly sediment collections in the Gulf of California to sea surface temperatures taken from satellite (AVHRR). Their results (Figure 6) show that the $U_{37}^{k'}$ index closely tracks changes in the satellite-derived SST, with little or no time offset. For most of the year, the temperatures estimated from the Prahl *et al.* (1988) calibration agree with SST. Somewhat lower than predicted $U_{37}^{k'}$ values were obtained during the warmest summer months (inferred SST >28 °C). These results could be rationalized by some combination of subsurface alkenone production during the summer thermal stratification, errors in the satellite SST measurements, and nonlinearity in the $U_{37}^{k'}$ index at the warm extreme of growth conditions. Muller and Fischer's time-series data from the Cap Blanc upwelling center off the coast of southwest Africa also show a strong seasonal cycle in $U_{37}^{k'}$ that is consistent with changes in SST. After removing a small temporal offset due to the sinking time of particles from the surface ocean to their sediment traps, the authors conclude that both the amplitude and absolute values of the $U_{37}^{k'}$ temperature estimates are in good agreement with weekly sea surface temperature estimates.

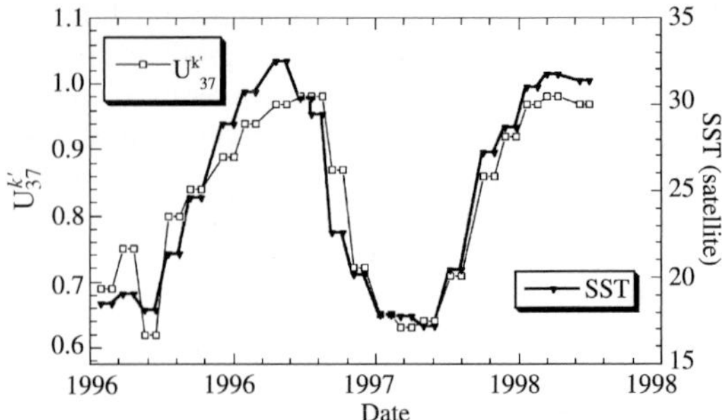

Figure 6 Comparison of Gulf of California sediment trap $U_{37}^{k'}$ time-series versus SST estimated from satellite measurements (Goni *et al.*, 2001). Note the rapid response of the $U_{37}^{k'}$ index to changes in SST. The $U_{37}^{k'}$ index may not record the warmest temperatures accurately; see text for discussion.

6.15.6.4 Core Tops

Surficial sediments should reflect the weighting function of alkenone production at all seasons and depths throughout the annual cycle—the integrated production temperature (IPT) concept of Conte *et al.* (1992). Core-top material also provides the benefit of temporal and spatial averaging of other factors, such as genetic variability and variations in growth rate that may influence alkenone systematics. This comes at costs: the time averaging varies with sedimentation rate and bioturbation, and much information that relates to the original production of alkenones in the surface ocean is lost. Furthermore, a sediment-based regression compares surficial material, representing centuries to millennia of ocean history to the short period of instrumental temperature data used by ocean atlases such as the Levitus (1994) global climatology.

Large data sets (Herbert *et al.*, 1998; Rosell-Mele *et al.*, 1995b; Sonzogni *et al.*, 1997) of core-top $U_{37}^{k'}$ show strong convergence with the original Prahl *et al.* (1988) temperature calibration, using mean annual surface temperature (MAST) (0–10 m) as the reference (Figure 7). A recent compilation by Muller *et al.* (1998) synthesized results of over 300 core-top analyses from the different ocean basins, determined by various laboratories. The Muller *et al.* (1998) data set encompasses the entire range of temperatures and biogeographic provinces of alkenone producers, but is biased toward continental margin sediments. Muller *et al.* noted that $U_{37}^{k'}$ in core tops also correlated highly to seasonal temperatures in the upper water column, as these covary with mean annual temperature. The correlation decreased significantly, however, if alkenone unsaturation was regressed against temperatures at 20 m and below. Sediments thus provide strong empirical evidence that alkenones synthesis occurs in the mixed layer in most areas of the ocean. Updated recently to 490 samples by P. Muller (P. Muller, personal communication), the core-top calibration of $U_{37}^{k'}$ to mean annual sea surface temperature does not differ statistically from the original Prahl *et al.* (1988) culture and water-column line. A single global, linear regression of $U_{37}^{k'}$ to MAST produces a standard error of estimate of 1.4 °C, of the same size as the Conte *et al.* (submitted) linear fit of particulate matter $U_{37}^{k'}$ to water-column measurements. A careful statistical comparison now needs to be performed to determine whether core-top and water-column regressions to temperature differ beyond the uncertainties of regression parameters.

The reproducibility of the $U_{37}^{k'}$ values in surface sediments from the same region can be quite extraordinary. Figure 8 displays two regions where our laboratory was able to obtain recent sediments from a number of box cores off the coast of California. The temperature estimates derived from $U_{37}^{k'}$ analysis not only agree closely with MAST using the Prahl/Muller equation, but they agree with each other to very nearly the analytical error.

Little support for the large range in alkenone parameters (ratios of C_{37} to C_{38} ketones, alkenones to alkenoates, etc.) observed in culture experiments comes from the sediment realm. As one example, I compared (Herbert, 2000) the frequency distribution of the ratio of total C_{37}

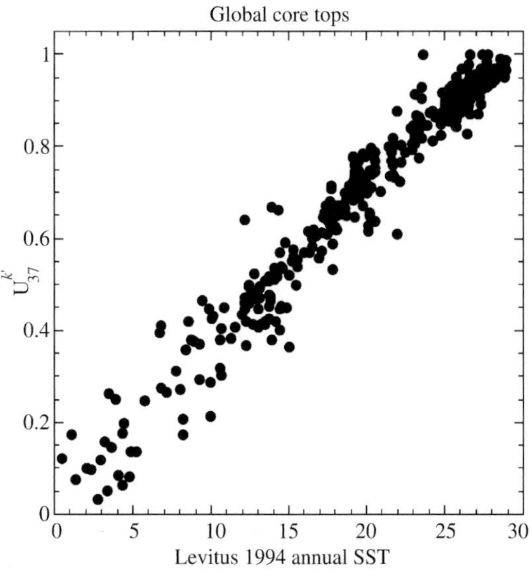

Figure 7 Global compilation of near-surface sediment $U_{37}^{k'}$ values (Muller *et al.*, 1998; Muller *et al.*, unpublished; Herbert, unpublished) plotted versus mean annual SST from the closest grid point of the Levitus (1994) World Ocean Atlas.

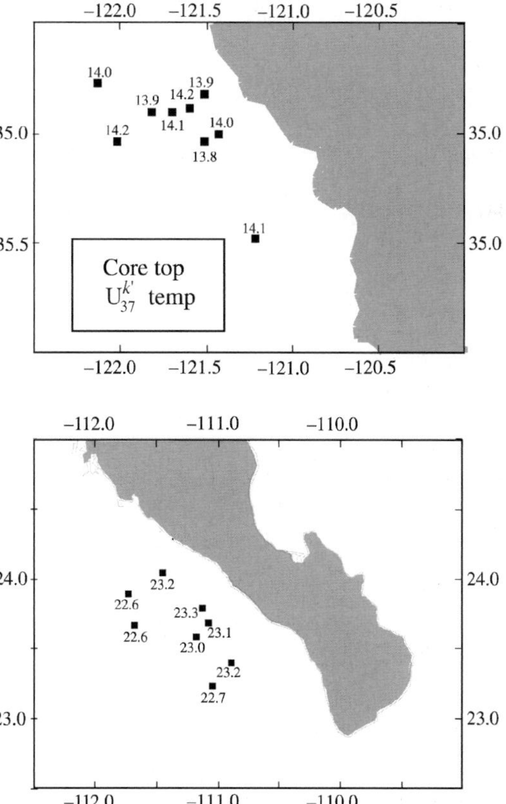

Figure 8 $U_{37}^{k'}$ temperature estimates taken from the top cm of box cores off the California Coast (Herbert *et al.*, 1998). Note that the temperature estimates are consistent to nearly the analytical error of the gas chromatographic technique.

ketones ($\sum C_{37}$) to total C_{38} ketones ($\sum C_{38}$) in a core-top data set generated by our laboratory. The data set is weighted to samples along the California margin, but also includes samples from the North Atlantic, equatorial Pacific, central Pacific gyre, Peru margin, and western Pacific. The tight cluster of core-top values around a mean of just over 1.0 contrasts with the extreme heterogeneity of culture results, but is in very good agreement with the average ratio of 1.04 from a large upper water-column data set off Bermuda (Conte *et al.*, 2001). The values we found are similar to the mean of ~1.2 determined by Rosell-Mele *et al.* (1994) in North Atlantic core top and by Sonzogni *et al.* (1997, 1998) in Indian Ocean sediments, and the average of ~1.15 reported by Sawada *et al.* (1996) from the Sea of Japan. Core-top data do not therefore encourage the idea that alkenone systematics can "fingerprint" regional variations in the fraction of production due to *E. huxleyi* and *G. oceanica* (cf. Volkman *et al.*, 1995; Sawada *et al.*, 1996).

The success of core-top temperature calibrations indicates that physiological state, genetic variability, and depth and seasonality of production play secondary roles to the control on the sedimentary $U_{37}^{k'}$ index exerted by mean annual near-surface temperature. In most cases, these factors produce errors at the level of 1.5 °C or less in the global core-top calibration. To this observer's opinion, core-top data cannot be reconciled with the large variations in the $U_{37}^{k'}$ index attributed to genetic or physiological factors by some culture studies. This does not indicate that the culture data are wrong in a technical sense, but that their results cannot always be extrapolated to the natural environment (Popp *et al.*, 1998).

Core-top data sets still leave significant room for improvement in several important regards. The important question of nonlinearity at the high and low ends of the $U_{37}^{k'}$–temperature relation remains unsettled. In particular, several studies suggest that the $U_{37}^{k'}$–temperature relationship flattens at temperatures above 26 °C or 27 °C (Sikes *et al.*, 1997; Sonzogni *et al.*, 1997; Bentaleb *et al.*, 2002). However, other studies of sediments underlying tropical waters find that the linear Prahl relationship seems to hold throughout the calibration range (Pelejero and Grimalt, 1997; Pelejero and Calvo, in press), and the global sediment compilation of Muller *et al.* (1998) provides no support for a nonlinear relationship. One would expect that calibrations on the extreme ends of the temperature range become difficult. Analytical difficulties grow as the di-unsaturated ketone becomes a minor peak at the low end of the index; similar difficulties pertain to the detection of the $C_{37:3}$ ketone in the face of chromatographic interferences in sediments under very warm ocean surface waters. In addition, it is also likely that seasonal production biases become large in high-latitude waters, emphasizing the need to combine modern-day ecological information with core-top data to better calibrate the $U_{37}^{k'}$ signal in cold waters. We also have little consensus on the temperature significance of the tetra-unsaturated C_{37} ketone. Originally included in the original $U_{37}^{k'}$ index by Brassell *et al.* (1986a), the $C_{37:4}$ ketone appears only in cold waters and in sediments underlying cold waters. Prahl *et al.* (1988) found that including the tetra-unsaturated ketone in a temperature equation did not improve the fit, and therefore omitted it from their $U_{37}^{k'}$ index. It remains to be explained why the $C_{37:4}$ ketone should be common in the high-latitude North Atlantic, where surface temperatures fall below 10 °C (Conte *et al.*, 1994a; Rosell-Mele *et al.*, 1994, 1995a; Rossell-Mele, 1998; Calvo *et al.*, 2002; Sicre *et al.*, 2002), but rare to absent in water-column particulates and sediments of the very cold waters of the Southern Ocean (Sikes and Volkman, 1993; Sikes *et al.*, 1997; Ternois *et al.*, 1998). Rosell-Mele and co-workers (Rosell-Mele, 1998; Rosell-Mele *et al.*, 2002) propose that the presence of large amounts of the $C_{37:4}$ ketone in lipid extracts may signal low salinity waters, at least in the North Atlantic.

6.15.7 SYNTHESIS OF CALIBRATION

The resemblance of core-top alkenone unsaturation data to both mean annual SST and the original Prahl *et al.* (1988) culture calibration is a quite remarkable result that is not completely understood. As the review above suggests, the calibration of a temperature proxy for paleoenvironmental analysis involves a host of steps, ranging from the physiology and genotype of the producing organisms, their ecology, and eventually the transport and degradation of particles in the water column and sediments. In the case of the alkenone thermometer, it is comforting to note that water-column and sediment calibrations come quite close. The consistency of the sediment regression to the original Prahl *et al.* (1988) linear relation is in some sense fortuitous. Other culture studies produce results that differ as much as 5 °C from the standard Prahl *et al.* (1988) calibration, and there is no inherent reason to prefer a linear calibration of unsaturation to growth temperature to a nonlinear one. Further, we know that alkenone producers do not operate at constant rates throughout the year in most regions of the ocean, and that they do not always live in the mixed layer. One should therefore keep in mind that the Prahl *et al.* (1988) and the identical Muller *et al.* (1998) relation of $U_{37}^{k'}$ with the mean annual SST are idealizations. We can note, however, that at least some of the caveats raised in the application of alkenone thermometry do not seem born out by sedimentary evidence. The hypothesis of Epstein *et al.* (1998) predicts that physiological factors would lead to a large offset in the $U^{k'}_{37}$ temperatures estimated from alkenones synthesized in upwelling and nonupwelling regions. Core-top data cover the oceans well enough to state that such an upwelling/nutrient bias must be no larger than the mean standard error of the entire regression (±1.4 °C), if it exists at all. Similarly, core-top data sets cover a large span in the proportions of strains of *E. huxleyi*, and in the contribution of *G. oceanica* to alkenone production, yet no statistical evidence emerges to treat the $U^{k'}_{37}$ of different oceanic biotic provinces differently.

The convergence of so much sedimentary data to a simple model must mean that an apparently fortuitous result (the good relation with mean annual SST) has an underlying predictability. One of the few efforts to quantify the consequences of seasonal variations in production and/or remineralization of alkenones is instructive. Conte *et al.* (in press) modeled the impact that such variations would have on the $U^{k'}_{37}$ in a time-averaged sediment. They found that seasonality produced only small offsets (<1 °C) from SST in the net (IPT) alkenone signal except at very high latitudes. Unless the season of the alkenone-producing bloom is restricted to the precise time of the coldest or warmest seasonal temperatures, it is not easy to cause the $U^{k'}_{37}$ delivered to sediments to depart much from the mean annual SST. This lesson was demonstrated by Sonzogni *et al.* (1997) and Muller *et al.* (1998), who used satellite chlorophyll estimates to make seasonally weighted flux estimates of alkenone production at core locations, assuming that alkenone fluxes correlate with bulk phytoplankton production. They found that the resulting flux-weighted temperature corrections to mean annual SST were negligible. It may also be that when subsurface production occurs, the temperature near the top of the seasonal thermocline is not colder by more than 1–2 °C from mean annual SST.

Alkenone production in oceanic gyres does appear to give evidence for a subsurface temperature bias. For example, Prahl *et al.* (1993), Doose *et al.* (1997), and Herbert *et al.* (1998) all report core-top $U^{k'}_{37}$ values lower than mean annual SST by 1–2 °C in gyre locations in the eastern North Pacific, consistent with subsurface fluorescence maxima (Prahl *et al.*, 1993), and maximal production during the late winter and early spring in the region. Ohkouchi *et al.* (1999) argued similarly for a subsurface gyre bias in a survey of core tops in the central North Pacific Ocean. One can therefore expect that a careful treatment of the core-top database may develop rules for how the $U^{k'}_{37}$ index will deviate systematically (~1 °C) according to distance from the nearest coastline, or some other simple proxy for gyre versus margin position. The influence of diagenesis on $U^{k'}_{37}$ values cannot be excluded, but its impact apparently ranges from negligible (most studies) to perhaps a warm bias ~1 °C (Hoefs *et al.*, 1998; Gong and Hollander, 1999).

Caution should probably be used in interpreting small down-core changes in $U^{k'}_{37}$ in high- and low-latitude regions. Here, one is in less reliable analytical territory (Grimalt *et al.*, 2001; Pelejero and Calvo, 2003), and the core-top and water-column $U^{k'}_{37}$ data can be modeled by either linear or polynomial fits (Sikes *et al.*, 1997; Sonzogni *et al.*, 1997). Even if evidence eventually conclusively supports nonlinear fits at the cold and/or warm extremes of the $U^{k'}_{37}$ range, interpreting small changes in these regions quantitatively is probably a losing game. Flattening of the $U^{k'}_{37}$ relationship with temperature would mean that the index loses sensitivity to temperature in very high and low latitudes. At the same time, the analytical error grows. One therefore needs to be wary of generating spuriously large temperature changes from $U^{k'}_{37}$ deviations of dubious reliability. As a practical limit, I suggest not interpreting $U^{k'}_{37}$ deviations quantitatively for paleotemperatures at values lower than ~0.20 or higher than ~0.96 (5 °C or 27 °C, respectively, according to the Prahl/Muller equations).

6.15.8 PALEOTEMPERATURE STUDIES USING THE ALKENONE METHOD

The rapidity and high precision of alkenone analysis makes the technique ideally suited to produce time series of past near-surface ocean temperatures. High signal-to-noise ratio can be demonstrated in a number of ways. The ocean-drilling program acquires offset holes at drilling sites to assure the continuity of the recovered sedimentary record. High-resolution analyses of offset holes that cover the same stratigraphic interval show that even small-scale changes in the $U^{k'}_{37}$ index are reproducible (Zhao *et al.*, 1993). Alkenone temperature estimates that cover the Late Holocene paint a picture of subdued SST changes, in accord with polar ice cores evidence that describes this period as quite stable in comparison to other recent intervals of Earth history. Alkenone-derived temperatures over the past 10 ka rarely deviate from modern values by more than 1–2 °C, even in very densely sampled records (e.g., Schulte *et al.*, 1999; Zhao *et al.*, 2000). In contrast to foraminiferal faunal and $\delta^{18}O$ estimates, the alkenone technique apparently has the reliability necessary to define Late Holocene cooling trends of only −0.27 °C kyr^{-1} to

$-0.15\,°C\,kyr^{-1}$ in a coherent manner in the northeast Atlantic and Mediterranean sea (Marchal *et al.*, 2002).

In the following sections, I divide alkenone paleotemperature studies roughly by the time span covered. While arbitrary to some degree, this should allow the reader to gauge the contributions of the alkenone technique to important paleoclimatic questions, which tend to be arrayed by the age and duration of Earth history studied. In addition, studies at different time resolutions will have different sets of supporting information, caveats, and questions to be addressed by future work. Nearly all of the studies cited rely on the standard Prahl *et al.* (1988) temperature scale. However, a few applications of nonstandard calibrations exist in the literature (Pelejero *et al.*, 1999a; Wang *et al.*, 1999; Calvo *et al.*, 2002).

6.15.8.1 Holocene High-resolution Studies

Because instrumental records rarely date back more than a century, the alkenone technique may play an important role in characterizing past ocean surface temperature variability on timescales of a few years to centuries. Alkenone analyses may thus complement information derived from geochemical analyses of corals, whose usefulness in studying past El Nino cycles and other phenomena is now well established. Unlike corals, alkenone-producing organisms range over nearly the entire ocean. Only certain locations will be favorable, however, for preserving alkenone signals for high-resolution analysis. These generally occur along continental margins, where sediment flux is high. Areas along various coastlines where anoxic, or dysaerobic bottom water contacts sediments have the additional potential to resolve yearly or even seasonal variations, through the deposition of laminated sediments. Such highly productive, highly preserving settings have elevated alkenone concentrations in the sediments. Because so little material (perhaps only 100 mg) is required for a reliable $U^{k'}_{37}$ determination, very high resolution sampling becomes feasible.

Two regions of the eastern Pacific have been tested by the alkenone method for details of past El Nino (ENSO) variability. A set of early papers by Farrington *et al.* (1988) and McCaffrey *et al.* (1990) targeted the Peru upwelling zone, which experiences large surface warmings during El Nino conditions. The authors used ^{210}Pb-dated box cores to study the alkenone record of the past few centuries at three locations along the Peru margin. Samples were taken at estimated 5–10 yr intervals, with one long record extending back to ~1680 AD (McCaffrey *et al.*, 1990). $U^{k'}_{37}$ temperature variations correlated in part with historical ENSO indices, but without one-to-one matches. McCaffrey *et al.* (1990) attributed the mismatches to a combination of dating uncertainties, the difficult in precisely resolving layers for sampling, and redistribution of the alkenone signal over a broader stratigraphic interval, so that ENSO anomalies are smoothed. They also noted that alkenone records may be biased away from ENSO warm events by the decline in phytoplankton production that accompanies El Nino anomalies.

Sediments from the Santa Barbara Basin off Southern California display layers that can be sampled at annual resolution. A pioneering study by Kennedy and Brassell (1992) produced annual-resolution data from the core top to an estimated basal age of 1915 AD, based on a varve chronology. The authors showed that SSTs varied by 1–2 °C on an annual basis in this time interval. Intervals of major historical El Nino-related warmings along the California Coast stood out in most cases as warm $U^{k'}_{37}$ anomalies. Our laboratory produced a slightly longer record that confirms most of the details of the Kennedy and Brassell (1992) record, although the absolute temperature estimates are offset due to interlaboratory differences. We found that the alkenone method detected 80–90% of the known El Nino warmings over the last century. We also found that most warm intervals had low alkenone abundances, which would be consistent with the historical association of decreased upwelling along the California margin during major El Nino events. Zhao *et al.* (2000) extended the Santa Barbara record to 1440 AD with approximately biannual sample resolution. They found that SST oscillated around its modern mean value of 15.5 °C with an amplitude of less than 3 °C. Their sampling recognized 5 of 12 very strong ENSO events over the period 1840–1920 for which ENSO anomalies in the equatorial Pacific have been defined. A significant trend to low SST around the turn of the twentieth century coincides with a known period of low SST along the California margin. Other approximately centennial-scale oscillations are evident in the time series as well (Zhao *et al.*, 2000). Significantly, the ~500 yr record shows no linear temperature trend over time, and no dip in SST during the Little Ice Age period of the late 1500s to early 1600s. If Zhao *et al.* (2000) alkenone record accurately reflects regional SST, then the Little Ice Age cooling of the North Atlantic may not have been expressed along the west coast of North America.

It is still too early to assess how well the alkenone method will resolve variations on the ENSO timescale. Issues of chronology are significant, since a varve counting error of only a few years will produce miscorrelations between alkenone records and other historical indices of the ENSO phenomenon. It is also not clear how reliable individual core records will be at recording SST at the annual scale. Patchiness of SST anomalies and surface productivity may impose noise on already small (1–3 °C) anomalies in mean annual SST. For ENSO-related alkenone work to progress, we will need more records gathered from the same basin, sampled with the best possible chronology. These will give us a more reliable sense of the success rate of the alkenone paleothermometer in resolving very short-lived SST variations in the late Holocene.

On somewhat longer timescales, several studies have used the alkenone index to resolve variations of SST within the Holocene (last 12 kyr). The Holocene represents an interesting mix of some of the factors that on longer timescales help to drive ice age cycles. Northern hemisphere summer insolation peaked at ~9 ka, and has declined continuously since that time as the Earth's orbital configuration shifted. The Holocene has not, however, seen any significant change in ice volume or atmospheric CO_2. A coherent long-term cooling of between 1 °C and 2 °C over the past 9–10 ka seems to have occurred in the Atlantic Ocean and the Mediterranean Sea, according to $U^{k'}_{37}$ time series (Zhao *et al.*, 1995; Bard *et al.*, 2000; Cacho *et al.*, 2001; Calvo *et al.*, 2002; Marchal *et al.*, 2002). The cooling appears amplified in some regions, such as the Mediterranean (Cacho *et al.*, 2002). In detail, the timing of peak Holocene temperatures progresses within the Mediterranean (Cacho *et al.*, 2002). Marchal *et al.* (2002) point out that the cooling trend exhibited by $U^{k'}_{37}$ in the North Atlantic is consistent with evidence for late Holocene glacial readvance in Iceland, with borehole temperature reconstructions from Greenland ice cores, and with pollen data in Europe and North America that indicate a southward migration of the cool spruce forest.

The cooling trend detected in the North Atlantic appears to be a regional, rather than global pattern. Alkenone data from the Indian Ocean (Bard *et al.*, 1997; Cayre and Bard, 1999), the South China Sea (Wang *et al.*, 1999; Kienast *et al.*, 2001; Steinke *et al.*, 2001), and the western tropical Atlantic (Ruhlemann *et al.*, 1999) in fact show very slight sea surface warming from the early Holocene toward the present. High-resolution sampling along the western margin of North America (Prahl *et al.*, 1995; Kienast and McKay, 2001; Herbert, unpublished data) shows no trend at all during the last 9 ka, although the alkenone data do suggest millennial oscillations of perhaps 1 °C amplitude. Variability within the Holocene also emerges from the Doose-Rolinski *et al.* (2001) high-resolution study of SST over the last 5 ka in the Arabian Sea. The authors sampled a high-deposition rate core at ~20 yr intervals, and measured the $\delta^{18}O$ of planktonic foraminifera in addition to alkenone unsaturation. Total variance within the 5 kyr period is ~0.6 °C (1σ), but temperature extremes of up to 3 °C are recorded (it should be noted that the authors used a nonstandard calibration whose slope would increase the estimated temperature changes by nearly 50% compared to the canonical (Prahl *et al.*, 1988 equation). Stronger than average northeast monsoon winds were inferred to result in cooler than average temperatures, with the reverse occurring when stronger south west monsoons dominated the system. Variance in SST apparently increased in the last 1,500 years, suggesting high variability of monsoonal winds during the latest Holocene in comparison to the mid-Holocene.

6.15.8.2 Millennial-scale Events of the Late Pleistocene and Last Glacial Termination

The discovery of very rapid climatic anomalies in the high-latitude North Atlantic region sparked the search for similar events in other regions. Among the prominent anomalies are the Younger Dryas event, Dansgaard–Oeschger (D/O) cycles, and Heinrich events. Pollen successions in northern Europe identified a brief interval during the last deglaciation when climate returned to very cold conditions. Spectacularly revealed in the isotopic record of Greenland ice cores (Dansgaard *et al.*, 1989; Johnsen *et al.*, 1992; Grootes *et al.*, 1993), the Younger Dryas interval lasted from ~11 ka to ~13 ka. Greenland ice cores also demonstrate a highly unstable climate during the last glacial interval (marine oxygen isotope stages 2 and 3). Isotopic changes equivalent to 6–8 °C changes in air temperature occur as rapid bursts known as the D/O cycles. These appear to be grouped in units of 3–4 cycles, which terminate in a longer period of unusually cold temperatures in Greenland. The D/O cycles can be recognized one-to-one in marine cores from the North Atlantic (Bond *et al.*, 1993). There, regionally coherent pulses of ice-rafted debris that originates from Canada, Iceland, and the Norwegian Sea, are termed Heinrich events. The most recent of these occurred during late glacial time at ~18 ka. Most authors correlate the Heinrich events with the terminations of bursts of D/O cycles (Bond and Lotti, 1995).

Regional amplitudes of the millennial-scale events, and their timing relative to the Greenland ice sheet template, constitute important pieces of the puzzle. Many theories propose that the millennial events originate from instabilities in the North Atlantic thermohaline circulation (Broecker, 1994) and propagate through the deep ocean "conveyor belt." It is also possible that the millennial-scale temperature changes originate outside the North Atlantic, and are merely well expressed there. Millennial anomalies could also spread by means other than the thermohaline circulation. It seems likely that good regional coverage of SST anomalies may in the end provide the "fingerprints" necessary to decide which models of millennial-scale variability during glacial times are the most plausible.

Alkenone SST reconstructions in the North Atlantic and Mediterranean show the Younger Dryas, D/O cycles, and Heinrich events very clearly (Eglinton *et al.*, 1992; Zhao *et al.*, 1993; Rosell-Mele *et al.*, 1997; Cacho *et al.*, 1999, 2002; Bard *et al.*, 2000; Calvo *et al.*, 2001). Millennial-scale temperature changes apparently are exported from the North Atlantic by the Canary Current (Zhao *et al.*, 1995; Bard *et al.*, 2000) or through the atmosphere (Cacho *et al.*, 2002). Recognizing that the Greenland anomalies propagate into the subtropical North Atlantic helps to explain initially puzzling features of high-resolution alkenone data taken during the glacial interval. Zhao *et al.* (1995) found that the time of maximum global ice volume (last glacial maximum (LGM), 21–23 ka) did not have the coldest SST of the glacial period off Northwest Africa. It has since become clearer that the interval of Heinrich event H2 (18 ka) produced the coldest temperatures of the late glacial period in the North Atlantic, and that this cooling is detected in numerous cores off northwest Africa (Zhao *et al.*, 1995). While the LGM period saw cold SST in the subtropical North Atlantic relative to the Holocene, it was sandwiched between even colder periods paced by millennial climate instability.

The alkenone records from the Mediterranean Sea by Cacho *et al.* (1999, 2002) show some of the most spectacular evidence for how millennial events pervade the regional climate of the North Atlantic and Western Europe (Figure 9). All of the millennial features of the Greenland ice cores for the last 50 ka are immediately recognizable as large SST changes, some as rapid as 6 °C per century (Cacho *et al.*, 2002). Intervals calibrated as colder than 12 °C by the $U_{37}^{k'}$ method also have pulses of the polar foraminifera *G. pachyderma* (left-coiling variety). The coldest events correspond in each case to the time of Heinrich periods of intense ice rafting in the high-latitude North Atlantic. Because the times of Heinrich events do not stand out as the coldest millennial periods in the Greenland ice core record, the SST data imply that these events may have a different spatial pattern and mode of propagation compared to the D/O anomalies. As found in cores off northwest Africa, the LGM in the Mediterranean did not have the coldest temperatures recorded during the last 50 ka. These occurred at ~16–18 ka, ~24 ka, ~30 ka, ~39 ka, and ~46 ka (Cacho *et al.*, 2002).

Details of millennial variability away from the North Atlantic are still emerging. A Younger Dryas cooling occurred along the northwest margin of North America (Kienast and McKay, 2001; Seki *et al.*, 2002). Its timing is identical to that of the Younger Dryas in the circum-North Atlantic region within the error of AMS ^{14}C dating. Off the coast of British Columbia, the Younger Dryas produced a cooling of ~4 °C from the Allerod–Bolling warm interval, and ended with a warming of almost 6 °C to peak Holocene temperatures at 10–11 ka (Kienast and McKay, 2001). The Younger Dryas event shows up along the central California coast in subdued form at

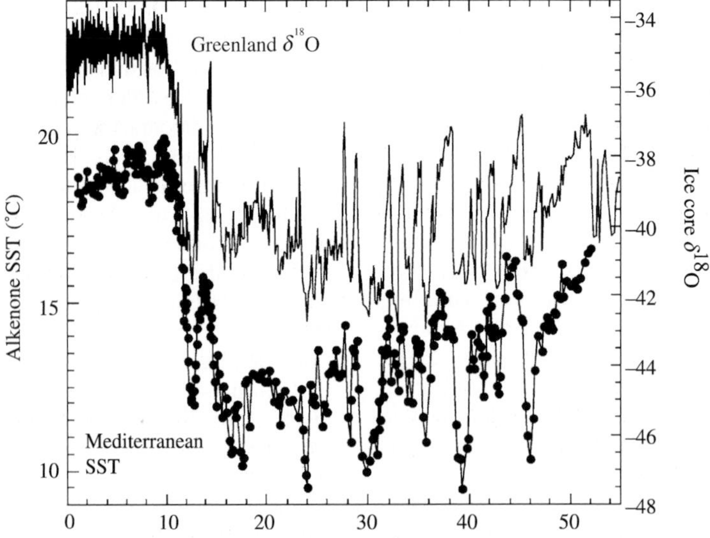

Figure 9 High resolution alkenone temperature estimates obtained by Cacho *et al.* (2002) from the Mediterranean Sea in relation to oscillations in temperature recorded by $\delta^{18}O$ in Greenland ice cores. Important millennial events (Younger Dryas = YD, H = Heinrich Event, D/O = Dansgaard/Oeschger Event) line up between the records to the precision of the independent chronologies.

ODP Site 1017 (Seki *et al.*, 2002). There, the cooling appears to be ~12 °C according to $U_{37}^{k'}$ estimates. The Seki *et al.* work also demonstrates that this part of the California margin felt the effects of D/O and Heinrich perturbations on SST. Substantial (3–4 °C) temperature anomalies correlate to the Greenland D/O events. The coldest temperatures of the 80 kyr record correspond to the intervals of North Atlantic Heinrich events (Seki *et al.*, 2002). In the tropics, the Younger Dryas period apparently led to small (0.5–1 °C) coolings in many locations in phase with the North Atlantic cooling. Well-dated alkenone evidence for a tropical expression of Younger Dryas cooling come from the South China Sea (Kienast *et al.*, 2001; Steinke *et al.*, 2001; Wang *et al.*, 1999), the Indian Ocean (Bard *et al.*, 1997; Cayre and Bard, 1999), and the South Atlantic (Kim *et al.*, 2002). Evidence for millennial-scale variability matching the Greenland ice core record is weaker—the very highly sampled 40 kyr record of Wang *et al.* (1999) in the South China Sea does not reveal significant changes in SST corresponding to either D/O or Heinrich events, but Schulte and Müller (2001) found small SST anomalies in the Arabian Sea at the same time as the North Atlantic millennial events.

Rapid temperature changes detected by alkenone paleothermometry may actually be anti-correlated to the North Atlantic pattern in some places. This effect is in fact predicted by the conveyor belt theory for the origin of millennial variability. If the North Atlantic thermohaline circulation drives millennial cycles in temperature, then the excess or deficit of heat transported to the high-latitude North Atlantic is compensated by adjustments in the cross-equatorial heat transport from the southern hemisphere. Records from the western tropical Atlantic and from the southern hemisphere might thus show warming at the same time as the coolings observed in Greenland ice cores. Resolving this question puts a premium on good dating, since the events in question lasted only centuries to one to two millennia. Ruhlemann *et al.* (1999) produced a densely dated (by AMS 14C) record of SST from a high deposition rate core off Grenada in the western tropical Atlantic. Both $\delta^{18}O$ of the planktonic foraminifer *G. ruber* and the $U_{37}^{k'}$ index show modest warming during the Younger Dryas and Heinrich event H1 times. The authors conclude that the apparent antiphasing of western tropical and North Atlantic SST supports the thermohaline model of millennial-scale events. Mazaud *et al.* (2002) attempted a similar analysis in the Southern Indian Ocean. In their case, they used a geomagnetic intensity signal in the sediments to attempt to synchronize alkenone and paleontological estimates of SST to the Greenland ice core data over the interval 32–50 ka. Alkenone analyses suggested that oscillations of 1–2 °C did occur at millennial timescales in this region. While several cold pulses appear to coincide with North Atlantic Heinrich events, longer coolings that may correlate to clusters of D/O cycles appear anti-correlated to the North Atlantic. The evidence thus favors a model that allows for different modes of producing and/or propagating different styles of millennial change of the surface temperature field.

6.15.8.3 Marine Temperatures during the LGM

Alkenone paleothermometry arrived at a time of controversy on the spatial pattern of cooling at the LGM. New estimates of tropical cooling derived from the Sr/Ca of fossil corals challenged the results of the seminal CLIMAP effort to map surface ocean temperatures at the last Ice Age (CLIMAP, 1976, 1981) CLIMAP employed a variety of planktonic microfossil groups, principally foraminifera, to derive seasonal anomalies of LGM temperatures relative to present day. One of the most significant conclusions of the studies was the poleward amplification of glacial cooling. The tropical ocean temperatures remained very near their present values, while temperatures decreased by as much as 8–10 °C in the North Atlantic, and 4–6 °C in the high latitude Southern Ocean (CLIMAP, 1976, 1981). Coral Sr/Ca thermometry placed tropical cooling in some places at 4–6 °C (Beck *et al.*, 1992; Guilderson *et al.*, 1994). Noble gases in continental groundwaters also suggested more cooling at low latitudes than evident in the CLIMAP reconstruction (Stute *et al.*, 1995).

The LGM is defined by the period during the last glacial cycle when ice volume reached its largest extent, at ~21–23 ka (EPILOG definition of Mix *et al.*, 2001). Since the definition relates to ice volume and carries a specific chronostratigraphic value, it does not necessarily correspond to the time of coldest ocean temperatures, as demonstrated in the high-resolution studies mentioned in the previous section. As reviewed by Mix *et al.* (2001), the best criterion for defining the LGM comes from multiple bracketing AMS ^{14}C dates. Oxygen isotopic data provide the next best tool for recognizing the LGM interval. The temperatures and temperature differences I review here follow the chronostratigraphic definition of the LGM. In most cases, the LGM is defined by isotopic data, rather than radiocarbon dates. The reader should also keep in mind that the LGM temperature anomaly depends on the reference frame. One gets somewhat different numbers if one uses only the latest Holocene alkenone SST points to define the contrast to LGM temperatures, as opposed to peak Holocene temperatures, because regional cooling and/or warming has occurred in the Holocene in many regions. All anomalies discussed here will be expressed as the differences between alkenone SST estimates at the stratigraphic level of the LGM and the latest Holocene samples, which I assume represent temperatures very close to modern day.

In this author's opinion, alkenone reconstructions favor rather decisively what might be termed a "modified CLIMAP" view of ocean surface cooling at 21 ka (Figure 10). Cases certainly occur where the $U_{37}^{k'}$ index indicates substantially more cooling at the LGM than the CLIMAP reconstruction (e.g., Jasper and Gagosian, 1989). The alkenone method typically favors more Ice Age tropical cooling than suggested by CLIMAP, but preserves the fundamental conclusion that the tropics cooled much less than did the high latitudes. In one particularly telling study, Bard and co-workers (Bard *et al.*, 1997; Sonzogni *et al.*, 1998) looked at an array of 20 cores in the Indian Ocean between 20° N and 20° S. $U_{37}^{k'}$ data showed more cooling (2–3 °C) away from the equator, but only ~1 °C cooling within 5° of the equator. Significantly, the alkenone temperature estimates weighed in with cooling in the northern Indian Ocean, where foraminiferal estimates had shown no cooling and even warming at the LGM (Bard *et al.*, 1997; Sonzogni *et al.*, 1998). The difference in results between faunal and alkenone techniques might be due to the superior signal-to-noise ratio of alkenone analysis, to the fact that foraminiferal faunas in the tropics respond to variables other than temperature, and to the observation that glacial faunas frequently have poor analogues in

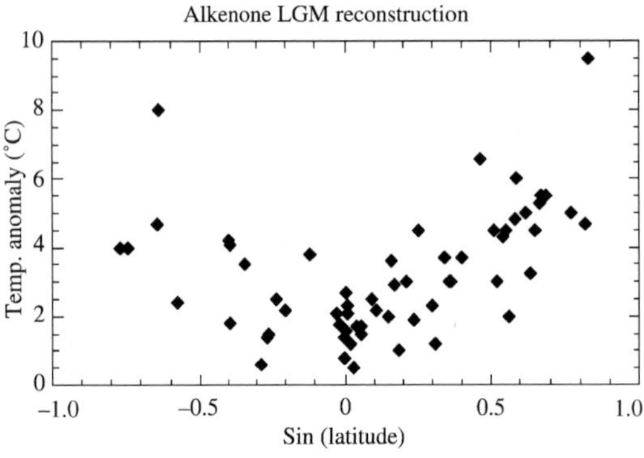

Figure 10 Compilation of all available alkenone estimates of cooling of ocean surface temperatures at the LGM relative to the late Holocene. The estimates have been projected onto the sine of latitude to approximately compensate for the distribution of ocean surface area from the equator to the poles. Note that the Ice Age anomalies are much stronger at mid and high latitudes than in the tropics. Scatter at any given latitude reflects variability in the quality of the chronological control used to assign the LGM level in alkenone time series, but also includes an important contribution from real heterogeneity of cooling at the LGM.

the modern ocean, making micropaleontological estimates more open to doubt. Other regions where the alkenone technique has been used to estimate tropical ocean temperatures at the LGM include the western equatorial Pacific (Ohkouchi *et al.*, 1994), the South China Sea (Wang *et al.*, 1999; Pelejero *et al.*, 1999a; Steinke *et al.*, 2001), and the tropical Atlantic (Zhao *et al.*, 1995; Schneider *et al.*, 1996; Sikes and Keigwin, 1994).

One should note that the impact of a few degrees of cooling in the tropical ocean is not climatically trivial. Since one-half of the Earth's surface lies between 30° of the equator, any cooling has a significant effect on the globally integrated surface cooling during the last Ice Age. It also seems likely that more $U_{37}^{k'}$ work will resolve spatial patterns of cooling that implicate dynamical adjustments of the tropical surface ocean to the changed sea level, orbital, and CO_2 boundary conditions at the LGM.

Evidence in hand already suggests that the ocean cooled differently according to region at the LGM. Although the details of how SST evolved during the last glaciation and through the glacial termination differ widely, one can distinguish two major patterns: those regions where the SST profile looks much like the oxygen isotope curve, and hence sea level and ice volume, and those in which SSTs depart significantly from the global pattern of ice volume. Many records show that SSTs ran in parallel with the ice volume cycle; e.g., in the high-latitude North Atlantic (Villanueva *et al.*, 1998; Calvo *et al.*, 2001), the South China Sea (Huang *et al.*, 1997; Pelejero *et al.*, 1999a; Steinke *et al.*, 2001), off Hawaii (Lee *et al.*, 2001), and parts of the South Atlantic (Schneider *et al.*, 1995, 1996; Kirst *et al.*, 1999; Sachs and Lehman, 2001). However, $U_{37}^{k'}$ data indicate that another common pattern was for SST to rise in advance of deglaciation. SST led ice volume temporally in the equatorial Pacific (Lyle *et al.*, 1992) and tropical Indian Ocean (Bard *et al.*, 1997). However, even within one ocean basin, there were strong regional differences in the timing of maximum cooling. Kirst *et al.* (1999) used three cores to form onshore–offshore transect along the Namibian upwelling margin in the South Atlantic. They demonstrated that both the $U_{37}^{k'}$ and the C_{37} total ketones (an index of productivity) varied systematically along the transect in the timing of maxima and minima. Warming began well before the LGM close to the African continental margin. At the same time that warming occurred, the concentration of C_{37} ketones declined steeply in the sediments. Offshore, temperatures followed ice volume closely. Kirst and colleagues ascribe the patterns they see to the impingement of warmer waters from the North along the coastal margin at the LGM, when wind systems were reorganized. The incursion of this warmer subtropical water coincided with less favorable conditions for upwelling along the Namibian continental margin. The offshore cores stayed distant enough from the front of equatorially derived water that they do not display the early warming, and instead sense the basin-wide cooling that reached its peak at the LGM. My own laboratory has found a similar pattern off the California margin, where core locations near the present-day southern boundary of the cold California Current show only small cooling at the LGM (Herbert *et al.*, 2001; Figure 11). As off Benguela, cores along the California margin some distance from the frontal boundaries display the close association between cold SST and the LGM that one would intuitively expect. Alkenone evidence thus suggests that several of the major eastern boundary current systems retracted poleward at the LGM.

A review of the current literature suggests that substantial gaps in coverage of the LGM surface ocean by the alkenone method still exist. Very few records have been obtained from the interior gyres of the major ocean basins. The sediments underlying these regions are less favorable for alkenone analysis, because they tend to have slow deposition rates and low biomarker contents due to low overlying productivity. Nevertheless, the gyres are very broad features in the modern ocean, and their thermal state at the LGM should

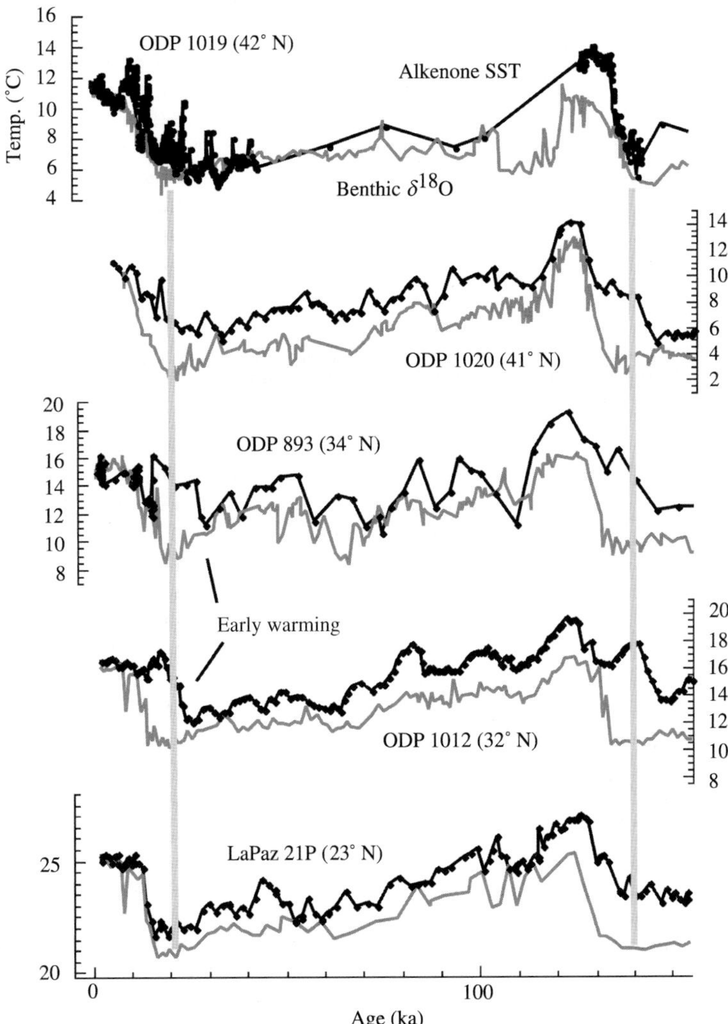

Figure 11 Regional variation in the timing of SST rise at the last two glacial terminations revealed by alkenone determinations along the California margin (Herbert *et al.*, 2001). Data are arranged from north to south to demonstrate the regional nature of SST response (dark lines). Benthic $\delta^{18}O$ data acquired on the same cores (lighter color) demonstrates unequivocally that SST rose early relative to the global sea level rises at each termination at many sites. The strongest anomalies occur in the southern California borderland region, which today represents the boundary between the cold, equatorward flowing California Current, and subtropical waters that flow northward seasonally (Herbert *et al.*, 2001).

not be neglected. In addition, the record of SSTs from the southern hemisphere badly lags the amount of data now available for much of the northern hemisphere. The Bard *et al.* (1997) study demonstrates nicely that SST rose in synchroneity in both hemispheres between 20° N and 20° S in the Indian Ocean beginning at ~15 ka. Such common phasing might indicate that forcing by CO_2 and/or water vapor link the surface ocean of both hemispheres during deglaciation (Bard *et al.*, 1997). More spatial coverage from other parts of the southern hemisphere is still needed. The temperature history of the warm pool of the western equatorial Pacific also awaits more investigation. However, this region is not well suited for the $U_{37}^{k'}$ proxy, as its temperature lies at or above the saturation of the $U_{37}^{k'}$ index at around 28 °C.

6.15.8.4 SST Records of the Late Pleistocene Ice Age Cycles

The rapidity and precision of alkenone analyses make them ideally suited for producing time series of the SST response to the late Pleistocene cycles in glaciation. However, legitimate questions arise about how the proxy will work on long time-scales. As with all biologically based proxies, the possibility exists that the response of the alkenone unsaturation could have changed over evolutionary time. That question is sharper in the case of the $U_{37}^{k'}$ proxy, because one of the principal modern producing species, *E. huxleyi*, first appeared in the geological record during marine oxygen isotope stage 8, at ~280 ka (Thierstein *et al.*, 1977). Indeed, micropaleontological examinations show that the rise to dominance of *E. huxleyi* occurred at different times during the last glacial cycle, depending on region (Jordan *et al.*, 1996). Gephyrocapsids dominated the coccolithophorid assemblage until the Holocene in regions such as the Benguela upwelling system (Summerhayes *et al.*, 1995). Several studies that document the sequence of coccolithophorids in late Pleistocene records fail to find any indication that the proportions of alkenone-producing species affect SST reconstructions by the $U_{37}^{k'}$ method. For example, Doose-Rolinski monitored the proportions of *Gephyrocapsa* to *Emiliania* at high resolution in the late Holocene in their Arabian Sea core. High-frequency SST changes did not correspond to changes in the ratio of these species. Jordan *et al.* (1996) also failed to find a relation between reconstructed SST changes and changes in the abundance of *E. huxleyi* in a 130 kyr record from the tropical Atlantic. Other work demonstrates that the proportions of C_{37} ketones to C_{38} ketones do not vary down-core over long time periods, consistent with the evolutionary conservatism of alkenone synthesis (Muller *et al.*, 1997; Rostek *et al.*, 1997, Yamamoto *et al.*, 2000).

Alkenone paleotemperatures appear to give reliable information on orbital-scale changes in SST over the past 500 ka. The 100 kyr cycle of glaciation dominates all long alkenone SST records (Eglinton *et al.*, 1992; Emeis *et al.*, 1995a; Schneider *et al.*, 1995, 1996; Sonzogni *et al.*, 1998; Villanueva *et al.*, 1998; Herbert *et al.*, 2001). There is no indication of significant linear trends in SST over the late Pleistocene, which might warn of systematic diagenetic or evolutionary artifacts. Rather, the results of alkenone work to date show remarkably consistent patterns of SST change in relation to oxygen isotope evidence for glacial–interglacial climate state. Alkenone data suggest that a number of previous interglacial intervals have been warmer by up to 3 °C from the late Holocene. In particular, alkenone estimates for temperatures at the peak of the previous interglacial period (oxygen isotope stage 5e), fall consistently above the late Holocene. This result is consistent with the higher northern hemisphere insolation at 125 ka as compared to the Holocene, and with some evidence for sea level higher than the modern during the last interglacial period. The alkenone results contrast with the CLIMAP reconstruction for isotope stage 5e, which showed no systematic departures in SST from modern-day values (CLIMAP, 1984). The alkenone estimates of a warm stage MIS 5e are consistent, however, with the warm temperatures derived from sparse Mg/Ca data from planktonic foraminifer (Lea *et al.*, 2000; Nuernberg *et al.*, 2000). One should note, however, that the magnitude of the MIS 5e anomaly declines if it is referenced to the Holocene insolation maximum at 9 ka, because many alkenone records show 1–2 °C warming in the early Holocene compared to the modern day.

Alkenone data do show characteristic regional patterns of SST change over the 100 kyr glacial cycles of the late Pleistocene. Records from the North Atlantic tend to show the classical "saw-tooth" pattern of long decline in SST, followed by a rapid rise at glacial terminations (Villanueva *et al.*, 1998; Calvo *et al.*, 2001). Similar asymmetrical cooling and warming also appears in the higher-latitude South Atlantic (Kirst *et al.*, 1999) and South China Sea (Pelejero *et al.*, 1999a). In many places, however, the ocean reached its coldest temperatures well before maximum glaciation. Thus, records from the Indian Ocean consistently show that the coldest temperatures in that region coincided with MIS 4 (Rostek *et al.*, 1993; Emeis *et al.*, 1995a; Bard *et al.*, 1997). Our own data from off the California margin show that the early warming (relative to deglaciation) detected at the end of the last Ice

Age recurred at each of the glacial terminations for the last five glacial cycles (Herbert *et al.*, 2001). In some cases, the positive relation between colder $U_{37}^{k'}$ estimates and increased concentrations of C_{37} ketones in the sediment implicates enhanced upwelling as the cause of cold Ice Age temperatures (Kirst *et al.*, 1999). However, in many locations the pattern of coccolithophorid productivity is completely decoupled from SST, or in inverse relation to the expected increase with colder temperatures if upwelling were the cause. In these cases, the temperature changes must involve changes in ocean heat transport that occur at a larger scale than the regional changes in wind field.

The spectral content of alkenone record has only been assessed in a few records. In one noteworthy example, Schneider *et al.* (1999) provide evidence that tropical Atlantic SST has a strong imprint from the precessional (~21 kyr) cycle of insolation. Precessional variability drives contrasts in summer and winter heating, particularly at low latitudes. An expected result is that precession regulates the monsoonal cycle of the tropics. This result is consistent with the strength of precession and the weakness of obliquity (41 kyr) components in the spectra of equatorial $U_{37}^{k'}$ records. Both Schneider *et al.* (1999) and our own unpublished work off the California margin also show that the 41 kyr component in SST is larger at higher latitudes in late Pleistocene time series than it is in low-latitude records.

One puzzle that remains is the persistent evidence from alkenones that the previous glacial period (MIS 6) was not nearly as cold as the most recent glacial maximum in many locations. Relatively warm estimates for MIS 6 typically come from tropical and subtropical regions (Rostek *et al.*, 1993, 1997; Emeis *et al.*, 1995a; Schneider *et al.*, 1996; Villanueva *et al.*, 1998). However, many other tropical locations provide SST estimates as cold as the most recent cycle for the previous glacial period (Pelejero *et al.*, 1999a; Wang *et al.*, 1999; Sicre *et al.*, 2000). At high latitudes, the temperatures recorded by alkenones for MIS 6 are in line with those of the last glacial period (Schneider *et al.*, 1996; Villanueva *et al.*, 1998; Kirst *et al.*, 1999; Calvo *et al.*, 2001; Herbert *et al.*, 2001 and unpublished). Furthermore, the stage-6 warm anomaly does not repeat itself in previous glacial intervals where $U_{37}^{k'}$ analysis has been performed. There is therefore no obvious reason to dismiss the alkenone result, other than that it does not square with oxygen isotope evidence for a glacial climate as extreme as the most recent episode. If it correctly interpreted, the alkenone data would show that the tropical ocean responded in different ways to glacial maxima over the course of time.

6.15.8.5 SST before the Late Pleistocene

What is the long-term prospect for using alkenones to deduce the large-scale thermal evolution of the oceans in the Cenozoic? There is certainly enormous potential. Unlike the $\delta^{18}O$ proxy, the alkenone unsaturation index depends in theory only on temperature, and thus isolates temperature changes from other influences such as salinity and global ice volume that mingle in the $\delta^{18}O$ record of both benthic and planktonic foraminifera. And unlike micropaleontologically based methods for SST reconstruction, which rely on knowledge of the present-day preferences of extant species, the alkenone paleothermometer may range well back in time, provided that it has behaved in an evolutionarily conservative manner, and provided that diagenesis does not overprint the original SST signal. Although one would clearly prefer that the modern calibration of the $U_{37}^{k'}$ index apply accurately throughout the span of alkenone records, all is not lost if this assumption is not correct. There should still be at a minimum a role for a proxy that can unambiguously record spatial and temporal *variability* of SST in long paleoceanographic time series.

It is one of the ironies of the paleoceanographic record that alkenones are frequently more abundant in sediments before the appearance of *E. huxleyi* than they are in the period in which this coccolithophorid dominates the floral assemblage. Three studies that relate the concentrations of C_{37} ketones to the abundance of *E. huxleyi* found that the ratio of alkenones to total organic carbon actually increased in sediments older than the *E. huxleyi* acme (Muller *et al.*, 1997; Rostek *et al.*, 1997; Sicre *et al.*, 2000). Down-core profiles of alkenone concentration often increase with age, a pattern opposite to what would be expected of long-term diagenetic control on their abundance (Muller *et al.*, 1997). At the same time, time series of alkenone parameters such as the chain length ratio ($\delta C_{37:}/\delta C_{37}$ ketones) show no change with time over the past few glacial cycles (Muller *et al.*, 1997; Rostek *et al.*, 1997). Our laboratory has produced longer records that also lead us to conclude that alkenones are very refractory, once incorporated into the sediment. Alkenone data from a 2.8 Myr record off the coast of Southern California (Z. Liu *et al.*, unpublished data) show higher concentrations in the older part of the record than in the late Pleistocene (Figure 3). Furthermore, there is no suggestion of a monotonic increase in SST going backward in time, as might be expected if diagenesis biased the unsaturation index. Changes in the dominant wavelength of the SST oscillations match the spectral evolution of oxygen isotopes from 41 kyr dominance in the early Pleistocene to the later rise of the 100 kyr cycle in the late Pleistocene (cf. Ruddiman *et al.*, 1986). We also note the conservatism of the chain length ratio over this 3 Myr interval.

Although reports of alkenones in sediments of Pliocene, Miocene, Oligocene, and even Eocene age are common (see Lichtfouse *et al.*, 1992; Brassell, 1993; van der Smissen and Rulkotter, 1996; Rinna *et al.*, 2002), the late Pliocene period provides the only example where data have been generated in enough continuity to see the evolution of SST as recorded by the $U_{37}^{k'}$ index. The alkenone method appears to resolve major cooling in both the North and South Atlantic during the onset of northern hemisphere glaciation (between ~3 Ma and ~2.5 Ma). My laboratory generated a low-resolution record (~50 kyr sampling) of alkenone temperatures from ~6 Ma to ~2.3 Ma in the subtropical North Atlantic (Herbert and Schuffert, 1998). We found warm temperatures before the period of time when sizable ice sheets grew in the northern hemisphere. Furthermore, variability was low (~1 °C). Larger-amplitude SST changes appeared at the time that northern hemisphere ice volume began to grow. The changes presumably record the onset of large, orbitally driven variations in ice volume. An even more substantial change in the pattern of SST over this critical paleoclimatic interval comes from the South Atlantic (Marlow *et al.*, 2000). This study followed SST off the Benguela margin from 4.5 Ma to the present, again at 40–50 kyr resolution. According to the alkenone index, a nearly 10 °C cooling occurred between ~3 Ma and ~1.8 Ma. The authors ascribe the cooling to a long-term increase in upwelling along the Benguela margin, but acknowledge that basin-wide changes in heat transport could also be involved. Put together, the two studies of late Pliocene SST demonstrate that the alkenone proxy will play a useful role in understanding the coupling between the growing amplitude of northern hemisphere glacial–interglacial variability and ocean dynamics in the late Pliocene and early Pleistocene.

6.15.8.6 Comparison with other Proxies: $\delta^{18}O$

It should be possible to test the fidelity of alkenone SST reconstructions with an independent paleothermometer such as $\delta^{18}O$. There are a number of reasons why this comparison is not straightforward, however. First, many planktonic foraminifera live below the mixed layer for part or all of their life cycle. Comparing their isotopic temperatures with the unsaturation index of the alkenone producers, which must live in the photic zone, may not yield consistent estimates of past temperature changes. Many foraminifera have seasonal cycles of production far more pronounced than those of *E. huxleyi* and *G. oceanica*, so one may expect offsets in paleotemperatures due to seasonal biases of either foraminiferal or alkenone production. The ideal comparison of foraminiferal isotopic and coccolithophorid $U_{37}^{k'}$ estimates should therefore come from the tropics, where the seasonal amplitude in temperature is small and where some species of planktonic foraminifera spend most of their life in the mixed layer. The ideal foraminiferal species for comparison should be *G. ruber*, which contains algal photosymbionts and consistently shows mixed-layer isotopic temperatures in recent sediments. *G. ruber* dominates many tropical assemblages. It becomes rare in waters colder than ~20 °C. The next best choice should be *G. sacculifer*, also a tropical species. Isotopic depth ranking shows, however, that *G. sacculifer* may often live below the mixed layer. In addition, *G. sacculifer* adds a calcite crust during gametogenesis that apparently forms at greater depth. Its isotopic temperatures are therefore likely to be somewhat cooler than the mixed layer.

Planktonic $\delta^{18}O$ values inherently contain more uncertainty as paleothermometers than do alkenone data. The analytical signal-to-noise ratio favors the $U_{37}^{k'}$ technique by four to five times (e.g., a typical analytical uncertainty of 0.1‰ in $\delta^{18}O$ corresponds to a temperature uncertainty of 0.4–0.05 °C). Planktonic foraminiferal isotopic values also incorporate the effects of changing global ice volume and regional evaporation/precipitation on the isotopic composition of the water in which the foraminifera grow. Indeed, one of the promises of SST proxies such as the $U_{37}^{k'}$ index is to constrain temperatures so that the other components of foraminiferal $\delta^{18}O$ can be studied.

The amplitudes of temperature change at the LGM inferred by both alkenones and planktonic $\delta^{18}O$ seem largely consistent. Broecker (1986) pointed out that the magnitude of tropical sea surface cooling at the LGM was strongly constrained by the modest amount of isotopic enrichment in ice age foraminifera beyond that required by the global ice volume effect. Estimating the ice volume effect precisely has been difficult. The canonical ice volume effect of 1.2‰ comes from a sea-level/$\delta^{18}O$ calibration (Fairbanks and Matthews, 1978). More recent estimates based on pore-water deconvolution put the ice volume effect at 0.9–1‰ (Schrag *et al.*, 1996). As discussed previously, alkenone SST estimates support tropical cooling of 1–2 °C in most locations. This cooling is enough to favor the recent smaller estimates for the global ice volume effect, because it would correct the typical glacial isotopic enrichment of 1–1.5‰ by several-tenths of a per mil (see Figure 12). Alternatively, one could infer a relative freshening of many tropical

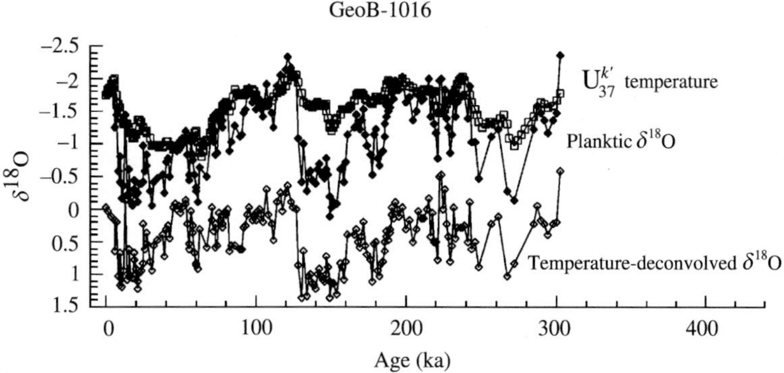

Figure 12 Alkenone data acquired with planktonic foraminiferal $\delta^{18}O$ can be used to remove the temperature effect from oxygen isotopic signals in order to assess global (ice volume) and regional (evaporation/precipitation balance) contributions to isotopic change. Data come from an equatorial Atlantic core studied by Schneider *et al.* (1996); the isotopic deconvolution was made by the present author. Note that the structure and amplitude of the temperature-corrected planktonic $\delta^{18}O$ record are consistent with current thinking on the global ice volume signal for the late Pleistocene 100 kyr cycles.

locations from the comparison of temperature and isotopic change. Pelejero *et al.* provide a regional data set of *G. ruber* isotopic values with corresponding alkenone SST estimates in the South China Sea (Pelejero *et al.*, 1999a). Cores in the northern part of the Sea display amplitudes in $\delta^{18}O$ of 1.4–1.6‰ between the late Holocene and LGM. The corresponding alkenone temperature anomalies are ~3.2–3.5 °C, equivalent to ~0.7‰ in $\delta^{18}O$ by the temperature effect. The lower latitude sites from the South China Sea produce smaller glacial–interglacial differences in $\delta^{18}O$ (1.3–1.4‰) and in $U^{k'}_{37}$ temperatures (2.3–2.6 °C). Cores collected off Hawaii yield glacial/interglacial isotopic differences from *G. ruber* of 1.1–1.3‰ and alkenone-based cooling of 2–3 °C (Lee *et al.*, 2001). However, Muller *et al.* compared isotopic and alkenone data in a tropical (11° S) core from the South Atlantic and found higher isotopic amplitude (1.9‰) accompanied by an ~3.5 °C temperature change deduced from the $U^{k'}_{37}$ index. Such evidence certainly suggests that regional changes in the $\delta^{18}O$ of seawater in the surface ocean at the LGM will have to be considered in addition to the global ice volume effect.

Several studies have in fact attempted to deduce regional changes in salinity by pairing foraminiferal $\delta^{18}O$ data with alkenone paleotemperature estimates. In a seminal paper, Rostek *et al.* (1993) subtracted the isotopic temperature effect from an Indian Ocean $\delta^{18}O$ curve obtained from *G. ruber* using the $U^{k'}_{37}$ index. Their reconstruction put salinity of the tropical Indian Ocean as 1–2 psu more saline than today, and also suggested that the salinities were fresher than modern during the previous interglacial peak at MIS 5e. Regional reconstructions of both $\delta^{18}O$ and $U^{k'}_{37}$ from the Mediterranean also appear to resolve glacial–interglacial shifts in salinity (Emeis *et al.*, 2000; Cacho *et al.*, 2001, 2002). The isotopic shift immediately outside the Mediterranean in the Gulf of Cadiz is modest (1.8‰), as is cooling (~3.5 °C). Glacial age isotopic anomalies grow progressively in the Mediterranean from 2.5‰ in the Alboran Sea to over 3‰ in the eastern Mediterranean (Emeis *et al.*, 2000; Cacho *et al.*, 2001, 2002). Much of the greater isotopic amplitude can be explained by intense surface-ocean cooling in the interior Mediterranean (5–8 °C, depending on location), away from the Atlantic connection. However, the residual isotopic enrichment within the Mediterranean indicates that this region was even more saline than today during the last glacial period, especially in the eastern basin. Another regionally coherent pattern of salinity change comes from the South China Sea (Wang *et al.*, 1999). Surface waters in this region, in contrast to the Mediterranean, apparently freshened during the LGM in comparison to the present day.

One danger, of course, for alkenone deconvolution of planktonic $\delta^{18}O$ records comes from the fact that the proxies originate from two different types of organisms. Some of the differences in amplitude and timing of the two signals may result from ecological changes, rather than changes in the $\delta^{18}O$ of the water the foraminifera grew in. Such ambiguities are present in all multiproxy paleo-environmental work; resolving them is one of the chief challenges of paleoceanography.

6.15.8.7 Comparison with other Proxies: Microfossils

CLIMAP (1976, CLIMAP, 1981) produced its reconstruction of temperatures at the LGM by using the modern-day ecological preferences of microfossil groups to estimate past SSTs. Since the CLIMAP effort, micropaleontologists have developed new methods of reconstructing SST in addition to the transfer-function method pioneered by CLIMAP investigators. Rather different sets of glacial–interglacial temperature anomalies can result from applying different microfossil calibrations to the same data set (see Mix *et al.*, 2001; Waelbroeck and Steinke, 2002). It is therefore simplistic to compare alkenone methods to a particular mode of micropaleontological reconstruction as a general test of either the geochemical or microfossil approach. Furthermore, there are relatively few cores in which the microfossil record (generally from planktonic foraminifera) has been directly compared with the $U_{37}^{k'}$ measurements. Experience shows that spatially sparse data sets can often produce apparent conflicts at the LGM that come simply from extrapolating results from one proxy over too great a distance (Broccoli and Marciniak, 1996).

Temperature estimates produced by alkenone and foraminiferal assemblage methods generally agree quite well for the Holocene (but see Marchal *et al.* (2002) for some discrepancies in the North Atlantic and Mediterranean). Alkenone temperatures usually fall in between the warm and cold season estimates of foraminifera (Chapman *et al.*, 1996; Wang *et al.*, 1999; Perez-Folgado *et al.*, 2003). Since both methods are in some sense optimized for late Holocene conditions, this is perhaps not surprising. Estimates of SST at the LGM more commonly diverge (Chapman *et al.*, 1996; Huang *et al.*, 1997; Perez-Folgado *et al.*, 2003). It is not evident how to reconcile changes in the $U_{37}^{k'}$ index relative to seasonal SST estimates derived from foraminiferal faunas. In one of the best-studied examples, Chapman *et al.* (1996) found that $U_{37}^{k'}$ estimates for glacial temperatures followed summer temperatures estimated from planktonic foraminifera at a subtropical north Atlantic site. Chapman *et al.* (1996) argued that the alkenone producers must have changed their season of production from late winter and spring (present-day) to summer during glacial and early Holocene times. This ecological switch, according to the authors, explains why the alkenone thermometer would underestimate glacial SST cooling at the study location. Chapman *et al.* (1996) do note that the temperatures inferred from alkenones match the isotopic temperatures derived from the planktonic foraminifer *G. bulloides* very well. Precisely the opposite temporal dichotomy between $U_{37}^{k'}$ and foraminiferal estimates was obtained from the South China Sea, where glacial alkenone temperatures are consistently colder than the cold season foraminiferal estimate (Huang *et al.*, 1997).

As noted earlier, most alkenone estimates from the tropics support more cooling than inferred by CLIMAP (see Bard *et al.*, 1997; Sonzogni *et al.*, 1998; Pelejero *et al.*, 1999; Lee *et al.*, 2001). It is not clear if the differences are larger than overlap of the standard error of the techniques. In most cases, the alkenone method has a clear advantage in signal-noise for resolving smaller temperature changes (see, e.g., Lee *et al.*, 2001).

6.15.8.8 Comparison with other Proxies: Mg/Ca

As with the $U_{37}^{k'}$ proxy, Mg/Ca measurements in foraminiferal calcite may yield paleotemperature measurements without the ambiguities of $\delta^{18}O$ or micropaleontological assemblage analyses. In contrast to the alkenone proxy, preservation markedly affects the Mg/Ca ratio in foraminiferal shells, and hence the paleotemperature estimate (Brown and Elderfield, 1996; Lea *et al.*, 1999, 2000; Rosenthal *et al.*, 2000). The alkenone temperature proxy may also have the advantage in several other aspects. Synthesis of alkenones is tied to the photic zone, unlike that of foraminiferal calcite. The signal-to-noise ratio of alkenone measurements is superior for the paleotemperature range 5–25 °C, where the exponential dependence of Mg/Ca results in a relatively low temperature sensitivity (Lea *et al.*, 1999, 2000; Elderfield and Ganssen, 2000). At higher temperatures, the Mg/Ca proxy may become superior because of the same exponential sensitivity to temperature. Furthermore, the Mg/Ca proxy should not lose sensitivity to temperature at the warm end of ocean temperatures, unlike the alkenone proxy. The Mg/Ca proxy should therefore be the tool of choice in determining past changes in SST in warm pool regions of the ocean.

Two direct comparisons of $U_{37}^{k'}$ and Mg/Ca estimates of glacial–interglacial SST changes come from the tropical Atlantic. Elderfield and Ganssen (2000) compared the Mg/Ca and $\delta^{18}O$ of 5 species of planktonic foraminifera at core site BOFS31K, the same core analyzed by Chapman *et al.* (1996) for foraminiferal SST estimates. Elderfield and Ganssen (2000) refer to Mg/Ca temperatures as "calcification temperatures" to account for the observation that many planktonic foraminifera calcify significantly below the mixed layer. Mg/Ca data produced two down-core clusters of species: a cold group (*G. inflata, N. pachyderma*, and *G. menardii*) and a warm group with Mg/Ca temperatures consistently 3–3.5 °C warmer (*G. ruber* and *G. bulloides*). Core-top temperatures estimated from the warm group fell ~2 °C colder than modern SST, while the alkenone estimate falls right at mean annual modern SST (Elderfield and Ganssen, 2000). Temperature changes estimated from *G. ruber* track alkenone temperatures very closely, while maintaining the ~2 °C offset toward colder temperatures. Mg/Ca measurements on *G. bulloides* show almost no cooling for most of the glacial period. Elderfield and Ganssen conclude that the foraminiferal Mg/Ca temperature estimates at the LGM follow the alkenone estimate (~2 °C) more closely than the foraminiferal transfer function estimate (cooling of ~5 °C). Their work also hints at the ecological uncertainties that enter into interpreting temperature records from different planktonic organisms that may have had diverging depth and seasonal habitats over the course of glacial to interglacial climate change.

Nuernberg *et al.* (2000) compared the Mg/Ca and alkenone approaches over three glacial–interglacial cycles at a cores at 1° S in the central South Atlantic. The foraminifer used was *G. sacculifer*. Core-top temperature estimates differed by ~2 °C. Once again, the Mg/Ca estimate fell consistently cooler than the alkenone record down-core. In most cases, the temperature curves were offset by 2–3 °C. Occasionally, as at the last glacial termination, the two temperature profiles coincided. The authors noted strong qualitative similarities in the two paleotemperature data sets. Both gave similar cooling at the LGM (3.5 °C for the $U_{37}^{k'}$ index, 3.4 °C for Mg/Ca), a warmer-than-Holocene stage 5e, and a warmer previous glacial cycle than the last glacial interval. Both geochemical proxies gave smaller SST changes at the LGM than planktonic foraminiferal transfer function estimates, which gave 9 °C cooling during the cold season and 5 °C cooling for the warm season. They also agreed on the greater warmth of MIS 5e, while the foraminiferal transfer function did not. But the correlation coefficients for the Mg/Ca and $U_{37}^{k'}$ proxies were not strong: 0.49 over the 270 kyr record, and 0.78 over the last 90 ka (Nuernberg *et al.*, 2000).

6.15.9 CONCLUSIONS

This review paints an optimistic picture of the ability of alkenone unsaturation indices recorded in sediments to capture past SST change on a wide variety of timescales. I believe this optimism is justified on a number of grounds. First, the analytical precision of the estimate is outstanding. One can anticipate that interlaboratory differences will diminish over time as a result of intercalibration experiments (e.g., Rosell-Mele *et al.*, 2001) and the maturation of analytical protocols. The precision of alkenone analyses is meaningful, in the sense that core-top sediments from the same region yield very similar values for $U_{37}^{k'}$, and very similar down-core estimates of cooling at time horizons such as the LGM. Unlike the SST proxies derived from foraminifera, the alkenone index survives extensive degradation in the water columns and sediments. And, it seems unlikely that the calibration of alkenone unsaturation to growth temperature will change with additional data.

The relationship observed in sediments closely mimics the relation measured between unsaturation in photic zone particulates (Conte *et al.* in press) and *in situ* temperature. We should note, however, that the calibration of sedimentary $U_{37}^{k'}$ to mean annual SST represents a statistical relationship and should not be interpreted in too simplistic a manner. We have seen above that this relation holds for sediments because alkenone synthesis generally occurs in the mixed layer, and because alkenone production by selected haptophyte algae is less seasonal than many other planktonic groups. Furthermore, temperatures during the season of maximal alkenone production generally come close to mean annual SST. The index, therefore, is a quite robust *proxy* for mean annual temperature. It should be kept in mind that regional variations in the factors that control the depth and seasonality of alkenone production can cause the index to deviate from mean annual SST.

As is the case for many other paleotemperature proxies, uncertainties grow at the extreme ends of the temperature range sampled by the $U_{37}^{k'}$ index. These come in part from analytical problems, and in part from the lack of a theoretical basis for preferring a linear or curvilinear calibration at very cold and warm temperatures. The behavior of alkenones as fine-grained particles may cause more significant problems for the proxy than does the calibration uncertainty. Recent evidence for advection of alkenones by deep currents (Benthien and Müller, 2000; Ohkouchi *et al.*, 2002), suggests that paleoceanographers ignore sedimentology at their peril. In particular, the spectacular evidence from $U_{37}^{k'}$ for rapid SST changes in some drift sediments (Sachs and Lehman, 1999) may need to be reinterpreted as a signal of pulses of input from high latitudes. New tools such as ^{14}C AMS dating of alkenones will lead us to a clearer picture of which sedimentary environments are the most favorable for applying the $U_{37}^{k'}$ index as a measure of past SST. The comforting news is that the very high correlation between sediment core-top $U_{37}^{k'}$ and mean annual SST indicates that advection cannot be a pervasive problem.

Alkenone paleotemperature estimates have already weighed in significantly on the controversy of tropical SST at the LGM. The "modified CLIMAP" view they support seems consistent with tropical cooling of 2–3 °C estimated from Mg/Ca of planktonic foraminifera (Lea *et al.*, 2000; Nuernberg *et al.*, 2000; Stott *et al.*, 2002; Visser *et al.*, 2003). SST change over deeper timescales awaits more investigation by the alkenone method. All indications point to the diagenetic stability of alkenones over millions of years. The long evolutionary history and apparent conservatism of alkenone synthesis suggest that the proxy will play a useful role in solving major paleoclimatic questions over most of the Cenozoic.

REFERENCES

Aksnes D. L., Egge J. K., Rosland R., and Heimdal B. R. (1994) Representation of *Emiliania-huxleyi* in phytoplankton simulation-models—a 1st approach. *Sarsia* **79**, 291–300.

Andruleit H. (1997) Coccolithophore fluxes in the Norwegian-Greenland sea: seasonality and assemblage alterations. *Mar. Micropaleontol.* **31**, 45–64.

Andruleit H., von Rad U., Bruns A., and Ittekkot V. (2000) Coccolithophore fluxes from sediment traps in the north-eastern Arabian sea off Pakistan. *Mar. Micropaleontol.* **38**, 285–308.

Antia A. N., Maaßen J., Herman P., Voß M., Scholten J., Groom S., and Miller P. (2001) Spatial and temporal variability of particle flux at the NW European continental margin. *Deep-Sea Res. II* **48**, 3083–3106.

Balch W. M., Holligan P. M., and Kilpatrick K. A. (1992) Calcification, photosynthesis and growth of the bloom-forming coccolithophore, *Emiliania huxleyi*. *Cont. Shelf Res.* **12**, 1353–1374.

Bard E., Rostek F., and Sonzogni C. (1997) Interhemispheric synchrony of the last deglaciation inferred from alkenone paleothermometry. *Nature* **385**, 707–710.

Bard E., Rostek F., Turon J.-L., and Gendreau S. (2000) Hydrological impact of Heinrich events in the subtropical northeast Atlantic. *Science* **289**, 1321–1324.

Beaufort L. and Heussner S. (1999) Coccolithophorids on the continental slope of the Bay of Biscay production, transport and contribution to mass fluxes. *Deep-Sea Res. II* **46**, 2147–2174.

Beaufort L. and Heussner S. (2001) Seasonal dynamics of calcareous nannoplankton on a West European continental margin: the Bay of Biscay. *Mar. Micropaleontol.* **43**, 27–55.

Beck J. W., Edwards R. L., Ito E., Taylor F. W., Recy J., Rougerie F., Joannot P., and Henin C. (1992) Sea-surface temperature from coral skeletal Strontium–Calcium ratios. *Science* **257**, 644–646.

Bell M. V. and Pond D. (1996) Lipid composition during growth of motile and coccolith forms of *Emiliania huxleyi*. *Phytochemistry* **41**, 465–471.

Bentaleb I., Grimalt J. O., Vidussi F., Marty J.-C., Martin V., Denis M., Hatte C., and Fontugne M. (1999) The C_{37} alkenone record of seawater temperatures during seasonal thermocline stratification. *Mar. Chem.* **64**, 301–313.

Bentaleb I., Fontugne M., and Beaufort L. (2002) Long-chain alkenones and $U_{37}^{k'}$ variability along a south–north transect in the western Pacific Ocean. *Global Planet. Change* **34**, 173–183.

Benthien A. and Müller P. J. (2000) Anomalously low alkenone temperatures caused by lateral particle and sediment transport in the Malvinas current region, western Argentine basin. *Deep-Sea Res. I* **47**, 2369–2393.

Bollman J. (1997) Morphology and biogeography of *Gephyrocapsa* coccoliths in Holocene sediments. *Mar. Micropaleontol.* **29**, 319–350.

Bond G. C. and Lotti R. (1995) Iceberg discharges into the North Atlantic on millennial timescales during the last glaciation. *Science* **267**, 1005–1010.

Bond G. C., Broecker W. S., Johnsen S., McManus J., Labeyrie L., Jouzel J., and Bonani G. (1993) Correlations between climate records from North Atlantic sediments and Greenland ice. *Nature* **366**, 552–554.

Boon J. J., van der Meer F. W., Schuyl P. J., de Leeuw J. W., Schenck P. A., and Burlingame A. L. (1978) *Organic Geochemical Analyses of Core Samples from Site 362, Walvis Ridge*. DSDP Leg 40, Initial Reports. DSDP Legs 38, 39, 40, and 41, Suppl., pp. 627–637.

Brand L. E. (1982) Genetic variability and spatial patterns of genetic differentiation in the reproductive rates of the marine coccolithophores *Emiliania huxleyi* and *Gephyrocapsa oceanica*. *Limmnol. Oceanogr.* **27**, 236–245.

Brand L. E. (1984) The salinity tolerance of forty-six marine phytoplankton isolates. *Estuar. Coast. Shelf Sci.* **18**, 543–556.

Brand L. E. (1991) Minimum iron requirements of marine phytoplankton and the implications for the biogeochemical control of new production. *Limnol. Oceanogr.* **36**, 1756–1771.

Brand L. E. (1994) Physiological ecology of marine coccolithophores. In *Coccolithophores* (eds. A. Winter and W. G. Siesser). Cambridge University Press, Cambridge, UK, pp. 39–49.

Brand L. E. and Guillard R. R. L. (1981) The effects of continuous light and light intensity on the reproduction rates of twenty-two species of marine phytoplankton. *J. Exp. Mar. Biol. Ecol.* **50**, 119–132.

Brassell S. C. (1993) Applications of biomarkers for delineating marine paleoclimate fluctuations during the Quaternary. In *Organic Geochemistry* (eds. M. H. Engel and S. A. Macko). Plenum, New York, pp. 669–738.

Brassell S. C., Eglinton G., Marlowe I. T., Pflaumann U., and Sarnthein M. (1986a) Molecular stratigraphy: a new tool for climatic assessment. *Nature* **320**, 129–133.

Brassell S. C., Brereton R. G., Eglinton G., Grimalt J. O., Liebezeit G., Marlowe I. T., Pflaumann U., and Sarnthein M. (1986b) Palaeoclimatic signals recognized by chemometric treatment of molecular stratigraphic data. *Org. Geochem.* **10**, 649–660.

Brassell S. C., Eglinton G., and Howell V. J. (1987) Palaeoenvironmental assessment for marine organic-rich sediments using molecular organic geochemistry. In *Marine Petroleum Source Rocks* (eds. A. J. Fleet and J. Brooks). Blackwell, London, pp. 79–98.

Broccoli A. J. and Marciniak E. P. (1996) Comparing simulated glacial climate and paleodata: a reexamination. *Paleoceanography* **11**, 3–14.

Broecker W. S. (1986) Oxygen isotope constraints on surface ocean temperatures. *Quat. Res.* **26**, 121–134.

Broecker W. S. (1994) Massive iceberg discharges as triggers for global climate change. *Nature* **372**, 421–424.

Broerse A. T. C., Ziveri P., and Honjo S. (2000a) Coccolithophore ($CaCO_3$) flux in the Sea of Okhotsk: seasonality, settling and alteration processes. *Mar. Micropaleontol.* **39**, 179–200.

Broerse A. T. C., Ziveri P., Van Hinte J. E., and Honjo S. (2000b) Coccolithophore export production, species composition, and coccolith-$CaCO_3$ fluxes in the NE Atlantic (34°N 21°W and 48°N 21°W). *Deep-Sea Res. II* **47**, 1877–1905.

Broerse A. T. C., Brummer G.-J. A., and Van Hinte J. E. (2000c) Coccolithophore export production in response to monsoonal upwelling off Somalia (northwestern Indian ocean). *Deep-Sea Res. II* **47**, 2179–2205.

Brown S. and Elderfield H. (1996) Variations in Mg/Ca and Sr/Ca ratios of planktonic foraminifera caused by post-depositional dissolution-evidence of shallow Mg-dependent dissolution. *Paleoceanography* **11**, 543–551.

Brown C. W. and Yoder J. A. (1994) Coccolithophorid blooms in the global ocean. *J. Geophys. Res.* **99**(C), 7467–7482.

Bukry D. (1974) Coccoliths as paleosalinity indicators-evidence from the Black sea. In *The Black Sea, its Geology, Chemistry and Geology* (eds. E. T. Degens and D. A. Ross). Am. Assoc. Petrol. Geol. Mem., Tulsa, OK, vol. 20, pp. 33–363.

Cacho I., Pelejero C., Grimalt J. O., Calfat A. M., and Canals M. (1999) C37 alkenone measurements of sea surface temperature in the Gulf of Lions (NW Mediterranean). *Org. Geochem.* **30**, 557–566.

Cacho I., Grimalt J. O., Canals M., Sbaffi L., Shackleton N. J., Schonfeld J., and Zahn R. (2001) Variability of the western Mediterranean sea surface temperature during the last 25, 000 years and its connection with the Northern Hemisphere climatic changes. *Paleoceanography* **16**, 40–52.

Cacho I., Grimalt J. O., and Canals M. (2002) Response of the western Mediterranean Sea to rapid climatic variability during the last 50,000 years: a molecular biomarker approach. *J. Mar. Sys.* **33–34**(C), 253–272.

Cadee G. C. (1985) Macroaggregates of *Emiliania huxleyi* in sediment traps. *Mar. Ecol. Prog. Ser.* **24**, 193–196.

Calvo E., Villanueva J., Grimalt J. O., Boelaert A., and Labeyrie L. (2001) New insights into the glacial latitudinal temperature gradients in the North Atlantic: results from $U_{37}^{k'}$ sea surface temperatures and terrigenous inputs. *Earth Planet. Sci. Lett.* **188**, 509–519.

Calvo E., Grimalt J., and Jansen E. (2002) High resolution U_{37}^{k} sea surface temperature reconstruction in the Norwegian Sea during the Holocene. *Quat. Sci. Rev.* **21**, 1385–1394.

Cayre O. and Bard E. (1999) Planktonic foraminiferal and alkenone records of the last deglaciation from the eastern Arabian Sea. *Quat. Res.* **52**, 337–342.

Chaler R., Grimalt J. O., Pelejero C., and Calvo E. (2000) Sensitivity effects in $U_{37}^{k'}$ paleotemperature estimation by chemical ionization mass spectrometry. *Anal. Chem.* **72**, 5892–5897.

Chapman M. R., Shackleton N. J., Zhao M., and Eglinton G. (1996) Faunal and alkenone reconstructions of subtropical North Atlantic surface hydrography and paleotemperature over the last 28 kyr. *Paleoceanography* **11**, 343–357.

CLIMAP (1976) The surface of the Ice-age earth. *Science* **191**, 1131–1137.

CLIMAP project members (1981) Seasonal reconstruction of the Earth's surface at the last glacial maximum GSA map and chart series MC-36. Geol. Soc. Amer., Boulder CO.

CLIMAP (1984) The last interglacial ocean. *Quat. Res.* **21**, 123–224.

Conte M., Volkman J. K., and Eglinton G. (1994a) Lipid biomarkers of the Haptophyta. In *The Haptophyte Algae* (eds. J. C. Green and B. S. C. Leadbeater). Clarendon Press, Oxford, pp. 351–377.

Conte M. H. and Eglinton G. (1993) Alkenone and alkenoate distributions within the euphotic zone of the eastern North Atlantic: correlation with production temperature. *Deep-Sea Res.* **40**, 1935–1961.

Conte M. H., Eglinton G., and Madureira L. A. S. (1992) Long-chain alkenones and alkyl alkenoates as palaeotemperature indicators: their production, flux, and early sedimentary diagenesis in the eastern North Atlantic. *Org. Geochem.* **19**, 287–298.

Conte M. H., Thompson A., and Eglinton G. (1994) Primary production of lipid biomarker compounds by *Emiliania huxleyi*: results from an experimental mesocosm study in fjords of southern Norway. *Sarsia* **79**, 319–332.

Conte M. H., Thompson A., Eglinton G., and Green J. C. (1995) Lipid biomarker diversity in the coccolithophorid *Emiliania huxleyi* (Prymniesiophyceae) and the related species *Gephyrocapsa oceanica*. *J. Phycol.* **31**, 272–282.

Conte M. H., Thompson A., Lesley D., and Harris R. P. (1998a) Genetic and physiological influences on the alkenone/alkenoate versus growth temperature relationship in *Emiliania huxleyi* and *Gephyrocapsa oceanica*. *Geochim. Cosmochim. Acta* **62**, 51–68.

Conte M. H., Weber J. C., and Ralph N. (1998b) Episodic particle flux in the deep Sargasso Sea an organic geochemical assessment. *Deep-Sea Res. I.: Ocean. Res.* **45**, 1819–1841.

Conte M. H., Weber J. C., King L. L., and Wakeham S. G. (2001) The alkenone temperature signal in the western North Atlantic surface waters. *Geochim. Cosmochim. Acta* **65**, 4275–4287.

Conte M. H., Sicre M. A., Weber J. C., and Shultz-Bull D. (xxxx) The global temperature calibration of the alkenone unsaturation index ($U_{37}^{k'}$) in surface waters and a comparison of the alkenone production temperature recorded by $U_{37}^{k'}$ in sediments with overlying sea surface temperature. *Paleoceanography* (in press).

Cortés M. Y., Bollmann J., and Thierstein H. R. (2001) Coccolithophore ecology at the HOT station ALOHA, Hawaii. *Deep-Sea Res. II* **48**, 1957–1981.

Cranwell P. A. (1985) Long-chain unsaturated ketones in recent lacustrine sediments. *Geochim. Cosmochim. Acta* **49**, 1545–1551.

Dansgaard W., White J. W. C., and Johnsen S. J. (1989) The abrupt termination of the Younger Dryas climate event. *Nature* **339**, 532–534.

De Leeuw J. W., Meer F. W. V. D., Rijpstra W. I. C., and Schencck P. A. (1980) On the occurrence and structural identification of long chain unsaturated ketones and hydrocarbons in sediments. In *Advances in Organic Geochemistry, 1979*, Phys. Chem. Earth, vol. 12 (eds. A. G. Douglas and J. R. Maxwell). Pergamon, Oxford, 211–217.

Doose H., Prahl F. G., and Lyle M. W. (1997) Biomarker temperature estimates from modern and last glacial surface waters of the California current system between 33° and 42°N. *Paleoceanography* **12**, 615–622.

Doose-Rolinski H., Rogalla U., Scheeder G., Luckge A., and von Rad U. (2001) High-resolution temperature and evaporation changes during the late Holocene in the northeastern Arabian Sea. *Paleoceanography* **16**, 358–367.

Edvardsen B., Eikrem W., Green J. C., Andersen R. A., Moon-van der Staay S. Y., and Medlin L. K. (2000) Phylogenetic reconstructions of the Haptophyta inferred from 18S ribosomal DNA sequences and available morphological data. *Phycologia* **39**, 19–35.

Eglinton G., Bradshaw S. A., Rosell A., Sarnthein M., Pflaumann U., and Tiedemann R. (1992) Molecular record of secular sea surface temperature changes on 100-year timescales for glacial terminations I, II, and IV. *Nature* **356**, 423–426.

Elderfield H. and Ganssen G. (2000) Past temperature and $\delta^{18}O$ of surface ocean waters inferred from foraminiferal Mg/Ca ratios. *Nature* **405**, 442–445.

Emeis K.-C., Anderson D. M., Doose H., Kroon D., and Schulz-Bull D. (1995a) Sea-surface temperatures and the history of monsoon upwelling in the northwest Arabian Sea during the last 500,000 years. *Quat. Res.* **43**, 355–361.

Emeis K.-C., Doose H., Mix A., and Schulz-Bull D. (1995b) Alkenone sea-surface temperatures and carbon burial at Site 846 (eastern equatorial Pacific Ocean): the last 1.3 MY. *Sci. Res. ODP* **138**, 605–613.

Emeis K.-C., Struck U., Schulz H.-M., Rosenberg R., Bernasconi S., Erlenkeuser H., Sakamoto T., and Martinez-Ruiz F. (2000) Temperature and salinity variations of Mediterranean Sea surface waters over the last 16,000 years from records of planktonic stable isotopes and alkenone unsaturation ratios. *Palaeogeogr. Paleaoeclimatol. Palaeoecol.* **158**, 259–280.

Epstein B., D'Hondt S., Quinn J. G., Zhang J., and Hargraves P. E. (1998) An effect of dissolved nutrient concentrations on alkenone-based temperature estimates. *Paleoceanography* **13**, 122–126.

Epstein B. L., D'Hondt S., and Hargraves P. E. (2001) The possible metabolic role of C37 alkenones in *Emiliania huxleyi*. *Org. Geochem.* **32**, 867–875.

Fairbanks R. G. and Matthews R. K. (1978) The marine oxygen isotopic record in Pleistocene coral, Barbados, West Indies. *Quat. Res.* **10**, 181–196.

Farrimond P. G., Eglinton P. G., and Brassell S. C. (1986) Alkenones in cretaceous black shales, Blake-Bahama basin, western North Atlantic. In *Advances in Organic Geochemistry 1985* (eds. D. Leytaheuser and J. Rullkotter). Pergamon, Oxford, vol. 10, pp. 897–903.

Farrington J. W., Davis A. C., Sulanowski J., McCaffrey M. A., McCarthy M., Clifford C. H., Dickinson D., and Volkman J. K. (1988) Biogeochemistry of lipids in surface sediments of the Peru upwelling area at 15° S. *Org. Geochem.* **19**, 277–285.

Ficken K. J. and Farrimond P. (1995) Sedimentary lipid geochemistry of Framvaren: impacts of a changing environment. *Mar. Chem.* **51**, 31–43.

Fisher N. S. and Honjo S. (1989) Interspecific differences in temperature and salinity responses in the coccolithophore *Emiliania huxleyi*. *Biol. Oceanogr.* **6**, 355–361.

Freeman K. H. and Wakeham S. G. (1992) Variation in the distribution and isotopic composition of alkenones in Black Sea particles and sediments. *Org. Geochem.* **19**, 277–285.

Fujiwara S., Tsuzuki M., Kawchi M., Minaka N., and Inouye I. (2001) Molecular phylogeny of the Haptophyta based on the *rbc*L gene and sequence variation in the spacer region of the rubisco operon. *J. Phycol.* **37**, 121–129.

Giraudeau J., Monteiro M. S., and Nikodemus K. (1993) Distribution and malformation of living coccolithophores in the northern Benguela upwelling system off Namibia. *Mar. Micropaleontol.* **22**, 93–110.

Gong C. and Hollander D. J. (1999) Evidence for differential degradation of alkenones under contrasting bottom water oxygen conditions: implication for paleotemperature reconstruction. *Geochim. Cosmochim. Acta* **63**, 405–411.

Goni M. A., Hartz D. M., Thunell R. C., and Tappa E. (2001) Oceanographic considerations in the application of the alkenone-based paleotemperature $U_{37}^{k'}$ index in the Gulf of California. *Geochim. Cosmochim. Acta* **65**, 545–557.

Grice K., Klein Breteler W. C. M., Schoten S., Grossi V., de Leeuw J. W., and Sinninge Damste J. S. (1998) Effects of zooplankton herbivory on biomarker proxy records. *Paleoceanography* **13**, 686–693.

Grimalt J. O., Rulkotter J., Sicre M.-A., Summons R., Farrington J., Harvey H. R., Goni M., and Sawada K. (2000) Modifications of the C_{37} alkenone and alkenoate composition in the water column and sediment: possible implications for sea surface temperature estimates in Paleoceanography. *Geochem. Geophys. Geosys. 1* (paper number 2000GC000053).

Grimalt J. O., Calvo E., and Pelejero C. (2001) Sea surface paleotemperature errors in $U_{37}^{k'}$ estimation due to alkenone measurements near the limit of detection. *Paleoceanography* **16**, 226–232.

Grootes P. M., Stuiver M., White J. W. C., Johnsen S., and Jouzel J. (1993) Comparison of oxygen isotope records from the GISP2 and GRIP Greenland ice cores. *Nature* **366**, 552–554.

Grossi V., Rapel D., Auber C., and Rontani J.-F. (2000) The effect of growth temperature on the long-chain alkenes composition in the marine coccolithophorid *Emiliania huxleyi*. *Phytochemistry* **54**, 393–399.

Guilderson T. P., Fairbanks R. G., and Rubenstone J. L. (1994) Tropical temperature variations since 20,000 years ago: modulating interhemispheric climate change. *Science* **263**, 663–665.

Haidar A. T. and Thierstein H. R. (2001) Coccolithophore dynamics off Bermuda (N. Atlantic). *Deep-Sea Res. II* **48**, 1925–1956.

Hamanaka J., Sawada K., and Tanoue E. (2000) Production rates of C37 alkenones determined by ^{13}C-labeling technique in the euphotic zone of Sagami Bay, Japan. *Org. Geochem.* **31**, 1095–1102.

Harada N., Handa N., Harada K., and Matsuoka H. (2001) Alkenones and particulate fluxes in sediment traps from the central equatorial Pacific. *Deep-Sea Res. I: Oceanogr. Res. Pap.* **48**, 891–907.

Herbert T. D. (2000) Review of alkenone calibrations (culture, water column, and sediments). *Geochem. Geophys. Geosys.* **1** [Paper no. 2000GC000055].

Herbert T. D. and Schuffert J. (1998) Alkenone unsaturation estimates of late Miocene through late Pliocene sea surface temperature changes, ODP Site 958. *Proc. Ocean Drill. Prog.: Sci. Results* **159T**, 17–22.

Herbert T. D., Schuffert J. D., Thomas D., Lange K., Weinheimer A., and Herguera J.-C. (1998) Depth and seasonality of alkenone production along the California margin inferred from a core-top transect. *Paleoceanography* **13**, 263–271.

Herbert T. D., Schuffert J. D., Andreasen D., Heusser L., Lyle M., Mix A., Ravelo A. C., Stott L. D., and Herguera J. C. (2001) Collapse of the California current during glacial maxima linked to climate change on land. *Science* **293**, 71–76.

Hoefs M. J. L., Versteegh G. J. M., Ripstra W. I. C., de Leeuw J. W., and Sinninghe Damste J. S. (1998) Postdepositional oxic degradational of alkenones: implications for the measurement of palaeo sea surface temperatures. *Paleoceanography* **13**, 42–59.

Houghton S. D. and Guptha M. V. S. (1991) Monsoonal and fertility controls on recent marginal sea and continental shelf coccolith assemblages from the western Pacific and the northern Indian oceans. *Mar. Geol.* **97**, 251–259.

Huang C. Y., Wu S. F., Zhao M., Chen M. T., Wang C. H., Tu X., and Yuan P. B. (1997) Surface ocean and monsoon climate variability in the South China sea since the last glaciation. *Mar. Micropaleontol.* **32**, 71–94.

Jasper J. and Gagosian R. B. (1989) Alkenone molecular stratigraphy in an oceanic environment affected by glacial freshwater events. *Paleoceanography* **4**, 603–614.

Jiang M. J. and Gartner S. (1984) Neogene and quaternary calcareous nannofossil biostratigraphy of the Walvis ridge. In *Init Rep. DSDP 74* (eds. T. C. Moore and E. Rabinowitz). US Govt. Printing Office, Washington, DC, pp. 561–595.

Johnsen S. J., Clausen H. B., Dansgaard W., Fuhrer K., Gundestrup N., Hammer C. U., Iversen P., Jouzel J., Stauffer B., and Steffensen J. P. (1992) Irregular glacial interstadials recorded in a new Greenland ice core. *Nature* **359**, 311–313.

Jordan R. W., Zhao M., Eglinton G., and Weaver P. P. E. (1996) Coccolith and alkenone stratigraphy at a NW African upwelling site (ODP 658C) over the last 130,000 years. In *Microfossils and Oceanic Environments*, British Micropaleontol. Soc. Series (eds. A. L. Moguilevsky and R. Whateley), pp. 111–130.

Kennedy J. and Brassell S. C. (1992) Molecular records of twentieth century El Nino events in laminated sediments from Santa Barbara basin. *Nature* **357**, 62–64.

Kienast S. and McKay J. L. (2001) Sea surface temperatures in the subarctic northeast Pacific reflect millennial-scale climate oscillations during the last 16 kyr. *Geophys. Res. Lett.* **28**, 1563–1566.

Kienast M., Steinke S., Stattegger K., and Calvert S. E. (2001) Synchronous tropical South China Sea SST change and Greenland warming during deglaciation. *Science* **291**, 2132–2134.

Kim J. H., Schneider R. R., Müller P. J., and Wefer G. (2002) Interhemispheric comparison of deglacial sea-surface temperature patterns in Atlantic eastern boundary currents. *Earth Planet. Sci. Lett.* **194**, 383–393.

Kirst G., Schneider R. R., Muller P. J., von Storch I., and Wefer G. (1999) Late quaternary temperature variability in the Benguela current system derived from Alkenones. *Quat. Res.* **52**, 92–103.

Koopmans M. P., Schaeffer-Reiss C., DeLeeuw J. W., Lean M. D., Maxwell J. R., Schaeffer P., and Sinninghe Damste J. S. (1997) Sulphur and oxygen sequestration of n-C37 and n-C38 unsaturated ketones in an immature kerogen and the release of their carbon skeletons during early stages of thermal maturation. *Geochim. Cosmochim. Acta* **61**, 2397–2408.

Lea D. W., Mashiotta T. A., and Spero H. J. (1999) Controls on magnesium and strontium uptake in planktonic foraminifera determined by live culturing. *Geochim. Cosmochim. Acta* **63**, 2369–2379.

Lea D. W., Pak D. K., and Spero H. J. (2000) Climate impact of late quaternary Equatorial Pacific sea surface temperature variations. *Science* **289**, 1719–1724.

Lee K. Y., Slowey N., and Herbert T. D. (2001) Glacial sea surface temperatures in the subtropical North Pacific: a comparison of $U_{37}^{k'}$, $\delta^{18}O$, and foraminiferal assemblage temperature estimates. *Paleoceanography* **16**, 268–279.

Levitus S. (1994) *Climatological Atlas of the World Ocean*, NOAA Prof. Pa.13. US Govt. Print. Off., Washington, DC, 173pp.

Li J., Philp R. P., Pu F., and Allen J. (1996) Long-chain alkenones in Qighai lake sediments. *Geochem. Cosmochim. Acta* **60**, 235–241.

Lichtfouse E., Littke R., Disko U., Willshc H., Ruldotter J., and Stein R. (1992) Geochemistry and petrology of organic matter in Miocene to quaternary deep sea sediments from the Japan Sea (Sites 798 and 799). *Sci. Res. ODP* **127/128**, 667–675.

Lyle M., Prahl F. G., and Sparrow M. A. (1992) Upwelling and productivity changes inferred from a temperature record in the central equatorial Pacific. *Nature* **355**, 812–815.

Madureira L. A. S., Conte M. H., and Eglinton G. (1995) The early diagenesis of lipid biomarker compounds in North Atlantic sediments. *Paleoceangraphy* **10**, 627–642.

Marchal O., Cacho I., Stocker T. F., Grimalt J. O., Calvo E., Martrat B., Shackleton N., Vautravers M., Cortijo E., van Kreveld S., Andersson C., Koc N., Chapman M., Sbaffi L., Duplessy J. C., Sarnthein M., Turon J. L., Duprat J., and Jansen E. (2002) Apparent long-term cooling of the sea surface in the northeast Atlantic and Mediterranean during the Holocene. *Quat. Sci. Rev.* **21**, 455–483.

Marlow J. R., Lange C. B., Wefer G., and Rosell-Mele A. (2000) Upwelling intensification as part of the Pliocene-pleistocene climate transition. *Science* **290**, 2288–2291.

Marlowe I. T., Green J. C., Neal A. C., Brassell S. C., Eglinton G., and Course P. A. (1984a) Long chain (n-C37–C39) alkenones in the prymnesiophyceae: distribution of alkenones and other lipids and their taxonomic significance. *British Phycol. J.* **19**, 203–216.

Marlowe I. T., Brassell S. C., Eglinton G., and Green J. C. (1984b) Long chain unsaturated ketones and esters in living algae and marine sediments. *Org. Geochem.* **6**, 135–141.

Marlowe I. T., Brassell S. C., Eglinton G., and Green J. C. (1990) Long-chain alkenones and alkyl alkenoates and the fossil coccolith record of marine sediments. *Chem. Geol.* **88**, 349–375.

Mayer B. and Schwark L. (1999) A 15,000-year stable isotope record from sediments of lake Steisslingen, southwest Germany. *Chem. Geol.* **161**, 315–337.

Mazaud A., Sicre M. A., Ezat U., Pichon J. J., Duprat J., Laj C., Kissel C., Beaufort L., Michel E., and Turon J. L. (2002) Geomagnetic-assisted stratigraphy and sea surface temperature changes in core MD94-103 (southern Indian ocean): possible implications for North-South climatic relationships around H4. *Earth Planet. Sci. Lett.* **201**, 159–170.

McCaffrey M. A., Farrington J. W., and Repeta D. J. (1990) The organic geochemistry of Peru margin surface sediments: 1. A comparison of the C37 alkenone and historical El Nino records. *Geochim. Cosmochim. Acta* **54**, 1671–1682.

McIntyre A. (1970) *Gephyrocapsa protohuxleyi* sp. n. as possible phyletic link and index fossil for the Pleistocene. *Deep-Sea Res.* **17**, 187–190.

Medlin L. K., Barker G. L. A., Campbell L., Green J. C., Hayes P. K., Marie D., Wrieden S., and Vaulot G. (1996) Genetic characterization of *Emiliania huxleyi* (Haptophyta). *J. Mar. Sys.* **9**, 13–31.

Mercer J. L., Zhao M., and Coleman S. M. (1999) Alkenone evidence of sudden changes in Chesapeake Bay conditions ca. 300 years ago. *EOS 80 Suppl.* S185.

Mix A. C., Morey A. E., Pisias N. G., and Hostetler S. W. (1999) Foraminiferal faunal estimates of paleotemperature: circumventing the no-analog problem yields cool ice age tropics. *Paleoceanography* **14**, 350–359.

Mix A. C., Bard E., and Schneider R. (2001) Environmental processes of the Ice age: land, ocean, glaciers (EPILOG). *Quat. Sci. Rev.* **20**, 627–657.

Mouzdahir A. v., Grossi Bakkas S., and Rontani J.-F. (2001) Visible light-dependent degradation of long-chained alkenes in killed cells of *Emiliania huxleyi* and *Nannocchlorpsis salina*. *Phytochemistry* **56**, 677–684.

Muller P. J. and Fischer G. (2001) A 4-year sediment trap record of alkenones from the filamentous upwelling region off Cape Blanc, NW: Africa and a comparison with distributions in underlying sediments. *Deep-Sea Res.* **48**, 1877–1903.

Muller P. J., Cepek M., Ruhland G., and Schneider R. R. (1997) Alkenone and coccolithophorid species changes in late quaternary sediments from the Walvis ridge: implications for the alkenone paleotemperature method. *Palaeogeogr. Palaeoclimatol. Palaeoecol.* **135**, 71–96.

Muller P. J., Kirst G., Ruhland G., von Storch I., and Rosell-Mele A. (1998) Calibration of the alkenone paleotemperature index $U_{37}^{k'}$ based on core-tops from the eastern South Atlantic and the global ocean (60° N–60° S). *Geochim. Cosmochim. Acta* **62**, 1757–1772.

Nanninga H. J. and Tyrell T. (1996) The importance of light for the formation of algal blooms by *Emiliania huxleyi*. *Mar. Ecol. Prog. Ser.* **136**, 195–203.

Nishida S. (1986) Nannoplankton flora in the southern Ocean, with special reference to siliceous varieties. *Mem. Nat. Inst. Polar. Res.* (spec. issue) **40**, 56–68.

Nuernberg D., Muller A., and Schneider R. R. (2000) Paleo-sea surface temperature calculations in the equatorial East Atlantic from Mg/Ca ratios in planktic foraminifera: a comparison to sea surface temperature estimates from $U_{37}^{k'}$, oxygen isotopes, and foraminiferal transfer functions. *Paleoceanography* **15**, 124–134.

Ohkouchi N., Kawamua K., Nakamura T., and Taira A. (1994) Small changes in the sea surface temperature during the last 20,000 years: molecular evidence from the western tropical Pacific. *Geophys. Res. Lett.* **21**, 2207–2210.

Ohkouchi N., Kawamura K., Kawahata H., and Okada H. (1999) Depth ranges of alkenone production in the central Pacific Ocean. *Global Biogeochem. Cycles* **13**, 695–704.

Ohkouchi N., Eglinton T. I., Keigwin L. D., and Hayes J. M. (2002) Spatial and temporal offsets between proxy records in a sediment drift. *Science* **298**, 1224–1227.

Okada H. and Honjo S. (1973) The distribution of oceanic coccolithophorids in the Pacific. *Deep-Sea Res.* **26**, 355–374.

Okada H. and McIntyre A. (1979) Seasonal distribution of modern coccolithophores in the western North Atlantic Ocean. *Mar. Biol.* **54**, 319–328.

Paasche E. (2002) A review of the coccolithophorid *Emiliania huxleyi* (Prymnesiophyceae) with particular reference to growth, coccolith formation, and calcification-photosynthesis interactions. *Phycologia* **40**, 503–529.

Pelejero C. and Grimalt J. O. (1997) The correlation between the $U_{37}^{k'}$ index and sea surface temperatures in the warm boundary: the South China Sea. *Geochim. Cosmochim. Acta* **61**, 4789–4797.

Pelejero C., Grimalt J. O., Heilig S., Kienast M., and Wang L. (1999a) High-resolution $U_{37}^{k'}$ temperature reconstructions in the South China Sea over the past 220 kyr. *Paleoceanography* **14**, 224–231.

Pelejero C., Kienast M., Wang L., and Grimalt J. O. (1999b) The flooding of Sundaland during the last deglaciation: imprints in hemipelagic sediments from the southern South China Sea. *Earth Planet. Sci. Lett.* **171**, 661–671.

Pelejero C. and Calvo E. (2003) The upper end of the $U_{37}^{k'}$ temperature calibration revisited. *Geochem. Geophys. Geosyst.* **4**, article 1014.

Perez-Folgado M., Sierro F. J., Flores J. A., Cacho I., Grimalt J. O., Zahn R., and Shackleton N. (2003) Western Mediterranean planktonic foraminifera events and millennial climatic variability during the last 70 kyr. *Mar. Micropaleontol.* **48**, 49–70.

Popp B. N., Kenig F., Wakeham S. G., and Bidigare R. R. (1998) Does growth rate affect ketone unsaturation and intracellular carbon isotopic variability in *Emiliania huxleyi*? *Paleoceanography* **13**, 35–41.

Prahl F. G. and Wakeham S. G. (1987) Calibration of unsaturation patterns in long-chain ketone compositions for paleotemperature assessment. *Nature* **330**, 367–369.

Prahl F. G., Muelhausen L. A., and Zahnle D. L. (1988) Further evaluation of long-chain alkenones as indicators of paleoceanography conditions. *Geochim. Cosmochim. Acta* **52**, 2303–2310.

Prahl F. G., de Lange G. J., Lyle M., and Sparrow M. A. (1989a) Post-depositional stability of long-chain alkenones under contrasting redox conditions. *Nature* **341**, 434–437.

Prahl F. G., Muelhausen L. A., and Lyle M. (1989b) An organic geochemical assessment of oceanographic conditions at MANOP site C over the past 26,000 years. *Paleoceanography* **5**, 495–510.

Prahl F. G., Collier R. B., Dymond J., Lyle M., and Sparrow M. A. (1993) A biomarker perspective on prymnesiophyte productivity in the northeast Pacific Ocean. *Deep-Sea Res.* **40**, 2061–2076.

Prahl F. G., Pisias N., Sparrow M. A., and Sabin A. (1995) Assessment of sea-surface temperature at 42° N in the California current over the last 30,000 years. *Paleoceanography* **10**, 763–773.

Prahl F. G., Dymond J., and Sparrow M. A. (2000) Annual biomarker record for export production in the central Arabian Sea. *Deep-Sea Res II* **47**, 1581–1604.

Prahl F. G., Pilskaln C. H., and Sparrow M. A. (2001) Seasonal record for alkenones in sedimentary particles from the Gulf of Maine. *Deep-Sea Res. I* **48**, 515–528.

Pujos A. (1987) Late Eocene to Pleistocene medium-sized and small-sized 'Reticulofenestrids'. *Abh. Geol. Bund. A* **39**, 239–277.

Rinna J., Warning B., Meyers P. A., Brumsack H.-J., and Rullkötter J. (2002) Combined organic and inorganic geochemical reconstruction of paleodepositional conditions of a Pliocene sapropel from the eastern Mediterranean Sea. *Geochim. Cosmochim. Acta* **66**, 1969–1986.

Rio D. (1982) The fossil distribution of coccolithophore genus gephyrocapsa kamptner and related plio-pleistocene chronostratigraphic problems. In *Init. Rep. DSDP* 68 (eds. W. L. Prell and J. V. Gardner). US Govt. Printing Office, Washington, DC, pp. 325–343.

Rontani F.-F., Cuny P., Gossi V., and Beker B. (1997) Stability of long-chain alkenones in senescing cells of *Emiliania huxleyi*: effect of photochemical and aerobic microbial degradation on the alkenone unsaturation ratio ($U_{37}^{k'}$). *Org. Geochem.* **26**, 503–509.

Rosell-Mele A. (1998) Interhemispheric appraisal of the value of alkenone indices as temperature and salinity proxies in high-latitude locations. *Paleoceanography* **13**, 694–703.

Rosell-Mele A., Carter J., and Eglinton G. (1994) Distribution of long-chain alkenones and alkyl alkenoates in marine surface sediments from the North East Atlantic. *Org. Geochem.* **22**, 501–509.

Rosell-Mele A., Carter J., Parry A. T., and Eglinton G. (1995a) Determination of the U_{37}^{k} index in geological samples. *Anal. Chem.* **67**, 1283–1289.

Rosell-Mele A., Eglinton G., Pflaumann U., and Sarnthein M. (1995b) Atlantic core top calibration of the U_{37}^{k} index as a sea-surface temperature indicator. *Geochim. Cosmochim. Acta* **59**, 3099–3107.

Rosell-Mele A., Maslin M. A., Maxwell J. R., and Schaeffer P. (1997) Biomarker evidence for Heinrich events. *Geochim. Cosmochim. Acta* **61**, 1671–1678.

Rosell-Mele A., Bard E., Emeis E.-C., Grimalt J. O., Muller P., Schneider R., Bouloubassi I., Epstein B., Fahl K., Fluegge A., Freeman K., Goni M., Guntner U., Hrtz D., Hellebust S., Herbert T., Ikehara M., Ishiwatari R., Kawamura K., Kenig F., de Leeuw J., Lehman S., Mejanelle L., Ohkouchi N., Pancost R. D., Pelejero C., Prahl F., Quinn J., Rontani J. F., Rostek F., Rulkotter J., Sachs J., Blanz T., Sawada K., Schulz-Bull D., Sikes E., Sonzogni C., Ternois Y., Versteegh G., Volkman J. K., and Wakeham S. (2001) Precision of the current methods to measure the alkenone proxy $U_{37}^{k'}$ and absolute alkenone abundance in sediments: results of an inter-laboratory comparison study. *Geochem. Geophys. Geosys.* **2**, paper 2000GC000141.

Rosell-Mele A., Jansen E., and Weinelt M. (2002) Appraisal of a molecular approach to infer variations in surface ocean freshwater inputs into the North Atlantic during the last glacial. *Global Planet. Change* **34**, 143–152.

Rosenthal Y., Lohmann G. P., Lohmann K. C., and Sherrell R. M. (2000) Incorporation and preservation of Mg in *Globigerinoides sacculifer*: implications for reconstructing the temperature and $^{18}O/^{16}O$ of seawater. *Paleoceanography* **15**, 135–145.

Rostek F., Ruhland G., Bassinot F. C., Muller P. J., Labeyrie L. D., Lancelot Y., and Bard E. (1993) Reconstructing sea surface temperature using $\delta^{18}O$ and alkenone records. *Nature* **364**, 319–321.

Rostek F., Bard E., Beaufort L., Sonzogni C., and Ganssen G. (1997) Sea surface temperature and productivity records for the past 240 kyr in the Arabian Sea. *Deep-Sea Res.* **44**, 1461–1480.

Roth P. H. (1994) Distribution of coccoliths in oceanic sediments. In *Coccolithophores* (eds. A. Winter and W. G. Siesser). Cambridge University Press, Cambridge, UK, pp. 199–218.

Ruddiman W. F., Raymo M., and McIntyre A. (1986) Matuyama 41000 year cycles: North Atlantic Ocean and northern hemisphere ice sheets. *Earth Planet. Sci. Lett.* **80**, 117–129.

Ruhlemann C., Mulitza S., Muller P. J., Wefer G., and Zahn R. (1999) Warming of the tropical Atlantic Ocean and slow-down of thermohaline circulation during the last deglaciation. *Nature* **402**, 511–514.

Sachs J. P. and Lehman S. J. (1999) Subtropical North Atlantic temperatures 60,000 to 30,000 years ago. *Science* **286**, 756–759.

Sachs J. and Lehman S. J. (2001) Glacial surface temperatures of the southeast Atlantic Ocean. *Science* **293**, 2077–2079.

Samtleben C. and Bickert T. (1990) Coccoliths in sediment traps from the Norwegian Sea. *Mar. Micropaleontol.* **16**, 39–64.

Sawada K., Handa N., Shiraiwa Y., Danbara A., and Montani S. (1996) Long-chain alkenones and alkyl alkenoates in the coastal and pelagic sediments of the northwest North Pacific, with special reference to the reconstruction of *Emiliania huxleyi* and *Geophyrocapsa oceanica* ratios. *Org. Geochem.* **24**, 751–764.

Sawada K., Handa N., and Nakatsuka T. (1998) Production and transport of long-chain alkenones and alkyl alkenoates in a sea water column in the northwestern Pacific off central Japan. *Mar. Chem.* **59**, 219–234.

Schneider R. R., Muller P. J., and Ruhland G. (1995) Late quaternary surface circulation in the east-equatorial Atlantic: evidence from alkenone sea surface temperatures. *Paleoceanography* **10**, 197–220.

Schneider R. R., Muller P. J., Ruhland G., Meinecke G., Schmidt H., and Wefer G. (1996) Late quaternary surface temperatures and productivity in the east-equatorial South Atlantic: response to changes in trade/monsoon wind forcing and surface water advection. In *The South Atlantic: Present and Past Circulation* (eds. G. Wefer, W. H. Berger, G. Siedler, and D. J. Webb). Springer, Berlin, pp. 527–551.

Schneider R. R., Muller P. J., and Acheson R. (1999) Atlantic alkenone sea surface temperature records. In *Reconstructing Ocean History: A Window into the Past* (eds. Abrantes and A. C. Mix). Academic Press, New York, pp. 33–55.

Schrag D. P., Hampt G., and Murray D. W. (1996) Pore fluid constraints on the temperature and oxygen isotopic composition of the Glacial Ocean. *Science* **272**, 1930–1932.

Schulte S. and Müller P. J. (2001) Variations in sea-surface temperature and primary productivity during Heinrich and

Dansgaard-Oeschger events in the northeastern Arabian Sea. *Geo-Mar. Lett.* **21**, 168–175.

Schulte S., Rostek F., Bard E., Rulkotter J., and Marchal O. (1999) Variations of oxygen-minimum and primary productivity recorded in sediments of the Arabian Sea. *Earth Planet. Sci. Lett.* **173**, 205–221.

Schulz H.-H., Schoener A., and Emeis K.-C. (2000) Long-chain alkenone patterns in the Baltic Sea an ocean-freshwater transition. *Geochim. Cosmochim. Acta* **64**, 469–477.

Seki O., Ishiwatari R., and Matsumoto K. (2002) Millennial climate oscillations in NE Pacific surface waters over the last 82 kyr: new evidence from alkenones. *Geophys. Res. Lett.* **29** 101029/2002GL015200.

Sicre M.-A., Ternois Y., Miquel J.-C., and Marty J.-C. (1999) Alkenones in the Mediterranean Sea: interannual variability and vertical transfer. *Geophys. Res. Lett.* **26**(12), 1735–1738.

Sicre M.-A., Ternois Y., Paterne M., Boireau P., Beaufort L., Martinez P., and Betrand P. (2000) Biomarker stratigraphic records over the last 150 Kyrs off the NW African coast at 25°N. *Org. Geochem.* **31**, 577–588.

Sicre M.-A., Bard E., Ezat U., and Rostek F. (2002) Alkenone distributions in the North Atlantic and Nordic Sea surface waters. *Geochem. Geophys. Geosys.* **3**, paper 2001GC000159.

Sikes C. S. and Fabry V. J. (1994) Photosynthesis, $CaCO_3$ deposition, coccolithophorids and the global carbon cycle. In *Regulation of Atmospheric CO_2 and O_2 by Photosynthetic Carbon Metabolism* (eds. N. E. Tolbert and J. Preiss). Oxford University Press, New York, pp. 217–233.

Sikes E. L. and Keigwin L. D. (1994) Equatorial Atlantic sea surface temperatures for the last 16 kyr: a comparison of U^{k}_{37}, $\delta^{18}O$, and foraminiferal assemblage temperature estimates. *Paleoceanography* **9**, 31–45.

Sikes E. L. and Volkman J. K. (1993) Calibration of alkenone unsaturation ratios (U^{k}_{37}) for paleotemperature estimation in cold polar waters. *Geochim. Cosmochim. Acta* **57**, 1883–1889.

Sikes E. L., Farrington J. W., and Keigwin L. D. (1991) Use of the alkenone unsaturation ratio U^{k}_{37} to determine past sea surface temperatures: core-top SST calibrations and methodology considerations. *Earth Planet. Sci. Lett.* **104**, 36–47.

Sikes E. L., Volkman J. K., Robertson L. G., and Pichon J.-J. (1997) Alkenones and alkenes in surface waters and sediments of the Southern Ocean: implications for paleotemperature estimation in polar regions. *Geochim. Cosmochim. Acta* **61**, 1495–1505.

Simoneit B. R. T., Prahl F. G., Leif R. N., and Mao S.-Z. (1994) Alkenones in sediments of middle valley. *Sci. Res. ODP* **139**, 479–484.

Sinninghe-Damsté J. S., Rijpstra W. I. C., and Reichart G.-J. (2002) The influence of oxic degradation on the sedimentary biomarker record: II. Evidence from Arabian Sea sediments. *Geochim. Cosmochim. Acta* **66**, 2737–2754.

Sonzogni C., Bard E., Rostek F., Dollfus D., Rosell-Mele A., and Eglinton G. (1997) Temperature and salinity effects on alkenone ratios measured in surface sediments from the Indian Ocean. *Quat. Res.* **47**, 344–355.

Sonzogni C., Bard E., and Rostek F. (1998) Tropical sea surface temperatures during the last glacial period: a view based on alkenones in Indian Ocean sediments. *Quat. Sci. Rev.* **17**, 1185–1201.

Sprengel C., Baumann K. H., and Neuer S. (2000) Seasonal and interannual variation of coccolithophore fluxes and species composition in sediment traps north of Gran Canaria (29 degrees N 15 degrees W). *Mar. Micropaleontol.* **39**, 157–178.

Sprengel C., Baumann K. H., Henderiks J., Henrich R., and Neuer S. (2002) Modern coccolithophore and carbonate sedimentation along a productivity gradient in the Canary Islands region: seasonal export production and surface accumulation rates. *Deep-Sea Res. II* **49**, 3577–3598.

Steinke S., Kienast M., Pflaumann U., Weinelt M., and Stattegger K. (2001) A high-resolution sea-surface temperature record from the tropical South China Sea (16,500–3,000 yr BP). *Quat. Res.* **55**, 352–362.

Stott L., Poulsen C., Lund S., and Thunell R. (2002) Super ENSO and global climate oscillations at millennial time-scales. *Science* **297**, 222–226.

Stute M., Forster M., Frischkorn H., Serejo A., Clark J. F., Schlosser P., Broecker W. S., and Bonani G. (1995) Cooling of tropical Brazil (5°C) during the last glacial maximum. *Science* **269**, 379–383.

Summerhayes C. P., Kroon D., Rosell-Mele A., Jordan R. W., Schrader H.-J., Hearn R., Villanueva J., Grimalt J. O., and Eglinton G. (1995) Variability in the benguela current upwelling system over the past 70,000 years. *Prog. Oceanogr.* **35**, 207–251.

Sun M.-Y. and Wakeham S. G. (1994) Molecular evidence for degradation and preservation of organic matter in the anoxic Black Sea basin. *Geochim. Cosmochim. Acta* **58**, 3395–3406.

Teece M. A., Getliff J. M., Leftley J. W., Parkes R. J., and Maxwell J. R. (1998) Microbial degradation of the marine prymnesiophyte *Emiliania huxleyi* under oxic and anoxic conditions as a model for early diagenesis: long chain alkadienes, alkenones, and alkyl alkenoates. *Org. Geochem.* **29**, 863–880.

Ternois Y., Sicre M.-A., Boireau A., Marty J.-C., and Miquel J.-C. (1996) Production pattern of alkenones in the Mediterranean Sea. *Geophys. Res. Lett.* **23**, 3171–3174.

Ternois Y., Sicre M.-A., Boireau A., Conte M. H., and Eglinton G. (1997) Evaluation of long-chain alkenones as paleotemperature indicators in the Mediterranean Sea. *Deep-Sea Res.* **44**, 271–286.

Ternois Y., Sicre M.-A., Boireau A., Beaufort L., Migquel J.-C., and Jeandel C. (1998) Hydrocarbons, sterols, and alkenones in sinking particles in the Indian Ocean sector of the Southern Ocean. *Org. Geochem.* **28**, 489–501.

Thiel V., Jenisch A., Landmann G., Reimer A., and Michaelis W. (1997) Unusual distributions of long-chain alkenones and tetrahymanol from the highly alkaline Lake Van, Turkey. *Geochim. Cosmochim. Acta* **61**, 2053–2064.

Thierstein H. R., Geitzenauer K. R., Molfino B., and Shackleton N. J. (1977) Global synchroneity of late Quaternary coccolith datum levels: validation by oxygen isotopes. *Geology* **5**, 400–404.

Thomsen C., Schulz-Bull D. E., Petrick G., and Duinker J. C. (1998) Seasonal variability of the long-chain alkenone flux and the effect on the $U^{k'}_{37}$ index in the Norwegian Sea. *Org. Geochem.* **28**, 311–323.

Thomsen H. A., Buck K. R., and Chavez F. P. (1994) Haptophytes as components of marine phytoplankton. In *The Haptophyte Algae* (eds. J. C. Green and B. S. C. Leadbeater). Clarendon Press, Oxford, pp. 187–208.

van der Smissen J. H. and Rulkotter J. (1996) Organofacies variations in sediments from the continental slope and rise of the New Jersey continental margin (Sites 903 and 905). *Proc. ODP Sci. Res.* **150**, 329–344.

Versteegh G. J. M., Riegman R., de Leew J. W., and Jansen J. H. F. (2001) $U^{k'}_{37}$ values for Isochrysis galbana as a function of culture temperature, light intensity and nutrient concentrations. *Org. Geochem.* **32**, 785–794.

Villanueva J. and Grimalt J. O. (1996) Pitfalls in the chromatographic determination of the alkenone $U^{k'}_{37}$ index for paleotemperature estimation. *J. Chromatogr. A* **723**, 285–291.

Villanueva J. and Grimalt J. O. (1997) Gas chromatographic tuning of the $U^{k'}_{37}$ paleothermometer. *Anal. Chem.* **69**, 3329–3332.

Villanueva J., Pelejero C., and Grimalt J. O. (1997) Clean-up procedures for the unbiased estimation of C_{37} alkenone sea surface temperatures and terrigenous n-alkane inputs in Paleoceanographyapy. *J. Chromatogr. A* **757**, 145–151.

Villanueva J., Grimalt J. O., Cortijo E., Vidal L., and Labeyrie L. (1998) Assessment of sea surface temperature variations in the central North Atlantic using the alkenone unsaturation index ($U_{37}^{k'}$). *Geochim. Cosmochim. Acta* **62**, 2421–2427.

Visser K., Thunell R., and Stott L. (2003) Magnitude and timing of temperature change in the Indo-Pacific warm pool during deglaciation. *Nature* **421**, 152–155.

Volkman J. K., Eglinton G., Corner E. D. S., and Sargent J. R. (1980a) Novel unsaturated straight-chain C37–C39 methyl and ethyl ketones in marine sediments and a coccolithophore *Emiliania huxleyi*. In *Advances in Organic Geochemistry 1979, Physics and Chemistry of the Earth* (eds. A. G. Douglas and J. R. Maxwell). Pergamon, Oxford, vol. 12, pp. 219–227.

Volkman J. K., Eglinton G., Corner E. D. S., and Forsberg T. E. V. (1980b) Long-chain alkenes and alkenones in the marine coccolithophorid *Emiliania huxleyi*. *Phytochemistry* **19**, 2619–2622.

Volkman J. K., Burton H. R., Everitt D. A., and Allen D. I. (1988) Pigment and lipid compositions of algal and bacterial communities in Ace Lake, Vestfold Hills, Antarctica. *Hydrobiologia* **165**, 41–57.

Volkman J. K., Jeffer S. W., Nichols P. D., Rogers G. I., and Garland C. D. (1989) Fatty acid and lipid composition of 10 species of microalgae used in mariculture. *J. Exp. Mar. Biol. Ecol.* **128**, 219–240.

Volkman J. K., Barrett S. M., Blackburn S. I., and Sikes E. L. (1995) Alkenones in *Gephyrocapsa oceanica*: implications for studies of paleoclimate. *Geochim. Cosmochim. Acta* **59**, 513–520.

Waelbroeck C. and Steinke S. (2002) Comment on Steinke *et al.* (2001). *Quat. Res.* **57**, 432–433.

Wang L., Sarnthein M., Erlenkeuser H., Grimalt J., Grootes P., Heilig S., Ivanova E., Kienast M., Pelejero C., and Pflaumann U. (1999) East Asian monsoon climate during the Late Pleistocene: high-resolution sediment records from the South China Sea. *Mar. Geol.* **156**, 245–284.

Weaver P. P. E., Chapman M. R., Eglinton G., Zhao M., Rutledge D., and Read G. (1999) Combined coccolith, foraminiferal, and biomarker reconstruction of paleoceanography conditions over the past 120 kyr in the northern North Atlantic (59° N, 23° W). *Paleoceanography* **14**, 336–349.

Wei W. C. (1993) Calibration of upper pliocene-lower pleistocene nannofossil events with oxygen isotope stratigraphy. *Paleoceanography* **8**, 85–99.

Winter A., Jordan R., and Roth P. (1994) Biogeography of living coccolithophores in ocean waters. In *Coccolithophores* (eds. A. Winter and W. G. Siesser). Cambridge University Press, Cambridge, UK, pp. 161–177.

Xu L., Reddy C. M., Farrington J. W., Frysinger G. S., Gaines R. B., Johnson C. G., Nelson R. K., and Eglinton T. I. (2001) Identification of a novel alkenone in Black Sea sediments. *Org. Geochem.* **32**, 633–645.

Yamamoto M., Ficken K., Baas M., Bosch H.-J., and de Leuw J. W. (1996) Molecular paleontology of the earliest danian at geulhemmerberg (the Netherlands). *Egol. en Mijnbouw* **75**, 255–267.

Yamamoto M., Shiraiwa Y., and Inouye I. (2000) Physiological responses of lipids in *Emiliania huxleyi* and *Gephyrocapsa oceanica* (Haptophyceae) to growth status and their implications for alkenone paleothermometry. *Org. Geochem.* **31**, 799–811.

Young J. R. and Westbroek P. (1991) Genotypic variation in the coccolithophorid species *Emiliania huxleyi*. *Mar. Micropaleontol.* **18**, 5–23.

Zhao M., Rosell A., and Eglinton G. (1993) Comparison of two U_{37}^{k} sea surface temperature records for the last climatic cycle at ODP site 658 from the sub-tropical Northeast Atlantic. *Palaeogeogr. Palaeoclimat. Palaeoecol.* **103**, 57–65.

Zhao M., Beveridge N. A. S., Shackleton N. J., Sarnthein M., and Eglinton G. (1995) Molecular stratigraphy of cores off northwest Africa: sea surface temperature history over the last 80 ka. *Paleoceanography* **10**, 661–675.

Zhao M., Eglinton, Read G. G., and Schimmelmann A. (2000) An alkenone ($U_{37}^{k'}$) quasi-annual sea surface temperature record (AD 1440 to 1940) using varved sediments from the Santa Barbara basin. *Org. Geochem.* **31**, 903–917.

Zink K. G., Leythaeuser D., Melkonian M., and Schwark L. (2001) Temperature dependency of long-chain alkenone distributions in recent to fossil limnic sediments and in lake waters. *Geochim. Cosmochim. Acta* **65**, 253–265.

Ziveri P. and Thunell R. (2000) Cocolithophore export in Guaymas Basin, Gulf of California: response to climate forcing. *Deep-Sea Res. II* **47**, 2073–2100.

Ziveri P., Thunell R., and Rio D. (1995) Export production of coccolithophore in an upwelling region: results from San Pedro basin, Southern California bight. *Mar. Micropaleontol.* **24**, 335–358.

Ziveri P., Broerse A. T. C., van Hinte J. E., Westbroek P., and Honjo S. (2000) The fate of coccoliths at 48° N 21° W, northeastern Atlantic. *Deep-Sea Res. II* **47**, 1853–1875.

6.16
Tracers of Past Ocean Circulation

J. Lynch-Stieglitz
Columbia University, Palisades, NY, USA

6.16.1 INTRODUCTION

Information about how the ocean circulated during the past is useful in understanding changes in ocean and atmospheric chemistry, changes in the fluxes of heat and freshwater between the ocean and atmosphere, and changes in global wind patterns. The circulation of surface waters in the ocean leaves an imprint on sea surface temperature, and is also inextricably linked to the patterns of oceanic productivity. Much valuable information about past ocean circulation has been inferred from reconstructions of surface ocean temperature and productivity, which are covered in separate chapters. Here the focus is on the geochemical tracers that are used to infer the flow patterns and mixing of subsurface water masses.

Several decades ago it was realized that chemistry of the shells of benthic foraminifera (carbon isotope and Cd/Ca ratios) carried an imprint of the nutrient content of deep-water masses (Shackleton, 1977; Broecker, 1982; Boyle, 1981). This led rapidly to the recognition that the water masses in the Atlantic Ocean were arrayed differently during the last glacial maximum than they are today, and the hypothesis that the glacial arrangement reflected a diminished contribution of low-nutrient North Atlantic deep water (NADW) (Curry and Lohmann, 1982; Boyle and Keigwin, 1982). More detailed spatial reconstructions indicated a shallow nutrient-depleted water mass overlying a more nutrient-rich water mass in the glacial Atlantic. These findings spurred advances not only in geochemistry but in oceanography and climatology, as workers in these fields attempted to simulate the inferred glacial circulation patterns and assess the vulnerability of the modern ocean circulation to changes such as observed for the last ice age.

While the nutrient distributions in the glacial Atlantic Ocean were consistent with a diminished flow of NADW, they also could have reflected an increase in inflow from the South Atlantic and/or a shallower yet undiminished deep-water mass. Clearly, tracers capable of giving information on deep-water flow rate, rather than nutrient content alone, were needed. Differences between surface water (measured on planktonic foraminifera) and deep-water (measured on coexisting benthic foraminifera) radiocarbon concentrations provided the first rate constraint (Broecker *et al.*, 1988; Shackleton *et al.*, 1988). Reduced amounts of protactinium relative to the more particle-reactive thorium in the glacial Atlantic suggested that deep water was exported from the Atlantic during glacial times (Yu *et al.*, 1996). More recently, density gradients in upper waters have been used to infer changes in the upper ocean return flow that compensates the deep-water export (Lynch-Stieglitz *et al.*, 1999b).

Many of these tracers of paleo-ocean flow have been applied to all of the ocean basins, and have been extended in time throughout the Neogene. Despite this progress, a consistent picture of the circulation of the ocean during even the last ice age has yet to emerge. While circulation tracers suggest a rearrangement of water masses in the Atlantic, there is still considerable disagreement about the water masses and circulation in the rest of the World Ocean. Some of this arises from still insufficient data coverage, but some is the result of conflicting information from the various deep circulation tracers. In this chapter, we examine in more detail the methods used to reconstruct past ocean circulation, which will illuminate the source of some of this conflicting information.

6.16.2 NUTRIENT WATER MASS TRACERS

6.16.2.1 Carbon Isotopes

Measuring the carbon isotope ratio in the calcite tests of bottom dwelling (benthic) foraminifera is perhaps the most widespread method for reconstructing the distribution and properties of deep-water masses. The carbon isotope ratios in the foraminifera reflect that ratio in seawater. In turn, the carbon isotope ratio in the deep sea is primarily controlled by the regeneration of ^{13}C-poor organic material and has a distribution much like a nutrient such as phosphate in the modern ocean. While a nutrient tracer reflects the increased contribution from the regeneration of organic material as the water mass ages, it also carries a signature of the initial nutrient content of the source regions. In today's Atlantic, the contrast between high-nutrient Southern Ocean source water and low-nutrient NADW is strong, and the residence time of deep water in the Atlantic Ocean is short, allowing little additional accumulation of nutrients from sinking organic material. In this case the nutrient tracers can be used to document the relative mixtures of Northern and Southern water masses.

6.16.2.1.1 *Controls on $\delta^{13}C$ of oceanic carbon*

During photosynthesis, organisms preferentially take up the lighter isotope of carbon (^{12}C) increasing surface ocean $\delta^{13}C$. When the ^{13}C-depleted organic matter is decomposed, this CO_2 release decreases the $\delta^{13}C$ of seawater DIC at depth (Deuser and Hunt, 1969; Craig, 1970; Kroopnick, 1985). If this process were the only one responsible for the distribution of $\delta^{13}C$ in the ocean, $\delta^{13}C$ of oceanic carbon would decrease by $\sim$1.1‰ for every 1 μmol kg^{-1} increase in oceanic PO_4 (Broecker and Maier-Reimer, 1992). Reflecting this biological cycling, as a deep water mass with a homogeneous source ages (such as in the deep Indian and Pacific oceans) the $\delta^{13}C$ and PO_4 increase in this biologically expected ratio (Figure 1). However, the discrimination against ^{13}C during photosynthesis increases from about 19‰ in the warm surface ocean to about 30‰ in the Antarctic, which means that the PO_4 and $\delta^{13}C$ changes associated with the uptake and regeneration of organic material can vary regionally.

Carbon isotope fractionation during air–sea exchange is also an important factor in determining the isotopic composition of seawater. If the CO_2 in the atmosphere were in isotopic equilibrium with the dissolved inorganic carbon (DIC) in the ocean, the DIC would be enriched in ^{13}C relative to the atmospheric CO_2 by ~8‰ at 20 °C (Inoue and Sugimura, 1985). This equilibrium fractionation depends on the temperature of equilibration, with the oceanic carbon more enriched relative to the atmospheric value by about 0.1‰ per degree of cooling (Mook *et al.*, 1974). If the surface ocean were everywhere in isotopic equilibrium with atmospheric CO_2, the 30 °C range in ocean temperatures would cause a 3‰ range in oceanic $\delta^{13}C$ with higher values where surface temperatures are cold, and lower values in warm surface waters. This range is similar to the magnitude of $\delta^{13}C$ change induced by biological processes. However, isotopic equilibrium is rarely reached, because the timescale for carbon isotope equilibration between the ocean

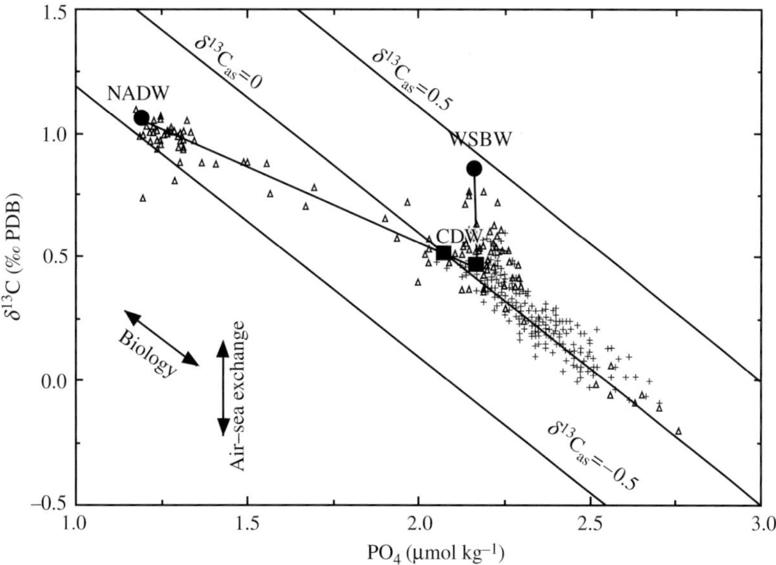

Figure 1 Deep water (>2,000 m) $\delta^{13}C$ and PO_4 data from the modern ocean after Lynch-Stieglitz *et al.* (1995). The lines indicate a "Redfield slope" and $\delta^{13}C_{as}$ indicates the deviation from the mean ocean Redfield slope for $\delta^{13}C$ and PO_4 (Broecker and Maier-Reimer, 1992). While the deep Indian and Pacific values follow the slope expected for biologic processes, deep waters which form in the North Atlantic and Southern Ocean have, respectively, lower and higher $\delta^{13}C$ due to air–sea exchange.

and atmosphere (~10 years for a 50 m deep mixed layer) is about 10 times longer than chemical equilibration between atmospheric CO_2 and DIC (Broecker and Peng, 1974; Tans, 1980). This isotopic equilibration time is significantly longer than surface waters stay in one place in the ocean.

The exchange of atmospheric CO_2 and surface ocean aqueous CO_2 has the potential to leave oceanic carbon depleted in ^{13}C in areas of CO_2 invasion (e.g., high northern latitudes, low-p_{CO_2} surface waters), and enriched in ^{13}C in areas of the ocean where CO_2 is outgassed (e.g., equatorial upwelling regions, high-p_{CO_2} surface waters) (Lynch-Stieglitz *et al.*, 1995). This is because the CO_2 gas that is dissolved in the ocean and exchanges between the ocean and atmosphere is isotopically lighter than the much larger pool of the DIC that consists mainly of bicarbonate and carbonate ions. This effect is observed in the ocean, because the timescale for carbon equilibrium between the ocean and atmosphere (which drives the air–sea exchange of the CO_2) is shorter than the timescale for isotopic equilibrium.

In general, surface-ocean $\delta^{13}C$ lies somewhere in between the value predicted if biological cycling alone controlled the distribution of $\delta^{13}C$ in the ocean and the value which would be at equilibrium with the atmosphere (Broecker and Peng, 1982; Broecker and Maier-Reimer, 1992). The departure from the biologically expected value can be expressed as an air–sea exchange signature ($\delta^{13}C_{as}=\delta^{13}C-(2.7-1.1\times PO_4)$) (Broecker and Maier-Reimer, 1992; Lynch-Stieglitz *et al.*, 1995). By definition, water with $\delta^{13}C_{as}$ of 0‰ has the same air–sea exchange signature as mean ocean deep water. A positive value of $\delta^{13}C_{as}$ means that the water has a higher $\delta^{13}C_{as}$ than mean ocean deep water (more of an influence of air–sea exchange at cold temperatures/less at warm temperatures), where as a negative $\delta^{13}C_{as}$ value implies less $\delta^{13}C$ enrichment due to air–sea exchange than for mean ocean deep water.

$\delta^{13}C_{as}$ in surface waters is highest in the Antarctic and the North Pacific due to exchange with the atmosphere at low temperatures. The lowest values of $\delta^{13}C_{as}$ are found in the centers of the subtropical gyres. Surface waters circulate in these anticyclonic gyres long enough to allow the oceanic dissolved inorganic carbon (DIC) to equilibrate isotopically with the atmosphere at high temperatures despite the relatively low rates of gas exchange. The situation in the North Atlantic seems slightly more complex. Air–sea exchange at cold temperatures will tend to raise the $\delta^{13}C_{as}$ of the northward flowing upper waters, but does not have time to overcome the effects of CO_2 invasion and the remnant low-$\delta^{13}C_{as}$ of the northward flowing surface water. Deep waters leaving the surface in the North Atlantic and the Antarctic reflect the low and high $\delta^{13}C_{as}$, respectively.

6.16.2.1.2 Carbon isotope ratios in benthic foraminifera

Belanger *et al.* (1981) demonstrated that the $\delta^{13}C$ in the benthic foraminifera *Planulina wuellerstorfi* decreased from highest values in the Norwegian-Greenland Sea, to lower values in the South Atlantic and lowest values on the East Pacific Rise. They argued that *P. wuellerstorfi* were reliable recorders of bottom-water conditions and not overly influenced by the incorporation of metabolic carbon. (While *wuellerstorfi* is variously assigned to the genera *Planulina, Cibicides, Cibicidoides*, and *Fontbotia* by various workers, there is no disagreement on the assignment of the species.) They also found that δ^{13}C in the genera *Oridorsalis* and *Pyrgo* did not follow the expected deep-water progression in δ^{13}C and argued that the shell chemistry reflected the lower δ^{13}C of their pore-water habitat. The same year, Graham *et al.* (1981) showed, with a larger set of core top data, that the δ^{13}C from the tests *P. wuellerstorfi* and *Cibicidoides kullenbergi* also followed the expected relationship with seawater chemistry. Graham *et al.* (1981) also found that these species show δ^{13}C values that are closest to what would be expected for precipitation in seawater. While they found that the δ^{13}C values in the *P. wuellerstorfi* and *C. kullenbergi* reflect the δ^{13}C of the ambient water, they are a full 1‰ lower than expected for calcite formed in equilibrium with this water (Woodruff *et al.*, 1980). However, carbon isotope equilibrium between calcite and dissolved carbon is not well known at the temperatures of the deep ocean (Grossman, 1984a; Romanek *et al.*, 1992). These studies paved the way for the widespread reconstruction of deep-water δ^{13}C that followed soon afterward (Curry and Lohmann, 1982; Boyle and Keigwin, 1982; Duplessy *et al.*, 1984, 1988; Curry *et al.*, 1988).

Shown in Figure 2 is a comparison of *Planulina* and *Cibicidoides* δ^{13}C with estimated water column δ^{13}C from Duplessy *et al.* (1984). At the time of this study, there was little quality water column δ^{13}C data with which to compare the core top values. However, in subsequent years the more extensive and improved data sets have confirmed that, on average, *Planulina* and *Cibicidoides* δ^{13}C reflect water column δ^{13}C with very little or no offset. Some of the scatter in Figure 2 may be attributed to errors in estimated oceanic δ^{13}C or core tops which are not Holocene in age. However, the more recent data sets also show that there can be significant offsets between the foraminifera and water column values at some locations even when living foraminifera are compared with water column δ^{13}C collected at the same location.

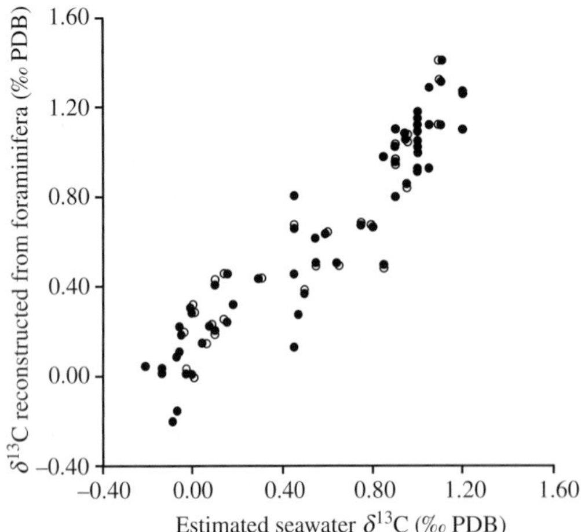

Figure 2 Carbon isotopic composition of core top benthic foraminifera in the genera *Cibicidoides* and *Planulina* versus estimated carbon isotopic composition of DIC in the water column above core location (after Duplessy *et al.*, 1984).

Perhaps it is not unexpected that benthic foraminifera do not always cleanly record the isotopic composition of bottom water. The fact that all of the benthic foraminifera have δ^{13}C that is significantly lighter than what would be expected for equilibrium with bottom water provides the first clue that additional processes may be at work. As postulated by Belanger *et al.* (1981), some foraminifera may live in the pore waters where δ^{13}C of the DIC is lower than in the overlying water column. Even so, the carbon isotopic composition of the calcite skeleton of most calcitic organisms does not reflect the composition of the ambient waters (Wefer and Berger, 1991) but is dominated by physiological or "vital" effects which usually involve the incorporation of isotopically light metabolic carbon or kinetic fractionation during the transport of carbon to the calcification site or during the calcification itself. The isotopic ratios of planktonic foraminifera are also dominated by these vital effects and are, in general, not in equilibrium with the seawater (e.g., Spero and Lea, 1996; Spero *et al.*, 1997).

A more detailed picture at how benthic foraminifera record bottom-water δ^{13}C has been gained through careful analysis of stained (live, or recently living) benthic foraminifera collected from surface sediments combined with measurements of both water column and pore-water δ^{13}C (Grossman, 1984a,b; Mackensen and Douglas, 1989; McCorkle *et al.*, 1990, 1997; Mackensen *et al.*, 1993; Rathburn *et al.*, 1996; Tachikawa and Elderfield, 2002). These studies confirm earlier notions that different species show different microhabitat preferences, with some species (including most *Cibicidoides* and *Planulina*) living on or near the sediment surface and others (including *Uvigerina*) living within the sediments. The deeper-dwelling species show depleted δ^{13}C values relative to bottom waters that, in general, reflect the magnitude of the pore-water depletions. These studies suggest that this microhabitat effect is the dominant control on benthic foraminiferal δ^{13}C, and the variations in the vital effects between species are small. While the number of

Cibicidoides analyzed has been small, the *Cibicidoides* species tend to have the smallest isotopic depletion relative to bottom waters. However, the isotopic depletion even for *Cibicidoides* can be significant, especially in areas where productivity, and thus the supply of organic material to the sediments, is high. Mackensen *et al.* (1993) showed that living and dead specimens of *P. wuellerstorfi* showed substantial depletions relative to bottom waters, particularly beneath high-productivity fronts in the Southern Ocean. While *P. wuellerstorfi* has been reported to live above the sediment surface (Lutze and Thiel, 1989), Mackensen *et al.* (1993) postulate that the foraminifera are calcifying within the organic fluff layer that can be found in high-productivity environments, with their low-δ^{13}C value reflecting either the low-δ^{13}C environment within the fluff layer or high growth rates of the foraminifera within these environments.

These relationships have been generally confirmed with down-core δ^{13}C studies. These studies show that the δ^{13}C from *Uvigerina* (infaunal) is always more negative than for *Cibicidoides*, and that this offset is variable in space and time, consistent with the microhabitat model. While many early isotopic studies were performed using *Uvigerina* because it was felt that it best reflected oxygen isotope equilibrium, most deep-water δ^{13}C reconstructions now use *Cibicidoides* or *Planulina* species. It is broadly acknowledged that these records, particularly in high-productivity areas, might be influenced by a microhabitat effect. However, by avoiding high-productivity areas and using supplemental information about changes in bottom-water characteristics and overlying productivity, this method has continued to yield regionally coherent and sensible patterns of deep-water δ^{13}C.

6.16.2.1.3 *Cadmium in benthic foraminifera*

The use of Cd/Ca measurements in the tests of benthic foraminifera for deep-water nutrient reconstructions was developed in parallel with the carbon isotope method. Cadmium concentrations in seawater follow a nutrient-like distribution, while calcium concentrations simply reflect variations in salinity. The benthic foraminifera incorporate the cadmium and calcium into their shells in proportion to their presence in seawater, which allows for the reconstruction of deep-water cadmium (and thus macronutrient) distributions.

6.16.2.1.4 *Cadmium in seawater*

Cadmium in seawater has a nutrient-like profile, with depleted values in warm surface waters, and the most enriched concentrations in deep waters (Boyle *et al.*, 1976). Many high-quality cadmium measurements in seawater since the early 1980s have confirmed this nutrient-like behavior, with cadmium showing an almost linear relationship with phosphorus in seawater. While it is still not clear whether cadmium is taken up as an essential nutrient or inadvertently, the cycling within the ocean is dominated by the incorporation into organic tissue in the surface ocean, and the regeneration of elemental cadmium as this material breaks down below the surface. The relationship with phosphorus is not quite linear, with a lower slope at lower phosphorus values (Figure 3). This structure may be due to the fact that as Cd/P in seawater decreases so will Cd/P in organic matter (Elderfield and Rickaby, 2000). That is, a fixed ("Redfield") ratio of Cd/P in organic matter should not be expected. Instead, if the cadmium is incorporated actively by organisms, the ecosystem finds a way to manage with less of this "nutrient," or less cadmium is incorporated inadvertently into particles, simply because there is less. Intermediate nutrient values can result from either mixing of a very low and high nutrient water mass (which would produce a linear Cd–P curve that passes through the origin and the Cd/P of deep water), or through the progressive drawdown of nutrients in surface water (which would produce a curved relationship). As a consequence, there are real differences in the Cd–P relationships in different ocean regions (Elderfield and Rickaby, 2000), which may of course change through time. Because

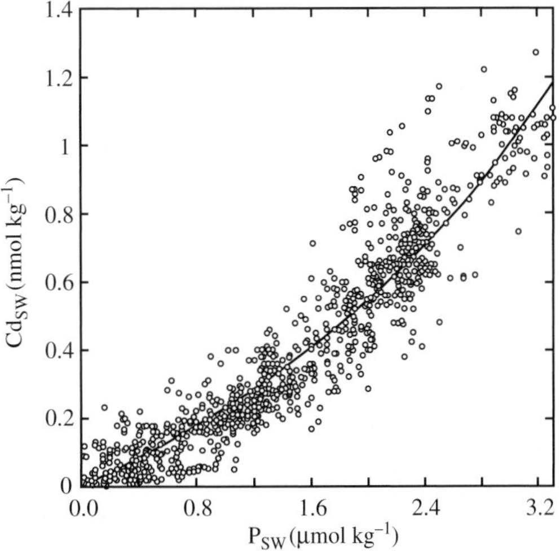

Figure 3 Seawater cadmium measurements from the modern ocean plotted versus PO_4 (after Elderfield and Rickaby, 2000). The curvature in this relationship is modeled by Elderfield and Rickaby (2000) (solid line) as resulting from the preferential uptake of cadmium in marine organic matter.

cadmium is removed from seawater via CdS precipitation in suboxic sediments, the global seawater cadmium inventory, and thus the global Cd–P relationship can change as the extent of suboxic sediments changes (Rosenthal *et al.*, 1995; van Geen *et al.*, 1995). The residence time for cadmium and phosphorous in seawater is sufficiently long that the cadmium concentration in the last glacial is unlikely to be lower than today by more than 5% (Boyle and Rosenthal, 1996), allowing us to make meaningful inferences about the cycling of the major nutrients via reconstructions of the cadmium concentration in seawater.

6.16.2.1.5 Cd/Ca in foraminifera

Boyle (1981) showed that with careful attention to cleaning, benthic foraminifera shells reflect the concentration of some trace metals (including cadmium) in seawater, with a preferential incorporation of cadmium into foraminifera. The empirical distribution coefficient (D, ratio of Cd/Ca in foraminifera to Cd/Ca in seawater) was shown to be 2.9 (Boyle, 1988). More core top measurements at all water depths have subsequently shown that this distribution coefficient increases with depth, starting at 1.0 at ~1 km and increasing until stabilizing at ~2.9 at a water depth of 3 km (Boyle, 1992). Boyle (1992) argues that the depth dependence is unlikely to be a temperature effect. However, once this depth-dependent distribution coefficient is applied, the reconstructed cadmium from foraminifera tests shows a linear relationship to cadmium estimated for the overlying water (Figure 4). Boyle *et al.* (1995) showed

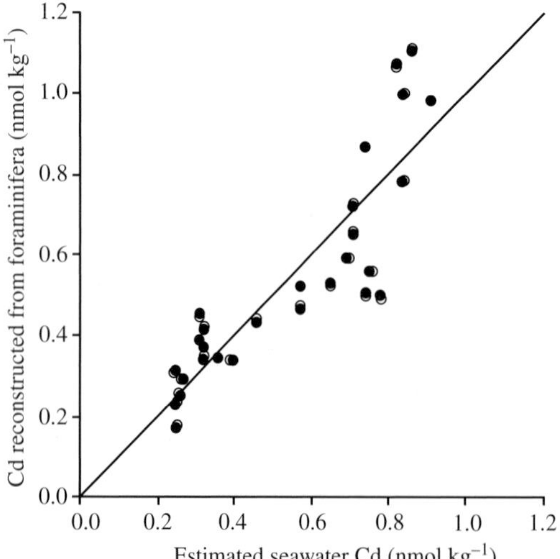

Figure 4 Seawater cadmium reconstructed from foraminifera using a depth-dependent empirical distribution coefficient versus estimated bottom-water cadmium (sources Boyle, 1988, 1992).

that for aragonitic foraminifera *Hoeglundina elegans* $D = 1$, which shows no depth dependence. McCorkle *et al.* (1995) observed an unexpected decrease in the apparent distribution coefficient of cadmium and other trace metals with depth between 2.5 km and 4 km in the deep Pacific, which suggested dissolution-driven loss of cadmium from the foraminiferal calcite after burial. Elderfield *et al.* (1996) favor the idea that the lower D in deep Pacific waters may be a calcification response to the low carbonate saturation state rather than representing post-depositional dissolution, a possibility also acknowledged by McCorkle *et al.* (1995). McCorkle *et al.* (1995) also observe a small decrease in the $\delta^{13}C$ content of foraminifera with depth, but this decrease is well within the other uncertainties related to the $\delta^{13}C$ tracer. There has been no systematic attempt to correct paleoceanographic cadmium reconstructions for this apparent dissolution (or low-carbonate saturation state) effect.

To the extent that the $\delta^{13}C$ values of infaunal benthic foraminifera reflect the low $\delta^{13}C$ of pore waters, it would be expected that the Cd/Ca will also reflect the interstitial, not bottom-water concentrations. However, while interstitial $\delta^{13}C$ values decrease with increasing depth in the sediment due to the regeneration of isotopically light organic matter, the distribution of cadmium in pore water is more complex. In shallow pore waters cadmium increases due to the regeneration of organic material, but in suboxic sediments, it then decreases due to the precipitation of CdS (McCorkle and Klinkhammer, 1991). As a consequence, unlike for $\delta^{13}C$ there is no systematic relationship between Cd/Ca of foraminifera that dwell on the surface and those below the sediment surface. Tachikawa and Elderfield (2002) showed that, for three sites in the North Atlantic pore waters were 2–4 times enriched in cadmium relative to bottom waters. They found that *P. wuellerstorfi*, whose $\delta^{13}C$ values suggested that they calcified in the bottom waters at the sediment surface, had a D of 3–4, consistent with culture studies (Havach *et al.*, 2001). The species which calcify within the pore water (infaunal species) showed lower D (~1) with respect to the pore-water cadmium concentrations. The net effect is that the Cd/Ca in *P. wuellerstorfi* and the infaunal species is quite similar. Further, Tachikawa and Elderfield argue that the depth-dependent D for foraminifera (which compared Cd/Ca ratios from primarily *Uvigerina* and *Cibicidioides/Planulina*) actually reflects a trend in the enrichment of pore-water cadmium relative to bottom-water cadmium. So, as with the $\delta^{13}C$ measurements, the Cd/Ca on *P. wuellerstorfi* or other surface dwelling foraminifera will give a more direct measure of bottom-water cadmium. The D for this species is not well constrained but is between 3 and 4. If infaunal foraminifera are used, the depth-dependent *D* of Boyle, which takes into account both the enrichment of pore-water cadmium relative to bottom water and the uptake of cadmium into the foraminifera, is more appropriate. As with $\delta^{13}C$, careful interpretation aided by considering supplemental information about changes in bottom-water characteristics has continued to yield regionally coherent and sensible patterns of deep-water cadmium (e.g., Boyle, 1992; Rickaby *et al.*, 2000).

6.16.2.2 Ba/Ca

Lea and Boyle (1989) showed that the barium content in foraminifera shells is, like cadmium, controlled by the barium content in bottom waters. Barium, in a broad sense, also cycles like a nutrient (depleted in surface waters, and higher in deep waters), but its regeneration occurs deeper than the organic matter. This results in a close correlation between barium and alkalinity in today's ocean (Lea, 1993). Like Cd/Ca, Ba/Ca in foraminifera has also been used, as a paleo-tracer of water masses (e.g., Lea and Boyle (1990), but suffers the same carbonate-saturation-state-linked effect as Cd/Ca (McCorkle *et al.*, 1995).

6.16.2.3 Zn/Ca

Benthic foraminiferal Zn/Ca is also controlled by bottom-water-dissolved zinc concentration (Marchitto *et al.*, 2000), which also follows a nutrient-type profile (Bruland *et al.*, 1978). Zinc is necessary nutrient for phytoplankton and, especially, diatom growth. This results in a very strong correlation between oceanic zinc and silicon concentrations. Like Cd/Ca, the apparent distribution coefficient for zinc decreases in waters which are more undersaturated with respect to calcium carbonate. This change in *D* is linearly related to the degree of undersaturation and, once this effect is taken into account, the Zn/Ca in the foraminifera reflect this well in the overlying water mass (Figure 5). Marchitto *et al.* (2000) suggest that because they respond to saturation state differently, Cd/Ca and Zn/Ca can be used together to assess both the deep-water mass properties and carbonate ion concentrations. Because of the close relationship between silicon and zinc, Zn/Ca in foraminifera is particularly useful for tracing the silicon-rich bottom water that originates in the Southern Ocean (Marchitto *et al.*, 2002).

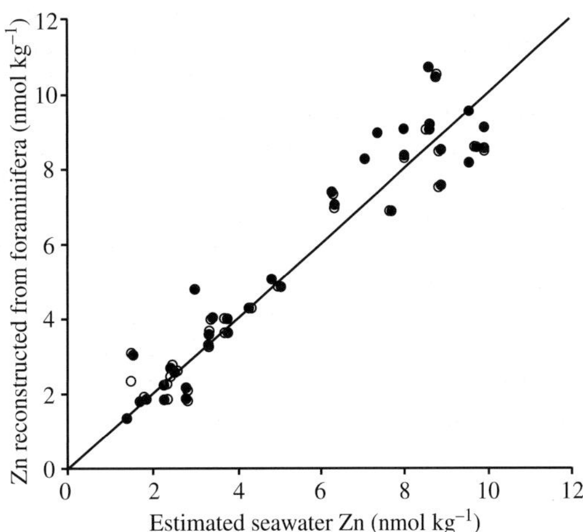

Figure 5 Seawater zinc reconstructed from foraminifera using an empirical distribution coefficient which depends on the degree of carbonate unsaturation versus estimated bottom-water zinc (after Marchitto *et al.*, 2000).

6.16.3 CONSERVATIVE WATER MASS TRACERS

In general, nutrient-type tracers have been used to infer the sources and mixing of deep-water masses in the past despite the complication that the nutrient content of deep water depends both on the nutrient content at the source of the deep water and the addition of nutrients via the regeneration of organic matter as the water mass ages. This is primarily due to the large and relatively readily measured contrasts in the nutrient-related tracers. Headway is now being made in the use of conservative tracers that reflect the mixing of source waters alone (such as temperature and salinity in the modern ocean).

6.16.3.1 Mg/Ca in Benthic Foraminifera

Mg/Ca in benthic foraminifera reflect the calcification temperature (Rosenthal *et al.*, 1997b; Rathburn and DeDeckker, 1997; Martin *et al.*, 2002). In principle, temperature can be used as a conservative tracer for subsurface water masses, but at this point the uncertainties in converting the Mg/Ca data into temperature data overwhelm the deep-water temperature gradients that distinguish water masses of different origins. Also, due to the nonlinear nature of the relationship between Mg/Ca and temperature, small errors in Mg/Ca can lead to large inferred temperature changes, especially at cold temperatures. However, development in this field is rapid, and there is promise that Mg/Ca could be used to assess the contributions to intermediate-water masses where the temperature signals are more easily distinguished.

6.16.3.2 Pore-water Chemistry

The $\delta^{18}O$ of seawater is set by processes that occur at the ocean surface (evaporation, precipitation, sea ice formation, and melt) (Craig and Gordon, 1965), and can be used as conservative tracers in the subsurface ocean. The $\delta^{18}O$ of glacial-age water has been reconstructed by measuring the $\delta^{18}O$ of pore water in deep-sea sediments (e.g., Schrag and DePaolo, 1993; Schrag *et al.*, 2002). The maximum contribution to $\delta^{18}O$ from the LGM appears at a depth of 20–60 m within the sediments. The actual $\delta^{18}O$ of the glacial-age deep water must be reconstructed by modeling the diffusion of water within the sediments. LGM estimates for the bottom-water temperature (also a conservative tracer) can then be made by using the benthic foraminiferal $\delta^{18}O$ with the information from pore fluids about the $\delta^{18}O$ of the water in which it calcified. Similarly, salinity can be reconstructed using pore fluid [Cl] (Adkins and Schrag, 2001). While the number of locations with such data are limited, Adkins *et al.* (2002) show the promise of using pore-water $\delta^{18}O$ and [Cl] to reconstruct the properties of deep-water masses. As more data are collected, these tracers will also be very useful for reconstructing the sources and mixing of deep-water masses.

6.16.3.3 Oxygen Isotopes in Benthic Foraminifera

The $\delta^{18}O$ of foraminiferal calcite reflects the $\delta^{18}O$ of the water in which it calcified and the temperature of calcification. Since both temperature and $\delta^{18}O$ of seawater are imprinted on a water mass as at the surface, the $\delta^{18}O$ of the foraminifera is a conservative tracer that can be used to determine the sources and mixing of water masses in the deep ocean. The use of $\delta^{18}O$ as a conservative tracer does not require the separation of the temperature and $\delta^{18}O$/salinity components of the foraminiferal $\delta^{18}O$ signal. Despite the fact that oxygen isotope data are routinely gathered simultaneously with the carbon isotope data, the use of $\delta^{18}O$ as a tracer for ocean circulation has lagged behind the use of $\delta^{13}C$. Zahn and Mix (1991) showed that the gradients in deep-water foraminiferal $\delta^{18}O$ were small, due to the fact that while NADW is warmer than Antarctic Bottom Water (which would lead to lower foraminiferal $\delta^{18}O$), the $\delta^{18}O$ of NADW is higher (counteracting the temperature effect). They also demonstrated that there were significant issues with interlaboratory and interspecies calibration that interfered with the effective use of this tracer. However, recent studies have shown that *Cibicidoides* and *Planulina* do a good job of tracking calcification temperature (Lynch-Stieglitz *et al.*, 1999a; Duplessy *et al.*, 2002) (Figure 6). While early studies showed that *Cibicidoides* $\delta^{18}O$ did not reflect equilibrium calcification (O'Neil *et al.*, 1969), later determinations of equilibrium calcification are consistent with the observed values for core-top *Cibicidoides* (Kim and O'Neil, 1997; Lynch-Stieglitz *et al.*, 1999a). Some of the interlaboratory calibration problems arise, because most mass spectrometry laboratories routinely calibrate using a carbonate standard (NBS-19) that can be up to 6‰ lower than benthic foraminifera from the deep ocean during glacial periods. Care must be taken to calibrate regularly with standards spanning a large range of isotopic ratios, and appropriate corrections must be made for many

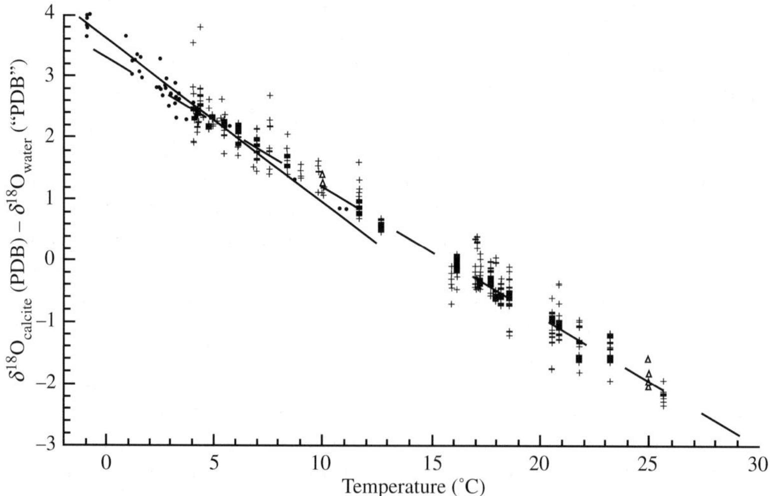

Figure 6 The isotopic fractionation, expressed as the difference between the $\delta^{18}O$ of foraminiferal calcite and the $\delta^{18}O$ of seawater (converted from SMOW to a "Peedee Belemnite (PDB)" like scale by subtracting 0.27 in accordance with paleoceanographic tradition) versus temperature of calcification. *Cibicidoides* and *Planulina* from Lynch-Stieglitz *et al.* (1999a) and (— +) and Duplessy *et al.* (2002) (— •) are shown along with the inorganic precipitation experiments of Kim and O'Neil (1997) (△).

mass spectrometers (e.g., Ostermann and Curry, 2000). Duplessy *et al.* (2002) showed coherent patterns of deep-ocean foraminiferal $\delta^{18}O$, which could be used to make inferences about deep-water sources. While the range of foraminiferal $\delta^{18}O$ in the deep ocean is small, the foraminifera seem to be able to record these differences well. Mackensen *et al.* (2001) and Matsumoto *et al.* (2001) also showed regionally coherent patterns of benthic foraminiferal $\delta^{18}O$ in the Southern Ocean. With careful attention to laboratory calibration, deep- and intermediate-water foraminiferal $\delta^{18}O$ reconstructions show great promise as a conservative tracer for ocean circulation.

6.16.3.4 Carbon Isotope Air–Sea Exchange Signature

Because the $\delta^{13}C$ in seawater is influenced by both biological cycling and air–sea exchange, one can use co-occurring nutrient data (influenced only by biological cycling) to isolate $\delta^{13}C$ differences that arise from air–sea exchange (Lynch-Stieglitz and Fairbanks, 1994). Because it is reset only at the surface, the air–sea exchange signature is a conservative tracer for ocean circulation. Today, newly formed NADW and Weddell Sea bottom water (WSBW) have $\delta^{13}C$ air–sea exchange signatures that differ by 0.8‰ (Lynch-Stieglitz *et al.*, 1995). However, the patterns in the air–sea exchange signature are only as good as the contributing $\delta^{13}C$ and Cd/Ca data, and areas where either dissolution or productivity overprints may be a problem must be avoided, thus limiting the use of this tracer.

6.16.4 NEODYMIUM ISOTOPE RATIOS

The deep-water masses of the ocean have distinctive neodymium isotope ratios that are recorded both in manganese nodules (Albarède and Goldstein, 1992; Albarède *et al.*, 1997) and dispersed manganese–iron oxides in deep-sea sediments (Rutberg *et al.*, 2000). Precisely how the water masses acquire their neodymium isotope ratios is still unresolved (Albarède *et al.*, 1997). This ratio is neither a conservative nor a nutrient water mass tracer. Neodymium is unaffected by biological cycling in the ocean, but there are both sources and sinks of neodymium beneath the sea surface, particularly at the sediment–water interface where it is mobilized from detrital material, and is precipitated in metallic crusts. The neodymium isotope ratios in NADW are high, reflecting the addition of neodymium from the predominantly old continentally derived detrital material, and the neodymium isotope ratios in the North Pacific are low, reflecting the influence of recent volcanic activity. In regions where there is active mixing between these water masses (the Atlantic Ocean and the Southern Ocean) the seawater (and manganese–iron precipitates) have intermediate values, as is the case for the nutrient and conservative water mass tracers (Rutberg *et al.*, 2000).

6.16.5 CIRCULATION RATE TRACERS

The water mass tracers discussed above can be used to assess the sources of subsurface water masses and the mixture of waters from different sources. However, this information can be used only to assess the relative rates of renewal of these deep-water masses from the various source regions (e.g., LeGrand and Wunsch, 1995). There are several geochemical techniques that can be used to assess the rates of ocean circulation.

6.16.5.1 Radiocarbon

Carbon-14 (radiocarbon) is produced in the atmosphere and decays with a half-life of 5,730 yr. Natural radiocarbon levels in the ocean are highest in the surface ocean, where they partially equilibrate with atmosphere. Deep-water concentrations are highest in newly formed NADW and decrease to the lowest values in the deep Pacific, where the deep waters have longest been out of contact with the atmosphere (Stuiver *et al.*, 1983). Benthic foraminifera record the radiocarbon concentrations of the deep water in which they grow, but the ^{14}C in the foraminifera will decay over time. In order to reconstruct the radiocarbon content at the time the foraminifera were living, radiocarbon measurements can be made on benthic and planktonic foraminifera from the same interval in the core (Broecker *et al.*, 1988, 1990; Shackleton *et al.*, 1988) in order to determine the age of the benthic foraminifera. However, the radiocarbon content of the planktonic foraminifera will reflect not only the age of the planktonic foraminifera, but the initial radiocarbon content of the near-surface water in which they grew. The initial radiocarbon in surface waters and in near-surface waters can be significantly less than expected for equilibrium with the atmosphere due to the relatively long time required for isotopic equilibrium between carbon in the surface ocean and atmosphere. One must also account for the fact that the radiocarbon content in the atmosphere changes with time, and will also impact the initial radiocarbon content of surface waters (Adkins and Boyle, 1997).

Also complicating the use of concurrent benthic and planktonic radiocarbon concentrations to assess water mass age is the requirement that the benthic and planktonic foraminifera measured at the same interval actually be the same age. If the peak abundance of planktonic and benthic foraminifera occurs at different depths in the core, bioturbation can introduce a spread in the planktonic–benthic ages in cases with low accumulation rate, which is unrelated to changes in bottom-water age (Peng *et al.*, 1977; Broecker *et al.*, 1999). While these problems can be overcome by using cores with high sedimentation rate (e.g., Keigwin and Schlegel, 2002), the widespread use of this techniques has been limited primarily by the identification of suitable cores with high enough foraminiferal abundances for accurate radiocarbon determinations. Sediments with high accumulation rates tend to either have low foraminiferal abundances (due to dilution by terrigenous material), or be in high-productivity regions of the ocean, where surface waters are expected to have variable and relatively low initial radiocarbon content.

Another way to measure the deep-water radiocarbon ventilation age in the past is by using deep-dwelling benthic corals (Adkins *et al.*, 1998; Mangini *et al.*, 1998; Goldstein *et al.*, 2001). Here, the age of the coral can be independently determined using uranium-series dating allowing the radiocarbon content of the coral to be used to determine the radiocarbon content of deep water at the time the coral grew. This method is unaffected by bioturbation, but is currently limited by the availability of deep-sea benthic corals.

Changes in the ventilation of the ocean can also be assessed by looking at how the ^{14}C content of the atmosphere has changed through time (e.g., Hughen *et al.*, 1998). The atmospheric ^{14}C content is controlled both by the rate of ^{14}C production in the atmosphere and the rate at which ^{14}C is transferred into the ocean. While a decrease in the ventilation of the ocean causes a buildup of radiocarbon in the atmosphere, one must also account for changes in production before interpreting the atmospheric radiocarbon changes in terms of ocean circulation (Marchal *et al.*, 2001).

6.16.5.2 $^{231}Pa/^{230}Th$ Ratio

Both protactinium and thorium are produced in the water column and scavenged by particles onto the seafloor. Because protactinium is scavenged less efficiently by marine particles (Anderson *et al.*, 1983), much of the protactinium produced in the Atlantic Ocean is transported to the Southern Ocean by deep waters before it is scavenged onto the seafloor. This produces a low (below production) $^{231}Pa/^{230}Th$ ratio in Atlantic Ocean and a high $^{231}Pa/^{230}Th$ ratio in Atlantic Ocean in the Southern Ocean (Yu *et al.*, 1996). Reconstruction of the $^{231}Pa/^{230}Th$ ratio suggests that this pattern of protactinium export persisted in glacial times, and a low deep-water residence time for the LGM Atlantic has been inferred (Yu *et al.*, 1996). However, the $^{231}Pa/^{230}Th$ ratio in sediments depends on the complex interplay between deep-water flow and the differential scavenging of the two elements, which depends both on the composition of the particles and the particle flux (Marchal *et al.*, 2000). Biogenic opal fractionates thorium from protactinium much less effectively than other particle types and the ratios in Southern Ocean sediments will be very sensitive to the particle composition (Luo and Ku, 1999; Yu *et al.*, 2001). While the Atlantic $^{231}Pa/^{230}Th$ fluxes are low and today predominantly reflect the short residence time of deep waters in this basin, in the Pacific Ocean $^{231}Pa/^{230}Th$ ratios are most sensitive to particle flux, with higher ratios where the scavenging is most intense at the boundaries (Yu *et al.*, 2001). Reconstructing circulation patterns from $^{231}Pa/^{230}Th$ ratios will, in general, require a detailed knowledge of particle fluxes and composition.

6.16.5.3 Geostrophic Shear Estimates from $\delta^{18}O$ in Benthic Foraminifera

Physical oceanographers routinely use the distribution of density in the ocean to help quantify ocean circulation. A balance between the pressure gradient force and Coriolis term (geostrophic balance) is assumed. Velocity can then be calculated from density information using either a measured or assumed velocity at a reference level, or by combining the geostrophic balance with other constraints (such as conservation of tracer flow across a section). This approach has also been applied to paleoceanography by using vertical density profiles reconstructed from the oxygen isotopic composition of foraminiferal calcite (Lynch-Stieglitz *et al.*, 1999a,b).

The $\delta^{18}O$ from the calcite tests of benthic foraminifera preserved in ocean sediments can be used to estimate upper-ocean density because both the $\delta^{18}O$ of calcite ($\delta^{18}O_{calcite}$) and density increase as a result of increasing salinity or decreasing temperature (Lynch-Stieglitz *et al.*, 1999a). The fractionation between calcite precipitated inorganically and the water in which it forms increases by ~0.2‰ for every 1 °C decrease in temperature (Kim and O'Neil, 1997). The relationship between $\delta^{18}O_{calcite}$ and salinity is more complex. The $\delta^{18}O_{calcite}$ reflects the $\delta^{18}O$ of the water in which the foraminifera grew. The $\delta^{18}O$ of seawater ($\delta^{18}O_{water}$) primarily reflects patterns of evaporation and freshwater influx to the surface of the ocean. Because salinity also reflects these same processes, salinity and $\delta^{18}O_{water}$ are often well correlated in the ocean.

Although the exact relationship varies in different areas of the surface ocean (Craig and Gordon, 1965), the vast majority of surface and warn subsurface waters in the ocean have salinity and $\delta^{18}O_{water}$ values that scatter around a linear trend. For times in the geologic past, the ability to reconstruct density from the $\delta^{18}O_{calcite}$ is most limited by the knowledge of the relationships between $\delta^{18}O_{water}$ and salinity, as well as the relationships between temperature and salinity. However, it is the gradients in density that are used for the geostrophic calculations, and the reconstructed density gradients are less affected by these limitations than the density itself.

This technique has been used to estimate and reconstruct the flow of the Gulf Stream in the Florida Straits (Lynch-Stieglitz *et al.*, 1999b) by reconstructing vertical density profiles on either side of this current. Perhaps more useful for the reconstruction of the large-scale ocean circulation is the fact that the difference in density between the eastern and western margins of the ocean basins reflects the strength of the geostrophic overturning circulation (Lynch-Stieglitz, 2001).

At this point, the vertical density structure for the past ocean can be reconstructed using benthic foraminifera only where the seafloor intersects the upper water column (ocean margins, islands, shallow seamounts). Although planktonic foraminifera calcify at various depth within the upper water column, we have no way to quantitatively reconstruct the depth at which they calcified. However, deep dwelling planktonic foraminifera can be used to reconstruct the spatial pattern of upper ocean flows (Matsumoto and Lynch-Stieglitz, 2003).

6.16.6 NON-GEOCHEMICAL TRACERS OF PAST OCEAN CIRCULATION

A review of tracers of past ocean circulation would not be complete without mentioning some of the sedimentological and paleontological approaches to reconstructing past ocean flow. Sediment grain size has been used to reconstruct current intensity since the 1970s (Ledbetter and Johnson, 1976). More recent grain-size studies emphasize the mean size of the "sortable silt" fraction (>10 μm), which is least likely to display a cohesive behavior (McCave *et al.*, 1995). Stronger currents show larger sortable silt size due to changes in both deposition and winnowing with current speed.

Benthic foraminifera assemblages have been related to distinct deep-water masses and have been used for tracking these water masses during the past (Streeter, 1973; Schnitker, 1974). In subsequent years, it has been determined that the deep-sea benthic assemblage responds strongly to the flux of organic material from overlying surface productivity (e.g., Miller and Lohmann, 1982; Loubere, 1991), limiting its use as a water mass tracer to low-productivity regions (Schnitker, 1994).

6.16.7 OCEAN CIRCULATION DURING THE LAST GLACIAL MAXIMUM

Some of these methods have been applied to examine ocean circulation throughout the Cenozoic era, and others to examine century scale changes in the very recent past. However, all of the methods have been used to look at the ocean circulation during the last glacial maximum (LGM, ca. $(1.9-2.3)\times 10^4$ yr ago) and have their maximum geographic coverage during this time. Here we will briefly examine the picture of ocean circulation that these techniques have produced for the LGM. Each of the paleoceanographic methods outlined above has its advantages and pitfalls, times and places where it should work well, and environments where variables other than ocean circulation will complicate the interpretations. There is no single ocean circulation scenario for the LGM that is consistent with all of the available data from all of the tracers. In the following description, regionally coherent signals are weighted more heavily than data from isolated sediment cores. Data in regions where the complications may overwhelm the signal (e.g., high-productivity regions for $\delta^{13}C$ data, the very deep ocean for cadmium data, etc.) are weighted less heavily. A quantitative description of past ocean circulation will require many more data, better understanding of the tracers, combined with better techniques for dealing with the kinds of uncertainties and multiple processes that affect the paleoceanographic proxies.

Both Cd/Ca and $\delta^{13}C$ distributions suggest a strong stratification in the North Atlantic Ocean, with a low nutrient, high $\delta^{13}C$ water mass (often called glacial North Atlantic intermediate water, GNAIW) occupying depths down to about 2,000 m, and a high nutrient, low $\delta^{13}C$ water mass of southern origin underneath (e.g., Boyle and Keigwin, 1987; Oppo and Lehman, 1993; Duplessy *et al.*, 1988) (Figure 7). These data are generally interpreted as supporting a shallower overturning circulation in the glacial Atlantic, similar to that observed in some models of LGM ocean circulation.

However, it is clear from many tracers that there was continued ventilation of deep waters in the North Atlantic as well. Boyle and Keigwin (1982) argued that the fact that reconstructed cadmium concentrations in the deep North Atlantic were still lower than average ocean values suggests that there was not a complete shutdown in NADW production. More recently, Rickaby *et al.* (2000) find that reconstructed glacial cadmium concentrations were lower in the western basin than the eastern basin in the North Atlantic below 2,500 m water depth, supporting a North Atlantic source of deep waters (Figure 8). The reconstructed $\delta^{13}C$ are also higher in the North Atlantic than in the South Atlantic during the LGM, suggesting continued ventilation of deep waters from the north (e.g., Matsumoto and Lynch-Stieglitz, 1999; Broecker, 2002). Barium (Lea and Boyle, 1990) and zinc (Marchitto *et al.*, 2002) data demand continued ventilation from the north as well.

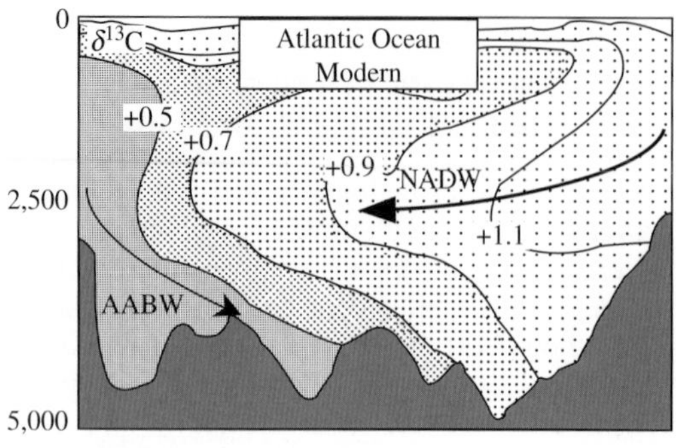

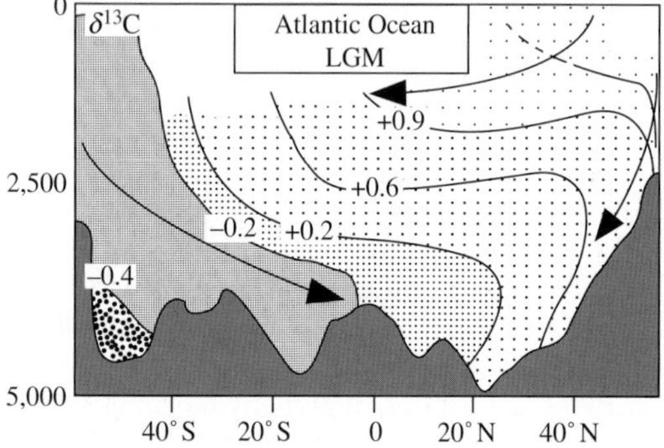

Figure 7 Modern and LGM distributions of $\delta^{13}C$ in the Atlantic Ocean inferred from benthic foraminifera (*Cibicidoides*) (Labeyrie, 1992) (reproduced by permission of Elsevier from *Quat. Sci. Rev.*, **1992**, *11*, 401–413).

Yu *et al.* (1996) find that the $^{231}Pa/^{230}Th$ ratio in the LGM Atlantic is lower than the production ratio, implying relatively short residence time for waters in the LGM Atlantic. However, this could be accomplished by rapid flushing of deep or intermediate waters either from the north or from the south, so does not demand strong ventilation of deep waters in the North Atlantic. Lynch-Stieglitz *et al.* (1999b) find a reduced vertical shear in the geostrophic velocity in the Florida Straits, consistent with a weaker overturning circulation. This does not contradict the idea of continued ventilation of deep- and intermediate-water masses in the North Atlantic during the LGM, but does suggest that this ventilation was either slower or accomplished by mechanisms that did not draw large quantities of surface water northward.

Radiocarbon reconstructions for the deep North Atlantic show older ventilation ages than today (Broecker *et al.*, 1990; Keigwin and Schlegel, 2002), but still younger than waters in the South Atlantic (Goldstein *et al.*, 2001), suggesting that while this deep water was ventilated from the North during the LGM, it

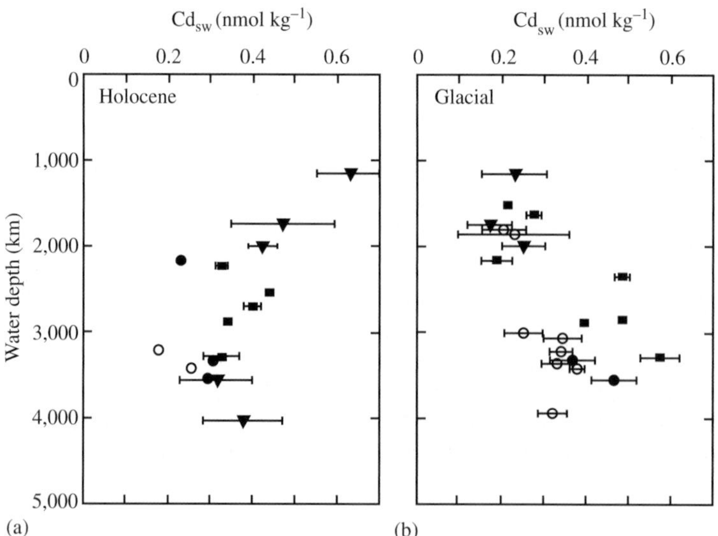

Figure 8 Cadmium reconstructions for the North Atlantic Ocean, showing the presence of a shallow, nutrient-depleted water mass overlying deeper, nutrient-rich waters (Rickaby *et al.*, 2000). Cores west of the Mid-Atlantic Ridge are plotted in open symbols, with lower nutrient values suggesting a source of new deep waters in the North Atlantic (reproduced by permission of Elsevier from *Geochim. Cosmochim. Acta.*, **2000**, *64*, 1229–1236).

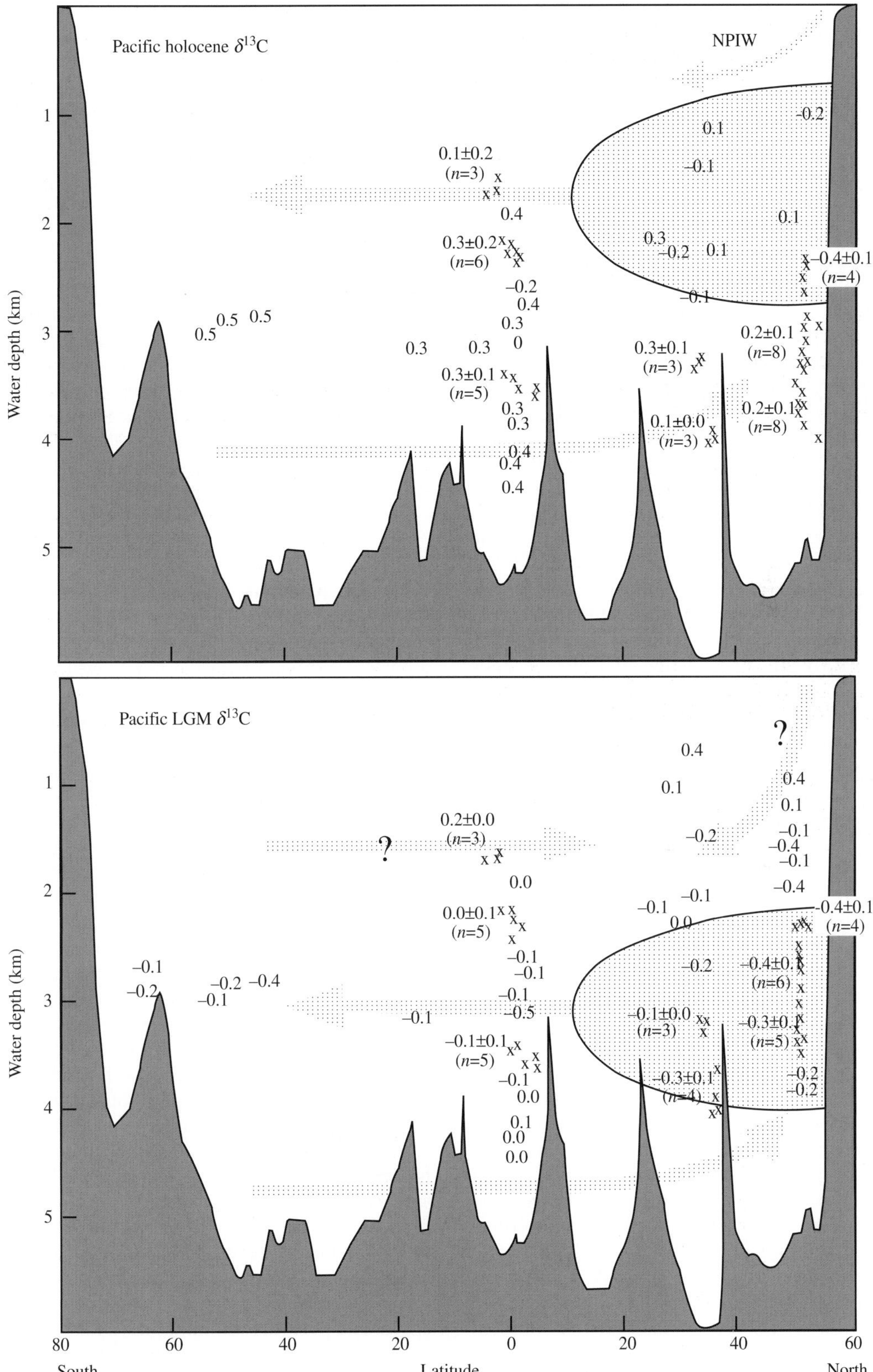

Figure 9 Modern and LGM distributions of benthic foraminiferal (*Cibicidoides*) $\delta^{13}C$ in the Pacific Ocean (after Matsumoto *et al.*, 2002).

was either ventilated more slowly than today or that the glacial NADW formed from surface waters which had radiocarbon concentrations that were less well equilibrated with the atmosphere than today.

The processes involved in the ventilation of the upper and lower deep waters in the North Atlantic were probably different for each water mass and are not well understood at this time. However, a simple shoaling (and possible southward shift) of the NADW overturning cell is probably a very simplistic view of the circulation of the LGM Atlantic.

Less is known about the state of the LGM Pacific Ocean, largely because the poor carbonate preservation in the Pacific Ocean limits the availability of cores with benthic foraminifera. The undersaturation with respect to calcium carbonate in the Pacific also complicates the interpretation of the trace-metal-based methods that suffer from dissolution or incorporation effects. However, like in the Atlantic, the carbon and oxygen isotope data suggest a deep hydrographic boundary at ~2,000 m water depth (Mix *et al.*, 1991; Herguera *et al.*, 1992; Keigwin, 1998; Matsumoto *et al.*, 2002), with a low-δ^{13}C, high-δ^{18}O water mass below this divide and a high-δ^{13}C, low-δ^{18}O water mass above this divide (Figure 9). The LGM Indian Ocean is also characterized by a benthic divide at ~2,000 m, with higher δ^{13}C and lower cadmium reconstructed for the upper water mass (Kallel *et al.*, 1988; Boyle *et al.*, 1995; McCorkle *et al*, 1998). This deep hydrographic boundary is also seen off Southern Australia (Lynch-Stieglitz *et al.*, 1996).

For both the upper and lower water masses, there is a general progression from highest δ^{13}C and Cd/Ca in the North Atlantic, to intermediate values in the South Atlantic and South Australia, to progressively lower δ^{13}C values in the North Pacific. Goldstein *et al.* (2001) also argue, based on limited glacial-age radiocarbon data, that the modern progression of youngest deep waters in the North Atlantic, intermediate values in the Southern Ocean, and oldest values in the deep Pacific was preserved during glacial times. The δ^{18}O values in the upper water mass are similar from North Atlantic to North Pacific, within the considerable scatter in the data. Whether or not the GNAIW reached the Pacific (after modification in the Southern Ocean) or there was another source nutrient poor upper deep water in the North Pacific remains a source of debate (Lynch-Stieglitz *et al.*, 1996; Keigwin, 1998; Oppo and Horowitz, 2000; Matsumoto *et al.*, 2002). Duplessy *et al.* (2002) (Figure 10) show a progression of oxygen isotope values in the deep (>2,000 m) water mass from highest

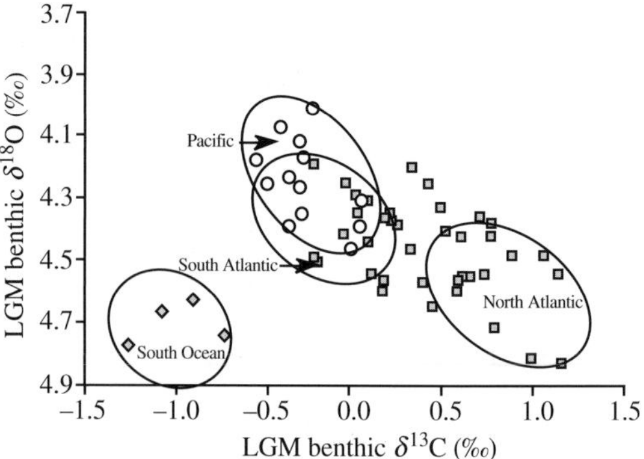

Figure 10 LGM distribution of benthic foraminiferal δ^{13}C and δ^{18}O in sediment cores from the deep ocean (mostly >2,000 m water depth) (Duplessy *et al.*, 2002). The very high δ^{18}O and low δ^{13}C in the Southern Ocean suggest a separate water mass from those filling the deep Atlantic and Pacific basins (Duplessey *et al.*, 2002) (reproduced by permission of Elsevier from *Quat. Sci. Rev.*, **2002**, *21*, 327).

values in the North Atlantic, to lowest in the North Pacific. This indicates either a distinct source of deep water to the Pacific Ocean, or progressive mixing with the overlying low-δ^{18}O water mass as the deep water ages.

Finally, there appears to be an even denser water mass that occupies the deepest portion of the Southern Ocean. This deep water has high-δ^{18}O values and very low δ^{13}C (e.g., Figure 10). This water mass is characterized by extremely low (about −1‰) δ^{13}C values and is observed in the Atlantic sector of the Southern Ocean (Figure 11). While there was some thought that the low-δ^{13}C values may reflect a productivity-related overprint (Mackensen *et al.*, 1993), a convincing case has been made that the δ^{13}C values are regionally coherent and unrelated to other measures of overlying productivity (Mackensen *et al.*, 2001; Ninnemann and Charles, 2002). One core within this deep-water mass shows LGM neodymium isotope values that have less of an Atlantic signature than today (Rutberg *et al.*, 2000). Although isolated low-δ^{13}C values have been found, there is currently insufficient spatial coverage, especially in deep waters, to document this deep-water mass in the Indian and Pacific sectors of the Southern Ocean. The Cd/Ca data are ever more limited than the δ^{13}C data in the Southern Ocean (due largely to low foraminiferal abundance, and potential saturation-state-related overprints), but suggest that the low-δ^{13}C values are not accompanied by significant increases in nutrients (Rosenthal *et al.*, 1997a,b). This suggests that air–sea exchange may have been responsible, at least in part for the lower δ^{13}C in the deep southern Ocean relative to the rest of the deep LGM ocean (Toggweiler, 1999).

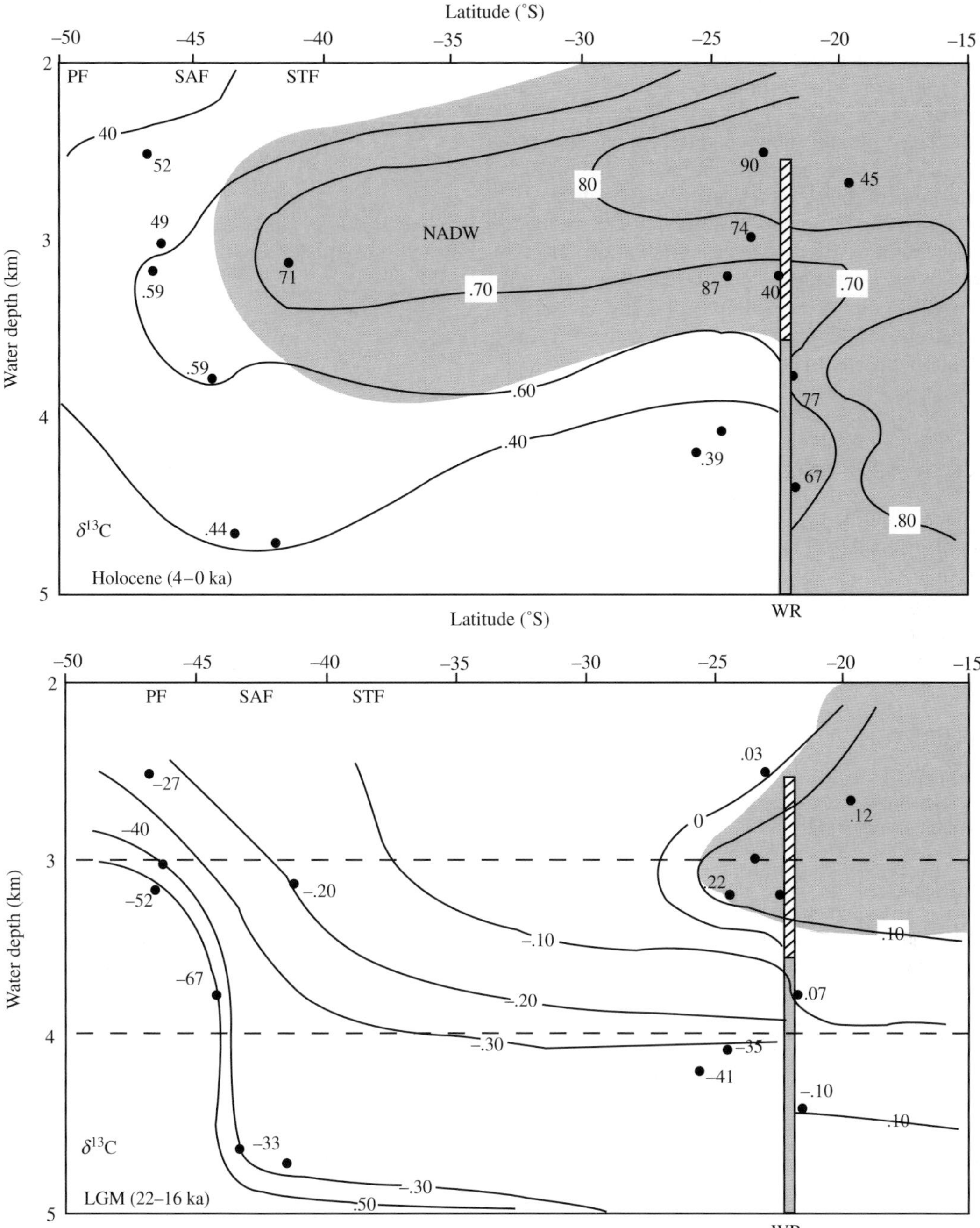

Figure 11 Modern and LGM distributions of benthic foraminiferal (*Cibicidoides*) $\delta^{13}C$ in the South Atlantic (Mackensen *et al.*, 2001). Data have been corrected for a photodetritus effect, but still show very low $\delta^{13}C$ (<-0.5 per mil) in the deepest portions of the Antarctic (reproduced by permission of Elsevier from *Global Planet. Change*, **2001**, *30*, 216–217).

Reconstructions of salinity from pore water [Cl] (Adkins *et al.*, 2002) suggest that this deep Southern Ocean water mass was considerably saltier than deep waters in the Atlantic and Pacific basins. Adkins *et al.* (2002) also argue that the pore-water $\delta^{18}O$ suggests that this dense, salty water is produced as a by-product of sea ice formation.

6.16.8 CONCLUSIONS

Considerable strides have been made since the early 1980s in developing methods to reconstruct past ocean circulation. There are now a number of different geochemical methods that can yield information on paleo-ocean flow. However, as the use of each method develops, it becomes clear that there are multiple processes affecting the tracers themselves, and the recording of the geochemical tracers into the sediments. We are still just beginning to understand most of these methods, and have limited amounts of data available. Learning how to fully appreciate and separate out these factors, combined with multiple measurements using methods affected by different processes, can ultimately lead to well-constrained reconstructions of past ocean circulation. However, even now these methods are capable of giving us a tantalizing glimpse of LGM ocean in which the water masses were arranged in a distinctly different pattern than today.

ACKNOWLEDGMENTS

I am grateful to Tom Marchitto for his help in preparing this chapter and Dan McCorkle for his thorough review. This work was supported by an NSF CAREER award to the author (OCE-9984989).

REFERENCES

Adkins J. F. and Boyle E. A. (1997) Changing atmospheric $\Delta^{14}C$ and the record of deep water paleoventilation ages. *Paleoceanography* **12**, 337–344.

Adkins J. F. and Schrag D. P. (2001) Pore fluid constraints on deep ocean temperature and salinity during the Last Glacial Maximum. *Geophys. Res. Lett.* **28**, 771–775.

Adkins J. F., Cheng H., Boyle E. A., Druffel E. R. M., and Edwards R. L. (1998) Deep-sea coral evidence for rapid change in ventilation of the deep North Atlantic 15,400 years ago. *Science* **280**, 725–728.

Adkins J. F., McIntyre K., and Schrag D. P. (2002) The salinity, temperature, and $\delta^{18}O$ of the glacial deep ocean. *Science* **298**, 1769–1773.

Albarède F. and Goldstein S. L. (1992) World map of Nd isotopes in sea-floor ferromanganese deposits. *Geology* **20**, 761–761.

Albarède F., Goldstein S. L., and Dautel D. (1997) The neodymium isotopic composition of manganese nodules from the Southern and Indian oceans, the global oceanic neodymium budget, and their bearing on deep ocean circulation. *Geochim. Cosmochim. Acta* **61**, 1277–1291.

Anderson R. F., Bacon M. P., and Brewer P. G. (1983) Removal of 231Pa and 230Th from the open ocean. *Earth Planet. Sci. Lett.* **62**, 7–23.

Belanger P. E., Curry W. B., and Matthews R. K. (1981) Core-top evaluation of benthic foraminiferal isotopic ratios for paleo-oceanographic interpretations. *Paleogeogr. Paleoclimatol. Paleoecol.* **33**, 205–220.

Boyle E. A. (1981) Cadmium, zinc, copper, and barium in foraminifera tests. *Earth Planet. Sci. Lett.* **53**, 11–35.

Boyle E. A. (1988) Cadmium: chemical tracer of deepwater paleoceanography. *Paleoceanography* **3**, 471–489.

Boyle E. A. (1992) Cadmium and $\delta^{13}C$ paleochemical ocean distributions during the stage 2 glacial maximum. *Ann. Rev. Earth Planet. Sci.* **20**, 245–287.

Boyle E. A. and Keigwin L. D. (1982) Deep circulation of the North Atlantic over the last 200,000 years: geochemical evidence. *Science* **218**, 784–787.

Boyle E. A. and Keigwin L. D. (1987) North Atlantic thermohaline circulation during the last 20,000 years linked to high latitude surface temperature. *Nature* **330**, 35–40.

Boyle E. A. and Rosenthal Y. (1996) Chemical hydrography of the South Atlantic during the last glacial maximum: Cd vs $\delta^{13}C$. In *The South Atlantic: Present and Past Circulation* (eds. G. Wefer, W. H. Berger, G. Siedler, and D. Webb). Springer, Berlin, pp. 423–443.

Boyle E. A., Sclater F. R., and Edmond J. M. (1976) On the marine geochemistry of cadmium. *Nature* **263**, 42–44.

Boyle E. A., Labeyrie L., and Duplessy J.-C. (1995) Calcitic foraminiferal data confirmed by cadmium in aragonitic *Hoeglundina*: application to the last glacial maximum in the northern Indian Ocean. *Paleoceanography* **10**, 881–900.

Broecker W. S. (1982) Glacial to interglacial changes in ocean chemistry. *Prog. Oceanogr.* **11**, 151–197.

Broecker W. S. (2002) Constraints on the glacial operation of the Atlantic Ocean's conveyor circulation. *Israel J. Chem.* **42**, 1–14.

Broecker W. S. and Peng T.-H. (1974) Gas exchange rates between air and sea. *Tellus* **26**, 21–35.

Broecker W. S. and Peng T.-H. (1982) *Tracers in the Sea.* Eldigio Press, Palisades, NY.

Broecker W. S. and Maier-Reimer E. (1992) The influence of air and sea exchange on the carbon isotope distribution in the sea. *Global Biogeochem. Cycles* **6**, 315–320.

Broecker W. S., Andree M., Bonani G., Wolfli W., Oeschger H., Klas M., Mix A., and Curry W. (1988) Preliminary estimates for the radiocarbon age of deep water in the glacial ocean. *Paleoceanography* **3**, 659–669.

Broecker W. S., Peng T.-H., Trumbore S., Bonani G., and Wolfli W. (1990) The distribution of radiocarbon in the glacial ocean. *Global Biogeochem. Cycles* **4**, 103–107.

Broecker W., Matsumoto K., Clark E., Hajdas I., and Bonani G. (1999) Radiocarbon are differences between coexisting foraminiferal species. *Paleoceanography* **14**, 431–436.

Bruland K. W., Knauer G. A., and Martin J. H. (1978) Zinc in north-east Pacific water. *Nature* **271**, 741–743.

Craig H. (1970) Abyssal carbon 13 in the south Pacific. *J. Geophys. Res.* **75**, 691–695.

Craig H. and Gordon L. I. (1965) Deuterium and oxygen 18 variations in the ocean and the marine atmosphere: stable isotopes in oceanographic studies and paleotemperatures. In *Proceedings of the 3rd Spoleto Conference, Spoleto, Italy* (ed. E. Tongiori). Sischi and Figli, Pisa, pp. 9–130.

Curry W. B. and Lohmann G. P. (1982) Carbon isotopic changes in benthic foraminifera from the western South Atlantic: reconstructions of glacial abyssal circulation patterns. *Quat. Res.* **18**, 218–235.

Curry W. B., Duplessy J.-C., Labeyrie L. D., and Shackleton N. J. (1988) Changes in the distribution of $\delta^{13}C$ of deep water $\sum CO_2$ between the last glaciation and the Holocene. *Paleoceanography* **3**, 317–341.

Deuser W. G. and Hunt J. M. (1969) Stable isotope ratios of dissolved inorganic carbon in the Atlantic. *Deep-Sea Res.* **16**, 221–225.

Duplessy J.-C., Shackleton N. J., Matthews R. K., Prell W., Ruddiman W. F., Caralp M., and Hendy C. H. (1984) ^{13}C record of benthic foraminifera in the last interglacial ocean: implications for the carbon cycle and the global deep water circulation. *Quat. Res.* **21**, 225–243.

Duplessy J.-C., Shackleton N. J., Fairbanks R. G., Labeyrie L., Oppo D., and Kallel N. (1988) Deepwater source variations during the last climatic cycle and their impact on the global deepwater circulation. *Paleoceanography* **3**, 343–360.

Duplessy J.-C., Labeyrie L., and Waelbroeck C. (2002) Constraints of the ocean oxygen isotopic enrichment between the Last Glacial Maximum and the Holocene: paleoceanographic implications. *Quat. Sci. Rev.* **21**, 315–330.

Elderfield H. and Rickaby R. E. M. (2000) Oceanic Cd/P ratio and nutrient utilization in the glacial Southern Ocean. *Nature* **405**, 305–310.

Elderfield H., Bertram C. J., and Erez J. (1996) A biomineralization model for the incorporation of trace elements into foraminiferal calcium carbonate. *Earth Planet. Sci. Lett.* **142**, 409–423.

Goldstein S. J., Lea D. W., Chakraborty S., Kashgarian M., and Murrell M. T. (2001) Uranium-series and radiocarbon geochronology of deep-sea corals: implications for Southern Ocean ventilation rates and the oceanic carbon cycle. *Earth Planet. Sci. Lett.* **193**, 167–182.

Graham D. W., Corliss B. H., Bender M. L., and Keigwin L. D. (1981) Carbon and oxygen isotopic disequilibria of recent deep-sea benthic foraminifera. *Mar. Micropaleontol.* **6**, 483–497.

Grossman E. L. (1984a) Carbon isotopic fractionation in live benthic foraminifera-comparison with inorganic precipitate studies. *Geochim. Cosmochim. Acta* **48**, 1505–1512.

Grossman E. L. (1948b) Stable isotope fractionation in live benthic foraminifera from the southern California borderland. *Palaeogeogr. Paleoclimatol. Palaeoecol.* **47**, 301–327.

Havach S. M., Chandler G. T., Wilson-Finelli A., and Shaw T. J. (2001) Experimental determination of trace element partition coefficients in cultured benthic foraminifera. *Geochim. Cosmochim. Acta* **65**, 1277–1283.

Herguera J. C., Jasen E., and Berger W. H. (1992) Evidence for a bathayal front at 2,000 m depth in the glacial Pacific, based on a depth transect on Ontong-Java Plateau. *Paleoceanography* **7**, 273–288.

Hughen K. A., Overpeck J. T., Lehman S. J., Kashgarian M., Southon J., Peterson L. C., Alley R., and Sigman D. M. (1998) Deglacial changes in ocean circulation from an extended radiocarbon calibration. *Nature* **391**, 65–68.

Inoue H. and Sugimura Y. (1985) Carbon isotopic fractionation during CO_2 exchange process between air and sea water under equilibrium and kinetic conditions. *Geochim. Cosmochim. Acta* **49**, 2453–2460.

Kallel N., Labeyrie L. D., Juillet-Leclerc A., and Duplessy J.-C. (1988) A deep hydrological front between intermediate and deep water masses in the glacial Indian Ocean. *Nature* **333**, 651–655.

Keigwin L. D. (1998) Glacial-age hydrography of the far northwest Pacific Ocean. *Paleoceanography* **13**, 323–339.

Keigwin L. D. and Schlegel M. A. (2002) Ocean ventilation and sedimentation since the glacial maximum at 3 km in the western North Atlantic. *Geochem. Geophys. Geosys.* **3**, 10.1029/2001GC000283.

Kim S.-T. and O'Neil J. (1997) Equilibrium and nonequilibrium oxygen isotope effects in synthetic carbonates. *Geochim. Cosmochim. Acta* **61**, 3461–3475.

Kroopnick P. M. (1985) The distribution of ^{13}C of $\sum CO_2$ in the world oceans. *Deep-Sea Res.* **32**, 57–84.

Labeyrie L. D. (1992) Changes in the vertical structure of the North Atlantic Ocean between glacial and modern times. *Quat. Sci. Rev.* **11**, 401–413.

Lea D. W. (1993) Constraints on the alkalinity and circulation of glacial circumpolar deep water from benthic foraminiferal barium. *Global Biogechem. Cycles* **7**, 695–710.

Lea D. W. and Boyle E. A. (1989) Barium content of benthic foraminifera controlled by bottom-water composition. *Nature* **338**, 751–753.

Lea D. W. and Boyle E. A. (1990) Foraminiferal reconstructions of barium distributions in water masses of the glacial oceans. *Paleoceanography* **5**, 719–742.

Ledbetter M. T. and Johnson D. A. (1976) Increased transport of Antarctic bottom water in the Vema Channel during the Last Ice Age. *Science* **194**, 837–839.

LeGrand P. and Wunsch C. (1995) Constraints from paleotracer data on the North Atlantic circulation during the last glacial maximum. *Paleoceanography* **10**, 1011–1045.

Loubere P. (1991) Deep-sea benthic foraminiferal assemblage response to a surface ocean productivity gradient: a test. *Paleoceanography* **6**, 193–204.

Luo S. and Ku T.-L. (1999) Oceanic 231Pa/230Th ratio influenced by particle composition and remineralization. *Earth Planet. Sci. Lett.* **167**, 183–195.

Lutze G. F. and Thiel H. (1989) Epibenthic foraminifera from elevated microhabitats: Cibicidoides wuellerstorfi and Planulina ariminensis. *J. Foraminiferal Res.* **19**, 153–158.

Lynch-Stieglitz J. (2001) Using ocean margin density to constrain ocean circulation and surface wind strength in the past. *Geochem. Geophys. Geosys.* **2**, 2001GC000208.

Lynch-Stieglitz J. and Fairbanks R. G. (1994) A conservative tracer for glacial ocean circulation from carbon isotope and paleonutrient measurements in benthic foraminifera. *Nature* **369**, 308–310.

Lynch-Stieglitz J., Stocker T. F., Broecker W. S., and Fairbanks R. G. (1995) The influence of air–sea exchange on the isotopic composition of oceanic carbon: observations and modeling. *Global Biogeochem. Cycles* **9**, 653–665.

Lynch-Stieglitz J., van Geen A., and Fairbanks R. G. (1996) Interocean exchange of Glacial North Atlantic Intermediate Water: evidence from sub-Antarctic Cd/Ca and carbon isotope measurements. *Paleoceanography* **11**, 191–201.

Lynch-Stieglitz J., Curry W. B., and Slowey N. (1999a) A geostrophic transport estimate for the Florida Current from the oxygen isotope composition of benthic foraminifera. *Paleoceanography* **14**, 360–373.

Lynch-Stieglitz J., Curry W. B., and Slowey N. (1999b) Weaker gulf stream in the Florida Straits during the last glacial maximum. *Nature* **402**, 644–648.

Mackensen A. and Douglas R. G. (1989) Down-core distribution of live and dead deep-water benthic foraminifera in box cores from the Weddell Sea and the California continental borderland. *Deep-Sea Res.* **36**, 879–900.

Mackensen A., Hubberten H.-W., Bickert T., Fischer G., and Fütterer D. K. (1993) The $\delta^{13}C$ in benthic foraminiferal tests of Fontbotia wuellerstorfi (Schwager) relative to the $\delta^{13}C$ of dissolved inorganic carbon in southern ocean deep water: implications for glacial ocean circulation models. *Paleoceanography* **8**, 587–610.

Mackensen A., Rudolph M., and Kuhn G. (2001) Late Pleistocene deep-water circulation in the sub-Antarctic eastern Atlantic. *Global Planet. Change* **30**, 197–229.

Mangini A., Lomitschka M., Eichstadter R., Frank N., Vogler S., Bonani G., Hajdas I., and Patzold J. (1998) Coral provides way to age deep water. *Nature* **392**, 347–348.

Marchal O., Francois R., Stocker T. F., and Joos F. (2000) Ocean thermohaline circulation and sedimentary $^{231}Pa/^{230}Th$ ratio. *Paleoceanography* **15**(6), 625–641.

Marchal O., Stocker T. F., and Muscheler R. (2001) Atmospheric radiocarbon during the Younger Dryas: Production, ventilation, or both? *Earth Planet. Sci. Lett.* **185**, 383–395.

Marchitto T. M., Curry W. B., and Oppo D. W. (2000) Zinc concentrations in benthic foraminifera reflect seawater chemistry. *Paleoceanography* **15**, 299–306.

Marchitto T. M., Oppo D. W., and Curry W. B. (2002) Paired benthic foraminiferal Cd/Ca and Zn/Ca evidence for a greatly increased presence of Southern Ocean water in the glacial North Atlantic. *Paleoceanography* **17**(3), art. no. 1038.

Martin P. A., Lea D. W., Rosenthal Y., Shackleton N. J., Sarnthien M., and Papenfuss T. (2002) Quaternary deep sea temperature histories derived from benthic foraminiferal Mg/Ca. *Earth Planet. Sci. Lett.* **198**, 193–209.

Matsumoto K. and Lynch-Stieglitz J. (1999) Similar glacial and Holocene deep water circulation inferred from southeast Pacific benthic foraminiferal carbon isotope composition. *Paleoceanography* **14**, 149–163.

Matsumoto K. and Lynch-Stieglitz J. (2003) Persistence of Gulf Stream separation during the Last Glacial Period:

implications for current separation theories. *J. Geophys. Res. Oceans* **108**(C6), art. no. 3174.

Matsumoto K., Lynch-Stieglitz J., and Anderson R. F. (2001) Similar glacial and Holocene Southern Ocean hydrography. *Paleoceanography* **16**, 445–454.

Matsumoto K., Oba T., Lynch-Stieglitz J., and Yamamoto H. (2002) Interior hydrography and circulation of the glacial Pacific Ocean. *Quat. Sci. Rev.* 1693–1704.

McCave I. N., Manighetti B., and Robinson S. G. (1995) Sortable silt and fine sediment size/composition slicing: parameters for palaeocurrent speed and palaeoceanography. *Paleoceanography* **10**, 593–610.

McCorkle D. C. and Klinkhammer G. P. (1991) Porewater cadmium geochemistry and the porewater cadmium–delta-C-13 relationship. *Geochim. Cosmochim. Acta* **55**, 161–168.

McCorkle D. C., Keigwin L. D., Corliss B. H., and Emerson S. R. (1990) The influence of microhabitats on the carbon isotopic composition of deep-sea benthic foraminifera. *Paleoceanography* **5**, 161–185.

McCorkle D. C., Martin P. A., Lea D. W., and Klinkhammer G. P. (1995) Evidence of a dissolution effect on benthic foraminiferal shell chemistry: $\delta^{13}C$, Cd/Ca, Ba/Ca, and Sr/Ca results from the Ontong Java Plateau. *Paleoceanography* **10**, 699–714.

McCorkle D. C., Corliss B. H., and Farnham C. A. (1997) Vertical distributions and stable isotopic compositions of live (stained) benthic foraminifera from the North Carolina and California continental margins. *Deep-Sea Res. I* **44**, 983–1024.

McCorkle D. C., Heggie D. T., and Veeh H. H. (1998) Glacial and Holocene stable isotope distributions in the southeastern Indian Ocean. *Paleoceanography* **13**, 20–34.

Miller K. G. and Lohmann G. P. (1982) Environmental distribution of recent benthic foraminifera on the northeast United States continental slope. *Geol. Soc. Am. Bull.* **93**, 200–206.

Mix A. C., Pisias N. G., Zahn R., Rugh W., Lopez C., and Nelson K. (1991) Carbon 13 in Pacific deep and intermediate waters, 0–370 ka: implications for ocean circulation and Pleistocene CO_2. *Paleoceanography* **6**, 205–226.

Mook W. G., Bommerson J. C., and Staverman W. H. (1974) Carbon isotope fractionation between dissolved bicarbonate and gaseous carbon dioxide. *Earth Planet. Sci. Lett.* **22**, 169–176.

Ninnemann U. S. and Charles C. D. (2002) Changes in the mode of Southern Ocean circulation over the last glacial cycle revealed by foraminiferal stable isotopic variability. *Earth Planet. Sci. Lett.* **201**, 383–396.

O'Neil J. R., Clayton R. N., and Mayeda T. K. (1969) Oxygen isotope fractionation in divalent metal carbonates. *J. Chem. Phys.* **5**, 5547–5558.

Oppo D. W. and Horowitz M. (2000) Glacial deep water geometry: South Atlantic benthic foraminiferal Cd/Ca and $\delta^{13}C$ evidence. *Paleoceanography* **15**, 147–160.

Oppo D. W. and Lehman S. J. (1993) Mid-depth circulation of the subpolar North Atlantic during the last glacial maximum. *Science* **259**, 1148–1152.

Ostermann D. R. and Curry W. B. (2000) Calibration of stable isotopic data: an enriched delta O-18 standard used for source gas mixing detection and correction. *Paleoceanography* **15**, 353–360.

Peng T.-H., Broecker W. S., Kipphut G., and Shackleton N. (1977) Benthic mixing in deep-sea cores as determined by ^{14}C dating and its implications regarding climate stratigraphy and the fate of fossil fuel CO_2. In *Fate of Fossil Fuel CO_2 in the Oceans* (eds. N. R. Anderson and A. Malahoff). Plenum, NY, pp. 355–373.

Rathburn A. E. and DeDeckker P. (1997) Magnesium and strontium compositions of recent benthic foraminifera from the Coral Sea, Australia and Prydz Bay, Antarctica. *Mar. Micropaleo.* **32**, 231–248.

Rathburn A. E., Corliss B. H., Tappa K. D., and Lohmann K. C. (1996) Comparisons of the ecology and stable isotopic compositions of living (stained) benthic foraminifera from the Sulu and South China seas. *Deep-Sea Res. I* **43**, 1617–1646.

Rickaby R. E. M., Greaves M. J., and Elderfield H. (2000) Cd in planktonic and benthic foraminiferal shells determined by thermal ionisation mass spectrometry. *Geochim. Cosmochim. Acta* **64**, 1229–1236.

Romanek C. S., Grossman E. L., and Morse J. W. (1992) Carbon isotopic fractionation in synthetic aragonite and calcite: effect of temperature and precipitation rate. *Geochim. Cosmochim. Acta* **56**, 419–430.

Rosenthal Y., Boyle E. A., Labeyrie L., and Oppo D. (1995) Glacial enrichments of authigenic Cd and U in sub-Antarctic sediments—a climatic control on the elements oceanic budget. *Paleoceanography* **10**, 395–413.

Rosenthal Y., Boyle E. A., and Labeyrie L. (1997a) Last glacial maximum paleochemistry and deepwater circulation in the Southern Ocean: evidence from foraminiferal cadmium. *Paleoceanography* **12**, 787–796.

Rosenthal Y., Boyle E. A., and Slowey N. (1997b) Temperature control on the incorporation of magnesium, strontium, fluorine, and cadmium into benthic foraminiferal shells from Little Bahama Bank: prospects for thermocline paleoceanography. *Geochim. Cosmochim. Acta* **61**, 3633–3643.

Rutberg R. L., Hemming S. R., and Goldstein S. L. (2000) Reduced North Atlantic deep water flux to the glacial Southern Ocean inferred from neodymium isotope ratios. *Nature* **405**, 935–938.

Schnitker D. (1974) West Atlantic abyssal circulation during the past 120,000 years. *Nature* **248**, 385–387.

Schnitker D. (1994) Deep-sea benthic foraminifers: food and bottom water masses. In *Carbon Cycling in the Glacial Ocean: Constraints on the Ocean's Role in Global Change*, NATO ASI Ser. I. 17. (eds. R. Zahn, T. F. Pedersen, M. A. Kaminiski, and L. Labeyrie). Springer, Berlin, pp. 539–554.

Schrag D. P. and DePaolo D. J. (1993) Determination of $\delta^{18}O$ of seawater in the deep ocean during the Last Glacial Maximum. *Paleoceanography* **8**, 1–6.

Schrag D. P., Adkins J. F., McIntyre K., Alexander J. L., Hodell D. A., Charles C. D., and McManus J. F. (2002) The oxygen isotopic composition of seawater during the Last Glacial Maximum. *Quat. Sci. Rev.* **21**, 331–342.

Shackleton N. J. (1977) Tropical rainforest history and the equatorial Pacific carbonate dissolution cycles. In *Fate of Fossil Fuel CO_2 in the Oceans* (eds. N. R. Anderson and A. Malahoff). Plenum, NY, pp. 401–427.

Shackleton N. J., Duplessy J. C., Arnold M., Maurice P., Hall M. A., and Cartlidge J. (1988) Radiocarbon age of last glacial Pacific deep water. *Nature* **335**, 708–711.

Spero H. J. and Lea D. W. (1996) Experimental determination of stable isotope variability in *Globigerina bulloides*: implications for paleoceanographic reconstructions. *Mar. Micropaleontol.* **28**, 231–246.

Spero H. J., Bijma J., Lea D. W., and Bemis B. E. (1997) Effect of seawater carbonate chemistry on planktonic foraminiferal carbon and oxygen isotope values. *Nature* **390**, 497–500.

Streeter S. S. (1974) Bottom water and benthonic foraminifera in the North Atlantic glacial-interglacial contrasts. *Quat. Res.* **3**, 131–141.

Stuiver M., Quay P. D., and Ostlund H. G. (1983) Abyssal water C-14 distribution and the age of the world oceans. *Science* **219**, 849.

Tachikawa K. and Elderfield H. (2002) Microhabitat effects on Cd/Ca and $\delta^{13}C$ of benthic foraminifera. *Earth Planet. Sci. Lett.* **202**, 607–624.

Tans P. P. (1980) On calculating the transfer of carbon-13 in reservoir models of the carbon cycle. *Tellus* **32**, 464–469.

Toggweiler J. R. (1999) Variation of atmospheric CO_2 by ventilation of the ocean's deepest water. *Paleoceanography* **14**, 571–588.

van Geen A., McCorkle D. C., and Klinkhammer G. P. (1995) Sensitivity of the phosphate-cadmium–carbon isotope relation in the ocean to cadmium removal by suboxic sediments. *Paleoceanography* **10**, 159–169.

Wefer G. and Berger W. H. (1991) Isotope paleontology: growth and composition of extant calcareous species. *Mar. Geol.* **100**, 207–248.

Woodruff F., Savin S. M., and Douglas R. G. (1980) Biological fractionation of oxygen and carbon isotopes by recent benthic foraminifera. *Mar. Micropaleo.* **5**, 3–11.

Yu E.-F., Francois R., and Bacon M. P. (1996) Similar rates of modern and last-glacial ocean thermohaline circulation inferred from radiochemical data. *Nature* **379**, 689–694.

Yu E. F., Francois R., Bacon M. P., *et al.* (2001) Fluxes of Th-230 and Pa-231 to the deep sea: implications for the interpretation of excess Th-230 and Pa-231/Th-230 profiles in sediments. *Earth Planet Sci. Lett.* **191**(3–4), 219–230.

Zahn R. and Mix A. C. (1991) Benthic foraminiferal $\delta^{18}C$ in the ocean's temperature–salinity–density field: constraints on Ice Age thermohaline circulation. *Paleoceanography* **6**, 1–20.

6.17
Long-lived Isotopic Tracers in Oceanography, Paleoceanography, and Ice-sheet Dynamics

S. L. Goldstein and S. R. Hemming
Columbia University, Palisades, NY, USA

6.17.1 INTRODUCTION

The decay products of the long-lived radioactive systems are important tools for tracing geological time and Earth processes. The main parent-daughter pairs used for studies in the Earth and Planetary sciences are Rb–Sr, Th–U–Pb, Sm–Nd, Lu–Hf, and Re–Os. Traditionally most practitioners have not focused their careers studying paleoceanography or paleoclimate, and the vast majority of investigations using these systems address issues in geochronology, igneous petrology, mantle geochemistry, and mantle and continental evolution. These tools are not currently considered as among the "conventional" tools for oceanographic, paleoceanographic, or paleoclimate studies. Nevertheless, these isotopic tracers have a long history of application as tracers of sediment provenance and ocean circulation. Their utility for oceanographic and paleoclimate studies is becoming increasingly recognized and they can be expected to play an important role in the future.

Frank (2002) published a general review of long-lived isotopic tracers in oceanography and authigenic Fe–Mn sediments. This chapter attempts to avoid duplication of that effort, rather, it focuses on the basis for using authigenic neodymium, lead, and hafnium isotopes in oceanography and paleoceanography. In particular, it reviews in detail the currently available data on neodymium isotopes in the oceans in order to evaluate its strengths and weaknesses as an oceanographic tracer and a proxy to investigate paleocirculation. Neodymium isotope ratios are highlighted because lead isotopes in the present-day oceans are contaminated by anthropogenic input, and dissolved hafnium thus far has not been measured. This chapter only gives a cursory summary of the results of studies on Fe–Mn crusts, which were covered extensively by Frank (2002). This chapter also summarizes the application of strontium, neodymium, and lead isotopes for tracing the sources of continental detritus brought to the oceans by icebergs and implications for the history of the North Atlantic ice sheets. This subject was not covered by Frank (2002), and in this case we summarize the major results. Dissolved strontium and osmium isotopes are treated in another chapter (see Chapter 6.20).

6.17.2 LONG-LIVED ISOTOPIC TRACERS AND THEIR APPLICATIONS

Long-lived systems are those with decay rates that are slow relative to the 4.56 Gyr age of the solar system, meaning effectively that the parent element still exists in nature. Therefore, the abundances ratios of the daughter products of decay are still increasing in rocks and water, but at rates slow enough that changes in the daughter-isotope ratios are negligible over short time periods. In this sense they contrast with short-lived radioactive systems, associated, for example, with cosmogenic nuclides and the intermediate products of uranium decay. The abundance of a nuclide formed by radioactive decay is referenced to a "stable" isotope (one that is not a decay product) for the same element. All of the long-lived radioactive decay systems have high atomic masses (from 87 amu to 208 amu) thus chemical and biological mass fractionation effects are small, unlike light stable isotopes such as those of carbon and oxygen. As a result, in a marine or ice-core sample the effects of biological or chemical mass fractionation are negligible and isotope ratios are conservative tracers, reflecting the sources of the individual elements. Transport processes on the Earth's surface generally sample a variety of continental sources, therefore in most cases the daughter isotope ratios represent mixtures from the different age terrains. If the elements are dissolved, then their isotope ratios will remain constant over any travel path as long as no additions from new sources with different isotope ratios are added. For particulates, sorting of minerals or particle sizes from a single source can lead to isotopic variability. Neodymium, lead, and hafnium are relatively insoluble elements, as a result their residence times in seawater are short relative to oceanic mixing times of ~1,500 yr (Broecker and Peng, 1982), and isotope ratios may vary both geographically and with depth in the oceans. The isotopic variability underlies their utilization as water-mass tracers. Thus far, this application has been particularly successful with neodymium.

6.17.3 SYSTEMATICS OF LONG-LIVED ISOTOPE SYSTEMS IN THE EARTH

Interpretation of the first-order isotopic variability of strontium, neodymium, hafnium, and lead in the Earth is founded on understanding the bulk chemical characteristics of the silicate Earth and how its gross differentiation has affected the parents and daughters. The best understood system among those under consideration here is $^{147}Sm \rightarrow {}^{143}Nd$. Both parent and daughter elements are rare earth elements (REEs) existing solely in the +3 oxidation state in natural systems. The rare earths are refractory during nebula condensation, partitioning into solids at high temperatures, thus the Sm/Nd and $^{143}Nd/^{144}Nd$ ratios of the Earth are considered to be the same as in chondritic meteorites, and there is a consensus on their values (Table 1).

Moreover, they are chemically similar. Both display very limited solubility in water, and differences in chemical behavior are mainly associated with atomic size. Lutetium and hafnium are also refractory elements during nebula condensation, and thus their relative abundance in the Earth are chondritic, however, at the time of writing the value of the bulk Earth $^{176}Hf/^{177}Hf$ ratio and the ^{176}Lu decay constant are being debated (Table 1). Neodymium- and hafnium-isotope ratios in this paper will be expressed as parts per 10^4 deviations from the bulk Earth values, viz., as ε_{Nd} and ε_{Hf}. For example,

$$\varepsilon_{Nd} = \left[\left\{\frac{(^{143}Nd/^{144}Nd)_{sample}}{(^{143}Nd/^{144}Nd)_{bulk\ Earth}}\right\} - 1\right] \times 10^4$$

The bulk Earth compositions of Rb–Sr and Th–U–Pb are less well known, complicated by the

Table 1 Accepted or commonly used parameters for long-lived decay systems.

Decay system	*Parent/daughter*	*(Present day)*	*Isotope ratio*	*(Present day)*	*Refs.*
$^{147}Sm \rightarrow {}^{143}Nd$	$^{147}Sm/^{144}Nd$	0.1966	$^{143}Nd/^{144}Nd$	0.512638	1
$^{87}Rb \rightarrow {}^{87}Sr$	$^{87}Rb/^{86}Sr$	0.09	$^{87}Sr/^{86}Sr$	0.7050	2
$^{176}Lu \rightarrow {}^{176}Hf$	$^{176}Lu/^{177}Hf$	0.0332	$^{176}Hf/^{177}Hf$	0.282772	3
$^{176}Lu \rightarrow {}^{176}Hf$	$^{176}Lu/^{177}Hf$	0.0334	$^{176}Hf/^{177}Hf$	0.28283	4
				(Initial solar system)	
$^{238}U \rightarrow {}^{206}Pb$			$^{206}Pb/^{204}Pb$	9.307	5
$^{235}U \rightarrow {}^{207}Pb$			$^{207}Pb/^{204}Pb$	10.294	5
$^{232}Th \rightarrow {}^{208}Pb$			$^{208}Pb/^{204}Pb$	29.476	5
Decay constant (yr^{-1})					
$\lambda\ ^{147}Sm$	6.54×10^{-12}				6
$\lambda\ ^{87}Rb$	1.42×10^{-11}				7, 13
$\lambda\ ^{176}Lu$	1.93×10^{-11}				8
$\lambda\ ^{176}Lu$	1.865×10^{-11}				9
$\lambda\ ^{176}Lu$	1.983×10^{-11}				10
$\lambda\ ^{238}U$	1.55125×10^{-10}				11, 13
$\lambda\ ^{235}U$	9.8485×10^{-10}				11, 13
$\lambda\ ^{232}Th$	4.9475×10^{-11}				12, 13

References are as follows: 1. Jacobsen and Wasserburg (1980); 2. O'Nions *et al.* (1977), DePaolo and Wasserburg (1976b); 3. Blichert- Toft and Albare(c)de (1997); 4. Tatsumoto *et al.* (1981) as adjusted by Vervoort *et al.* (1996); 5. Tatsumoto *et al.* (1973); 6. Lugmair and Scheinin (1974); 7. Neumann and Huster (1976); 8. Sguigna *et al.* (1982); 9. Scherer *et al.* (2001); 10. Bizzarro *et al.* (2003); 11. Jaffey *et al.* (1971); 12. Le Roux and Glendenin (1963); and 13. Steiger and Jäger (1977). For Pb isotopes the initial values for the solar system are known much better than the present-day bulk Earth, and these are given. For the Lu–Hf system, both the decay constant and the bulk Earth value are currently in dispute. The isotope ratios of Nd and Hf in the text are given as deviations from the bulk Earth value in parts per 10^4.

volatility of rubidium and lead during nebular condensation combined with the variable depletion of volatile elements in the Earth. Of these systems, there is a consensus nevertheless for the Rb/Sr and $^{87}Sr/^{86}Sr$ ratios of the bulk Earth (DePaolo and Wasserburg, 1976b; O'Nions *et al.*, 1977). For the Th–U–Pb, the refractory behavior of thorium and uranium puts strong constraints on the $^{208}Pb/^{206}Pb$ of the Earth, but the Th/Pb and U/Pb ratios are not as well constrained. The coupled nature of U–Pb isotopic evolution (two elements, and two decay systems) means that systematics of closed-system evolution of the solar system is well constrained such that planetary bodies should lie on a ~4.56 Ga isochron line in $^{206}Pb/^{204}Pb-^{207}Pb/^{204}Pb$ space. However, balancing the bulk Earth composition to that line has been problematic. Details of these complications are beyond the scope of this contribution, however, there are extensive discussions in the literature.

Long-lived isotopes are useful tracers for paleoceanography and paleoclimate studies because they vary over the surface of the Earth and within. Variability between the continental crust and mantle is primarily a result of (i) the gross chemical differentiation of the Earth, associated with mantle melting and plate tectonics to form continental crust, and the different behaviors of the parent and daughter elements during magma formation, and (ii) the age of the continental crust. Among the systems under consideration, Rb–Sr, Sm–Nd, and Lu–Hf isotopic variability on the Earth's surface follows general expectations from melting behavior, at least in a gross sense. Rubidium, neodymium, and hafnium are more likely to enter magma than their partner elements, and as a result the continents have higher Rb/Sr and lower Sm/Nd and Lu/Hf ratios than the Earth's mantle. During the course of time this has resulted in the continents having high $^{87}Sr/^{86}Sr$ and low $^{143}Nd/^{144}Nd$ and $^{176}Hf/^{177}Hf$ ratios compared to the mantle and bulk Earth (i.e., negative ε_{Nd} and ε_{Hf} values), as illustrated by Figure 1. Mid-ocean ridge and ocean island basalts (MORB and OIB), derived from melting of the mantle, have low $^{87}Sr/^{86}Sr$ and high $^{143}Nd/^{144}Nd$ and $^{176}Hf/^{177}Hf$ ratios (positive ε_{Nd} and ε_{Hf} values). Convergent-margin volcanics have similar isotopic characteristics as oceanic basalts as long as they are not situated on old continental crust or contain large amounts of subducted sediment.

The distinction between continental- and mantle-isotope ratios increases with the age of the continental terrain. Because the age of the continental crust is geographically variable, the continents are isotopically heterogeneous on regional scales and this heterogeneity forms the basis for tracing sources and transport. As a reflection of melting behavior and continental age, isotopic variations between some of these isotope systems are systematic in continental rocks and oceanic basalts. Nd–Hf isotopes are usually positively correlated while Nd–Sr and Hf–Sr isotopes are usually negatively correlated. However, these are gross features of the decay systems

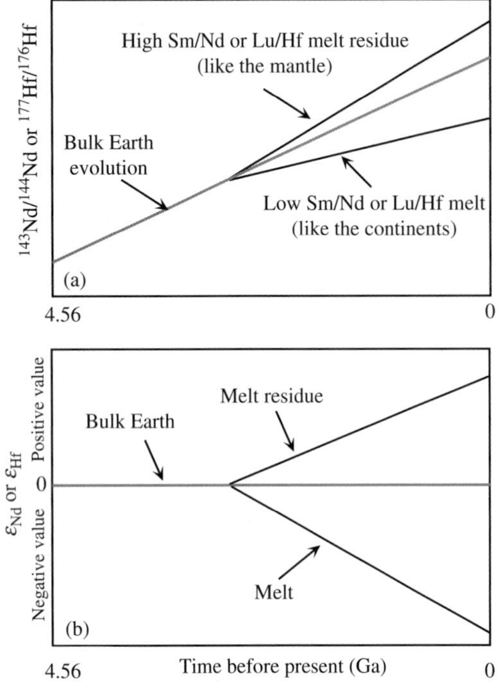

Figure 1 Systematics of Nd- and Hf-isotopic evolution in the bulk Earth, continental crust, and mantle. Daughter elements Nd and Hf are more incompatible during mantle melting (more likely to go into a partial melt of mantle rock) than Sm and Lu, respectively. As a result, the continental crust has a lower Sm/Nd and Lu/Hf ratio than the mantle, and lower Nd- and Hf-isotope ratios. Young continental crust has isotope ratios similar to the mantle, and the older the continental terrain, the lower the Nd- and Hf-isotope ratios. Rb–Sr behaves in the opposite sense, such that the parent element Rb is more incompatible than the daughter element Sr. (a) Schematic example of the evolution of Nd- and Hf-isotope ratios of a melt and the melt residue from a melting event around the middle of Earth history from a source with the composition of the bulk Earth. (b) The same scenario as in (a), but with the isotope ratios plotted as ε_{Nd} and ε_{Hf}. The "bulk Earth" value throughout geological time is defined as ε_{Nd} and $\varepsilon_{Hf} = 0$, and ε-value of a sample is the parts per 10^4 deviation from the bulk Earth value.

and there are some complicating features. For example, a significant portion of continental hafnium is located in the mineral zircon, a heavy mineral that can be separated by sedimentary processes. Therefore, in sediments where zircon has been separated by sorting, the Lu/Hf and $^{176}Hf/^{177}Hf$ ratios should be higher than expected from the crustal age. As discussed below, this appears to have important effects on the isotope ratio of dissolved hafnium in the oceans. The Rb–Sr system is more likely than neodymium and hafnium to be disturbed by weathering and metamorphism due to higher solubility of rubidium and strontium in water. In addition, minerals such as micas generally have high Rb/Sr ratios, and feldspars have lower Rb/Sr ratios; therefore, crystal sorting of a sample during transport by currents can result in significant strontium-isotopic heterogeneity of samples from the same source. These processes affecting hafnium and strontium isotopes can degrade correlations with neodymium isotopes in marine or eolian samples.

Lead-isotope ratios in continental rocks and oceanic basalts often do not display strong correlations with Nd–Sr–Hf isotope ratios, even though the melting behavior of thorium and uranium relative to lead is similar to Rb–Sr (Hofmann *et al.*, 1986). $^{206}Pb/^{204}Pb$ ratios usually show substantial overlap in oceanic basalts and continental rocks. In $^{206}Pb/^{204}Pb$–$^{207}Pb/^{204}Pb$ x–y plots, continental rocks generally fall above oceanic basalts, reflecting higher $^{207}Pb/^{204}Pb$ for a given $^{206}Pb/^{204}Pb$ ratio, a consequence of the short half-life of ^{235}U compared to ^{238}U (Table 1) coupled with a higher U/Pb ratio in the continents than the mantle over the first half of Earth history. As a result, continent-derived lead can often be distinguished from mantle-derived lead.

6.17.3.1 Early Applications to the Oceans

Investigations involving the oceans and marine sediments were among the earliest applications of long-lived radioactive systems. The first major lead isotopic study by Patterson *et al.* (1953) reported data on a Pacific manganese nodule and a red clay. This was followed by Patterson's (1956) classic work on the age of the Earth, where lead-isotope ratios of marine sediments were used to estimate the average lead-isotope composition of the Earth. Although subsequent work has shown the approach to be flawed, this work still is credited for defining the genetic link between the formation of meteorites and the Earth. Chow and Patterson's (1959) classic paper on lead-isotope ratios in manganese nodules can be considered as the work that sets the stage for all future applications of long-lived radioactive isotopes to the oceans. They wrote "the oceans serve as collecting and mixing reservoirs for leads which are derived form vast land areas, thus provide samples of leads … (reflecting) large continental segments. Lead isotopes can also be used as tracers for the study of circulation and mixing patterns of the oceans." This and their subsequent study (Chow and Patterson, 1962) outlined the geographical variability of lead in manganese nodules and pelagic clays in the oceans. They showed, for example, that different regions have characteristic lead-isotope ratios, and that they are lowest in the northwest Pacific and highest in the northeast Atlantic (Figure 2). They further

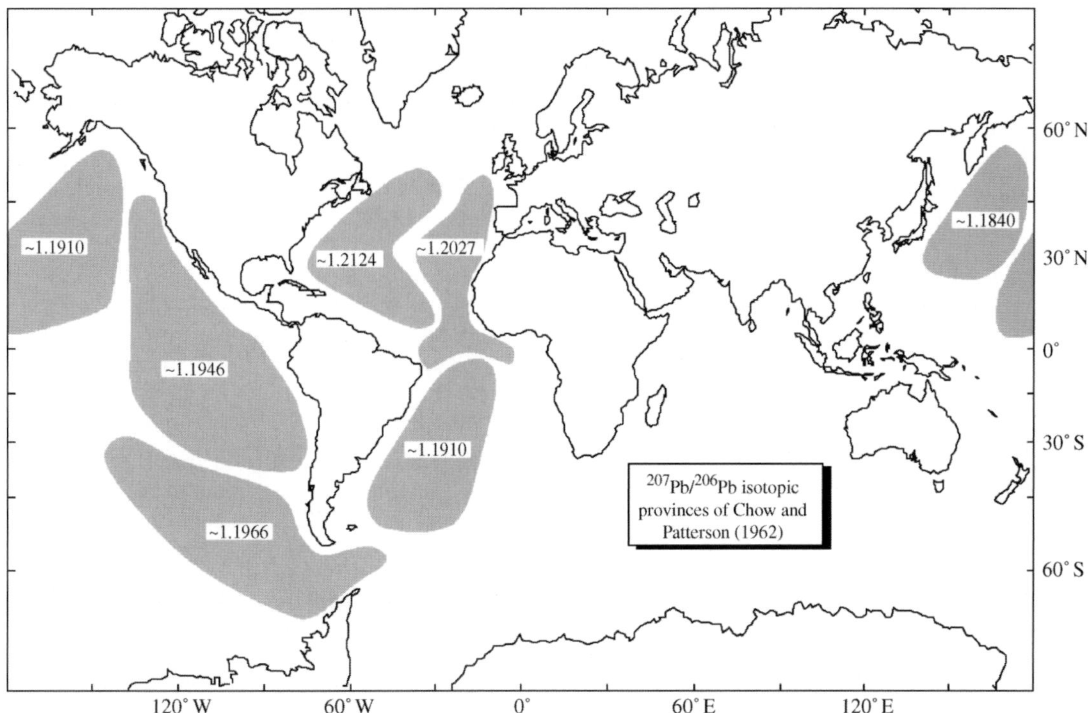

Figure 2 Marine Pb-isotopic provinces defined by Chow and Patterson (1962). The Pb-isotope ratios are given as $^{207}Pb/^{206}Pb$ ratios rather than $^{207}Pb/^{204}Pb$ and $^{206}Pb/^{204}Pb$ because of easier measurement and fractionation control. The map shows that major provinces in the oceans were defined by this early work. Among the implications is that North Atlantic Pb has an old continental provenance, defined by high $^{207}Pb/^{206}Pb$ ratios.

identified the southwest Atlantic as a province having relatively low lead-isotope ratios, like the north Pacific. They concluded that the source of marine lead is the continental crust, but without knowledge of isotopic compositions of mantle lead. Nearly a decade later, Reynolds and Dasch (1971) and Dasch *et al.* (1971) confirmed a continental source for the lead in manganese nodules and clays far from ocean ridges, but showed that manganese-rich metalliferous sediments deposited near mid-ocean ridges contain mantle-derived lead, extracted by hydrothermal processes near ridges.

The first neodymium-isotopic analyses aimed at characterizing the oceans closely followed the initial development of neodymium isotopes as a chronometer and tracer (Richard *et al.*, 1976; DePaolo and Wasserburg, 1976a; O'Nions *et al.*, 1977). O'Nions *et al.* (1978) was the first to report neodymium (along with lead and strontium) isotopes in manganese nodules and hydrothermal sediments. They confirmed the distinction between continental and mantle provenances of lead in hydrogenous and hydrothermal manganese sediments, respectively. Consistent with the previous studies, they found that strontium in these deposits is derived from seawater. All of the neodymium-isotope ratios in their samples from the Pacific were similar and lower than the bulk Earth value ($\varepsilon_{Nd} = -1.5$ to -4.3). They concluded correctly that neodymium in the oceans is mainly derived from the continents, and incorrectly that neodymium-isotope ratios in the oceans are like strontium isotopes in the sense that they are globally well-mixed and display this restricted range. In order to reach this conclusion they ignored a sample from the Indian Ocean having high strontium-isotope ratios and a lower, even more continent-like neodymium-isotope ratio of $\varepsilon_{Nd} = -8.5$.

The conclusion of O'Nions *et al.* (1978) that neodymium is relatively well mixed in the oceans was soon shown to be incorrect by studies of Fe–Mn sediments (Piepgras *et al.*, 1979; Goldstein and O'Nions, 1981). These studies showed that the Pacific, Indian, and Atlantic oceans have distinct and characteristic neodymium-isotopic signatures. The North Atlantic has the lowest (old continent-like) values ($\varepsilon_{Nd} = -10$ to -14), the Pacific the highest ($\varepsilon_{Nd} = 0$ to -5), and the Indian is intermediate ($\varepsilon_{Nd} = -7$ to -10). The data, in addition, indicated some systematic geographic variability within ocean provinces. For example, the lowest values in the Atlantic were found in the far north, and some western Indian Ocean samples from locations near southern Africa had Atlantic-type values. In the Pacific, the highest values were found in hydrothermal crusts or in the far northwest. The inter-ocean differences were attributed to the variability in the age of the surrounding continental crust, or significant contributions of volcanism-derived neodymium in the Pacific. During the period 1978–1981 two additional Pacific manganese nodule-analyses were published by Elderfield *et al.* (1981). Goldstein and O'Nions (1981) also showed that neodymium-isotope ratios can be distinguished in the leachable (precipitated) component of sediments and the solid-clay residue. The first direct measurements on Atlantic and Pacific seawater by Piepgras and Wasserburg (1980) showed substantial vertical variability, and general agreement with the Fe–Mn nodule and crust data for waters at similar depths. In addition, they suggested that North Atlantic Deep Water may have a distinct neodymium-isotopic signature, and that neodymium isotopes may have significant applications to paleoceanography.

In contrast to the application of lead and neodymium isotopes to the oceans, the first hafnium-isotopic study was not published until 1986. White *et al.* (1986) showed that six manganese nodules from the Atlantic, Indian, and Pacific oceans have a restricted range of hafnium-isotope ratios, with positive ε_{Hf} values (+0.4 to +0.9). Thus, hafnium isotopes neither show the distinct continental signature observed for marine neodymium-isotope ratios, nor the large variability of neodymium-isotope ratios in the different oceans. Therefore, although Sm–Nd and Lu–Hf systems have evolved coherently in the continents and mantle, showing positive covariations that reflect the fractionation of Sm/Nd and Lu/Hf ratios during mantle melting and continent generation (Figure 1), manganese nodules fall off this "crust–mantle correlation" to high hafnium-isotope ratios for a given neodymium-isotope ratio. White *et al.* (1986) suggested that a significant portion of marine hafnium is derived from the mantle through hydrothermal activity, or subaerial or subaqueous weathering of volcanics. Furthermore, they suggested a low continental flux of hafnium due to retention in zircon. Despite the interest generated by surprisingly high hafnium-isotope ratios, the next study was not published for 11 yr (Godfrey *et al.*, 1997). This delay was mainly due to the difficulties of analyzing hafnium-isotope ratios by thermal-ionization mass spectrometry, a consequence of the high first-ionization potential for hafnium. This situation changed in the mid-1990s with the development of multiple collector inductively coupled plasma-source mass spectrometers.

All of the early investigators, with the exception of Harry Elderfield, were mainly mantle or planetary geochemists, and these isotopic studies had relatively small impact on the oceanography community. Nevertheless, the stage was set for future studies. The gross characteristics of the marine lead-isotopic variability was outlined by the mid-1970s based on Fe–Mn sediments as seawater proxies. The gross characteristics of the global geographic neodymium-isotopic variability in the oceans (Albare(c)de and Goldstein, 1992), and shown in Figure 3, was outlined by the early 1980s.

6.17.4 NEODYMUIM ISOTOPES IN THE OCEANS

6.17.4.1 REEs in Seawater

Neodymium is a valuable tracer for oceanographic studies, as a product of radioactive decay and as an REE. They behave as a coherent group of elements whose chemical behavior is defined by filling of the 4*f* electron shell. Conventionally, REE abundances are shown in the sequence of increasing atomic number. In studies of igneous rocks, measured REE abundances are normalized to average chondritic-meteorite values, and in marine studies they are usually normalized to "average shale" (cf. Taylor and McLennan, 1985). As mentioned above, they exist in the Earth in the +3 oxidation state and display similar chemical behavior. Their relatively small chemical behavioral differences are mainly due to the effects of decreased ionic radius with higher atomic number. Two exceptions are europium and cerium. In high-oxidation environments like the oceans, cerium can exist as Ce^{+4}, in which case it forms an insoluble oxide. In low-oxidation environments such as within the Earth, europium can exist as Eu^{+2}. This rarely occurs at the Earth's surface, but it has important effects in magma chambers where Eu^{+2} is preferentially incorporated into feldspars. Magmatic fractionation of feldspar leads to a marked negative europium anomaly in convergent-margin volcanic rocks, and is a distinctive characteristic of the composition of the average upper-continental crust (e.g. Taylor and McLennan, 1985; Wedepohl, 1995; Rudnick and Fountain, 1995; Gao *et al.*, 1998). Dissolved REEs in seawater are stabilized as carbonate complexes (Cantrell and Byrne, 1987). The fraction of each REE that exists as a carbonate complex increases with increasing atomic number.

As a result of these factors, the REE pattern of seawater is distinctive. The seawater pattern is heavy REE enriched (e.g., Elderfield and Greaves, 1982). In addition, cerium shows a marked depletion compared with its neighbors, lanthanum and praseodymium. Because average shale contains a negative europium anomaly compared to chondrites, the practice of normalizing seawater to

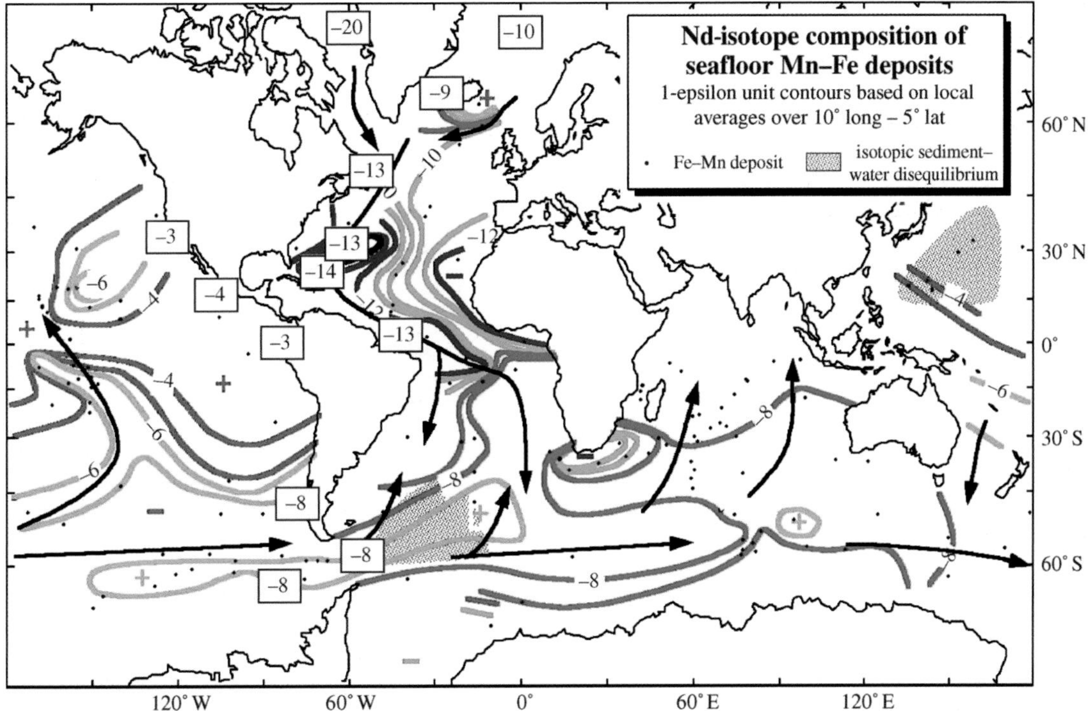

Figure 3 *Map of Nd-isotope variability in ferromanganese deposits.* The map shows systematic geographic variability with lowest values in the North Atlantic, highest values in the Pacific, and intermediate values elsewhere. Arrows illustrate general movement of deep water, and show that the contours generally follow deep-water flow. Shaded fields delineate regions where the Fe–Mn and deep seawater data differ by $>2\varepsilon_{Nd}$ units (after Albare(c)de and Goldstein, 1992).

shale mutes the magnitude of the europium anomaly in patterns normalized to shale, such as published seawater REE diagrams, as compared to patterns normalized to chondrites. Nevertheless, variability in the magnitude of europium anomalies in different source rocks causes Eu/REE ratios in seawater to vary more than the trivalent REEs. The most recent general review of REEs in the oceans is by Elderfield (1988).

Abundances of REE's in depth profiles, with the exception of cerium, generally show depletions in surface waters and enrichments in deep water. Moreover, concentrations are generally lower in the North Atlantic and higher in the Pacific. With respect to both of these characteristics they are similar to SiO_2, and REE abundances of intermediate and deep waters tend to correlate with silicate (Elderfield and Greaves, 1982; Debaar *et al.*, 1983, 1985). Nevertheless, despite similarities in the behavior of SiO_2 and the REE, there are important differences, with the Atlantic generally having higher REE/SiO_2 than the Pacific (cf. Elderfield, 1988, and references therein). Despite these inter-ocean differences, silicate is often used as a proxy for REE behavior in the oceans. As shown below, neodymium-isotope ratios place important constraints on the mode of REE cycling in the oceans.

6.17.4.2 Neodymium-isotope Ratios in Seawater

There are only a small number of published papers, fewer than 20, reporting neodymium-isotope ratios in seawater. The early work was entirely carried out in the lab of Professor Gerald Wasserburg at Caltech (Piepgras and Wasserburg, 1980, 1982, 1983, 1985, 1987; Stordal and Wasserburg, 1986; Spivack and Wasserburg, 1988), which deserves credit for mapping the primary variability in seawater. All of the other published data are from Cambridge, Harvard, Lamont, Tokyo, and Toulouse. A map of the locations of seawater data is shown in Figure 4. The reason for the small number of studies is that the task is analytically challenging, due to the very low abundance of neodymium in seawater, which generally ranges ~15–45 pmol kg^{-1} (~2–7 ng L^{-1}). Most isotope laboratories measure neodymium as a positive metal ion (Nd^+), and are comfortable measuring 80–100 ng of neodymium, which would require processing of 10–40 L of seawater per sample. For this reason most of the seawater neodymium-isotope ratios in the literature were measured as a positive metal oxide (NdO^+), which affords higher efficiency of ion transmission and allows measurements of smaller samples. Even so,

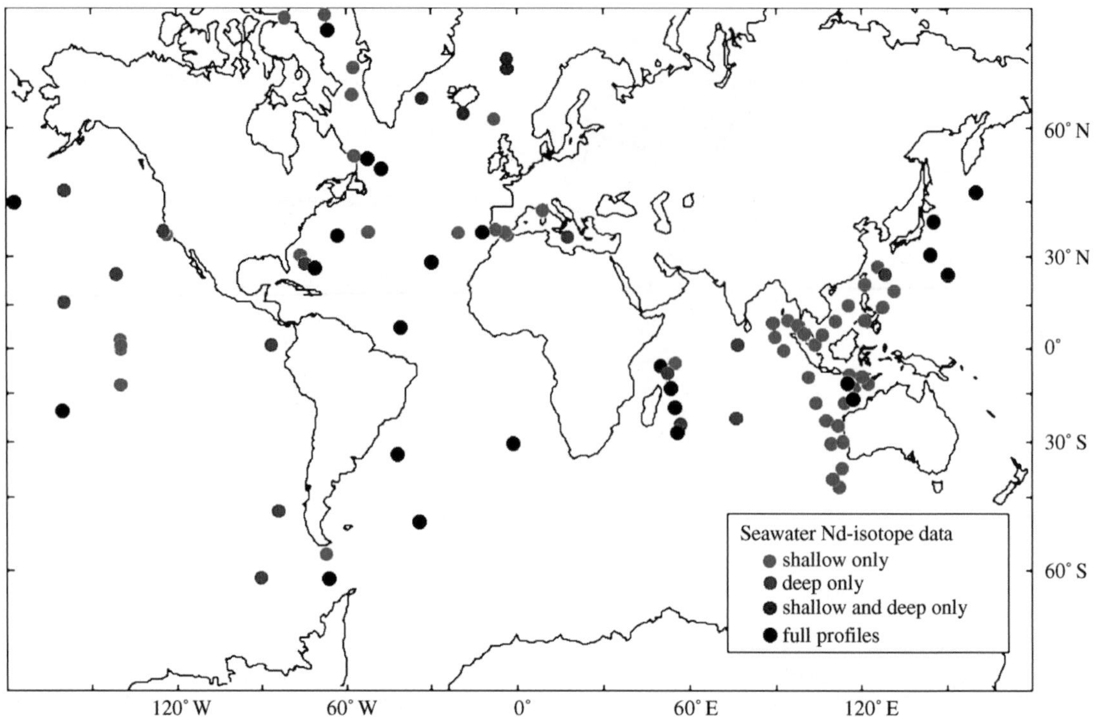

Figure 4 *Map of locations of published seawater Nd-isotope data.* Locations are distinguished by shallow-only data, deep-only data, deep- and shallow-only data, and full profiles, where "shallow" is <2,000 mb sl. All publications are cited in the text except Henry *et al.* (1994), which reports the two central Mediterranean sites.

measurements of ~15–20 ng of neodymium still require processing of 3–10 L of seawater per sample.

The variability of neodymium isotopes in the oceans is closely related to global ocean circulation. The present-day mode of thermohaline circulation has been likened to a "great ocean conveyor" (Broecker and Peng, 1982; Broecker and Denton, 1989) in which deep water formed in the North Atlantic flows southward where it enters the circum-Antarctic. Circumpolar water flows to the Indian and Pacific, where it upwells. The North Atlantic is recharged by the surface-return flow from the Pacific through the Indonesian Straits and the Indian Ocean, and by northward movement of intermediate and deep water in the Atlantic from the circum-Antarctic (Rintoul, 1991). In the latter case, southward flowing NADW (North Atlantic Deep Water) is "sandwiched" by northward flowing Antarctic bottom water (AABW) and Antarctic Intermediate Water (AAIW). The principal observations on the variability of neodymium-isotope ratios globally are summarized below, with the publications that made the main discoveries.

Neodymium-isotope ratios in deep seawater show the same general geographical pattern as in Fe–Mn nodules and crusts. The global geographical pattern is illustrated in the Fe–Mn nodule-crust map (Figure 3), and by some depth profiles (Figure 5). The North Atlantic is characterized by low values, the Pacific by high values, and the Indian Ocean is intermediate (Piepgras and Wasserburg, 1980, 1987; Bertram and Elderfield, 1993). In addition, neodymium-isotope ratios in the circum-Antarctic are also intermediate to the North Atlantic and Pacific (Piepgras and Wasserburg, 1982). Typical values are in the range of Fe–Mn nodules and crusts as noted above. The same studies show that neodymium concentrations are generally highest in the Pacific and lowest in the Atlantic (Figure 6).

NADW has a narrow range of $\varepsilon_{Nd} = -13$ to -14 (Figure 5), which defines most of the mid-range and deep water in the North Atlantic south of ~55° N (Piepgras and Wasserburg, 1980, 1987). The uniformity is perhaps surprising considering the variability of NADW sources. NADW is a mixture of sources displaying two very different neodymium-isotope components. Baffin Bay between the Labrador and Greenland receives its neodymium from old continental crust with very low neodymium-isotope ratios of $\varepsilon_{Nd} < -20$. NADW sources from the Norwegian and Greenland Sea sources east of Greenland have much higher neodymium-isotope ratios of $\varepsilon_{Nd} = -7$ to -10, and these higher values are observed in the Denmark Straits and Faeroe Channel overflow (Stordal and Wasserburg, 1986).

Surface waters show more variable neodymium-isotope ratios than intermediate and deep waters (Figure 5). In contrast to the dominance

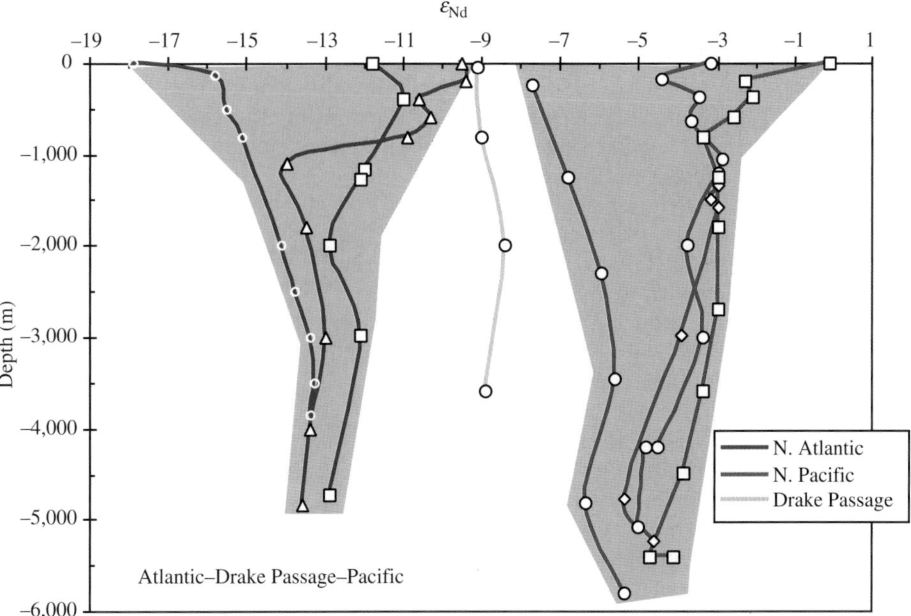

Figure 5 *Representative depth profiles of Nd-isotope ratios from the North Atlantic and Pacific oceans and the Drake Passage.* The Atlantic and Pacific profiles were chosen to encompass the range of values for deep water in those oceans. Symbols show the data. The diagram illustrates the differences between the oceans and the greater variability of Nd isotopes in shallow versus deep waters (sources Piepgras and Wasserburg, 1982, 1983, 1987; Spivack and Wasserburg, 1988; Piepgras and Jacobsen, 1988; Shimizu *et al.*, 1994).

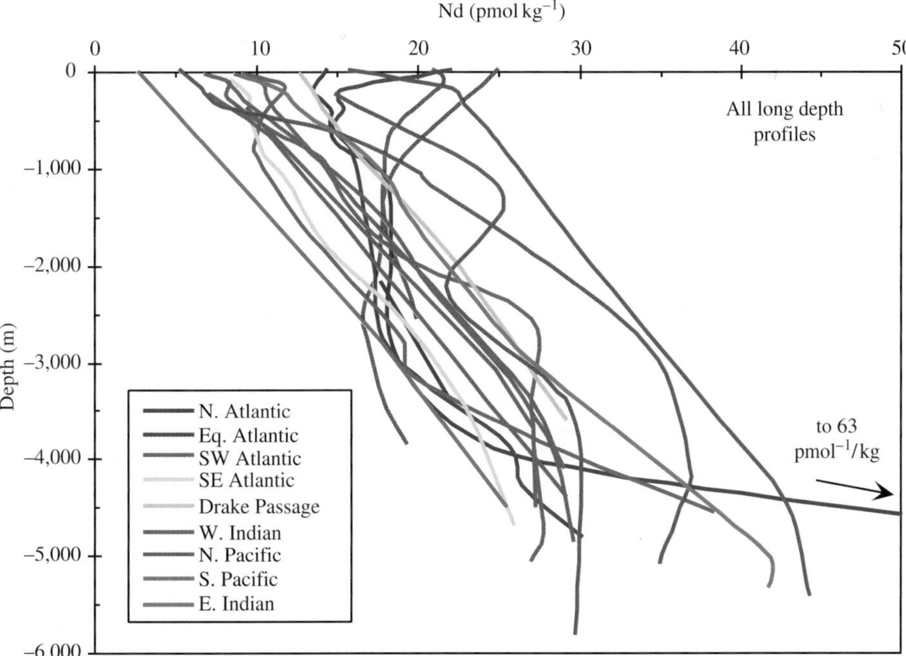

Figure 6 *Nd concentrations versus depth for all long depth profiles in the literature.* The data show that higher concentrations with depth are a general feature in the oceans, although this is not necessarily the case in the North Atlantic. Deep waters generally have the highest abundances of Nd in the North Pacific, and the lowest in the North Atlantic, but there are several exceptions (sources Piepgras and Wasserburg, 1980, 1982, 1983, 1987; Spivack and Wasserburg, 1988; Piepgras and Jacobsen, 1988; Bertram and Elderfield, 1993; Jeandel, 1993; Shimizu *et al.*, 1994; Jeandel *et al.*, 1998).

of NADW values ($\varepsilon_{Nd} = -13$ to -14) in the intermediate and deep water of the North Atlantic, surface waters are variable, ranging from $\varepsilon_{Nd} = -8$ to -26 (Piepgras and Wasserburg, 1980, 1983, 1987; Spivack and Wasserburg, 1988; Stordal and Wasserburg, 1986). The greater variability of surface waters is observed throughout the oceans. Like silicate, neodymium abundances are generally depleted in surface waters compared to intermediate and deep water (Figure 6).

The limited seawater data from the circum-Antarctic display uniform neodymium-isotope ratios intermediate to the Atlantic and Pacific ($\varepsilon_{Nd} = -8$ to -9). The only neodymium-isotope data available are from two profiles in the Drake Passage and one in the eastern Pacific sector (Piepgras and Wasserburg, 1982).

Where depth profiles sample different water masses, neodymium-isotope ratios vary with the water mass. In South Atlantic intermediate and deep water, neodymium-isotope ratios are generally intermediate to the NADW and Drake Passage values. Depth profiles show a zig-zag pattern (Figure 7), higher at intermediate depths dominated by AAIW, lower at greater depths dominated by NADW, and higher at deepest levels dominated by AABW (Piepgras and Wasserburg, 1987; Jeandel, 1993). A published diagram from von Blanckenburg (1999) overlaying neodymium-isotopic profiles and salinity (Figure 8) elegantly shows that the general characteristics of neodymium-isotope ratios vary with salinity in southward and northward flowing-water masses in the Atlantic.

Pacific intermediate and deep water dominantly display high values, of $\varepsilon_{Nd} = -2$ to -4 (Piepgras and Wasserburg, 1982; Piepgras and Jacobsen, 1988; Shimizu *et al.*, 1994). Near southern Chile, and in the west-central Pacific, where circumpolar deep water moves northward, neodymium-isotope ratios display lower values closer to those typical of circumpolar seawater. In the central Pacific the circumpolar neodymium-isotope signal can be detected as far north as ~40° N (Shimizu *et al.*, 1994).

Neodymium-isotope ratios of intermediate and deep water in the Indian Ocean are intermediate to the Atlantic and Pacific. They generally fall between $\varepsilon_{Nd} = -7$ to -9, and are consistent with domination by northward flowing circumpolar water (Bertram and Elderfield, 1993; Jeandel *et al.*, 1998). A depth profile east of southern Africa (Figure 7) displays the same zig-zag pattern as South Atlantic intermediate and deep water, reflecting advection of NADW into the western Indian Ocean (Bertram and Elderfield, 1993).

The broadest characteristics can be concisely summarized. Neodymium-isotope ratios in seawater vary systematically with location throughout the oceans, high in the Pacific, low in the Atlantic, and intermediate in the Indian and circum-antarctic. In addition, they are variable in depth profiles that contain different water masses, and neodymium-isotope ratios associated with end-members of water masses are conserved over long advective pathways.

6.17.4.3 Where does Seawater Neodymium Come From?

The first-order question arising from the global variability is, where does the neodymium in seawater come from? The negative ε_{Nd} values clearly show that throughout the oceans the dissolved neodymium is dominantly derived from the continents (cf. Figure 1). The characteristic values in the Atlantic and Pacific indicate at first glance that they reflect the age of the surrounding continental crust. The North Atlantic, surrounded by old continental crust, has the lowest values, and the Pacific, surrounded by orogenic belts has the highest values. However, the relationships are not simple ones.

In the North Atlantic, the present-day neodymium-isotope ratio of NADW of $\varepsilon_{Nd} \approx -13.5$ is determined by efficient mixing of two disparate sources to the west and east of Greenland. The Baffin Bay source with $\varepsilon_{Nd} < -20$ is easily explained by the surrounding early and middle Precambrian continental crust. The high ε_{Nd} value of -7 to -9 of the Denmark Straits–Norwegian Sea sources are less easy to explain. Both Greenland and Norway are primarily surrounded by Precambrian continental crust, whose early to mid-Proterozoic average age suggests an average ε_{Nd} much more negative. The only substantial sources of higher neodymium-isotope ratios in the region are Iceland and the flood basalts of the British Tertiary Province, although these cover a minor portion of the surrounding land area. Unless there is a significant input into the Denmark Straits–Norwegian Sea derived from the Pacific and leakage into these regions through the Arctic, the high neodymium-isotope ratios almost certainly show substantial input from these local volcanic sources. As mentioned previously, the value of $\varepsilon_{Nd} \approx -13.5$ that characterizes NADW is the signature imparted to most of the intermediate and deep water throughout the North Atlantic south of ~55° N. The observation that the Baffin Bay–Denmark Straits–Norwegian Sea sources are so different implies that this distinctive present-day NADW signature is inherently unstable and may easily be subject to change with time.

In the Pacific, the mechanism through which seawater obtains its high neodymium-isotope ratios is not well identified. The source is clearly not from the surrounding continental masses. The major Pacific sediment sources from the surrounding

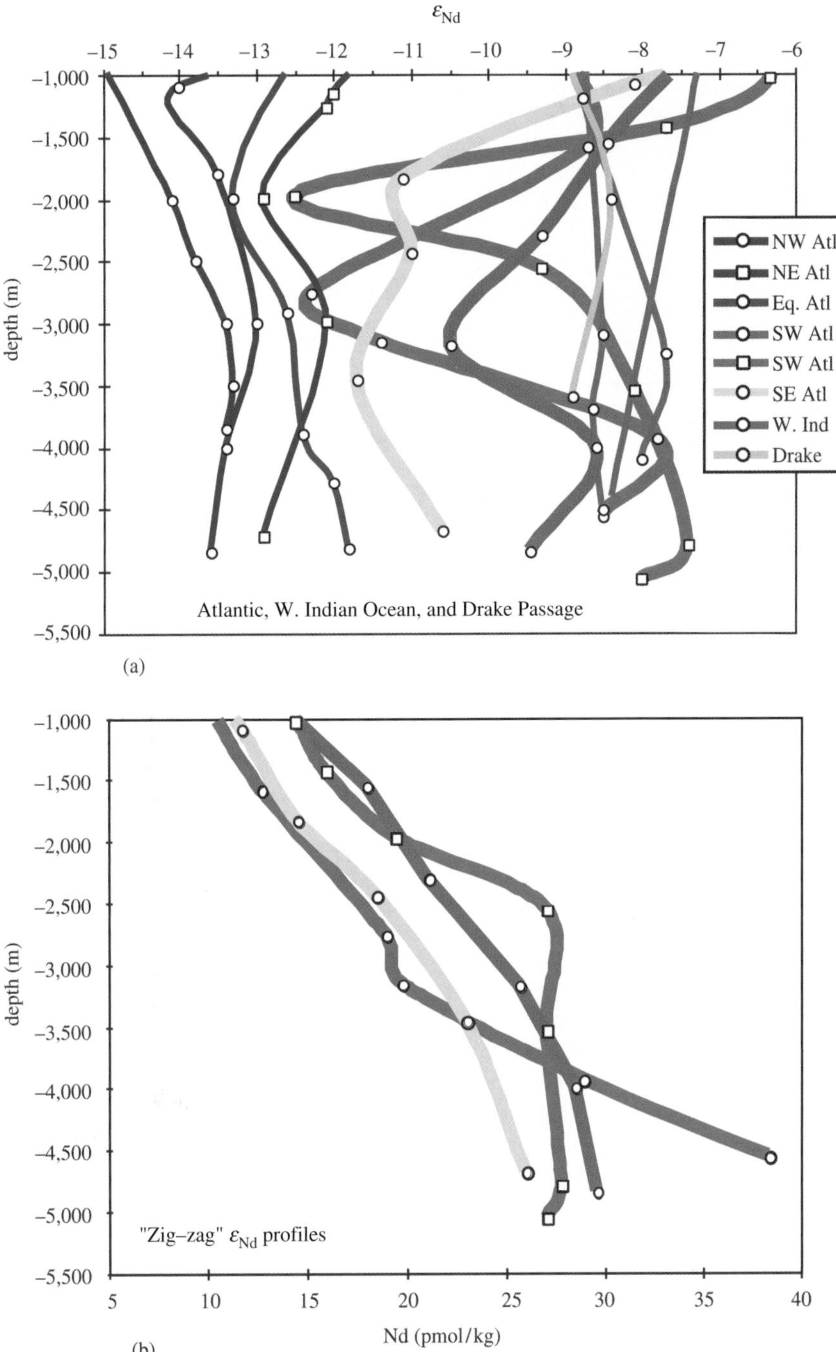

Figure 7 *Nd-isotope ratios and concentrations versus depth in the Atlantic and western Indian Oceans, and the Drake Passage.* (a) Only full profiles are shown. Western Indian Ocean data are similar to the single profile from the Drake Passage, but at slightly higher Nd-isotope ratios, which may indicate that the Drake Passage profile does not give the upper limit for the circum-Antarctic. Profiles from the South Atlantic and one from the West Indian show large variations with depth, which can range from near North Atlantic values, reflecting NADW, to near circumpolar-West Indian values, reflecting AABW and AAIW in individual profiles. (b) Nd concentrations of those profiles showing zig-zag ε_{Nd} patterns. The zig-zag profiles show that the Nd-isotopic signature of water masses is conserved over the transport path in the Atlantic (sources Piepgras and Wasserburg, 1982; 1987; Spivack and Wasserburg, 1988; Bertram and Elderfield, 1993; Jeandel, 1993).

continental crust, including all of the major rivers as well as wind-blown dust from China that is the main detritus source for the central Pacific, have much lower neodymium-isotope ratios than the typical Pacific seawater value of −4 (e.g. Goldstein *et al.*, 1984; Goldstein and Jacobsen, 1988; Nakai *et al.*, 1993; Jones *et al.*, 1994). Among Pacific land masses, the only major

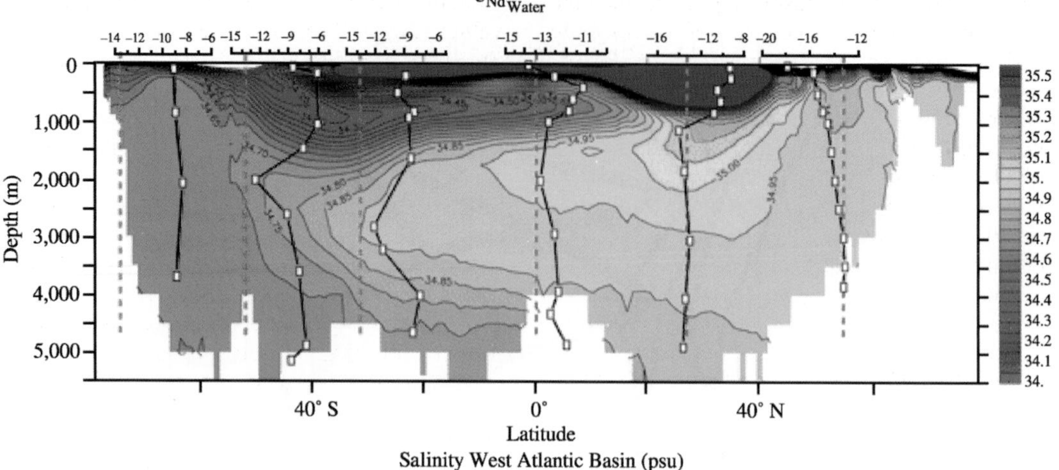

Figure 8 *Nd-isotope ratio and salinity profiles in the Atlantic.* Nd-isotope ratios with depth are consistent with salinity in the Atlantic, further showing that the Nd-isotopic signatures of water masses are conserved with advective transport (source von Blanckenburg, 1999).

sediment source with comparable neodymium-isotope ratios is Taiwan, and this has been suggested as the source of the high neodymium-isotope ratios of Pacific seawater, owing to its anomalously high erosion rate (Goldstein and Jacobsen, 1988). However, it is noteworthy that the detrital component of western and central Pacific sediments is $\varepsilon_{Nd} \approx -10$ to -12 (Nakai *et al.*, 1993) and this signature is neither imparted to coexisting seawater or Pacific Fe–Mn oxide precipitates. Indeed, if the continents surrounding the Pacific were the major source of dissolved neodymium, Nd-isotope ratios in the Pacific and Atlantic would be at most only marginally different. Early studies of the REEs (Michard *et al.*, 1983) and neodymium-isotope ratios in hydrothermal solutions (Piepgras and Wasserburg, 1985) from Pacific seafloor spreading axes showed that ridge volcanism cannot be a significant source of Pacific seawater neodymium. The most likely source of high neodymium isotope ratios in the Pacific is circum-Pacific volcanism. Volcanic ash is widely dispersed in the Pacific as small highly reactive particles with large surface area, and Pacific sediments have large numbers of ash layers. The Pacific Ocean is surrounded by convergent plate margins, and especially in the northwest Pacific the volcanoes lie upwind. In addition, the Pacific contains a large abundance of intraplate volcanic islands. Neodymium-isotope ratios of arc volcanics (probably the largest source) and ocean islands typically have high values of $\varepsilon_{Nd} \approx +7$ to $+10$. The effects of exchange between volcanic par ticles and seawater have been inferred from studies of neodymium isotopes in waters passing through the Indonesian Straits (Jeandel *et al.*, 1998; Amakawa *et al.*, 2000), and Papua New Guinea (Lacan and Jeandel, 2001). In the cases, increases in neodymium-isotope ratios along water advection pathways are attributed to local volcanic sources. Thus, it is likely that addition of neodymium from volcanic particles imparts the distinctive neodymium-isotopic fingerprint to Pacific seawater, which is ultimately intermediate to values of recent volcanic and old continental sources.

The inefficiency of neodymium exchange between continental detritus and seawater is clearly seen in the Indian Ocean. The major sources of sediment to the Indian Ocean are the Ganges–Brahmaputra and Indus river systems, with $\varepsilon_{Nd} \approx -10$ to -12 (Goldstein *et al.*, 1984), whose neodymium-isotope ratios are reflected in the Bengal Fan (e.g. Bouquillon *et al.*, 1990; Galy *et al.*, 1996; Pierson-Wickmann *et al.*, 2001). The neodymium-isotope ratio of Indian seawater is $\varepsilon_{Nd} \approx -8$, significantly higher than these sediment sources, even in Indian water close to the Indus and Bengal fans. This difference exists despite the high abundance of neodymium in continental sediment (~30 ppm) compared to Indian seawater (~0.003 ng g^{-1}), which shows that a small amount of exchange in regions with high sedimentation rates like these two Fans should have a major local effect on seawater neodymium. The fact that the large difference persists between the fan sediments and proximal Indian seawater shows that the neodymium in terrigenous detritus is tightly bound to the sediment. The neodymium-isotope ratios of Indian seawater are the same as the circum-Antarctic, as observed in direct seawater measurements and in Fe–Mn oxides. This is consistent with Indian Ocean hydrography, which shows that Indian intermediate and deep seawater is primarily fed by northward advecting water from the circum-Antarctic.

These observations, taken together, indicate that the neodymium-isotope ratios of intermediate and deep seawater are imprinted mainly in the Atlantic and Pacific oceans. The circum-Antarctic is fed by both and its intermediate neodymium-isotope ratio reflects those of the Atlantic and Pacific water sources. Indian Ocean intermediate and deep water is fed primarily by the circum-Antarctic and tends to retain a circum-Antarctic neodymium-isotope ratio.

6.17.4.4 Neodymium Isotopes as Water-mass Tracers

How well do neodymium-isotope ratios trace water masses? The apparent answer is that neodymium isotopes trace them very well. Where there is a significant neodymium-isotopic contrast between water masses with depth at a single location, as in the South Atlantic, neodymium-isotope ratios vary coherently with the water mass (Figures 7 and 8). The North Atlantic profiles show uniform ε_{Nd} values between −12 and −14 at depths >2,500 m. The Drake Passage profile shows a uniform ε_{Nd} value of ∼−9. In this context, Equatorial Atlantic neodymium-isotope ratios are slightly higher than those in the North Atlantic, especially in deep water, consistent with addition of some AABW. In the South Atlantic, profiles show a strong zig-zag pattern. The lowest values nearly reach those of NADW, strongly indicating that NADW keeps its neodymium signature as it travels southward in the Atlantic. In the SE Atlantic, the NADW wedge can be identified in Figures 7 and 8, shallower than in the SW Atlantic. At shallower depths than the NADW "wedge," neodymium-isotope ratios increase toward circumpolar values in northward flowing AAIW. At deeper depths than the NADW "wedge," the neodymium-isotope ratios increase toward circumpolar values in northward flowing AABW.

The highest neodymium-isotope ratios in the South Atlantic at depths greater than 2,500 m reach ε_{Nd} ∼−7.5, which are higher than the Drake Passage values (Figure 7). This may indicate a source of neodymium with high isotope ratios in the South Atlantic. However, it is premature to conclude that deep South Atlantic neodymium-isotope ratios "overstep" the Southern Ocean values, for the following reasons. The maxima for all of the deep South Atlantic waters are between $\varepsilon_{Nd} = -7$ to −9, more variable than presently available data from the Drake Passage but still quite similar. This range is also similar to circumpolar Fe–Mn sediments (Albare(c)de *et al.*, 1997). Depth profiles from the western Indian Ocean near southern Africa are similar to the South Atlantic and Drake Passage, but like the South Atlantic they slightly exceed the Drake Passage values (Figure 7). Of the three long profiles in the literature (Bertram and Elderfield, 1993), two are relatively constant ε_{Nd} with depth, while the third shows a similar zig-zag as the South Atlantic, and reflects the flow of NADW around southern Africa. It is noteworthy that the Drake Passage profile (from Piepgras and Wasserburg, 1982), is the only one in the literature from the true circum-Antarctic, and its deepest sample is from a relatively shallow depth of ∼3,500 m. Either Drake passage seawater has more variable Nd isotope ratios than indicated by presently available data, or there is some addition of Nd to the circumpolar Atlantic sector from a high ε_{Nd} source.

In the global ocean, neodymium-isotope ratios in deep water (here considered to be >2,500 m) show remarkably good covariations with salinity and silicate (Figure 9), which provides key evidence of its value as a water-mass tracer. In the Atlantic and western Indian oceans, deep seawater faithfully records mixing between northern- and southern-derived water masses (Figure 10(a)). With northern waters characterized by low ε_{Nd} and high salinity, and southern waters by high ε_{Nd} and low salinity, the data show a very good covariation between the end-members and fall within a mixing envelope. Of particular interest are the data from the southwest Atlantic, where the high salinity waters fall well within the range of the north Atlantic data, and the low salinity waters within the range of South Atlantic–Drake Passage–West Indian Ocean data. A similar relationship also holds between SiO_2 and ε_{Nd} (Figure 10(b)), where northern source deep waters have low SiO_2 and ε_{Nd}, and southern source waters high SiO_2 and ε_{Nd}. Here again, most of the Atlantic data fall within a mixing envelope between the two end-members.

The Pacific is more of a homogenous pool of water than the Atlantic, and relationships between neodymium-isotope ratios and water masses are not as clearcut. Nevertheless, there is strong evidence that neodymium isotopes behave here too as conservative water-mass tracers. To the east of New Zealand there is a tongue of deep circumpolar water that moves northward. In depth profiles from the south central Pacific, the neodymium-isotope ratios of deep water reaches circumpolar values, while the intermediate water is more Pacific-like. More compelling evidence for the mainly conservative nature of neodymium in deep water is illustrated by comparing neodymium isotopes and silicate in Pacific and circumpolar waters (Figure 11). They show a good positive correlation, consistent with mixing between Pacific and circumpolar waters. In the Pacific region salinity cannot be used as a water-mass

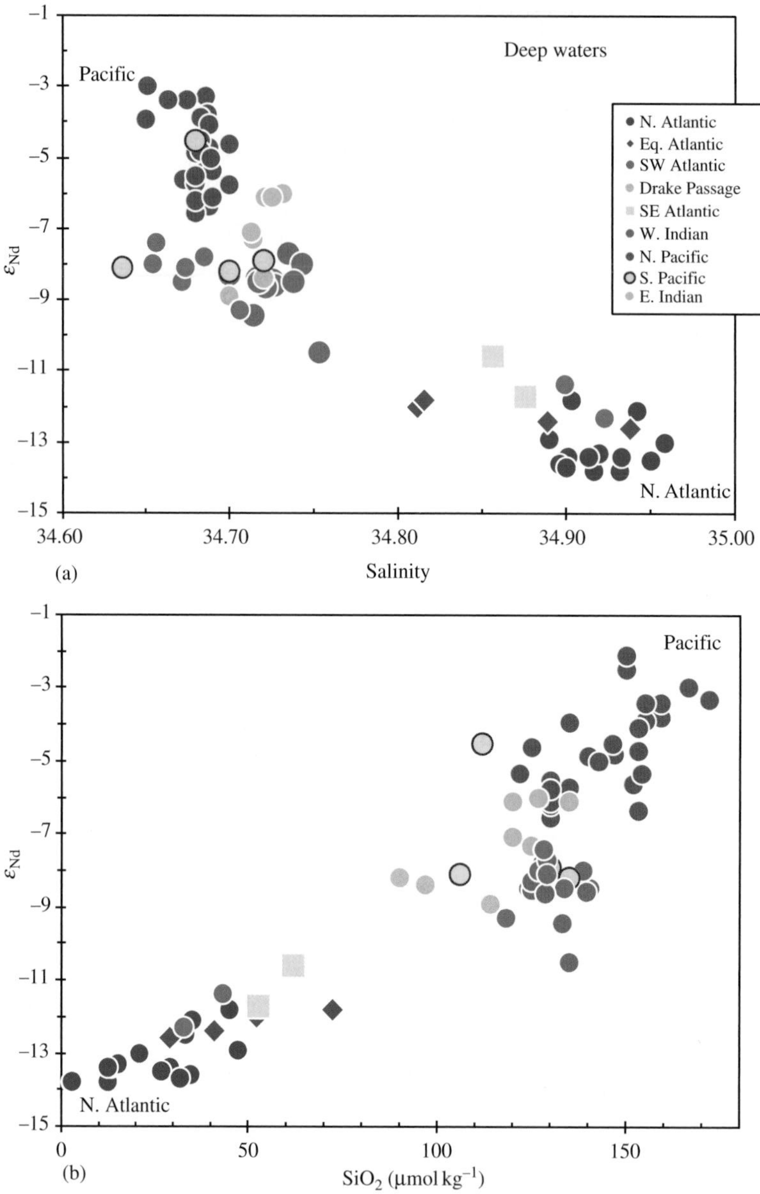

Figure 9 *Nd-isotope ratios versus salinity and silicate in deep seawater.* (a) Seawater Nd-isotope ratios display a good covariation with salinity, but the correlation breaks down somewhat in the Pacific, because its salinity is similar to the circum-Antarctic and South Atlantic. (b) Throughout the oceans there is a very good covariation with SiO_2. In both diagrams, the South Atlantic data responsible for the zig-zag patterns in profile (Figure 8) are also consistent with salinity and silicate, with high and low ε_{Nd} reflecting AABW and NADW, respectively. Thus, the Nd-isotope ratios trace the water masses of the present-day deep oceans very well. Plotted data are from >2,500 mb sl, except two Drake Passage data from 1,900 m and 2,000 m (Nd data sources: Piepgras and Wasserburg, 1980, 1982, 1983, 1987; Spivack and Wasserburg, 1988; Piepgras and Jacobsen, 1988; Bertram and Elderfield, 1993; Jeandel, 1993; Shimizu *et al.*, 1994; Jeandel *et al.*, 1998). Where salinity or silicate were not available in the publication, they were estimated from Levitus (1994), using the location and depth of the water sample.

proxy as in the Atlantic because salinity is similar in Pacific and circum-Antarctic waters (Figure 9(a)).

Taking all of these considerations into account, the available data strongly indicate that neodymium-isotope ratios in deep water are conservative tracers of water masses. While the relationships are most clear in the Atlantic, where a large portion of the basin consists of different water masses at different depths from northern and southern sources, the relationships also hold well in the Pacific. An important further implication is

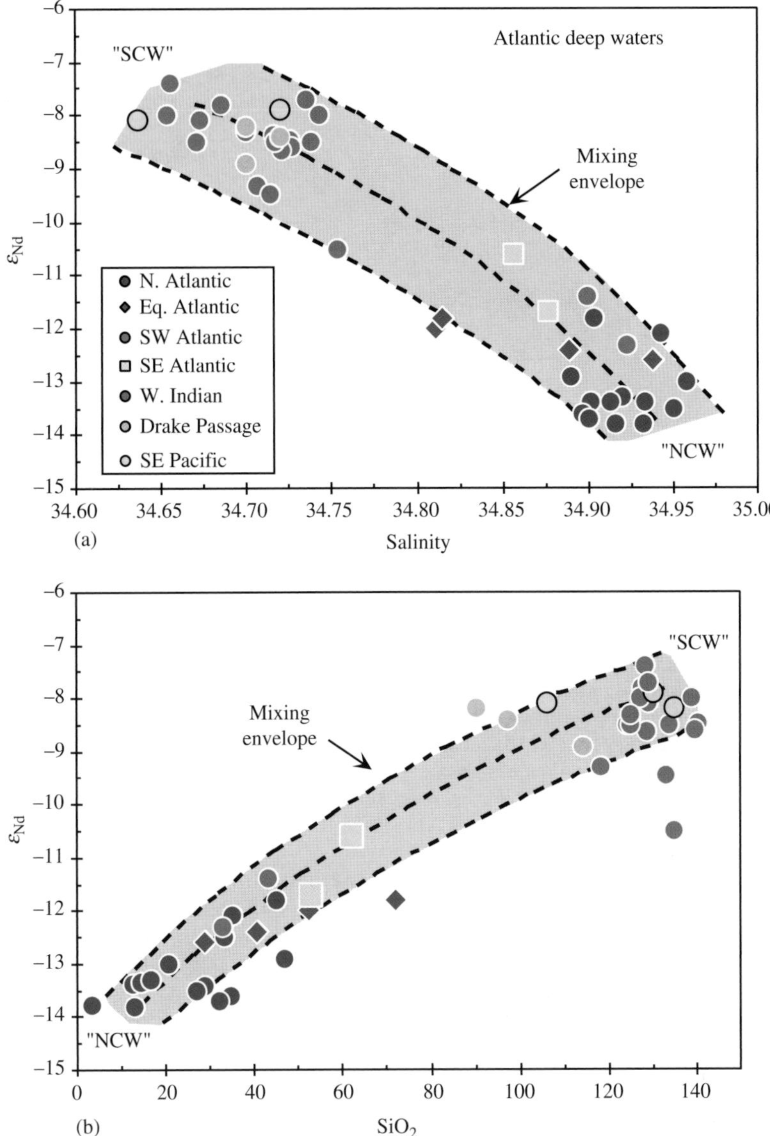

Figure 10 *Nd-isotope ratios versus salinity and silicate in Atlantic, West Indian, and Southern Ocean deep waters.* Mixing lines between northern and southern component end-member compositions (NCW and SCW, respectively) are shown, since they are nonlinear. Salinity, silicate, and Nd-isotope end-members are shown; Nd concentrations range from 16 nmol kg^{-1} to 20 nmol kg^{-1} in NCW and 27 nmol kg^{-1} to 30 nmol kg^{-1} in SCW, approximating reasonable ranges for NADW and AABW, respectively. In both (a) and (b) the data generally fall within the "mixing envelopes," showing that the Nd-isotope ratios of Atlantic deep waters reflect mixing between the northern and southern component waters. Plotted data are from >2,500 mb sl, except two Drake Passage data from 1,900 m and 2,000 m. (Nd data sources: Piepgras and Wasserburg, 1980, 1982, 1983, 1987; Spivack and Wasserburg, 1988; Piepgras and Jacobsen, 1988; Bertram and Elderfield, 1993; Jeandel, 1993). Where salinity or silicate were not available in the publication, they were estimated from Levitus (1994), using location and depth.

that neodymium isotopes should have great potential as tracers of past ocean circulation.

6.17.4.5 The "Nd Paradox"

The previous section demonstrated that neodymium-isotope ratios in the deep-ocean trace mixing between Southern and Northern Ocean waters in the Atlantic and Pacific. However, the variability of neodymium concentrations in seawater was not discussed. As noted in the discussion of REE in seawater (Section 6.17.4.1 and Figure 6), abundances generally show depletions in surface waters and enrichments in deep water, and are lower in the North Atlantic and higher in the Pacific. With respect to these characteristics the REEs are similar to SiO_2. It has already been

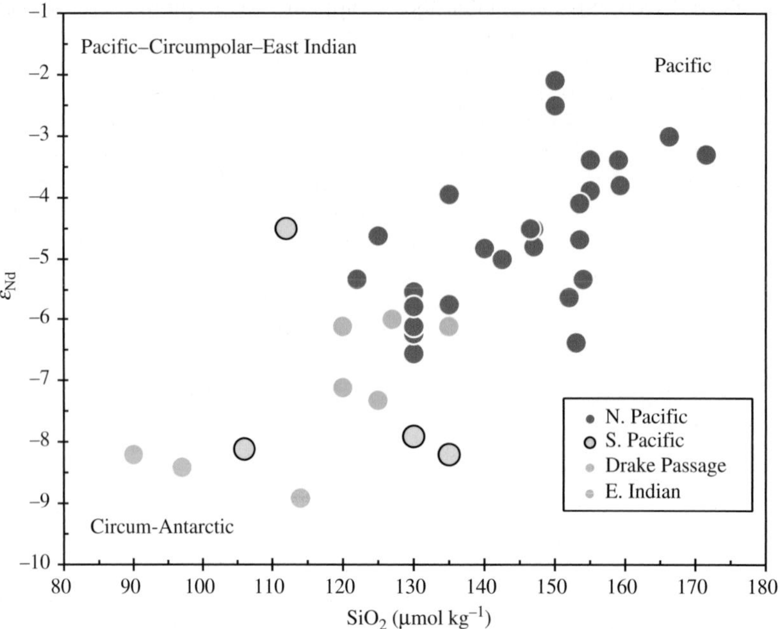

Figure 11 *Nd-isotope ratios versus silicate in Pacific, Indian, and Southern Ocean deep waters.* The positive correlation shows that Nd-isotope ratios trace mixing of deep waters from the circum-Antarctic and Pacific. Plotted data are from >2,500 mb sl, except two Drake Passage data from 1,900 m and 2,000 m (Nd data sources: Piepgras and Wasserburg, 1980, 1982; Piepgras and Jacobsen, 1988; Bertram and Elderfield, 1993; Shimizu *et al.*, 1994; Jeandel *et al.*, 1998). Where salinity or silicate were not available in the publication, they were estimated from Levitus (1994), using location and depth.

shown that neodymium-isotope ratios and silicate display an excellent global covariation (Figure 9(b)), and silicate is often used as a proxy for REE behavior in the oceans (e.g. Elderfield, 1988). This has been taken as evidence that silicate and Nd display a similar behavior in the oceans.

Silicate is depleted from surface waters by biological processes and remineralized in the deep water. Moreover, silicate tends to accumulate in water masses as they age (cf. Broecker and Peng, 1982). These processes account for both the increasing concentration of silicate with depth, and its increasing concentrations from the North Atlantic to the circum-Antarctic to the Pacific. With a few exceptions (notably the North Atlantic), neodymium abundances show a smooth increase with depth, and in deep water they are highest in the Pacific, lowest in the Atlantic, and intermediate in the Indian (Figure 6).

Considered separately, the neodymium-isotope ratios and the neodymium concentrations have very different implications. (i) Neodymium-isotope ratios are variable in the different oceans and within an ocean they fingerprint the advective paths of water masses. This requires that the residence time of neodymium is shorter than the ocean mixing time of $\sim 10^3$ yr (e.g., Broecker and Peng, 1982). (ii) Neodymium concentrations appear to mimic silicate, implying orders of magnitude longer residence times of $\sim 10^4$ yr and indicating increased addition through dissolution along the advected path of water masses. These observations were first addressed in detail by Bertram and Elderfield (1993), and further emphasized by Jeandel and colleagues (Jeandel *et al.*, 1995, 1998; Tachikawa *et al.*, 1997a,b; Lacan and Jeandel, 2001), who termed this the "Nd paradox." Both groups have suggested a general model of neodymium cycling in the oceans, that treats neodymium and the other REE as an analog to silicate. Neodymium is introduced into the surface ocean through partial dissolution of atmospheric input. For example, Jeandel *et al.* (1995) estimate that half of the global atmospheric input is dissolved when entering seawater. The cycling model suggests that in the deep-water source regions such as the North Atlantic the neodymium reaches the deep ocean through water mass subduction; elsewhere the neodymium is scavenged by organisms, vertically transported to deep water as the organisms settle in the water column, and added to the deep ocean through particulate-water exchange. Throughout the oceans neodymium is advected along with water masses. Thus, the inheritance of neodymium-isotope ratios of deep-water sources along with increasing neodymium concentrations as water ages are explained by a combination of lateral advection and vertical cycling.

It was argued above (Section 6.17.4.4) that the neodymium-isotope ratios, are conservative tracers of the advective movement and mixing of water masses (e.g., Figures 7, 10, and 11), that distinguish the southward flow of NADW and northward flow of AABW and AAIW in the Atlantic, as well as exchange between Pacific and circum-Antarctic waters. To the extent that neodymium-isotope ratios in the Atlantic and Pacific trace exchange of water between these oceans and the circum-Antarctic, they are potentially powerful oceanographic and paleoceanographic tracers. However, this implies that neodymium-isotopic signatures are imparted to deep water primarily at two locations, the North Atlantic and the Pacific, and that mixing yields intermediate neodymium-isotope ratios in the circum-Antarctic. If vertical transport from surface waters and neodymium exchange in deep waters is an important process throughout the oceans, then neodymium that is extraneous to the deep water masses must be added along the transport path. Unless the added neodymium fortuitously has the same isotope ratio as the water mass, the value of neodymium-isotope ratios as a water-mass tracer will be compromised. The extent that neodymium isotopes are compromised as water-mass tracers depends on how much the deep-water neodymium-isotopic fingerprint is changed by the addition or exchange.

A test against a significant addition of shallow neodymium to deep waters is to show that deep and shallow waters at the same location have markedly different neodymium-isotope ratios. While most published depth profiles are ambiguous in this regard, there are several published profiles in which the shallow and deep water have significantly different neodymium-isotopic signatures (Figure 12(a)). These neodymium-isotopic profiles are from all of the ocean basins; therefore, it is reasonable to infer that this relationship is a general feature of the oceans. The contrasts between deep and shallow water can be quite large, $\sim 4\varepsilon_{Nd}$ units in the Atlantic, and $\sim 8\varepsilon_{Nd}$ units in the South Pacific. Concentration profiles of the same samples show smoothly increasing neodymium abundances with increasing depth (in all cases but the North Atlantic), with deep-water concentrations generally more than a factor of 2 higher than shallow waters (Figure 12(b)). Without the constraints from the neodymium-isotope ratios, the concentration profiles are consistent with shallow scavenging of neodymium and remineralization at increasing depths. However, taken together, the differences between deep and shallow water neodymium-isotope ratios and concentrations preclude the addition of significant quantities of neodymium scavenged at shallow levels to the deep water. This means that vertical cycling at these sites does not appear to explain the higher neodymium concentrations in the deep water.

Can the Nd paradox be resolved by water-mass mixing? Vertical cycling does not appear to explain both the neodymium-isotope ratios and Nd concentrations in the oceans. Neodymium-isotope covariations with salinity and silicate in the Atlantic and Pacific are broadly consistent with mixing between northern- and southern-source waters in both basins (Figures 10 and 11). However, using the analogy to silicate, the increasing concentrations of neodymium from the Atlantic, to the circum-Antarctic/Indian, to the Pacific, have been explained in terms of increased addition of neodymium with the aging of deep water. If the high neodymium-isotope ratios of Pacific water are a result of reaction of volcanic ash with seawater, then can this could be the source of the higher neodymium concentrations in the Pacific compared to the Atlantic? Can the "Nd paradox" be explained by addition of Nd to deep water in the North Atlantic and the Pacific combined with water mass exchange between the North Atlantic–circum-Antarctic–Pacific oceans?

In terms of the global ocean, the intermediate neodymium-isotope ratios and neodymium concentrations of the circum-Antarctic/Indian oceans are not explicable through simple mixing of North Atlantic and Pacific source waters. In the circum-Antarctic and Indian oceans, neodymium concentrations are too low (Figure 13). This is reasonable, and could be a reflection of neodymium scavenging in the water column during advective transport between the North Atlantic and circum-Antarctic on the one hand, and the Pacific and circum-Antarctic on the other.

The North Atlantic and the Pacific are not directly linked, rather their linkage is modulated by the circum-Antarctic. Therefore, a more direct question is whether the neodymium-isotope ratios and concentrations are explicable in terms of binary mixing between Northern and Southern component waters within each basin. In the Pacific, it was pointed out by Piepgras and Jacobsen (1988) that neodymium-isotope ratios appear to follow simple mixing between the North Pacific and the Southern Ocean (illustrated in Figure 11) but concentrations do not. They attributed the apparent coherence of the neodymium isotopes and its absence in the neodymium concentrations to the loss of similar proportions of neodymium by scavenging from waters derived from the Pacific and circum-Antarctic during advective transport. The losses would not change the neodymium-isotope ratios, and the appearance of binary mixing systematics would be conserved.

In the Atlantic, like the Pacific, neodymium-isotope ratios are consistent with mixing of

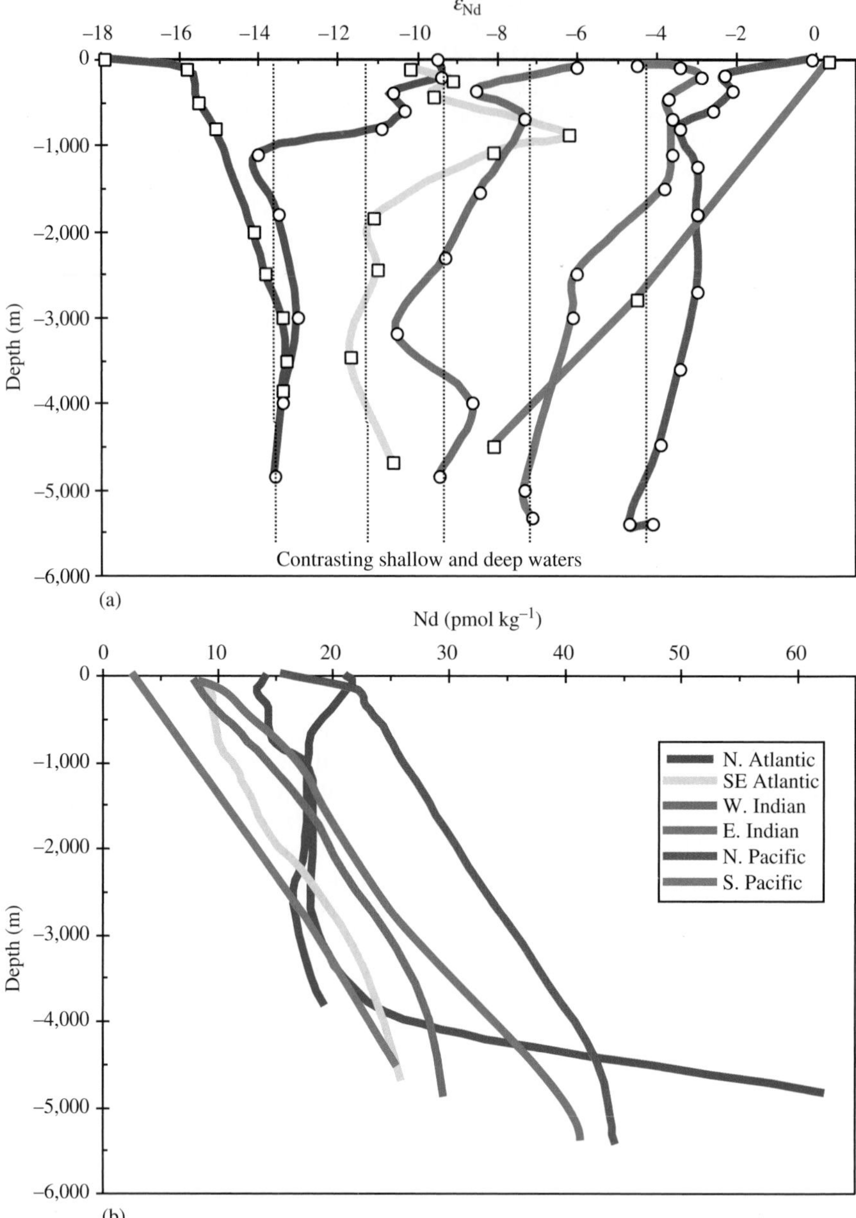

Figure 12 *Comparison of Nd-isotopes and abundances in shallow and deep waters.* (a) These examples, from all the ocean basins, highlight profiles where the Nd-isotope ratios of shallow and deep waters show large differences, as high as $\sim 4\varepsilon_{Nd}$ units in the Atlantic and $\sim 8\varepsilon_{Nd}$ units in the South Pacific. Dashed vertical lines emphasize the differences between the shallow and deep waters. (b) Concentration profiles show smoothly increasing Nd abundances with increasing depth in all cases but the North Atlantic. Deep-water concentrations are generally greater than twice as high as shallow waters. The abundance profiles are consistent with shallow scavenging of Nd and addition at increasing depths, but the differences between deep and shallow water Nd-isotope ratios preclude significant addition Nd scavenged at shallow levels. (Nd data sources: Piepgras and Wasserburg, 1982, 1987; Piepgras and Jacobsen, 1988; Bertram and Elderfield, 1993; Jeandel, 1993; Jeandel *et al.*, 1998).

Northern and Southern end-members using salinity or silicate as conservative water-mass mixing proxies (Figure 10). Comparing neodymium-isotope ratios and concentrations, the compositions of many samples are also consistent with North–South water-mass mixing (Figure 14(a)). However, a substantial portion of the data is outside any reasonable mixing envelope. Whereas in the Pacific–circum-Antarctic case above, the non-conservative behavior of the concentrations can be explained by loss during advection, in the Atlantic the neodymium concentrations of the samples outside the mixing envelope are too high. Moreover, those that are too high tend to be

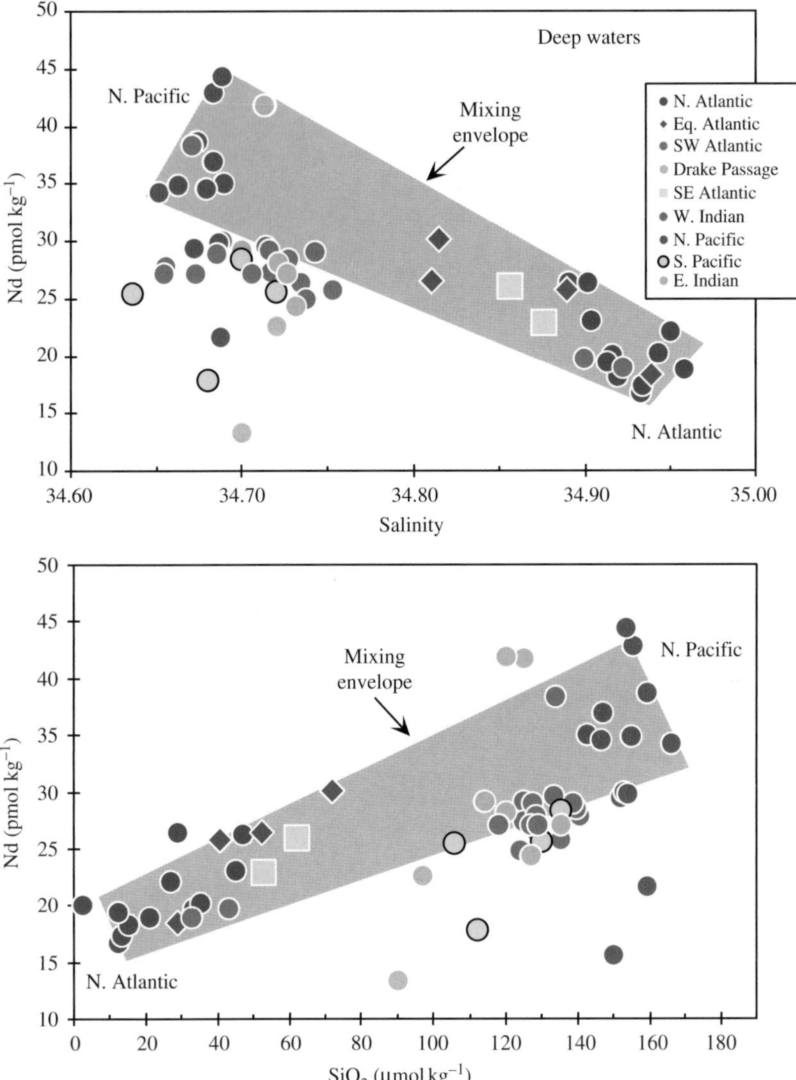

Figure 13 *Nd abundance versus (a) salinity and (b) silicate in deep seawater.* Nd concentrations do not show the same well-behaved characteristics as Nd-isotope ratios with salinity and silicate in global deep water. Mixing envelopes are shown between North Atlantic and Pacific end-members, and the circum-Antarctic, South Atlantic, and Indian Ocean samples fall outside of it. Plotted data are from >2,500 mb sl, except two Drake Passage data from 1,900 m and 2,000 m (Nd data sources: Piepgras and Wasserburg, 1980, 1982, 1983, 1987; Spivack and Wasserburg, 1988; Piepgras and Jacobsen, 1988; Bertram and Elderfield, 1993; Jeandel, 1993; Shimizu *et al.*, 1994; Jeandel *et al.*, 1998). Where salinity or silicate were not available in the publication, they were estimated from Levitus (1994), using location and depth.

the deepest samples (not shown in Figure 14 but this can be inferred from Figure 6 combined with Figure 9). The neodymium concentrations of the available data set in the Atlantic, therefore, point to an overabundance of neodymium in the deepest waters, as noted by Bertram and Elderfield (1993), and Jeandel and colleagues (Jeandel *et al.*, 1995, 1998; Tachikawa *et al.*, 1997, 1999a,b; Lacan and Jeandel, 2001). The neodymium-isotope ratios, however, place important constraints on the addition process. Figure 14(a) shows that the samples lying outside of the mixing envelope from the North Atlantic and the Equatorial Atlantic have neodymium-isotope ratios that are the same as samples within the mixing envelope. Comparison of neodymium concentrations with salinity (Figure 14(b)) shows that the same samples fall outside the simple mixing envelope. Therefore, the neodymium concentrations are enriched, but the isotope ratios are normal for the location. It was shown in Figures12(a) and (b) that depth profiles exist from all the oceans showing that high neodymium concentrations in deep water cannot be explained by addition of neodymium from

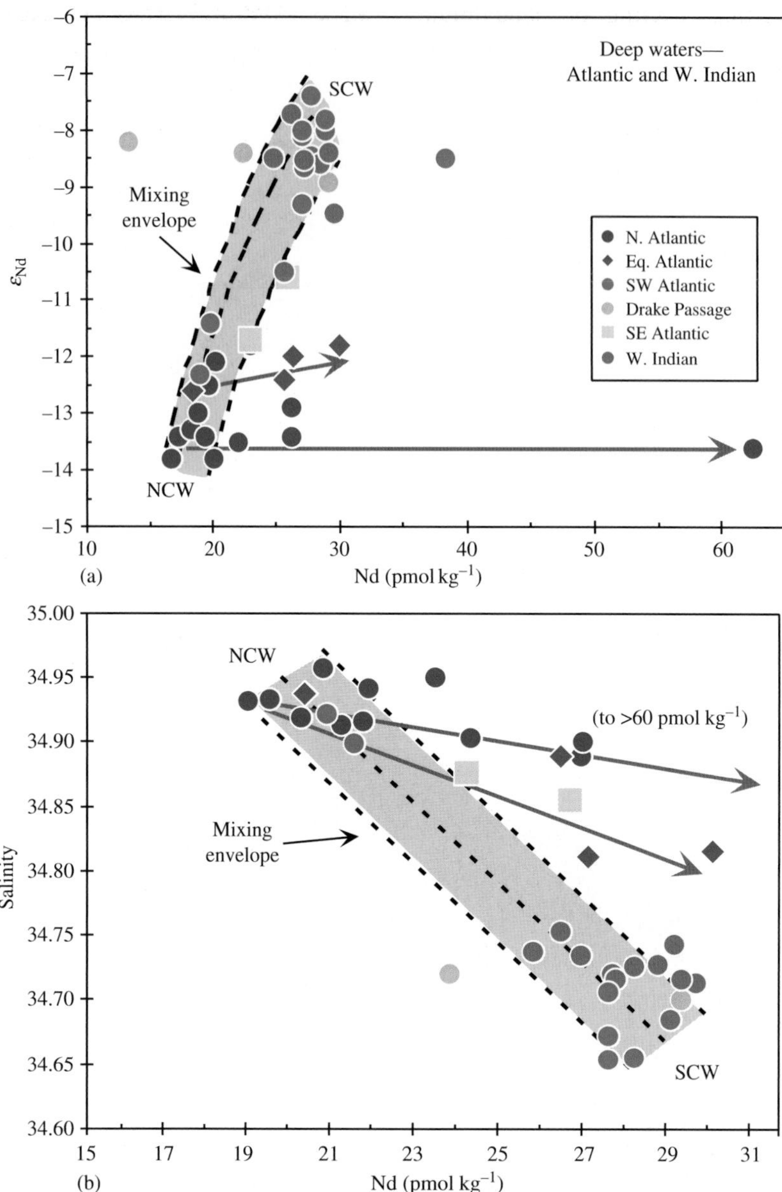

Figure 14 *Nd abundance versus salinity and silicate in deep Atlantic region seawater.* Mixing lines between northern and southern component waters (NCW and SCW) are shown in (a) because they are nonlinear. In both (a) and (b) much of the data fall within the "mixing envelopes," but many samples show high concentrations of Nd. These are generally the deepest samples. Arrows show that large increases in Nd concentrations in a region or profile are not accompanied by significant changes in Nd-isotope ratio (Nd data sources: Piepgras and Wasserburg, 1980, 1982, 1983, 1987; Spivack and Wasserburg, 1988; Bertram and Elderfield, 1993; Jeandel, 1993). Where salinity or silicate were not available in the publication, they were estimated from Levitus (1994), using location and depth.

surface waters. Profiles of ε_{Nd} and neodymium versus depth from the South Atlantic and western Indian oceans are shown in Figure 7. The concentration profiles shown in Figure 7(b) are those with zig-zag ε_{Nd} patterns. In all these cases, the neodymium concentrations increase smoothly with depth, despite the zig-zag pattern of the neodymium-isotope ratios. The neodymium-isotope ratios reflect the deep-water masses, while the neodymium concentrations appear to be decoupled.

The increasing concentrations of neodymium with depth indicate that neodymium is added to deep water as water masses laterally advect in the Atlantic, however, the neodymium that enriches deep waters along the advective path in the Atlantic has isotope ratios expected for the respective water masses. The isotope ratios appear to preclude vertical cycling of neodymium through scavenging at shallow levels, followed by addition at deeper levels, as a primary cause of increasing neodymium concentrations with depth. The "Nd paradox" still stands.

6.17.4.6 Implications of Nd Isotopes and Concentrations in Seawater

The forgoing discussion shows that neodymium-isotope ratios are excellent conservative tracers of water masses throughout the oceans. However the processes that are controlling neodymium concentrations are still not well understood. An important avenue of further research will be to compare ε_{Nd} of authigenic and terrigenous sediments in key areas. Nevertheless, the close relationship between neodymium-isotope ratios and water masses, coupled with the range of global variability in the oceans, indicate that it can be an effective water mass tracer in the present day and shows great potential as a tracer of past ocean circulation.

6.17.5 APPLICATIONS TO PALEOCLIMATE

6.17.5.1 Radiogenic Isotopes in Authigenic Ferromanganese Oxides

The discussions (Section 6.17.4) above showed that the primary controls on the composition of neodymium-isotope ratios in seawater in crusts are the provenance (possibly modified by weathering, e.g., von Blanckenburg and Nagler, 2001) and ocean circulation. In order for a tracer to be valuable for paleoceanographic studies, its modern distribution should follow water masses in a simple way, and neodymium isotopes clearly fulfill this criterion. A world map of neodymium-isotope ratios in the outer portions of Fe–Mn nodules and crusts (Figure 3) shows that they display systematic geographical variations. In most regions they are the same as the local bottom water. A spectacular indication of the veracity of neodymium isotopes in manganese crusts and nodules as water-mass tracers is the comparison of the pattern of ^{14}C ages of bottom waters in the Pacific Ocean (Figure 15;

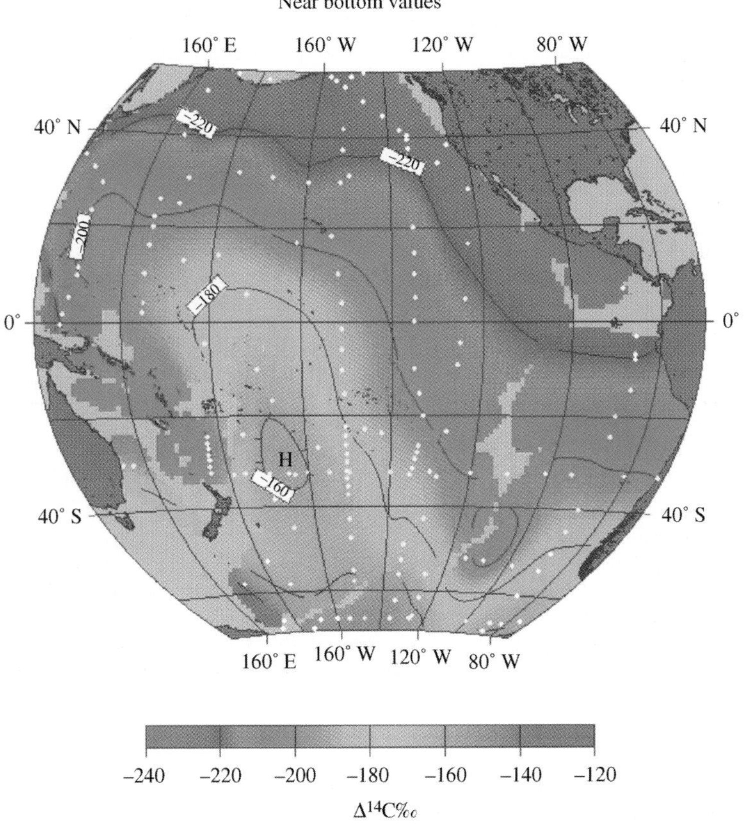

Figure 15 *$\Delta^{14}C$ in deep Pacific seawater.* The tongue of high $\Delta^{14}C$ shows the present-day path of circumpolar water entering the Pacific in the present day. The location and shape of the tongue of high ^{14}C water is similar to the "tongue" of low Nd isotope ratios in Fe–Mn deposits shown in Figure 3. In the Fe–Mn samples the low Nd isotope "tongue" is an integrated signal over 10^5–10^6 yr. The comparison shows that the Nd isotope signal in the Fe–Mn samples track the path of circumpolar water into the Pacific, and that on average the pathway has remained the same over the last million years (reproduced by permission of R. Key, Princeton University from *Ocean Circulation and Climate: Observing and Modelling the Global Ocean*, **2001**, pp. 431–452).

Schlosser *et al.*, 2001) with the world map neodymium-isotope ratios of Fe–Mn nodules and crusts (Figure 3). Manganese crusts and nodules grow at a rate of a few millimeters per million years and thus each sample generally integrates several glacial–interglacial cycles. Thus, the strong correlation with modern water-mass characteristics is a testament to the general stability of the modern ocean circulation through the Pleistocene.

However, Albare(c)de and Goldstein (1992) noted that there are some notable deviations between modern water values and crusts. Of particular note is the southwestern Atlantic region where the crust data have higher neodymium-isotope ratios than modern deep water (Figure 3). Data from disseminated authigenic neodymium from marine sediment core RC11-83, from the Cape Basin in the southeast Atlantic (Rutberg *et al.*, 2000), provides a resolution of this apparent conflict. This study traced the export of NADW to the Southern Ocean through the last glacial period to ~70 ka, and was the first to track deep-ocean circulation using neodymium-isotope ratios on glacial–interglacial timescales. It was found that the Holocene and marine-isotope stage 3 (MIS 3) record is like that predicted from the modern water column data, while the MIS 2 and 4 records are more like Pacific seawater (Figure 16). The integrated signal over the last 4 MISs is comparable to the outer portions of manganese crusts and nodules of that region. This further emphasizes that neodymium-isotope ratios in marine authigenic Fe–Mn oxides have great potential as paleoceanographic water-mass proxies.

Although lead-isotope ratios in the present-day oceans cannot be calibrated due to anthropogenic contamination, as previously discussed, one of the first major long-lived radiogenic isotope studies showed that the pre-industrial distribution can be determined using authigenic Fe–Mn sediments (Chow and Patterson, 1959, 1962). Further studies (Abouchami and Goldstein, 1995; von Blanckenburg *et al.*, 1996) have shown that lead-isotope ratios from the surface layers of Fe–Mn nodules and crusts have similar patterns to neodymium isotopes. That is, they are consistent with an older continental provenance in the Atlantic and a younger one in the Pacific, and the present-day geographical distribution is consistent with dispersion of the signals by ocean circulation.

The primary limitation of using Fe–Mn crusts for paleoceanographic studies derives from

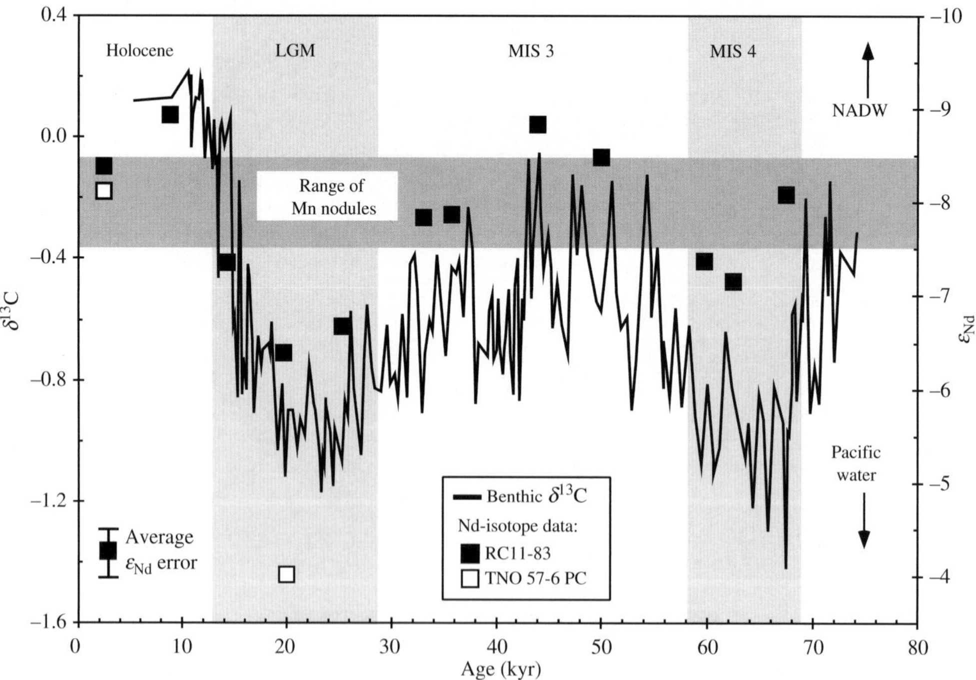

Figure 16 *Nd-isotope ratios of Fe–Mn leachates and benthic foraminiferal $\delta^{13}C$ versus age in southeast Atlantic cores.* The ε_{Nd} axis is reversed in order to facilitate comparison. The $\delta^{13}C$ variations were previously interpreted as reflecting changes in thermohaline circulation intensity over the Holocene and the last ice age (Charles and Fairbanks, 1992; Charles *et al.*, 1996). The down-core Nd-isotope ratios are consistent with stronger and weaker thermohaline circulation intensity during warm and cold marine isotope stages (MIS 1,3 and 2,4), respectively. In TNO57-6, located shallower and farther south than RC11-83, the glacial Nd-isotope value is like the Pacific, and may indicate a shutdown of NADW export to the circum-Antarctic. Horizontal shaded region shows the range of Nd-isotope ratios in circumpolar Fe–Mn nodules (Albare(c)de *et al.*, 1997) (after Rutberg *et al.*, 2000).

the slow growth rates, which limit the possible time resolution. The only Fe–Mn crust with an age record allowing a time resolution of a few thousand years is sample VA13-2 (cf. Segl *et al.*, 1984), from the central Pacific, where significant changes on glacial and subglacial scales are not expected. A study by Abouchami *et al.* (1997) established the constancy of neodymium-isotope ratios in the Pacific through the last several glacial stages. Fe–Mn encrustations have been used extensively over the last several years to investigate long-term changes in the sources of neodymium and lead at different localities in the oceans (cf. Frank, 2002, and references therein). For higher time resolution, the record on disseminated Fe–Mn oxides in South Atlantic core RC11-83 by Rutberg *et al.* (2000) shows great future potential for addressing issues associated with changes of ocean circulation (Figure 16). In addition, Vance and Burton (1999) have concluded that they can trace changes in the provenance of neodymium in surface waters on the basis of analyses of carefully cleaned foraminifera.

6.17.5.2 Long-term Time Series in Fe–Mn Crusts

The current state of the long-term time series data from Fe–Mn crusts long-term has been addressed in detail in the review by Frank (2002). Here we only present a general summary.

On large timescales over millions of years, neodymium and lead isotopes in Fe–Mn crusts have the potential for addressing changes in weathering contributions to the ocean, major circulation changes controlled by tectonic movements of the continents, and opening or closing of important ocean passages. The isotopic signature of the input to the oceans may vary for a variety of reasons, including changes in contributions to the oceans from continental terrains of different ages, changes in runoff and elevation, and even changes in the degree of incongruent weathering. As a result, applications on long timescales are intriguing but complicated.

The review by Frank (2002) discusses the long-term time series neodymium- and lead-isotope data from 19 Mn crusts, and some clear patterns are evident. The North Atlantic and Pacific record the extreme old and young continental provenance, respectively, throughout the time span of the crusts of up to ~60 Myr, with the Indian Ocean remaining intermediate throughout. In addition, the initiation of northern hemisphere glaciation is accompanied by a dramatic trend towards older continental sources in North Atlantic crusts. The data show that the glacial erosion process yielded an increased contribution of ancient continental products to the dissolved load of the Atlantic, and possibly a higher concentration as well. Moreover, the Southern Ocean does not show a correlative change. This has been interpreted to indicate that the time-integrated (over the $\sim 10^5 - 10^6$ yr time resolution of a single Fe–Mn crust sample) contribution of NADW to the Southern Ocean decreased with the onset of northern hemisphere glaciation (Frank *et al.*, 2002). Several papers have raised questions about incongruent chemical weathering imprints on the isotope composition of dissolved neodymium (von Blanckenburg and Nagler, 2001) and hafnium and lead isotopes (Piotrowski *et al.*, 2000; van de Flierdt *et al.*, 2002).

6.17.5.3 Hf–Nd Isotope Trends in the Oceans

Among long-lived isotope systems, the best correlated in continental rocks and oceanic basalts are neodymium- and hafnium-isotope ratios. They form a very coherent positive linear trend which is taken as the "mantle–crust array" (Figure 17(a)), reflecting the effects of coupled fractionation of Sm/Nd and Lu/Hf ratios through the history of the differentiation of the silicate earth (e.g. Vervoort *et al.*, 1999). However, marine Hf–Nd isotopic variations lie on a slope that diverges from the crust–mantle array. While the range of ε_{Hf} values in continental rocks is about twice that of ε_{Nd}, the variability of hafnium isotopes in Fe–Mn crusts and nodules is considerably smaller than that of neodymium isotopes (White *et al.*, 1986; Piotrowski *et al.*, 2000; Godfrey *et al.*, 1997; Albare(c)de *et al.*, 1998; van de Flierdt *et al.*, 2002). As a result, Nd–Hf isotope ratios of Fe–Mn crusts and nodules fall off the Nd–Hf "mantle–crust array" (Figure 17(b)), where relative to the array, the hafnium-isotope ratio is too high for a given neodymium-isotope ratio.

This observation has been interpreted in various ways. (i) The residence time of hafnium in the oceans is longer than neodymium (White *et al.*, 1986). (ii) Hydrothermal systems associated with ridge volcanism make a significant contribution to the hafnium in the oceans (White *et al.*, 1986), unlike neodymium (Michard *et al.*, 1983; Piepgras and Wasserburg, 1985). (iii) Hafnium isotopes in seawater reflect incongruent weathering of continental rocks. This last possibility could have a large effect because hafnium is a major element in the mineral zircon, comprising a few to several percent by weight. Moreover, the high hafnium abundance in zircon means that the Lu/Hf ratio is close to zero. As a result, hafnium isotopes in zircons do not change markedly with time due to radioactive decay. Due to the high abundance,

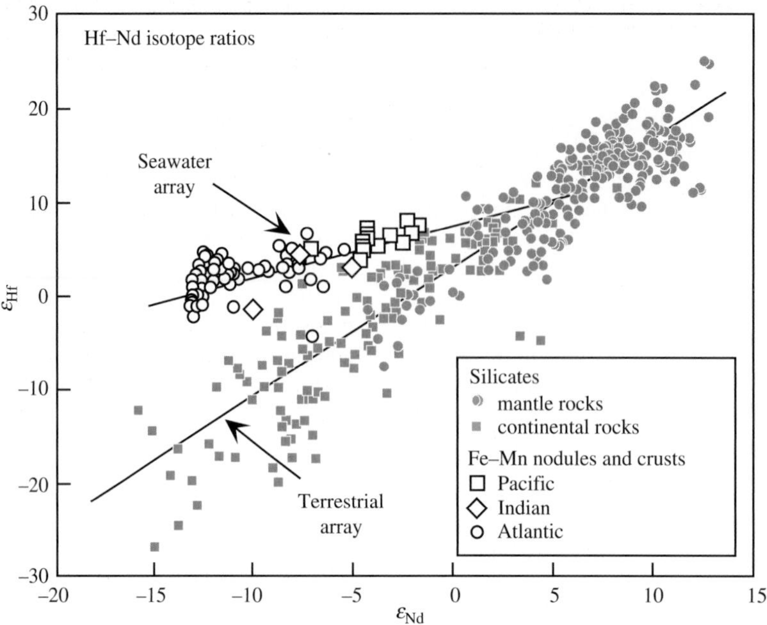

Figure 17 *Nd–Hf isotope ratios in rocks and Fe–Mn deposits.* In mantle and continental rocks these two isotope ratios a positive covariation, reflecting correlated fractionations of Sm/Nd and Lu/Hf during mantle melting and continent formation over Earth history. Seafloor Fe–Mn deposits are distinct, and lie off the continent-mantle "array," at high Hf isotope ratios for a given Nd isotope ratio. Possible explanations are given in the text. The figure is modified from van de Flierdt *et al.* (2002). The crust–mantle "array" is based on Vervoort *et al.* (1999). (Hf isotope data on Fe–Mn samples: White *et al.*, 1986; Godfrey *et al.*, 1997; Albare(c)de *et al.*, 1998; Burton *et al.*, 1999; Piotrowski *et al.*, 2000; David *et al.*, 2001; van de Flierdt *et al.*, 2002).

a significant fraction of the hafnium budget of the continental crust resides in zircon. Finally, zircon is highly resistant to weathering. As a result, hafnium may be preferentially weathered from non-zircon portion of a rock, with a significantly higher Lu/Hf and hafnium-isotope ratio (depending on the geological age) than the bulk rock (White *et al.*, 1986; Piotrowski *et al.*, 2000; Godfrey *et al.*, 1997; van de Flierdt *et al.*, 2002).

Of course, all of these factors may be partly acting to produce the marine Hf–Nd isotope trend. At first glance a longish apparent residence time for hafnium is puzzling. However, White *et al.* (1986) note that the flux of continental-derived hafnium to the oceans is likely to be quite small due to the insoluble nature of zircon and hafnium in general. They refer to a model of scavenging by particulate matter of Balistrieri *et al.* (1981) and find that if they use the relevant constants for hafnium they get scavenging residence times of the REEs of 10^2 yr, compared to 10^{12} yr for hafnium! They note that these predictions are qualitatively consistent with the observation that hafnium is only slightly enriched in deepsea red clays while rare earths are strongly enriched (Patchett *et al.*, 1984). Van de Flierdt *et al.* (2002) report a high resolution study of some North Atlantic manganese crusts, and find that at about the time of onset of northern hemisphere glaciation, the marine Hf–Nd isotope array moved steeply downward in the direction of silicate reservoirs. They further showed a systematic negative correlation between lead and hafnium isotopes. The hafnium-isotope ratios decrease and lead-isotopes increase at the time of the northern hemisphere glaciation, which is consistent with addition of both elements through greater weathering of ancient zircon. Accordingly, they provide evidence that in some cases the normally inert zircon can be ground finely enough to be partly dissolved.

Thus, it is possible that all of the three controls are acting on the hafnium contribution to seawater. Despite the low concentrations of neodymium in seawater, the high Nd/Hf ratios in Fe–Mn crusts indicate relatively lower hafnium abundances. Because zircon is not a crystallizing phase in basalts, hafnium is not sequestered in zircon in the oceanic crust and it may be more available for dissolution due to hydrothermal processes compared to continental rocks. As a result the flux of hafnium from the oceanic crust into seawater, relative to the continental flux, may be higher than for neodymium, making it more visible.

6.17.6 LONG-LIVED RADIOGENIC TRACERS AND ICE-SHEET DYNAMICS

The previous discussions focused on utilizing long-lived radioactive systems, either dissolved in seawater or in authigenic precipitates. Here we discuss its utility in silicate detritus or individual minerals in marine sediments to follow the history of northern hemisphere ice sheets during the last glacial period. This is merely one example of the powerful potential of radiogenic isotopes as provenance tracers wih paleoceanographic implication.

Ice rafting was an important depositional process in the North Atlantic during Pleistocene glacial cycles (e.g., Ruddiman, 1977). The variable geologic age and history of land covered by ice sheets allows the potential for understanding their growth and evolution from millions to hundreds of year timescales through tracking of their detritus into marine sediments. As a result, the distribution of ice rafted detritus in the North Atlantic has been used to infer the major sources of icebergs and the general pattern of surface circulation. There has been a similar pattern of ice rafted deposition during glacial and interglacial times, with a southward shift in the locus of melting related to colder surface water in glacial times. Detailed studies of ice rafting in conjunction with other evidence for changes in the ocean–atmosphere–ice system in the North Atlantic provide important insights into processes that control climate conditions. A key element for understanding ocean–ice sheet–atmosphere interactions is identifying the major sources of ice-rafted detritus. In the following sections we give some examples of the application of isotopic provenance studies to constrain ice-rafted detritus (and thus iceberg) sources in the North Atlantic region.

6.17.6.1 Heinrich Events

"Heinrich events" of the North Atlantic are found as prominent layers, rich in ice-rafted detritus, within marine sediment cores (Heinrich, 1988). Heinrich layers are important as they may be related to extreme climate events worldwide (e.g. Broecker, 1994; Broecker and Hemming, 2001). The six Heinrich layers of the last glacial cycle can be divided into two groups based on the flux of ice-rafted grains and on the concentration of detrital carbonate in the coarse fraction (Bond *et al.*, 1992; Broecker *et al.*, 1992; Gwiazda *et al.*, 1996a; McManus *et al.*, 1998). Heinrich events are named in numerical order with H1 being the most recent. All six Heinrich layers are characterized by high percentages of ice-rafted detritus. However, in the case of H3 and H6 the flux of ice-rafted detritus, as indicated by number of lithic grains per gram or by $^{230}Th_{xs}$ measurements (McManus *et al.*, 1998), is not greatly increased relative to the background. Instead the high ice-rafted detritus percentage appears to be related to low foraminifera abundances. The array of data that have been collected on Heinrich layer provenance reveal a remarkably complete story of the geological history of the Heinrich layers' source. The four prominent Heinrich layers, H1, H2, H4, and H5, appear to have formed from massive discharges of icebergs from Hudson Strait. The provenance of these Heinrich layers is very distinctive within the IRD belt and hence can be mapped by any number of geochemical measurements. Isotopic studies to date have examined Heinrich layer provenance using K/Ar, Nd, Sr, and Pb isotopic techniques.

6.17.6.2 K/Ar ages of Heinrich Event Detritus

The first geochemical provenance measurement of the Heinrich layers was the K/Ar apparent age of <2 μm and 2–16 μm fine fractions (Jantschik and Huon, 1992). Ambient North Atlantic sediments have apparent K/Ar ages of ~400 Myr (Hurley *et al.*, 1963; Huon and Ruch, 1992; Jantschik and Huon, 1992), whereas the sediments from Heinrich layers H1, H2, H4, and H5 yielded apparent ages of ~1 Gyr. Variation in the potassium concentration is small, and thus the K/Ar age signal is a product of the radiogenic $^{40}Ar^*$ concentration (Hemming *et al.*, 2002a). Hemming *et al.* (2002a) showed that the $^{40}Ar^*$ is quite uniform in eastern North Atlantic cores and that the K/Ar age and $^{40}Ar/^{39}Ar$ spectra of <2 μm terrigenous sediment from Heinrich layer H2 in the eastern North Atlantic and from Orphan Knoll (southern Labrador Sea/western Atlantic) are indistinguishable. Taken together these results imply the entire fine fraction of Heinrich layer H2 was derived from sources bordering the Labrador Sea. The same pattern most likely characterizes H1, H4, and H5, because their K/Ar ages and $^{40}Ar^*$ concentrations are similar to H2 in eastern North Atlantic cores.

6.17.6.3 Nd–Sr–Pb Isotopes in Terrigenous Sediments

Strontium-isotope ratios are not particularly diagnostic of source terrain ages because Rb/Sr ratios are easily disturbed. However, Sm/Nd ratios tend to be conservative and neodymium-isotope ratios are effective tracers of the mean crustal age of sediment sources. Neodymium-isotope ratios of Heinrich layers H1, H2, H4, and H5 are consistent with derivation from a source with Archean heritage, and Grousset *et al.* (1993)

suggested sources surrounding the Labrador Sea or Baffin Bay (Figure 18). Strontium isotopes can be useful, however, when coupled with neodymium isotopes to trace the combined Sr–Nd characteristics of source terrains. Neodymium- and strontium-isotope ratios for different grain-size fractions from North Atlantic sediments suggest that the total terrigenous sediment load within these Heinrich layers in the IRD belt may be derived from the same limited range of sources (Revel *et al.*, 1996; Hemming *et al.*, 1998; Snoeckx *et al.*, 1999; Grousset *et al.*, 2000, 2001). The University of Colorado group has extensively characterized the composition of potential source areas in the vicinity of the Hudson Strait, Baffin Bay, other regions along the western Labrador coast, and the Gulf of St. Lawrence (Barber, 2001; Farmer *et al.*, 2003). Their data are consistent with a Hudson Strait provenance for Heinrich layers H1, H2, H4, and H5, and demonstrate an absence of substantial southeastern Laurentide ice sheet sources within pure Heinrich intervals. Lead-isotope ratios of the fine terrigenous fraction of Heinrich layers are distinctive (Hemming *et al.*, 1998), and are also consistent with derivation from the Hudson Strait region. New results from Farmer *et al.* (2003) may allow further distinction of fine-grained sediment sources with lead isotopes.

6.17.6.4 Isotopic and Geochronologic Measurements on Individual Mineral Grains

In addition to the bulk geochemical methods, several studies have examined individual grains or composite samples of feldspar grains for their lead-isotope ratios (Gwiazda *et al.*, 1996a,b; Hemming *et al.*, 1998), or individual grains of hornblende for their $^{40}Ar/^{39}Ar$ ages (Gwiazda *et al.*, 1996c; Hemming *et al.*, 1998, 2000b; Hemming and Hajdas, 2003). These studies provide remarkable insights into the geologic history of Heinrich layers that allow further refinement of the interpretations based on bulk isotopic analyses. Feldspars have high lead abundance and very low uranium and thorium abundance and thus the lead-isotope ratios of feldspar approximate the initial lead-isotope ratios of its source (e.g. Hemming *et al.*, 1994 1996, 2000b). Lead-isotope data from Heinrich layer H1, H2, H4, and H5 feldspar grains form a linear trend that indicates an Archean (~2.7 Ga)

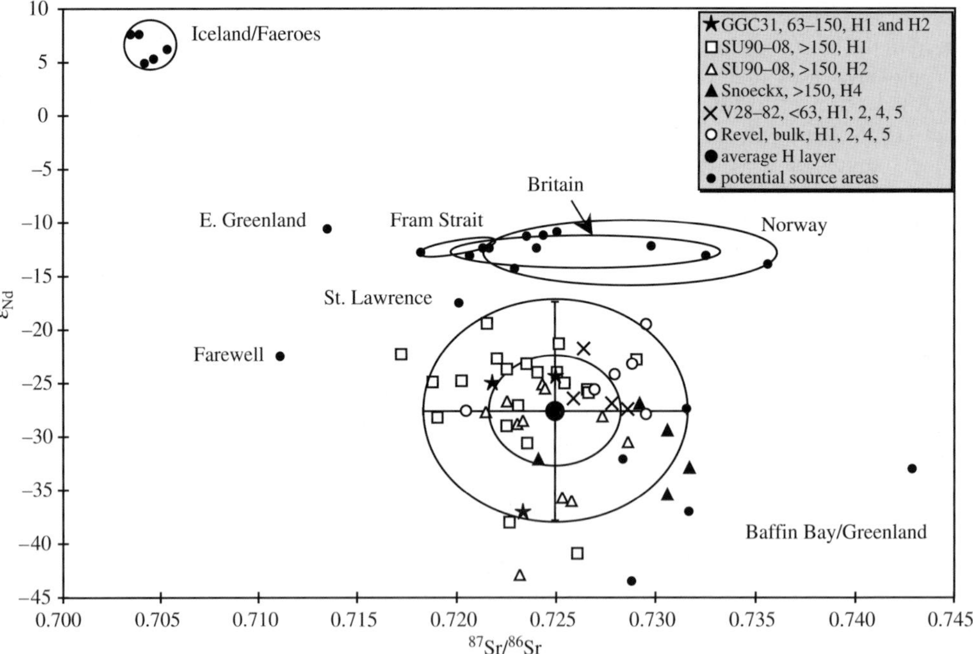

Figure 18 *Nd–Sr-isotope ratios of terrigenous clastic components of Heinrich layers H1, H2, H4, and H5.* Shown for reference are the average and 1 and 2 sigma range for the data published on these Heinrich layers and reported compositions of potential source areas of ice rafted detritus (Grousset *et al.*, 2001, and references therein). Note that the data are from different size fractions in different publications, but there is not evidence that there is a substantial bias even in the $^{87}Sr/^{86}Sr$ where one might be expected. This is most likely due to the large fraction of glacial flour with approximately the same composition in the fine as the coarse fraction in contrast to the way most sedimentary grain size variations are produced (Sources Grousset *et al.*, 2001; Snoeck *et al.*, 1999; Hemming *et al.*, 1998; Hemming, unpublished; Revel *et al.*, 1996).

heritage and a Paleoproterozoic (~1.8 Ga) metamorphic event (Figure 19(b)). Heinrich layer grains are similar in composition to H2 from Hudson Strait-proximal core HU87-009 and to feldspar grains from Baffin Island till (Hemming *et al.*, 2000b). However, they are distinctly different from feldspar grains from Gulf of St. Lawrence core V17-203, where Appalachian (Paleozoic) and Grenville (~1 Ga) sources are found. $^{40}Ar/^{39}Ar$ ages of individual hornblende grains from Heinrich layers H1, H2, H4, and H5 cluster around the implied Paleoproterozoic metamorphic events from the feldspar lead-isotope data (Gwiazda *et al.*, 1996c; Hemming *et al.*, 1998, 2000a; Hemming and Hajdas, 2003), and consistent with hornblende grains from Baffin Island tills (Hemming *et al.*, 2000b).

6.17.6.5 Contrasting Provenance of H3 and H6

Events H3 and H6 do not appear to be derived from the same sources as H1, H2, H4, and H5 (Figures 19(c) and 20). Using lead-isotope compositions of composite feldspar samples, Gwiazda *et al.* (1996a) found that H3 and H6 resemble the ambient sediment in V28-82, suggesting a large European source contribution, which agrees with the conclusion of Grousset *et al.* (1993). Lead-isotope data from composites of 75 to 300 grains from Gwiazda *et al.* (1996a) are shown in Figure 19(c). As mentioned above, H3 and H6 seem to be low-foraminifera intervals rather than ice-rafting events. These conclusions are consistent with other observations around the North Atlantic. Although Heinrich layer H3 appears to be a Hudson Strait event (Grousset *et al.*, 1993; Bond and Lotti, 1995; Rashid *et al.*, 2003, and Hemming, unpublished lead-isotope and $^{40}Ar/^{39}Ar$ hornblende data from Orphan Knoll), it does not spread Hudson Strait-derived IRD as far to the east as the other Heinrich events (Grousset *et al.*, 1993; Figure 19). Additionally, Kirby and Andrews (1999) proposed that H3 (and the Younger Dryas) represent across Strait (also modeled by Pfeffer *et al.*, 1997) rather than along Strait flow as inferred for H1, H2, H4, and H5. The strontium-isotope map of H3 shows a striking pattern of decrease in $^{87}Sr/^{86}Sr$ ratios nearly perpendicular to the IRD belt (Figure 21), consistent with a mixture of sediments from the Labrador Sea icebergs with those derived from icebergs from eastern Greenland, Iceland, and Europe. H6 has not been studied, but the composition of organic material in H3 does not stand out prominently in the studies mentioned above (Madureira *et al.*, 1997; Rosell-Mele *et al.*, 1997; Huon *et al.*, 2002).

6.17.6.6 Summary of Geochemical Provenance Analysis of Heinrich Layers

Heinrich layers H1, H2, H4, and H5 have several distinctive characteristics that distinguish them from ambient IRD, and they are derived from a mix of provenance components that are all consistent with derivation from a small region near Hudson Strait. Heinrich layers H3 and H6 have different sources, at least in the eastern North Atlantic. Less is known about H6, but H3 appears to have a Hudson Strait source in the southern Labrador Sea and western Atlantic, consistent with a similar but weaker event compared to the big four. Important related questions are as follows: How many types (provenance, flux, etc.) of Heinrich layers are there? Are IRD events in previous glacial intervals akin to the six in the last glacial period?

6.17.6.7 Trough Mouth Fans as Archives of Major IRD Sources

Trough mouth fans are major glacial-marine sediment fans that form when ice streams occupied glacial troughs that extend to the shelf-slope boundary (e.g., Vorren and Laberg, 1997). They are a significant resource and archive for understanding the potential compositional characteristics of sediment-laden icebergs, because they contain source-specific detrital material in highly concentrated accumulations. An important consideration that emphasizes the value of this archive is the recognition that the debris flows mimic the tills in the source area and, therefore, also the composition of the IRD in the calving icebergs. Furthermore, most icebergs experience substantial melting close to the ice-sheet margin, and the sediments they carry tend to melt out before the iceberg has reached the open ocean (Syvitski *et al.*, 1996; Andrews, 2000). Sediments deposited on the trough mouth fans are not only good representative point sources of the glacial drainage area; they are also located in positions of high iceberg production, that increase the likelihood of an iceberg's being carried to deep-marine environments in the surface-ocean currents before loosing its sediment load. Locations of documented trough mouth fans in the northern hemisphere are shown in Figure 22.

Although there is substantial overlap in the geological histories of the continental sources around the North Atlantic, there are also some systematic variations that can allow distinction of different sources. Realization of the full potential awaits characterizing the sources with multiple tracers as well as sedimentological studies, and documenting the geographic pattern of dispersal in marine sediments within small time-windows.

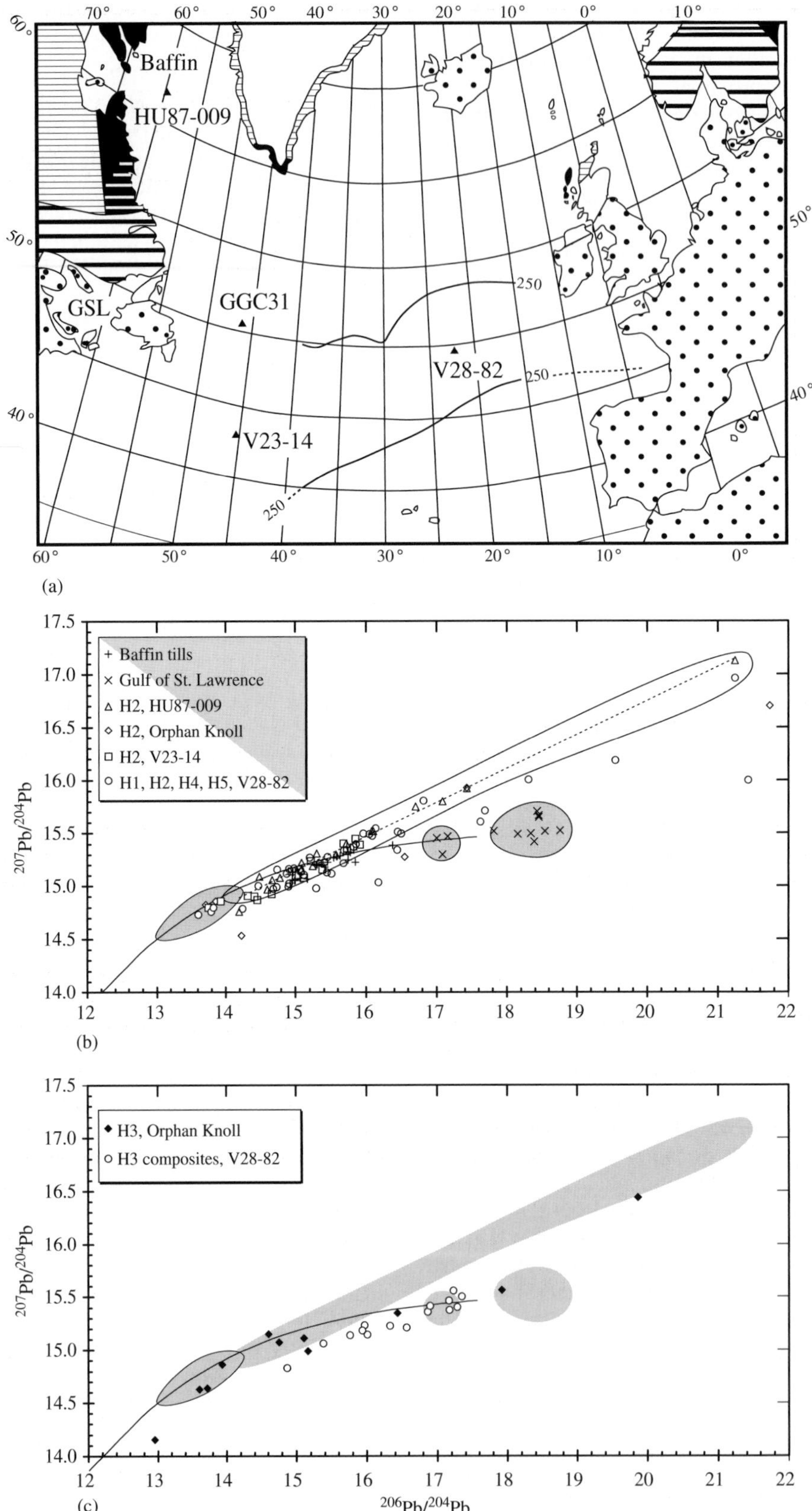
Baffin
HU87-009
GSL
GGC31
V28-82
V23-14
250
Baffin tills
Gulf of St. Lawrence
H2, HU87-009
H2, Orphan Knoll
H2, V23-14
H1, H2, H4, H5, V28-82
H3, Orphan Knoll
H3 composites, V28-82
$^{207}Pb/^{204}Pb$
$^{206}Pb/^{204}Pb$
(a)
(b)
(c)

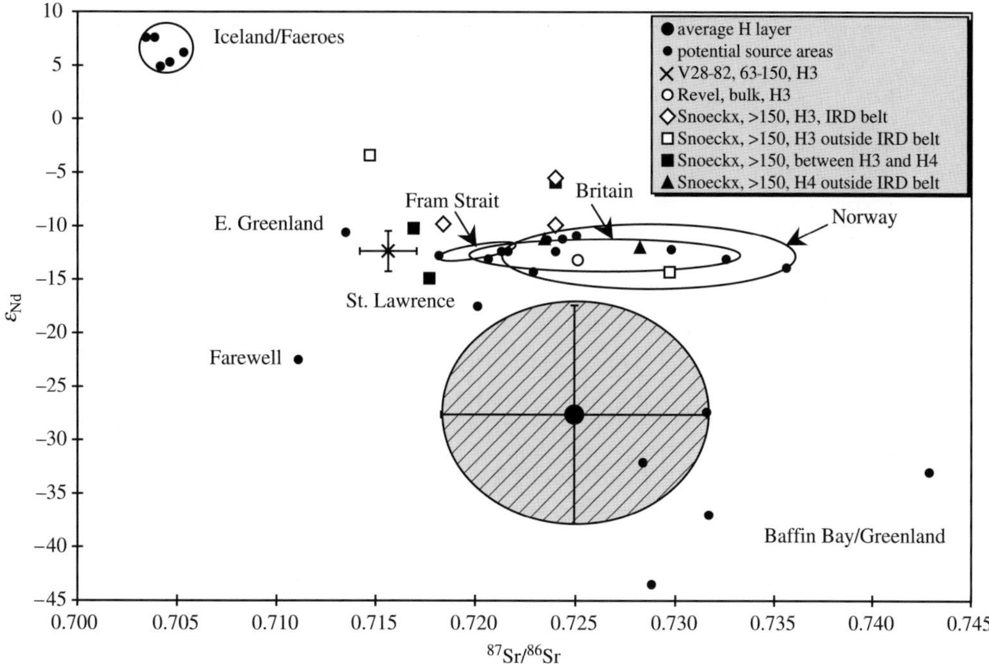

Figure 20 *Nd–Sr-isotope ratios of terrigenous clastic components of Heinrich layer H3.* Shown for reference are the average and 2 sigma range for the data published on Heinrich layers H1, H2, H4, and H5, and reported compositions of potential source areas of ice rafted detritus (Grousset *et al.*, 2001, and references therein). Also included is the average of 5 unpublished analyses across H3 from V28-82, where the error bars represent the range of values measured (A. Jost and S. Hemming) (sources Grousset *et al.*, 1993; Revel *et al.*, 1996; Snoeckx *et al.*, 1999).

A first step towards the goal of characterizing IRD sources in trough mouth fans was made by Hemming *et al.* (2002b) using $^{40}Ar/^{39}Ar$ data from populations of individual hornblende grains. The results conform closely to expectations based on the mapped ages of rocks in the region of the grains analyzed. For example, samples from the mid-Norwegian margin and from the northeastern margin of Greenland provide Caledonian ages. Samples from the Bear Island trough mouth fan have little hornblende indicating a dominant sedimentary source, and the hornblendes that are found yield a dominantly Paleoproterozoic spectrum of ages. Samples from near the southern tip of Greenland exhibit a mixture of Paleoproterozoic and a distinctive ~1.2 Ga population that is inferred to have derived from the alkaline Gardar complex. Samples from the Gulf of St. Lawrence have a virtually pure Grenville population, and samples from the Hudson Strait have a dominant Paleoproterozoic population with a small Archean contribution.

6.17.6.8 $^{40}Ar/^{39}Ar$ Hornblende Evidence for History of the Laurentide Ice Sheet During the Last Glacial Cycle

The history of the northern hemisphere ice sheets is an important aspect of the paleoclimate system. The layered record of IRD and other climate indicators preserved in deepsea sediment cores provides the potential to unravel the sequences of events surrounding important intervals during the last glacial cycle. Results of an extensive campaign to analyze the $^{40}Ar/^{39}Ar$ ages of multiple individual hornblende grains in core

Figure 19 *Pb isotopes in feldspar grains from Heinrich layers.* (a) Map showing locations of cores analyzed with geology and IRD belt for reference. (b) Data from Heinrich layers H1, H2, H4, and H5 from several North Atlantic and Labrador Sea cores. Also shown are data from Gulf of St. Lawrence core V17-203 (Hemming, unpublished data) and from Baffin Island tills (Hemming *et al.*, 2002b). Reference fields are Superior province (Gariépy and Alle(c)gre, 1985), Labrador Sea reference line (LSRL, Gwiazda *et al.*, 1996b), Grenville (DeWolf and Mezger, 1994) and Appalachian (Ayuso and Bevier, 1991). Data sources are H2 from HU87-009, V23-14, and V28-82 (Gwiazda *et al.*, 1996b), Orphan Knoll core GGC31 (Hemming, unpublished), H1, H4, and H5 from V28-82 (Hemming *et al.*, 1998). (c) Data from Heinrich layer H3 with reference fields from (b). Data sources are H3 from V28-82 (Gwiazda *et al.*, 1996a), H3 from Orphan Knoll core GGC31 (Hemming, unpublished data).

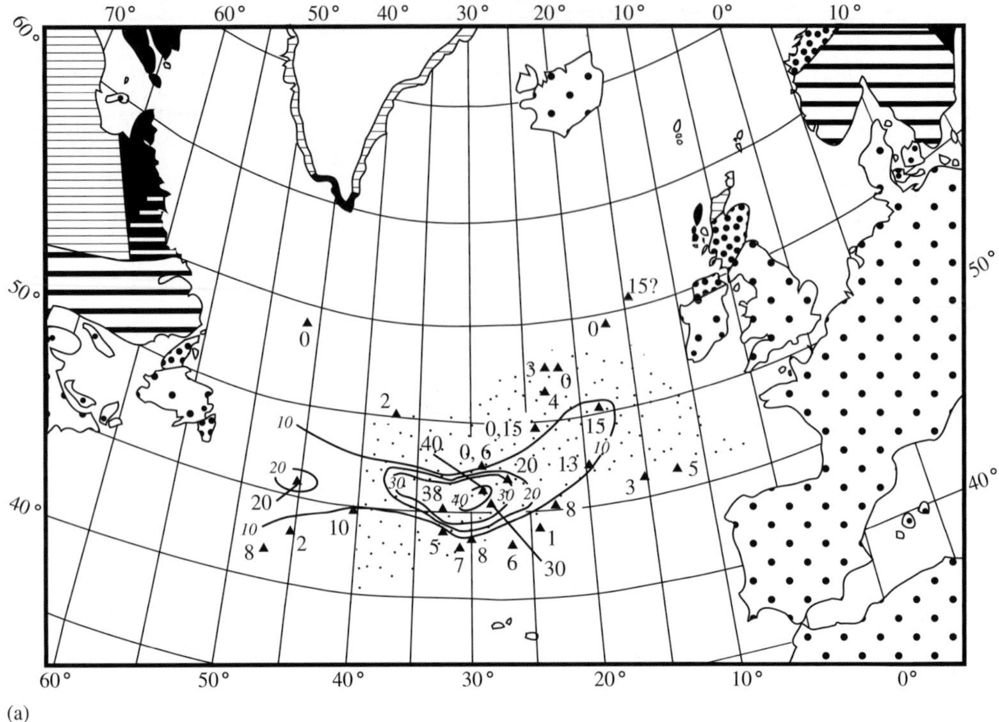

(a)

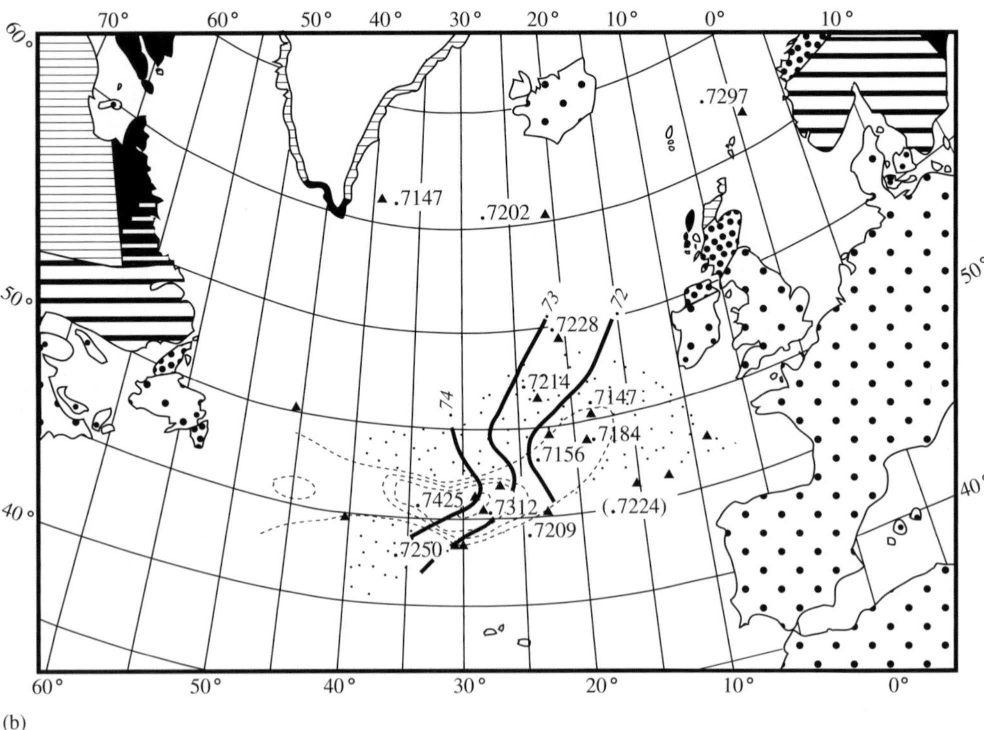

(b)

Figure 21 *Maps of data from Heinrich layer H3.* General geology and the Ruddiman (1977) IRD belt are shown. (a) Isopach map with 10 cm contour interval. (b) $^{87}Sr/^{86}Sr$ values for siliciclastic detritus in H3. Isopachs are shown in light dashed lines for reference. (c) 25–23 ka 250 mg cm^{-2} kyr^{-1} contours defining the approximately E–W IRD belt (Ruddiman, 1977), contours of 10, 50, and 100 sand-sized ash shards per cm^{2} defining the approximately N–S trajectory of currents bringing Icelandic detritus into the North Atlantic (Ruddiman and Glover, 1982), and light dashed lines of 10 cm thickness intervals for H3.

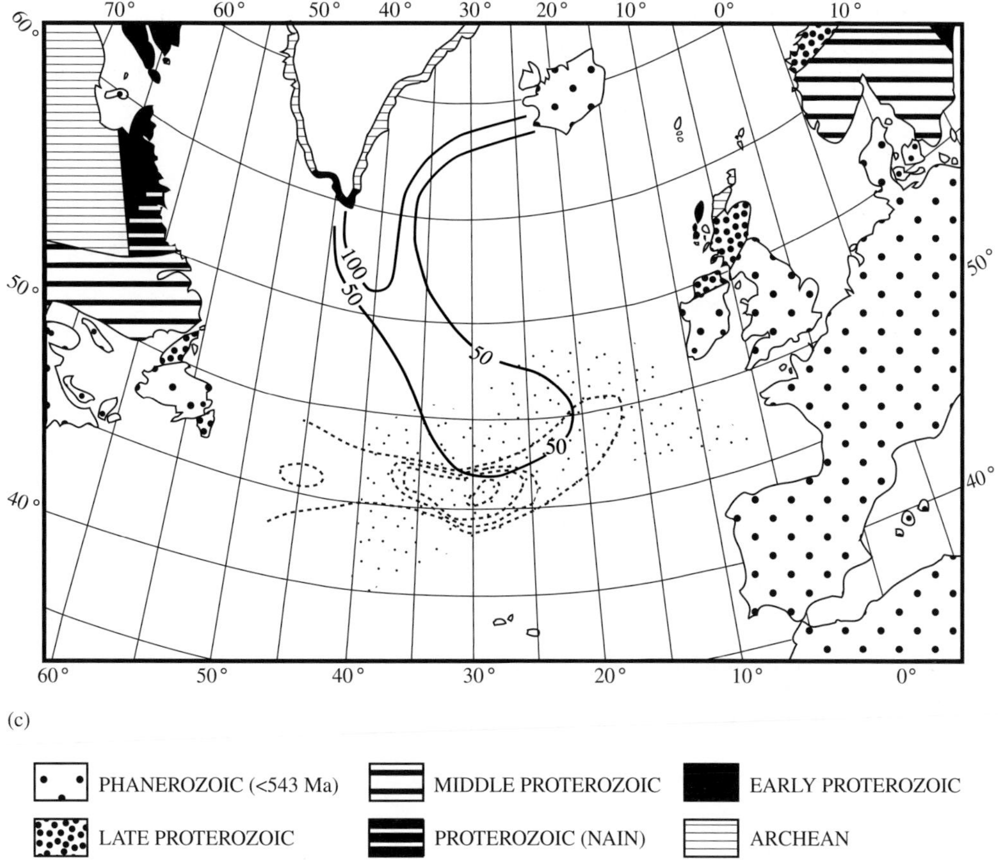

Figure 21 (continued).

V23-14 (Hemming and Hajdas, 2003) are presented as an example of the potential of this approach.

Core V23-14 is located within the thickest part of Ruddiman's (1977) ice-rafted detritus belt, and directly downstream of any Gulf of St. Lawrence region contributions (Figure 19(a)). Thus, the provenance of IRD from this core provides constraints on the evolution of the Laurentide ice sheet or smaller satellite ice sheets (e.g., Stea *et al.*, 1998) and their iceberg contributions to the northwest Atlantic margin. The hornblende data are binned according to the age brackets based on known geological ages in the North Atlantic region. Data from within the Heinrich layers are lumped into one interval with an assumed duration of 0.1 kyr. There is some disagreement about the duration of Heinrich layers, ranging from estimates that they were virtually instantaneous, to estimates of 1,000 yr or more. Currently available published results of dating and flux methods used do not allow clearly constraining these estimates, although a best average estimate of duration is taken to be 500+/−250 yr. By lumping the Heinrich layer samples into a small interval, and plotting the cumulative fraction data against estimated age, we emphasize the ambient evolution of the Laurentide ice sheet (Figure 23).

In the period 42.5–26 ka (by ^{14}C dating), there is little evidence of iceberg contributions from Laurentian sources south of 55° N, which should provide dominantly Grenville and Appalachian ages. In contrast, in the period 26–14 ^{14}C ka, there is abundant evidence of contributions from this sector. Finally, in the period 14–6 ^{14}C ka, Grenville and Appalachian ages are again absent. These observations are consistent with the results of Hemming *et al.* (2000b) from Orphan Knoll in the interval above H2. They are also consistent with the observations of Ruddiman (1977), who presented ice-rafted detritus fluxes across the North Atlantic in several time intervals, including 40–25 ka, and 25–13 ka. The average flux appears to be approximately double the 40–25 ka flux in the 25–13 ka interval, and there is an overall correspondence between the flux of ice-rafted detritus and the extent of the northern hemisphere ice sheets (Ruddiman, 1977).

Results from studies of the ice-rafted detritus population from core V23-14 provide insights into the evolution of ice sheets of the northwest Atlantic margin since 43 ka. Between Heinrich layers H5 and H3 (MIS 3), it appears that the

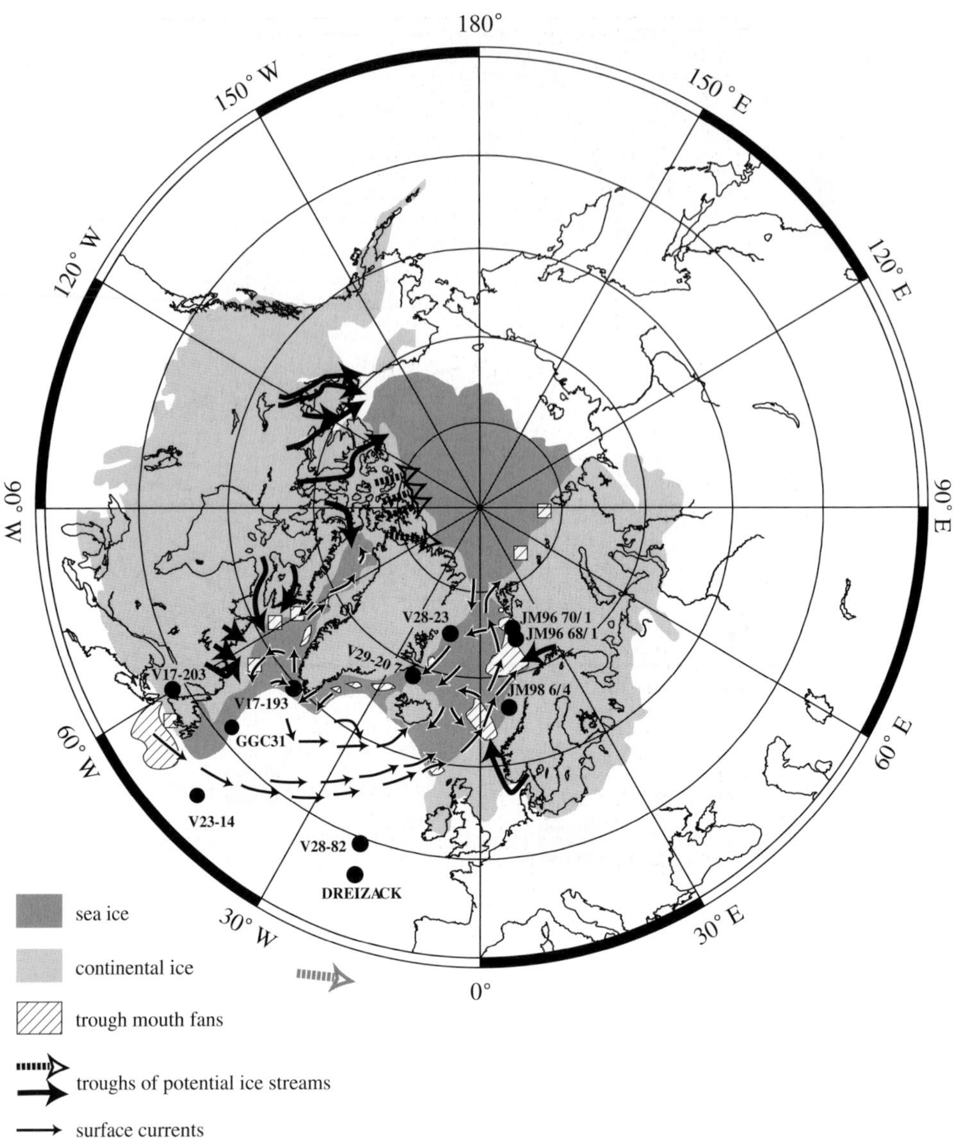

Figure 22 Polar projection from 40° to 90° showing the extents of continental and sea ice in the northern hemisphere, and cores presented and/or discussed here. The locations of known trough mouth fans are shown as well as inferred ice streams of the Laurentide ice sheet. Trough mouth fans (TMF) of the Nordic Seas are from Vorren and Laberg (1997), those around Greenland are from Funder *et al.* (1989), and along the western Labrador Sea are from Hesse and Khodabakhsh (1998). Arrows marking troughs across the Laurentide margin are based on mapping of J. Kleman. Arrow marking large trough feeding the North Sea TMF and the Barents Sea arrow indicating a large trough feeding Bear Island TMF is from Stokes and Clark (2001).

ice sheet (or sheets) did not extend far enough southeast to drop iceberg deposits with Grenville or Appalachian derivation into the North Atlantic. Between H3 and H1 (MIS 2), and outside the Heinrich layers, significant portions of the hornblende grains have ages indicating derivation from the southeastern sector, and thus indicate a significant expansion of the sheets at ~26 ^{14}C ka. After H1 (MIS 1), no detritus attributable to the southeastern sector is found, and judging from the present day situation, most of the ice-rafted detritus was likely derived from the Greenland ice sheet in the Holocene interval.

6.17.7 FINAL THOUGHTS

This chapter has summarized the basis for the application of long-lived isotopic tracers in oceanography and paleoceanography, focusing on neodymium isotopes in the oceans. It also summarized the current state of knowledge on the applications of these tracers to delineate the

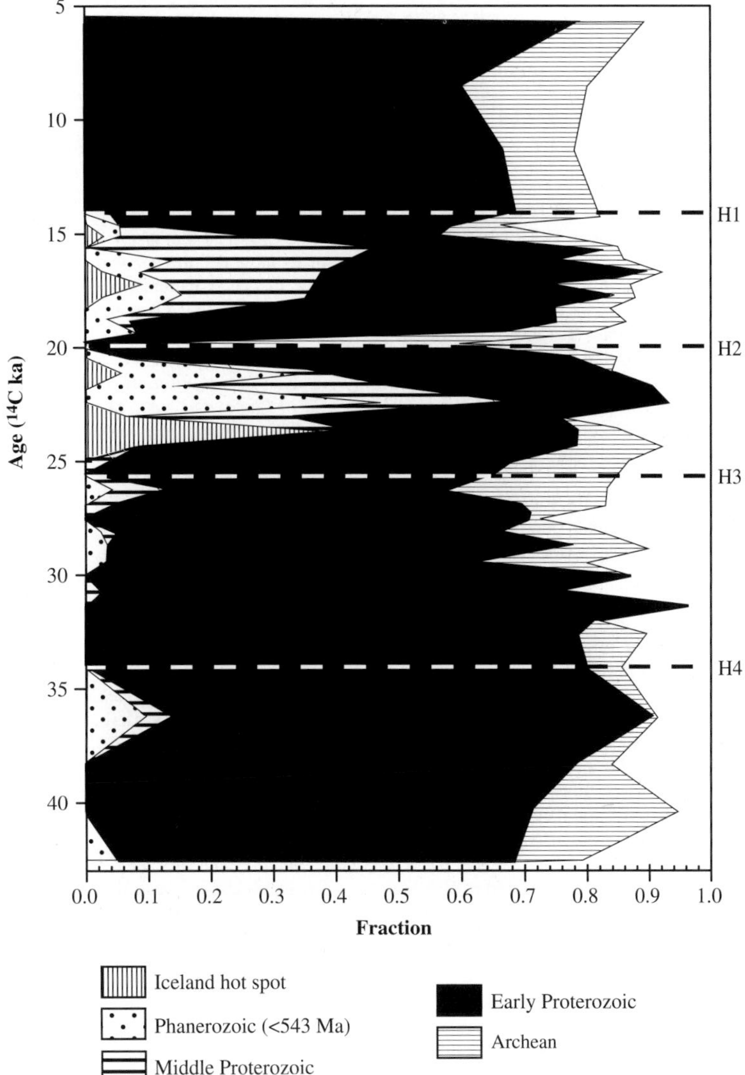

Figure 23 *Down-core plot of the hornblende Ar age populations from core V23-14.* Heinrich layers are indicated by the dashed lines (Source Hemming and Hajdas, 2003).

ice-sheet sources of North Atlantic Heinrich events. Especially since the early 1990s in particular, long-lived isotopic tracers have matured as paleoclimatic tools, as their potential value is becoming increasingly recognized and used by an growing number of investigators. Over the next decade they are certain to be among the primary methods used to generate new discoveries about the Earth's climatic history.

ACKNOWLEDGMENTS

This chapter is dedicated to Professors Gerald J. Wasserburg of Caltech, whose retirement roughly coincides with the publication of the Treatise, and Wally Broecker of LDEO. Gerry and colleagues characterized against all odds the neodymium isotopic variations in the global oceans, and the resulting body of work forms the basis of subsequent studies. Throughout his career Wally has greatly advanced the field of paleoceanography and global climate change, and has embraced the application of radiogenic isotope tracers. His contribution to emphasizing the importance of Heinrich layers is particularly relevant to this chapter. This is LDEO Contribution #6498.

REFERENCES

Abouchami W. and Goldstein S. L. (1995) A lead isotopic study of circum-antarctic manganese nodules. *Geochim. Cosmochim. Acta* **59**(9), 1809–1820.

Abouchami W., Goldstein S. L., Galer S. J. G., Eisenhauer A., and Mangini A. (1997) Secular changes of lead and neodymium in central Pacific seawater recorded by a Fe–Mn crust. *Geochim. Cosmochim. Acta* **61**(18), 3957–3974.

Albare(c)de F. and Goldstein S. L. (1992) World map of Nd isotopes in sea-floor ferromanganese deposits. *Geology* **20**(8), 761–763.

Albare(c)de F., Goldstein S. L., and Dautel D. (1997) The neodymium isotopic composition of manganese nodules from the southern and Indian oceans, the global oceanic neodymium budget, and their bearing on deep ocean circulation. *Geochim. Cosmochim. Acta* **61**(6), 1277–1291.

Albare(c)de F., Simonetti A., Vervoort J. D., Blichert-Toft J., and Abouchami W. (1998) A Hf–Nd isotopic correlation in ferromanganese nodules. *Geophys. Res. Lett.* **25**(20), 3895–3898.

Amakawa H., Alibo D. S., and Nozaki Y. (2000) Nd isotopic composition and REE pattern in the surface waters of the eastern Indian Ocean and its adjacent seas. *Geochim. Cosmochim. Acta* **64**(10), 1715–1727.

Andrews J. T. (2000) Icebergs and iceerg rafted detritus (IRD) in the North Atlantic: facts and assumptions. *Oceanography* **13**, 100–108.

Ayuso R. A. and Bevier M. L. (1991) Regional differences in Pb isotopic compositions of feldspars in plutonic rocks of the northern appalachian mountains, USA and Canada: a geochrmical method of terrane correlation. *Tectonics* **10**, 191–212.

Balistrieri L., Brewer P. G., and Murray J. W. (1981) Scavenging residence times of trace-metals and surface-chemistry of sinking particles in the deep ocean. *Deep-sea Res. Part A: Oceanogr. Res. Pap.* **28**(2), 101–121.

Barber D. (2001) Laurentide ice sheet dynamics from 35 to 7 ka: Sr–Nd–Pb isotopic provenance of northwest Atlantic margin sediments. PhD, University of Colorado.

Bertram C. J. and Elderfield H. (1993) The geochemical balance of the rare-earth elements and neodymium isotopes in the oceans. *Geochim. Cosmochim. Acta* **57**(9), 1957–1986.

Bizzarro M., Baker J. A., Haack H., Ulfbeck D., and Rosing M. (2003) Early history of Earth's crust–mantle system inferred from hafnium isotopes in chondrites. *Nature* **421**(6926), 931–933.

Blichert-Toft J. and Albare(c)de F. (1997) The Lu–Hf isotope geochemistry of chondrites and the evolution of the mantle–crust system. *Earth Planet. Sci. Lett.* **148**(1–2), 243–258.

Bond G., Heinrich H., Broecker W., Labeyrie L., McManus J., Andrews J., Huon S., Jantschik R., Clasen S., Simet C., Tedesco K., Klas M., Bonani G., and Ivy S. (1992) Evidence for massive discharges of icebergs into the North-Atlantic Ocean during the last glacial period. *Nature* **360**(6401), 245–249.

Bond G. C. and Lotti R. (1995) Iceberg discharges into the North-Atlantic on millennial time scales during the last glaciation. *Science* **267**(5200), 1005–1010.

Bouquillon A., France-Lanord C., Michard A., and Tiercelin J. J. (1990) Sedimentology and isotopic chemistry of the Bengal fan sediments: the denudation of the Himalaya. *Proc. ODP Sci. Res.* **116**, 43–53.

Broecker W. S. (1994) Massive iceberg discharges as triggers for global climate-change. *Nature* **372**(6505), 421–424.

Broecker W. S. and Denton G. H. (1989) The role of ocean-atmosphere reorganizations in glacial cycles. *Geochim. Cosmochim. Acta* **53**, 2465–2501.

Broecker W. S. and Hemming S. (2001) Paleoclimate-climate swings come into focus. *Science* **294**(5550), 2308–2309.

Broecker W. S. and Peng T.-H. (1982) *Tracers in the Sea.* Eldigio Press, Palisades, NY.

Broecker W. S., Bond G. C., Klas M., Clark E., and McManus J. (1992) Origin of the Northern Atlantic's Heinrich events. *Clim. Dyn.* **6**, 265–293.

Burton K. W., Lee D. C., Christensen J. N., Halliday A. N., and Hein J. R. (1999) Actual timing of neodymium isotopic variations recorded by Fe–Mn crusts in the western North Atlantic. *Earth Planet. Sci. Lett.* **171**(1), 149–156.

Cantrell K. J. and Byrne R. H. (1987) Rare earth element complexation by carbonate and oxalate ions. *Geochim. Cosmochim. Acta* **51**, 597–605.

Charles C. D. and Fairbanks R. G. (1992) Evidence from Southern Ocean sediments for the effect of North Atlantic deep-water flux on climate. *Nature* **355**, 416–419.

Charles C. D., Lynch-Stieglitz J., Ninnemann U. S., and Fairbanks R. G. (1996) Climate connections between the hemisphere revealed by deep-sea sediment core/ice core correlations. *Earth Planet. Sci. Lett.* **142**, 19–27.

Chow T. J. and Patterson C. C. (1959) Lead isotopes in manganese nodules. *Geochim. Cosmochim. Acta* **17**, 21–31.

Chow T. J. and Patterson C. C. (1962) The occurence and significance of lead isotopes in pelagic sediments. *Geochim. Cosmochim. Acta* **26**, 263–308.

Dasch E. J., Dymond J. R., and Heath G. R. (1971) Isotopic analysis of metalliferous sediment from the east Pacific rise. *Earth Planet. Sci. Lett.* **13**, 175–180.

David K., Frank M., O'Nions R. K., Belshaw N. S., and Arden J. W. (2001) The Hf isotope composition of global seawater and the evolution of Hf isotopes in the deep Pacific Ocean from Fe–Mn crusts. *Chem. Geol.* **178**(1–4), 23–42.

Debaar H. J. W., Bacon M. P., and Brewer P. G. (1983) Rare-earth distributions with a positive Ce anomaly in the western North-Atlantic Ocean. *Nature* **301**(5898), 324–327.

Debaar H. J. W., Bacon M. P., Brewer P. G., and Bruland K. W. (1985) Rare-earth elements in the Pacific and Atlantic oceans. *Geochim. Cosmochim. Acta* **49**(9), 1943–1959.

DePaolo D. J. and Wasserburg G. J. (1976a) Nd isotopic variations and petrogenetic models. *Geophys. Res. Lett.* **3**, 249–252.

DePaolo D. J. and Wasserburg G. J. (1976b) Inferences about magma sources and mantle structure from variations of Nd-143–Nd-144. *Geophys. Res. Lett.* **3**(12), 743–746.

DeWolf C. P. and Mezger K. (1994) Lead isotope analyses of leached feldspars—constraints on the early crustal history of the Grenville Orogen. *Geochim. Cosmochim. Acta* **58**(24), 5537–5550.

Elderfield H. (1988) The oceanic chemistry of the rare-earth elements. *Phil. Trans. Roy. Soc. London Ser. A: Math. Phys. Eng. Sci.* **325**(1583), 105–126.

Elderfield H. and Greaves M. J. (1982) The rare earth elements in seawater. *Nature* **296**, 214–219.

Elderfield H., Hawkesworth C. J., Greaves M. J., and Caalvert S. E. (1981) Rare-earth element geochemistry of oceanic ferromanganese nodules and associated sediments. *Geochim. Cosmochim. Acta* **45**, 513–528.

Farmer G. L., Barber D., and Andrews J. (2003) Provenance of late quaternary ice-proximal sediments in the North Atlantic: Nd, Sr, and Pb isotopic evidence. *Earth Planet. Sci. Lett.* **209**(1–2), 227–243.

Frank M. (2002) Radiogenic isotopes: tracers of past ocean circulation and erosional input. *Rev. Geophys.* **40**(1) article no. 1001.

Frank M., Whiteley N., Kasten S., Hein J. R., and O'Nions K. (2002) North Atlantic deep water export to the southern ocean over the past 14 Myr: evidence from Nd and Pb isotopes in ferromanganese crusts. *Paleoceanography* **17**(2) article no. 1002.

Funder S., Larsen H. C., and Fredskild B. (1989) Quaternary geology of the ice-free areas and adjacent shelves of Greenland. In *Quaternary Geology of Canada and Greenland: K-1* (ed. R. J. Fulten). Geological Survey of Canada, Boulder, CO, pp. 741–792.

Galy A., FranceLanord C., and Derry L. A. (1996) The Late Oligocene Early Miocene Himalayan belt: constraints deduced from isotopic compositions of Early Miocene turbidites in the Bengal fan. *Tectonophysics* **260**(1–3), 109–118.

Gao S., Luo T. C., Zhang B. R., Zhang H. F., Han Y. W., Zhao Z. D., and Hu Y. K. (1998) Chemical composition of the continental crust as revealed by studies in East China. *Geochim. Cosmochim. Acta* **62**(11), 1959–1975.

Gariépy C. and Alle(c)gre C. J. (1985) The lead isotope geochemistry and geochronology of late-kinematic intrucives from the Abitibi Greenstone belt, and the implications for late Archaean crustal evolution. *Geochim. Cosmochim. Acta* **49**, 2371–2383.

Godfrey L. V., Lee D. C., Sangrey W. F., Halliday A. N., Salters V. J. M., Hein J. R., and White W. M. (1997) The Hf isotopic composition of ferromanganese nodules and crusts and hydrothermal manganese deposits: implications for seawater Hf. *Earth Planet. Sci. Lett.* **151**(1–2), 91–105.

Goldstein S. J. and Jacobsen S. B. (1988) Nd and Sr isotopic systematics of river water suspended material—implications for crustal evolution. *Earth Planet. Sci. Lett.* **87**(3), 249–265.

Goldstein S. L. and O'Nions R. K. (1981) Nd and Sr isotopic relationships in pelagic clays and ferromanganese deposits. *Nature* **292**, 324–327.

Goldstein S. L., Onions R. K., and Hamilton P. J. (1984) A Sm–Nd isotopic study of atmospheric dusts and particulates from major river systems. *Earth Planet. Sci. Lett.* **70**(2), 221–236.

Grousset F. E., Labeyrie L., Sinko J. A., Cremer M., Bond G., Duprat J., Cortijo E., and Huon S. (1993) Patterns of ice-rafted detritus in the glacial North-Atlantic (40-degrees-55-degrees-N). *Paleoceanography* **8**(2), 175–192.

Grousset F. E., Pujol C., Labeyrie L., Auffret G., and Boelaert A. (2000) Were the North Atlantic Heinrich events triggered by the behavior of the European ice sheets? *Geology* **28**(2), 123–126.

Grousset F. E., Cortijo E., Huon S., Herve L., Richter T., Burdloff D., Duprat J., and Weber O. (2001) Zooming in on Heinrich layers. *Paleoceanography* **16**(3), 240–259.

Gwiazda R. H., Hemming S. R., and Broecker W. S. (1996a) Provenance of icebergs during Heinrich event 3 and the contrast to their sources during other Heinrich episodes. *Paleoceanography* **11**(4), 371–378.

Gwiazda R. H., Hemming S. R., and Broecker W. S. (1996b) Tracking the sources of icebergs with lead isotopes: the provenance of ice-rafted debris in Heinrich layer 2. *Paleoceanography* **11**(1), 77–93.

Gwiazda R. H., Hemming S. R., Broecker W. S., Onsttot T., and Mueller C. (1996c) Evidence from Ar-40/Ar-39 ages for a Churchill province source of ice-rafted amphiboles in Heinrich layer 2. *J. Glaciol.* **42**(142), 440–446.

Heinrich H. (1988) Origin and consequences of cyclic ice rafting in the northeast Atlantic Ocean during the past 130, 000 years. *Quat. Res.* **29**(2), 142–152.

Hemming S. R. and Hajdas I. (2003) Ice-rafted detritus evidence from Ar-40/Ar-39 ages of individual hornblende grains for evolution of the eastern margin of the Laurentide ice sheet since 43 C-14 ky. *Quat. Int.* **99**, 29–43.

Hemming S. R., McLennan S. M., and Hanson G. N. (1994) Lead isotopes as a provenance tool for quartz—examples from Plutons and Quartzite, northeastern Minnesota, USA. *Geochim. Cosmochim. Acta* **58**(20), 4455–4464.

Hemming S. R., McDaniel D. K., McLennan S. M., and Hanson G. N. (1996) Pb isotope constraints on the provenance and diagenesis of detrital feldspars from the Sudbury basin, Canada. *Earth Planet. Sci. Lett.* **142**(3–4), 501–512.

Hemming S. R., Broecker W. S., Sharp W. D., Bond G. C., Gwiazda R. H., McManus J. F., Klas M., and Hajdas I. (1998) Provenance of Heinrich layers in core V28-82, northeastern Atlantic: Ar-40/Ar-39 ages of ice-rafted hornblende, Pb isotopes in feldspar grains, and Nd–Sr–Pb isotopes in the fine sediment fraction. *Earth Planet. Sci. Lett.* **164**(1–2), 317–333.

Hemming S. R., Bond G. C., Broecker W. S., Sharp W. D., and Klas-Mendelson M. (2000a) Evidence from 40Ar–39Ar ages of individual hornblende grains for varying Laurenide sources of iceberg discharges 22,000 to 10,500 yr BP. *Quat. Res.* **54**, 372–373.

Hemming S. R., Gwiazda R. H., Andrews J. T., Broecker W. S., Jennings A. E., and Onstott T. C. (2000b) Ar-40/Ar-39 and Pb–Pb study of individual hornblende and feldspar grains from southeastern Baffin Island glacial sediments: implications for the provenance of the Heinrich layers. *Can. J. Earth Sci.* **37**(6), 879–890.

Hemming S. R., Hall C. M., Biscaye P. E., Higgins S. M., Bond G. C., McManus J. F., Barber D. C., Andrews J. T., and Broecker W. S. (2002a) Ar-40/Ar-39 ages and Ar-40* concentrations of fine-grained sediment fractions from North Atlantic Heinrich layers. *Chem. Geol.* **182**(2–4), 583–603.

Hemming S. R., Vorren T. O., and Kleman J. (2002b) Provinciality of ice rafting in the North Atlantic: application of Ar-40/Ar-39 dating of individual ice rafted hornblende grains. *Quat. Int.* **95–96**, 75–85.

Henry F., Jeandel C., Dupre B., and Minster J. F. (1994) Particulate and dissolved Nd in the western Mediteranean Sea: sources, fate and budget. *Mar. Chem.* **45**, 283–305.

Hesse R. and Khodabakhsh S. (1998) Depositional facies of late Pleistocene Heinrich events in the Labrador Sea. *Geology* **26**(2), 103–106.

Hofmann A. W., Jochum K. P., Seufert M., and White W. M. (1986) Nb and Pb in oceanic basalts: new constraints on mantle evolution. *Earth Planet. Sci. Lett.* **79**, 33–45.

Huon S. and Ruch P. (1992) Mineralogical, K–Ar and Sr-87/Sr-86 isotope studies of Holocene and late glacial sediments in a deep-sea core from the northeast Atlantic Ocean. *Mar. Geol.* **107**(4), 275–282.

Huon S., Grousset F. E., Burdloff D., Bardoux G., and Mariotti A. (2002) Sources of fine-sized organic matter in North Atlantic Heinrich layers: delta C-13 and delta N-15 tracers. *Geochim. Cosmochim. Acta* **66**(2), 223–239.

Hurley P. M., Heezen B. C., Pinson W. H., and Fairbairn H. W. (1963) K–Ar values in pelagic sediments of the North Atlantic. *Geochim. Cosmochim. Acta* **27**, 393–399.

Jacobsen S. B. and Wasserburg G. J. (1980) Sm–Nd isotopic evolution of chondrites. *Earth Planet. Sci. Lett.* (50), 139–155.

Jaffey A. H., Flynn K. F., Glendeni Le., Bentley W. C., and Essling A. M. (1971) Precision measurement of half-lives and specific activities of U-235 and U-238. *Phys. Rev. C* **4**(5), 1889ff.

Jantschik R. and Huon S. (1992) Detrital silicates in Northeast Atlantic deep-sea sediments during the Late Quaternary—mineralogical and K–Ar isotopic data. *Eclogae Geologicae Helvetiae* **85**(1), 195–212.

Jeandel C. (1993) Concentration and isotopic composition of Nd in the South-Atlantic Ocean. *Earth Planet. Sci. Lett.* **117**(3–4), 581–591.

Jeandel C., Bishop J. K., and Zindler A. (1995) Exchange of neodymium and its isotopes between seawater and small and large particles in the Sargasso Sea. *Geochim. Cosmochim. Acta* **59**(3), 535–547.

Jeandel C., Thouron D., and Fieux M. (1998) Concentrations and isotopic compositions of neodymium in the eastern Indian Ocean and Indonesian straits. *Geochim. Cosmochim. Acta* **62**(15), 2597–2607.

Jones C. E., Halliday A. N., Rea D. K., and Owen R. M. (1994) Neodymium isotopic variations in North Pacific modern silicate sediment and the insignificance of detrital ree contributions to seawater. *Earth Planet. Sci. Lett.* **127**(1–4), 55–66.

Kirby M. E. and Andrews J. T. (1999) Mid-Wisconsin Laurentide ice sheet growth and decay: implications for Heinrich events 3 and 4. *Paleoceanography* **14**(2), 211–223.

Lacan F. and Jeandel C. (2001) Tracing Papua New Guinea imprint on the central Equatorial Pacific Ocean using neodymium isotopic compositions and rare earth element patterns. *Earth Planet. Sci. Lett.* **186**(3–4), 497–512.

Le Roux L. J. and Glendenin L. E. (1963) Half-life of ^{232}Th. *Proc. Natl. Meet. Nuclear Energy*, 83–94.

Levitus S. (1994) *World Ocean Atlas*. US Department of Commerce, Boulder, CO.

Lugmair G. and Scheinin N. B. (1974) Sm–Nd ages: a new dating method. *Meteoritics* **19**, 369.

Madureira L. A. S., vanKreveld S. A., Eglinton G., Conte M., Ganssen G., vanHinte J. E., and Ottens J. J. (1997) Late quaternary high-resolution biomarker and other sedimentary climate proxies in a northeast Atlantic core. *Paleoceanography* **12**(2), 255–269.

McManus J. F., Anderson R. F., Broecker W. S., Fleisher M. Q., and Higgins S. M. (1998) Radiometrically determined sedimentary fluxes in the sub-polar North Atlantic during the last 140,000 years. *Earth Planet. Sci. Lett.* **155**(1–2), 29–43.

Michard A., Albarede F., Michard G., Minster J. F., and Charlou J. L. (1983) Rare-earth elements and uranium in high-temperature solutions from East Pacific Rise hydrothermal vent field (13-degrees-N). *Nature* **303**(5920), 795–797.

Nakai S., Halliday A. N., and Rea D. K. (1993) Provenance of dust in the Pacific Ocean. *Earth Planet. Sci. Lett.* **119**(1–2), 143–157.

Neumann W. and Huster E. (1976) Discussion of Rb-87 half-life determined by absolute counting. *Earth Planet. Sci. Lett.* **33**(2), 277–288.

O'Nions R. K., Hamilton P. J., and Evensen N. M. (1977) Variations in ^{143}Nd/^{144}Nd and ^{87}Sr/^{86}Sr ratios in oceanic basalts. *Earth Planet. Sci. Lett.* **34**, 13–22.

O'Nions R. K., Carter S. R., Cohen R. S., Evensen N. M., and Hamilton P. J. (1978) Nd and Sr isotopes in oceanic ferromanganese deposits and ocean floor basalts. *Nature* **273**, 435–438.

Patchett P. J., White W. M., Feldmann H., Kielinczuk S., and Hofmann A. W. (1984) Hafnium rare-earth element fractionation in the sedimentary system and crustal recycling into the earths mantle. *Earth Planet. Sci. Lett.* **69**(2), 365–378.

Patterson C. C. (1956) The age of meteorites and the earth. *Geochim. Cosmochim. Acta* **10**, 230–237.

Patterson C. C., Goldberg E. D., and Inghram M. G. (1953) Isotopic compositions of quaternary leads from the Pacific Ocean. *Bull. Geol. Soc. Am.* **64**, 1387–1388.

Pfeffer W. T., Dyurgerov M., Kaplan M., Dwyer J., Sassolas C., Jennings A., Raup B., and Manley W. (1997) Numerical modeling of late Glacial Laurentide advance of ice across Hudson Strait: insights into terrestrial and marine geology, mass balance, and calving flux. *Paleoceanography* **12**(1), 97–110.

Piepgras D. J. and Jacobsen S. B. (1988) The isotopic composition of neodymium in the North Pacific. *Geochim. Cosmochim. Acta* **52**(6), 1373–1381.

Piepgras D. J. and Wasserburg G. J. (1980) Neodymium isotopic variations in seawater. *Earth Planet. Sci. Lett.* **50**, 128–138.

Piepgras D. J. and Wasserburg G. J. (1982) Isotopic composition of neodymium in waters from the Drake Passage. *Science* **217**(4556), 207–214.

Piepgras D. J. and Wasserburg G. J. (1983) Influence of the Mediterranean outflow on the isotopic composition of neodymium in waters of the North-Atlantic. *J. Geophys. Res.: Oceans Atmos.* **88**(NC10), 5997–6006.

Piepgras D. J. and Wasserburg G. J. (1985) Strontium and neodymium isotopes in hot springs on the East Pacific Rise and Guaymas Basin. *Earth Planet. Sci. Lett.* **72**(4), 341–356.

Piepgras D. J. and Wasserburg G. J. (1987) Rare-earth element transport in the western North-Atlantic inferred from Nd isotopic observations. *Geochim. Cosmochim. Acta* **51**(5), 1257–1271.

Piepgras D. J., Wasserburg G. J., and Dasch E. J. (1979) The isotopic composition of Nd in different ocean masses. *Earth Planet. Sci. Lett.* **45**, 223–236.

Pierson-Wickmann A. C., Reisberg L., France-Lanord C., and Kudrass H. R. (2001) Os–Sr–Nd results from sediments in the Bay of Bengal: implications for sediment transport and the marine Os record. *Paleoceanography* **16**(4), 435–444.

Piotrowski A. M., Lee D. C., Christensen J. N., Burton K. W., Halliday A. N., Hein J. R., and Gunther D. (2000) Changes in erosion and ocean circulation recorded in the Hf isotopic compositions of North Atlantic and Indian Ocean ferromanganese crusts. *Earth Planet. Sci. Lett.* **181**(3), 315–325.

Rashid H., Hesse R., and Piper D. J. W. (2003) Distribution, thickness and origin of Heinrich layer 3 in the Labrador Sea. *Earth Planet. Sci. Lett.* **205**(3–4), 281–293.

Revel M., Sinko J. A., Grousset F. E., and Biscaye P. E. (1996) Sr and Nd isotopes as tracers of North Atlantic lithic particles: paleoclimatic implications. *Paleoceanography* **11**(1), 95–113.

Reynolds P. H. and Dasch E. J. (1971) Lead isotopes in marine manganese nodules and the ore-lead growth curve. *J. Geophys. Res.* **76**, 5124–5129.

Richard P., Shimizu N., and Alle(c)gre C. J. (1976) ^{143}Nd/^{146}Nd, a natural tracer: an application to oceanic basalts. *Earth Planet. Sci. Lett.* **31**, 269–278.

Rintoul S. R. (1991) South Atlantic interbasin exchange. *J. Geophys. Res.* **96**, 2675–2692.

Rosell-Mele A., Maslin M. A., Maxwell J. R., and Schaeffer P. (1997) Biomarker evidence for "Heinrich" events. *Geochim. Cosmochim. Acta* **61**(8), 1671–1678.

Ruddiman W. F. (1977) Late quaternary deposition of ice-rafted sand in subpolar North-Atlantic (Lat 40-degrees to 65-degrees-N). *Geol. Soc. Am. Bull.* **88**(12), 1813–1827.

Ruddiman W. F. and Glover L. K. (1982) Mixing of volcanic ash zones in subpolar North Atlantic sediments. In *The Ocean Floor, Bruce C. Heezen Memorial Volume* (eds. R. A. Scrutton and M. Talwani). Wiley, Chichester, NY, pp. 37–60.

Rudnick R. L. and Fountain D. M. (1995) Nature and composition of the continental-crust—a lower crustal perspective. *Rev. Geophys.* **33**(3), 267–309.

Rutberg R. L., Hemming S. R., and Goldstein S. L. (2000) Reduced north Atlantic deep water flux to the glacial southern ocean inferred from neodymium isotope ratios. *Nature* **405**(6789), 935–938.

Scherer E., Munker C., and Mezger K. (2001) Calibration of the lutetium–hafnium clock. *Science* **293**(5530), 683–687.

Schlosser P., Bullister J. L., Fine R., Jenkim W. J., Key R., Lupton J., Roether W., and Smethie W. M. Jr. (2001) Transformation and age of water masses. In *Ocean circulation and climate: Observing and Modelling the Global Ocean* (eds G. Siedler, J. Church, and J. Gould). Academic Press, pp. 431–452.

Segl M., Mangini A., Bonani G., Hofmann H. J., Nessi M., Suter M., Wölfli W., Friedrich G., Plüger W. L., Wiechowski A., and Beer J. (1984) ^{10}Be-dating of a manganese crust from central North Pacific and implications for ocean palaeocirculation. *Nature* **309**, 540–543.

Sguigna A. P., Larabee A. J., and Waddington J. C. (1982) The half-life of Lu-176 by a gamma–gamma coincidence measurement. *Can. J. Phys.* **60**(3), 361–364.

Shimizu H., Tachikawa K., Masuda A., and Nozaki Y. (1994) Cerium and neodymium isotope ratios and ree patterns in seawater from the North Pacific Ocean. *Geochim. Cosmochim. Acta* **58**(1), 323–333.

Snoeckx H., Grousset F., Revel M., and Boelaert A. (1999) European contribution of ice-rafted sand to Heinrich layers H3 and H4. *Mar. Geol.* **158**(1–4), 197–208.

Spivack A. J. and Wasserburg G. J. (1988) Neodymium isotopic composition of the Mediterranean outflow and the eastern North-Atlantic. *Geochim. Cosmochim. Acta* **52**(12), 2767–2773.

Stea R. R., Piper D. J. W., Fader G. B. J., and Boyd R. (1998) Wisconsinan glacial and sea-level history of Maritime

Canada and the adjacent continental shelf: a correlation of land and sea events. *Geol. Soc. Am. Bull.* **110**(7), 821–845.

Steiger R. H. and Jäger E. (1977) Subcommission on geochronology—convention on use of decay constants in geochronology and cosmochronology. *Earth Planet. Sci. Lett.* **36**(3), 359–362.

Stokes C. R. and Clark C. D. (2001) Palaeo-ice streams. *Quat. Sci. Rev.* **20**(13), 1437–1457.

Stordal M. C. and Wasserburg G. J. (1986) Neodymium isotopic study of Baffin-bay water—sources of Ree from very old terranes. *Earth Planet. Sci. Lett.* **77**(3–4), 259–272.

Syvitski J. P. M., Andrews J. T., and Dowdeswell J. A. (1996) Sediment deposition in an iceberg-dominated glacimarine environment, East Greenland: basin fill implications. *Global Planet. Change* **12**(1–4), 251–270.

Tachikawa K., Jeandel C., and Dupre B. (1997) Distribution of rare earth elements and neodymium isotopes in settling particulate material of the tropical Atlantic Ocean (EUMELI site). *Deep-Sea Res. I: Oceanogr. Res. Pap.* **44**(11), 1769–1792.

Tachikawa K., Jeandel C., and Roy-Barman M. (1999a) A new approach to the Nd residence time in the ocean: the role of atmospheric inputs. *Earth Planet. Sci. Lett.* **170**(4), 433–446.

Tachikawa K., Jeandel C., Vangriesheim A., and Dupre B. (1999b) Distribution of rare earth elements and neodymium isotopes in suspended particles of the tropical Atlantic Ocean (EUMELI site). *Deep-Sea Res. I: Oceanogr. Res. Pap.* **46**(5), 733–755.

Tatsumoto M., Knight R. J., and Allegre C. J. (1973) Time differences in formation of meteorites as determined from ratio of Pb-207 to Pb-206. *Science* **180**(4092), 1279–1283.

Tatsumoto M., Unruh D. M., and Patchett P. J. (1981) U-Pb nd Lu–;Hf systematics of Antarctic meteorites. *Proc. 6th Symp. Antarctic Meteorit.*, 237–249.

Taylor S. R. and McLennan S. M. (1985) *The Continental Crust: Its Composition and Evolution*. Blackwell, Oxford.

van de Flierdt T., Frank M., Lee D. C., and Halliday A. N. (2002) Glacial weathering and the hafnium isotope composition of seawater. *Earth Planet. Sci. Lett.* **198**(1–2), 167–175.

Vance D. and Burton K. (1999) Neodymium isotopes in planktonic foraminifera: a record of the response of continental weathering and ocean circulation rates to climate change. *Earth Planet. Sci. Lett.* **173**(4), 365–379.

Vervoort J. D., Patchett P. J., Gehrels G. E., and Nutman A. P. (1996) Constraints on early Earth differentiation from hafnium and neodymium isotopes. *Nature* **379**(6566), 624–627.

Vervoort J. D., Patchett P. J., Blichert-Toft J., and Albarede F. (1999) Relationships between Lu–Hf and Sm–Nd isotopic systems in the global sedimentary system. *Earth Planet. Sci. Lett.* **168**(1–2), 79–99.

von Blanckenburg F. (1999) Perspectives: paleoceanography—tracing past ocean circulation? *Science* **286**(5446), 1862–1863.

von Blanckenburg F. and Nagler T. F. (2001) Weathering versus circulation-controlled changes in radiogenic isotope tracer composition of the Labrador Sea and North Atlantic deep water. *Paleoceanography* **16**(4), 424–434.

von Blanckenburg F., O'Nions R. K., and Hein J. R. (1996) Distribution and sources of pre-anthropogenic lead isotopes in deep ocean water from Fe–Mn crusts. *Geochim. Cosmochim. Acta* **60**(24), 4957–4963.

Vorren T. O. and Laberg J. S. (1997) Trough mouth fans—Palaeoclimate and ice-sheet monitors. *Quat. Sci. Rev.* **16**(8), 865–881.

Wedepohl K. H. (1995) The composition of the continental-crust. *Geochim. Cosmochim. Acta* **59**(7), 1217–1232.

White W. M., Patchett J., and Ben Othman D. (1986) Hf isotope ratios of marine sediments and Mn nodules: evidence for a mantle source of Hf in seawater. *Earth Planet. Sci. Lett.* **79**, 46–54.

6.18
The Biological Pump in the Past

D. M. Sigman
Princeton University, NJ, USA

and

G. H. Haug
Geoforschungszentrum Potsdam, Germany

6.18.1 INTRODUCTION

It is easy to imagine that the terrestrial biosphere sequesters atmospheric carbon dioxide; the form and quantity of the sequestered carbon, living or dead organic matter, are striking. In the ocean, there are no aggregations of biomass comparable to the forests on land. Yet biological productivity in the ocean plays a central role in the sequestration of atmospheric carbon dioxide, typically overshadowing the effects of terrestrial biospheric carbon storage on timescales longer than a few centuries. In an effort to communicate the ocean's role in the regulation of atmospheric carbon dioxide, marine scientists frequently refer to the ocean's biologically driven sequestration of carbon as the "biological pump." The original and strict definition of the biological (or "soft-tissue") pump is actually more specific: the sequestration of carbon dioxide in the ocean interior by the biogenic flux of organic matter out of surface waters and into the deep sea prior to decomposition of that organic matter back to carbon dioxide (Volk and Hoffert, 1985) (Figure 1). The biological pump extracts carbon from the "surface skin" of the ocean that interacts with the atmosphere, presenting a lower partial pressure of carbon dioxide (CO_2) to the atmosphere and thus lowering its CO_2 content.

The place of the biological pump in the global carbon cycle is illustrated in Figure 2. The atmosphere exchanges carbon with essentially three reservoirs: the ocean, the terrestrial biosphere, and the geosphere. The ocean holds ~50 times as much carbon as does the atmosphere, and

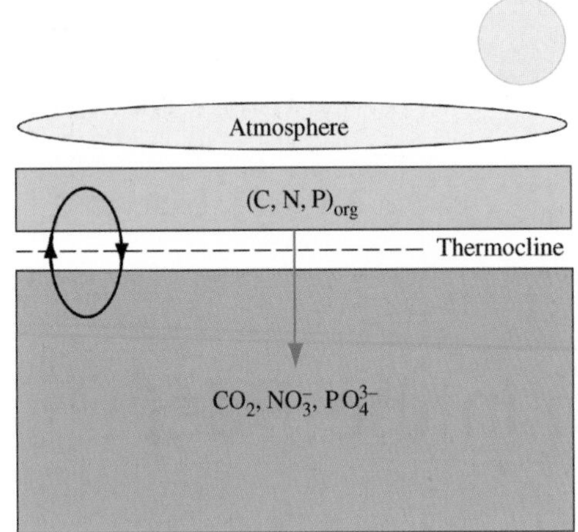

Figure 1 A schematic of the ocean's "biological pump," the sequestration of carbon and nutrients (nitrogen and phosphorus) in the ocean interior (lower dark blue box) by the growth of phytoplankton (floating unicellular algae) in the sunlit surface ocean (upper light blue box), the downward rain of organic matter out of the surface ocean and into the deep ocean (green downward arrow), and the subsequent breakdown of this organic matter back to carbon dioxide (CO_2), nitrate (NO_3^-), and phosphate (PO_4^{3-}). The nutrients and CO_2 are reintroduced to the surface ocean by mixing and upwelling (the circling arrows at left). The biological pump lowers the CO_2 content of the atmosphere by extracting it from the surface ocean (which exchanges CO_2 with the atmosphere) and sequestering it in the isolated waters of the ocean interior. In most of the low- and mid-latitude ocean, the surface is isolated from the deep sea by a temperature-driven density gradient, or "thermocline," keeping nutrient supply low and leading to essentially complete consumption of NO_3^- and PO_4^{3-} by phytoplankton at the surface. At the higher latitudes, where there is no permanent thermocline, more rapid communication with the deep sea leads to incomplete consumption of NO_3^- and PO_4^{3-}.

~20 times as much as the terrestrial biosphere. Thus, on timescales that are adequately long to allow the deep sea to communicate with the surface ocean and atmosphere ($\gtrsim$500 yr), carbon storage in the ocean all but sets the concentration of CO_2 in the atmosphere. The effect of the biological pump is not permanent on this timescale of ocean circulation; the downward transport of carbon is balanced by the net upward flux from the CO_2-rich waters of the deep sea, which, in the absence of the biological pump, would work to homogenize the carbon chemistry of the ocean, raising atmospheric CO_2 in the process.

The dynamics of marine organic carbon can be described as a set of three nested cycles in which the biological pump is the cycle with flux and reservoir of intermediate magnitude (Figure 3). The cycle with the shortest timescale, which operates within the surface ocean, is composed of net primary production by phytoplankton (their photosynthesis less their respiration) and heterotrophic respiration by zooplankton and bacteria that oxidize most of the net primary production back to CO_2. This cycle is by far the greatest in terms of the flux of carbon, but the reservoir of sequestered carbon that accumulates in surface waters (phytoplankton biomass, dead organic particles, and dissolved organic carbon) is small relative to the atmospheric reservoir of CO_2, and it has a short residence time (less than a year). A small imbalance in this cycle, between net primary production and heterotrophic respiration, feeds the next cycle in the form of organic carbon that sinks (or is mixed) into the deep sea. This "export production" drives the biological pump (Figure 3).

At the other extreme, a small fraction of the organic matter exported from the surface ocean escapes remineralization in the water column and sediments and is thus buried in the sediments, removing carbon from the ocean/atmosphere system (Figure 3). On the timescale of geologic processes, this carbon removal is balanced by the oxidation of the organic matter when it is exposed at the earth surface by uplift and weathering, or when it is released by metamorphism and volcanism. While the fluxes involved in this cycle are small, the reservoir is large, so that its importance increases with timescale, becoming clearly relevant on the timescale of millions of years.

The biological pump, in the strict sense, does not include the burial of organic matter in marine sediments. There are several related reasons for this exclusion. First, as described below, the variations in atmospheric carbon dioxide that are correlated with the waxing and waning of ice ages have driven much of the thinking about the role of organic carbon production on atmospheric composition. Only a very small fraction of the organic matter exported out of surface waters is preserved and buried in the underlying sediment, so that this process cannot sequester a significant amount of carbon on the timescale of millennia and thus is not a candidate process for the major carbon dioxide variations over glacial cycles. Second, on the timescale for which organic matter burial is relevant, i.e., over millions of years, it is thought to be only one of several mechanisms by which atmospheric CO_2 is regulated. Most hypotheses regarding the history of CO_2 over geological time involve weathering of rocks on land and the precipitation of carbonates in the ocean, due to both the larger fluxes and larger reservoirs involved in the geological trapping of CO_2 as carbonate as opposed to organic carbon (Figure 2). Finally, the importance of biological productivity in determining the burial rate of organic carbon is not at all clear. While some examples of high organic carbon burial appear to be due to high

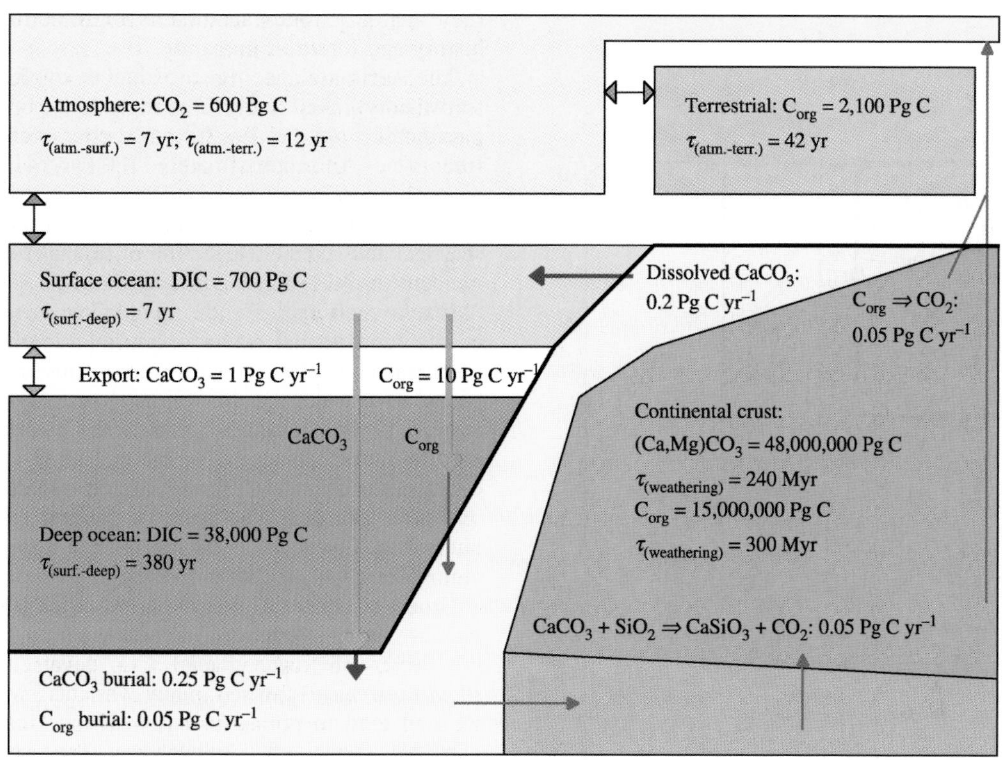

Figure 2 A simplified view of the Holocene (pre-industrial) carbon cycle (largely based on Holmen (1992)). Units of carbon are petagrams (Pg, 10^{15} g). The fluxes are colored according to the residence time of carbon in the reservoirs they involve, from shortest to longest, as follows: red, orange, purple, blue. Exchanges of the atmosphere with the surface ocean and terrestrial biosphere are relatively rapid, such that changes in the fluxes alter atmospheric CO_2 on the timescale of years and decades. Exchange between the surface ocean and the deep ocean is such that centuries to millennia are required for a change to yield a new steady state. Because the deep-ocean carbon reservoir is large relative to the surface ocean, terrestrial biosphere, and atmosphere, its interactions with the surface ocean, given thousands of years, determine the total amount of carbon to be partitioned among those three reservoirs. In a lifeless ocean, the only cause for gradients in CO_2 between the deep ocean and the surface ocean would be temperature (i.e., the "solubility pump" (Volk and Hoffert, 1985))—deep waters are formed from especially cold surface waters, and CO_2 is more soluble in cold water. The biological pump, represented by the right downward arrow, greatly enhances the surface–deep CO_2 gradient, through the rain of organic matter ("C_{org}") out of the surface ocean. The "carbonate pump," represented by the left downward arrow, is the downward rain of calcium carbonate microfossils out of the surface ocean. Its effect is to actually raise the p_{CO_2} of the surface ocean; this involves the alkalinity of seawater and is discussed in Chapters 6.04, 6.10, 6.19, and 6.20. Almost all of the organic matter raining out of the surface ocean is degraded back to CO_2 and inorganic nutrients as it rains to the seafloor or once it is incorporated into the shallow sediments; only less than 1% (~0.05 Pg out of ~10 Pg) is removed from the ocean/atmosphere system by burial (the downward blue arrow). This is in contrast to the calcium carbonate rain out of the surface (the downward purple arrow), ~25% of which is buried. In parallel, the weathering rate of calcium carbonate on land is significant on millennial timescales (the left-pointing purple arrow), while the release of geologically sequestered C_{org} occurs more slowly, on a similar timescale as the release of CO_2 from the metamorphism of calcium carbonate to silicate minerals.

oceanic productivity (Vincent and Berger, 1985), others involve diverse additional processes, such as the delivery of sediments to the ocean (FranceLanord and Derry, 1997), which influences the ease with which the organic matter rain enters the sedimentary record (Canfield, 1994). In the latter case, the distribution of oceanic productivity is of secondary importance to the effect of organic carbon burial on atmospheric CO_2.

The most direct evidence for natural variations in atmospheric CO_2 comes from the air that is trapped in Antarctic glacial ice. Records from Antarctic ice cores indicate that the concentration of CO_2 in the atmosphere has varied in step with the waxing and waning of ice ages (Barnola *et al.*, 1987; Petit *et al.*, 1999) (Figure 4). During interglacial times, such as the Holocene (roughly the past 10^4 years), the atmospheric partial pressure of CO_2 (p_{CO_2}) is typically near 280 ppm by volume (ppmv). During peak glacial times, such as the last glacial maximum ~1.8×10^4 yr ago, atmospheric p_{CO_2}

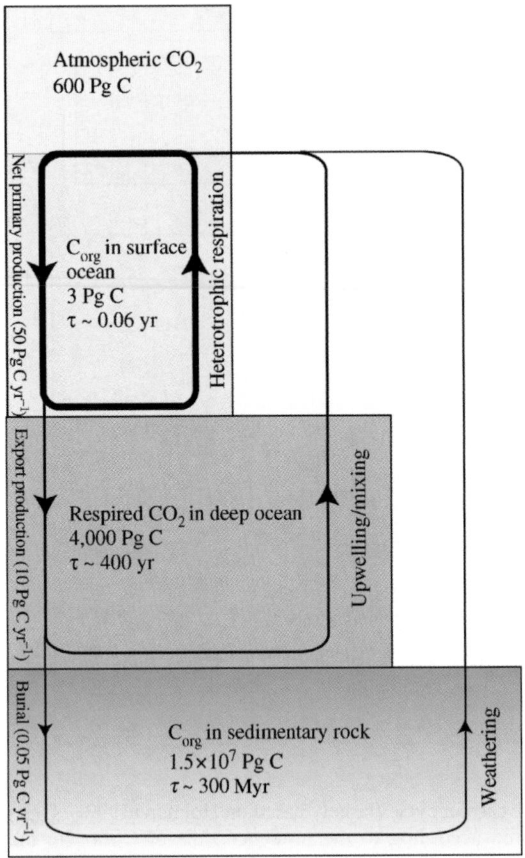

Figure 3 The nested cycles of marine organic carbon, including (i) net primary production and heterotrophic respiration in the surface ocean, (ii) export production from the surface ocean and respiration in the deep sea followed by upwelling or mixing of the respired CO_2 back to the surface ocean (the biological pump, the subject of this chapter), and (iii) burial of sedimentary organic carbon and its release to the atmosphere by weathering. The slight imbalance of each cycle fuels the longer-timescale cycles. The carbon reservoirs shown refer solely to the fraction affected by organic carbon cycling; for instance, the deep-ocean value shown is of CO_2 produced by respiration, not the total dissolved inorganic carbon reservoir.

is 180–200 ppmv, or ~80–100 ppmv lower. CO_2 is a greenhouse gas, and model calculations suggest that its changes play a significant role in the energetics of glacial/interglacial climate change (Weaver *et al.*, 1998; Webb *et al.*, 1997). However, we have not yet identified the cause of these variations in CO_2. The explanation of glacial/interglacial cycles in CO_2 is a major motivation in the paleoceanographic and paleoclimatological communities, for three reasons: first, these CO_2 changes are the most certain of any in our record of the past (excluding the anthropogenic rise in CO_2); second, they show a remarkably strong connection to climate, inspiring considerations of both cause and effect; and third, they are of a timescale that is relevant to the history and future of humanity.

The ultimate pacing of glacial cycles is statistically linked to cyclic changes in the orbital parameters of the Earth, with characteristic frequencies of approximately 100 kyr, 41 kyr, and 23 kyr (Berger *et al.*, 1984; Hays *et al.*, 1976). These orbitally driven variations in the seasonal and spatial distribution of solar radiation incident on the Earth surface, known as the "Milankovitch cycles", are thought by many to be the fundamental driver of glacial/interglacial oscillations in climate. However, positive feedbacks within the Earth's climate system must amplify orbital forcing to produce the amplitude and temporal structure of glacial/interglacial climate variations. Carbon dioxide represents one such feedback. The point of greatest uncertainty in this feedback is the mechanism by which atmospheric CO_2 is driven to change.

Broecker (1982a,b) first hypothesized a glacial increase in the strength of the biological pump as the driver of lower glacial CO_2 levels. This suggestion has spawned many variants, which we will tend to phrase in terms of the "major" nutrients, nitrogen and phosphorus. These nutrients are relatively scarce in surface ocean waters but are required in large quantities to build all algal biomass. In terms of the major nutrients, the biological pump hypotheses fall into two groups: (i) changes in the low and mid-latitude surface ocean, where the major nutrients appear to limit the extraction of CO_2 by biological production, and (ii) changes in the polar and subpolar ocean regions, where the major nutrients are currently not completely consumed and not limiting. In both cases, the central biological process is "export production," the organic matter produced by phytoplankton that is exported from the surface ocean, resulting in the sequestration of its degradation products (inorganic carbon and nutrients) in the ocean interior. Here, we review the concepts, tools, and observations relevant to variability in the biological pump on the millenial timescale, with a strong focus on its potential to explain glacial/interglacial CO_2 change.

The biogenic rain to the deep sea has important mineral components: calcium carbonate, mostly from coccolithophorids (phytoplankton) and foraminifera (zooplankton), and opal, mostly from diatoms (phytoplankton) and radiolaria (zooplankton). The calcium carbonate ($CaCO_3$) component is important for atmospheric carbon dioxide in its own right (Figure 2). In contrast to organic carbon, the production, sinking, and burial of $CaCO_3$ acts to raise atmospheric carbon dioxide concentrations. This is unintuitive, in that $CaCO_3$, like organic carbon, represents a repository for inorganic carbon and a vector for the removal of

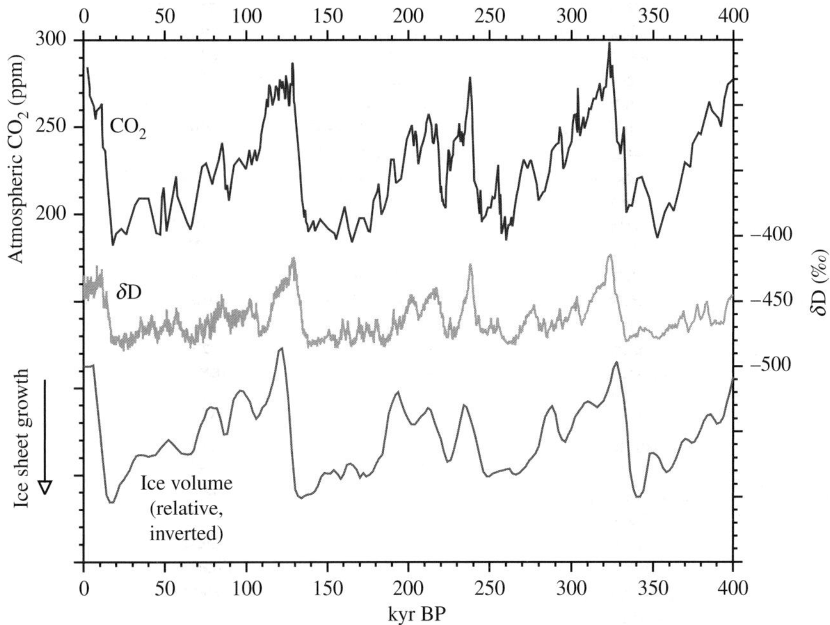

Figure 4 The history of atmospheric CO_2 back to 400 kyr as recorded by the gas content in the Vostok ice core from Antarctica (Petit *et al.*, 1999). The ratio of deuterium to hydrogen in ice (expressed as the term δD) provides a record of air temperature over Antarctica, with more negative δD values corresponding to colder conditions. The history of global ice volume based on benthic foraminiferal oxygen isotope data from deep-sea sediment cores (Bassinot *et al.*, 1994) is plotted as relative sea level, so that ice ages (peaks in continental ice volume) appear as sea-level minima, with a full glacial/interglacial amplitude for sea level change of ~130 m (Fairbanks, 1989). During peak glacial periods, atmospheric CO_2 is 80–100 ppmv lower than during peak interglacial periods, with upper and lower limits that are reproduced in each of the 100 kyr cycles. Ice cores records, including the Vostok record shown here, indicate that atmospheric CO_2 was among the first parameters to change at the termination of glacial maxima, roughly in step with southern hemisphere warming and preceding the decline in northern hemisphere ice volume.

this carbon from the surface ocean and atmosphere. The difference involves ocean "alkalinity." Alkalinity is the acid-titrating capacity of the ocean. As it increases, the pH of seawater rises (i.e., the concentration of H^+, or protons, decreases), and an increasing amount of CO_2 is stored in the ionic forms bicarbonate (HCO_3^-) and carbonate (CO_3^{2-}). This "storage" is achieved by the loss of H^+ from carbonic acid (H_2CO_3), which is itself formed by the combination of dissolved CO_2 with its host H_2O. Thus, an increase in ocean alkalinity lowers the p_{CO_2} of surface waters and thus the CO_2 concentration of the overlying atmosphere. The carbonate ion holds two equivalents of alkalinity for every mole of carbon: CO_3^{2-} must absorb two protons before it can leave the ocean as CO_2 gas. In the precipitation of $CaCO_3$, CO_3^{2-} is removed from the ocean, lowering the alkalinity of the ocean water and thus raising its p_{CO_2}. We can describe this in terms of a chain of reactions: the CO_3^{2-} that was lost to precipitation is replaced by the deprotonation of a HCO_3^-, generating a proton. This proton then combines with another HCO_3^- to produce H_2CO_3, which dissociates to form CO_2 and H_2O, yielding the summary reaction: $Ca^{2+} + 2HCO_3^- \rightarrow CaCO_3 + CO_2 + H_2O$. Thus, the biological precipitation of $CaCO_3$ raises the p_{CO_2} of the water in which it occurs.

The biological formation of $CaCO_3$ affects atmospheric CO_2 from two perspectives. First, the precipitation of $CaCO_3$ in surface waters and its sinking to the seafloor drives a surface-to-deep alkalinity gradient, which raises the p_{CO_2} of surface waters in a way that is analogous to the p_{CO_2} decrease due to the biological pump; this might be referred to as the "carbonate pump." Just as with the CO_2 produced in deep water by organic matter degradation, the chemical products of the deep redissolution of the $CaCO_3$ rain are eventually mixed up to the surface again, undoing the effect of their temporary (~1,000 yr) sequestration in the abyss. Second, ~25% of the $CaCO_3$ escapes dissolution and is buried, thus sequestering carbon and alkalinity in the geosphere, on a timescale of thousands to millions of years (Figure 2). An excess in the loss of alkalinity by calcium carbonate burial rate relative to the input of alkalinity by continental weathering will drive an increase in the p_{CO_2} of the whole ocean on the timescale of thousands of years. In summary, $CaCO_3$ precipitation can alter atmospheric p_{CO_2} by generating a surface-to-deep gradient in seawater alkalinity (the carbonate pump) and by changing the total amount of alkalinity in the ocean.

The $CaCO_3$ cycle is a central part of the effect of biological productivity on atmospheric CO_2. However, it is not within the strict definition of the biological pump, which deals specifically with organic carbon. Moreover, the effect of the $CaCO_3$ rain is determined not only by the actual magnitude of the rain to the seafloor but also by its degree of preservation and burial at the seafloor, a relatively involved subject that is treated elsewhere in this volume. Thus, in our discussions below, we try as much as possible to limit ourselves to the geochemical effects of the biogenic rain of organic matter, bringing $CaCO_3$ into the discussion only when absolutely necessary and then trying to focus on its biological production and not its seafloor preservation.

6.18.2 CONCEPTS

At the coarsest scale, phytoplankton abundance (Figure 5(a)) and nutrient availability (Figures 5(b)–(d)) are correlated across the global surface ocean, indicating that the supply of major nutrients is a dominant control on productivity. In this context, two types of environments emerge from the global ocean distributions (Figure 5): (i) the tropical and subtropical ocean, where productivity is low and limited by the supply of nitrate and phosphate, and (ii) the subpolar and polar regions, where the supply of nitrate and phosphate is high enough so as not to limit phytoplankton growth. This distinction frames much of the conceptual discussion that follows. One important result will at this point seem counterintuitive: the global biological pump is driven from the low-latitude regions, where the biological pump is working at maximal efficiency with respect to the major nutrient supply, and weakened by the high-latitude regions, which are not maximally efficient with respect to the major nutrients and which are thus allowing deeply sequestered CO_2 to escape back to the atmosphere.

The relative importance of the low versus high latitudes in controlling atmospheric CO_2 is currently a matter of debate, with the lines drawn between the two types of models that are used to study the biological pump: geochemical box models of the ocean, the first tools available for quantifying the role of the biological pump, and the newer ocean general circulation models. Relative to box models, at least some general circulation models express a greater sensitivity of atmospheric CO_2 to the low-latitude ocean and a lesser sensitivity to the high-latitude ocean (Archer *et al.*, 2000b). This difference in sensitivity has major implications for the feasibility of different oceanic changes to explain glacial/interglacial CO_2 change. While some authors have argued that the discrepancy can be framed in terms of the amount of vertical mixing in the low versus high-latitude ocean (Broecker *et al.*, 1999), we are not convinced by this argument, as such differences should be captured by the deep-ocean temperatures in the two types of models. Rather, we favor the arguments of Toggweiler *et al.* (2003a) that the difference is mostly in the degree of air/sea equilibration of CO_2 that is allowed in the high-latitude regions of the models. If this is so, there is good reason to believe that the high latitudes are as important in the biological pump as the box models suggest (Toggweiler *et al.*, 2003a). Regardless of our view, it must be recognized that the quantitative arguments made below are based largely on our experience with box models (Hughen *et al.*, 1997; Sigman *et al.*, 1998, 1999b, 2003), and that some workers would find different areas of emphasis, although the same basic processes would probably be described (e.g., see Archer *et al.*, 2000a).

While the global correlation between nutrients and productivity is strong, it is far from perfect. For instance, the chlorophyll levels among the high-latitude regions do not correlate with their surface nutrient concentration, nor does such a correlation hold particularly well within a single high-latitude region (Figure 5). These discrepancies provide a starting point for addressing what controls polar productivity and how the different polar regions impact the biological pump. We also look to the current differences among the polar regions to understand how the high-latitude leaks in the biological pump might have changed in the past.

The general link between nutrient concentration and productivity is also broken in some low-latitude regions. For instance, coastal productivity is high even where there is no apparent elevation in major nutrients. While part of this discrepancy is probably due to nutrient inputs from the continents that are not captured in Figure 5 because they are sporadic and/or compositionally complex, the discrepancy also reflects the very rapid cycling of organic matter in the shallow ocean. We will see that the coastal regions, despite their extremely high productivity, are not central to the biological pump, illustrating the distinction between productivity and the biological pump.

While this comparison between "nutrients" and productivity is useful, the surface distributions of the nutrients are themselves not all identical.

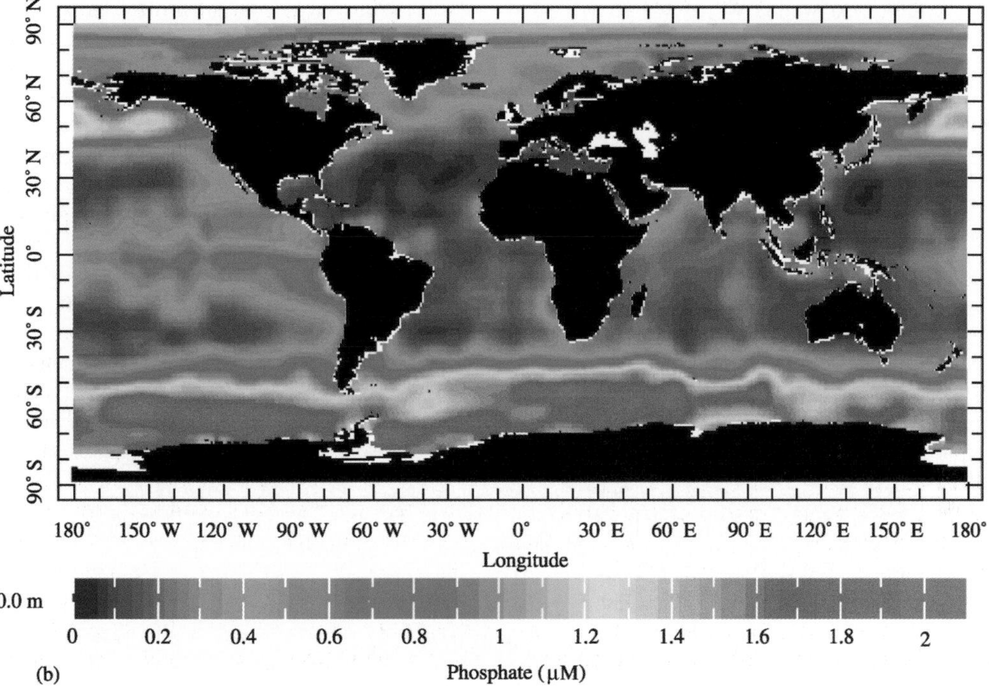

Figure 5 The concentrations in the surface ocean of chlorophyll (a qualitative index of phytoplankton abundance and thus net primary productivity, (a)), the "major nutrients" phosphate (PO_4^{3-}, (b)) and nitrate (NO_3^-, (c)), and the nutrient silicate (SiO_4^{4-}, (d)), which is required mostly by diatoms, a group of phytoplankton that builds opal tests. The chlorophyll map in (a) is from composite data from the NASA SeaWiFs satellite project collected during 2001 (http://seawifs.gsfc.nasa.gov/SEAWIFS.html). The nutrient data in (b), (c), and (d) are from 1994 World Ocean Atlas (Conkright *et al.*, 1994), and the maps are generated from the International Research Institute for Climate Prediction Data Library (http://ingrid.ldeo.columbia.edu). Comparison of (a) with (b), (c), and (d) demonstrates that phytoplankton abundance, at the coarsest scale, is driven by the availability of these nutrients, and these nutrients are generally most available in the polar ocean and along the equatorial upwellings where nutrient-rich deep water mixes more easily to the surface. However, among the nutrient-bearing high-latitude regions, there is not a good correlation between nutrient concentration and chlorophyll, indicating that other parameters come to limit productivity in these regions. The high-latitude ocean of the southern hemisphere, which has the highest nutrient concentrations, is known as the Southern Ocean. It is composed of the more polar Antarctic zone, where the silicate concentration is high and diatom productivity is extensive, and the more equatorward Subantarctic zone, where nitrate and phosphate remain high but silicate is scarce. The strong global correlation between phosphate and nitrate, originally recognized by Redfield (Redfield, 1942, 1958; Redfield *et al.*, 1963), allows us to group the two nutrients together when considering their internal cycling within the ocean, as they are exhausted in more or less the same regions. However, their input/output budgets may change over hundreds and thousands of years, and the relationship between these two nutrients may have been different in the past.

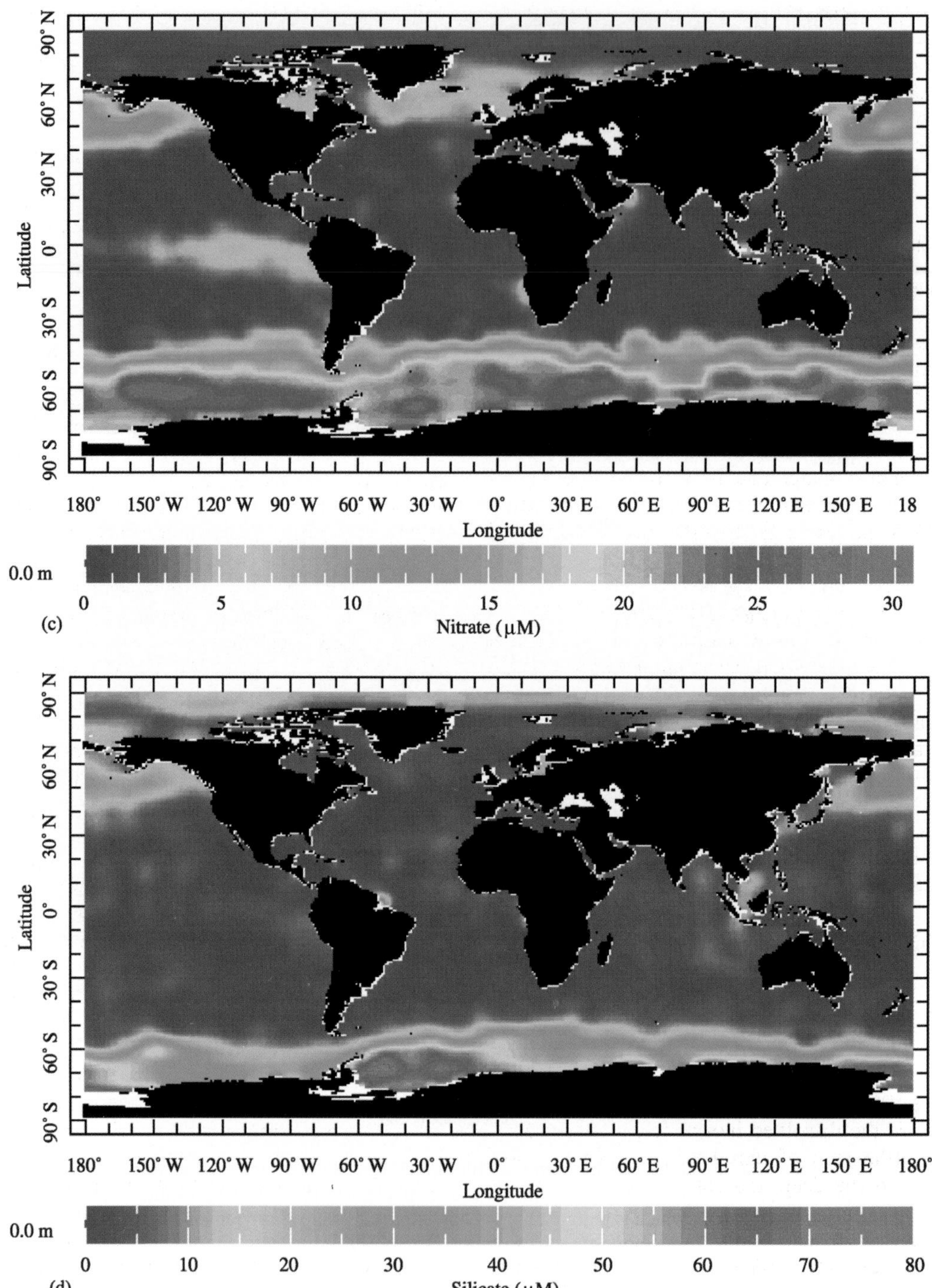

Figure 5 (continued).

For instance, nitrate and phosphate penetrate much further north than does silicate in the Southern Ocean, the polar ocean surrounding Antarctica (Figures 5(b)–(d)). As silicate is needed by diatoms but not other types of phytoplankton, this difference has major implications for the ecology of the Southern Ocean, present and past. Moreover, since surface water from 40–50° S is fed into the thermocline that supplies the low-latitude surface ocean with nutrients, the nutrient conditions of the Southern Ocean affect the nutrient supply and ecology of lower latitudes (Brzezinski *et al.*, 2002; Matsumoto *et al.*, 2002b). With regard to nitrate and phosphate (Figures 5(b) and (c)), a more careful comparison of their concentration variations would show lower nitrate-to-phosphate ratios in the Pacific than elsewhere, indicative of the nitrate loss that is occurring in that basin (Deutsch *et al.*, 2001; Gruber and Sarmiento, 1997). The input/output budget of the marine nitrogen cycle and its affect on the ocean nitrate reservoir arises as a matter of great importance in the section that follows.

6.18.2.1 Low- and Mid-latitude Ocean

We can describe the global biological pump in terms of a balance of fluxes between the sunlit surface ocean and the cold, dark, voluminous deep ocean (Figure 1). Phytoplankton consume CO_2 and nutrients (in particular, the major nutrients nitrogen and phosphorus) in the surface, and subsequent processing drives a downward rain of this organic matter that is degraded after it has sunk into the ocean interior, effectively extracting the CO_2 and nutrients from the surface ocean and lowering atmospheric CO_2. Balancing this downward flux of particles is the net upward flux of dissolved inorganic carbon and nutrients into the surface, due to a combination of diffuse vertical mixing over the entire ocean and focused vertical motion (upwelling and downwelling) in specific regions.

This steady state can be described by the following expressions (Broecker and Peng, 1982), which indicate a balance between the gross upward (left-hand side) and gross downward (right-hand side) transport of phosphorus (Equation (1)) and carbon (Equation (2)):

$$[PO_4^{3-}]_{deep} \times Q = [PO_4^{3-}]_{surface} \times Q + EP_P \quad (1)$$

$$[DIC]_{deep} \times Q = [DIC]_{surface} \times Q + (EP_P \times C/P_{EP}) \quad (2)$$

where $[PO_4^{3-}]$ is the phosphate concentration and [DIC] is the dissolved inorganic carbon concentration (the sum of dissolved CO_2, HCO_3^-, and CO_3^{2-}) in the deep and surface layers, Q is the exchange rate of water between the deep and surface, EP_P is the export production in terms of phosphorus, and C/P_{EP} is the carbon to phosphorus ratio of the export production. In these expressions, phosphorus is intended to represent the major nutrients in general; the distinctions between phosphorus and nitrogen will be considered later.

The biological pump is the biologically driven gradient in [DIC] between the surface and the deep sea ($[DIC]_{deep} - [DIC]_{surface}$). Because the deep ocean is a very large reservoir, increasing this gradient essentially means decreasing $[DIC]_{surface}$. Because the surface ocean equilibrates with the gases of the atmosphere, an increase in the gradient lowers atmospheric CO_2 as well. Solving Equation (2) for $[DIC]_{deep} - [DIC]_{surface}$ gives

$$[DIC]_{deep} - [DIC]_{surface} = (EP_P \times C/P_{EP})/Q \quad (3)$$

Solving Equation (1) for EP_P gives:

$$EP_P = ([PO_4^{3-}]_{deep} - [PO_4^{3-}]_{surface}) \times Q \quad (4)$$

Substitution of Equation (4) into Equation (3) for EP_P gives the following expression for $[DIC]_{deep} - [DIC]_{surface}$:

$$[DIC]_{deep} - [DIC]_{surface} = ([PO_4^{3-}]_{deep} - [PO_4^{3-}]_{surface}) \times C/P_{EP} \quad (5)$$

Thus, given the assumptions above, the surface–deep gradient in inorganic carbon concentration (the "strength" of the biological pump) is determined by (i) the C/P of the organic matter that is exported from the surface ocean, and (ii) the nutrient concentration ($[PO_4^{3-}]$) gradient between the surface and deep ocean.

In the low- and mid-latitude ocean, this balance is simplified by the nearly complete extraction of $[PO_4^{3-}]$ and $[NO_3^-]$ from the surface ocean (Figures 5(b) and (c)). The warm sunlit surface layer is separated by a strong temperature-driven density gradient ("thermocline") from the cold deep ocean, preventing rapid communication between the low-latitude ocean surface and the ocean interior. As a result, nutrient supply from below occurs slowly, and phytoplankton completely strip surface waters of available nitrogen and phosphorus; i.e., low- and mid-latitude phytoplankton growth is limited by the major nutrients nitrogen and phosphorus. If the low-latitude surface ocean is so strongly limited by the major nutrients that $[PO_4^{3-}]_{surface}$ would not be raised above zero for any imaginable increase in Q (i.e., if surface ocean plant productivity remains major-nutrient limited over a broad range of surface–deep mixing rates), then changes in upwelling and vertical mixing are balanced by equivalent changes in biological export production. In this case, changes in Q could cause changes in productivity over time, but these would not change the surface–deep gradient in [DIC] maintained by the biological pump.

If this is an adequate description of the real low-latitude ocean, then the strength of its biological pump (the amplitude of the surface–deep gradient in inorganic carbon concentration) is controlled solely by: (i) the carbon/nutrient ratio of the organic matter, and (ii) the nutrient content of the deep ocean (see Equation (5)). If the C/P ratio of export production increased or the $[PO_4^{3-}]$ of the deep ocean increased, then biological production at the surface would drive an increase in the downward flux of carbon that is not at first matched by any change in the upward flux of carbon dioxide associated with surface–deep mixing. A new balance between the downward and upward fluxes of carbon would eventually be reached as the [DIC] gradient between the surface and the deep increases, so that a given amount of mixing transports more CO_2 back into the surface ocean. This increase in surface–deep [DIC] gradient would be mostly due to a drop in dissolved inorganic carbon concentration of the surface ocean, which would lower the p_{CO_2} of the surface ocean and drive an associated decrease in atmospheric CO_2.

Biological production in the ocean tends to incorporate carbon, nitrogen, and phosphorus into biomass in the ratios of 106/16/1 (Redfield, 1942, 1958; Redfield *et al.*, 1963). These ratios determine the amount of inorganic carbon that is sequestered in the deep sea when the supply of nutrients to the surface is completely converted to export production, as is the case in the modern low- and mid-latitude ocean. It is unclear at this point how variable the C/N/P ratios of export production might be through time. When Broecker (1982a) first hypothesized a glacial increase in the strength of the biological pump as the driver of lower glacial CO_2 levels, he considered the possibility (suggested to him by Peter Weyl) that the C/P export production changes between ice ages and interglacials, with a higher C/P ratio during the last ice age, but he could find neither a mechanism nor direct paleoceanographic evidence for such a change.

Roughly the same situation persists today. We remain exceedingly uncertain about the robustness of the "Redfield ratios" through time and space. However, we do not yet have any mechanistic understanding of their variability (Falkowski, 2000), neither have we recognized a paleoceanographic archive that would help us to assess this variability. The organic matter preserved in the sediment record represents a minute and compositionally distinct fraction of the organic matter exported from the surface ocean. We have no reason to believe that variations in the elemental ratios of this residual sedimentary organic matter will reflect that of export production, especially in the face of changing seafloor conditions.

One of the persistent views of geochemists, to the frustration of their more biologically oriented colleagues, is that the complex recycling of elements in the low-latitude surface ocean has no major significance for the low-latitude biological pump. The argument is that, given the nutrient limitation that predominates in the low-latitude surface, the amount of carbon to be exported is determined by the supply of the major nutrients from the subsurface and the C/N/P uptake ratios of phytoplankton, regardless of the upper ocean recycling of a given nutrient element before export. However, the C/N/P ratios of export production may be altered by ecological effects on the relationship between phytoplankton biomass and the organic matter actually exported from the surface ocean, even without changes in the elemental requirements of the phytoplankton themselves. As a result, the ecology of the surface ocean could drive variations in the strength of the biological pump, for instance, by changing the C/P ratio of export production independent of the C/P ratio of phytoplankton, so as to change the amount of carbon exported for a given phosphorus supply. It was with this reasoning that Weyl originally justified the idea of glacial/interglacial changes in the C/P ratio of export production (Broecker, 1982a). Terrestrial biogeochemists and limnologists are actively seeking a quantitative theory for the ecological constraints imposed by elemental cycling (Elser *et al.*, 2000); oceanographers may benefit from a similar focus on the controls on elemental ratios, in particular, in export production. In any case, while changes in the chemical composition of export production may play a role in the variability of the ocean's biological pump, we are not yet in a position to posit specific hypotheses about such changes or to recognize them in the sedimentary record.

Broecker (1982b) and McElroy (1983) initially hypothesized that the oceanic reservoirs of phosphate and nitrate, respectively, might increase during glacial times, which would allow enhanced low-latitude biological production to lower atmospheric CO_2. However, it was recognized that this basic mechanism would require large changes in nutrient reservoirs to produce the entire observed amplitude of CO_2 change. According to a box-model calculation, the immediate effect of an increase in the nutrient reservoir is to extract CO_2 from the surface ocean and atmosphere, sequestering it in the deep sea, with a 30% increase in the oceanic nutrient reservoir driving a 30–40 ppmv CO_2 decrease (Table 1, see "closed system effects") (Sigman *et al.*, 1998). The longer-term effect on CO_2 from the ocean alkalinity balance depends greatly on whether $CaCO_3$ export increases in step with C_{org} export. It is not known whether a global increase in low-latitude production would increase the $CaCO_3$ rain in step with the C_{org} rain—this would probably depend on the circumstances of

Table 1 Atmospheric CO_2 effects of a 30% increase in the ocean nutrient inventory.

CO_2 changes (ppm)	*$CaCO_3$ production*	
	Proportional to C_{org}	*Constant*
Closed system effects	−34	−43
Open system effects	+17	−3
Total CO_2 change	−17	−46

the increase—so we should consider both possibilities.

If $CaCO_3$ export increases proportionately with C_{org} export (see "$CaCO_3$ production proportional to C_{org}" in Table 1), the increase in the $CaCO_3$ flux to the seafloor causes a gradual loss of alkalinity from the ocean, which works to raise atmospheric CO_2 (see "open system effects" in Table 1). As a result, only a 15–25 ppmv drop in CO_2 results from a 30% increase in the oceanic major nutrient reservoir (Table 1). However, if $CaCO_3$ production remains constant as C_{org} production increases (see "$CaCO_3$ production constant" in Table 1), the effects on the whole ocean reservoir of alkalinity are minimal (see "open system effects" in Table 1), and the increase in the oceanic nutrient reservoir is much more effective at lowering atmospheric CO_2. In this case, the box model calculation suggests that a 30% increase in the oceanic nutrient reservoir lowers atmospheric CO_2 by ~50 ppmv (Table 1), while some general circulation models would suggest declines in CO_2 of as much as 80 ppm (Archer *et al.*, 2000a). The required increase in the ocean nutrient reservoir could perhaps be lowered by a coincident decrease in $CaCO_3$ export that was modest enough not to violate observations of the glacial-age sedimentary $CaCO_3$ distribution (Archer and Maier-Reimer, 1994; Brzezinski *et al.*, 2002; Matsumoto *et al.*, 2002b; Sigman *et al.*, 1998). Nevertheless, a 30% increase in oceanic nutrients would require a dramatic change in ocean biogeochemistry.

Given a residence time of >6 thousand years for oceanic phosphate and the nature of the input/output terms in the oceanic phosphorus budget (Froelich *et al.*, 1982; Ruttenberg, 1993), it is difficult to imagine how oceanic phosphate could vary so as to cause the observed amplitude, rate, and phasing of atmospheric CO_2 change (Ganeshram *et al.*, 2002). Entering an ice age, Broecker (1982a) considered the possibility of a large phosphorus input from the weathering of shelf sediments, as ocean water is captured in ice sheets and sea level drops. Ice-core studies dispatched with this admittedly unlikely scenario by demonstrating that, going into an interglacial period, CO_2 begins to rise before sea level rises (Sowers and Bender, 1995), so that the phosphorus inventory decrease would occur too late to explain the rise of CO_2 into interglacials.

There is growing evidence that the nitrogen cycle is adequately dynamic to allow for a large change in the oceanic nitrate reservoir on glacial/interglacial timescales (Codispoti, 1995; Gruber and Sarmiento, 1997). Denitrification, the heterotrophic reduction of nitrate to N_2 gas, is the dominant loss term in the oceanic budget of "fixed" (or bioavailable) nitrogen. Nitrogen isotope studies in currently active regions of denitrification provide compelling evidence that water column denitrification was reduced in these regions during glacial periods (Altabet *et al.*, 1995; Ganeshram *et al.*, 1995, 2002; Pride *et al.*, 1999). In addition, it has been suggested that N_2 fixation, the synthesis of new fixed nitrogen from N_2 by cyanobacterial phytoplankton, was greater during glacial periods because of increased atmospheric supply of iron to the open ocean (iron being a central requirement of the enzymes for N_2 fixation) (Falkowski, 1997). Both of these changes, a decrease in water column denitrification and an increase in N_2 fixation, would have increased the oceanic nitrate reservoir. It has been suggested that such a nitrate reservoir increase would lead to significantly higher export production in the open ocean during glacial periods, potentially explaining glacial/interglacial atmospheric CO_2 changes (Broecker and Henderson, 1998; Falkowski, 1997; McElroy, 1983).

The major conceptual questions associated with this hypothesis involve the feedback of the nitrogen cycle and the strictness with which marine organisms must adhere to the Redfield ratios, both N/P and C/N. Oceanic N_2 fixation provides a mechanism by which the phytoplankton community can add fixed nitrogen to the ocean when conditions favor it. By contrast, riverine input of phosphorus, the major source of phosphate to the ocean, is not directly controlled by marine phytoplankton. Because phytoplankton have no way to produce biologically available phosphorus when there is none available, geochemists have traditionally considered phosphorus to be the fundamentally limiting major nutrient on glacial/interglacial timescales (Broecker and Peng, 1982). For a nitrate reservoir increase alone to drive an increase in low-latitude biological production, the nutrient requirements of the upper-ocean biota must be able to deviate from Redfield stoichiometry to adjust to changes in the N/P ratio of nutrient supply, so as to fully utilize the added nitrate in the absence of added phosphate. Alternatively, if this compensatory shift in biomass composition is incomplete, the surface ocean will tend to shift toward phosphate limitation, and an increase in export production will be prevented.

It was originally suggested by Redfield (Redfield *et al.*, 1963), based on analogy with freshwater systems, that N_2 fixers in the open ocean will enjoy greater competitive success under conditions of nitrate limitation but will be discouraged in the case of phosphate limitation. Having posited this sensitivity, Redfield hypothesized that N_2 fixation acts as a negative feedback on the nitrate reservoir, varying so as to prevent large changes in the nitrate reservoir that are not associated with a coincident change in the phosphate reservoir (Redfield *et al.*, 1963).

On the grand scale, this feedback must exist and must put some bounds on the degree to which the oceanic nitrate reservoir can vary independently from phosphorus. However, the quantitative constraint that this negative feedback places on the global nitrate reservoir is not yet known, with one suggested possibility being that iron is so important to N_2 fixers that changes in its supply can overpower the nitrate/phosphate sensitivity described by Redfield, leading to significant variations in the nitrate reservoir over time (Broecker and Henderson, 1998; Falkowski, 1997).

In summary, there are two components of the paradigm of phosphorus's control on the nitrate reservoir and the low-latitude biological pump, both stemming from Redfield's work: (i) oceanic N_2 fixation proceeds in phosphorus-bearing, nitrogen-poor environments, but not otherwise, and (ii) export production has a consistent C/N/P stoichiometry. The first statement would have to be inaccurate for the nitrate reservoir to vary independently of the phosphate reservoir, and the second statement would have to be inaccurate for such an independent change in the nitrate reservoir to actually cause a change in the biological pump. Partly as a reaction to studies that have considered the influence of other parameters, in particular, the supply of iron to open ocean N_2 fixation (Falkowski, 1997), some papers have restated the traditional Redfield-based paradigm (Ganeshram *et al.*, 2002; Tyrrell, 1999). However, the real challenge before us is to test and quantify the rigidity of the Redfield constraints relative to the influence of other environmental parameters. Biogeochemical studies of the modern ocean (Karl *et al.*, 1997; Sanudo-Wilhelmy *et al.*, 2001) and the paleoceanographic record (Haug *et al.*, 1998) are both likely to play a role in the evaluation of these questions regarding Redfield stoichiometry and N_2 fixation, which are, in turn, critical for the hypothesis of the low latitude biological pump as the driver of glacial/interglacial CO_2 change.

The mean concentration of O_2 in the ocean interior would be lowered in the glacial ocean by an increase in low-latitude carbon export, regardless of whether it is driven by an increase in the ocean phosphorus reservoir, the ocean nitrogen reservoir, or a change in the carbon/nutrient ratio of sinking organic matter. As described below, this makes deep-ocean $[O_2]$ a possible constraint on the strength and efficiency of the biological pump in the past. Here, however, we focus on this sensitivity of deep-ocean $[O_2]$ as part of an additional negative feedback in the ocean nitrogen cycle. This feedback may restrict the variability of the oceanic nitrate reservoir, even in the case that Redfield's N_2 fixation-based feedback on the ocean nitrogen cycle is weak and ineffective.

Denitrification occurs in environments that are deficient in O_2. An increase in the ocean's nitrate content will drive higher export production (neglecting, for the moment, the possibility of phosphate limitation). When this increased export production is oxidized in the ocean interior, it will drive more extensive O_2 deficiency. This will lead to a higher global rate of denitrification, which, in turn, will lower the ocean's nitrate content. Thus, the sensitivity of denitrification to the O_2 content of the ocean interior generates a hypothetical negative feedback that may, like the N_2 fixation-based feedback, work to stabilize the nitrate content of the ocean (Toggweiler and Carson, 1995).

While it is generally true that the major nutrients nitrate and phosphate are absent in low- and mid-latitude surface ocean, there are important exceptions (Figures 5(b) and (c)). Wind-driven upwelling leads to nonzero nutrient concentrations and high biological productivity at the surface along the equator and the eastern margins of the ocean basins (e.g., off of Peru, California, and western Africa). Why have we given such short shrift to these biologically dynamic areas in our discussion of the low- and mid-latitude biological pump?

While critically important for ocean ecosystems and potentially important for interannual variations in atmospheric CO_2, the nutrient status in equatorial and coastal upwelling systems does not greatly affect atmospheric CO_2 on centennial or millennial timescales. Above, we demonstrated that greater vertical exchange between the nutrient-poor surface ocean and the deep sea causes higher export production but does not affect the [DIC] gradient between the surface and the deep sea, so that it would not drive a change in atmospheric CO_2. Following this same reasoning, low-latitude upwellings may generate much export production, but they are not millennial-scale sinks for atmospheric CO_2 because the upwelling brings up both nutrients and respiratory CO_2 from the ocean interior. Neither do the upwellings drive a net loss of CO_2 from the ocean, because the high-nutrient surface waters of the low-latitude upwellings do not contribute appreciably to the ventilation of the ocean subsurface. Put another way, the nutrients supplied to the surface are eventually consumed as the nutrient-bearing surface water flows away from the site of upwelling, before the water has an opportunity to descend back into the ocean interior. This pattern is evident in surface ocean p_{CO_2} data (Takahashi *et al.*, 1997). For instance, in the core of equatorial Pacific upwelling, p_{CO_2} is high both because the water is warming and because deeply sequestered metabolic CO_2 is brought to the surface and evades back to the atmosphere. Adjacent to the equatorial upwelling, the tropical and subtropical Pacific is a region of low p_{CO_2} which is associated with the consumption of the excess surface nutrients that have escaped consumption at the site of upwelling. It does not matter for atmospheric CO_2 whether the upwelled nutrients are converted to export production on-axis of the upwelling system or further off-axis. Thus, while changes in upwelling, biological production, and nutrient status are central to the history of the low-latitude ocean, they have limited importance for atmospheric CO_2 changes on glacial/interglacial timescales. Increased export production driven by higher rates of vertical mixing or upwelling would only play a role in lowering atmospheric CO_2 if it caused a change in the chemical composition of the exported organic material, such as a decrease in its $CaCO_3/C_{org}$ ratio or an increase in the mean depth at which the organic rain is metabolized in the ocean interior (e.g., Boyle, 1988b; Dymond and Lyle, 1985).

The surface chlorophyll distribution shows that coastal environments, even in regions without upwelling, are among the most highly productive in the ocean (Figure 5(a)). However, it is believed that coastal ocean productivity does not have a dominant effect on atmospheric CO_2 over centuries and millennia. As biomass does not accumulate significantly in the ocean, any long-term excess in net primary production relative to heterotrophy must lead to export of organic matter out of the surface layer, in particular, as a downward rain of biogenic particles. In the case of the coastal ocean, this export is stopped by the shallow seafloor, frequently within the depth range that mixes actively with the surface ocean. The high net primary productivity of the coastal ocean owes much to the presence of the shallow seafloor, as the nutrients released from the degradation of settled organic matter are available to the phytoplankton in the sunlit surface. Yet, just as the seafloor prevents the loss of nutrients from the coastal surface ocean, so too does it prevent the export of carbon. Almost as soon as the settled organic carbon is respired, the CO_2 product is free to diffuse back into the atmosphere. Were the same export to have occurred over the open ocean, it would have effectively sequestered the carbon in that biogenic rain within the ocean interior for roughly a thousand years. Thus, the shallow seafloor of the ocean margin leads to high primary productivity while at the same time limiting its effect on the carbon cycle over the timescale of centuries and millennia. It should be noted, however, that the ocean margins are critically important in the global carbon cycle on the timescale of millions of years, due to their role in organic carbon burial.

It has been hypothesized that a significant fraction of the organic matter raining onto the shallow margin is swept over the shelf/slope break, thus transporting organic matter into the ocean interior (Walsh *et al.*, 1981). This has been interpreted as rendering ocean margin productivity important to atmospheric CO_2 on millennial timescales, in that the organic matter being swept off the shelf would be an important route by which CO_2 is shuttled into the deep sea. However, this view should be considered carefully. The ocean margins have high net primary production because of nutrient recycling, but the recycled nutrients were produced by the decomposition of organic matter that would have otherwise been exported out of the system. Mass balance dictates that export of organic carbon from the coastal surface ocean, like export production in the open ocean surface, is set by the net supply of nutrients to the system. If a low-latitude ocean margin receives most of its net nutrient supply from the deep sea, then it is no different than the neighboring open ocean in the quantity of carbon that it can export. There is one caveat to this argument: if organic matter deposited on the shelf can have its nutrients stripped out without oxidizing the carbon back to CO_2 and this "depleted" organic carbon is then transported off the shelf and into the ocean interior, then ocean margin productivity, having been freed from the Redfield constraint on the elemental composition of its export production, could drive a greater amount of organic carbon export for a given amount of nutrient supply.

6.18.2.2 High-latitude Ocean

Broecker's (Broecker, 1981, 1982a,b; Broecker and Peng, 1982) initial ideas about the biological pump revolutionized the field of chemical oceanography. However, soon after their description, several groups demonstrated that his focus on the low-latitude ocean missed important aspects of the ocean carbon cycle. The thermocline outcrops at subpolar latitudes, allowing deep waters more ready access to the surface in the polar regions. In these regions, the nutrient-rich and CO_2-charged waters of the deep ocean are exposed to the atmosphere and returned to the subsurface before the available nutrients are fully utilized by phytoplankton for carbon fixation. This incomplete utilization of the major nutrients allows for the leakage of deeply sequestered CO_2 back into the atmosphere, raising the atmospheric p_{CO_2}. Work is ongoing to understand what limits phytoplankton growth in these high-latitude regions. Both light and trace metals such as iron are scarce commodities in these regions and together probably represent the dominant controls on polar productivity, with light increasing in importance toward the poles due to the combined effects of low irradiance, sea ice coverage, and deep vertical mixing (see Chapters 6.02, 6.04, and 6.10). The Southern Ocean, which surrounds Antarctica, holds the largest amount of unused surface nutrients (Figures 5(b)–(d)), yet the surface chlorophyll suggests that it is perhaps the least productive of the polar oceans; iron and light probably both play a role in explaining this pattern (Martin *et al.*, 1990; Mitchell *et al.*, 1991).

The efficiency of the global biological pump with respect to the ocean's major nutrient content is determined by the nutrient status of the polar regions that ventilate the deep sea, or more specifically, the nutrient concentration of the new water that enters the deep sea from polar regions (Figure 6). Deep water is enriched in nutrients and CO_2 because of the low- and mid-latitude biological pump, which sequesters both nutrients and inorganic carbon in subsurface. In high-latitude regions such as the Antarctic, the exposure of this CO_2-rich deep water at the surface releases this sequestered CO_2 to the atmosphere. However, the net uptake of nutrients and CO_2 in the formation of phytoplankton biomass and the eventual export of organic matter subsequently lowers the p_{CO_2} of surface waters, causing the surface layer to reabsorb a portion of the CO_2 that was originally lost from the upwelled water. Thus, the ratio between export production and the ventilation of CO_2-rich subsurface water, not the absolute rate of either process alone, controls the exchange of CO_2 between the atmosphere and high-latitude surface ocean. The nutrient concentration of water at the time that it enters the subsurface is referred to as its "preformed" nutrient concentration. The preformed nutrient concentration of subsurface water provides an indicator of the cumulative nutrient utilization that it underwent while at the surface, with a higher preformed nutrient concentration indicating lower cumulative nutrient utilization at the surface and thus a greater leak in the biological carbon pump. This picture overlooks a number of important facts. For instance, ocean/atmosphere CO_2 exchange is not instantaneous, so water can sink before it has reached CO_2 saturation with respect to the atmosphere. In polar regions, this typically means that deep water may leave the surface before it has lost as much CO_2 to the atmosphere as it might have (Stephens and Keeling, 2000).

The global efficiency of the biological pump can be evaluated in terms of a global mean preformed nutrient concentration for surface water that is folded into the ocean interior. In the calculation of this mean preformed nutrient concentration, each subsurface water formation term is weighted according to the volume of the ocean that it ventilates because this corresponds to the amount of CO_2 that could be sequestered in a given portion of the ocean interior. As a result, the nutrient status of surface ocean regions that directly ventilate the ocean subsurface (deep, intermediate, or thermocline waters) have special importance for atmospheric CO_2.

The high-latitude North Atlantic and the Southern Ocean are the two regions that appear to dominate the ventilation of the modern deep ocean. Through the preformed nutrient concentrations of the newly formed deep waters, their competition to fill the deep ocean largely determines the net efficiency of the global biological pump (Toggweiler *et al.*, 2003b). Of these two regions, the North Atlantic has a low preformed nutrient concentration and thus drives the ocean toward a high efficiency for the biological pump. In contrast, the preformed nutrient concentration of Southern Ocean-sourced deep water is high and thus drives the global biological pump toward a low efficiency.

Within the polar oceans involved in deep water formation, certain regions are more important than others. The Antarctic Zone, the most polar region in the Southern Ocean, is involved in the formation of both deep and intermediate-depth waters, making this region important to the atmosphere/ocean CO_2 balance. The quantitative effect of the Subantarctic Zone on atmospheric CO_2 is less certain, depending on the degree to which the nutrient status of the Subantarctic surface influences the preformed nutrient concentration of newly formed subsurface water (Antarctic Intermediate Water and Subantarctic mode water), but its significance is probably much less than that of the Antarctic.

Questions regarding the most critical regions extend to even smaller spatial scales. For instance, with regard to deep-ocean ventilation in the Antarctic, it is uncertain as to whether the entire Antarctic surface plays an important role in the CO_2 balance, or whether instead only the specific locations of active deep water formation (e.g., the Weddell Sea shelf) are relevant. This question hinges largely on whether the surface water in the region of deep water formation exchanges with the open Antarctic surface, or whether it instead comes directly from the shallow subsurface and remains isolated until sinking. If the latter is true, then the open Antarctic may be irrelevant to the chemistry of the newly formed deep water. This would be problematic for paleoceanographers, since essentially all of the reliable paleoceanographic records come from outside these rather special environments. There is, however, a new twist to this question. While deep water formation in the modern Antarctic has long been thought of as restricted to the shelves, efforts to reproduce

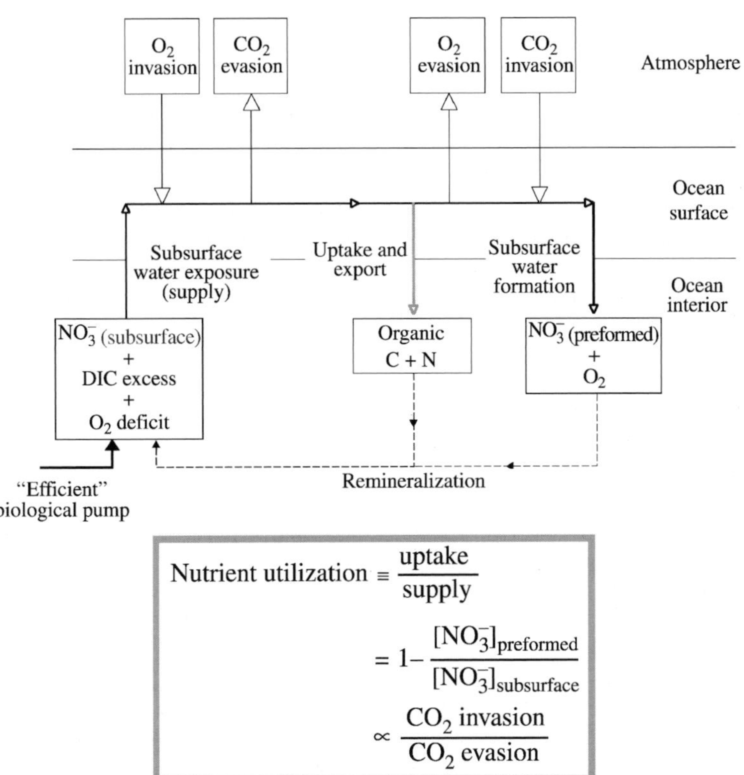

$$\text{Nutrient utilization} \equiv \frac{\text{uptake}}{\text{supply}} = 1 - \frac{[NO_3^-]_{preformed}}{[NO_3^-]_{subsurface}} \propto \frac{CO_2\ \text{invasion}}{CO_2\ \text{evasion}}$$

Figure 6 The effect on atmospheric CO_2 of the biological pump in a region of deep-ocean ventilation where the major nutrients are not completely consumed, such as the Antarctic zone of the Southern Ocean. In this figure, NO_3^- is intended to represent the major nutrients and thus should not be distinguished conceptually from PO_4^{3-}. The biological pump causes deep water to have dissolved inorganic carbon (DIC) in excess of its "preformed" [DIC] (the concentration of DIC in the water when it left the surface; see lower left box, which represents water in the ocean interior). The low- and mid-latitudes (and a few polar regions, such as the North Atlantic) house a biological pump that is "efficient" in that it consumes all or most of the major nutrient supply. As a result, it drives the DIC excess toward a concentration equivalent to the deep $[NO_3^-]$ multiplied by the C/N ratio of exported organic matter (left lower arrow represents this influence on deep-water chemistry). The respiration of organic matter in the ocean interior also leads deep water to have an O_2 deficit relative to saturation. In high-latitude regions such as the Antarctic, the exposure of this deep water at the surface releases this sequestered CO_2 to the atmosphere and leads to the uptake of O_2 from the atmosphere (upper left). However, the net uptake of nutrients and CO_2 in the formation of phytoplankton biomass and the eventual export of organic matter ("uptake and export") subsequently lowers the p_{CO_2} of surface waters, causing the surface layer to reabsorb a portion of the CO_2 that was originally lost from the upwelled water (upper right). The net excess in photosynthesis to respiration that drives this export out of the surface also produces O_2, some of which evades to the atmosphere, partially offsetting the initial O_2 uptake by the surface ocean. While the down-going water has lost its DIC excess by exchange with the atmosphere, a portion of the original DIC excess will be reintroduced in the subsurface when the exported organic matter is remineralized (the more complete the nutrient consumption in the surface, the greater the DIC excess that is reintroduced in the subsurface). The ratio between export and the ventilation of CO_2-rich subsurface water, which controls the net flux of CO_2 between the atmosphere and surface ocean in this region, can be related to the nutrients. Nutrient utilization, defined as the ratio of the rate of nutrient uptake (and export) to the rate of nutrient supply, expresses the difference between nutrient concentration of rising and sinking water. Nutrient utilization relates directly (through the Redfield ratios) to the ratio of CO_2 invasion (driven by organic matter export) relative to evasion (driven by deep-water exposure at the surface). The lower the nutrient utilization, the lower the ratio of invasion to evasion and thus the more this region represents a leak in the biological pump. This diagram overlooks a number of important facts. For instance, ocean/atmosphere CO_2 exchange is not instantaneous, so water can sink before it has reached CO_2 saturation with respect to the atmosphere.

deep-sea nutrient chemistry and radiochemistry raise the possibility that the deep ocean may also be ventilating in a more diffuse mode throughout the open Antarctic (Broecker *et al.*, 1998; Peacock, 2001). If this is the case, the significance of the open Antarctic in the CO_2 balance is less susceptible to uncertainties about lateral exchange with the shelf sites of deep water formation. In the discussion that follows, we assume that the open Antarctic does affect the CO_2 exchanges associated with deep water formation, either by active exchange of surface water with the specific sites of

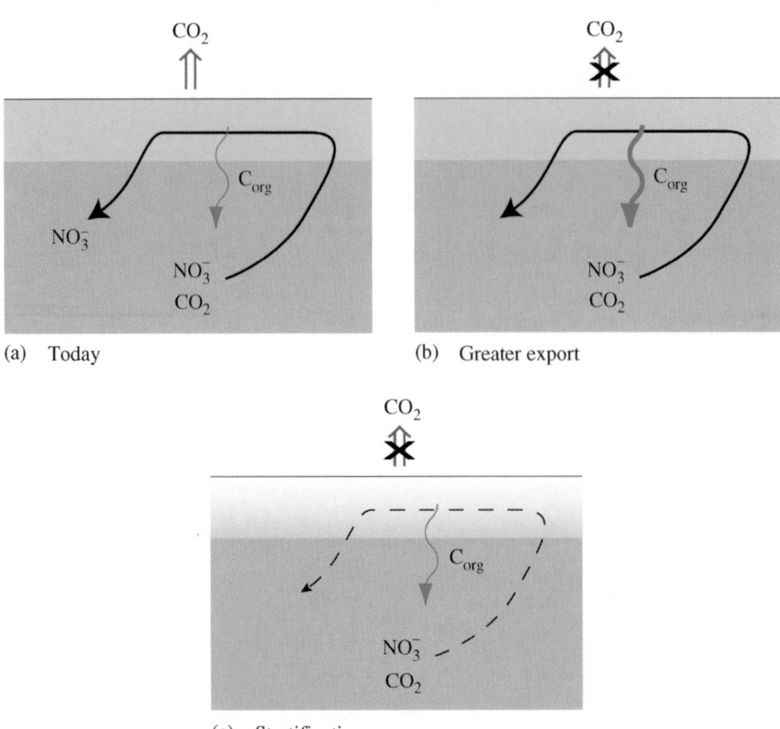

Figure 7 Schematic illustrations of the CO_2 leak from the Southern Ocean to the atmospheric that exists today and of two alternative scenarios by which this leak might have been stopped during ice ages. The incomplete consumption of nutrients in the modern Southern Ocean causes waters that come to the surface in the Southern Ocean to release CO_2 to the atmosphere because not all of the nutrient supply to the surface drives a downward rain of organic material, allowing CO_2 sequestered by the lower latitude biological pump to escape (a). During ice ages, atmospheric CO_2 could have been reduced by increasing the nutrient efficiency of the biological pump in the Southern Ocean (b and c). This could have been driven by an increase in biological productivity (b), which would have actively increased the downward flux of carbon, or by density "stratificaton" of the upper ocean, allowing less nutrient- and CO_2-rich water to the Southern Ocean surface (c). As described in Figure 8, because of the existence of other regions of deep water formation, the evasion of CO_2 is reduced upon stratification (c) even if the degree of nutrient consumption in the surface does not change.

deep water formation or by its actual participation in the ventilation process. In any case, we must improve our understanding of the modern physical and chemical oceanography of the polar regions if we are to determine their roles in atmospheric CO_2 change.

In their hypothesis of a glacial decrease in Southern Ocean CO_2 leak, early workers considered two causes: (i) an increase in biological export production and (ii) a decrease in the exposure rate of deep waters at the polar surface (Figure 7). One can imagine processes that would have reduced the evasion of CO_2 from the Southern Ocean by either of the two mechanisms mentioned above. For instance, an increase in the input of dust and its associated trace metals to the Southern Ocean might have driven an increase in the rate of nutrient and carbon uptake by phytoplankton (Martin, 1990). Alternatively, an increase in the salinity-driven stratification of the Antarctic and/or a decrease in wind-driven upwelling could have lowered the rate of nutrient and carbon dioxide supply to the surface (discussed further in Section 6.18.3) (Francois *et al.*, 1997; Keeling and Visbeck, 2001; Sigman and Boyle, 2000, 2001; Toggweiler and Samuels, 1995). Under most conditions, these two hypothesized changes do not have the same quantitative effect on atmospheric CO_2. For instance, stratification can cause a significantly larger amount of CO_2 decrease than does increased export production for a given amount of nutrient drawdown in Southern Ocean surface waters (Sigman *et al.*, 1999b).

These differences can be understood in the context of the mean preformed nutrient concentration of the ocean interior (Figure 8). There are two Southern Ocean mechanisms by which the global efficiency of the biological pump may be increased during glacial times: (i) decreasing the preformed nutrient concentration of Southern-Ocean-sourced deep water (Figures 8(c) and (d)), and (ii) decreasing the importance of Southern-Ocean-sourced deep waters in the ventilation of the deep ocean, relative to the North Atlantic or some other source of deep water with low

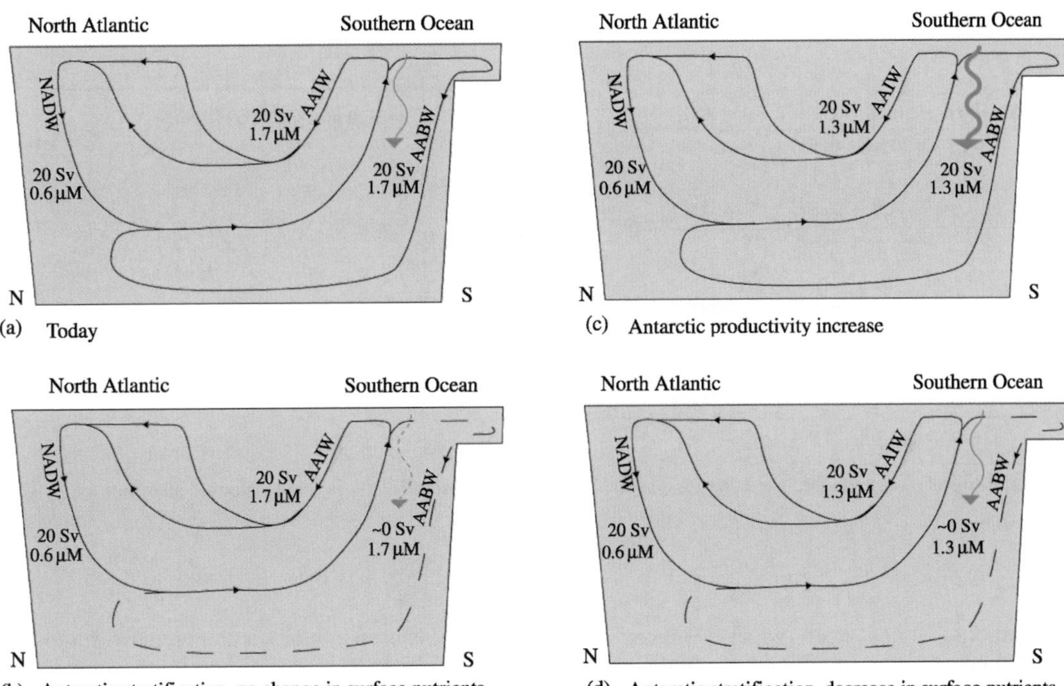

Figure 8 Cartoons depicting the ventilation of the ocean interior by the North Atlantic and Southern Ocean and the preformed [PO_4^{3-}] of the different ventilation terms, for today (a) and three hypothesized ice-age conditions ((b), (c), and (d)); calculation of the mean preformed [PO_4^{3-}] of the ocean interior for each of these cases (e); and the effects of the hypothesized glacial changes (b), (c), and (d) on atmospheric CO_2 compared with the change in mean preformed [PO_4^{3-}], as calculated by the CYCLOPS ocean geochemistry box model (f) (Keir, 1988; Sigman *et al.*, 1998). In the interglacial case (a), there are three sources of subsurface water shown: North Atlantic Deep Water (NADW), Antarctic Intermediate Water (AAIW), and Antarctic Bottom Water (AABW). In this simple picture, NADW and AAIW formation are related; NADW is drawn into the Antarctic surface by the wind-driven ("Ekman") divergence of surface waters and then forms AAIW because of surface-water convergence further north (Gordon *et al.*, 1977; Toggweiler and Samuels, 1995). The rates of formation (in Sverdrups, $10^6\ m^3\ s^{-1}$) and preformed [PO_4^{3-}] (in μM) are from the CYCLOPS model. In (b), stratification is assumed to shut off AABW formation. Because we do not change NADW, we are forced to leave AAIW unchanged as well. In (b), export production in the Antarctic decreases proportionally with the decrease in upwelling to the surface, such that surface [PO_4^{3-}] remains at 1.7 μM. This would be the expected outcome if, for instance, the Antarctic is iron limited and most of the iron comes from upwelling, during both interglacials and ice ages. In (d), the same circulation change occurs as in (b), but in this case export production does not decrease as upwelling decreases, so that surface [PO_4^{3-}] drops to ~1.3 μM. This would be the expected outcome if, for instance, the Antarctic iron is supplemented by a high dust input during ice ages. In (c), circulation remains constant but Antarctic export production increases (as indicated by the increase in the downward green arrow). This would be the expected outcome if, for instance, the combined input of iron from dust supply and deep water is greater during ice ages (Martin, 1990; Watson *et al.*, 2000). In (e), the mean preformed [PO_4^{3-}] of the ocean interior is calculated as the total phosphate input to the ocean interior divided by the ventilation rate (North Atlantic terms on the left, Antarctic terms on the right). In (f), the calculated change in atmospheric p_{CO_2} (relative to the interglacial case (a)) is plotted against the mean preformed [PO_4^{3-}] of each case, as calculated in (e). The strong correlation between these two parameters (a 170 ppm decrease in CO_2 for a 1 μM decrease in [PO_4^{3-}]) indicates that the mean preformed [PO_4^{3-}] is an excellent predictor for the effect of a given change on the strength of the biological pump. In these experiments, the mean gas exchange temperature for the ocean interior was held constant, so that no change in the "solubility pump" was allowed (see text). In addition, calcium carbonate dynamics are included in these model experiments, so that a fraction of the CO_2 response (~25%) is not the result of the biological pump, in the most strict sense. Finally, none of these experiments address the evidence that NADW formation was reduced during glacial times (Boyle and Keigwin, 1982).

preformed nutrients (Figures 8(b) and (d)). Increasing surface productivity would lower the preformed nutrient concentration of the deep sea solely by reducing the preformed nutrient concentration of the Southern Ocean contribution to deep water (Figure 8(c)). Stratification of the Southern Ocean surface would reduce the fraction of the deep ocean ventilated by the Southern Ocean, allowing the ocean interior to migrate toward the lower preformed nutrient concentration

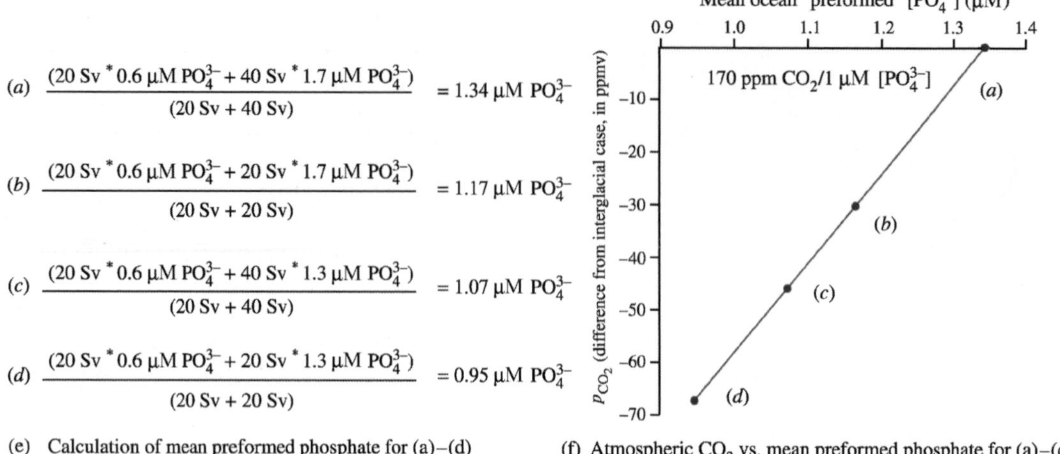

Figure 8 (continued).

of northern-sourced deep water (Figure 8(b)). In addition, if upon stratification phytoplankton growth rate does not decrease as much as does the supply of nutrients to the surface, then stratification would also lower the preformed nutrient concentration of southern-sourced deep waters (Figure 8(d)). In this case, stratification lowers the preformed nutrient concentration by both (i) and (ii) above, leading to a lower net preformed nutrient concentration for the ocean interior and thus making it a more potent mechanism for lowering atmospheric CO_2. Of course, there are important additional considerations. First, a decrease in Southern Ocean deep water might allow the deep-sea temperature to rise, which would push CO_2 back into the atmosphere. This apparently did not occur during glacial times (Schrag *et al.*, 1996), suggesting that all glacial-age deep-water sources were very cold. Second, and more importantly, North Atlantic deep water formation may have been weaker during the last ice age (Boyle and Keigwin, 1982) (see Chapter 6.16). As this deep-water source is the low preformed nutrient end-member today, its reduction would have worked to increase the preformed nutrient concentration of the deep ocean during the last ice age, potentially offsetting the effects of a decrease in Southern-sourced deep water. These questions about actual events aside, the mean preformed nutrient concentration of the deep ocean provides an important predictive index for the effect of changes in the polar ocean on atmospheric CO_2.

Some attention has focused on the prevention of CO_2 release from the Antarctic surface by sea ice cover as a potential driver of glacial/interglacial CO_2 change (Stephens and Keeling, 2000). Though the mechanism is essentially physical, it nevertheless would cause an increase in the global efficiency of the biological pump by removing its polar leak. This is an important reminder that the biological pump is not solely a biological process, but rather arises from the interaction between the biology and physics of the ocean. For prevention of CO_2 release by sea ice cover to be the sole mechanism for glacial/interglacial CO_2 change, nearly complete year-round ice coverage of the Antarctic would be required. This seems unlikely to most investigators (Crosta *et al.*, 1998). As a result, a hybrid hypothesis has been proposed for the glacial Antarctic, in which intense surface stratification and nutrient consumption during the summer was followed by wintertime prevention of gas exchange due to ice cover (Moore *et al.*, 2000). While promising, this mechanism faces a discrepancy with the evidence for lower export production in the glacial Antarctic (see below). In the modern Antarctic, nutrients are supplied to the surface largely by wintertime vertical mixing. Thus, without year-round stratification, higher annual export production would have been required to lower the surface nutrient concentration below current summertime levels, even for a brief summer period. An alternative combined gas exchange limitation/stratification mechanism that does not violate the productivity constraint is that of spatial segregation of the two mechanisms within the Antarctic (Figure 9). Near the Antarctic continental margin, the formation of thick ice cover and its associated salt rejection could drive overturning, but the near-complete ice cover in that region could prevent CO_2 release. The ice formed along the coast would eventually drift into the open Antarctic, where it would melt, thus stratifying the water column and preventing the release of CO_2 from open Antarctic as well. In any case, ice cover provides an alternative or

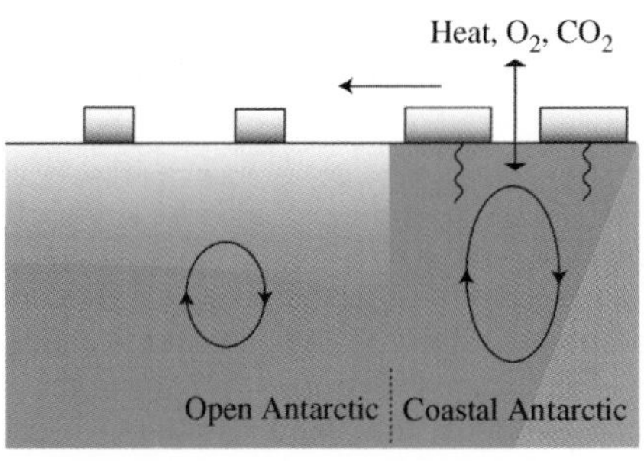

(a) Today

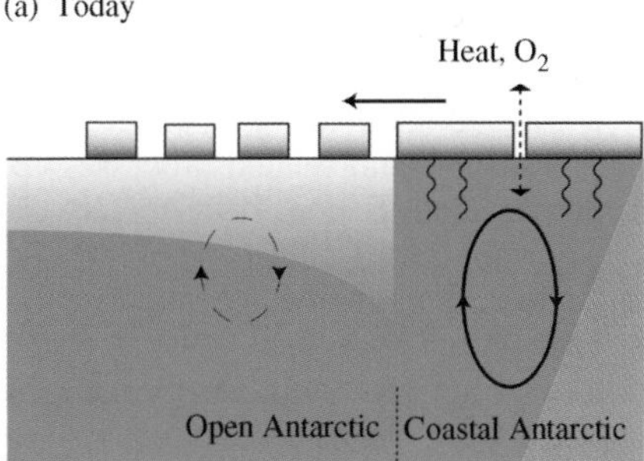

(b) Hybrid sea ice/stratification hypothesis

Figure 9 North-South (left-to-right) depth section of the shallow Antarctic, (a) under modern conditions, and (b) under glacial conditions as posited by a hybrid hypothesis involving both limitation of gas exchange by sea ice near the Antarctic continent and sea ice melt-driven stratification of the open Antarctic. During glacial times, sea ice formation was likely vigorous in the Antarctic, and the salt rejection during freezing would have worked to increase the density of the surface, perhaps driving vigorous overturning (bold loop close to the Antarctic continent) (Keeling and Stephens, 2001). While this alone would act to ventilate the deep sea and release CO_2 to the atmosphere, the ice coverage may have been adequately dense to prevent CO_2 outgassing, which occurs more slowly than the exchange of heat and O_2 (Stephens and Keeling, 2000). Here, we posit that this region of ice formation and ventilation was limited to the waters close to the Antarctic continent. The winds in the Antarctic would have drawn the ice northward, and some of it would have melted in the open Antarctic. This would have worked to strengthen the vertical stratification in the open Antarctic, which may have reduced CO_2 loss from the ocean in this region (Francois *et al.*, 1997). Depending on the predominant route of deep-ocean ventilation in the modern Antarctic (coastal or open ocean, see text), either the coastal sea ice coverage or the open Antarctic stratification may be taken as the most important limitation on CO_2 release. This hybrid hypothesis has a significant list of requirements in order to be feasible. For instance, the surface waters in the hypothesized region of convection must not mix laterally with the surface waters of the open Antarctic.

additional potential mechanism for preventing the release of biologically sequestered CO_2 from the Antarctic, and future paleoceanographic work must try to provide constraints on its actual importance.

6.18.3 TOOLS

While there are many aspects of biological productivity that one might hope to reconstruct through time, there are two parameters that are most fundamental to the biological pump: (i) export production and (ii) nutrient status. In regions where a deviation from major-nutrient limitation is highly unlikely, in particular, the tropical and subtropical ocean, export production is the major regional constraint that local paleoceanographic data can provide. For instance, if export production in the tropics was higher some time in the past, then there are at least two plausible explanations: (i) there was more rapid water exchange between the surface and the nutrient-bearing subsurface water, or (ii) there was an increase in the nutrient content of the subsurface water (and/or in N_2 fixation in the surface; see discussion above). While physically driven changes in nutrient supply (e.g., changes in upwelling intensity) have limited significance for the biological pump, changes in deep-ocean nutrient content could drive a strengthening of the pump and thus an associated decline in atmospheric CO_2 (see discussion above). Thus, it is critical for studies of the biological pump not only to reconstruct changes in low-latitude export production but also to determine what caused the changes.

In the polar ocean, both export production and nutrient status must be known to determine the impact on atmospheric CO_2, because both of these terms are needed to determine the ratio of CO_2 supply from deep water to CO_2 sequestration by export production (Figure 7). For instance, a decrease in export production associated with an increase in surface nutrients would imply an increased leak in the biological pump, whereas a decrease in export production associated with reduced nutrient availability would imply a smaller leak in the pump. In low-latitude regions of upwelling, nutrient status is less important from the perspective of the global biological pump. Nevertheless, since nutrient status is potentially variable in these environments, it must be constrained to develop a tractable list of explanations for an observed change in productivity.

A great many approaches have been taken to study the biological pump in the past. While there are approaches that take advantage of virtually every aspect of the available geologic archives, we focus here on tools to reconstruct export production and nutrient status and limit ourselves to those that are largely based on geochemical measurement. We refer the interested reader to Fischer and Wefer (1999) for additional information and references.

6.18.3.1 Export Production

Export production, as defined above, is the flux of organic carbon that sinks (or is mixed) out of the surface ocean and into the deep sea. While it is extremely difficult to imagine how paleoceanographic measurements can provide direct constraints on this highly specific parameter, approaches exist for the reconstruction of the flux of biogenic debris to the seafloor. To the degree that sinking biogenic debris has a predictable chemical composition and that its export out of the surface ocean is correlated with its rain rate to the seafloor, these approaches can provide insight into export production variations (Muller and Suess, 1979; Ruhlemann *et al.*, 1999; Sarnthein *et al.*, 1988) (see Chapter 6.04).

Only a very small fraction of the organic carbon exported out of the surface ocean accumulates in deep-sea sediments, with loss occurring both in the water column and at the seafloor. Moreover, various environmental conditions may influence the fraction of export production preserved in the sedimentary record (see Chapter 6.11). The mineral components of the biogenic rain (calcium carbonate, opal) do not represent a source of chemical energy to the benthos and thus may be preserved in a more predictable fashion in deep-sea sediments. Thus, paleoceanographers sometimes hope to reconstruct export production (the flux of organic carbon out of the surface ocean) from the biogenic rain of opal or carbonate to the seafloor (Ruhlemann *et al.*, 1999). We know that systematic variations in the ratio of opal or calcium carbonate to organic carbon in export production do occur, so there are many cases where changes in the mineral flux can be interpreted either as a change in the rate of export production or in its composition. Despite these limitations, we are better off with this information than without it.

The most basic strategy to reconstruct the biogenic rain to the seafloor is to measure accumulation rate in the sediments. This approach is appropriate for materials that accumulate without loss in the sediments or that are preserved to a constant or predictable degree. For the biogenic components of interest, organic carbon, opal, and calcium carbonate, the degree of preservation varies with diverse environmental variables, detracting from the usefulness of accumulation rate for the reconstruction of their rain to the seafloor. Nevertheless, because the environmental variables controlling preservation in the sediments are often linked to rain rate to the seafloor, accumulation rate can sometimes be a sound basis for at least the qualitative reconstruction of changes in some components of the biogenic rain. For instance, the fraction of the opal rain that is preserved in the sediment appears to be higher in opal-rich environments (Broecker and Peng, 1982). As a result, an observed increase in sedimentary opal accumulation rate may have been partially due to an increase in preservation. However, an increase in opal rain (and thus diatom-driven export) would have been needed to increase the sedimentary opal content in the first place.

If an age model can be developed for a deep-sea sediment core, then the accumulation histories of the various sediment components can be reconstructed. Assuming that the age model is correct, the main uncertainty in the environmental significance of the reconstruction is then the potential for lateral sediment transport. Sediments can be winnowed from or focused to a given site on the seafloor.

The geochemistry of radiogenic thorium provides a way to evaluate the effect of lateral sediment transport and related processes (Bacon, 1984; Suman and Bacon, 1989) (see Chapter 6.09). Thorium-230 (^{230}Th) is produced at a constant rate throughout the oceanic water column by the decay of dissolved ^{234}U. As ^{230}Th is produced, it is almost completely scavenged onto particles at the site of its production. As a result, the accumulation rate of ^{230}Th in deep-sea sediments should match its integrated production in the overlying water column. If the accumulation of ^{230}Th over a time period, as defined by a sediment age model, is more or less than should have been produced in the overlying water column during that period, then sediment is being focused or winnowed, respectively.

Thorium-230 is also of great use as an independent constraint on the flux of biogenic material to the seafloor (Bacon, 1984; Suman and Bacon, 1989). Because the production rate of ^{230}Th in the water column is essentially constant over space and time, its concentration in sinking particles is diluted in high-flux environments with a large biogenic flux to the seafloor, yielding lower sedimentary concentrations of ^{230}Th in these environments. One limitation on the use of ^{230}Th is the radioactive decay of this isotope, which has a half-life of 7.5×10^4 yr. Helium-3 can apparently be used in similar ways to ^{230}Th and has the advantage that it is a stable isotope (Marcantonio *et al.*, 1996); however, it has been studied less than ^{230}Th.

The addition of other scavenged elements further enhances the utility of ^{230}Th. ^{231}Pa, like ^{230}Th, is produced throughout the ocean water column by radioactive decay of a uranium isotope (^{235}U in the case of ^{231}Pa). However, protactinium is somewhat less particle-reactive than thorium. As a result, the downward flux of ^{231}Pa to the seafloor varies across the ocean. Protactinium from low-flux regions mixes into high-flux regions to be scavenged, resulting in higher ^{231}Pa fluxes in environments with high biogenic rain rates. Thus, the ^{231}Pa/^{230}Th ratio of sediments has been studied as an index of the particle flux through the water column. This index is potentially complementary to thorium-normalized accumulation rates in that the ^{231}Pa/^{230}Th ratio should not be affected by losses during sedimentary diagenesis. Limitations of this approach include (i) the tendency for the less easily scavenged isotope (^{231}Pa) to adsorb preferentially to certain types of particles (e.g., diatom opal), which can make it difficult to distinguish a change in the flux of particles from a compositional change, and (ii) the scavenging of protactinium on basin margins, which can vary in importance as ocean circulation changes (Chase *et al.*, 2003b; Walter *et al.*, 1999; Yu *et al.*, 2001). Beryllium-10 may be used in similar ways to ^{231}Pa but has been studied less (Anderson *et al.*, 1990).

The flux of barium to the seafloor appears to be strongly related to the rain of organic matter out of the surface ocean (Dehairs *et al.*, 1991). Apparently, the oxidation of organic sulfur produces microsites within sinking particles that become supersaturated with respect to the mineral barite ($BaSO_4$). On this basis, barium accumulation has been investigated and applied as a measure of export production in the past, representing a more durable sedimentary signal of the organic carbon sinking flux than sedimentary organic carbon itself (Dymond *et al.*, 1992; Gingele *et al.*, 1999; Paytan *et al.*, 1996). While debates continue on aspects of the biogenic barium flux, some problems with preservation are broadly recognized. In sedimentary environments with low bottom water O_2 and/or high organic carbon rain rates, active sulfate reduction in the shallow sediments can cause the barite flux to dissolve. In addition, there is some low level of barite dissolution under all conditions; if the biogenic barium flux is low, a large fraction of the barite can dissolve at the seafloor. Thus, barium accumulation studies appear to be most applicable to environments of intermediate productivity.

The rapidly growing field of organic geochemistry promises new approaches for the study of biological productivity in the past. By studying specific chemical components of the organic matter found in marine sediments, uncertainties associated with carbon source can be removed, and a richer understanding of past surface conditions can be developed (e.g., Hinrichs *et al.*, 1999; Martinez *et al.*, 1996). To some degree, organic geochemical approaches may allow us to circumvent the thorny problem of breakdown and alteration at the seafloor (Sachs and Repeta, 1999).

6.18.3.2 Nutrient Status

The measurable geochemical parameters currently available for addressing the nutrient status of the surface ocean include (i) the Cd/Ca ratio (Boyle, 1988a) and $^{13}C/^{12}C$ of planktonic (surface-dwelling) foraminiferal carbonate (Shackleton *et al.*, 1983), which are intended to record the concentration of cadmium and the $^{13}C/^{12}C$ of DIC in surface water, (ii) the nitrogen isotopic composition of bulk sedimentary organic matter (Altabet and Francois, 1994a) and microfossil-bound organic matter (Shemesh *et al.*, 1993; Sigman *et al.*, 1999b), which may record the degree of nitrate utilization by phytoplankton in surface water, (iii) the silicon isotopic composition (De La Rocha *et al.*, 1998) and Ge/Si ratio (Froelich and Andreae, 1981) of diatom microfossils, which may record the degree of silicate utilization in surface water (but see Bareille *et al.* (1998)), and (iv) the carbon isotopic composition of organic matter in sediments (Rau *et al.*, 1989) and diatom microfossils (Rosenthal *et al.*, 2000; Shemesh *et al.*, 1993), which may record the dissolved CO_2 concentration of surface water and/or the carbon uptake rate by phytoplankton.

The concentration of dissolved cadmium is strongly correlated with the major nutrients throughout the global deep ocean, and the Cd/Ca ratio of benthic (seafloor-swelling) foraminifera shells records the cadmium concentration of seawater (Boyle, 1988a). As a result, benthic foraminiferal Cd/Ca measurements allow for the reconstruction of deep-ocean nutrient concentration gradients over glacial/interglacial cycles. Planktonic foraminiferal Cd/Ca measurements in surface water provide an analogous approach to reconstruct surface ocean nutrient concentration, although this application has been less intensively used and studied.

A number of factors appear to complicate the link between planktonic foraminiferal Cd/Ca and surface nutrient concentrations in specific regions. First, temperature may have a major effect on the Cd/Ca ratio of planktonic foraminifera (Elderfield and Rickaby, 2000; Rickaby and Elderfield, 1999). Second, planktonic foraminiferal shell growth continues below the surface layer (Bauch *et al.*, 1997; Kohfeld *et al.*, 1996) and probably integrates the Cd/Ca ratio of surface and shallow subsurface waters. This is of greatest concern in polar regions such as the Antarctic, where there are very sharp vertical gradients within the depth zone in which planktonic foraminferal calcification occurs. Third, carbonate geochemistry on the deep seafloor may affect the Cd/Ca ratio of foraminifera preserved in deep-sea sediments (McCorkle *et al.*, 1995). Finally, the link between cadmium and phosphate concentration in surface waters is not as tight as it is in deeper waters (Frew and Hunter, 1992).

Because of isotopic fractionation during carbon uptake by phytoplankton, there is a strong correlation between the $^{13}C/^{12}C$ of DIC and nutrient concentration in deep waters; as a result, the $^{13}C/^{12}C$ of benthic foraminiferal fossils is a central tool in paleoceanography. The $^{13}C/^{12}C$ of planktonic foraminiferal calcite has been used as a tool to study the strength of the biological pump (Shackleton *et al.*, 1983). However, the exchange of CO_2 with the atmosphere leads to a complicated relationship between the $^{13}C/^{12}C$ of DIC and the nutrient concentration in surface waters (Broecker and Maier-Reimer, 1992), such that even a perfect reconstruction of surface water $^{13}C/^{12}C$ would not provide direct information of surface-ocean nutrient status. In addition, the $^{13}C/^{12}C$ of planktonic foraminiferal fossils found in surface sediments appears to be an imperfect recorder of the $^{13}C/^{12}C$ of DIC in modern surface waters, for a variety of reasons (Spero *et al.*, 1997; Spero and Lea, 1993; Spero and Williams, 1988). Finally, the same concerns about the calcification depth noted for Cd/Ca also apply to carbon isotopes or, for that matter, any geochemical signal in planktonic foraminiferal calcite.

During nitrate assimilation, phytoplankton preferentially consume ^{14}N-nitrate relative to ^{15}N-nitrate (Montoya and McCarthy, 1995; Waser *et al.*, 1998), leaving the surface nitrate pool enriched in ^{15}N (Sigman *et al.*, 1999a). This results in a correlation between the $^{15}N/^{14}N$ ratio of organic nitrogen and the degree of nitrate utilization by phytoplankton in surface waters (Altabet and Francois, 1994a,b). There are major uncertainties in the use of this correlation as the basis for paleoceanographic reconstruction of nutrient status, which include (i) the isotopic composition of deep-ocean nitrate through time, (ii) temporal variations in the relationship between nitrate utilization and the nitrogen isotopes in the surface ocean (i.e., the "isotope effect" of nitrate assimilation), and (iii) the survival of the isotope signal of sinking organic matter into the sedimentary record (e.g., Lourey *et al.*, in press). The nitrogen isotope analysis of microfossil-bound organic matter (Sigman *et al.*, 1999b) and of specific compound classes such as chlorophyll-degradation products (Sachs and Repeta, 1999) promises to provide tools to evaluate and circumvent the effect of diagenesis. However, it remains to be seen whether selective nitrogen pools such as microfossil-bound nitrogen are tightly linked to the nitrogen isotope ratio of the integrated sinking flux, the parameter that theoretically relates most directly to the degree of nitrate utilization in surface waters (Altabet and Francois, 1994a).

The isotopic composition of silicon in diatom opal has been investigated as a proxy for the degree of silicate utilization by diatoms, based on the fact that diatoms fractionate the silicon isotopes (^{30}Si and ^{28}Si) during uptake (De La Rocha *et al.*, 1997, 1998). This application is analogous to the use of nitrogen isotopes to study nitrate utilization, but with important differences. On the one hand, the upper ocean cycle is arguably simpler for silica than bio-available nitrogen, which bodes well for the silicon isotope system. On the other hand, there are very few regions of the surface ocean that maintain high dissolved silicate concentrations (Figure 4(d)). As a result, in regions of strong silicate gradients, the link between silicate utilization and silicon isotopic composition may be compromised by mixing processes in surface waters.

The carbon isotopic composition of sedimentary organic matter was originally developed as a paleoceanographic proxy for the aqueous CO_2 concentration of Southern Ocean surface waters (Rau *et al.*, 1989). The aqueous CO_2 concentration is a nearly ideal constraint for understanding a region's effect on the biological pump, as it would provide an indication of its tendency to release or absorb carbon dioxide. However, it has become clear that growth rate and related parameters are as important as the concentration of aqueous CO_2 for setting the $^{13}C/^{12}C$ of phytoplankton biomass and the sinking organic matter that it yields (Popp *et al.*, 1998). In addition, active carbon acquisition by phytoplankton is also probably important for phytoplankton $^{13}C/^{12}C$ in at least some environments (Keller and Morel, 1999). Thus, the $^{13}C/^{12}C$ of organic carbon is a useful paleoceanographic constraint (Rosenthal *et al.*, 2000), but one that is currently difficult to interpret in isolation.

6.18.3.3 Integrative Constraints on the Biological Pump

Above, we have focused on approaches to reconstruct the role of a specific region of the surface ocean on the global biological pump. However, if our goal is to explain the global net effect of ocean biology on the carbon cycle, we must also search for less local, more integrative constraints on the biological pump. This is possible because the atmosphere, surface ocean and deep sea are each relatively homogenous geochemical reservoirs, while being distinct from one another. There are a number of global scale geochemical parameters that may provide important constraints on the biological pump; we describe several of these below.

6.18.3.3.1 Carbon isotope distribution of the ocean and atmosphere

As described above, the biological pump tends to sequester ^{12}C-rich carbon in the ocean interior. All else being equal, the stronger the global biological pump, the higher will be the $^{13}C/^{12}C$ of dissolved inorganic carbon in the surface ocean and of carbon dioxide in the atmosphere. Broecker (1982a,b) and Shackleton (see Shackleton *et al.*, 1983) compared sediment core records of the $^{13}C/^{12}C$ of calcite precipitated by planktonic and benthic foraminfera, the goal being to reconstruct the $^{13}C/^{12}C$ difference in DIC between the surface ocean and the deep sea, a measure of the strength of the global ocean's biological pump. Indeed, this work was the first suggestion that the biological pump was stronger during ice ages, thus potentially explaining the lower CO_2 levels of glacial times. Our view of these results is now more complicated (e.g., Spero *et al.*, 1997); however, the basic conclusion remains defensible (Hofmann *et al.*, 1999). Reliable measurement of the $^{13}C/^{12}C$ of atmospheric CO_2 has proven challenging (Leuenberger *et al.*, 1992; Marino and McElroy, 1991; Smith *et al.*, 1999). Moreover, there are additional modifiers of the $^{13}C/^{12}C$ of atmospheric CO_2, such as the temperature of gas exchange. Nevertheless, these data also seem consistent with the biological pump hypothesis for glacial/interglacial CO_2 change (Smith *et al.*, 1999).

6.18.3.3.2 Deep-ocean oxygen content

The atmosphere/ocean partitioning of diatomic oxygen (O_2) is a potentially important constraint on the strength of the biological pump; the stronger the pump, the more O_2 will be shuttled from the deep ocean to the surface ocean and atmosphere. The rain of organic detritus into the deep ocean drives an O_2 demand by the deep-ocean benthos as it sequesters carbon dioxide in the deep sea, while exposure of deep waters at the surface recharges them with O_2 as it allows deep waters to degas excess CO_2 to the atmosphere (Figure 6). A decrease in atmospheric CO_2 due to the biological pump should, therefore, be accompanied by a decrease in the O_2 content of the ocean subsurface.

The concentration of dissolved O_2 in the ocean interior has long been a target for paleoceanographic reconstruction (the atmospheric change in O_2 content would be minute). Sediments underlying waters with nearly no O_2 tend to lack burrowing organisms, so that sediments in these regions are undisturbed by "bioturbation" and can be laminated; this is perhaps our most reliable paleoceanographic indicator of deep-water O_2 content. Arguments have been made for surface area-normalized sedimentary organic carbon content as an index of O_2 content in some settings (Keil and Cowie, 1999). It remains to be seen whether this is complicated by the potential for changes in the rain rate of organic matter to the sediments. There are a number of redox-sensitive metals, the accumulation of which gives information on the O_2 content of the pore water in shallow sediments (Anderson *et al.*, 1989; Crusius *et al.*, 1996; Crusius and Thomson, 2000). Unfortunately, the O_2 content of the sediment pore waters can vary due to organic matter supply to the sediments as well as the O_2 content of the bottom water bathing the seafloor, so that these two parameters can be difficult to separate (a situation that is analogous to that for sedimentary organic carbon content). While interesting data and arguments have been put forward in support of various approaches (Hastings *et al.*, 1996; Russell *et al.*, 1996), it seems fair to argue that the paleoceanographic community still lacks a reliable set of methods for the global reconstruction of deep ocean dissolved O_2 content, especially in environments far from complete O_2 consumption.

Initial model results suggested that a biological pump mechanism for the glacial/interglacial CO_2 change would have rendered the ocean subsurface so O_2 deficient as to prevent the presence of burrowing organisms and oxic respiration over large expanses of the seafloor, which should leave some tell-tale signs in the sediment record. However, observations have changed this story significantly. For instance, the O_2 minimum, which is at intermediate depths in the modern ocean, may have migrated downward into the abyssal ocean during the last ice age (Berger and Lange, 1998; Boyle, 1988c; Herguera *et al.*, 1992; Marchitto *et al.*, 1998), so that the O_2 decrease was apparently focused in waters which are currently relatively rich in O_2, perhaps avoiding widespread anoxia at any given depth (Boyle, 1988c). For this reason, testing the biological pump hypothesis by reconstructing deep ocean O_2 will require that we do more than simply search for extensive deep-sea anoxia; rather, it will probably require a somewhat quantitative indicator of dissolved O_2 that works at intermediate O_2 concentrations.

6.18.3.3.3 Phasing

With adequate dating and temporal resolution in paleoceanographic and paleoclimatic records, the sequence of past events and changes can be reconstructed, providing among the most direct evidence for cause and effect. There is much information on phasing that is relevant to the biological pump and its role in glacial/interglacial CO_2 change. We limit ourselves here to one example that arises largely from ice core records: the timing of CO_2 change relative to temperature and glaciation (Broecker and Henderson, 1998 and references therein). Near the end of ice ages, atmospheric CO_2 rises well before most of the deglaciation (Monnin *et al.*, 2001; Sowers and Bender, 1995); this was referred to above as strong evidence against Broecker's shelf phosphorus hypothesis for CO_2 change (see Section 6.18.2.1). While the phasing of temperature is still debated, it appears that most of the warming in the high-latitude southern hemisphere preceded most of the warming in the high-latitude northern hemisphere, and that the atmospheric CO_2 rise lagged only slightly behind the southern hemisphere warming. This latter observation is roughly consistent with a variety of hypotheses for changes in the biological pump, but appears inconsistent with alternative hypotheses that rely solely on changes in the oceanic calcium carbonate budget, which operates on a longer timescale than the biological pump (Archer and Maier-Reimer, 1994; Opdyke and Walker, 1992). With continued work, the detailed timing of CO_2 change may provide quantitative constraints on changes in the oceanic calcium carbonate budget as a partial contributor to CO_2 change; calcium carbonate plays a role in many of the biological pump hypotheses (Archer *et al.*, 2000a; Sigman *et al.*, 1998; Toggweiler, 1999).

6.18.4 OBSERVATIONS

To this point, a number of central observations and concepts have already been described to motivate a search for changes in the biological pump over glacial/interglacial cycles. First, there are carbon dioxide variations over glacial/interglacial cycles that are of the right magnitude and temporal structure to be caused by changes in the biological pump. Second, carbon isotope data for carbon dioxide and foraminiferal carbonate appear consistent with a biological-pump mechanism. Third, changes in the nitrogen cycle have been recognized that would tend to increase the oceanic nitrate reservoir during glacial times, although an actual increase in this reservoir has not been demonstrated; such a reservoir change might be expected to strengthen the low-latitude pump during glacial times. Finally, observations about phytoplankton in polar regions, in particular, their incomplete consumption of the major nutrients and their tendency toward iron limitation, suggest that simple changes in either the iron supply to the Antarctic surface ocean or in polar ocean circulation could lead to an increase in the efficiency of the high-latitude biological pump during glacial times. These observations, together with other ideas about the operation of the ocean carbon cycle, warrant that we consider the basic regional observations on biological productivity and nutrient status over glacial/interglacial cycles.

6.18.4.1 Low- and Mid-latitude Ocean

There have been many studies of the response of coastal and equatorial upwelling systems to glacial/interglacial climate change, and this overview cannot do justice to the entire body of work. The coastal upwelling regions along the western continental margins show an Atlantic/Pacific difference in their response to glacial/interglacial climate change. Studies along the western coast of Africa suggest an increase in productivity during glacial times (Summerhayes *et al.*, 1995; Wefer *et al.*, 1996). However, in the eastern Pacific, the coastal upwelling zones off of California and Mexico in the north and off of Peru in the south were less productive during the last glacial period (Dean *et al.*, 1997; Ganeshram and Pedersen, 1998; Ganeshram *et al.*, 2000; Heinze and Wefer, 1992). This response has generally been explained as the effect of continental cooling (and a large North American ice sheet in the case of the California Current) on the winds that currently drive coastal upwelling in the eastern tropical Pacific (Ganeshram and Pedersen, 1998; Herbert *et al.*, 2001).

Early paleoceanographic studies in the equatorial Pacific found that export production was greater in the equatorial Pacific during ice ages (Lyle, 1988; Pedersen, 1983; Pedersen *et al.*, 1991; Rea *et al.*, 1991), and some subsequent studies have supported this conclusion (Herguera and Berger, 1991; Murray *et al.*, 1993; Paytan *et al.*, 1996; Perks *et al.*, 2000). However, this interpretation has been put in question by reconstructions using ^{230}Th- and ^{3}He-normalized accumulation rates, ^{231}Pa/^{230}Th ratios (Marcantonio *et al.*, 2001b; Schwarz *et al.*, 1996; Stephens and Kadko *et al.*, 1997) and other evidence (Loubere, 1999). From studies along the eastern equatorial Atlantic, the consensus is for higher productivity during the last ice age than the Holocene; because of the limited extent of the Atlantic basin, it is difficult to differentiate this change from the glacial-age increase in productivity in African coastal upwelling (Lyle *et al.*, 1988; Martinez *et al.*, 1996; Moreno *et al.*, 2002; Rutsch *et al.*, 1995; Schneider *et al.*, 1996). Proxies for nutrient utilization (Altabet, 2001; Farrell *et al.*, 1995; Holmes *et al.*, 1997), pH (Sanyal *et al.*, 1997), and wind strength (Stutt *et al.*, 2002) would suggest that any export production increases that did occur in the low-latitude upwelling systems of the Pacific and Atlantic during the last glacial maximum were due to higher wind-driven upwelling, which increased the nutrient supply to the surface.

Upwelling associated with monsoonal circulation has apparently behaved predictably since the last ice age (Duplessy, 1982; Prell *et al.*, 1980). With less summertime warming of the air over Asia during the last glacial maximum, there was a weakening of the southwesterly winds of the summertime monsoon that drive coastal upwelling off the horn of Africa and Saudi Arabia, leading to a glacial decrease in export production in that region (Altabet *et al.*, 1995; Anderson and Prell, 1993; Marcantonio *et al.*, 2001a; Prell *et al.*, 1980). To the east and in the more open Indian Ocean, where the (southward) wintertime monsoonal winds drive upwelling or vertical mixing, productivity was apparently higher during the last glacial maximum, again consistent with the expected effect of a cooler Eurasian climate on the monsoon cycle (Beaufort, 2000; Duplessy, 1982; Fontugne and Duplessy, 1986). In the South China Sea, where the winter monsoon (seaward flow) is responsible for the exposure of nutrient-rich deep waters, export production was apparently higher during the last glacial maximum, again suggesting a stronger winter monsoon associated with colder winter conditions over Eurasia (Huang *et al.*, 1997; Thunell *et al.*, 1992). The effect of the wintertime monsoon winds may also have impacted the western equatorial Pacific (Herguera, 1992; Kawahata *et al.*, 1998).

The low-productivity regions of the low- and mid-latitude ocean are an important source of information on the strength of the low-latitude biological pump, as they may be less susceptible to wind-driven changes in nutrient supply than are the upwelling regions discussed above, so that changes in productivity would be more closely tied to changes in the oceanic nutrient reservoir (and/or in N_2 fixation). As mentioned above, studies in the western equatorial Pacific indicate that the sedimentary C_{org} was higher in these regions during glacial times (Figure 10; Perks *et al.* (2000)); however, studies from the more southern western Pacific appear to suggest lower productivity (Kawahata *et al.*, 1999). Studies in the western equatorial and tropical Atlantic indicate lower $CaCO_3$ and C_{org} burial rates during glacial times (François *et al.*, 1990; Ruhlemann *et al.*, 1996). In the tropical Indian Ocean, away from the Arabian Sea, productivity was apparently higher during glacial times, but this has again been explained as the result stronger wind-driven mixing during the wintertime monsoon (Beaufort, 2000; Fontugne and Duplessy, 1986). Discrete sedimentation events of massive mat-forming diatoms are found in glacial-age sediments from the equatorial Indian; their significance is unclear (Broecker *et al.*, 2000).

As described in Section 6.18.2, there is substantial evidence for a change in the spatial pattern of denitrification across glacial/interglacial transitions (Altabet *et al.*, 1995, 2002; Emmer and Thunell, 2000; Ganeshram *et al.*, 1995; Pride *et al.*, 1999), and it seems likely that this led to a net global decrease in the loss rate of nitrate from the ocean during the last glacial maximum. Depending on the changes in N_2 fixation over glacial/interglacial cycles and the role of phosphorus in these changes (see Section 6.18.2), low-latitude productivity may or may not have responded so as to strengthen the biological pump. As discussed above, increased low-latitude productivity due to increased upwelling or vertical mixing has limited significance for the strength of the biological pump. It is a challenge to develop an approach to study past productivity changes that can distinguish between a change in the physical rate of vertical exchange and a change in the nutrient content of the subsurface. Thus, it is difficult to test the hypothesis that the ocean nutrient reservoir drove a significant glacial increase in the low- and mid-latitude biological pump, so as to explain glacial/interglacial CO_2 changes. Nevertheless, it is notable that no direct support for the hypothesis has arisen without a focused search. Based on the overview above, the evidence for a net global increase in low-latitude productivity during the last ice age is not compelling. Moreover, the evidence that does exist for regional increases in low-latitude productivity seems easily explained in terms of changes in the wind-driven upwelling. Comparison of accumulation records for organic carbon and opal (Herguera and Berger, 1994) and of records from opposite sides of the South Atlantic basin (Ruhlemann *et al.*, 1999) has led to the suggestion that the nutrient content of the thermocline was lower during ice ages. If such was indeed the case, it could be associated with a tendency for nutrients to shift into the deeper ocean during ice ages (Boyle, 1988b,c); nevertheless, it provides little encouragement for hypotheses of a larger oceanic nutrient reservoir during glacial times.

While it is tempting to phrase orbitally driven climate variability in terms of "ice ages" and "interglacials," this simplification breaks down frequently in the lower-latitude ocean (Clemens *et al.*, 1991; McIntyre and Molfino, 1996; Sachs *et al.*, 2001). The precession component of the Earth's orbital variations is by far the most directly important for the energy budget of the tropics. The Pleistocene variability of productivity in some tropical regions (Pailler *et al.*, 2002) and coastal and equatorial upwellings (Moreno *et al.*, 2002; Rutsch *et al.*, 1995; Summerhayes *et al.*, 1995) is dominated by the precession ($\sim$23 kyr) cycle, with the result that our focus on the Last Glacial Maximum/Holocene comparison can be misleading about the real timing of change. New data make this case very strongly for equatorial Pacific productivity (Perks *et al.*, 2000). As the variations in atmospheric CO_2, glacial ice volume, and polar temperature all have less variability at the precession frequency, one would surmise that the low-latitude productivity changes dominated by precession are not major drivers of variability in the biological pump or in atmospheric CO_2. This is consistent with the conceptual point that wind-driven changes in low-latitude productivity

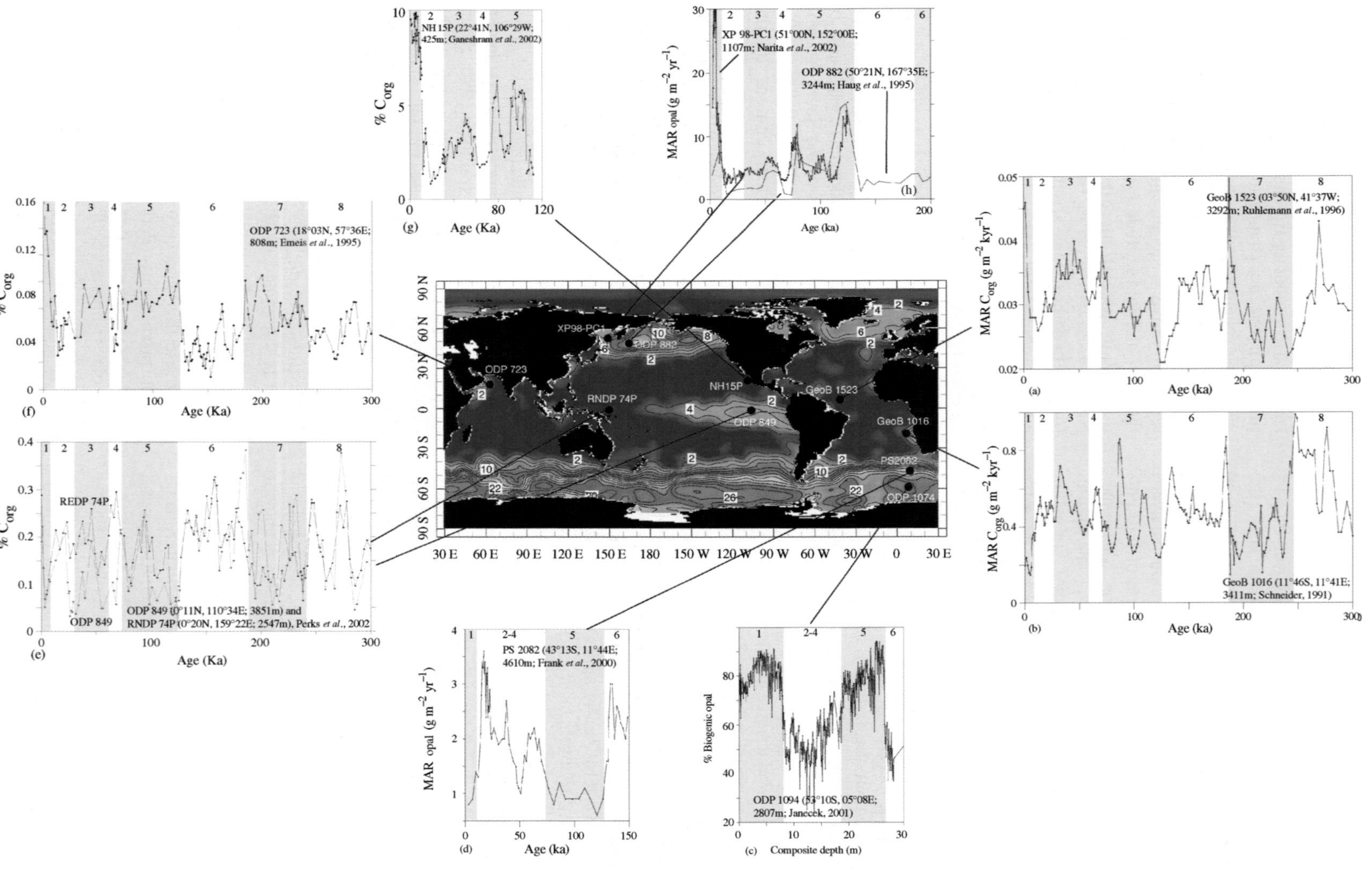

NH 15P (22°41N, 106°29W; 425m; Ganeshram et al., 2002)
% C_org
Age (Ka)
(g)
XP 98-PC1 (51°00N, 152°00E; 1107m; Narita et al., 2002)
ODP 882 (50°21N, 167°35E; 3244m; Haug et al., 1995)
MAR opal (g m^-2 yr^-1)
Age (ka)
(h)
ODP 723 (18°03N, 57°36E; 808m; Emeis et al., 1995)
% C_org
Age (Ka)
(f)
REDP 74P
ODP 849
ODP 849 (0°11N, 110°34E; 3851m) and RNDP 74P (0°20N, 159°22E; 2547m), Perks et al., 2002
% C_org
Age (Ka)
(e)
GeoB 1523 (03°50N, 41°37W; 3292m; Ruhlemann et al., 1996)
MAR C_org (g m^-2 kyr^-1)
Age (ka)
(a)
GeoB 1016 (11°46S, 11°41E; 3411m; Schneider, 1991)
MAR C_org (g m^-2 kyr^-1)
Age (ka)
(b)
PS 2082 (43°13S, 11°44E; 4610m; Frank et al., 2000)
MAR opal (g m^-2 yr^-1)
Age (ka)
(d)
ODP 1094 (53°10S, 05°08E; 2807m; Janecek, 2001)
% Biogenic opal
Composite depth (m)
(c)
XP98-PC1
ODP 882
ODP 723
NH15P
GeoB 1523
RNDP 74P
ODP 849
GeoB 1016
PS2082
ODP 1074

are of limited importance to CO_2 transfer between the atmosphere and ocean.

6.18.4.2 High-latitude Ocean

The global high-latitude open ocean includes the Arctic, the North Atlantic, the Southern Ocean, and the Subarctic North Pacific (Figure 5). We address each of these in turn.

The modern open Arctic appears to be relatively unproductive. Annually averaged solar insolation is low and sea ice cover is extensive, so that light limitation of phytoplankton growth is likely. Surface nitrate is also quite low in the Arctic surface (Figure 5(c)), at least partially due to a strong permanent "halocline," or vertical salinity gradient, that drives a strong vertical density gradient in the upper Arctic ocean, isolating fresh surface waters from saltier and thus denser nutrient-rich deep water. However, as the major nutrients are not completely absent in the surface, the major nutrients are probably not the dominant limitation for phytoplankton in the open Arctic. The few paleoproductivity studies of the open Arctic suggest that it was less productive during the last ice age (Schubert and Stein, 1996; Wollenburg *et al.*, 2001). A range of ice age conditions may have contributed to this decrease, including greater ice cover and thus more extreme light limitation as well as sea-level-driven loss of the nutrient supply from the Pacific that currently enters from the Bering Strait. Our current understanding of the Arctic would suggest that it is not in itself a central part of ocean/atmosphere CO_2 partitioning, so little energy has been expended to understand its glacial/interglacial cycle.

The high-latitude North Atlantic has a productive spring bloom but tends towards major-nutrient limitation in the summer. This tendency toward nutrient limitation is due at least partially to the formation of deep water in the high-latitude North Atlantic. The Gulf Stream extension provides salty, warm, low-nutrient water to the high-latitude surface, and the cooling of this water eventually leads to the formation of deep waters (see Chapter 6.16). As a result of this steady state, neither surface waters nor underlying deep waters

Figure 10 Paleoceanographic records relevant to the history of export production from various regions of the global ocean, overlain on a map of surface nitrate concentration (in μM; Conkright *et al.*, 1994). The records are chosen to be somewhat representative as a crude first-order paleoproductivity record for their oceanic regime; the extreme uncertainty associated with some of these reconstructions is discussed at length in Section 6.18.2. The records are of the sedimentary concentration or accumulation of organic carbon or biogenic opal. While the concentration records are complicated by potential changes in dilution by other sediment components, there are some circumstances where accumulation rate estimates are also problematic. The referenced sources should be consulted for information on these concerns. The biogeochemical regime of the surface ocean is roughly indicated by surface nitrate concentration (see Figure 5), which indicates whether a record is from a nutrient-poor, low-latitude region, a nutrient-bearing low-latitude upwelling region, or a nutrient-bearing polar region. Records are plotted versus time (except core ODP 1094 in the Southern Ocean) and glacial–interglacial stages are numbered, with a gray shading during interglacials. (a) The C_{org} accumulation record from core GeoB 1523 for the last 300 ka (Ruhlemann *et al.*, 1996) conforms with the general finding that productivity was lower in the oligotrophic tropical Atlantic during the last ice age. (b) The C_{org} accumulation record from core GeoB 1016 for the last 300 ka (Schneider, 1991) is among the large body of evidence indicating that the Benguela upwelling system and the eastern equatorial Atlantic were more productive during the last ice age. However, all of the Atlantic records show a strong precessional response, indicating only a loose connection with ice volume. Ruhlemann *et al.* (1999) have noted an anti-correlation between eastern and western tropical Atlantic records. (c) Biogenic opal concentrations of ODP Site 1094 in the Antarctic zone of Southern Ocean (Janecek, 2001) conform with the large body of data indicating lower opal accumulation rates during ice ages. The upper 30 m of the composite sediment sequence represent approximately the last 135 ka. (d) Biogenic opal accumulation rate in core PS2062 (Frank *et al.*, 2000) over the last 150 ka conforms with a large body of data indicating higher opal and organic carbon accumulation in the Subantarctic zone of the Southern Ocean during ice ages, opposite to the temporal pattern observed in the Antarctic. (e) The C_{org} concentration in RNDP 74P from the western equatorial Pacific and in ODP Site 849 from the eastern equatorial Pacific over the last 300 ka (Perks *et al.*, 2000) suggests higher productivity across the entire equatorial upwelling during the last ice age but shows its dominant response to precession, not polar ice volume. (f) The C_{org} concentration in ODP Site 723 over the last 300 ka (Emeis *et al.*, 1995) conforms with the large body of data suggesting lower productivity in the Oman upwelling system during ice ages because of a weaker summer monsoon (and thus weaker upwelling) when Eurasia is cold. (g) The C_{org} concentration in core NH15P during the last 120 ka in the coastal upwelling zone off northwestern Mexico (Ganeshram and Pedersen, 1998) conforms with a large body of data indicating lower productivity during the last ice age because the North American ice sheet caused a reduction in wind-driven coastal upwelling. (h) Biogenic opal accumulation rates in the Subarctic North Pacific (ODP Site 882; Haug *et al.* (1995)) and the Okhotsk Sea (Core XP98-PC1 Narita *et al.* (2002)) suggest lower diatom export during glacials, perhaps because of stronger salinity stratification of the upper water column and thus reduced nutrient supply. See text for more references.

can supply large amounts of nutrients to the surface, with the largest nutrient source being the Antarctic Intermediate Water that penetrates from the south. The equivocal evidence for changes in North Atlantic productivity over glacial/interglacial cycles appears to suggest lower export production during the last ice age (Manighetti and McCave, 1995; Thomas *et al.*, 1995). This is consistent with the persistent formation of a low-nutrient subsurface water in the glacial North Atlantic (Oppo and Lehman, 1993), even though this water mass was distinct from modern North Atlantic Deep Water (again, see Chapter 6.16). If, during ice ages, the North Atlantic continued to import lower-latitude surface waters and produce a subsurface water mass with low preformed nutrients, then it would have played roughly the same role in the biological pump as it does today. However, a glacial weakening of the North Atlantic source would have had implications for the global mean preformed nutrient concentration of the deep ocean (Section 6.18.2.2 and Figure 8).

Paleoceanographic work in the Southern Ocean indicates a strong glacial/interglacial oscillation, but with very different responses in the (poleward) Antarctic than in the (equatorward) Subantarctic. In the modern Antarctic, biological export production is dominated by diatoms, the group of phytoplankton that precipitate tests of hydrated silica ("opal"), and these microfossils are frequently the dominant component of the underlying sediments. The accumulation of diatomaceous sediments in the Antarctic was clearly lower during the last ice age (Mortlock *et al.*, 1991), and other paleoceanographic indicators suggest that the biological export of carbon was also lower (Francois *et al.*, 1997; Kumar *et al.*, 1995; Rosenthal *et al.*, 2000). However, given an apparent bias of at least some of these proxies to preferentially record the carbon export closely associated with diatom microfossils (Chase *et al.*, 2003b; Yu *et al.*, 2001), it remains possible that the biological export of carbon from the glacial Antarctic surface was as high or higher than during interglacial times but that it occurred in a form that was poorly preserved in the sediment record (Arrigo *et al.*, 1999; Boyle, 1998; Brzezinski *et al.*, 2002; DiTullio *et al.*, 2000; Moore *et al.*, 2000; Tortell *et al.*, 2002).

In the Atlantic and Indian sectors of the Subantarctic, there is very strong evidence for higher export production during glacial times and an associated increase in the relative importance of diatom production (Chase *et al.*, 2001; Francois *et al.*, 1997; Kumar *et al.*, 1995; Mortlock *et al.*, 1991; Rosenthal *et al.*, 1995). This high glacial productivity has not yet been recognized in the Pacific sector (Chase *et al.*, 2003a). If this productivity increase was indeed absent in the Pacific sector of the Subantarctic, it provides an important constraint on its cause. For instance, this finding may be consistent with iron fertilization as the dominant driver of the glacial increase in Subantarctic productivity, as the increase in iron input may have been extremely limited in parts of the Pacific sector (Chase *et al.*, 2003a).

While these changes in export production are critical to the workings of the ice age Southern Ocean, they do not address directly the effect of the Southern Ocean on atmospheric CO_2, which is largely determined by the competition between the upwelling of respiratory CO_2 and nutrients to the surface and export of organic matter out of the surface (Section 6.18.2). In the Antarctic, nitrogen isotope data suggest that the fraction of the nitrate consumed during the last ice age was higher than its current value (Francois *et al.*, 1997; Sigman *et al.*, 1999b). These result seem to support the long-standing hypothesis that nutrient utilization changes in Antarctic waters are a fundamental driver of glacial/interglacial changes in the atmospheric concentration of CO_2; box model calculations predict the 25–40% higher nitrate utilization (i.e., 50–65% in the glacial ocean compared to 25% during the present interglacial) could lower atmospheric CO_2 by the full glacial/interglacial amplitude under the relevant conditions (Sigman *et al.*, 1999b). Since paleoceanographic proxy data suggest that Antarctic export production was lower during the last ice age, one would infer that more complete nitrate utilization in the Antarctic was due to a lower rate of nitrate supply from the subsurface, implying that the fundamental driver of the CO_2 change was an ice age decrease in the ventilation of deep waters at the surface of the Antarctic (François *et al.*, 1997). Such a change in the Antarctic would also help to explain observations in the low-latitude ocean. For instance, the extraction of nutrients in the Antarctic should lower the nutrient content of the waters that subducted into the intermediate-depth ocean and thermocline at the equatorward margin of the Antarctic. This would work to prevent the transfer of nutrients from cold, deep ocean into the warmer, upper ocean, potentially explaining an apparent transfer of nutrients from the mid-depths to the deep ocean during glacial times (Boyle, 1988b; Herguera *et al.*, 1992; Keir, 1988; Matsumoto *et al.*, 2002a; Sigman and Boyle, 2000).

Two possible causes for Antartic stratification have been discussed. First, the southern hemisphere westerlies winds apparently shifted northward during glacial times (Hebbeln *et al.*, 2002; McCulloch *et al.*, 2000), which should have reduced Ekman-driven upwelling in the Antarctic (Toggweiler *et al.*, 1999) ("wind-shift" mechanism). Second, the vertical gradient in density, as determined jointly by the vertical gradients in temperature and salinity, may have been stronger in the glacial Antarctic ("density" mechanism).

However, both of these mechanisms are currently incomplete as explanations for the glacial reduction in atmospheric CO_2 (Keeling and Visbeck, 2001; Sigman and Boyle, 2001). As an example, we consider the "density" mechanism in more detail. In the modern Antarctic, as in most polar regions, the vertical salinity gradient tends to stratify the upper ocean, while temperature tends to destabilize it. Thus, either an increase in the relative importance of salinity or a decrease in the importance of temperature could be the fundamental trigger for the development of strong stratification in the glacial Antarctic; once stratification begins, it will be reinforced by the accumulation of freshwater in the upper layer. Given the abundance of sea ice in the glacial Antarctic (Crosta *et al.*, 1998) and the typical association of sea ice with freshwater release, it is tempting to call upon sea ice as the initial driver of stratification. However, sea ice does not represent a source of freshwater *per se*, but rather a mechanism for transporting freshwater from one region of the Antarctic to another, with net ice formation in the coastal Antarctic and net melting in the open Antarctic. As a result, stratification of the open Antarctic by the melting of sea ice might occur at the expense of the coastal Antarctic, where sea ice formation would make surface waters more saline and thus more dense, leading to overturning and the accompanying release of CO_2 to the atmosphere. Thus, we are more encouraged by hypotheses that involve changes in temperature (i.e., cooling of the deep ocean) as the trigger for stratification (Gildor and Tziperman, 2001; Gildor *et al.*, 2002), although these have been tested only in simple physical models. In any case, if the evidence for stratification could be taken as overwhelming, we would still need to determine what caused it and how much it lowered atmospheric CO_2 during glacial times.

However, the evidence for Antarctic stratification is not overwhelming; indeed, many investigators are not convinced (e.g., Anderson *et al.*, 1998; Elderfield and Rickaby, 2000). The interpretation from nitrogen isotopes of higher nitrate utilization in the Antarctic faces a number of apparent disagreements with other proxies of nutrient status, in particular, the Cd/Ca (Boyle, 1988a; Keigwin and Boyle, 1989) and $^{13}C/^{12}C$ (Charles and Fairbanks, 1988) of planktonic foraminiferal calcite. Measurements of these ratios in planktonic foraminifera indicate no clear decrease in the nutrient concentration of Antarctic surface water, while such a decrease would have been expected on the basis of the nitrogen isotope data. Each of the paleochemical proxies has significant uncertainties (see Section 6.18.3.2). Nevertheless, the need to define new interpretations of a variety of measurements to fit the stratification hypothesis does not inspire confidence.

Dissolved silicate is a major nutrient for diatom growth because of the silica tests that these phytoplankton precipitate. Much like nitrate and phosphate, silicate is nearly completely depleted in the low-latitude surface ocean but is found at relatively high concentrations in the modern Antarctic (Figure 5(d)). The silicon isotopic composition of diatom microfossils implies that there was a reduction in Antarctic silicate utilization during the last ice age (De La Rocha *et al.*, 1998), in contrast to the evidence for enhanced nitrate utilization. While this may indicate a disagreement among proxies, the alternative is that the difference signals a real oceanographic change. Field observations, incubations and culture studies (Franck *et al.*, 2000; Hutchins and Bruland, 1998; Takeda, 1998) indicate that iron-replete conditions (such as may result from the dustiness of the ice age atmosphere (Mahowald *et al.*, 1999)) favor a higher nitrate/silicate ratio in diatoms, and phytoplankton species composition changes may have reinforced this shift (Tortell *et al.*, 2002). Indeed, taking both the nitrogen and silicon isotope data at face value implies that the nitrate/silicate uptake ratio of Antarctic phytoplankton was higher during the last ice age, leading to lower nitrate but higher silicate concentrations in the glacial Antarctic relative to modern times (Brzezinski *et al.*, 2002). From the perspective of our efforts to reconstruct the history of export production, this possibility is problematic in that the rain of biogenic silica out of the surface ocean would have decreased independently from the rain of organic carbon.

If the change in nitrate/silicate uptake ratio was adequately great, it may have actually removed the tendency for preferential depletion of silicate relative to nitrate that is observed in the modern Antarctic, possibly increasing the silicate/nitrate ratio of Southern-source water that supplies nutrients to the low-latitude surface (Brzezinski *et al.*, 2002). A greater supply of silicate to the low-latitude ocean could have driven an increase in the importance of silica-secreting phytoplankton (diatoms) relative to $CaCO_3$-precipitating phytoplankton (coccolithophorids) (Matsumoto *et al.*, 2002b); this may explain a shift toward higher opal accumulation rates in some equatorial regions (Broecker *et al.*, 2000). If increased diatom productivity at low latitudes in the southern hemisphere occurred at the expense of coccolithophores, the high sinking rates of diatom-derived particulate matter may have facilitated the transport of organic matter through intermediate waters to the deep ocean (Boyle, 1988b), again consistent with evidence for the "nutrient deepening" during glacial times (Boyle, 1988c, 1992; Boyle *et al.*, 1995; Herguera *et al.*, 1992). In addition, a floral shift away from coccolithophores to diatoms would have also lowered the $CaCO_3$/organic carbon rain ratio, weakening the ocean carbonate pump. Both the increase in the remineralization depth of organic carbon and a decrease in the rain ratio would have worked to lower-atmospheric CO_2 (Archer and Maier-Reimer, 1994; Berger and Keir, 1984; Boyle, 1988b; Broecker and Peng, 1987; Brzezinski *et al.*, 2002; Dymond and Lyle, 1985; Keir and Berger, 1983; Matsumoto *et al.*, 2002b; Sigman *et al.*, 1998).

In the Subantarctic, planktonic foraminiferal Cd/Ca and the $^{13}C/^{12}C$ of diatom-bound organic matter are consistent with an ice age state of higher nutrient utilization (Rosenthal *et al.*, 1997, 2000). However, the consideration of proxy complications argue against such a change (Elderfield and Rickaby, 2000). The significance of the nitrogen isotope data for the glacial Subantarctic is uncertain (François *et al.*, 1997; Sigman *et al.*, 1999a). Thus, while the Subantarctic was certainly more productive during glacial times, the history of its nutrient status is unclear and deserves further investigation. Given our current understanding of the role that the Subantarctic plays in the carbon cycle, the Subantarctic is less important in itself than in what it indicates about the nutrient status of the Antarctic zone at its (up-stream) southern end. The Subantarctic also represents the gateway by which the Antarctic affects the low-latitude ocean and is thus central in questions regarding the impact of Antarctic nutrient changes on low-latitude productivity (Brzezinski *et al.*, 2002; Matsumoto *et al.*, 2002b).

The Subarctic Pacific, like the Antarctic, is characterized by year-round nonzero concentrations of the major nutrients, nitrate and phosphate. However, the Subarctic Pacific maintains a higher degree of nutrient utilization (lower surface nutrient concentrations) than the Antarctic. In the open Subarctic Pacific, summer nitrate concentration frequently falls below 8 μM in surface water, despite nitrate concentrations of greater than 35 μM in the upwelling deep water below the permanent halocline. One important difference between the Subarctic Pacific and the Antarctic is the much stronger halocline in the former (Reid, 1969, Talley, 1993; Warren, 1983), where the extremely low salinity of the upper 400 m limits the exposure of nutrient- and CO_2-charged subsurface waters at the surface. It is roughly this salinity-driven stratification that has been proposed for the glacial Antarctic. Thus, the modern Subarctic Pacific provides something of a modern analogue for the glacial Antarctic stratification hypothesis (Haug *et al.*, 1999; Morley and Hays, 1983).

The Subarctic Pacific undergoes changes in productivity and nutrient regime over glacial/interglacial cycles, although these changes are of smaller amplitude than is observed for the Southern Ocean. Opal accumulation was apparently lower during the last glacial maximum in the western Subarctic Pacific and its marginal seas (Gorbarenko, 1996; Haug *et al.*, 1995; Narita *et al.*, 2002). This has been interpreted, in much the same way as the Antarctic changes, as indicating intensified stratification during glacial times (Narita *et al.*, 2002). In this light, the smaller glacial/interglacial signal of this region relative to the Antarctic is explained by the fact the modern Subarctic Pacific is already strongly stratified. However, the dominant feature in most sediment records from western Subarctic Pacific and Bering Sea is of a high-productivity event upon the transition from the last ice age to the Holocene (Keigwin *et al.*, 1992; Nakatsuka *et al.*, 1995), for which there is as yet no well-supported explanation. Moreover, the eastern Subarctic Pacific seems to lack a glacial/interglacial signal, instead being characterized by abrupt events of opal deposition that are not clearly linked to global climate change (McDonald *et al.*, 1999).

6.18.5 SUMMARY AND CURRENT OPINION

While the early focus of hypotheses for glacial/interglacial CO_2 change was on the biological pump (Broecker, 1982a,b), a number of alternative mechanisms, mostly involving the calcium carbonate budget, attracted increasing attention during the late 1980s and the 1990s (Archer and Maier-Reimer, 1994; Boyle, 1988b; Broecker and Peng, 1987; Opdyke and Walker, 1992). As problems have been recognized with these mechanisms as the sole driver of glacial/interglacial CO_2 change (Archer and Maier-Raimer, 1994; Sigman *et al.*, 1998), the biological pump is receiving attention once again. Hypotheses exist for both low- and high-latitude changes in the biological pump. Their strengths, weaknesses, and central questions are now fairly clear.

Currently, the most popular low-latitude hypotheses depend largely on whether the nitrogen cycle alone can drive a large-scale change in low-latitude export production. As of the early 2000s, while this question has not yet been directly addressed, no new observations have arisen that would overturn the traditional view that the nitrogen cycle could have had only a limited effect without cooperation from the phosphorus cycle and that the phosphorus cycle could not have been adequately dynamic over glacial/interglacial cycles (e.g., Froelich *et al.*, 1982; Haug *et al.*, 1998; Redfield, 1942, 1958; Redfield *et al.*, 1963; Ruttenberg, 1993; Sanudo-Wilhelmy *et al.*, 2001). Thus, while the low-latitude biological pump has by no means been eliminated as the driver of glacial/interglacial cycles, our attention is focused on the polar ocean, the Antarctic in particular.

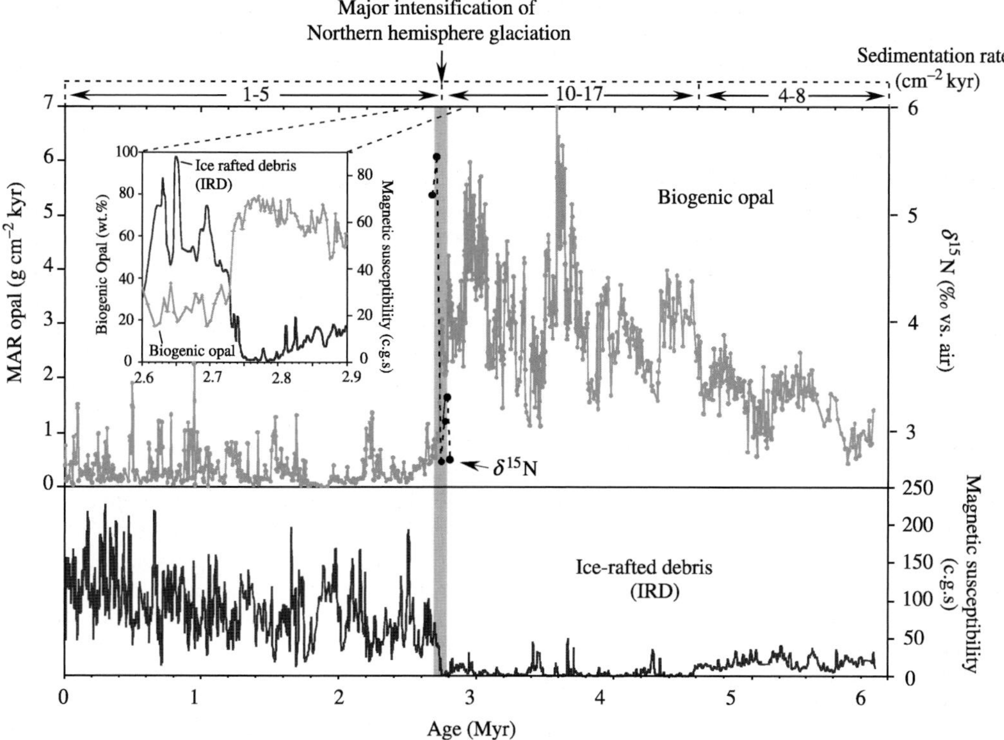

Figure 11 A 6 Myr biogenic opal record from Ocean Drilling Program Site 882 in the western Subarctic Pacific (50°21′ N, 167°35′ E; water depth 3,244 m; figure from Haug *et al.* (1999)), showing an approximately fourfold decrease in opal accumulation rate at 2.73 Myr ago (top panel). This abrupt drop in opal occurred at isotope stage G6 (the former isotope stage 110) synchronously with the massive onset of ice rafted debris, as indicated by the increase in magnetic susceptibility (bottom panel). Maxima in opal accumulation during the last 2.73 Myr are generally linked to interglacial times or deglaciations, as reported by Keigwin *et al.* (1992) for the last deglaciation. Since silicate is supplied by the exposure of nutrient-rich deep water, the high opal flux rates of the mid-Pliocene Subarctic Pacific require a rate of deep-water exposure of a magnitude similar to that observed in the modern Antarctic. Because silicate consumption is nearly complete in the modern Subarctic Pacific (Figure 5(d)), the sharp drop in opal flux at 2.73 Myr ago cannot be attributed simply to a decrease in the completeness of silicate consumption. Rather, it must record a decrease in the rate of exposure of silicate-rich deep water at the Subarctic Pacific surface. Sediment $\delta^{15}N$ ($^{15}N/^{14}N_{sample}/^{15}N/^{14}N_{air} - 1) \times 1{,}000$) increases concurrently with the sharp drop in opal accumulation at the event 2.73 Myr ago (from ~3‰ to 5‰, upper panel). This shift toward higher $\delta^{15}N$ suggests an increase in nitrate utilization (Altabet and Francois, 1994a), providing additional evidence that the decrease in opal flux resulted from a decrease in the exposure of nutrient-rich deep water, which lowered the nitrate supply and thus forced more complete nitrate consumption while decreasing the silicate available for opal production (reproduced by permission of Nature Publishing Group from *Nature 401*, 779–782).

We have discussed three rough categories of Antarctic change that might drive a decline in CO_2 during ice ages: (i) increased biological productivity, as in response to increased iron supply, (ii) vertical stratification of the upper ocean, and (iii) gas exchange limitation by ice cover. The productivity hypothesis is not supported by data, which suggest a smaller biogenic rain in the glacial Antarctic, although it remains possible that our proxies are telling us only about a reduction in the rain of bulk biogenic material (i.e., diatom-derived opal) and are leading us to overlook an increase in "export production," the rain of organic carbon out of the surface ocean. The gas exchange hypothesis remains possible but is difficult to test and would seem to require extreme conditions to be the sole cause for the observed amplitude of CO_2 change. The stratification hypothesis is supported by some data but conflicts with other data, or at least with the traditional interpretations of those data. While there are major concerns about each of the polar hypotheses, these hypotheses have recently been strengthened by the realization that they would have impacted the lower-latitude ocean and the $CaCO_3$ budget in ways that should have strengthened their overall effect on atmospheric CO_2 (e.g., Brzezinski *et al.*, 2002; Matsumoto *et al.*, 2002b; Toggweiler, 1999).

As is clear from the treatment above, the stratification hypothesis has particular resonance with the authors of this chapter. This is driven largely by the conceptual and observational links we observe between two nutrient-bearing polar ocean regions, the Antarctic and the Subarctic Pacific. As described above, there is a strong analogy between the Subarctic Pacific that we observe today and the glacial-age Antarctic that is posed by the stratification hypothesis. Moreover, at a major cooling event 2.7 Myr ago, it appears that the Subarctic Pacific underwent the transition to its current salinity-stratified condition (Figure 11) (Haug *et al.*, 1999), just as stratification has been hypothesized for the Antarctic on the transitions to ice age conditions. Ongoing work on other intervals of the climate record seems to support the existence of a strong generalized link between climate cooling and polar ocean stratification. The physical mechanism for this link, however, is still unresolved.

To these authors, there is no infallible recipe for the investigation of the biological pump in the past. Nevertheless, there are glaring uncertainties that require our attention, both in our concepts and our tools. The quantitative effects of the low latitude and polar ocean on atmospheric CO_2 are still a matter of debate (Broecker *et al.*, 1999; Toggweiler *et al.*, 2003a,b). Until our understanding of the modern ocean and carbon cycle has progressed to the degree that such major conceptual questions can be resolved, it will be extremely difficult to generate any consensus on the role of the biological pump in driving carbon dioxide change. After optimistic discussion of "proxy calibration" in the 1990s, we now recognize that the sedimentary and geochemical parameters we use to study the history of the biological pump are the product of multiple environmental variables, requiring that we think deeply about their significance in each case that they are applied. The quantity of work on the history of the biological pump has increased markedly over the last decade; a concerted effort to review, evaluate, and synthesize this information would greatly improve its usefulness. At the same time, we must continue to support a vanguard in search of untapped constraints. We have emphasized above that the biological pump is not strictly a biological phenomenon but rather results from the interaction of ocean biology, chemistry, and physics; as such, progress in its study is tied to our understanding of the history of the ocean in general, including its physical circulation and fundamental conditions (Adkins *et al.*, 2002).

ACKNOWLEDGMENTS

DMS is supported by the US NSF through grants OCE-9981479, OCE-0081686, DEB-0083566 (to Simon Levin), and OCE-0136449, and by British Petroleum and Ford Motor Company through the Princeton Carbon Mitigation Initiative. This chapter is dedicated to the memory of David S. Sigman, the only biochemist who would have tried to read it.

REFERENCES

Adkins J. F., McIntyre K., and Schrag D. P. (2002) The salinity, temperature, and delta O-18 of the glacial deep ocean. *Science* **298**(5599), 1769–1773.

Altabet M. A. (2001) Nitrogen isotopic evidence for micronutrient control of fractional NO_3^- utilization in the equatorial Pacific. *Limnol. Oceanogr.* **46**(2), 368–380.

Altabet M. A. and Francois R. (1994a) Sedimentary nitrogen isotopic ratio as a recorder for surface ocean nitrate utilization. *Global Biogeochem. Cycles* **8**(1), 103–116.

Altabet M. A. and Francois R. (1994b) The use of nitrogen isotopic ratio for reconstruction of past changes in surface ocean nutrient utilization. In *Carbon Cycling in the Glacial Ocean: Constraints on the Ocean's Role in Global Change* (eds. R. Zhan, M. Kaminski, L. Labeyrie, and T. F. Pederson). Springer, Berlin, Heidelberg, New York vol. 17, pp. 281–306.

Altabet M. A., Francois R., Murray D. W., and Prell W. L. (1995) Climate-related variations in denitrification in the Arabian Sea from sediment $^{15}N/^{14}N$ ratios. *Nature* **373**, 506–509.

Altabet M. A., Higginson M. J., and Murray D. W. (2002) The effect of millennial-scale changes in Arabian Sea denitrification on atmospheric CO_2. *Nature* **415**(6868), 159–162.

Anderson D. M. and Prell W. L. (1993) A 300 kyr record of upwelling off Oman during the Late Quaternary: evidence of the Asian southwest monsoon. *Paleoceanography* **8**, 193–208.

Anderson R., Kumar N., Mortlock R., Froelich P., Kubik P., Dittrich-Hannen B., and Suter M. (1998) Late-Quaternary changes in productivity of the Southern Ocean. *J. Mar. Sys.* **17**, 497–514.

Anderson R. F., Lehuray A. P., Fleisher M. Q., and Murray J. W. (1989) Uranium deposition in Saanich Inlet sediments, Vancouver Island. *Geochim. Cosmochim. Acta* **53**(9), 2205–2213.

Anderson R. F., Lao Y., Broecker W. S., Trumbore S. E., Hofmann H. J., and Wolfi W. (1990) Boundary scavenging in the Pacific Ocean: a comparison of ^{10}Be and ^{231}Pa. *Earth Planet. Sci. Lett.* **96**, 287–304.

Archer D. and Maier-Reimer E. (1994) Effect of deep-sea sedimentary calcite preservation on atmospheric CO_2 concentration. *Nature* **367**, 260–263.

Archer D., Winguth A., Lea D., and Mahowald N. (2000a) What caused the glacial/interglacial atmospheric pCO_2 cycles? *Rev. Geophys.* **38**(2), 159–189.

Archer D. E., Eshel G., Winguth A., Broecker W. S., Pierrehumbert R., Tobis M., and Jacob R. (2000b) Atmospheric pCO_2 sensitivity to the biological pump in the ocean. *Global Biogeochem. Cycles* **14**(4), 1219–1230.

Arrigo K. R., Robinson D. H., Worthen D. L., Dunbar R. B., DiTullio G. R., VanWoert M., and Lizotte M. P. (1999) Phytoplankton community structure and the drawdown of nutrients and CO_2 in the Southern Ocean. *Science* **283**(5400), 365–367.

Bacon M. P. (1984) Glacial to interglacial changes in carbonate and clay sedimentation in the Atlantic Ocean estimated from Th-230 measurements. *Isotope Geosci.* **2**, 97–111.

Bareille G., Labracherie M., Mortlock R. A., Maier-Reimer E., and Froelich P. N. (1998) A test of $(Ge/Si)_{opal}$ as a paleorecorder of $(Ge/Si)_{seawater}$. *Geology* **26**(2), 179–182.

Barnola J. M., Raynaud D., Korotkevich Y. S., and Lorius C. (1987) Vostok ice core provides 160,000-year record of atmospheric CO_2. *Nature* **329**, 408414.

Bassinot F. C., Labeyrie L. D., Vincent E., Quidellerur X., Shackleton N. J., and Lancelot Y. (1994) The astronomical theory of climate and the age of the Brunhes-Matuyama magnetic reversal. *Earth Planet. Sci. Lett.* **126**, 91–108.

Bauch D., Carstens J., and Wefer G. (1997) Oxygen isotope composition of living *Neogloboquadrina pachyderma* (sin.) in the Arctic Ocean. *Earth Planet. Sci. Lett.* **146**(1–2), 47–58.

Beaufort L. (2000) Dynamics of the monsoon in the equatorial Indian Ocean over the last 260,000 years. *Quat Int.* **31**, 13–18.

Berger A., Imbrie J., Hays J., Kukla G., and Saltzman B. (1984). Milankovitch and Climate, Reidel, Boston vol. 1, 493pp.

Berger W. H. and Keir R. S. (1984) Glacial-Holocene changes in atmospheric CO_2 and the deep-sea record. In *Climate Processes and Climate Sensitivity*. Geophyisical Monograph 29 (eds. J. E. Hansen and T. Takahashi). American Geophysical Union, Washington DC, pp. 337–351.

Berger W. H. and Lange C. B. (1998) Silica depletion in the thermocline of the glacial North Pacific: corollaries and implications. *Deep-Sea Res. II* **45**(8–9), 1885–1904.

Boyle E. A. (1988a) Cadmium: chemical tracer of deepwater paleoceanography. *Paleoceanography* **3**(4), 471–489.

Boyle E. A. (1988b) The role of vertical chemical fractionation in controlling late quaternary atmospheric carbon dioxide. *J. Geophys. Res.* **93**(C12), 15701–15714.

Boyle E. A. (1988c) Vertical oceanic nutrient fractionation and glacial/interglacial CO_2 cycles. *Nature* **331**, 55–56.

Boyle E. A. (1992) Cadmium and $\delta^{13}C$ paleochemical ocean distributions during the stage 2 glacial maximum. *Ann. Rev. Earth Planet. Sci.* **20**, 245–287.

Boyle E. A. (1998) Pumping iron makes thinner diatoms. *Nature* **393**, 733–734.

Boyle E. A. and Keigwin L. D. (1982) Deep circulation of the North Atlantic for the last 200,000 years: geochemical evidence. *Science* **218**, 784–787.

Boyle E. A., Labeyrie L., and Duplessy J.-C. (1995) Calcitic foraminiferal data confirmed by cadmium in aragonitic *Hoeglundina*: application to the last glacial maximum in the northern Indian Ocean. *Paleoceanography* **10**(5), 881–900.

Broecker W. S. (1981) Glacial to interglacial changes in ocean and atmospheric chemistry. In *Climatic Variations and Variability: Facts and Theory* (ed. A. Berger). Kluwer, Boston, pp. 111–121.

Broecker W. S. (1982a) Glacial to interglacial changes in ocean chemistry. *Progr. Oceanogr.* **2**, 151–197.

Broecker W. S. (1982b) Ocean chemistry during glacial time. *Geochim. Cosmochim. Acta* **46**, 1689–1706.

Broecker W. S. and Henderson G. M. (1998) The sequence of events surrounding Termination II and their implications for the cause of glacial–interglacial CO_2 changes. *Paleoceanography* **13**(4), 352–364.

Broecker W. S. and Maier-Reimer E. (1992) The influence of air and sea exchange on the carbon isotope distribution in the sea. *Global Biogeochem. Cycles* **6**(3), 315–320.

Broecker W. S. and Peng T.-H. (1982) *Tracers in the Sea*. Eldigio Press, Palisades, New York.

Broecker W. S. and Peng T.-H. (1987) The role of $CaCO_3$ compensation in the glacial to interglacial atmospheric CO_2 change. *Global Biogeochem. Cycles* **1**(1), 15–29.

Broecker W. S., Peacock S. L., Walker S., Weiss R., Fahrbach E., Schroeder M., Mikolajewicz U., Heinze C., Key R., Peng T. H., and Rubin S. (1998) How much deep water is formed in the Southern Ocean? *J. Geophys. Res.: Oceans* **103**(C8), 15833–15843.

Broecker W. S., Lynch-Stieglitz J., Archer D., Hofmann M., Maier-Reimer E., Marchal O., Stocker T., and Gruber N. (1999) How strong is the Harvardton-Bear constraint? *Global Biogeochem. Cycles* **13**(4), 817–820.

Broecker W. S., Clark E., Stiegllitz J. L., Beck W., Stott L. D., Hajdas I., and Bonani G. (2000) Late glacial diatom accumulation at 9° S in the Indian Ocean. *Paleoceanography* **15**, 348–352.

Brzezinski M. A., Pride C. J., Franck V. M., Sigman D. M., Sarmiento J. L., Matsumoto K., Gruber N., Rau G. H., and Coale K. H. (2002) A switch from $Si(OH)_4$ to NO_3^- depletion in the glacial Southern Ocean. *Geophys. Res. Lett.* **29**(12), 10.1029/12001GL014349.

Canfield D. E. (1994) Factors influencing organic carbon preservation in marine sediments. *Deep-Sea Res.* **114**, 315–329.

Charles C. D. and Fairbanks R. G. (1988) Glacial to interglacial changes in the isotopic gradients of southern Ocean surface water. In *Geological History of the Polar Oceans: Arctic versus Antarctic* (eds. U. Bleil and J. Thiede). Kluwer Academic, pp. 519–538.

Chase Z., Anderson R. F., and Fleisher M. Q. (2001) Evidence from authigenic uranium for increased productivity of the glacial Subantarctic Ocean. *Paleoceanography* **16**(5), 468–478.

Chase Z., Anderson R. F., Fleisher M. Q., and Kubik P. W. (2003a) Accumulation of biogenic and lithogenic material in the Pacific sector of the Southern Ocean during the past 40,000 years. *Deep-Sea Res. II* **50**(3–4), 739–768.

Chase Z., Anderson R. F., Fleisher M. Q., and Kubik P. W. (2003b) Scavenging of ^{230}Th, ^{231}Pa and ^{10}Be in the Southern Ocean (SW Pacific sector): the importance of particle flux, particle composition and advection. *Deep-Sea Res. II* **50** (3–4), 739–768.

Clemens S. C., Prell W. L., Murray D. W., Shimmield G., and Weedon G. (1991) Forcing mechanisms of the Indian Ocean monsoon. *Nature* **353**, 720–725.

Codispoti L. A. (1995) Biogeochemical cycles: Is the ocean losing nitrate? *Nature* **376**(6543), 724.

Conkright M., Levitus S., and Boyer T. (1994) *World Ocean Atlas 1994*, **vol. 1**: Nutrients. US Department of Commerce, Washington DC, pp. 16.

Crosta X., Pichon J.-J., and Burckle L. H. (1998) Application of modern analogue technique to marine Antarctic diatoms: reconstruction of maximum sea-ice extent at the Last Glacial Maximum. *Paleoceanography* **13**(3), 284–297.

Crusius J. and Thomson J. (2000) Comparative behavior of authigenic Re, U, and Mo during reoxidation and subsequent long-term burial in marine sediments. *Geochim. Cosmochim. Acta* **64**(13), 2233–2242.

Crusius J., Calvert S., Pedersen T., and Sage D. (1996) Rhenium and molybdenum enrichnments in sediments as indicators of oxic, suboxic and sulfidic conditions of deposition. *Earth Planet. Sci. Lett.* **145**(1–4), 65–78.

Dean W. E., Gardner J. V., and Piper D. Z. (1997) Inorganic geochemical indicators of glacial–interglacial changes in productivity and anoxia on the California continental margin. *Geochim. Cosmochim. Acta* **61**(21), 4507–4518.

Dehairs F., Stroobants N., and Goeyens L. (1991) Suspended barite as a tracer of biological activity in the Southern Ocean. *Mar. Chem.* **35**, 399–410.

De La Rocha C. L., Brzezinski M. A., and DeNiro M. J. (1997) Fractionation of silicon isotopes by marine diatoms during biogenic silica formation. *Geochim. Cosmochim. Acta* **61**, 5051–5056.

De La Rocha C. L., Brzezinski M. A., DeNiro M. J., and Shemesh A. (1998) Silicon-isotope composition of diatoms as an indicator of past oceanic change. *Nature* **395**(6703), 680–683.

Deutsch C., Gruber N., Key R. M., Sarmiento J. L., and Ganaschaud A. (2001) Denitrification and N_2 fixation in the Pacific Ocean. *Global Biogeochem. Cycles* **15**(2), 483–506.

DiTullio G. R., Grebmeier J. M., Arrigo K. R., Lizotte M. P., Robinson D. H., Leventer A., Barry J. P., VanWoert M. L., and Dunbar R. B. (2000) Rapid and early export of *Phaeocystis antarctica* blooms in the Ross Sea, Antarctica. *Nature* **404**, 595–598.

Duplessy J. C. (1982) Glacial to interglacial contrasts in the northern Indian Ocean. *Nature* **295**, 494–498.

Dymond J. and Lyle M. (1985) Flux comparisons between sediments and sediment traps in the eastern tropical Pacific:

implications for atmospheric CO_2 variations during the Pleistocene. *Limnol. Oceanogr.* **30**(4), 699–712.

Dymond J., Suess E., and Lyle M. (1992) Barium in deep-sea sediment: a geochemical proxy for paleoproductivity. *Paleoceanography* **7**(2), 163–181.

Elderfield H. and Rickaby R. E. M. (2000) Oceanic Cd/P ratio and nutrient utilization in the glacial Southern Ocean. *Nature* **405**, 305–310.

Elser J. J., Sterner R. W., Gorokhova E., Fagan W. F., Markow T. A., Cotner J. B., Harrison J. F., Hobbie S. E., Odell G. M., and Weider L. J. (2000) Biological stoichiometry from genes to ecosystems. *Ecol. Lett.* **3**, 540–550.

Emeis K.-C., Anderson D. M., Doose H., Kroon D., and Schulz-Bull D. E. (1995) Sea-surface temperatures and the history of monsoon upwelling in the NW Arabian Sea during the last 500,000 yr. *Quat. Res.* **43**, 355–361.

Emmer E. and Thunell R. C. (2000) Nitrogen isotope variations in Santa Barbara Basin sediments: implications for denitrification in the eastern tropical North Pacific during the last 50,000 years. *Paleoceanography* **15**(4), 377–387.

Fairbanks R. G. (1989) A 17,000-year glacio-eustatic sea level record: influence of glacial melting rates on the Younger Dryas event and deep-ocean circulation. *Nature* **342**, 637–642.

Falkowski P. G. (1997) Evolution of the nitrogen cycle and its influence on the biological sequestration of CO_2 in the ocean. *Nature* **387**(6630), 272–275.

Falkowski P. G. (2000) Rationalizing elemental ratios in unicellular algae. *J. Phycol.* **36**, 3–6.

Farrell J. W., Pedersen T. F., Calvert S. E., and Nielsen B. (1995) Glacial–interglacial changes in nutrient utilization in the equatorial Pacific Ocean. *Nature* **377**(6549), 514–517.

Fischer G. and Wefer G. (1999) *Use of Proxies in Paleoceanography: Examples from the South Atlantic* (eds. G. Fischer and G. Wefer). Springer, Berlin, 727pp.

Fontugne M. R. and Duplessy J.-C. (1986) Variations of the monsoon regime during the upper Quaternary: evidence from carbon isotopic record of organic matter in North Indian Ocean sediment cores. *Palaeogeogr. Paleoclimatol. Palaeoecol.* **56**, 69–88.

FranceLanord C. and Derry L. A. (1997) Organic carbon burial forcing of the carbon cycle from Himalayan erosion. *Nature* **390**(6655), 65–67.

Franck V. M., Brzezinski M. A., Coale K. H., and Nelson D. M. (2000) Iron and silicic acid availability regulate Si uptake in the Pacific Sector of the Southern Ocean. *Deep-Sea Res. II* **47**, 3315–3338.

François R., Bacon M. P., and Suman D. O. (1990) Thorium-230 profiling in deep-sea sediments: high-resolution records of flux and dissolution of carbonate in the equatorial Atlantic during the last 24,000 years. *Paleoceanography* **5**, 761–787.

François R. F., Altabet M. A., Yu E.-F., Sigman D. M., Bacon M. P., Frank M., Bohrmann G., Bareille G., and Labeyrie L. D. (1997) Water column stratification in the Southern Ocean contributed to the lowering of glacial atmospheric CO_2. *Nature* **389**, 929–935.

Frank M., Gersonde R., van der Loeff M.R., Bohrmann G., Nurnberg C. C., Kubik, P. W., Suter M., and Mangini A. (2000) Similar glacial and interglacial export bioproductivity in the Atlantic sector of the Southern Ocean: multiproxy evidence and implications for glacial atmospheric CO_2. *Paleoceanography* **15**(6), 642–658, 2000PA000497.

Frew R. D. and Hunter K. A. (1992) Influence of Southern Ocean waters on the cadmium-phosphate properties of the global ocean. *Nature* **360**(6400), 144–146.

Froelich P. N. and Andreae M. O. (1981) The marine geochemistry of germanium: ekasilicon. *Science* **213**, 205–207.

Froelich P. N., Bender M. L., Luedtke N. A., Heath G. R., and DeVries T. (1982) The marine phosphorus cycle. *Am. J. Sci.* **282**, 474–511.

Ganeshram R. S. and Pedersen T. F. (1998) Glacial–interglacial variability in upwelling and bioproductivity off NW Mexico: implications for quaternary paleoclimate. *Paleoceanography* **13**(6), 634–645.

Ganeshram R. S., Pedersen T. F., Calvert S. E., and Murray J. W. (1995) Large changes in oceanic nutrient inventories from glacial to interglacial periods. *Nature* **376**(6543), 755–758.

Ganeshram R. S., Pedersen T. F., Calvert S. E., McNeill G. W., and Fontugne M. R. (2000) Glacial–interglacial variability in denitrification in the world's oceans: causes and consequences. *Paleoceanography* **15**(4), 361–376.

Ganeshram R. S., Pedersen T. F., Calvert S. E., and Francois R. (2002) Reduced nitrogen fixation in the glacial ocean inferred from changes in marine nitrogen and phosphorus inventories. *Nature* **415**, 156–159.

Gildor H. and Tziperman E. (2001) Physical mechanisms behind biogeochemical glacial–interglacial CO_2 variations. *Geophys. Res. Lett.* **28**, 2421–2424.

Gildor H., Tziperman E., and Toggweiler J. R. (2002) Sea ice switch mechanism and glacial–interglacial CO_2 variations. *Global Biogeochem. Cycles* **16**(3) 10.1029/2001GB001446.

Gingele F. X., Zabel M., Kasten S., Bonn W. J., and Nurnberg C. C. (1999) Biogenic barium as a proxy for paleoproductivity: methods and limitations of application. In *Use of Proxies in Paleoceanography: Examples from the South Atlantic* (eds. G. Fischer and G. Wefer). Springer, Berlin, pp. 345–364.

Gorbarenko S. A. (1996) Stable isotope and lithologic evidence of late-Glacial and Holocene oceanography of the northwestern Pacific and its marginal seas. *Quat. Res.* **46**, 230–250.

Gordon A. L., Taylor H. W., and Georgi D. T. (1977) Antarctic oceanographic zonation. In *Polar Oceans* (ed. M. J. Dunbar). Arctic Institute of North America, pp. 219–225.

Gruber N. and Sarmiento J. L. (1997) Global patterns of marine nitrogen fixation and denitrification. *Global Biogeochem. Cycles* **11**, 235–266.

Hastings D. W., Emerson S. R., and Mix A. C. (1996) Vanadium in foraminiferal calcite as a tracer for changes in the areal extent of reducing conditions. *Paleoceanography* **11**(6), 666–678.

Haug G. H., Maslin M. A., Sarnthein M., Stax R., and Tiedemann R. (1995) Evolution of northwest Pacific sedimentation patterns since 6 Ma: site 882. *Proc. ODP. Sci. Results* **145**, 293–315.

Haug G. H., Pedersen T. F., Sigman D. M., Calvert S. E., Nielsen B., and Peterson L. C. (1998) Glacial/interglacial variations in productivity and nitrogen fixation in the Cariaco Basin during the last 550 ka. *Paleoceanography* **13**(5), 427–432.

Haug G. H., Sigman D. M., Tiedemann R., Pedersen T. F., and Sarnthein M. (1999) Onset of permanent stratification in the Subarctic Pacific Ocean. *Nature* **40**, 779–782.

Hays J. D., Imbrie J., and Shackleton N. J. (1976) Variations in the Earth's orbit: pacemaker of the ice ages. *Science* **194**, 1121–1132.

Hebbeln D., Marchant M., and Wefer G. (2002) Paleoproductivity in the southern Peru-Chile Current through the last 33 000 yr. *Mar. Geol.* **186**(3–4), 487–504.

Heinze H.-M. and Wefer G. (1992) The history of coastal upwelling off Peru over the past 650,000 years. In *Upwelling Systems: Evolution Since the Early Miocene* (eds. C. P. Summerhayes, W. L. Prell, and K.-C. Emeis). Geological Society of London, vol. 64, pp. 451–462.

Herbert T. D., Schuffert J. D., Andreasen D., Heusser L., Lyle M., Mix A., Ravelo A. C., Stott L. D., and Herguera J. C. (2001) Collapse of the California current during glacial maxima linked to climate change on land. *Science* **293**, 71–76.

Herguera J. C. (1992) Deep-sea benthic foraminifera and biogenic opal—glacial to postglacial productivity changes in the western equatorial Pacific. *Mar. Micropaleonthol.* **19**, 79–98.

Herguera J. C. and Berger W. H. (1991) Paleoproductivity from benthic foraminifera abundance—glacial to postglacial change in the west-equatorial Pacific. *Geology* **19**(12), 1173–1176.

Herguera J. C. and Berger W. H. (1994) Glacial to postglacial drop in productivity in the western equatorial Pacific—mixing rate versus nutrient concentrations. *Geology* **22**(7), 629–632.

Herguera J. C., Jansen E., and Berger W. H. (1992) Evidence for a bathyal front at 2,000 m depth in the glacial Pacific, based on a depth transect on Ontong Java Plateau. *Paleoceanography* **7**(3), 273–288.

Hinrichs K. U., Schneider R. R., Muller P. J., and Rullkotter J. (1999) A biomarker perspective on paleoproductivity variations in two Late Quaternary sediment sections from the Southeast Atlantic Ocean. *Org. Geochem.* **30**(5), 341–366.

Hofmann M., Broecker W. S., and Lynch-Stieglitz J. (1999) Influence of a [CO_2(aq)] dependent biological C-isotope fractionation on glacial C-13/C-12 ratios in the ocean. *Global Biogeochem. Cycles* **13**(4), 873–883.

Holmen K. (1992) The global carbon cycle. In *Global Biogeochemical Cycles* (eds. S. S. Butcher, R. J. Charlson, G. H. Orians, and G. V. Wolfe). Academic Press, New York, pp. 239–262.

Holmes M. E., Schneider R. R., Muller P. J., Segl M., and Wefer G. (1997) Reconstruction of past nutrient utilization in the eastern Angola Basin based on sedimentary $^{15}N/^{14}N$ ratios. *Paleoceanography* **12**(4), 604–614.

Huang C.-Y., Liew P.-M., Zhao M., Chang T.-C., Kuo C.-M., Chen M.-T., Wang C.-H., and Zheng L.-F. (1997) Deep sea and lake records of the Southeast Asian paleomonsoons for the last 25 thousand years. *Earth Planet. Sci. Lett.* **146**, 59–72.

Hutchins D. A. and Bruland K. W. (1998) Iron-limited diatom growth and Si : N uptake ratios in a coastal upwelling regime. *Nature* **393**, 561–564.

Janecek T. R. (2001) Data report: Late Pleistocene biogenic opal data for Leg 177 Sites 1093 and 1094. *Proc. Ocean Drilling Program, Sci. Results* **177**, 1–5.

Karl D., Letelier R., Tupas L., Dore J., Christian J., and Hebel D. (1997) The role of nitrogen fixation in biogeochemical cycling in the subtropical North Pacific Ocean. *Nature* **388**(6642), 533–538.

Kawahata H., Suzuki A., and Ahagon N. (1998) Biogenic sediments in the West Caroline Basin and the western equatorial Pacific during the last 330,000 years. *Mar. Geol.* **149**, 155–176.

Kawahata H., Ohkushi K. I., and Hatakeyama Y. (1999) Comparative Late Pleistocene paleoceanographic changes in the mid latitude Boreal and Austral Western Pacific. *J. Oceanogr.* **55**, 747–761.

Keeling R. F. and Stephens B. B. (2001) Antarctic sea ice and the control of Pleistocene climate instability. *Paleoceanography* **16**(1), 112–131.

Keeling R. F. and Visbeck M. (2001) Palaeoceanography: Antarctic stratification and glacial CO_2. *Nature* **412**, 605–606.

Keigwin L. D. and Boyle E. A. (1989) Late quaternary paleochemistry of high-latitude surface waters. *Paleogeogr. Palaeoclimatol. Paleoecol.* **73**, 85–106.

Keigwin L. D., Jones G. A., and Froelich P. N. (1992) A 15,000 year paleoenvironmental record from Meiji Seamount, far northwestern Pacific. *Earth Planet. Sci. Lett.* **111**, 425–440.

Keil R. G. and Cowie G. L. (1999) Organic matter preservation through the oxygen-deficient zone of the NE Arabian Sea as discerned by organic carbon: mineral surface area ratios. *Mar. Geol.* **161**(1), 13–22.

Keir R. and Berger W. (1983) Atmospheric CO_2 content in the last 120,000 years: the phosphate extraction model. *J. Geophys. Res.* **88**, 6027–6038.

Keir R. S. (1988) On the late Pleistocene ocean geochemistry and circulation. *Paleoceanography* **3**, 413–445.

Keller K. and Morel F. M. M. (1999) A model of carbon isotopic fractionation and active carbon uptake in phytoplankton. *Mar. Ecol. Prog. Ser.* **182**, 295–298.

Kohfeld K. E., Fairbanks R. G., Smith S. L., and Walsh I. D. (1996) *Neogloboquadrina pachyderma* (sinistral coiling) as paleoceanographic tracers in polar oceans: evidence from northeast water polynya plankton tows, sediment traps, and surface sediments. *Paleoceanography* **11**(6), 679–699.

Kumar N., Anderson R. F., Mortlock R. A., Froelich P. N., Kubik P., Dittrich-Hannen B., and Suter M. (1995) Increased biological productivity and export production in the glacial Southern Ocean. *Nature* **378**, 675–680.

Leuenberger M., Siegenthaler U., and Langway C. C. (1992) Carbon isotope composition of atmospheric CO_2 during the last Ice Age from an Antarctic ice core. *Nature* **357**(6378), 488–490.

Loubere P. (1999) A multiproxy reconstruction of biological productivity and oceanography in the eastern equatorial Pacific for the past 30,000 years. *Mar. Micropaleontol.* **37**, 173–198.

Lourey M. J., Trull T. W., and Sigman D. M. (2003) An unexpected decrease of δ^{15} N of surface and deep organic nitrogen accompanying Southern Ocean seasonal nitrate depletion. *Global Biogeochem. Cycles* (in press).

Lyle M. (1988) Climatically forced organic carbon burial in the equatorial Atlantic and Pacific Oceans. *Nature* **335**, 529–532.

Lyle M., Murray D., Finney B., Dymond J., Robbins J., and Brookforce K. (1988) The record of Last Pleistocene biogenic sedimentation in the eastern tropical Pacific Ocean. *Paleoceanography* **3**, 39–59.

Mahowald N., Kohfeld K., Hansson M., Balkanski Y., Harrison S. P., Prentice I. C., Schulz M., and Rodhe H. (1999) Dust sources and deposition during the last glacial maximum and current climate: a comparison of model results with paleodata from ice cores and marine sediments. *J. Geophys. Res.: Atmos.* **104**, 15895–15916.

Manighetti B. and McCave I. (1995) Depositional fluxes and paleoproductivity and ice rafting in the NE Atlantic over the past 30 ka. *Paleoceanography* **10**, 579–592.

Marcantonio F., Anderson R. F., Stute M., Kumar N., Schlosser P., and A M. (1996) Extraterrestrial He-3 as a tracer of marine sediment transport and accumulation. *Nature* **383**(6602), 705–707.

Marcantonio F., Anderson R., Higgins S., Fleisher M., Stute M., and Schlosser P. (2001a) Abrupt intensification of the SW Indian Ocean monsoon during the last delaciation: constraints from Th, Pa and He isotopes. *Earth Planet. Sci. Lett.* **184**, 505–514.

Marcantonio F., Anderson R., Higgins S., Stute M., and Schlosser P. (2001b) Sediment focussing in the central equatorial Pacific Ocean. *Paleoceanography* **16**, 260–267.

Marchitto T. M., Curry W. B., and Oppo D. W. (1998) Millennial-scale changes in North Atlantic circulation since the last glaciation. *Nature* **393**(6685), 557–561.

Marino B. D. and McElroy M. B. (1991) Isotopic composition fo atmospheric CO_2 inferred from carbon in C4 plant cellulose. *Nature* **349**, 127–131.

Martin J. H. (1990) Glacial–interglacial CO_2 change: the iron hypothesis. *Paleoceanography* **5**, 1–13.

Martin J. H., Fitzwater S. E., and Gordon R. M. (1990) Iron deficiency limits growth in Antarctic waters. *Global Biogeochem. Cycles* **4**, 5–12.

Martinez P. H., Bertrand P. H., Bouloubassi I., Bareille G., Shimmield G., Vautravers B., Grousset F., Guichard S., Ternois Y., and Sicre M.-A. (1996) An integrated view of inorganic and organic biogeochemical indicators of paleoproductivity changes in a coastal upwelling area. *Org. Geochem.* **24**, 411–420.

Matsumoto K., Oba T., Lynch-Stieglitz J., and Yamamoto H. (2002a) Interior hydrography and circulation of the glacial Pacific Ocean. *Quat. Sci. Rev.* **21**, 1693–1704.

Matsumoto K., Sarmiento J. L., and Brzezinski M. A. (2002b) Silicic acid "leakage" from the Southern Ocean as a possible mechanism for explaining glacial atmospheric pCO_2. *Global Biogeochem. Cycles* **16** 10.1029/2001GB001442.

McCorkle D. C., Martin P. A., Lea D. W., and Klinkhammer G. P. (1995) Evidence of a dissolution effect on benthic foraminiferal shell chemistry: delta C-13, Cd/Ca, Ba/Ca, and Sr/Ca results from the Ontong Java Plateau. *Paleoceanography* **10**(4), 699–714.

McCulloch R. D., Bentley M. J., Purves R. S., Hulton N. R. J., Sugden D. E., and Clapperton C. M. (2000) Climate inferences from glacial and palaeoecological evidence at the last glacial termination, southern South America. *J. Quat. Sci.* **15**, 409–417.

McDonald D., Pederson T., and Crusius J. (1999) Multiple Late Quaternary episodes of exceptional diatom production in the Gulf of Alaska. *Deep-Sea Res. II* **46**, 2993–3017.

McElroy M. B. (1983) Marine biological controls on atmospheric CO_2 and climate. *Nature* **302**, 328–329.

McIntyre A. and Molfino B. (1996) Forcing of Atlantic equatorial and subpolar millennial cycles by precession. *Science* **274**(5294), 1867–1870.

Mitchell B. G., Brody E. A., Holm-Hansen O., McClain C., and Bishop J. (1991) Light limitation of phytoplankton biomass and macronutrient utilization in the Southern Ocean. *Limnol. Oceanogr.* **36**(8), 1662–1677.

Monnin E., Indermuhle A., Dallenbach A., Fluckiger J., Stauffer B., Stocker T. F., Raynaud D., and Barnola J.-M. (2001) Atmospheric CO_2 concentrations over the last glacial termination. *Science* **291**, 112–114.

Montoya J. P. and McCarthy J. J. (1995) Isotopic fractionation during nitrate uptake by marine phytoplankton grown in continuous culture. *J. Plankton Res.* **17**(3), 439–464.

Moore J. K., Abbott M. R., Richman J. G., and Nelson D. M. (2000) The Southern Ocean at the last glacial maximum: a strong sink for atmospheric carbon dioxide. *Global Biogeochem. Cycles* **14**, 455–475.

Moreno A., Nave S., Kuhlmann H., Canals M., Targarona J., Freudenthal T., and Abrantes F. (2002) Productivity response in the North Canary Basin to climate changes during the last 250,000 yr: a multi-proxy approach. *Earth Planet. Sci. Lett.* **196**, 147–159.

Morley J. J. and Hays J. D. (1983) Oceanographic conditions associated with high abundances of the radiolarian *Cycladophora davisiana. Earth Planet. Sci. Lett.* **66**, 63–72.

Mortlock R. A., Charles C. D., Froelich P. N., Zibello M. A., Saltzman J., Hyas J. D., and Burckle L. H. (1991) Evidence for lower productivity in the Antarctic during the last glaciation. *Nature* **351**, 220–223.

Muller P. J. and Suess E. (1979) Productivity, sedimentation rate, and sedimentary organic matter in the oceans: I. Organic carbon preservation. *Deep-Sea Res. I* **26**, 1347–1362.

Murray R. W., Leinen M., and Isern A. R. (1993) Biogenic flux of Al to sediment in the central equatorial Pacific Ocean: evidence for increased productivity during glacial periods. *Paleoceanography* **8**(5), 651–670.

Nakatsuka T., Watanabe K., Handa N., Matsumoto E., and Wada E. (1995) Glacial to interglacial surface nutrient variation of the Bering deep basins recorded by delta-C-13 and delta-N-15 of sedimentary organic matter. *Paleoceanography* **10**(6), 1047–1061.

Narita H., Sato M., Tsunogai S., Murayama M., Nakatsuka T., Wakatsuchi M., Harada N., and Ujiié Y. (2002) Biogenic opal indicating less productive northwestern North Pacific during the glacial ages. *Geophys. Res. Lett.* **29**(15) 10.1029 22.1–22.4.

Opdyke B. N. and Walker J. C. G. (1992) Return of the coral reef hypothesis: basin to shelf potitioning of $CaCO_3$ and its effect on atmospheric CO_2. *Geology* **20**, 733–736.

Oppo D. W. and Lehman S. J. (1993) Mid-depth circulation of the subpolar North Atlantic during the last glacial maximum. *Science* **259**, 1148–1152.

Pailler D., Bard E., Rostek F., Zheng Y., Mortlock R., and van Geen A. (2002) Burial of redox-sensitive metals and organic matter in the equatorial Indian Ocean linked to precession. *Geochim. Cosmochim. Acta* **66**(5), 849–865.

Paytan A., Kastner M., Martin E. E., Macdougall J. D., and Herbert T. (1996) Glacial to interglacial fluctuations in productivity in the equatorial Pacific as indicated by marine barite. *Science* **274**(5291), 1355–1357.

Peacock S. (2001) Use of tracers to constrain time-averaged fluxes in the ocean. PhD, Columbia University.

Pedersen T. F. (1983) Increased productivity in the eastern equatorial Pacific during the last glacial maximum (19,000 to 14,000 yr B.P.). *Geology* **11**, 16–19.

Pedersen T. F., Nielsen B., and Pickering M. (1991) Timing of Late Quaternary productivity pulses in the Panama Basin and implications for atmospheric CO_2. *Paleoceanography* **6**, 657–677.

Perks H. M., Charles C. D., and Keeling R. F. (2000) Precessionally forced productivity variations across the equatorial Pacific. *Paleoceanography* **17**(3) 10.1029/2000PA000603.

Petit J. R., Jouzel J., Raynaud D., Barkov N. I., Barnola J.-M., Basile I., Bender M., Chappellaz J., Davis M., Delaygue G., Delmotte M., Kotlyakov V. M., Legrand M., Lipenkov V. Y., Lrius C., Pepin L., Ritz C., Saltzman E., and Stievenard M. (1999) Climate and atmospheric history of the past 420,000 years from the Vostok ice core, Antarctica. *Nature* **399**, 429–436.

Popp B. N., Laws E. A., Bidigare R. R., Dore J. R., Hanson K. L., and Wakeham S. G. (1998) Effect of phytoplankton cell geometry on carbon isotopic fractionation. *Geochim. Cosmochim. Acta* **62**(1), 69–77.

Prell W. L., Hutson W. H., Williams D. F., Be A. W. H., Geitzenauer K., and Molfino B. (1980) Surface circulation of the Indian Ocean during the Last Glacial maximum, approximately 18,000 yr BP. *Quat. Res.* **14**(3), 309–336.

Pride C., Thunell R., Sigman D., Keigwin L., Altabet M., and Tappa E. (1999) Nitrogen isotopic variations in the Gulf of California since the last deglaciation: response to global climate change. *Paleoceanography* **14**(3), 397–409.

Rau G. H., Tkahashi T., and Des Marais D. J. (1989) Latitudinal variations in plankton $\delta^{13}C$: implications for CO_2 and productivity in past oceans. *Nature* **341**, 516–518.

Rea D., Pisias N., and Newberry T. (1991) Late Pleistocene paleoclimatology of the central equatorial Pacific: flux patterns of biogenic sediments. *Paleoceanography* **6**, 227–244.

Redfield A. C. (1942) The processes determining the concentration of oxygen, phosphate and other organic derivatives within the depths of the Atlantic ocean. *Pap. Phys. Oceanogr. Meteorol.* **IX**.

Redfield A. C. (1958) The biological control of chemical factors in the environment. *Am. Sci.* **46**, 205–221.

Redfield A. C., Ketchum B. H., and Richards F. A. (1963) The influence of organisms on the composition of seawater. In *The Sea* (ed. M. N. Hill). Interscience, Vol. 2, pp. 26–77.

Reid J. L. (1969) Sea surface temperature, salinity, and density of the Pacific Ocean in summer and in winter. *Deep-Sea Res. I* **16**(suppl.), 215–224.

Rickaby R. E. M. and Elderfield H. (1999) Planktonic foraminiferal Cd/Ca: Paleonutrients or paleotemperature? *Paleoceanography* **14**(3), 293–303.

Rosenthal Y., Boyle E. A., Labeyrie L., and Oppo D. (1995) Glacial enrichments of authigenic Cd and U in Subantarctic sediments—a climatic control on the elements oceanic budget. *Paleoceanography* **10**(3), 395–413.

Rosenthal Y., Boyle E. A., and Labeyrie L. (1997) Last glacial maximum paleochemistry and deepwater circulation in the Southern Ocean: evidence from foraminiferal Cadmium. *Paleoceanography* **12**(6), 787–796.

Rosenthal Y., Dahan M., and Shemesh A. (2000) Southern Ocean contributions to glacial-interglacial changes of atmospheric CO_2: evidence from carbon isotope records in diatoms. *Paleoceanography* **15**(1), 65–75.

Ruhlemann C., Muller P. J., and Schneider R. R. (1999) Organic carbon and carbonate as paleoproductivity proxies: examples from high and low productivity areas of the tropical Atlantic. In *Use of Proxies in Paleoceanography: Examples from the South Atlantic* (eds. G. Fischer and G. Wefer). Springer, Berlin, pp. 315–344.

Ruhlemann M., Frank M., Hale W., Mangini A., Multiza P., and Wefer G. (1996) Late Quaternary productivity changes in the western equatorial Atlantic: evidence from ^{230}Th-normalized carbonate and organic carbon accumulation rates. *Mar. Geol.* **135**, 127–152.

Russell A. D., Emerson S., Mix A. C., and Peterson L. C. (1996) The use of foraminiferal uranium/calcium ratios as an indicator of changes in seawater uranium content. *Paleoceanography* **11**(6), 649–663.

Rutsch H.-J., Mangini A., Bonani G., Dittrich-Hannen B., Kubik P., Suter M., and Segl M. (1995) ^{10}Be and Ba concentrations in West African sediments trace productivity in the past. *Earth Planet. Sci. Lett.* **133**, 129–143.

Ruttenberg K. C. (1993) Reassessment of the oceanic residence time of phosphorous. *Chem. Geol.* **107**, 405–409.

Sachs J. P. and Repeta D. J. (1999) Oligotrophy and nitrogen fixation during Eastern Mediterranean sapropel events. *Science* **286**, 2485–2488.

Sachs J. P., Anderson R. F., and Lehman S. J. (2001) Glacial surface temperatures of the southeast Atlantic Ocean. *Science* **293**(5537), 2077–2079.

Sanudo-Wilhelmy S., Kustka A., Gobler C., Hutchins D., Yang M., Lwiza K., Burns J., Capone D., raven J., and Carpenter E. (2001) Phosphorus limitation of nitrogen fixation by *Trichodesmium* in the central Atlantic Ocean. *Nature* **411**(6833), 66–69.

Sanyal A., Hemming N. G., Broecker W. S., and Hanson G. N. (1997) changes in *p*H in the eastern equatorial Pacific across stage 5–6 boundary based on boron isotopes in foraminifera. *Global Biogeochem. Cycles* **11**(1), 125–133.

Sarnthein M., Winn K., Duplessy J.-C., and Fontugne M. R. (1988) Global variations of surface ocean productivity in low and mid latitudes: influence on CO_2 reservoirs of the deep ocean and atmosphere during the last 21,000 years. *Paleoceanography* **3**(3), 361–399.

Schneider R. R. (1991) Spätquartäre Produktivitätsänderungen im östlichen Anfola Becken: Reaktion auf Variationen im Passat-Monsun Windsystem und der Advektion des Benguela-Küstenstroms. *Ber. Fachbereich Geowiss., Univ. Bremen* **21**, 1–198.

Schneider R. R., Muller P. J., Ruhland G., Meinecke G., Schmidt H., and Wefer G. (1996) Late Quaternary surface temperatures and productivity in the East-equatorial South Atlantic: Response to changes in Trade/Monsoon wind forcing and surface water advection. In *The South Atlantic: Present and Past Circulation* (eds. G. Wefer, W. H. Berger, G. Siedler, and D. J. Webb). Springer, Berlin, pp. 527–551.

Schrag D., Hampt G., and Murray D. (1996) Pore fluid constraints on the temperature and oxygen isotopic composition of the glacial ocean. *Science* **272**(5270), 1930–1932.

Schubert C. J. and Stein R. (1996) Deposition of organic carbon in Arctic sediments: terrigenous supply versus marine productivity. *Org. Geochem.* **24**, 421–436.

Schwarz B., Mangini A., and Segl M. (1996) Geochemistry of a piston core from Ontong Java Plateau (western equatorial Pacific): evidence for sediment redistribution and changes in paleoproductivity. *Geol. Rundsch.* **85**, 536–545.

Shackelton N. J., Hall M. A., Line J., and Cang S. (1983) Carbon isotope data in core V19-30 confirm reduced carbon dioxide concentration of the ice age atmosphere. *Nature* **306**, 319–322.

Shemesh A., Macko S. A., Charles C. D., and Rau G. H. (1993) Isotopic evidence for reduced productivity in the glacial southern ocean. *Science* **262**, 407–410.

Sigman D. M. and Boyle E. A. (2000) Glacial/interglacial variations in atmospheric carbon dioxide. *Nature* **407**(6806), 859–869.

Sigman D. M. and Boyle E. A. (2001) Palaeoceanography: Antarctic stratification and glacial CO_2. *Nature* **412**, 606.

Sigman D. M., McCorkle D. C., and Martin W. R. (1998) The calcite lysocline as a constraint on glacial/interglacial low-latitude production changes. *Global Biogeochem. Cycles* **12**(3), 409–427.

Sigman D. M., Altabet M. A., Francois R., McCorkle D. C., and Fischer G. (1999a) The δ^{15}N of nitrate in the Southern Ocean: consumption of nitrate in surface waters. *Global Biogeochem. Cycles* **13**(4), 1149–1166.

Sigman D. M., Altabet M. A., Francois R., McCorkle D. C., and Gaillard J.-F. (1999b) The isotopic compositon of diatom-bound nitrogen in Southern Ocean sediments. *Paleoceanography* **14**(2), 118–134.

Sigman D. M., Lehman S. J., and Oppo D. W. (2003) Evaluating mechanisms of nutrient depletion and ^{13}C enrichment in the intermediate-depth Atlantic during the last ice age. *Paleoceanography* **18**(3), 1072, doi: 10.1029/2002PA000818.

Smith H. J., Fischer H., Wahlen M., Mastroianni D., and Deck B. (1999) Dual modes of the carbon cycle since the Last Glacial Maximum. *Nature* **400**, 248–250.

Sowers T. and Bender M. L. (1995) Climate records covering the last deglaciation. *Science* **269**, 210–214.

Spero H. J. and Lea D. W. (1993) Intraspecific stable isotope variability in the planktic foraminifera *Globigerinoides sacculifer*: Results from laboratory experiments. *Mar. Micropaleontol.* **22**, 221–234.

Spero H. J. and Williams D. F. (1988) Extracting environmental information from planktonic foraminiferal delta 13-C. *Nature* **335**, 717–719.

Spero H. J., Bijma J., Lea D. W., and Bemis B. E. (1997) Effect of seawater carbonate concentration on foraminiferal carbon and oxygen isotopes. *Nature* **390**, 497–499.

Stephens B. B. and Keeling R. F. (2000) The influence of Antarctic sea ice on glacial–interglacial CO_2 variations. *Nature* **404**, 171–174.

Stephens M. and Kadko D. (1997) Glacial-Holocene calcium carbonate dissolution at the central equatorial Pacific seafloor. *Paleoceanography* **12**, 797–804.

Stutt J. B. W., Prins M. A., Schneider R. R., Weltje G. J., Jansen J. H. F., and Postma G. (2002) A 300 kyr record of aridity and windstrength in southwestern Africa; inferences from grain-size distributions of sediments on Walvis Ridge, SE Atlantic. *Mar. Geol.* **180**(1–4), 221–233.

Suman D. O. and Bacon M. P. (1989) Variations in Holocene sedimentation in the North-American Basin determined from Th-230 measurements. *Deep-Sea Res. A* **36**(6), 869–878.

Summerhayes C. P., Kroon D., Rosell-Mele A., Jordan R. W., Schrader H.-J., Hearn R., Villanueva J., Grimalt J. O., and Eglinton G. (1995) Variability in the Benguela Current upwelling system over the past 70,000 years. *Progr. Oceanogr.* **35**, 207–251.

Takahashi T., Feely R. A., Weiss R. F., Wanninkohf R. H., Chipman D. W., Sutherland S. C., and Takahashi T. T. (1997) Global air-sea flux of CO_2: an estimate based on measurements of sea-air pCO(2) difference. *Proc. Natl. Acad. Sci. USA* **94**(16), 8292–8299.

Takeda S. (1998) Influence of iron availability on nutrient consumption ratio of diatoms in oceanic waters. *Nature* **393**, 774–777.

Talley L. D. (1993) Distribution and formation of North Pacific intermediate water. *J. Phys. Oceanogr.* **23**, 517–537.

Thomas E., Booth L., Maslin M., and Shackleton N. J. (1995) Northeastern Atlantic benthic foraminifera during the last 45,000 years—changes in productivity seen from the bottom up. *Paleoceanography* **10**(3), 545–562.

Thunell R. C., Miao Q., Calvert S. E., and Pederson T. F. (1992) Glacial–Holocene biogenic sedimentation patters in the South China Sea: productivity variations and surface water pCO_2. *Paleoceanography* **7**, 143–162.

Toggweiler J. R. (1999) Variations in atmospheric CO_2 driven by ventilation of the ocean's deepest water. *Paleoceanography* **14**(5), 571–588.

Toggweiler J. R. and Carson S. (1995) What are the upwelling systems contributing to the ocean's caron and nutrient budgets? In *Upwelling in the Ocean: Modern Processes and Ancient Records* (eds. C. P. Summerhayes, K.-C. Emeis, M. V. Angel, R. L. Smith, and B. Zeitzschel). Wiley, New York.

Toggweiler J. R. and Samuels B. (1995) Effect of Drake Passage on the global thermohaline circulation. *Deep-Sea Res. I* **42**(4), 477–500.

Toggweiler J. R., Carson S., and Bjornsson H. (1999) Response of the ACC and the Antarctic pycnocline to a meridional shift in the southern hemisphere westerlies. *EOS, Trans. AGU* **80**(49), OS286.

Toggweiler J. R., Gnanadesikan A., Carson S., Murnane R., and Sarmiento J. L. (2003a) Representation of the carbon cycle in box models and GCMs: 1. Solubility pump. *Global Biogeochem. Cycles* **17**(1) 0.1029/2001GB001401.

Toggweiler J. R., Murnane R., Carson S., Gnanadesikan A., and Sarmiento J. L. (2003b) Representation of the carbon cycle in box models and GCMs: 2. Organic pump. *Global Biogeochem. Cycles* **17**(1) 10.1029/2001GB001841.

Tortell P. D., DiTullio G. R., Sigman D. M., and Morel F. M. M. (2002) CO_2 effects on taxonomic composition and nutrient utilization in an equatorial Pacific phytoplankton assemblage. *Mar. Ecol. Progr. Ser.* **236**, 37–43.

Tyrrell T. (1999) The relative influences of nitrogen and phosphorus on oceanic primary production. *Nature* **400**, 525–531.

Vincent E. and Berger W. H. (1985) Carbon dioxide and polar cooling in the Miocene: the Monterey hypothesis. In *The Carbon Cycle and Atmospheric CO_2: Natural Variations Archean to Present* (eds. E. T. Sundquist and W. S. Broecker). American Geophysical Union, Washington DC, pp. 455–468.

Volk T. and Hoffert M. I. (1985) Ocean carbon pumps: analysis of relative strengths and efficiencies in ocean-driven atmospheric CO_2 changes. In *The Carbon Cycle and Atmospheric CO_2: Natural Variations Archean to Present* (eds. E. T. Sundquist and W. S. Broecker). American Geophysical Union, Washington DC, pp. 99–110.

Walsh J. J., Rowe G. T., Iverson R. L., and McRoy C. P. (1981) Biological export of shelf carbon is a neglected sink of the global CO_2 cycle. *Nature* **291**, 196–201.

Walter H.-J., Rutgers van der Loeff M. M., and Francois R. (1999) Reliability of the $^{231}Pa/^{230}Th$ activity ratio as a tracer for bioproductivity of the ocean. In *Use of Proxies in Paleoceanography: Examples from the South Atlantic* (eds. G. Fischer and G. Wefer). Springer, Berlin, pp. 393–408.

Warren B. (1983) Why is no deep water formed in the North Pacific? *J. Mar. Res.* **41**, 327–347.

Waser N. A. D., Turpin D. H., Harrison P. J., Nielsen B., and Calvert S. E. (1998) Nitrogen isotope fractionation during the uptake and assimilation of nitrate, nitrite, and urea by a marine diatom. *Limnol. Oceanogr.* **43**(2), 215–224.

Watson A. J., Bakker D. C. E., Ridgewell A. J., Boyd P. W., and Law C. S. (2000) Effect of iron supply on Southern Ocean CO_2 uptake and implications for atmospheric CO_2. *Nature* **407**(6805), 730–733.

Weaver A. J., Eby M., Fanning A. F., and Wiebe E. C. (1998) Simulated influence of carbon dioxide, orbital forcing and ice sheets on the climate of the Last Glacial Maximum. *Nature* **394**, 847–853.

Webb R. S., Lehman S. J., Rind D. H., Healy R. J., and Sigman D. M. (1997) Influence of ocean heat transport on the climate of the Last Glacial Maximum. *Nature* **385**(6618), 695–699.

Wefer G., Berger W. H., Siedler G., and Webb D. J. (1996) The South Atlantic: present and past circulation. Springer, Berlin.

Wollenburg J., Kuhnt W., and Mackensen A. (2001) Changes in Arctic Ocean paleproductivity and hydrography during the last 145 kyr: the benthic foraminiferal record. *Paleoceanography* **16**, 65–77.

Yu E. F., Francois R., Bacon M. P., Honjo S., Fleer A. P., Manganini S. J., van der Loeff M. M. R., and Ittekot V. (2001) Trapping efficiency of bottom-tethered sediment traps estimated from the intercepted fluxes of Th-230 and Pa-231. *Deep Sea Res. I* **48**(3), 865–889.

6.19
The Oceanic $CaCO_3$ Cycle

W. S. Broecker
Columbia University, Palisades, NY, USA

6.19.1 INTRODUCTION

Along with the silicate debris carried to the sea by rivers and wind, the calcitic hard parts manufactured by marine organisms constitute the most prominent constituent of deep-sea sediments. On high-standing open-ocean ridges and plateaus, these calcitic remains dominate. Only in the deepest portions of the ocean floor, where dissolution takes its toll, are sediments calcite-free. The foraminifera shells preserved in marine sediments are the primary carriers of paleoceanographic information. Mg/Ca ratios in these shells record past surface water temperatures; temperature corrected $^{18}O/^{16}O$ ratios record the volume of continental ice; $^{13}C/^{12}C$ ratios yield information about the strength of the ocean's biological pump and the amount of carbon stored as terrestrial biomass; the cadmium and zinc concentrations serve, respectively, as proxies for the distribution of dissolved phosphate and dissolved silica in the sea. While these isotopic ratios and trace element concentrations constitute the workhorses of the field of paleoceanography, the state of preservation of the calcitic material itself has an important story to tell. It is this story with which this chapter is concerned.

In all regions of the ocean, plots of sediment composition against water depth have a characteristic shape. Sediments from mid-depth are rich in $CaCO_3$ and those from abyssal depths are devoid of $CaCO_3$. These two realms are separated by a transition zone spanning several hundreds of meters in water depth over which the $CaCO_3$ content drops toward zero from the 85–95% values which characterize mid-depth sediment. The upper bound of this transition zone has been termed the "lysocline" and signifies the depth at which dissolution impacts become noticeable. The lower bound is termed the "compensation depth" and signifies the depth at which the $CaCO_3$ content is reduced to 10%. While widely used (and misused), both of these terms suffer from ambiguities. My recommendation is that they be abandoned in favor of the term "transition zone." Where quantification is appropriate, the depth of the mid-point of $CaCO_3$ decline should be given. While the width of the zone is also of interest, its definition suffers from the same problems associated with the use of the terms "lysocline" and "compensation depth," namely, the boundaries are gradual rather than sharp.

While determinations of sediment $CaCO_3$ content as a function of water depth in today's ocean or at any specific time in the past constitute a potentially useful index of the extent of dissolution, it must be kept in mind that this relationship is highly nonlinear. Consider, for example, an area where the rain rate of calcite to the seafloor is 9 times that of noncarbonate material. In such a situation, were 50% of the calcite to be dissolved, the $CaCO_3$ content would drop only from 90% to only 82%, and were 75% dissolved away, it would drop only to 69% (see Figure 1). One might counter by saying that as the $CaCO_3$ content can be measured to an accuracy of ±0.5% or better, one could still use $CaCO_3$ content as a dissolution index. The problem is that in order to obtain a set of sediment samples covering an appreciable range of water depth, topographic gradients dictate that the cores would have to be collected over an area covering several degrees. It is unlikely that the ratio of the rain rate of calcite to that of noncalcite would be exactly the same at all the coring sites. Hence, higher accuracy is not the answer.

6.19.2 DEPTH OF TRANSITION ZONE

As in most parts of today's deep ocean the concentrations of Ca^{2+} and of CO_3^{2-} are nearly constant with water depth, profiles of $CaCO_3$ content with depth reflect mainly the increase in the solubility of the mineral calcite with pressure (see Figure 2). This increase occurs because the volume occupied by the Ca^{2+} and CO_3^{2-} ions dissolved in seawater is smaller than when they are combined in the mineral calcite. Unfortunately, a sizable uncertainty exists in the magnitude of this volume difference. The mid-depth waters in the ocean are everywhere supersaturated with respect to calcite. Because of the pressure dependence of solubility, the extent of supersaturation decreases with depth until the saturation horizon is reached. Below this depth, the waters are undersaturated with respect to calcite. While it is tempting to conclude that the saturation horizon corresponds to the top of the transition zone, as we shall see, respiration CO_2 released to the pore waters complicates the situation by inducing calcite dissolution above the saturation horizon.

One might ask what controls the depth of the transition zone. The answer lies in chemical economics. In today's ocean, marine organisms manufacture calcitic hard parts at a rate several times faster than CO_2 is being added to the ocean–atmosphere system (via planetary outgassing and weathering of continental rocks) (see Figure 3). While the state of saturation in the ocean is set by

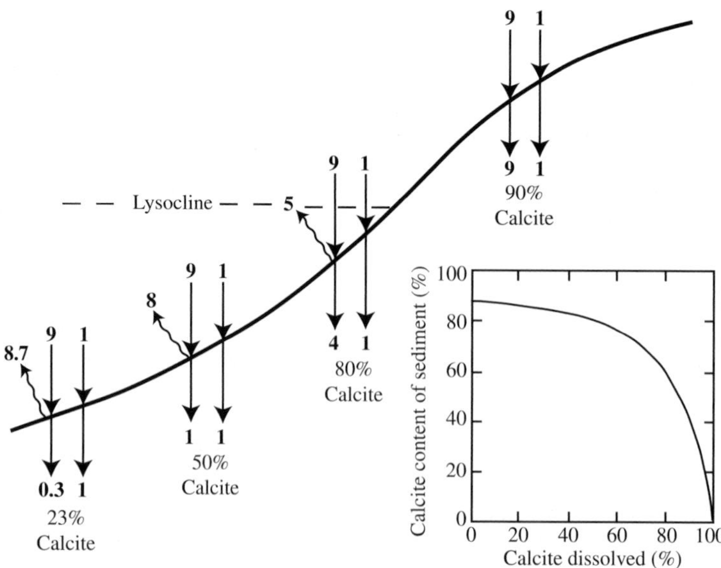

Figure 1 A diagrammatic view of how the extent of dissolution impacts the percent calcite in the sediment. In each example, the right-hand vertical arrows give the rain rate and accumulation rate of non-$CaCO_3$ debris and the left-hand vertical arrows the rain rate and accumulation rate of calcite. The wavy arrows represent the dissolution rates of calcite. As can be seen from the graph on the lower right, the percent of calcite in the sediment gives a misleading view of the fraction of the raining calcite which has dissolved, for large amounts of dissolution are required before the calcite content of the sediment drops significantly.

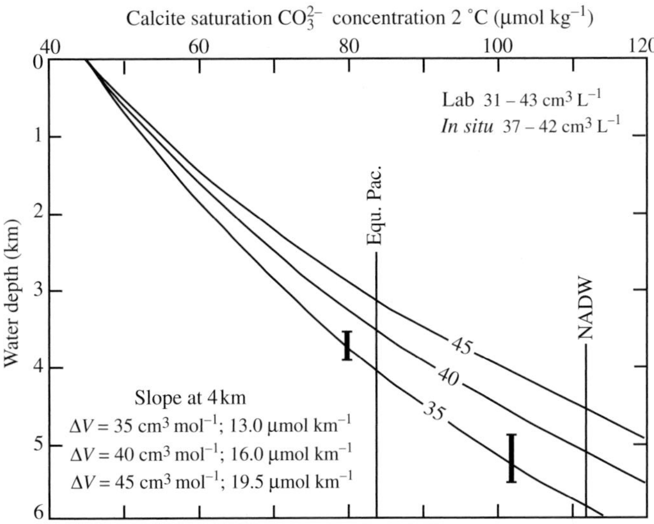

Figure 2 Saturation carbonate ion concentration for calcite at deep-water temperatures as a function of water depth (i.e., pressure). Curves are drawn for three choices of ΔV between Ca^{2+} and CO_3^{2-} ions when combined in calcite and when dissolved in seawater. The vertical black lines indicate the average CO_3^{2-} concentration in the deep equatorial Pacific and in NADW. The bold segments indicate the onset of dissolution as indicated by *in situ* experiments carried out in the deep North Pacific (Peterson, 1966) and in the deep North Atlantic (Honjo and Erez, 1978). As indicated in the upper right-hand corner, estimates of ΔV cover a wide range.

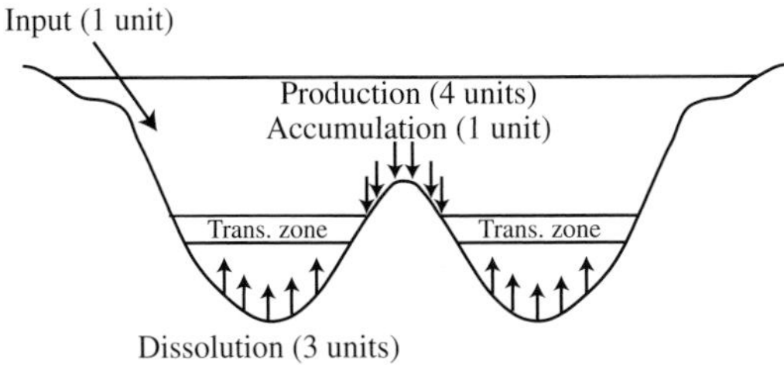

Figure 3 Marine organisms produce calcite at ~4 times the rate at which the ingredients for this mineral are supplied to the sea by continental weathering and planetary outgassing. A transition zone separates the mid-depth ocean floor where calcite is largely preserved from the abyssal ocean floor where calcite is largely dissolved.

the product of the Ca^{2+} and CO_3^{2-} concentrations, calcium has such a long residence (10^6 yr) that, at least on the timescale of a single glacial cycle (~10^5 yr), its concentration can be assumed to have remained unchanged. Further, its concentration in seawater is so high that $CaCO_3$ cycling within the sea does not create significant gradients. In contrast, the dissolved inorganic carbon (i.e., CO_2, HCO_3^-, and CO_3^{2-}) in the ocean is replaced on a timescale roughly equal to that of the major glacial to interglacial cycle (10^5 yr). But, since in the deep sea CO_3^{2-} ion makes up only ~5% of the total dissolved inorganic carbon, its adjustment time turns out to be only about one-twentieth that for dissolved inorganic carbon or ~5,000 yr. Hence, the concentration of CO_3^{2-} has gradients within the sea and likely has undergone climate-induced changes.

Hence, at least on glacial to interglacial timescales, attention is focused on distribution of CO_3^{2-} concentration in the deep sea for it alone sets the depth of the transition zone. Thus, it is temporal changes in the concentration of carbonate ion which have captured the attention of those paleoceanographers interested in glacial to interglacial changes in ocean operation. These changes involve both the carbonate ion concentration averaged over the entire deep ocean and its distribution with respect to water depth and geographic location. Of course, it is the global average carbonate ion concentration in the deep sea that adjusts in order to assure that burial of $CaCO_3$ in the sediments matches the input of CO_2 to the ocean atmosphere system (or, more precisely, the input minus the fraction destined to be buried as organic residues). For example, were some anomaly to cause the burial of $CaCO_3$ in seafloor sediments to exceed supply, then the CO_3^{2-} concentration would be drawn down. This drawdown would continue until a balance between removal and supply was restored. As already mentioned, the time constant for this adjustment is on the order of 5,000 yr.

6.19.3 DISTRIBUTION OF CO_3^{2-} ION IN TODAY'S DEEP OCEAN

As part of the GEOSECS, TTO, SAVE and WOCE ocean surveys, $\sum CO_2$ and alkalinity measurements were made on water samples captured at various water depths in Niskin bottles. Given the depth, temperature and salinity for these samples, it is possible to compute the *in situ* carbonate ion concentrations. LDEO's Taro Takahashi played a key role not only in the measurement programs, but also in converting the measurements to *in situ* carbonate ion concentrations. Because of his efforts and, of course, those of many others involved in these expeditions, we now have a complete picture of the distribution of CO_3^{2-} ion concentrations in the deep sea.

Below 1,500 m in the world ocean, the distribution of carbonate ion concentration is remarkably simple (see Broecker and Sutherland, 2000 for summary). For the most part, waters in the Pacific, Indian, and Southern Oceans have concentrations confined to the range $83 \pm 8\ \mu mol\ kg^{-1}$. The exception is the northern Pacific, where the values drop to as low as $60\ \mu mol\ kg^{-1}$. In contrast, much of the deep water in the Atlantic has concentrations in the range $112 \pm 5\ \mu mol\ kg^{-1}$. The principal exception is the deepest portion of the western basin where Antarctic bottom water (AABW) intrudes.

As shown by Broecker *et al.*, the deep waters of the ocean can be characterized as a mixture of two end members, i.e., deep water formed in the northern Atlantic and deep water formed in the Southern Ocean. These end members are characterized by quite different values of a quasi-conservative property, PO_4^* (i.e., $PO_4 - 1.95 + O_2/175$). Although these two deep-water sources have similar initial O_2 contents, those formed in the northern Atlantic have only roughly half the PO_4 concentration of the deep waters descending in the Southern Ocean. Thus, the northern end member is characterized by a PO_4^* value of 0.73 ± 0.03, while the southern end member is characterized by a value of $1.95 \pm 0.05\ \mu mol\ kg^{-1}$. In Figure 4 is shown a plot of carbonate ion concentration for waters deeper than 1,700 m as a function of PO_4^*. The points are color coded according to O_2 content. As can be seen, the high O_2 waters with northern Atlantic PO_4^* values have carbonate ion concentrations of $\sim 120\ \mu mol\ kg^{-1}$, while those formed in Weddell Sea and Ross Sea have values closer to $90\ \mu mol\ kg^{-1}$.

The sense of the between-ocean difference in carbonate ion concentration is consistent with the PO_4^*-based estimate that Atlantic deep water (i.e., North Atlantic deep water (NADW)) is a mixture of about 85% deep water formed in the northern Atlantic and 15% deep water formed in Southern Ocean, while the remainder of the deep ocean is flooded with a roughly 50–50 mixture of these two source waters (Broecker, 1991). The interocean difference in carbonate ion concentration relates to the fact that deep

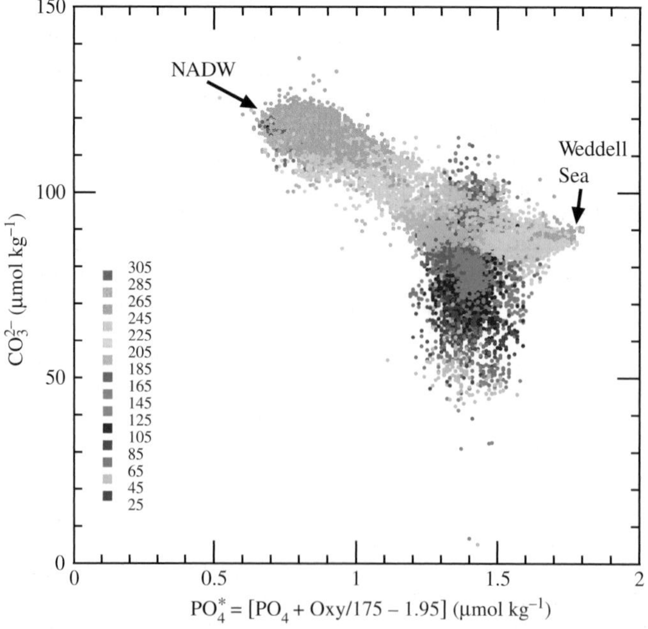

Figure 4 Plot of carbonate ion concentration as a function of PO_4^* for waters deeper than 1,700 m in the world ocean. The points are coded according to O_2 content (in $\mu mol\ kg^{-1}$). This plot was provided by LDEO'S Stew Sutherland and Taro Takahashi. It is based on measurements made as part of the GEOSECS, TTO, SAVE and WOCE surveys.

water formed in the northern Atlantic has a higher CO_3^{2-} concentration than that produced in the Southern Ocean. The transition zone between NADW and the remainder of the deep ocean is centered in the western South Atlantic and extends around Africa into the Indian Ocean (fading out as NADW mixes into the ambient circumpolar deep water).

The difference in carbonate ion concentration between NADW and the rest of the deep ocean is related to the difference in PO_4 concentration. NADW has only about half the concentration of PO_4 as does, for example, deep water in equatorial Pacific. This is important because, for each mole of phosphorus released during respiration, ~120 mol of CO_2 are also produced. This excess CO_2 reacts with CO_3^{2-} ion to form two HCO_3^- ions. Were PO_4 content the only factor influencing the interocean difference in carbonate ion concentration, then it would be expected to be more like 90 μmol kg^{-1} rather than the observed 30 μmol kg^{-1}. So, something else must be involved.

This something is CO_2 transfer through the atmosphere (Broecker and Peng, 1993). The high-phosphate-content waters upwelling in the Southern Ocean lose part of their excess CO_2 to the atmosphere. This results in an increase in their CO_3^{2-} ion content. In contrast, the low-PO_4-content waters reaching in the northern Atlantic have CO_2 partial pressures well below that in the atmosphere and hence they absorb CO_2. This reduces their CO_3^{2-} concentration. Hence, it is the transfer of CO_2 from surface waters in the Southern Ocean to surface waters in the northern Atlantic reduces the contrast in carbonate ion concentration between deep waters in the deep Atlantic and those in the remainder of the deep ocean.

One other factor expected to have an impact on the carbonate ion concentration in deep Pacific Ocean and Indian Ocean turns out to be less important. Much of the floor of these two oceans lies below the transition zone. Hence, most of the $CaCO_3$ falling into the deep Pacific Ocean and Indian Ocean dissolves. One would expect then that the older the water (as indicated by lower $^{14}C/C$ ratios), the higher its CO_3^{2-} ion concentration would be. While to some extent this is true, the trend is much smaller than expected. The reason is that in the South Pacific Ocean and South Indian Ocean an almost perfect chemical titration is being conducted, i.e., for each mole of respiration CO_2 released to the deep ocean, roughly one mole of $CaCO_3$ dissolves (Broecker and Sutherland, 2000). So indeed, the older the water, the higher its $\sum CO_2$ content. But, due to $CaCO_3$ dissolution, there is a compensating increase in alkalinity such that the carbonate ion concentration remains largely unchanged. Only in the northern reaches of these oceans does the release of metabolic CO_2 overwhelm the supply of $CaCO_3$ allowing the CO_3 concentration to drop.

As in the depth range of transition zone, the solubility of $CaCO_3$ increases by ~14 μmol kg^{-1} km^{-1} increase in water depth, the 30 μmol kg^{-1} higher CO_3^{2-} concentration in NADW should (other things being equal) lead to a 2 km deeper transition zone in the Atlantic than in the Pacific Ocean and the Indian Ocean. In fact, this is more or less what is observed.

6.19.4 DEPTH OF SATURATION HORIZON

A number of attempts have been made to establish the exact depth of the calcite saturation horizon. The most direct way to do this is to suspend preweighed calcite entities at various water depths on a deep-sea mooring, then months later, recover the mooring and determine the extent of weight loss (see Figure 5). Peterson (1966) performed such an experiment at 19 °N in the Pacific Ocean using polished calcite spheres and observed a pronounced depth-dependent increase in weight loss that commenced at ~3,900 m. Honjo and Erez (1978) performed a similar experiment at 32 °N in the Atlantic and found that coccoliths, foraminifera shells and reagent calcite experienced a 25–60% weight loss at 5,500 m but no measurable weight loss at 4,900 m. Thus the North Atlantic–North Pacific depth difference in the depth of the onset of dissolution is more or less consistent with expectation. Broecker and Takahashi (1978) used a combination of the depth of the onset of sedimentary $CaCO_3$ content decline and the results of a technique referred to as *in situ* saturometry (Ben-Yaakov and Kaplan, 1971) to define the depth dependence of solubility. While fraught with caveats, these results are broadly consistent with those from the mooring experiment. By measuring the composition of pore waters extracted *in situ* from sediments at various water depths, Sayles (1985) was able to calculate what he assumed to be saturation CO_3^{2-} concentrations. Finally, several investigators have performed laboratory equilibrations of calcite and seawater as a function of confining pressure. But, as each approach is subject to biases, more research is needed before the exact pressure dependence of the solubility of calcite can be pinned down.

6.19.5 DISSOLUTION MECHANISMS

Three possible dissolution processes come to mind. The first of these is termed water column dissolution. As foraminifera shells fall quite rapidly and as they encounter calcite undersaturated water only at great depth, it might be concluded that dissolution during fall is unimportant. But it has been suggested that

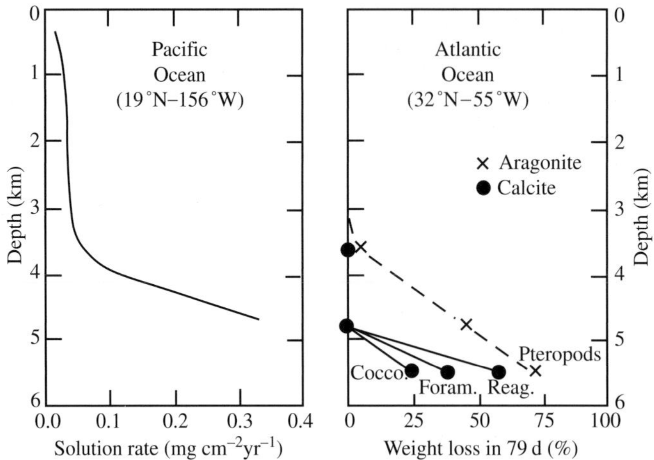

Figure 5 Results of *in situ* dissolution experiments. Peterson (1966) re-weighed polished calcite spheres after a 250 d deployment on a mooring in the North Pacific. Honjo and Erez (1978) observed the weight loss for calcitic samples (coccoliths, foraminifera and reagent calcite) and an aragonitic sample (pteropods) held at depth for a period of 79 d. While Peterson hung his spheres directly in seawater, the Honjo–Erez samples were held in containers through which water was pumped. The results suggest that the calcite saturation horizon lies at 4,800 ± 200 m in the North Atlantic and at about 3,800 ± 200 m in the North Pacific. For aragonite, which is 1.4 times more soluble than calcite, the saturation horizon in the North Atlantic is estimated to be in the range 3,400 ± 200 m.

organisms feeding on falling debris ingest and partially dissolve calcite entities (Milliman *et al.*, 1999). Because of their small size, coccoliths are presumed to be the most vulnerable in this regard. But little quantitive information is available to permit quantification of this mode of dissolution.

The other two processes involve dissolution of calcite after it reaches the seafloor. A distinction is made between dissolution that occurs before burial (i.e., interface dissolution) and dissolution that takes place after burial (i.e., pore-water dissolution). The former presumably occurs only at water depths greater than that of the saturation horizon. But the latter has been documented to occur above the calcite saturation horizon. It is driven by respiration CO_2 released to the pore waters.

Following the suggestion of Emerson and Bender (1981) that the release of respiration CO_2 in pore waters likely drives calcite dissolution above the saturation horizon, a number of investigators took the bait and set out to explore this possibility. David Archer, as part of his PhD thesis research with Emerson, developed pH microelectrodes that could be slowly ratcheted into the upper few centimeters of the sediment from a bottom lander. He deployed these pH microelectrodes along with the O_2 microelectrodes and was able to show that the release of respiration CO_2 (as indicated by a reduction in pore-water O_2) was accompanied by a drop in pH (and hence also of CO_3^{2-} ion concentration). Through modeling the combined results, Archer *et al.* (1989) showed that much of the CO_2 released by respiration reacted with $CaCO_3$ before it had a chance to escape (by molecular diffusion) into the overlying bottom water. As part of his PhD research, Burke Hales, a second Emerson student, improved Archer's electrode system and made measurements on the Ceara Rise in the western equatorial Atlantic (Hales and Emerson, 1997) and on the Ontong–Java Plateau in the western equatorial Pacific (Hales and Emerson, 1996) (see Figure 6). Taken together, these two studies strongly support the proposal that dissolution in pore waters of sediments leads to substantial dissolution of calcite. This approach has been improved upon by the addition of an LIX electrode to measure CO_2 itself and a micro-optode to measure Ca^{2+} (Wenzhöfer *et al.*, 2001).

In another study designed to confirm that most of the CO_2 released into the upper few centimeters of the sediments reacts with $CaCO_3$ before escaping to the overlying bottom water, Martin and Sayles (1996) deployed a very clever device that permitted the *in situ* collection of closely spaced pore-water samples in the upper few centimeters of the sediment column. Measurements of $\sum CO_2$ and alkalinity on these pore-water samples revealed that the gradient of $\sum CO_2$ (μmol km^{-1}) with depth is close to that of alkalinity (μequiv. kg^{-1}). This can only be the case if much of the respiration CO_2 reacts with $CaCO_3$ to form a Ca^{2+} and two HCO_3^- ions.

Dan McCorkle of Woods Hole Oceanographic Institution conceived of yet another way to confirm that pore-water respiration CO_2 was largely neutralized by reaction with $CaCO_3$. As summarized in Figure 7, he made $^{13}C/^{12}C$ ratio measurements on $\sum CO_2$ from pore-water profiles and found that the trend of $\delta^{13}C$ with

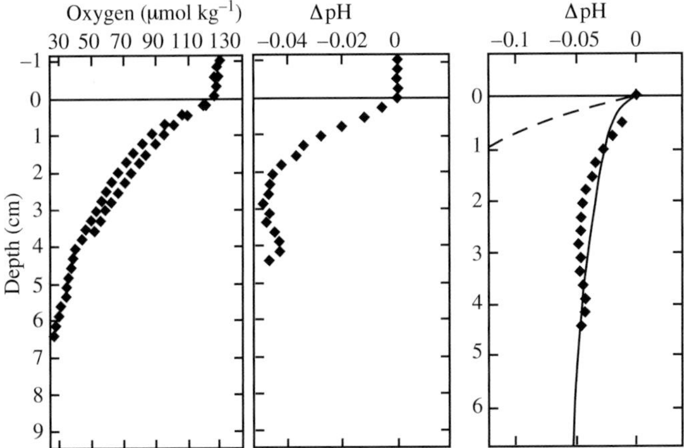

Figure 6 Microelectrode profiles of dissolved O_2 and ΔpH obtained by Hales and Emerson (1996) at 2.3 km depth on the Ontong–Java Plateau in the western equatorial Pacific. On the right are model curves showing the pH trend expected if none of the CO_2 released during the consumption of the O_2 was neutralized by reaction with sediment $CaCO_3$ (dashed curve) and a best model fit to the measured ΔpH trend (solid curve). The latter requires that much of the respiration CO_2 reacts with $CaCO_3$ before it escapes into the overlying bottom water.

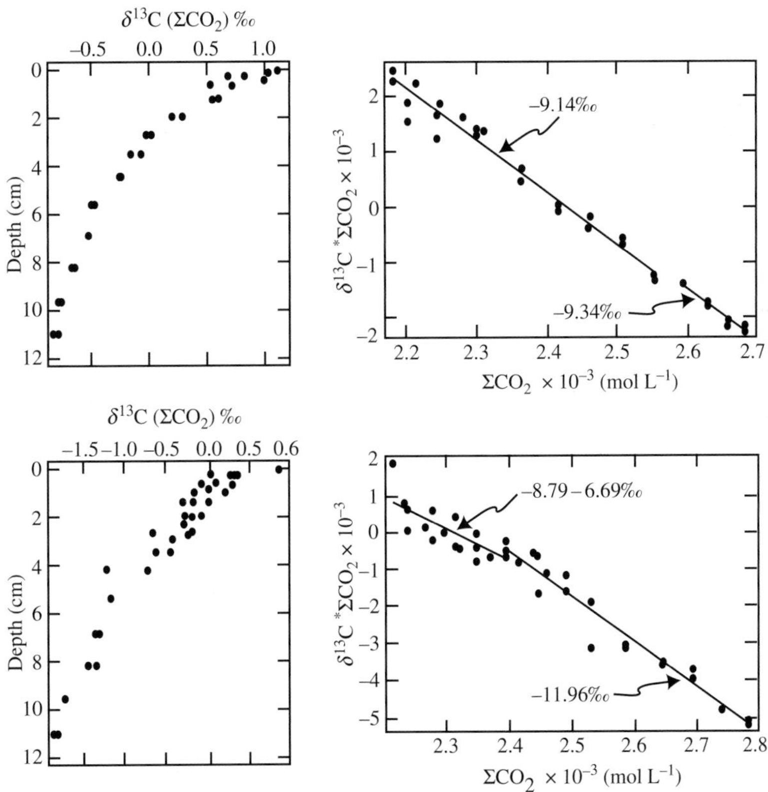

Figure 7 On the left are shown plots versus depth in the sediment of $\delta^{13}C$ in pore water total dissolved inorganic carbon (i.e., $\sum CO_2$) for two sites on the Ceara Rise (5 °S in the western Atlantic) (Martin *et al.*, 2000). On the right are plots of $\delta^{13}C$ versus $\sum CO_2$. The slopes yield the isotopic composition of the excess CO_2. As can be seen, it requires that the respiration CO_2 be diluted with a comparable contribution from dissolved $CaCO_3$.

excess $\sum CO_2$ is consistent with a 50–50 mixture of carbon derived from marine organic matter (−20‰) and that derived from marine calcite (+1‰) (Martin *et al.*, 2000). Again, these results require that a large fraction of the metabolic CO_2 reacts with $CaCO_3$.

There is, however, a fly in the ointment. Benthic flux measurements made by deploying chambers

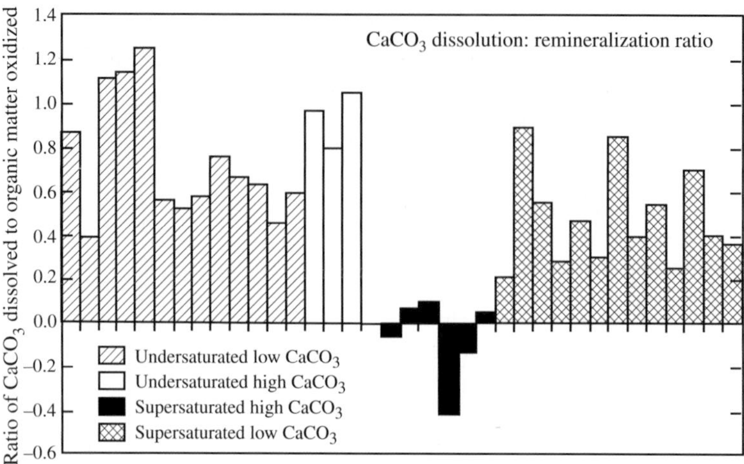

Figure 8 Summary of the ratio of $CaCO_3$ dissolved and organic material oxidized for bottom chamber deployments in the northeastern Pacific, Ontong–Java Plateau, Ceara Rise, Cape Verde Plateau, northwestern Atlantic continental rise and California borderland basins (R. A. Jahnke and D. B. Jahnke, 2002). The absence of measurable alkalinity fluxes from high-$CaCO_3$ sites bathed in supersaturated bottom water appears to be inconsistent with observations (see text).

on the seafloor reveal a curious pattern (see Figure 8). R. A. Jahnke and D. B. Jahnke (2002) found that alkalinity and calcium fluxes from sediments (both high and low in $CaCO_3$ content) below the calcite saturation horizon and on low-$CaCO_3$-content sediments from above the saturation horizon yield more or less the expected fluxes. However, chambers deployed on high-$CaCO_3$-content sediments from above the saturation horizon yield no measurable alkalinity flux. Yet pore-water profiles and electrode measurements for these same sediments suggest that calcite is dissolving. Whole foraminifera shell weight and $CaCO_3$ size index measurements (see below) agree with conclusion of these authors that calcite dissolution is not taking place. R. A. Jahnke and D. B. Jahnke (2002) propose that impure $CaCO_3$ coatings formed on the surfaces of calcite grains are redissolved in contact with respiration CO_2-rich pore waters and that the products of this dissolution diffuse back to sediment–water interface. Based on this scenario, the reason that the these authors record no calcium or alkalinity flux is that the ingredients for upward diffusion of calcium and alkalinity are being advected downward bound to the surfaces of calcite grains. Hence, there is no net flux of either property into their benthic chamber. That such coatings form was demonstrated long ago by Weyl (1965), who showed that when exposed in the laboratory to supersaturated seawater it was the calcite crystal surfaces that achieved saturation equilibrium with seawater rather than vice versa.

Broecker and Clark (2003) fortify the mechanism proposed by R.A. Jahnke and D. B. Jahnke, providing additional evidence by proposing that it must be coatings rather than the biogenic calcite itself that dissolve. As shown in Figure 9, while on the Ontong–Java Plateau, there is a progressive decrease in shell weight and $CaCO_3$ size index with water depth; on the Ceara Rise, neither of these indices shows a significant decrease above a water depth of 4,100 m. This is consistent with the conclusion that no significant dissolution occurs at the depth of 3,270 m where the pore-water and chamber measurements were made.

Although Berelson *et al.* (1994) report chamber-based alkalinity fluxes from high-calcite sediment, the sites at which their studies were performed are very likely bathed in calcite-undersaturated bottom water. If so, coatings would not be expected to form.

One other observation, i.e., core-top radiocarbon ages, appears to be at odds with pore-water dissolution. The problem is as follows. To the extent that respiration CO_2-driven dissolution occurring in the core-top bioturbated zone is homogeneous (i.e., all calcite entities lose the same fraction of their weight in a unit of time), then the core-top radiocarbon age should decrease slowly with increasing extent of dissolution. The reason is that dissolution reduces the time of residence of $CaCO_3$ entities in the core-top mixed layer, and hence also their apparent ^{14}C age. But, as shown by Broecker *et al.* (1999), core-top radiocarbon ages on Ontong–Java Plateau cores from a range of water depths reveal an increase rather than a decrease with water depth (see Figure 10). This increase is likely the result of dissolution that occurs on the seafloor in calcite-undersaturated bottom waters before the calcite is incorporated into the core-top mixed layer. In this case, the reduction of $CaCO_3$ input to the sediment leads to an increase in the average residence time

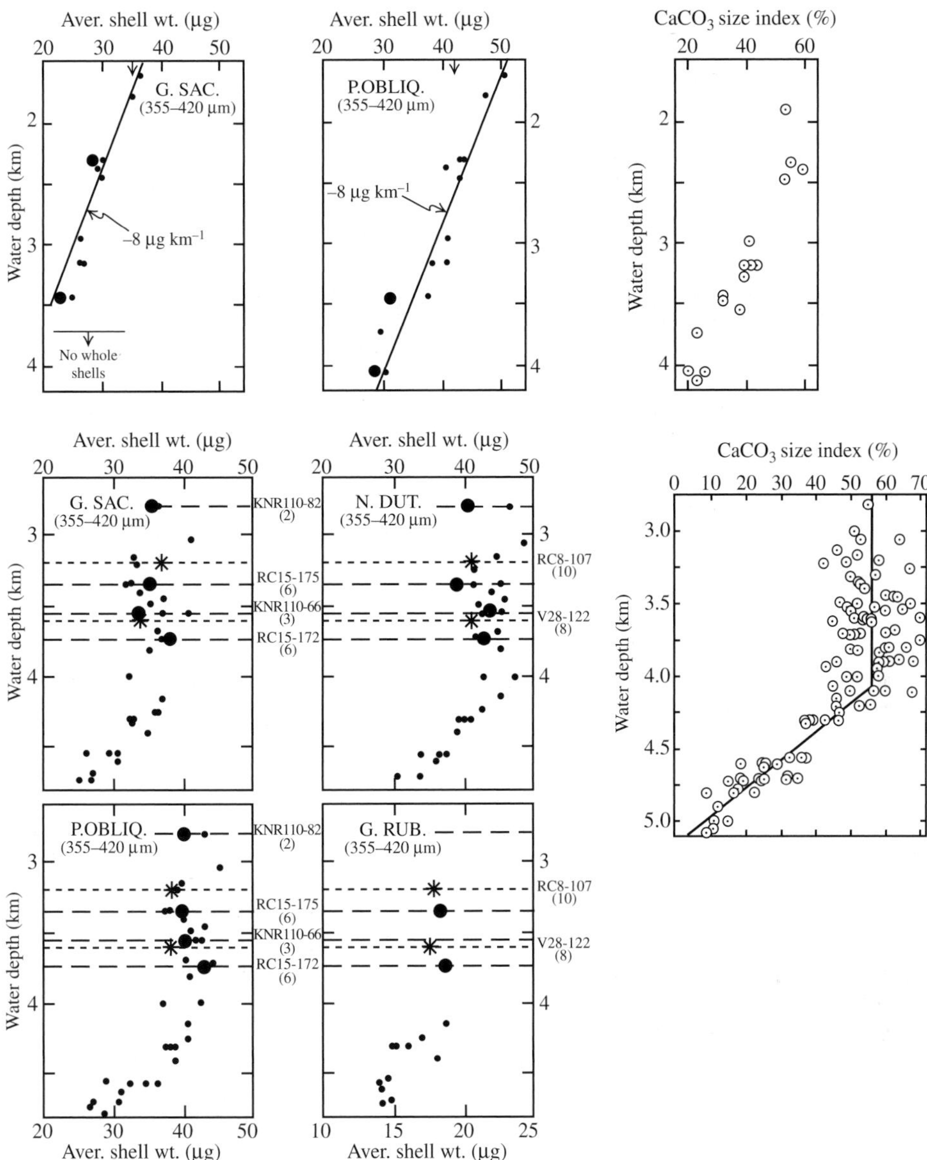

Figure 9 The upper panel shows shell weight and $CaCO_3$ size fraction results from core top covering a range of water depth on the Ontong–Java Plateau. The lower panel shows shell weight results from Ceara Rise and $CaCO_3$ size fraction results from the equatorial Atlantic.

of calcite in the bioturbated layer. It may be that competition between pore-water dissolution and seafloor dissolution changes with depth. As shown in Figure 3, down to about 3 km pore-water dissolution appears to have the upper hand (and hence the ^{14}C ages becomes progressively younger with water depth). Below 3 km, the situation switches and seafloor dissolution dominates (hence, the ^{14}C ages become progressively older with increasing water depth).

6.19.6 DISSOLUTION IN THE PAST

One of the consequences of dissolution of $CaCO_3$ in pore waters is that it creates an ambiguity in all of the sediment-based methods for reconstructing past carbonate ion distributions in the deep sea. By "sediment-based" methods, one means methods involving some measure of the preservation of the $CaCO_3$ contained in deep-sea sediments. The ambiguity involves the magnitude of the offset between the bottom-water and the pore-water carbonate ion concentrations. The results obtained using any such methods can be applied to time trends in bottom-water carbonate ion concentration only if the pore-water–bottom-water offset is assumed to have remained nearly constant.

Fortunately, two methods have been proposed for which this ambiguity does not exist. One

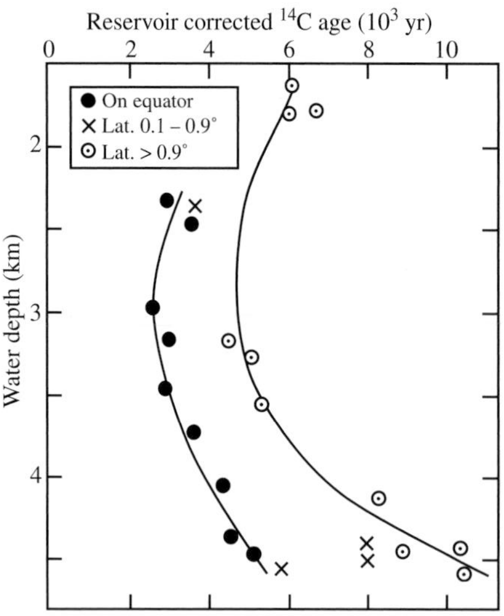

Figure 10 ^{14}C ages (reservoir corrected by 400 yr) as a function of water depth for core-top samples from the Ontong–Java Plateau (Broecker *et al.*, 1999). As can be seen, the ages for cores taken on the equator are systematically younger than those for cores taken a degree or so off the equator. The reason is that the sedimentation rates are twice as high on, than off the equator, while the depth of bioturbation is roughly the same. The onset of the increase in core-top age occurs at a depth of ~3 km. If this onset can be assumed to represent the depth of the saturation horizon (see text), then those results suggest a value ΔV of $\sim 45\ cm^3\ mol^-$ for the reaction $CO_3^{2-} + Ca^{2+} \Leftrightarrow CaCO_3$ (calcite). On the other hand, this depth may represent the horizon where interface dissolution just matches pore-water dissolution.

involves measurements of boron isotope ratios in benthic foraminifera (Sanyal *et al.*, 1995) and the other Zn/Cd ratios in benthic foraminifera (Marchitto *et al.*, 2000). Unfortunately, as of early 2003, neither of these methods has received wide enough application to allow its utility to be proven (see below). Until this has been done, we are left with the ambiguity as to whether sediment-based methods reflect mainly changes in bottom-water CO_3^{2-} or as proposed by Archer and Maier-Reimer (1994) in the pore-water-bottom-water CO_3^{2-} offset.

6.19.7 SEDIMENT-BASED PROXIES

A number of schemes have been proposed by which changes in the carbonate ion concentration in the deep sea might be reconstructed. The most obvious of these is the record of the $CaCO_3$ content of the sediment. Unfortunately, as already discussed, the $CaCO_3$ content depends on the ratio of the rain rate of $CaCO_3$ to that of silicate debris as well as on the extent of dissolution of the calcite. Unless quite large, changes in the extent of dissolution cannot be reliably isolated from changes in the composition of the raining debris. Other schemes focus on the state of preservation of the calcite entities. One involves the ratio of dissolution-prone to dissolution-resistant planktonic foraminifera shells (Ruddiman and Heezen, 1967; Berger, 1970). The idea is that the lower this ratio, the greater the extent of dissolution. A variant on this approach is to measure the ratio of foraminifera fragments to whole shells (Peterson and Prell, 1985; Wu and Berger, 1989). The idea behind both approaches is that as dissolution proceeds, the foraminifera shells break into pieces. These methods suffer, however, from two important drawbacks. First, any method involving entity counting is highly labor-intensive. Second, the results depend on the initial makeup of the foraminifera population in the sediment.

Furthermore, neither of these methods has yet been calibrated against present-day pressure-normalized carbonate ion concentration nor has either one been widely applied. At one point, the author became enamored with a simplified version of the fragment method. Instead of counting fragments (a labor-intensive task), the ratio of $CaCO_3$ in the greater than 63 μm fraction to the total $CaCO_3$ was measured, the idea being that as dissolution proceeded, calcite entities larger than 63 μm would break down to entities smaller than 63 μm. This method was calibrated by conducting measurements on core-top samples from low-latitude sediments spanning a range of water depth in all three oceans (Broecker and Clark, 1999). While these results were promising, when the method was extended to glacial sediment, it was found that the core-top calibration relationship did not apply (Broecker and Clark, 2001a). A possible reason is that the ratio of the fine (coccolith) to coarse (foraminifera) $CaCO_3$ grains in the initial material was higher during glacial time than during the Holocene.

Despite their drawbacks, these methods have led to several important findings. First, it was clearly demonstrated that during glacial time the mean depth of the transition zone did not differ greatly from today's. This finding is important because it eliminates one of the hypotheses which have been put forward to explain the lower glacial atmospheric CO_2 content, namely, the coral reef hypothesis (Berger, 1982). According to this idea, shallow-water carbonates (mainly coral and coralline algae) formed during the high-sea stands of periods of interglaciation would be eroded and subsequently dissolved during the low-sea stands of periods of glaciation, alternately reducing and increasing the sea's CO_3^{2-} concentration. But in order for this hypothesis to be viable, the transition zone would have to have been displaced downward by several kilometers during glacial time. Rather, the reconstructions suggest that the displacement was no more than a few hundred meters.

Two other findings stand out. First, as shown by Farrell and Prell (1989), at water depths in the 4 km range in the eastern equatorial Pacific, the impact of dissolution was greater during interglacials than during glacials (i.e., the transition zone was deeper during glacial time). Second, fragment-to-whole foraminifera ratios measured on a series of cores from various depths in the Caribbean Sea clearly demonstrate better preservation during glacials than interglacials (Imbrie, 1992). These findings have been confirmed by several investigators using a range of methods. Taken together, these findings gave rise to the conclusion that the difference between the depth of the transition zone in the Atlantic from that in the Pacific was somewhat smaller than now during glacial time. In addition, the existence of a pronounced dissolution event in the Atlantic Ocean at the onset of the last glacial cycle has been documented (Curry and Lohmann, 1986).

6.19.8 SHELL WEIGHTS

An ingenious approach to the reconstruction of the carbonate ion concentration in the deep sea was developed by WHOI's Pat Lohmann (1995). Instead of focusing on ratios of one entity to another, he developed a way to assess the extent of dissolution experienced by shells of a given species of planktonic foraminifera. He did this by carefully cleaning and sonification of the greater than 63 μm material sieved from a sediment sample. He then picked and weighed 75 whole shells of a given species isolated in a narrow size fraction range (usually 355–420 μm). In so doing, he obtained a measure of the average shell wall thickness. By obtaining shell weights for a given species from core-top samples spanning a range in water depth, Lohmann was able to show that the lower the pressure-normalized carbonate ion concentration, the smaller the whole shell weight (and hence the thinner the shell walls) (see Figure 11).

Lohmann's method seemingly has the advantage over those used previously in that no assumptions need to be made about the initial composition of the sediment. However, Barker and Elderfield (2002) make a strong case that the thickness of the foraminifera shell walls varies with growth conditions. They did so by weighing shells of temperate foraminifera from core tops from a number of locales in the North Atlantic. They found strong correlations between shell weight and both water temperature and carbonate ion concentration, the warmer the water and the higher its carbonate ion concentration, the thicker the shells. If, as Barker and Elderfield (2002) content, it is the carbonate ion concentration that drives the change in initial wall thickness, then glacial-age shells should have formed with thicker shells than do their Late Holocene counterparts. Fortunately, the ice-core-based atmospheric CO_2 record allows the carbonate ion concentration in the glacial surface waters to be reconstructed and hence presumably also the growth weight of glacial foraminifera. At this point, however, several questions remain unanswered. For example, does the dependence of shell weight on surface water carbonate ion concentration established for temperate species apply to tropical species? Perhaps the shell weight dependence flattens as the high carbonate ion concentrations characteristic of tropical surface waters are approached. Is carbonate ion concentration the only environmental parameter on which initial shell weights depends? As discussed below, there

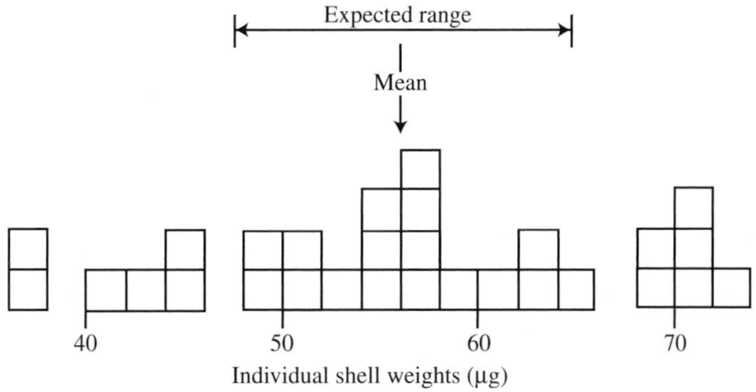

Figure 11 Weights of 29 individual *P. obliquiloculata* shells picked from the 355–420 μm size fraction. If all the shells had the same wall thickness, a spread in weight of 17 μg would be expected (assuming that shell weight varies with the square of size). Clearly, this indicates that shells of the same size must have a range in wall thickness. As can be seen, the observed range in weight is twice the expected range.

Table 1 The calcite-saturation carbonate ion concentration in cold seawater and the slope of this solubility as a function of water depth based on a 1 atm solubility of 45 μmol CO_3^{2-} kg^{-1} and a ΔV of 40 cm^3 mol^{-1}.

Water depth (km)	*Calcite sat.* (μmol CO_3^{2-} kg^{-1})	*Sol. slope* (μmol kg^{-1} km^{-1})	CO_3^{2-} *versus shell wt.* (μmol kg^{-1} μg^{-1})
1.5	58.8	10.3	1.4
2.0	64.2	11.4	1.5
2.5	70.2	12.5	1.6
3.0	76.7	13.6	1.8
3.5	83.8	14.9	2.0
4.0	91.6	16.3	2.1
4.5	100.1	17.9	2.3

Also shown is the slope of the shell-weight loss–carbonate ion concentration relationship for various water depths. The 0.7 μmol kg^{-1} km^{-1} increase in carbonate ion concentration in the Ontong–Java Plateau deep-water column is taken into account.

is reason to believe that the situation is perhaps more complicated.

Lohmann's method has other drawbacks. Along with all sediment-based approaches, it suffers from an inability to distinguish changes in bottom-water carbonate ion concentration from changes in bottom-water to pore-water concentration offset. Shell thickness may also depend on growth rate and hence nutrient availability. Finally, a bias is likely introduced when dissolution becomes sufficiently intense to cause shell breakup, in which case the shells with the thickest walls are likely to be the last to break up. Nevertheless, Lohmann's method opens up a realm of new opportunities.

The sensitivity of shell-weight to pressure-normalized carbonate ion concentration (i.e., after correction for the increase in the solubility of calcite with water depth) was explored by determining the weight of Late Holocene shells from various water depths in the western equatorial Atlantic (Ceara Rise) and western equatorial Pacific (Ontong–Java Plateau). This strategy takes advantage of the contrast in carbonate ion concentration between the Atlantic and Pacific deep waters. As shown in Figure 9, shell weights for Ontong–Java Plateau samples do decrease with water depth and hence with decreasing pressure-normalized carbonate ion concentrations (Broecker and Clark, 2001a). However, the surprise is that there is no evidence of either weight loss or shell break at depths less than 4,200 m for Ceara Rise core-top samples. Rather, weight loss and shell breakup is evident only for samples from deeper than 4,200 m. This observation is in agreement with the benthic chamber results of R. A. Jahnke and D. B. Jahnke and hence supports the hypothesis that above the calcite saturation horizon the gradients in pore-water composition are fueled primarily by the dissolution of "Weyl" (1965) coatings rather than of the biogenic calcite itself.

The Ontong–Java results yield a weight loss of ~8 μg for each kilometer increase in water depth. In order to convert this to a dependence on pressure-normalized carbonate ion concentration, it is necessary to take into account the change in *in situ* carbonate ion concentration in the water column over the Ontong–Java Plateau water column (i.e., $CO_3^{2-} = 72 + 3(z - 2)$ μmol kg^{-1}, where z is the water depth in km) and the pressure dependence of the saturation carbonate ion concentration. The latter depends on the difference in volume between Ca^{2+} and CO_3^{2-} ions when in solution and when they are bound into calcite. The relationship is as follows:

$$(CO_{3\ \mathrm{sat}}^{2-})^z = (CO_{3\ \mathrm{sat}}^{2-})^0 e^{PV/RT}$$

where the units of z are km, of ΔV, L mol^{-1}, of R, L atm, and T, K. If ΔV is re-expressed as cm^3 mol^{-1}, the relationship becomes

$$(CO_{3\ \mathrm{sat}}^{2-})^z = (CO_{3\ \mathrm{sat}}^{2-})^0 e^{z\Delta V/225}$$

where ΔV is the volume of the ions when bound into calcite minus that when they are dissolved in seawater. $(CO_3^{2-})^0$ is 45 mol kg^{-1} and while the exact value of ΔV remains uncertain, 40 cm^3 mol^{-1} fits most ocean observations (Peterson, 1966; Honjo and Erez, 1978; Ben-Yaakov and Kaplan, 1971; Ben-Yaakov *et al.*, 1974). Listed in Table 1 are the saturation concentrations based on this ΔV and the slope of the solubility as a function of water depth. Also given are estimates of the weight loss for foraminifera shells per unit decrease in carbonate ion concentration.

6.19.9 THE BORON ISOTOPE PALEO pH METHOD

Theoretical calculations by Kakihana *et al.* (1977) suggested that the uncharged species of dissolved borate ($B(OH)_3$) should have a 21 per mil higher $^{11}B/^{10}B$ ratio than that for the charged species ($B(OH)_4^-$). Hemming and Hanson (1992) demonstrated that this offset might be harnessed as a paleo pH proxy. Their reasoning was as follows. As the residence time of borate in seawater is tens of millions of years, on the timescale of glacial cycles the isotope composition of oceanic borate could not have changed. They further reasoned that it must be the charged borate species that is incorporated into marine $CaCO_3$ and hence marine calcite should have an isotope composition close to that of the charged species in seawater. This is important because as shown in Figure 12 the isotopic composition of the charged species must depend on the pH of the seawater. The higher the pH, the larger the fraction of the borate in the charged form and hence the closer its isotopic composition will be to that for bulk seawater borate. In contrast, for pH values the isotopic composition of the residual amount of charged borate must approach a value 21 per mil lower than that for bulk seawater borate. Working with a graduate student, Abhijit Sanyal, Hemming applied his method to foraminifera shells and demonstrated that indeed foraminifera shells record pH (Sanyal *et al.*, 1995). Benthic foraminifera had the expected offset from planktonics. Glacial age *G. sacculifer*, as dictated by the lower glacial atmospheric CO_2 content, recorded a pH about 0.15 units higher than that for Holocene shells. Sanyal went on to grow planktonic foraminifera shells at a range of pH values (Sanyal *et al.*, 1996, 2001). He also precipitated inorganic $CaCO_3$ at a range of pH values

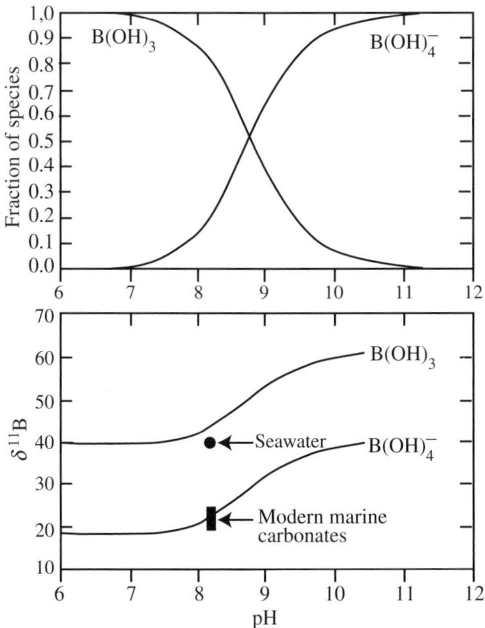

Figure 12 Speciation of borate in seawater as a function of pH (upper panel). Isotopic composition of the uncharged ($B(OH)_3$) and charged ($B(OH)_4^-$) species as a function of pH(lower panel) (Hemming and Hanson, 1992). As marine carbonates incorporate only the charged species, their isotopic composition is close to that of $B(OH)_4^-$.

(Sanyal *et al.*, 2000). These results yielded the expected pH dependence of boron isotope composition. However, they also revealed sizable species-to-species offsets (as do the carbon and oxygen isotopic compositions).

The waterloo of this method came when glacial-age benthic foraminifera were analyzed. The results suggested that the pH of the glacial deep ocean was 0.3 units greater than today (Sanyal *et al.*, 1995). This corresponds to a whopping 90 μmol kg^{-1} increase in carbonate ion concentration. The result was exciting because, if correct, the lowering of the CO_2 content of the glacial atmosphere would be explained by a whole ocean carbonate ion concentration change. But this result was clearly at odds with reconstructions of the depth of the glacial transition zone. Such a large increase in deep-water carbonate ion concentration would require that it deepened by several kilometers. Clearly, it did not. Archer and Maier-Reimer (1994) proposed a means by which this apparent disagreement might be explained. They postulated that if during glacial time the release of metabolic CO_2 to sediment pore waters (relative to the input of $CaCO_3$) was larger than today's, this would have caused a shoaling of the transition zone and thereby thrown the ocean's $CaCO_3$ budget out of kilter. Far too little $CaCO_3$ would have been buried relative to the ingredient input. The result would be a steady increase in the ocean's carbonate ion inventory (see Figure 13) and a consequent progressive deepening of the transition zone. This deepening would have continued until a balance between input and loss was once again achieved. In so doing, a several kilometer offset between the depth of the saturation horizon and the depth of the transition zone would have been created. However, this explanation raised three problems so serious that the boron isotope-based deep-water pH change has fallen into disrepute. First, it required that the change in glacial ecology responsible for the increase in the rain of organic matter be globally uniform. Otherwise, there would have been very large "wrinkles" in the depth of the glacial transition zone. No such wrinkles have been documented. Second, at the close of each glacial period when the flux of excess organic matter was shut down, there must have been a prominent global preservation event. In order to restore the saturation horizon to its interglacial position, an excess over ambient $CaCO_3$ accumulation of ~3 g cm^{-2} would have to have occurred over the entire seafloor. It would be surprising if some residue from this layer were not to be found in sediments lining the abyssal plains. It has not. These sediments have no more than 0.2% by weight $CaCO_3$. In other words, of the 3 g cm^{-2} deposited during the course of the carbonate ion drawdown,

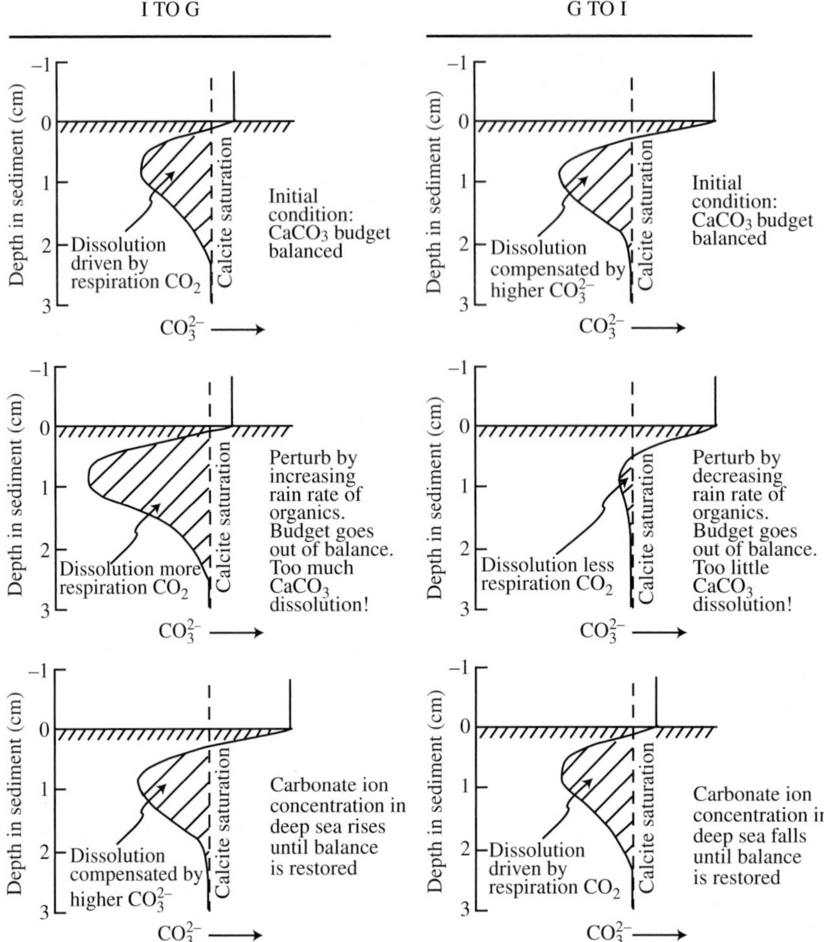

Figure 13 Shown on the left is the sequence of events envisioned by Archer and Maier-Reimer (1994) for the transition from interglacial (I) to glacial (G) conditions. An increase in respiration CO_2 release to the sediment pore waters enhances calcite dissolution, thereby unbalancing the $CaCO_3$ budget. This imbalance leads to a buildup in CO_3^{2-} ion concentration in the deep sea until it compensates for the extra respiration CO_2. On the right is the sequence of events envisioned for the transition from G to I conditions. The input of excess respiration CO_2 to the sediments ceases, thereby reducing the rate of calcite dissolution. This leads to an excess accumulation of $CaCO_3$ on the seafloor and hence to a reduction in carbonate ion concentration which continues until steady state is reestablished.

almost nothing remains. Finally, based on model simulations, Sigman *et al.* (1998) have shown that it is not possible to maintain for tens of thousands of years a several-kilometer separation between the saturation horizon and the transition zone.

This "waterloo" was unfortunate for the author considers the boron method to be basically sound and potentially extremely powerful. The answer to the benthic enigma may lie in species-to-species differences in the boron isotope "vital" effect for benthic foraminifera. The measurement method use by Sanyal *et al.* (1996) required a large number of benthic shells in order to get enough boron to analyze. This created a problem because, as benthics are rare among foraminifera shells, mixed benthics rather than a single species were analyzed. If the boron isotope pH proxy is to become an aid to deep-ocean studies, then techniques requiring smaller amounts of boron will have to be created. There also appears to be a problem associated with variable isotopic fractionation of boron during thermal ionization. As this fractionation depends on the ribbon temperature and perhaps other factors, it may introduce biases in the results for any particular sample. Hopefully, a more reproducible means of ionizing boron will be found.

6.19.10 Zn/Cd RATIOS

The other bottom-water CO_3^{2-} ion concentration proxy is based on the Zn/Cd ratio in benthic foraminifera shells. As shown by Marchitto *et al.* (2000), the distribution coefficient of zinc between shell and seawater depends on CO_3^{2-} ion concentration, such that the lower the carbonate ion concentration, the large the Zn/Cd ratio in the foraminifera shell. Assuming that the Zn/Cd ratio in seawater was the same during the past as it is today, the ratio of these two trace elements should serve as a paleo carbonate ion proxy. However, there are problems to be overcome. For example, in today's ocean, zinc correlates with silica and cadmium with phosphorus. As silica is 10-fold enriched in deep Pacific water relative to deep Atlantic water while phosphorus is only twofold enriched, differential redistribution of silica and phosphorus in the glacial ocean poses a potential bias. However, as at high carbonate supersaturation the distribution coefficient for zinc flattens out, it may be possible to use measurements on benthic foraminifera from sediments bathed in highly supersaturated waters to sort this out. But, as is the case for the boron isotope proxy, much research will be required before reconstructions based on Zn/Cd ratios can be taken at face value.

6.19.11 DISSOLUTION AND PRESERVATION EVENTS

There are several mechanisms that might lead to carbonate ion concentration transients at the beginning and end of glacial periods. One such instigator is changes in terrestrial biomass. Shackleton (1997) was the first to suggest that the mass of carbon stored as terrestrial biomass was smaller during glacial than during interglacial periods. He reached this conclusion based on the fact that measurements on glacial-age benthic foraminifera yielded lower $\delta^{13}C$ values than those for their interglacial counterparts. Subsequent studies confirmed that this was indeed the case and when benthic foraminifera ^{13}C results were averaged over the entire deep sea, it was found that the ocean's dissolved inorganic carbon had a $^{13}C/^{12}C$ ratio 0.35 ± 0.10 per mil lower during glacial time than during the Holocene (Curry *et al.*, 1988). If this decrease is attributed to a lower inventory of wood and humus, then the magnitude of the glacial biomass decrease would have been 500 ± 150 Gt of carbon. The destruction of this amount of organic material at the onset of a glacial period would create a 20 μmol kg^{-1} drop in the ocean's CO_3^{2-} concentration and hence produce a calcite dissolution event. Correspondingly, the removal of this amount of CO_2 from the ocean–atmosphere reservoir at the onset of an interglacial period would raise the carbonate ion concentration by 20 μmol kg^{-1} and hence produce a calcite preservation event. This assumes that the time over which the biomass increase occurred was short compared to the CO_3^{2-}response (i.e., ~5,000 yr). If this is not the case, then the magnitude of the carbonate ion changes would be correspondingly smaller.

Another possible instigator of such transients was proposed by Archer and Maier-Reimer (1994). Their goal was to create a scenario by which the lower CO_2 content of the glacial atmosphere might be explained. As already mentioned, it involved a higher ratio of organic carbon to $CaCO_3$ carbon in the material raining to the deep-sea floor during glacial times than during interglacial times, and hence an intensification of pore-water dissolution. As in the case for the terrestrial biomass change, such an increase would have thrown the ocean's carbon budget temporarily out of kilter. The imbalance would have been remedied by a buildup of carbonate ion concentration at the onset of glacials and a drawdown of carbonate ion concentration at the onset of interglacials (see Figure 14). Hence, it would also lead to a dissolution event at the onset of glacial episodes and a preservation event at the onset of interglacial episodes. Were the changes in organic to $CaCO_3$ rain proposed by Archer and Maier-Reimer to have explained the entire glacial to interglacial CO_2 change, then the magnitude of the transients would have been ~4 times larger than that resulting from 500 Gt C changes in terrestrial biomass.

Regardless of their origin, these dissolution events and preservation events would be short-lived. As they would disrupt the balance between

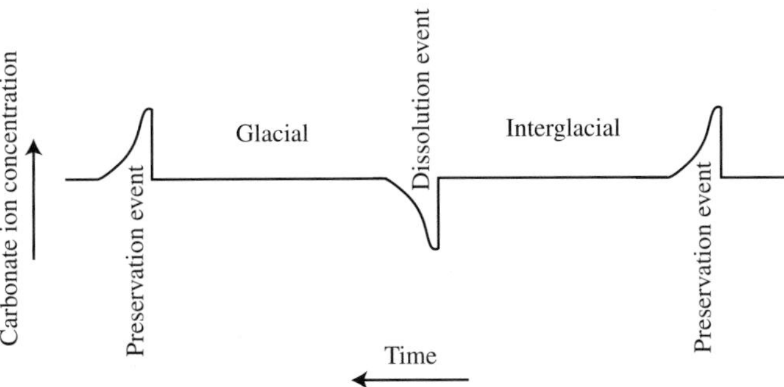

Figure 14 Idealized scenario for carbonate ion concentration changes associated with dissolution and preservation events.

burial and supply, they would be compensated by either decreased or increased burial of $CaCO_3$ and the balance would be restored with a constant time of ~5,000 yr (see Figure 14).

Clear evidence for the compensation for an early Holocene preservation event is seen in shell weight results form a core from 4.04 km depth on the Ontong–Java Plateau in the western equatorial Pacific (see Figure 15). A drop in the weight of *P. obliquiloculata* shells of 11 μg between about 7,500 yr ago and the core-top bioturbated zone (average age 4,000 yr) requires a decrease in carbonate ion concentration between 7,500 y ago and today (see Table 1). This Late Holocene CO_3^{2-} ion concentration drop is characterized by an up-water column decrease in magnitude becoming imperceptible at 2.31 km. It is interesting to note that during the peak of the preservation event, the shell weights showed only a small decrease with water depth (see Figure 15), suggesting either that the pressure effect on calcite solubility was largely compensated by an increase with depth in the *in situ* carbonate ion concentration or that the entire water column was supersaturated with respect to calcite.

In the equatorial Atlantic only in the deepest core (i.e., that from 5.20 km) is the Late Holocene intensification of dissolution strongly imprinted.

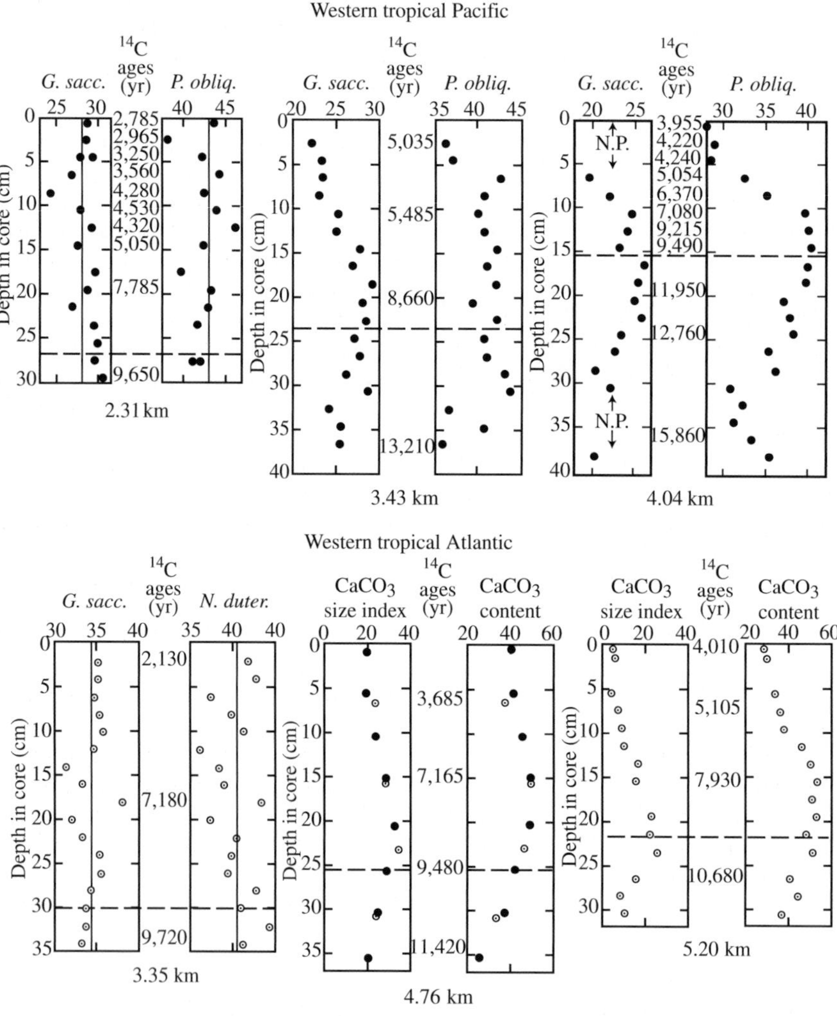

Figure 15 The Early Holocene preservation event. Records for six cores are shown: three from Ontong–Java Plateau in western tropical Pacific (BC36:0° 158 °E, 2.31 km; BC51: 0° 161 °E, 3.43 km and BC56: 0° 162 °E, 4.04 km) and three from the western tropical Atlantic (RC15-175: 4 °N, 47 °W, 3.35 km; RC16-55: 10 °N, 45 °W, 4.76 km and RC17-30: 11 °N, 41 °W, 5.20 km). In the upper panel, N.P. indicates that there are no whole shells present. In the lower panel, the open circles represent measurements made on trigger weight cores and the closed circles, measurements made on piston cores. Where measurements were made on both the depth scale is that for the piston core, and the trigger weight sample depths have been multiplied by a factor of 1.5 to compensate for foreshortening (Broecker *et al.*, 1993). The shell weights are in μg, the size index is the percentage of the $CaCO_3$ contained in the >63 μm fraction and the calcium carbonate content is in percent. The dashed lines show the depth of the 9,500 B.P. radiocarbon-age horizon.

As in this core no whole shells are preserved, the evidence for an Early Holocene preservation event is based on $CaCO_3$ size-index and $CaCO_3$ content measurements. As can be seen in Figure 15, both show a dramatic decrease starting about 7,500 yr ago. As for the Pacific, the magnitude of the imprint decreases up-water column, becoming imperceptible at 3.35 km.

If either the biomass or the respiration CO_2 mechanisms are called upon, the magnitude of the Early Holocene CO_3^{2-} maximum must have been uniform throughout the deep ocean. The most straightforward explanation for the up-water column reduction in the magnitude of the preservation event is that at mid-depths; the sediment pore waters are presently close to saturation with respect to calcite. Hence, the Early Holocene maximum in deep-sea CO_3 ion concentration pushed them into the realm of supersaturation. If so, there is no need to call on a depth dependence for the magnitude of the preservation event.

The post-8,000-year-ago decrease in CO_3^{2-} ion concentration of 23 μmol kg^{-1} required to explain the Late Holocene 11 μg decrease in *P. obliquiloculata* shell weights observed in the deepest Ontong–Java Plateau core is twice too large to be consistent with the 20 ppm increase in atmospheric CO_2 content over this time interval (Indermühle *et al.*, 1999). The significance of this remains unknown.

In the equatorial Atlantic $CaCO_3$ content, $CaCO_3$ size-index and shell-weight measurements reveal three major dissolution events, one during marine isotope stage 5d, one during 5b and one during stage 4 (see Figures 16(a) and (b)). As these events are only weakly imprinted on Pacific sediments, it appears that a major fraction of the carbonate ion reduction was the result of enhanced penetration into the deep Atlantic of low carbonate ion concentration Southern Ocean water. If this conclusion proves to be correct, then it suggests that the balance between the density of deep waters formed in the northern Atlantic and those formed in the Southern Ocean is modulated by the strength of northern hemisphere summer insolation (i.e., by Milankovitch cycles).

6.19.12 GLACIAL TO INTERGLACIAL CARBONATE ION CHANGE

In addition to the preservation and dissolution event transients, there were likely carbonate ion concentration changes that persisted during the entire glacial period. These changes could be placed in two categories. One involves a change in the average CO_3^{2-} concentration of the entire deep sea necessary to compensate for a change in the ratio of calcite production by marine organisms to ingredient supply. The other involves a redistribution of carbonate ion within the deep sea due to a redistribution of phosphate (and hence also of respiration CO_2) and/or to a change in the magnitude of the flux of CO_2 through the atmosphere from the Southern Ocean to the northern Atlantic.

Based on shell-weight measurements, Broecker and Clark (2001c) attempted to reconstruct the depth distribution of carbonate ion concentration during late glacial time for the deep equatorial Atlantic Ocean and Pacific Ocean. At the time their paper was published, these authors were unaware of the dependence of initial shell weight on carbonate ion concentration in surface water established by Barker and Elderfield (2002) for temperate species. Since during the peak glacial time the atmosphere's p_{CO_2} was ~80 ppm lower than during the Late Holocene, the carbonate ion concentration in tropical surface waters must have been 40–50 μmol kg^{-1} higher at that time. Based on the Barker and Elderfield (2002) trend of ~1 μg increase in shell weight per 9 μmol kg^{-1} increase in carbonate ion concentration, this translates to an 8 μg heavier initial shell weights during glacial time. Figure 17 shows, while this correction does not change the depth dependence or interocean concentration difference, it does greatly alter the magnitude of the change. In fact, were the correction made, it would require that the carbonate ion concentration in virtually the entire glacial deep ocean was lower during glacial time than during interglacial time. For the deep Pacific Ocean and the Indian Ocean, this flies in the face of all previous studies which conclude that dissolution was less intense during periods of glaciation than during periods of interglaciation. However, as the Broecker and Clark study concentrates on the Late Holocene while earlier studies concentrate on previous periods of interglaciation, it is possible that the full extent of the interglacial decrease in carbonate ion during the present interglacial has not yet been achieved. Of course, it is also possible that significant thickening of foraminifera shells during glacial time did not occur. Until this matter can be cleared up, reconstruction of glacial-age deep-sea carbonate ion concentrations must remain on hold.

6.19.13 NEUTRALIZATION OF FOSSIL FUEL CO_2

The ultimate fate of much of the CO_2 released to the atmosphere through the burning of coal, oil, and natural gas will be to react with the $CaCO_3$ stored in marine sediments (Broecker and Takahashi, 1977; Sundquist, 1990; Archer *et al.*, 1997). The amount of $CaCO_3$ available for dissolution at any given place on the seafloor depends on the calcite content in the sediment

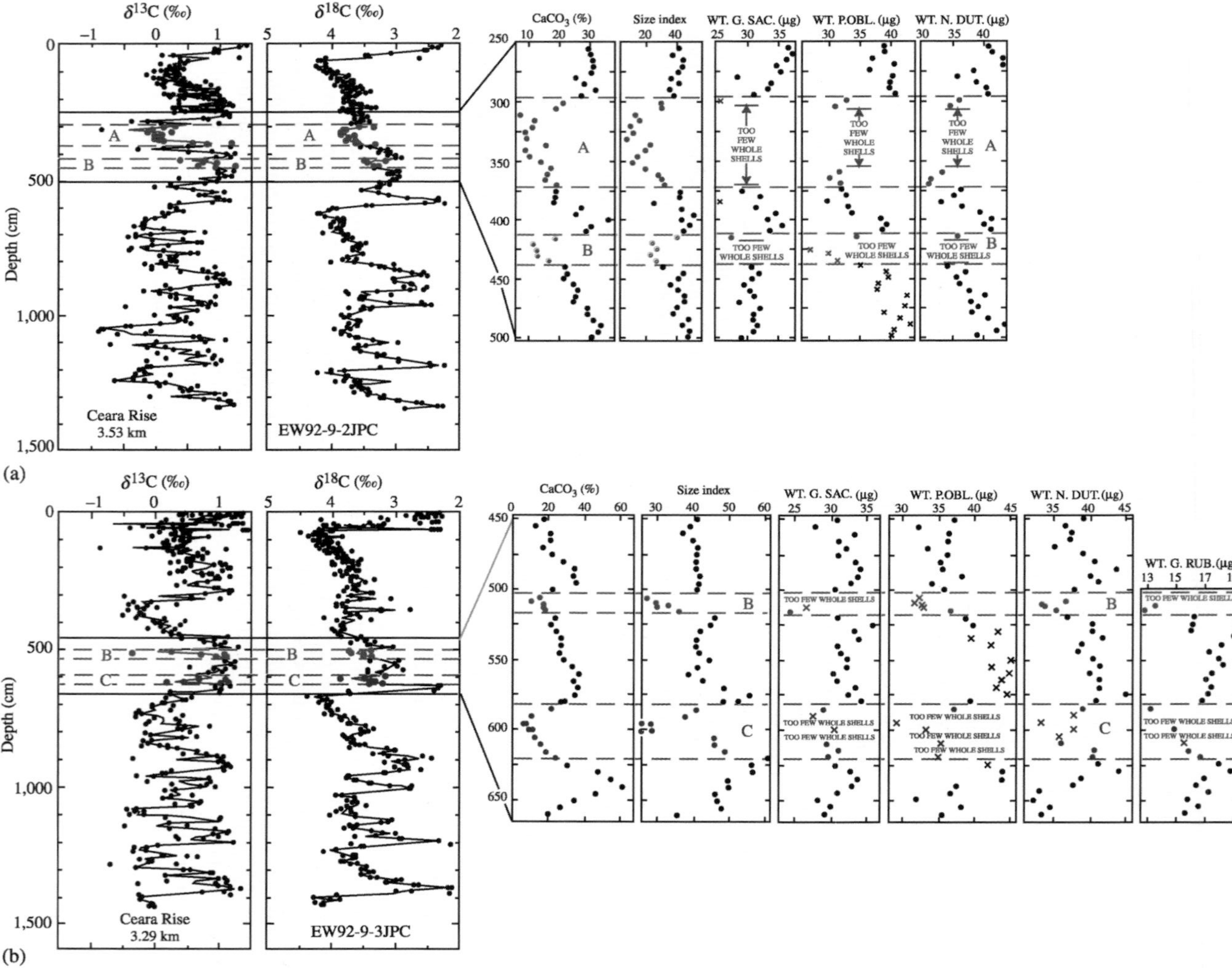

Figure 16 (A) $CaCO_3$, size index and shell weight results for a portion of jumbo pistoncore EW92-9-2 (a) and EW92-9-3 (b) from the northern flank of the Ceara Rise. The results in (a) document dissolution events A (stage 4) and B (stage 5b). (B) The results in (b) document dissolution events B (stage 5b) and C (stage 5d). The ^{18}O and ^{13}C records for benthic foraminifera are from Curry (1996). The *x*s represent samples in which only 15–30 whole shells were found.

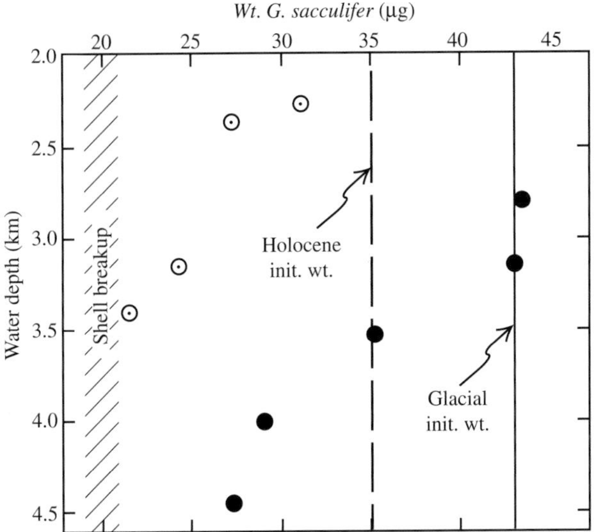

Figure 17 Average shell weights for whole *G. sacculifer* shells of glacial age as a function water depth. The open circles are for cores from the Ontong–Java Plateau, and the closed circles are for cores from the equatorial Atlantic. The vertical dashed line is the initial *G. sacculifer* weight obtained from measurements on Holocene samples from the Ceara Rise. The solid vertical line is the estimated initial weight for glacial-age *G. sacculifer* based on the assumption that the carbonate ion concentration dependence of initial shell weight established by Barker and Elderfield (2002) applies to tropical species.

and the depth to which sediments are stirred by organisms. The former is now well mapped and the latter has been documented in many places by radiocarbon measurements. The amount of $CaCO_3$ available for dissolution at any given site is given by

$$\sum CaCO_3 = \frac{h\rho f_c}{1 - f_c}$$

where h is the depth of bioturbation, ρ is the water-free sediment density, and f_c is the weight-fraction calcite. The high-$CaCO_3$ sediments that drape the oceans' ridges and plateaus typically have ~90% $CaCO_3$ and a water-free density of 1 g cm^{-3}. The bioturbation depth in these sediments averages 8 cm. Hence, the upper limit on amount of $CaCO_3$ available for dissolution in such a sediment is 72 g cm^{-2}. As roughly one quarter of the seafloor is covered with calcite-rich sediments, this corresponds to $\sim 6.3 \times 10^{19}$ g $CaCO_3$ (i.e., 7,560 Gt C). This amount could neutralize 6.3×10^{17} mol of fossil fuel CO_2. This amount exceeds the combined oceanic inventory of dissolved CO_3^{2-} (1.6×10^{17} mol) and of dissolved HBO_3^- (0.8×10^{17} mol). It is comparable to the amount of recoverable fossil fuel carbon.

I say 'upper limit' because once this amount of $CaCO_3$ has been dissolved, the upper 8 cm of the sediment would consist entirely of a noncarbonate residue. As molecular diffusion through such a thick residue would be extremely slow, the rate of dissolution of $CaCO_3$ stored beneath this $CaCO_3$-free cap would be minuscule, and further neutralization would be confined to the fall to the seafloor of newly formed $CaCO_3$.

The rate of this dissolution of the $CaCO_3$ stored in the uppermost sediment will depend not only on the magnitude of the reduction of the deep ocean's CO_3^{2-} content, but also on the rate at which the insoluble residue is stirred into the sediment. This bioturbation not only homogenizes the mixed layer, but is also exhumes $CaCO_3$ from beneath the mixed layer.

REFERENCES

Archer D. and Maier-Reimer E. (1994) Effect of deep-sea sedimentary calcite preservation on atmospheric CO_2 concentration. *Nature* **367**, 260–263.

Archer D., Emerson S., and Reimers C. (1989) Dissolution of calcite in deep-sea sediments: pH and O_2 microelectrode results. *Geochim. Cosmochim. Acta* **53**, 2831–2845.

Archer D., Kheshgi H., and Maier-Reimer E. (1997) Multiple timescales for neutralization of fossil fuel CO_2. *Geophys. Res. Lett.* **24**, 405–408.

Archer D., Kheshgi H., and Maier-Riemer E. (1998) Dynamics of fossil fuel CO_2 neutralization by marine $CaCO_3$. *Global Biogeochem. Cycles* **12**, 259–276.

Barker S. and Elderfield H. (2002) Response of foraminiferal calcification to glacial–interglacial changes in atmospheric carbon dioxide. *Science* **297**, 833–836.

Ben-Yaakov S. and Kaplan I. R. (1971) Deep sea *in situ* calcium carbonate saturometry. *J. Geophys. Res.* **76**, 772–781.

Ben-Yaakov S., Ruth E., and Kaplan I. R. (1974) Carbonate compensation depth: relation to carbonate solubility in ocean waters. *Science* **184**, 982–984.

Berelson W. M., Hammond D. E., and Cutter G. A. (1990) *In situ* measurements of calcium carbonate dissolution rates in deep-sea sediments. *Geochim. Cosmochim. Acta* **54**, 3013–3020.

Berelson W. M., Hammond D. E., McManus J., and Kilgore T. E. (1994) Dissolution kinetics of calcium carbonate in equatorial Pacific sediments. *Global Biogeochem. Cycles* **8**, 219–235.

Berger W. H. (1970) Planktonic foraminifera: selective solution and the lysocline. *Mar. Geol.* **8**, 111–138.

Berger W. H. (1982) Increase of carbon dioxide in the atmosphere during deglaciation: the coral reef hypothesis. *Naturwissenschaften* **69**, 87–88.

Broecker W. S. (1991) The great ocean conveyor. *Oceanography* **4**, 79–89.

Broecker W. S. and Clark E. (1999) $CaCO_3$ size distribution: a paleo carbonate ion proxy. *Paleoceanography* **14**, 596–604.

Broecker W. S. and Clark E. (2001a) Reevaluation of the $CaCO_3$ size index paleocarbonate ion proxy. *Paleoceanography* **16**, 669–771.

Broecker W. S. and Clark E. (2001b) A dramatic Atlantic dissolution event at the onset of the last glaciation. *Geochem. Geophys. Geosys.* **2**, 2001GC000185, Nov. 2.

Broecker W. S. and Clark E. (2001c) Glacial to Holocene redistribution of carbonate ion in the deep sea. *Science* **294**, 2152–2155.

Broecker W. S. and Clark E. (2001d) An evaluation of Lohmann's foraminifera-weight index. *Paleoceanography* **16**, 531–534.

Broecker W. S. and Clark E. (2002) A major dissolution event at the close of MIS 5e in the western equatorial Atlantic. *Geochem. Geophys. Geosys.* **3**(2) 10.1029/2001GC000210.

Broecker W. S. and Clark E. (2003) Pseudo-dissolution of marine calcite. *Earth Planet. Sci. Lett.* **208**, 291–296.

Broecker W. S. and Peng T.-H. (1993) Interhemispheric transport of $\sum CO_2$ through the ocean. In *The Global Carbon Cycle*. NATO SI Series (ed. M. Heimann). Springer, vol. 115, pp. 551–570.

Broecker W. S. and Sutherland S. (2000) The distribution of carbonate ion in the deep ocean: support for a post-Little Ice Age change in Southern Ocean ventilation. *Geochem. Geophys. Geosys.* **1**, 2000GC000039, July 10.

Broecker W. S. and Takahashi T. (1977) Neutralization of fossil fuel CO_2 by marine calcium carbonate. In *The Fate of Fossil Fuel CO_2 in the Oceans* (eds. N. R. Andersen and A. Malahoff). Plenum, New York, pp. 213–248.

Broecker W. S. and Takahashi T. (1978) The relationship between lysocline depth and *in situ* carbonate ion concentration. *Deep-Sea Res.* **25**, 65–95.

Broecker W. S., Lao Y., Klas M., Clark E., Bonani G., Ivy S., and Chen C. (1993) A search for an early Holocene $CaCO_3$ preservation event. *Paleoceanography* **8**, 333–339.

Broecker W. S., Clark E., Hajdas I., Bonani G., and McCorkle D. (1999) Core-top ^{14}C ages as a function of water depth on the Ontong-Java Plateau. *Paleoceanography* **14**, 13–22.

Curry W. B. (1996) Late Quaternary deep circulation in the western equatorial Atlantic. In *The South Atlantic: Present and Past Circulation* (eds. G. Wefer, W. H. Berger, G. Siedler, and D. J. Webb). Springer, New York, pp. 577–598.

Curry W. B. and Lohmann G. P. (1986) Late Quaternary carbonate sedimentation at the Sierra Leone rise (eastern equatorial Atlantic Ocean). *Mar. Geol.* **70**, 223–250.

Curry W. B., Duplessy J. C., Labeyrie L. D., and Shackleton N. J. (1988) Changes in the distribution of $\delta^{13}C$ of deep water $\sum CO_2$ between the last glaciation and the Holocene. *Paleoceanography* **3**, 317–341.

Emerson S. and Bender M. (1981) Carbon fluxes at the sediment–water interface of the deep-sea: calcium carbonate preservation. *J. Mar. Res.* **39**, 139–162.

Farrell J. W. and Prell W. L. (1989) Climatic change and $CaCO_3$ preservation: an 800,000 year bathymetric reconstruction from the central equatorial Pacific Ocean. *Paleoceanography* **4**, 447–466.

Hales B. and Emerson S. (1996) Calcite dissolution in sediments of the Ontong-Java Plateau: *in situ* measurements of porewater O_2 and pH. *Global Biogeochem. Cycles* **5**, 529–543.

Hales B. and Emerson S. (1997) Calcite dissolution in sediments of the Ceara Rise: *in situ* measurements of porewater O_2, pH and $CO_{2(aq)}$. *Geochim. Cosmochim. Acta* **61**, 501–514.

Hales B., Emerson S., and Archer D. (1994) Respiration and dissolution in the sediments of the western North Atlantic: estimates from models of *in situ* microelectrode measurements of porewater oxygen and pH. *Deep-Sea Res.* **41**, 695–719.

Hemming N. G. and Hanson G. N. (1992) Boron isotopic composition and concentration in modern marine carbonates. *Geochim. Cosmochim. Acta* **56**, 537–543.

Honjo S. and Erez J. (1978) Dissolution rates of calcium carbonate in the deep ocean: an *in situ* experiment in the North Atlantic. *Earth Planet. Sci. Lett.* **40**, 226–234.

Imbrie J. (1992) On the structure and origin of major glaciation cycles: I. Linear responses to Milankovitch forcing. *Paleoceanography* **7**, 701–738.

Indermühle A., Stocker T. F., Joos F., Fischer H., Smith H. J., Wahlen M., Deck B., Mastroianni D., Techumi J., Blunier T., Meyer R., and Stauffer B. (1999) Holocene carbon-cycle dynamics based on CO_2 trapped in ice at Taylor Dome, Antarctica. *Nature* **398**, 121–126.

Jahnke R. A. and Jahnke D. B. (2002) Calcium carbonate dissolution in deep-sea sediments: implications of bottom water saturation state and sediment composition. *Geochim. Cosmochim. Acta* (submitted for publication, November, 2001).

Kakihana H., Kotaka M., Satoh S., Nomura M., and Okamoto M. (1977) Fundamental studies on the ion exchange separation of boron isotopes. *Bull. Chem. Soc. Japan* **50**, 158–163.

Lohmann G. P. (1995) A model for variation in the chemistry of planktonic foraminifera due to secondary calcification and selective dissolution. *Paleoceanography* **10**, 445–457.

Marchitto T. M., Jr., Curry W. B., and Oppo D. W. (2000) Zinc concentrations in benthic foraminifera reflect seawater chemistry. *Paleoceanography* **15**, 299–306.

Martin W. R. and Sayles F. L. (1996) $CaCO_3$ dissolution in sediments of the Ceara Rise, western equatorial Atlantic. *Geochim. Cosmochim. Acta* **60**, 243–263.

Martin W. R., McNichol A. P., and McCorkle D. C. (2000) The radiocarbon age of calcite dissolving at the sea floor: estimates from pore water data. *Geochim. Cosmochim. Acta* **64**, 1391–1404.

Milliman J. D., Troy P. J., Balch W. M., Adams A. K., Li Y.-H., and Mackenzie F. T. (1999) Biologically mediated dissolution of calcium carbonate above the chemical lysocline? *Deep-Sea Res.* **46**, 1653–1669.

Peterson M. N. A. (1966) Calcite: rates of dissolution in a vertical profile in the central Pacific. *Science* **154**, 1542–1544.

Peterson L. C. and Prell W. L. (1985) Carbonate preservation and rates of climatic change: an 800 kyr record from the Indian Ocean. In *The Carbon Cycle and Atmospheric CO_2: Natural Variations Archean to Present,* Geophys. Monogr. Ser. 32 (eds. E. T. Sundquist and W. S. Broecker) pp. 251–270.

Reimers C. E., Jahnke R. A., and Thomsen L. (2001) In situ sampling in the benthic boundary layer. In *The Benthic Boundary Layer: Transport Processes and Biogeochemistry* (eds. B. P. Boudreau and B. B. Jørgensen). Oxford University Press, Oxford, pp. 245–268.

Ruddiman W. F. and Heezen B. C. (1967) Differential solution of planktonic foraminifera. *Deep-Sea Res.* **14**, 801–808.

Sanyal A., Hemming N. G., Hanson G. N., and Broecker W. S. (1995) Evidence for a higher pH in the glacial ocean from boron isotopes in foraminifera. *Nature* **373**, 234–236.

Sanyal A., Hemming N. G., Broecker W. S., Lea D. W., Spero H. J., and Hanson G. N. (1996) Oceanic pH control on the boron isotopic composition of foraminifera: evidence from culture experiments. *Paleoceanography* **11**, 513–517.

Sanyal A., Nugent M., Reeder R. J., and Bijma J. (2000) Seawater pH control on the boron isotopic composition of calcite: evidence from inorganic calcite precipitation experiments. *Geochim. Cosmochim. Acta* **64**, 1551–1555.

Sanyal A., Bijma J., Spero H., and Lea D. W. (2001) Empirical relationship between pH and the boron isotopic composition of *Globigerinoides sacculifer*: implications for the boron isotope paleo-pH proxy. *Paleoceanography* **16**, 515–519.

Sayles F. L. (1985) $CaCO_3$ solubility in marine sediments: evidence for equilibrium and non-equilibrium behavior. *Geochim. Cosmochim. Acta.* **49**, 877–888.

Shackleton N. J. (1977) Tropical rainforest history and the equatorial Pacific carbonate dissolution cycles. In *The Fate of Fossil Fuel CO_2 in the Oceans* (eds. N. R. Anderson and A. Malahoff). Plenum, New York, pp. 401–428.

Sigman D. M., McCorkle D. C., and Martin W. R. (1998) The calcite lysocline as a constraint on glacial/interglacial low-latitude production changes. *Global Biogeochem. Cycles* **12**, 409–427.

Sundquist E. T. (1990) Long-term aspects of future atmospheric CO_2 and sea-level changes. In *Sea-level Change*. National Research Council Studies in Geophysics (ed. R. Revelle). National Academy Press, Washington, pp. 193–207.

Wenzhöfer F., Adler M., Kohls O., Hensen C., Strotmann B., Boehme S., and Schulz H. D. (2001) Calcite dissolution driven by benthic mineralization in the deep-sea: *in situ* measurements. *Geochim. Cosmochim. Acta* **65**, 2677–2690.

Weyl P.K. (1965) The solution behavior of carbonate materials in seawater. In *Proc. Int. Conf. Tropical Oceanography, Miami Beach, Florida*, pp. 178–228.

Wu G. and Berger W. H. (1989) Planktonic foraminifera: differential dissolution and the quaternary stable isotope record in the west equatorial Pacific. *Paleoceanography* **4**, 181–198.

6.20
Records of Cenozoic Ocean Chemistry

G. E. Ravizza

University of Hawaii, Manoa, HI, USA

and

J. C. Zachos

University of California, Santa Cruz, CA, USA

6.20.1 INTRODUCTION

Numerous lines of evidence show that there have been dramatic changes in the marine realm during the last 65 Myr. These changes occur over varying timescales. Some are relatively abrupt, occurring on timescales of thousands to tens of thousands of years. Others occur more gradually, over million-year timescales. Many of the most valuable monitors of past changes in ocean chemistry, such as the $\delta^{13}C$ and $\delta^{18}O$ of foraminiferal calcite are subject to high-frequency variations that must be smoothed out if long-term, secular, trends are to be recognized clearly. Conversely, other records of past seawater chemistry, such as the marine strontium isotope record, respond only slowly to high-frequency external forcing and are incapable of recording it with fidelity. Nevertheless, it is likely that high-frequency forcing related to glacial erosion and shifts in the hydrologic cycle play an important role in shaping the marine strontium isotope record. Therefore, even though the focus of this review is on records of Cenozoic ocean chemistry that emphasize long-term changes, the different timescales on which Cenozoic ocean chemistry changes are not fully separable.

In this review emphasis is placed on isotopic records of ocean chemistry. In general terms, a conscious decision was made to emphasize those records that document long-term changes in the chemical and physical properties of the global ocean over the course of the Cenozoic. For example, while reconstructions of burial fluxes of barium or phosphorus may place valuable constraints on paleo-productivity in a specific setting, making extrapolations to infer globally integrated trends from these data sets is very difficult because of sparse data coverage in space and time. Similarly, we have also chosen to exclude discussion of short-residence-time tracers like lead and neodymium isotopes that can yield important information about changing patterns of ocean circulation and regional shifts in oceanic inputs. A recent discussion of these tracer systems is available elsewhere (Frank, 2002). The geological records that are emphasized include those of the stable carbon and oxygen isotopes preserved in benthic foraminiferal calcite, and marine strontium and osmium isotopes. These four records clearly manifest significant changes in global ocean chemistry. At any given time, the stable carbon and oxygen isotope records include an important component of spatial variability; however, these are not so large as to obscure the pattern of temporal variation preserved in the sediment record.

Boron isotopes as a paleo-pH proxy and Mg/Ca as a paleo-temperature proxy are also discussed, because information provided by these relatively new proxies has important implications for the better-established records mentioned in the preceding paragraph. The application of boron isotopes to reconstructing surface water pH provides a means of estimating past atmospheric CO_2 levels. Much of the discussion of the marine strontium isotope record in the past has been linked to implicit assumptions about Cenozoic variations in atmospheric CO_2 levels. New boron isotope results indicate that previous assumptions about Cenozoic atmospheric CO_2 levels need to be reconsidered. Mg/Ca ratio variations in benthic foraminiferal calcite are discussed because of the promise that combined Mg/Ca and $\delta^{18}O$ studies hold for resolving the dual influence of ice volume and deep-water temperature on benthic foraminiferal $\delta^{18}O$ records.

There are several other topics that would be equally appropriate to consider in a review of this type. Some of these, such as the $\delta^{34}S$ record of seawater and the history of the calcium carbonate compensation depth (CCD), are mentioned briefly as they relate to records that are discussed in greater detail. Other topics have been omitted. We hope readers will recognize that the topics covered here are determined not only by the scientific interests of the authors, but also by the practical limits of what can be covered in a single review.

6.20.2 CENOZOIC DEEP-SEA STABLE ISOTOPE RECORD

Much of what is currently understood about the Cenozoic history, of deep-sea temperature, carbon chemistry, and global ice volume, has been gleaned from the stable isotope ratios of benthic foraminifera. Benthic foraminifera extract carbonate and other ions from seawater to construct their tests. In many species, this is achieved near carbon and oxygen isotopic equilibrium. Kinetic fractionation effects tend to be small and constant (Grossman, 1984, 1987). As a result, shell $\delta^{13}C$ and $\delta^{18}O$ strongly covary with the isotopic composition of seawater and dissolved inorganic carbon (DIC). For both carbon and oxygen isotopes, there is also a temperature-dependent fractionation effect. For oxygen isotopes, the effect is relatively large, 0.25‰ $(°C)^{-1}$, about an order of magnitude larger than that for carbon isotopes. As such, in addition to monitoring changes in seawater isotope ratios, the oxygen isotopes can be used to evaluate temperature. Moreover, the tests of benthic foraminifera are relatively resistant to dissolution and other diagenetic processes, making them ideal archives of ocean history.

The latest "global" deep-sea stable isotope record for the Cenozoic, presented in Figure 1, is based on benthic foraminifera oxygen and carbon isotope records compiled from over 40 pelagic sediment cores (Zachos *et al.*, 2001). The raw data were smoothed using a five-point running mean, and curve-fitted with a weighted (2%) running mean. For the carbon isotope record, the curve fit was terminated at the middle–late Miocene boundary because of a marked increase in basin-to-basin carbon isotope fractionation at that time (Wright *et al.*, 1991). This record provides a fairly

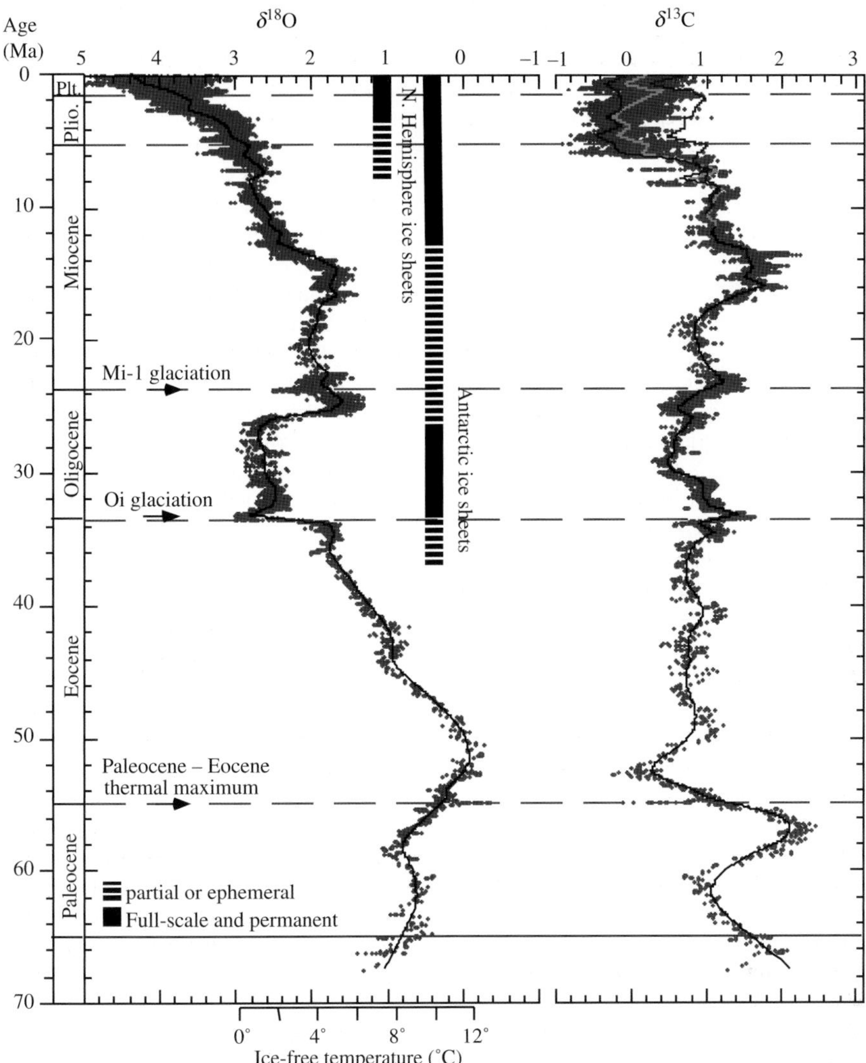

Figure 1 The global deep-sea isotope record based on data compiled from 40 DSDP and ODP sites. These data were derived from pelagic deep-sea cores (e.g., from depths greater than 1,000 m) with lithologies that are predominantly fine-grained, carbonate-rich (>50%) oozes or chalks. Most of the data are derived from analyses of two common and long-lived benthic taxa, *Cibicidoides* and *Nuttallides*. The absolute ages are relative to the standard Geomagnetic Polarity Timescale (GPTS) for the Cenozoic (Berggren *et al.*, 1995). To minimize biases related to inconsistencies in sampling density in space and time, the raw data were smoothed using a five-point running mean, and curve-fitted with a locally weighted mean. The smoothing results in a loss of detail that is undetectable from the long-timescale perspective. For the carbon isotope record, the global curve fit was terminated just before the Late Miocene and replaced with separate Atlantic and Pacific curves because of a marked increase in basin-to-basin carbon isotope fractionation (Wright and Miller, 1993). The temperature scale represents mean deep-sea temperature for the period of time preceding the onset of large-scale glaciation on Antarctica (~35 Ma). After this time, much of the variability in the δ^{18}O record reflects on changes in global ice volume on Antarctica and in the N. Hemisphere. The vertical bars provide a semiquantitative representation of ice volume in each hemisphere relative to the LGM with the dashed bar representing periods of minimal ice coverage (~<50%), and the full bar representing close to maximum ice-coverage (>50% of present). In the more recent portion of the carbon isotope record, separate curve fits were derived for the Atlantic and Pacific to illustrate the increase in basin-to-basin fractionation which exceeds ~1.0‰ in some intervals. Prior to 15 Ma, interbasin gradients are on the order of a few tenths of a per mil or less (source Zachos *et al.*, 2001).

coarse perspective on the long-term variations in Cenozoic climate and ocean chemistry. For the finer-scale variations that tend to be masked in this long-term perspective, individual high-resolution records are preferable. Several high-resolution records spanning "critical" intervals are plotted in Figure 2.

6.20.2.1 Oxygen Isotopes and Climate

The conservative physical and chemical characteristics of bottom waters are fixed at the site of deep-water formation, which for much of the Cenozoic appears to have been the high-latitude polar seas. As a consequence, the long-term oxygen isotope variations recorded by benthic foraminifera reflect largely on changes in high-latitude sea surface temperature (Shackleton *et al.*, 1985). In general, a $\delta^{18}O$ increase of 1.0‰ is equivalent to roughly 4 °C of cooling. Foraminifera oxygen isotopes also record change in seawater $\delta^{18}O$ which, on the million-year timescale, is controlled primarily by changes in the volume of continental ice sheets which are isotopically depleted (−30‰ to −40‰) relative to seawater. For example, melting the present-day ice sheets would decrease mean ocean $\delta^{18}O$ by more than 1.0‰ while raising sea level by 100 m (~0.01‰ m^{-1}). It is assumed that the exchange of isotopes with the crust has negligible effects on seawater $\delta^{18}O$ on the timescale of the Cenozoic.

Separating the relative contributions of these two variables to the deep-sea oxygen isotope record is a challenging exercise. Consideration of additional factors including the lower boundaries of seawater temperature (e.g., freezing) can place some limits on the temperature effect (Miller *et al.*, 1987, 1991b; Zachos *et al.*, 1993, 1994). However, only with independent measures of temperature that are unaffected by salinity, can one effectively isolate the component related to ice volume. To this end, the degree of saturation of alkenones of autotrophs or Mg/Ca of benthic foraminifera (Lear *et al.*, 2000; Billups and Schrag, 2002) can be used to constrain temperature.

Over the long term, the Cenozoic deep-sea oxygen isotope record is dominated by two important features that relate to major shifts in mean climatic state. The first is a rise in values from 53 to 35. This trend, which is mostly gradual but punctuated by several steps, is an expression of the Eocene transition from greenhouse to ice-house conditions. In the Early Eocene (~53 Ma) the deep sea was relatively warm, ~7 °C warmer than present, and there were no ice sheets. Over the next 20 Myr. the ocean cools, and the first large ice sheets appear on Antarctica. The latter event is reflected by the relatively sharp 1.2‰ increase in $\delta^{18}O$ at 33.4 Ma. This pattern reverses slightly toward the end of the Oligocene, but by middle Miocene, ice sheets begin to expand slowly, eventually covering most of Antarctica as reflected in $\delta^{18}O$. The second significant step in the Cenozoic is associated with the gradual buildup of northern hemisphere ice sheets between 3.5 Ma and 2.5 Ma.

On short timescales, the $\delta^{18}O$ record reveals considerable variability. Most of this is concentrated in the Milankovitch bands, and therefore reflects on orbitally modulated changes in ice volume and/or deep-sea temperatures. The largest amplitude oscillations occur in the Quaternary when ice sheets are present on North America. Prior to the Quaternary, the signal amplitude in $\delta^{18}O$ is about one-third to one-half, mostly reflecting variations in the volume of Antarctic ice sheets (Figure 2(a)). These lower-amplitude oscillations persist through much of the Neogene and the Oligocene with most of the variance concentrated in the obliquity bands (Tiedemann *et al.*, 1994; Shackleton and Crowhurst, 1997; Flower *et al.*, 1997; Zachos *et al.*, 2001).

A small component of the short-term variability falls under the category of anomalies or transients. This includes a short-lived but abrupt negative excursion 55 Myr ago (Figure 2(b)). The magnitude of the $\delta^{18}O$ change implies a 4–5 °C transient warming of the deep sea. This event, referred to as the Paleocene–Eocene thermal maximum, is by far the most extreme of the rapid climatic changes inferred from the oxygen isotope records. Other isotope anomalies representing brief climatic excursions have been documented in the earliest Oligocene (~33.4 Ma), at the O/M boundary (23.0 Ma), and in the middle Miocene (~14 Ma) (Miller *et al.*, 1991a; Zachos *et al.*, 1996).

Despite efforts to produce a globally averaged record, gradients in deep-sea temperature combined with the uneven distribution of deep-sea cores can introduce subtle biases into the global $\delta^{18}O$ record. To start, deep-sea temperature is not uniform: the upper ocean is several degrees warmer than the deep ocean. Gradients also exist spatially, especially at intermediate depths with proximity to the polar and tropical oceans. As a result, global compilations utilizing data from sites located at different water depths and basins can be somewhat misleading. For example, the older, Early Paleogene, portions of cores tend to be biased toward the upper ocean because of long-term subsidence and seafloor subduction (Zachos *et al.*, 2001). Moreover, because no single record spans the entire Cenozoic, shifts in $\delta^{18}O$ can be artificially produced through splicing. One example is the negative shift in $\delta^{18}O$ observed in the Late Oligocene (~27 Ma), which in the compilation is larger than recorded in any individual record. A number of deep-sea records

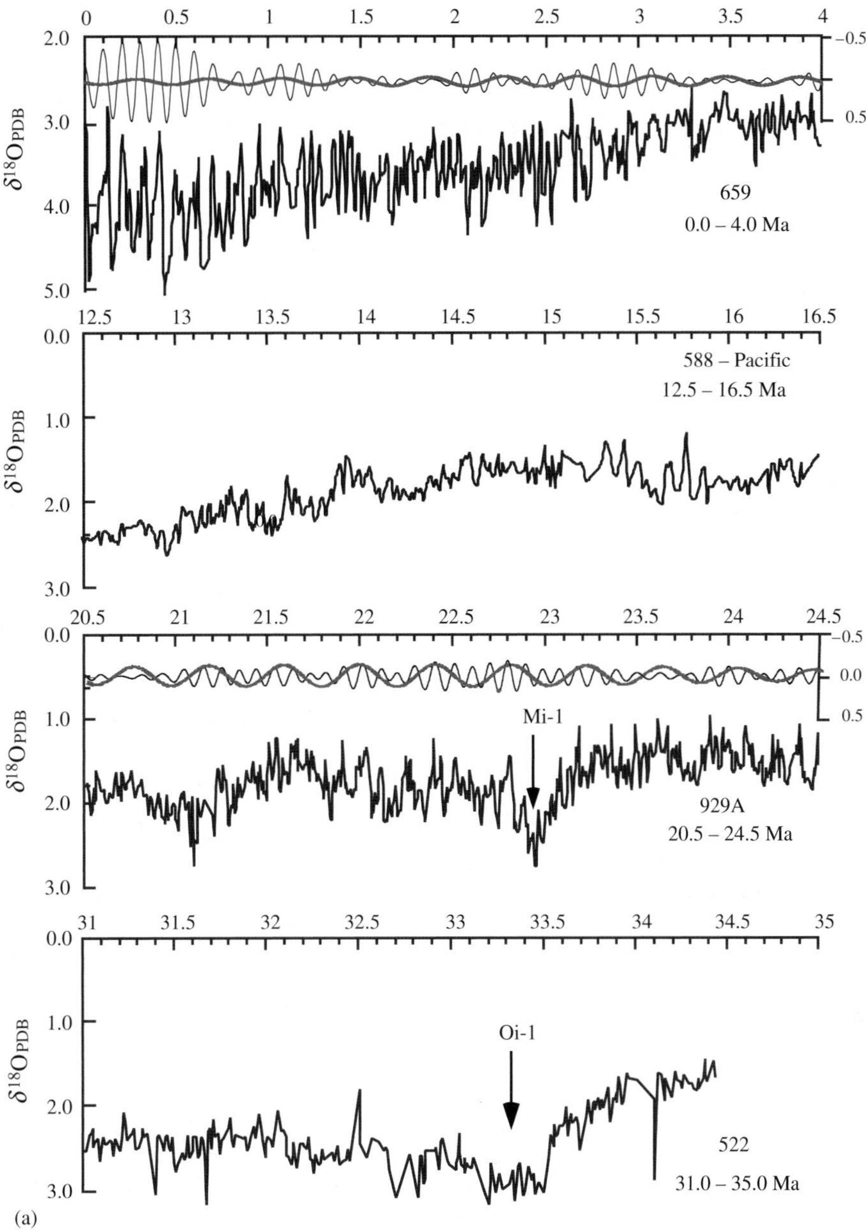

Figure 2 (a) High-resolution 4 Myr long δ^{18}O time series representing four intervals of the Cenozoic (Zachos *et al.*, 2001). The data are from sites 659, eastern equatorial Atlantic (Tiedemann *et al.*, 1994); 588, Southwest Pacific (Flower and Kennett, 1993); 929, western equatorial Atlantic (Paul *et al.*, 2000); 522, South Atlantic (Zachos *et al.*, 1996) and 689 Southern Ocean (Diester-Haass, 1996). Sampling intervals range from 3 kyr, to 10 kyr. Note that the δ^{18}O axes on all plots are set to the same scale (3.0‰) although different ranges to accommodate the change in mean ocean temperature/ice volume with time. The upper curves in each panel represent Gaussian band-pass filters designed to isolate variance associated with the 400 kyr and 100 kyr eccentricity cycles. The 400 kyr filter has a central frequency (cf) = 0.0025 and a bandwidth (bw) = 0.0002; the 100 kyr cf = 0.01 and bw = 0.002. (b) Multiple benthic isotope records characterizing the Late Paleocene thermal maximum event at 55 Ma (Thomas and Shackleton, 1996; Bralower *et al.*, 1995). The timing of the beginning of the event is placed at 54.95 Ma (Röhl *et al.*, 2000). Data are plotted on the timescale of these two papers combined with the cyclostratigraphy for site 690 (Cramer, in press); data from other sites are correlated to the 690 record. The apparent late initiation of the events at 690 is the result of the lack of specimens of *N. truempyi* during the first part of the event at that site; note low-isotope values of *Bulimina ovula* in the gap in the *N. truempyi* data. The oxygen isotope data indicate an abrupt 4–6 °C warming of the deep ocean in a period of ~10 kyr, by far the most rapid rate of warming of the last 65 Myr. The negative carbon isotope excursion is thought to represent the influx of methane from dissociation of gas hydrates.

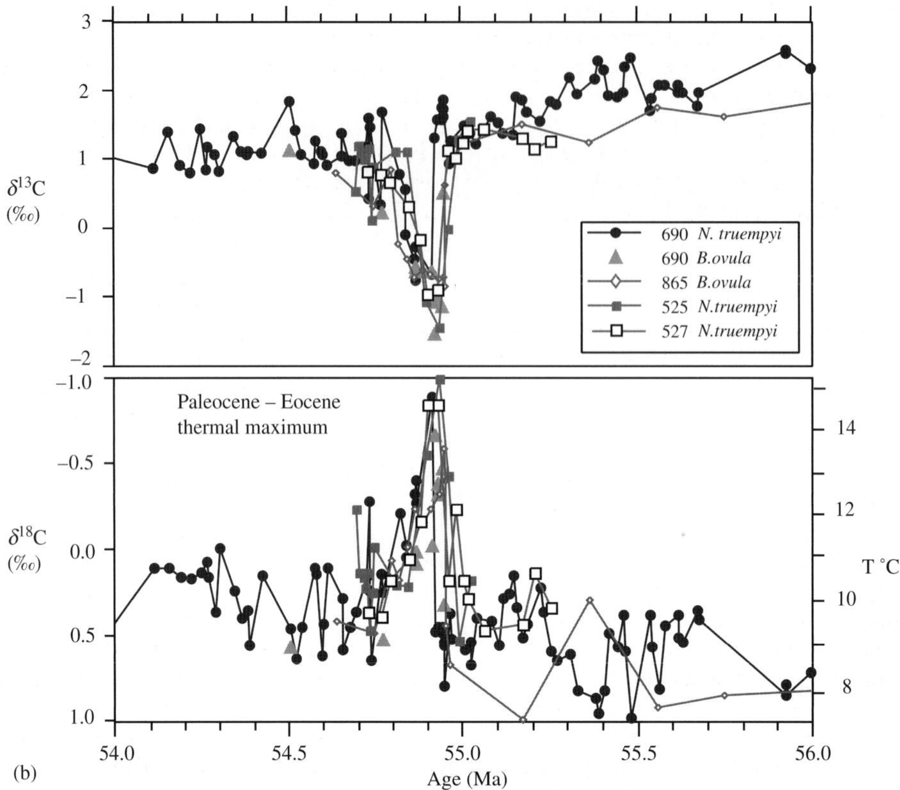

Figure 2 (continued).

either terminate or begin at this point. The few that span this interval show a shift, but with a magnitude about half (~0.5‰) that represented in the global compilation.

6.20.2.2 Carbon Isotopes and Ocean Carbon Chemistry

The distribution of carbon isotopes within the ocean is dependent on two processes, ocean circulation and export production. The $\delta^{13}C_{DIC}$ of deep-water masses are initially set at the site of sinking (via equilibrium exchange with the atmosphere), but tend to progressively change as they slowly migrate through the basins. The respiration of organic matter delivered by export production releases isotopically depleted CO_2 ($\alpha_{T_{CO_2},CH_2O} = 1.021$) to the DIC pool. Thus, as deep waters "age" and their DIC and nutrient contents increase, $\delta^{13}C$ decreases. Initial offsets in deep-water chemical characteristics, however, are retained to the extent that $\delta^{13}C$ can be used to distinguish water masses deriving from different sources. Moreover, the relationship between dissolved nutrients, primarily PO_4, and $\delta^{13}C_{DIC}$, is nearly linear such that the latter has served as a proxy for estimating the distribution of the former, at least for short timescales.

The mean value of ocean $\delta^{13}C_{DIC}$ is not stationary. It changes in response to variations in the fluxes of carbon between the ocean, and the major sources and sinks of carbon (Kump and Arthur, 1999). The major sources are volcanic and metamorphic outgassing of CO_2, and rock (organic and inorganic) weathering. The major sinks are organic and inorganic carbon burial as represented by the following equation from Kump and Arthur (1999):

$$\frac{d}{dt}(M_o\delta_{carb}) = F_w\delta_w + F_{volc}\delta_{volc} - F_{b,carb}\delta_{carb} - F_{borg}(\delta_{carb} + \Delta_B) \quad (1)$$

where M_o is the total dissolved carbon in the ocean, F_w and F_b are the fluxes for weathering and burial, and δ_{carb}, δ_{volc}, and δ_w are the average carbon isotopic compositions of marine carbonates, volcanic CO_2 (crustal + mantle), and weathered or riverine carbon, respectively. δ_{carb} is effectively equal to the mean ocean $\delta^{13}C_{DIC} + 1.0$‰. For F_{borg}, δ is based on a fractionation factor Δ_B which is relatively large, on the order of ~20‰. Because of the large range in isotope values for these sources and sinks, on timescales of >10^5 yr even small changes in the fluxes to or from these reservoirs can impart noticeable changes in mean ocean $\delta^{13}C_{DIC}$.

Two aspects of the Cenozoic marine $\delta^{13}C$ record merit discussion. The first is the spatial distribution of $\delta^{13}C$ between ocean basins, which is relatively insignificant from 65 Ma to 8 Ma (Figure 1). The lack of gradients between basins is generally viewed as evidence of a single dominant deep-water source during the Cenozoic, Antarctica (Wright and Miller, 1993). Deep waters from other regions (e.g., evaporitic, marginal seas), while important locally, are effectively negligible on a global scale. Over the last 8 Myr, the flux of North Atlantic deep water has been large enough to dominate the Atlantic, at least on a periodic basis (i.e., northern hemisphere interglacials). This, coupled with increased isolation of the deep Atlanticas the Tethys and Panamanian gateways close, allows for the development of the large carbon isotope gradient with the Pacific (i.e., basin-to-basin fractionation). This illustrates how the distribution of carbon isotopes is influenced by circulation patterns.

The second observation concerns the long- and short-term trends. Much has been said about the long-term patterns. In particular, the gradual increases in $\delta^{13}C$ over the Late Paleocene and middle Miocene have been attributed to increased rates of organic carbon burial, possibly brought about by tectonic factors that created or destroyed basins, or to changes in mantle outgassing rates (Berner *et al.*, 1983; Shackleton, 1985). The Late Neogene decline in $\delta^{13}C$ is similarly viewed as a signal of changes in the size of the organic carbon reservoir and/or burial rates (Shackleton, 1985; Raymo, 1997). The ability to isolate the process(es) responsible for these long-term trends (i.e., $>10^5$ yr), however, is limited by the ability to constrain other aspects of the global carbon and related geochemical cycles. For example, the burial of reduced sulfur is tightly coupled to organic carbon cycling as represented by Equation (2):

$$2CH_2O + SO_4^{2-} + H^+ \Leftrightarrow 2CO_2 + HS^- + H_2O \quad (2)$$

This microbe mediated reaction is accompanied by significant isotopic fractionation of the sulfur ($\alpha_{SO_4\text{-}HS} = 1.025$) such that large increases in the rate of sulfate reduction can significantly increase the $\delta^{34}S$ of the remaining sulfate reservoir (residence time of sulfur in the ocean would require timescales of 10^5 yr and greater). As a result, a sustained shift in the rate of marine organic carbon production and/or burial as proposed for the Late Paleocene or Neogene should be accompanied by gradual changes in seawater $\delta^{34}S$. Only recently, however, has a record of seawater $\delta^{34}S$ became available (Paytan *et al.*, 1998) to test these hypotheses. This record shows relatively constant values over much of the Cenozoic with two notable exceptions, the Late Paleocene to Early Eocene and the Pleistocene.

The short-term trends, particularly the excursions, merit special consideration. The most prominent occurs at 55 Ma, approximately the Paleocene–Eocene (P–E) boundary. It is characterized by an abrupt 3.0‰ decrease in benthic foraminiferal $\delta^{13}C$ as well as a 4–5 °C global warming, the P–E thermal maximum (Figure 2(b)). Such a rapid and large decrease in mean ocean $\delta^{13}C$ can only be achieved with the addition of a large quantity of ^{12}C enriched carbon. The mean $\delta^{13}C$ of carbon derived from volcanic outgassing is −7‰, whereas the mean for methane ranges from −40‰ to − 60‰, the latter dependent upon whether it is thermogenically or bacterially produced. The rate of change associated with the P–E excursion is more readily achieved through the addition of methane (Dickens *et al.*, 1997). The largest reservoir of methane near the Earth's surface is the marine hydrate reservoir. Computations show that dissociation and oxidation of 2,000 Gt of hydrate methane would be sufficient to produce an excursion of this magnitude. It is likely that other smaller shifts in ocean $\delta^{13}C$ might have similar origins (Dickens, 2001).

The benthic $\delta^{13}C$ record is marked by other excursions in the Cenozoic at the Eocene–Oligocene and Oligocene–Miocene boundaries. These anomalies, however, are positive and more gradual. The direction of change indicates perturbations in one or more fluxes of the global carbon cycle, possibly the burial rates of reduced carbon, which are inferred to increase during each of these events.

6.20.3 THE MARINE STRONTIUM AND OSMIUM ISOTOPE RECORDS

The marine strontium isotope record is the proxy record most commonly used to constrain the geologic history of chemical weathering. However, in recent years it has been widely criticized as a proxy indicator of past silicate weathering rates. The osmium isotope record is analogous to the strontium record in many respects, and can help to constrain interpretations of the marine strontium isotope record. In this section the geochemical factors that influence the osmium and strontium isotope compositions of seawater are reviewed, and the structure of these two records of Cenozoic ocean chemistry is discussed.

6.20.3.1 Globally Integrated Records of Inputs to the Ocean

There are two critically important similarities between osmium and strontium isotopes as paleoceanographic tracers. The first is the sense of parent–daughter fractionation during mantle melting. In the Rb–Sr and Re–Os systems, the radioactive parents, rubidium and rhenium, are partitioned into the melt preferentially to the daughter elements strontium and osmium. Continental crust, a product of mantle differentiation, is characterized by higher Rb/Sr and Re/Os ratios relative to the deep in the Earth. Given the significant mean age of average upper continental crust, ~2 Gyr, *in situ* decay of ^{87}Rb and ^{187}Re has produced significant amounts of ^{87}Sr and ^{187}Os. Thus, in both systems, more radiogenic isotope signatures (higher $^{87}Sr/^{86}Sr$ and $^{187}Os/^{188}Os$) characterize old-crustal rocks relative to recent mantle-derived rocks. The second, and equally important, similarity is the relative isotopic homogeneity of both dissolved strontium and osmium in seawater. This was only recently confirmed for osmium by direct analyses of seawater (Sharma *et al.*, 1997; Levasseur *et al.*, 1998; Woodhouse *et al.*, 1999). The fact that both osmium and strontium are isotopically well mixed in the modern ocean suggests that temporal variations in both isotope records can provide a globally integrated history of oceanic inputs. These two attributes provide a basis for using the marine strontium and osmium isotope records to constrain changes in the riverine solute flux through time.

Mixing of multiple sources of isotopically distinct inputs provides a useful framework for interpreting temporal variations in seawater $^{87}Sr/^{86}Sr$ and $^{187}Os/^{188}Os$. Early efforts to understand the marine strontium isotope record recognized changing proportions of strontium derived from weathering of three end-members: old felsic crust, young basalts, and marine carbonates as the primary influence on the strontium isotopic composition of seawater (Brass, 1976). Although our understanding of the marine strontium cycle has advanced significantly, these rock reservoirs remain fundamentally important. Hydrothermal alteration of basalt at mid-ocean ridges provides a continuous supply of relatively unradiogenic strontium to the ocean that is assumed to be proportional to oceanic crustal production rate. The supply of strontium from weathering of rocks exposed on land is titrated against this unradiogenic strontium input from oceanic crust. Marine carbonates have strontium concentrations that range from several hundred to more than 1,000 ppm, and represent a third important source of strontium to the oceans. The low Rb/Sr of most marine carbonates render production of ^{87}Sr by *in situ* decay in these rocks unimportant. Consequently, strontium derived from weathering of marine carbonates is generally assumed to be similar in isotopic composition to seawater. This is strictly true of strontium released to pore waters during early diagenetic recrystallization of biogenic carbonate (Gieskes *et al.*, 1986). In the context of these three mixing end-members, the marine strontium isotope record can be interpreted as the result of changes in the relative proportions of strontium delivered to the oceans from weathering of old continental crust and young basalts, as buffered by rapid recycling of strontium associated with marine carbonates. This simplified description of the marine strontium cycle can be represented by the equation

$$\frac{\mathrm{d}R_{\mathrm{SW}}}{\mathrm{d}t} = t^{-1}(f_{\mathrm{R}}(R_{\mathrm{R}} - R_{\mathrm{SW}}) + f_{\mathrm{HT}}(R_{\mathrm{HT}} - R_{\mathrm{SW}}) + f_{\mathrm{D}}(R_{\mathrm{D}} - R_{\mathrm{SW}})) \tag{3}$$

where R_i is $^{87}Sr/^{86}Sr$ and f_i is the ratio of the strontium flux from source i to the ocean to the total flux of strontium to the ocean, and t is the marine residence time of strontium (~2 Myr). The subscripts SW, R, HT, and D correspond to seawater, rivers, hydrothermal, and diagenetic terms, and $\mathrm{d}R_{\mathrm{SW}}/\mathrm{d}t$ is the slope of the seawater strontium curve. A summary of representative values for the present-day marine strontium budget is given in Table 1.

An analogous expression can be written for the seawater osmium isotope system but the diagenetic term representing recrystallization of biogenic carbonate would be eliminated. Although the marine osmium cycle is less well documented than the strontium cycle, first-order estimates for most key parameters are available (Table 1). The short marine-residence time of osmium (~10–40 kyr: see Peucker-Ehrenbrink and Ravizza, 2001) and the low sample density in Cenozoic osmium isotope record precludes time-dependent modeling at this time. In addition, it is uncertain if the HT term for osmium can strictly be linked to oceanic crustal production. Instead, it is perhaps better represented by a generic unradiogenic flux that includes both hydrothermal and cosmic inputs (see below for additional discussion). Available data (Table 1) suggest that similar proportions of osmium and strontium are derived from the radiogenic "continental" end-member. This provides some support for a simple "two-component" approach to the marine strontium and osmium isotope records. To the extent that globally averaged riverine input is the product of chemical weathering of average upper crust, the isotopic compositions of strontium and osmium are expected to be relatively invariant. If so, then the gross structure of the marine osmium isotope record should mimic that of the strontium record and changes in the strontium and osmium isotope composition of seawater should be representative of changing solute flux to the global ocean.

Table 1 Summary of the present-day marine Sr and Os budgets.

Isotope ratios	$^{87}Sr/^{86}Sr$	$^{187}Os/^{188}Os$
Seawater	0.70916	1.06
Average upper crust	1.26–1.40	0.716
Average riverine input	0.7119	1.4
Diagenetic flux	0.7084	Unknown
Fractional contribution from riverine flux	0.67	0.73
Elemental fluxes	Sr (mol yr^{-1})	Os (mol yr^{-1})
Riverine flux	3.3×10^{10}	1,850
Diagenetic flux	0.34×10^{10}	Unknown
Calculated hydrothermal/unradiogenic flux	1.6×10^{10}	680
Seawater concentration	590 μM	50 fm
Seawater inventory	1.3×10^{17} mol	7.4×10^{7} mol
Calculated residence time	2.4×10^{6} yr	30,000 yr

Data sources: Sr budget as compiled by Elderfield and Schultz (1996); Os budget as compiled by Peucker-Ehrenbrink and Ravizza (2000); Average upper crust: Peucker-Ehrenbrink and Jahn (2001); Goldstein and Jacobsen (1988).
Notes: Fractional riverine contribution calculated from isotope mass balance. Calculated hydrothermal/unradiogenic flux terms are calculated assuming the present-day ocean is at steady state with respect to Sr and Os isotope mass balance.

6.20.3.2 Osmium–Strontium Decoupling

A more careful examination of the geochemistry of the Rb–Sr and Re–Os systems reveals significant differences between these two isotope systems—differences that have the potential to decouple the marine strontium and osmium isotope records from one another. As monovalent and divalent cations, rubidium and strontium partition readily into major rock-forming minerals in both high- and low-temperature environments. As noted above the affinity of Sr^{+2} for carbonate minerals is a particularly important aspect of the surficial strontium cycle. In contrast, rhenium and osmium, both third-series transition metals, are redox active and strongly siderophile and chalcophile. As a result, rhenium and osmium tend to be associated with trace phases like sulfides and metal oxides. These very different geochemical affinities suggest that the fluxes and isotopic composition of dissolved strontium and osmium carried by individual rivers need not be well correlated in the modern Earth system. It is also possible, though less likely, that the globally averaged fluxes and isotopic composition of dissolved strontium and osmium can vary with time in an uncorrelated manner. If this is the case, then the Cenozoic marine strontium and osmium isotope records are not expected to resemble one another. Below we outline several specific aspects of the geochemical cycles of the Rb–Sr and Re–Os systems that have the potential to decouple oceanic inputs of osmium and strontium.

6.20.3.2.1 Decoupled riverine fluxes of strontium and osmium?

The association of strontium with carbonates, and rhenium and osmium with sedimentary organic matter may effectively decouple the riverine fluxes of strontium and osmium. It is well known that calcium carbonate weathers much more rapidly than silicate minerals and that estimates of silicate weathering fluxes based on strontium isotope data are complicated by this phenomenon (Palmer and Edmond, 1992; see also Jacobson *et al.* (2002a) for a discussion). In general, though not always (see below), terrestrial weathering of carbonates increases strontium flux and lowers $^{87}Sr/^{86}Sr$ relative to an equivalent carbonate-free catchment. The lower $^{87}Sr/^{86}Sr$ of average river flux (0.712) relative to eroding upper crust (0.716) is the global manifestation of the buffering influence that carbonate dissolution exerts on riverine strontium isotope composition.

In contrast to Rb–Sr system, there is no analogue to the buffering influence of carbonate weathering on seawater osmium isotope variations. Instead, it has been suggested that chemical weathering of old organic-rich sediments may actually cause large amplitude changes in the seawater $^{187}Os/^{188}Os$, because they are enriched in both rhenium and osmium and have unusually large Re/Os ratios (Ravizza, 1993). Recent studies demonstrate that rhenium and osmium are efficiently mobilized during black shale weathering (Peucker-Ehrenbrink and Hannigan, 2000; Jaffe *et al.*, 2002; Pierson-Wickmann *et al.*, 2002). If the Os/C_{org} ratios of recent Black Sea sediments (Ravizza *et al.*, 1991) are characteristic of the sedimentary organic carbon reservoir, in general, as much as 5–10% of the continental crustal inventory of osmium may be associated with sedimentary organic matter. This is comparable to the fraction of continental crustal strontium associated with marine carbonates. While the potential importance of weathering old organic-rich shales for causing increases in the seawater

$^{187}Os/^{188}Os$ has been recognized in many studies (Pegram *et al.*, 1992; Ravizza, 1993; Peucker-Ehrenbrink *et al.*, 1995; Singh *et al.*, 1999; Pierson-Wickmann *et al.*, 2000), the potential buffering influence of rapid recycling of osmium rich, reducing marine sediments on continental margins has received little attention. Although the connection between erosion of sedimentary organic matter and the marine osmium isotope record is not well understood, it does provide a clear example of the contrasting aqueous geochemistry of the Rb–Sr and Re–Os systems. These differences can make the marine strontium and osmium isotope records responsive to different aspects of continental weathering.

6.20.3.2.2 *Decoupled unradiogenic fluxes?*

Dissimilarities between osmium and strontium also exist for the unradiogenic inputs to the ocean, providing additional ways in which these two paleoceanographic records could be decoupled. While it is well established that hydrothermal alteration of mid-ocean ridge basalts (MORBs) is the primary source of unradiogenic strontium to the ocean, it is unlikely that this process can balance the osmium isotope budget. Initial investigation of osmium in high-temperature vent fluids suggests that additional source(s) of unradiogenic osmium to the ocean must exist (Sharma *et al.*, 2000). Again, the geochemistry of osmium suggests other likely unradiogenic sources to the ocean. Core formation and the compatible behavior of osmium during mantle melting are responsible for producing the 10^4-fold depletion in typical crustal rocks relative to undifferentiated extraterrestrial material. Although it is unlikely that changes in cosmic dust flux through time are responsible for structure in the Cenozoic osmium isotope record (Peucker-Ehrenbrink, 1996), the flux of unradiogenic osmium contributed to the modern ocean by cosmic dust dissolution remains uncertain (see Peucker-Ehrenbrink and Ravizza, 2000) and some workers argue that it may be significant (Sharma *et al.*, 2000). The compatible behavior of osmium during partial melting of the mantle yields concentrations in MORBs that are ~100 times lower than in ultramafic rocks. Consequently, submarine alteration of ultramafic rocks exposed on the seafloor is commonly invoked as a potentially important source of unradiogenic osmium to the oceans (Palmer and Turekian, 1986; Martin, 1991; Snow and Reisberg, 1995; Peucker-Ehrenbrink, 1996; Levasseur *et al.*, 1999). The recent discovery of ultramafic hosted hydrothermal systems (Douville *et al.*, 2002; Kelly *et al.*, 2001) has intensified interest in this potential source of osmium to seawater. Low-temperature hydrothermal activity (Ravizza *et al.*, 1996; Sharma *et al.*, 2000) has also been suggested as a significant source of unradiogenic osmium to seawater. This source is conceptually appealing because the low sulfide concentration of these fluids may allow osmium concentration to reach substantially higher than that in high-temperature, sulfide-rich fluids. A single analysis of a low-temperature fluid lends some support to this notion (Sharma *et al.*, 2000), but additional data are required to document the influence of low-temperature hydrothermal activity on the marine osmium isotope balance.

Ultramafic hydrothermal alteration is the most interesting of the potential mechanisms for decoupling unradiogenic inputs of osmium and strontium to the ocean. If this proves to be the major source of unradiogenic osmium to seawater, then unradiogenic osmium flux could be anticorrelated with oceanic crustal production rates because ultramafic rocks are preferentially exposed on slow spreading ridges. Alternatively, if low-temperature hydrothermal osmium flux proves to be important, it is unclear that this would undermine the assumption that unradiogenic osmium flux is proportional to oceanic crustal production rate. While major impact events, like the K–T boundary, can clearly decouple the strontium and osmium isotope records, the short-lived nature of this perturbation makes it relatively unimportant for the Cenozoic evolution of ocean chemistry.

6.20.3.3 Reconstructing Seawater Isotope Composition from Sediments

Reconstructing the isotopic composition of ancient seawater is more problematic for osmium than for strontium. Analyses of well-preserved microfossils, typically cleaned foraminifera, yield a fairly robust record of the $^{87}Sr/^{86}Sr$ of ancient seawater because Sr^{+2} is lattice bound in biogenic carbonates. Moreover, strontium isotopic analyses of bulk carbonate and associated pore waters provide the basis for quantitative models of strontium diagenesis in carbonates demonstrating that diagenetic artifacts, even in bulk carbonate analyses, are relatively modest in nearly pure carbonate sequences (Richter and DePaolo, 1988; Richter and Liang, 1993). By comparison with strontium, osmium burial in marine sediments is both more complicated and less-well understood. Osmium burial is more complicated because authigenic osmium enrichment occurs in a variety of depositional settings. These include manganese nodules (Luck and Turekian, 1983; Burton *et al.*, 1999), slowly accumulating pelagic clays (Esser and Turekian, 1988), organic-rich marine sediments (Ravizza and Turekian, 1989, 1992), and metalliferous sediments accumulating near mid-ocean ridges (Ravizza and McMurtry, 1993; Ravizza *et al.*, 1996). This pattern of enrichment demonstrates that osmium is effectively removed from seawater to the solid phase under both oxidizing and reducing conditions. By analogy with better-studied trace metals such as vanadium and molybdenum, it is likely that sedimentary organic matter and iron and manganese oxides play important roles in sequestering osmium from seawater under reducing and oxidizing conditions, respectively. The finely dispersed nature of these phases in the sediment complicates reconstructing the marine osmium isotope record, because there is no simple means of physically isolating material that contains exclusively seawater-derived osmium.

Two different strategies are employed to reconstruct the $^{187}Os/^{188}Os$ of ancient seawater from the sediment record. Analyses of bulk sediments are used in organic-rich (Ravizza, 1998; Cohen *et al.*, 1999) and metalliferous sediments (Ravizza, 1993) where hydrogenous osmium dominates the total sediment osmium budget. In slowly accumulating pelagic clays where osmium associated with cosmic dust and detrital material accounts for more than 50% of the total osmium inventory, leaching methods must be used to selectively liberate hydrogenous (seawater-derived osmium) from pelagic clays (Pegram *et al.*, 1992; Pegram and Turekian, 1999). Although modern calibration studies have vindicated both approaches, neither is completely convincing because both rely on the assumptions about either the selectivity of chemical leaching methods (see Peucker-Ehrenbrink *et al.* (1995) and Pegram and Turekian (1999) for discussions of potential artifacts), or the osmium concentration of local detrital material. As a result of the lack of compelling geochemical arguments, demonstrating the integrity of the marine osmium isotope record, a stratigraphic approach is used in which coeval sediment records from differing locations and depositional setting are compared to one another. Similar records of temporal variations $^{187}Os/^{188}Os$ preserved in widely separated sediment sequences provide empirical evidence suggesting that both records are accurately recording seawater $^{187}Os/^{188}Os$.

6.20.3.4 Cenozoic Strontium and Osmium Isotope Records

6.20.3.4.1 Overview of the Cenozoic marine strontium isotope record

As the result of several decades of investigation the Cenozoic history of seawater $^{87}Sr/^{86}Sr$ variation has become a well-established record of changing ocean chemistry. Peterman *et al.* (1970) presented the first Phanerozoic record of changes in the $^{87}Sr/^{86}Sr$ ratio of seawater derived from analyses of biogenic carbonate. The development of high-precision strontium isotope stratigraphy established the marine strontium isotope record as a valuable stratigraphic tool, and laid the foundation for developing a detailed composite record of Cenozoic seawater strontium isotope variations (DePaolo and Ingram, 1985). Although this record continues to be refined, its major features are well established (Figure 3). A recent compilation of marine strontium isotope record, based on a subset of all published data, can be found in McArthur *et al.* (2001).

The most striking aspect of the Cenozoic strontium record is the nearly monotonic, and relatively rapid rise in seawater $^{87}Sr/^{86}Sr$ ratio during the last 40 Myr, as compared to the small

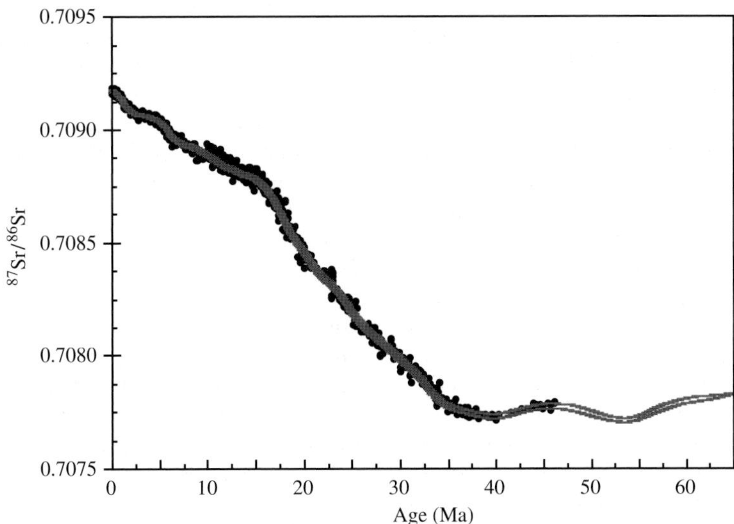

Figure 3 Composite marine strontium isotope record (sources Miller *et al.*, 1991a; Oslick *et al.*, 1994; Hodell and Woodruff, 1994; Mead and Hodell, 1995; Farrell *et al.*, 1995; Martin *et al.*, 1999; Reilly *et al.*, 2002). For the Early Cenozoic where data coverage is sparse the best fit of McArthur *et al.* (2001) is plotted.

amplitude variations that characterize the Early Cenozoic record (65–40 Ma). This difference between the Early and Late Cenozoic portions of strontium record is accompanied by a dramatic contrast in data density. The latter part of the Cenozoic marine strontium isotope record is heavily sampled for two reasons. First the steep slope of the $^{87}Sr/^{86}Sr$ versus age curve allows absolute age dating of marine carbonates with a precision that can be better than ± 1 Ma. Second, the hypothesis that the rapid increase in seawater $^{87}Sr/^{86}Sr$ reflects a significant and systematic increase in alkalinity flux to the ocean associated with accelerated rates of chemical weathering driven by Himalayan uplift (Raymo and Ruddiman, 1992) provided additional impetus to refine the Late Cenozoic strontium isotope record. As a result of these efforts, the fine structure of the last 40 Ma of the Cenozoic strontium isotope record is well documented. Samples are commonly analyzed at 100 kyr intervals in multiple records and independent age control is based on magnetostratigraphy, biostratigraphy, and, more recently, orbital tuning (Farrell *et al.*, 1995; Martin *et al.*, 1999). By comparison the details of seawater $^{87}Sr/^{86}Sr$ ratio variations during the first 25 Ma of the Cenozoic are only poorly constrained.

6.20.3.4.2 Overview of the Cenozoic marine osmium isotope record

The history of Cenozoic variations in the osmium isotopic composition of seawater is also preserved in the marine sediment record. This record is only poorly documented compared to the strontium record. In large part, this reflects the fact that the osmium isotope system is a fairly new paleoceanographic tracer. The first report of temporal changes in the osmium isotope composition of seawater was made only in 1992 (Pegram *et al.*, 1992). Initially, changes in osmium isotope composition produced by decay of ^{187}Re (half-life ~42 Gyr) to ^{187}Os were reported as $^{187}Os/^{186}Os$ ratio variations. Subsequently, the convention of reporting $^{187}Os/^{188}Os$ ratios was adopted because the decay of ^{190}Pt can produce small but measurable amounts of ^{186}Os (Walker *et al.*, 1997). In this review all data initially reported as $^{187}Os/^{186}Os$ ratios have been converted to $^{187}Os/^{188}Os$. Both the present-day marine osmium budget and the marine osmium isotope record are the subject of a recent review (Peucker-Ehrenbrink and Ravizza, 2000).

The Cenozoic marine osmium isotope record is characterized by a shift from generally unradiogenic values in the Early Cenozoic to more radiogenic values in the Neogene. This trend culminates with higher present-day seawater $^{187}Os/^{188}Os$ ratios than at any other time during the Cenozoic (Figure 4). It is clear that the marine strontium and osmium curves do not closely resemble one another. Based on a comparison of these two isotope records, the Cenozoic can be subdivided into three different intervals. (i) From the present to the mid-Miocene (0–15 Ma) the marine strontium and osmium isotope records are broadly correlated and are found to be rising toward more radiogenic isotope ratios. (ii) From

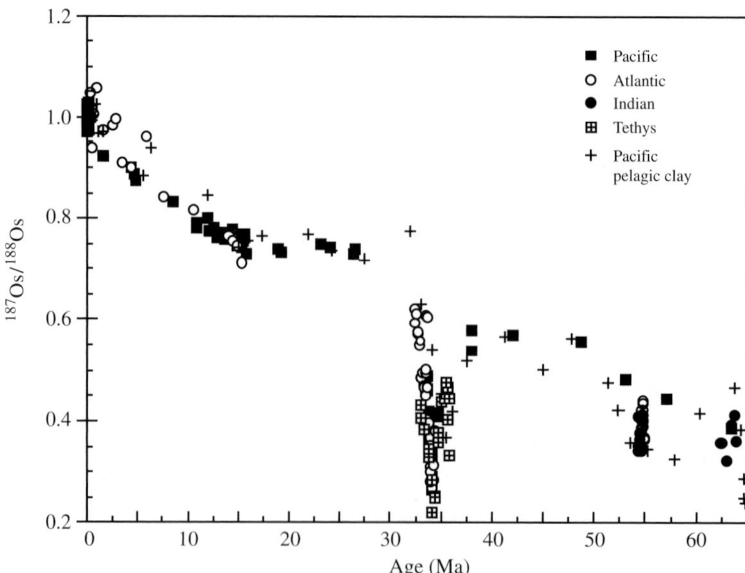

Figure 4 Composite record of seawater $^{187}Os/^{188}Os$ variations during the Cenozoic (sources Pegram *et al.*, 1992; Ravizza, 1993, 1998; Peucker-Ehrenbrink *et al.*, 1995; Reusch *et al.*, 1998; Oxburgh, 1998; Pegram and Turekian, 1999; Ravizza *et al.*, 2001; Ravizza and Peucker-Ehrenbrink, 2003).

the mid-Miocene to the Eocene–Oligocene transition (15–35 Ma) seawater $^{87}Sr/^{86}Sr$ ratio is found to be rising fairly rapidly. In contrast, $^{187}Os/^{188}Os$ ratio remains relatively constant for the majority of this time interval, with the notable exception of the Eocene–Oligocene transition itself. During the Paleocene and Eocene (35–65 Ma), it appears that the osmium record exhibits considerably more variability than does the strontium record. Neither record shows clear evidence of systematic change to more or less radiogenic isotope compositions during this early part of the Cenozoic.

Recently, a substantial body of data constraining the isotope composition and flux of dissolved osmium carried by rivers has been reported (Sharma *et al.*, 1999; Levasseur *et al.*, 1999; Martin *et al.*, 2001). These new data show that while the strontium and osmium isotope compositions of present-day seawater conform with a two-component mixing model fairly well (Table 1), the strontium and osmium isotope compositions of Paleogene seawater do not (Figure 5). This requires either decoupling of strontium and osmium of a magnitude that is not observed in the Neogene marine isotope balances, or a substantial shift in the strontium and osmium isotope composition of riverine influx. By assuming that the mole fractions of osmium and strontium contributing to seawater from riverine input have remained the same as in the modern, a model-based isotopic composition of Paleogene riverine input can be calculated (Figure 5). While the calculated $^{87}Sr/^{86}Sr$ ratio is well within the range measured for modern rivers, the calculated $^{187}Os/^{188}Os$ ratio is not, being lower than average modern riverine input by a factor of ~3. A coupled Sr–C isotope model was designed to partition all seawater strontium isotope variation into changing isotope composition of river flux and the $^{87}Sr/^{86}Sr$ ratio of the silicate portion of riverine flux (Kump and Arthur, 1997). These results indicate a large increase in the $^{87}Sr/^{86}Sr$ ratio and the silicate portion of riverine strontium flux from Paleogene ratio of 0.7095 to a present-day ratio of 0.716. Thus, the marine osmium isotope record may reveal substantial changes in the composition of weathered silicate rock that are partially obscured in the marine strontium isotope record by the buffering effect of carbonate weathering.

6.20.3.4.3 Significance of uplift and weathering of the Himalayan–Tibetan Plateau (HTP)

Among the many recent papers that discuss the marine strontium isotope record, the influence of uplift and weathering of the HTP is pre-eminent as a potential cause of the increasingly radiogenic, or "continental," character during the Cenozoic. Several workers argue that HTP weathering is

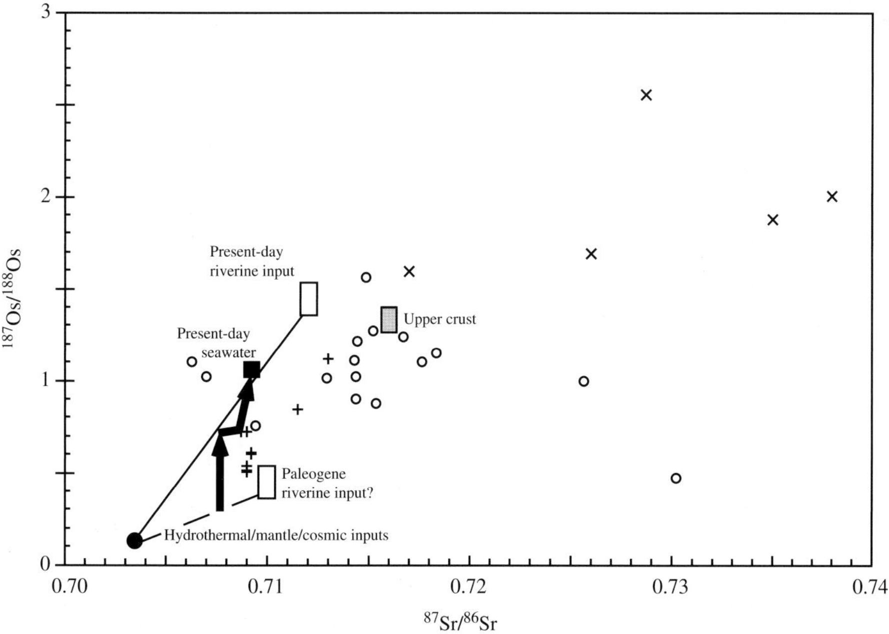

Figure 5 A cross-plot of $^{187}Os/^{188}Os$ versus $^{87}Sr/^{86}Sr$ illustrating the isotopic ratios associated with various oceanic inputs. Solid black arrows schematically illustrate the temporal evolution of seawater during the Cenozoic. Data sources (see Table 1 also) are: (–) Cenozoic seawater; (○) Loess deposits, Peucker-Ehrenbrink and Jahn (2001); (+) Indus paleosols, Chesley *et al.* (2000); and (×) Ganges paleosols, Chesley *et al.* (2000).

the primary cause of the post-40 Ma rise in seawater $^{87}Sr/^{86}Sr$ ratio (Richter *et al.*, 1992; Edmond, 1992; Raymo and Ruddiman, 1992). Several lines of evidence have been used to support this argument. The $^{87}Sr/^{86}Sr$ of Ganges–Brahmaputra (G–B) river waters are unusually high compared to that of global average river flux (Krishnaswami *et al.*, 1992; Palmer and Edmond, 1992). The most rapid increase in the seawater strontium curve coincides with rapid HTP uplift (Richter *et al.*, 1992; Hodell and Woodruff, 1994; see also Figure 6). Calculations of the influence of modern HTP rivers on the present-day seawater $^{87}Sr/^{86}Sr$ ratio have also been used to argue for the importance of HTP weathering (Hodell *et al.*, 1990).

The importance of HTP orogenesis as a cause of the Cenozoic rise in seawater $^{87}Sr/^{86}Sr$ has likely been overstated by many workers. Krishnaswami *et al.* (1992) estimated that G–B flow could account for only one-third of the post-40 Ma rise in seawater $^{87}Sr/^{86}Sr$ ratio. The most recent data indicate that the G–B river system contributes 2% of the global riverine strontium flux with $^{87}Sr/^{86}Sr$ ratio of 0.73 (Galy *et al.*, 1999). Using Equation (3) and the modern marine strontium budget to calculate hypothetical steady-state $^{87}Sr/^{86}Sr$ seawater ratio in the absence of G–B inflow yields a value between 0.7090 and 0.7089, depending on the $^{87}Sr/^{86}Sr$ assumed for the diagenetic strontium flux. This corresponds to ~20% of the post-40 Ma rise in seawater $^{87}Sr/^{86}Sr$ ratio. Some argue that all rivers draining the HTP should be considered, increasing the strontium contribution to ~25% of the global strontium flux (Richter *et al.*, 1992). However, among these rivers only the G–B has $^{87}Sr/^{86}Sr$ ratio substantially larger than average global river input (Palmer and Edmond, 1989, 1992). This implies that strontium supplied by other HTP rivers would contribute to rising

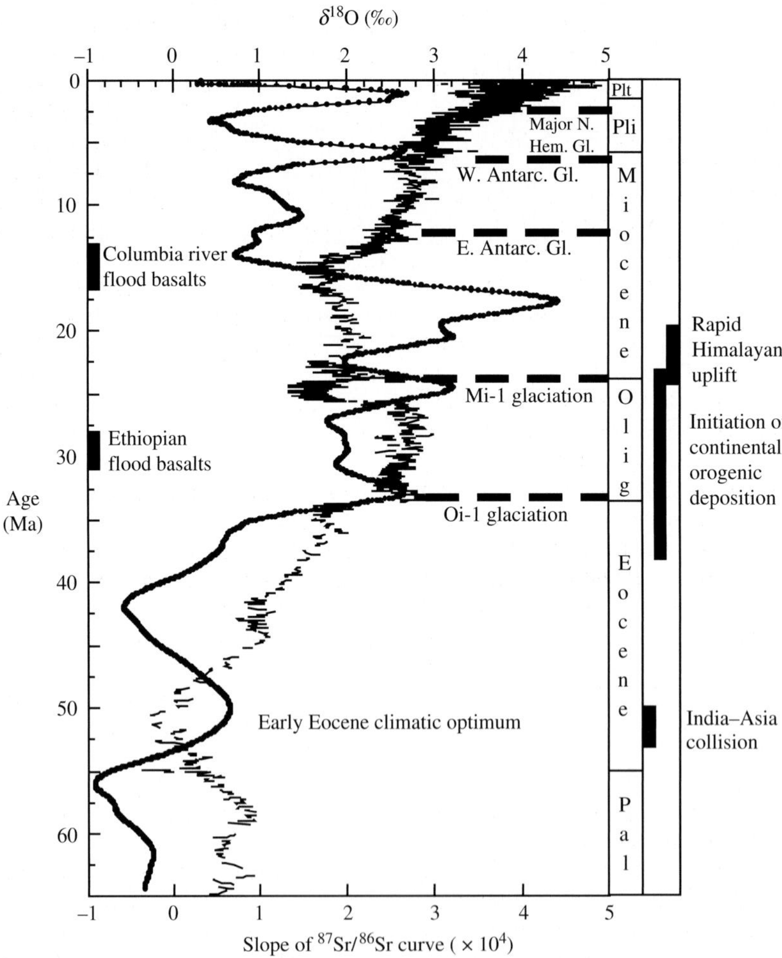

Figure 6 Comparison of the slope of the marine $^{87}Sr/^{86}Sr$ curve (McArthur *et al.*, 2001) to the composite benthic foraminiferal $\delta^{18}O$ record (Zachos *et al.*, 2001). Age ranges for flood basalt eruption are from Wignall (2001). Timing of glacial expansion events is after Zachos *et al.* (2001). Himalayan uplift and erosion history after Najman *et al.* (2000, 2001).

seawater $^{87}Sr/^{86}Sr$ ratio only if their flow represented a net increase in river water flux to the ocean, and not simply a redistribution from another part of the globe. Although it is not possible to preclude a causal relationship between rising seawater $^{87}Sr/^{86}Sr$ ratio and HTP uplift, the detailed changes of the strontium isotope record are difficult to reconcile with the uplift history of the HTP (Figure 6). This is particularly true of the early phases of orogenesis because they remain poorly dated (see Najman *et al.*, 2001).

Early interpretations of the marine osmium isotope record also emphasized a Himalayan influence (Pegram *et al.*, 1992), particularly during the last 15 Ma when both $^{87}Sr/^{86}Sr$ and $^{187}Os/^{188}Os$ rise in concert (Peucker-Ehrenbrink *et al.*, 1995). The modern river data do not preclude this possibility. Recent analyses of dissolved osmium in the G–B water (Sharma *et al.*, 1999; Levasseur *et al.*, 1999) indicate that this river system contributes very roughly 2% of the total riverine osmium flux, similar to that estimated for strontium. The $^{187}Os/^{188}Os$ ratio associated with these analyses are highly variable ranging from 2.9 in the Ganges, twice that of global average river input, to 1.07 in the Brahmaputra. Additional osmium data from the G–B river system are needed to determine if the few available data are representative of seasonally averaged fluxes.

Weathering of Cambrian to Precambrian black shales from the Lesser Himalaya (LH) may be important in causing the rapid rise in seawater $^{187}Os/^{188}Os$ ratio during the last 15 Myr. These rocks appear to be the primary source of rhenium, and unusually radiogenic osmium to the modern G–B river system (Singh *et al.*, 1999; Pierson-Wickmann *et al.*, 2000; Dalai *et al.*, 2002). Analyses of paleosols constrain the $^{187}Os/^{188}Os$ ratio of the Ganges (1.6–2.6) during the last 18 Myr, with the highest ratio in the youngest (3 Ma) sample (Chesley *et al.*, 2000; see Figure 5). This range is similar to the more radiogenic measurements of modern Ganges river water (Levasseur *et al.*, 1999) and Ganges bed load (Pierson-Wickmann *et al.*, 2000). Mass-balance considerations suggest that total amount of ^{187}Os available from the LH black shales is sufficient to influence the osmium isotope composition of the global ocean (Singh *et al.*, 1999).

Given that some workers argue that the LH carbonates are also the source of the unusually radiogenic strontium in the modern G–B system (English *et al.*, 2000), it is tempting to link the rising strontium and osmium isotopic composition of global seawater from 15 Ma to the present to weathering of the lesser Himalaya. However, not all workers agree on the importance of LH carbonates in the G–B strontium budget (see Galy *et al.*, 1999; Singh *et al.*, 1998, and references therein). In addition, the onset of coupled increase of seawater strontium and osmium isotope ratios predates the estimated time of LH exposure to erosion at 11 Ma (Chesley *et al.*, 2000). If input from the G–B is responsible for driving seawater strontium and osmium to more radiogenic isotope composition, the strontium and osmium fluxes provided by these rivers must have been substantially higher in the past (Singh *et al.*, 1999; Chesley *et al.*, 2000).

6.20.3.4.4 Glaciation and the marine strontium and osmium isotope records

Glaciation as a means of enhancing the flux of radiogenic strontium to the ocean is a second recurring theme in interpretations of the Cenozoic seawater strontium isotope record (Armstrong, 1971; Miller *et al.*, 1988; Hodell *et al.*, 1990; Capo and DePaolo, 1990). By comparison to the constraints on the timing of uplift and erosion in the HTP, the timings of major glacial events during the Cenozoic are very well established. Direct comparison of changes in slope of the marine strontium isotope record to oxygen isotope records that constrain changes in climate illustrates this clearly (Figure 6). For example, it has recently been argued that the initial rise in the Cenozoic $^{87}Sr/^{86}Sr$ record may be causally linked to the growth and decay of ice sheets in the Late Eocene and Early Oligocene (Zachos *et al.*, 1999). Although changes in the slope of the $^{87}Sr/^{86}Sr$ record do occur close to the time of many other major glacial events, there is no systematic lead–lag relationship between the two records. For example, a local maximum in the slope of the $^{87}Sr/^{86}Sr$ record is nearly coincident with the first major glaciation of the Oligocene (Oi-1), but a similar local maximum clearly predates the Mi-1 glaciation (Figure 6). Nevertheless mass-balance calculations (Hodell *et al.*, 1990; Blum, 1997) suggest that glacial enhancement of ^{87}Sr flux to seawater indicates that this is a plausible mechanism for driving at least part of the increasing seawater $^{87}Sr/^{86}Sr$ during the last several million years. Coupled studies of base cation flux and strontium isotopes demonstrate that silicate weathering rates are substantially increased, perhaps threefold compared to old saprolitic soils (Blum, 1997), and that the youngest moraines preferentially release ^{87}Sr (Blum and Erel, 1995). However, in glaciated regions with trace carbonate phases distributed throughout bedrock, rapid carbonate weathering can mute the $^{87}Sr/^{86}Sr$ signal of increased weathering rate (Jacobson *et al.*, 2003). Thus, while glacially enhanced silicate weathering may be an effective agent of CO_2 drawdown, quantitatively estimating this effect based on the marine strontium isotope record is not a promising endeavor. Any such effort is further complicated by the fact that the seawater strontium isotope record cannot capture high-frequency glacial forcing because of its long marine-residence time (Richter and Turekian, 1993; Henderson *et al.*, 1994).

The short marine-residence time of osmium allows the isotope composition of seawater to change on glacial–interglacial timescales. Records spanning recent glacial events indicate substantial shifts, ~10%, toward lower-seawater $^{187}Os/^{188}Os$ ratio during peak glacial conditions that are interpreted as diminished river flux to the ocean (Oxburgh, 1998). However, investigation of osmium release from glacial soil sequences shows that ^{187}Os release is enhanced in a manner similar to that documented for ^{87}Sr (Peucker-Ehrenbrink and Blum, 1998), suggesting that the net effect of glaciations should be to drive seawater $^{187}Os/^{188}Os$ to higher values. This conclusion is supported by a recent investigation of the Eocene–Oligocene transition that demonstrates that the abrupt and permanent increase in seawater $^{187}Os/^{188}Os$ ratio is contemporaneous with major Antarctic glaciation (Ravizza and Peucker-Ehrenbrink, 2003; see also Figure 7). Together, these studies provide evidence that the marine osmium isotope record is influenced by glacial cycles. This in turn suggests that the large amplitude, high-frequency glacial cycles play an important role in causing seawater osmium and strontium isotope compositions to shift to more radiogenic values in the latter part of the Cenozoic.

6.20.3.5 Variations in the Strontium and Osmium Isotope Composition of Riverine Input

The most widely recognized ambiguity associated with interpreting the marine strontium and osmium isotope records is the inability to distinguish between temporal changes in the isotopic composition of riverine input and temporal changes in the total flux of strontium and osmium to the ocean. Relatively early models recognized the significance of this uncertainty (Richter *et al.*, 1992; Berner and Rye, 1992). Subsequent analyses of strontium isotopes in paleosol carbonates (Quade *et al.*, 1997) and pedogenic minerals in deep-sea clays (Derry and France-Lanord, 1996) yielded compelling evidence for $^{87}Sr/^{86}Sr$ ratio of river input as high as 0.755 from the G–B river system 7.5 Myr ago. Prior to this time, riverine $^{87}Sr/^{86}Sr$ ratio was 0.712, similar to global average riverine flux. Complementary analyses of osmium isotopes (Chesley *et al.*, 2000) demonstrated coupled strontium and osmium isotope variations in rivers draining the Himalaya during the last 18 Myr (Figure 5). The studies are critically important because they demonstrate that significant temporal variations in riverine strontium and osmium isotope compositions can occur in a large river system. Given that the G–B river system is unusually radiogenic among modern rivers and yet still delivers only ~2% of

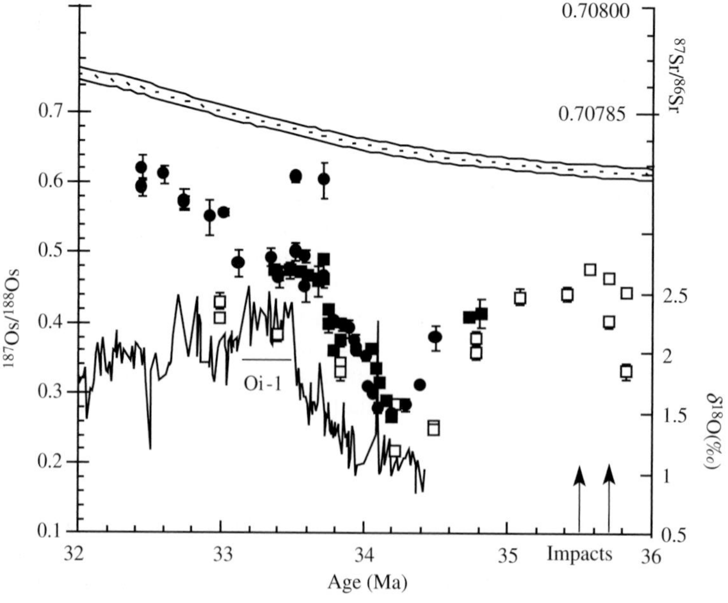

Figure 7 The marine osmium isotope record of the Eocene–Oligocene transition (Ravizza and Peucker-Ehrenbrink, 2003). The benthic foraminiferal oxygen isotope record from site 522 (Zachos *et al.*, 1996) and a fit to the marine strontium isotope record are shown for comparison. Note that the rise of "ice house" conditions in the Early Oligocene is closely associated with an abrupt and permanent increase in seawater $^{187}Os/^{188}Os$. This provides evidence supporting the hypothesis that glaciation contributes to rising seawater $^{187}Os/^{188}Os$ and $^{87}Sr/^{86}Sr$ ratios (( )95% C. L. from McArthur *et al.*, 2001).

the global riverine strontium flux, it is much less likely that such large and rapid changes in the strontium and osmium isotope composition of global average river flux can occur.

The combined osmium and strontium records are consistent with a shift to more radiogenic riverine isotope compositions during the Cenozoic (Figure 5). This suggests that erosion and weathering of unradiogenic lithologies were more important in the past. Several studies suggest that plateaus in the $^{87}Sr/^{86}Sr$ record are caused by erosion of volcanic deposits characterized by low $^{87}Sr/^{86}Sr$ values, either flood basalt provinces (Taylor and Lasaga, 1999; Dessert *et al.*, 2001; McArthur *et al.*, 2001) or island arc volcanics (Reusch and Maasch, 1998). The rapid decline in the slope of the $^{87}Sr/^{86}Sr$ curve between 17 Ma and 14 Ma to near zero provides a specific Cenozoic example (Figure 6). This feature has been attributed to weathering of the Columbia River flood basalts (Hodell *et al.*, 1990; Taylor and Lasaga, 1999), and also to island arc weathering accelerated by the collision of New Guinea with the Australian continent (Reusch and Maasch, 1998). However, emplacement of Ethiopian Traps the Early Oligocene coincides with a much less pronounced decrease in slope of the $^{87}Sr/^{86}Sr$ curve (Figure 6), even though it is estimated to be more than 3 times larger than the CRFB Province (Wignall, 2001). Thus, it seems unlikely that the plateau in the marine strontium isotope record between 17 Ma and 14 Ma can be attributed entirely to flood basalt weathering.

The influence of island arc and flood basalt weathering on the marine osmium isotope record is even less well understood. Analyses of modern rivers draining Papua New Guinea indicate that arc weathering represents an important source of unradiogenic osmium to seawater (Martin *et al.*, 2000, 2001). Though detailed investigations of modern flood basalt weathering have not been made, the marine osmium isotope record across the Triassic–Jurassic boundary is consistent with the hypothesis that emplacement of the Central Atlantic magmatic province is responsible for a shift to lower $^{187}Os/^{188}Os$ ratios (Cohen and Coe, 2002). However, available Cenozoic data do not yield evidence of a significant influence of either CRFB volcanism or arc-continent collision on the marine osmium isotope record between 14 Ma and 17 Ma (Reusch *et al.*, 1998). The fact that the marine osmium isotope record continues to rise during the plateau in the strontium record supports the ideas outlined above that the unradiogenic sources of osmium and strontium are not tightly coupled. The pronounced excursion to unradiogenic $^{187}Os/^{188}Os$ values in the Late Eocene and at the K–T boundary provides additional evidence of Sr–Os decoupling (Figure 4). For the case of the K–T boundary extraterrestrial osmium associated with the K–T impact event is clearly an important factor; in the Late Eocene the influence of extraterrestrial osmium input is possible though less certain (Ravizza and Peucker-Ehrenbrink, 2003).

6.20.3.6 Osmium and Strontium Isotopes as Chemical Weathering Proxies

The need to constrain the geologic history of chemical weathering motivated the detailed reconstruction of past changes in the strontium and osmium isotope composition of seawater. While these combined records remain the best available indicators of changes in patterns of Cenozoic weathering, neither can be used to quantitatively reconstruct CO_2 consumption rates associated with chemical weathering. Silicate weathering rate is only one of the many factors that influences the flux of ^{87}Sr to the ocean. Detailed comparisons of strontium isotope data and river solute chemistry in modern rivers demonstrate that other factors, particularly the influence of strontium released by rapidly weathering carbonates, can obscure the contribution from silicate weathering (Palmer and Edmond, 1989; Jacobson *et al.*, 2002a,b, 2003). Similarly for osmium, silicate-weathering contributions to the ^{187}Os flux are likely convoluted with those from weathering of organic-rich sediments. The former process represents a CO_2 sink in the weathering cycle, while the latter represents a CO_2 source. In addition, the $^{187}Os/^{188}Os$ of osmium contributed by weathering of sedimentary organic matter is strongly influenced by the age of the sediment. This further complicates separating the various sources contributing to riverine osmium flux. Thus, for osmium, like strontium, there is no simple connection between seawater isotope composition and silicate weathering rates.

The marine strontium and osmium isotope records remain important records of ocean chemistry, because they do reflect the changing influence of continental inputs through time. In this regard the marine osmium isotope record has considerable unexplored potential because of its ability to respond to high-frequency external forcing, allowing detailed correlation with carbon and oxygen isotope records. For example, the excursion to more radiogenic seawater $^{187}Os/^{188}Os$ ratio during the unusual warmth of the Paleocene–Eocene thermal maximum (PETM) provides the best available evidence supporting the operation of feedback between global climate and chemical weathering rates (Figure 8). The combined marine strontium (Figure 6) and osmium (Figure 7) isotope records are consistent with the hypothesis that the first major Oligocene glaciation (Oi-1) enhanced silicate weathering rates, providing a positive feedback for additional

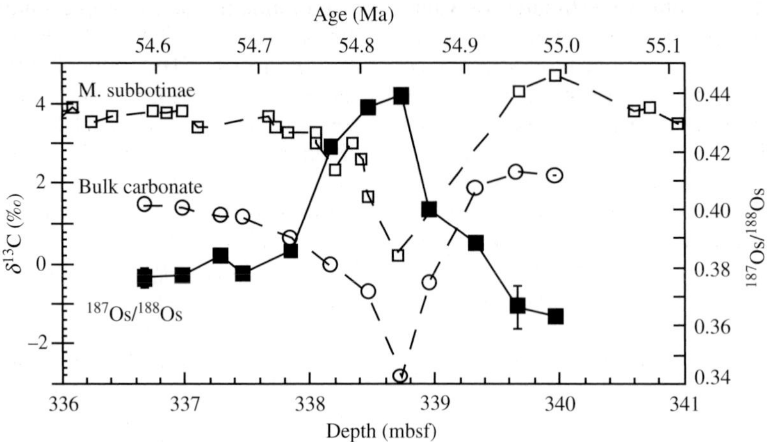

Figure 8 The osmium isotope excursion documented at site 549 from Ravizza *et al.* (2001) is coincident with the carbon isotope excursion that marks the PETM. This association has been interpreted as evidence of accelerated global average weathering rates in response to this global warming event and provides a clear example of the utility of coupling osmium isotope data with conventional stable isotope records of ocean chemistry.

cooling. Contemporaneous changes in the carbon isotopic composition of deep water (Figure 1) and a substantial deepening of the calcium carbonate compensation depth suggest that changes in the ocean–atmosphere carbon cycle were intimately associated with this climatic event. Considered in isolation, the marine strontium and osmium isotope records are ambiguous in that there is not a single unique interpretation of the underlying causes of these documented changes in global ocean chemistry. However, these records can place important constraints on the models of changing ocean chemistry when they are carefully integrated with other records of global climate and ocean chemistry.

6.20.4 Mg/Ca RECORDS FROM BENTHIC FORAMINIFERA

Over the past several years, Mg/Ca variations preserved in benthic foraminifera have received attention as a record of deep-ocean paleotemperatures (Lear *et al.*, 2000; Billups and Schrag, 2002). This work is part of a large and vigorous effort to exploit the potential of Mg/Ca ratio as a paleo-temperature proxy in both benthic and planktonic foraminifera on a variety of timescales. From a historical perspective, it is noteworthy that a positive correlation between the Mg/Ca ratio of biogenic calcite and precipitation temperature was recognized long ago (Chave, 1954). Other factors such as taxa, growth rate, and saturation state are also known to influence the Mg/Ca ratio of biogenic calcite (see Mackenzie *et al.* (1983) for a summary of this early work). These complicating factors detracted from early notions that Mg/Ca ratio of biogenic carbonates might be useful for reconstructing paleo-temperatures. Renewed interest in Mg/Ca ratio as a paleo-temperature proxy differs from these early efforts in many respects. For the purpose of this review some of the more important differences include improved analytical precision and accuracy, the emphasis on microfossils, and the abundance of complementary oxygen isotope data (see also Chapter 6.14).

6.20.4.1 Coupling Benthic Foraminiferal Mg/Ca and Oxygen Isotope Records

The Mg/Ca and $\delta^{18}O$ of benthic foraminifers are both influenced by the temperature of calcification. This link is the reason that a discussion of Mg/Ca is included in this review. As outlined above, the oxygen isotope composition of biogenic calcite is a function of the temperature and the oxygen isotope composition of the water in which the organism calcified. For benthic foraminiferal records it is widely assumed that mainly the ice volume changes drive the temporal variations in the $\delta^{18}O$ of seawater. Thus, benthic foraminiferal $\delta^{18}O$ records have the dual influence of changing ice volume and deep-water temperatures embedded in them. Mg/Ca paleo-thermometry applied to benthic foraminifera offers the possibility of determining absolute deep-water temperatures. These temperature estimates can be used with the measured $\delta^{18}O$ of the same benthic foraminifera to calculate temporal changes in $\delta^{18}O$ of seawater. Reconstructions of the $\delta^{18}O$ of seawater are of interest because variance in this parameter is strongly influenced by the growth and decay of ice sheets. Note however, that even with benefit of a well-constrained $\delta^{18}O$ of seawater record, quantitative estimates of Cenozoic ice volume still require assumptions about the average $\delta^{18}O$ of the global ice reservoir. Still the possibility of accurately determining deep-sea paleo-temperatures and simultaneously better constraining changing ice volume throughout the Cenozoic has motivated a great deal of recent interest in Mg/Ca variations in benthic foraminifera. Lear *et al.* (2000) first applied this approach to a composite record spanning much of the Cenozoic. More recently, this initial effort was supplemented by additional data from a Southern Ocean site spanning roughly the last 28 Myr (Billups and Schrag, 2002).

6.20.4.2 Calibration of the Mg/Ca Thermometer

Before discussing Cenozoic benthic foraminiferal Mg/Ca records, we present an overview of how Mg/Ca ratios measured in benthic foraminifera are used to calculate paleo-temperatures. The most recent calibration efforts postdate the most recent work on the Cenozoic Mg/Ca ratio. Revisions to the calibration equation and other aspects of core-top data sets have important implications for how Cenozoic benthic Mg–Ca records are interpreted. The influence of temperature on Mg/Ca ratio of mixed Cibicidoides reported by Lear *et al.* (2002; Figure 9(a)) is: $(\mathrm{Mg/Ca})_{\mathrm{foram}} = 0.867\mathrm{e}^{0.109T}$, where T represents bottom-water temperature. Also, because these data are largely from core-top calibration studies, this formulation implicitly assumes that calcification occurs from modern seawater with a Mg/Ca ratio of approximately 5.2 mol mol^{-1}. Explicitly including the possibility of variable seawater Mg/Ca ratio results in a modified equation: $(\mathrm{Mg/Ca})_{\mathrm{foram}} = R(0.867\mathrm{e}^{0.109T})$, where R is the Mg/Ca ratio of seawater at some time in the past divided by the present-day seawater Mg/Ca ratio. Other terms are the same as in the previous equation. The problems this possibility poses for Mg/Ca paleo-thermometry are conceptually similar to those presented by $\delta^{18}O$ of seawater variations in oxygen isotope paleo-thermometry, because shell composition depends both on the temperature and the composition of seawater. Variation in seawater Mg/Ca ratio is likely to be less directly coupled to climate change than $\delta^{18}O$ of seawater because changing ice volume should not directly influence seawater Mg/Ca ratio. Presumably seawater Mg/Ca ratio is less variable on short timescales, because the marine residence times of magnesium and calcium are relatively long. Current estimates of the oceanic residence times of calcium in seawater are ~1 Myr while estimates of magnesium residence times are substantially longer. Possible secular trends in seawater Mg/Ca ratio are discussed below.

Examination of core-top calibration data at low temperature (<5 °C), particularly those reported by Martin *et al.* (2002), reveals a distinctly stronger temperature dependence, and a more nearly linear relationship, than is indicated by the full data set (Figure 9(b)). If magnesium incorporation into biogenic calcite behaves as a true solid solution, equilibrium thermodynamics predicts that Mg/Ca ratio should depend exponentially on temperature. The deviation of the low-temperature data from the exponential best fit can be interpreted as evidence that factors other than temperature exert an increasingly important influence on the Mg/Ca ratio of foraminiferal calcite at these low temperatures (Lear *et al.*, 2002). This calibration uncertainty is less important than the potential influence of factors other than temperature on Early Cenozoic benthic foraminiferal Mg/Ca records, because warm deep-water temperatures amplify the temperature component of Mg/Ca variation. Lear *et al.* (2002) include vital effects, dissolution artifacts, growth rate effects, and carbonate saturation state in a list of factors that may exert a secondary control on the Mg/Ca ratios of benthic foraminifera. Some of these same parameters have been recognized as important influences on the Mg/Ca ratio of larger calcifying organisms. For example, Mg/Ca ratios in coralline algae and pelecypods are influenced by growth rate, which is likely to be influenced by temperature (Moberly, 1968). A similar scheme of "nested" influences on Mg/Ca records of benthic foraminifera could be invoked to explain the unusually strong apparent influence of temperature on Mg/Ca ratio in the study by Martin *et al.* (2002) (see Figure 9(b)).

6.20.4.3 Cenozoic Benthic Foraminiferal Mg/Ca Records

Although recent studies of benthic foraminiferal Mg/Ca ratio variations do not span the entire Cenozoic, one study (Lear *et al.*, 2000) does extend back to approximately 50 Ma capturing the unusual warmth of the Eocene climatic optimum. A second study (Billups and Schrag, 2002) reports additional data spanning the last 28 Myr. Results from these two studies are grossly similar in that both show generally decreasing Mg/Ca ratios in benthic foraminifera with time (Figure 10). This is consistent with cooling deep-water temperatures. However, direct comparison of the two data sets is made complicated by several factors. Lear *et al.* (2000) compiled a composite record that includes several different species of benthic foraminifera. Differences between measured Mg/Ca ratios of coexisting species were interpreted as evidence of vital effects, and data were adjusted by empirical correction factors to yield a record that is effectively normalized to *O. umbonatus*.

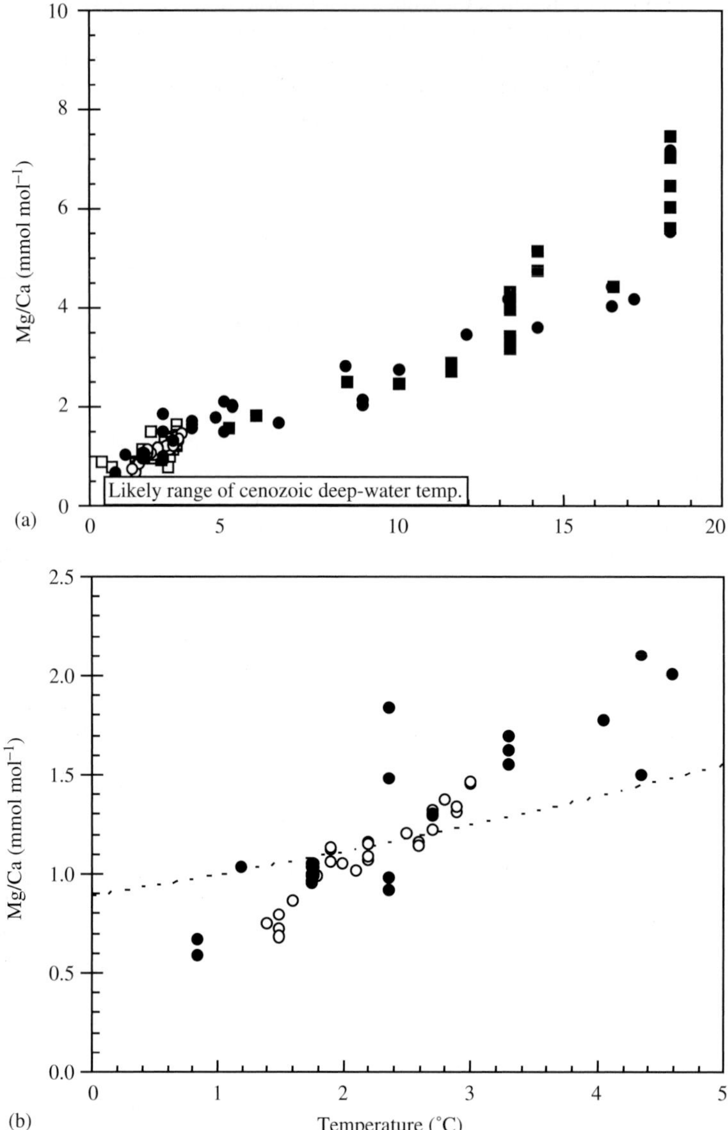

Figure 9 Summary of available data used to calibrate the influence of calcification temperature on the Mg/Ca of benthic foraminifera, after Lear *et al.* (2002). Data are from multiple sources. Data included both core-top samples and analyses of cultured foraminifera. Panel (a) displays the full data set used to obtain the exponential calibration given in the text ((■) Rosenthal *et al.*, 1997 (corrected)). Panel (b) displays a subset of the core-top data that suggest a steeper slope than the full data set. This complicating aspect of the calibration is particularly important to paleo-temperature reconstructions in the Late Cenozoic when deep-water temperatures fall in this low-temperature range ((○) Martin *et al.*, 2002; (●) Lear *et al.*, 2002). See text for further discussion.

In addition, this record was smoothed in order to facilitate comparison to composite benthic foraminiferal $\delta^{18}O$ records. In contrast, the Billups and Schrag (2002) record is based predominately on a single species, *C. mundulus*, with some additional data from *C. wuellerstorfi*. Smoothing is not required for calculating $\delta^{18}O$ of seawater in this record because paired Mg/Ca and $\delta^{18}O$ data are available for the majority of the samples. This approach is desirable because it provides a framework for examining higher-frequency variations in deep-water temperature and $\delta^{18}O$ of seawater. Comparison of these two data sets, in the context of the recent calibration efforts discussed above, indicates that application of the Mg/Ca paleo-temperature proxy to benthic foraminifera is more complicated than initially assumed.

Potential analytical bias, vital effects, and calibration issues all complicate interpretation of the records shown in Figure 10. Measured benthic foraminiferal Mg/Ca ratios in site 747 (Billups and Schrag, 2002) are in general higher than in coeval samples from the composite record of

Lear *et al.* (2000) implying warmer temperatures. The sense of this offset is unexpected given the high latitude of site 747. A systematic analytical bias between the two data sets is possible because different methods were used to make Mg/Ca measurements. Note that applying corrections used by Lear *et al.* (2000) to normalize *C. mundulus* (+0.15) and *C. wuellerstorfi* (+0.45) to normalize Mg/Ca data to *O. umbonatus* tend to amplify this difference. It is noteworthy that core-top analyses discussed above (Lear *et al.*, 2002) did not substantiate a systematic offset between *O. umbonatus* and *C. wuellerstorfi*, highlighting the fact that the general question of vital effects remains open. Although the best available calibration (Figure 9(a)) is based on a mixed assemblage of *Cibicidoides*, *C. mundulus* is not included. Note that not all extant *Cibicidoides* species conform to the calibration shown in Figure 9(a) (Lear *et al.*, 2002). Thus, even the nearly monospecific data set of Billups and Schrag (2002) does not entirely avoid potential problems associated with vital effects. Uncertainty in the functionality of the appropriate Mg/Ca calibration also deserves mention. Billups and Schrag (2002) used a linear calibration curve linking Mg/Ca to temperature rather than an exponential fit. The slope of the calibration line is essentially identical to that shown by the Martin *et al.* (2002) data (Figure 9(b)), but as noted above there is an offset in absolute Mg/Ca ratio. Applying the more recent exponential calibration (Lear *et al.*, 2002) to the Billups and Schrag (2002) data set amplifies the amplitude of calculated temperature variations and increases the average temperature by approximately 3 °C (Figure 11). These differences propagate directly into calculated records of $\delta^{18}O$ of seawater. Given that modern calibration studies can yield linear Mg/Ca versus temperature responses at low temperature (Figure 9(b)),

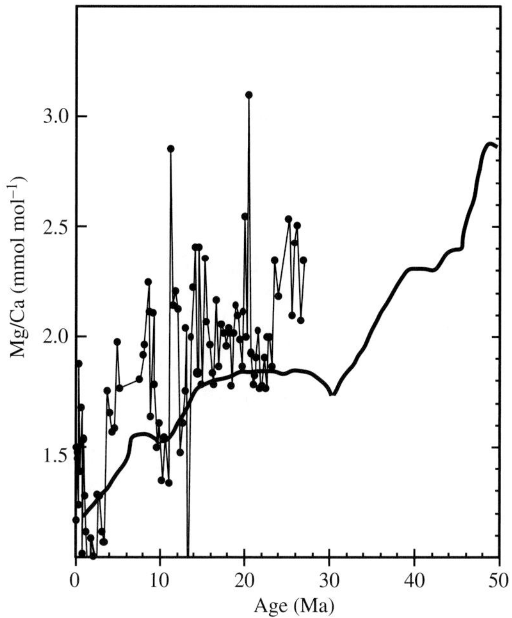

Figure 10 Comparison of Mg/Ca ratios measured in benthic foraminifera from a smoothed, composite record (thick solid curve: Lear *et al.*, 2000), and from site 747 (filled circles: Billups and Schrag, 2002). Both records display an overall trend of declining Mg/Ca ratio with decreasing age. This trend is indicative of cooling deep-water temperatures during the Cenozoic.

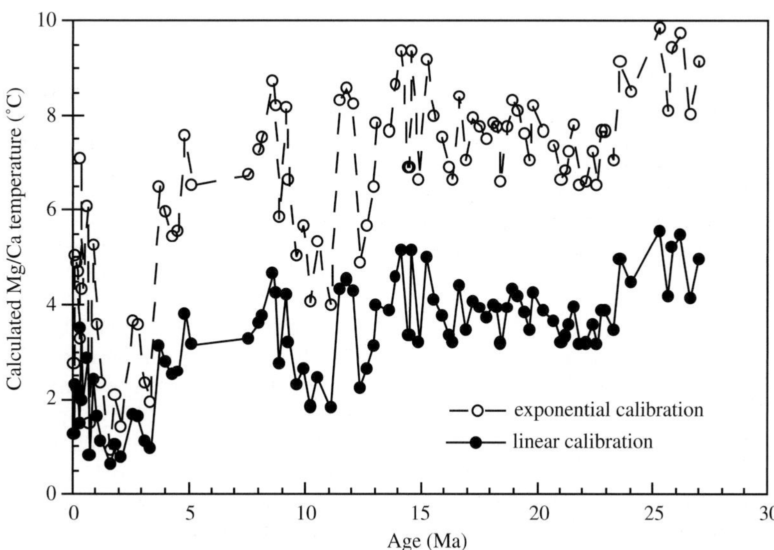

Figure 11 Comparison of deep-water temperatures calculated from the Mg/Ca data of Billups and Schrag (2002) using the linear calibration originally used in this publication and the more recent exponential calibration reported by Lear *et al.* (2002).

careful consideration of the appropriate calibration approach is warranted. This is particularly true in the Late Cenozoic where deep-water temperature estimates tend to be lower.

6.20.4.4 Changing Seawater Mg/Ca Ratio

Possible variations in the Mg/Ca ratio of seawater during the Cenozoic also contribute uncertainty to temperature estimates based on Mg/Ca ratios measured in benthic foraminifera (see Billups and Schrag (2002) for a discussion of model based estimates of the Cenozoic evolution of seawater Mg/Ca ratios; see also Chapter 6.21). Analyses of fluid inclusions in evaporites (Zimmermann, 2000) are consistent with declining seawater Mg/Ca ratio during the Cenozoic. Taken at face value these data suggest a 12% seawater magnesium depletion 5 Myr ago, and further declining to 30–40% of modern value by the Late Eocene (37 Myr ago). However, Zimmermann (2000) notes that dolomitization of calcium carbonate during evaporite formation could create the false impression of lower-seawater magnesium concentrations in the fluid inclusion data. On Phanerozoic timescales, seawater Mg/Ca estimates based on fluid inclusions (Lowenstein *et al.*, 2001) and fossil echinoderms (Dickson, 2002) yield similar records. This lends some credibility to estimates of Cenozoic seawater Mg/Ca ratio based on fluid inclusions. Based on benthic $\delta^{18}O$ data, the assumption of ice-free conditions, and the revised Mg/Ca temperature calibration, Lear *et al.* (2002) calculated that the maximum likely depletion in seawater Mg/Ca ratio was approximately 35%. Even though the magnitude of this estimate is similar to fluid inclusion-based estimates, the timing of these changes remains ill-constrained. While substantial uncertainty remains in converting foraminiferal Mg/Ca data to absolute temperature, it is clear that this proxy has great potential. Likely the most immediate progress will be made in studies that emphasize high temporal resolution and couple Mg/Ca and stable oxygen isotope data to deconvolve the dual influence of changing deep-water temperatures and ice volume during events of rapid growth and decay of ice sheets. On these short timescales abrupt changes in seawater Mg/Ca proportions are unexpected.

6.20.5 BORON ISOTOPES, PALEO-pH, AND ATMOSPHERIC CO_2

The boron isotopic composition of calcite precipitated by foraminifera has been used to reconstruct the pH of the ancient seawater. In appropriate oceanographic settings, pH estimates derived from boron isotope analyses of planktonic foraminifera, in conjunction with additional assumptions about the dissolved inorganic carbon inventory in the surface ocean, can be used to calculate atmospheric CO_2 concentrations. Declining levels of atmospheric CO_2 have been invoked commonly (Raymo and Ruddiman, 1992) as a potentially important causative factor in long-term cooling and the expansion of polar ice that is indicated by the Cenozoic benthic $\delta^{18}O$ record (Figure 1). In this context, the use of boron isotopes as a paleo-pH proxy represents an important new record of changing ocean chemistry during the Cenozoic.

6.20.5.1 The pH Dependence of Boron Isotope Fractionation

There are two stable isotopes of boron, ^{10}B and ^{11}B. ^{10}B is the more abundant of the two, accounting for ~80% of all boron atoms. Variations in boron isotope abundance are reported relative to an NIST boric acid standard as $\delta^{11}B$. In seawater there are two different species of boron. The relative proportion of the two species is a function of pH as indicated by the reaction

$$B(OH)_3 + H_2O = B(OH)_4^- + H^+$$

The p*K*a of boric acid is ~8.7 so that in the pH range relevant to seawater, 7.2–8.2, these two forms of boron coexist. Isotope exchange equilibrium with respect to boron isotopes occurs rapidly such that ^{10}B is preferentially partitioned into the tetrahedral $B(OH)_4^-$ species relative to trigonal boric acid species. At isotopic equilibrium the difference between the $\delta^{11}B$ of these two species is close to 20 per mil. This isotope fractionation factor appears to be largely insensitive to other factors such as temperature and pressure. Given the constant contrast in $\delta^{11}B$ between $B(OH)_3$ and $B(OH)_4^-$, mass-balance considerations require that the $\delta^{11}B$ of the two species vary systematically with pH. The $\delta^{11}B$ of each species can be calculated explicitly once the $\delta^{11}B$ of seawater, the p*K*a of boric acid, and the isotope fractionation are specified (Figure 12). It is important to note that because the p*K*a of boric acid is temperature dependent, calcification temperature can also exert an important influence on the $\delta^{11}B$ of the $B(OH)_4^-$ species (Palmer *et al.*, 1998). It is also noteworthy that at the low end of the seawater pH range boron isotopes lose sensitivity as a pH proxy, because the slope of the $\delta^{11}B$ versus pH curve (Figure 12) becomes nearly flat for both the boron species. This reflects the fact that boron occurs dominantly as boric acid at these lower pH values with only roughly 2% of $B(OH)_4^-$.

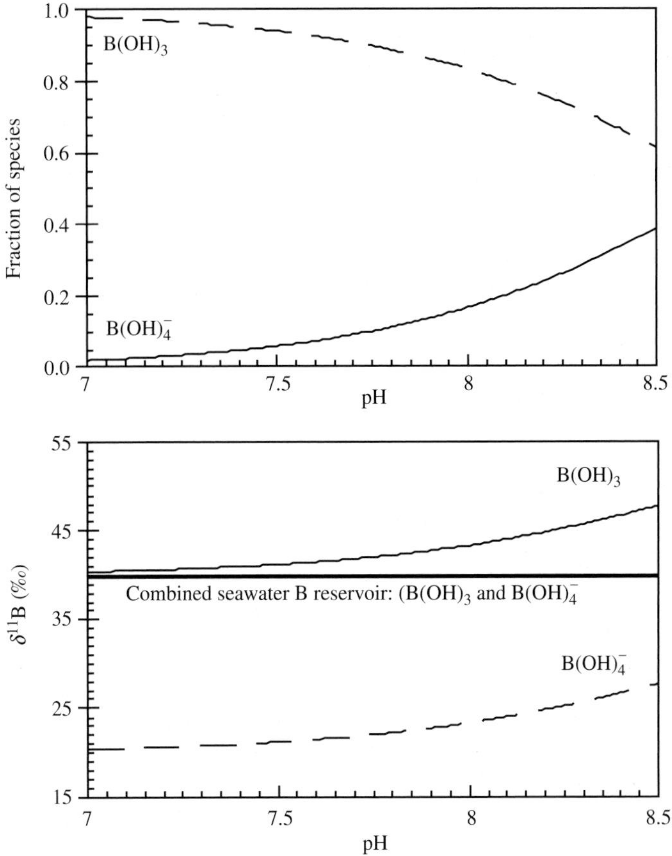

Figure 12 Schematic illustration of isotopic contrast between coexisting boron species as a function of pH. The range of pH shown encompasses all pH values relevant to seawater. Note that at low-pH boric acid dominates the boron inventory and $\delta^{11}B$ becomes relatively insensitive to small pH changes. These curves were calculated using a p*K*a of 8.7 for boric acid, a constant isotopic fractionation of 20 per mil between the two boron species, and $\delta^{11}B$ of 39.6 per mil for the total seawater boron reservoir.

6.20.5.2 Boron Partitioning into Calcite

Both inorganic precipitation experiments (Sanyal *et al.*, 2000) and culture experiments (Sanyal *et al.*, 1996, 2001) have shown that the isotopic composition of boron incorporated into calcite is a function of pH (Figure 13). The measured $\delta^{11}B$ of calcite precipitated over a range of pH are similar to that predicted for $B(OH)_4^-$ indicating that the anionic species of boron is selectively partitioned into calcite as suggested in earlier studies (Vengosh *et al.*, 1991; Hemming and Hanson, 1992). The selective partitioning of a single boron species into calcite effectively records the pH of the water from which the calcite precipitated. Culture studies exhibit a systematic offset in the $\delta^{11}B$ measured in different species of foraminifera indicating that vital effects can influence the absolute $\delta^{11}B$ of foraminiferal calcite. Detailed comparison to theoretical predictions of boron isotope fractionation is hampered by uncertainties in both the isotope fractionation factor and the p*K*a of boric acid. Both vital effects in the biogenic material and kinetic factors in the inorganic system may contribute to the variable offset between the $\delta^{11}B$ of biogenic and inorganic calcite (Figure 13). In the case of planktonic foraminifera that host photosymbionts, daytime drawdown of CO_2 within a microenvironment may contribute to the documented vital effect (Sanyal *et al.*, 2001). Correlated variations in $\delta^{11}B$ and $\delta^{13}C$ in modern coral support such an interpretation (Hemming *et al.*, 1998a). Investigations of both core-top and cultured specimens of *G. sacculifer* and *O. universa* show a similar $\delta^{11}B$ offset relative to one another providing strong evidence that species-dependent vital effects do influence the $\delta^{11}B$ of planktonic foraminiferal calcite (Sanyal *et al.*, 2001; Figure 13).

The few analyses of boron concentration variations in calcite as a function of precipitation pH that have been done in culture (Sanyal *et al.*, 1996) and inorganic precipitation (Sanyal *et al.*, 2000) experiments indicate that boron concentrations increase with increasing precipitation pH. This observation is grossly consistent with

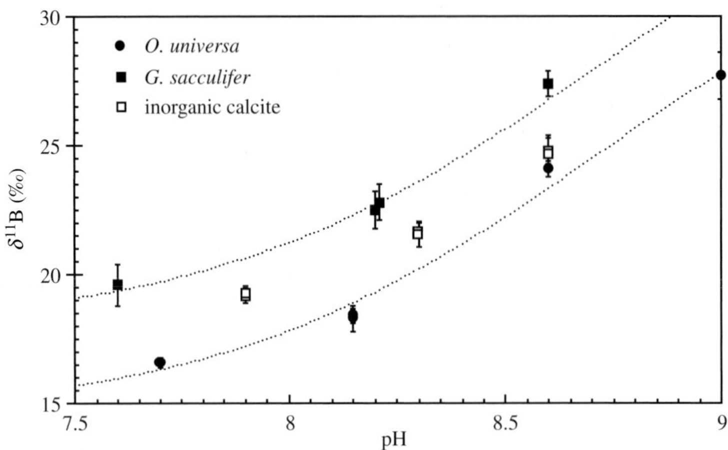

Figure 13 Summary of efforts to calibrate δ^{11}B variations in calcite as a function of the pH of precipitation. Results are replotted from Sanyal *et al.* (1995, 1996, 2000, 2001). The shape of the dashed curves is that predicted by the pH control on the δ^{11}B of $B(OH)_4^-$. The position of these curves was adjusted to fit the *O. universa* and *G. sacculifer* data, highlighting the nearly constant boron isotopic offset between these two species of planktonic foraminifera. The upper curve closely approximates the calculated δ^{11}B of $B(OH)_4^-$ (see Figure 12). Note that the *O. universa* and *G. sacculifer* data plotted include both cultured foraminifera and core-top samples.

the increasing concentration of $B(OH)_4^-$ with increasing pH and a simple proportionality between the $B(OH)_4^-$ concentration in seawater and the boron concentration in calcite. Although there is some scatter in the available data, these results suggest that boron concentration analyses may be complementary to δ^{11}B analyses in paleo-pH reconstructions. However, it is important to note that subtle surface structural effects can give rise to sixfold variations in boron concentration on small spatial scales within a single crystal (Hemming *et al.*, 1998b). Thus, it could be argued that δ^{11}B could accurately record precipitation pH, even though boron concentrations do not exhibit the systematic variation expected based on simple models of element partitioning. While much remains to be learned about the details of boron incorporation into inorganic and biogenic calcite, available data yield strong evidence that calcite δ^{11}B can record the pH of the solution from which it precipitated.

6.20.5.3 Paleo-pH and Atmospheric CO_2 Reconstruction

Several efforts to reconstruct the pH of ancient seawater have been made, starting with the work of Spivack *et al.* (1993). Efforts to quantify changes in surface and deep-water pH during recent glacial–interglacial cycles followed (Sanyal *et al.*, 1995, 1997). The efforts most relevant to this chapter are those of Palmer and Pearson presented in a series of recent papers (Palmer *et al.*, 1998; Pearson and Palmer, 1999, 2000). This work focuses on reconstructing the pH of surface waters in the tropical Pacific Ocean over Cenozoic timescales as a means of constraining atmospheric CO_2 variations. In the modern ocean pH exhibits substantial spatial variability unrelated to atmospheric CO_2 levels. For example, in highly productive regions pH can vary by as much as 0.7 pH units in the upper 1,000 m to values below 7.5. This variability is driven largely by the production of CO_2 during oxidation of sinking organic matter. By concentrating their efforts on a region of the ocean that is likely to have remained thermally stratified, and exhibited only modest productivity variations, Palmer *et al.* (1998) argued that surface-dwelling foraminifera are likely to have grown in waters close to exchange equilibrium with respect to atmospheric CO_2. Analyses of multiple species of foraminifera believed to calcify over distinct depth ranges exhibit trends of δ^{11}B that become less positive with increasing depth in the water column (Palmer *et al.*, 1998). This is consistent with the expected trend of decreasing pH with increasing depth, resulting from organic matter oxidation in the water column. By applying a similar approach to an assemblage of Eocene foraminifera, Pearson and Palmer (1999) estimated the δ^{11}B of Eocene seawater to fall between 38 per mil and 41 per mil, with a most likely value of 40.5 per mil. Values greater than 41 per mil imply a very steep vertical pH gradient and absolute pH values so low as to interfere with calcification. Seawater δ^{11}B lighter than 38‰ imply only small pH changes with depth, and these authors argued that the low productivity implied by such a shallow pH gradient was unlikely. Only gradual changes in δ^{11}B of seawater are likely, because the marine residence time of boron is on the order of 10 Myr. Detailed paleo-pH profiles at a few time intervals distributed throughout the Cenozoic (Palmer *et al.*, 1998; Pearson and Palmer, 1999) provide a framework constraining seawater $\delta^{11}B$ during the Cenozoic. In a follow-on study Pearson and Palmer (2000) combined data from analyses of planktonic foraminifera that calcified in the mixed layer with their model-based constraints on seawater $\delta^{11}B$ to make 35 individual surface water pH estimates distributed over the past 60 Myr (Figure 14). To estimate atmospheric CO_2 from this surface ocean pH requires additional assumptions. As noted above the surface waters at this site are assumed to remain close to equilibrium with respect to atmospheric CO_2 throughout the Cenozoic. Second, an independent estimate of either total alkalinity or total DIC is required to calculate the partial pressure of CO_2 from pH. To make this estimate, Pearson and Palmer (2000) used reconstructions of CCD, and assumed the depth of calcite saturation relative to the CCD remained invariant with time. They further assumed that the vertical gradient in alkalinity remained constant with time and that the calcium concentration of seawater varied in concert with alkalinity.

Although the atmospheric CO_2 record resulting from the Pearson and Palmer (2000) study is subject to substantial uncertainty, several features of the record are of interest. Although Figure 14 does indicate that CO_2 levels were roughly 10 times higher than present-day levels during the Early Eocene, the record as a whole does not, however, suggest a gradual decline in CO_2 levels throughout the Cenozoic. This is best illustrated by the variability in the middle Eocene CO_2 estimates ranging over nearly an order of magnitude on relatively short timescales. Note that this time is characterized by relatively light benthic $\delta^{18}O$ values indicative of warm deep-water temperatures and little or no polar ice (Figure 1). The last 25 Myr of the CO_2 reconstruction show smaller amplitude variations with the majority of data suggesting values between 200 ppm and 400 ppm. Reconstructions of CO_2 between 16 Ma and 4 Ma based on carbon isotope variations in alkenones yield similar values ranging between 180 ppm and 320 ppm (Pagani *et al.*, 1999). As these two methods are largely independent of one another, the similarity of the two data during the time interval where they overlap lends credibility to the combined data sets. Both records suggest a component of high-frequency variation in atmospheric CO_2 levels, but detailed comparison of the two records is hampered by the low-temporal resolution of the two records. For example, the boron isotope record suggests that high atmospheric CO_2 levels may have been associated with emplacement of the Columbia River flood basalts at ~17 Ma, while the expansion of the East Antarctic ice sheet at 15 Ma may be associated with a local minimum in CO_2 (Pearson and Palmer, 2000). In contrast, alkenone-based CO_2 estimates do not indicate a pronounced CO_2 decrease during expansion of the east Antarctic ice sheet (Pagani *et al.*, 1999). Based on available data it is unclear if these two records contradict one another or if aliasing of high-frequency variation simply creates this impression.

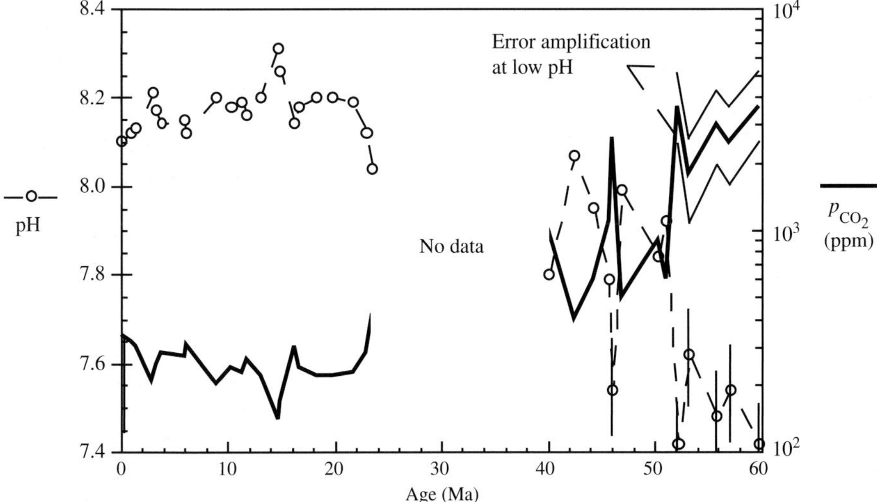

Figure 14 Estimates of mixed-layer pH and atmospheric CO_2 derived from boron isotope analyses of planktonic foraminifera recovered from drill cores taken in the western equatorial Pacific. Results are replotted from Pearson and Palmer (2000). Note that uncertainty associated with both pH and CO_2 estimates amplifies in the Early Cenozoic. This reflects the intrinsic insensitivity of the boron isotope pH proxy at low pH (see Figure 12). Atmospheric CO_2 estimates require additional assumptions about the size of the DIC pool in surface waters.

6.20.5.4 Outstanding Questions about Paleo-pH Reconstructions

Several important questions related to the use of boron isotopes as a paleo-pH proxy remain unanswered. While progress has been made documenting the presence of vital effects on the $\delta^{11}B$ of biogenic calcite (Sanyal *et al.*, 2001; Hemming *et al.*, 1998a), further work on this issue is required. Reconstruction of Early Cenozoic pH requires working with planktonic foraminifera that are no longer extant and thus vital effects cannot be well constrained without a substantial additional effort to work on multiple species with overlapping age ranges. Dissolution effects and diagenetic alteration are known to adversely affect the application of other paleoceanographic proxies, and the influence of this phenomenon on $\delta^{11}B$ is yet to be investigated. Given that tropical planktonic foraminifera are particularly vulnerable to diagenetic alteration on the seafloor (Pearson *et al.*, 2001), the influence of this process on $\delta^{11}B$ of foraminiferal calcite is important to evaluate. Recent efforts to estimate glacial–interglacial pH changes in deep waters that are based on preservation of carbonate microfossils (Anderson and Archer, 2002) have yielded results that are at odds with boron-isotope-based estimates (Sanyal *et al.*, 1995), also highlighting the need for additional work. The potential for temporal variations in the $\delta^{11}B$ of seawater also warrants additional investigation. Model calculations based on a refined version of the present-day marine boron cycle, and assumptions about oceanic crustal production and weathering rates (Lemarchand *et al.*, 2000) suggest that temporal changes in the $\delta^{11}B$ of seawater may be substantially larger than estimated by Pearson and Palmer (2000) using the pH profile approach. The model calculations of Lemarchand *et al.* (2000) suggest that seawater $\delta^{11}B$ could have been as low as 36 per mil. If true, this would imply that the high CO_2 levels estimated in the Early Cenozoic (Figure 14) are incorrect and that levels similar to modern values are more likely. Thus, independent data constraining the $\delta^{11}B$ of seawater would also represent an important contribution to advancing the use of boron isotopes as a paleo-pH proxy. Both continuing refinements to application of boron isotopes to paleo-pH reconstructions and coupled application of independent means of estimating paleo-p_{CO_2} levels are important areas for future work.

6.20.6 CLOSING SYNTHESIS: DOES OROGENESIS LEAD TO COOLING?

Recently much debate has focused on the hypothesis that increased silicate weathering rates, associated with Himalayan uplift, played a causative role in long-term Cenozoic cooling by reducing atmospheric carbon dioxide levels (Raymo *et al.*, 1988; Raymo and Ruddiman, 1992). Indeed, it seems appropriate to revisit this idea at the close of this review because the idea fueled much of the effort to refine and integrate the paleoceanographic records discussed above. What is the current status of this idea? Two key problems with the "orogenesis leads to cooling" hypothesis as initially articulated by Raymo and co-workers have been raised. First, simple mass-balance considerations preclude the possibility of accelerated rates of CO_2 drawdown by chemical weathering throughout the Cenozoic without a comparable increase in the flux of CO_2 to the atmosphere from volcanic and metamorphic sources (Caldeira *et al.*, 1993; Volk *et al.*, 1993; Berner and Caldeira, 1997; Broecker and Sanyal, 1998). Second, recent efforts to determine the history of atmospheric CO_2 levels suggest that cooling and ice growth during the Neogene are not associated with systematic decreases in atmospheric CO_2 (Pagani *et al.*, 1999; Pearson and Palmer, 2000), also undermine this hypothesis.

In spite of the critiques outlined in the preceding paragraph, it is inappropriate to discount entirely the role of orogenesis in Cenozoic cooling entirely. The mass-balance problems outlined above can be addressed. Kump and Arthur (1997) presented a conceptual framework in which steady-state levels of atmospheric CO_2 could be reduced without changes in global weathering flux by increasing "weatherability." In essence they argued that tectonic and climatic factors could act to allow weathering fluxes to remain high enough to balance CO_2 input in spite of significantly cooler global climate. This concept provides a defensible framework for a causative link between Himalayan uplift and global cooling, because physical weathering can be reasonably argued to enhance "weatherability" (Kump *et al.*, 2000). It also provides a basis for arguing that a limited decoupling of average global temperatures and silicate weathering rates played an important role in Cenozoic cooling. Both of these are key components of the "orogenesis leads to cooling" hypothesis.

The available data constraining the Cenozoic history of atmospheric CO_2 suggest nearly 10-fold higher concentration during the Paleocene and Early Eocene, relative to the Neogene (Pearson and Palmer, 2000). Thus, the gross contrast between the unusual warmth of the Early Paleogene and the "ice house" conditions of the Neogene can still be interpreted as the result of tectonically driven changes in weathering, even though the many features of Neogene climate change cannot. It is likely that factors other than orogenesis contributed to enhanced weatherability during the Cenozoic. For example, accelerated physical erosion due to glaciation may greatly increase the area of fresh mineral surfaces available to chemical weathering (Blum, 1997; Zachos *et al.*, 1999). Many factors, in addition to atmospheric CO_2 concentration, may influence Earth's climate over the course of the Cenozoic. Thus while the "orogenesis leads to cooling" hypothesis is not dead, it does not provide a single unified explanation of the established record of Cenozoic climatic variation.

The mass-balance argument that CO_2 consumption by silicate weathering cannot outpace CO_2 production by volcanic and metamorphic degassing also has important implications for interpretation of heavy isotope records. If, to first approximation, the average long-term rate of silicate weathering has remained constant during the Cenozoic, and this can be extended to the total flux of strontium and osmium, then the variations in the marine $^{87}Sr/^{86}Sr$ and $^{187}Os/^{188}Os$ records are better interpreted as the result of changing isotopic composition of riverine flux (Kump and Arthur, 1997). Under this interpretation, the Cenozoic strontium and osmium isotope record imply a secular change in the age and/or composition of weathered material. Both records suggest that the Cenozoic is characterized by an increasing contribution from older material with a cratonic affinity and radiogenic strontium and osmium characteristics relative to recently mantle-derived material (Figure 5). How such a shift in the nature of weathered material might influence "weatherability," or if it would at all, has not been considered in detail.

The efforts of many researchers have contributed to a greatly improved history of Cenozoic ocean chemistry and climate history. These efforts have revealed a great deal of fine structure in these records that is global in nature, ranging from episodes of unusual warmth such as the PETM to episodes of rapid glacial expansion in the Early Oligocene and Early Miocene. It is possible that short-term fluctuations in atmospheric CO_2 played an important causative role in some of these events, even though available data argue against a simple causal link between atmospheric CO_2 and long-term Cenozoic cooling. A refined vision of a global thermostat originally proposed by Walker *et al.* (1981) that includes secular changes in global "weatherability" (Kump and Arthur, 1997) can be invoked to explain how a negative feedback between global temperature and silicate weathering rate can stabilize Earth's climate system in the wake of large perturbations.

The causes of events that perturb Earth's climate system are less well understood and more varied in nature. Understanding these phenomena likely requires studies that focus on specific climatic events rather than long-term climate evolution. Careful integration of the various paleo-proxies discussed in this review offers the potential for the first time to establish the lead–lag relationships between ice growth, deep-water temperature change, continental weathering, and atmospheric CO_2 during important climate transitions and on timescales similar to the residence of inorganic carbon in the ocean-atmosphere system. These types of studies will represent a major step forward in our ability to test fundamental hypotheses about the underlying causes of the many climate changes documented during the Cenozoic. This integrated multiproxy approach may also lay a foundation for exploring the possibility that these short-term climate events can influence Earth's long-term climate evolution.

REFERENCES

Anderson D. M. and Archer D. (2002) Glacial-interglacial stability of ocean pH inferred from foraminifer dissolution rates. *Nature* **416**(6876), 70–73.

Armstrong R. L. (1971) Glacial erosion and the variable isotopic composition of strontium in sea water. *Nature; Phys. Sci.* **230**(14), 132–133.

Berggren W. A., Kent D. V., Swisher C. C., III, and Aubry M.-P. (1995) A revised Cenozoic geochronology and chronostratigraphy Geochronology, time scales and global stratigraphic correlation. *Special Publication—SEPM (Society for Sedimentary Geology).* **54**, 129–212.

Berner R. A. and Caldeira K. (1997) The need for mass balance and feedback in the geochemical carbon cycle. *Geology* 955–956.

Berner R. A. and Rye D. M. (1992) Calculation of the Phanerozoic strontium isotope record of the oceans from a carbon cycle model. *Am. J. Sci.* **292**, 136–148.

Berner R. A., Lasaga A. C., and Garrels R. M. (1983) The carbon-silicate geochemical cycle and its effect on atmospheric carbon dioxide over the past 100 million years. *Am. J. Sci.* **283**, 641–683.

Billups K. and Schrag D. P. (2002) Paleotemperatures and ice volume of the past 27 Myr revisited with paired Mg/Ca and $^{18}O/^{16}O$ measurements on benthic foraminifera. *Paleoceanography* **17**, 3-1–3-11.

Blum J. D. (1997) The effect of late Cenozoic glaciation and tectonic uplift on silicate weathering rates and the marine $^{87}Sr/^{86}Sr$ record. In *Tectonic uplift and climate change* (ed. W. F. e. Ruddiman). Plenum, New York, pp. 259–288.

Blum J. D. and Erel Y. (1995) A silicate weathering mechanism linking increases in marine $^{87}Sr/^{86}Sr$ with global glaciation. *Nature* **373**, 415–418.

Bralower T. J., Zachos J. C., Thomas E., Parrow M., Paull C. K., Kelly D. C., Premoli Silva I., Sliter W. V., and Lohmann K. C. (1995) Late Paleocene to Eocene paleoceanography of the equatorial Pacific Ocean: stable isotopes recorded at Ocean Drilling Program Site 865, Allison Guyot. *Paleoceanography* **10**(4), 841–865.

Brass G. W. (1976) The variation of the marine $^{87}Sr/^{86}Sr$ ratio during Phanerozoic time: interpretation using a flux model. *Geochim. Cosmochim. Acta* **40**, 721–730.

Broecker W. S. and Sanyal A. (1998) Does atmospheric CO_2 police the rate of chemical weathering? *Global Biogeochem. Cycles* **12**(3), 403–408.

Burton K. W., Bourdon B., Birck J.-L., Allegre C. J., and Hein J. R. (1999) Osmium isotope variations in the oceans recorded by Fe–Mn crusts. *Earth Planet. Sci. Lett.* **171**(1), 185–197.

Caldeira K., Arthur M. A., Berner R. A., and Lasaga A. C. (1993) Cooling in the late Cenozoic. *Nature* **361**, 123–124.

Capo R. C. and DePaolo D. J. (1990) Seawater strontium isotopic variations from 2.5 million years ago to the present. *Science* **249**, 51–55.

Chave K. E. (1954) Aspects of the biogeochemistry of magnesium: [Part] 1. Calcareous marine organisms; [Part] 2. Calcareous sediments and rocks. *J. Geol.* **62**(3), 266–283.

Chesley J. T., Quade J., and Ruiz J. (2000) The Os and Sr isotopic record of Himalayan paleorivers: Himalayan tectonics and influence on ocean chemistry. *Earth Planet. Sci. Lett.* **179**(1), 115–124.

Cohen A. S. and Coe A. L. (2002) New geochemical evidence for the onset of volcanism in the Central Atlantic magmatic province and environmental change at the Triassic–Jurassic boundary. *Geology* **30**(3), 267–270.

Cohen A. S., Coe A. L., Bartlett J. M., and Hawkesworth C. J. (1999) Precise Re-Os ages of organic-rich mudrocks and the Os isotope composition of Jurassic seawater. *Earth Planet. Sci. Lett.* **167**, 159–173.

Dalai T. K., Singh S. K., Trivedi J. R., and Krishnaswami S. (2002) Dissolved rhenium in the Yamuna River system and the Ganga in the Himalaya: role of black shale weathering on the budgets of Re, Os, and U in rivers and CO_2 in the atmosphere. *Geochim. Cosmochim. Acta* **66**(1), 29–43.

DePaolo D. J. and Ingram B. L. (1985) High resolution stratigraphy with Sr isotopes. *Science* **227**, 938–941.

Derry L. A. and France-Lanord C. (1996) Neogene Himalayan weathering history and river $^{87}Sr/^{86}Sr$: impact on the marine Sr record. *Earth Planet. Sci. Lett.* **142**, 59–74.

Dessert C., Dupre B., Francois L. M., Schott J., Gaillardet J., Chakrapani G., and Bajpai S. (2001) Erosion of Deccan traps determined by river geochemistry: impact on the global climate and the $^{87}Sr/^{86}Sr$ ratio of seawater. *Earth Planet. Sci. Lett.* **188**(3–4), 459–474.

Dickens G. (2001) On the fate of past gas: what happens to methane released from a bacterially mediated gas hydrate capacitor? *Geochem. Geophys. Geosys.-G3* **2**, 2000GC000131.

Dickens G. R., Castillo M. M., and Walker J. C. G. (1997) A blast of gas in the latest Paleocene: simulating first-order effects of massive dissociation of oceanic methane hydrate. *Geology* **25**(3), 259–262.

Dickson J. A. D. (2002) Fossil echinoderms as monitor of the Mg/Ca ratios of Phanerozoic oceans. *Science* **298**, 1222–1224.

Diester-Haass L. (1996) Late Eocene–Oligocene paleoceanography in the southern Indian Ocean (ODP Site 744). *Marine Geol.* **130**(1–2), 99–119.

Douville E., Charlou J. L., Oelkers E. H., Bienvenu P., Jove Colon C. F., Donval J. P., Fouquet Y., Prieur D., and Appriou P. (2002) The Rainbow Vent fluids (36 degrees 14′N, MAR); the influence of ultramafic rocks and phase separation on trace metal content in Mid-Atlantic Ridge hydrothermal fluids. *Chem. Geol.* **184**(1–2), 37–48.

Edmond J. M. (1992) Himalayan tectonics, weathering processes, and the strontium isotope record in marine limestones. *Science* **258**, 1594–1597.

Elderfield H. and Schultz A. (1996) Mid-ocean ridge hydrothermal fluxes and the chemical composition of the ocean. *Ann. Rev. Earth Planet. Sci.* **24**, 191–224.

English N. B., Quade J., DeCelles P. G., and Garzione C. N. (2000) Geologic control of Sr and major element chemistry in Himalayan rivers, Nepal. *Geochim. Cosmochim. Acta* **64**(15), 2549–2566.

Esser B. K. and Turekian K. K. (1988) Accretion rate of extraterrestrial particles determined from osmium isotope systematics of Pacific pelagic clay and manganese nodules. *Geochim. Cosmochim. Acta* **52**, 1383–1388.

Farrell J. W., Clemens S. C., and Gromet L. P. (1995) Improved chronostratigraphic reference curve of late Neogene seawater $^{87}Sr/^{86}Sr$. *Geology* **23**, 403–406.

Flower B. P. and Kennett J. P. (1993) Middle Miocene ocean-climate transition; high-resolution oxygen and carbon isotopic records from Deep Sea Drilling Project Site 588A, Southwest Pacific. *Paleoceanography* **8**(6), 811–843.

Flower B. P., Zachos J. C., and Paul H. (1997) Milankovitch-scale climate variability recorded near the Oligocene/Miocene boundary. *Proc. Ocean Drill. Prog. Sci. Results* **154**, 433–439.

Frank M. (2002) Radiogenic isotopes: tracers of past ocean circulation and erosional input. *Rev. Geophys.* **40**(1), 1.1–1.38.

Galy A., France-Lanord C., and Derry L. A. (1999) The strontium isotopic budget of Himalayan rivers in Nepal and Bangladesh. *Geochim. Cosmochim. Acta* **63**(13–14), 1905–1925.

Gieskes J. M., Elderfield H., and Palmer M. R. (1986) Strontium and its isotopic composition in interstitial waters of marine carbonate sediments. *Earth Planet. Sci. Lett.* **77**(2), 229–235.

Goldstein S. J. and Jacobsen S. B. (1988) Nd and Sr isotopic systematics of river water suspended material: implications for crustal evolution. *Earth Planet. Sci. Lett.* **87**, 249–265.

Grossman E. L. (1984) Stable isotope fractionation in live benthic foraminifera from the Southern California Borderland. *Palaeogeogr. Palaeoclimatol. Palaeoecol.* **47**(3–4), 301–327.

Grossman E. L. (1987) Stable isotopes in modern benthic foraminifera: a study of vital effect. *J. Foraminiferal Res.* **17**(1), 48–61.

Hemming N. G. and Hanson G. N. (1992) Boron isotopic composition and concentration in modern marine carbonates. *Geochim. Cosmochim. Acta* **56**(1), 537–543.

Hemming N. G., Guilderson T. P., and Fairbanks R. G. (1998a) Seasonal variations in the boron isotopic composition of coral; a productivity signal? *Global Biogeochem. Cycles* **12**(4), 581–586.

Hemming N. G., Reeder R. J., and Hart S. R. (1998b) Growth-step-selective incorporation of boron on the calcite surface. *Geochim. Cosmochim. Acta* **62**(17), 2915–2922.

Henderson G. M., Martel D. J., O'Nions K., and Shackleton N. J. (1994) Evolution of seawater $^{87}Sr/^{86}Sr$ over the last 400 ka: the absence of glacial/interglacial cycles. *Earth Planet. Sci. Lett.* **128**, 643–651.

Hodell D. A. and Woodruff F. (1994) Variations in the strontium isotopic ratio of seawater during the Miocene: stratigraphic and geochemical implications. *Paleoceanography* **9**, 405–426.

Hodell D. A., Mead G. A., and Mueller P. A. (1990) Variation in the strontium isotopic composition of seawater (8 Ma to present): implications for chemical weathering rates and dissolved fluxes to the oceans. *Chem. Geol.; Isotope Geosci. Sect.* **80**(4), 291–307.

Jacobson A. D., Blum J. D., Chamberlain C. P., Poage M. A., and Sloan V. F. (2002a) Ca/Sr and Sr isotope systematics of a Himalayan glacial chronosequence: carbonate versus silicate weathering rates as a function of landscape surface age. *Geochim. Cosmochim. Acta* **66**(1), 13–27.

Jacobson A. D., Blum J. D., and Walter L. M. (2002b) Reconciling the elemental and Sr isotope composition of Himalayan weathering fluxes: insights from the carbonate geochemistry of stream waters. *Geochim. Cosmochim. Acta* **66**(19), 3417–3429.

Jacobson A. D., Blum J. D., Chamberlain C. P., Craw D., and Koons P. O. (2003) Climatic and tectonic controls on chemical weathering in the New Zealand Southern Alps. *Geochim. Cosmochim. Acta* **67**(1), 29–46.

Jaffe L. A., Peucker-Ehrenbrink B., and Petsch S. T. (2002) Mobility of rhenium, platinum group elements and organic carbon during black shale weathering. *Earth Planet. Sci. Lett.* **198**(3–4), 339–353.

Kelly D. S., Karson J. A., Blackman D. K., Frueh-Green G. L., Butterfield D. A., Lilley M. D., Olson E. J., Schrenk M. O., Roe K. K., Lebon G. T., and Rivizzigno P. (2001) An off-axis hydrothermal vent field near the Mid-Atlantic Ridge at 30 degrees N. *Nature* **412**(6843), 145–149.

Krishnaswami S., Trivedi J. R., Sarin M. M., Ramesh R., and Sharma K. K. (1992) Strontium isotopes and rubidium in the Ganga–Brahmaputra River system: weathering in the Himalaya, fluxes to the Bay of Bengal and contributions to the evolution of oceanic $^{87}Sr/^{86}Sr$. *Earth Planet. Sci. Lett.* **109**(1–2), 243–253.

Kump L. and Arthur M. A. (1997) Global chemical erosion during the Cenozoic: weatherability balances the budgets. In *Tectonic Uplift and Climate Change* (ed. W. F. Ruddiman). Plenum, New York, pp. 400–424.

Kump L. R. and Arthur M. A. (1999) Interpreting carbon-isotope excursions: carbonates and organic matter. *Chem. Geol.* **161**(1–3), 181–198.

Kump L. R., Brantley S. L., and Arthur M. A. (2000) Chemical weathering, atmospheric CO_2, and climate. *Ann. Rev. Earth Planet. Sci.* **28**, 611–667.

Lear C. H., Elderfield H., and Wilson P. A. (2000) Cenozoic deep-sea temperatures and global ice volumes from Mg/Ca in benthic foraminiferal calcite. *Science* **287**(5451), 269–272.

Lear C. H., Rosenthal Y., and Slowey N. (2002) Benthic foraminiferal Mg/Ca-paleothermometry: a revised core-top calibration. *Geochim. Cosmochim. Acta* **66**(19), 3375–3387.

Lemarchand D., Gaillardet J., Lewin E., and Allegre C. J. (2000) The influence of rivers on marine boron isotopes and implications for reconstructing past ocean pH. *Nature* **408**(6815), 951–954.

Levasseur S., Birk J.-L., and Allegre C. J. (1998) Direct measurement of fetomoles of osmium and the $^{187}Os/^{186}Os$ ratio in seawater. *Science* **282**, 272–274.

Levasseur S., Birck J. L., and Allegre C. J. (1999) The osmium riverine flux and the oceanic mass balance of osmium. *Earth Planet. Sci. Lett.* **174**(1–2), 7–23.

Lowenstein T. K., Timofeeff M. N., Brennan S. T., Hardie L. A., and Demicco R. V. (2001) Oscillations in Phanerozoic seawater chemistry: evidence from fluid inclusions. *Science* **294**(5544), 1086–1088.

Luck J. M. and Turekian K. K. (1983) Osmium-187/osmium-186 in manganese nodules and the Cretaceous-Tertiary boundary. *Science* **222**(4624), 613–615.

Mackenzie F. T., Bischoff W. D., Bishop F. C., Loijens M., Schoonmaker J., and Wollast R. (1983) Magnesian calcites: low-temperature occurrence, solubility and solid-solution behavior. In *Carbonates: Mineralogy and Chemistry. Rev. Mineral.* **11**, 97–144.

Martin C. E. (1991) Os isotopic characteristics of mantle derived rocks. *Geochim. Cosmochim. Acta* **55**, 1421–1434.

Martin C. E., Peucker-Ehrenbrink B., Brunskill G. J., and Szymczak R. (2000) Sources and sinks of unradiogenic osmium runoff from Papua New Guinea. *Earth Planet. Sci. Lett.* **183**(1–2), 261–274.

Martin C. E., Peucker-Ehrenbrink B., Brunskill G., and Szymczak R. (2001) Osmium isotope geochemistry of a tropical estuary. *Geochim. Cosmochim. Acta* **65**(19), 3193–3200.

Martin E. E., Shackleton N. J., Zachos J. C., and Flower B. P. (1999) Orbitally-tuned Sr isotope chemostratigraphy for the late middle to late Miocene. *Paleoceanography* **14**(1), 74–83.

Martin P. A., Lea D. W., Rosenthal Y., Shackleton N. J., Sarnthein M., and Papenfuss T. (2002) Quaternary deep sea temperature histories derived from benthic foraminiferal Mg/Ca. *Earth Planet. Sci. Lett.* **198**(1–2), 193–209.

McArthur J. M., Howarth R. J., and Bailey T. R. (2001) Strontium isotope stratigraphy; LOWESS Version 3; best fit to the marine Sr-isotope curve for 0–509 Ma and accompanying look-up table for deriving numerical age. *J. Geol.* **109**(2), 155–170.

Mead G. A., and Hodell D. A. (1995) Controls on the (super 87) Sr/(super 86) Sr composition of seawater from the middle Eocene to Oligocene; Hole 689B, Maud Rise, Antarctica. *Paleoceanography* **10**(2), 327–346.

Miller K. G., Fairbanks R. G., and Mountain G. S. (1987) Tertiary oxygen isotope synthesis, sea level history, and continental margin erosion. *Paleoceanography* **2**, 1–19.

Miller K. G., Feigenson M. D., Kent D. V., and Olsson R. K. (1988) Upper Eocene to Oligocene isotope ($^{87}Sr/^{86}Sr$, $\delta^{18}O$, $\delta^{13}C$) standard section, Deep Sea Drilling Project Site 522. *Paleoceanography* **3**(2), 223–233.

Miller K. G., Feigenson M. D., Wright J. D., and Clement B. M. (1991a) Miocene isotope reference section, Deep Sea Drilling Project Site 608: an evaluation of isotope and biostratigraphic resolution. *Paleoceanography* **6**(1), 33–52.

Miller K. G., Wright J. D., and Fairbanks R. G. (1991b) Unlocking the ice house: Oligocene–Miocene oxygen isotopes, eustasy and margin erosion. *J. Geophys. Res.* **96**, 6829–6848.

Moberly R., Jr. (1968) Composition of magnesian calcites of algae and pelecypods by electron microprobe analysis. *Sedimentology* **11**, 61–82.

Najman Y., Bickle M., and Chapman H. (2000) Early Himalayan exhumation: isotopic constraints from the Indian foreland basin. *Terra Nova* **12**(1), 29–34.

Najman Y., Pringle M., Godin L., and Oliver G. (2001) Dating of the oldest continental sediments from the Himalayan foreland basin. *Nature* **410**(6825), 194–197.

Oslick J. S., Miller K. G., Feigenson M. D., Wright J. D. (1994) Oligocene-Miocene strontium isotopes: stratigraphic revisions and correlations to an inferred glacioeustatic record. *Paleoceanography* **9**(3), 427–443.

Oxburgh R. (1998) Variations in the osmium isotope composition of seawater over the past 200,000 years. *Earth Planet. Sci. Lett.* **159**, 183–191.

Pagani M., Freeman K. H., and Arthur M. A. (1999) Late Miocene atmospheric CO_2 concentrations and the expansion of C_4 gasses. *Science* **285**(5429), 876–879.

Palmer M. R. and Edmond J. M. (1989) The strontium isotope budget of the modern ocean. *Earth Planet. Sci. Lett.* **92**, 11–26.

Palmer M. R. and Edmond J. M. (1992) Controls over the strontium isotope composition of river water. *Geochim. Cosmochim. Acta* **56**, 2099–2111.

Palmer M. R. and Turekian K. K. (1986) $^{187}Os/^{186}Os$ in marine manganese nodules and the constraints on the crustal geochemistries of rhenium and osmium. *Nature* **319**, 216–220.

Palmer M. R., Pearson P. N., and Cobb S. J. (1998) Reconstructing past ocean pH-depth profiles. *Science* **282**(5393), 1468–1471.

Paul H. A., Zachos J. C., Flower B. P., and Tripati A. (2000) Orbitally induced climate and geochemical variability across the Oligocene/Miocene boundary. *Paleoceanography* **15**(5), 471–485.

Paytan A., Kastner M., Campbell D., and Thiemens M. H. (1998) Sulfur isotopic composition of Cenozoic seawater sulfate. *Science* **282**(5393), 1459–1462.

Pearson P. N. and Palmer M. R. (1999) Middle Eocene seawater pH and atmospheric carbon dioxide concentrations. *Science* **284**(5421), 1824–1826.

Pearson P. N. and Palmer M. R. (2000) Atmospheric carbon dioxide concentrations over the past 60 million years. *Nature* **406**, 695–699.

Pearson P. N., Ditchfield P. W., Singano J., Harcourt-Brown K. G., Nicholas C. J., Olsson R. K., Shackleton N. J., and Hall M. A. (2001) Warm tropical sea surface temperatures in the Late Cretaceous and Eocene epochs. *Nature* **413**(6855), 481–487.

Pegram W. J. and Turekian K. K. (1999) The osmium isotopic composition change of Cenozoic sea water as inferred from a deep-sea core corrected for meteoritic contributions. *Geochim. Cosmochim. Acta* **63**(23–24), 4053–4058.

Pegram W. J., Krishnaswami S., Ravizza G., and Turekian K. K. (1992) The record of seawater $^{187}Os/^{186}Os$

variation through the Cenozoic. *Earth Planet. Sci. Lett.* **113**, 569–576.

Peterman Z. E., Hedge C. E., and Tourtelot H. A. (1970) Isotopic composition of strontium in sea water throughout Phanerozoic time. *Geochim. Cosmochim. Acta* **34**(1), 105–108.

Peucker-Ehrenbrink B. (1996) Accretion of extraterrestrial matter during the last 80 million years and its effect on the marine osmium isotope record. *Geochim. Cosmochim. Acta* **60**, 3187–3196.

Peucker-Ehrenbrink B. and Blum J. D. (1998) The effects of global glaciation on the osmium isotopic composition of continental runoff and seawater. *Geochim. Cosmochim. Acta* **62**, 3193–3203.

Peucker-Ehrenbrink B. and Hannigan R. E. (2000) Effects of black shale weathering on the mobility of rhenium and platinum group elements. *Geology* **28**(5), 475–478.

Peucker-Ehrenbrink B. and Jahn B.-J. (2001) Rhenium–osmium isotope systematics and platinum group element concentrations: loess and the upper continental crust. *Geochem. Geophys. Geosys.* #2001GC000172.

Peucker-Ehrenbrink B. and Ravizza G. (2000) The marine osmium isotope record. *Terra Nova* **12**, 205–219.

Peucker-Ehrenbrink B., Ravizza G., and Hofmann A. W. (1995) The marine $^{187}Os/^{186}Os$ record of the past 80 million years. *Earth Planet. Sci. Lett.* **130**, 155–167.

Pierson-Wickmann A.-C., Reisberg L., and France-Lanord C. (2000) The Os isotopic composition of Himalayan river bedloads and bedrocks: importance of black shales. *Earth Planet. Sci. Lett.* **176**(2), 203–218.

Pierson-Wickmann A.-C., Reisberg L., and France-Lanord C. (2002) Behavior of Re and Os during low-temperature alteration: results from Himalayan soils and altered black shales. *Geochim. Cosmochim. Acta* **66**(9), 1539–1548.

Quade J., Roe L., DeCelles P. G., and Ojha T. P. (1997) The late Neogene $^{87}Sr/^{86}Sr$ record of lowland Himalayan rivers. *Science* **276**, 1828–1831.

Ravizza G. (1993) Variations of the $^{187}Os/^{186}Os$ ratio of seawater over the past 28 million years as inferred from metalliferous carbonates. *Earth Planet. Sci. Lett.* **118**, 335–348.

Ravizza G. (1998) Osmium-isotope geochemistry of Site 959: implications for Re–Os sedimentary geochronology and reconstruction of past variations in the Os-isotopic composition of seawater. *Proc. Ocean Drill. Prog., Sci. Res.* **159**, 181–186.

Ravizza G. and McMurtry G. M. (1993) Osmium isotopic variations in metalliferous sediments from the East Pacific Rise and Bauer Basin. *Geochim. Cosmochim. Acta* **57**, 4301–4310.

Ravizza G. and Peucker-Ehrenbrink B. (2003) The marine $^{187}Os/^{188}Os$ record of the Eocene–Oligocene transition: the interplay of weathering and glaciation. *Earth Planet. Sci. Lett.* **210**, 151–165.

Ravizza G. and Turekian K. K. (1989) Application of the ^{187}Re-^{187}Os system to black shale chronology. *Geochim. Cosmochim. Acta* **53**, 3257–3262.

Ravizza G. and Turekian K. K. (1992) The osmium isotopic composition of organic-rich marine sediments. *Earth Planet. Sci. Lett.* **110**, 1–6.

Ravizza G., Turekian K. K., and Hay B. J. (1991) The geochemistry of rhenium and osmium in recent sediment from the Black Sea. *Geochim. Cosmochim. Acta* **55**, 3741–3752.

Ravizza G., Martin C. E., German C. R., and Thompson G. (1996) Os isotopes as tracers in seafloor hydrothermal systems: metalliferous deposits from the TAG hydrothermal area, 26 degrees N Mid-Atlantic Ridge. *Earth Planet. Sci. Lett.* **138**, 105–119.

Ravizza G., Norris R. N., Blusztajn J., and Aubry M. P. (2001) An osmium isotope excursion associated with the Late Paleocene thermal maximum: evidence of intensified chemical weathering. *Paleoceanography* **16**(2), 155–163.

Raymo M. E. (1997) Carbon cycle models: how strong are the constraints? In *Tectonic Uplift and Climate Change* (ed. W. F. Ruddiman). Plenum, New York, pp. 367–381.

Raymo M. E. and Ruddiman W. F. (1992) Tectonic forcing of late Cenozoic climate. *Nature* **359**, 117–122.

Raymo M. E., Ruddiman W. F., and Froelich P. N. (1988) The influence of late Cenozoic mountain building on ocean geochemical cycles. *Geology* **16**, 649–653.

Reilly T. J., Miller K. G., and Feigenson M. D. (2002) Latest Eocene–earliest Miocene Sr isotopic reference section, Site 522, eastern South Atlantic. *Paleoceanography* **17**(3), 9.

Reusch D. N. and Maasch K. A. (1998) The transition from arc volcanism to exhumation, weathering of young Ca, Mg, Sr silicates, and CO_2 drawdown. In *Tectonic Boundary Conditions for Climate Reconstructions*, Oxford Monographs on Geology and Geophysics (eds. T. J. Crowley and K. C. Burke), vol. 39, pp. 261–276.

Reusch D. N., Ravizza G., Maasch K. A., and Wright J. D. (1998) Miocene seawater $^{187}Os/^{188}Os$ ratios inferred from metalliferous carbonates. *Earth Planet. Sci. Lett.* **160**(1–2), 163–178.

Richter F. M. and DePaolo D. J. (1988) Diagenesis and Sr isotope evolution of seawater using data from DSDP 590B and 575. *Earth Planet. Sci. Lett.* **90**, 382–394.

Richter F. M. and Liang Y. (1993) The rate and consequences of Sr diagenesis in deep-sea carbonates. *Earth Planet. Sci. Lett.* **117**, 553–565.

Richter F. M. and Turekian K. K. (1993) Simple models for the geochemical response of the ocean to climatic and tectonic forcing. *Earth Planet. Sci. Lett.* **119**, 121–131.

Richter F. M., Rowley D. B., and DePaolo D. J. (1992) Sr isotope evolution of seawater: the role of tectonics. *Earth Planet. Sci. Lett.* **109**, 11–23.

Röhl U., Bralower T. J., Norris R. D., and Wefer G. (2000) New chronology for the Late Paleocene thermal maximum and its environmental implications. *Geology* **28**, 927–930.

Rosenthal Y., Boyle E. A., and Slowey N. (1997) Temperature control on the incorporation of magnesium, strontium, fluorine, and cadmium into benthic foraminiferal shells from Little Bahama Bank: prospects for thermocline paleoceanography. *Geochim. Cosmochim. Acta* **61**, 3633–3643.

Sanyal A., Hemming N. G., Hanson G. N., and Broecker W. S. (1995) Evidence for a higher pH in the glacial ocean from boron isotopes in foraminifera. *Nature* **373**(6511), 234–236.

Sanyal A., Hemming N. G., Broecker W. S., Lea D. W., Spero H. J., and Hanson G. N. (1996) Oceanic pH control on the boron isotopic composition of Foraminifera: evidence from culture experiments. *Paleoceanography* **11**(5), 513–517.

Sanyal A., Hemming N. G., Broecker W. S., and Hanson G. N. (1997) Changes in pH in the eastern equatorial Pacific across stage 5–6 boundary based on boron isotopes in Foraminifera. *Global Biogeochem. Cycles* **11**(1), 125–133.

Sanyal A., Nugent M., Reeder R. L., and Bijma J. (2000) Seawater pH control on the boron isotopic composition of calcite: evidence from inorganic calcite precipitation experiments. *Geochim. Cosmochim. Acta* **64**(9), 1551–1555.

Sanyal A., Bijma J., Spero H., and Lea D. W. (2001) Empirical relationship between pH and the boron isotopic composition of *Globigerinoides sacculifer*: implications for the boron isotope paleo-pH proxy. *Paleoceanography* **16**(5), 515–519.

Shackleton N. J. (1985) Oceanic carbon isotope constraints on oxygen and carbon dioxide in the Cenozoic atmosphere. In *The Carbon Cycle and Atmospheric CO_2: Natural Variations—Archean to Present*, Chapman Conference on Natural Variations in Carbon Dioxide and the Carbon Cycle. *AGU Monogr.* **32**, 412–417.

Shackleton N. J. and Crowhurst S. (1997) Sediment fluxes based on an orbitally tuned time scale 5 Ma to 14 Ma, Site 926. *Proc. Ocean Drill. Prog., Sci. Results* **54**, 69–82.

Sharma M., Papanastassiou D. A., and Wasserburg G. J. (1997) The concentration and isotopic composition of osmium in the oceans. *Geochim. Cosmochim. Acta* **61**, 3287–3299.

Sharma M., Wasserburg G. J., and Hofmann A. W. (1999) Himalayan uplift and osmium isotopes in oceans and rivers. *Geochim. Cosmochim. Acta* **63**(23–24), 4005–4012.

Sharma M., Wasserburg G. J., Hofmann A. W., and Butterfield D. A. (2000) Osmium isotopes in hydrothermal fluids from the Juan de Fuca Ridge. *Earth Planet. Sci. Lett.* **179**(1), 139–152.

Singh S. K., Trivedi J. R., Pande K., Ramesh R., and Krishnaswami S. (1998) Chemical and strontium, oxygen, and carbon isotopic compositions of carbonates from the Lesser Himalaya: implications to the strontium isotope composition of the source waters of the Ganga, Ghaghara, and the Indus rivers. *Geochim. Cosmochim. Acta* **62**(5), 743–755.

Singh S. K., Trivedi J. R., and Krishnaswami S. (1999) Re–Os isotope systematics in black shales from the Lesser Himalaya: their chronology and role in the $^{187}Os/^{188}Os$ evolution of seawater. *Geochim. Cosmochim. Acta* **63**(16), 2381–2392.

Snow J. E. and Reisberg L. (1995) Os isotopic systematics of the MORB mantle: results from altered abyssal peridotites. *Earth Planet. Sci. Lett.* **136**, 723–733.

Spivack A. J., You C.-F., and Smith H. J. (1993) Foraminiferal boron isotope ratios as a proxy for surface ocean pH over the past 21 Myr. *Nature* **363**(6425), 149–151.

Taylor A. S. and Lasaga A. C. (1999) The role of basalt weathering in the Sr isotope budget of the oceans. *Chem. Geol.* **161**(1–3), 199–214.

Thomas E. and Shackleton N. J. (1996) The Paleocene–Eocene benthic foraminiferal extinction and stable isotope anomalies. *Geol. Soc. Spec. Publ.* **101**, 401–441.

Tiedemann R., Sarnthein M., and Shackleton N. J. (1994) Astronomic timescale for the Pliocene Atlantic delta (super 18) O and dust flux records of Ocean Drilling Program Site 659. *Paleoceanography* **9**(4), 619–638.

Vengosh A., Kolodny Y., Starinsky A., Chivas A. R., and McCulloch M. T. (1991) Coprecipitation and isotopic fractionation of boron in modern biogenic carbonates. *Geochim. Cosmochim. Acta* **55**(10), 2901–2910.

Volk T., Caldeira K., Arthur M. A., Berner R. A., Lasaga A. C., Raymo M. E., and Ruddiman W. (1993) Cooling in the late Cenozoic: discussions and reply. *Nature* **361**(6408), 123–124.

Walker J. C. G., Hays P. B., and Kasting J. F. (1981) A negative feedback mechanism for the long term stabilization of Earth's surface temperature. *J. Geophys. Res.* **86**, 9776–9782.

Walker R. J., Morgan J. W., Beary E. S., Smoliar M. I., Czamanske G. K., and Horan M. F. (1997) Applications of the $^{190}Pt-^{186}Os$ isotope system to geochemistry and cosmochemistry. *Geochim. Cosmochim. Acta* **61**, 4799–4807.

Wignall P. B. (2001) Large igneous provinces and mass extinctions. *Earth Sci. Rev.* **53**(1–2), 1–33.

Woodhouse O. B., Ravizza G., Falkner K. K., Statham P. J., and Peucker-Ehrenbrink B. (1999) Osmium in seawater: vertical profiles of concentration and isotopic composition in the eastern Pacific Ocean. *Earth Planet. Sci. Lett.* **173**(3), 223–233.

Wright J. D. and Miller K. G. (1993) Southern Ocean influences on Late Eocene to Miocene deepwater circulation. The Antarctic paleoenvironment: a perspective on global change: Part two. *Antarct. Res. Ser.* **60**, 1–25.

Wright J. D., Miller K. G., and Fairbanks R. G. (1991) Evolution of modern deepwater circulation: evidence from the Late Miocene Southern Ocean. *Paleoceanography* **6**(2), 275–290.

Zachos J. C., Lohmann K. C., Walker J. C. G., Wise S. W., and Anonymous (1993) Abrupt climate changes and transient climates during the Paleogene: a marine perspective. *J. Geol.* **101**(2), 191–213.

Zachos J. C., Stott L. D., and Lohmann K. C. (1994) Evolution of early Cenozoic marine temperatures. *Paleoceanography* **9**(2), 353–387.

Zachos J. C., Quinn T. M., and Salamy K. A. (1996) High-resolution (10^4 years) deep-sea foraminiferal stable isotope records of the Eocene–Oligocene climate transition. *Paleoceanography* **11**(3), 251–266.

Zachos J. C., Opdyke B. N., Quinn T. M., Jones C. E., and Halliday A. N. (1999) Early Cenozoic glaciation, Antarctic weathering, and seawater $^{87}Sr/^{87}Sr$: is there a link? *Chem. Geol.* **161**(1–3), 165–180.

Zachos J., Pagani M., Sloan L., Thomas E., Billups K., Smith J. P., and Uppenbrink J. P. (2001) Trends, rhythms, and aberrations in global climate 65 Ma to present. *Science* **292**(5517), 686–693.

Zimmermann H. (2000) Tertiary seawater chemistry: implications from primary fluid inclusions in marine halite. *Am. J. Sci.* **300**(10), 723–767.

6.21
The Geologic History of Seawater

H. D. Holland
Harvard University, Cambridge, MA, USA

6.21.1 INTRODUCTION

Aristotle proposed that the saltness of the sea was due to the effect of sunlight on water. Robert Boyle took strong exception to this view and—in the manner of the Royal Society—laid out a program of research in the opening paragraph of his *Observations and Experiments about the Saltness of the Sea* (1674) (Figure 1):

> The Cause of the Saltness of the Sea appears by *Aristotle's* Writings to have busied the Curiosity of Naturalists before his time; since which, his Authority, perhaps much more than his Reasons, did for divers Ages make the Schools and the generality of Naturalists of his Opinion, till towards the end of the last Century, and the beginning of ours, some Learned Men took the boldness to question the common Opinion; since when the Controversie has been kept on foot, and, for ought I know, will be so, as long as 'tis argued on both sides but by Dialectical Arguments, which may be probable on both sides, but are not convincing on either. Wherefore I shall here briefly deliver some particulars about the Saltness of the Sea, obtained by my own trials, where I was able; and where I was not, by the best Relations I could procure, especially from Navigators.

Boyle measured and compiled a considerable set of data for variations in the saltness of surface seawater. He also designed an improved piece of

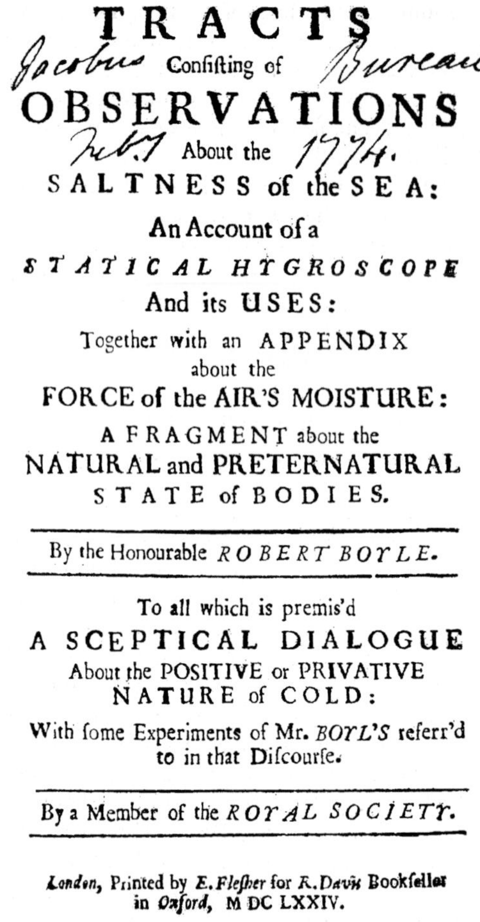

TRACTS
Confifting of
OBSERVATIONS
About the
SALTNESS of the SEA:
An Account of a
STATICAL HYGROSCOPE
And its USES:
Together with an APPENDIX
about the
FORCE of the AIR'S MOISTURE:
A FRAGMENT about the
NATURAL and PRETERNATURAL
STATE of BODIES.

By the Honourable ROBERT BOYLE.

To all which is premis'd
A SCEPTICAL DIALOGUE
About the POSITIVE or PRIVATIVE
NATURE of COLD:
With fome Experiments of Mr. BOYL'S referr'd
to in that Difcourfe.

By a Member of the ROYAL SOCIETY.

London, Printed by E. Flesher for R. Davis Bookfeller
in Oxford, M DC LXXIV.

Figure 1 Title page of Robert Boyle's Tracts consisting of Observations about the Saltness of the Sea and other essays (1674).

equipment for sampling seawater at depth, but the depths at which it was used were modest: 30 m with his own instrument, 80 m with another, similar sampler. However, the younger John Winthrop (1606–1676), an early member of the Royal Society, an important Governor of Connecticut, and a benefactor of Harvard College, was asked to collect seawater from the bottom of the Atlantic Ocean during his crossing from England to New England in the spring of 1663. The minutes of the Royal Society's meeting on July 20, 1663, give the following account of his unsuccessful attempt to do so (Birch, 1756; Black, 1966):

> Mr. Winthrop's letter written from Boston to Mr. Oldenburg was read, giving an account of the trials made by him at sea with the instrument for sounding of depths without a line, and with the vessel for drawing water from the bottom of the sea; both which proved successless, the former by reason of too much wind at the time of making soundings; the latter, on account of the leaking of the vessel. Capt. Taylor being to go soon to Virginia, and offering himself to make the same experiments, the society recommended to him the trying of the one in calm weather, and of the other with a stanch vessel.
>
> Mr. Hooke mentioning, that a better way might be suggested to make the experiment above-mentioned, was desired to think farther upon it, and to bring in an account thereof at the next meeting.

A little more than one hundred years later, in the 1780s, John Walker (1966) lectured at Edinburgh on the saltness of the oceans. He marshaled all of the available data and concluded that "these reasons seem all to point to this, that the water of the ocean in respect to saltness is pretty much what it ever has been."

In this opinion he disagreed with Halley (1715), who suggested that the salinity of the oceans has increased with time, and that the ratio of the total salt content of the oceans to the rate at which rivers deliver salt to the sea could be used to ascertain the age of the Earth. The first really serious attempt to measure geologic time by this method was made by Joly (1899). His calculations were refined by Clarke (1911), who inferred that the age of the ocean, since the Earth assumed its present form, is somewhat less than 100 Ma. He concluded, however, that "the problem cannot be regarded as definitely solved until all available methods of estimation shall have converged on one common conclusion." There was little appreciation in his approach for the magnitude of: (i) the outputs of salt from the oceans, (ii) geochemical cycles, and (iii) the notion of a steady-state ocean. In fact, Clarke's "age" of the ocean turns out to be surprisingly close to the oceanic residence time of Na^+ and Cl^-.

The modern era of inquiry into the history of seawater can be said to have begun with the work of Conway (1943, 1945), Rubey (1951), and Barth (1952). Much of the progress that was made between the appearance of these publications and the early 1980s was summarized by Holland (1984). This chapter describes a good deal of the progress that has been made since then.

6.21.2 THE HADEAN (4.5–4.0 Ga)

The broad outlines of Earth history during the Hadean are starting to become visible. The solar system originated 4.57 Ga (Allègre *et al.*, 1995). The accretion of small bodies in the solar nebula occurred within ~10 Myr of the birth of the solar system (Lugmaier and Shukolyukov, 1998). The Earth reached its present mass between 4.51 Ga and 4.45 Ga (Halliday, 2000; Sasaki and Nakazawa, 1986; Porcelli *et al.*, 1998). The core formed in <30 Ma (Yin *et al.*, 2002; Kleine *et al.*, 2002). The early Earth was covered by a magma ocean, but this must have cooled quickly at the end of the accretion process, and the first primitive crust must have formed shortly thereafter.

At present we do not have any rocks older than 4.03 Ga (Bowring and Williams, 1999). There are, however, zircons older than 4.03 Ga which were weathered out of their parent rocks and incorporated in 3 Ga quartzitic rocks in the Murchison District of Western Australia (Froude *et al.*, 1983; Compston and Pidgeon, 1986; Nutman *et al.*, 1991; Nelson *et al.*, 2000). A considerable number of 4.2–4.3 Ga zircon grains have been found in the Murchison District. A single 4.40 Ga zircon grain has been described by Wilde *et al.* (2001). The oxygen of these zircons has apparently retained its original isotopic composition. Mojzsis *et al.* (2001), Wilde *et al.* (2001), and Peck *et al.* (2001) have shown that the $\delta^{18}O$ values of the zircons which they have analyzed are significantly more positive than those of zircons which have crystallized from mantle magmas (see Figure 2). The most likely explanation for this difference is that the melts from which the zircons crystallized contained a significant fraction of material enriched in ^{18}O. This component was probably a part of the pre-4.0 Ga crust. Its enrichment in ^{18}O was almost certainly the result of subaerial weathering, which generates ^{18}O-enriched clay minerals (see, e.g., Holland, 1984, pp. 241–251). If this interpretation is correct, the data imply the presence of an active hydrologic cycle, a significant quantity of water at the Earth surface, and an early continental crust. However, as Halliday (2001) has pointed out, inferring the existence of entire continents from zircon grains in a single area requires quite a leap of the imagination. We need more zircon data from many areas to confirm the inferences drawn from the small amount of available data. Nevertheless, the inferences themselves are reasonable and fit into a coherent model of the Hadean Earth.

At present, little can be said with any degree of confidence about the composition of the proposed Hadean ocean, but it was probably not very different from that of the Early Archean ocean, about which a good deal can be inferred.

6.21.3 THE ARCHEAN (4.0–2.5 Ga)

6.21.3.1 The Isua Supracrustal Belt, Greenland

The Itsaq Gneiss complex of southern West Greenland contains the best preserved occurrences of ≥3.6 Ga crust. The gneiss complex had a complicated early history. It was added to and modified during several events starting ca. 3.9 Ga (Nutman *et al.*, 1996). Supracrustal, mafic, and ultramafic rocks comprise ~10% of the complex; these range in age from ≥3.87 Ga to ca. 3.6 Ga. A large portion of the Isua supracrustal belt (Figure 3) contains rocks that may be felsic volcanics and volcaniclastics, and abundant, diverse chemical sediments (Nutman *et al.*, 1997). The rocks are deformed, and many are substantially altered by metasomatism. However, transitional stages can be seen from units with relatively well-preserved primary volcanic and sedimentary features to schists in which all primary features have been

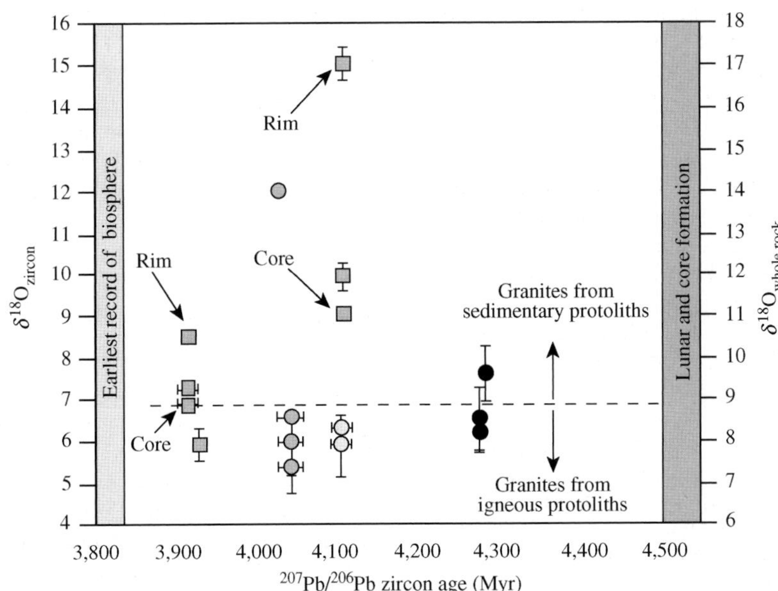

Figure 2 Ion microprobe $\delta^{18}O$ data for individual zircon spot analyses versus $^{207}Pb/^{206}Pb$ zircon age. The right vertical axis shows the estimated $\delta^{18}O$ data for the whole rock ($\delta^{18}O_{WR}$) from which the zircon crystallized (source Mojzsis *et al.*, 2001).

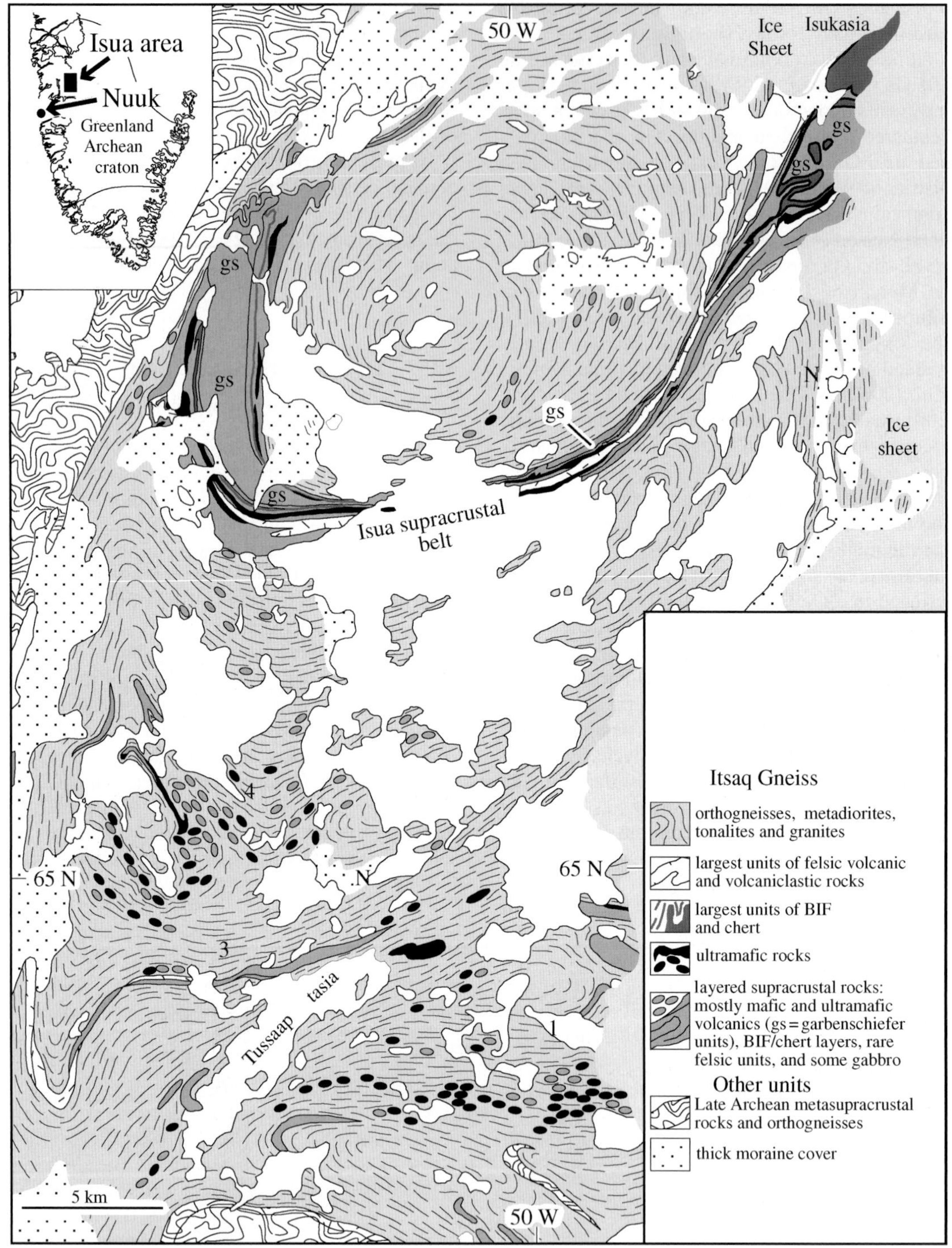

Figure 3 Map of the northern part of the Itsaq Gneiss complex (source Nutman *et al.*, 2002).

obliterated. Most of the Isua greenstone belt consists of fault-bounded rock packages mainly derived from basaltic and high-manganese basaltic pillow lava and pillow lava breccia, chert-banded iron formation (chert-BIF), and a minor component of clastic sedimentary rocks derived from chert and basaltic volcanic rocks (Myers, 2001). The Isua sequence, as we now see it, resembles deep-sea sequences rather than platform deposits. The Isua rocks could have been deposited in a purely oceanic environment without a significant sialic detrital component, and intruded by the felsic gneisses during or after tectonic emplacement into the Amitsoq protocontinent (Rosing *et al.*, 1996).

The most compelling evidence for an ocean during the deposition of the Isua greenstone belt is provided by the BIF deposits which occur in this sequence. They are highly metamorphosed. Boak and Dymek (1982) have shown that the pelitic rocks in the sequence were exposed to temperatures of ~550 °C at pressures of ~5 kbar. Temperatures during metamorphism could not have exceeded ~600 °C. The mineralogy and the chemistry of the iron formations have obviously been altered by metamorphism, but their major features are still preserved. Magnetite-quartz BIFs are particularly common. Iron enrichment can be extensive, as shown by the presence of a two-billion ton body at the very north-easternmost limit of the supracrustal belt (Bridgewater *et al.*, 1976).

In addition to the quartz-magnetite iron formation, Dymek and Klein (1988) described magnesian iron formation, aluminous iron formation, graphitic iron formation, and carbonate rich iron formation. They pointed out that the composition of the iron formation as a whole is very similar to that of other Archean and Proterozoic iron formations that have been metamorphosed to the amphibolite facies.

The source of the iron and probably much of the silica in the BIF was almost certainly seawater that had cycled through oceanic crust at temperatures of several hundred degree celsius. The low concentration of sulfur and of the base metals that are always present in modern solutions of this type indicates that these elements were removed, probably as sulfides of the base metals before the deposition of the iron oxides and silicates of the Isua iron formations.

The molar ratio of Fe_2O_3/FeO in the BIFs is less than 1.0 in all but one of the 28 analyses of Isua iron formation reported by Dymek and Klein (1988). In the one exception the ratio is 1.17. Unless the values of this ratio were reduced significantly during metamorphism, the analyses indicate that magnetite was the dominant iron oxide, and that hematite was absent or very minor in these iron formations. This, in turn, shows that some of the hydrothermal Fe^{2+} was oxidized to Fe^{3+} prior to deposition, but that not enough was oxidized to lead to the precipitation of Fe_2O_3 and/or Fe^{3+} oxyhydroxide precursors, or that these phases were subsequently replaced by magnetite.

The process(es) by which the precursor(s) of magnetite in these iron formations precipitated is not well understood. A possible explanation involves the oxidation of Fe^{2+} by reaction with seawater to produce Fe^{3+} and H_2 followed by the precipitation of "green rust"—a solid solution of $Fe(OH)_2$ and $Fe(OH)_3$—and finally by the dehydration of green rust to magnetite (but see below). Figure 4 shows that the boundary between the stability field of $Fe(OH)_2$ and amorphous $Fe(OH)_3$ is at a rather low value of $p\varepsilon$. The field of green rust probably straddles this boundary. Saturation of solutions with siderite along this

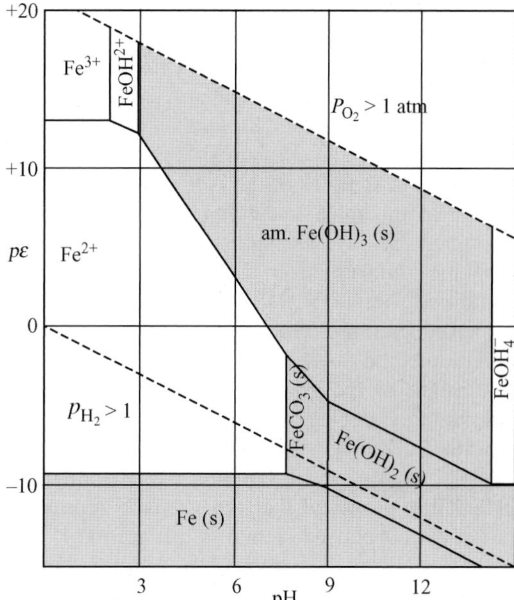

Figure 4 Diagram $p\varepsilon$ versus pH for the system $Fe-CO_2-H_2O$. The solid phases are $Fe(OH)_3$ (amorphous), $FeCO_3$ (siderite), $Fe(OH)_2$(s), and Fe(s); $C_T = 10^{-3}$ M. Lines are calculated for Fe(II) and Fe(III) $= 10^{-5}$ M at 25 °C. The possible conversion of carbonate to methane at low $p\varepsilon$ values was ignored (source Stumm and Morgan, 1996).

boundary at a total carbon concentration of 10^{-3} M and a total Fe concentration of 10^{-5} M lies within a reasonable range for the pH of Archean seawater. However, the sequence of mineral deposition was probably more complex (see below).

The reaction of Fe^{2+} with H_2O is greatly accelerated by solar UV (see, e.g., Braterman *et al.*, 1984), and it has been suggested that solar UV played an important role in the deposition of oxide facies iron formation (Cairns-Smith, 1978; Braterman *et al.*, 1983; Sloper, 1983; François, 1986, 1987; Anbar and Holland, 1992). However, it now appears that solar UV played no more than a minor role in the deposition of oxide facies BIF. The reasons for this conclusion are detailed later in the chapter. The absence of significant quantities of hematite indicates that seawater from which the precursor of the magnetite-quartz iron formation was deposited was mildly reducing. This is corroborated by the mineralogy of the silicates, in which iron is present exclusively or nearly so in the divalent state. The only possible indication of a relatively high oxidation state is the presence of cerium anomalies reported in some of Dymek and Klein's (1988) rare earth element (REE) analyses. However, the validity of these anomalies is somewhat uncertain, because the analyses do not include praseodymium, and because neodymium measurements are lacking in a number of cases. Fryer (1983) had earlier observed that no significantly anomalous behavior of cerium has been found in any Archean iron formations, and neither Appel (1983) and Shimizu *et al.* (1990), nor Bau and Möller (1993) were able to detect cerium anomalies in samples of the Isua iron formation.

One of the most intriguing sedimentary units described by Dymek and Klein (1988) is the graphitic iron formation. Four samples of this unit contained between 0.70% and 2.98% finely dispersed graphite. Quartz, magnetite, and cummingtonite are their main mineral constituents. The origin of the graphite has been a matter of considerable debate. Schidlowski *et al.* (1979) reported a range of −5.9‰ to −22.2‰ for the $\delta^{13}C$ value of 13 samples of graphite from Isua. They proposed that the graphite represents the metamorphosed remains of primary Isua organisms, that the isotopically light carbon in some of their graphite samples reflects the isotopic composition of these organisms, and that the isotopically heavy carbon in some of their graphite samples reflects the redistribution of carbon isotopes between organic and carbonate carbon during amphibolite grade metamorphism. Perry and Ahmad (1977) found $\delta^{13}C$ values between −9.3‰ and −16.3‰ in Isua supracrustal rocks and pointed out that the fractionation of the carbon isotopes between siderite and graphite in their samples is consistent with inorganic equilibrium of these phases at ~400–500 °C. Oehler and Smith (1977) found graphite with $\delta^{13}C$ values of −11.3‰ to −17.4‰ in Isua metapelites (?) containing 150–4,800 ppm reduced carbon, and graphite with a $\delta^{13}C$ range between −21.4‰ and −26.9‰ in metasediments from the Isua iron formation which contain only trace amounts of graphite (4–56 ppm). The carbon in the latter samples is thought to be due to postdepositional contamination.

Since then Mojzsis *et al.* (1996) have used *in situ* ion microprobe techniques to measure the isotopic composition of carbon in Isua BIF and in a unit from the nearby Akilia Island that may be BIF. $\delta^{13}C$ in carbonaceous inclusions in the BIF ranged from −23‰ to −34‰; those in carbon inclusions occluded in apatite micrograins from the Akilia Island BIF ranged from −21‰ to −49‰. Since the carbon grains embedded in apatite were small and irregular, the precision and accuracy of the individual $\delta^{13}C$ measurements was typically ±5‰ (1σ). These measurements tend to confirm the Schidlowski *et al.* (1979) interpretation, and suggest that the graphite in the Isua and Akilia BIFs could well be the metamorphosed remains of primitive organisms. However, other interpretations have been advanced very forcefully (see, e.g., Holland, 1997; Van Zuilen *et al.*, 2002).

Naraoka *et al.* (1996) have added additional $\delta^{13}C$ measurements, which fall in the same range as those reported previously. They emphasize that graphite with $\delta^{13}C$ values around −12‰ was probably formed by an inorganic, rather than by a biological process. Rosing (1999) reported $\delta^{13}C$ values ranging from −11.4‰ to −20.2‰ in 2–5 μm graphite globules in turbiditic and pelagic sedimentary rocks from the Isua supracrustal belt. He suggests that the reduced carbon in these samples represents biogenic detritus, which was perhaps derived from planktonic organisms.

This rather large database certainly suggests that life was present in the 3.7–3.9 Ga oceans, but it is probably best to treat the proposition as likely rather than as proven. One of the arguments against the presence of life before 3.8 Ga is based on the likelihood of large extraterrestrial impacts during a late heavy bombardment (LHB). The craters of the moon record an intense bombardment by large bodies, ending abruptly ca. 3.85 Ga (Dalrymple and Ryder, 1996; Hartmann *et al.*, 2000; Ryder, 1990). The Earth was probably impacted at least as severely as the moon, and there is a high probability that impacts large enough to vaporize the ocean's photic zone occurred as late as 3.8 Ga (Sleep *et al.*, 1989). The environment of the early Earth, therefore, may have been extremely challenging to life (Chyba, 1993; Appel and Moorbath, 1999). There has, however, been little direct examination of the Earth's surface environment during this period. The metasediments at Isua and on Akilia Island supply a small window on the effects of extraterrestrial bombardment between ca. 3.8 Ga and 3.9 Ga. Anbar *et al.* (2002) have determined the concentration of iridium and platinum in three samples of a ~5 m thick BIF chert unit and in three samples of mafic–ultramafic flows interposed with BIF in a relatively undeformed section on Akilia Island. The iridium and platinum concentrations in the Akilia metasediments are all extremely low. The iridium content of only one of the BIF/chert samples is above 3 ppt, the detection limit of the ID–ICP–MS techniques. The concentration of platinum is below the detection limit (40 ppt) in nearly all of the BIF/chert samples. Both elements were readily detected in the mafic–ultramafic samples. The extremely low concentration of iridium and platinum in the BIF/chert samples shows that their composition was not significantly affected by extraterrestrial impacts. It is difficult, however, to extrapolate from these few analyses to other environments between 3.8 Ga and 3.9 Ga, especially because it has been proposed that these rocks are not sedimentary (Fedo and Whitehouse, 2002). As Anbar *et al.* (2002) have pointed out, the large time gaps between the large, life-threatening impact events require extensive sampling of the rock record for the detection of the impact events. The slim evidence supplied by the Anbar *et al.* (2002) study encourages the view that conditions were sufficiently benign for the existence of life on Earth during the deposition of the sediments at Isua and on Akilia, but optimism on this point must surely be tempered, and it is premature to speculate on the effects of the potential biosphere on the state of the oceans 3.8–3.9 Ga.

6.21.3.2 The Mesoarchean Period (3.7–3.0 Ga)

Evidence for the presence of life is abundant during the later part of the Mesoarchean period. Unmetamorphosed carbonaceous shales with $\delta^{13}C$ values that are consistent with a biological origin of the contained carbon are reasonably common (see, e.g., Schidlowski *et al.*, 1983; Strauss *et al.*, 1992). The nature of the organisms that populated the Mesoarchean oceans is still hotly disputed. The description of microfossils in the 3.45 Ga Warrawoona Group of Western Australia by Schopf and Packer (1987) and Schopf (1983) suggested that some of these microfossils were probably the remains of cyanobacteria. If so, oxygenic photosynthesis is at least as old as 3.45 Ga. Brasier *et al.* (2002) have re-examined the type sections of the material described by Schidlowski *et al.* (1983) and have reinterpreted all of his 11 holotypes as artifacts formed from amorphous graphite within multiple generations of metalliferous hydrothermal vein chert and volcanic glass. However, Schopf *et al.* (2002) maintain that the laser Raman imagery of this material not only establishes the biogenicity of the fossils which they have studied, but also provides insight into the chemical changes that accompanied the metamorphism of these organics. Whatever the outcome of this debate, it is most likely that the oceans contained an upper, photic zone, that ~20% of the volcanic CO_2 added to the atmosphere was reduced and buried as a constituent of organic water, and that the remaining 80% were buried as a constituent of marine carbonates (see, e.g., Holland, 2002).

The influence of these processes on the composition of Mesoarchean seawater is still unclear. De Ronde *et al.* (1997) have studied the fluid chemistry of what they believe are Archean seafloor hydrothermal vents, and have explored the implications of their analyses for the composition of contemporary seawater. They estimate that seawater contained 920 mmol L^{-1} Cl, 2.25 mmol L^{-1} Br, 2.3 mmol L^{-1} SO_4, 0.037 mmol L^{-1} I, 789 mmol L^{-1} Na, 5.1 mmol L^{-1} NH_4, 18.9 mmol L^{-1} K, 50.9 mmol L^{-1} Mg, 232 mmol L^{-1} Ca, and 4.52 mmol L^{-1} Sr. This composition, if correct, implies that Archean seawater was rather similar to modern seawater. Unfortunately, there is considerable doubt about the correctness of these concentrations. First, the composition of seawater is altered significantly during passage through the oceanic crust, and the reconstruction of the composition of seawater from that of hydrothermal fluids is not straightforward. Second the charge balance of the proposed seawater is quite poor. Third, the quartz which contained the fluid inclusions analyzed by De Ronde *et al.* (1997) is intimately associated with hematite and goethite. The former mineral is most unusual as an ocean floor mineral at 3.2 Ga. The latter is also unusual, because these sediments passed through a metamorphic event at 2.7 Ga during which the temperature rose to >200 °C (De Ronde *et al.*, 1994). It is not unlikely that the inclusion fluids analyzed by De Ronde *et al.* were trapped more recently than 3.2 Ga, and that they are not samples of 3.2 Ga seawater (Lowe, personal communication, 2002).

The direct evidence for the composition of Mesoarchean seawater is, therefore, quite weak. For the time being it seems best to rely on indirect evidence derived from the mineralogy of sediments from this period, the composition of these minerals, and the isotopic composition of their contained elements. The carbonate minerals in Archean sediments are particularly instructive (see, e.g., Holland, 1984, chapter 5). Calcite, aragonite, and dolomite were the dominant carbonate minerals. Siderite was only a common constituent of BIFs. These observations imply that the Archean oceans were saturated or, more likely, supersaturated with respect to $CaCO_3$ and $CaMg(CO_3)_2$. Translating this observation into values for the concentration of Ca^{2+}, Mg^{2+}, HCO_3^-, and CO_3^{2-} in seawater is difficult in the absence of other information, but it can be shown that for the likely range of values of atmospheric P_{CO_2} ($\leq$0.03 atm; Rye *et al.*, 1995), the pH of seawater was probably $\geq$6.5. At saturation with respect to calcite and dolomite at 25 °C, the ratio $m_{Mg^{2+}}/m_{Ca^{2+}}$ in solutions is close to 1.0. This does not have to be the value of the ratio in Archean seawater. In Phanerozoic seawater (see below), the Mg^{2+}/Ca^{2+} ratio varied considerably from values as low as 1 up to its present value of 5.3 (Lowenstein *et al.*, 2001; Horita *et al.*, 2002).

A rough upper limit to the Fe^{2+}/Ca^{2+} ratio in Archean seawater can be derived from the scarcity of siderite except as a constituent of carbonate iron formations. At saturation with respect to siderite and calcite, the ratio $m_{Fe^{2+}}/m_{Ca^{2+}}$ is approximately equal to the ratio of the solubility product of siderite (Bruno *et al.*, 1992) and calcite (Plummer and Busenberg, 1982):

$$\frac{m_{Fe^{2+}}}{m_{Ca^{2+}}} \approx \frac{K_{sid}}{K_{cal}} = \frac{10^{-10.8}}{10^{-8.4}} = 4 \times 10^{-3} \qquad (1)$$

The absence of siderite from normal Archean carbonate sequences indicates that this is a reasonable upper limit for the Fe^{2+}/Ca^{2+} ratio in normal Archean seawater. An approximate lower limit can be set by the Fe^{2+}content of limestones and dolomites. These contain significantly more Fe^{2+} and Mn^{2+} than their Phanerozoic counterparts (see, e.g., Veizer *et al.*, 1989), a finding that is consistent with a much lower O_2 content in the atmosphere and in near-surface seawater than today.

The strongest evidence for no more than a few ppm O_2 in the Archean and Early Paleoproterozoic atmosphere is the evidence for mass-independent fractionation (MIF) of the sulfur isotopes in pre-2.47 Ga sulfides and sulfates (Farquhar *et al.*, 2000, 2001; Pavlov and Kasting, 2002; Bekker *et al.*, 2002). In the absence of O_2, solar UV interacts with SO_2 and generates MIF of the sulfur isotopes in the reaction products. The MIF signal is probably preserved in the sedimentary record, because this signal in elemental sulfur produced by this process differs from that of the gaseous products. The fate of the elemental sulfur and sulfur gases is not well understood, but S^0 may well be deposited largely as a constituent of sulfide minerals and the sulfur that is present as a constituent of sulfur gases in part as a constituent of sulfates. Today very little of the MIF signal is preserved, because all of the products are gases that become isotopically well mixed before the burial of their contained sulfur.

It is not surprising that the geochemical cycle of sulfur during the low-O_2 Archean differed from that of the present day. As shown in Figure 5, the mass-dependent fractionation of the sulfur isotopes in sedimentary sulfides was smaller prior to 2.7 Ga than in more recent times. Several explanations have been advanced for this observation. The absence of microbial sulfate reduction is one. However, the presence of microscopic sulfides in ca. 3.47 Ga barites from north pole, Australia with a maximum sulfur fractionation of 21.1% and a mean of 11.6% clearly indicates that microbial sulfate reduction was active during the deposition of these, probably evaporitic sediments (Shen *et al.*, 2001). A second explanation involves high temperatures in the pre-2.7 Ga oceans (Ohmoto *et al.*, 1993; Kakegawa *et al.*, 1998). However, Canfield *et al.* (2000) have shown that at both high and low temperatures large fractionations are expected during microbial sulfate reduction in the presence of abundant sulfate. A third possibility is that the concentration of sulfate in the Mesoarchean oceans was very much smaller than its current value of 28 mm. At present it appears that, until ca. 2.3 Ga, $m_{SO_4^{2-}}$ was $\leq 200\ \mu mol\ L^{-1}$, the concentration below which isotopic fractionation during sulfate reduction is greatly reduced (Harrison and Thode, 1958; Habicht *et al.*, 2002). Such a low sulfate concentration in seawater prior to 2.3 Ga is quite reasonable. In the absence of atmospheric O_2, sulfide minerals would not have been oxidized during weathering, and this source of river SO_4^{2-} would have been extremely small. The other major source of river SO_4^{2-}, the solution of evaporite minerals, would also have been minimal. Constructing a convincing, quantitative model of the Archean sulfur cycle is still, however, very difficult. Volcanic SO_2 probably disproportionated, at least partially, into H_2S and H_2SO_4 by reacting with H_2O at temperatures below 400 °C. SO_4^{2-} from this source must have cycled through the biosphere. Some of it was probably lost during passage through the oceanic crust at hydrothermal temperatures. Some was lost as a constituent of sulfides (mainly pyrite) and relatively rare sulfates (mainly barite).

Another potentially major loss of SO_4^{2-} may well have been the anaerobic oxidation of

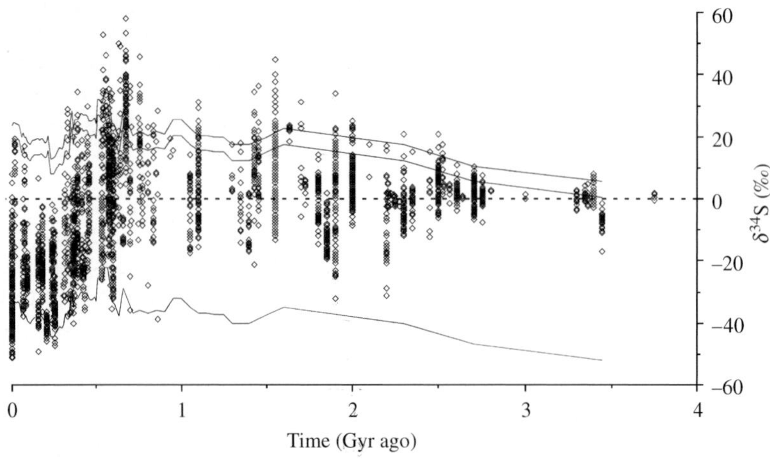

Figure 5 The isotopic composition of sedimentary sulfides over geologic time (sources Canfield and Raiswell, 1999).

methane via the overall reaction

$$CH_4 + SO_4^{2-} \rightarrow HCO_3^- + HS^- + H_2O \quad (2)$$

(Iversen and Jørgensen, 1985; Hoehler and Alperin, 1996; Orphan *et al.*, 2001). It seems likely that the reaction is accomplished in part by a consortium of Archaea growing in dense aggregates of ~100 cells, which are surrounded by sulfate-reducing bacteria (Boetius *et al.*, 2000; De Long, 2000). In sediments rich in organic matter SO_4^{2-} is depleted rapidly. Below the zone of SO_4^{2-} depletion CH_4 is produced. The gas diffuses upward and is destroyed, largely in the transition zone, where the concentration of SO_4^{2-} in the interstitial water is in the range of 0.1–1 mmol kg^{-1} (Iversen and Jørgensen, 1985). The rate of CH_4 oxidation is highest where its concentration is equal to that of SO_4^{2-}. In two stations studied by Iversen and Jørgensen, the total anaerobic methane oxidation was close to 1 mmol m^{-2} d^{-1}, of which 96% occurred in the sulfate–methane transition zone. If this rate were characteristic of the ocean floor as a whole, SO_4^{2-} reduction within marine sediments would occur at a rate of $\sim(1\times10^{14})$ mol yr^{-1}, which exceeds the present-day input of volcanic SO_2 by ca. two orders of magnitude. The process is, therefore, potentially important for the global geochemistry of sulfur and carbon.

The anaerobic oxidation of CH_4 also occurs in anoxic water masses. In the Black Sea only ~2% of the CH_4 which escapes from sediments reaches the atmosphere. The remainder is largely lost by sulfate reduction in the anoxic parts of the water column (Reeburgh *et al.*, 1991). In an ocean containing ≤1 mmol L^{-1} SO_4^{2-} the rate of CH_4 loss in the water column would almost certainly be smaller than in the Black Sea today, and the flux of CH_4 to the atmosphere would almost certainly be greater than today. Pavlov *et al.* (2000) have shown that the residence time of CH_4 in an anoxic atmosphere is $\sim 3\times10^4$ yr, i.e., some 1,000 times longer than today. The combination of a higher rate of CH_4 input and a longer residence time in the atmosphere virtually assures that the partial pressure of CH_4 was much higher in the Archean atmosphere than its present value of $\sim(1\times10^{-6})$ atm. A CH_4 pressure of 10^{-4}–10^{-3} atm is not unlikely (Catling *et al.*, 2001). At these levels CH_4 generates a very significant greenhouse warming, enough to overcome the likely lower luminosity of the Sun during the early part of Earth history (Kasting *et al.*, 2001). The recent discovery of microbial reefs in the Black Sea fueled by the anaerobic oxidation of methane by SO_4^{2-} (Michaelis *et al.*, 2002) suggests that this process was important during the Archean, and that it can account for some of the organic matter generated in the oceans before the rise of atmospheric O_2.

6.21.3.3 The Neoarchean (3.0–2.5 Ga)

The spread in the $\delta^{34}S$ value of sulfides and sulfates increased significantly between 3.0 Ga and 2.5 Ga. The first major increase in the $\delta^{34}S$ range occurred ~2.7 Ga (see Figure 5). However, sulfate concentrations probably stayed well below the present value of 28 mmol kg^{-1} until the Neoproterozoic. Grotzinger (1989) has reviewed the mineralogy of Precambrian evaporites and has shown that calcium sulfate minerals (or their pseudomorphs) are scarce before ~1.7–1.6 Ga. Bedded or massive gypsum/anhydrite formed in evaporitic environments is absent in the Archean and Paleoproterozoic record. A low concentration of SO_4^{2-} in the pre-1.7 Ga oceans is the most reasonable explanation for these observations (Grotzinger and Kasting, 1993).

Rather interestingly, the oldest usable biomarkers in carbonaceous shales date from the Neoarchean. Molecular fossils extracted from 2.5 Ga to 2.7 Ga shales of the Fortescue and Hamersley groups in the Pilbara Craton, Western Australia, indicate that the photic zone of the water column in the areas where these shales were deposited was probably weakly oxygenated, and that cyanobacteria were part of the microbial biota (Brocks *et al.*, 1999, 2002; Summons *et al.*, 1999). The similarity of the timing of the rise in the range of $\delta^{34}S$ in sediments and the earliest evidence for the presence of cyanobacteria may, however, be coincidental, because to date no sediments older than 2.7 Ga have been found that contain usable biomarker molecules (Brocks, personal communication, 2002).

Despite the biomarker evidence for the generation of O_2 at 2.7 Ga, the atmosphere seems to have contained very little or no O_2, and much of the ocean appears to have been anoxic. Pyrite, uraninite, gersdorffite, and, locally, siderite occur as unequivocally detrital constituents in 3,250–2,750 Ma fluvial siliciclastic sediments in the Pilbara Craton in Australia (Rasmussen and Buick, 1999). These sediments have never undergone hydrothermal alteration. Some grains of siderite display evidence of several episodes of erosion, rounding, and subsequent authigenic overgrowth (see Figure 6). Their frequent survival after prolonged transport in well-mixed and, therefore, well-aerated Archean rivers that contained little organic matter strongly implies that the contemporary atmosphere was much less oxidizing than at present. The paper by Rasmussen and Buick (1999) was criticized by Ohmoto (1999), but staunchly defended by Rasmussen *et al.* (1999).

These observations complement those made since the early 1990s on the gold–uranium ores of the Witwatersrand Basin in South Africa and on the uranium ores of the Elliot Lake District in

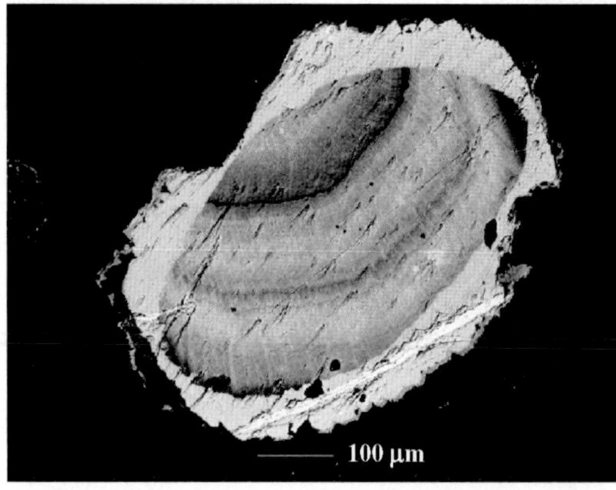

Figure 6 Rounded siderite grain, with core of compositionally banded siderite and gray to black syntaxial overgrowths (source Rasmussen and Buick, 1999).

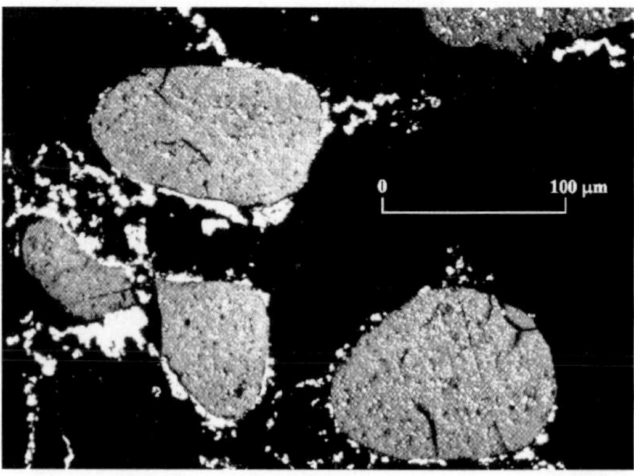

Figure 7 Detrital grains of uraninite with characteristic dusting of galena, partly surrounded by PbS overgrowths. The big grain displays a typical "muffin shape." Basal Reef, footwall; Loraine Gold Mines, South Africa oil immersion; 375× (source Schidlowski, 1966).

Canada. The origin of these ores has been hotly debated (see, e.g., Phillips *et al.*, 2001). The rounded shape of many of the pyrite and uraninite grains (see Figures 7 and 8) are in a geologic setting appropriate for the placer accumulation of heavy minerals. Figure 7 shows some of the muffin-shaped uraninite grains described by Schidlowski (1966), and Figure 8 shows rounded grains of pyrite described by Ramdohr (1958). It is clear that some of the rounded pyrite grains are replacements of magnetite, ilmenite, and other minerals. The origin of any specific rounded pyrite grain if based on textural evidence alone is, therefore, somewhat ambiguous. However, the Re–Os age of some pyrite grains indicates that they are older than the depositional age of the sediments (Kirk *et al.*, 2001). The detrital origin of the uraninite muffins is essentially established by their chemical composition. As shown in Table 1 these contain significant concentrations of ThO_2, which are characteristic of uraninite derived from pegmatites but not of hydrothermal pitchblende. It is, therefore, very difficult to assign anything but a detrital origin to the uraninite in the Witwatersrand ores (see, e.g., Hallbauer, 1986).

Experiments by Grandstaff (1976, 1980) and more recently by Ono (2002) on the oxidation and dissolution or uraninite can be used to set a rough upper limit of 10^{-2}–10^{-3} atm on the O_2 content of the atmosphere during the formation of the Au–U deposits of the Witwatersrand Basin (Holland, 1984, chapter 7). This maximum O_2 pressure is much greater than that permitted by the presence of MIF of sulfur isotopes during the last 0.5 Ga of the Archean; the observations do, however, complement each other.

The chemical composition of soils developed during the Late Archean and during the Paleoproterozoic also fit the pattern of a low- or no-O_2 atmosphere. During weathering on such an Earth, elements which are oxidized in a high-O_2 atmosphere remain in their lower valence states and behave differently within soils, in groundwaters, and in rivers. The theory connecting this qualitative statement to the expected behavior of redox sensitive elements has been developed in papers by Holland and Zbinden (1988), Pinto and Holland (1988), and Yang and Holland (2003). The available data for the chemical evolution of paleosols have been summarized by Rye and Holland (1998) and by Yang and Holland (2003). The composition of paleosols is consistent with a change from a low- or no-O_2 atmosphere to a highly oxygenated atmosphere between 2.3 Ga and 2.0 Ga; a different interpretation of the

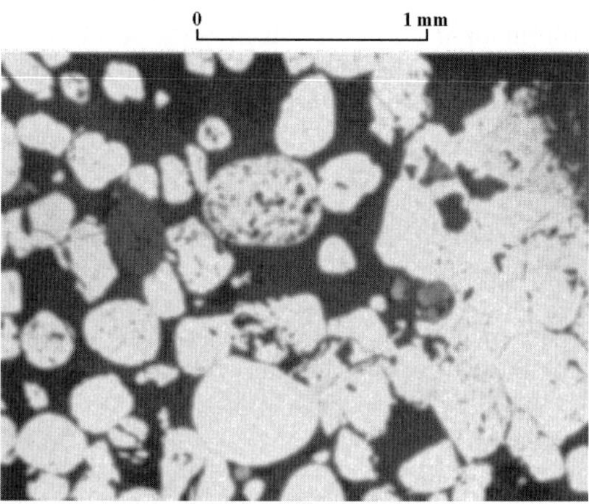

Figure 8 Conglomerate consisting of several types of pyrite together with zircon, chromite, and other heavy minerals. The large pyrite grain in the right part of the figure is a complex assemblage of older pyrite grains which have been cemented by younger pyrite (source Ramdohr, 1958).

Table 1 Electron microprobe analyses of uraninite in some Witwatersrand ores.

Source	*Grain no.*	*UO_2* (%)	*ThO_2* (%)	*PbO_2* (%)	*FeO* (%)	*TiO_2* (%)	*CaO* (%)	*Total* (%)	*UO_2/ThO_2*
Cristaalkop Reef (171) (Vaal Reefs South Mine)	T21	65.8	5.3	23.3	0.8	<0.01	0.6	95.8	12.4
	T22	66.5	6.1	21.9	0.5	0.02	0.7	95.7	10.9
	T23	70.5	1.7	23.8	0.5	0.04	0.8	97.3	41.5
	T23	62.4	6.1	24.7	0.4	<0.01	1.1	94.7	10.2
	T25	63.0	8.0	23.0	0.5	<0.01	1.0	95.5	7.9
	T27	66.6	3.2	23.3	1.0	0.04	0.9	95.0	20.8
	T28	66.3	1.4	28.0	0.6	0.02	0.9	97.2	47.4
	T29	63.5	3.9	28.5	0.6	<0.01	0.8	97.3	16.3
	T30	65.7	10.2	18.9	0.7	0.08	1.1	96.7	6.4
	Average	65.6	5.1	23.9	0.6	0.02	0.9	96.1	12.9
Carbon Leader (135) (Western Deep Levels Mine)	T13	69.6	2.7	26.1	0.2	0.08	1.0	99.7	25.8
	T14	66.4	2.5	27.8	0.2	0.12	0.7	97.7	26.6
	T15	63.8	2.0	30.3	0.2	0.06	0.7	97.1	31.9
	T16	62.6	7.0	24.5	0.2	0.04	0.8	95.1	8.9
	T17	67.1	5.2	21.1	0.2	0.06	0.7	94.4	12.9
	T18	69.2	2.1	28.0	0.2	0.10	0.7	100.3	33.0
	T19	67.9	7.2	18.3	0.2	0.06	0.6	94.3	9.4
	Average	66.7	4.1	25.2	0.2	0.07	0.7	97.0	16.3
Carbon Leader (167) (West Driefontein Mine)	B43	61.1	5.4	27.9	1.1	0.25	0.4	96.2	11.3
	B44	71.3	2.6	24.2	0.6	0.25	0.6	99.6	27.4
	B46	68.2	4.3	28.2	0.4	0.10	0.5	101.7	15.9
	B47	70.0	2.1	23.6	0.4	0.10	0.6	96.8	33.3
	B48	68.2	3.5	24.7	0.5	0.12	0.3	97.3	19.5
	B49	67.3	6.4	21.6	0.4	0.16	0.5	96.4	10.5
	B52	67.6	9.2	22.8	0.9	0.14	0.3	100.9	7.3
	Average	68.1	4.2	24.7	0.6	0.16	0.5	98.3	16.2
Main Reef (151) (SA Lands Mine)	B31	68.7	3.3	14.7	2.3	1.20	0.4	90.6	20.8
	B32	69.1	2.1	21.0	1.1	0.50	0.4	94.2	32.9
	B36	67.3	1.5	24.4	0.9	0.19	0.3	94.6	44.9
	B37	63.5	2.9	17.3	5.0	2.03	0.5	91.2	21.9
	Average	67.2	2.5	19.4	2.3	0.98	0.4	92.8	26.9
Basal Reef (184) (Welkom Mine)	U4	68.2	6.3	19.6	2.4	<0.1	0.4	96.9	10.8
	U5	70.5	3.3	16.8	4.1	0.2	0.4	95.3	21.4
	U7	70.1	5.2	24.8	1.1	0.1	0.5	101.8	13.5
	U8	66.6	4.5	25.2	0.9	<0.1	0.3	97.5	14.8
	U9	72.5	2.6	26.1	1.6	<0.1	0.5	103.3	27.9
	U10	64.8	3.3	23.7	5.2	0.2	0.5	97.7	19.6
	Average	68.8	4.2	22.7	2.6	0.1	0.4	98.7	16.4
Overall average:		67.2	3.9	23.6	1.0	0.16	0.6	93.8	17.2

Source: Feather (1980).

available data has been proposed by Ohmoto (1996) and by Beukes *et al.* (2002a,b).

One consequence of the proposed great oxidation event (GOE) of the atmosphere between 2.3 Ga and 2.0 Ga is that trace elements such as molybdenum, rhenium, and uranium, which are mobile during weathering in an oxidized environment, would have been essentially immobile before 2.3 Ga. Their concentration in seawater would then have been very much lower than today, and their enrichment in organic carbon-richshales would have been minimal. This agrees with the currently available data (Bekker *et al.*, 2002; Yang and Holland, 2002). Carbonaceous shales older than ca. 2.3 Ga are not enriched in molybdenum, rhenium, and uranium. A transition to highly enriched shales occurs ~2.1 Ga; by 1.6 Ga the enrichment of carbonaceous shales in these elements was comparable to that in their Phanerozoic counterparts (see, e.g., Werne *et al.*, 2002).

The data for the mineralogy of BIFs tell much the same story. These sediments provide strong evidence for the view that the deep oceans were anoxic throughout Archean time (James, 1992). Evidence regarding the oxidation state of the shallow parts of the Archean oceans is still very fragmentary. The shallow water facies of the 2.49 ± 0.03 Griquatown iron formation (Nelson *et al.*, 1999) in the Transvaal Supergroup of South Africa (Beukes, 1978 , 1983; Beukes and Klein, 1990) were deposited on the ~800 km × 800 km shelf shown in the somewhat schematic Figure 9. The stratigraphic relations are illustrated in the south–north cross-section of Figure 10. Several of the units in the Danielskuil Member of the Griquatown iron formation can be traced across the shallow platform from the subtidal, low-energy epeiric sea, through the high-energy zone of the shelf, into the lagoonal, near-shore parts of the platform. In the deeper parts of the shelf, siderite and iron silicates dominate the mineralogy of the iron formation. Siderite and minor (<10%) hematite dominate the sediments of the high-energy zone. Greenalite and siderite lutites are most common in the platform lagoonal zone.

The dominance of Fe^{2+} minerals in even the shallowest part of the platform can only be explained if the O_2 content of the ambient atmosphere was very low. The half-life of Fe^{2+} oxidation in the Gulf Stream and in Biscayne Bay, Florida is only a few minutes (Millero *et al.*, 1987). The half-life of Fe^{2+} oxidation is similar in the North Sea, the Sargasso Sea, Narragansett Bay, and Puget Sound (for summary see Millero *et al.*, 1987).

The half-life of Fe^{2+} in the solutions from which the Griquatown iron formation was deposited was obviously many orders of magnitude longer than this, and it is useful to inquire into the cause for the difference. Stumm and Lee (1961) have shown that the rate of oxidation of Fe^{2+} in aqueous solutions is governed by the equation

$$-\mathrm{d}m_{Fe^{2+}}/\mathrm{d}t = km^2_{OH^-}m_{O_2}m_{Fe^{2+}} \qquad (3)$$

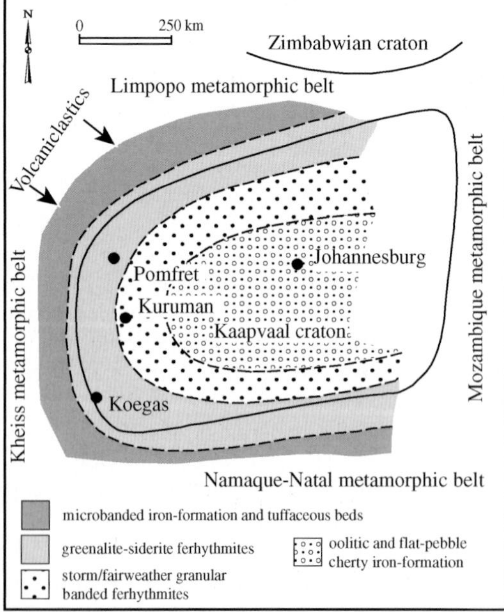

Figure 9 Depositional model for the Kuruman–Griquatown transition zone in a plan view, illustrating lithofacies distribution during drowning of the Kaapvaal craton (source Beukes and Klein, 1990).

where

$$m_{OH^-} = \text{concentration of free } OH^-$$

$$m_{O_2} = \text{concentration of dissolved } O_2$$

Integration of Equation (3) yields

$$\ln m_{Fe^{2+}}/_0m_{Fe^{2+}} = -km^2_{OH^-}m_{O_2}t \qquad (4)$$

where $_0m_{Fe^{2+}}$ is the initial concentration of Fe^{2+} in the solution. If we substitute a_{OH^-} for m_{OH^-}, the value of k for modern seawater is $\sim 0.9 \times 10^{15}\ \mathrm{min}^{-1}$ (Millero *et al.*, 1987). An upper limit for $\ln m_{Fe^{2+}}/_0m_{Fe^{2+}}$ can be obtained from the field data for the Griquatown iron formation on the Campbellrand platform. Hematite accounts for $\leq 10\%$ of the iron in the near-shore iron formation. If all of the iron that was oxidized to Fe^{3+} was precipitated as a constituent of Fe_2O_3 during the passage of seawater across the platform,

$$m_{Fe^{2+}}/_0m_{Fe^{2+}} \geq 0.9 \qquad (5)$$

The time, t, required for the passage of water across the Campbellrand platform is uncertain. The modern Bahamas are probably a reasonable analogue for the Campbellrand platform. On the Bahama Banks tidal currents of 25 cm s^{-1} are common, and velocities of 1 m s^{-1} have been recorded in channels (Sellwood, 1986). At a rate of 25 cm s^{-1} it would have taken seawater ~1 month (4×10^4 min) to traverse the ~800 km diameter of the Campbellrand platform. This period is much longer than the time required to precipitate Fe^{3+} oxyhydroxide after the oxidation of Fe^{2+} to Fe^{3+} (Grundl and Delwiche, 1993). The best estimate of the residence time of seawater on the Grand Bahama Bank is ~1 yr (Morse *et al.*, 1984; Millero, personal communication). The pH of the solutions from which the iron formations were deposited was probably less than that of seawater today, but probably not lower than 7.0.

If we combine all of these rather uncertain values for the terms in Equation (4), we obtain

$$m_{O_2} \sim (<0.10)/0.9 \times 10^{15} \times (\geq 10^{-14.0}) \times 4 \times 10^4\ \mathrm{mol\ kg^{-1}} < 3 \times 10^{-7}\ \mathrm{mol\ kg^{-1}} \qquad (6)$$

In an atmosphere in equilibrium with seawater containing this concentration of dissolved O_2,

$$P_{O_2} = 2.4 \times 10^{-4}\ \mathrm{atm}$$

The maximum value of atmospheric P_{O_2} estimated in this manner is consistent with inferences from the MIF of the sulfur isotopes during the deposition of the Griquatown iron formation that $P_{O_2} \leq 1 \times 10^{-5}$ PAL.

A rather curious observation in the light of these observations is that in many unmetamorphosed

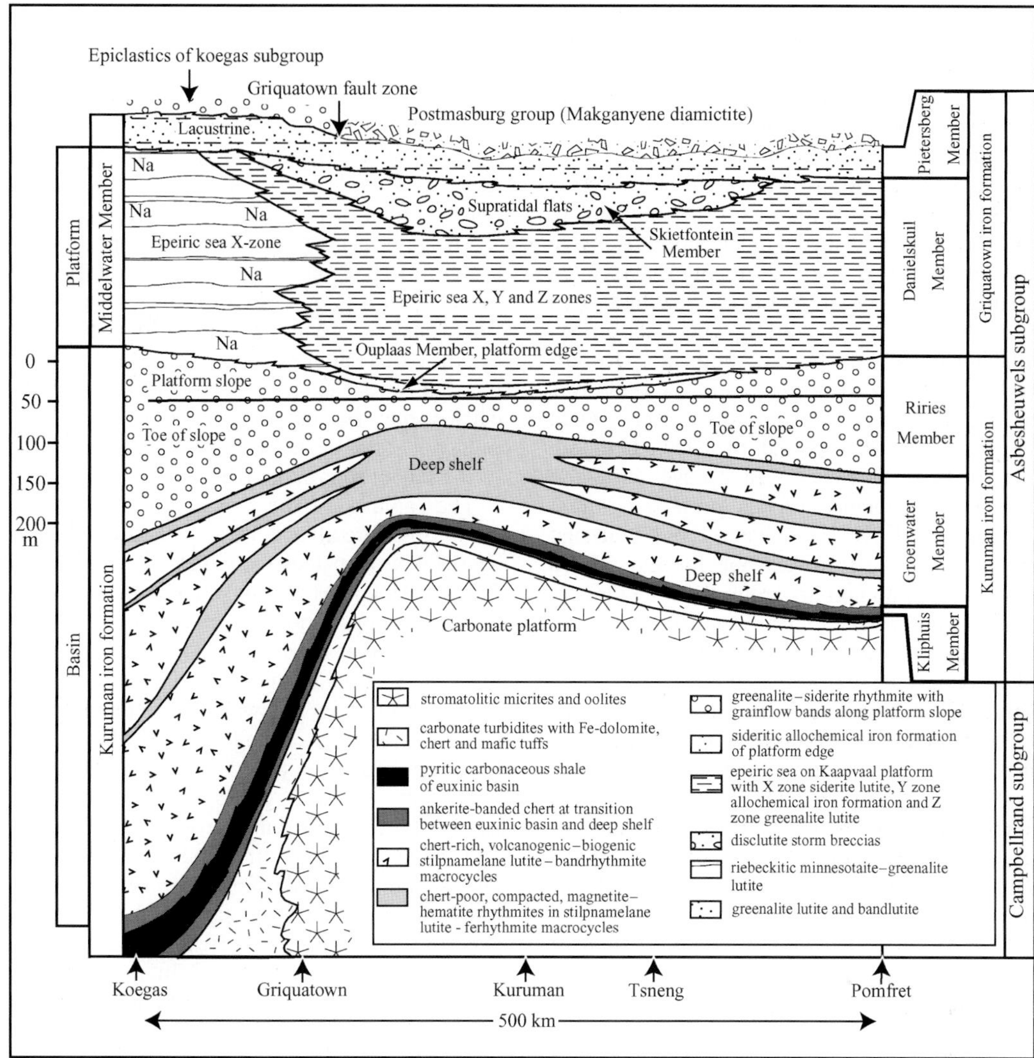

Figure 10 Longitudinal cross-section illustrating stratigraphic relationships and inferred palaeodepositional environments of the Asbesheuwels Subgroup in the Griqualand West basin (after Beukes, 1978).

oxide facies BIFs, the first iron oxide mineral precipitated was frequently hematite (Han, 1982, 1988). This phase was later replaced by magnetite. Klein and Beukes (1989) have reported the presence of hematite as a minor component in iron formations of the Paleoproterozoic Transvaal Supergroup, South Africa. The hematite occurs in two forms: as fine hematite dust and as very fine grained specularite. The former could well be a very early phase in these BIFs. The early deposition of hematite in BIFs followed by large-scale replacement by magnetite could simply be the result of reactions in mixtures of hydrothermal vent fluids with ambient O_2-free seawater. High-temperature hydrothermal vent fluids are strongly undersaturated with respect to hematite. However, on mixing with ambient O_2-free seawater their pH rises, and path calculations indicate that they can become supersaturated with respect to hematite. During hematite precipitation P_{H_2} would increase, and the early hematite could well be replaced by magnetite during early diagenesis. The sequence of mineral deposition from such solutions depends not only on the composition of the vent fluids and that of the ambient seawater, but also on the kinetics of the precipitation mechanisms. These, in turn, could have been influenced by bacterial processes (Konhauser *et al.*, 2002). A thorough study of the effects of these parameters remains to be done.

6.21.4 THE PROTEROZOIC

6.21.4.1 The Paleoproterozoic (2.5–1.8 Ga)

In 1962 the author divided the evolution of the atmosphere into three stages (Holland, 1962). On the basis of rather scant evidence from the mineralogy of Precambrian sedimentary uranium deposits, it was suggested that free oxygen was not present in appreciable amounts until ca. 1.8 Ga, but that by the end of the Paleozoic the O_2 content of the atmosphere was already a large fraction of its present value. In a similar vein, Cloud (1968) proposed that the atmosphere before 1.8–2.0 Ga could have contained little or no free oxygen. In 1984 the author published a much more extensive analysis of the rise of atmospheric oxygen (Holland, 1984). Progress since 1962 had been rather modest. After reading the section on atmospheric O_2 in the Precambrian, Robert Garrels commented that this part of the book was very long but rather short on conclusions. Much more progress was reported in 1994 (Holland, 1994), and the last few years have shown a widespread acceptance of his proposed "great oxidation event" (GOE) between ca. 2.3 Ga and 2.0 Ga. This acceptance has not, however, been universal. Ohmoto and his group have steadfastly maintained (Ohmoto, 1996, 1999) that the level of atmospheric O_2 has been close to its present level during the past 3.5–4.0 Ga.

During the past few years the most exciting new development bearing on this question has been the discovery of the presence of mass-independent fractionation (MIF) of the sulfur isotopes in sulfides and sulfates older than ca. 2.47 Ga. Figure 11 summarizes the available data for the degree of MIF ($\Delta^{33}S$) in sulfides and sulfates during the past 3.8 Ga. The presence of values of $\Delta^{33}S > 0.5$‰ in sulfides and sulfates older than ca. 2.47 Ga indicates that the O_2 content of the atmosphere was $<10^{-5}$ PAL prior to 2.47 Ga (Farquhar *et al.*, 2001; Pavlov and Kasting, 2002). The absence of significant MIF in sulfides and sulfates $\leq$2.32 Ga (Bekker *et al.*, 2002) is indicative of O_2 levels $>10^{-5}$ PAL. There are no data to decide when between 2.32 Ga and 2.47 Ga the level of atmospheric P_{O_2} rose, and whether the rise was a single event, or whether P_{O_2} oscillated before a final rise by 2.3 Ga.

The presence of O_2 in the atmosphere–ocean system at 2.32 Ga is supported strongly by the presence of a large body of shallow-water hematitic ironstone ore in the Timeball Hill Formation, South Africa (Beukes *et al.*, 2002a,b). Apparently, the shallow oceans have been oxidized ever since. The deeper oceans may have continued in a reduced state at least until the disappearance of the Paleoproterozoic BIFs ca. 1.7 Ga.

The rapidity of the rise of the O_2 content of the atmosphere after 2.3 Ga is a matter of dispute. The Hekpoort paleosols, which developed on the 2.25 Ga Hekpoort Basalt, consist of an oxidized hematitic upper portion and a reduced lower portion. Beukes *et al.* (2002a,b) have pointed out that the section through the Hekpoort paleosol near Gaborone in Botswana is similar to modern tropical laterites. Yang and Holland (2003) have remarked on the differences between the chemistry and the geology of the Hekpoort paleosols and Tertiary groundwater laterites, and have proposed that the O_2 level in the atmosphere during the formation of the Hekpoort paleosols was between ca. 2.5×10^{-4} atm and 9×10^{-3} atm, i.e., considerably lower than at present. Paleosols developed in the Griqualand Basin on the Ongeluk Basalt, which is of the same

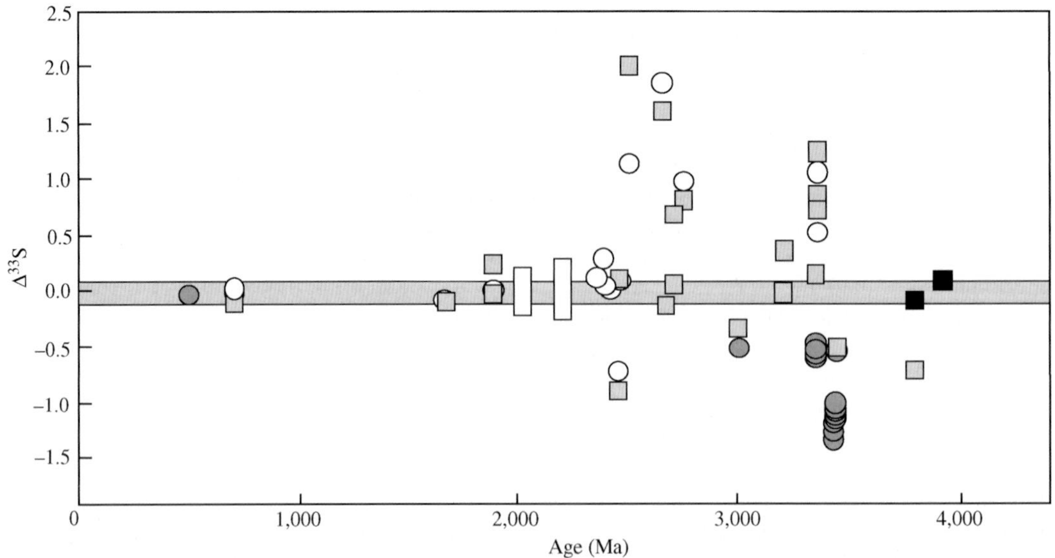

Figure 11 Summary of data for the degree of MIF of the sulfur isotopes in sulfides and sulfates. Data from Farquhar *et al.* (2000) (chemically defined sulfur minerals (□) sulfides, (■) total sulfur, and (○) sulfates; (●) macroscopic sulfate minerals) with updated ages and from Bekker *et al.* (2002) ((□) range of values for pyrites in black shales). The gray band at $\Delta^{33}S \sim 0$ represents the mean and 1 SD of recent sulfides and sulfates from Farquhar *et al.* (2000).

age as the Hekpoort Basalt, are highly oxidized. The difference between their oxidation state and that of the Hekpoort paleosols may be due to a slightly younger age of the paleosols in the Griqualand Basin. They probably formed during the large, worldwide positive variation of Figure 12 in the $\delta^{13}C$ value of marine carbonates (Karhu and Holland, 1996). This excursion is best interpreted as a signal of the production of a large quantity of O_2 between 2.22 Ga and 2.06 Ga. Estimates of this quantity are perforce very rough. The average isotopic composition of organic carbon during the excursion is somewhat uncertain, as is the total rate of carbon burial with organic matter and carbonate sediments during the $\delta^{13}C$ excursion, and there is the possibility that the $\delta^{13}C$ excursion in Figure 12 was preceded by another, shorter excursion (Bekker *et al.*, 2001; Young, 1969). It seems likely, however, that the total excess quantity of O_2 produced during the ca. 160 Ma of the $\delta^{13}C$ excursion was ~12–22 times the inventory of atmospheric oxygen (Karhu, 1993; Karhu and Holland, 1996). This large amount of O_2 must somehow have disappeared into the sedimentary record. The most likely sinks are crustal iron and sulfur. As indicated by the mineralogy of marine evaporites and the isotopic composition of sulfur in black shales, the SO_4^{2-} concentration of seawater probably remained very modest until well beyond 2.0 Ga (Grotzinger and Kasting, 1993; Shen *et al.*, 2002). Iron is, therefore, the most likely major sink for the O_2 produced during the $\delta^{13}C$ excursion between 2.22 Ga and 2.06 Ga. This is not unreasonable. Before the rise of atmospheric O_2, FeO was not oxidized to Fe_2O_3 during weathering, as shown by the record of the Fe_2O_3/FeO ratio in pre-2.3 Ga sedimentary rocks (Bekker *et al.*, 2003). The appearance of extensive red beds ca. 2.3 Ga indicates that a major increase in the Fe_2O_3/FeO ratio of sediments and sedimentary rocks occurred at that time. Shales before the GOE contained, on average, ~6.5% FeO and ~1.3% Fe_2O_3. Shales deposited between 2.3 Ga and 2.1 Ga contain, on average, ~4.1% FeO and

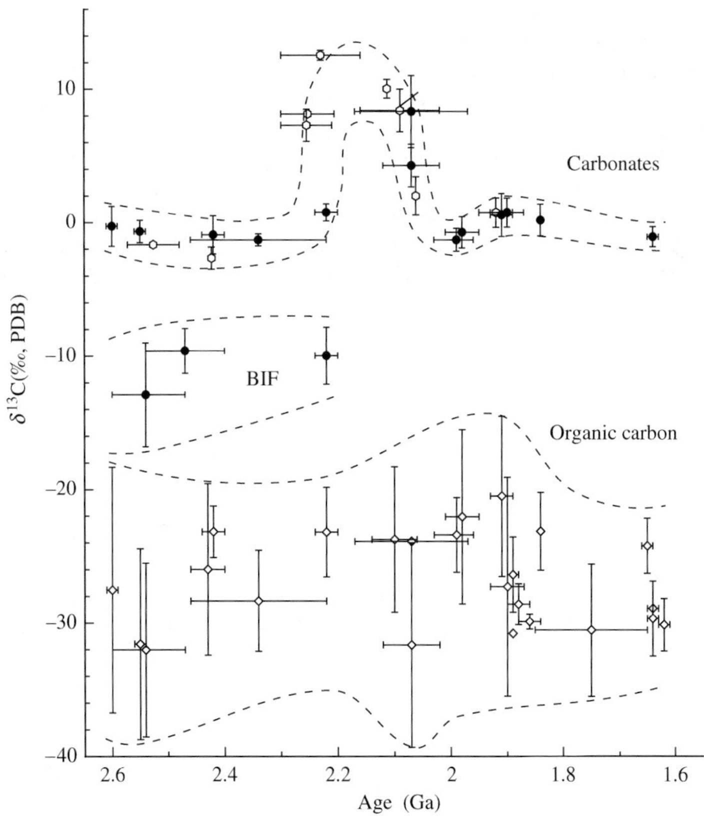

Figure 12 Variation in isotopic composition of carbon in sedimentary carbonates and organic matter during Paleoproperozoic time. Mean $\delta^{13}C$ values of carbonates from Fennoscandian Shield from Karhu (1993) are indicated by open circles. Vertical bars represent ±1 SD of $\delta^{13}C$ values, and horizontal bars indicate uncertainty in age of each stratigraphic unit. Arrows combine dated formations that are either preceded or followed by major $\delta^{13}C$ shift. BIF denotes field for iron and manganese formations. Note that uncertainties given for ages do not necessarily cover uncertainties in entire depositional periods of sample groups. PDB—Peedee belemnite (source Karhu and Holland, 1996).

2.5% Fe_2O_3. There seems to have been little change between 2.1 Ga and 1.0 Ga (Bekker *et al.*, 2003). Approximately 2% of the FeO in pre-GOE rocks seem to have been converted to Fe_2O_3 during weathering in the course of the GOE. Since each mole of FeO requires 0.25 mol O_2 for conversion to Fe_2O_3, ~0.08 mol O_2 was used during the weathering of each kilogram of rock. If weathering rates during the GOE were comparable to current rates, some 1.6×10^{12} mol O_2 were used annually to convert FeO to Fe_2O_3 during weathering in the course of the $\delta^{13}C$ excursion. The total O_2 use was, therefore, $\sim 2.6 \times 10^{20}$ mol, i.e., ~6 times the present atmospheric O_2 inventory. Some of the excess O_2 was probably used to increase the redox state of the crustal sulfur cycle. The duration of the $\delta^{13}C$ excursion is roughly equal to the half-life of sedimentary rocks at 2 Ga. The Fe_2O_3/FeO ratio of rocks subjected to weathering at the end of the $\delta^{13}C$ excursion was, therefore, greater than at its beginning and was approaching a value typical of Mesoproterozoic sedimentary rocks. Post-$\delta^{13}C$ excursion weathering of sediments produced during the $\delta^{13}C$ excursion, therefore, required much less additional O_2 than the weathering of pre-GOE rocks.

Although this is a likely explanation for the fate of most of the "extra" O_2 generated during the $\delta^{13}C$ excursion between ca. 2.22 Ga and 2.06 Ga, it does not account for the GOE itself or for the cause of the $\delta^{13}C$ excursion. The appearance of O_2 in the atmosphere between ca. 2.47 Ga and 2.32 Ga could be explained easily if cyanobacteria evolved at that time. However, this explanation has been rendered very unlikely by the discovery of biomarkers that are characteristic of cyanobacteria and eukaryotes in 2.5–2.7 Ga sedimentary rocks (Brocks *et al.*, 1999). An alternative explanation involves a change in the redox state of volcanic gases as the trigger for the change of the oxidation state of the atmosphere (Kasting *et al.*, 1993). These authors pointed out that the loss of H_2 from the top of a reducing atmosphere into interplanetary space would have increased the overall oxidation state of the Earth as a whole, and almost certainly that of the mantle. This, in turn, would have led to an increase in the f_{O_2} of volcanic gases and to a change in the redox state of the atmosphere.

In a more detailed analysis of this mechanism, Holland (2002) showed that the change in the average f_{O_2} of volcanic gases required for the transition of the atmosphere from an anoxygenic to an oxygenic state is quite small. There is no inconsistency between the required change in f_{O_2} and the limits set on such changes by the data of Delano (2001) and Canil (1997, 1999, 2002) for the evolution of the redox state of the upper mantle during the past 4.0 Ga. The estimated changes in f_{O_2} due to H_2 loss are consistent with the likely changes in the redox state of the upper mantle if the major control on that state is exerted by the Fe_2O_3/FeO buffer. In this explanation the average composition of volcanic gases before the GOE was such that 20% of their contained CO_2 could be reduced to CH_2O, and all of the sulfur gases to FeS_2. Excess H_2 present in the gases would have escaped from the atmosphere, possibly via the decomposition of CH_4 in the upper atmosphere. The loss of H_2 would have produced an irreversible oxidation of the early Earth (Catling *et al.*, 2001). The GOE began when the composition of volcanic gases had changed, so that not enough H_2 was present to convert 20% of the contained CO_2 to CH_2O and all of the sulfur gases to FeS_2. Before the GOE the only, or nearly the only, sulfate mineral deposited in sediments seems to have been barite. Since barium is a trace element, its precipitation as $BaSO_4$ accounted for only a small fraction of the atmospheric input of volcanic sulfur. After the GOE a fraction of volcanic sulfur began to leave the atmosphere–ocean system as a constituent of other sulfate minerals as well, largely as gypsum ($CaSO_4 \cdot 2H_2O$) and anhydrite ($CaSO_4$).

During the Phanerozoic close to half of the volcanic sulfur in volcanic gases has been removed as a constituent of FeS_2, the other half as a constituent of gypsum and anhydrite (see, e.g., Holland, 2002). The shift from the essentially complete removal of volcanic sulfur as a constituent of FeS_2 to the present state was gradual (see below). It was probably controlled by a feedback mechanism involving an increase in the sulfur content of volcanic gases. This was probably the result of an increase in the rate of subduction of $CaSO_4$ added to the oceanic crust by the cycling of sea water at temperatures above ca. 200 °C.

The burial of excess organic matter during the $\delta^{13}C$ excursion between 2.22 Ga and 2.06 Ga almost certainly required an excess of PO_4^{3-}. It seems likely that this excess was released from rocks during weathering due to the lower pH of soil waters related to the generation of H_2SO_4 that accompanied the oxidative weathering of sulfides. Toward the end of the $\delta^{13}C$ excursion, this excess PO_4^{3-} was probably removed by adsorption on the Fe^{3+} hydroxides and oxyhydroxides produced by the oxidative weathering of Fe^{2+} minerals (Colman and Holland, 2000). Although this sequence of events is reasonable, and although some parts of it can be checked semiquantitatively, the proposed process by which the anoxygenic atmosphere became converted to an oxygenic state should be treated with caution. Too many pieces of the puzzle are still either missing or of questionable shape.

A most interesting and geochemically significant change in the oceans may have occurred ca. 1.7 Ga. BIFs ceased to be deposited. They are apparently absent from the geologic record until their reappearance 1 Ga later in association with the very large Neoproterozoic ice ages (Beukes and Klein, 1992). Three explanations have been advanced for the hiatus in BIF deposition between 1.7 Ga and 0.7 Ga. The first proposes that the deposition of BIF ended when the deep waters of the oceans became aerobic (Cloud, 1972; Holland, 1984). After 1.7 Ga, Fe^{2+} from hydrothermal vents was oxidized to Fe^{3+} close to the vents and was precipitated as Fe^{3+} oxides and/or oxyhydroxides on the floor of the oceans. The second explanation proposes that anoxic bottom waters persisted until well after the deposition of BIFs ceased, and that an increase in the concentration of H_2S rather than the advent of oxygen was responsible for removing iron from deep ocean water (Canfield, 1998). The sulfur isotope record indicates that the concentration of oceanic sulfate began to increase ~2.3 Ga leading to increasing rates of sulfide production by bacterial sulfate reduction. Canfield (1998) has suggested that sulfide production became sufficiently intense ~1.7 Ga to precipitate the total hydrothermal flux of iron as a constituent of pyrite in the deep oceans. As a basis for this contention, he points out that the generation of aerobic deep ocean water would have required levels of atmospheric O_2 within a factor of 2 or 3 of the present level, a level which he believes was not attained until the Neoproterozoic. However, Canfield's (1998) analysis of his three-box model of the oceans assumes that the rate of sinking of organic matter into the deep ocean was the same during the Paleoproterozoic as at present. This is unlikely. Organic matter requires ballast to make it sink. Today most of the ballast is supplied by siliceous and carbonate tests (Logan *et al.*, 1995; Armstrong *et al.*, 2002; Iglesias-Rodriguez *et al.*, 2002; Sarmiento *et al.*, 2002). Clays and dust seem to be minor constituents of the ballast, although they may have been more important before the advent of soil-binding plants. There is no evidence for the production of siliceous or calcareous tests in the Paleoproterozoic oceans. Inorganically precipitated SiO_2 and/or $CaCO_3$ could have been important, but precipitation of these phases probably occurred mainly in shallow-water evaporitic settings. It is, therefore, likely that ballast was much scarcer during the Paleoproterozoic than today, and that the quantity of particulate organic matter (POC) transported annually from shallow water into the deep oceans was much smaller than today. This, in turn, implies that the amount of dissolved O_2 that was required to oxidize the rain of POC was much smaller than today. Evidence from paleosols suggests that atmospheric O_2 levels ca. 2.2 Ga were ≥15% PAL (Holland and Beukes, 1990). This implies that the proposal for the end of BIF deposition based on the development of oxygenated bottom waters ca. 1.7 Ga is quite reasonable. It does not, of course, prove that the proposal is correct. For one thing, too little is known about the mixing time of the Paleoproterozoic oceans. Data for the oxidation state of the deep ocean since 1.7 Ga are needed to settle the issue. The third explanation posits that no large hydrothermal inputs such as are required to produce BIFs occurred between 1.7 Ga and 0.7 Ga. This seems unlikely but not impossible.

6.21.4.2 The Mesoproterozoic (1.8–1.2 Ga)

Sedimentary rocks of the McArthur Basin in Northern Australia provide one of the best windows on the chemistry of the Mesoproterozoic ocean. Some 10 km of 1.6–1.7 Ga sediments accumulated in this intracratonic basin (Southgate *et al.*, 2000). In certain intervals, they contain giant strata-bound Pb–Zn–Ag mineral deposits (Jackson *et al.*, 1987; Jackson and Raiswell, 1991; Crick, 1992). The sediments have experienced only low grades of metamorphism.

Shen *et al.* (2002) have reported data for the isotopic composition of sulfur in carbonaceous shales of the lower part of the 1.72–1.73 Ga Wollogorang Formation and in the lower part of the 1.63–1.64 Ga Reward Formation of the McArthur Basin. These shales were probably deposited in a euxinic intracratonic basin with connection to the open ocean. The $\delta^{34}S$ of pyrite in black shales of the Wollogorang Formation ranges from −1‰ to +6.3‰ with a mean and SD of 4.0 ± 1.9‰ ($n = 14$). Donnelly and Jackson (1988) reported similar values. The $\delta^{34}S$ values of pyrite in the lower Reward Formation range from +18.2‰ to +23.4‰ with an average and SD of 18.4 ± 1.8‰ ($n = 10$). The spread of $\delta^{34}S$ values within each formation is relatively small. The sulfur is quite ^{34}S-enriched compared to compositions expected from the reduction of seawater sulfate with a $\delta^{34}S$ of 20–25‰ (Strauss, 1993). This is especially true of the sulfides in the Reward Formation. Shen *et al.* (2002) propose that the Reward data are best explained if the concentration of sulfate in the contemporary seawater was between 0.5 mmol kg^{-1} and 2.4 mmol kg^{-1}. Sulfate concentrations in the Mesoproterozoic ocean well below those of the present oceans have also been proposed on the basis of the rapid change in the value of $\delta^{34}S$ in carbonate associated sulfate of the 1.2 Ga Bylot Supergroup of northeastern Canada (Lyons *et al.*, 2002). However, the value of $m_{SO_4^{2-}}$ in Mesoproterozoic seawater is still rather uncertain.

Somewhat of a cross-check on the SO_4^{2-} concentration of seawater can be obtained from the evaporite relics in the McArthur Group

(Walker *et al.*, 1977). Up to 40% of the measured sections of the Amelia Dolomite consist of such relics in the form of carbonate pseudomorphs after a variety of morphologies of gypsum and anhydrite crystals, chert pseudomorphs after anhydrite nodules, halite casts, and microscopic remnants of original, unaltered sulfate minerals. Muir (1979) and Jackson *et al.* (1987) have pointed out the similarity of this formation to the recent sabkhas along the Persian Gulf coast. The pseudomorphs crosscut sedimentary features such as bedding and laminated microbial mats, suggesting that the original sulfate minerals crystallized in the host sediments during diagenesis.

Pseudomorphs after halite are common throughout the McArthur Group. The halite appears to have formed by almost complete evaporation of seawater in shallow marine environments and probably represents ephemeral salt crusts. The general lack of association of halite and calcium sulfate minerals in these sediments probably resulted in part from the dissolution of previously deposited halite during surface flooding, but also indicates that evaporation did not always proceed beyond the calcium sulfate facies.

This observation allows a rough check on the reasonableness of the Shen *et al.* (2002) estimate of the sulfate concentration in seawater during the deposition of the McArthur Group. On evaporating modern seawater, gypsum begins to precipitate when the degree of evaporation is ~3.8. As shown in Figure 13, the onset of gypsum and/or anhydrite precipitation occurs at progressively greater degrees of evaporation as the product $m_{Ca^{2+}}m_{SO_4^{2-}}$ in seawater decreases. Today $m_{Ca^{2+}}m_{SO_4^{2-}} = 280$ (mmol kg^{-1})^2. If this product is reduced to 23 (mmol kg^{-1})^2, anhydrite begins to precipitate simultaneously with halite at a degree of evaporation of 10.8. The presence of gypsum casts without halite in the sediments of the McArthur Group indicates that in seawater at that time $m_{Ca^{2+}}m_{SO_4^{2-}} > 23$ (mmol kg^{-1})^2 provided the salinity of seawater was the same as today. If $m_{SO_4^{2-}}$ was 2.4 mmol kg^{-1}, the upper limit suggested by Shen *et al.* (2002), $m_{Ca^{2+}}$, must then have been >10 mmol kg^{-1}, the concentration of Ca^{2+} in modern seawater. An SO_4^{2-} concentration of 2.4 mmol kg^{-1} is, therefore, permissible. Sulfate concentrations as low as 0.5 mmol kg^{-1} require what are probably unreasonably high concentrations of Ca^{2+} in seawater to account for the precipitation of gypsum before halite in the McArthur Group sediments.

The common occurrence of dolomite in the McArthur Group indicates that the $m_{Mg^{2+}}/m_{Ca^{2+}}$ ratio in seawater was >1 (see below). This is also indicated by the common occurrence of aragonite as the major primary $CaCO_3$ phase of sediments on Archean and Proterozoic carbonate platforms (Grotzinger, 1989; Winefield, 2000). Although these hints regarding the composition of Mesoproterozoic seawater are welcome, they need to be confirmed and expanded by analyses of fluid inclusions in calcite cements.

Perhaps the most interesting implication of the close association of gypsum, anhydrite, and halite relics in the McArthur Group is that the temperature during the deposition of these minerals was not much above 18 °C, the temperature at which gypsum, anhydrite, and halite are stable together (Hardie, 1967). At higher temperatures anhydrite is the stable calcium sulfate mineral in equilibrium with halite. The coexistence of gypsum and anhydrite with halite suggests that the temperature

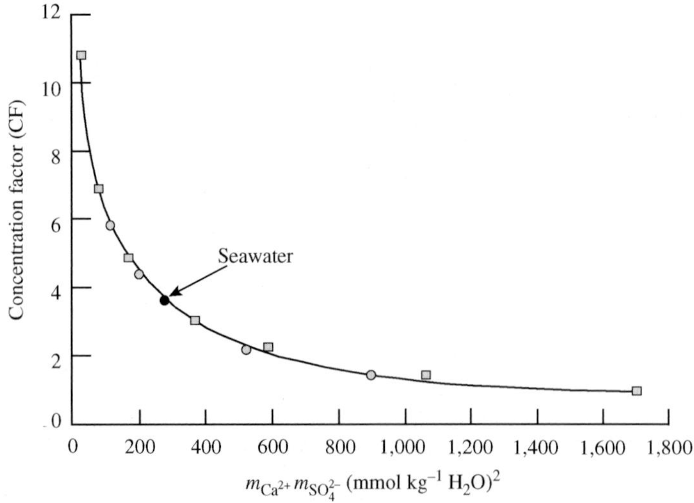

Figure 13 The relationship between the value of the product $m_{Ca^{2+}}$ $m_{SO_4^{2-}}$ in seawater and the concentration factor at which seawater becomes saturated with respect to gypsum at 25 °C and 1 atm (source Holland, 1984).

during their deposition was possibly lower but probably no higher than in the modern sabkhas of the Persian Gulf, where anhydrite is the dominant calcium sulfate mineral in association with halite (Kinsman, 1966).

In their paper on the carbonaceous shales of the McArthur Basin, Shen *et al.* (2002) comment that euxinic conditions were common in marine-connected basins during the Mesoproterozoic, and they suggest that low concentrations of seawater sulfate and reduced levels of atmospheric oxygen at this time are compatible with euxinic deep ocean waters. Anbar and Knoll (2002) echo this sentiment. They point out that biologically important trace metals would then have been scarce in most marine environments, potentially restricting the nitrogen cycle, affecting primary productivity, and limiting the ecological distribution of eukaryotic algae. However, some of the presently available evidence does not support the notion of a Mesoproterozoic euxinic ocean floor. Figure 14 shows that the redox sensitive elements molybdenum, uranium, and rhenium are well correlated with the organic carbon content of carbonaceous shales in the McArthur Basin. The slope of the correlation lines is close to that in many Phanerozoic black shales, suggesting that the concentration of these elements in McArthur Basin seawater was comparable to their concentration in modern seawater. Preliminary data for the isotopic composition of molybdenum in the Wollogorang Formation of the McArthur Basin (Arnold *et al.*, 2002) suggest somewhat more extensive sulfidic deposition of molybdenum in the Mesoproterozoic than in the modern oceans. Their data may, however, reflect a greater extent of shallow water euxinic basins rather than an entirely euxinic ocean floor. The good correlation of the concentration of sulfur and total iron in the McArthur Basin shales (Shen *et al.*, 2002) confirms the euxinic nature of the Basin; the large value of the ratio of sulfur to total iron indicates that this basin cannot have been typical of the oceans as a whole. Additional data for the concentration of redox sensitive elements in carbonaceous shales and more data for the isotopic composition of molybdenum and perhaps of copper in carbonaceous shales will probably clarify and perhaps settle the questions surrounding the redox state of the deep ocean during the Mesoproterozoic.

6.21.4.3 The Neoproterozoic (1.2–0.54 Ga)

After what appears to have been a relatively calm and uneventful climatic, atmospheric, and marine history during the Mesoproterozoic, the

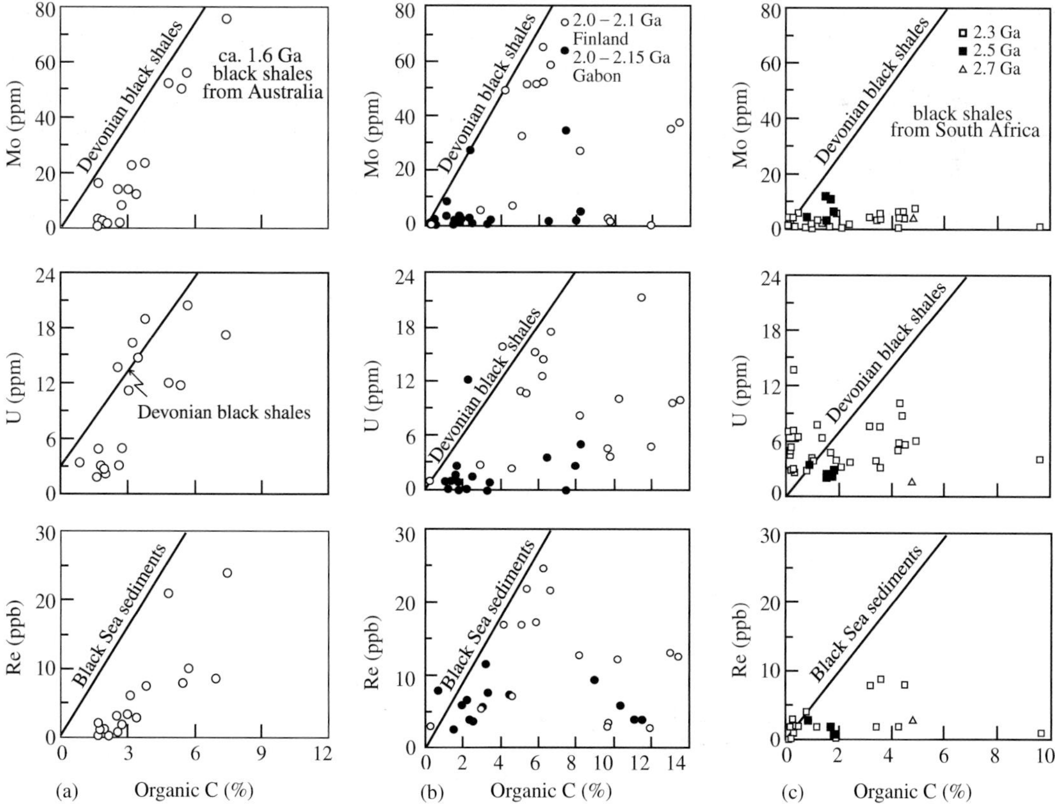

Figure 14 The concentration of Mo, U, and Re in carbonaceous shales: (a) McArthur Basin, Australia, 1.6 Ga; (b) Finland and Gabon, 2.0–2.15 Ga; and (c) South Africa, ≥2.3 Ga.

Neoproterozoic returned to the turbulence of the Paleoproterozoic era. The last 300 Ma of the Proterozoic were times of extraordinary global environmental and biological change. Major swings in the $\delta^{13}C$ value of marine carbonates were accompanied by several very large glaciations, the sulfate content of seawater rose to values comparable to that of the modern oceans (Horita *et al.*, 2002), and the level of atmospheric O_2 probably attained modern values by the time of the biological explosion at the end of the Precambrian and the beginning of the Paleozoic. A great deal of research has been done on the last few hundred million years of the Proterozoic, stimulated in part by the discovery of the extensive glacial episodes of this period. Nevertheless, many major questions remain unanswered. The description of the major events and particularly their causes are still quite incomplete.

Figure 15 is a recent compilation of measurements of the $\delta^{13}C$ values of marine carbonates between 800 Ma and 500 Ma (Jacobsen and Kaufman, 1999). The $\delta^{13}C$ values experienced a major positive excursion interrupted by sharp negative spikes. The details of these spikes are still quite obscure (see, e.g., Melezhik *et al.*, 2001), but the negative excursions associated with major glaciations between ca. 720 Ma and 750 Ma and between 570 Ma and 600 Ma were almost certainly separated by a 100 Ma plateau of very strongly positive $\delta^{13}C$ values (see, e.g., Walter *et al.*, 2000; Shields and Veizer, 2002; Halverson, 2003). In some ways the Late Neoproterozoic positive $\delta^{13}C$ excursion is reminiscent of the Paleoproterozoic excursion between 2.22 Ga and 2.06 Ga. Their duration and magnitude are similar. However, the Paleoproterozoic ice ages preceded the large positive $\delta^{13}C$ excursion, whereas in the Neoproterozoic they occur close to the beginning and close to the end of the excursion. Figure 15 can be used to estimate the "excess O_2" produced during the Neoproterozoic excursion. The average value of $\delta^{13}C$ between 700 Ma and 800 Ma was +3.0‰; between 595 Ma and 700 Ma it was +8.0‰, and between 540 Ma and 595 Ma it was −0.4‰. The average $\delta^{13}C$ value for the period between 595 Ma and 800 Ma was +5.6‰. The variation of $\delta^{13}C$ during this time interval has been defined much more precisely by Halverson (2003). Although his $\delta^{13}C$ curve differs considerably from that of Jacobsen and Kaufman (1999), Halverson's (2003) average value of $\delta^{13}C$ between 595 Ma and 800 Ma is very similar to that of Jacobsen and Kaufman (1999). This indicates that ~40% of the carbon in the sediments of this period were deposited as a constituent of organic matter.

The rate of excess O_2 generation was probably $\sim 6 \times 10^{12}$ mol yr^{-1}, and the total excess O_2 produced between 600 Ma and 800 Ma was $\sim 12 \times 10^{20}$ mol. This quantity is ~30 times the O_2 content of the present atmosphere, 0.4×10^{20} mol. O_2 buildup in the atmosphere could, therefore, have been only a small part of the effect of the large $\delta^{13}C$ excursion. The excess O_2

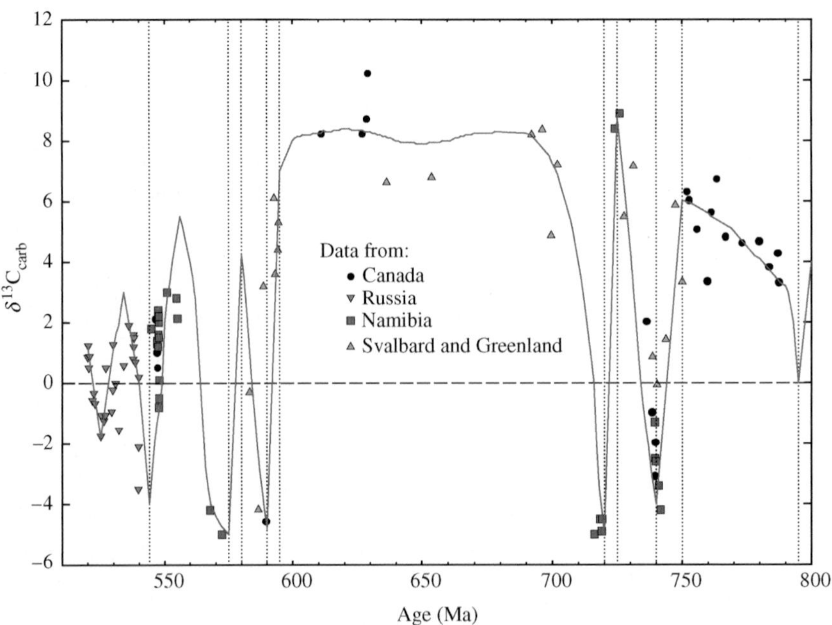

Figure 15 Temporal variations in $\delta^{13}C$ values of marine carbonates between 800 Ma and 500 Ma (source Jacobsen and Kaufman, 1999).

is also much larger than 0.8×10^{20} mol, the quantity required to raise the SO_4^{2-} concentration of seawater from zero to its present value by oxidizing sulfide. Fortunately, additional sulfate sinks are available to account for the estimated excess O_2: $CaSO_4$ and $CaSO_4 \cdot 2H_2O$ in evaporites, $CaSO_4$ precipitated in the oceanic crust close to MORs during the cycling of seawater at hydrothermal temperatures, and an increase in the Fe_2O_3/FeO ratio in sedimentary rocks. These sinks seem to be of the right order of magnitude to account for the use of the excess O_2. The magnitude of the $CaSO_4$ reservoir in sedimentary rocks during the last part of the Neoproterozoic has been estimated on the basis of models based on sulfur isotope data to be $(2 \pm 0.5) \times 10^{20}$ mol (Holser *et al.*, 1989). The conversion of this quantity of sulfur from sulfide to sulfate requires $(4 \pm 1) \times 10^{20}$ mol O_2.

At present the loss of $CaSO_4$ from seawater to the oceanic crust seems to be $\sim 1.0 \times 10^{12}$ mol yr^{-1} (Holland, 2002). At this rate the loss of SO_4^{2-} to the oceanic crust between 600 Ma and 800 Ma would have been 2×10^{20} mol. The total O_2 sinks due to the sulfur cycle during this period might, therefore, have amounted to $\sim 6 \times 10^{20}$ mol. The increase in the Fe_2O_3/FeO ratio in sedimentary rocks probably required $\sim 1 \times 10^{20}$ mol O_2. Given all the rather large uncertainties and somewhat shaky assumptions which have been made in this mass balance calculation, the agreement between the estimated quantity of excess O_2 and the estimated quantity of O_2 required to convert the sulfur cycle from its pre-1,200 Ma state to its state at the beginning of the Paleozoic is quite reasonable. The logic behind the change is also compelling. Carbon, iron, and sulfur are the three elements which dominate the redox state of the near-surface system. The carbon cycle seems to have been locked into its present state quite early in Earth history, probably by its linkage to the geochemical cycle of phosphorus. The iron cycle took on a more modern cast during the positive Paleoproterozoic $\delta^{13}C$ excursion. It is not unreasonable to propose that the positive $\delta^{13}C$ excursion during the Neoproterozoic was responsible for converting the sulfur cycle to its modern mode and for generating a further increase in the Fe_2O_3/FeO ratio.

Two questions now come to mind: (i) What triggered the Neoproterozoic $\delta^{13}C$ excursion? and (2) Are the strong negative excursions due to instabilities inherent in the long positive excursion? The answers that have been given to both questions are still speculative, but it seems worthwhile to attempt a synthesis. The $\delta^{13}C$ excursion was accompanied by the reappearance of BIFs (Klein and Beukes, 1993), which are related to glacial periods but in a somewhat irregular manner (Young, 1976; James, 1983). They are widely distributed, and their tonnage is significant. Their reappearance virtually demands that the deeper parts of the oceans were anoxic. If, as suggested earlier, the deep oceans were oxidized during the Mesoproterozoic, they returned to their pre-1.7 Ga state during the last part of the Neoproterozoic. One possible cause for this return is the appearance of organisms which secreted SiO_2 or $CaCO_3$, that could serve as ballast for particulate organic matter. Recently discovered vase-shaped microfossils (VSMs) in the Chuar Group of the Grand Canyon could be members of one of these groups. The fossils appear to be testate amoebae (Porter and Knoll, 2000; Porter *et al.*, 2003). The structure and composition of testate amoebae tests are similar to those inferred for the VSMs. A number of testate amoebae have agglutinated tests; others have tests in which internally synthesized 1 μm to >10 μm scales of silica are arrayed in a regular pattern (Figure 16). The age of the VSM fossils in the Grand Canyon and in the Mackenzie Mountain Supergroup, NWT is between 742 ± 7 Ma and ca. 778 Ma. Their presence at this time suggests that they or other SiO_2-secreting organisms could have supplied ballast for the transport of particulate organic matter into the deep ocean near the beginning of the Neoproterozoic $\delta^{13}C$ excursion. If the O_2 content of the atmosphere at that time was still significantly less than today, the flux of organic matter required to make the deep oceans anoxic would only need to have been a small fraction of the present-day flux. An increase in the flux of organic matter to the deep ocean probably followed the Cambrian explosion (Logan *et al.*, 1995).

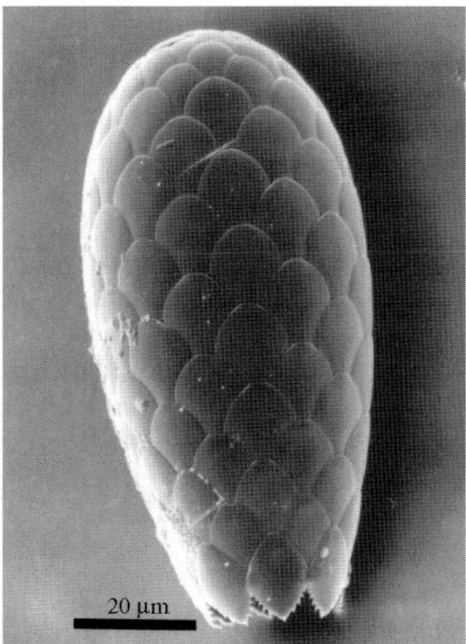

Figure 16 The test of *Euglypha tuberculata*. Note regularly arranged siliceous scales: scale bar, 20 μm (courtesy of Ralf Meisterfeld).

Arguments can be raised against the importance of the evolution of SiO_2-secreting organisms as ballast for organic matter. The VSM organisms were shallow water and benthic, not open ocean and planktonic, and the remains of SiO_2-secreting organisms have not been found in Neoproterozoic deep-sea sediments. Alternative explanations have been offered for the burial of "excess" organic carbon between 800 Ma and 600 Ma. Knoll (1992) and Hoffman *et al.* (1998) pointed out that the Late Proterozoic was a time of unusual, if not unique, formation of rapidly subsiding extensional basins flooded by marine waters. Organic matter buried with rapidly deposited sediments is preserved more readily than in slowly deposited sediments (Suess, 1980). However, the consequences of this effect on the total rate of burial of organic matter are small unless additional PO_4^{3-} becomes available. Anoxia would tend to provide the required addition (Colman and Holland, 2000). However, the evidence for anoxia is still limited. The reappearance of BIFs demands deep-ocean anoxia during their formation, but the Neoproterozoic BIFs are associated with the major glaciations, which may indicate—but surely does not prove—that deep-water anoxia between 800 Ma and 550 Ma was restricted to these cold periods. Other indications of deep-water anoxia are needed to define the oxidation state of deep water and the causes of anoxia during the Neoproterozoic.

In a low-sulfate ocean, anoxia would probably have increased the rate of methane production in marine sediments. Some of this methane probably escaped into the water column. In the deeper, O_2-free parts of the oceans, methane was oxidized in part by organisms using SO_4^{2-} as the oxidant. In the upper parts of the oceans, it was partly oxidized by O_2. The remainder escaped into the atmosphere. There it was in part decomposed and oxidized inorganically to CO_2; in part it was returned to the ocean in nonmethanogenic areas and was oxidized there biologically. Some methane generated in marine sediments was probably sequestered at least temporarily in methane clathrates. The methane concentration in the Neoproterozoic atmosphere could have been as high as 100–300 ppm (Pavlov *et al.*, 2003). If methane concentrations were as high as this, it would have been a very significant greenhouse gas. At a concentration of 100 ppm, methane could well have increased the surface temperature by 12 °C (Pavlov *et al.*, 2003).

It seems strange, therefore, that the Late Neoproterozoic should have been a time of severe glaciations (Kirschvink, 1992). In their review of the snowball Earth hypothesis, Hoffman and Schrag (2002) point out that in some sections a steep decline in $\delta^{13}C$ values by 10–15‰ preceded any physical evidence of glaciation or sea-level fall. The most likely explanation for such a sharp drop involves a major decrease in the burial rate of organic carbon. The reason(s) for this are still obscure. It is intriguing that major periods of phosphogenesis coincided roughly with the glaciations between 750–800 Ma and ca. 620 Ma (Cook and McElhinny, 1979). The first of these appears to coincide with the appearance of Rapitan-type iron ores. The second seems to have commenced shortly after the latest Neoproteroizc glaciation (Hambrey and Harland, 1981), reached a peak during the Early Cambrian, then declined rapidly, and finally ended in the Late Mid-Cambrian (Cook and Shergold, 1984, 1986). However, the time relationship between the glacial and the phosphogenic events is still not entirely resolved. The rapid removal of phosphate from the oceans may have begun before the onset of glaciation. Perhaps continental extension and rifting during the Neoproterozoic and the creation of many shallow epicontinental seaways at low paleolatitudes created environments that were particularly favorable for the deposition of phosphorites (Donnelly *et al.*, 1990).

Some of the Neoproterozoic BIFs and associated rock units are also quite enriched in P_2O_5. For instance, the P_2O_5 content of the Rapitan iron formation in Canada and its associated hematitic mudstones ranges from 0.49% to 2.16% (Klein and Beukes, 1993). It can readily be shown, however, that the phosphate output from the oceans into the known phosphorites and BIFs is a small fraction of the phosphate metabolism of the oceans as a whole between 800 Ma and 600 Ma. This observation does not eliminate the possibility that phosphate removal into other sedimentary rocks was abnormally rapid during the two phosphogenic periods between 800 Ma and 600 Ma. Such abnormally rapid phosphate removal as a constituent of apatite would have decreased the availability of phosphate for deposition with organic matter. This would have produced a decrease in the $\delta^{13}C$ value of carbonates as observed before the onset of the snowball Earth glaciations. It would probably also have reduced the rate of methane generation, the methane concentration in the atmosphere, and the global temperature. If this, in turn, led to the onset of glaciation, the decrease in the rate of weathering would have further restricted the riverine flow of phosphate and thence to a decrease in $\delta^{13}C_{carb}$ and the global temperature. This scenario is highly speculative, but it does seem to account at least for the onset of the glaciations.

It seems very likely that the continents were largely ice covered during the Neoproterozoic glaciations. The state of the oceans is still a matter of debate. The Hoffman–Schrag Snowball Earth hypothesis posits that the oceans were completely, or nearly completely, ice covered. This seems unlikely. Leather *et al.* (2002) have pointed out that the sedimentology and stratigraphy of the Neoproterozoic glacials of Arabia were more like those of the familiar oscillatory glaciations of the Pleistocene than those required by the Snowball Earth hypothesis. Similarly, Condon *et al.* (2002), who studied the stratigraphy and sedimentology of six Neoproterozoic glaciomarine successions, concluded that the Neoproterozoic seas were not totally frozen, and that the hydrologic cycle was functioning during the major glaciations. This suggests that the tropical oceans were ice free or only partially ice covered. Perhaps an Earth in such a state might be called a frostball, rather than a snowball. Chemical weathering on the continents in this state would have been very minor. CO_2 released from volcanoes would have built up in the atmosphere and in the oceans until its partial pressure was high enough to overcome the low albedo of the Earth at the height of the glacial episodes. In the Hoffman–Schrag model, some 10 Ma of CO_2 buildup are needed to raise the atmospheric CO_2 pressure sufficiently to overcome the low albedo of a completely ice-covered Earth. The only test of this timescale has been provided by the Bowring *et al.* (2003) data for the duration of the Gaskiers glacial deposits in Newfoundland. This unit is often described as a Varanger-age glaciomarine deposit. It is locally overlain by a thin cap carbonate bed with a highly negative carbon isotopic signature. The U–Pb geochronology of zircons separated from ash beds below, within, and above the glacial deposits indicates that these glacial deposits accumulated in less than 1 Ma. The short duration of this episode may be more consistent with a frostball than with a snowball Earth.

At the very low stand of sea level during the heights of these glaciations, the release of methane from clathrates might have contributed significantly to the subsequent warming and to the negative value of $\delta^{13}C_{CARB}$ in the ocean–atmosphere system (Kennedy *et al.*, 2001). Intense weathering after the retreat of the glaciers would—among other products—have released large quantities of phosphate. This might have speeded the recovery of photosynthesis and the return of the $\delta^{13}C$ of marine carbonates to their large positive values along the course of the 800–600 Ma positive $\delta^{13}C$ excursion.

Between the end of the positive $\delta^{13}C$ excursion and the beginning of the Cambrian, several large marine evaporites were deposited. These are still preserved and offer the earliest opportunity to use the mineralogy of marine evaporites and the composition of fluid inclusions in halite to reconstruct the composition of the contemporaneous seawater. Horita *et al.* (2002) have used this approach to show that the sulfate concentration in latest Neoproterozoic seawater was ~ 23 mmol kg^{-1}, i.e., only slightly less than in modern seawater, and significantly greater than during some parts of the Phanerozoic. By the latest Proterozoic, a new sulfur regimen had been installed. The cycling of seawater through MORs had probably reached present-day levels, the S/C ratio in volcanic gases had therefore risen, and the proportion of volcanic sulfur converted to constituents of sulfides and sulfates had approached unity as demanded by the composition of average Phanerozoic volcanic gases (Holland, 2002). The conversion of the composition of sedimentary rocks from their pre-GOE to their modern composition had been nearly completed.

The level of atmospheric O_2 seems to have been the last of the redox parameters to approach modern values. The evidence for this comes from the changes in the biota that occurred between the latest Proterozoic and the middle of the Cambrian period. These changes are discussed in the next section. Although a good deal of progress has been made since 1980 in our understanding of the atmosphere and oceans during the Proterozoic Era, we are still woefully ignorant of even the most basic oceanographic data for Precambrian seawater.

6.21.5 THE PHANEROZOIC

6.21.5.1 Evidence from Marine Evaporites

Our understanding of the chemical evolution of Phanerozoic seawater has increased enormously since the end of World War II. Rubey's (1951) presidential address to the Geological Society of America was aptly entitled "The geologic history of seawater, an attempt to state the problem." During the following year, Barth (1952) introduced the concept of the characteristic time in his analysis of the chemistry of the oceans, and this can, perhaps, be considered the beginning of the application of systems analysis to marine geochemistry. Attempts were made by the Swedish physical chemist Lars Gunnar Sillén to apply equilibrium thermodynamics to define the chemical history of seawater, but, as he pointed out, "practically everything that interests us in and around the sea is a symptom of nonequilibrium… What we can hope is that an equilibrium model may give a useful first approximation to the real system, and that the deviations of the real system may be treated as disturbances" (Sillén, 1967). Similar sentiments were expressed by Mackenzie and Garrels (1966) and by Garrels and Mackenzie (1971). They proposed that there has been little change in seawater composition since 1.5–2 Ga, although they were concerned by the discovery by Ault and Kulp (1959), Thode *et al.* (1961),

Thode and Monster (1965), and Holser and Kaplan (1966) of very significant fluctuations in the isotopic composition of sulfur in seawater during the Phanerozoic.

A good deal of optimism regarding the constancy or near constancy of the composition of seawater during the Phanerozoic was, however, permitted by the Holser (1963) discovery that the Mg/Cl and Br/Cl ratios in brines extracted from fluid inclusions in Permian halite from Hutchinson, Kansas were close to those of modern brines. Holland (1972) published his analysis of the constraints placed by the constancy of the early mineral sequence in marine evaporites during the Phanerozoic. This paper showed that most of the seawater compositions permitted by the precipitation sequence $CaCO_3$–$CaSO_4$–NaCl in marine evaporites fall within roughly twice and half of the concentration of the major ions in seawater today. These calculations were extended by Harvie *et al.* (1980) and Hardie (1991) to include the later, more complex mineral assemblages of marine evaporites. Their calculations explained the mineral sequence in modern evaporites, and confirmed that the composition of Permian seawater was similar to that of modern seawater.

The development of a method to extract brines from fluid inclusions in halite and to obtain quantitative analyses by means of ion chromatography (Lazar and Holland, 1988) led to a study of the composition of trapped brines in halite from several Permian marine evaporites (Horita *et al.*, 1991). Their results together with those of Stein and Krumhansl (1988) confirmed that the composition of modern seawater is similar to that of Permian seawater, and suggested that the composition of seawater has been quite conservative during the Phanerozoic. This suggestion, however, has turned out to be far off the mark. Analyses of fluid inclusions in halite of the Late Silurian Salina Group of the Michigan Basin (Das *et al.*, 1990), the Middle Devonian Prairie Formation in the Saskatchewan Basin (Horita *et al.*, 1996), and a growing number of other marine evaporites (for a summary of the results of other groups, see Horita *et al.* (2002)) have shown that the similarity between Permian and modern seawater is the exception rather than the rule. Only the fluid inclusions in halite of the latest Neoproterozoic Ara Formation in Oman (Horita *et al.*, 2002) are similar to their Permian and modern equivalents. All of the other fluid inclusions contain brines which are very significantly depleted in Mg^{2+} and SO_4^{2-} relative to modern seawater. Their composition is consistent with the mineralogy of the associated evaporites. The difference between these inclusion fluids and their modern counterparts is either due to significant differences in the composition of the seawater from which they were derived or to reactions which depleted the Mg^{2+} and SO_4^{2-} content of the brines along their evaporation path. The Silurian and Devonian evaporites which we studied are associated with large carbonate platforms. The dolomitization of $CaCO_3$ followed by the deposition of gypsum and/or anhydrite during the passage of seawater across such platforms can deplete the evaporating brines in Mg^{2+} and SO_4^{2-}. Their composition can then become similar to that of the brines in the Silurian and Devonian halites. We opted for this interpretation of the fluid inclusion data, wrongly as it turned out. An obvious test of the proposition that the differences were due to dolomitization and $CaSO_4$ precipitation was to analyze fluid inclusions in halite from marine evaporites which are not associated with extended carbonate platforms. Zimmermann's (2000) work on Tertiary evaporites did just that. Her analyses showed that in progressively older Tertiary evaporites the composition of the seawater from which the brines in these evaporites developed was progressively more depleted in Mg^{2+} and SO_4^{2-} (Figure 17). As shown in Figure 18, this trend continued into the Cretaceous and was not reversed until the Triassic or latest Permian. The composition of fluid inclusion brines in marine halite is, therefore, a reasonably good guide to the composition of their parent seawater. However, changes due to the reaction of evaporating seawater with the sediments across which it passes en route to trapping have almost certainly occurred and cannot be neglected.

6.21.5.2 The Mineralogy of Marine Oölites

The proposed trend of the Mg/Ca ratio of seawater (Figure 17) during the Tertiary is supported by two independent lines of evidence: the mineralogy of marine oölites and the magnesium content of foraminifera. Sandberg (1983, 1985) discovered that the mineralogy of marine oölites has alternated several times between dominantly calcitic and dominantly aragonitic (see Figure 18). On this basis he divided the Phanerozoic into periods of calcitic and aragonitic seas. He suggested that the changes in mineralogy were related to changes in atmospheric P_{CO_2} or to changes in the Mg/Ca ratio of seawater. The experiments by Morse *et al.* (1997) have shown that changes in the Mg/Ca ratio are the most likely cause of the changes in oölite mineralogy. The most recent switch in oölite mineralogy occurred near the base of the Tertiary (see Figures 17 and 18). Unfortunately, the Mg/Ca ratio of seawater at the time of this switch can be defined only roughly, because the change in oölite mineralogy from calcite to aragonite depends on temperature as well as on the Mg/Ca ratio of

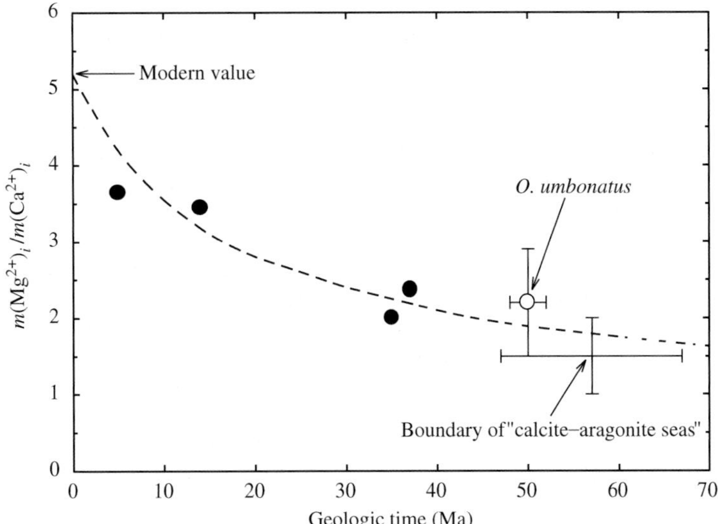

Figure 17 $m(Mg^{2+})_i/m(Ca^{2+})_i$ ratio in seawater during the Tertiary based on analyses of fluid inclusions in marine halite (● and dashed line), compared with data based on the Mg/Ca ratio of *O. umbonatus* (○) (Lear *et al.*, 2000) and the boundary of "aragonite-calcite seas" of Sandberg (1985) (source Horita *et al.*, 2002).

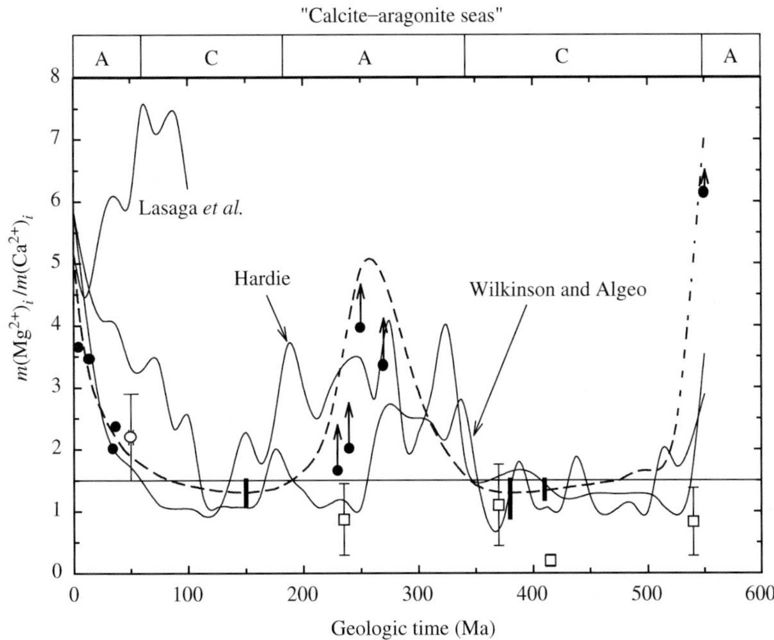

Figure 18 $m(Mg^{2+})_i/m(Ca^{2+})_i$ ratio in seawater during the Phanerozoic based on analyses of fluid inclusions in marine halite (solid symbols and dashed line), compared with the data based on Mg/Ca of *O. umbonatus* (○) (Lear *et al.*, 2000) and of abiogenic marine carbonate cements (□) (Cicero and Lohmann, 2001). Also shown are the results of modeling by Lasaga *et al.* (1985), Wilkinson and Algeo (1989), and Hardie (1996). "A" and "C" at the top indicate "aragonite seas" and "calcite seas" of Sandberg (1985) (source Horita *et al.*, 2002).

seawater, and probably also on other compositional and kinetic factors.

6.21.5.3 The Magnesium Content of Foraminifera

The magnesium content of foraminifera supplies another line of evidence in support of a low Mg/Ca ratio in Early Tertiary seawater. Lear *et al.* (2000) found that the magnesium content of *O. umbonatus* decreased progressively with increasing age during the Tertiary. This change in composition is due in large part to changes in seawater temperature and in the Mg/Ca ratio of seawater, but it can also be overprinted by diagenetic processes. Much more data are needed to define the course of the Mg/Ca ratio of seawater on the basis of paleontologic data, but it is encouraging that its rather uncertain course during the Tertiary is consistent with that derived from the two other lines of evidence.

The correlation between the mineralogy of marine oölites and the Mg/Ca ratio of seawater as inferred from the composition of fluid inclusion brines extends throughout the Phanerozoic. This is also true for the correlation of the temporal distribution of the taxa of major calcite and aragonite reef builders and the temporal distribution of KCl-rich and $MgSO_4$-rich marine evaporites (Figure 19; Stanley and Hardie, 1998). There is no reason, therefore, to doubt that the composition of seawater has changed significantly during the Phanerozoic. As shown in Figures 20–23, only the concentration of potassium seems to have remained essentially constant. Lowenstein *et al.* (2001) have confirmed these trends and have shown that ancient inclusion fluids in halite had somewhat lower Na^+ concentrations and higher Cl^- concentrations during halite precipitation than present-day halite-saturated seawater brines.

Figures 20–23 compare the changes in the composition of seawater during the Phanerozoic that have been proposed by various authors since the early 1980s. Some of these differences are sizable. The Berner–Lasaga–Garrels (BLAG) box model developed by Berner *et al.* (1983) and Lasaga *et al.* (1985) included most of the major geochemical processes that affect the composition of seawater. Their changes in the concentration of Mg^{2+} and Ca^{2+} during the past 100 Ma are, however, much smaller than the estimates of Horita *et al.* (2002). The differences are due, in part, to the absence of dolomite as a major sink of Mg^{2+} in the BLAG model. However, Wallmann (2001) proposed concentrations of Ca^{2+} much higher than those of Horita *et al.* (2002), because he assumed, as an initial value in his model, that at 150 Ma the Ca^{2+} concentration in seawater was twice that of modern seawater.

The effect of penecontemporaneous dolomite deposition on the chemical evolution of seawater was included in the model published by Wilkinson and Algeo (1989), which was based on Given and Wilkinson's (1987) compilation of the distribution of limestones and dolomites in Phanerozoic sediments. Their calculations suggest that the concentration of Ca^{2+} in seawater remained relatively constant (±20%) during the Phanerozoic, but that the concentration of Mg^{2+} changed significantly, largely in phase with their proposed dolomite-age curve. The changes that they proposed for the concentration of both elements differ significantly from those of

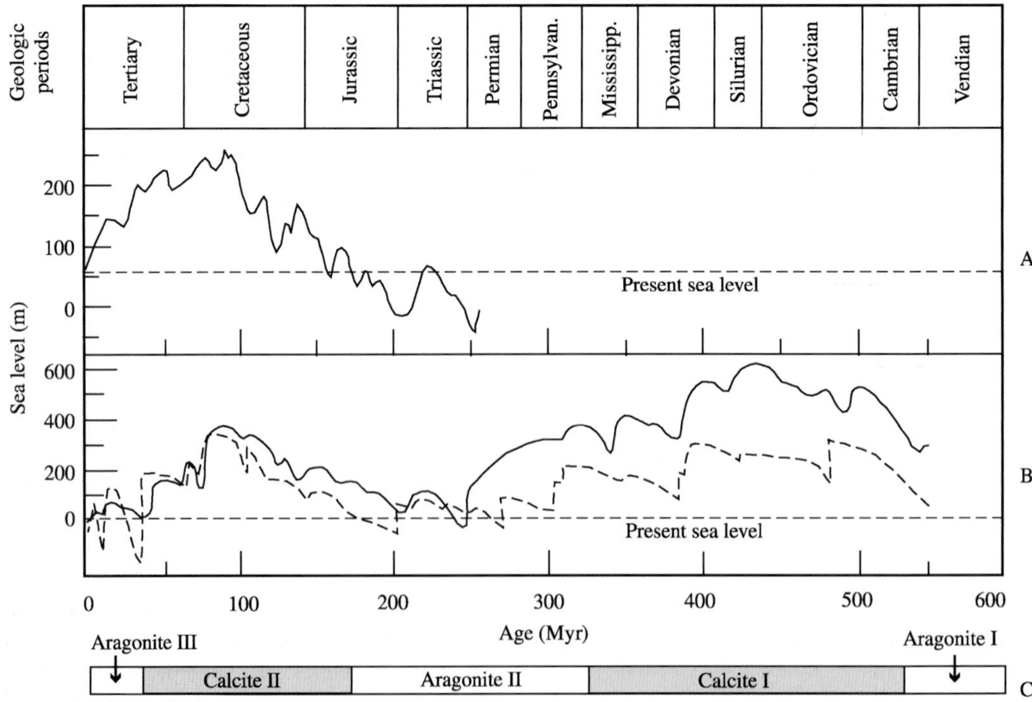

Figure 19 Sea-level changes and secular variations in the mineralogy of marine carbonates (Holland and Zimmermann, 2000): A—mean sea level during the Mesozoic and Cenozoic (Haq *et al.*, 1987); B—mean sea level during the Phanerozoic ((– –) Vail *et al.*, 1977; (—) Hallam, 1984); and C—secular variation in the mineralogy of Phanerozoic nonskeletal marine carbonates (source Stanley and Hardie, 1998).

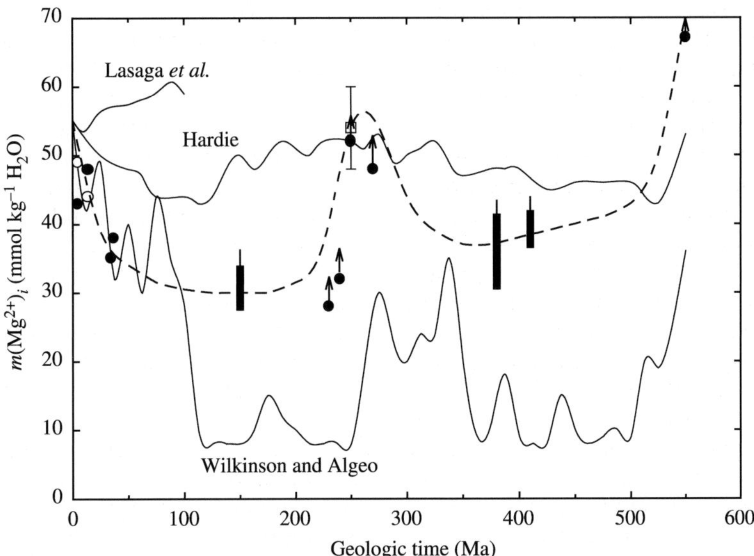

Figure 20 Concentration of Mg^{2+} in seawater during the Phanerozoic based on analyses of fluid inclusions in marine halite (solid symbols): thick and thin vertical bars are based on the assumption of different values for $m(Ca^{2+})_i/m(SO_4^{2-})_i$. Dashed line is our best estimate of age curve: (□)—Horita *et al.* (1991) and (○)—Zimmermann (2000). Also shown are the results of modeling by Lasaga *et al.* (1985), Wilkinson and Algeo (1989), and Hardie (1996) (source Horita *et al.*, 2002).

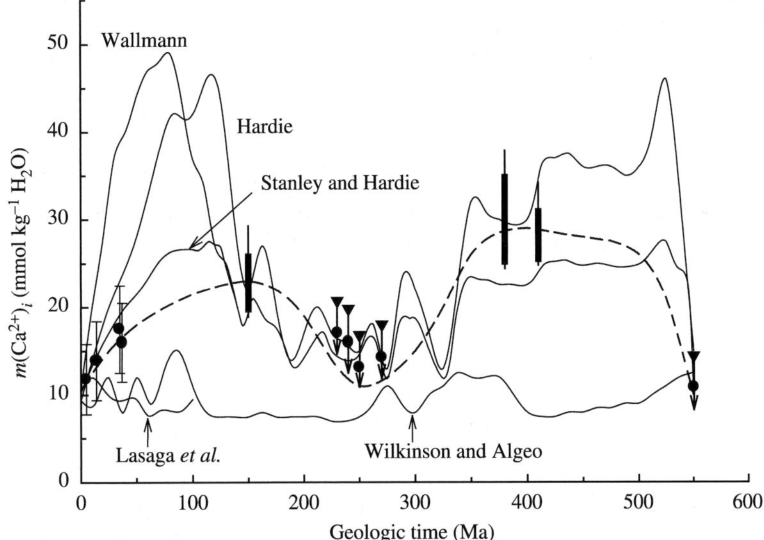

Figure 21 Concentration of Ca^{2+} in seawater during the Phanerozoic based on analyses of fluid inclusions in marine halite (solid symbols): circles–triangles and thick–thin vertical bars are based on the assumption of different values for $m(Ca^{2+})_i m(SO_4^{2-})_i$. Dashed line is our best estimate of age curve. Also shown are the results of modeling by Lasaga *et al.* (1985), Wilkinson and Algeo (1989), Hardie (1996), Stanley and Hardie (1998), and Wallmann (2001) (source Horita *et al.*, 2002).

Horita *et al.* (2002), in part because of their incomplete compilation of Phanerozoic carbonate rocks (Holland and Zimmermann, 2000).

6.21.5.4 The Spencer–Hardie Model

A very different approach to estimating the composition of seawater during the Phanerozoic was taken by Spencer and Hardie (1990) and Hardie (1996). These authors proposed that the composition of seawater was determined by the mixing ratio of river water and mid-ocean ridge solutions coupled with the precipitation of solid $CaCO_3$ and SiO_2 phases. They accepted Gaffin's (1987) curve for the secular variation of ocean crust production, which is based on the Exxon first-order

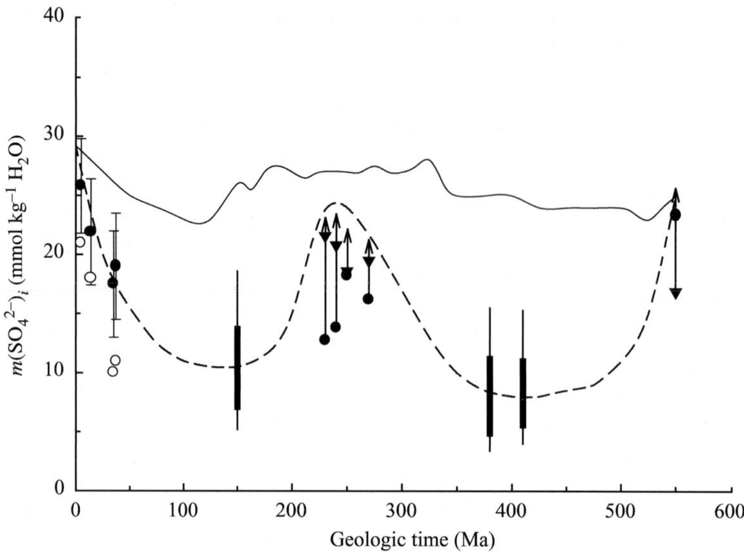

Figure 22 Concentration of SO_4^{2-} in seawater during the Phanerozoic based on analyses of fluid inclusions in marine halite (solid symbols): circles–triangles and thick–thin vertical bars are based on the assumption of different values for $m(Ca^{2+})_i m(SO_4^{2-})_i$. Dashed line is our best estimate of age curve. Data in open and filled circles are from Zimmermann (2000). Also shown are the results of modeling by Hardie (1996) (source Horita *et al.*, 2002).

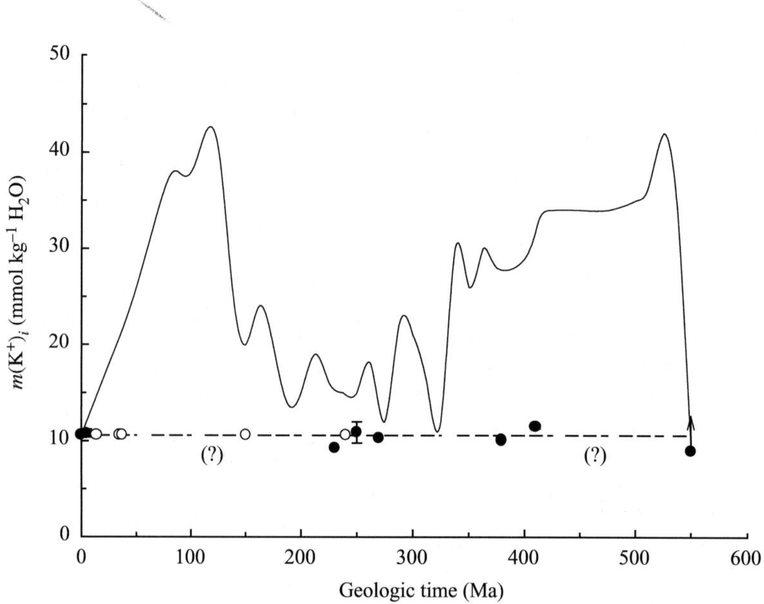

Figure 23 Concentration of K^+ in seawater during the Phanerozoic based on analyses of fluid inclusions in marine halite (solid symbols): circles–triangles and thick–thin vertical bars are based on the assumption of different values for $m(Ca^{2+})_i m(SO_4^{2-})_i$. Dashed line is our best estimate of age curve. Open circles are from Zimmermann (2000). Also shown are the results of modeling by Hardie (1996) (source Horita *et al.*, 2002).

global sea-level curve (Vail *et al.*, 1977). They assumed that, as a first approximation, the MOR flux of hydrothermal brines has scaled linearly with the rate of ocean crust production as estimated from the sea-level curves. Spencer and Hardie (1990) and Hardie (1996) then applied their mixing model to estimate the course of the composition of seawater during the Phanerozoic.

The sea-level curves in Figure 19 have two maxima. In the Hardie (1996) model these maxima are taken to coincide with maxima in the rate of seawater cycling through MORs and hence to minima in the concentration of Mg^{2+} and SO_4^{2-} in the contemporaneous seawater. To that extent the predictions of the Hardie (1996) model agree well with the fluid inclusion data and represented a distinct advance in understanding the chemical evolution of seawater during the Phanerozoic.

There are, however, rather serious discrepancies between the Hardie (1996) model and the fluid inclusion data of Horita *et al.* (2002). The most glaring is the difference between the large variations in the K^+ concentration predicted by Hardie (1996) and the essentially constant value of the K^+ concentration indicated by the fluid inclusion data. The reasons for the near constancy of the K^+ concentration in Phanerozoic seawater are not completely understood. They must, however, be related to the mechanisms by which K^+ is removed from seawater. The most important of these are almost certainly the uptake of K^+ by riverine clays and by silicate phases produced during the alteration of oceanic crust by seawater at temperatures below 100 °C. Neither process is included in the Hardie (1996) model.

The Hardie (1996) model has also been criticized on several other grounds (Holland *et al.*, 1996; Holland and Zimmermann, 1998). At present it is, perhaps, best regarded as a rough, first-order approximation. Currently the fluid inclusion data for brines in marine halite are probably the best indicators of the composition of seawater during the Phanerozoic. However, these data are much in need of improvement. The coverage of the Phanerozoic era is still quite spotty, and a large number of additional measurements are needed before anything more can be claimed than a preliminary outline of the Phanerozoic history of seawater. Even the presently available data are not above suspicion. The assumptions that underlie the data points in Figures 17–23 were detailed by Horita *et al.* (2002). They are reasonable but not necessarily correct. The composition of brines trapped in halite is far removed from that of their parent seawater. Reconstructing the evolution of these brines is a considerable challenge, given the complexity of the precipitation, dissolution, reprecipitation, and mixing processes in evaporite basins.

6.21.5.5 The Analysis of Unevaporated Seawater in Fluid Inclusions

Analyses of unevaporated seawater are probably essential for defining precisely the evolution of seawater during the Phanerozoic. Johnson and Goldstein (1993) have described single-phase fluid inclusions in low-magnesium calcite cement of the Wilberns Formation in Texas. These almost certainly contain Cambro–Ordovician seawater. The salinity of the inclusion fluids ranges from 31‰ to 47‰. This is essentially identical to the range of seawater salinity observed in shallow-water marine settings today, and is consistent with the precipitation of the calcite cements within slightly restricted environments. Banner has reported a similar salinity range for fluid inclusions in low-magnesium calcite cements in the Devonian Canning Basin of Australia. The fluid inclusions in both areas have diameters $\leq 30\ \mu m$. They are large enough for making heating–freezing measurements but too small until now to serve for quantitative chemical analysis. The recent development of analytical techniques based on ICPMS technology brings the analysis of these fluid inclusions within reach. It will be important to determine whether their composition agrees with the composition of seawater that has been inferred from studies of the composition of inclusion fluids in halite from marine evaporites.

6.21.5.6 The Role of the Stand of Sea Level

The correlation between the composition of seawater and the sea-level curves in Figure 19 is quite striking. The reasons for the correlation are not entirely clear. Hardie (1996) suggested that the stand of sea level reflects the rate of ocean crust formation, that this determines the rate of seawater cycling through MORs, and hence the mixing ratio of hydrothermally altered seawater with average river water. However, the rate of seafloor spreading has apparently not changed significantly during the last 40 Ma (Lithgow-Bertelloni *et al.*, 1993), and Rowley's (2002) analysis of the rate of plate creation and destruction indicates that the rate of seafloor spreading has not varied significantly during the past 180 Ma. If this is correct, the rate of ridge production has been essentially constant since the Early Jurassic. Since the rate of seawater cycling through MORs is probably proportional to this rate, other changes in the Earth system have been responsible for the changes in sea level and in the composition of seawater during the past 180 Ma. Holland and Zimmermann (2000) have pointed out that marine carbonate sediments deposited during the past 40 Ma contain, on average, less dolomite than Proterozoic and Paleozoic carbonates. The lower dolomite content of the more recent carbonate sediments is due to the increase in the deposition of $CaCO_3$ in the deep sea, where dolomitization only takes place in unusual circumstances. The decrease in the rate of Mg^{2+} output from the oceans due to dolomite formation has been balanced by an increase in the output of oceanic Mg^{2+} by the reaction of seawater with clay minerals and with ocean-floor basalts, mainly at MORs. The increase in the output of Mg^{2+} into these reservoirs has been brought about by an increase in the Mg^{2+} concentration of seawater. A simple quantitative model of these processes (Holland and Zimmermann, 2000) can readily account for the observed increase in the Mg^{2+} and SO_4^{2-} concentration of seawater during the Tertiary.

This explanation cannot account for the changes in seawater chemistry before the development of abundant open-ocean $CaCO_3$ secreting organisms. Coccolithophores first appeared in the Jurassic and diversified tremendously during the Cretaceous. The foraminifera radiated explosively during the Jurassic and Cretaceous. Prior to the evolution of coccoliths and planktonic foraminifera, carbonate sediments were largely or entirely deposited on the continents and in shallow-water marine settings. It is likely, therefore, that changes in seawater composition before ca. 150 Ma were related either to changes in the rate of seafloor spreading or to the mineralogy of continental and near-shore carbonate sediments. There are not enough data to rule out the first alternative. However, the apparent near-constancy of the rate of ocean crust formation during the past 180 Ma (Rowley, 2002) is not kind to the notion of major changes in this rate during the first part of the Phanerozoic. The second alternative is more attractive. Flooding of the continents was extensive during high stands of sea level, and dolomitization, which is favored in warm, shallow evaporative settings, must have been widespread. During low stands of sea level, carbonate deposition on the continents was probably replaced by deposition on rims along continental margins, where dolomitization was kinetically less favored. One can imagine that this is the major reason for the correlation of the Mg^{2+} and SO_4^{2-} concentration of seawater and the stand of sea level. The relationship may, however, be more complicated. During the Tertiary, sea level fell, yet the shallow water, near-shore carbonates deposited during the last 40 Ma are strongly and extensively dolomitized. It is not clear why this should not have happened equally enthusiastically during the Permian and Late Neoproterozoic low stands of sea level.

6.21.5.7 Trace Elements in Marine Carbonates

Many attempts have been made to relate the trace element distribution in marine minerals, particularly in carbonates, to the rate of seawater cycling through MORs, and to tectonics in general. The concentration of lithium and the concentration and isotopic composition of strontium have been studied particularly intensively. In their paper on the lithium and strontium content of foraminiferal shells, Delaney and Boyle (1986) proposed that the Li/Ca ratio has varied rather little during the past 116 Ma, but that the Sr/Ca ratio increased significantly during this time period. More recent measurements by Lear *et al.* (2003) on a very large number of foraminifera have shown that the Sr/Ca ratio of benthic foraminifera has had a more complicated history during the past 75 Ma. Relating the Sr/Ca ratio of foraminiferal shells to the Sr/Ca ratio in seawater is complicated, because the Sr/Ca ratio in foraminifera is a rather strong function of temperature, shell size, and pressure, and because it is species specific (Elderfield *et al.*, 2000, 2002). Lear *et al.* (2003) have taken all these factors into account, have combined their data for a variety of benthic foraminifera, and have proposed the course shown in Figure 24 for the Sr/Ca ratio in seawater during the last 75 Ma. The ratio decreased rapidly between 75 Ma and 40 Ma; during the last 40 Ma it climbed significantly, but the increase was interrupted by a decrease between 15 Ma and 7 Ma.

The data in Figure 24 can be combined with those in Figure 21 for the course of the calcium concentration in seawater to yield the course of the strontium concentration of seawater during the last 75 Ma. At 75 Ma, the strontium concentration was $\sim 22 \times 10^{-5}$ mol kg^{-1} H_2O. At 40 Ma, it was $\sim 11 \times 10^{-5}$ mol kg^{-1} H_2O. At present it is 8.5×10^{-5} mol kg^{-1} H_2O. The 60% decrease in the strontium concentration during the last 75 Ma exceeds the decrease of the calcium concentration (45%).

Reconstructing the course of the Sr/Ca concentration in seawater from the composition of fossils is limited by the effects of diagenesis. These are particularly disturbing in carbonates older than 100 Ma. Figure 25 shows Steuber and Veizer's (2002) data for the strontium content of Phanerozoic biological low-magnesium calcites. The averages of the strontium concentrations indicate a course similar to that of the changes in sea level during the Phanerozoic; but the scatter in their data is so large that the significance of their average curve is somewhat in doubt. Lear *et al.* (2003) have attempted to interpret the changes in the concentration and the isotopic composition of strontium in seawater during the last 75 Ma. The rise of the Himalayas and perhaps of other major mountain chains, the lowering of sea level, and the transfer of a significant fraction of marine $CaCO_3$ deposition from shallow to deep waters have all played a role in the changes in seawater composition during the Tertiary. A quantitative treatment of the available data for strontium is

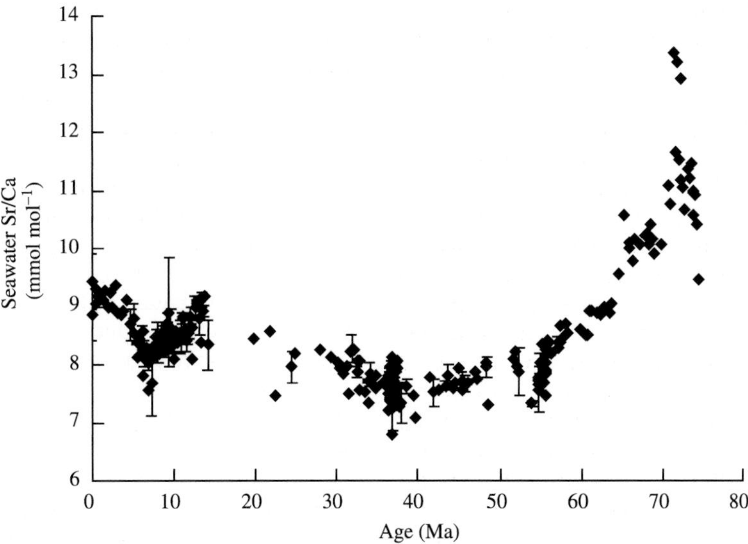

Figure 24 Seawater Sr/Ca record for the Cenozoic (source Lear *et al.*, 2003).

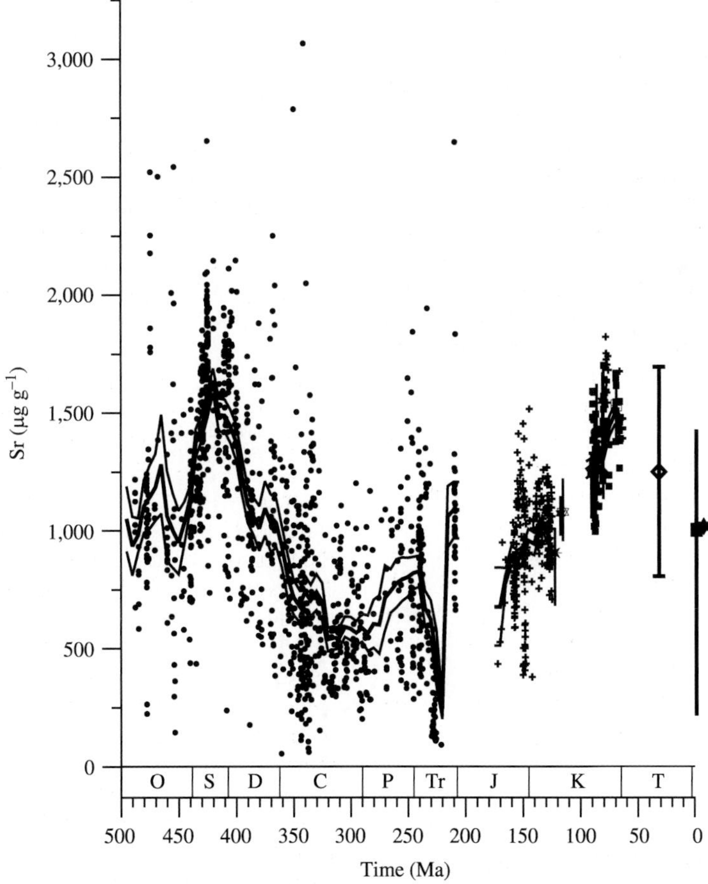

Figure 25 Sr concentrations in biological low-Mg calcite of brachiopods (dots), belemnites (crosses) and rudist bivalve (boxes). Mean values (bold curve) and two standard errors (thin curves) were calculated by moving a 20 Myr window in 5 Myr intervals across the data set. Ranges (vertical lines) and mean values (stars) of intrashell variations for concentrations in single rudist shells are also shown, but were not used in calculation of running means (source Steuber and Veizer, 2002).

still quite difficult; there are still too many poorly defined parameters in the controlling equations.

6.21.5.8 The Isotopic Composition of Boron in Marine Carbonates

The isotopic composition of several elements in marine mineral phases is a much better indicator of oceanic conditions than their concentration in these phases. The $^{11}B/^{10}B$ ratio of living planktonic foraminifera is related to the pH of seawater (Sanyal *et al.*, 1996). This relationship has opened the possibility of using the $^{11}B/^{10}B$ ratio in foraminifera to infer the pH of seawater in the past (Spivack *et al.*, 1993) and thence the course of past P_{CO_2}. This approach has been applied by Pearson and Palmer (1999, 2000) and by Palmer *et al.* (2000) to estimate the pH of seawater and P_{CO_2} during the Cenozoic. The results are, however, somewhat uncertain, because they depend rather heavily on the assumed value of the isotopic composition of boron in Cenozoic seawater, and because the fractionation of the boron isotopes during the uptake of this element in foraminifera is somewhat species specific (Sanyal *et al.*, 1996).

6.21.5.9 The Isotopic Composition of Strontium in Marine Carbonates

A very important contribution to paleoceanography has been made by measurements of the isotopic composition of strontium in Phanerozoic carbonates. The Burke *et al.* (1982) compilation of the isotopic composition of strontium in 744 Phanerozoic marine carbonates has now been expanded by a factor of ~6. Figure 26 is taken from the summary of these data by Veizer *et al.* (1999). The major features of the Burke curve have survived, and many of its features have been sharpened considerably. The curve in Figure 26 is quite robust, and our view of variations in the $^{87}Sr/^{86}Sr$ ratio of seawater during the Phanerozoic is unlikely to change significantly. The $^{87}Sr/^{86}Sr$ ratio of seawater has fluctuated quite significantly. The overall decrease from its high value during the Cambrian to a minimum in the Jurassic and the return to its Early Paleozoic value during the Tertiary does not mirror the two megacycles of the sea-level curve. There is an obvious second-order correlation with orogenies, but their effect on the $^{87}Sr/^{86}Sr$ ratio of seawater has been overshadowed by changes in the isotopic composition and the flux of river strontium to the oceans. The very rapid rise of the $^{87}Sr/^{86}Sr$ ratio in seawater during the Tertiary must be related to the weathering of rocks of very high $^{87}Sr/^{86}Sr$ ratios, and changes in the $^{87}Sr/^{86}Sr$ ratio of rocks undergoing weathering have probably played a major role in determining the fluctuations in the $^{87}Sr/^{86}Sr$ ratio of seawater during the entire Phanerozoic.

6.21.5.10 The Isotopic Composition of Osmium in Seawater

Large changes in the composition of rocks undergoing weathering must also be invoked to

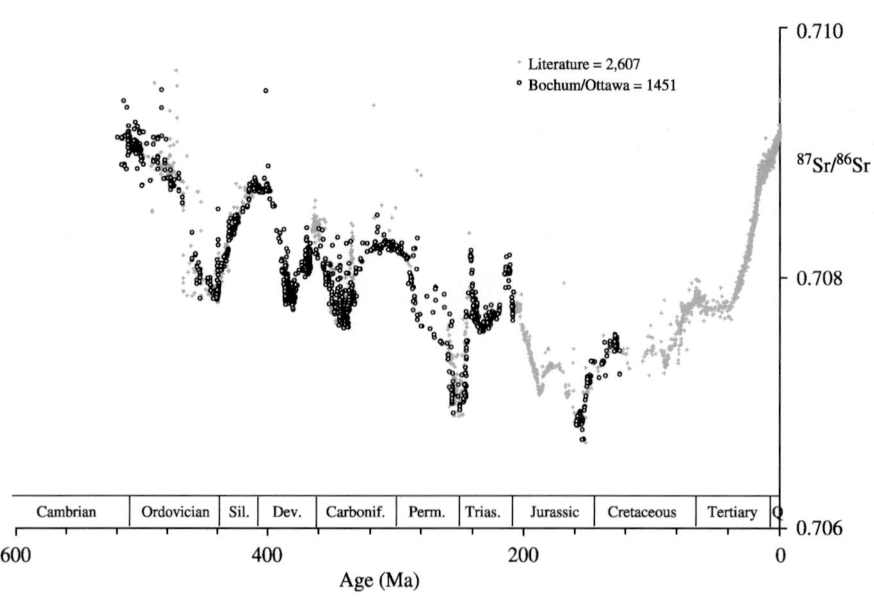

Figure 26 $^{87}Sr/^{86}Sr$ variations for the Phanerozoic based on 4,055 samples of brachiopods ("secondary" layer only for the new Bochum/Ottawa measurements), belemnites, and conodonts (source Veizer *et al.*, 1999).

explain the major changes in the $^{187}Os/^{186}Os$ ratio of seawater during the Cenozoic (see Figure 27) (Pegram and Turekian, 1999; Peucker-Ehrenbrink *et al.*, 1995).

6.21.5.11 The Isotopic Composition of Sulfur and Carbon in Seawater

Interestingly, the first-order variation of the isotopic composition of sulfur and carbon in seawater during the Phanerozoic is qualitatively similar to that of strontium. The value of $\delta^{34}S$ at the base of the Phanerozoic is highly positive. It drops to a minimum in the Permian and then rises again to its present, intermediate level (Figures 28 and 29). The variation of $\delta^{13}C$ during the Phanerozoic is nearly the inverse of the $\delta^{34}S$ curve (Figure 30). Both describe a half-cycle rather than a two-cycle path. The inverse variation of $\delta^{34}S$ and $\delta^{13}C$ strongly suggests that the geochemical cycles of the two elements are closely linked, as suggested

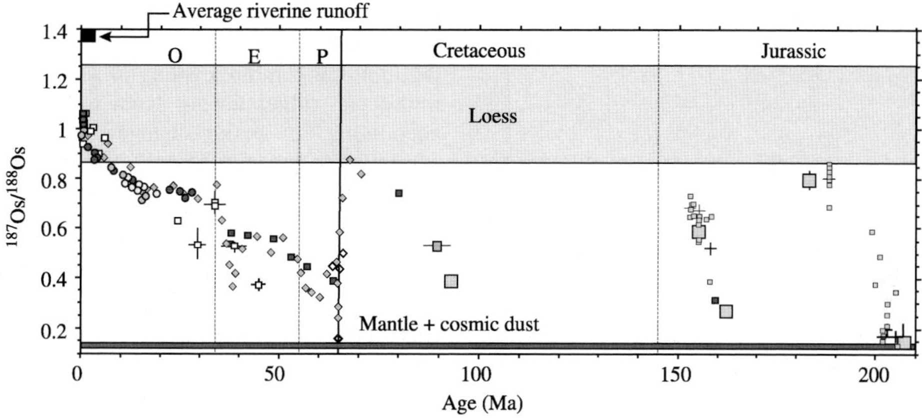

Figure 27 The marine Os isotope record during the last 200 Ma (source Peucker-Ehrenbrink and Ravizza, 2000).

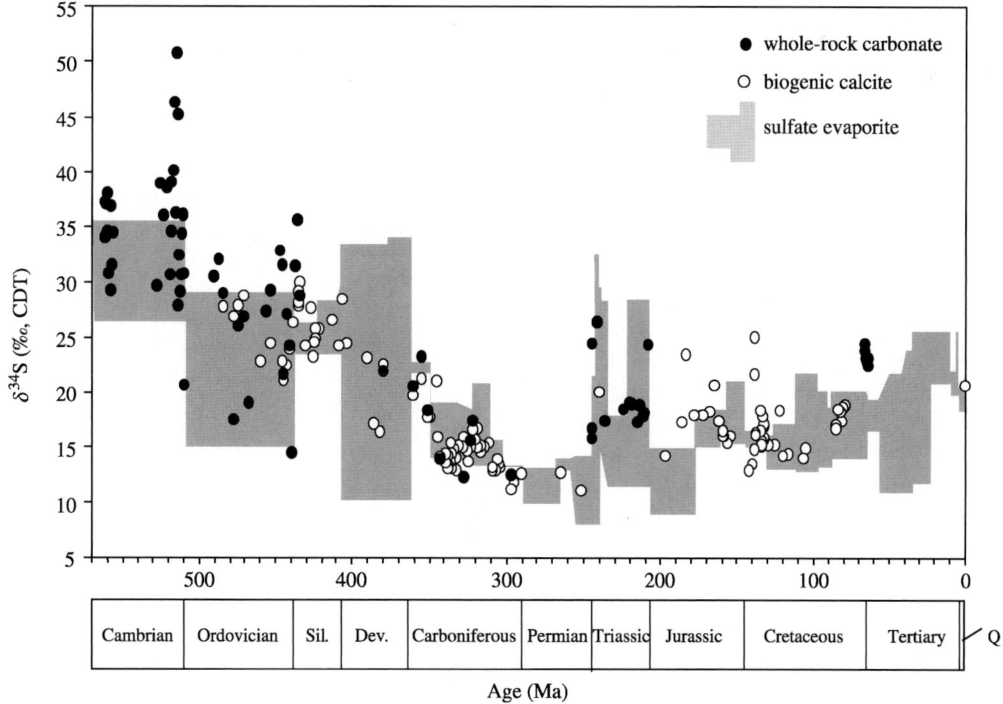

Figure 28 The sulfur isotopic composition of Phanerozoic seawater sulfate based on the analysis of structurally substituted sulfate in carbonates (Kampschulte and Strauss, in press) and evaporite based δ^{34} data (source Strauss, 1999).

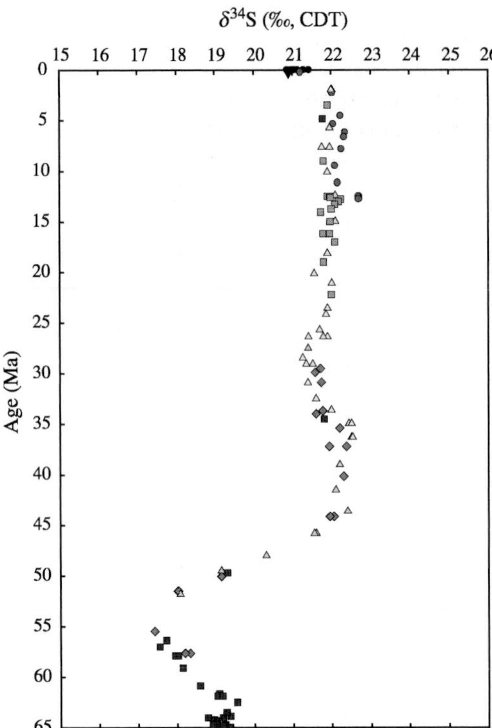

Figure 29 Isotopic composition of seawater sulfate during the past 65 Ma (Paytan *et al.*, 1998, 2002) with ages modified somewhat (sources Paytan, personal communication, 2003).

by Holland (1973) and Garrels and Perry (1974). Since that time the database for assessing the variation of the isotopic composition of both elements has been enlarged very considerably, and the literature dealing with the linkage between their geochemistry has grown apace (e.g., Veizer *et al.*, 1980; Holser *et al.*, 1988, 1989; Berner, 1989; Kump, 1993; Carpenter and Lohmann, 1997; Petsch and Berner, 1998). Nevertheless, a truly quantitative understanding of the linkage has not yet been achieved. The inputs of sulfur and carbon to the oceans as well as the functional relationships that relate the composition of these reservoirs to their outputs into the rock record are still not very well defined. The large difference between the residence time of HCO_3^- and SO_4^{2-} in the oceans has invited deviations from steady state both in their concentration in seawater and in the isotopic composition of their constituents.

6.21.5.12 The Isotopic Composition of Oxygen in Seawater

The use of the isotopic composition of oxygen in marine carbonates as a means of reconstructing the course of the $\delta^{18}O$ value of seawater during the Phanerozoic has been a matter of considerable contention. The $\delta^{18}O$ value of carbonates and cherts tends to become more negative with increasing geologic age (see, e.g., Figure 31). In many instances this decrease is clearly due to diagenetic changes involving reactions with isotopically light water. However, the $\delta^{18}O$ of carbonates which have been chosen carefully to avoid overprinting by diagenetic alteration also tends to decrease with increasing geologic age.

The strongest argument in favor of a near-constant $\delta^{18}O$ value of seawater during much of Earth history is based on an analysis of the effects of seawater cycling through MORs at high temperatures (Muehlenbachs and Clayton, 1976; Holland, 1984; Muehlenbachs, 1986). Changes in the high-temperature cycling of seawater through mid-ocean ridges do not seem to be capable of accounting for major changes in the $\delta^{18}O$ of seawater. At present the only promising mechanism for explaining large changes in the $\delta^{18}O$ of seawater is a major change in the ratio of low-temperature to high-temperature alteration of the oceanic crust (Lohmann and Walker, 1989). The $\delta^{18}O$ of high-temperature vent fluids (200–400 °C) scatter from +0.2‰ to +2.15‰ with an average value of +1.0‰ with respect to the entering seawater (Bach and Humphris, 1999). $\delta^{18}O$ data for basement fluids at temperatures below 100 °C are few. Elderfield *et al.* (1999) have shown that thermally driven seawater in an 80 km transect across the eastern flank of the Juan de Fuca Ridge at 48° latitude has temperatures between 15.5 °C and 62.8 °C and $\delta^{18}O$ values, on average, −1.05‰ with respect to the entering seawater. About 5% of the global ridge flank heat flux would have to be associated with exchange of $\delta^{18}O$ in the ridge flanks to balance the enrichment in seawater $\delta^{18}O$ due to high-temperature hydrothermal activity. The estimate of 5% is similar to that proposed by Mottl and Wheat (1994) (see also Mottl, 2003). It remains to be seen whether the large changes in the balance between low-temperature and high-temperature alteration of the oceanic crust that seem to be required to shift the $\delta^{18}O$ values of seawater to −8‰ have actually occurred. Even if they have not, the Earth's total hydrologic system is so complex that major changes in the $\delta^{18}O$ of seawater during the history of the planet should not be dismissed out of hand.

6.21.6 A BRIEF SUMMARY

The large amount of research that has been completed since the early 1980s has done much to clarify our understanding of the chemical evolution of seawater. In the Precambrian, major advances have centered on the effects of the rise

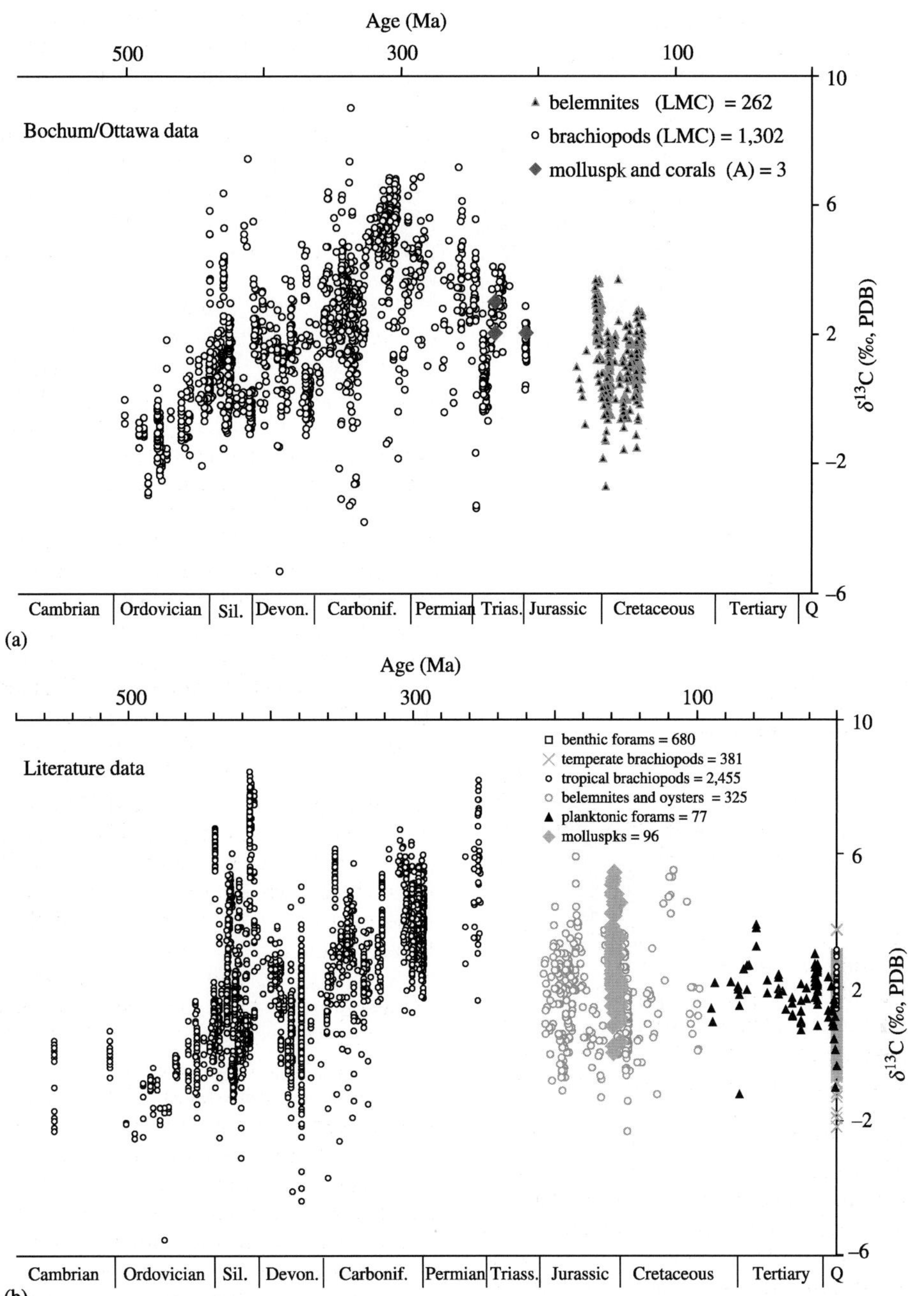

Figure 30 Carbon isotope composition of Phanerozoic low-Mg calcitic shells (source Veizer *et al.*, 1999).

of oxygen on the marine geochemistry of sulfur and the redox sensitive elements. In the Phanerozoic the definition, albeit imprecise, of the concentration of the major constituents of seawater and of the relationship between changes in their concentration, isotopic composition, tectonics, and biological evolution represent a major advance.

Despite this progress our knowledge of the ancient oceans is miniscule compared to our knowledge of the modern ocean. Even the course of the major element concentrations in Phanerozoic seawater is still rather poorly defined, and estimates of their concentration in Precambrian seawater are little more than guesswork. The most promising avenue to a more satisfactory paleoceanography is probably the analysis of fluid inclusions containing unevaporated seawater. This presents formidable analytical problems. If they can be solved, and if a sufficient number of such inclusions spanning a large fraction of Earth history are analyzed, the

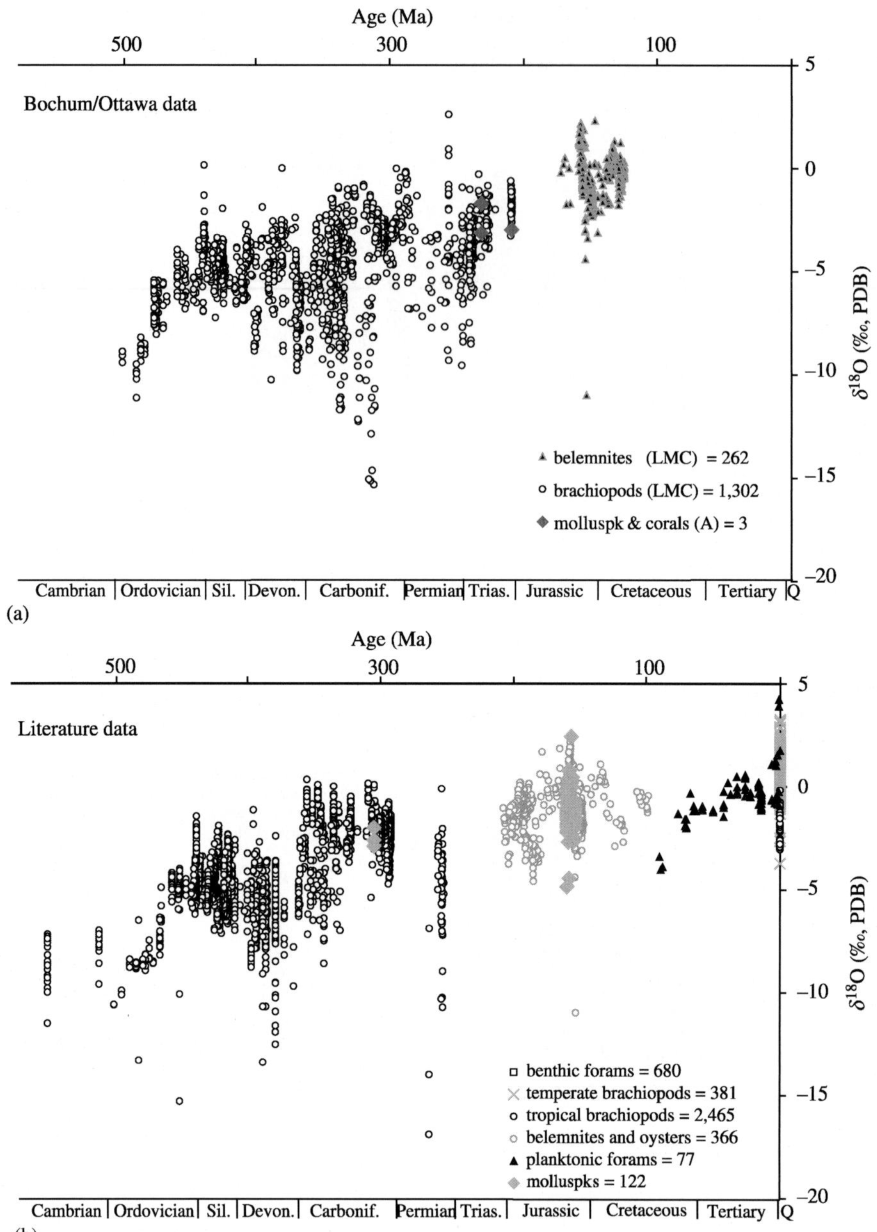

Figure 31 Oxygen isotope composition of Phanerozoic low-Mg calcitic shells (source Veizer *et al.*, 1999).

state of paleoceanography will be improved dramatically.

ACKNOWLEDGMENTS

I would like to express my deep gratitude to Andrey Bekker, Roger Buick, Paul Hoffman, Stein Jacobsen, Jim Kasting, Andrew Knoll, Dan Schrag, Yanan Shen, Roger Summons, and Jan Veizer for taking the time to review this chapter. Their comments ranged from thoroughly enthusiastic to highly critical. All of the reviews were helpful; some were absolutely vital. Andrey Bekker kindly supervised the assembly of the figures. I also wish to acknowledge financial support from NASA Grant NCC2-1053 144167 for the work of the Harvard-MIT NASA Astrobiology Institute.

REFERENCES

Allègre C., Manhès G., and Göpel C. (1995) The age of the Earth. *Geochim. Cosmochim. Acta* **59**, 1445–1456.

Anbar A. D. and Holland H. D. (1992) The photochemistry of manganese and the origin of banded iron formations. *Geochim. Cosmochim. Acta* **56**, 2595–2603.

Anbar A. D. and Knoll A. H. (2002) Proterozoic ocean chemistry and evolution: a bioinorganic bridge? *Science* **297**, 1137–1142.

Anbar A. D., Zahnle K. J., Arnold G. L., and Mojzsis S. J. (2002) Extraterrestrial iridium, sediment accumulation and the habitability of the early Earth's surface. *J. Geophys. Res.* **106**, 3219–3236.

Appel P. W. U. (1983) Rare earth elements in the Early Archean Isua iron-formation, West Greenland. *Precamb. Res.* **20**, 243–258.

Appel P. W. U. and Moorbath S. (1999) Exploring Earth's oldest geological record in Greenland. *EOS, Trans., AGU* **80**, 257, 261, 264.

Armstrong R. A., Lee G. F., Hedges J. I., Honjo S., and Wakeham A. (2002) A new mechanistic model for organic carbon fluxes in the ocean based on the quantitative association of POC with ballast minerals. *Deep-Sea Res.* **49**, 219–236.

Arnold G. L., Anbar A. D., and Barling J. (2002) Oxygenation of Proterozoic oceans: insight from molybdenum isotopes (abstr.). *Geochim. Cosmochim. Acta* **66** (15A), A30 (abstracts of the 12th V. M. Goldschmidt Conference, Davos, Switzerland, August 18–23, 2002).

Ault W. U. and Kulp J. L. (1959) Isotopic geochemistry of sulfur. *Geochim. Cosmochim. Acta* **16**, 201–235.

Bach W. and Humphris S. E. (1999) Relationship between the Sr and O isotope compositions of hydrothermal fluids and the spreading and magma-supply rates at oceanic spreading centers. *Geology* **27**, 1067–1070.

Barth T. F. W. (1952) *Theoretical Petrology*. Wiley, New York.

Bau M. and Möller P. (1993) Rare earth element systematics of the chemically precipitated component in Early Precambrian iron formations and the evolution of the terrestrial atmosphere–hydrosphere–lithosphere system. *Geochim. Cosmochim. Acta* **57**, 2239–2249.

Bekker A., Kaufman A. J., Karhu J. A., Beukes N. J., Swart Q. D., Coetzee L. L., and Eriksson K. A. (2001) Chemostratigraphy of the Paleoproterozic Duitschland Formation, South Africa: implications for coupled climate change and carbon cycling. *Am. J. Sci.* **301**, 261–285.

Bekker A., Holland H. D., Rumble D., Yang W., Wang P.-L., and Coetzee L. L. (2002) MIF of S, oölitic ironstones, redox sensitive elements in shales, and the rise of atmospheric oxygen (abstr.). *Geochim. Cosmochim. Acta* **66**, A64.

Bekker A., Holland H. D., Young G. M., and Nesbitt H. W. (2003) The fate of oxygen during the early Paleoproterozoic carbon isotope excursion. *Astrobiology* **2**(4), 477.

Berner R. A. (1989) Biogeochemical cycles of carbon and sulfur and their effect on atmospheric oxygen over Phanerozoic time. *Paleogeogr. Paleoclimatol. Paleoecol.* **75**, 97–122.

Berner R. A., Lasaga A. C., and Garrels R. M. (1983) The carbonate-silicate geochemical cycle and its effect on atmospheric carbon dioxide. *Am. J. Sci.* **283**, 641–683.

Beukes N. J. (1978) Die Karbonaatgesteentes en Ysterformasies van die Ghaap-Groep van die Transvaal-Supergroep in Noord-Kaapland. PhD Thesis, Rand Afrikaans University.

Beukes N. J. (1983) Palaeoenvironmental setting of iron-formations in the depositional basin of the Transvaal Supergroup, South Africa. In *Iron Formation: Facts and Problems* (eds. A. F. Trendall and R. C. Morris). Elsevier, Amsterdam, pp. 131–209.

Beukes N. J. and Klein C. (1990) Geochemistry and sedimentology of a facies transition from micro-banded to granular iron-formation in the early Proterozoic Transvaal Supergroup, South Africa. *Precamb. Res.* **47**, 99–139.

Beukes N. J. and Klein C. (1992) Time distribution, stratigraphy, and sedimentologic setting, and geochemistry of Precambrian iron-formations. In *The Proterozoic Biosphere* (eds. J. W. Schopf and C. Klein). Cambridge University Press, Cambridge, pp. 139–146.

Beukes N. J., Dorland H., and Gutzmer J. (2002a) Pisolitic ironstone and ferricrete in the 2.22–2.4 Ga Timeball Hill Formation, Transvaal Supergroup: implications for the history of atmospheric oxygen (abstr.). *2002 Annual Meeting of the Geological Society of America,* 283p.

Beukes N. J., Dorland H., Gutzmer J., Nedachi M., and Ohmoto H. (2002b) Tropical laterites, life on land, and the history of atmospheric oxygen in the Paleoproterozoic. *Geology* **30**, 491–494.

Birch T. (1756) *A History of the Royal Society of London* (ed. A. Millar). London, vol. 1.

Black R. C. (1966) *The Younger John Winthrop*. Columbia University Press, New York.

Boak J. L. and Dymek R. F. (1982) Metamorphism of the ca. 3, 800 Ma supercrustal rocks at Isua, West Greenland: implications for early Archaean crustal evolution. *Earth Planet. Sci. Lett.* **59**, 155–176.

Boetius A., Ravenschlag K., Schubert C. J., Rickert D., Widdel F., Gieseke A., Amann R., Jørgensen B. B., Witte U., and Pfannkuche O. (2000) A marine microbial consortium mediating anaerobic oxidation of methane. *Nature* **407**, 623–626.

Bowring S. A. and Williams I. S. (1999) Priscoan (4.00–4.03) orthogneiss from northwestern Canada. *Contrib. Mineral. Petrol.* **134**, 3–16.

Bowring S. A., Landing E., Myrow P., and Ramenzavi J. (2003) Geochronological constraints on the terminal Neoproterozoic events and the rise of Metazoans. *Astrobiology* **2**(4), 457–458.

Brasier M. D., Green O. R., Steele A., Van Kranendonk M., Jephcoat A. P., Kleppe A. K., Lindsay J. F., and Grassineau N. V. (2002) Questioning the evidence for Earth's oldest fossils. In *Astrobiology Science Conference, NASA Ames Research Center, April 7–11, 2002.*

Braterman P. S., Cairns-Smith A. G., and Sloper R. W. (1983) Photo-oxidation of hydrated Fe^{+2}—significance for banded iron formations. *Nature* **303**, 163–164.

Braterman P. S., Cairns-Smith A. G., Sloper R. W., Truscott T. G., and Craw M. (1984) Photooxidation of iron (II) in water between pH 7.4 and 4.0. *J. Chem. Soc.: Dalton Trans.*, 1441–1445.

Bridgewater D., Keto L., McGregor V. R., and Myers J. S. (1976) Archaean gneiss complex in Greenland. In *Geology of Greenland*, Geological Survey of Greenland (eds. A. Escher and W. S. Watt). Copenhagen, pp. 20–75.

Brocks J. J., Logan G. A., Buick R., and Summons R. E. (1999) Archean molecular fossils and the early rise of Eukaryotes. *Science* **285**, 1033–1036.

Brocks J. J., Summons R. E., Logan G. A., and Buick R. (2002) Molecular fossils in Archean Rocks: constraints on the oxygenation of the upper water column (abstr.). In *Astrobiology Science Conference, NASA Ames Research Center, April 7–11, 2002.*

Bruno J., Wersin P., and Stumm W. (1992) On the influence of carbonate in mineral dissolution: II. The stability of $FeCO_3$(s) at 25° and 1 atm. total pressure. *Geochim. Cosmochim. Acta* **56**, 1149–1155.

Burke W. H., Denison R. E., Heatherington E. A., Koepnick R. B., Nelson H. F., and Otto J. B. (1982) Variation of seawater $^{87}Sr/^{86}Sr$ throughout Phanerozoic time. *Geology* **10**, 516–519.

Cairns-Smith A. G. (1978) Precambrian solution photochemistry-inverse segregation and banded iron formations. *Nature* **276**, 807–808.

Canfield D. E. (1998) A new model for Proterozoic ocean chemistry. *Nature* **396**, 450–452.

Canfield D. E. (2001) Biogeochemistry of Sulfur Isotopes. In *Stable Isotope Geochemistry*, Reviews in Mineralogy and Geochemistry (eds. J. W. Valley and D. R. Cole). Washington, DC, vol. 43, pp. 607–636.

Canfield D. E. and Raiswell R. (1999) The evolution of the sulfur cycle. *Am. J. Sci.* **299**, 697–729.

Canfield D. E., Habicht K., and Thamdrup B. (2000) The Archean sulfur cycle and the early history of atmospheric oxygen. *Science* **288**, 658–661.

Canil D. (1997) Vanadium partitioning and the oxidation state of Archaean Komatiite magmas. *Nature* **389**, 842–845.

Canil D. (1999) Vanadium partitioning between orthopyroxene, spinel and silicate melt and the redox states of mantle source regions for primary magmas. *Geochim. Cosmochim. Acta* **63**, 557–572.

Canil D. (2002) Vanadium in peridotites, mantle redox and tectonic environments: Archean to present. *Earth Planet. Sci. Lett.* **195**, 75–90.

Carpenter S. J. and Lohmann K. C. (1997) Carbon isotope ratios of Phanerozoic marine cements: re-evaluating the global carbon and sulfur systems. *Geochim. Cosmochim. Acta* **61**, 4831–4846.

Catling D. C., Zahnle K. J., and McKay C. P. (2001) Biogenic methane, hydrogen escape, and the irreversible oxidation of early Earth. *Science* **293**, 839–843.

Chyba C. F. (1993) The violent emergence of the origin of life: progress and uncertainties. *Geochim. Cosmochim. Acta* **57**, 3351–3358.

Cicero A. D. and Lohmann K. C. (2001) Sr/Mg variation during rock-water interaction: implications for secular changes in the elemental chemistry of ancient seawater. *Geochim. Cosmochim. Acta* **65**, 741–761.

Clarke F. W. (1911) *The Data of Geochemistry,* 2nd edn. The Government Printing Office, Washington, DC.

Cloud P. E., Jr. (1968) Atmospheric and hydrospheric evolution on the primitive Earth. *Science* **160**, 729–736.

Cloud P. E., Jr. (1972) A working model for the primitive Earth. *Am. J. Sci.* **272**, 537–548.

Colman A. S. and Holland H. D. (2000) The global diagenetic flux of phosphorus from marine sediments to the oceans: redox sensitivity and the control of atmospheric oxygen levels. In *Marine Authigenesis: From Global to Microbiol*, SEPM Special Publication No. 66 (eds. C. R. Glenn, L. Prévôt-Lucas, and J. Lucas). Tulsa, UK, pp. 53–75.

Compston W. and Pidgeon R. T. (1986) Jack Hills, evidence of more very old detrital zircons in Western Australia. *Nature* **321**, 766–769.

Condon D. J., Prave A. R., and Benn D. I. (2002) Neoproterozoic glacial-rainout intervals: observations and implications. *Geology* **30**, 35–38.

Conway E. J. (1943) The chemical evolution of the ocean. *Proc. Roy. Irish Acad.* **48B**(8), 161–212.

Conway E. J. (1945) Mean losses of Na, Ca, etc. in one weathering cycle and potassium removal from the ocean. *Am. J. Sci.* **243**, 583–605.

Cook P. J. and McElhinny M. W. (1979) A reevaluation of the spatial and temporal distribution of sedimentary phosphate deposits in the light of plate tectonics. *Econ. Geol.* **74**, 315–330.

Cook P. J. and Shergold J. H. (1984) Phosphorus, phosphorites and skeletal evolution at the Precambrian–Cambrian boundary. *Nature* **308**, 231–236.

Cook P. J. and Shergold J. H. (1986) Proterozoic and Cambrian phosphorites—nature and origins. In *Phosphate Deposits of the World: Volume 1. Proterozoic and Cambrian Phosphorites* (eds. P. J. Cook and J. H. Shergold). Cambridge University Press, Cambridge, pp. 369–386.

Crick I. H. (1992) Petrological and maturation characteristics of organic matter from the Middle Proterozoic McArthur Basin, Australia. *Austral. J. Earth Sci.* **39**, 501–519.

Dalrymple G. B. and Ryder G. (1996) Argon-40/argon-39 age spectra of Apollo 17 highlands breccia samples by laser step heating and the age of the Serenitatis basin. *J. Geophys. Res.* **101**, 26069–26084.

Das N., Horita J., and Holland H. D. (1990) Chemistry of fluid inclusions in halite from the Salina Group of the Michigan Basin: implications for Late Silurian seawater and the origin of sedimentary brines. *Geochim. Cosmochim. Acta* **54**, 319–327.

Delaney M. L. and Boyle F. A. (1986) Lithium in foraminiferal shells: implications for high-temperature hydrothermal circulation fluxes and oceanic crustal generation rates. *Earth Planet. Sci. Lett.* **80**, 91–105.

Delano J. W. (2001) Redox history of the Earth's interior: implications for the origin of life. *Origins Life Evol. Biosphere* **31**, 311–334.

De Long E. F. (2000) Resolving a methane mystery. *Nature* **407**, 577–579.

De Ronde C. E. J., de Wit M. J., and Spooner E. T. C. (1994) Early Archean ($>$3.2 Ga) iron-oxide-rich hydrothermal discharge vents in the Barberton greenstone belt. *South Africa Geol. Soc. Am. Bull.* **106**, 86–104.

De Ronde C. E. J., Channer R. M., Faure de R., Bray C. J., and Spooner E. T. C. (1997) Fluid chemistry of Archean seafloor hydrothermal vents: implications for the composition of circa 3.2 Ga seawater. *Geochim. Cosmochim. Acta* **61**, 4025–4042.

Donnelly T. H. and Jackson M. J. (1988) Sedimentology and geochemistry of a mid-Proterozoic lacustrine unit from Northern Australia. *Sedim. Geol.* **58**, 145–169.

Donnelly T. H., Shergold J. H., Southgate P. N., and Barnes C. J. (1990) Events leading to global phosphogenesis around the Proterozoic–Cambrian boundary. In *Phosphorite Research and Development*, Geological Society Special Publication No. 52 (eds. A. J. G. Notholt and I. Jarvis). London, pp. 273–287.

Dymek R. F. and Klein C. (1988) Chemistry, petrology and origin of banded iron-formation lithologies from the 3,800 Ma Isua supracrustal belt. *West Greenland. Precamb. Res.* **39**, 247–302.

Elderfield H., Wheat C. G., Mottle M. J., Monnin C., and Spiro B. (1999) Fluid and geochemical transport through oceanic crust: a transect across the eastern flank of the Juan de Fuca Ridge. *Earth Planet. Sci.* **172**, 151–165.

Elderfield H., Cooper M., and Ganssen G. (2000) Sr/Ca in multiple species of planktonic foraminifera: implications for reconstructions of seawater Sr/Ca. *Geochem. Geophys. Geosys.* **1**, paper number 1999GC000031.

Elderfield H., Vautravers M., and Cooper M. (2002) The relationship between cell size and Mg/Ca, Sr/Ca, $\delta^{18}O$ and $\delta^{13}C$ of species of planktonic foraminifera. *Geochem. Geophys. Geosys.* **3**(8), paper number 10.1029/2001GC000194.

Farquhar J., Bao H., and Thiemens M. (2000) Atmospheric influence of Earth's earliest sulfur cycle. *Science* **289**, 756–758.

Farquhar J., Savarino J., Airieau S., and Thiemens M. (2001) Observation of wavelength-sensitive mass-independent sulfur isotope effect during SO_2 photolysis: implications for the early atmosphere. *J. Geophys. Res.* **106**, 32829–32839.

Feather C. E. (1980) Some aspects of Witwatersrand mineralization with special reference to uranium minerals: Prof. Paper. *US Geol. Surv.*, 1161-M.

Fedo C. M. and Whitehouse M. J. (2002) Metasomatic origin of quartz-pyroxene rock, Akilia, Greenland, and implications for Earth's earliest life. *Science* **296**, 1448–1452.

François L. M. (1986) Extensive deposition of banded iron formations was possible without photosynthesis. *Nature* **320**, 352–354.

François L. M. (1987) Reducing power of ferrous iron in the Archean ocean: 2. Role of $Fe(OH)^+$ photo-oxidation. *Paleoceanography* **2**, 395–408.

Froude D. O. (1983) Ion microprobe identification of 4,100–4,200 Myr old terrestrial zircons. *Nature* **304**, 616–618.

Fryer B. J. (1983) Rare-earth elements. In *Iron Formation: Facts and Problems* (eds. A. F. Trendall and R. C. Morris). Elsevier, Amsterdam, pp. 345–358.

Gaffin S. (1987) Ridge volume dependence on sea floor generation rate and inversion using long-term sea level change. *Am. J. Sci.* **287**, 596–611.

Garrels R. M. and Mackenzie F. T. (1971) *Evolution of Sedimentary Rocks*. W.W. Norton, New York.

Garrels R. M. and Perry E. C., Jr. (1974) Cycling of carbon, sulfur and oxygen through geologic time. In *The Sea* (ed. E. D. Goldberg). Wiley, New York, vol. 5, pp. 303–336.

Given R. K. and Wilkinson B. H. (1987) Dolomite abundance and stratigraphic age: constraints on rates and mechanisms of Phanerozoic dolostone formation. *J. Sedim. Petrol.* **57**, 1068–1078.

Grandstaff D. E. (1976) A kinetic study of the dissolution of uraninite. *Econ. Geol.* **71**, 1493–1506.

Grandstaff D. E. (1980) Origin of uraniferous conglomerates at Elliott Lake, Canada and Witwatersrand, South Africa: implications for oxygen in the Precambrian atmosphere. *Precamb. Res.* **13**, 1–26.

Grotzinger J. P. (1989) Facies and evolution of Precambrian carbonate depositional systems: emergence of the modern platform archetype. In *Controls on Carbonate Platform and Basin Development*, Special Publication No. 44 (eds. P. D. Crevello, J. L. Wilson, J. F. Sarg, and J. F. Read). The Society of Economic Paleontologists and Mineralogists, Tulsa, UK, pp. 79–106.

Grotzinger J. P. and Kasting J. F. (1993) New constraints on Precambrian ocean composition. *J. Geol.* **101**, 235–243.

Grundl T. and Delwiche J. (1993) Kinetics of ferric oxyhydroxide precipitation. *J. Contamin. Hydrol.* **14**, 71–87.

Habicht K. S., Gade M., Thamdrup B., Berg P., and Canfield D. E. (2002) Calibration of sulfate levels in Archean ocean. *Science* **298**, 2372–2374.

Hallam A. (1984) Pre-quaternary sea level changes. *Ann. Rev. Earth Planet. Sci.* **12**, 205–243.

Hallbauer D. K. (1986) The mineralogy and geochemistry of Witwatersrand pyrite, gold, uranium and carbonaceous matter. In *Mineral Deposits of Southern Africa* (eds. C. R. Anhaeusser and S. Maske). Geological Society of South Africa, Johannesburg, SA, pp. 731–752.

Halley E. (1715) A short account of the cause of the saltness of the ocean, and of the several lakes that emit no rivers; with a proposal, by help thereof, to discover the Age of the World. *Phil. Trans. Roy. Soc. London* **29**, 296–300.

Halliday A. N. (2000) Terrestrial accretion rates and the origin of the Moon. *Earth Planet. Sci. Lett.* **176**, 17–30.

Halliday A. N. (2001) In the beginning. *Nature* **409**, 144–145.

Halverson G. P. (2003) Towards an integrated stratigraphic and carbon isotopic record for the Neoproterozoic. Doctoral Dissertation, Department of Earth and Planetary Sciences, Harvard University, Cambridge, MA.

Hambrey M. J. and Harland W. B. (1981) *Earth's Pre-Pleistocene Glacial Record.* Cambridge University Press, Cambridge.

Han T.-M. (1982) Iron formations of Precambrian age: hematite–magnetite relationships in some Proterozoic iron deposits—a microscopic observation. In *Ore Genesis—The State of the Art* (eds. G. C. Amstutz, A. E. Goresy, G. Frenzel, C. Kluth, G. Moh, A. Wauschkuhn, and R. A. Zimmermann). Springer, Berlin, pp. 451–459.

Han T.-M. (1988) Origin of magnetite in iron-formations of low metamorphic grade. In *Proceedings of the 7th Quadrennial IAGOD Symposium*, pp. 641–656.

Haq B. U., Hardenbol J., and Vail P. R. (1987) Chronology of fluctuating sea levels since the Triassic. *Science* **235**, 1156–1167.

Hardie L. A. (1967) The gypsum-anhydrite equilibrium at one atmosphere pressure. *Am. Mineral.* **52**, 171–200.

Hardie L. A. (1991) On the significance of evaporites. *Ann. Rev. Earth Planet. Sci.* **19**, 131–168.

Hardie L. A. (1996) Secular variation in seawater chemistry: an explanation for the coupled variation in the mineralogies of marine limestones and potash evaporites over the past 600 my. *Geology* **24**, 279–283.

Harrison A. G. and Thode H. G. (1958) Mechanisms of the bacterial reduction of sulfate from isotope fractionation studies. *Trans. Faraday Soc.* **53**, 84–92.

Hartmann W. K., Ryder G., Grinspoon D., and Dones L. (2000) The time-dependent intense bombardment of the primordial Earth/Moon system. In *Origin of the Earth and Moon* (eds. R. M. Canup and K. Righter). University of Arizona Press, Tucson, pp. 493–512.

Harvie C. E., Weare J. H., Hardie L. A., and Eugster H. P. (1980) Evaporation of seawater: calculated mineral sequences. *Science* **208**, 498–500.

Hoehler T. M. and Alperin M. J. (1996) Anaerobic methane oxidation by a methanogen-sulfate reducer consortium: geochemical evidence and biochemical considerations. In *Microbial Growth in C1 Compounds* (eds. M. E. Lindstrom and F. R. Tabita). Kluwer Academic, San Diego.

Hoffman P. F. and Schrag D. P. (2002) The snowball earth hypothesis: testing the limits of global change. *Terra Nova* **14**, 129–155.

Hoffman P., Kaufman A. J., Halverson G. P., and Schrag D. P. (1998) A Neoproterozoic snowball Earth. *Science* **281**, 146–1342.

Holland H. D. (1962) Model for the evolution of the Earth's atmosphere. In *Petrologic Studies: A Volume to Honor A. F. Buddington* (eds. A. E. J. Engel, H. L. James, and B. F. Leonard). Geological Society of America, pp. 447–477.

Holland H. D. (1972) The geologic history of seawater: an attempt to solve the problem. *Geochim. Cosmochim. Acta* **36**, 637–651.

Holland H. D. (1973) Systematics of the isotopic composition of sulfur in the oceans during the Phanerozoic and its implications for atmospheric oxygen. *Geochim. Cosmochim. Acta* **37**, 2605–2616.

Holland H. D. (1984) *The Chemical Evolution of the Atmosphere and Oceans.* Princeton University Press, Princeton, NJ, 582p.

Holland H. D. (1994) Early Proterozoic atmospheric change. In *Early Life on Earth*, Nobel Symposium 84 (ed. S. Bengtson). Columbia University Press, New York, pp. 237–244.

Holland H. D. (1997) Evidence for life on Earth more than 3, 850 million years ago. *Science* **275**, 38–39.

Holland H. D. (2002) Volcanic gases, black smokers, and the Great Oxidation Event. *Geochim. Cosmochim. Acta* **66**, 3811–3826.

Holland H. D. and Beukes N. J. (1990) A paleoweathering profile from Griqualand West, South Africa: evidence for a dramatic rise in atmospheric oxygen between 2.2 and 1.9 BYBP. *Am. J. Sci.* **290**, 1–34.

Holland H. D. and Zbinden E. A. (1988) Paleosols and the evolution of the atmosphere: Part I. In *Physical and Chemical Weathering in Geochemical Cycles* (eds. A. Lerman and M. Meybeck). Kluwer Academic, San Diego, pp. 61–82.

Holland H. D. and Zimmermann H. (1998) On the secular variations in the composition of Phanerozoic marine potash evaporites: Comment and reply. *Geology* **26**, 91–92.

Holland H. D. and Zimmermann H. (2000) The dolomite problem revisited. *Int. Geol. Rev.* **42**, 481–490.

Holland H. D., Horita J., and Seyfried W. E. (1996) On the secular variations in the composition of Phanerozoic marine potash evaporites. *Geology* **24**, 993–996.

Holser W. T. (1963) Chemistry of brine inclusions in Permian salt from Hutchinson, Kansas. In *Symposium on Salt (First)* (ed. A. C. Bersticker). Northern Ohio Geol. Soc., Cleveland, OH, pp. 86–95.

Holser W. T. and Kaplan I. R. (1966) Isotope geochemistry of sedimentary sulfates. *Chem. Geol.* **1**, 93–135.

Holser W. T., Schidlowski M., Mackenzie F. T., and Maynard J. B. (1988) Biogeochemical cycles of carbon and sulfur. In *Chemical Cycles in the Evolution of the Earth* (eds. C. B. Gregor, R. M. Garrels, F. T. Mackenzie, and J. B. Maynard). Wiley-Interscience, New York, pp. 105–173.

Holser W. T., Maynard J. B., and Cruikshank K. M. (1989) Modelling the natural cycle of sulphur through Phanerozoic time. In *Evolution of the Global Biogeochemical Sulphur Cycle* (eds. P. Brimblecombe and A. Y. Lein). Wiley, New York, chap 2, pp. 21–56.

Horita J., Friedman T. J., Lazar B., and Holland H. D. (1991) The composition of Permian seawater. *Geochim. Cosmochim. Acta* **55**, 417–432.

Horita J., Weinberg A., Das N., and Holland H. D. (1996) Brine inclusions in halite and the origin of the Middle Devonian Prairie Evaporites of Western Canada. *J. Sedim. Res.* **66**, 956–964.

Horita J., Zimmermann H., and Holland H. D. (2002) The chemical evolution of seawater during the Phanerozoic: implications from the record of marine evaporites. *Geochim. Cosmochim. Acta* **66**, 3733–3756.

Iglesias-Rodriguez M. D., Armstrong R. A., Feely R., Hood R., Kleypas J., Milliman J. D., Sabine C., and Sarmiento J. L. (2002) Progress made in study of ocean's calcium carbonate budget. *EOS* **83**, 365, 374–375.

Iversen N. and Jørgensen B. B. (1985) Anaerobic methane oxidation rates at the sulfate-methane transition in marine sediments from Kattegat and Skagerrak (Denmark). *Limnol. Oceanogr.* **30**, 944–955.

Jackson M. J. and Raiswell R. (1991) Sedimentology and carbon–sulfur geochemistry of the Velkerry Formation, a mid-Proterozoic potential oil source in Northern Australia. *Precamb. Res.* **54**, 81–108.

Jackson M. J., Muir M. D., and Plumb K. A. (1987) Geology of the southern McArthur Basin, Northern Territory. *Bureau of Mineral Resources, Geology and Geophysics*. Bulletin 220.

Jacobsen S. B. and Kaufman A. J. (1999) The Sr, C, and O isotopic evolution of Neoproterozoic seawater. *Chem. Geol.* **161**, 37–57.

James H. L. (1983) Distribution of banded iron-formation in space and time. In *Iron Formation: Facts and Problems* (eds. A. F. Trendall and R. C. Morris). Elsevier, Amsterdam, pp. 471–490.

James H. L. (1992) Precambrian iron-formations: nature, origin, and mineralogical evolution from sedimentation to metamorphism. In *Diagenesis III: Developments in Sedimentology* (eds. K. H. Wolf and G. V. Chilingarian). Elsevier, Amsterdam, pp. 543–589.

Johnson W. J. and Goldstein R. H. (1993) Cambrian seawater preserved as inclusions in marine low-magnesium calcite cement. *Nature* **362**, 335–337.

Joly J. (1899) An estimate of the geological age of the Earth. *Sci. Trans. Roy. Dublin Soc.* **7**(II), 23–66.

Kakegawa T., Kawai H., and Ohmoto H. (1998) Origins of pyrite in the .5 Ga Mt. McRae Shale, the Hamersley District, Western Australia. *Geochim. Cosmochim. Acta* **62**, 3205–3220.

Karhu J. A. (1993) Paleoproterozoic evolution of the carbon isotope ratios of sedimentary carbonates in the Fennoscandian Shield. *Geol. Soc. Finland Bull.* **371**, 87.

Karhu J. A. and Holland H. D. (1996) Carbon isotopes and the rise of atmospheric oxygen. *Geology* **24**, 867–870.

Kasting J. F., Eggler D. H., and Raeburn S. P. (1993) Mantle redox evolution and the state of the Archean atmosphere. *J. Geol.* **101**, 245–257.

Kasting J. F., Pavlov A. A., and Siefert J. L. (2001) A coupled ecosystem-climate model for predicting the methane concentration in the Archean atmosphere. *Origin Life Evol. Biosphere* **31**, 271–285.

Kennedy M. J., Christie-Blick N., and Sohl L. E. (2001) Are Proterozoic cap carbonates and isotopic excursions a record of gas hydrate destabilization following Earth's coldest intervals? *Geology* **29**, 443–446.

Kinsman D. J. J. (1966) Gypsum and anhydrite of Recent age, Trucial Coast, Persian Gulf. In *2nd Symposium on Salt* (ed. J. L. Rau). The Northern Ohio Geological Society, Cleveland, OH, vol. 1, pp. 302–326.

Kirk J., Ruiz J., Chesley J., Titley S., and Walshe J. (2001) A detrital model for the origin of gold and sulfides in the Witwatersrand basin based on Re–Os isotopes. *Geochim. Cosmochim. Acta* **65**, 2149–2159.

Kirschvink J. L. (1992) Late Proterozoic low-latitude glaciation: the snowball earth. In *The Proterozoic Biosphere* (eds. J. W. Schopf and C. Klein). Cambridge University Press, Cambridge, pp. 51–52.

Klein C. and Beukes N. J. (1989) Geochemistry and sedimentology of a facies transition from limestone to iron-formation deposition in the Early Proterozoic Transvaal Supergroup, South Africa. *Econ. Geol.* **84**, 1733–1774.

Klein C. and Beukes N. J. (1993) Sedimentology and geochemistry of the glaciogenic Late Proterozoic Rapitan iron-formation in Canada. *Econ. Geol.* **88**, 542–565.

Kleine T., Münker C., Mezger K., and Palme H. (2002) Rapid accretion and early core formation on asteroids and the terrestrial planets from Hf–W chronometry. *Nature* **418**, 952–955.

Knoll A. H. (1992) Biological and biogeochemical preludes to the Edeacaran radiation. In *Origin and Early Evolution of the Metazoa* (eds. J. H. Lipps and P. W. Signor). Plenum, New York, chap. 4.

Konhauser K. O., Hamade T., Raiswell R., Morris R. C., Ferris F. G., Southam G., and Canfield D. E. (2002) Could bacteria have formed the Precambrain banded iron formations? *Geology* **30**, 1079–1082.

Kump L. R. (1993) The coupling of the carbon and sulfur biogeochemical cycles over Phanerozoic time. In *Interactions of C, N, P, and S Biogeochemical Cycles and Global Change* (eds. R. Wollast, F. T. Mackenzie, and L. Chou). Springer, pp. 475–490.

Lasaga A. C., Berner R. A., and Garrels R. M. (1985) An improved geochemical model of atmospheric CO_2 fluctuations over the past 100 million years. In *The Carbon Cycle and Atmospheric CO_2, Natural Variations Archean to Present*, Geophysical Monograph 32 (eds. E. T. Sundquist and W. S. Broecker). American Geophysical Union, Washington, DC, pp. 397–411.

Lazar B. and Holland H. D. (1988) The analysis of fluid inclusions in halite. *Geochim. Cosmochim. Acta* **52**, 485–490.

Lear C. H., Elderfield H., and Wilson P. A. (2000) Cenozoic deep-sea temperatures and global ice volumes from Mg/Ca in benthic foraminiferal calcite. *Science* **287**, 269–272.

Lear C. H., Elderfield H., and Wilson P. A. (2003) A Cenozoic seawater Sr/Ca record from benthic foraminiferal calcite and its application in determining global weathering fluxes. *Earth Planet. Sci. Lett.* **208**, 69–84.

Leather J., Allen P. A., Brasier M. D., and Cozzi A. (2002) Neoproterozoic snowball Earth under scrutiny: evidence from the Fig glaciation of Oman. *Geology* **30**, 891–894.

Lithgow-Bertelloni C., Richards M. A., Ricard Y., O'Connell R. J., and Engebretson D. C. (1993) Toroidal–poloidal partitioning of plate motions since 120 Ma. *Geophys. Res. Lett.* **20**, 375–378.

Logan G. A., Hayes J. M., Hieshima G. B., and Summons R. E. (1995) Terminal Proterozoic reorganization of biogeochemical cycles. *Nature* **376**, 53–56.

Lohmann K. C. and Walker C. G. (1989) The $\delta^{18}O$ record of Phanerozoic abiotic marine calcite cements. *Geophys. Res. Lett.* **16**, 319–322.

Lowenstein T. K., Timofeeff M. N., Brennan S. T., Hardie L. A., and Demicco R. V. (2001) Oscillations in Phanerozoic seawater chemistry: evidence from fluid inclusions in salt deposits. *Science* **294**, 1086–1088.

Lugmaier G. W. and Shukolyukov A. (1998) Early solar system timescales according to $^{53}Mn–^{53}Cr$ systematics. *Geochim. Cosmochim. Acta* **62**, 2863–2886.

Lyons T. W., Gellatly A. M., and Kah L. C. (2002) Paleoenvironmental significance of trace sulfate in sedimentary carbonates. In *Abstracts Volume, 6th International Symposium on the Geochemistry of the Earth's*

Surface, May 20–24, 2002, Honolulu, Hawaii, pp. 162–165.

Mackenzie F. T. and Garrels R. M. (1966) Chemical mass balance between rivers and oceans. *Am. J. Sci.* **264**, 507–525.

Melezhik V. A., Gorokhov I. M., Kuznetsov A. B., and Fallick A. E. (2001) Chemostratigraphy of neoproterozoic carbonates: implications for "blind" dating. *Terra Nova* **13**, 1–11.

Michaelis W., Seifert R., Nauhaus K., Treude T., Thiel V., Blumenberg M., Knittel K., Gieseke A., Peterknecht K., Pape T., Boetius A., Amann R., Jørgensen B. B., Widdel F., Peckmann J., Pimenov N., and Gulin M. (2002) Microbial reefs in the Black Sea fueled by anaerobic oxidation of methane. *Science* **297**, 1013–1015.

Millero F. J., Sotolongo S., and Izaguirre M. (1987) The oxidation kinetics of Fe(II) in seawater. *Geochim. Cosmochim. Acta* **51**, 793–801.

Mojzsis S. J., Arrhenius G., McKeegan K. D., Harrison T. M., Nutman A. P., and Friend C. R. L. (1996) Evidence for life on Earth before 3,800 million years ago. *Nature* **384**, 55–59.

Mojzsis S. J., Harrison T. M., and Pidgeon R. T. (2001) Oxygen-isotope evidence from ancient zircons for liquid water at the Earth's surface 4,300 Myr ago. *Nature* **409**, 178–181.

Morse J., Millero F. J., Thurmond V., Brown E., and Ostlund H. G. (1984) The carbonate chemistry of Grand Bahama Bank waters: after 18 years another look. *J. Geophys. Res.* **89**, 3604–3614.

Morse J. W., Wang Q., and Tsio M.-Y. (1997) Influences of temperature and Mg : Ca ratio on $CaCO_3$ precipitates from seawater. *Geology* **25**, 85–87.

Mottl M. J. (2003) Partitioning of energy and mass fluxes between mid-ocean ridge axes and flanks at high and low temperature. In *Energy and Mass Transfer in Marine Hydrothermal Systems* (eds. P. E. Halbach, V. Tunnicliffe, and J. R. Hein). Dahlem University Press, pp. 271–286.

Mottl M. J. and Wheat C. G. (1994) Hydrothermal circulation through mid-ocean ridge flanks: fluxes of heat and magnesium. *Geochim. Cosmochim. Acta* **58**, 2225–2237.

Muehlenbachs K. (1986) Alteration of the oceanic crust and the ^{18}O history of seawater. In *Stable Isotopes in High Temperature Geological Processes*, Reviews in Mineralogy (eds. J. W. Valley, H. P. Taylor, Jr., and J. R. O'Neil). Mining Society of America, Chelsea, MI, vol. 16, chap. 12, pp. 425–444.

Muehlenbachs K. and Clayton R. N. (1976) Oxygen isotope composition of the oceanic crust and its bearing on seawater. *J. Geophys. Res.* **81**, 4365–4369.

Muir M. D. (1979) A sabkha model for deposition of part of the Proterozoic McArthur Group of the Northern Territory, and implications for mineralization. *BMR J. Austral. Geol. Geophys.* **4**, 149–162.

Myers J. S. (2001) Protoliths of the 3.8–3.7 Ga Isua greenstone belt, West Greenland. *Precamb. Res.* **105**, 129–141.

Naraoka H., Ohtake M., Maruyama S., and Ohmoto H. (1996) Non-biogenic graphite in 3.8-Ga metamorphic rocks from the Isua district, Greenland. *Chem. Geol.* **133**, 251–260.

Nelson D. R., Trendall A. F., and Altermann W. (1999) Chronological correlations between the Pilbara and Kaapvaal cratons. *Precamb. Res.* **97**, 165–189.

Nelson D. R., Robinson B. W., and Myers J. S. (2000) Complex geological histories extending for $\geq$4.0 Ga deciphered from xenocryst zircon microstructures. *Earth Planet. Sci.* **181**, 89–102.

Nutman A. P., Kinny P. D., Compston W., and Williams J. S. (1991) Shrimp U–Pb zircon geochronology of the Narryer Gneiss Complex, Western Australia. *Precamb. Res.* **52**, 275–300.

Nutman A. P., McGregor V. R., Friend C. R. L., Bennett V. C., and Kinny P. D. (1996) The Itsaq Gneiss complex of southern West Greenland: the world's most extensive record of early crustal evolution (3,900–3,600 Ma). *Precamb. Res.* **78**, 1–39.

Nutman A. P., Bennett V. C., Friend C. R. L., and Rosing M. T. (1997) 3,710 and >3,790 Ma volcanic sequences in the Isua (Greenland) supracrustal belt: structural and Nd isotope implications. *Chem. Geol. (Isotope Geosci.)* **141**, 271–287.

Nutman A. P., Friend C. R. L., and Bennett V. (2002) Evidence for 3,650–3,600 Ma assembly of the northern end of the Itsaq Gneiss Comples, Greenland: implication for early Archaean tectonics. *Tectonics* **21**, 1–28.

Oehler D. Z. and Smith J. W. (1977) Isotopic composition of reduced and oxidized carbon in early Archaean rocks from Isua, Greenland. *Precamb. Res.* **5**, 221–228.

Ohmoto H. (1996) Evidence in pre-2.2 Ga paleosols for the early evolution of the atmospheric oxygen and terrestrial biota. *Geology* **24**, 1135–1138.

Ohmoto H. (1999) Redox state of the Archean atmosphere: evidence from detrital heavy minerals in ca. 3,250–2, 750 Ma sandstones from the Pilbara Craton, Australia: Comment. *Geology* **27**, 1151–1152.

Ohmoto H., Kakegawa T., and Lowe D. R. (1993) 3.4-billion-year-old biogenic pyrites from Barberton, South Africa: sulfur isotope evidence. *Science* **262**, 555–557.

Ono S. (2002) Detrital uraninite and the early Earth's atmosphere: SIMS analyses of uraninite in the Elliot Lake district and the dissolution kinetics of natural uraninite. Doctoral Dissertation, Pennsylvania State University, State College, PA.

Orphan V. J., House C. H., Hinrichs K. U., McKeegan K. D., and DeLong E. F. (2001) Methane-consuming archaea revealed by directly coupled isotopic and phylogenetic analysis. *Science* **293**, 484–487.

Palmer H. R., Pearson P. N., and Cobb S. J. (2000) Reconstructing past ocean pH-depth profiles. *Science* **282**, 1468–1471.

Pavlov A. A. and Kasting J. F. (2002) Mass-independent fractionation of sulfur isotopes in Archean sediments: strong evidence for an anoxic Archean atmosphere. *Astrobiology* **2**, 27–41.

Pavlov A. A., Kasting J. F., Brown L. L., Rages K. A., and Freedman R. (2000) Greenhouse warming by CH_4 in the atmosphere of early Earth. *J. Geophys. Res.* **105**, 11981–11990.

Pavlov A. A., Hurtgen M., Kasting J. F., and Arthur M. A. (2003) A methane-rich Proterozoic atmosphere? *Geology* **31**, 87–90.

Paytan A., Kastner M., Campbell D., and Thiemens M. (1998) Sulfur isotopic composition of Cenozoic seawater sulfate. *Science* **282**, 1459–1462.

Paytan A., Mearon S., Cobb K., and Kastner M. (2002) Origin of marine barite deposits: Sr and S characterization. *Geology* **30**, 747–750.

Pearson P. N. and Palmer M. R. (1999) Middle Eocene seawater pH and atmospheric carbon dioxide concentrations. *Science* **284**, 1824–1826.

Pearson P. N. and Palmer M. R. (2000) Atmospheric carbon dioxide concentrations over the past 60 million years. *Nature* **406**, 695–699.

Peck W. H., Valley J. W., Wilde S. A., and Geraham C. M. (2001) Oxygen isotope ratios and rare earth elements in 3.3 to 4.4 Ga zircons: ion microprobe evidence for high $\delta^{18}O$ continental crust and oceans in the early Archean. *Geochim. Cosmochim. Acta* **65**, 4215–4229.

Pegram W. J. and Turekian K. K. (1999) The Osmium isotopic composition change of Cenozoic seawater as inferred from a deep-sea core corrected for meteoritic contributions. *Geochim. Cosmochim. Acta* **63**, 4053–4058.

Perry E. C., Jr. and Ahmad S. N. (1977) Carbon isotope composition of graphite and carbonate minerals from the 3.8-AE metamorphosed sediments, Isukasia, Greenland. *Earth Planet. Sci. Lett.* **36**, 281–284.

Petsch S. T. and Berner R. A. (1998) Coupling the geochemical cycles of C, P, Fe, and S: the effect on atmospheric O_2 and the isotopic records of carbon and sulfur. *Am. J. Sci.* **298**, 246–262.

Peucker-Ehrenbrink B. and Ravizza G. (2000) The marine osmium isotope record. *Terra Nova* **12**, 205–219.

Peucker-Ehrenbrink B., Ravizza G., and Hofmann A. W. (1995) The marine $^{187}Os/^{186}Os$ record of the past 80 million years. *Earth Planet. Sci. Lett.* **130**, 155–167.

Phillips G. N., Law J. D. M., and Myers R. E. (2001) Is the redox state of the Archean atmosphere constrained? *Soc. Econ. Geologists SEG Newslett.* **47**(1), 9–18.

Pinto J. P. and Holland H. D. (1988) Paleosols and the evolution of the atmosphere: Part II. In *Paleosols and Weathering Through Geologic Time*, Geol. Soc. Am. Spec. Pap. 216 (eds. J. Reinhardt and W. Sigleo), pp. 21–34.

Plummer L. N. and Busenberg E. (1982) The solubilities of calcite, aragonite and vaterite in CO_2–H_2O solutions between 0° and 90°, and an evaluation of the aqueous model for the system $CaCO_3$–CO_2–H_2O. *Geochim. Cosmochim. Acta* **46**, 1011–1040.

Porcelli D., Cassen P., Woolum D., and Wasserburg G. J. (1998) Acquisition and early losses of rare gases from the deep Earth. In *Origin of the Earth and Moon: Lunar Planetary Institute Contribution, Report 957*, pp. 35–36.

Porter S. M. and Knoll A. H. (2000) Testate amoebae in the Neoproterozoic era: evidence from vase-shaped microfossils in the Chuar Group, Grand Canyon. *Paleobiology* **26**, 360–385.

Porter S. M., Meisterfeld R., and Knoll, A. H. (2003) Vase-shaped microfossils from the Neoproterozoic Chuar Group, Grand Canyon: a classification guided by modern Testate amoebae. *J. Paleontol.* **77**, 205–255.

Ramdohr P. (1958) Die Uran-und Goldlagerstätten Witwatersrand-Blind River District-Dominion Reef-Serra de Jacobina: erzmikroskopische Untersuchungen und ein geologischer Vergleich. *Abh. Deutschen Akad. Wiss Berlin* **3**, 1–35.

Rasmussen B. and Buick R. (1999) Redox state of the Archean atmosphere: evidence from detrital heavy minerals in ca. 3, 250–2,750 Ma sandstones from the Pilbara Craton, Australia. *Geology* **27**, 115–118.

Rasmussen B., Buick R., and Holland H. D. (1999) Redox state of the Archean atmosphere: evidence from detrital heavy minerals in ca. 3,250–2,750 Ma sandstones from the Pilbara Craton, Australia: Reply. *Geology* **27**, 1152.

Reeburgh W. S., Ward B. B., Whalen S. C., Sandbeck K. A., Kilpatrick K. A., and Kerkhof L. J. (1991) Black Sea methane geochemistry. *Deep-Sea Res.* **38**(suppl. 2), 1189–1210.

Rosing M. T. (1999) ^{13}C-depleted carbon microparticles in >3,700-Ma sea-floor sedimentary rocks from West Greenland. *Science* **283**, 674–676.

Rosing M. T., Rose N. M., Bridgewater D., and Thomsen H. S. (1996) Earliest part of Earth's stratigraphic record: a reappraisal of the >3.7 Ga Isua (Greenland) supracrustal sequence. *Geology* **24**, 43–46.

Rowley D. B. (2002) Rate of plate creation and destruction: 180 Ma to present. *Geol. Soc. Am. Bull.* **114**, 927–933.

Rubey W. W. (1951) Geologic history of seawater, an attempt to state the problem. *Bull. Geol. Soc. Am.* **62**, 1111–1147.

Ryder G. (1990) Lunar samples, lunar accretion and the early bombardment of the Moon. *EOS, Trans., AGU* **71**(10), 313, 322–323.

Rye R. and Holland H. D. (1998) Paleosols and the evolution of the atmosphere: a critical review. *Am. J. Sci.* **298**, 621–672.

Rye R., Kuo P. H., and Holland H. D. (1995) Atmospheric carbon dioxide concentration before 2.2 billion years ago. *Nature* **378**, 603–605.

Sandberg P. A. (1983) An oscillating trend in Phanerozoic non-skeletal carbonate mineralogy. *Nature* **305**, 19–22.

Sandberg P. A. (1985) Nonskeletal aragonite and pCO_2 in the Phanerozoic and Proterozoic. In *The Carbon Cycle and Atmospheric CO_2, Natural Variations Archean to Present*, Geophysical Monograph 32 (eds. E. T. Sundquist and W. S. Broecker) American Geophysical Union, Washington, DC, pp. 585–594.

Sanyal A., Hemming N. G., Broecker W. S., Lea D. W., Spero H. J., and Hanson G. N. (1996) Oceanic pH control on the boron isotopic composition of foraminifera: evidence from culture experiments. *Paleoceanography* **11**, 513–517.

Sarmiento J. L., Dunne J., Gnanadesikan A., Key R. M., Matsumoto K., and Slater R. (2002) A new estimate of the $CaCO_3$ to organic carbon export ratio. *Global Biogeochem. Cycles* **16**(4), 54-1–54-12.

Sasaki S. and Nakazawa K. J. (1986) Metal–silicate fractionation in the growing Earth: energy source for the terrestrial Magma Ocean. *J. Geophys. Res.* **91**, B9231–B9238.

Schidlowski M. (1966) Beiträge zur Kenntnis der radioactiven Bestandteile der Witwatersrand-Konglomerate. I Uranpecherz in den Konglomeraten des Oranje-Freistaat-Goldfeldes. *N. Jb. Miner Abh.* **105**, 183–202.

Schidlowski M., Appel P. W. U., Eichmann R., and Junge C. E. (1979) Carbon isotope geochemistry of the 3.7×10^9 yr-old Isua sediments, West Greenland: implications for the Archaean carbon and oxygen cycles. *Geochim. Cosmochim. Acta* **43**, 189–199.

Schidlowski M., Hayes J. M., and Kaplan I. R. (1983) Isotopic inferences of ancient biochemistries: carbon, sulfur, hydrogen, and nitrogen. In *Earth's Earliest Biosphere, Its Origin and Evolution* (ed. J. W. Schopf). Princeton University Press, Princeton, NJ, chap. 7, pp. 149–186.

Schopf J. W. (1983) Microfossils of the Early Archean Apex Chert: new evidence for the antiquity of life. *Science* **260**, 640–646.

Schopf J. W. and Packer B. M. (1987) Early Archean (3.3-billion to 3.5-billion-year-old) microfossils from Warrawoona Group, Australia. *Science* **237**, 70–73.

Schopf J. W., Kudryatsev A. B., Agresti D. G., Czaja A. D., and Widowiak T. J. (2002) Laser-Raman imagery of the oldest fossils on Earth. In *Astrobiology Science Conference, NASA Ames Research Center, April 7–11, 2002*.

Sellwood B. W. (1986) Shallow marine carbonate environments. In *Sedimentary Environments and Facies* (ed. H. G. Reading). Blackwell, UK, chap. 10, pp. 283–342.

Shen Y., Buick R., and Canfield D. E. (2001) Isotopic evidence for microbial sulphate reduction in the early Archaean era. *Nature* **410**, 77–81.

Shen Y., Canfield D. E., and Knoll A. H. (2002) Middle Proterozoic ocean chemistry: evidence from the McArthur Basin, Northern Australia. *Am. J. Sci.* **302**, 81–109.

Shields G. and Veizer J. (2002) Precambrian marine carbonate isotope database: version 1.1. *Geochem. Geophys. Geosys.* **3**, doi: 10.1029/2001GC000266.

Shimizu H., Umemoto N., Masuda A., and Appel P. W. U. (1990) Sources of iron-formations in the Archean Isua and Malene supracrustals, West Greenland: evidence from La–Ce and Sm–Nd isotopic data and REE abundances. *Geochim. Cosmochim. Acta* **54**, 1147–1154.

Sillén L. G. (1967) The ocean as a chemical system. *Science* **156**, 1189–1197.

Sleep N. H., Zahnle K. J., Kasting J. F., and Morowitz H. J. (1989) Annihilation of ecosystems by large asteroid impacts on the early Earth. *Nature* **342**, 139–142.

Southgate P. N., Bradshaw B. E., Domagala J., Jackson M. J., Idnurm M., Krassay A. A., Page R. W., Sami T. T., Scott D. L., Lindsay J. F., McConachie B. A., and Tarlowski C. (2000) Chronostratigraphic basin framework for Paleoproterozoic rocks (1,730–1,575 Ma) in northern Australia and implications for base-metal mineralization. *Austral. J. Earth Sci.* **47**, 461–483.

Spencer R. J. and Hardie L. A. (1990) Control of seawater composition by mixing of river waters and mid-ocean ridge hydrothermal brines. In *Fluid–Mineral Interactions: A Tribute to H. P. Eugster*. Special Publication 2 (eds. R. J. Spencer and I.-M. Chou). Geochemical Society, San Antonio, TX, pp. 409–419.

Spivack A. J., You C.-F., and Smith H. J. (1993) Foraminiferal boron isotope ratios as a proxy for surface ocean pH over the past 21 Myr. *Nature* **363**, 149–151.

Stanley S. M. and Hardie L. A. (1998) Secular oscillations in the carbonate mineralogy of reef-building and sediment-producing organisms driven by tectonically forced shifts in seawater chemistry. *Paleogeogr. Paleoclimatol. Paleoecol.* **144**, 3–19.

Stein C. L. and Krumhansl J. L. (1988) A model for the evolution of brines in salt from the lower Salado Formation, southeastern New Mexico. *Geochim. Cosmochim. Acta* **52**, 1037–1046.

Steuber T. and Veizer J. (2002) Phanerozoic record of plate tectonic control of seawater chemistry and carbonate sedimentation. *Geology* **30**, 1123–1126.

Strauss H. (1993) The sulfur isotopic record of Precambrian sulfates: new data and a critical evaluation of the existing record. *Precamb. Res.* **63**, 225–246.

Strauss H. (1999) Geological evolution from isotope proxy signals-sulfur. *Chem. Geol.* **161**, 89–101.

Strauss H., DesMarais D. J., Hayes J. M., and Summons R. E. (1992) The carbon-isotopic record. In *The Proterozoic Biosphere* (eds. J. W. Schopf and C. Klein). Cambridge University Press, Cambridge, chap. 3, pp. 117–127.

Stumm W. and Lee G. F. (1961) Oxygenation of ferrous iron. *Ind. Eng. Chem.* **53**, 143–146.

Stumm W. and Morgan J. J. (1996) *Aquatic Chemistry: Chemical Equilibria and Rates in Natural Waters*. Wiley-Interscience, New York.

Suess E. (1980) Particulate organic carbon flux in the oceans—surface productivity and oxygen utilization. *Nature* **288**, 260–263.

Summons R. E., Jahnke L. L., Hope J. M., and Logan G. A. (1999) 2-methylhopanoids as biomarkers for cyanobacterial oxygenic photosynthesis. *Nature* **400**, 554–557.

Thode H. G. and Monster J. (1965) Sulfur isotope geochemistry of petroleum, evaporites, and ancient seas. In *Fluids in Subsurface Environments, Mem. 4*. Am. Assoc. Petrol. Geol., Tulsa, OK, pp. 367–377.

Thode H. G., Monster J., and Sunford H. B. (1961) Sulfur isotope geochemistry. *Geochim. Cosmochim. Acta* **25**, 159–174.

Vail P. R., Mitchum R. W., and Thompson S. (1977) Seismic stratigraphy and global changes of sea level 4, Global cycles of relative changes of sea level. *AAPG Memoirs* **26**, 82–97.

Van Zuilen M. A., Lepland A., and Arrhenius G. (2002) Reassessing the evidence for the earliest traces of life. *Nature* **418**, 627–630.

Veizer J., Holser W. T., and Wilgus C. K. (1980) Correlation of $^{13}C/^{12}C$ and $^{34}S/^{32}S$ secular variations. *Geochim. Cosmochim. Acta* **44**, 579–587.

Veizer J., Hoefs J., Lowe D. R., and Thurston P. C. (1989) Geochemistry of Precambrian carbonates: II. Archean greenstone belts and Archean seawater. *Geochim. Cosmochim. Acta* **53**, 859–871.

Veizer J., Ala D., Azmy K., Bruckschen P., Buhl D., Bruhn F., Carden G. A. F., Diener A., Ebneth S., Godderis Y., Jasper T., Korte C., Pawellek F., Podlaha O. G., and Strauss H. (1999) $^{87}Sr/^{86}Sr$, $\delta^{13}C$ and $\delta^{18}O$ evolution of Phanerozoic seawater. *Chem. Geol.* **161**, 59–88.

Walker J. (1966) *Lectures on Geology*. The University of Chicago Press, Chicago, IL.

Walker R. N., Muir M. D., Diver W. L., Williams N., and Wilkins N. (1977) Evidence of major sulphate evaporite deposits in the Proterozoic McArthur Group, Northern Teritory, Australia. *Nature* **265**, 526–529.

Wallmann K. (2001) Controls on the Cretaceous and Cenozoic evolution of seawater composition, atmospheric CO_2 and climate. *Geochim. Cosmochim. Acta* **65**, 3005–3025.

Walter M. R., Veevers J. J., Calver C. R., Gorjan P., and Hill A. C. (2000) Dating the 840–544 Neoproterozoic interval by isotopes of strontium, carbon, and sulfur in seawater and some interpretative models. *Precamb. Res.* **100**, 371–433.

Werne J. P., Sageman B. B., Lyons T. W., and Hollander D. J. (2002) An intergrated assessment of a "type euxinic" deposit: evidence for multiple controls on black shale deposition in the Middle Devonian Oatka Creek formation. *Am. J. Sci.* **302**, 110–143.

Wilde S. A., Valley J. W., Peck W. H., and Graham C. M. (2001) Evidence from detrital zircons for the existence of continental crust and oceans on the Earth 4.4 Gyr ago. *Nature* **409**, 175–178.

Wilkinson B. H. and Algeo T. J. (1989) Sedimentary carbonate record of calcium-magnesium cycling. *Am. J. Sci.* **289**, 1158–1194.

Winefield P. R. (2000) Development of late Paleoproterozoic aragonitic sea floor cements in the McArthur Group, Northern Australia. In *Carbonate Sedimentation and Diagenesis in the Evolving Precambrian World*, SEPM Special Publication 67 (eds. J. P. Grotzinger and N. P. James). pp. 145–159.

Yang W. and Holland H. D. (2002) The redox sensitive trace elements Mo, U, and Re in Precambrian carbonaceous shales: indicators of the Great Oxidation Event (abstr.). *Geol. Soc. Am. Ann. Mtng.* **34**, 382.

Yang W. and Holland H. D. (2003) The Hekpoort paleosol profile in Strata 1 at Gaborone, Botswana: soil formation during the Great Oxidation Event. *Am. J. Sci.* **303**, 187–220.

Yin Q., Jacobsen S. B., Yamashita K., Blichert-Toft J., Télouk P., and Albarède F. (2002) A short timescale for terrestrial planet formation from Hf–W chronometry of meteorites. *Nature* **418**, 852–949.

Young G. M. (1969) Geochemistry of Early Proterozoic tillites and argillites of the Gowganda Formation, Ontario, Canada. *Geochim. Cosmochim. Acta* **33**, 483–492.

Young G. M. (1976) Iron-formation and glaciogenic rocks of the Rapitan Group, Northwest Territories, Canada. *Precamb. Res.* **3**, 137–158.

Zimmermann H. (2000) Tertiary seawater chemistry—implications from fluid inclusions in primary marine halite. *Am. J. Sci.* **300**, 723–767.

Volume Subject Index

The index is in letter-by-letter order, whereby hyphens and spaces within index headings are ignored in the alphabetization (e.g. Arabian–Nubian Shield precedes Arabian Sea). Terms in parentheses are excluded from the initial alphabetization. In line with normal materials science practice, compound names are not inverted but are filed under substituent prefixes.

The index is arranged in set-out style, with a maximum of three levels of heading. Location references refer to the page number. Major discussion of a subject is indicated by bold page numbers. Page numbers suffixed by *f* or *t* refer to figures or tables.